Geology Underfoot in Central Nevada

Richard L. Orndorff
Robert W. Wieder
Harry F. Filkorn

Mountain Press Publishing Company
Missoula, Montana
2001

All photographs by the authors unless otherwise credited

Cover painting © 2001 by John Megahan
Spire at Cathedral Gorge

All mileages are from the *Official Highway Map of Nevada* (Nevada State Department of Transportation) and from *Nevada Atlas and Gazetteer* (Delorme).

Library of Congress Cataloging-in-Publication Data

Orndorff, Richard L.
Geology underfoot in central Nevada / Richard L. Orndorff, Robert W. Wieder, Harry F. Filkorn.
p. cm.
Includes bibliographical references and index.
ISBN 0-87842-418-0 (alk. paper)
1. Geology—Nevada—Guidebooks. 2. Nevada—Guidebooks. I. Wieder, Robert W., 1958– II. Filkorn, Harry F. III. Title.

QE137.O76 2000
557.93—dc21

00-048047

PRINTED IN THE UNITED STATES OF AMERICA

Mountain Press Publishing Company
P.O. Box 2399
Missoula, Montana 59806
406-728-1900

Abandoned gold miner's cabin sits next to U.S. 50 near Eureka, Nevada. —Dave Futey photo

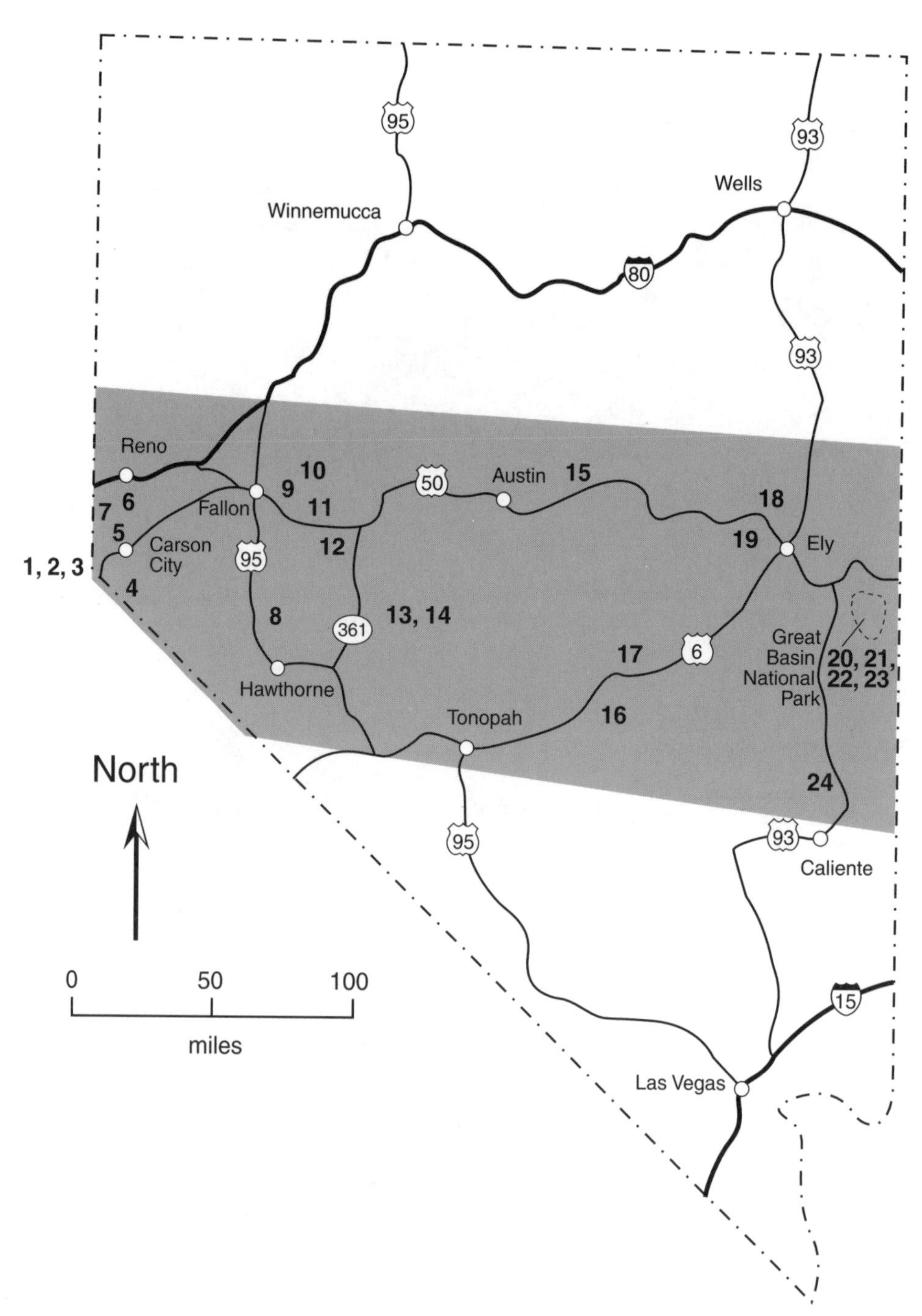

Sites featured in this book. Numbers correspond with vignette number.

Contents

Preface *vii*

Geology of Central Nevada *1*

1. **Tahoe's Recycled Sandbox**
   Sand Harbor *11*
2. **Vulcan's Throat**
   Cave Rock *23*
3. **Grinding Glaciers and Sliding Slopes**
   Emerald Bay *29*
4. **Finding Fault**
   Genoa Fault Scarp *39*
5. **Take a Walk on the Wild Slide**
   Slide Mountain *49*
6. **Fire Down Below**
   Steamboat Hot Springs *61*
7. **Of Lava and Lahars**
   Mount Rose *71*
8. **Nevada's Greatest Lake**
   Lake Lahontan *83*
9. **Varnished Rock in a Desert Gallery**
   Grimes Point *97*
10. **Deciphering Time's Muck Heaps**
    Hidden Cave *107*
11. **One Grain at a Time**
    Sand Mountain *121*
12. **Whole Lotta Shakin' Going On**
    Fairview Peak *131*
13. **Boomtown Ghosts and the Hunt for Gold**
    Berlin Ghost Town *145*
14. **Mesozoic Monsters of the Deep**
    Nevada's Ichthyosaurs *155*

15. **Remnants of a Burning Wasteland**
Hickison Summit *167*

16. **Explosions in an Alien Landscape**
Lunar Crater Volcanic Field *177*

17. **Message from the Mantle**
Black Rock Lava Flow *185*

18. **Desert Gemstones**
Garnet Hill *193*

19. **Open Pits and Altered Rock**
Robinson Mining District *203*

20. **Island in the Desert**
Wheeler Peak *213*

21. **Beauty Beneath the Surface**
Lehman Caves *229*

22. **Gnarled Elders of the Snake Range**
Bristlecone Pines *241*

23. **Nevada's Only Glacier**
Glacial Features in
Great Basin National Park *253*

24. **Hoodoos and Badlands**
Cathedral Gorge *265*

Glossary *273*

Sources of More Information *285*

Index *291*

Preface

Many people perceive geologists as quirky. Admittedly, we are frequently caught acting in ways that may seem bizarre when considered out of context. We lick rocks. We skid to a halt at roadcuts to see faults, folds, unusual rocks, or sometimes just for the chance of finding something new. We argue about oblique slip and punctuated equilibrium over pitchers of beer, rarely noticing the looks other bar patrons shoot our way. Our dream vehicles have four-wheel drive and a short wheelbase—the better to navigate fissures, washes, sagebrush, and the occasional range cow. The perfect Christmas gift for a geologist is a new pair of boots, although we seem to never discard the old ones. We have holes in our jeans pockets from leaking acid vials, and we experience anxiety attacks when we wander more than 100 feet from rock hammer and hand lens. Perhaps the strongest single characteristic of geologists is that we like to be outdoors looking at rocks. And if we're lucky, we communicate that sense of pleasure in this book.

Geology takes us places we wouldn't ordinarily visit and allows us to meet people we wouldn't ordinarily meet. For example, while writing this book, we looked for outcrops we read about in texts, some published almost one hundred years ago. One such search led us far down a rough dirt road south of Eureka. We were looking for trilobites but found that mining had long since destroyed the outcrop in question. However, during our search, we discovered a Peruvian sheepherder living high up in the mountains with his two dogs—a black-and-white border collie and a medium-size, dark brown mutt. The mutt had a curious strip of shiny, silver duct tape on his head. The herder spoke no English, and we spoke no Spanish, but we surmised that the dog was difficult to see in the distance, hence the reflective tape. We gratefully credit the herder for the idea. Dave Futey, who we thank for shooting many of the photos in this book, is an avid outdoorsman and photographer. His camera and the quest for the perfect photo often lead him astray, though. Until we slapped some duct tape on his forehead, we spent an inordinate amount of time looking for him in the mountains and deserts of central Nevada. We spent roughly an hour with the herder, communicating with the few words we all understood, gestures, and pictures drawn in the dirt. We understood that he felt at home here in Nevada, just as he would in his native Andes. We felt at home, too, with him and with the landscape.

We thank our families and friends for putting up with us—first while we poked around central Nevada and then while we put words on paper. We thank David Weide for reviewing the manuscript and providing useful insights and helpful suggestions. Any errors here are solely the fault of the authors. The people at Mountain Press have helped us every step of the way. If we write again, it will be because they made it such an enjoyable experience.

Stokes Castle in Austin, Nevada, stand as a tribute to Austin's early days as a gold-mining boomtown. —Dave Futey photo

Geology of Central Nevada

Nevada is a mystery to most people. Mention the state and images of giant casinos filled with clanging slot machines come to mind. These diversions exist only in populated areas, however, and Nevada is largely unpopulated. Most of the state is a wild and rugged landscape, either dry desert or daunting mountain. Its very ruggedness speaks volumes about its turbulent geologic history. Nevada hosts more than two hundred mountain ranges separated from one another by arid desert valleys. It harbors perennially snow-covered peaks and saline lakes, hot springs and rushing streams, volcanoes and faults, and great fractures in the earth that continue to produce earthquakes today. Groves of aspen crown alpine meadows, and oceans of fragrant sagebrush fill valleys. Nevada is a wonderland, a geologic epic that we hope to share with you.

When we look at the geologic framework of a landscape, we are observing the end result of a sequence of events, each superimposed upon the other. These events have occurred over geologic time—that is, the interval of time from the formation of the earth to the present. To put geologic time in perspective, Don Eicher, in his book *Geologic Time,* compressed the 4.6-billion-year history of the earth into a single year. The oldest rocks we know of, and thus our first clues to the story of the earth's past, date to mid-March. Living organisms first arose in the sea in May, and land plants and animals in late November. Dinosaurs dominated the earth in mid-December but vanished on December 26, the same day the Rocky Mountains formed. Human ancestors first walked upright during the evening of December 31, and the most recent continental ice sheets receded from the upper Midwest about 1 minute and 15 seconds before midnight. Rome rose and fell between 11:59:45 and 11:59:50, and Christopher Columbus set foot in the New World at 11:59:57. James Hutton, the founder of the theory of uniformitarianism and modern geology, published his work 1 second before midnight. The last 200 years of history squeezed into that final second before the new year.

Geologists formally divide geologic time into eons, eras, periods, and epochs. Significant events in earth history mark each of these intervals. The Phanerozoic eon, for example, begins about 570 million years ago and encompasses only the last 15 percent of geologic time. Phanerozoic means "apparent life," and this eon starts with the appearance of multicelled organisms with hard parts, such as shells and teeth, preserved in rock as fossils. The Phanerozoic eon contains the Paleozoic, Mesozoic, and Cenozoic eras, meaning "ancient life," "middle life," and "new life," respectively. Paleozoic time witnessed a succession of dominant organisms, from invertebrates

ERA	PERIOD	EPOCH	TIME
Cenozoic	Quaternary	Holocene Pleistocene	present
			1.6 million years
	Tertiary	Pliocene Miocene Oligocene Eocene Paleocene	
			67 million years
Mesozoic	Cretaceous Jurassic Triassic		
			245 million years
Paleozoic	Permian Pennsylvanian Mississippian Devonian Silurian Ordovician Cambrian		
			570 million years
Precambrian			4.6 billion years

Geologic time scale

to fish to amphibians. Reptiles first evolved in Paleozoic time but did not dominate the world until Mesozoic time. Although dinosaurs flourished during Mesozoic time, they became extinct at its boundary with the Cenozoic era, the age of mammals. During this last era, mammals emerged as the dominant animals.

The fossils in rock allow us to establish a sequence of ages. We can say "This rock is older than that one" based on the fossils each contain, but fossils alone provide only a relative age. To assign a numeric value to the age of a rock we must look elsewhere.

We can determine the absolute ages of rocks—and thus assign dates to the divisions of earth history—using radiometric dating. This method relies on the presence of naturally occurring radioactive elements. Many rocks form with small but detectable quantities of such elements, which decay naturally over time into what are called stable daughter products—that is, a different element that is not radioactive. The decay proceeds at a constant rate known as a half-life—the time required for one-half of the radioactive parent element to decay to a stable daughter product. That process of radioactive decay provides us with an atomic clock we can use to determine the age of a rock. For instance, assume that a lava flow containing the radioactive element potassium-40 cools and solidifies into basalt, a dark volcanic rock. Potassium-40 decays to the stable daughter product argon-40. The half-life for this process is 1.3 billion years. The rock initially contains no argon-40, but over time quantities build up as the potassium-40 decays. If we measure the amounts of potassium-40 and argon-40 in the lab and discover that these quantities are the same, we know the absolute age of the rock is exactly one half-life, or 1.3 billion years, because one-half of the radioactive parent potassium-40 has transformed into stable daughter product argon-40.

Nevada sits in the center of the Basin and Range, a region of north-south trending mountain ranges separated by deep valleys. This region's distinctive topography is the product of plate tectonics; that is, the interaction between large, slow-moving plates of the earth's outer skin. The earth consists of a core, mantle, and crust. The core, at the center of the earth, contains mostly the very heavy elements iron and nickel and generates the earth's magnetic field. The mantle is the dense and thick layer between the core and the earth's very thin outer layer, or crust. Relatively low-density rock forms the crust. The widely accepted theory of plate tectonics introduces two additional components, the lithosphere and the asthenosphere. The lithosphere contains the uppermost mantle and crust, which together behave as a

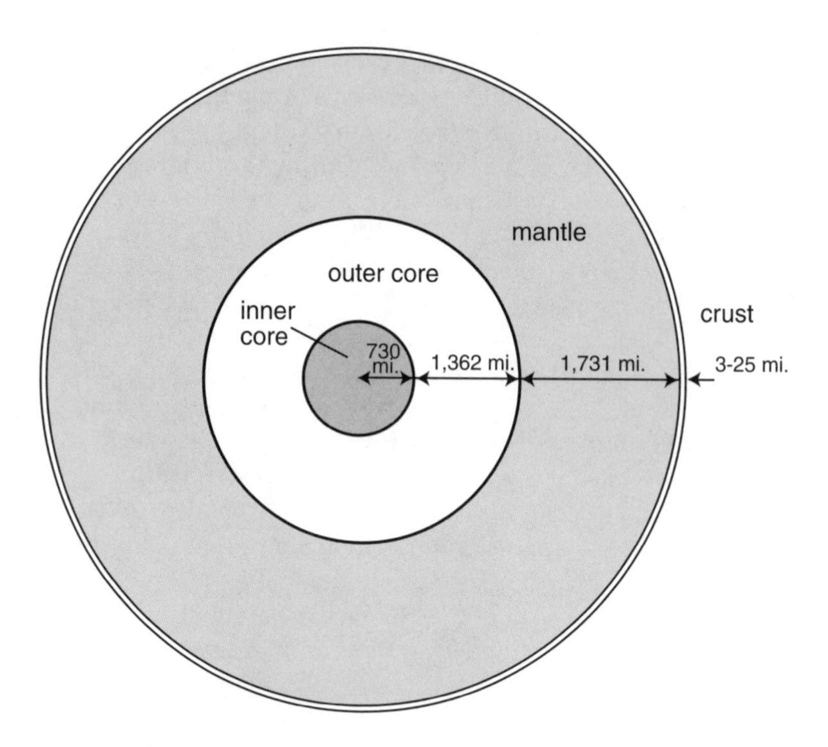

The earth consists of the core, the mantle, and the crust.

brittle solid. The lithosphere is broken into a number of plates that float on the ductile upper mantle—the asthenosphere. The flow of material within this region of the mantle drives movement of the numerous plates of the lithosphere. Plates interact with each other at their boundaries by moving together, moving apart, or sliding past one another. These interactions control the location of most of the world's volcanoes and the occurrence of earthquakes.

The Basin and Range lies within the North American plate. Semi-molten rock in the asthenosphere is welling up beneath the Basin and Range, causing the lithosphere to dome and thin over a large area. The upwelling asthenosphere then drags the crust to the east and west, literally pulling the continent apart. As the crust extends, pieces of the lithosphere fracture and fall, creating the desert valleys that alternate with mountains in the Basin and Range. The spreading of the continent poses no immediate danger to inhabitants because the rates of plate movement average only inches per year. If this process continues for millions of years, a great rift valley will form, deepening over time until it fills with water to create a new ocean. We can

The asthenosphere wells up beneath continental lithosphere.

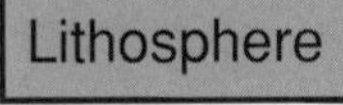

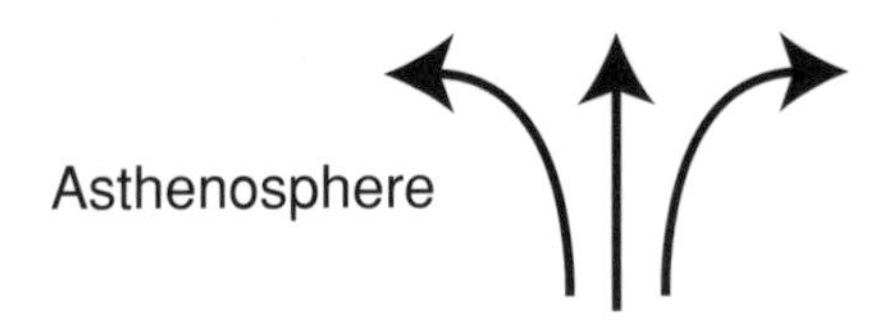

The lithosphere domes broadly and thins.

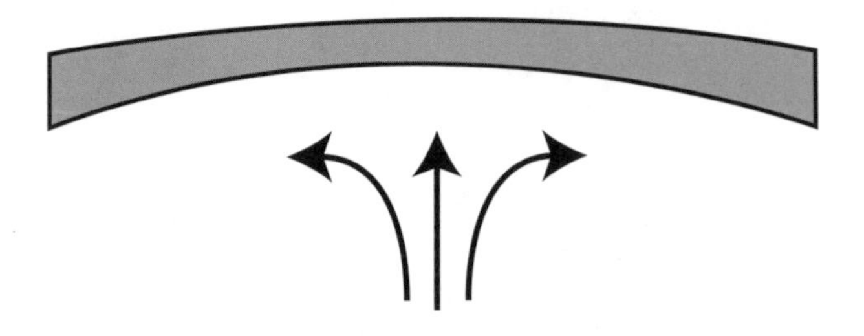

The lithosphere extends, and faults create alternating mountains and valleys.

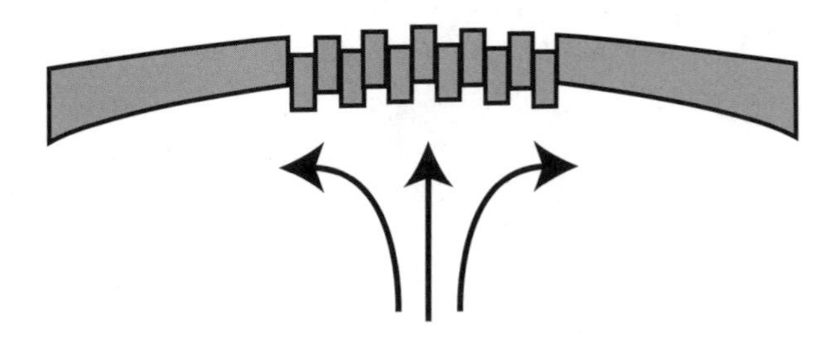

How the Basin and Range evolved

see Nevada's future elsewhere in the world. The East African Rift Valley is a spreading plate boundary in which the center has literally fallen out. The Red Sea is a young ocean that has filled a rift valley.

Nevada also sits within the Great Basin. Hydrologic, rather than tectonic, features define this province. Most river systems eventually empty into the sea. The Amazon, Nile, and Mississippi—the three largest rivers in the world—flow into the Atlantic Ocean, Mediterranean Sea, and Gulf of Mexico, respectively. Rivers within the Great Basin behave differently. This central part of the southwestern United

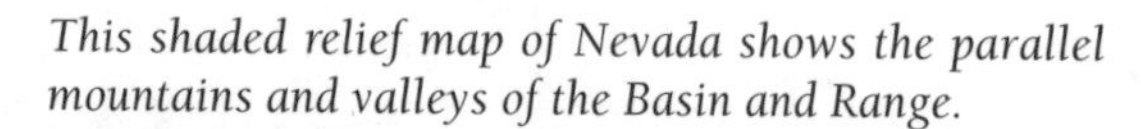
This shaded relief map of Nevada shows the parallel mountains and valleys of the Basin and Range.

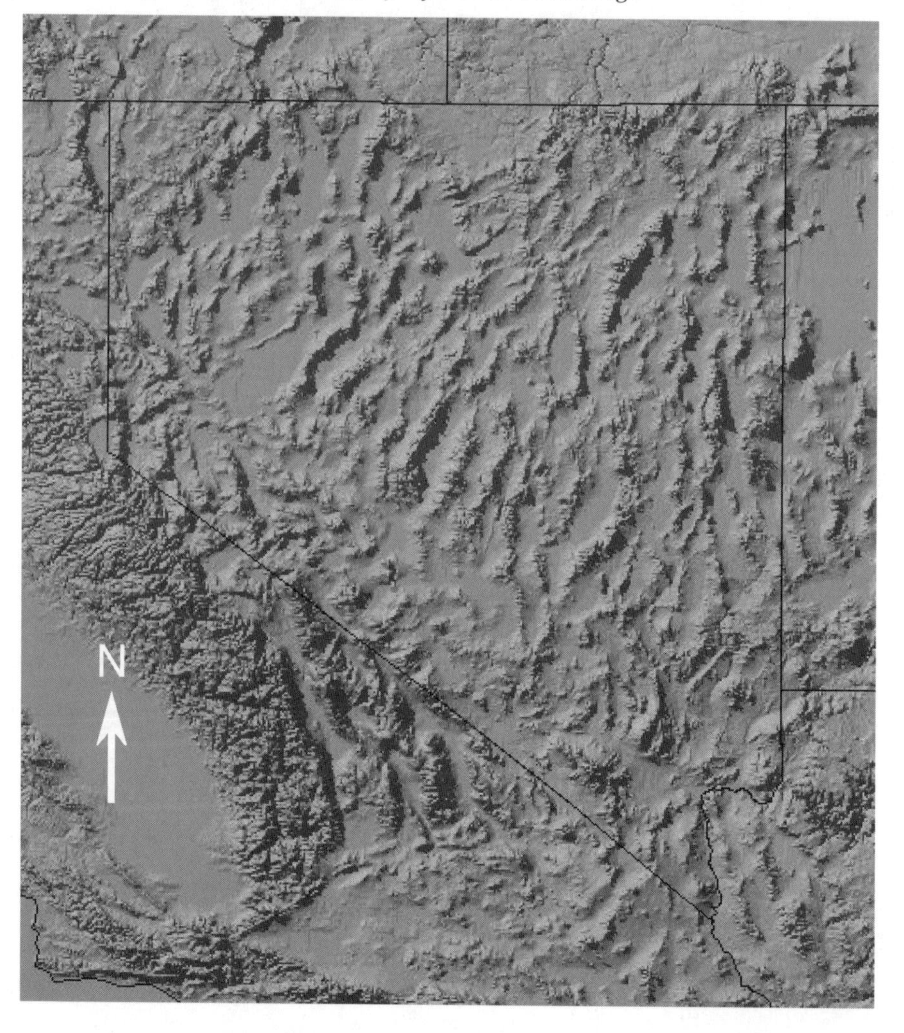

States is a closed basin—no rivers exit to the sea. The Truckee River flows from Lake Tahoe into Pyramid Lake, and no farther. The Humboldt River flows west across Nevada to the Humboldt Sink, a broad salty basin that holds water during part of the year. Highlands to the north and south prevent the escape of water in those directions, and the Great Basin reaches from the Sierra Nevada in the west to the Rocky Mountains in the east. These mountain barriers effectively create an immense bowl from which water escapes only by evaporation.

The earth's crust consists of three types of rock: igneous, sedimentary, and metamorphic. Igneous rock cools and solidifies from molten rock, which is called magma if it remains in the earth's interior and lava if it spills out onto the surface. Magma that intrudes beneath the earth's surface becomes intrusive igneous rock, such as granite or granodiorite. Intrusive rock cools slowly, allowing large mineral crystals, such as those within granite, to grow. Lava erupts onto the earth's surface during volcanism and cools rapidly, becoming extrusive igneous rock. Such rapid cooling produces rock that has very small

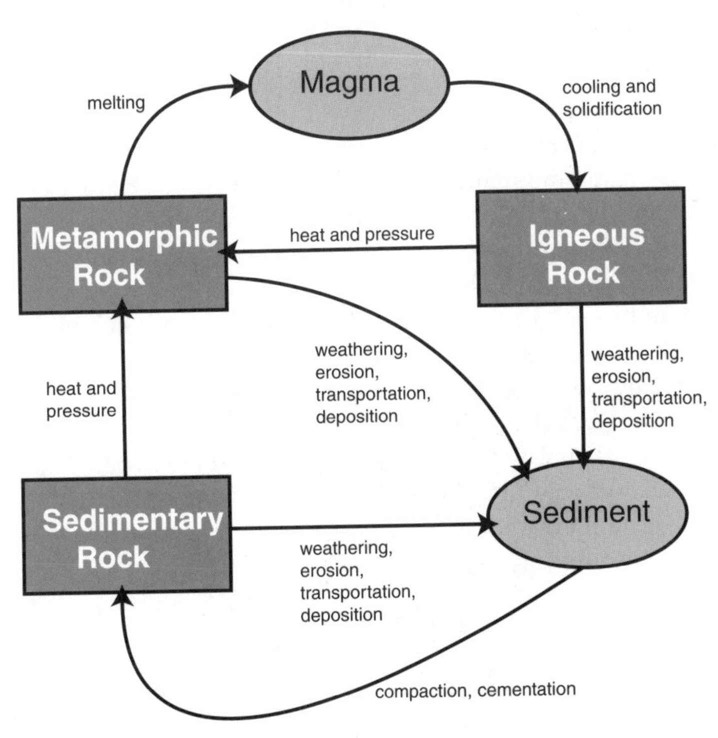

The rock cycle shows how the earth constantly recycles itself.

mineral crystals, such as basalt. Sedimentary rock consists of the weathered remains of pre-existing rock. Wind, water, or ice erode, pick up, and transport sediment, then deposit it. The sediment compacts and cements into new rock—sandstone, for example. Metamorphic rock forms when heat or pressure alter pre-existing rock without melting it. Quartzite is the metamorphic version of sandstone—the original sedimentary grains have crushed together and partially recrystallized to produce a rock that is denser and stronger than the original sedimentary rock. Gneiss (say "nice") and schist are other examples of metamorphic rock.

The earth is an ever-changing system in which rocks are constantly changing from one type to another. The rock cycle illustrates the interactions between the three rock types. We will encounter all three types in our journey across central Nevada.

In the chapters that follow, we visit sites that reflect the diverse geologic processes that have shaped Nevada over millions of years. Each geological vignette tells the story of a specific feature, process, or event in Nevada's geologic history. Each vignette stands alone, independent of the others, but together they weave a rich tapestry of Basin and Range geology.

Our main route is U.S. 50, the loneliest highway in America, which will carry us from west to east as we traverse the core of both the Basin and Range and the Great Basin. We will look at Paleozoic rock in Great Basin National Park and fossil remnants of giant sea creatures that roamed the oceans during Mesozoic time. We will observe the effects of recent glaciation on Wheeler Peak, look at evidence for huge lakes that once filled today's desert valleys, and walk across volcanoes and lava flows. We will stand on recent fault scarps, next to steaming hot springs, and beneath huge landslides. We will also look at gold- and copper-bearing rock and associated historic mining activities in Nevada.

We hope to share with you an appreciation of the varied landscapes in Nevada. We also hope to instill a sense of the dynamic nature of geology, an understanding that the earth is not a finished story but instead a continuing tale with an ever-changing plot.

U.S. 50, also known as "the loneliest road in America," travels east-west across central Nevada. —Dave Futey photo

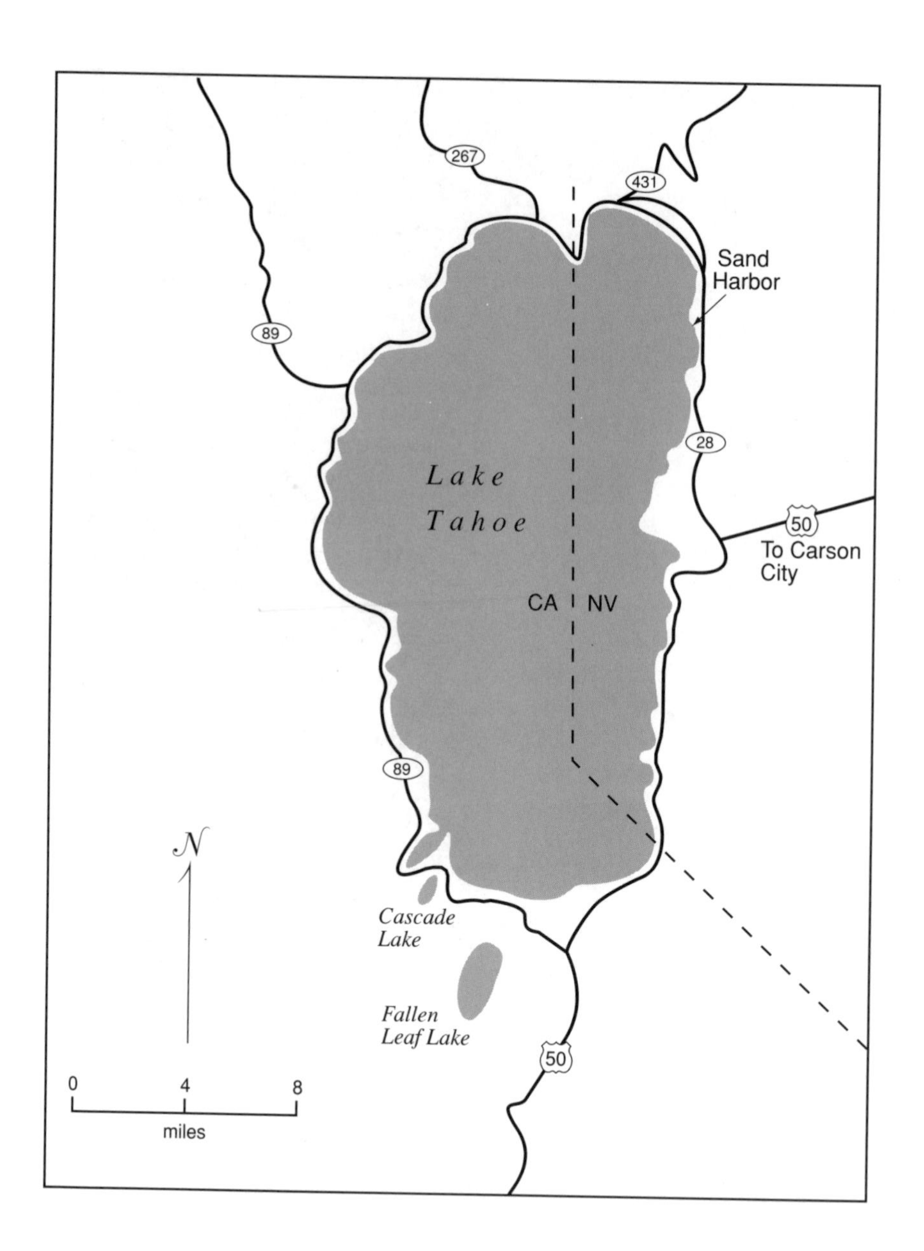

▶ GETTING THERE

You can easily access Sand Harbor Beach State Recreation Area along Nevada 28, about 5.8 miles south of the intersection with Nevada 431. From the south, it is about 8.5 miles north of the intersection of Nevada 28 and U.S. 50. Turn west into the main parking area (south of the boat launch). The park is open year-round and for a nominal fee, one can swim, picnic, camp, and study geology.

1

Tahoe's Recycled Sandbox

SAND HARBOR

The Sierra Nevada stretches for 400 miles along the eastern side of California. For most of its length, a single crest divides the drainage between the gentle western slope and the much steeper eastern slope. Near the southernmost headwaters of the Truckee River, however, the Sierra crest divides. The main crest continues its northwestern trend, while a second crest, the Carson Range, branches off to the north. Nestled within the valley separating these two lofty ranges lies Lake Tahoe. Often called the jewel of the Sierras, Lake Tahoe is about 22 miles long and 12 miles wide. At 1,645 feet deep, it is the deepest lake in North America.

Tahoe's unusual beauty attracts its share of sightseers and recreationists. They come from all over the world to enjoy the deep, clear water and the high mountain setting. One of the nicest places to enjoy the Lake Tahoe experience is from the beach at Sand Harbor, along Nevada 28 about 5.8 miles south of the junction with Nevada 431. A seemingly unlimited supply of soft, fine-grained sand makes this one of the best beaches on the lake. You might wonder how all of this sand ended up here. The answer to this question lies in the rocks here and the processes that affect them.

On foot, exit the parking lot at its northeast end and begin walking along the boardwalk interpretive trail. This is a short, pleasant walk among the pines and cliffrose, the latter adding a sweet smell to the air in early summer. Interpretive signs along the way tell of the natural history of the area. You will soon come to a few places where you can exit the boardwalk and walk out on one of the promontories of strange, rounded rock. Choose one, walk out to one of the odd-looking boulders, and look it over. First we'll figure out what kind of rock this is, then we can see if it plays a role in the sand supply system.

Look at the texture of your chosen boulder. Without much trouble you should be able to pick out individual grains, some light and some

A group of granodiorite boulders accessible from the boardwalk

darker. It's a pretty safe bet to say that a rock like this, in which you can easily discern individual mineral crystals, is a plutonic rock—that is, an igneous rock that cooled slowly at depth. The magma that eventually became this rock intruded beneath miles of overlying rock that insulated it. Magma at depth may take thousands of years to completely solidify, and because it cools slowly, the crystals have time to grow to a visible size.

Now look at the mineral grains that make up this rock. Find a relatively small piece that you can bring up to your nose instead of having to take your nose down to the rock. You may notice that some of the grains look clear and glassy. These are grains of quartz. Other grains are milky white. If you have found a hand sample, move it around in the sunlight, and you will see that the light reflects off of some of the milky white crystals. These are plagioclase feldspar crystals. The atoms in this mineral align in discrete planes, some of which can act as zones of weakness. The crystal breaks easily along these weakened zones in a process called cleavage, and the cleavage surfaces reflect light. The darker minerals are mainly elongate, rectangular hornblende and platy, shiny biotite. Coarse-grained igneous rocks that contain primarily quartz and feldspar with lesser amounts of dark minerals are said to be granitic. The light and dark minerals give the

Close-up of a block of granodiorite from Sand Harbor. The light minerals are quartz and feldspar. The dark elongate crystals are hornblende. The dark hexagonal crystals, like the one to the left of the date on the coin, are biotite.

rock a salt-and-pepper look. In a true granite, the feldspar present is mostly potassium feldspar. In this rock, the feldspar is primarily plagioclase, so geologists call it a granodiorite. The rock here at Sand Harbor, and most of the rock that underlies the Lake Tahoe basin, is part of the Sierran batholith, a group of hundreds of discrete igneous intrusions, or plutons, that make up most of the Sierra Nevada.

When these plutons first solidified, they were hot and under a lot of pressure beneath miles of rock. Gradually, the rocks began to cool, and faults like the one exposed near Genoa (vignette 4) brought them closer to the surface. Over time, erosion wore away miles of overlying rock, dramatically decreasing the pressure upon the granodiorite—a process called unloading. This change in temperature and pressure caused the rock to crack. These cracks tended to align themselves along discrete planes known as joints. A group of more or less parallel joints is called a joint set, and two or more different joint sets crisscross many rocks in the Sierra Nevada.

The action of water weathers granitic rocks in characteristic patterns, either by mechanical means such as frost wedging, or by

Pluton intrudes at high pressure deep below the earth's surface.

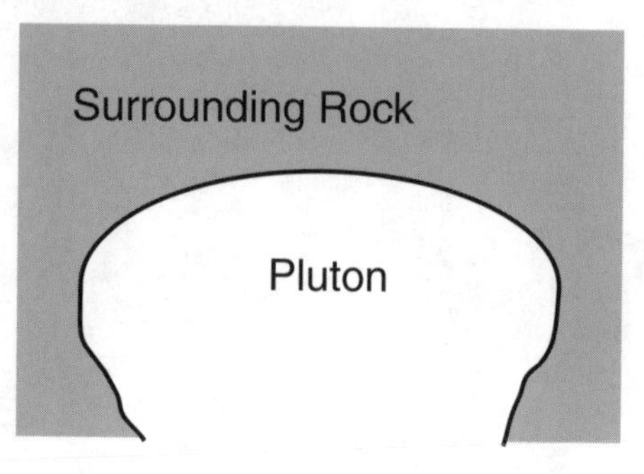

Uplift and erosion expose the pluton to the surface. Decreased pressure causes the pluton to expand and the outer surface cracks.

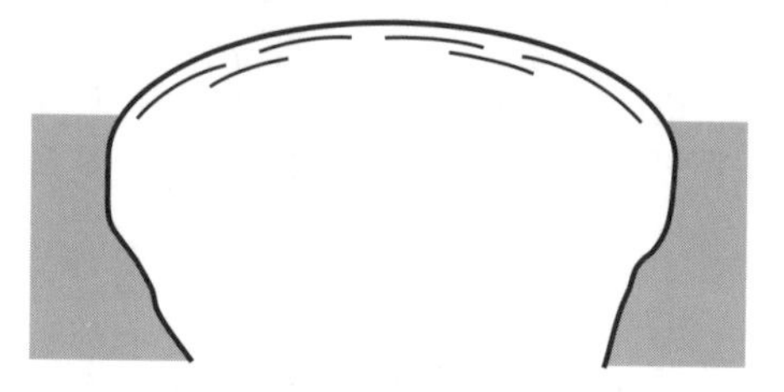

Mass wasting and erosion remove layers, causing more parallel fractures to form.

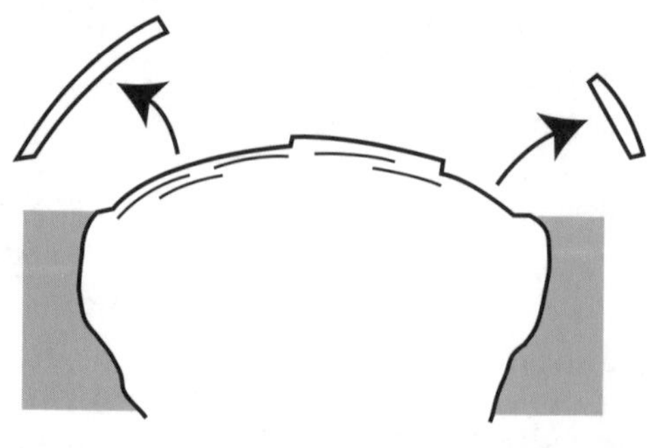

When weathering and erosion unload the miles of overlying rock, this igneous intrusion, or pluton, expands, developing cracks parallel to the exposed surface of the pluton.

Granodiorite along Nevada 28 south of Sand Harbor State Park shows parallel surface fractures characteristic of unloading and exfoliation.

chemical weathering. In frost wedging, water seeps into the small joint fractures in the rock and later freezes when the temperature drops. Water, unlike most fluids, expands when it freezes. This expansion pushes on the fracture and wedges it open a little wider and deeper, eventually breaking the rock apart. This simple process, repeated over and over for many thousands of years, can transform a once mighty mountain peak into a shattered mass of broken rock.

In chemical weathering, water and organic acids seep into the tiny fractures and chemically alter the black mineral biotite into the green mineral chlorite. This alteration process causes a slight amount of expansion, which pries apart other grains, allowing more water to seep into the rock. The dilute acid solutions also alter the feldspars into clay, a weak mineral that tends to fall apart easily and destabilize the other grains in the rock. These grains, mostly quartz and unweathered feldspar, break off the surface and accumulate as a coarse sand at the base of the outcrop.

The way in which various sets of joints intersect each other has a lot to do with how a rock weathers. Where they intersect at nearly right angles, they create sharp edges and corners. These corners provide excellent areas for the weathering process to begin, because a

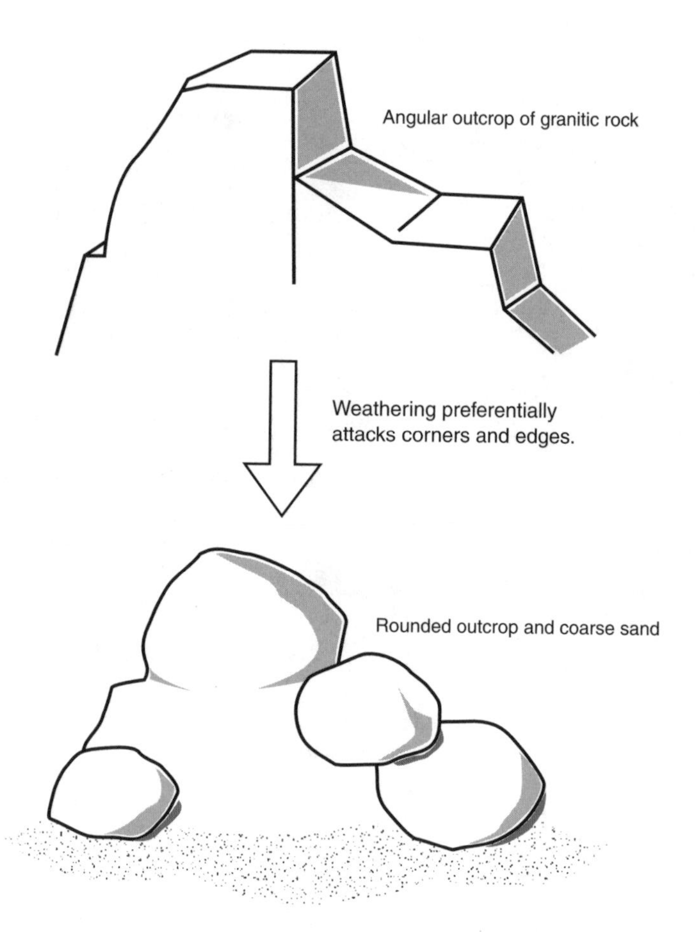

Spheroidal weathering produces a rounded outcrop from an angular one by preferentially attacking corners and edges.

Spheroidal weathering of granodiorite along Nevada 28 near Sand Harbor State Park. Preferential weathering of corners and edges rounds outcrops.

sharp corner provides more surface area to attack than a flat area. The result is that the sharp corners wear away faster than the rest of the rock, eventually leaving just rounded knobs such as the ones you see here at Sand Harbor.

Now walk down to the north end of the boardwalk, and look at the granodiorite promontory to the north across the small embayment. In the outcrop to the north, you can see widened cracks and flat surfaces that are the remnants of the joint system in these rocks. Most joints begin as tight fractures, but over time mechanical and chemical weathering processes widen them, as you see here.

Head back toward the boardwalk. Near where you re-enter the boardwalk you'll find some fine examples of deeply weathered pieces of granodiorite surrounded by coarse sand. Pick up a handful of this sand for a closer look. See if you can pick out the individual mineral grains using the same criteria you used when identifying them in the rock. The clear, glassy grains are quartz, the milky white ones are feldspar, and most of the dark ones are hornblende.

We seem to be getting somewhere now in our quest for the source of this sand. Weathering breaks down the granodiorite into a rubble of quartz, feldspar, and hornblende. Many of the mountains that sur-

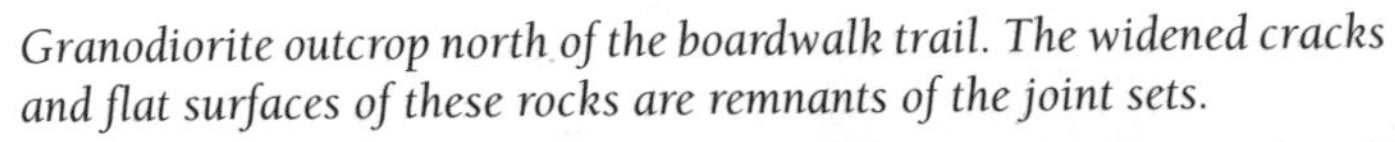

Granodiorite outcrop north of the boardwalk trail. The widened cracks and flat surfaces of these rocks are remnants of the joint sets.

This coarse sand from disintegrating granodiorite has not yet weathered to individual mineral grains.

This granodiorite sand has broken down to individual grains of quartz, feldspar, and hornblende.

The banging and jostling of waves has reduced the coarse-grained sand to this fine-grained sand.

round Lake Tahoe also consist of granodiorite, so it looks like we have a big enough source to make a beach like this. But is that the end of the story?

Walk out on to one of the beaches, and pick up a handful of sand. You may notice that this sand has a much finer texture than that which we examined earlier. Look around and you will see dunes composed almost entirely of this fine-grained sand. It seems that something more than just the decomposition of granodiorite is going on here.

The back-and-forth action of waves jostles the sand particles, causing them to collide and rub against each other. Over time, this process further breaks down the individual grains into smaller pieces, so any sand particles caught up in this nearshore environment will eventually wear to a finer texture. The level of Lake Tahoe has fluctuated in the past, with dry periods significantly lowering the lake. During such a time, when the lake level was very low, winds blew the fine sand from the dry lake bottom up onto the shore. The granodiorite headlands here at Sand Harbor acted like a baffle, breaking the wind and causing it to drop its load in dunes here—much the way a snow fence causes blowing snow to drift rather than blow across a highway.

We can see evidence of sand on the move here today as well. Many of the rounded boulders and knobs have dark lichens on their surface. Lichens grow extremely slowly, so their presence indicates that the rocks have been sitting here in this position for a long time. But the lower portion of some of these rocks lacks lichens and is much lighter in color. Sand probably buried the rock up to this line and then, more recently, the wind removed some sand from around the rock's base, exposing the lighter area that lichens have not yet claimed. Thus, the processes that formed this beach are still at work modifying it.

Stroll out onto one of the granodiorite promontories. Note how you can see the shape of the rocks far below the surface of the water. Lake Tahoe is very special in many ways, and water clarity is one of the most impressive. Scientists classify lakes based on diversity and productivity. Diversity refers to the number of different organisms that occupy a lake, and productivity refers to the total number of plants and animals within it. There are essentially two extremes for lakes.

Oligotrophic lakes have high diversity and low productivity, while eutrophic lakes demonstrate low diversity and high productivity.

Windblown sand once buried this boulder up to the level of the dark lichens.

Healthy alpine lakes are oligotrophic; they feature very cold, clear water with many different types of organisms. The total number of organisms, though, is fairly low.

Humans have greatly harmed some alpine lakes through human-caused eutrophication, a process that is presently degrading Lake Tahoe. Fertilizer from lawns and unburned engine fuel from motorboats enter the lake. Aerobic bacteria use dissolved oxygen from the water to break these substances down. Nutrients from the breakdown of fertilizer and fuel feed lake algae, which, along with organic debris, begin to cloud the water. Decreased dissolved oxygen kills fish and other organisms, which adds to the nutrient load and further decreases oxygen as the dead creatures decay. As cloudiness increases, bottom-dwelling plants receive less sunlight, so they die, too. Nutrient levels continue to increase, oxygen decreases, and algae dominates the system. Eventually the once-clear oligotrophic lake becomes eutrophic, supporting only algae whose blooms spread across the lake surface.

In the case of Lake Tahoe, such a process would destroy a national treasure. Unfortunately, it has begun. Already visibility has decreased and algae has flourished along the shoreline. In the last thirty years, the depth to which a person can see in the lake has dropped more than 30 percent, from 102 feet to about 70 feet. This problem is beginning to attract the attention of residents here. Jet skis are now prohibited because their simple engines released a large quantity of unburned fuel into the water. Further steps will protect this lake nestled between mountains of granodiorite, the jewel that is Lake Tahoe.

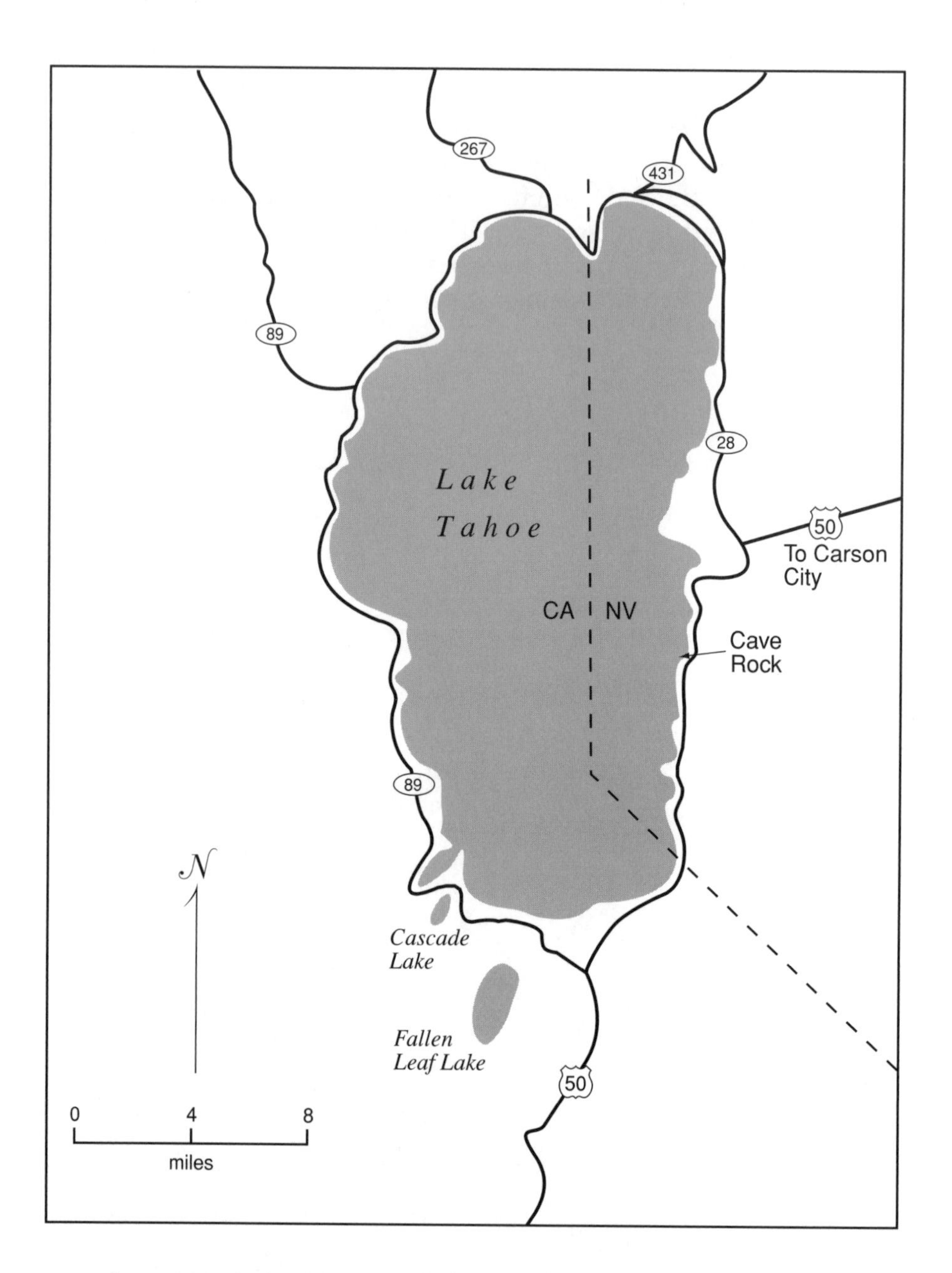

▶GETTING THERE

Cave Rock is hard to miss if you are traveling along the east side of Lake Tahoe on U.S. 50—tunnels take you right through it. The best place to stop and explore this structure is the parking lot at the small park and boat launch just south of the tunnels on the west side of the highway. Proceed down the incline and park. The park collects a small fee for the day, and a minimal fee for a 15-minute photo stop.

2

VULCAN'S THROAT

CAVE ROCK

Anyone traveling down U.S. 50 along the east side of Lake Tahoe can hardly help but notice the strange structure known as Cave Rock jutting into the lake from the west side of the Carson Range. As you approach, there seems no way around it, and, sure enough, tunnels cut through it to accommodate the highway. You might think that these tunnels gave Cave Rock its name, but if you pull into the parking lot of the small state park just south of it, you will see remnants of its true namesake just above the southbound tunnel. Park your car,

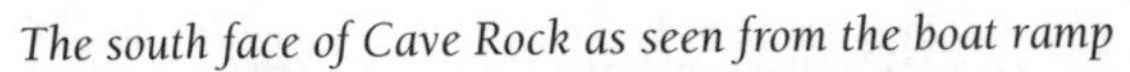

The south face of Cave Rock as seen from the boat ramp

and walk to the north end of the park near the boat launch for the best view of Cave Rock. U.S. 50 is too busy to safely cross for a closer look, but from this vantage you can still obtain a lot of information. What is this structure, and who or what created the caves?

When approaching a geologic problem, it is always best to begin with the basics, one of which is to figure out what type of rock we are dealing with. Along the east side of the paved area leading up to the boat launch, you can see a hill with many blocks of light-gray rock of various sizes strewn about it. These blocks have broken off Cave Rock and moved downslope. Such blocks are called talus, and when you can't reach the outcrop, the next best method to determine the rock type is to look at the talus that came from it.

Pick up a piece of this rock and look at it closely. Notice the crystalline texture of this igneous rock, which solidified from a magma. The size of the crystals in an igneous rock depend on how fast the magma cooled. The slower it cooled, the larger the crystals. This rock has many large crystals, some light and some dark, embedded in a much finer matrix. Geologists call this texture porphyritic. The light crystals are a type of feldspar called sanidine, and the dark crystals are hornblende. Compositionally, that makes the rock a latite, so we can describe it as a latite porphyry. The larger crystals grew while the magma was still deep underground, cooling slowly. Then, before completely crystallizing, the magma erupted onto the surface, and the remainder of it cooled quickly as the fine-grained matrix. The

Close-up of the latite porphyry of Cave Rock. The light-colored crystals are feldspar and the dark, needlelike crystals are hornblende.

process of figuring out what type of rock this is, then, has given us some important information about the outcrop that it came to be.

Movement along faults of the Sierra Nevada frontal fault zone (vignette 4) created the basin that holds Lake Tahoe. This valley, known as a graben, dropped between faults. It resembles the Carson Valley and the other valleys to the east in the Basin and Range. But while all these basins are similar in structure, something must be different here, because the valleys of the Basin and Range are dry, but the Tahoe basin holds the deepest lake in North America.

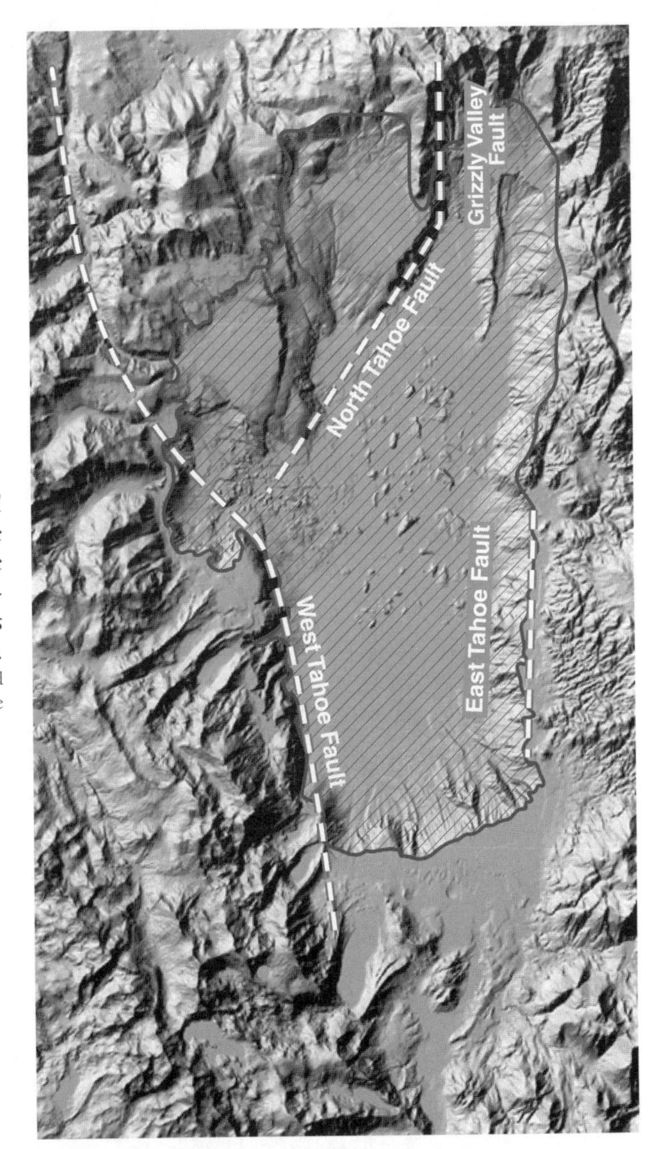

Faults control the shape of the Lake Tahoe basin. Cross-hatching marks lake's location. —U.S. Geological Survey image

One of the differences is the amount of precipitation that the Tahoe area receives. A tall mountain range like the Sierra Nevada acts like a wall, blocking moisture-laden clouds from moving farther east. As the air rises to pass over the range, it cools, and the moisture within it condenses, falling to the ground as rain or snow. This rain-shadow effect (vignette 3) causes the area to the east to receive only a fraction of the precipitation that falls in the Sierra Nevada. So, now we have a water source for the lake, but why does it collect here in such a spectacular fashion?

The answer to this question brings us to a unique characteristic of the Tahoe basin. The northern outlet is blocked by a huge pile of volcanic rock. Volcanic eruptions between 22 million and 2 million years ago produced this dam. This volcanism took the form of massive, blocky mudflows called lahars that slid down the flanks of volcanoes (vignette 7), as well as andesitic lava flows that poured out of numerous vents around the lake and farther east. Many of these vents have yet to be located and may be buried under later lava flows, but a few of them have been identified.

Cave Rock is the remnant of a former volcano. A volcano is simply a conduit through which magma, ash, and associated gasses reach the earth's surface. We commonly think of a volcano as having a conical shape, but that is not always the case. Sometimes flows pour out of cracks in the ground, called fissures, and blanket large areas with relatively flat sheets of lava. The familiar conical shape results from the accumulation of material from previous eruptions around the vent. As more lava, ash, and broken rock pour out of the volcano and onto its flanks, the cone builds higher and wider. Eventually eruptions cease and the cone stops growing. At this point, erosion begins wearing it down, but not all rocks erode at the same rate. Composition, jointing, and overall strength make some rocks more resistant to weathering than others. The harder, more resistant rocks weather much more slowly than the softer, less resistant rocks. This process, called differential weathering, often results in an uneven land surface, where the more resistant rocks make up the high points and the softer rocks erode to leave the low areas. Lahar deposits erode easily. The andesite flows better resist weathering, but they tend to be jointed, which gives the forces of mechanical weathering, such as frost wedging, a good foothold. Eventually these rocks break down, too.

When a volcanic vent stops erupting, lava, ash, and rock fragments remain inside the vent and in the feeder tubes leading up to it—just as water stays in the pipes after you turn off the faucet. This remaining plug of magma cools and solidifies into rock that tends to resist

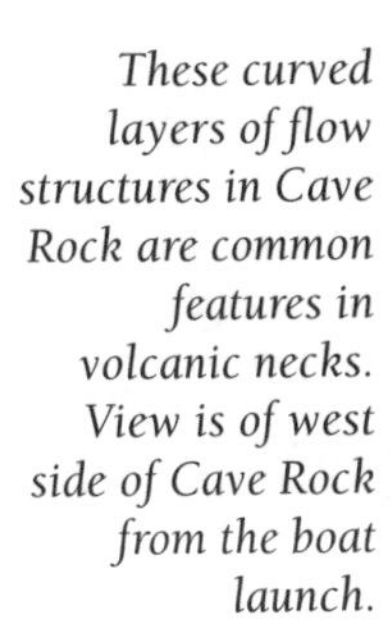

These curved layers of flow structures in Cave Rock are common features in volcanic necks. View is of west side of Cave Rock from the boat launch.

erosion. Over time, the less resistant lahar deposits and flows surrounding this plug erode away, leaving the harder rock of the plug standing alone. This type of structure is called a volcanic neck, and that's what Cave Rock is.

Look at the area above the caves. The rock looks layered, and the layers appear to be coming out at you. If you look to the left of the caves, along the western edge, you can see that these layers are nearly vertical and that they bend and curve. Although it may be hard to see at first, this layered structure makes up many parts of Cave Rock, and the layers splay in many different directions. These bending and curving layers, or flow structures, represent separate injections of lava and show that flow was occurring as the rock was cooling. Flow structures are common characteristics of many volcanic necks.

Now that we know a little about the geologic structure of Cave Rock, let's turn our attention back to the caves themselves and ponder for a moment how they came to be. Are they also related to the volcanic activity? Or was their formation due to completely different processes?

Look at the relationship of the caves to the rock surrounding them—binoculars will help. If the caves developed as part of the volcanic activity, you would expect the flow structures to show this, by bending around them, for instance. This is not the case, though. The rock around the caves looks eroded and broken off. These caves, it appears, were cut into the rock after it was formed.

A dam at the outlet to the Truckee River keeps the lake level constant now, but the Tahoe shoreline has changed its position many times in the past. Wave-cut benches known as strandlines mark the slopes as high as 600 feet above the present lake level. Ice dams that blocked the outlet during glacial advances caused most of the fluctuations. A significant strandline is present nearly 200 feet above the present lake level, approximately the same level as the caves. When the lake stood at this higher level, wave action eroded these caves into this rock face.

Some quarried granodiorite blocks form a wall to the west of the base of Cave Rock. This is all that is left of the Bigler Toll Road, which workers built when logging operations here cut timber for the Comstock mines. At around the time of the Civil War, the caves presented a more impressive spectacle, a major sight to see when traveling around Tahoe. Unfortunately, the construction of the tunnel below it destroyed part of the cave. What we see today is but a remnant of the caves that a traveler one hundred years ago would have seen, but even this remnant sufficiently piques the interest of many a twenty-first-century traveler.

3

Grinding Glaciers and Sliding Slopes

EMERALD BAY

Many of the mountain peaks around Lake Tahoe held glaciers multiple times over the past 2 million years. These glaciers played a huge role in shaping the terrain around the lake. Although ice never filled the basin itself, a few of the glaciers did reach the shoreline. One place where the ice came right down to lake level is at Emerald Bay. We will begin our exploration of the glacially sculpted landscape in the parking area at the Emerald Bay overlook.

You are standing in California, within spittin' distance (as geologists like to say) of the Nevada state line. Processes that we'll discuss here affected both the Nevada and California sections of the Sierra Nevada, and from Emerald Bay we can see Nevada mountains to the north and east. We have snuck into California because the landforms at this site are much more accessible than similar ones in Nevada. Consider this a visit of convenience.

Rockslide scar along California 89 southwest of the Emerald Bay overlook parking lot

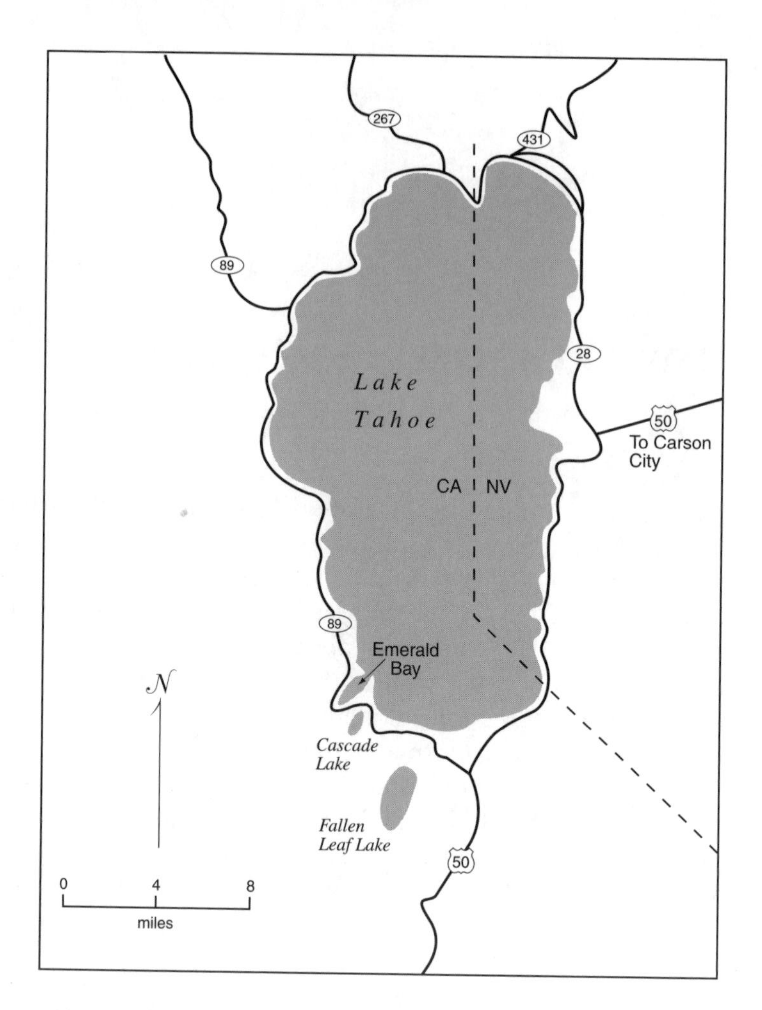

▶GETTING THERE

To reach Emerald Bay from Carson City, drive 26 miles west on U.S. 50 to the California state line. Continue west on U.S. 50 for 5 miles through South Lake Tahoe to the intersection of California 89 and turn north (right). The parking area at Emerald Bay is on the northeast (right) side of the road about 9 miles north of this intersection.

From Reno, take U.S. 395 south to Nevada 431, which leads southwest across Mount Rose Summit to Incline Village. Turn south onto Nevada 28 and drive 13 miles to U.S. 50. Drive west on U.S. 50 for 17 miles, crossing the California state line, and turn north on California 89. The parking area at Emerald Bay is on the northeast (right) side of the road about 9 miles north of this junction.

From there you will have a good view of the bay to the north and the rockslide scar across the road to the southwest. Glacial striations are evident on a granodiorite outcrop to the southeast. To get there, exit the parking area on foot and walk east along the paved sidewalk that follows the road for several hundred feet.

From the parking lot, look across the highway to the southwest at a large, light patch of rocks along the otherwise tree-covered slope. This is the scar of the Emerald Bay landslide, not a glacial feature but a conspicuous one. This slope first slid in the winter of 1953, probably triggered by earlier road construction that undercut and loosened the jointed blocks of granodiorite. Precipitation and snowmelt then lubricated these joints, some of which run nearly parallel to the slope, and released a moderate slide that damaged the highway. During the winter of 1955–1956, a season of higher than normal precipitation, the slope failed again. This slide was much larger than the earlier one and obliterated several hundred yards of the highway. The slope continues to slide a bit almost every year as precipitation and snowmelt lubricate fractures. The prohibitive cost of removing all the potential slide material from above makes this active slide a continual hazard to traffic on the road below.

Walk out to the point and look northeast across the lake to the Carson Range. Note how rounded those peaks appear. Now look to the peaks of the main Sierra Nevada in the southeast. These mountains are much more jagged. The reason for this difference lies in the glacial history of the two ranges. Although glaciers existed on peaks all around the lake, they were much more extensive on the west side. What could be the reason for this?

An easy answer would be that the western range is higher and colder than the more eastern Carson Range, so glaciers were more apt to form on the western peaks. This, however, is not the case. The mountains on both sides of the range reach similar heights, so another reason must account for this discrepancy.

The answer to this puzzle lies in the wind pattern and its effect on precipitation. As moisture-bearing air moves from west to east, it rises over the western crest of the Sierra Nevada. As it rises, the air cools and the moisture condenses, falling to the ground as rain or snow. The air mass continues moving east after having lost most of its moisture. The effect of this is that the annual average precipitation steadily decreases as you move eastward across the basin. This rain-shadow effect causes the Carson Range to receive only about 20 inches of precipitation per year, much less than the Sierra Nevada's 90 or so inches per year. Glaciers, then, could grow only on the shaded sides of the highest peaks in the Carson Range; for instance, on the northeastern slope of Mount Rose.

In today's climate, mountain precipitation falls as rain in the summer and snow in the winter. The snow may pile up in some winters, but it never gets so deep that the warm summer air cannot melt most of it. This was not always the case. In the recent past, geologically speaking, the climate in North America was substantially cooler (vignette 23). So much snow fell during the long winter that it did not all melt during the short, cool summer. A year-round snowpack began to accumulate on the peaks. New layers of snow buried the older snow, compressing it until it eventually recrystallized as ice. As this mass of snow and ice thickened, it began to flow under its own weight, picking up bits of rock and sediment as it moved. These pieces of rock and sediment ground and scraped the surfaces they rode over. In the Sierra Nevada the extensive glaciation scraped away all the loose, weathered rock and left behind fresh, jagged peaks on which weathering could begin all over again. In the Carson Range, where glacial ice accumulated on only a few peaks, chemical and mechanical weathering has proceeded uninterrupted, so the peaks are more rounded.

An aerial view of the southwest shoreline of Lake Tahoe. Emerald Bay, a glacially carved valley, filled with water as the lake level rose following the retreat of alpine glaciers in the Sierra Nevada.

As long as the amount of snow that falls exceeds the amount that melts, a glacier will grow and move forward, or advance. Alpine glaciers, which form on mountains, tend to move downslope following pre-existing valleys. A major glacial advance, the Tahoe stage, occurred between about 160,000 and 130,000 years ago. Another advance, the Tioga stage of about 100,000 to 20,000 years ago, was somewhat less extensive but still left many recognizable features.

A number of times, advancing glaciers blocked the outlet of the lake, and the water rose as much as 600 feet above its present level. The density of ice is nine-tenths that of water, so when the level of the water reached nine-tenths the height of the ice dam, it began to float, releasing catastrophic floods down the Truckee River canyon.

As the glaciers moved down the valleys, they picked up and pushed huge volumes of rock and sediment. Eventually the climate began to warm again, and more snow melted in the summer than fell in the winter. The glaciers began to retreat and left piles of sediment behind. From the observation point, look down into Emerald Bay at the two ridges that project from the shore and nearly cut the bay off from the rest of the lake. These ridges are piles of rocks and sand, called moraines, that a retreating glacier dumped here. This moraine marks the farthest advance of this particular glacier, so we call it a terminal moraine. Originally, these ridges connected, and Emerald Bay was a separate lake, much like Fallen Leaf Lake to the southwest. Erosion eventually breached this moraine, creating Emerald Bay.

Depositional features, such as moraines, tell only part of the glacial history here at Emerald Bay. Advancing glaciers carried some rock and sediment on top of the ice and some within the ice. However, they carried the vast majority of the rock debris along the sides and bottom of the glacier. As this material moved along under the weight of a thousand feet of ice, it ground away at the bedrock it overrode, sculpting the landscape.

A number of factors affect the ability of a glacier to abrade the bedrock. First and foremost, of course, is the amount of debris embedded in the basal layer of ice. Clean ice does not abrade solid rock, while ice with lots of sand and rock fragments abrades very effectively. The sliding velocity of the basal ice also contributes to a glacier's scouring power. Suppose two glaciers contain a similar amount of rock and sand in their basal layers, but one moves 50 feet per year and the other moves 100 feet per year. It stands to reason that the surface under the second glacier is subject to twice as much abrasion as the surface under the first glacier during the same amount of time.

Particle characteristics affect abrasion. Very hard, angular pieces abrade a surface more readily than softer, flat ones.

A glacier scrapes most effectively if it constantly renews its supply of abrasive rock particles. If no renewal takes place, the abrading particles soon become flat and smooth, losing their effectiveness. New fragments come to the fore through a combination of friction, heat from the earth, and the pressure of the overlying ice, which melts the bottom of the glacier, allowing the particles to move down toward the underlying rock surface. This gives the glacier a constant supply of new tools to wear away the bedrock.

All this scraping and grinding tends to leave behind distinctive erosional features that can tell us not only that a glacier once existed here, but also the direction in which the ice flowed. Fannette Island, the small island out in the bay, is one such feature. Note the overall shape of the island. It is a bit longer than it is wide, somewhat rounded, and the rock surface looks relatively smooth, especially on the southwest end. This is a glacially sculpted knob of bedrock known as a roche moutonnée, literally a "sheepback rock." Its long axis, southwest to northeast, parallels the direction of ice flow. A roche moutonnée begins as a small, nondescript hill. As the glacier rides

View of Emerald Bay from the overlook. The two ridges that encircle the bay are remnants of a terminal glacial moraine that erosion later breached. Fannette Island is a roche moutonnée, a bedrock knob that glacial ice streamlined.

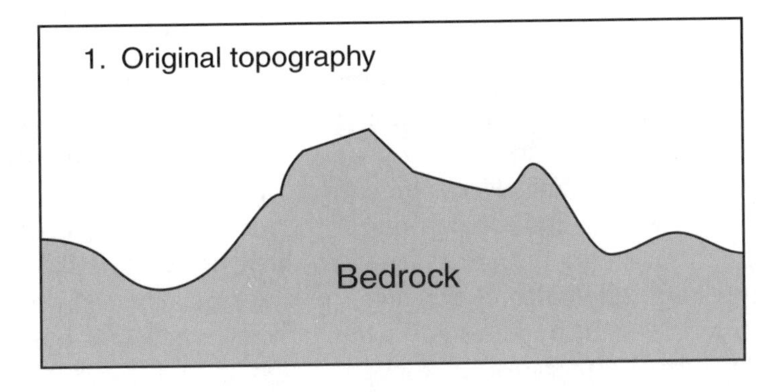
1. Original topography
Bedrock

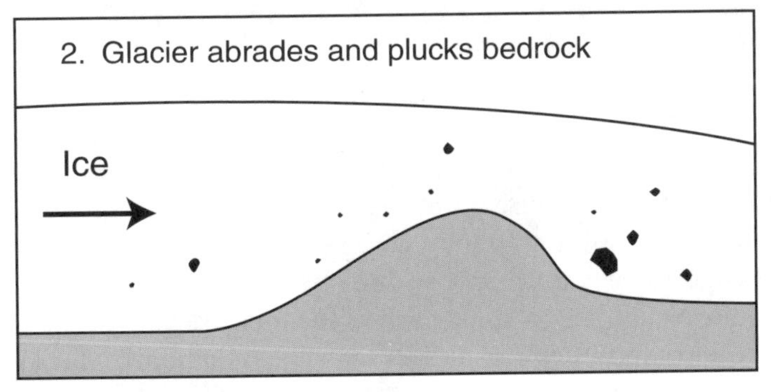
2. Glacier abrades and plucks bedrock
Ice

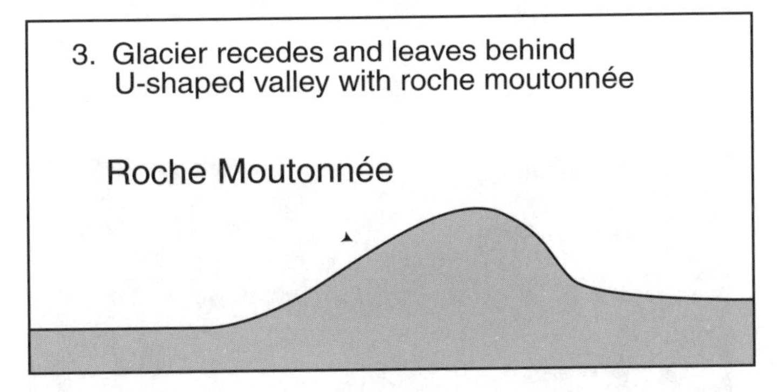
3. Glacier recedes and leaves behind
U-shaped valley with roche moutonnée
Roche Moutonnée

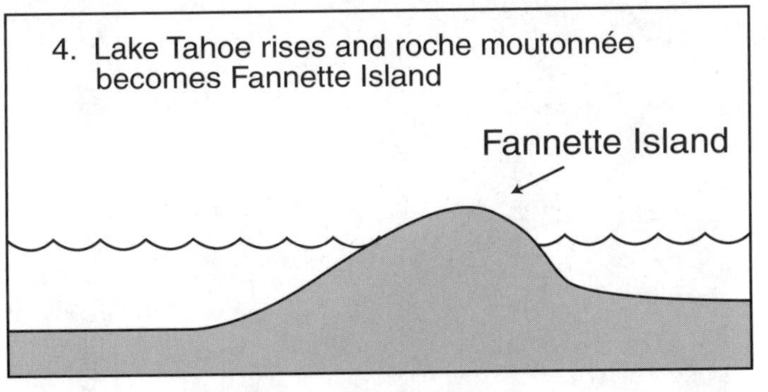
4. Lake Tahoe rises and roche moutonnée
becomes Fannette Island
Fannette Island

Development of a roche moutonnée

over the hill, the ice scrapes the upstream side into a smooth, gentle incline. As the ice passes over the crest of the hill, the pressure it exerts on the rock decreases and instead of abrading and smoothing the rock, it breaks off pieces and carries them away. This so-called plucking gives the downstream side of the roche moutonnée a steep, craggy form. In granitic rocks, such as those at Emerald Bay, joint fractures typically enhance the effects of plucking. The ice works its way into the small fractures and widens them, eventually loosening and breaking off pieces of the rock.

From the overlook you can easily see that Fannette Island takes this general form. Abrasion has left the upstream end, closest to you, smooth, rounded, and gently sloping. The downstream side has the steep, rough shape associated with plucking.

To discuss other erosional features around Emerald Bay, let's move southeast to Eagle Falls. From the southwest corner of the parking area, take the trail that follows along the highway to the falls. Cross the bridge to the south side of the stream, and walk out onto the granodiorite. Note how some parts of the rock are polished to a shine. Feel how smooth they are, almost as if a giant lapidary wheel ground

Glacial striations on a polished granodiorite surface on the east side of Eagle Falls at Emerald Bay.

Glacial striations in a granitic outcrop to the south of the Emerald Bay parking area

them. As the glacier passed over this bedrock, the fine sediment embedded within the ice buffed the surface down to this smooth polish.

If you look carefully at these polished surfaces, you will see sets of parallel scratches within some of them. These scratches, or striations, are further evidence of abrasion where the glacial ice met bedrock. Try to find a couple of different sets of striations. You may notice that they all point in the same general direction. Striations offer key evidence to the direction the ice was moving. Here at Emerald Bay, the scenario seems simple—the glacier originated high on the mountain and followed this valley down to the lake. We don't need the striations to figure this out. But things are not always so straightforward. If you were trying to figure out the direction of flow for a huge continental ice sheet that spread out over a large, flat expanse, you might welcome the clues these small scratches provide. After exploring the area around Eagle Falls to your satisfaction, follow the trail back to the parking lot. In this age of global warming, the ice ages of Pleistocene time seem very remote. We are finding, though, that it doesn't take much to significantly shift the climate cooler or warmer. Is the ice age over, or are we just between ice ages? The debate goes on. Look out over Emerald Bay and try to imagine ice filling it. Imagine huge blocks of ice breaking off the glacier and floating away across the lake. It has only been 10,000 years since such a scene existed here. Who is to say when it may happen again?

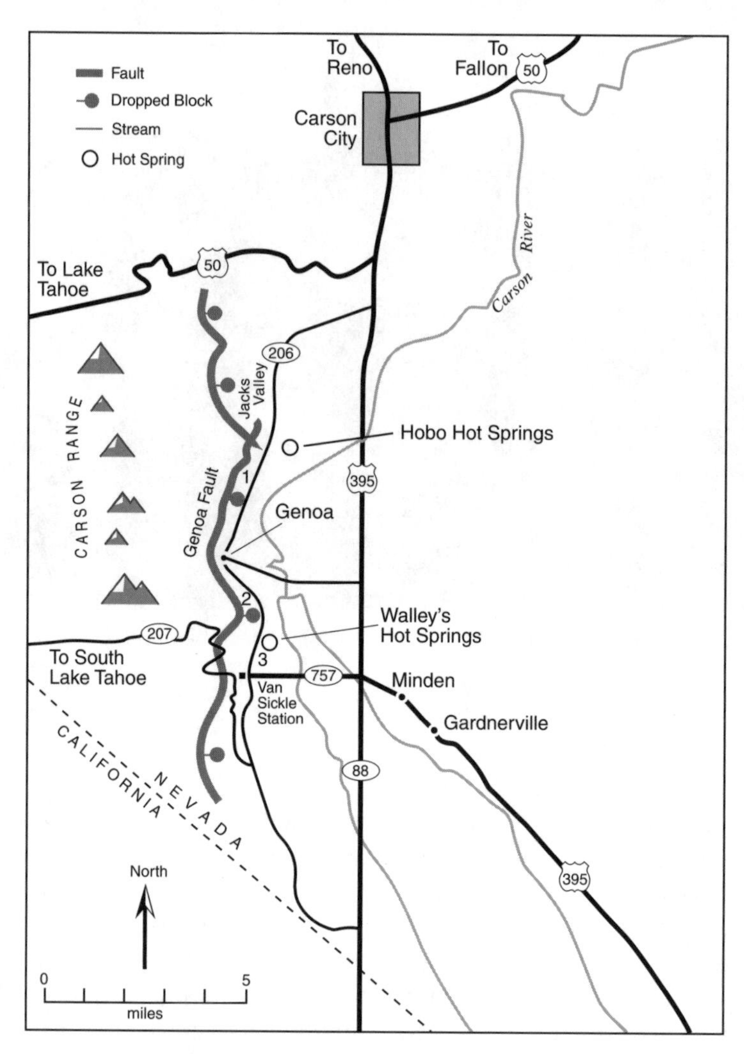

▶ GETTING THERE

Our study of the Genoa fault begins about 5 miles south of Carson City at the intersection of U.S. 395 and Nevada 206. Set your odometer to zero and proceed southwest on Nevada 206 over a small rise and into Jacks Valley. As you proceed south, the mountain range and the road converge, and at mile 5.5 you have a good view of the scarp cutting across alluvium. This is stop 1. The shoulder is wide, and there is a small pullout.

Continue south on Nevada 206 through the town of Genoa to stop 2 at the gravel pit, 1.2 miles south of town and 10.2 miles from the intersection with U.S. 395. Turn right into the gravel pit and park. After thoroughly exploring this magnificent exposure of the fault scarp, you may want to continue south on Nevada 206 to another good exposure on the right side of the road, 0.8 miles south of the gravel pit. From there you may be able to see steam rising from the water of hot springs on the east side of the road. The hot springs percolate through surface fractures associated with this fault system.

4

FINDING FAULT
GENOA FAULT SCARP

Many people think geology exists only as history, so it's always exciting to view processes that are active and observable. The area along the Sierra frontal fault zone affords many opportunities to see geology in action. The Genoa fault scarp is one spectacular example. This fault displays direct evidence that a mountain range is still rising.

The Sierra Nevada has a distinctly asymmetric form—the steep, rocky eastern face sharply contrasts with the relatively gentle western slope. Movement along a system of steep normal faults on the eastern face of the range is responsible for the asymmetry. This zone of faults, known as the Sierra frontal fault zone, forms the boundary between the Basin and Range and the Sierra Nevada geomorphic provinces.

The Genoa fault is the local expression of the frontal fault zone in this part of Nevada, extending from the California-Nevada border north to Jacks Valley, just west of where this vignette begins. The Carson Valley, east of the Genoa fault, is sliding down the fault relative to the Carson Range, a north-trending extension of the Sierra Nevada. Faults along the base of a mountain range, such as the Genoa fault, are common throughout the Basin and Range province.

The Sierra frontal fault system has not formed one long, continuous fault scarp—the break in topography caused by surface faulting. Rather, it consists of many smaller segments. As we begin this vignette at the intersection of U.S. 395 and Nevada 206, we are near the southern terminus of the Carson Range segment of this fault system. Travel west on Nevada 206 for 1.7 miles from the intersection. Here we cross one of this system's faults. It is difficult to pinpoint the fault trace, but movement along this fault has lifted the hill that the road is about to traverse. After climbing this small rise, follow the road as it curves south, and proceed along the east side of Jacks Valley. Look west across the valley at the Carson Range. All the relief you see there

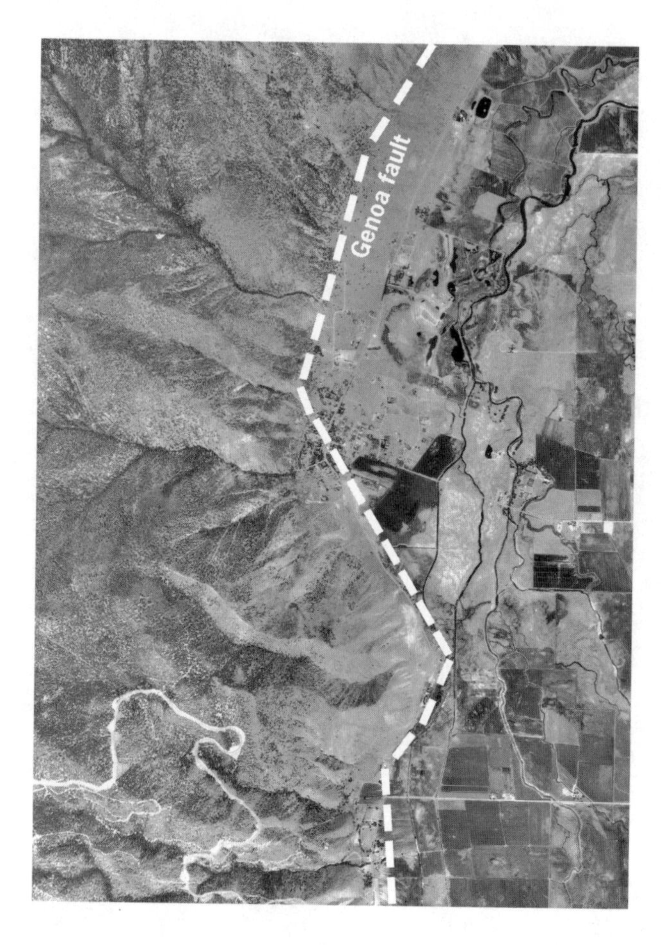

Aerial photograph of the Genoa fault scarp

is the result of movement along the Genoa fault. The base of this range marks the fault trace.

To the southeast, at the faulted base of the hill, the water of Hobo Hot Springs emerges. Springs commonly surface along faults, and if there is an available heat source, such as a shallow magma chamber, this water is often hot. The mechanism is simple. It begins when precipitation seeps into the ground and becomes part of the groundwater reservoir. In tectonically active areas such as this, it is not unusual to find a rather shallow heat source, which then heats the groundwater. If this groundwater comes in contact with a fault, it can travel along it to the surface, where it bubbles forth as a hot spring. This process is so common that a series of springs, especially hot springs, is one of the common criteria geologists use to infer the presence of a fault.

The afternoon sun accentuates the scarp, which appears as a dark line at the base of the range, in this view at the southern end of Jacks Valley (stop 1).

Continue south on Nevada 206 to mile 5.5, our first stop. An excellent view of the fault scarp presents itself on the west side of the road. Note the obvious change in slope marking the position of the scarp as it cuts across the unconsolidated, rocky debris, or alluvium, piled up at the base of the range. You can best see the scarp here in the afternoon or evening, when the low sun angle makes it appear as a dark line at the base of the range.

Sharp, well-preserved fault scarps like this one indicate relatively recent activity. Approximately 40 feet of displacement is displayed in this area. This section of the fault probably moved most recently less than 600 years ago. From here follow the road south into Genoa.

The town of Genoa is Nevada's oldest settlement, located along the California Emigrant Trail. Mormons built a trading station, stockade, and corral at this site in 1850. Emigrant travel soon became heavy, and by 1852 a permanent settlement began to grow around the trading post. Known then as Mormon Station, it was designated the seat of government for Carson County in 1854. In 1855 the name was changed to Genoa. With the discovery and subsequent development of the Comstock mines, northeast of Genoa near Virgina City, the town grew substantially, but as was the case with so many similar towns, when the mines played out, the population moved on. In 1910, half of the business district burned down, including the original fort

The morning sun shines brilliantly on the surface of the Genoa fault in the old gravel pit south of town.

and the courthouse. Genoa still exists. It has a small historic district, a courthouse museum, and a state historic park, as well as a few stores and restaurants.

About 1.2 miles south of town, you can see a spectacular exposure of the fault scarp in a small, abandoned gravel pit. Turn right into the pit, park, and look to the southwest at a sheer rock face approximately 40 feet high. As movement along this fault has raised the Carson Range, the loose, granular alluvium that often covers faults has sloughed off, leaving this stark, imposing fault scarp for all to admire.

Faults are not always as obvious as the one before you. Often, geologists must look a little deeper into the structure of the area before determining if a fault is present. Geologists look for certain characteristics that help them identify a fault. If you are lucky, a fault will display many of these features, but more typically you will find only a few. Let's discuss some of these distinguishing characteristics, and then see if we can identify any of them on the outcrop in front of us.

If the rocks in question retain their original depositional layering, or bedding, then omission or repetition of the beds may indicate faulting. A second clue is the discontinuity of stuctures. In this case, strata may abruptly end against a different type of rock or, in igneous rocks, a dike may suddenly end as if sliced off by a knife. Also, faults often act as avenues for migrating fluids. These solutions can either alter the rock or leave behind deposits of various minerals, such as silica

Contact between alluvium and bedrock along the fault scarp in the Genoa gravel pit

or calcium carbonate. This process, known as mineralization, appears as an encrusted mass along the fault trace and is another clue that a fault is present.

Now focus your attention on the outcrop. This coarsely crystalline rock is igneous, so there are no layers or bedding to help us identify it as a fault. A quick scan of the surface also reveals no significant dikes to be offset or truncated. We also see no evidence of mineralization. So far, we have been unsuccessful in finding definitive evidence of a fault, but let's keep looking.

Walk up to the outcrop and feel it. Note that some areas feel smooth and polished. Upon closer inspection, you'll notice thin scratches, or striations, on these polished areas. Geologists call this type of polished and grooved surface slickensides. As rocks on one side of the fault plane slide past rocks on the other side, they grind together. This grinding action tends to smooth the rock surface along the fault plane, resulting in polished grooves such as the ones you see here. Now we are getting somewhere, but let's look for more evidence.

Often the landscape offers hints that can help us determine if we really are looking at a fault. Of these features, the fault scarp, that telltale break in topography, is the most recognizable. A scarp, an abbreviation of the term *escarpment,* is a steep slope that separates two more gently sloping surfaces. Scarps can form in a number of ways, such as differential erosion (erosional scarps) or wave action (beach scarps). A fault scarp, however, owes its relief directly to movement along a fault. This outcrop in the gravel pit is a wonderful example of a fault scarp. Fault movement created the steep slope—the slickensides tell us this—that breaks the topography between the gentle slope you are standing on and another somewhat gentler slope above it. Fault scarps are not always this easy to see. As they age, they begin to erode, and unconsolidated sediment and vegetation covers them. Hence, it is always a pleasure to find a stark example such as this.

The series of springs aligned along the base of a mountain range also hints that a fault cuts through here. We will see more of this type of evidence down the road.

Now is a good time to discuss the direction of strike and dip of a fault plane. Strike is generally defined as the direction of a hypo-

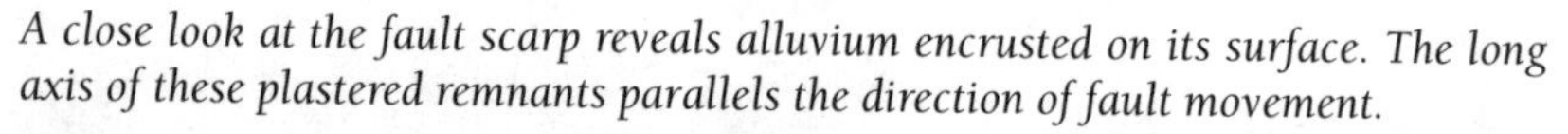

A close look at the fault scarp reveals alluvium encrusted on its surface. The long axis of these plastered remnants parallels the direction of fault movement.

thetical line formed at the intersection of a surface, in this case the fault plane, and a horizontal plane. The Genoa fault generally parallels the Carson Range, which trends north-south, so the strike of the fault is approximately north-south. The dip, measured perpendicular to the strike, describes the angle from horizontal that the fault plane slopes. A simple way to determine the direction of dip is to pour a little water on the surface in question. The water will flow in the direction of dip, in this case to the east.

A quick review of the information we have collected so far tells us that we have a break in slope along the base of a mountain range, with numerous springs, some of them hot, aligned along it, and with slickensides on its surface. This is pretty good evidence that we are looking at a fault. The next step is to test our skills at determining what type of fault it is. First we must determine the direction of movement along the fault. Was it vertical, horizontal, or a little of both? The key to determining this lies with taking a close look at the slickensides, which develop parallel to the direction of movement. In this case, the striations are oriented almost straight up and down. So the latest movement along this fault was vertical. Sharp eyes will be able to uncover more evidence of the direction of movement here. If you look closely you will see a few small areas, mainly on the south end of the outcrop near the bottom, where some shreds of alluvium cling to the fault surface. These remnants are a series of narrow, elongate mounds, up to about 8 inches across, of fine, mudlike material plastered onto the rock wall. The long axis of these encrustations parallels the striations, again telling us that the direction of the latest movement was vertical. This is known as a dip-slip fault because the movement, or slip, is directly up or down the dip of the fault plane.

So, now we have a fault plane that slopes relatively steeply to the east, on which the movement has been vertical. But which fault block has moved down and which has moved up? An interesting way to check the relative movement along a fault like this is to move your fingers up and down its surface parallel to the direction of the striations. You may notice that the surface feels smooth in one direction and slightly rough in the other. The direction in which the surface feels smooth is the direction that the dropped block moved. See if this little test tells you that the block above the fault, the hanging wall, has moved down relative to the block below the fault, the foot wall. A fault in which the hanging wall moves down relative to the foot wall is called a normal fault. Normal faults are the most common type of fault in the Basin and Range.

By studying the cross sections of the nearby river and stream terraces, geologists found that this fault scarp has a vertical displacement of about 34 feet and that it slid in a series of seismic events rather than a single large event. This fault moved most recently less than 1,000 years ago and maybe only a few hundred years ago. Over time, thousands of these seismic events have combined to produce the steep eastern face of the Carson Range in front of you. Total vertical offset along the Genoa fault may be as much as 8,000 feet. All along the eastern front of the Sierra Nevada, from the Mojave Desert to the southern Cascade Range, faults such as this one have worked independently, with the overall effect of raising this range to its present lofty height.

From the gravel pit, drive about one-half mile farther south on Nevada 206 to Walley's Hot Springs. Just past the hot springs, look for another good exposure of the Genoa fault scarp on the west side of the road, and pull into the turnout. Here, too, you can see slickensides, although not on such a grand scale as the last stop. The dark

Slickensides exposed in an outcrop on the west side of Nevada 206, near Walley's Hot Springs (camera lens cap for scale)

View northwest from Nevada 757 at the Carson Range. Cottonwood trees at the base of the range mark areas where springs surface along the Genoa fault trace.

rocks are metamorphosed Mesozoic volcanic rocks. Across the road, to the east, steam rises from the water. The hydrothermal water flows from hot springs that rise along the fault.

Drive another half mile to Van Sickle Station. Sagebrush, rabbitbrush, and other common plants of the high desert cover most of Jacks Valley and nearby Carson Valley. But, here and there along the fault, such as around Van Sickle Station, grow pockets of lush vegetation that include cottonwood trees. The unexpected plants in these lush oases flourish where more springs surface along faults.

To wrap up your tour of the Genoa fault, you might want to now take one of the nearby roads, such as Nevada 757, a mile or so east and then look back to get a much bigger view of the entire picture. From here, the Carson Range towers above you, and you know that a very young and active fault extends all along its base. You can see pockets of lush vegetation flourishing around the springs that surface along the fault's trace. And you can view the steplike pattern that cycles of uplift and erosion have sculpted throughout the history of the Genoa fault.

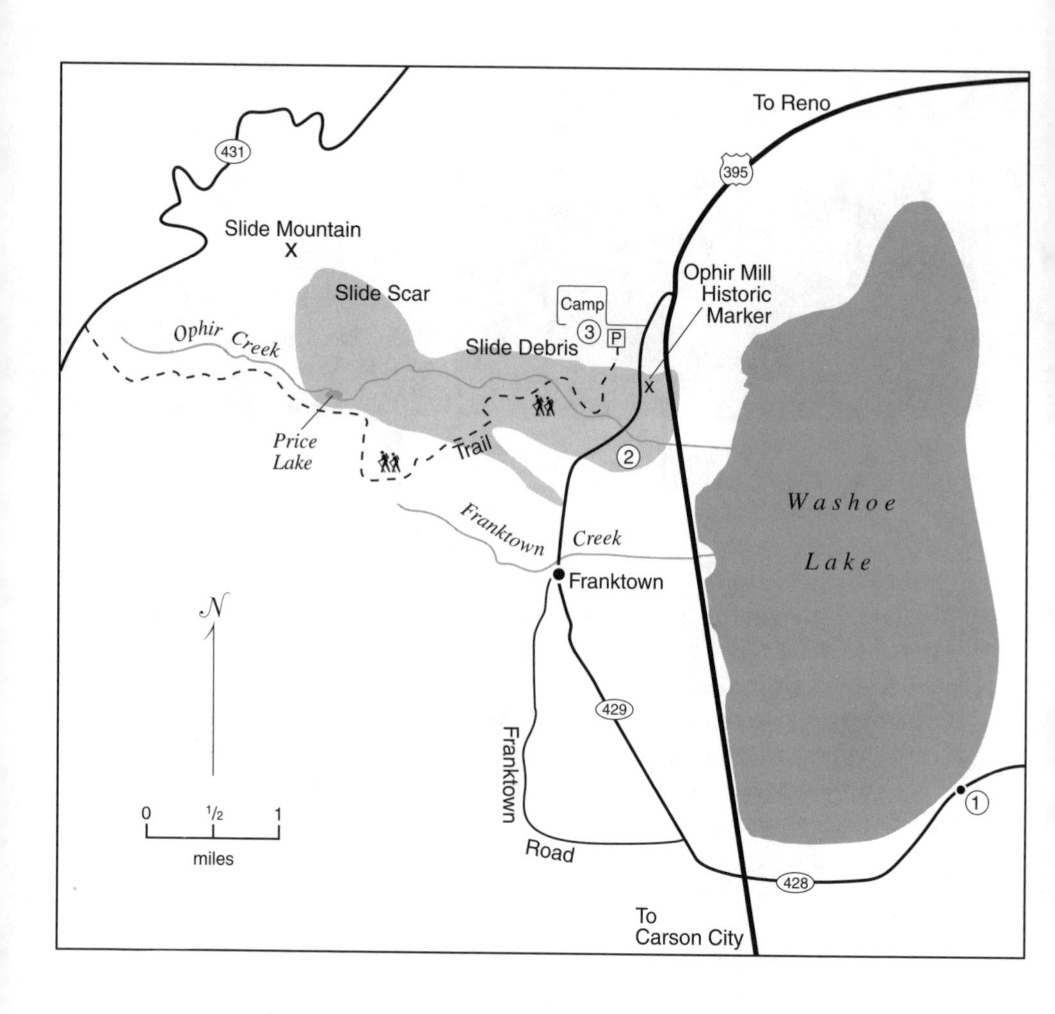

▶GETTING THERE

Our first view of Slide Mountain comes from the eastern shore of Washoe Lake. To reach stop 1, take the Eastlake Blvd. exit (Nevada 428) off U.S. 395. Head north 1.8 miles around the east side of the lake to the South Beach section of Washoe Lake State Park. Turn west and park. To reach stop 2, turn south (right) out of the parking lot and travel back along Nevada 428. Cross under U.S. 395 and proceed north on Nevada 429 for 5.7 miles to the Ophir Creek crossing just north of Bowers Mansion State Park. Park on the shoulder. To reach stop 3, continue north on Nevada 429 another 0.8 mile to Davis Creek State Park. Turn west (left) and follow the entrance road to a parking area on the south (left) at the trailhead.

5

Take a Walk on the Wild Slide

SLIDE MOUNTAIN

On May 30, 1983, a large mass of debris—rocks and mud mixed with water—slid down the steep southeast-facing slope of Slide Mountain. Much of this slurrylike debris flow moved down upper Ophir Creek canyon and into Upper Price Lake. Upon entering the lake, it displaced a great volume of the lake's water, which then flooded downstream into Lower Price Lake. The combined contents of both lakes then rushed down Ophir Creek, passing over and picking up rocks and mud from previous landslides. When the leading edge of this

Rockslide scar on the southeastern flank of Slide Mountain, with debris lobes from multiple slide events in the foreground

debris flow reached the mouth of Ophir Creek canyon, it measured about 30 feet high and was traveling at about 15 miles per hour. It then fanned out, covering a section of Nevada 429 with as much as 9 feet of muck. Silty run-off continued downstream to the edge of U.S. 395. In all, the churning mass of mud and rock destroyed two houses, and it overtook the five people running to escape it, injuring three and killing one. One person escaped relatively unscathed.

This was only the latest in a series of similar events here. Our vantage point at stop 1 on the east side of Washoe Lake affords a good view of the massive slide scar—brilliantly white on sunny days—on the southeastern flank of the mountain. Ophir Creek canyon occupies the area below the scar. Note how the canyon appears to be filled with small hills or mounds. This so-called hummocky topography is common in areas that have been subjected to repeated debris flows. Also note the steplike pattern of later flows deposited on previous ones.

The native occupants of this area had a name for Slide Mountain that translates roughly to "mountain that fell in upon itself." Historical accounts report large slides in late 1852, in April 1862, and in July 1890. The 1890 event apparently resembled the 1983 flow, sending a 30-foot wall of water and debris down Ophir Creek that destroyed

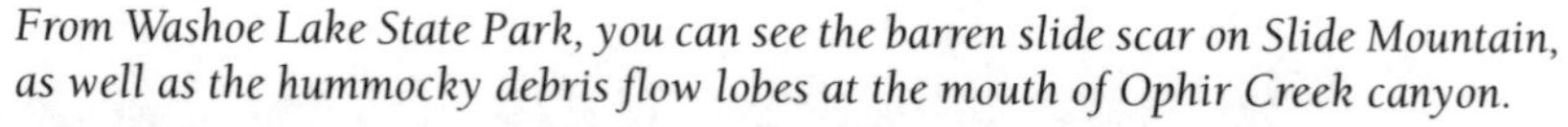

From Washoe Lake State Park, you can see the barren slide scar on Slide Mountain, as well as the hummocky debris flow lobes at the mouth of Ophir Creek canyon.

Debris lobes from multiple slides, viewed from Slide Mountain above the scar

the only house in the area and flooded about 200 feet of the V&T Railroad track. According to eyewitness accounts, a rock slide from the mountain probably triggered the 1890 flow.

What is it about this particular mountain that spawns repeated rock slides? Other high peaks in the area, such as Mount Rose or Marlette Peak, don't recurrently send masses of debris down their flanks. What is unique about the geology of Slide Mountain?

Rocks tend to slide along zones of weakness, so we should first look for reasons why the southeast face of Slide Mountain might be a weakened zone. In sections of stratified rock, a weak layer may lie between two stronger layers—for example, a layer of weak, flaky shale between layers of massive sandstone. If movement along faults tilted these strata sufficiently and water saturated the shale layer, you would have a setup for slope failure. Might this be the case on Slide Mountain?

To check this hypothesis, we must proceed 5.7 miles north on Nevada 429 to the Ophir Creek crossing (stop 2). Carefully pull off onto the wide shoulder and look to the west at the large hill com-

Aerial photograph of the southeastern flank of Slide Mountain, with rockslide debris extending eastward into Washoe Valley

Toe of debris flow lobe at Ophir Creek crossing. Note the unsorted texture characteristic of these flows. To the left is a higher, more forested lobe. The upper slide scar is visible in the background.

posed of boulders and unconsolidated sediment. This is one lobe of the Ophir Creek debris tongue. Cross the road and pick up a few of the cobble-size rock fragments and examine them. Note that they are all of similar composition and texture, and are crystalline in nature. These are igneous rocks, not sedimentary. The relatively coarse-grained texture of this rock tells us that it cooled slowly, probably at a substantial depth. The lesser amount of dark minerals, such as biotite and hornblende, set within the more abundant light minerals, mainly quartz and feldspar, give this rock the characteristic salt-and-pepper look of granitic rock. Look around and you will notice that almost every rock, from pebble to boulder, shares this granitic composition. A quick look at a geologic map will confirm that Slide Mountain, like many of the peaks in the Carson Range, consists almost entirely of the igneous rock granodiorite and not stratified sedimentary rocks such as shale and sandstone. So, a weak layer of sedimentary rock probably is not the mechanism we are searching for.

Granitic rocks typically crack in smooth fractures called joints. The length of such fractures usually ranges from a few feet to hundreds of feet. Generally, when joints first form, the space between the joint faces is minimal. However, over time the processes of weathering can combine with the ever-present effects of gravity to greatly widen these spaces.

Geologists classify joints either geometrically or genetically. A geometrical classification describes how joint surfaces relate to the bedding or other similar structure in the rocks. A genetic classification, on the other hand, is somewhat more significant in that it takes into account the origin of the joints.

Several mechanisms may ultimately cause jointing in rocks. One common type of jointing is that produced by contraction in cooling igneous rocks—columnar jointing in basalt flows is a classic example. Joints also form in response to stresses of folding and faulting, which create the fractures during tectonic activity. Slide Mountain sits within a very active seismic area, and this faulting and uplift may have played a role in fracturing these rocks.

Folding and faulting do not always fully relieve tectonic stresses but leave residual stress within the rocks. At some later time, when conditions allow, these stresses can cause jointing. A specific type of residual stress often plays a major role in the formation of joints in massive granitic rocks. These rocks normally cool and solidify under high pressure at depths of several miles. As erosion removes the overlying rock, the granitic rock gradually nears the earth's surface. This process, known as unloading, decreases the pressure on the granitic rock and allows it to expand and fracture. As we discussed earlier, the relatively large grain size of the granodiorite at Slide Mountain tells us that it was probably formed at depth. This rock is well jointed. Geologists believe that unloading played a role in the formation of these joints.

We've seen that western Nevada is an area of intense faulting. The mountains around you are still rising along major fault zones (vignette 4). The jointing of the granodiorite, whether a result of unloading, stresses associated with faulting, or both, may play a significant role in the repeated slides down this mountain—but how?

Joints are not randomly distributed in rocks, but instead are grouped together in sets that are more or less parallel. Their orientation can greatly affect the strength of a rock body. On Slide Mountain, one of the joint sets dips at an angle nearly parallel to the mountain's southeast slope. This configuration greatly decreases the stability of the slope and sets the stage for a rockslide. Prolonged chemical weathering and frost wedging combine to widen these joints, which exacerbates the slope's instability. Finally, water seeps into the joints and

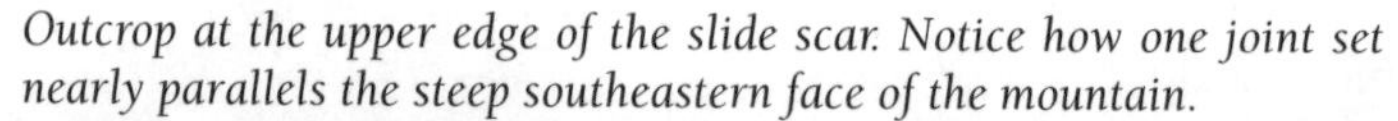

Outcrop at the upper edge of the slide scar. Notice how one joint set nearly parallels the steep southeastern face of the mountain.

A debris flow picks up and incorporates everything in its path—rocks, sediment, water, and even large trees.

lubricates the surface. Years of abnormally high precipitation magnify this effect. This combination of factors by itself may be enough to start a slide—as apparently was the case in May 1983. But, because Slide Mountain sits within a seismically active area, an earthquake could also trigger a rock slide. Gravity does the rest.

The story does not end here. Still at Ophir Creek crossing (stop 2), look up the canyon of Ophir Creek. The jumbled mass of debris clogging the canyon is known as the Ophir Creek debris tongue. Numerous debris flows of at least two different types contributed to these piles of rock and mud. In the first type, the rockslide enters the canyon of Ophir Creek, mixes with the water from the creek as well as that from the lakes, and becomes a swiftly flowing mass of rock, mud, and vegetation. Fluid flows such as this contain much material coarser than sand and are called wet debris flows. Wet debris flows can travel at speeds up to 15 miles per hour and carry with them everything in their path. Once the flow reaches the mouth of the canyon, it slows considerably, drops the coarser material, and continues down the slope as a broad sheet of finer material. The event of 1983 was a wet debris flow.

Very large rockfalls can cause a different type of flow, known as rockfall avalanches. These slides can flow long distances at fantastic

Large, angular blocks of granodiorite in slide debris. Note the absence of fine material between the boulders.

rates of speed, supported by trapped air and huge quantities of crushed rock and dust. Large quantities of water need not be present. Debris flows that form from these large rockfall avalanches commonly travel in excess of 60 miles per hour. The large lobes of debris in the drainage of Ophir Creek most likely were this type of flow. In some places, though not visible from our vantage point, these debris lobes flowed uphill. This would have required the high rate of speed typical of rockfall avalanches. The large size of the lobes also points to this type of flow.

At stop 2, Nevada 429 crosses the edge of the coarse debris from one of these rockfall avalanche deposits. A look to the west, up the canyon, gives you a good view of a drainage clogged with the debris of numerous flows. Note the great difference in size of the material in the deposit, which is one characteristic of a debris flow. A look to the east, down near U.S. 395, shows the area where the flow splayed out as it left the canyon, spreading a layer of mud and debris across the landscape. To the east, you can also see some houses precariously placed, considering the history of this drainage.

By studying the degree of weathering within a debris flow as well as the overlapping and cross-cutting relationships between the various debris flows, geologists have counted at least nine major slides that have occurred in the last 50,000 years. The large debris lobe to the south (left) of Ophir Creek is one of the younger ones. This deposit is typical of debris flows, consisting of angular blocks of fresh granodiorite in poorly sorted silty sand. As catastrophic as the event of 1983 may have seemed to those trying to outrun it, that wet debris flow, confined only to the immediate drainage of Ophir Creek, was insignificant when compared to the flows that deposited large lobes of debris like this one.

Nevada 429 crosses the toe of a debris flow, here viewed from the south.

Angular, slightly weathered granodiorite cobbles in the young debris flow where Nevada 429 crosses Ophir Creek

Proceed north (0.4 mile) on Nevada 429 to the historical marker designating the Ophir Mill site. To the west, you can see the steep face of another lobe of this debris tongue. The relatively unweathered granodirite blocks tell us that this lobe is of a similar age to the one south of Ophir Creek. About 0.1 mile farther north, the road cuts through another lobe with a slightly different character. Here the larger granodiorite blocks are more extensively weathered, and most of the smaller pieces are completely decomposed. The soil is dark gray to orange. The degree of weathering here indicates that this debris flow is much older than the others we have seen.

If you want to get out and walk over some of the features we've discussed, continue north on Nevada 429 about 0.4 mile to Davis Creek County Park. Turn left into the park, and then make another left into the parking area for the Ophir Creek trailhead. The first mile

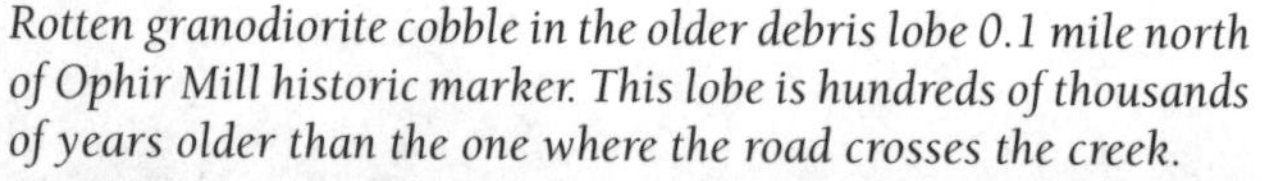

Rotten granodiorite cobble in the older debris lobe 0.1 mile north of Ophir Mill historic marker. This lobe is hundreds of thousands of years older than the one where the road crosses the creek.

Ophir Creek has incised into the slide debris, creating the canyon seen here from the hiking trail.

of this trail has a relatively gentle grade that most hikers should be able to negotiate. This part of the trail traverses some granitic bedrock and some debris flows. As you walk, see if you can tell the difference between where the trail is on the weathered, rounded blocks of bedrock and where it crosses the angular blocks and hummocky terrain of debris lobes.

After walking almost exactly 1.0 mile, you will come to an overlook with a view into the Ophir Creek gorge. This is a good place to stop. Look across the gorge to the south at the steplike pattern where younger debris flows have overridden older debris lobes. If you look down into the gorge, you will see that the lower stretch is extremely steep, a result of the debris flow of 1983 cutting into the canyon wall and carrying the material away.

From here, the trail becomes quite steep and treacherous, suitable for only the hardiest of hikers. If you wish to go farther, you will see more debris lobes as well as the slide scars and boulder fields produced by rockfall avalanches. The views are nice, but maybe you would rather relax, peer into the gorge, and reflect upon the history of this very active area. Then, you may get a slightly uneasy feeling as your thoughts turn to the future and the high probability of catastrophic debris flows triggered by rockslides from Slide Mountain.

View of the slide scar from above Upper Price Lake

The hummocky nature of the Slide Mountain debris flow deposits stands out in the view from the east side of U.S. 395. The slide scar is visible in the background.

6

Fire Down Below

STEAMBOAT HOT SPRINGS

Geothermal areas have always drawn both geologists and non-geologists alike—after all, who doesn't like hot springs? In western Nevada, hot water and steam surface in a number of places, the most spectacular of which is Steamboat Hot Springs, about 10 miles south of Reno. Steamboat's geothermal potential has attracted attention for well over one hundred years. Beginning in the 1860s, entrepreneurs developed the springs as spas and resorts to attend to the needs of the miners on the nearby Comstock Lode. In the 1950s and 1960s, a few energy companies drilled some exploratory wells in an attempt to locate a steam reservoir to generate electricity. These early efforts were unsuccessful, but later drilling did reach an adequate reservoir. In 1979, Phillips Petroleum Company achieved the first such success. From U.S. 395 just south of the intersection with Nevada 431, you can see a power-generating plant that taps this geothermal potential.

The Steamboat Springs geothermal area lies at the northeastern reach of the Steamboat Hills. By looking at the bedrock around Steamboat Hot Springs and the adjacent Steamboat Hills, we can

Geothermal power plant as seen from U.S. 395

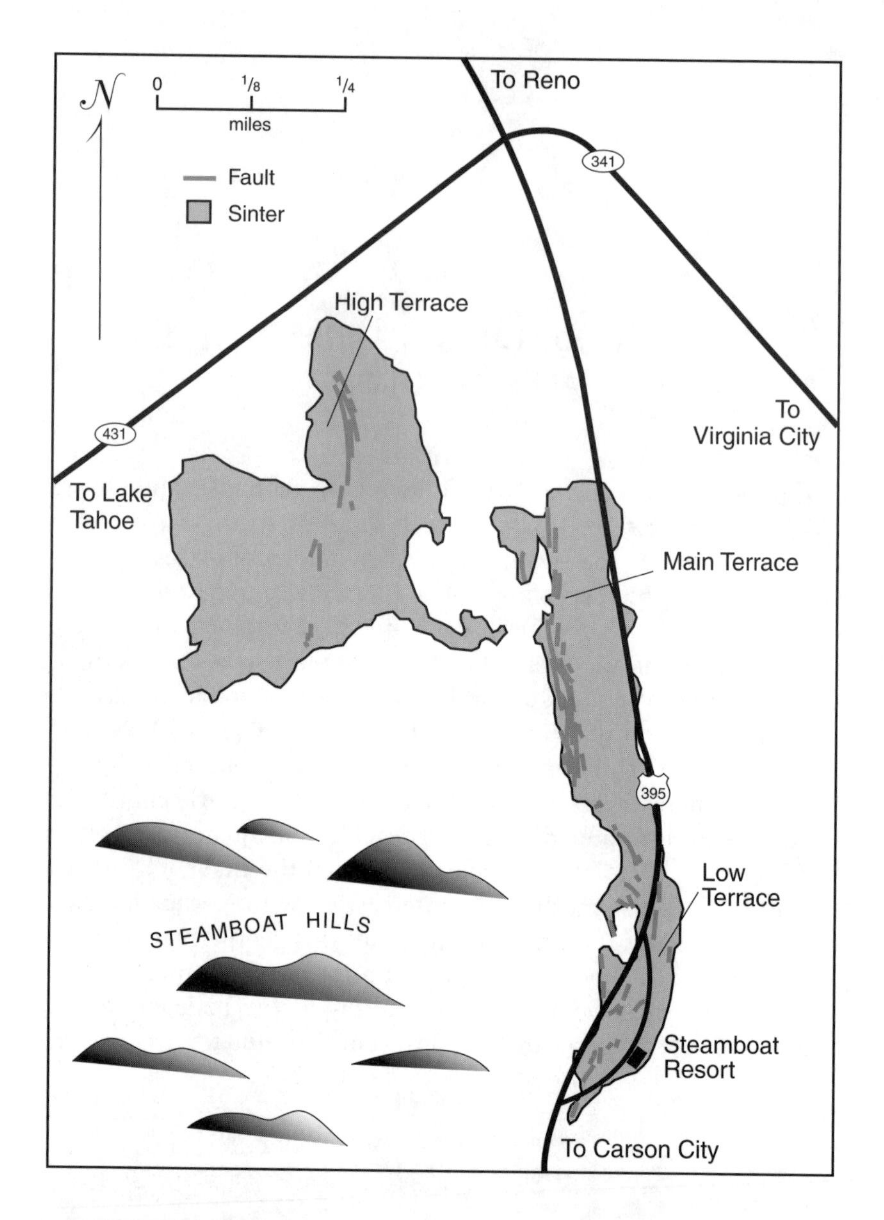

▶GETTING THERE

You can easily access the Steamboat Resort area, about 10 miles south of Reno, from U.S. 395. The road to the resort is on the east side of U.S. 395, about 1.2 miles south of the intersection with Nevada 431. Turn onto this small access road and proceed south about 200 feet, then pull off to the right. Here you can explore some hydrothermal features. Several fissures break the surface on the west side of the road, and some of them are active. To the west lie the Steamboat Hills, from which plumes of steam rise on cool mornings. The flat area to the northwest is the Main Terrace. Just to the south is a large artesian well that supplies water to the resort.

reconstruct a brief geologic history of this area and possibly pinpoint a source for the heat that drives this hydrothermal system.

The oldest rocks that crop out in the Steamboat Hills are metamorphic rocks more than 70 million years old and 65- to 70-million-year-old plutonic rocks that may be related to the Sierra Nevada batholith. Altered and weathered granodiorite from this last group is one of the most common rocks in the thermal region. Beginning in late Oligocene time, about 24 million years ago, and continuing through early Pleistocene time, about 1.1 million years ago, a series of lava flows of varying composition erupted in the Steamboat area. The Miocene Alta and Kate Peak formations, 12.5 to 20 million years old, were two of the most extensive of these flows, covering much of this area of

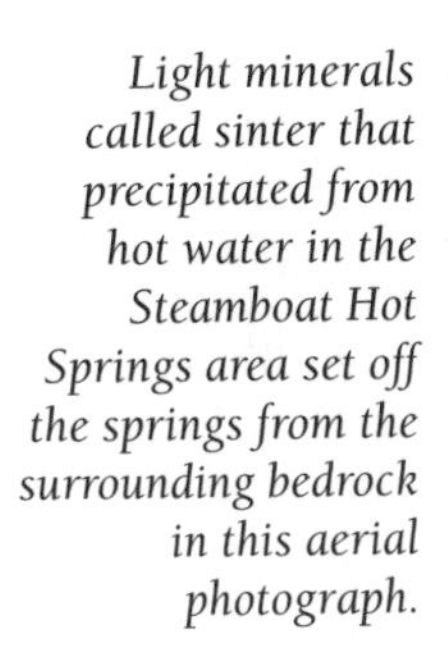
Light minerals called sinter that precipitated from hot water in the Steamboat Hot Springs area set off the springs from the surrounding bedrock in this aerial photograph.

After a thorough exploration of this area, proceed south to the main resort building, where you can stop in for a quick history lesson and some fine hospitality. The Steamboat Villa Hot Springs Spa, as it is called, offers hot mineral baths and a geothermal steam room, as well as various massage therapies.

The roads to the Main Terrace, the High Terrace, and Pine Basin are off-limits to the public. You can view a small grove of Jeffrey pines, as well as other aspects of the Steamboat Hills area, looking south from Nevada 431.

A small grove of Jeffrey pine stands in the foreground while steam rises in the background in this view from Nevada 431 southwest toward Steamboat Hills.

western Nevada and some of eastern California. The Steamboat Hills are part of a northeast-trending line of four rhyolite domes that erupted between 3 and 1 million years ago, toward the end of this episode of volcanism. One of these domes lies about 3 miles southwest of the thermal area; the others are to the northeast. The magma chamber that extruded these flows provides the heat for the Steamboat hydrothermal system.

While hydrothermal activity generally centers in regions with volcanic histories, not all volcanic areas develop hydrothermal systems. It takes more than a heat source. Many other areas of relatively recent volcanic activity, such as Lunar Crater (vignette 16) and Sunset Crater in Arizona, lack hydrothermal systems. What is different here at Steamboat Springs?

To get a closer look at the hydrothermal features, follow the directions in "Getting There" to the area north of the Steamboat Resort, and park. Get out and walk around the area west of the road. You will notice a couple of things right away. One is the strange, relatively flat lunar-style landscape almost devoid of vegetation that surrounds you. Then there are the cracks or fractures—known as fissures—in this surface.

Look into one of the fissures and inspect the white mineral deposits encrusting the walls. This material is called siliceous sinter—hard silica deposits left behind as hot spring waters evaporate. Travertine, a form of calcium carbonate, is another common mineral deposit found around many hot springs, but at Steamboat Springs, siliceous sinter predominates. The young sinter found in active areas such as this is porous and is composed mostly of opal. As later deposits bury it, heat, pressure, and time squeeze the sinter, fill the pores, and transform the opal into another type of quartz, known as chalcedony.

Note the porous texture of this opaline sinter from the Low Terrace.

Young, flat accumulations of sinter are called hot spring terraces and are common features of active hot springs. The area around the steamboat resort, including where you are standing, is known as the Low Terrace. If you look to the northwest across U.S. 395, you will see another flat area of the same pale sinter deposits. That is the Main Terrace. Over time, erosion and tectonic processes modify the sinter terraces. Faults fracture and raise them, and streams dissect them, until they bear little or no resemblance to their original form.

In the past, activity at Steamboat was considerably greater and included pools of standing hot water and occasional geysers. The Low Terrace and the Main Terrace, north of the resort, are still hydrothermally active. During the Ice Age, when groundwater levels were higher, the High Terrace was also active. Although still considered thermally active, the High Terrace has probably not discharged water on the surface within the past 30,000 years. Although none of the fissures on the Low Terrace presently contain standing water, many do release steam.

Walk around the Low Terrace and locate a fissure, preferably an active one emitting steam. Be careful, because the footing around these cracks can be a little treacherous. The fissures vary somewhat in size—try to find one with at least a few inches of separation so you can observe its structure.

Cracks in the earth's surface can form in a number of ways. They can be shrinkage fractures that open as a lava flow cools. They can simply be joints that erosion has widened. They can also have a

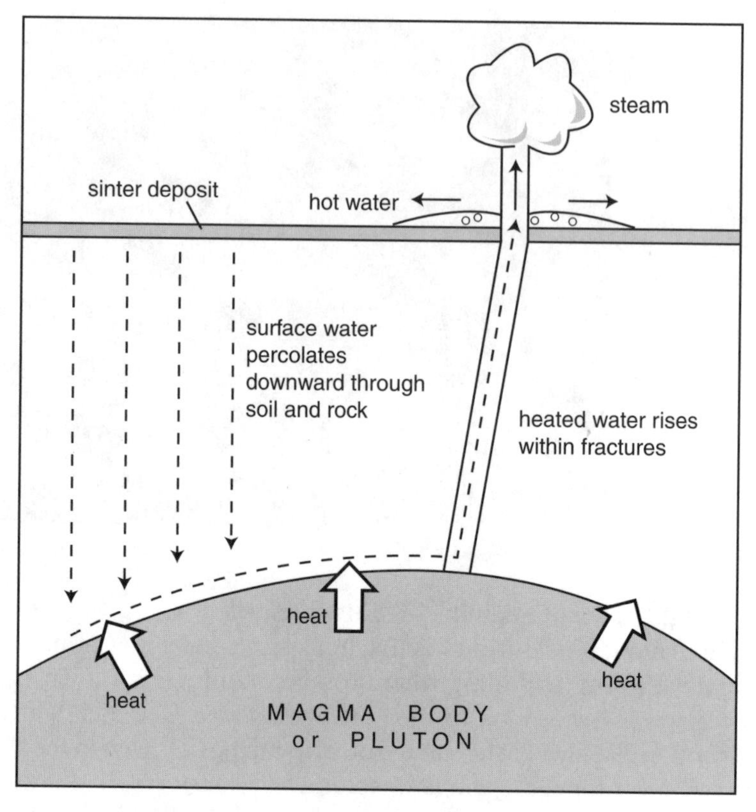

Infiltrating groundwater and a subterranean heat source interact and send steam to the surface.

tectonic origin, such as the surface expression of a fault or cracks forced open by a rising magma body deforming the surface above it.

The fact that the fractures on the Low Terrace extend deep enough to tap into a powerful subsurface heat source—large enough to heat groundwater to boiling and then send it to the surface—probably rules out an origin as shrinkage fractures or widened joints. This leaves a tectonic origin as the most likely scenario.

Geologists recognize three distinct fault systems in the Steamboat area. One trends northeast, parallel to the axis of the hills; one trends northwest; and a third trends approximately north. This latter system is most prominent in the thermal area and shows evidence of being most active recently. This fault system probably provides the structural control for the whole Steamboat hydrothermal system. Surface water seeps down along these faults, reaches the heat source, and subsequently rises back to the surface as hot water and steam.

But how do these surface fractures relate to the fault system? Are they simply a surface expression of the faults, or are other processes involved? Knowing many faults slice through this area, we might initially think that these cracks result from the rocks splitting apart in response to fault movement. Let's look for evidence.

Return to the fissure you picked out earlier. If you look closely, you will notice that some areas are open while sinter clogs other areas. Studies done over time have documented abrupt closure and reopening of some of the fissures here at Steamboat. This is not behavior you would normally expect if these fissures simply responded to movement along a fault. What could explain this behavior?

As the heated water rises, the pressure on it decreases and the water releases gases. One of these gases is hydrogen sulfide. Near the earth's surface hydrogen sulfide reacts with atmospheric oxygen to form sulfuric acid. This acid attacks the sinter as well as the bedrock, disintegrating rock adjacent to the fissures. So, although the fracture system of the sinter terraces is of tectonic origin, sulfuric acid's corrosive properties keep the channels open near the surface. You can see this

A long fissure in the Lower Terrace on the west side of the resort road. Note the corroded edges of the fissure and the sinter rubble that clogs it in the foreground.

process in individual fissures. Note that some parts of the fissures are open, allowing vapors to escape, while sinter rubble clogs other parts.

Much of the sinter of the Steamboat Springs area contains detectable amounts of gold, silver, mercury, arsenic, and antimony. Relatively high concentrations of gold and silver have been detected in the siliceous mud deposited in the springs. Across the street on the Main Terrace, which is closed to the public, the sinter deposits that surround some of the springs are pink from the presence of cinnabar. Other sinter on the Main Terrace is reddish orange, indicating the presence of metastibnite, an ore of antimony. Magmatic water, which accounts for less than 10 percent of the total water in the Steamboat hydrothermal system, rises from deep within the earth and deposits these minerals at or near the surface. This system closely resembles many hydrothermal ores deposited during Tertiary time in the Great Basin, such as the Comstock Lode. The terraces at Steamboat Springs likely are a modern analog to those deposits.

Previously, we discussed how hydrogen sulfide released by the hot, rising water reacts with atmospheric oxygen to form sulfuric acid. That reaction takes place near the surface. As the acidic water seeps

Steam escapes through an open fissure on the Low Terrace northwest of the Steamboat Villa Hot Springs Spa. Note how sinter rubble clogs the fissure in the foreground. A plume of steam rises from the Steamboat Hills.

back into the ground, it chemically leaches the bedrock, replacing the original silicate minerals of the igneous rocks with opal. This process, called acid-sulfate alteration, has altered much of the bedrock around the hydrothermal area and also affects the type of vegetation growing nearby. In an area known as the Pine Basin, a stand of Jeffrey pine grows at a much lower elevation than their lowest limit in the nearby Carson Range. Another, smaller stand of pines is observable from Nevada 431 west of U.S. 395. Sagebrush and other vegetation that normally inhabit this life zone cannot tolerate the acidic soil in the acid-sulfate alteration zone. The pines, however, can tolerate it and consequently have no competition for the limited moisture

Acid-tolerant Jeffrey pines grow on lighter-colored, heavily altered rock to the left. Darker, less altered rock to the right supports piñon pine and sagebrush.

View west at the Main Terrace of Steamboat Springs. A plume of steam rises from the Steamboat Hills in the left background; the Carson Range appears in the background to the right.

available. Rocks in parts of the Virginia Range to the east and the Carson Range to the west have also undergone acid-sulfate alteration, although oxidation of the iron-sulfide mineral pyrite is the culprit. In those places, stands of Jeffrey pines also grow at lower elevations than normal.

How long has Steamboat Springs been hydrothermally active? From the relationships between the sinter deposits, alluvial deposits, and volcanic rocks for which age dates are available, we can draw a general picture of this area's hydrothermal history. At Sinter Hill, some chalcedonic sinter rubble lies below the base of a basaltic andesite flow 2.5 million years old. This tells us that the hot springs have been active for more than 2.5 million years, and probably close to 3 million years. This is a long time for a hydrothermal system to remain continuously active. Such longevity would require a substantial magma body to supply the heat. More likely, the thermal activity has been intermittent throughout this time span. A lack of specific ages for some of the rock units leaves inconclusive the exact relationship between active and inactive episodes. One thing is clear, though: the Steamboat Springs area has one of the longest and most complex histories of any geothermal system in the world.

If you have time, wrap up your exploration of Steamboat Springs with a dip in the hot water at the resort. And while you're soaking, contemplate the geologic events that make this simple pleasure possible.

Steamboat Villa Spa and Resort on the Low Terrace

7

Of Lava and Lahars

MOUNT ROSE

From the large parking area at the summit of Nevada 431 you have a nice view of Mount Rose to the northwest. Standing at a height of 10,776 feet above sea level, it is one of the highest peaks in the Carson Range, a north-trending extension of the Sierra Nevada. The Sierra Nevada is a batholith, a huge chunk of igneous rock that cooled and crystallized deep underground more than 80 million years ago. Faults later lifted the batholith and erosion removed the overlying rock, exposing the batholith at the surface. Mt. Rose is unique within this mountain range, and not only because it looks different than the

Brown andesite produced during violent volcanic eruptions caps Mt. Rose.

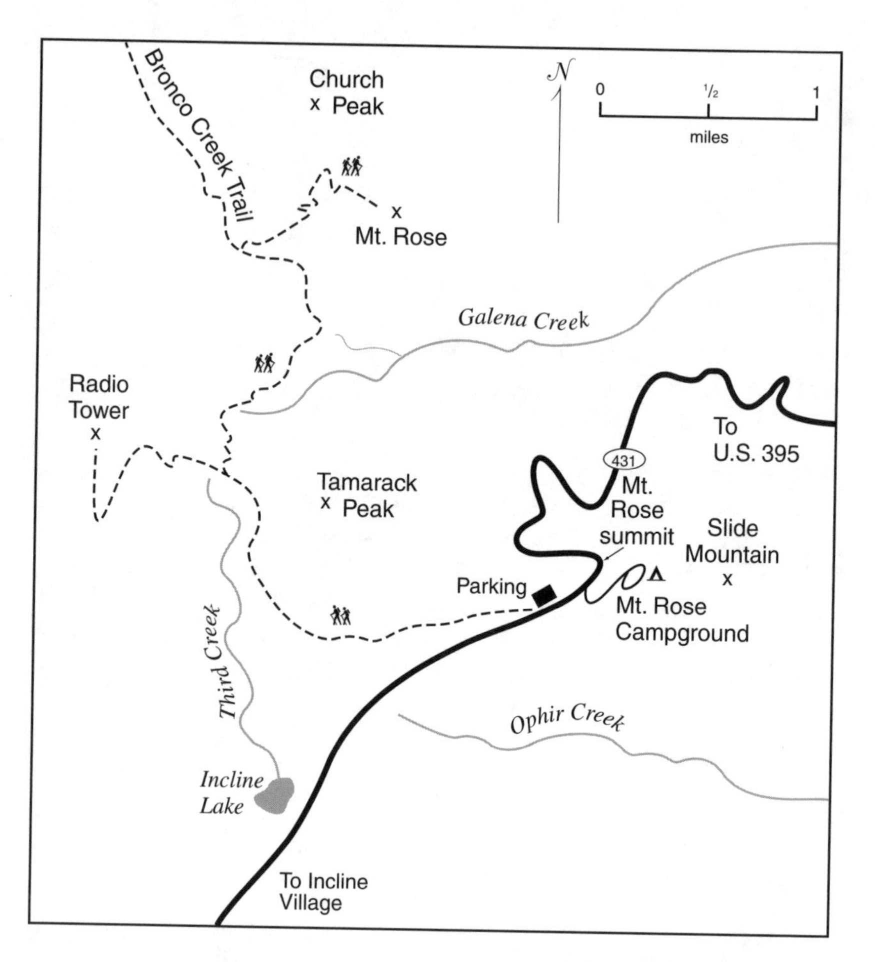

GETTING THERE

To reach Mount Rose from the Reno area, drive south on U.S. 395 about 8 miles to the junction with Nevada 431 and turn west (right). For the first few miles, the road climbs the Mt. Rose fan and glacial outwash complex. The outwash gets progressively younger as you proceed west. Soon the route begins to wind up the eastern face of the Carson Range, passing through outcrops of granodiorite, andesite, and mudflow. As you climb higher, take in the good views of the Pleasant and Washoe Valleys and the Virginia Range to the east, and Mt. Rose to the northwest. You will reach Mt. Rose summit 16.5 miles from the junction with U.S. 395. Park on the shoulder near the small building. This is a popular hike on summer weekends, when many vehicles may park along this section of highway, so we advise an early start. The trail begins at the dirt road on your right, 0.2 miles west of the summit.

To reach Mt. Rose from the Lake Tahoe area, take Nevada 431 east out of Incline Village 7.8 miles to Mt. Rose summit and the trailhead, which begins at the dirt road on your left near the small building.

You may wish to bring binoculars and a magnifying glass or hand lens.

surrounding terrain. It is a remnant of a later period of violent volcanism, and its flanks bear witness to a variety of processes that once laid this landscape to waste.

Walk over to the outcrop of light gray rock on the west side of the parking area and examine it closely. Check out its coarse-grained crystalline texture. The rock contains abundant light minerals, mainly clear quartz and white to pink feldspar, as well as lesser amounts of dark minerals, such as platy biotite and rodlike hornblende. In igneous rocks, crystal size tells us how rapidly the rock cooled. It takes time for dispersed atoms in magma to fit themselves into complex crystal lattices. Large crystals tell of slow cooling deep beneath the earth's surface that allowed plenty of time for atoms to migrate. As rock cools rapidly on the earth's surface, however, only small crystals grow; the rock cools before dispersed atoms can organize themselves into large crystal lattices. The rock here clearly cooled and crystallized slowly. Look on the ground around the outcrop, and you'll notice that this rock decomposes to a coarse-grained sand consisting mainly of BB-size particles of quartz and feldspar. This rock should look familiar to you if you have traveled throughout the Sierra Nevada. You see it in Yosemite, Kings Canyon, and Sequoia National Parks and in the Lake Tahoe basin. The rock exposed here in the parking area, like that found virtually everywhere within the Sierra Nevada, is granodiorite. Slide Mountain to the east consists entirely of this rock.

Now shift your attention to the northwest, and take a look at Mt. Rose. Binoculars help but are not absolutely necessary. Even at this distance, you can see that the mountain is significantly darker than the granodiorite here in the parking area. The rocks that constitute Mt. Rose appear to be something other than the typical granitic rock that makes up most of this range. Let's proceed to the Mt. Rose trailhead, 0.2 mile west on Nevada 431. We'll walk partway up this trail to see if we can discover why Mt. Rose looks so different from its neighbors.

After finding suitable parking along the shoulder of the road, walk to the trailhead near the small concrete building and examine the rocks exposed there. Here, as in the parking area at the summit, we find granodiorite. Find some large rounded boulders near the trail register. If you look at the ground beneath them, you will see the ubiquitous coarse-grained sand that granodiorite leaves behind as it weathers. Note that the trail itself consists almost exclusively of this sand.

Rounded granodiorite boulders at the Mt. Rose trailhead

Let's start walking up the trail. This section of the trail doubles as a service road to a radio relay station on Relay Ridge to the west, and it is wide and well graded. As you begin hiking, look upslope to the right where more rounded knobs of weathered granodiorite lie scattered among the trees—no hints of anything different so far.

After you have walked about 0.5 mile, the pebbles and cobbles that scatter the slopes around the trail change markedly. They are darker than the granodiorite that we left behind. Pick up a piece of this dark rock and examine it closely. You see large, light crystals embedded in a darker fine-grained material. Also note that the soil itself has changed to a slightly darker brown. Sage and other smaller shrubs dominate the vegetation here, whereas conifers grew at the trailhead. The change in soil and the consequent change in vegetation hint that the subsurface bedrock has changed. Let's walk a little farther and see if we can figure out what is going on. As you stroll this section of the trail, take in the spectacular view of Lake Tahoe to the south. This lake occupies a fault-bounded area, or graben, dropped between two raised blocks of granodiorite—the Sierra Nevada on the west and the Carson Range on the east. You can easily see this relationship from your high perch on the south flank of Mt. Rose.

A 2-foot-wide boulder of andesite in the lahar deposit about 0.8 mile from the trailhead

About 0.8 mile into the hike, you reach a large outcrop on your right. The western end of this outcrop is rugged and blocky. Stop here and study it. The first feature you may notice are the large blocks of rock sticking out of the face of the outcrop. Upon closer inspection, you will see that these broken pieces of rock range in size from pebbles to boulders more than 3 feet across. Most of the blocks have sharp corners and edges, not rounded ones like the boulders of granodiorite we saw at the trailhead. Now turn your attention to the fine-grained material that holds these blocks together like mortar holds bricks in a wall. Rocks like this that consists of large, angular blocks held together by a fine-grained matrix are called breccias (say BRETCH-yuhz).

Pick out one of the nicely exposed angular rock fragments and look at its texture—a magnifying glass or hand lens will come in handy here. The rock contains many large, light crystals embedded in a mass of very small, dark crystals. The angular chunks here seem to have characteristics of both intrusive and extrusive igneous rocks. The fine-grained rock, or groundmass, indicates rapid cooling, whereas the larger crystals must have cooled more slowly. In this case, the magma source for these rocks began cooling slowly at depth, which allowed the early growth of large plagioclase feldspar crystals. Then, before any other minerals could crystallize, the magma erupted onto the surface, where it cooled rapidly into a groundmass of very small crystals. The composition of this volcanic rock, with lots of plagioclase feldspar, lesser amounts of pyroxene and hornblende, and very little quartz, makes it an andesite.

Next, examine the matrix between the angular blocks. Some of it contains fine particles of mud, but if you look closely you will see that much of it, too, appears to consist of larger crystals in a fine-grained groundmass. Both the larger blocks and much of the matrix are andesite. Apparently a mixture of mud and molten andesite transported the solid blocks.

Volcanic eruptions take on many different styles. Some produce rivers of red-hot lava. Others send out slowly advancing piles of crusty lava that burn everything in their path. Still other, more explosive, eruptions shoot huge clouds of ash and debris into the air, blanketing the surrounding area with thick lava flows and hot ash deposits (vignette 15). The chemical composition of the magma and the amount of gas and steam present affect the type of eruption. Andesitic magmas tend to be thick and flow slowly. They also tend to contain a lot

The fine-grained rock in the foreground is tuff. The coarse, blocky rock overlying the tuff is the remnant of a lahar mudflow. This outcrop lies on the Mt. Rose trail about 1 mile from the trailhead.

of dissolved gas. These conditions promote explosive eruptions. One such violent eruption produced the rugged breccia here on the south flank of Mt. Rose. From 22 million to 2 million years ago, western Nevada was the scene of much volcanic activity. Explosive eruptions poured sheets of lava over the landscape and sent billowing clouds of ash and steam into the air.

Geologists have invented many terms to describe rocks such as the ones before you. The umbrella term *mudflow* describes a layer of jumbled rock in which a fine-grained, muddy matrix supports blocks of many sizes. Because we know that a volcanic eruption produced this mudflow, we can more specifically name it. A mudflow strictly of volcanic origin is called a lahar, an Indonesian term.

From the static, blocky look of this outcrop, it seems unlikely that the mess ever flowed quickly and fluidly—but looks can be deceiving. Lahars, like all mudflows, can race downslope. The speed varies with the steepness of the slope and the proportion of solid material to fluids. Lahars can travel from 10 miles per hour to more than 50 miles per hour. A cohesive slurry of fine ash particles, mud, and water fills the spaces between the larger rock fragments, buoying them up. In these flows, it is not really the water that carries the particles

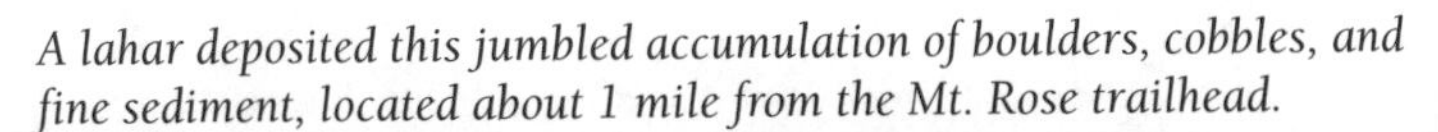

A lahar deposited this jumbled accumulation of boulders, cobbles, and fine sediment, located about 1 mile from the Mt. Rose trailhead.

along, but the particles that carry the water along. The water just lubricates the whole mess, allowing it to flow. It doesn't take a whole lot of water to get one of these flows moving, and once one starts you'd do best to keep your distance. Even a relatively small eruption can trigger devastating lahars. A single lahar on the flanks of Nevado del Ruiz volcano in Colombia killed 25,000 people in 1984.

A lahar can be either hot or cold depending on its origin. In a hot lahar, volcanic ash and lava mix with any water present on the mountain, releasing a high-speed flow that picks up additional debris on its way down the mountain. The water may come from snowpack, an alpine lake, or just the groundwater present at the time of eruption. In a cold lahar, torrential rains or excessive snowmelt drench unstable slopes of ash and send this gray slurry rushing downslope. It is not always easy to tell if prehistoric lahars such as this one were hot or cold. The fact that portions of the matrix as well as the blocks are andesitic indicates that this lahar flowed during a volcanic eruption. It was most likely a hot lahar.

In this case, an explosive eruption spewed out lava and ash, which moved out of the vent and down the side of the volcano, incorporating water and debris as it moved. Some lava cooled and solidified, but the flow tore off and swallowed blocks of the lava as the hotter

Deeply weathered andesite flows along the Mt. Rose trail

Exposure of glacial till along the trail

material behind flowed over it. These broken pieces are the large blocks of andesite you see protruding from the face of the outcrop. Eventually, the whole steaming mass came to rest in much the same form as we see it today.

Resume walking up the trail, and in about 0.1 mile you will come to an outcrop of an andesite lava flow. Solid rock here looks different from the breccias we saw earlier. Note how the light plagioclase feldspar crystals are all nearly the same size. This is a common characteristic of the andesite flows on Mount Rose.

Soon the trail bends to the northwest, and we lose the view of Lake Tahoe. About this time, the forest thickens noticeably as we enter a more protected drainage area. The draw offers some protection from the wind, and the trees provide welcome relief from the baking sun. Snowpack may pile a little deeper and last a little longer—an obvious observation if you take this hike any earlier than late June—thus providing more moisture here. When added together, these factors can change the vegetation of an area, like the change we see here from sagebrush to trees. But there is another important control as well. The thick soil here is part of a deposit left behind as the glacier that filled this valley receded about 140,000 years ago. It contains abundant silt-size particles that hold moisture better than the thin mantle of sandy soil we saw earlier. The soil's ability to hold water contributes greatly to the change in plant life.

We end our walk about 2.5 miles from the trailhead, where the Mt. Rose trail splits off from the service road to Relay Ridge. From this vantage point, you have another nice view of Mt. Rose and a greater appreciation of why this mountain stands out from the surrounding peaks. Look at the rocks strewn about the surface here at the trail junction. They are almost exclusively blocks of andesite similar to the lava flow we examined earlier. A glacier transported these blocks to this spot from higher on Mt. Rose. The rocks give us a good indication that the same type of volcanic rock that we studied along the trail makes up the mountain's higher slopes. This andesite gives the mountain its dark appearance, in contrast to the granodiorite that is so common in this range.

If you are a hardy hiker, you may want to continue up the trail. But be warned, it is 3.5 steep miles to the peak. Along the trail you will see more lahar deposits. Some of the lahars contain very large angular blocks that give you a good feel for the awesome power of these flows. A set of switchbacks across blocks of andesitic talus lead to the andesite lava flow that makes up the peak itself.

Andesite outcrop near the summit of Mt. Rose

The view from the top is a full 360-degree panorama that includes Lake Tahoe to the southwest, the Basin and Range to the east, and Reno to the north. To the south, you can see the high peaks of the Carson Range, including Slide Mountain, Jobs Peak, and Jobs Sister. Here on top of the world, you can consider the violent events that built this mountain while enjoying both a wonderful view and a cooling breeze.

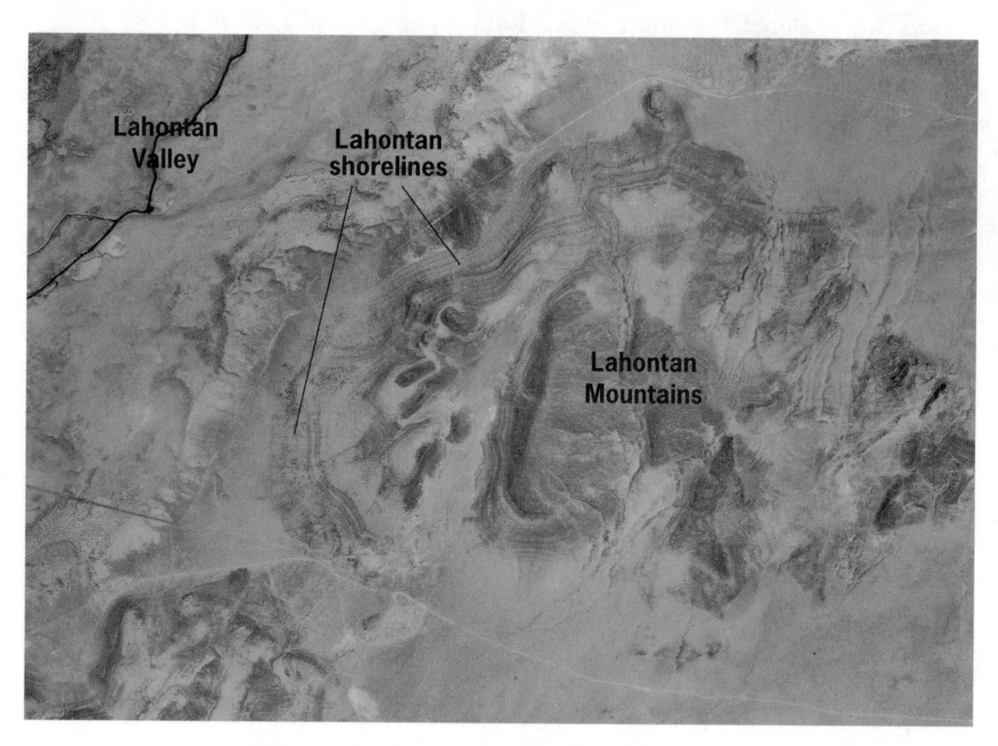

Wave-cut shorelines on the west side of the Lahontan Mountains stand out in this aerial view.

8

Nevada's Greatest Lake

LAKE LAHONTAN

Walker Lake seems strangely out of place. The blue water sparkles in the sun as gentle breezes send small waves toward shore. Anybody who has driven through Nevada understands that water is a scarce commodity indeed, but Nevada's arid desert valleys have not always been so dry. At times during the Quaternary period, roughly the last 2 million years of earth history, many of these dry valleys filled with water in the form of great interconnected lakes. Even the driest desert in North America, California's Death Valley, once held a deep lake that scientists call Lake Manley. Nevada had its share of such lakes as well, the largest of which was Lake Lahontan. Desert lakes like these are called pluvial lakes, bodies of water that are strongly sensitive to climate change. Walker Lake is one of the few freshwater remnants of pluvial Lake Lahontan. True to its nature, Lahontan expanded and shrank as climate changed, achieving its greatest size as continental and mountain glaciers swelled during the Ice Age, and finally shrinking to disconnected remnant lakes and dusty flats under today's hotter, drier climate.

Here in the Walker Lake Recreation Area, we are standing in one of the seven drainage basins that filled with water and joined to form Lake Lahontan 15,000 years ago. All but the Honey Lake basin, which lies mostly in California, sit entirely within the state of Nevada. Of the five bodies of water that exist today, two are saline—Black Rock Playa and the Humboldt-Carson Sink. Pyramid and Walker Lakes contain fresh water, as does Lake Tahoe. While the first four fit our mold of a desert lake, Lake Tahoe does not, because a river—the Truckee—originates from it. As the Truckee River leaves the lake it flows eastward through Reno before turning north and entering Pyramid Lake, which is part of the much larger Lake Lahontan basin. Six rivers terminate in the Lahontan subbasins. Four of these—the Truckee, Carson, Walker, and Humboldt Rivers—contribute 96 percent of the total surface water influx.

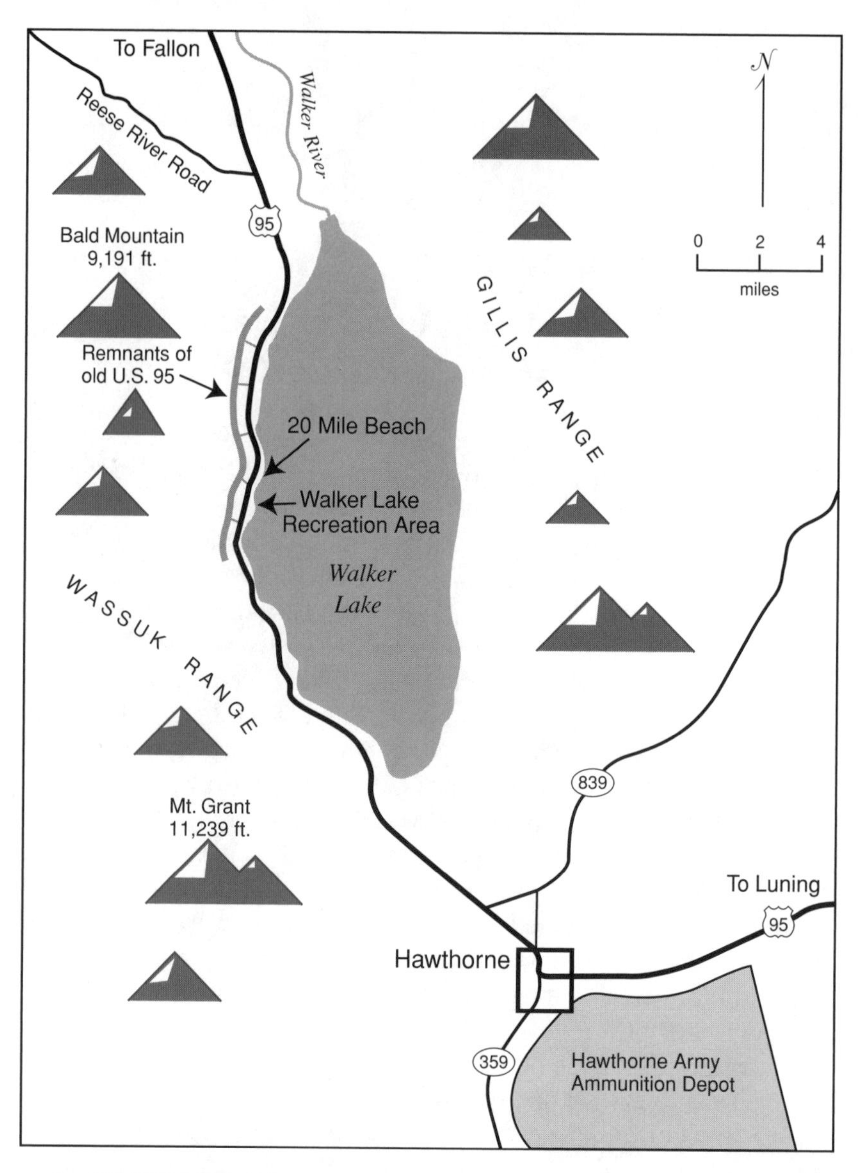

▶ GETTING THERE

We are going to visit two locations, about 70 miles apart, while we discuss Lake Lahontan. First we will look at wave-cut shorelines at Walker Lake, one of the few freshwater lakes in Nevada. We will then discuss playas at Grimes Point, which overlooks the Carson Sink.

The Walker Lake Recreation Area, a well-marked site on the west shore of Walker Lake, lies 60 miles south of Fallon and 12.5 miles north of Hawthorne on U.S. 95. We will move from there to 20 Mile Beach, about 1 mile north of the Walker Lake Recreation Area. Our last stop at Walker Lake lies along old U.S. 95, a piece of paved road 0.25 mile west of the

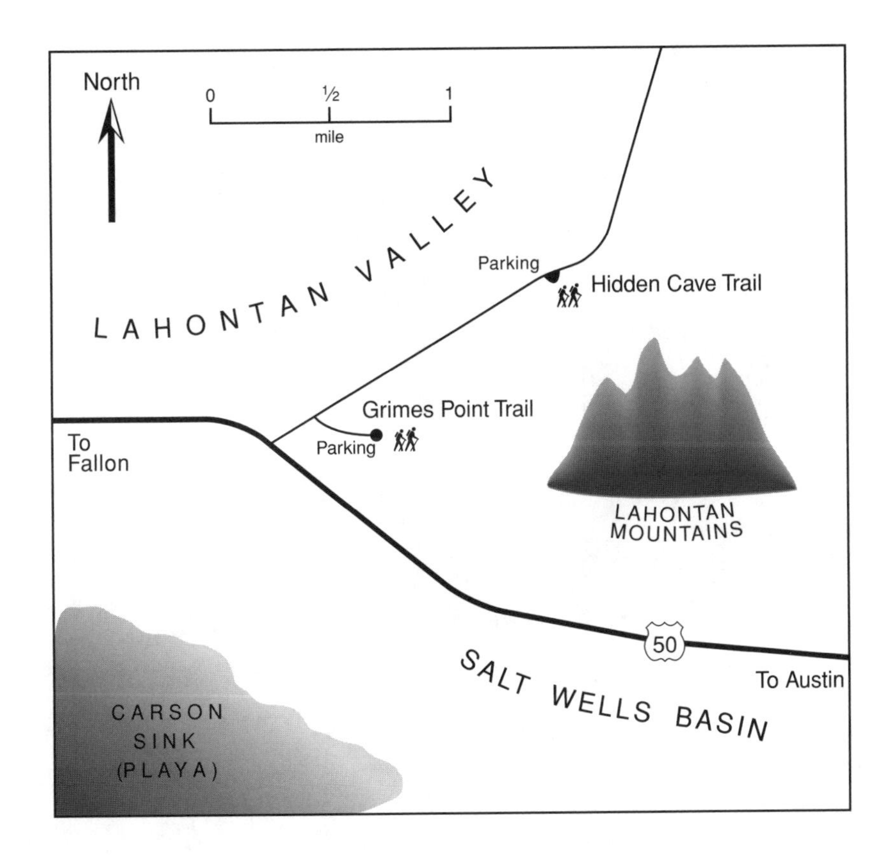

new road. To reach it, drive north of 20 Mile Beach and take one of the many small dirt roads that lead west from the new road to the old one.

Our second stop, Grimes Point, lies east of Fallon. From Walker Lake, drive 60 miles north on U.S. 95 to Fallon, then 11 miles east on U.S. 50. The turnoff is well marked, with a paved parking lot just north of the highway. On U.S. 50, this site is 72 miles east of Carson City and 100 miles west of Austin. From Reno, drive east on Interstate 80 to alternate U.S. 50, which connects to U.S. 50, for a total distance of 68 miles. Adjacent to the parking lot is a small picnic area with restroom facilities. North and east are the Lahontan Mountains, and southwest is the Carson Sink, a marshy remnant of pluvial Lake Lahontan.

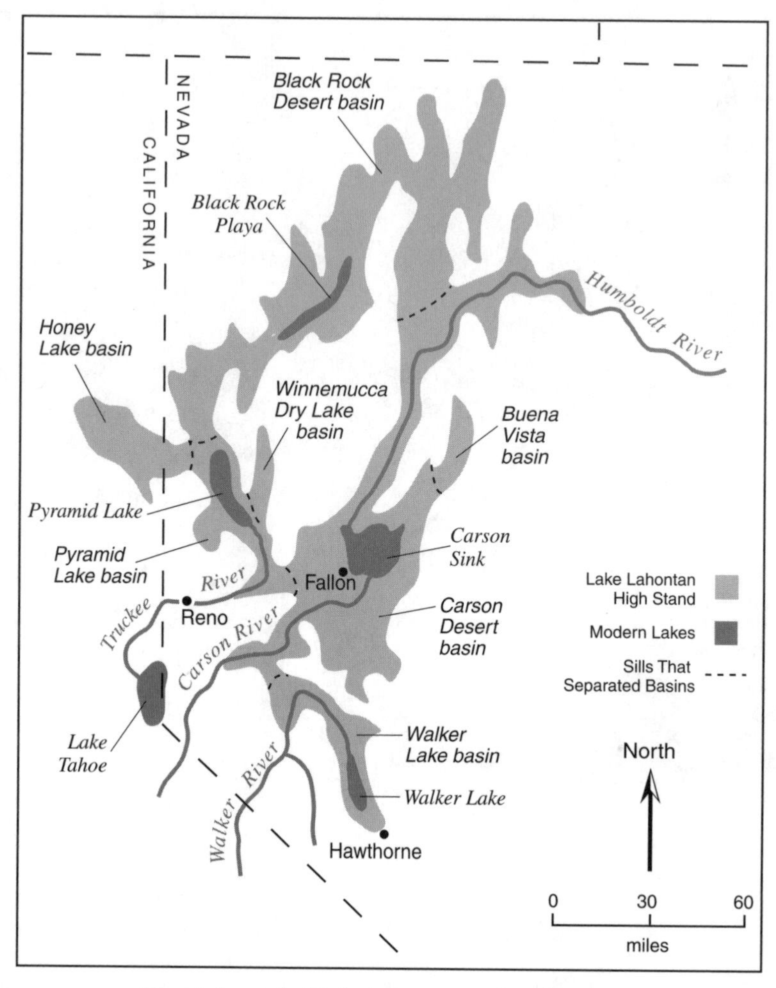

The high stand of Lake Lahontan and modern lakes within the Ice Age lake's subbasins

Many clues point to the existence of a large lake, or series of lakes, in these basins. If you look north from the Walker Lake Recreation Area, you can see a number of small steps—old shorelines—between the water and U.S. 95. A large body of water will experience proportionally large waves—essentially surface features transmitting energy that winds impart to the water. At the shoreline, waves erode bedrock and sediment, eventually carving flat surfaces called wave-cut terraces. Terraces represent times of relative stability when the lake level remained constant for an extended period, and they provide impor-

Waves from Lake Lahontan carved this low terrace in the middle ground on the west side of the Walker Lake basin.

tant evidence about the spatial extent of ancient lake systems. The longer a lake's water level remains stable, the flatter and broader its terraces will be.

Beach bars often develop near terraces. They are ridges of sand and gravel on the lake's floor just beyond the zone of breaking waves. Nearshore currents deposit the sand and gravel that make bars. Most of these ridges run parallel or roughly parallel to the shoreline, but they can also form perpendicular to it. Bars, while not pinning down a shoreline, do indicate that the shoreline was nearby. The terraces you can see here are rather small, remnants of lakes that filled only the Walker Lake subbasin.

Drive north about 1 mile to 20 Mile Beach, and pull off the road into the parking lot. West of U.S. 95, you can see the Wassuk Range, and a much larger terrace inscribed along its base. Waves cut this shoreline when the Walker Lake basin had joined with other basins as Lake Lahontan. It is broader and stands at a much higher elevation than the low terraces near the present lake surface.

Now drive north and take one of the many unnamed dirt roads that lead 0.25 mile west of U.S. 95 to a paved remnant of the older road. There are interesting ridges of material just east of the road. Look closely—they consist of rounded gravel cemented together by a lumpy, coral-like material. Breaking waves smoothed and sorted the gravel you see here; this was a Lahontan shoreline. The cement is tufa.

Tufa is a type of limestone, or calcium carbonate, that precipitates when calcium-rich water seeps into an alkaline lake. Calcium carbonate, unlike many other water-soluble materials such as salt and sugar, is more soluble in cold water than in hot. Cold mountain rain,

Vegetation clearly delineates ancient shorelines in the recreation areas on the west side of Walker Lake.

Sharen Orndorff studies tufa-cemented beach gravel in this ancient Walker Lake shoreline.

snowmelt, and ice melt percolated into soil and rock, eventually flowing into one of the many pluvial lakes. The lakes themselves, standing at lower elevations, would have been much warmer than the water entering them as springs. This rapid shift in temperature—as well as the presence of algae, which decreases solubility of calcium carbonate—caused calcium carbonate to precipitate as tufa. Tall towers, low mounds, or isolated pods of tufa lie scattered throughout many pluvial lake basins in Nevada. These accumulations stand in desert valleys today as remnants of former lakes.

Fossil remains of once-living organisms also help geologists determine the extent of pluvial lakes. As you look closely at the beach gravels here, you may pick out some rounded shell fragments as well. Breaking waves fracture and abrade shells rapidly, so there is little chance of finding well-preserved specimens in wave-cut terraces. Fossils of freshwater gastropods and pelecypods—both bivalves, animals with two-part shells—are present in Nevada's desert valleys, telling us that lakes once filled these basins. They tell us something else, too. The calcium carbonate that makes up their shells contains carbon-14. So, radiocarbon dating of these shells can provide ages for the fossils, and therefore for the lakes. Other materials, including wood, tufa, bone, and charcoal, also incorporate carbon-14, and so can yield radiocarbon dates.

Based on these and other clues, we can reconstruct both the maximum size of Lake Lahontan and its history of growth and decline over the last 50,000 years. Determining the maximum size of the lake is relatively easy; just identify the highest-elevation terrace, then find all the land surface that lies below this elevation within the connected subbasins. At its greatest extent, Lake Lahontan covered 8,570 square miles, more than the 7,540-square-mile surface area of Lake Ontario. With a maximum depth of 900 feet in Pyramid basin, Lake Lahontan was deeper than all the Great Lakes except Lake Superior, whose maximum depth is 1,333 feet.

Tufa entirely encrusts this cobble.

Numerous small terraces above the present water level and below the road mark the west side of the Walker Lake shoreline. These terraces and several others above the road are shorelines that waves cut during higher lake stands.

Determining a history of the lake level's rises and falls is more problematic. Key information for our analysis comes from both land and aquatic environments, but these clues are only helpful if they can supply temporal data as well. This requirement restricts us to materials that can provide radiocarbon dates. For example, suppose we excavate a trench in central Nevada and discover a tree trunk with roots embedded within a fossil soil at an elevation of 5,000 feet above sea level. Carbon-14 dating of the wood tells us that this tree is 10,000 years old. We know that soils develop and trees grow on land rather than under water. So, we deduce that 10,000 years ago the lake surface elevation was below this location. Less than a mile from our tree trunk, we dig another trench and discover fossil gastropod shells within clay at an elevation of 4,950 feet above sea level. Radiocarbon analysis of the gastropod shells gives an age of 10,000 years before present, making them contemporaneous with the excavated tree. We conclude that 10,000 years ago the lake surface covered this spot. Our combined data allows us to place the lake surface elevation between 4,950 and 5,000 feet above sea level 10,000 years ago. An accurate depiction of lake growth and decline therefore depends on the quality of the field data. Researchers base the reconstruction of Lake Lahontan over the last 50,000 years on large quantities of widespread and well-dated terrestrial and aquatic material from Nevada and California.

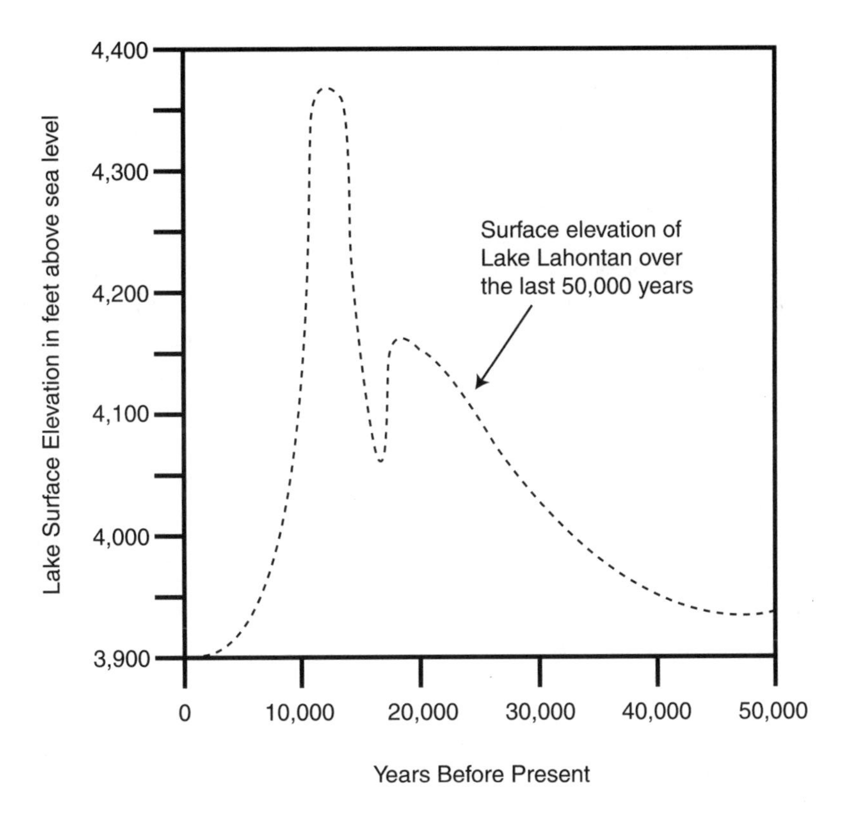

Variations in Lake Lahontan's surface elevation over the past 50,000 years

Not all the Lahontan subbasins respond to changing climate in the same way. Small and relatively narrow subbasins, such as the Walker Lake basin, produce a lake that is very sensitive to small changes in climate. Slight decreases in temperature or increases in precipitation noticeably deepen lakes in such basins. Increases in temperature or decreases in precipitation, on the other hand, rapidly lower lake levels. These basins therefore tend to record frequent subtle climate changes. Large and broad basins, such as the Pyramid Lake subbasin, produce lakes that are less sensitive to changing climate. Surface elevations of these lakes vary only with large changes in temperature and precipitation.

The earliest evidence of a lake in the Lahontan basin dates to approximately 45,000 years ago. At that time a lake with a surface elevation of 3,940 feet above sea level filled the interconnected western basins—the Black Rock Desert, Pyramid Lake, Winnemucca Dry Lake, and Honey Lake basins. A sill, the lowest point on a divide separating one basin from another, marks the overflow point where one full basin spills into its neighbor. As lakes grow in adjoining basins they can join into one connected lake that submerges the sill formerly separating them. This is what happened as the lake in the western basins grew and eventually united. Between 45,000 and 20,000 years ago, lake level rose to 4,150 feet above sea level, the elevation of the sill between the western subbasins and the Carson Desert. Continental ice sheets developed in Canada and the northern United States at this same time. Between 20,000 and 16,500 years ago, the lake level in the western basins remained constant as water spilled into the Carson Desert. The surface elevation decreased to 4,070 feet above sea level in the western basins by 16,000 years ago, probably because the climate warmed. From 15,000 to 13,500 years ago, the lake level grew rapidly, inundating the sills connecting all subbasins of Lake Lahontan. By 14,000 years ago, Lake Lahontan was a single body of fresh water. The lake surface reached its maximum elevation of 4,360 feet above sea level 13,500 years ago and remained at that level for the next 1,000 years. About 12,500 years ago, the lake surface fell rapidly to its modern elevation of about 3,900 feet above sea level.

The Walker Lake subbasin has experienced several puzzling dry periods that seem unrelated to climate change. Walker River supplies water to Walker Lake, much like the Truckee River supplies water to Pyramid Lake. Both rivers originate in the Sierra Nevada; thus, both rivers derive water from the same climate zone. Modern discharges for the Truckee and Walker Rivers generally correlate, meaning that changes in one river's flow mirror changes in the other's. Pyramid Lake has never completely dried up during the last 45,000 years, but Walker Lake has completely desiccated during three extended periods: from 40,000 to 15,000 years ago, from 12,500 to 4,700 years ago, and from 2,800 to 2,000 years ago. With the similarities that exist between the two lakes' modern water supplies, we would expect that if one lake were to dry up, the other would, too. Scientists think these dry periods relate not to the amount of water in the Walker River but instead to the river's path. The Walker River must have been periodically diverted, possibly by fault movement or sediment deposition, into the Carson Desert, depleting Walker Lake of its primary source of water. As the lake in the Carson Desert grew, Walker Lake shrank and finally dried up completely. Walker Lake returned only when the river returned to its former—and present—bed.

Why did the size of Lake Lahontan change so radically from its maximum extent about 13,500 years ago to the present-day leftovers in scattered subbasins? The answer, scientists think, lies in global windfield patterns. The jet stream is a zone of high-velocity wind that carries moist air from the Pacific Ocean into the northwestern United States and Canada. This pattern of wind motion strongly influences climate in coastal Washington and Oregon, making it quite different from that in the southwestern United States. During periods of continental glaciation, when ice sheets covered all of Canada and parts of the northern United States, the topographic influence of the very thick ice sheet split the jet stream. The temperature of the ice sheet also affected wind motion, establishing a strong region of high pressure—called an anticyclone—over the glaciated zone. This anticyclone drove part of the jet stream north and the other part south, a deflection of about 3 degrees of latitude during the maximum extent of Lake Lahontan. This detour brought the jet stream and its moisture-laden air south into the previously—and presently—arid Great Basin. This is a very good example of feedback in a natural system. Climate change caused continental glaciation, which in turn changed the path of the jet stream, thus further altering climate within the Great Basin.

Let's move north now to Grimes Point, about 70 miles from Walker Lake. As you drive through the adjacent valleys, notice the many wave-cut shorelines at the valley margins. In this part of Nevada, virtually every valley carries marks of Lake Lahontan. Pull into the Grimes Point parking lot, stretch, and look north. You can clearly

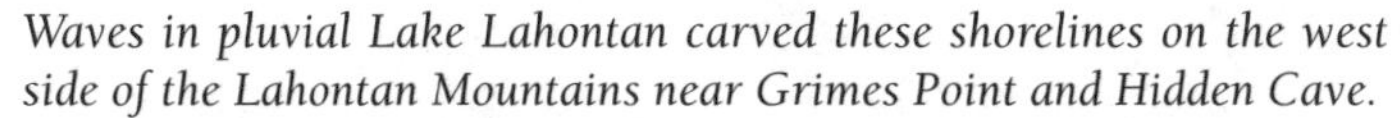

Waves in pluvial Lake Lahontan carved these shorelines on the west side of the Lahontan Mountains near Grimes Point and Hidden Cave.

see wave-cut shorelines along the base of the Lahontan Mountains. Grimes Point was once a peninsula jutting into Lake Lahontan, which filled the broad depressions to the south.

The Carson Sink lies to the southwest and Eight Mile Flat to the southeast. Eight Mile Flat is a playa—a salt flat devoid of vegetation. Playas owe their existence to the large lake that once filled this basin. As Lake Lahontan evaporated, it deposited formerly dissolved minerals, the last vestige of which was the suite of salts that gives the surface its brilliant white color. Playas exist throughout the Great Basin and come in a variety of sizes. Some playas, such as Carson Sink, hold water—albeit unpalatable, salty water—through part or all of the year. Others hold no water and haven't for a very long time.

Whitish salts that precipitated in an evaporating lake fill the central playa in this aerial view of the Carson Sink. Wave-cut shorelines are visible to the south.

As you drive by many of the playas within the Great Basin you will observe ongoing mining operations. A few miles east of us on U.S. 50, a salt mine sits in the middle of Four Mile Flat. Miners extract mineral deposits in the evaporites that desiccating pluvial lakes left behind. These economically valuable compounds within evaporites include sodium carbonate, sodium sulfate, borax, lithium, phosphate, bromine, and halite (table salt). To produce evaporite deposits in inland lakes, an environment must meet four general conditions.

First, the environment must have the proper topographic and climatic conditions necessary to produce large lakes. The Great Basin itself is essentially one gigantic closed basin made up of many smaller closed basins. The cooler and wetter climate during the last glaciation, roughly 20,000 years ago, favored the accumulation of water within these closed basins.

Second, a source of salts must exist within the contributing drainage basin. Typically, dissolved minerals come from weathering of exposed rock, but other processes contributed as well within the Great

Valuable salts are mined in the center of Four Mile Flat, a remnant of Lake Lahontan.

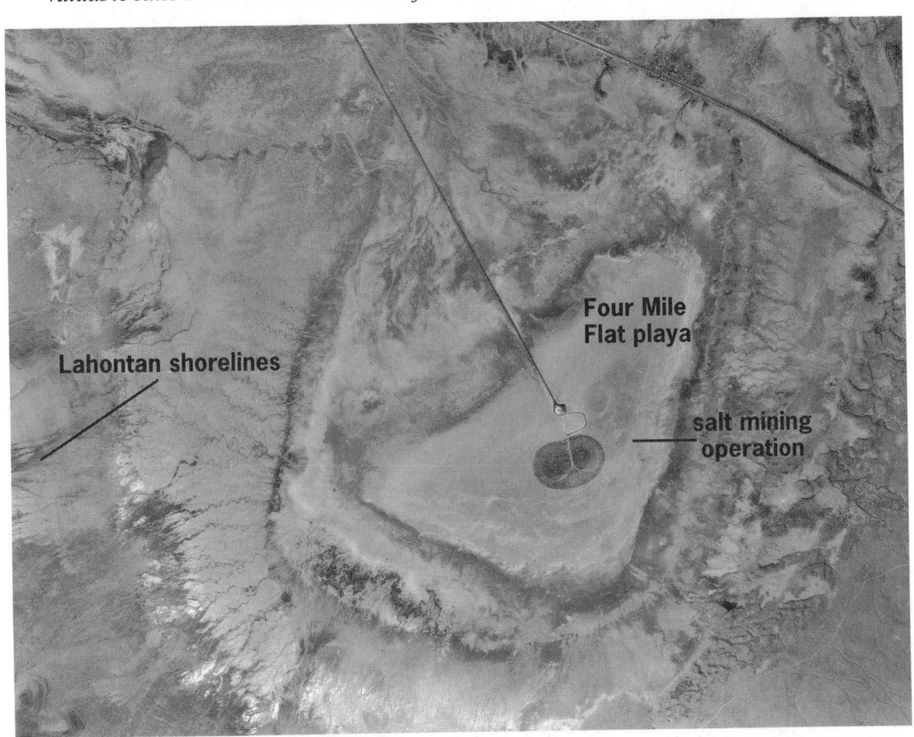

Basin. This region played host to much recent volcanism that brought an abundance of molten rock—the source for lava—near the surface. We also know the Great Basin is rife with faults. As molten rock solidifies it releases magmatic fluids, very hot water full of dissolved minerals. The many faults act as conduits, carrying this hot water to the surface as mineral springs. These springs supplied lots of dissolved salts to the growing pluvial lakes, enhancing their potential to eventually produce valuable evaporite deposits.

The third condition relates to the ability of a lake to concentrate dissolved minerals. If a lake overflows, it sends both water and dissolved minerals into another basin, thus preventing an increase in salinity in the overflowing lake. However, if a lake within a closed basin lake does not overflow, water leaves the basin only through evaporation, a route that dissolved materials cannot take. So, the concentration of minerals in the water increases over time.

Finally, climate must change in such a way as to either increase the rate of evaporation or decrease the quantity of basinwide run-off such that the lake completely evaporates. As the water evaporates, increasing salinity causes a sequence of minerals to be deposited, the order of which depends on temperature, pressure, and mineral concentration. What the drying lake leaves behind, then, is a playa, which, if the four conditions are met, can be mined for valuable mineral deposits.

Most of Nevada sits in the Great Basin, an inwardly draining region of the southwestern United States. Rain that falls here does not flow eventually into the sea; instead it collects in the many closed basins that make up the Great Basin. With our present hot and dry climate, evaporation rapidly sends most of that water back into the atmosphere. Some perennial freshwater lakes, such as Walker Lake and Pyramid Lake, survive in Nevada. These derive their water primarily from alpine run-off in the Sierra Nevada. They are beautiful pools in an otherwise arid region. Imagine for a moment the cooler, wetter climate that existed here 10,000 or 20,000 years ago and a sea of blue filling the Carson Sink and Eight Mile Flat. That was Lake Lahontan, a giant body of water that was truly Nevada's greatest lake.

9

Varnished Rock in a Desert Gallery

GRIMES POINT

Grimes Point, on the southwest edge of the Lahontan Mountains, is one of the oldest petroglyph, or rock carving, sites in the Great Basin. With more than nine hundred basalt boulders exhibiting more than three thousand graphics, it may also be the Great Basin's largest rock art site. The Lahontan Mountains contain abundant evidence of prehistoric cultures. Archeologists know relatively little about the different cultures that may have interacted at Grimes Point in the distant past, so they refer to all these people as Paleo-Indians. At least nine other sites with rock paintings, or pictographs, and petroglyphs lie in the vicinity of Grimes Point, as do numerous rock shelters and burial sites. Evidence from these locales suggest that humans have occupied the Lahontan Mountains for more than 10,000 years.

That long ago, Grimes Point would have looked much different than it does today. The point was a peninsula jutting out into pluvial Lake Lahontan (vignette 8). The shoreline perhaps would have been marshy, with stands of reeds and grass. The lake dried up during the last 10,000 years, but it did so gradually, leaving behind wetlands that would have provided food for the hunter-gatherers who lived here.

Many Great Basin petroglyph sites appear to be related to hunting, perhaps as a magical aid. Most sites are adjacent to game trails, potential ambush areas, or lines of upright boulders positioned as fences. Grimes Point fits this mold, as animals would certainly have visited the verdant lakeshore for food and water. On the crest of the Grimes Point peninsula are the remains of a fence—small mounds of stone running east to west and separated by 25 feet. These piles have hollow centers that may have once held wooden posts. Archeologists infer that early inhabitants drove game into and along this fence as part of their hunting strategy, "corralling" the animals for easier killing.

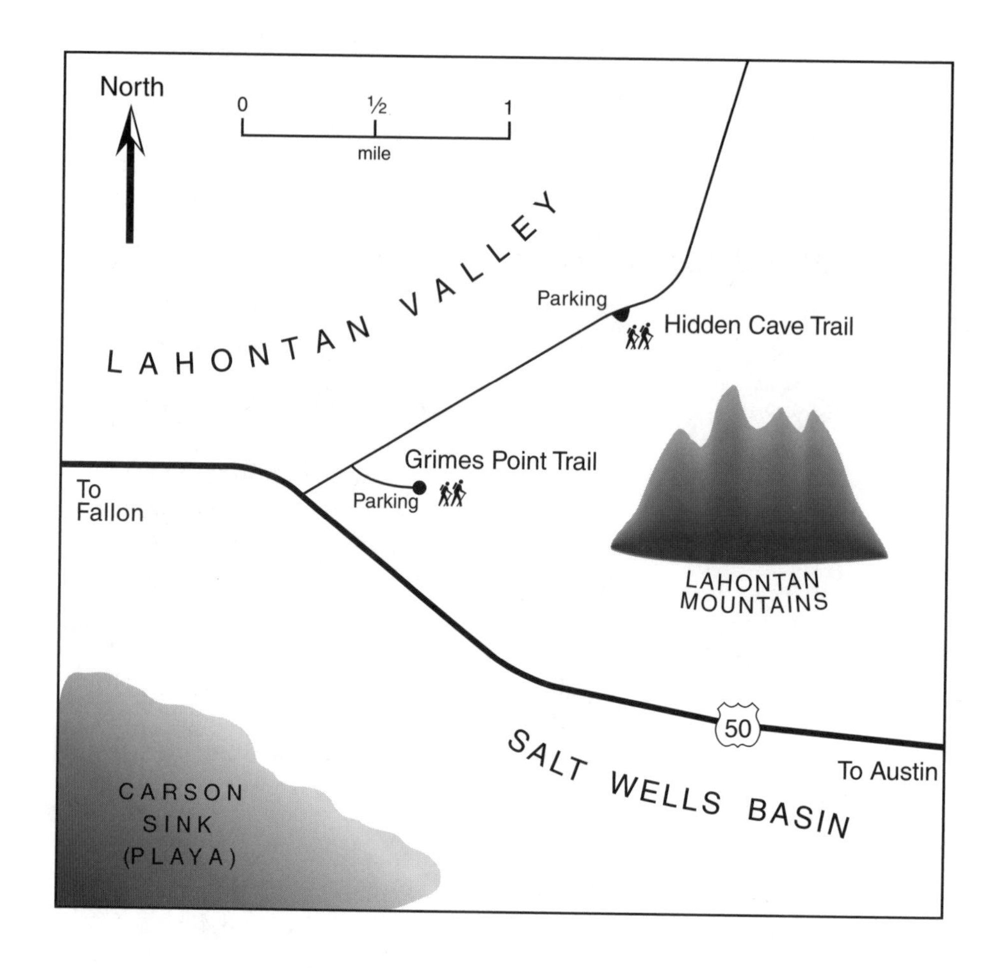

GETTING THERE

Grimes Point lies on U.S. 50 about 11 miles east of Fallon, 72 miles east of Carson City, and 100 miles west of Austin. From Reno, drive east on Interstate 80 to alternate U.S. 50 (which connects to U.S. 50) for a total distance of 68 miles. The turnoff to Grimes Point is well marked, with a paved parking lot just north of the highway. A small picnic area adjacent to the parking lot offers signs describing the area, as well as restroom facilities. North and east are the Lahontan Mountains, and southwest is the Carson Sink, a marshy remnant of pluvial Lake Lahontan. The 0.5-mile trail is well maintained, with only gentle slopes, and presents interpretations of the rock art it passes.

Combining your trip to Grimes Point with a visit to Hidden Cave will enhance your appreciation of both the geology and human history of this part of Nevada. Hidden Cave lies only about 1 mile northeast of Grimes Point on a well-graded dirt road.

The view southwest into the Carson Sink, a remnant of Pleistocene Lake Lahontan. Grimes Point was once a peninsula in that great lake. The boulder in the foreground features a curvilinear design on the left with representational elements on the right.

The Grimes Point Archeological Area represents a relatively recent attempt to preserve the site. In the 1930s, miners worked the caves here for gravel, as well as guano for fertilizer, leaving gaping scars on the surface. Grimes Point has since been used as an unofficial shooting range, then as a dump for the city of Fallon. Vandals defaced many panels of rock art, chipping rock and painting graffiti over petroglyphs. Pot hunters scoured the surface for artifacts. In the mid-1970s, personnel from the Bureau of Land Management reversed the trend of neglect that was gradually destroying this rich site. The BLM cleaned up the trash, built fences, and prohibited further mining. More importantly, the BLM and the Youth Conservation Corps constructed an interpretive trail to provide access for and information to the interested public. Our trip to Grimes Point highlights the stories told on the surfaces of the rocks along this trail.

The Grimes Point interpretive trail winds upward from the parking lot onto the ancient peninsula. Behind you, to the southwest, lies the flat valley floor that Lake Lahontan once covered. Ahead are the Lahontan Mountains, from which Grimes Point projects. Probably

Fresh and patinated surfaces on boulders at the edge of Grimes Point, with several wave-cut shorelines from Pleistocene Lake Lahontan in the background

the first thing you notice when walking the trail at Grimes Point is the glossy, black patina on the surfaces of basalt boulders. Fractured surfaces of rock reveal that the rock itself is a dull gray that is lighter than the surface patina. How do we explain this difference?

The surface coating that we see here is desert varnish, a ubiquitous feature in arid lands. A combination of manganese and iron oxides as well as clay minerals gradually covers exposed rock, forming a dark coating. Scientists in the early and middle 1900s believed that desert varnish formed from materials that were leached from the rock itself, but sophisticated chemical tests showed that elements found within the surface coating were absent from the rock. Varnish depths vary from 0.0004 to 0.02 inch thick but are typically between 0.0004 and 0.012 inch thick. The thickest desert varnish layers are roughly one-fifth the thickness of an average sheet of typing paper, while most are much thinner. Varnish varies from orange to black depending on the relative amounts of iron and manganese oxides within the layer. The more manganese the varnish has, the deeper black the color. More iron oxide produces a brown or orange varnish.

A surface patina rich in manganese blackens the surface of the boulders scattered about at Grimes Point. As you travel across Nevada and the rest of the southwestern United States, you will encounter a wide variety of varnish colors, from orange to red to brown to black, reflecting the relative dominance of iron versus manganese oxides.

The manganese in desert varnish is the first mystery we encounter here. It is concentrated at up to one hundred times the level found in the immediate environment. How can that be? If the chemical constituents of desert varnish were simply airborne materials, similar to blowing dust, we would expect the particles to randomly adhere to rock. That is not the case at all. The process at work preferentially takes up ambient manganese and deposits it, along with iron, as an oxide in desert varnish. It turns out that the agent responsible for this enhancement of manganese is biological. Mixotrophic bacteria, microorganisms that are able to subsist on a diet of both organic and inorganic material, live on the surface of most rock but are most abundant in arid and semiarid environments that are inhospitable to competing organisms. These bacteria favor an environment that has little organic material, near-neutral pH, periodic wetting followed by complete desiccation, and specific types of clay minerals. Mixotrophic bacteria consume ambient manganese, then secrete manganese oxide as a waste product. Manganese oxide cements airborne clay minerals to the rock, thus providing shelter for the microbes. Laboratory

Scanning electron microscope image of an unlaminated desert varnish —John Van Hoesen SEM photo

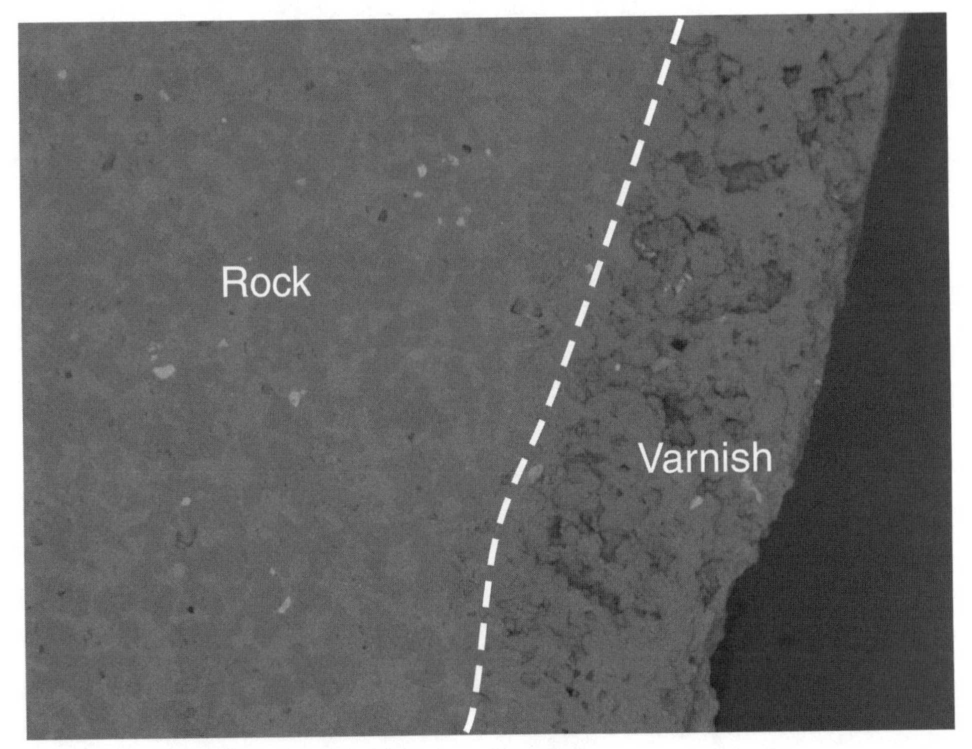

experiments have verified the ability of manganese-oxidizing microorganisms to generate varnish on rock.

The varnish these organisms create has a characteristic form and structure, or morphology, that varies from botryoidal to lamellate. Microscopic bumps and rounded knobs cover the surface of botryoidal varnish, while lamellate varnish appears as a flat layer with only very subtle bumps. These features are far too small to see with the naked eye.

The two natural morphologies seem tied to environmental moisture. Lamellate varnish contains a relative abundance of clay. Clay minerals are shaped like flat platelets, hence the flat structure of the resulting varnish layers. Botryoidal varnish is much higher in manganese oxide and contains relatively little clay. Heightened moisture levels in the atmosphere lead to a greater abundance of microbes. This larger population concentrates much more manganese on the rock surface than does the smaller population typical during arid periods. The botryoidal structure is essentially microscopic mounds of bacterial waste that the larger population of microbes during moist times excretes onto the rock.

Heightened aridity, on the other hand, reduces microbial activity and results in a lamellate varnish. Even under dry conditions enough microbes are present to cement the available clay minerals to the rock surface. Aridity, however, leads to more clay cementation relative to

A rectilinear petroglyph decorates a varnished boulder. Weathering has weakened and removed the surface layer, revealing the true color of the rock.
—Dave Futey photo

manganese, so lamellate varnish has a flatter microtopography. Aridity may influence the color of a varnish layer as well. Some researchers think manganese-poor orange varnishes indicate an alkaline environment, a result of very dry conditions. However, the debate continues about this relationship.

Almost all the basalt boulders at Grimes Point have a surface sheen, or polish. You can best observe this sheen in the morning or afternoon, when sunlight reflects from the boulders at an oblique angle. What has polished these rocks? When rockhounds polish stones, they place rough rocks in a tumbler with water and coarse grit. As the movement of the tumbler rolls the stones around, the grit gradually rounds them. After a while, the rockhound replaces this grit mixture with fresh water and finer grit. Four or five cycles in the tumbler with ever-finer grit produces a high polish and a stone that can be used for jewelry. A natural version of that process polished the rocks at Grimes Point. Desert varnish contains much manganese, a metal that holds a high polish. Instead of a tumbler filled with water and grit, we have persistent winds carrying dust from the playas that lie south and west of Grimes Point. Thousands of years of abrasion have polished the varnished surface to the sheen we see today.

The study of rock art and desert varnish have been intertwined at Grimes Point. Variations in the degree of desert varnish development establish a relative chronology in rock art styles here. In turn, rock

Cupule and Groove petroglyphs cover this boulder. Compare the fresh surface (upper left) *with the heavily varnished rest of the rock. Varnish has completely repatinated the cupules in the surface, attesting to their antiquity.*

Paleo-Indians carved the representational figure on the left more recently than the grooves on the right, as evidenced by the darker patination of the older glyphs.

art styles have helped researchers to use desert varnish in establishing absolute ages for rock exposures.

Archeologists were quick to note that the degree of varnish in petroglyphs at Grimes Point varies. Some petroglyphs are as dark as the surrounding rock while others appear much lighter. When first carved, all the graphics would have been the dull gray of the rock's fresh surfaces. The desert varnish coated the carved surfaces over the ensuing thousands of years. Over a long span of time, Paleo-Indians worked and reworked many boulders at Grimes Point. The larger and flatter rocks are a good place to look for degrees of patination in the different rock art styles. Based on this patination, archeologists established a sequence of artistic styles.

As you walk the interpretive trail at Grimes Point, try to discern the differences in style and patination in the carved basalt. The most highly varnished rock carvings at Grimes Point belong to the style called Pit and Groove, or Cupule and Groove. The names come from this style's characteristic uniform, circular depressions and grooves carved into the boulders. Conical depressions indent surfaces of all orientation, from horizontal to vertical, and do not appear to have been used for grinding or storage. Somewhat lighter in color than Pit and Groove petroglyphs are those of the Great Basin Curvilinear style. These are abstract shapes with contorted curves and circular themes. Lighter yet, but still varnished, are Great Basin Rectilinear petroglyphs. In rock art of this style we see crosses and sharp angles instead of circles and flowing curves. The least patinated carvings belong to the Great Basin Representational style. These glyphs are obviously patterned after lizards, deer, and humans, as well as other organisms

A unique serrated boulder falls into the Cupule and Groove category, one of the most ancient petroglyph styles found here.

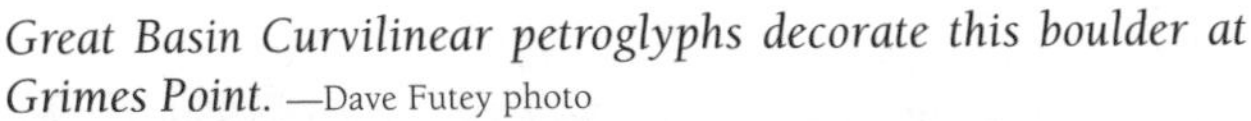

Great Basin Curvilinear petroglyphs decorate this boulder at Grimes Point. —Dave Futey photo

and objects from the daily lives of these early Americans. Yet another style of rock carving, called Stillwater Faceted, adorns boulders elsewhere in the Lahontan Mountains, northeast of Grimes Point near Fish Cave. Rocks shaped in this style, which is not present at Grimes Point, have faces that have been rubbed smooth. In some cases, intersecting faces have been ground flat to form a sharp edge. In all cases, rocks exhibiting the Stillwater Faceted style have been heavily repatinated, leading archeologists to believe that this style is even older than Pit and Groove petroglyphs. Do you see the differences in patination and style that allowed archeologists to develop the relative chronology?

If you wander off the trail and among the boulders, be careful not to touch the petroglyphs themselves. Oils from human skin will damage the artifacts. These carvings date to some of the earliest civilizations in North America, so please respect them for the national treasure that they are.

Some researchers suggest that desert varnish can be used to determine an absolute age for a particular rock surface. The basis for these claims lies in the presence of cations (pronounced KAT-eye-onnz) within desert varnish. Cations are positively charged atoms that may be mobile—that is, easily leached by water—or immobile and stable. Titanium is the most immobile of the cations, while calcium and potassium are two of the more mobile ones. The titanium, calcium, and potassium come from clay and other windblown dust that adhere to the varnish. Using ion particle accelerators, researchers can calculate the ratio of calcium and potassium to titanium in desert varnish. The working hypothesis is that, over time, more mobile cations are regularly leached from desert varnish, preventing their levels in the varnish from increasing, while immobile cations keep accumulating. This means that the ratio of mobile to immobile cations decreases over time. The ratio could hypothetically supply an absolute age for a varnished rock face. This research is controversial but offers some hope for age determination at sites like Grimes Point.

The boulders at Grimes Point hold the accumulated "writings" of thousands of years of Paleo-Indian civilizations. These peoples were some of the earliest inhabitants of the Great Basin and perhaps North America as a whole. If you wish to learn more about them, please continue your investigation at Hidden Cave (vignette 10), a very early habitation site in the Lahontan Mountains. The Hidden Cave trail is less than 1 mile northeast of Grimes Point.

10

Deciphering Time's Muck Heaps

HIDDEN CAVE

Twentieth-century Americans first visited Hidden Cave in the 1920s. Groups of archeologists interested in prehistoric humans in the area excavated the cave in 1940, 1951, and 1979. The cave has provided a wealth of artifacts and information, not only about ancient populations but also about changing climate in central Nevada. Hidden Cave, named for its small and difficult to locate opening, is not the only point of interest here. As we stroll counterclockwise around the 1-mile loop trail on the northwest flank of Eetza Mountain in the Lahontan Mountains, we uncover clues that allow us to imagine this region 10,000 years ago when the climate was much wetter. At that time, the small playalike depressions northwest of the trail were marshes filled with plant and animal life. Under such conditions hunter-gatherers would have thrived. The adjacent caves and rock shelters provided these people with convenient habitation and storage facilities. Remnants of such use in turn provide us with information about their lives.

Looking west at the broad flat near Hidden Cave. When Paleo-Indians lived here, roughly 10,000 years ago, this was a fertile wetland. —Dave Futey photo

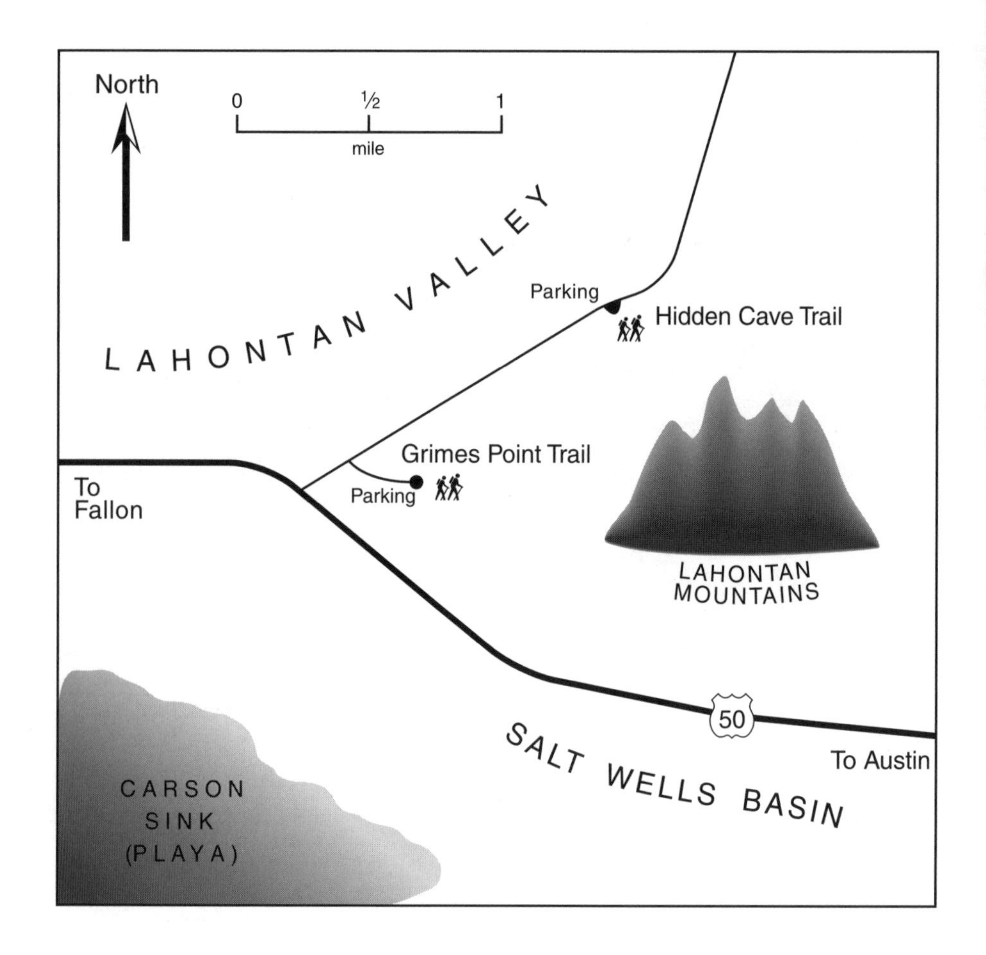

GETTING THERE

The turnoff to Hidden Cave is the same as that for Grimes Point and is well marked. It lies 11 miles east of Fallon, 72 miles east of Carson City, and 100 miles west of Austin, all on U.S. 50. From Reno, drive east on Interstate 80 to alternate U.S. 50, which connects to U.S. 50, for a total distance of 68 miles. Drive past the parking area (on the right) for Grimes Point and continue 1 mile north to a small parking lot on the east (right side) of the road. The 1-mile loop trail leaves from and returns to this lot.

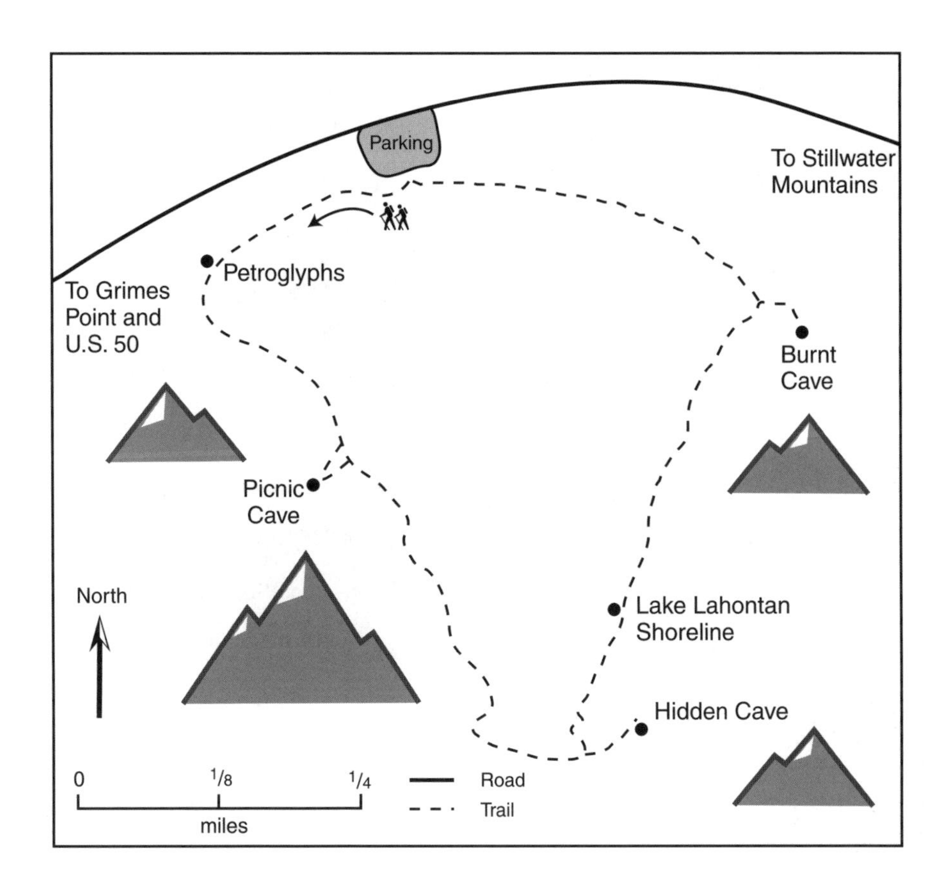

The trail gains about 100 feet in elevation before dropping again. As you walk the trail, please keep an eye out for loose rock, steep slopes, and rattlesnakes—who are more than happy to be left alone. Access to Hidden Cave is restricted to public tours, which the Churchill County Museum in Fallon, Nevada, offers several times each month. Check with the museum for the tour schedule. The tours afford visitors a close-up view of the stratigraphy and archeology of the cave.

Representational petroglyph of a horned toad near the beginning of the Hidden Cave trail

Our first stop is a prominent boulder next to the trail about 300 feet from the trailhead. Well-preserved petroglyphs—prehistoric carvings in rock—adorn it. Early inhabitants of this area inscribed these images in desert varnish, a surface patina of manganese, iron, and clay that gradually coats rock in arid regions (vignette 9). These particular petroglyphs date from between 2,000 and 500 years ago. Archeologists do not know the cultural significance of such drawings, but we can certainly enjoy them as representational and abstract art.

Picnic Cave, about 500 feet upslope from the petroglyph boulder, offers us a look at the structure of these rock shelters. Picnic Cave is a broad indentation that slopes down toward its mouth, which prevents debris from accumulating on the cave floor. On the cave's walls and ceiling, you can clearly see a lumpy, white encrustation that glues individual rocks together into a single unit. The roof of the cave consists of gravel and boulders that Ice Age Lake Lahontan deposited at one of its high stands (vignette 8). As dissolved minerals accumulated in the lake, calcium carbonate in the form of tufa began to precipitate. Tufa forms in brackish, mineral-rich water, perhaps with the

Tufa, a form of precipitated calcium carbonate, cements boulders together to make up the roof of Picnic Cave. —Dave Futey photo

aid of algae. Here, tufa has acted as mortar, binding lake gravel and other rock together into a single, coherent unit. Beneath the tufa layer, on the walls of the cave, look for a residual soil rich in reddish gravel. This soil developed on the surface of a once-exposed basalt flow, an iron-rich volcanic rock, that lake sediment later covered. When the lake's waters fell beneath the level the caves now occupy, the impact and abrasion of waves pounding on the shoreline carved the shelters we see here today. The caves in this valley formed about 21,000 years ago.

The next stop, about 0.3 mile from Picnic Cave, is Hidden Cave itself. Join a public tour, which the Churchill County Museum in Fallon, Nevada, offers several times each month, to enter the cave and look closely at its stratigraphy. A heavy steel door protects the cave's entrance to prevent vandalism. Excavators in 1940 used dynamite to enlarge the entrance to what we see here.

A local legend dating from the 1800s tells of a stagecoach robber who, when caught, admitted that he had hidden his ill-gotten gains

Richard Orndorff inspects tufa-encrusted boulders on the roof of Picnic Cave.

in one of the caves in the Grimes Point area. Four youths, searching for this treasure, stumbled upon Hidden Cave in the 1920s. During a boisterous rock fight, one of the boys hunkered down into a small crevice and discovered that it led into the mountainside. Afraid that mountain lions might enter the cave, he and his pals initially covered the entrance with rocks. Six months later they returned and explored the cave. The boys wriggled into the opening, barely large enough for them, and found that after some twists and turns it widened into a proper cave. They found a cavern with walls and floors blackened by guano and with dozens of bats dozing on the ceiling as its only inhabitants. The boys kept the cave a secret for many years thereafter.

In the mid-1930s, an entrepreneur by the name of Mr. McRiley mined Hidden Cave and many other caves in the area for bat guano, a material rich in nitrogen and used as high-grade fertilizer. Mr. McRiley was overheard commenting to the local postmaster that the digging would go much more quickly "if it weren't for all that Indian

The surface of tufa has a characteristic texture that forms when calcium carbonate precipitates in a shallow lake.
—Dave Futey photo

junk." Someone repeated this conversation to Fallon resident Margaret Wheat, who in 1938 petitioned the Southwest Museum to visit the cave.

The 1940 excavation, headed by the husband and wife team of S. M. and G. N. Wheeler, began by blasting a larger opening into the cave. In addition to their work at Hidden Cave, this couple also unearthed burial sites nearby in the Lahontan Mountains, in Spirit Cave and Fish Cave. The Nevada State Museum in Carson City houses the artifacts uncovered in 1940. A 1951 excavation headed by N. Roust and G. Grosscup, both from the University of California, and R. Morrison of the U.S. Geological Survey followed the Wheeler excavation. The goals of the second group were to learn more about both the inhabitants of Hidden Cave and the history of Lake Lahontan. The collection of the Lowie Museum at University of California at Berkeley includes artifacts this group uncovered. In 1979 and 1980, D. H. Thomas of the American Museum of Natural History and B.

A black-collared lizard perches on tufa-cemented beach gravels making up an ancient Lake Lahontan shoreline. —Dave Futey photo

Hatoff, archeologist for the Carson City District, led the latest excavation of Hidden Cave. Their study answered new questions about early civilizations and applied modern scientific techniques to this rich site.

Cultural artifacts that excavators have found within Hidden Cave include pendants, beads, cord, netting, projectile points, grinding tools, and a spear-throwing device called an atlatl. Plant matter and animal bones are also abundant in Hidden Cave. Finding cultural and other artifacts was not the only point of these investigations. The timing of human occupation and changing conditions of the inhabitants' environment are important to any archeological study. For this information scientists at Hidden Cave turned to stratigraphy.

Stratigraphy is the study of sedimentary rocks, including their composition, temporal and spatial relationships, and environments of deposition. Archeologists dug trenches in Hidden Cave, revealing a sequence of layered deposits that provided a wealth of information about its history. The stratigraphy of Hidden Cave records high and low stands of Lake Lahontan, wet and dry climates, volcanic eruptions, and even human habitation.

Radiocarbon analysis of organic material and carbonate rock in Hidden Cave provided dates for many of the deposits. The volcanic ash layers also provided discrete benchmarks for dating. When certain types of volcanoes erupt they emit a great deal of ash that carries

The entrance to Hidden Cave has been fitted with a door to prevent vandalism. Note the tufa-cemented gravels and boulders surrounding the cave mouth.

the signature chemistry of the magma source. If we know the timing of the eruption from radiometric dating, the distinctive chemistry of the emitted ash allows these layers to become high-resolution timelines. Two such layers lie within Hidden Cave.

The first thing to note as you enter Hidden Cave is that the floor and walls resemble those of Picnic Cave—red gravel and tufa-cemented boulders and lake gravel, respectively. Again, notice the bumpy, almost coral-like, appearance of the tufa. Your guide on the cave tour will discuss artifacts found within the cave, but he or she will also talk about cave stratigraphy. Some layers are conspicuous, while others feature subtle gradations that require careful observation to pick out.

Sediment in Hidden Cave records the history of climate change and lake growth and decline in west-central Nevada. Streams carry sediment into lakes, and as the moving water enters a lake it slows, dropping coarse sediment near the shoreline. Fine sediment, such as clay, settles out much more slowly, so it accumulates farther from shore. Little or no sediment accumulates in the very deepest, central parts of a large lake, as even fine material settles to the bottom before reaching these sections.

Sediment may pile up in a cave even when lake levels drop below the cave's mouth. In a very wet climate, mud and gravel from the cave walls may slump, creating a layer of mixed fine and coarse sediment on the cave floor. Under dry conditions, cave walls no longer collapse. Instead, episodic floods erode silt, fine-grained and uniformly light tan, from the slopes surrounding the cave mouth and carry it into the cave. Sedimentary structures within these silt deposits include ripple crossbedding from rapidly moving water, mud cracks that indicate wetting and drying cycles, and guano rafts, bat feces that detach from the floor and float during flooding. During a thunderstorm in August 1979, researchers observed rainwater entering Hidden Cave as a turbid stream of water and sediment. This inflow lasted for ten minutes and deposited a fine layer of silt on the cave floor. An exceptionally arid climate precludes floods, so sedimentation within the cave temporarily ceases during very dry times.

As you walk through Hidden Cave, pay attention to changes in color and grain size in the various strata that you see. These are important clues to the history of the cave. On the tour, you may not be able to pick out all the characteristics of each layer, but you will certainly be able to discern transitions, some of which represent drastic changes in the lake or surface environment.

Hidden Cave stratigraphy begins 21,000 years ago with deposition of sand and clay, sediment that tells of a large, deep lake. This continued until 18,000 years ago, when the lake surface dropped below the cave mouth, and sand and pebbles that washed from the walls accumulated on the cave floor. A layer of oolitic tufa lies on top of the pebbles and sand. Ooids are small, rounded grains of finely laminated calcium carbonate, about the size of fish eggs. They develop in shallow, alkaline water where gentle waves roll shell fragments, some microscopic, back and forth. Calcium carbonate precipitates in fine layers around the fragments, gradually enlarging them. The cave lay in the shallows of a mineral lake until 15,000 years ago, when geologists hypothesize that a lake of such great depth immersed the cave that no sediment accumulated there for 5,000 years.

Ten thousand years ago, the cave mouth lay within the zone of breaking waves. The waves deposited clean beach gravel in the cave. Then the lake level dropped below the cave mouth for good. Relatively moist conditions prevailed from 10,000 to 7,500 years ago, weakening the cave's walls, which collapsed, adding mudflow gravels to the stratigraphy. Increasing aridity prevented further mudflows from

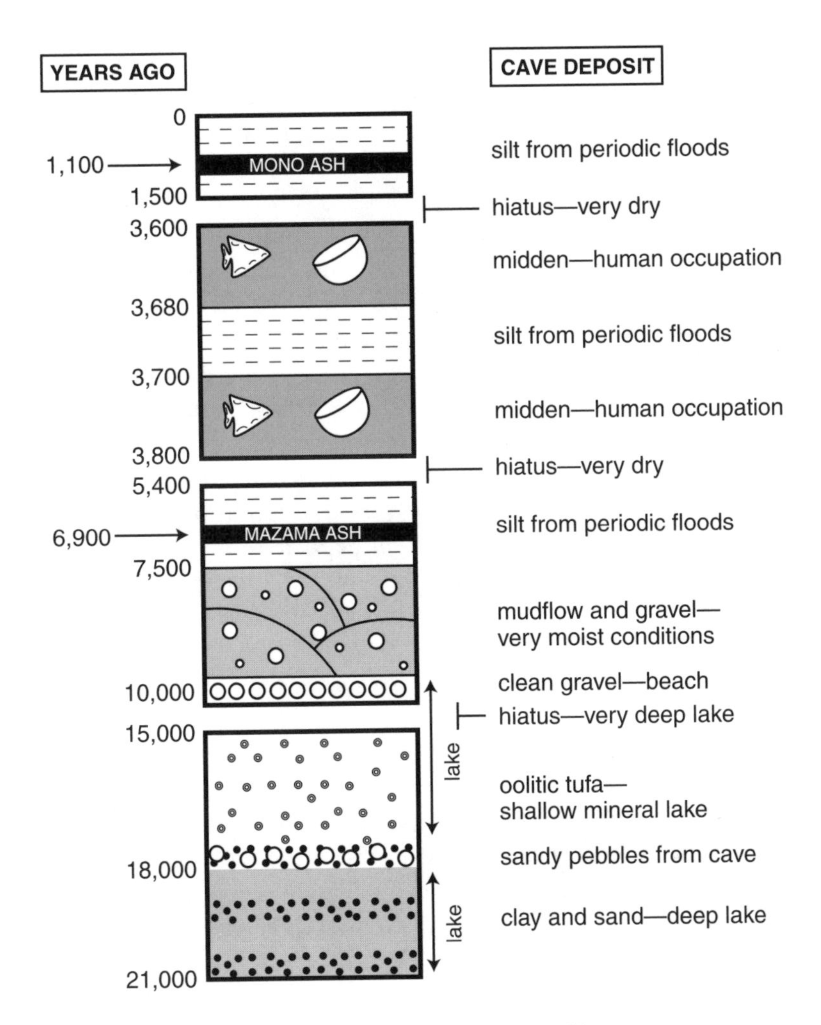

Stratigraphy of the deposits within Hidden Cave

7,500 to 5,400 years ago. Instead, episodic floods brought in silt from adjacent slopes. We find our first volcanic ash marker within this layer of silt. Mount Mazama, a volcano that sat at the site of modern-day Crater Lake, Oregon, exploded 6,900 years ago, blanketing the western United States with ash. A thin layer of Mazama ash lies within the silt. The characteristic chemistry of its obsidian fragments pinpoints the ash's origin. The absence of sediment in Hidden Cave from 5,400 to 3,800 years ago prevents us from drawing absolute conclusions about climatic conditions. We think that this period was so dry that even storm run-off was inadequate to transport sediment into the cave.

Records of human occupation are obviously very important to archeologists. Middens, from the Danish word *mogdynge* meaning "muck heap," are layers of primarily organic material that ancient humans transported into the cave. They are the primary source of artifacts found here. Rich in plant fibers and animal bones, these layers represent periods of intensive occupation and use. The presence of charcoal from cooking fires renders them a characteristic dark gray. The oldest of two middens piled up from 3,800 years to 3,700 years ago. Humans apparently abandoned the cave for 20 years, during which time silt deposition resumed. The second and more recent midden dates from 3,680 to 3,600 years ago. Once again, humans left behind a rich layer of plant and animal debris.

No material accumulated in Hidden Cave during the ensuing 2,100 years. Again scientists hypothesize very dry conditions. From 1,500 years ago to the present, storm run-off deposited silt within the cave. The Mono Craters in eastern California erupted 1,100 years ago, adding to the cave's stratigraphy a characteristic marker bed encased within the silt.

The last stop on the loop trail is Burnt Cave, a shelter that hunter-gatherers occasionally used. What makes this cave somewhat unusual are red pictographs that modern vandalism now partially obscure. A pictograph is a painting on rock, distinguishing it from a petroglyph, which is carved into the rock. Pictographs are generally more delicate than petroglyphs, as pigments chemically degrade and flake off the surface over time. Dark red on a charcoal black background, these rock drawings are difficult to see and nearly impossible to photograph. You will see them more clearly after your eyes adjust to the dim light within the cave.

Ancient people used three essential ingredients to create the paint for these images: pigment, to give the pictograph its particular color; binder, to hold the pigment particles together and glue them to the rock; and a vehicle, to make the mix fluid enough to apply readily. Pigments are the most indispensable ingredients in the mix. They may come from native ore or mineral clay, which are durable materials that weather well, or a vegetable dye, which degrades over time and loses its color. Some typical mineral pigments in pictographs are hematite for red and brown tones, malachite for green, goethite for yellow, gypsum for white, graphite for black, and azurite for blue. The unique chemical structure of each of these minerals reflects light in a characteristic manner, lending each its signature color. The red pictographs on the walls of Burnt Cave probably get their color from

This outcrop of tufa-cemented boulders and beach gravel from pluvial Lake Lahontan lies on the western edge of the Lahontan Mountains.

hematite, a form of iron oxide that is common in central Nevada. Common binders in pictographs are blood, egg white and yolk, seed oil, plant resin, milk, and honey. Vehicles include plant juice, water, animal oil, and urine. Artists mixed these three ingredients together using grinding tools and paint pots, artifacts that researchers have excavated here in Burnt Cave. Then artists applied the resulting paint with fingers, plant fibers, frayed twigs, yucca spines, or blowpipes.

Hidden Cave and the other caves on the loop trail represent only some of the many artifact-rich sites in this area. The 1940 Wheeler group also was the first to study two nearby caves, Spirit Cave and Fish Cave. These two sites yielded human remains that have become very important in North American archeology. Researchers unearthed a human mummy wrapped in a finely twined, woven mat in Spirit Cave. They found this so-called Spirit Cave Man close to two cremation sites and fragments of other remains. Radiocarbon analysis of the mat, hair, and bone from the mummy yielded dates ranging from 9,460 to 9,430 years before present, making this one of the oldest human remains found in North America. Other artifacts in Spirit Cave date from 9,300 to 1,700 years before present, indicating an extensive period of habitation and use. The very old dates associated with Spirit Cave are not isolated examples. Plaited mats in Grimes Burial

Cave and Crypt Cave, within walking distance of Hidden Cave, date to 9,470 and 9,120 years before present, respectively. Human bone found on Wizard Beach along Pyramid Lake dates to 9,225 years before present. Fish Cave held the skeleton of a 50-year-old woman with a complete set of healthy teeth—which says something important about our modern diet—as well as isolated fragments from other remains.

The Hidden Cave loop represents an interplay between geologic and human history. Inland lakes created the caves here by cementing gravel and boulders together, then eroding depressions at the shoreline. Humans came much later, drawn to the fertile wetlands along the shoreline of the retreating lake. They found the caves to be hospitable shelters and left evidence of their habitation superimposed on sediment deposited by a variety of lake and land processes. We have learned much from study of Hidden Cave stratigraphy, but as always, there is still a great deal more to learn. As you walk the loop or tour Hidden Cave keep in mind the antiquity of this site. Please help preserve it for others to experience.

11

One Grain at a Time

SAND MOUNTAIN

The Hollywood depiction of deserts is very often a sweeping expanse of sand. Who can forget camels charging over cresting dunes in the classic movie *Lawrence of Arabia*? Or Clint Eastwood staggering through a waterless wasteland of blowing sand in Sergio Leone's *The Good, the Bad and the Ugly*? The truth is, though, that sand dunes make up only about 20 percent of the surface area of the world's deserts, and they are particularly scarce within the Great Basin.

The desert surface within the Great Basin is predominantly bedrock and gravel, with isolated exposures of fine-grained sediment. What's lacking, of course, is the lush vegetation of humid regions. Plants do live in deserts, but those plants that thrive there do so because of special adaptations that allow them to find and use water very efficiently. And although it seems nonsensical in regions with so little available water, most desert landforms are shaped by moving water. Precipitation is rare indeed, but when it comes it typically does so with great intensity. The arid climate and lack of widespread vegetation, which protects soil from erosion, allows wind to be a more active agent of erosion and deposition in deserts than it is elsewhere. Wind is the force behind the formation of the huge accumulation of sand at Sand Mountain.

Sand Mountain is a dune, a deposit of windblown sand. Dunes are dynamic features that migrate downwind, some slowly and others rapidly. In many areas of the world, people must clear dunes from roads much as we clear winter snow from roads in northern North America.

As you drive north from U.S. 50, look at Sand Mountain and also the valley in which it sits. Sand Mountain measures 1.6 miles long, with a maximum width of about 0.3 mile. It stands 370 feet above the valley floor at its highest point. Sand Mountain has a sinuous S shape when viewed from above. To the west and north, widely scattered dune fields indicate active sand transport. Four Mile Flat, a dry rem-

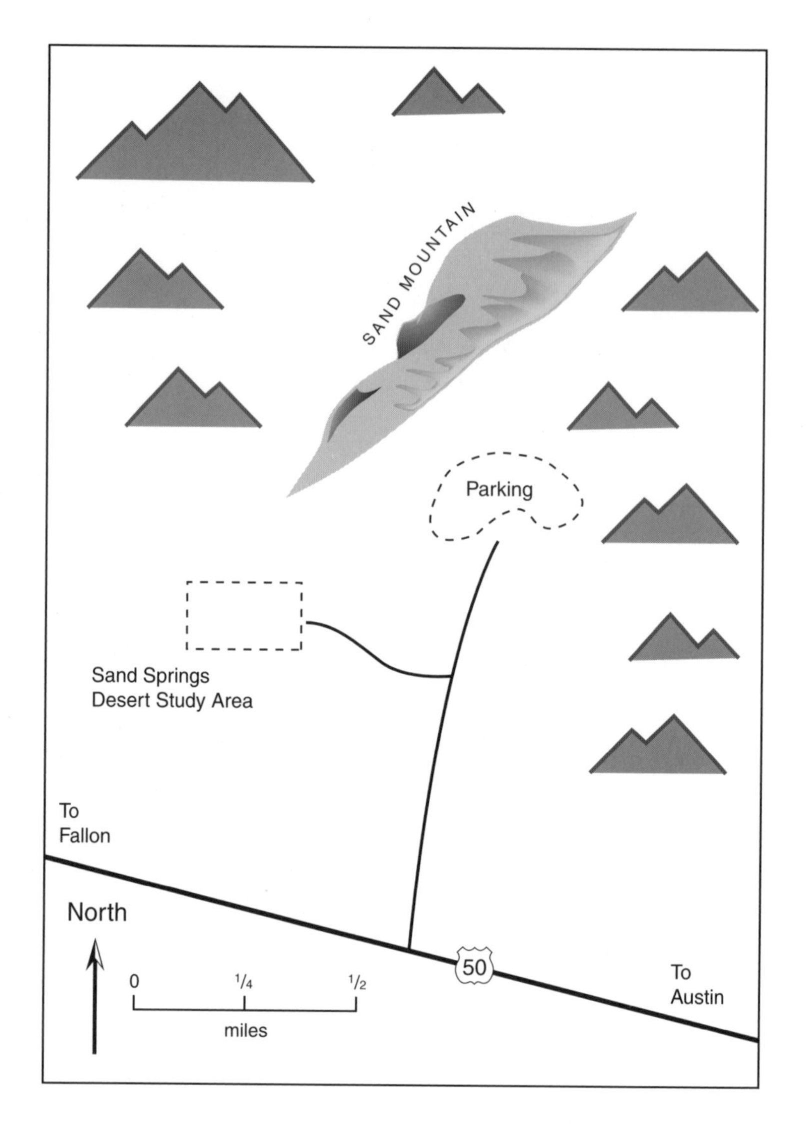

▶ GETTING THERE

On U.S. 50, the turnoff to Sand Mountain lies 26 miles east of Fallon, 87 miles east of Carson City, and 84 miles west of Austin. From Reno, take Interstate 80 to alternate U.S. 50, which connects to U.S. 50, for a total distance of 83 miles. Turn north at the signpost for the Sand Mountain Recreation Area. Drive 0.5 mile before turning west to the parking lot for the Sand Springs Desert Study Area. We will begin here, then drive the remaining 0.25 mile north to Sand Mountain itself to look at dunes. At Sand Mountain, park anywhere within the broad flat, but watch out for depressions with soft sand that may mire your vehicle.

After visiting Sand Mountain, you may want to walk the nature trail at Sand Springs Desert Study Area. This is a protected site with a maintained trail, interpretive signs, and a well-preserved Pony Express station.

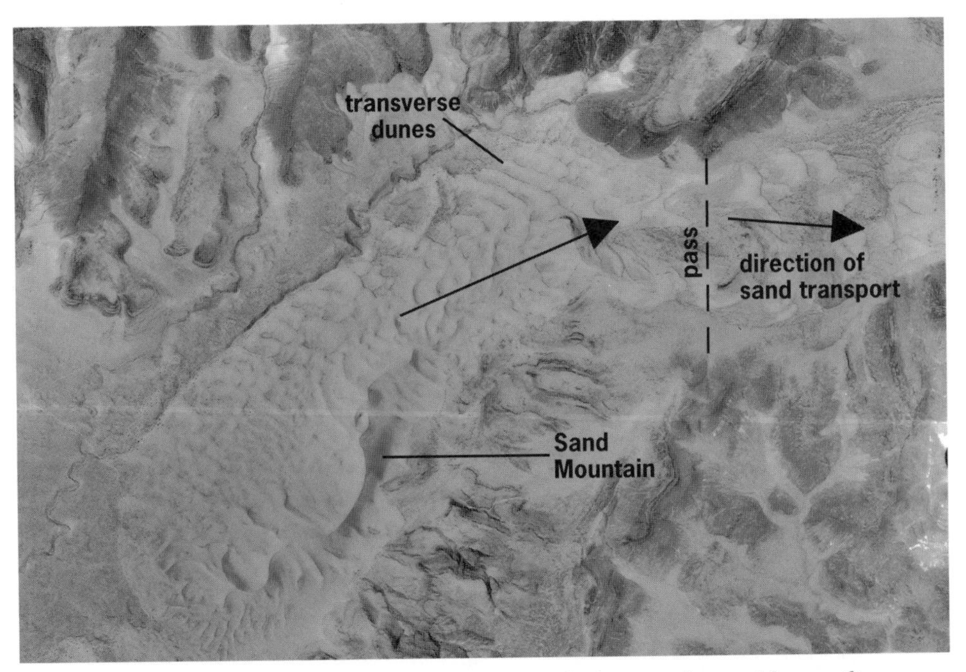

Sand Mountain is one part of a field of dunes. Sand moves northeast with prevailing winds over a mountain pass, then down into the neighboring valley. Note that the ridges of the transverse dunes are perpendicular to the wind direction.

nant of Lake Lahontan—an enormous body of water during Pleistocene time (vignette 8)—lies to the south. Prevailing winds, sand supply, and the orientation of the bounding mountain range have all influenced the morphology of Sand Mountain.

Turn west and drive to the Sand Springs Desert Study Area parking lot. This is a good place to begin our discussion of sand transport and dune formation. Sand will have accumulated in pockets near the parking area. Pick some up and take a close look at it. The sand grains are all roughly the same size, but color variation indicates a variety of mineral grains. Wind, due to the very low density of air, erodes less powerfully than water or ice and can only move relatively small pieces of sediment. Wind can transport dustlike silt and slightly larger sand particles, though the mode of transportation differs for the two grain size ranges. Silt grains are small enough that wind can actually sweep them into the atmosphere, suspend them there, and carry them far from their source. Sand grains, on the other hand, are too large to be suspended in a column of air. They move instead by saltation, the

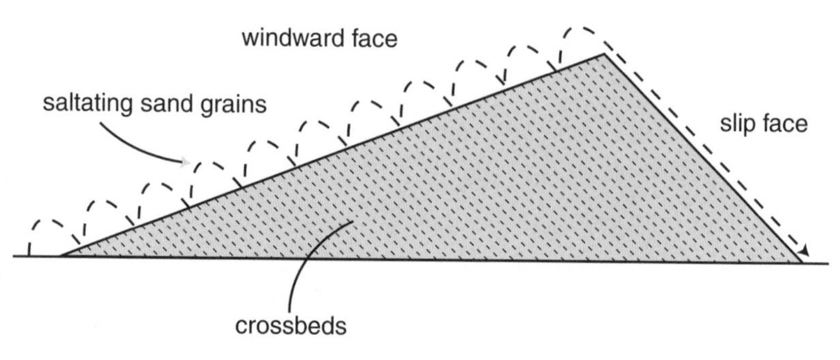

Cross section of a typical desert dune

North of the Sand Springs Desert Study Area, gravel and cobbles cover the desert floor. Larger rocks in the foreground are ventifacts that windblown sand has abraded and polished. —Dave Futey photo

root of which is the French word *sauter*, "to jump." Wind shear lifts sand grains, carries them a short distance, then drops them again to the desert surface. Grains that move by saltation bombard larger rocks, abrading and polishing them to form features called ventifacts. When winds that come predominantly from one direction abrade ventifacts, flat surfaces called facets can develop. Because the wind direction shifts here, facets are rare. The impact of saltating sand grains may slowly push some of the surface gravel, a process called surface creep.

Moving sand has polished large cobbles and boulders north of the Sand Springs Desert Study Area parking lot. Their surface sheen stands out best in the morning or late afternoon sun.

Ripples mark the surfaces of low mounds of sand just north of the Sand Springs Desert Study Area parking lot. These ripples can form anytime sand moves, whether propelled by wind, stream current, or beach surf.

When wind rapidly removes fine material, it leaves behind bowl-shaped depressions called blowouts. Preferential removal of fine material from the soil layer, combined with infrequent wetting and drying, often produces a surface layer of gravel on the desert surface. Over time, wind removes most of the fine sediment—silt and sand—leaving a tightly interlocking fabric of gravel and rocks at the surface. This so-called desert pavement effectively slows or even prevents continued blowouts. Look for desert pavement covering the surface where sand is not accumulating.

Let's proceed to the parking area east of Sand Mountain. Keep an eye out for off-road-vehicle enthusiasts as you walk around the dunes. If you could look at the longitudinal cross section of an isolated dune, you would can see that it is asymmetrical. The windward face of the dune slopes more gently than the leeward face. Saltating grains bounce up the windward face to the crest of the dune and then cascade down

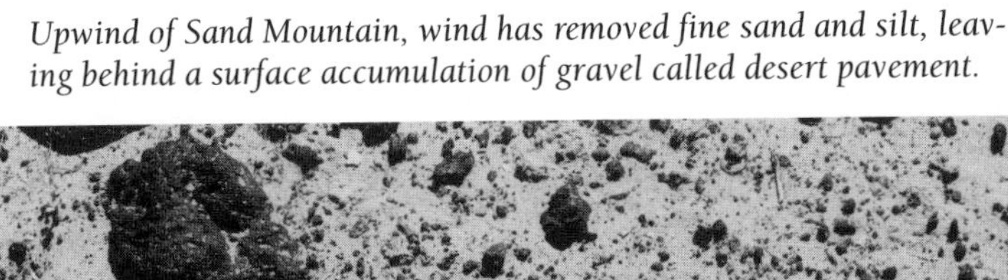

Upwind of Sand Mountain, wind has removed fine sand and silt, leaving behind a surface accumulation of gravel called desert pavement.

Windblown sand forms ripple marks on dune surfaces near Sand Mountain. These ripples are asymmetric, with gently sloping windward faces and steeper lee slopes. —Dave Futey photo

the lee face, also called the slip face. Small avalanches commonly slump down the barely stable slip face of a dune. You can see such slumps as you investigate the various types of dunes near Sand Mountain. If you could slice a dune in half—not a practical proposition without a backhoe—you would discover that the interior of the dune consists of fine, angled layers called crossbeds that record the movement and deposition of sand on the slip face. Each crossbedded layer represents a new slip face created as the dune migrated. The net result of the movement of individual sand grains is the large-scale movement of the entire dune.

Blowing sand builds different types of dunes depending on both the sand supply and the prevailing winds. Barchan dunes are shaped like crescents with tips that point downwind. These solitary features form with a limited supply of sand and a constant wind direction. They migrate across flat stretches of land that have relatively hard surfaces and little vegetation. The orientation of barchan dunes southwest of Sand Mountain tells us that the dunes are moving toward Sand Mountain and will eventually become part of this larger dune. Barchan dunes also occupy the southeast flanks of Sand Mountain.

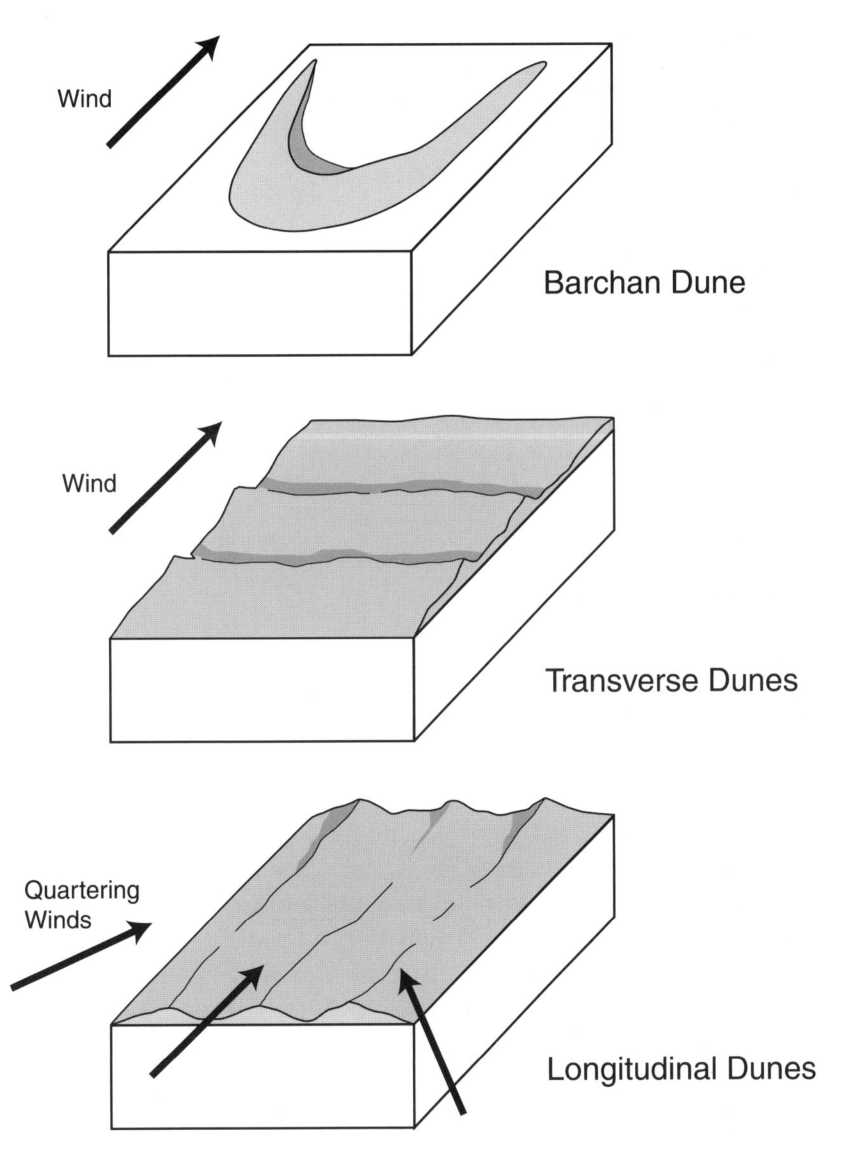

Three types of dunes found near Sand Mountain

These are easier to approach and investigate than more abundant, but more distant, dunes to the southwest. Where sand is more plentiful, barchan dunes may coalesce into barchanoid dunes, scalloped landforms whose curved crests run roughly perpendicular to prevailing winds. An even greater sand supply may produce transverse dunes,

Shadows on a cloudy afternoon delineate the crescent arms of barchan dunes on the east flank of Sand Mountain.

with linear crests and troughs oriented perpendicular to the wind direction. Winds have constructed transverse and barchanoid dunes in the broad plain to the west of Sand Mountain. You can also see both barchanoid and transverse dunes on the slope downwind of Sand Mountain. Morning and evening light allows clear viewing of the dune crests.

Sand Mountain itself is a longitudinal dune. The crests of such dunes parallel the dominant wind direction. Quartering winds—those that vary slightly in direction but remain within the same compass quadrant—shape longitudinal dunes. Longitudinal dunes typically form as a field of such features. It is less common to find an isolated example like Sand Mountain, which owes its existence to a feature called a wind gap.

Sand Mountain lies in a protected valley that slopes up to the northeast from the Salt Wells Basin. The Stillwater Mountains bound the valley on the north, west, and east, leaving the southern corridor open for inward migration of sand. The bordering mountains to the north and east stand 900 to 1,200 feet above the valley floor. A small pass, though, breaks this wall about 380 feet below the neighboring peaks. This pass acts as a wind gap that funnels the winds that have created Sand Mountain. Winds at this latitude blow predominantly from the southwest to the northeast. As winds from the southwest move into and through the valley, they accelerate as the valley narrows leading up to the pass. Once through the constricted, unnamed pass, these winds sink and spread out into Fairview Valley, and their velocity slows.

Sand Mountain is a linear dune, with a sinuous crest that parallels the dominant wind direction, southwest to northeast. Low barchan dunes sit as isolated features at the margins of Sand Mountain.

Even high-velocity wind has a limited capacity to transport sand. Downwind of Sand Mountain, rising slopes steepen considerably, making continued transport a difficult proposition. Sand Mountain sits at an equilibrium point in the narrowing valley, where wind can still transport sand upslope. Beyond Sand Mountain, wind cannot carry much sand up the steepening slopes, even though wind speed increases into the pass. Dunes within and beyond the wind gap indicate that the wind transports some, but not much, sand over the pass. Will winds push Sand Mountain itself up and through the wind gap? The relatively small quantities of sand downwind of the gap mean Sand Mountain is essentially a stationary dune, with little sand moving farther upslope.

Where does Sand Mountain's sand come from? The valleys of central Nevada were once inundated by Lake Lahontan, a Pleistocene lake that rose to its latest full stage about 10,000 to 20,000 years ago. Large lakes, like oceans, often build sand beaches with sediment that inflowing streams supply. Currents deposit the coarse sand and gravel near shore and carry the finer sediment to deeper water before depositing it. Wave action then keeps the sand near the shore, while currents move the sand parallel to shore, shaping it into a beach. When Lake Lahontan dried up, it left distinctive lake-related structures, including beaches, behind that mark its passing. Winds later relatively

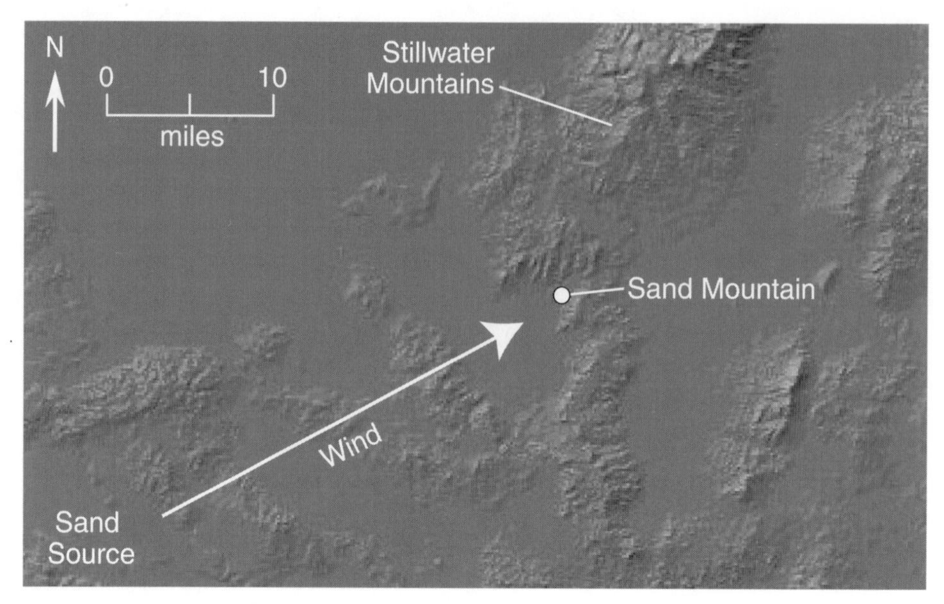

Wind carries sand from southwest to northeast to create the Sand Mountain dune field.

easily dispersed these beaches. Some of Lake Lahontan's sandy beaches existed in Simpson Pass, 10 miles west-southwest of Sand Mountain. Samples of Simpson Pass sediment from the surface and just below the surface strikingly resemble the sediment that makes up Sand Mountain. Composition of light and dark mineral grains are the same. The main difference is that the Simpson Pass sediment contains a much higher percentage of fine material than does Sand Mountain. This is to be expected, because the wind sorts sediment while it transports it. Fine materials remain in suspension for great distances, but sand grains bounce slowly along the surface and travel shorter distances. The wind would have rapidly segregated Simpson Pass sediment into suspended fine material and saltating sand grains. It is this sand that moved into the Stillwater Mountains and built Sand Mountain, an uncommon dune in the Great Basin.

12

Whole Lotta Shakin' Going On

FAIRVIEW PEAK

It is one thing to study and map the faults of a given area and quite another thing to be rudely awakened and shaken out of bed by an earthquake along one of them. Anyone in the Los Angeles area whom the Northridge earthquake jarred awake on January 17, 1994, can attest to that. A large earthquake can give one renewed appreciation for the power of nature.

In the early morning hours of December 16, 1954, two major earthquakes rocked eastern Churchill County, awakening residents throughout much of central Nevada. These shocks struck four minutes apart and had magnitudes of 7.1 and 6.9, respectively. The epicenter of the first shock was near a system of faults along the east side of Fairview Peak. The epicenter of the second shock was along the west side of Dixie Valley at the base of the Stillwater Range. The combined length of the surface fractures these two earthquakes produced measures about 60 miles. Thanks to the sparse population in the area, the temblors resulted in no deaths or injuries. But what exactly are earthquakes, and why do they occur?

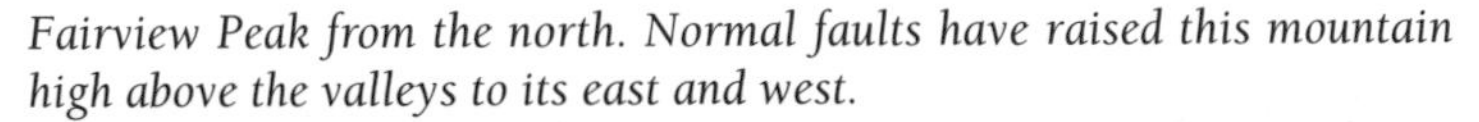

Fairview Peak from the north. Normal faults have raised this mountain high above the valleys to its east and west.

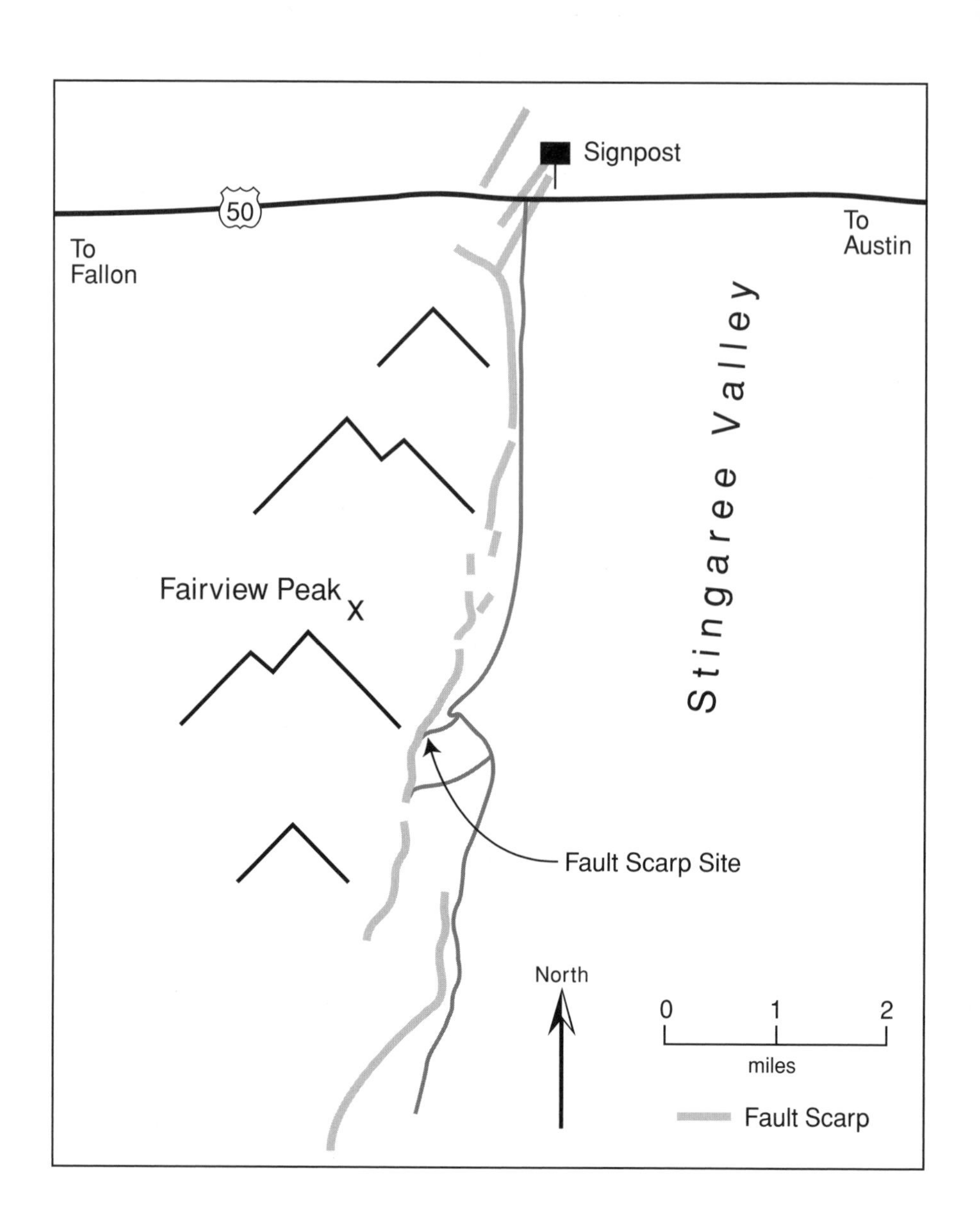

GETTING THERE

The Fairview Peak fault scarp is exposed along the east side of Fairview Peak, about 40 miles east of Fallon. If you are traveling from the west, the dirt road that leads to the site heads south from U.S. 50 about 1.7 miles east of the intersection with Nevada 121, just past Drum Summit. If you are traveling from the east, the road heads off about 5.5 miles west of the intersection with Nevada 361. The turnoff is well marked, and an interpretive sign 0.3 mile south of U.S. 50 will let you know that you are on the right road. As you travel along this road, look to your right to see a distinct break in slope where the fault scarp cuts along the base of the mountain

If you looked west from time to time as you drove south along the eastern face of Fairview Peak, you saw the slope break that corresponds to the fault scarp next to which you are parked. The scarp itself is an unvegetated slope steeper than both the slopes above and below it. Fairview Peak is not a particularly large mountain by Nevada standards, but like many mountains in this state, it is growing. And these peaks do not grow gradually; they burst upward, accompanied by a release of energy called an earthquake.

An earthquake is a response to stresses that build up within the earth's crust. These stresses build where crustal plates slide past each other, collide with each other, or spread apart. In the Basin and Range, these stresses result from crustal extension or spreading. As the stress slowly builds, the rocks begin to deform. Rock deforms in two different ways. Let's consider two dowels, one of wood and the other of metal. If you bend the wooden dowel slightly its shape changes. If you release it, it returns to its original shape. That's elastic strain—it springs back like an elastic band. If you bend it farther, it eventually breaks, and when that happens, you feel a jolt in your hands as the dowel releases the elastic strain it had stored. If you bend the metal dowel, it deforms, but when you release the stress, it does not return to its original shape. That's plastic deformation. Rock at the earth's surface is brittle and undergoes elastic strain—as the wooden dowel does—in response to applied stress. Rock deep beneath the surface deforms plastically.

As stress slowly builds, rocks near the surface begin to deform; the more the rocks deform, the more energy they store. This straining cannot go on indefinitely, and sooner or later the rocks fracture and slip past each other. The physical point where this rupture begins is

range. Proceed south for about 5 miles, then turn right onto another dirt road that heads southwest. This road is a little rough and somewhat steep, but most passenger cars should be able to negotiate it. Follow this road 0.7 mile to the end, and park. There is another interpretive sign at the parking area. From here, you can trace the well-exposed fault scarp a good distance to the north and south. You may also want to return to the main dirt road and explore the area to the south, where you will find other fine exposures of this fault scarp. Access these exposures from any one of the several unnamed dirt roads leading west.

called the focus of an earthquake. The quake's epicenter is the point on the earth's surface directly above the focus. The plane along which the rocks slip past each other is the fault. When the fault ruptures, the rocks release some of their stored energy, partly as heat and partly as waves that transmit energy away from the fault. These waves are what we feel as the earthquake.

Earthquake waves are not all the same. There are three types of waves: P waves, S waves, and surface waves. P waves are the fastest waves; S waves are slower than P waves but faster than surface waves. The body of the earth transmits P and S waves but not surface waves, which, as their name indicates, travel only along the earth's surface. Surface waves cause most of the earthquake damage to buildings and other structures because they produce large-scale up-and-down and side-to-side shaking.

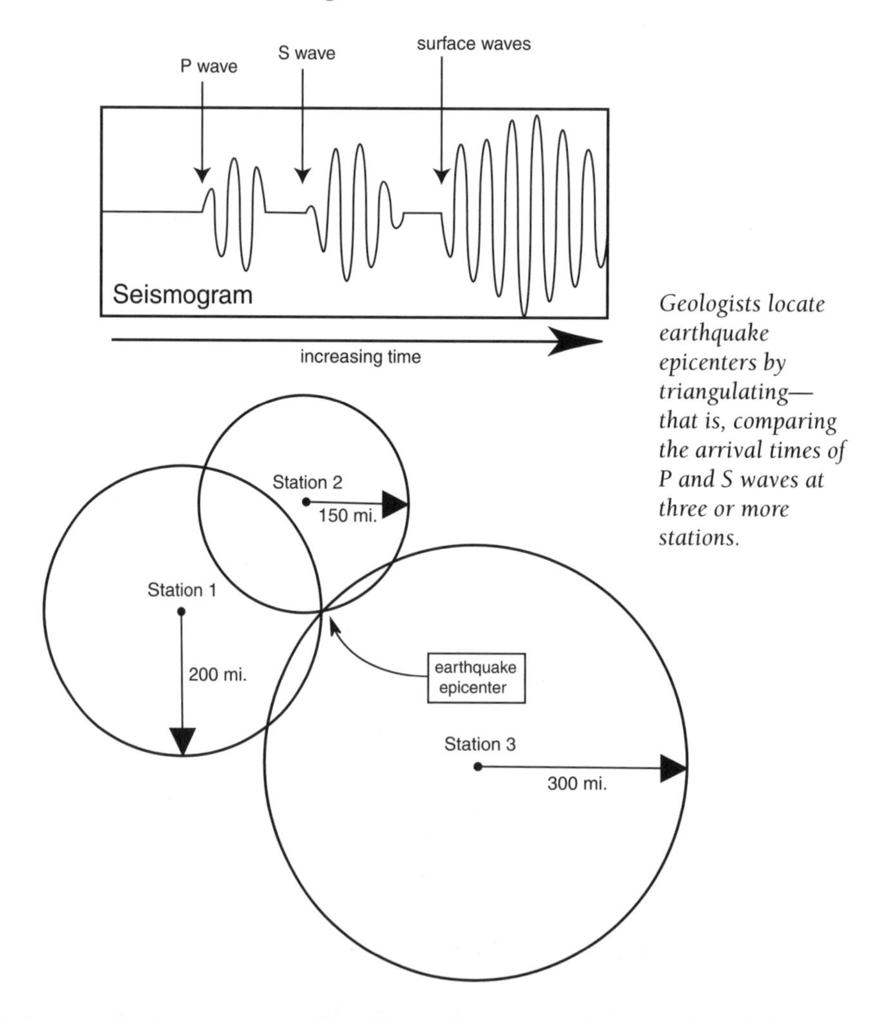

Geologists locate earthquake epicenters by triangulating—that is, comparing the arrival times of P and S waves at three or more stations.

Geologists use P and S waves to locate the epicenter of an earthquake in a process called triangulation. Consider an analogy. In the interest of science, a middle-aged geologist has agreed to run a series of races against an Olympic middle-distance runner. The geologist is more an ambler than a runner; the last time he ran was when someone shouted "Free beer!"—but even then he was a little too late. As you might expect, the Olympic athlete is in superb shape, and she feels confident of victory at any distance. The first race is 100 meters, which the runner wins by 8 seconds. They next run 400 meters; she finishes in 59 seconds while he shuffles across the finish line in 3 minutes. She wins the mile by 6 minutes. The final race is the marathon, which the Olympic athlete finishes in 2 hours and 30 minutes. The geologist still hasn't reached the finish line after two days, but race organizers aren't really concerned. They figure he stopped to look at a roadcut and got distracted. He'll show eventually; they promised him free beer, after all.

How does this story relate to earthquakes? The key is that when one runner is consistently faster than another, the elapsed time between their finishes increases as a function of the distance they've run. The P wave is the conditioned athlete, while the S wave is the geologist. P waves always arrive before S waves, and the difference in their respective arrival times tells seismologists how far their particular station is from the epicenter. If two stations share data, they can narrow the location of the epicenter to two points. With three stations, they can locate the epicenter exactly, hence the term *triangulation*.

Along the east side of Fairview Peak, a spectacular fault scarp remains as the surface expression of the first of the 1954 earthquakes. Walk around and examine the fault scarp. Note how fresh and youthful it looks. Even though the earthquake struck more than forty-five years ago, the scarp looks as if it formed only yesterday—a testament to its remarkable preservation in the arid climate of this region.

Here, the scarp cuts through alluvium shed from the mountain and measures about 15 feet high. Scarp heights of from 5 to 20 feet are common in this area, although true displacements are on the order of 5 to 13 feet. Alluvium has slid and settled into the base of the fault in areas, forming what are called gravity grabens that exaggerate the apparent offset. See if you can find places along the scarp where the land surface to its east appears to slope down and into the fault. Those "back slopes" at the base of the scarp are the gravity grabens. Earthfalls at the top of the scarp also tend to accentuate offset. Look for slumping and retreat of the upper face that has occurred since the

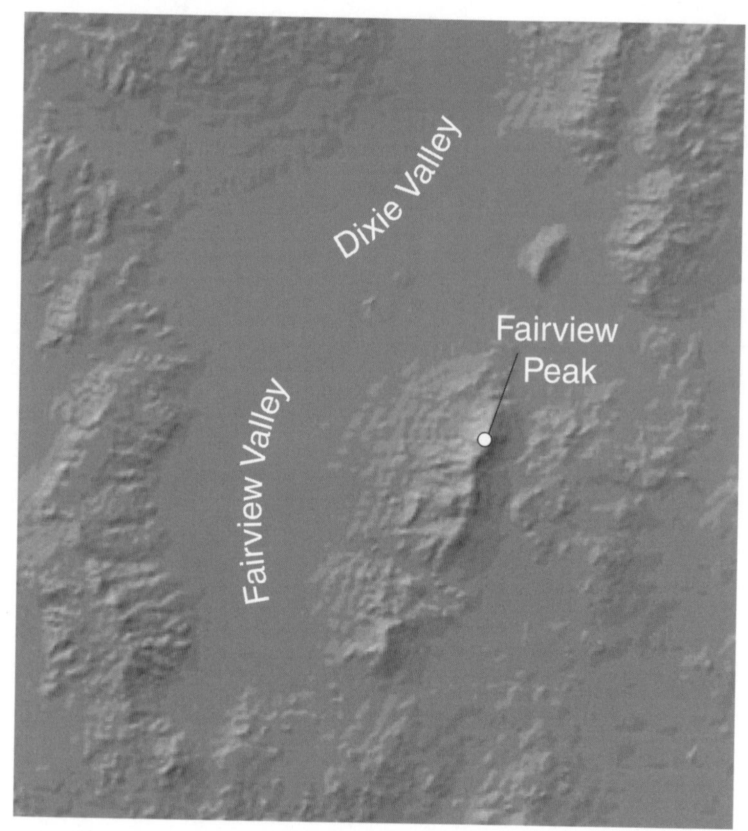

Asymmetric faulting along the base of the Fairview Range has produced a steeper eastern slope and a shallower western slope.

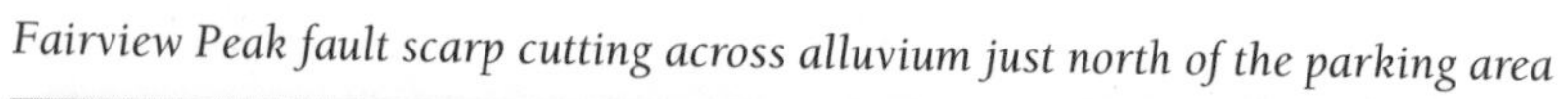

Fairview Peak fault scarp cutting across alluvium just north of the parking area

The Fairview Peak fault scarp near the parking area shows offset of about 10 feet. —Dave Futey photo

1954 earthquake. The more the slope fails along the scarp, the longer the unvegetated face becomes and the greater the fault's offset appears.

The two fault blocks moved up or down relative to one another, but they also moved side-by-side past each other. As you look west toward the scarp, you may be able to identify small offsets in gullies that demonstrate that the block across from you slid to the right as it moved upward. This type of offset is most apparent immediately after fault motion. Erosion smooths the land surface over time and evens drainages back out.

Look closely at some of the rocks that litter the scarp. Many of these angular pieces exhibit linear scratches called slickensides. These slickensides formed as the rocks ground against other rocks when the fault moved. These features are more conspicuous on larger rocks than on smaller rocks.

For the local residents abruptly shaken out of bed early that December morning in 1954, one of the first questions to come to mind after the ground stopped shaking might have been, "How big was it?" Seismologists determine the size of an earthquake in two ways. The first, intensity, is a qualitative measurement of the amount of shaking. Intensity rates the degree of damage to structures, the amount of disturbance to the surface of the earth, and the extent of animal reac-

Slickensides, such as these in a rock at the Fairview Peak fault scarp, are grooves and scratches cut into rock during fault movement. They parallel the direction of offset. —Dave Futey photo

tion to the shaking. Before instruments capable of quantitatively measuring magnitude became available, this was the only method to compare the sizes of various earthquakes. The most widely accepted scale of earthquake intensity is a twelve-degree scale based on the one developed by Giuseppe Mercalli in 1902. The Modified Mercalli Scale ranges from I, felt by few people, to XII, total damage. Scientists developed this scale to fit the conditions in California. Other countries, including Japan, have developed intensity scales to fit conditions in those areas. However, intensity ratings are somewhat subjective and depend on the population density and the type of building construction in the area.

According to the accounts of local residents, the 1954 earthquakes did not extensively damage the wooden structures in the area, but they shook many residents out of their beds, knocked most items off shelves, moved heavy furniture across the floor, and destroyed chimneys. They also opened substantial breaks in the ground, a fact that is obvious from our present vantage point. These surface fault breaks are large enough to justify a Modified Mercalli rating in the IX to X range, although a rating based solely on property damage would be somewhat less. If these earthquakes had struck in a more densely populated area, they probably would have wrought significantly more property damage, which would have more accurately reflected the intensity rating based on surface fractures. So, here we have a good example of the difficulty in applying this scale.

When instruments capable of quantitatively measuring the energy an earthquake releases became available, seismologists developed a standard quantitative scale to compare earthquakes worldwide. Originally introduced in Japan and later developed by Charles Richter in 1935, this scale used the wave heights, or amplitudes, measured by a seismograph to calculate the magnitude of the earthquake. Because the sizes of earthquakes vary enormously, the amplitudes of the ground motions differ by factors of thousands from one event to another. Therefore, Richter used a logarithmic scale in which each unit of magnitude corresponds to a tenfold increase in wave amplitude. So, an earthquake of magnitude 7 is ten times more powerful than one of magnitude 6. This trick has the effect of compressing the range of wave amplitudes. In the 1954 earthquakes, the magnitude 7.1 quake was 1.6 times more powerful than the 6.9 quake. The Richter scale of magnitude has no upper or lower limit, although crustal rocks certainly have limits to their strength. Rock can store energy as strain until the strain exceeds the rock's strength. At that point, the rock ruptures and releases that stored energy. Even the strongest rocks can only store a given amount of energy before they fracture.

Different types of forces within the earth's crust create different types of faults. Tensional forces pull the crust apart and lead to normal faults, where the block above the fault, or hanging wall, moves down relative to the block below the fault, the footwall. On the other hand, compressional forces created where plates converge tend to form reverse faults, where the hanging wall moves up relative to the footwall. A third type of force, shear, can create strike-slip faults in which blocks of rock slide horizontally past each other. This type of fault may create a visible scarp similar to that of a normal fault. Geologists determine the amount of movement along a strike-slip fault by measuring the offset of certain features, such as streams or dikes, that cross the fault. If you are looking across the fault from one side and the other side appears to have moved laterally to the right, you are looking at a right-lateral strike-slip fault. If the block across the fault from you has moved laterally to the left, the movement is left-lateral.

Observe the orientation of the fault plane and fault blocks here. If you are standing below and to the east of the scarp, the fault plane slopes steeply toward you. You are standing on the hanging wall, which has moved down relative to the footwall. This appears to be a normal fault, one that forms where tension pulls the earth's crust apart, a

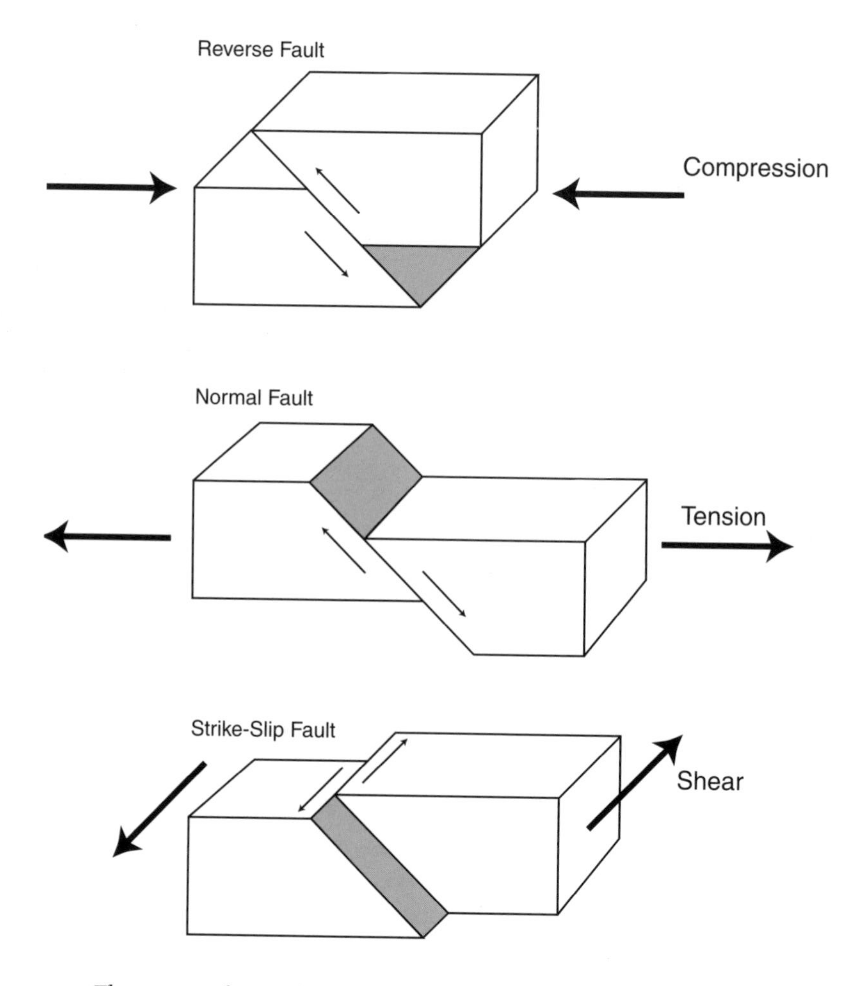

Three types of crustal stress result in three types of earthquake faults.

process currently in progress in the Basin and Range. Rising magma, bowing and stretching the crust, commonly causes this type of crustal extension.

The Fairview Peak fault, and the Dixie Valley fault to the north, are part of the Central Nevada Seismic Belt. Several other earthquakes with magnitudes greater than 6 have shaken this part of Nevada in the recent past. These include the 1915 Pleasant Valley earthquake, the 1932 Cedar Mountain earthquake, and the Rainbow Mountain and Stillwater earthquakes of August 1954. In the Cedar Mountain event, to the south of Fairview Peak, the fault moved primarily as a

View west from the road to the parking area. The light line at the base of the hill is the trace of the Fairview Peak fault.

strike-slip fault with right-lateral movement. This fault lies within a northwest-trending zone called the Walker Lane belt, where shear causes right-lateral movement to dominate. In the Pleasant Valley quake to the north, the fault blocks moved mainly up or down the fault plane in response to tension. Close inspection of the Fairview Peak and Dixie Valley faults in the years since the 1954 earthquakes revealed that much of the Fairview Peak fault shows significant right-lateral offset, whereas the Dixie Valley fault offset rocks up or down the slope of the fault. This tells us that the Fairview Peak earthquake may have been far enough south to feel the influence of the shearing within the Walker Lane belt. The Dixie Valley earthquake, being slightly north, instead displayed the up and down motion more typical of the central Basin and Range, where crustal extension dominates.

So, what does all this mean? For one thing, the block on which Fairview Peak lies has moved up about 10 feet in this area relative to the block on which we are standing. This may not seem that significant, considering the relief that surrounds you, but if you remember that a thousand such earthquakes may have struck over the last few million years, a picture of how the topography of the Basin and Range has evolved may come into focus. Each earthquake has resulted in normal faulting, which moves mountains up and valleys down. The Basin and Range probably began as a fairly flat surface. Continued east-west extension has produced a landscape of north-south trending mountain ranges separated by deep valleys. As long as tension continues to control deformation in Nevada, as elsewhere in the

The steep slope from bottom left to upper right is the fault scarp. It represents offset of about 10 feet created during a single earthquake. —Dave Futey photo

Fault scarp as it appears 0.5 mile south of the parking area. Displacement here is approximately 10 feet.

Basin and Range, normal faulting will dominate and mountains will grow taller.

You'll find other good exposures of the Fairview Peak fault both north and south of the area we have been studying. Take the time to look around and investigate this fault scarp in different places, and compare the scarp heights from one place to another. Displacement is not constant along a fault—some lengths are offset more than others as rock releases some energy stored as elastic deformation. Later processes, such as landslides and erosion, will create the impression of greater offset than really exists. Look for slumps and rills that seem to accentuate displacement of the scarp surface.

The 1954 shocks, the most recent large earthquakes near Fairview Peak, released a huge amount of energy that shook the land surface and incrementally lifted the range to our west. The release was only temporary, however. Tension continues here, and even now bedrock beneath our feet is deforming and storing energy. We don't know when the next quake will rattle central Nevada, but we know it will come. That's the nature of geology here in the dynamic Basin and Range.

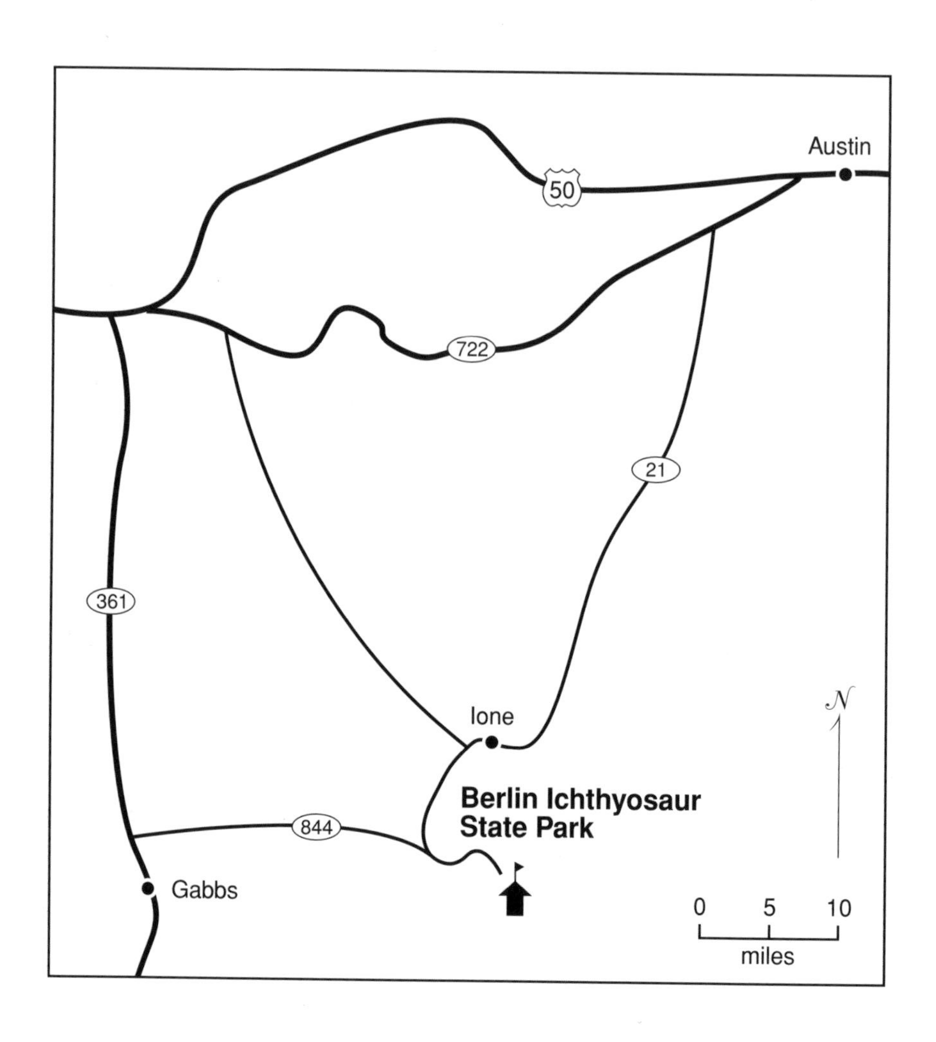

GETTING THERE

To reach Berlin Ghost Town from Fallon, drive east on U.S. 50 for 47 miles. Turn south (right) onto Nevada 361 and continue for 29 miles. Just before reaching the town of Gabbs, turn east (left) onto Nevada 844 and drive 20 miles to the state park entrance. To reach Berlin Ghost Town from Austin, head west on U.S. 50 for 62 miles and then drive south on Nevada 361 for 29 miles. Just before reaching the town of Gabbs, turn east (left) onto Nevada 844 and drive 20 miles to the state park entrance. If you are traveling from Tonopah, head west on U.S. 6 for 41 miles. Turn north (right) onto U.S. 95 and travel 39 miles. Turn north (right) onto Nevada 361 and drive 32 miles, through Gabbs, before turning east (right) on Nevada 844. Continue 20 miles to the park entrance. Park your vehicle in the small lot near the entrance and explore Berlin on foot following the roads and trails on the included map. Berlin is an important piece of Nevada mining history, so please don't remove or otherwise disturb any artifacts, and be especially careful about fire.

13

Boomtown Ghosts and the Hunt for Gold

BERLIN GHOST TOWN

The history of Berlin is, in a sense, a microcosm of the history of gold mining in Nevada and the West as a whole. A plot of annual mineral production in Nevada shows not a steady increase over time but rather a series of dramatic peaks and valleys that correspond to the natural cycle of activity in the mining industry. This cycle controlled more than just mineral wealth. Population in Nevada grew and shrank dramatically in response to mineral discovery and production. Population growth associated with the discovery of the Comstock gold and silver lode in 1859 led Abraham Lincoln to declare Nevada a state in 1864 (it was formerly part of the Utah Territory). Many towns appeared out of nowhere then vanished just as rapidly several years later.

Berlin is exceptional only in its state of preservation. As gold prospectors moved from one boomtown to another, they naturally took all their possessions with them. In much of Nevada, wood for construction is relatively rare, so miners often tore down their dwellings and transported those building materials as well. In Berlin this did not happen, so we are left with an intact example of a Nevada mining town dating from the second half of the nineteenth century.

Cycles of activity in boomtowns follow four phases: discovery, prosperity, decline, and depression. The initial discovery of gold or another valuable mineral and the establishment of mines brings a great influx of prospectors to the area. During the prosperity phase, establishment of new mines slows as production of gold from established mines increases. The duration of prosperity depends on the quantity of gold ore available and the ability of the miners to successfully excavate it. Decline follows as mining depletes the ore body and mines

An abandoned ore car sits on the rail that once led from the mine hoist, in the background, to the mill. —Dave Futey photo

begin to close. Decline ends in depression and the abandonment of the once-prosperous town.

Berlin, part of the Union Mining District, blossomed during the early cycle of mining in Nevada. Discovery during this cycle extended from 1849 to 1868, prosperity from 1869 to 1880, decline from 1881 to 1891, and depression from 1892 to 1899. Prospectors discovered precious metals here a bit later than elsewhere in the district, so activity in Berlin lagged a bit behind this general cycle. In Union Canyon, south of Berlin, prospectors first discovered silver in 1863. The initial assay of material from Berlin Canyon was reported in 1869. The Berlin mine was established in 1896 and operated until 1908. The town declined rapidly from 1908 to 1911, at which point it emptied and became the ghost town we see today.

Nevada's turbulent geologic history enriched the state in mineral wealth. Volcanic activity with abundant magma sources has led directly to the long history of successful gold mining in the state. Each volcano connects through a vent to a magma body, many of which contain the element gold. To produce an ore body, this magmatic gold must be concentrated into an economically viable deposit. As magma cools, minerals begin to crystallize into rock. When most of the magma has solidified, what remains is a mineral-rich fluid called a hydrothermal solution. Mostly water, this hot liquid is rich in dissolved quartz and also may contain valuable minerals such as gold, silver,

The mine hoist brought ore to the surface from deep shafts that followed veins of gold. Miners dumped the ore into rail cars for transport to the mill for processing.

platinum, and copper. Owing to their fluidity, hydrothermal solutions can move through fractures and pore spaces in rock. As the liquid moves away from the hot magma, it cools, which allows minerals to precipitate. Quartz veins with inclusions of gold crystallize within fractures and faults to form lode ore. Microscopic crystals of gold accumulate in pore spaces within rock to form invisible but valuable disseminated ore bodies. If you stroll around the foothills bordering Berlin, you will see white streaks in the bedrock. These are veins of quartz like those that miners followed in their search for gold here.

Prospectors in Berlin, and elsewhere in Nevada first, discovered gold not as a lode deposit but as weathered and eroded remains within a stream channel. Moving water transports light materials more easily than heavy ones. Gold has a specific gravity of 19.3, meaning it is 19.3 times more dense than water. So, water preferentially concentrates weathered gold grains within point and channel bars in streams, while it carries other mineral grains farther downstream. Such concentrations of gold and other heavy minerals that streams or waves create are called placer deposits.

Panning is the simplest method of prospecting and mining for placer gold. Panners scoop sediment into a shallow pan with gently sloping sides, then swirl the pan back and forth in the stream with the top edge barely submerged, allowing water to remove lighter mineral grains. If gold is present, it will remain in the pan after other materials have been swept away. Panning is a slow process, and miners quickly developed more sophisticated methods of collecting placer gold.

Miners built tools called cradles to speed the hydraulic separation process. The most important component of the cradle is a riffle box, a flat surface with raised bars across the floor to trap gold, and sides that prevent the escape of sediment. The riffle box was mounted on rockers. One miner dumped sediment and water through a mesh screen onto the riffle box. Another miner rocked the whole contraption back and forth, allowing water to carry light sediment over the riffles and out of the cradle. The workers periodically scooped up with a spoon any gold captured by the bars. As long as miners continued to collect gold in the stream, they built ever-larger hydraulic

Cradles like the one shown here allowed miners to separate gold from stream sediment more efficiently than by panning.

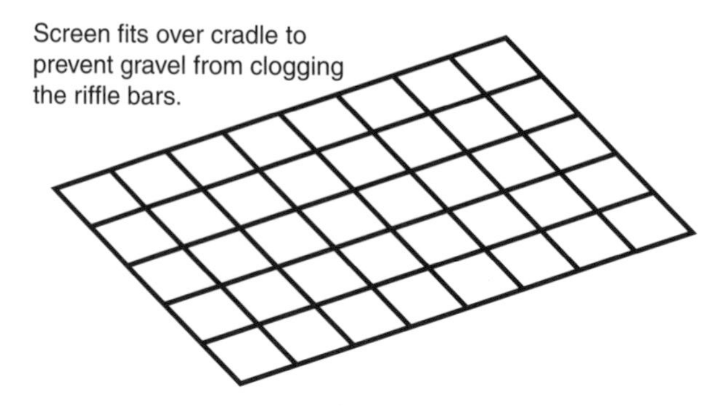
Screen fits over cradle to prevent gravel from clogging the riffle bars.

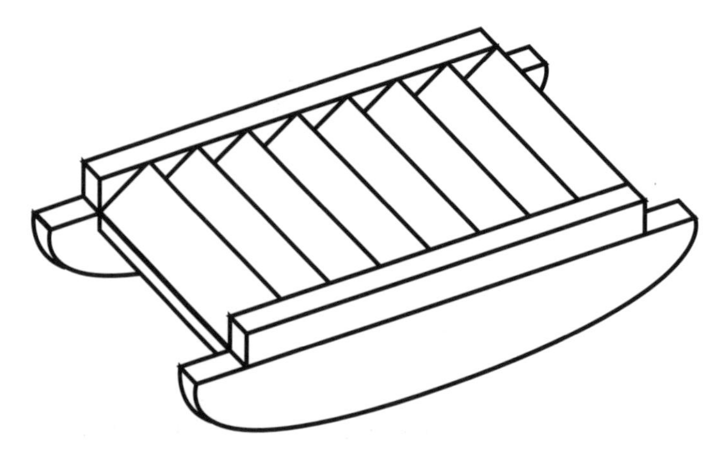
Miners rock the cradle back and forth in the flowing stream while shoveling sediment through the screen onto the riffle bars.
Riffle bars trap heavy minerals, such as gold, while allowing light minerals to flow downstream.

machines. But in Berlin, as in most other mining boomtowns, miners quickly identified the bedrock source of the placer gold and focused their attention there.

Mining for lode ore in bedrock is a more difficult proposition than placer mining. Gold in quartz veins invariably led into solid rock, necessitating removal by the backbreaking and extremely dangerous construction of an underground horizontal tunnel or vertical shaft. This was the main type of mining practiced in Berlin. Miners excavated an eight-level mine, which they accessed along an inclined shaft to remove gold ore.

The buildings within Berlin formed the mine headworks—all the structures devoted to the removal and processing of ore. These include the hoist, mill, assay office, machine shop, warehouse, and living quarters. The original hoist probably consisted of a winchlike

Layout of buildings and other features within the ghost town of Berlin

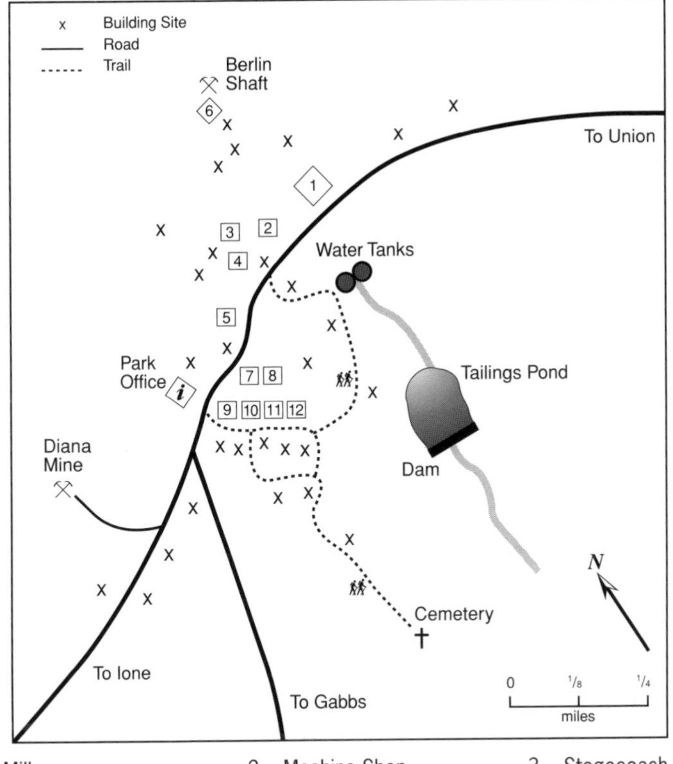

1	Mill	2	Machine Shop	3	Stagecoach Shop
4	Assay Office	5	Warehouse	6	Hoist Building
7	Bachelor Quarters	8	Stevenson Home	9	Foreman's Home
10	Watson Home	11	Blacksmith Shop	12	Phillips Home

windlass and bucket with which a single miner could raise material to the surface. As the mine grew, the steam-driven engine hoist that we see remnants of today was built. This enabled more efficient movement of men and ore. The hoist brought ore up through a vertical shaft to the surface, where it was loaded into rail cars that took it to the mill. Berlin's hoist is labeled number 6 on the site map.

The work of the mine was conducted in the drifts, or working tunnels. Drifts in small mines typically wandered mazelike through the earth following the hot lead, the concentrated ore deposit. In larger mines, however, engineers attempted to discern the size and shape of the ore body and planned organized tunnels and shafts to remove it. We don't know if trained engineers worked in Berlin, but probably they did not. Miners dug spacious caverns, called stopes, where the lode enlarged into ore pockets. In these rooms, workers left pillars of rock to support the roof, a practice learned from coal mining. A wooden frame shored tunnels and stopes in weak bedrock; square-set timbering, developed in the Comstock mines of northern Nevada, is one such method of support. Wood for shoring no doubt came from adjacent mountains, where cooler temperatures and higher precipitation supported pine forests. If you take a tour of the Diana Mine, just west of the park office, you will have the opportunity to move through shored tunnels, observe vertical connecting shafts, and see tools that the miners used.

Any combination of buckets, rail cars, mule train, or hoist transported the mined material to the surface. Workmen at the hoist separated ore from waste rock. Numerous waste heaps surround the town of Berlin. Ongoing assays ascertained that workers weren't sending valuable material to these heaps. Assays are chemical tests to determine the concentration, and hence the value, of gold and other metals in rock. An assayer performed these tests in the assay office, building number 4 on the site map.

Berlin's assay office kept busy with samples from the large mine and from prospectors exploring the surrounding mountains. The assayer determined the type and quantity of metallic minerals in each sample brought in. A typical assay office was furnished with a small crusher, a furnace, a worktable, crucibles, glassware, and a balance. An apprentice pulverized samples to dust-size grains and screened them for uniformity. The assayer weighed a portion of the dust, then mixed it with lead oxide, borax to catalyze the alloy process, and wheat flour to make a paste before placing it in the furnace. Shutting the furnace door reduced the oxygen supply, creating an environment that transformed lead oxide to molten lead. The molten lead descended through the mix and collected whatever valuable minerals were present. When this mass cooled, the assayer released the metal nugget from the glassy slag with a hammer blow. He then melted the lead alloy in an oxygen-rich environment to once again form lead oxide. Melting took place in a small bowl made of bone ash, which has the

The mine hoist carted gold ore to the crushing mill (right). *The machine shop* (left) *serviced the rail cars, crushers, and other mining equipment.* —Dave Futey photo

Miners performed chemical tests to assess the gold content of ore in the assay office in Berlin.

unique ability to absorb lead oxide, thus leaving behind a bead of doré, an alloy of gold and silver. The assayer weighed this bead, then heated it gently in nitric acid, which dissolved silver but left gold to be dried and weighed. The weights of the original sample, the doré bead, and the final gold allowed the assayer to determine both the gold and silver concentrations of the assayed ore.

Return to the hoist, then follow the rail line to the rear of the mill, the largest building in Berlin and number 1 on the site map. Once workers separated the ore from the waste, high-sided wagons with a three- to four-ton capacity transported the ore to the mill. There, crushers—miners with sledges—pounded it to the size of apples. They shoveled these chunks into a jaw crusher that reduced the ore to pea-size pieces. The cast-iron Blake crusher was the favorite of western miners due to its simplicity and durability, and it was probably the type used in Berlin. Crushed ore then moved to a gravity stamp mill such as the one we see in Berlin, in the same building as the crusher. Stamp mills consisted of five or more steam-driven pillars—the Berlin mill had thirty—whose stamps were 1,000-pound cylinders of iron. The stamps stomped on the ore to pulverize it. Miners moved ore laterally from stamp to stamp, until the pulverized rock finally moved through a fine screen. This dust commonly traveled in water down a bumper table, a type of riffle box, to concentrate ore.

In the amalgamation process that followed, mill workers mixed ore with water and salt in heated barrels or on an amalgamation table

This rail line, entering the upper floor of the crushing mill here, carried gold ore from the mine hoist.

Gold mining in Berlin demanded skilled carpentry in the construction of the mill, seen here, and other structures. —Dave Futey photo

to form a paste, to which mercury was added. After twenty-four hours of mixing, amalgamated gold and mercury formed a heavy liquid that workers separated from the paste in the barrel. Mill workers poured the paste, known as pulp, over a mercury-coated copper plate. Gold and silver adhered to the mercury while other material continued to flow over the surface. The final process, retorting, involved the heating of amalgam, the remaining mixture of mercury and gold, to vaporize the mercury and form gold bullion. Much, but not all, of the mercury was trapped and recycled for later use. Workers shipped the gold bullion by wagon and train to city banks in the West. Pulverized ore, waste from which miners had removed the gold, was piled near the mill. You can distinguish such tailings waste from the other waste heaps by its very small grain size. The tailings pond, just downstream of the water tanks, holds an accumulation of this fine-grained waste.

From 1898 until 1908, years of prosperity for the Union Mining District, Berlin supported some 250 people, including miners, blacksmiths, woodcutters, charcoal makers, a doctor, a nurse, and a prostitute. By 1911, the city was empty—after producing an estimated $850,000 worth of gold at $20 per ounce. As you stroll through Berlin's abandoned streets try to imagine the vibrant and raucous mining town that once was.

A frieze outside the fossil shelter shows a reconstruction of an ichthyosaur based on the remains inside. Note the large eye and long snout with abundant sharp teeth. —Dave Futey photo

14

Mesozoic Monsters of the Deep

NEVADA'S ICHTHYOSAURS

Berlin-Ichthyosaur State Park encompasses two world-class attractions: Berlin Ghost Town, an excellent example of an abandoned turn-of-the-twentieth-century frontier mining settlement, and one of the world's largest concentrations of exposed fossil ichthyosaurs—an extinct group of marine reptiles. Located in west-central Nevada, the park sits on the western flank of the Shoshone Mountains at an elevation of about 7,000 feet above sea level. A seemingly endless expanse of sagebrush covers the relatively featureless landscape of the adjacent valley. The rocky flanks of the Shoshone Mountains support scattered sage, juniper, and piñon pine. This sparsely populated part of Nevada has changed little since the first settlers arrived more than one hundred years ago. But it has changed much since ichthyosaurs lived here more than 200 million years ago.

Outside the Ichthyosaur Fossil Shelter, you will see a sculptured wall adorned with the life-size image of an ichthyosaur. Its name comes from a combination of the Greek words *ichthys*, meaning "fish," and *sauros*, meaning "lizard," because they superficially resemble fish. In spite of their fishlike appearance, they were air-breathing reptiles that had to surface for air or drown. Ichthyosaurs did not lay eggs as most other fish and reptiles do. Instead, they gave birth to live offspring. Their fossil bones are common in the Luning formation of Nevada, which consists mainly of sedimentary rocks of marine origin that are about 230 million years old, or late Triassic in age. The Luning formation of central Nevada has yielded bones of the largest ichthyosaurs found anywhere in the world.

The species of ichthyosaur preserved at Berlin-Ichthyosaur State Park is called *Shonisaurus popularis*. Members of this species were very large; the frieze outside the shelter shows a creature with an overall length of 56 feet and an estimated weight of 40 tons. The head alone is 10 feet long. These extinct sea creatures compare roughly to the size of a modern sperm whale, or nearly twice the size of a killer

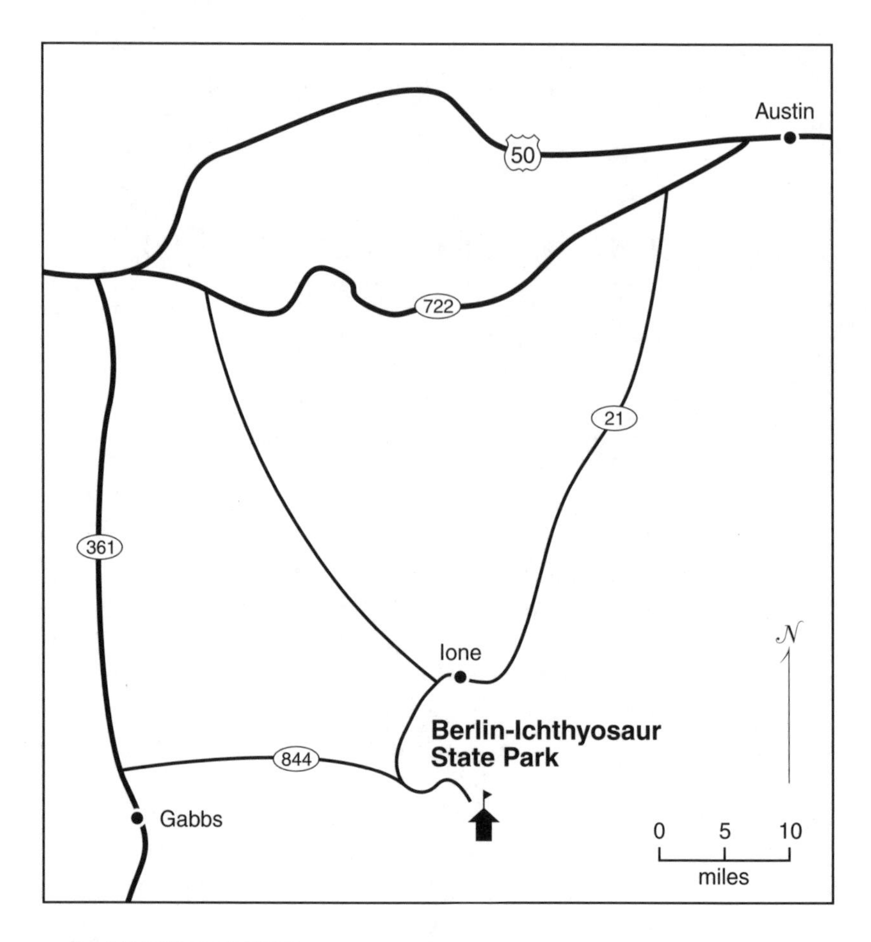

▶GETTING THERE

To reach Berlin-Ichthyosaur State Park from Fallon, drive east on U.S. 50 for 47 miles. Turn south onto Nevada 361 and continue for 29 miles. Just before reaching the town of Gabbs, turn east (left) onto Nevada 844, and drive 20 miles to the state park entrance. To reach the park from Austin, head west on U.S. 50 for 62 miles before driving south on Nevada 361 for 29 miles. Just before reaching the town of Gabbs, turn east onto Nevada 844 and drive 20 miles to the state park entrance. If you are traveling from Tonopah, head west on U.S. 6 for 41 miles. Turn north onto U.S. 95 and travel 39 miles. Turn north onto Nevada 361 and drive 32 miles, through Gabbs, before turning east on Nevada 844. Continue 20 miles to the park entrance.

Drive through Berlin Ghost Town, following signs to the Ichthyosaur Fossil Shelter, and park in the lot adjacent to the building. A short and easy nature trail leads north from the campground to the fossil shelter.

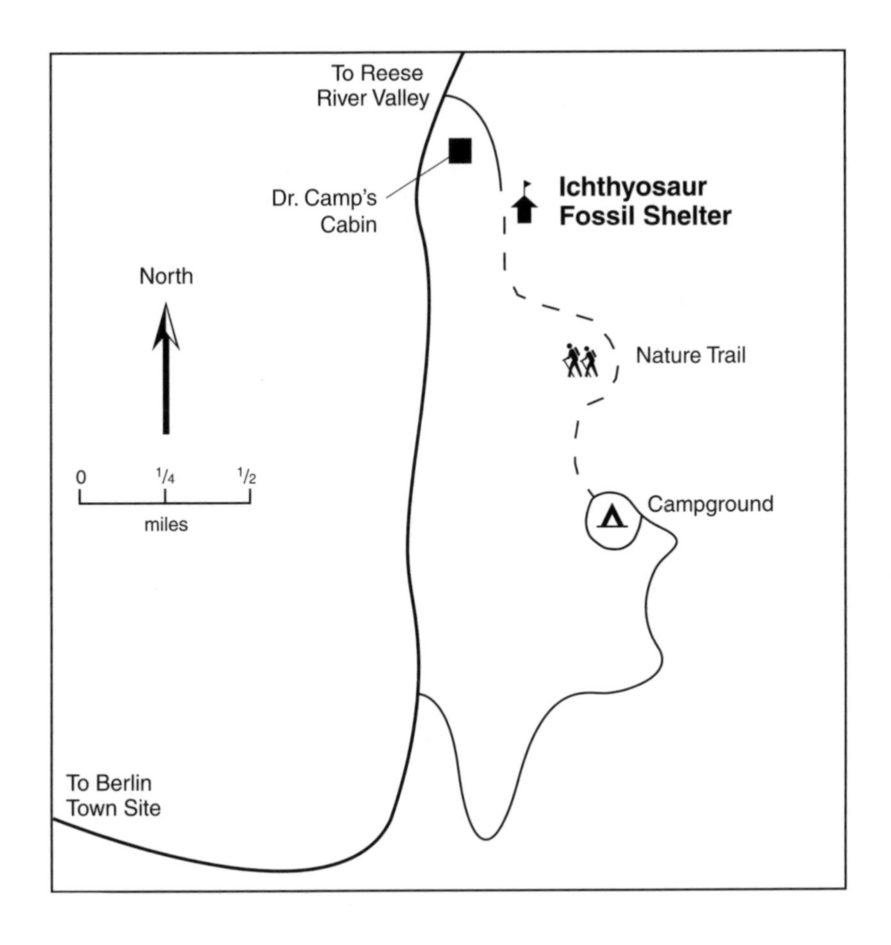

The park offers several guided tours of the Fossil Shelter daily from Memorial Day to Labor Day and on Saturdays and Sundays during the spring and fall. Check with the park for current tour schedules. Tours of the quarry are also available by request from November 14 to March 15.

Camping is available at the park campground. Each of the campsites has a covered picnic table, fire ring, and grill. Drinking water is provided at the campground from mid-April to October. The park has a day-use picnic area and an extensive system of hiking trails. A very easy hiking trail leads from the northern end of the campground to the Fossil Shelter, a one-way distance of about one-half mile. The higher elevations and shade from juniper and piñon pines make temperatures at the campground comfortable in the summer.

You may wish to combine a visit to the ichthyosaur shelter with a stroll through the ghost town of Berlin (vignette 13), whose residents first discovered ichthyosaur bones while mining for gold in Union Canyon.

whale. The ichthyosaur's highly streamlined body, well fitted to life as an active predator in a marine habitat, resembles the form of some of the fastest and most active swimmers in today's oceans, such as swordfish, marlin, and tuna. Estimations of ichthyosaur swimming speeds indicate that they were probably the fastest swimmers of the marine reptiles. The elongate mouth and strong jaws hold rows of pointed, conical teeth, similar in shape to those of the modern porpoise. A circular set of overlapped, bony plates internally reinforced the reptiles' proportionately large eyes and compensated for changes in water pressure when the animals dove or surfaced. This feature enabled them to consistently maintain their highly developed sense of vision at all swimming depths. The dimensions illustrated here make *Shonisaurus popularis* the largest ichthyosaur in the world and the largest animal known from the Triassic period.

Before entering the shelter, let's spend some time talking about history. Prospectors from the nearby mining town of Berlin discovered the first ichthyosaur bones in Union Canyon during the late 1800s, in what is now Berlin-Ichthyosaur State Park. At that time, the bones were a curious novelty. Miners used some of the bones to decorate the fireplace hearths of their homes in Berlin. They used larger vertebrae, with their flattened, disklike shape, circular outline, and slightly

The fossil shelter houses a preserved excavation of ichthyosaur bones from Mesozoic time.

Dr. Camp studied both ichthyosaur bones and rock units in his cabin in Union Canyon. His observations led to the first paleogeographic and skeletal reconstructions of Nevada ichthyosaurs.

concave sides, as dinner plates. The discovery of these ichthyosaur fossils remained unknown to the scientific community until a professor of geology at Stanford University, Siemon M. Muller, visited Union Canyon in 1928 and recognized that the bones belonged to an ichthyosaur. Due to the remote location of the site, the large size of the creatures, and the hardness of the fossil-bearing rock, Muller decided not to attempt to excavate the fossils at that time. Excavation did not commence until 25 years later. Margaret Wheat, an amateur fossil collector and friend of Muller, visited the site in 1952 and marveled at the fossils of these immense creatures. The following year, she returned with Berkeley paleontologist Charles L. Camp, who recognized the paleontological significance of the abundant large ichthyosaurs here. Camp and fellow paleontologist Samuel P. Welles organized a field expedition and returned to Union Canyon in 1954 to begin excavation, a project that continued for the next three years. Work at the site halted temporarily at the end of 1957, but resumed six years later. The excavations then continued from 1963 through 1965. A stroll through Union Canyon takes you by Dr. Camp's cabin, where for many years scientists worked diligently to piece together the puzzle that ichthyosaurs here represented.

The Nevada state legislature designated Union Canyon as the Ichthyosaur Paleontological State Monument in 1955. The monument constructed an A-frame building, the fossil shelter, over the main quarry in 1966 to provide a place for visitors to view the exposed fossils and to protect the fossils from weathering. Later, in 1970, the boundaries of the monument were enlarged to encompass the nearby ghost town of Berlin, much of which is still in excellent condition. Ichthyosaur State Park became a Registered Natural Landmark in 1975, and two years later Nevada claimed the ichthyosaur as the official state fossil.

Altogether, fieldwork at the Union Canyon locality in Berlin-Ichthyosaur State Park has yielded at least thirty-seven mostly complete ichthyosaur skeletons. Charles Camp studied the numerous bones uncovered at the site and determined that they were unlike any other ichthyosaur known to science at that time. Therefore, he named these newly discovered ichthyosaurs *Shonisaurus*, after the Shoshone Mountains. Based upon differences in skeletal anatomy, he assigned them to three different species: *Shonisaurus popularis, Shonisaurus silberlingi,* and *Shonisaurus mulleri*. Researchers recently have questioned the differences between these three species, and now it seems more likely that all the remains from the Union Canyon locality represent just one species, *Shonisaurus popularis*.

Upon entering the shelter, you will see ichthyosaur bones laid out on an irregular surface with a surrounding walkway. The ichthyosaur remains exposed at the fossil shelter were excavated and then left in place. This exhibit consists of nine fairly complete skeletons in close proximity to one another. It may take some time to distinguish between fossil and rock, so be patient. The most recognizable fossilized bones are probably the disk-shaped vertebrae and long ribs. Your tour guide will point out other features, including skulls and smaller bones.

Paleontology is the study of ancient life, and paleontologists piece together remnant bits and pieces of fossilized organisms. But they are also interested in the relationship between that organism and its environment. So there are really two puzzles: what environment did this animal live in, and how did its morphology, its physical structure, enable it to live there successfully? We will focus first on the question of environment; this is the field of paleogeography. Paleogeography is a difficult science, especially with old rocks such as these that have experienced many changes since deposition of the original sediment. Sometimes researchers must base their hypotheses about environments on sparse clues, some of which may seem contradictory.

Take a look at the rock upon which the bones rest. After leaving the shelter, you can get a much better view of the local bedrock, which is similar to what we see here in the shelter, simply by wandering the

Vertebrae and ribs lie exposed within the fossil shelter. These bones remain in the same position in which excavators found them.

slopes of Union Canyon. Many clues indicate that the ichthyosaur bones were deposited in a marine environment. Perhaps one of the most convincing pieces of evidence is the fact that the rocks that enclose the articulated ichthyosaur skeletons consist of very fine-grained sediments. As rivers carry sediment into the ocean, speed of the river water decreases rapidly, which causes the coarse sediment to drop to the bottom near the shore. Fine sediment remains in suspension much longer and travels far out to sea, where it settles slowly to the ocean bottom. The fine-grained sediment in the rocks here indicates that they probably accumulated in the deeper areas of the continental shelf and adjacent ocean basin. During the time that the Union Canyon ichthyosaurs lived, this region of central Nevada was farther south, probably in the tropics, and was situated along the western coast of what is now North America.

Other fossils found with the ichthyosaur bones also support an interpretation of a deep-water environment. Most of these other fossils are from swimming organisms, such as ammonoid and nautiloid cephalopod mollusks, that lived just above the deep seafloor. The general lack of fossils of scavenging marine organisms also fits well with this interpretation, because scavengers characteristically avoid deep marine waters. Also, the rocks here typically lack remains of organisms that lived either on or in the seafloor in shallow marine settings. In contrast, some of the other strata of the Luning formation contain an abundant and diverse assemblage of such organisms, including mollusks, brachiopods, corals, echinoderms, and sponges—a definite indication of a nearshore environment.

Most of the bones of each of the ichthyosaur skeletons are in the correct anatomical position relative to each other. Since the bones of the skeletons are articulated, or not scattered from each other, the ichthyosaurs must have arrived at their final resting place before they had much time to decay, which would have allowed the bones to separate. Strong currents probably would have moved at least some of the carcasses during their decomposition, and the bones of the skeletons would have been scattered during their transport. So, ocean currents must have been fairly weak, as they typically are in deeper water. The relative completeness of the skeletons also indicates that sediment buried them rapidly enough to allow the bones to settle onto the seafloor as the flesh and ligaments decomposed and the skeletal structure collapsed. If the skeletons had remained exposed on

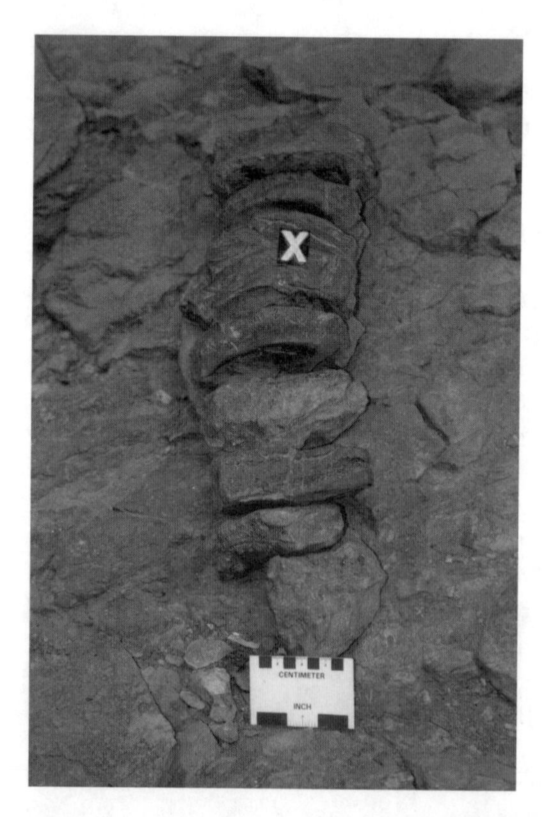

A vertebral column of an ichthyosaur retains its proper orientation even after 100 million years.
—Dave Futey photo

the seafloor for a long time, the bones would have disaggregated and scattered.

Take a look at the orientation of the ichthyosaur skeletons. Most of the skeletons lie roughly parallel to each other lengthwise, and in a northerly direction. Some researchers suggest that the ocean currents along the seafloor, while generally so weak that they did not transport the carcasses during their decay, nonetheless were strong enough to orient the bodies parallel to the current. If you balk at the idea that weak currents could reorient a carcass weighing tens of tons, think of a dead fish floating on a lake. A carcass traps gas as the flesh decomposes. That gas makes the carcass bouyant, so the fish floats during that stage of decomposition. Such bouyancy may have helped "unweight" the ichthyosaur carcasses, allowing weak currents to gently reorient them.

The surfaces of some bones are corroded and pitted by dissolution, evidence that, although they were buried fairly rapidly, they lay exposed on the seafloor before sediment interred them. Scientists see this same type of submarine corrosion in the exposed bones of modern whale skeletons found in the deep Santa Catalina Basin of the Pacific Ocean. The presence of brachiopod shells attached on the upper surfaces of some bones also indicates that the ichthyosaur bones lay exposed on the seafloor. These organisms cement themselves to a hard surface in order to live. This seems contradictory; earlier we stated that sediment here lacks bottom-dwelling organisms, which supports the hypothesis of a deep-water environment. The sediment surrounding the skeletons indeed lacks fossils of burrowing scavengers. The brachiopods represent an ephemeral population that temporarily anchored on the ichthyosaur skeletons until fine sediment subsequently buried them. The bones lay on the surface long enough for mollusks to anchor on them, but fine sediment buried them before they could decompose.

Onward to the creatures themselves! Take some time to look at the bones laid out in front of you. Stroll the walkway and view them from different angles. Is it obvious to you that there are nine individuals here? Is it obvious exactly how the bones fit together? Is it even obvious that these are bones? With so many specimens of *Shonisaurus popularis* collected from this region, scientists know more about the skeleton of this species of ichthyosaur than any other from late Triassic time. But skeletal reconstruction is a tricky business. Since the early excavations here, paleontologists have worked hard to understand how these puzzle pieces fit together into a functioning carnivore. This has often involved revising the work of earlier scientists.

A broken skull lies on the surface within the fossil shelter. Ichthyosaur skulls proved to be the most delicate of the bones found here; none has survived intact. —Dave Futey photo

Over many years, paleontologists have readjusted their estimates of the size of the head and paddles, the length of the tail, and the position and size of the dorsal fin. These anatomical misproportions most likely resulted from the fact that earlier skeletal reconstructions built a picture of a single complete skeleton from the study of portions of several different individuals. This is known as a composite skeletal reconstruction. Although scientists prefer to reconstruct the skeleton of any species from a single individual, this is not possible here. Even in the most complete specimens, certain skeletal parts are not preserved. For example, no entirely complete specimen of the skull of *Shonisaurus popularis* is known. Imagine putting together a three-dimensional puzzle. You don't know what picture belongs on the front, you must take pieces from a number of different boxes, and many of the pieces have broken into smaller pieces. The scientists working with these bones faced that task. It was only many years of work that allowed them to go from what you see before you to the frieze outside, from broken bone to living organism.

One major unresolved problem concerns the explanation for exactly how a large number of articulated ichthyosaur skeletons came to be preserved so close to each other. Let's explore two plausible explanations for this occurrence. Perhaps the ichthyosaurs died singly over an extended period of time; their carcasses sank to the bottom and weak currents naturally concentrated them into a depression or submarine valley in the seafloor. Or maybe this deposit of multiple ichthyosaur skeletons records a massive die-off. In this scenario, something caused all of the ichthyosaurs preserved here to die at about the same time. Several different explanations, based mainly upon studies of modern marine vertebrates, attempt to account for the possibility of an ichthyosaur mass mortality event. These include rapid changes in the temperature or chemistry of the seawater, volcanic eruptions, large storms, specific behaviors (such as spawning, stranding, or coastal foraging), and poisoning by toxic plankton blooms or red tides.

Changes in the physical or chemical conditions of the seawater that were drastic enough to kill ichthyosaurs also would have killed other marine organisms. However, no other fossil evidence exists along with the ichthyosaur skeletons to indicate any such mass die-off. Volcanic eruptions could possibly kill many ichthyosaurs, but we should see some evidence, such as a volcanic ash layer, preserved in the sediments. None is present in the rock record. Similarly, the stratigraphic record lacks evidence of a severe storm, such as a coarser-grained layer of sediment. The notion that some ichthyosaurian behavioral attribute caused mass mortality also is improbable. Studies of other fossils prove that ichthyosaurs bore live offspring in much the same manner that modern whales do. So, we can rule out a mass mortality caused from spawning behavior because ichthyosaurs did not spawn. And while mass mortality by stranding or beaching previously held support as the cause for the large concentration of ichthyosaur skeletons here, researchers now consider that scenario unlikely. After all, our examination of the sedimentary layers containing the fossils indicate a deep marine setting, not the shallow coastal setting necessary to support a case for stranding.

One of the more likely possibilities is that the ichthyosaurs may have eaten fish or shellfish tainted with a neurotoxin that would have paralyzed them. Paralytic shellfish poisoning in humans also results from a marine neurotoxin. The neurotoxin may have originated at the base of the marine food chain, in oceanic plankton, and then become concentrated in the tissues of the animals that ate the plankton. The ichthyosaurs, consumers higher in the food chain, then ate these plankton-feeding organisms. Researchers presume that ichthyosaurs were negatively buoyant, so the creatures would have sunk to the bottom after death. This type of poisoning mechanism has triggered some of the mass kills of modern whales along the eastern coast of New England. However, the whales were positively buoyant, and they subsequently became stranded on the beach.

In paleontology, it seems the more we learn, the more mysterious the past becomes. Have we solved the mystery of the monsters of the deep? No. Our inquiry has simply led us to ask more questions, questions about the environment in which they lived, what they looked like, and how they died. Each new generation of scientists builds on the work of the generation before, providing some answers but invariably providing many more questions. As you continue to investigate the fossils within the shelter and examine the rock containing the fossils, consider these questions. And ask some of your own.

15

REMNANTS OF A BURNING WASTELAND

HICKISON SUMMIT

The Monitor Valley of central Nevada may not look like much from U.S. 50 at 60 miles per hour, but a little careful observation can turn up clues to a very interesting past. Prehistoric sites in the area, such as those at Hickison Summit, have yielded numerous artifacts and petroglyphs dated at more than 10,000 years before present. At Hickison Summit you'll find many easily accessible petroglyph panels carved into the rock. These carved figures tell a story. They speak of a time when the climate here was much wetter than it is today, and many of the valleys held large lakes whose shores offered comfortable places to make camp. The rocks in which ancient peoples carved these figures tell a story, too. They relate to about 23 million years ago, during Miocene time, when violent volcanic eruptions would have made much of central Nevada a very uncomfortable, even deadly, place to live.

At the Hickison Summit Petroglyph Recreation Area, pick up a trail brochure, and begin the walk to the various petroglyph panels. They may be hard to identify at first, a fact that tells us something about the material they have been carved into.

The pinkish, crumbly rock unit into which ancient people carved the petroglyphs is part of the Bates Mountain tuff. It is a type of volcanic rock known as an ignimbrite, or ash-flow tuff, born of a particularly violent style of volcanic eruption. The eruptions resemble the 1980 eruption at Mount St. Helens, but on a much grander scale. In 1980, Mount St. Helens ejected about 0.15 cubic miles of magma, whereas the eruption that spewed out the Bates Mountain tuff produced more than 250 cubic miles of magma and volcanic ash—almost 2,000 times the volume of the Mt. St. Helens eruption.

The amount of gas within a batch of magma determines how explosive the eruption will be. When the magma is deep beneath the earth's surface, the gas is compressed under a lot of pressure. As the

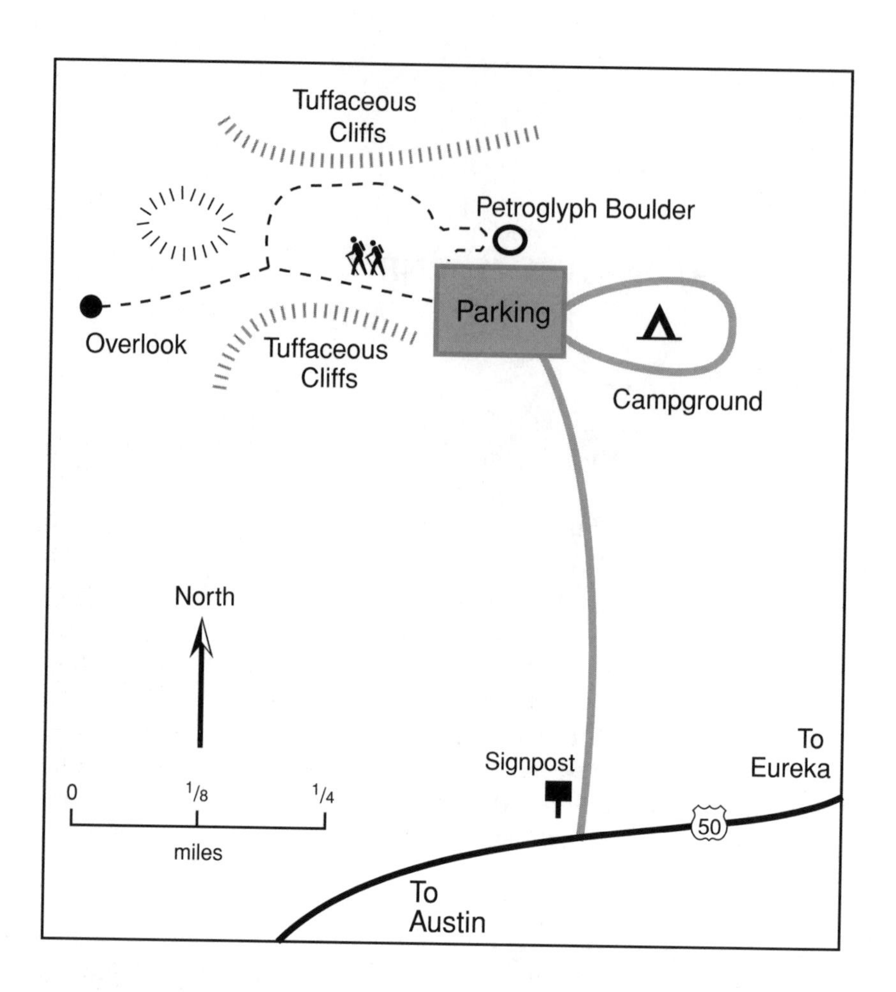

▶ GETTING THERE

Hickison Summit Petroglyph Recreation Area is located along U.S. 50 approximately 25 miles east of Austin and 45 miles west of Eureka. A sign at the entrance road directs you to the parking area and campground. A short trail will lead you to the petroglyph sites and out to an overlook for a good view of the flow units. Please do not touch or deface the petroglyphs.

Early inhabitants inscribed petroglyphs in the soft nonwelded tuff at Hickison Summit. A light desert varnish coats this relatively stable face.

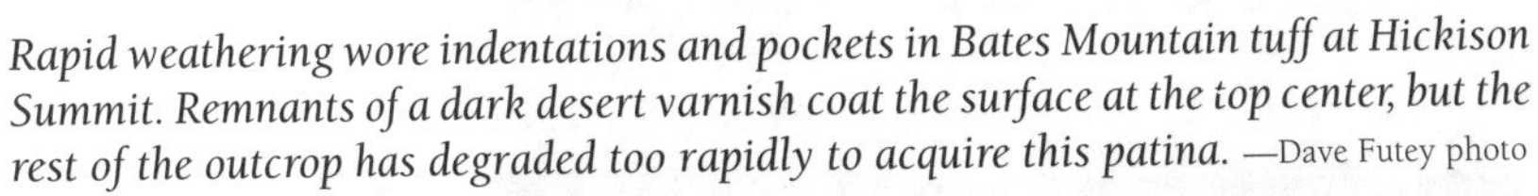

Rapid weathering wore indentations and pockets in Bates Mountain tuff at Hickison Summit. Remnants of a dark desert varnish coat the surface at the top center, but the rest of the outcrop has degraded too rapidly to acquire this patina. —Dave Futey photo

magma rises toward the surface, the gas expands. If the magma contains a lot of gas, it erupts explosively, sending forth hot, glowing, churning clouds of volcanic rock fragments and ash that can blanket a whole landscape. Magmas that form ignimbrites are very gas-rich and hence, very explosive. The ignimbrites spread out laterally across the terrain at high speeds, so they tend to be thickest in pre-existing valleys, and form only a thin veneer on ridges. This contrasts with typical ash-fall deposits that fall like snow, covering the landscape with a similar thickness of ash everywhere.

Geologists typically view ignimbrites in a vertical section, such as a cliff face, and divide them into what are called flow units. Each flow unit represents a single flow. Sometimes flow units erupt in such close succession that they cool as a single unit, called a compound cooling unit. In the area around Hickison Summit, two or three flow units represent the Bates Mountain tuff.

Pumice, a bubble-filled, glassy rock that floats in water, makes up much of an ignimbrite. When glass gets hot, it gets soft, and soft glass deforms easily. In many ignimbrites, these hot pumice fragments, or clasts—usually fist size or smaller—flatten out under the weight of overlying volcanic deposits and weld together. This forms a dense,

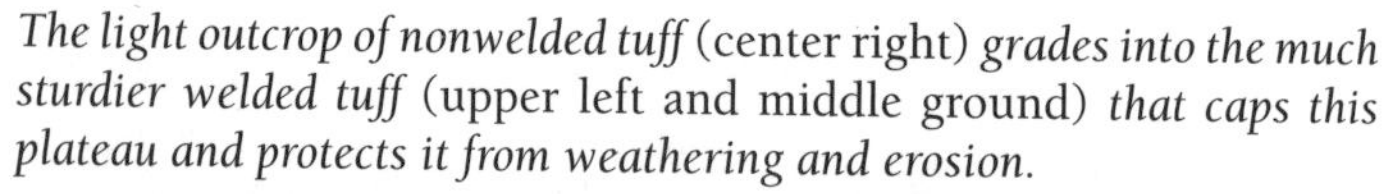

The light outcrop of nonwelded tuff (center right) *grades into the much sturdier welded tuff* (upper left and middle ground) *that caps this plateau and protects it from weathering and erosion.*

more coherent rock called welded tuff. The most densely welded ignimbrites consist exclusively of glass and crystals. The welded tuff at Hickison Summit is deep brown and very hard. More commonly, these cooling units weld only partially. This happens in the following way. As the hot (greater than 1,000 degrees Fahrenheit), foaming mass flows, it breaks apart into pumice fragments and small shards of glass. When the still very hot flow finally comes to a stop, it settles and compacts. At this point the pumice fragments, still hot and soft, collapse and flatten under the weight of the overlying rock mass, then fuse with the glass shards into a welded tuff. The resulting banded or streaked texture exists only in ignimbrites and is a valuable feature used to identify this rock type.

A typical cooling unit of the Bates Mountain tuff measures from 50 to 150 feet thick and consists of a nonwelded basal portion that grades upward into a cliff of densely welded tuff. Look at the outcrop at stop 2 on the trail map brochure. The rock is generally pinkish with many dark brown to black chunks supported in a finer matrix. Some of these chunks have the bubbled texture of pumice, and others do not. The chunks without bubbles are pieces of the rock that surrounded the vent of the volcano and blew out during the eruption. The fine-grained matrix is ash, glass shards, and small crystals, mainly of quartz and feldspar. The rock has a texture and feel somewhat reminiscent

The texture of nonwelded tuff at Hickison Summit resembles that of papier-mâche.

of papier-mâche. The petroglyphs hold up poorly in this crumbly and soft material. This then, is the nonwelded portion of one of the flows. Next, walk toward stop 6 on the trail map brochure and look back and to the right. A cliffy layer of rock caps the ridge. It is the welded portion of the flow unit. If you hike up to one of these cliffs, you will see that this is a much harder rock than the nonwelded tuff. You will also be able to see the characteristic flattened pumice fragments in the welded tuff.

Ignimbrites such as the Bates Mountain tuff can cover vast areas very rapidly. Between 40 and 20 million years ago large volcanic flows similar to the Bates Mountain tuff erupted from huge calderas and blanketed much of central and western Nevada. When the Bates Mountain tuff exploded, many such eruptive centers existed throughout Nevada and the West, including the Long Valley caldera in eastern California and the Yellowstone caldera of Wyoming.

A mistaken idea sometimes surfaces that a caldera is simply a volcano that blew its top off. However, calderas form in a somewhat more complex process. Generally, a caldera-forming eruption consists of two distinct phases. In the first phase, a single vent sends a column of ash high into the air. As the eruption continues, the vent widens and the volcano ejects an ever-larger column of ash. Eventually the ash column becomes unstable and begins to collapse on itself. The ash and pumice spread out laterally as a pyroclastic flow.

Layering in some nonwelded portions of the Bates Mountain tuff

The column does not collapse all at once, but over a long period, as the vent spews more material that then falls back to the earth as part of the pyroclastic flow.

Before long, so much material has erupted out of the magma chamber that its roof begins to cave in. The collapsed roof of the magma chamber becomes the caldera floor. Now phase two begins. As the roof caves in, a number of vents open up along ring fractures that form the boundary of the caldera. These vents discharge more pyroclastic flows. Some of the pyroclastic flows pool deeply in the caldera. Others travel far beyond the caldera, spreading as layers of hot

The stages in the development of a resurgent caldera

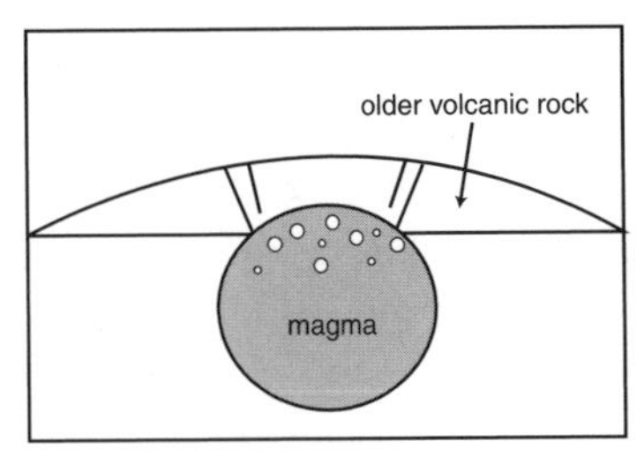

1. Magma with dissolved gas cracks the overlying rock into concentric ring fractures.

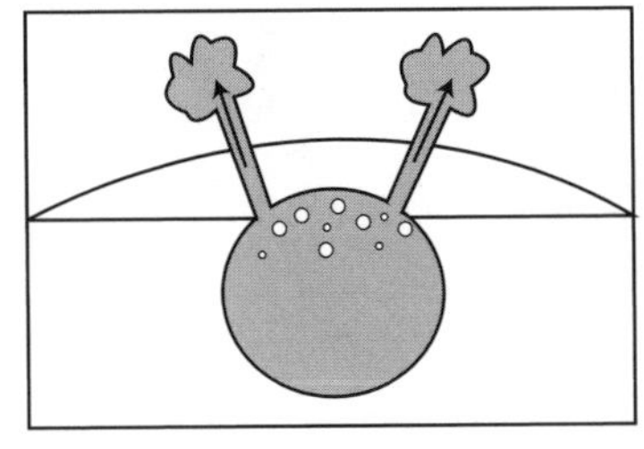

2. Initial burst of ash and gas through fractures

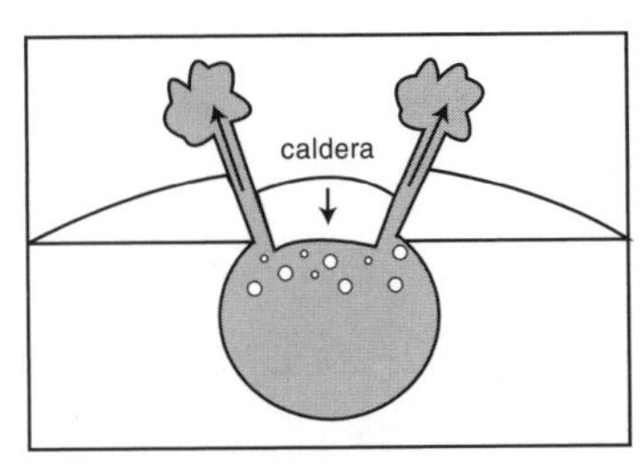

3. Rock overlying magma chamber collapses.

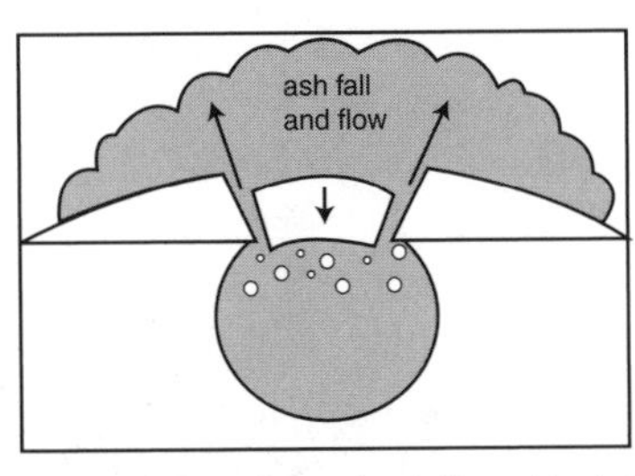

4. Ash falls and pyroclastic flows blanket the landscape.

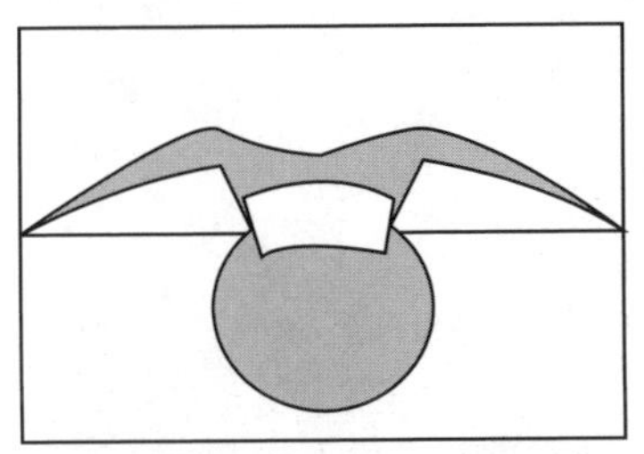

5. Hot ash flows compact and become welded tuffs.

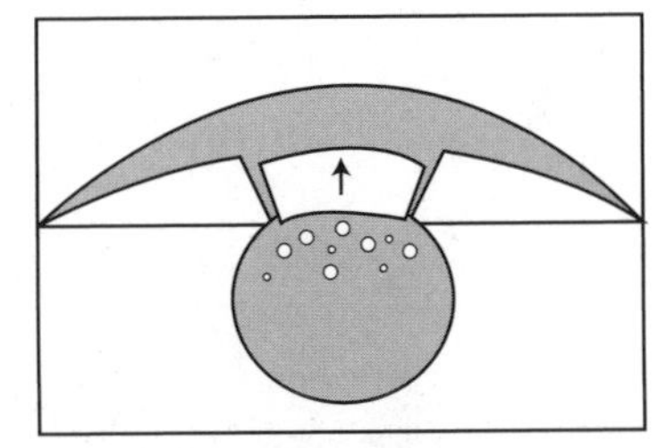

6. Resurgent volcanism domes volcanic rock.

pumice and ash across the countryside. Geologists refer to this type of caldera as a Crater Lake type, after the classic Crater Lake caldera in Oregon, created when the roof of erupting Mt. Mazama collapsed.

The resurgent caldera, the granddaddy of all volcanic structures, follows the same two phases of development as a Crater Lake type caldera. However, after the explosive eruption ceases, the floor of a resurgent caldera domes, or resurges. The main eruption released a huge quantity of magma from the magma chamber, removing the weight of overlying molten rock from deeper magma. The deeper magma then rises in the same way an iceberg rises in the water when its top breaks off. The rising magma domes the floor of the resurgent caldera. Sheer size is one of the definitive characteristics of resurgent calderas—some measure more than 50 miles across. The eruption at Yellowstone 600,000 years ago created a resurgent caldera. This eruption spewed huge sheets of ignimbrite and sent ash as far away as Louisiana.

Explosive, caldera-forming volcanic eruptions were commonplace throughout much of Nevada from about 40 million to 20 million years ago, during Tertiary time. One of these caldera complexes lies to the

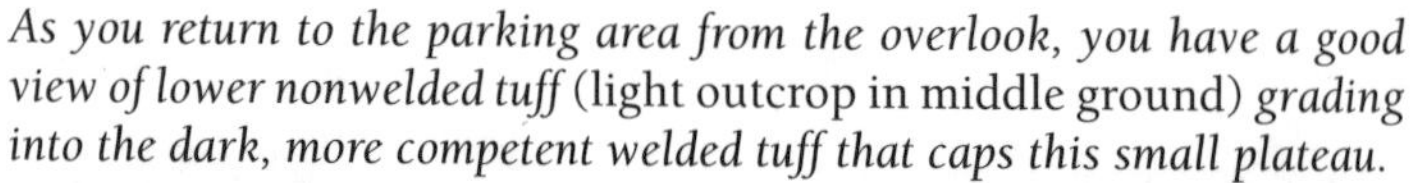

As you return to the parking area from the overlook, you have a good view of lower nonwelded tuff (light outcrop in middle ground) *grading into the dark, more competent welded tuff that caps this small plateau.*

south in the nearby Toquima Range. This caldera complex was active between 27 million and 24 million years ago, but it was not the source of the Bates Mountain tuff. The flow units of these three calderas in the Toquima Range are slightly older, and all were much smaller, accumulating mainly within the calderas themselves. The Bates Mountain tuff, on the other hand, covers an estimated area of almost 3,500 square miles. Unfortunately, the location of the caldera that served as the source of this massive unit remains a mystery. With regular, violent eruptions like this one, what a truly inhospitable place Nevada was during Tertiary time.

When humans inhabited central Nevada about 10,000 years ago, all that remained of this violent past were the eroded remnants of ash-flow sheets, lifted and dropped along Basin and Range block faults. This faulting left basins that the wet climate of 10,000 years ago transformed into lakes supporting abundant plant and animal life. The Basin and Range faulting also created high areas above the lakes, perfect vantage points for observing and ambushing prey. Life was good here—maybe good enough to while away some time carving figures into an attractive rock with a violent past.

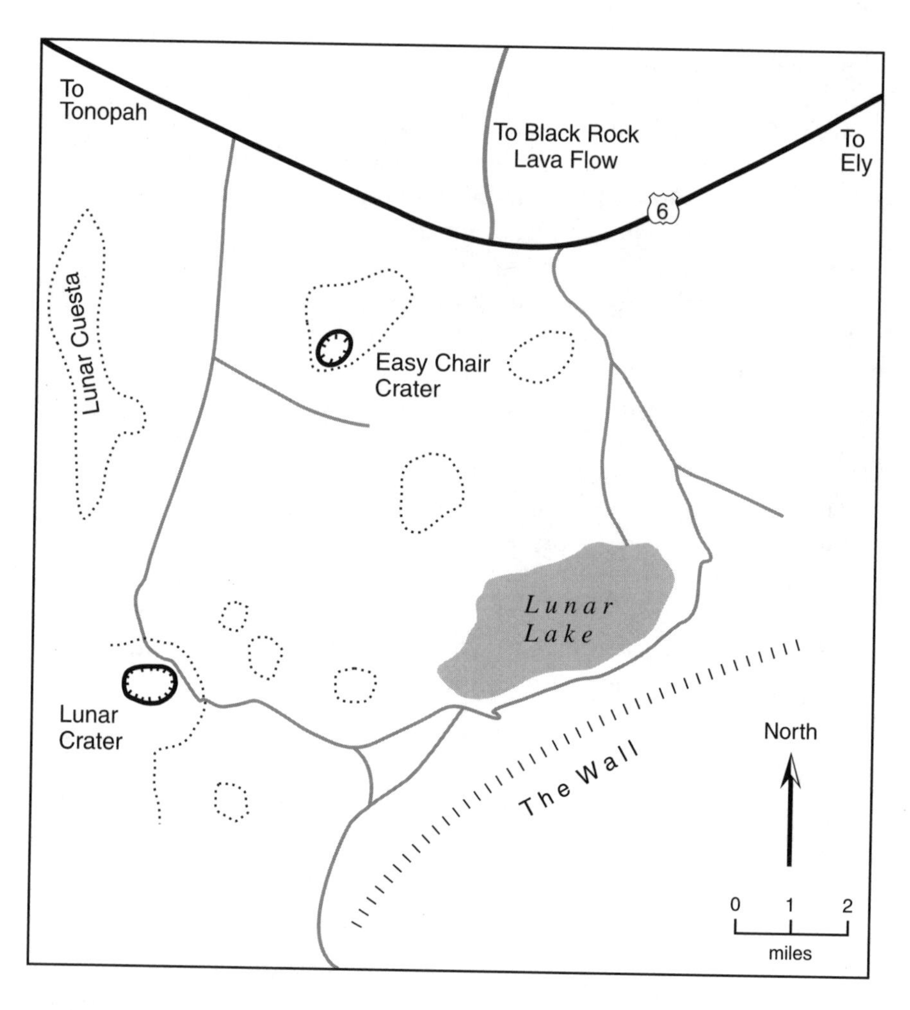

▶ GETTING THERE

From Las Vegas, drive 207 miles north on U.S. 95 to Tonopah. Follow U.S. 6 east for 74 miles to the signpost that leads you south onto a graded dirt road to Lunar Crater. From Ely, travel 93 miles west on U.S. 6 to the turnoff to Lunar Crater. The graded dirt roadway loops through part of the volcanic field. It heads south, passing the marked turnoff to Easy Chair Crater to the east, and reaches Lunar Crater about 7 miles south of the paved road. The dirt road then winds east past Lunar Lake, a playa, before heading north to U.S. 6. This scenic loop provides visitors close-up views of some of the most interesting volcanic features found here, including Lunar Crater itself. More adventuresome visitors can easily explore areas adjacent to the scenic loop on foot. There are no services in the volcanic field, so all visitors, whether in a car or on foot, should take plenty of water.

16

Explosions in an Alien Landscape

LUNAR CRATER VOLCANIC FIELD

Lunar Crater volcanic field sits at the southern end of the Pancake Range, one of Nevada's more unusually named ranges. The volcanic field is an excellent example of a young and pristine volcanic terrain. A history of explosive eruptions and lava flows gives the landscape its otherworldly topography, which scientists and engineers have used to test equipment for interplanetary exploration.

Centrally located in the Great Basin, Lunar Crater volcanic field has an arid, continental climate. Temperatures here in the high desert span the extremes; the mean minimum in January is -17 degrees Fahrenheit, and the mean maximum in July is 90 degrees Fahrenheit. Mean annual precipitation measures less than 5 inches, usually evenly distributed over the entire year. Elevations here are more than a mile above sea level. The summits of most of the cinder cones are between 6,000 and 7,000 feet, and even at the bottom of Lunar Crater, the

Lunar Crater, with a number of cinder cone volcanoes in the background
—Dave Futey photo

lowest point in the volcanic field, the elevation is 5,570 feet. The high elevation and harsh climate support sparse vegetation, including perennial grasses such as Indian Ricegrass and shrubs such as saltbrush, greasewood, and sagebrush.

The dry climate and rugged terrain here discourage permanent human occupation. Early Native Americans, including the ancestral Shoshone, probably realized that the features of this landscape were very unusual and possibly regarded the region as a place of spiritual significance. The legendary mountain man Jedediah Smith was most likely the first pioneer to travel through the Lunar Crater volcanic field. In the summer of 1827, he was returning from his first trip to California, and his route likely crossed the region near the present location of U.S. 6. Years later, in 1869, a group of men from an expedition led by George Wheeler explored and mapped some of the region.

The rugged topography here has served an interesting and unusual purpose. Scientists studied Lunar Crater volcanic field during the early years of space exploration, in preparation for the first missions to the moon. The many well-preserved craters and other volcanic features here resemble those on the moon's surface. More recently this region has become the subject of detailed studies of newer technologies, such as remote sensing and spectral analyses of satellite images, in preparation for current and future missions to Mars.

To begin our discussion of the geologic history of Lunar Crater volcanic field, let's first stop at Easy Chair Crater, accessible via a maintained gravel road leading eastward from the loop road, about four miles south of U.S. 6. The parking area lies at the southern base of this cinder cone. From there, it is a relatively short, easy walk up to the rim of the crater. Volcanic activity in this area began about 3 million years ago and has continued intermittently until as recently as 20,000 years ago. Volcanism started when a fault zone, or fissure, broke the earth's brittle crust. This weakened zone then provided an avenue of escape for molten rock, or magma, that was present deep beneath the earth's surface. The crust may have fractured from the pressure the molten rock itself exerted as it forced its way toward the surface. Or, forces from regional deformation, such as the crustal extension that created the Southwest's Basin and Range, may have created the fault zone.

At least three major episodes of volcanism, each superimposed over the preceding one, created the volcanic landscape before you. Each period of volcanic activity began as localized eruptions through short segments of fissures called vents. In this phase of volcanism, dissolved gas drove these volcanic eruptions, which sprayed molten rock and

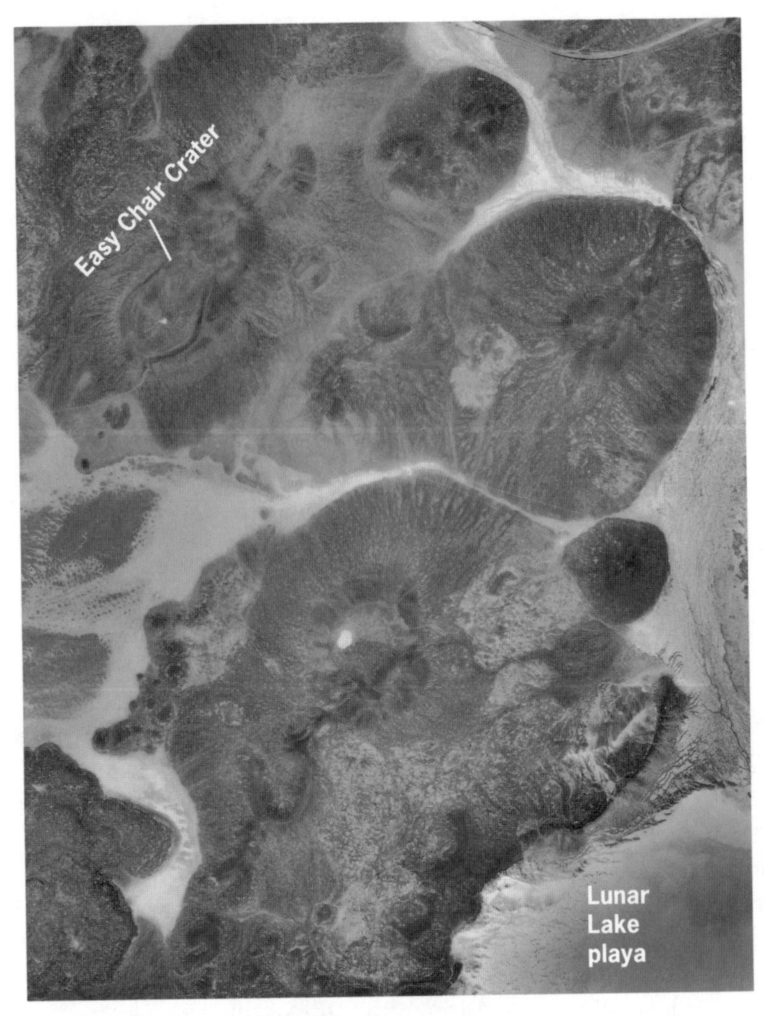

An explosive eruption of Easy Chair Crater, a cinder cone volcano, blew an asymmetric crater in the cone's southwest flank. Numerous other craters and lava flows decorate the landscape.

gas into the air. All magma bodies contain such gases, and like the magma body as a whole, they are initially under very high pressure. So long as the confining pressure is maintained, the gases remain dissolved in the molten magma, but when this pressure is released, they can escape explosively. The opening of a soda can provides a good analogy for an eruption. The can keeps the liquid within under pressure and maintains gaseous carbon dioxide in solution. When you open the can, the liquid within experiences ambient atmospheric pressure, which is much lower than that of the initially pressurized can. The decrease in pressure allows some of the dissolved carbon dioxide gas to escape, which you clearly hear as a hiss. As magma bodies approach the earth's surface, they too respond to decreased pressure. At some point, a fracture develops from the surface to the magma body, exposing the magma to surface atmospheric pressure. The eruption begins as the volatile gases escape.

Magma is much more viscous than water, so volatile gases cannot escape magma as easily as they do a can of soda. Instead, they carry molten rock to the surface as they leave. The eruptions at Easy Chair Crater ejected blobs of molten rock that cooled and solidified while still airborne. The resulting gravel-size particles, known as cinders, fell back to the ground and settled around the perimeter of the vent. Repeated eruptions here eventually built mounds of rock around the many vents, creating cinder cone volcanoes such as Easy Chair Crater.

As you walk up to the rim of the crater, examine the predominantly gravel-size cinders underfoot. They are black with a deep red to brown staining on the surface. This is basalt, fine-grained volcanic rock that is high in iron and magnesium. Iron in basalt oxidizes, or combines with oxygen, when exposed at the surface, creating iron oxide, or rust. Those of you who have lived in northern or northeastern states probably know this process—you have observed it on your vehicles. The primary constituent of a steel car body is iron. Abundant surface water and the lavish use of salt to melt snow on roadways causes vehicle bodies to rust rapidly. Water and salt act as catalysts for the oxidation process, that is, they accelerate the creation of iron oxide (rust) without themselves being changed during the chemical reaction.

An explosive eruption through the side of the cinder cone created the crater itself, an event that we will discuss when we get to Lunar Crater. The eruption directed its force southward, blasting away part of the southern rim. With the crater's southern rim lower than the northern rim, the landform resembles a large, cushiony chair. In reality, though, cinders are anything but soft and comfortable. Take care when exploring this feature, especially if you decide to climb the higher slopes.

The magma already lost much of its volatile gas during the initial cinder-spewing phase of volcanism, so the second eruptive phase mainly produced molten rock. With less gas to force it out, hot lava did not spray into the air. Instead, it eventually filled the vent until it overflowed and poured onto the surrounding landscape. Because this molten rock was very hot and fluid, the lava flows sometimes spread

Easy Chair Crater as viewed from the south. The asymmetric crater formed after the original cinder cone accumulated.

over large areas before cooling and solidifying. We will further discuss lava flows in the Lunar Crater volcanic field when we visit the Black Rock lava flow in vignette 17.

Our next stop is Lunar Crater, a National Natural Landmark since 1973, located seven miles south of the highway on the graded loop road. A parking area along the edge of the crater's rim gives visitors direct views into this immense volcanic depression. The crater itself is about 3,800 feet in diameter and 430 feet deep, and it encompasses more than 400 acres. A large explosion created this crater when groundwater that occupied pore spaces in rock or sediment came in contact with hot rocks or magma very near the earth's surface. Liquid water is much denser than water vapor; if you heat water to the boiling point the steam takes up greater volume than the original liquid.

Expanding gas is the fundamental process behind explosions. Conventional bombs utilize chemical reactions, such as the oxidation, or burning, of gunpowder, that create rapidly expanding gases within a confined system. Pressure builds to enormous levels until the gas blasts free of its container, sending shrapnel in every direction. A similar process occurred at Lunar Crater. Either a magma body or newly formed igneous rock heated groundwater that was confined in tiny pore spaces in rock. The temperature of the fresh igneous rock would have been about 1,000 degrees Fahrenheit, almost five times the boiling point of water. The groundwater flashed into steam upon contacting the magma or hot rock, generating very high pressure. The

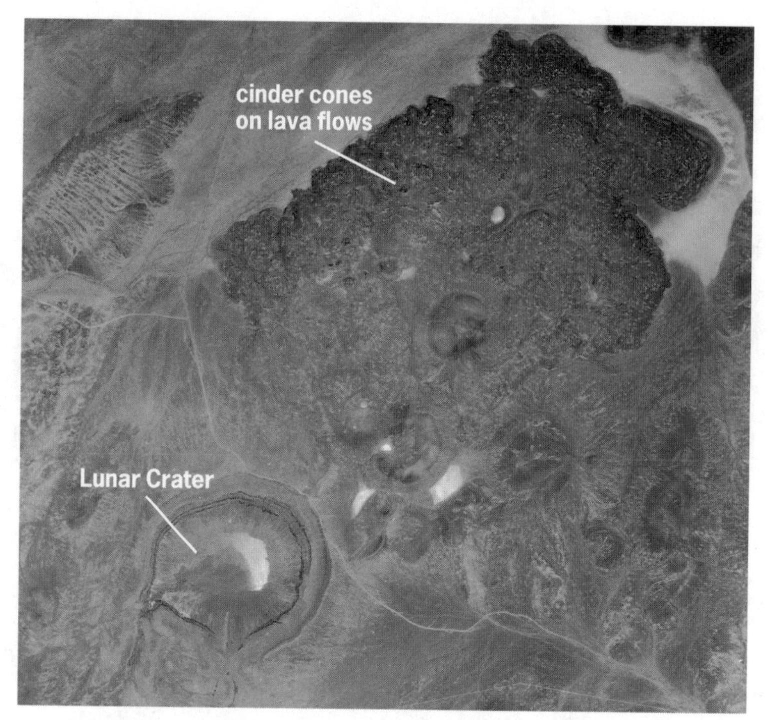

An explosive eruption blasted Lunar Crater into the surface of a pre-existing lava flow. The margins of the dark flow stand out in the upper part of this aerial photograph.

overlying rocks contained the pressurized steam until the gas suddenly burst out in a catastrophic explosion called a phreatic eruption.

Lunar Crater is a maar, a landform produced by the explosive heating of groundwater. The explosion that created this feature broke through nearly horizontal layers of earlier flows and cinder deposits, so the rim of the crater does not project far above the surrounding landscape, and it blends into the surrounding topography. The floor of the crater is more than 400 feet below the rim. The asymmetric depression in Easy Chair Crater, the cinder cone volcano we visited earlier, formed by this same process. That explosive eruption blew southward through the side of the pre-existing cinder cone.

If you continue along the loop road, you will encounter The Wall, an arcuate, northwest-facing cliff at the eastern edge of the volcanic field. Aptly named, The Wall is vertical in some places, and the upper ridges rise nearly a thousand feet above the adjacent lowland. The

The Wall rises over the east side of Lunar Crater volcanic field. The toe of a more recent flow is in the middle ground.

Wall is the eroded edge of a type of lava flow called an ash flow tuff (vignette 15). Ash flow tuffs erupt as frothy, floodlike surges of molten rock, fiery ash, and hot gasses that can blanket a landscape in a matter of seconds. When the flow ceases, the hot gases escape from between the particles of molten rock, allowing the glowing-hot rock and ash to settle, compress, and weld together. Two ash flow tuffs, the Shingle Pass tuff and the tuff of Lunar Cuesta, form most of the vertical faces of The Wall. These two ash flow tuffs erupted about 10 million to 20 million years ago during the Miocene epoch, a period of widespread volcanic activity in the American west. Younger basalts later flowed over sections of The Wall.

As you drive north along The Wall, Lunar Lake, a light-colored flat, lies to the west. This broad depression measures a little more than two square miles in area and typically contains either very shallow water at its southern end or, more often, none at all. This type of ephemeral desert lake is known as a playa (vignette 8). It collects water from a relatively small part of the surrounding landscape, primarily the areas to the north and northeast. Because the region is arid, the lakebed only contains water after storms. The typically hard, dry, and dusty surface is devoid of obvious life.

The Lunar Crater volcanic field is a wonderful place to explore a variety of volcanic landforms. Its lunar landscape graphically illustrates the explosive outcome when water and magma meet. Combine your visit here with a stop at the Black Rock Lava Flow, just north of U.S. 6. There we will see what this lava flow has to tell us about where it came from and what modern geologic processes act on it today.

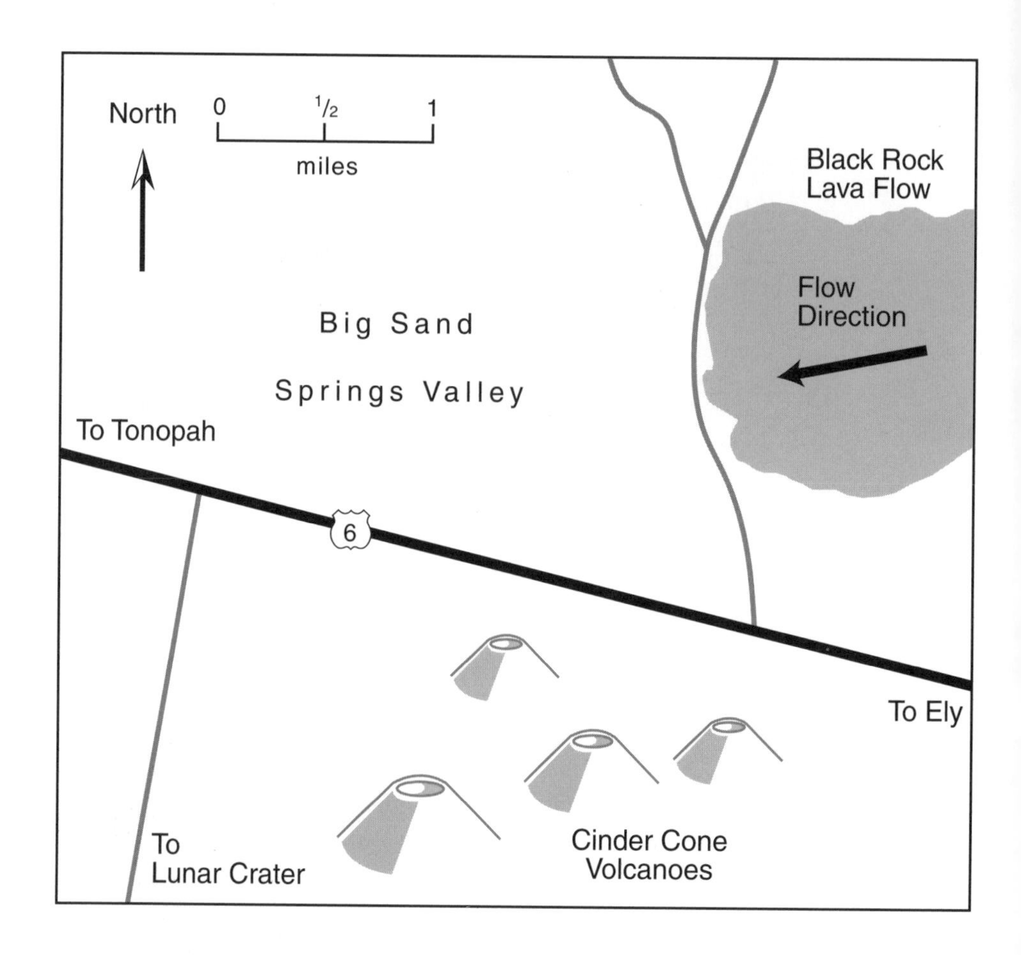

▶ GETTING THERE

From Las Vegas, drive 207 miles north on U.S. 95 to Tonopah. Follow U.S. 6 east for 79 miles to a signpost that leads you north onto a dirt road to the Black Rock lava flow. From Ely, travel 88 miles west on U.S. 6 to the turnoff to the lava flow. The turnoff is about 5 miles east of the marked road that leads south to Lunar Crater.

Drive north on the graded dirt road for about 2 miles, at which point the road winds along the margin of the flow. There is plenty of room to park at the toe of the flow and explore it on foot. Take care walking across the flow's blocky, rugged surface.

17

Message from the Mantle

BLACK ROCK LAVA FLOW

The Black Rock lava flow sits within the Lunar Crater volcanic field, a terrain shaped by eruptions that began about 3 million years ago and continued until as recently as 20,000 years ago. The Black Rock flow is the youngest lava flow in this area. It erupted so recently that you can still see surface features that formed during its eruption. Perhaps more intriguing, the flow also holds pieces of rock from deep within the earth's mantle. A careful hike up onto the lava provides an excellent view of its desolate and very rugged surface.

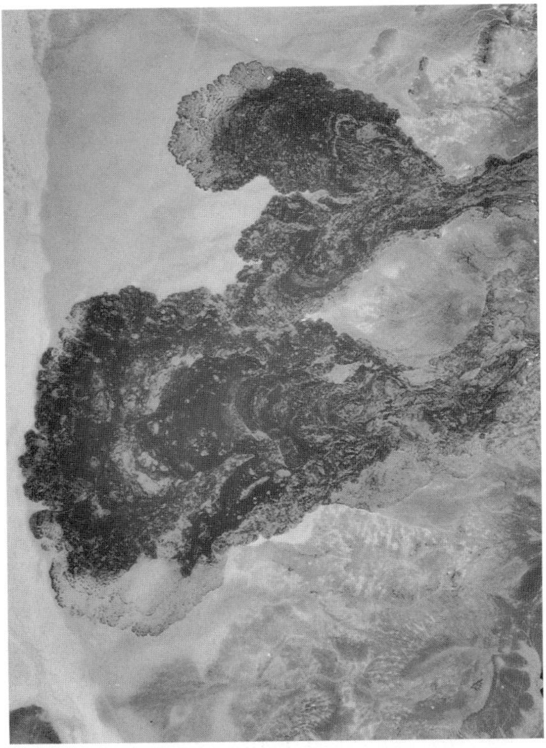

The dark Black Rock lava flow sharply contrasts with the much lighter evaporite deposits and alluvium on the desert floor.

The flow is a thick, tongue-shaped mass of basaltic lava. Its blocky surface stands fairly high above the desert floor. The flow extends downward and westward from Black Rock Crater, a cinder cone volcano at the eastern edge of the Lunar Crater volcanic field. In almost any direction from the edge of the Black Rock lava flow, you can see multiple volcanic cones and flows. The volcanic landforms in this panorama formed not during a single eruptive event but in at least three major episodes of volcanism, each having two eruptive phases. Each period of volcanic activity began with localized eruptions in which molten rock and gas sprayed into the air through short segments of bedrock fissures. In this phase of volcanic activity, Black Rock Crater and other volcanoes here sent blobs of molten rock flying through the air, where they cooled and solidified. The resulting coarse-grained, gravel-size cinders fell back to the ground around the perimeter of the vent. Repeated eruptions eventually built mounds of rock around the many vents here, creating cinder cone volcanoes such as Black Rock Crater.

During these localized eruptions, the magma lost much of its gas, whose escape was the driving force behind the flying blobs of lava. So, the second eruptive phase mainly produced molten rock. With

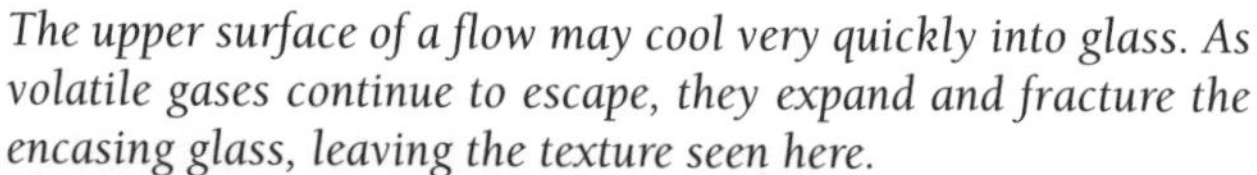

The upper surface of a flow may cool very quickly into glass. As volatile gases continue to escape, they expand and fracture the encasing glass, leaving the texture seen here.

less gas to drive it out, the magma did not spray into the air. Instead, lava flowed from the vent into the crater until it overflowed onto the surrounding landscape. These very hot and fluid flows sometimes spread over large areas before cooling and solidifying. The molten basalt must have been at least 1,800 degrees Fahrenheit as it poured onto the earth's surface forming the Black Rock lava flows and others in this region.

Look closely at the rock that makes up the Black Rock flow. It is typically gray to nearly black. A hand lens or magnifying glass will help you see the texture more clearly. Under magnification, you can clearly identify very small, interlocking crystals. The gray minerals are plagioclase, and the black ones are pyroxene. In some places, you will also find larger crystals of these minerals, called megacrysts, loosely dispersed within the basalt. You will see occasional vesicles—small gas bubbles that were preserved as the lava solidified. Basalts that have many of these bubbles are called vesicular basalts.

Look around until you find some basalt that contains small pieces of very different rock. Particularly noticeable are bright green chunks of the mineral olivine. These "alien" bits of rock within the basalt are called inclusions. As molten magma rises from deep beneath the earth's surface, it tears off chunks of the rock through which it moves. If these rocks have a higher melting point than the magma's temperature, they remain as solid rocks within the magma as it spills onto the surface, cools, and crystallizes. By examining the composition of inclusions, including olivine, in the basalt, geologists conclude that this

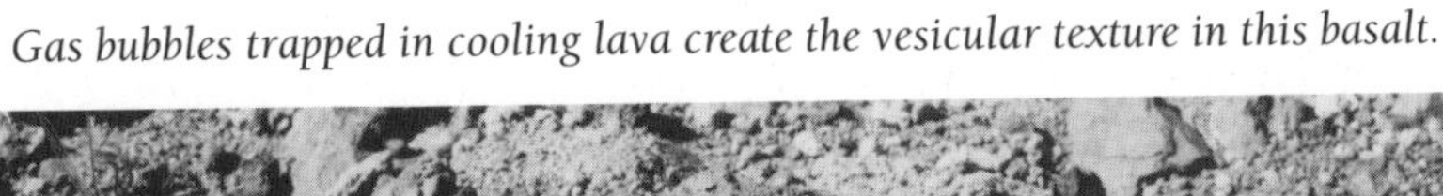

Gas bubbles trapped in cooling lava create the vesicular texture in this basalt.

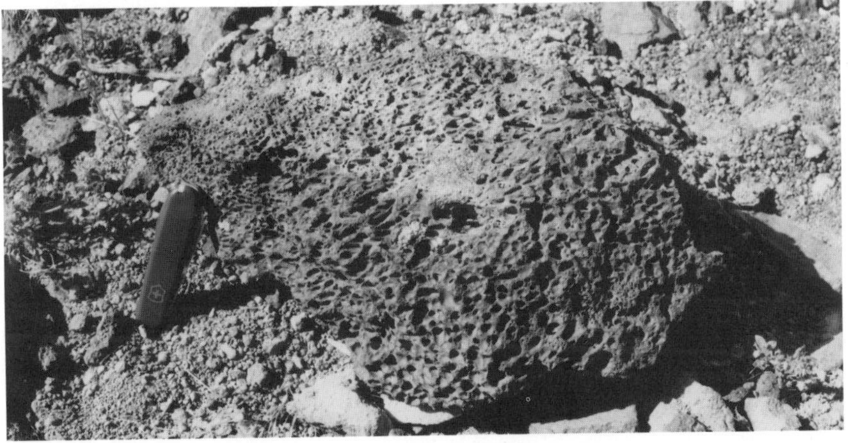

Inclusion of mantle rock in basalt of the Black Rock lava flow

basalt derived from melting of rock deep beneath the earth's crust, in the region called the upper mantle. The inclusions in the Black Rock lava flow have the same composition as the rocks in the upper mantle, indicating their region of origin.

Lava flows come in two basic shapes: elongate or equant. Elongate lava flows, as the name implies, are relatively long and thin. The largest elongate lava flow in the Lunar Crater volcanic field is about 3.8 miles long, 1 mile wide, and 10 to 16 feet thick. In contrast, equant flows are relatively short, with lengths less than three times their width. Equant flows also are commonly much thicker than elongate flows, up to about 80 feet thick. Variability in temperature, rate of extrusion, and composition of the lava, along with the slope of the underlying surface, probably control the different appearances of the flows. The Black Rock flow is an equant lava flow.

The surface at the toe of the flow has a predominantly uneven and blocky texture called aa (say AH-ah). This name comes to us from the Hawaiians, who live on volcanic islands. The rugged aa texture forms when the upper portion of a lava flow cools rapidly, solidifying and fracturing into blocks, while the underlying, still-hot lava continues to flow out from the vent. Carefully climb onto the flow surface. As you inspect the basalt, you will begin to notice variations in the surface structure. The blocky aa texture gives way to a ropy flow texture in some areas of the flow. This more fluid surface texture is called pahoehoe (say Pah-HOY-hoy), another Hawaiian term. Very hot, fluid basalt moves rapidly and smoothly across the landscape,

The angular, blocky surface of an aa flow on the southwestern edge of the Black Rock lava flow is easily accessible from U.S. 6.

creating a smooth, ropy surface such as the one here. The surface of the Black Rock lava flow is largely aa with only localized pahoehoe. In some places, you can still see original flow structures in the lava. Large, solidifying, surface blocks of lava riding on the molten material below jammed up against each other to create pressure ridges—fractured, linear mounds of basalt, typically several feet tall and 5 to 20 feet long. As the surface of the flow cooled, molten basalt continued to flow beneath its thin crust, jamming pieces of the solid surface together. Converging chunks rode up where they collided, forming pressure ridges. Blocky chunks of basalt mounded on the sides of flow channels that flowed persistently. Where the flow edge extended outward, the lava formed basalt fingers that look like large lobes of black honey frozen in time. These are clearly visible on the air photo of the flow.

The surfaces of the flows in the Lunar Crater volcanic field tell us something about the relative ages of the flows and the natural processes that have acted on these surfaces through time. As you investigate the Black Rock flow, you will find that sand and silt are filling the characteristic fractures and depressions on the surface of the aa flow. Plants have gained a foothold in some of these spots. A layer of windblown silt and sand, called an eolian mantle, is in the process of covering surfaces of younger flows—those less than a quarter of a million years old. These accumulations of fine-grained sediment on the surfaces of the lava enhance the formation of soil, which in turn further thickens the eolian mantle. Eventually, the blanket of sediment

Hot, fluid lava, called pahoehoe, moves easily across the land surface and develops a ropy structure characteristic of pahoehoe flows.

and soil smooths the surfaces of the flows, and erosion crumbles the higher protrusions on the lava's surface, flattening the surfaces of the flows. So, unlike the surface of the Black Rock flow, the surfaces of the older flows—those from one-quarter to three-quarters of a million years old—look flat and smooth. If you drive to Lunar Crater (vignette 16), you will see several of these older flows with smooth upper surfaces. The competing processes of erosion and accumulation of windblown debris are in balance, so these flat flow surfaces are relatively stable. Through time, surface run-off from infrequent storms erodes small gullies in the eolian mantle, and stream erosion becomes the dominant surface process. This condition marks flows older than three-quarters of a million years old. Stream erosion eventually will remove the eolian mantle and expose the underlying lava itself to the forces of erosion. The overall aridity of the central Nevada climate makes the processes of accretion and erosion proceed at a much slower pace than they would in a more humid climate.

This volcanic field encompasses a large area, more than 100 square miles. In fact, the section of the volcanic field near Lunar Crater is only the western part of a linear chain of volcanic features that continues northeastward, across the western flank and crest of the Pancake Range, for about 15 miles. Altogether, Lunar Crater volcanic field encompasses more than one hundred volcanic vents and at least thirty-five lava flows. However, the entire Lunar Crater volcanic field is just the youngest and easternmost portion of an even larger trend of regional volcanism. This regional trend originated in southern California, near Death Valley, and migrated northeastward over time, across the central Great Basin and eventually into the Lunar Crater area.

Volcanic activity can migrate in two ways. Either the source of magma can move beneath the crust, or the crust itself can move over a stationary magma source. The Hawaiian Islands, for example, formed as the Pacific plate moved northwest over a stationary magma source, or hot spot. The North American plate is moving generally westward, and this movement may have contributed slightly to the apparent northeasterly movement of regional volcanism. But, the source of the magma itself is also moving. Why and how, we don't really know, but it certainly relates to the broad doming and fracturing of the earth's crust that has created the Basin and Range topography. The ages of the volcanic rocks along this trend indicate that the deep source of the magmas has moved northeastward from southern California at a rate of about one-half inch per year. For example, the volcanic rocks in the Reveille Range, southwest of the Lunar Crater area, are about 4 million years old, whereas those of the Lunar Crater field are less than 3 million years old.

Combine your visit here with stops at Easy Chair Crater and Lunar Crater (vignette 16), where we explore cinder cone volcanoes and the products of explosive eruptions triggered when groundwater meets red-hot magma.

The Black Rock lava flow, seen here from Easy Chair Crater, issued from the asymmetric cone to the right and then flowed westward (toward the left in the photo). *It flowed both north and south around the cinder cone on the left.*

18

Desert Gemstones

GARNET HILL

Garnet Fields Rockhound Area, locally known as Garnet Hill, is one of the best-known and most easily accessible sites for collecting garnets in Nevada. Although rockhounds have collected garnets from this area for more than one hundred years, the crystals still abound. Visitors commonly find a number of well-shaped gemstones in an hour or less. These garnets have a unique history. Unlike most other garnets, these minerals crystallized directly from volcanic gases during the late stages of an eruption.

Juniper and piñon pines grow on weathered rhyolite that litters the flanks of Garnet Hill. Deep red garnets line some voids within the rhyolite.
—Dave Futey photo

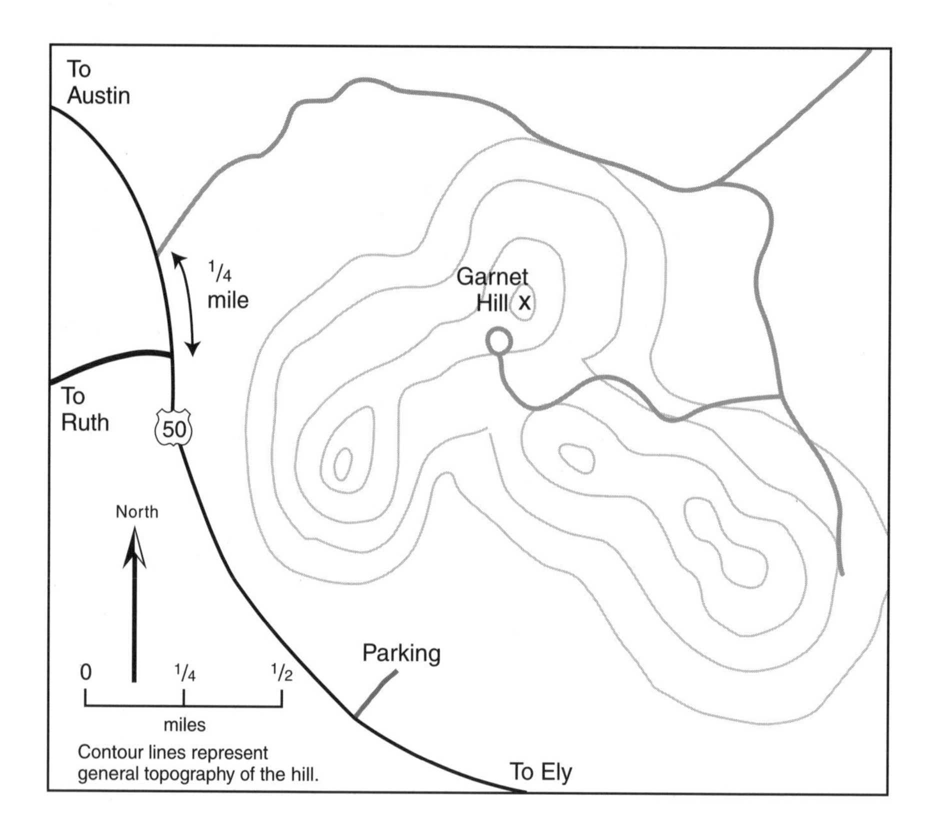

▶GETTING THERE

Garnet Hill is approximately 5.25 miles west of Ely on U.S. 50. The turnoff to the collection site is 0.25 mile west of the turnoff to Ruth. A sign leads you onto a graded dirt road north of U.S. 50. Follow this road 3.1 miles, then turn right and drive an additional 1.7 miles. Turn right again on another dirt road and follow it 1.4 miles to the small parking area beneath Garnet Hill.

You'll find several picnic tables, a group campsite, and a restroom facility next to the parking lot. Tent camping is permitted, and the lot will accommodate small recreational vehicles. The entire area is between about 6,300 and 7,400 feet above sea level and within the piñon-juniper ecologic zone. Summer temperatures at these elevations usually are mild, but it is just as dry as at lower elevations. Carry plenty of water, as none is available at the parking area.

You may forcibly remove garnets from bedrock with a hammer and chisel, a difficult proposition, or you may look for loose garnets on the slopes and within the many small channels on the hill's flanks. A geologist's hammer will come in handy—ordinary carpentry hammers are not up to the task of breaking rhyolite. When breaking rock, always wear eye protection. A hand lens or a magnifying glass will be useful for looking closely at crystals you find at Garnet Hill.

Garnets are semiprecious stones that lack the inherent value of much rarer emeralds, sapphires, and diamonds. Garnets of gem quality, suitable for use in jewelry, are rare, but collectors have found some here, particularly on the summit at the end of the ridge that trends southwest from Garnet Hill. However, many of the partially formed garnets also make nice earrings, rings, or other jewelry because the mounting can hide the unfaceted side.

We recognize gemstones in nature by their color, translucence, and crystal faces. The number and shape of crystal faces reflect the interior arrangement of atoms in a mineral. The elements that make up a mineral are bound together with mathematical regularity; that is, the angles between specific types of atoms remain the same wherever they exist in that mineral. The angles between a crystal's faces, then, reflect that internal regularity. And, while two minerals may contain the same elements, how the atoms tie together gives each mineral its unique qualities.

The two minerals diamond and graphite offer a good example of this phenomenon. Both diamond and graphite contain only carbon atoms, but while diamonds are often clear and white, graphite is gray with a metallic sheen. Diamond is the hardest of minerals. Graphite is very soft, so soft that we use it as pencil "lead." A complicated three-dimensional lattice ties together a diamond's carbon atoms and gives the gem its well-defined crystal faces. The types and arrangements of bonds within graphite differ greatly from those in diamonds. Very weak bonds between the carbon atoms in graphite allow the carbon to easily break off in flat sheets. When you press your pencil to paper, you break carbon atoms from the mineral, leaving a dark line of graphite. The color, hardness, and value differences between diamond and graphite reflect nothing less than the internal arrangement of atoms, even though they are all carbon atoms.

The particular type of garnet at Garnet Hill is a beautiful dark red variety called almandine. Almandine consists primarily of the elements iron, aluminum, silicon, and oxygen bound together into a complex three-dimensional network. The element manganese sometimes takes the place of iron in the crystal structure, shifting the composition toward the manganese-rich variety of garnet called spessartine. Some garnets at Garnet Hill contain both iron and manganese in varying proportions. Mineralogists call them almandine-spessartine garnets. As you collect these faceted stones, observe the variation in color between them, from deep red to orange red, that reflects the influence of manganese. Garnets as a mineral group vary greatly in color, from white to green, brown, red, and purple.

Richard Orndorff holds a small handful of faceted garnets that he found in gullies on the hillside. These specimens weathered out of the rhyolite and washed downhill.

The garnets at Garnet Hill typically exhibit well-developed crystal faces, or natural facets, on half of their surfaces. The most common shape, or the crystal habit, of the faceted garnets is a trapezohedron. If you look closely at many different Garnet Hill specimens, you will probably see different shapes as well, including simple dodecahedrons and complicated hexoctahedrons. Compare your finds with the illustration of crystal habits for garnets.

The majority of garnets here are fairly small, usually about 0.2 inch or less in diameter. Larger gemstones obviously are much less common than smaller ones—that enhances the value of the larger ones. The major requirement for the production of large crystals is time. When molten rock cools fairly rapidly, as it does on the earth's surface, only small crystals grow. Dispersed atoms do not have time before the magma hardens to adopt the large, orderly arrangements that make up large crystals. If molten rock cools extremely rapidly, no crystals at all form. Instead, lava hardens into volcanic glass called obsidian. However, if molten magma cools very slowly, atoms have time to join the developing crystal lattices, and large crystals grow. Large crystals typically grow in rock that forms deep underground, where high temperatures keep molten rock from cooling too rapidly.

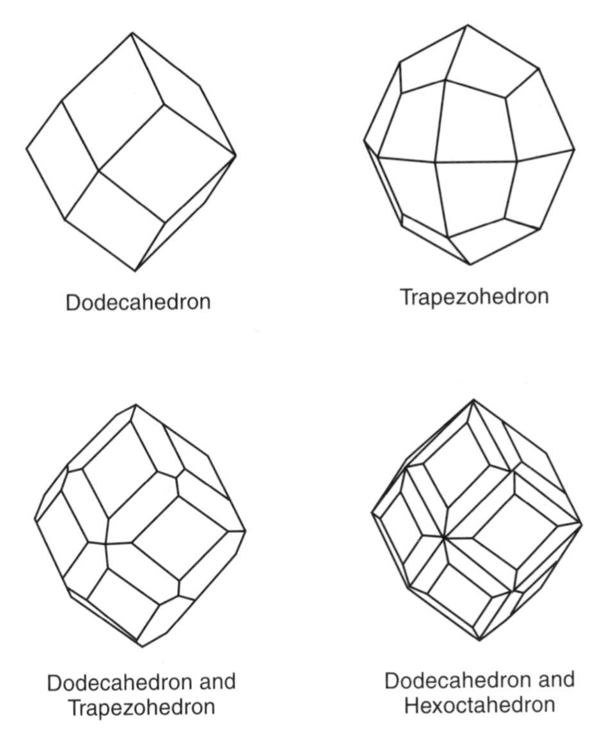

Crystal shapes you might see in garnets from Garnet Hill

Most of the garnets at Garnet Hill crystallized within small holes, called vesicles, in a light grayish pink volcanic rock called rhyolite. The vesicles are the remains of gas bubbles, left when pockets of gas escaped as the magma cooled. As you stroll up the hill from the parking area, take a look at the fine-grained, light tan to pink rhyolite that crops out here. This rhyolite is relatively young, geologically speaking. Although the rhyolite at Garnet Hill has not been dated, the radiometric age of a similar rhyolite outcrop about 2 miles southwest of Garnet Hill places it at about 37 million years old, or late Eocene in age.

Garnet most commonly forms in metamorphic rocks or in intrusive igneous rocks that cool and solidify deep within the earth. To find garnets as we do here in volcanic rock, which cooled quickly at the surface, is quite unusual. These garnets, then, are not the product of metamorphism, nor did they crystallize directly from molten rock. They crystallized in a remarkable fashion, forming directly from hot vapor as it escaped from the cooling rhyolite lava.

A small excavation made by an ambitious garnet miner. Surface rock has been picked through, so experienced rockhounds search for gemstones in buried, unweathered rhyolite.

The Garnet Hill rhyolite complex developed in two stages. First, an explosive, gas-rich eruption of lava sent hot ash into the atmosphere. The ash eventually settled and solidified into a rock called tuff. Later, a mass of molten rock, also rhyolitic in composition, intruded the ash flow tuff from below. This event was much less violent than the initial eruption. The rhyolite that intruded beneath the tuff contained dissolved volatile gases that were released as very hot vapor when the rhyolite began to cool and solidify. The vapor escaped upward through fractures and interconnected vesicles in the volcanic rock. As the vapor ascended through the rock, it rapidly cooled below its solidifying temperature. Minerals crystallized directly from the gas onto the walls of the passageways in much the way that frost crystallizes on cold objects on winter nights.

Two stages of crystallization produced two different sets of minerals within the fractures and vesicles. During the early, high-temperature stage, garnet, as well as the common rock-forming minerals quartz and feldspar, crystallized from the vapor. The garnets probably started to grow as soon as the lava began to solidify. The temperature of the rhyolite at that time probably was about 950 degrees Fahrenheit. As

Scanning electron microscope image of a garnet crystal from Garnet Hill
—John Van Hoesen SEM photo

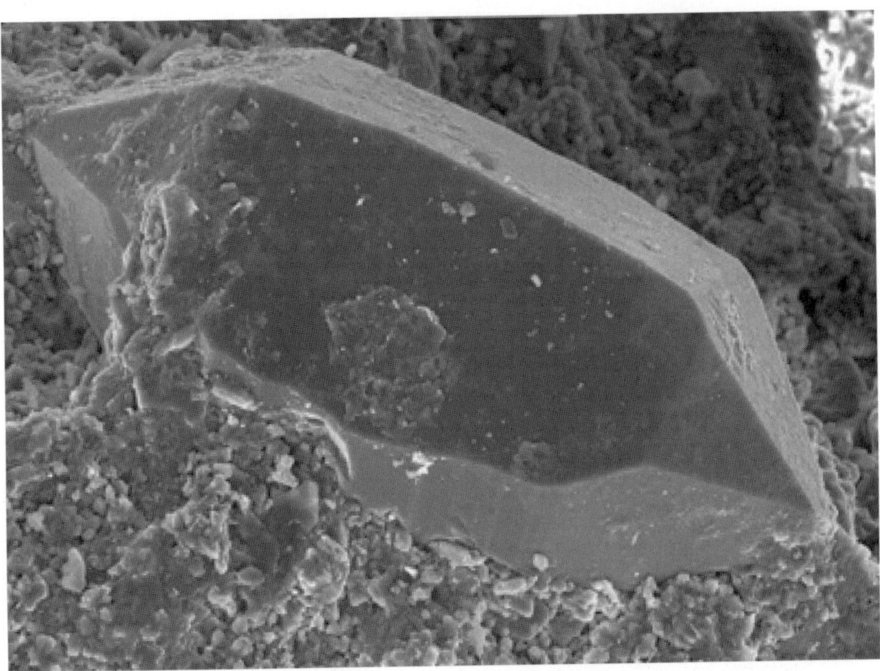

Scanning electron microscope image of a quartz crystal from Garnet Hill
—John Van Hoesen SEM photo

One way to find garnets is to split the rhyolite apart with hammer and chisel. This method yields fine specimens but can be arduous work.
—Dave Futey photo

the vapor cooled below 600 degrees Fahrenheit, small amounts of the common minerals calcite and stilbite crystallized.

As you pry open blocks of rhyolite on the slopes here, notice the clear and white crystals that surround the garnets. These are the other minerals that crystallized from the hot vapor along with their more colorful companions. Crystals that form at the same time often intergrow, while crystals that form at different times grow only on the surfaces of the earlier crystals. You can observe the relationships with a good magnifying glass. When you look at the mineral assemblage through a magnifying glass, you can make out slight differences in color and crystal form. Clear quartz crystals are long and thin with pointed ends, while feldspar and calcite take on a blockier shape. These differences again reflect differences in the atomic bonds that control each mineral's structure.

At Garnet Hill, the tuff that once covered the rhyolite has eroded away, exposing the garnet-bearing rhyolite flow at the surface. There are two ways to collect garnets at Garnet Hill, and either method can produce good results. You might want to look for rocks with small quartz-lined cavities exposed on their outer surfaces. Because the quartz crystals grew during the same stage of mineralization as garnet, these minerals occur together. Break these rocks open with a geologic hammer, exercising a certain amount of finesse. Even though garnet is a fairly hard mineral, you can shatter the crystals if you strike too hard. You can further expose or entirely remove the garnets from the host rock by either breaking away the surrounding rock or prying the crystal out with a chisel. To avoid damaging nice garnets, try not to pry against or directly hit the crystals.

If breaking rhyolite sounds like too arduous a task beneath a hot central Nevada sun, let nature do the work for you: scan the ground for crystals that the natural forces of weathering have removed from the rhyolite. This method works well, because rhyolite is a relatively weak rock and weathers rapidly, while garnets strongly resist weathering. So, the garnets remain on the surface after the rhyolite that once held them has disintegrated. The dark red, faceted garnets stand out against the light brown soils on the slopes. With the sun at your back, walk slowly and scan the ground, watching for garnet facets to

Searching a small gully for garnets that weathered out of the rhyolite and washed downslope

flash as they reflect sunlight. Small concentrations of garnets typically accumulate in drainages, so searching the many smaller ravines and gullies on the slopes can prove particularly productive, especially soon after a summer rainstorm. This method of collecting usually yields garnets that are nearly entirely free of the rhyolite host rock. It is less labor intensive than hammering away at the rhyolite and may seem like nothing more than a stroll through piñon pines and juniper trees that shade these slopes. You can explore a large area and take in the beautiful scenery.

The higher elevations of Garnet Hill also provide a good vantage point to view the mining operations to the southwest near Ruth, Nevada. You can combine your trip to Garnet Hill with a visit to the Liberty Pit, a giant open-pit copper mine that lies just minutes away. A public overlook provides views into one of the giant excavations, and displays offer information about the geologic history of the copper ore.

19

Open Pits and Altered Rock

ROBINSON MINING DISTRICT

The mine overlook in Ruth, Nevada, sits on the eastern edge of a gaping hole in the ground called Liberty Pit, one of five such features in our immediate neighborhood. From west to east, they are the Veteran Pit, Tripp Pit, Liberty Pit, Ruth Pit, and Kimberly Pit. You can see all of the Liberty Pit and parts of some of the others from the overlook. These stupendous excavations grew to their present size as miners followed gold- and copper-rich ore through bedrock in Robinson Canyon, the valley in which Ruth sits.

About 100 million years ago, magma from deep within the earth intruded existing sedimentary rock. This magma brought valuable metals into the region. These metals exist not only within igneous rock but also within surrounding sedimentary rock that the magma

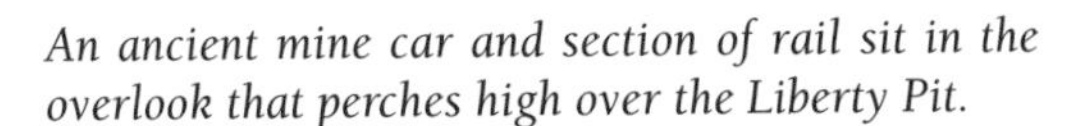
An ancient mine car and section of rail sit in the overlook that perches high over the Liberty Pit.

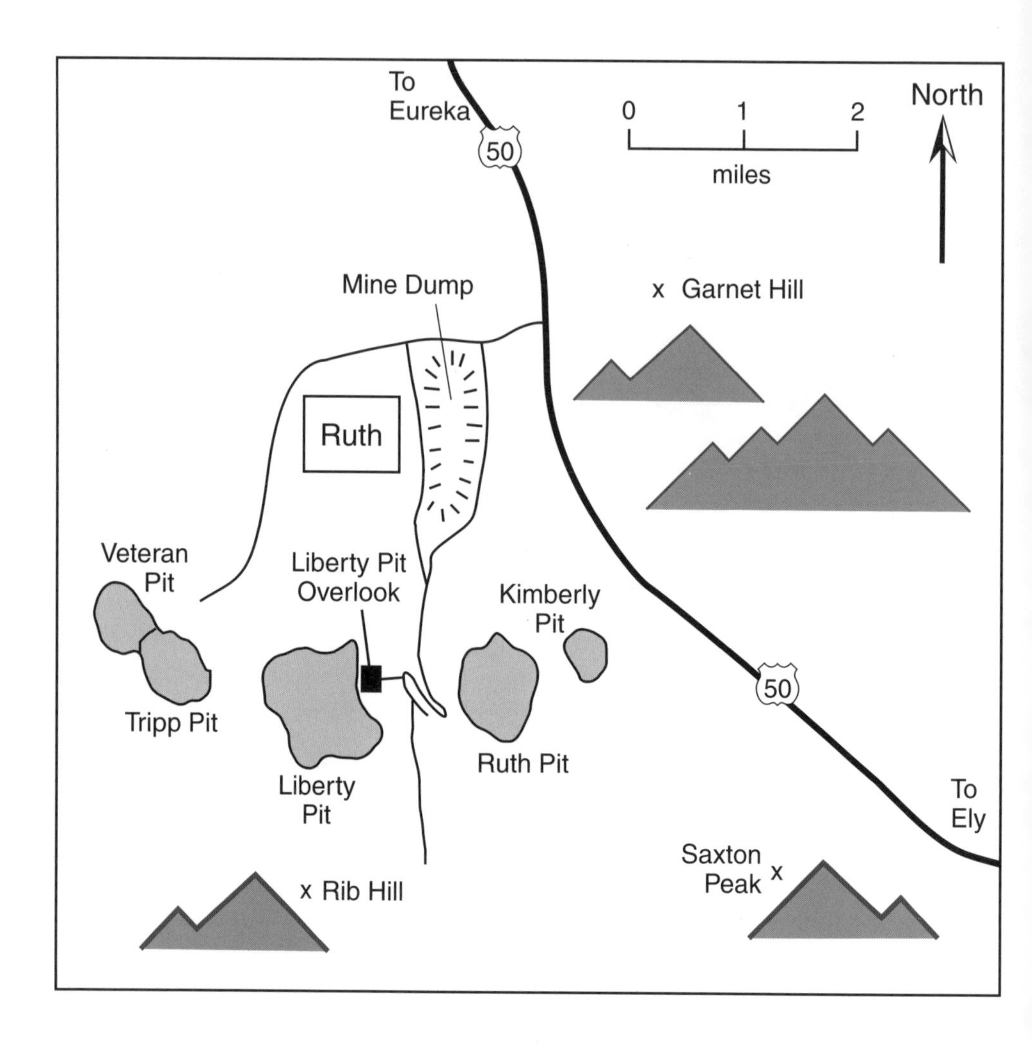

GETTING THERE

The Robinson Mining District is in the small town of Ruth, about 5 miles west of Ely on U.S. 50. Turn west toward Ruth on the marked road, continue for approximately 0.5 mile, then turn south (left) following the sign to the Liberty Pit Overlook. To your right, you will see a large pile of mine tailings, waste products from the ore separation process. The road begins to climb and winds back and forth at about 2.5 miles. Near the top, a sign instructs you to turn right and pull into a large parking area that overlooks the Liberty Pit. Within the parking lot you will find interpretive signs, rock and ore samples, and examples of antique mining equipment.

chemically and thermally altered when it intruded. We will discuss the geologic history of Robinson Canyon, including the process of hydrothermal alteration, but first let's look at the history of mining here.

Late in the 1860s, prospectors discovered gold and silver ore in the Egan Range, home to Robinson Canyon. Shortly thereafter, in the 1870s, miners found copper as well. Copper proved to be much more abundant than either gold or silver, so commercial copper production from underground shafts and open-pit mines commenced in 1908. Visitors here can see horizontal shafts in the steep walls of Liberty Pit. Some of them were probably underground shafts that existed before the pit reached its present size; others are exploratory shafts that geologists dug to direct surface mining operations toward the richest ore. Mining in Robinson Canyon has proceeded in fits and starts in response to prevailing economic conditions. When the value of copper rises, it is more profitable to mine, so operations intensify. When the value drops, mining slows or even ceases for a while. Mines of all types experience similar cyclical activity.

It is difficult to get a real sense of scale when looking into Liberty Pit. It helps to see ongoing mining operations; trucks moving into and out of the pit allow for a comparative sense of depth. Remember,

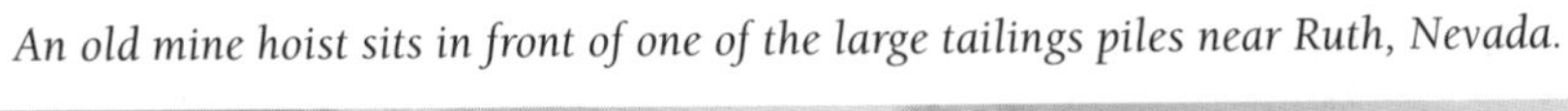

An old mine hoist sits in front of one of the large tailings piles near Ruth, Nevada.

Terracing within Liberty Pit allows huge mining trucks to carry loads of copper ore from the base to the processing center.

though, that earth-moving vehicles used in mining are much larger than trucks you see on the road. This further distorts your sense of scale and can make the pit appear smaller than it actually is.

Very large boulders border the parking area. They are samples of bedrock units that have played important roles in the geologic history of Robinson Canyon. As we go through the history of events that shaped this region, we will direct your attention to relevant rock units in this display. These rocks include unaltered and altered limestone and shale, as well as the igneous rock ultimately responsible for the genesis of the ore.

Scientists have worked out the convoluted geologic history of this region primarily based on the idea of superposition. Superposition states that if a rock unit or structure, such as a fault, cuts across another unit or structure, it is younger than the feature it cuts. We say that younger units or structures are superposed over older ones. If, for example, a fault offsets a sandstone bed, we know that the sandstone formed first and the cross-cutting fault followed. This is a very simple example. We commonly see complicated arrangements of rock layers and structures that resulted from multiple events over hun-

Various bedrock units on display at the edge of Liberty Pit. Samples here include altered and unaltered sedimentary rock as well as the quartz monzonite porphyry that was ultimately responsible for copper mineralization here.

dreds of millions of years. The challenge for geologists, then, is to decipher the sequence of events that tell a coherent geologic story. The Robinson Mining District posed just such a puzzle. Let's piece together the sequence of major events that created the bedrock and other structures found here.

Oceans covered most of Nevada during Paleozoic and early Mesozoic time, about 500 to 200 million years ago. A thick stack of marine sediment accumulated, then compacted and cemented into sedimentary rock. The primary host rocks for the mines here are the Chainman shale and the Ely limestone, both sedimentary rocks from the late Paleozoic era. Take a look at the labeled samples of these two rock units in the parking area. Shales are typically gray with fine laminations, or layers, while limestones can vary in color from white to black. Notice that the limestone shows other colors, especially blues and greens. A later period of mineralization added these colored minerals to the sedimentary rocks.

East-west compression folded and faulted the regional bedrock in middle Mesozoic time. Faults lifted the marine sedimentary units and exposed them to weathering at the earth's surface. Sediment is deposited in flat layers, so folded sedimentary rock shows the effects of stresses that occurred after compaction and cementation. Look on the walls of the pit for numerous signs that rock layers in this region no longer lie horizontally.

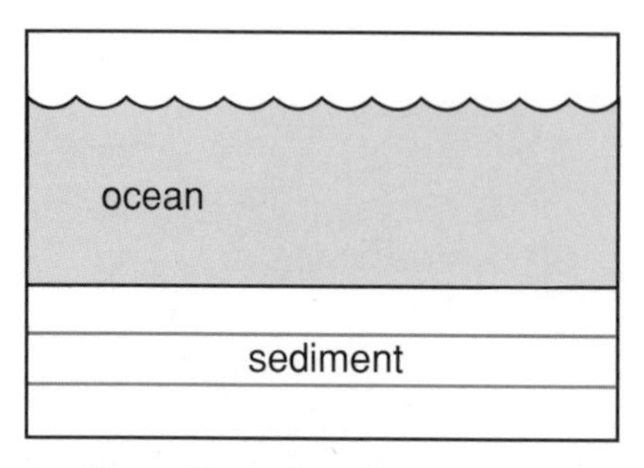

1. Deposition of sediment in a marine environment

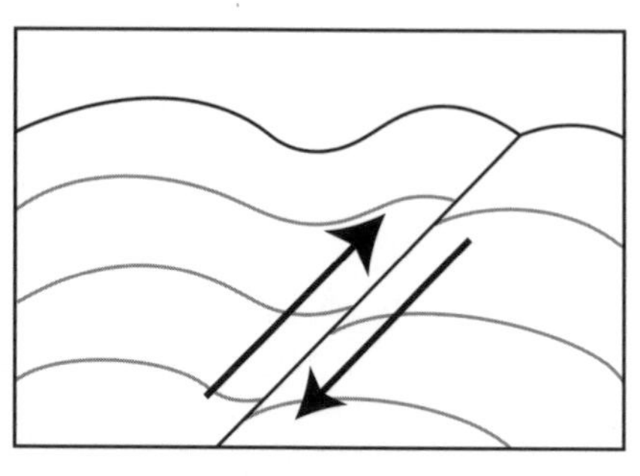

2. Uplift and compression fold and fault sedimentary rock

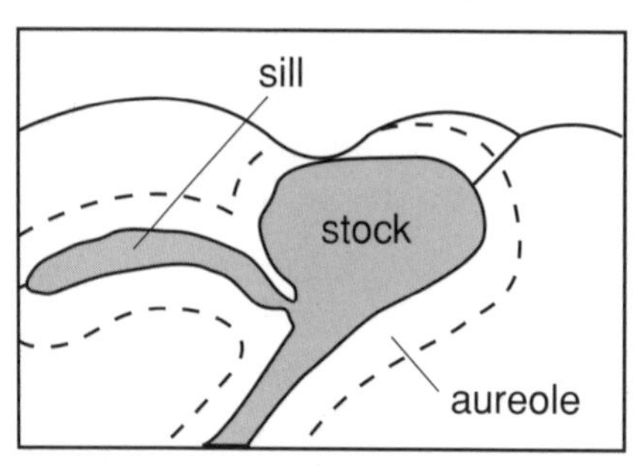

3. Magma intrudes deformed rock along planes of weakness, such as faults and bedding.

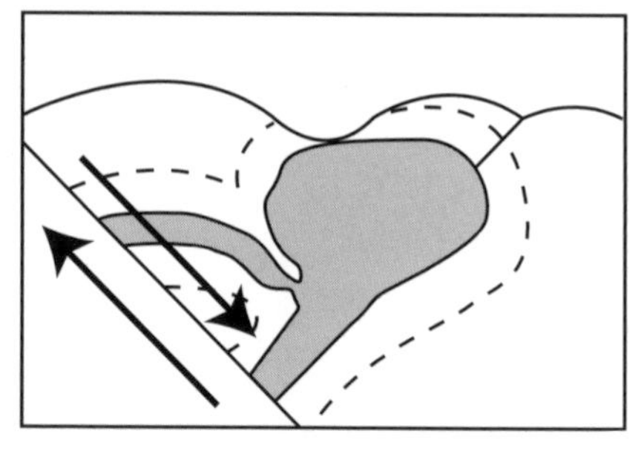

4. Fault offset creates "rootless" intrusions.

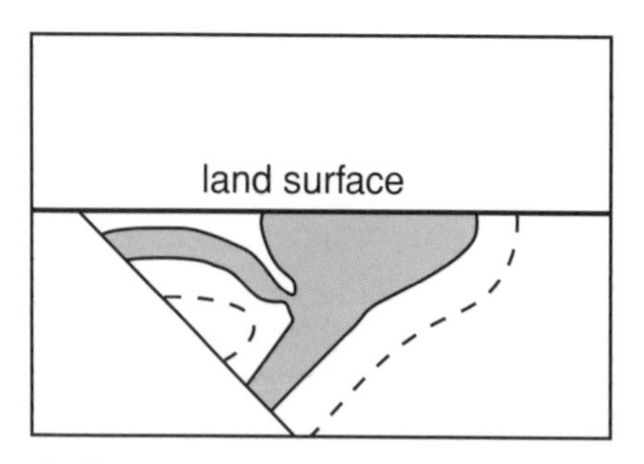

5. Erosion removes upper layer of intrusions and surrounding rock.

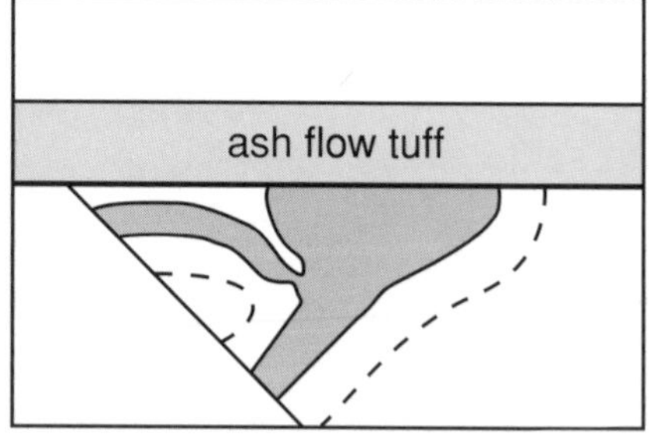

6. Tertiary volcanic rock blankets surface.

Geologic history of the Robinson Mining District's ore deposit

Late in Mesozoic time, about 110 million years ago, magma penetrated the deformed sedimentary bedrock, including the shale and limestone. It intruded along pre-existing pathways of least resistance, such as faults and fractures. Intrusions like these are called plutons, and their more specific names reflect their geometric relation to the beds they intrude. Magma did not move uniformly into the bedrock. Instead it formed isolated and distinct igneous bodies called stocks and sills. A stock is a medium-size pluton, less than 40 square miles in area when viewed from above, and more or less uniform in width, depth, and breadth. A sill, on the other hand, is a thin, tabular intrusion that lies parallel to bedding in the host rock. The Arcturus limestone of Permian age is the youngest formation that the magma intruded. The stocks and sills cut across this and older rock units, so the intrusions are younger than the deformed sedimentary rock.

Geologists call the igneous rock here quartz monzonite porphyry. Take a close look at the labeled sample. Because it formed beneath the surface and cooled slowly, you can easily discern the mineral crystals. The flecks of color you see in the rock represent the variety of minerals there. The major constituents of quartz monzonite are clear quartz, white to pink feldspar, and black hornblende. Some especially large pink feldspar crystals surrounded by smaller crystals of the other minerals also mark the surface of the rock. When we have larger crystals, called phenocrysts, set into a matrix of smaller crystals, we say that the rock is a porphyry. The geologic name, therefore, tells us about the composition—quartz monzonite—and the rock texture—porphyry.

After the plutons moved into the sedimentary rock and began to cool, mineral-rich hydrothermal fluids from the magma altered both the plutons and surrounding country rock. When magma cools, minerals crystallize in order of their individual melting points. If magma cools below the melting point of a particular mineral, it solidifies. When magma has mostly solidified, some liquid remains—very hot water with abundant dissolved minerals. In Robinson Canyon, this hydrothermal fluid contained dissolved gold, silver, and lots of copper. As the fluid cooled, elements in solution precipitated to form minerals. Some of the hydrothermal minerals here, such as the glossy, black, platy biotite, hold no particular value. But others obviously do.

Metals precipitated into tiny pore spaces in the plutons as disseminated deposits, ore that disperses throughout the rock and is difficult to recognize with the naked eye. If you look very closely at the quartz

Robinson Mining District tailings piles as viewed from the parking area on Garnet Hill

monzonite porphyry, you can see gold-colored metallic specks. These are small crystals of the mineral chalcopyrite, which contains copper. Sometimes hydrothermal fluids move out of igneous rock and fill adjacent faults and fractures with precipitated minerals. These so-called vein deposits contain large crystals of gold, silver, or copper that a person can see clearly. The major part of the Comstock ore and the ore at Berlin (vignette 13) were vein deposits.

Hydrothermal fluids not only changed the plutons, but they also migrated outward through tiny pore spaces in the neighboring sedimentary units. As these fluids traveled, they cooled, allowing minerals to solidify. Part of the economically viable ore deposit in Robinson Canyon is altered sedimentary rock surrounding plutons. Alteration, and the concentration of associated valuable minerals, is greatest nearest the plutons and decreases with increasing distance. The zone of hydrothermal alteration is called an aureole. The blue and green minerals within the Ely limestone sample are copper oxides that hydrothermal fluids deposited. Many copper oxides—like azurite, malachite, and turquoise—are highly valued for their vibrant colors. You can clearly see blue azurite and green malachite in the Ely limestone.

Abandoned headworks from the early days of mining in Robinson Canyon

Extensive regional faulting displaced both bedrock and plutons. Faults cut off some igneous bodies from their deeper source rock. Geologists call such amputated plutons rootless intrusions. Over time, weathering and erosion removed surface bedrock and exhumed sections of the plutons. A weathering rind, exhibiting physical and chemical changes wrought by exposure at the earth's surface, identifies these surfaces.

Weathering and erosion brought a unique type of ore enrichment to the altered sedimentary rocks. Much of the hydrothermally altered limestone surrounding the now-rootless plutons contained very low concentrations of valuable minerals because they were on the outer fringes of the alteration aureole. However, acidic water percolated from the surface downward through exposed rock. The acid allowed the water to dissolve and transport metals from the upper layers and redeposit them in lower layers, greatly enriching these deeper layers. This process increased the concentration of copper in some of this rock to the point that it became a viable ore deposit.

Violent volcanism in middle Tertiary time, about 20 million years ago, blanketed the region with volcanic rock. Layers of volcanic ash, called ash flow tuffs, covered the plutons that had been exposed at the surface. Erosion later carved channels through the ash flows, revealing the metal-rich ore bodies that miners found in the 1800s.

Several of the buildings that make up the headworks at the copper mine, with tailings piles to the left and right

Stroll along the edge of Liberty Pit, paying attention to the variety of colors and shapes of rock layers within as well as to the boulders displayed at the rim. These rocks tell a complex geologic tale in both their composition and their structural relationship to one another. To early geologists, they were a puzzle to solve, and the solution offered immediate economic rewards. Understanding what happened allowed astute geologists and miners to locate the richest ore. If the idea of searching for valuable minerals appeals to you, stop by Garnet Hill (vignette 18) just to our north. The geologic history there ties into that of Robinson Canyon, but instead of copper, gold, and silver, we find an accumulation of semiprecious gems called garnets.

20

Island in the Desert

WHEELER PEAK

Wheeler Peak, in Great Basin National Park, stands 13,063 feet above sea level and approximately 8,000 feet above Spring Valley, west of the Snake Range. It is the second tallest mountain in Nevada, the tallest being 13,143-foot Boundary Peak in the White Mountains on the Nevada-California border. Mountains are no rarity in Nevada. More than two hundred mountain ranges occupy the state, looking from the air like so many caterpillars headed north and south. Though intervening desert valleys isolate each mountain range from the others, they all share a common and complex geologic history. And Wheeler Peak is a fine place to explore that shared ancestry.

The Wheeler Peak trail winds through pine forest before moving into alpine tundra at treeline. Wheeler Peak is the flat-topped mountain on the right. —Dave Futey photo

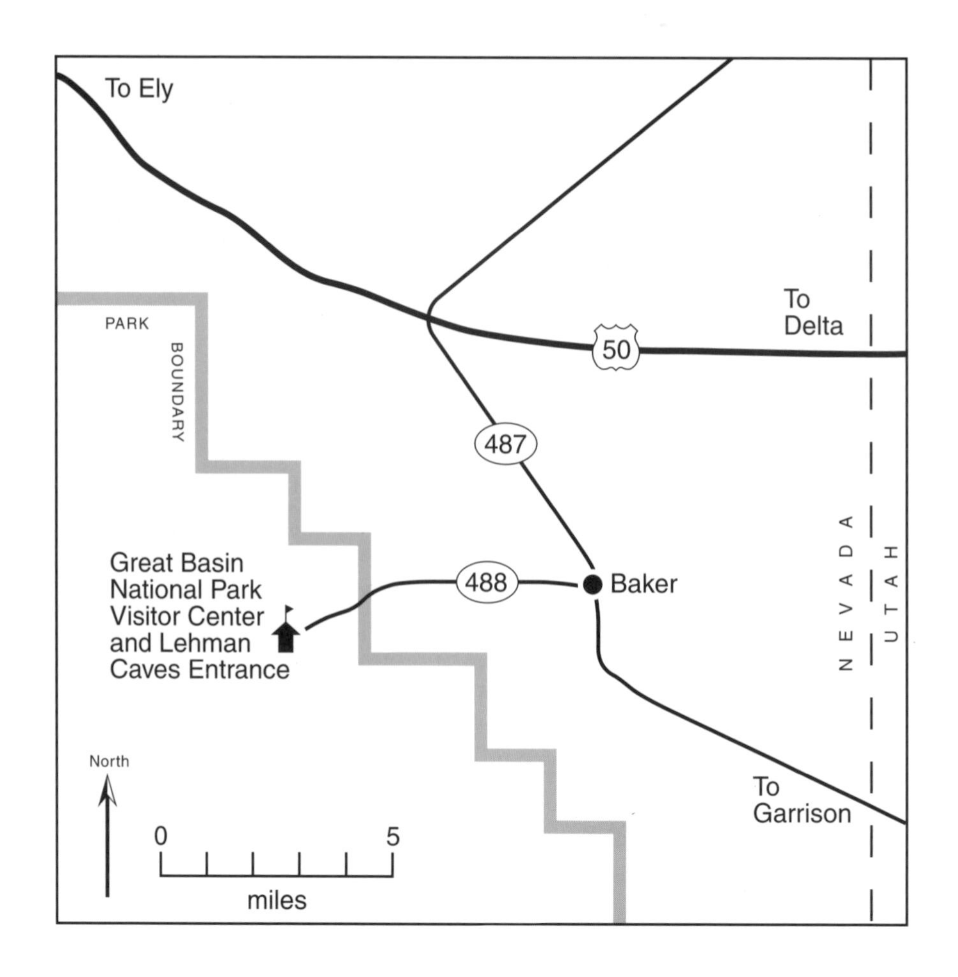

▶GETTING THERE

Great Basin National Park lies in eastern Nevada near the town of Baker. From Las Vegas, drive 20 miles north on Interstate 15, then 234 miles north on U.S. 93. Turn east on U.S. 50 and continue 37 miles before turning south on Nevada 487. Travel 5 miles into Baker, then go west on Nevada 488. The park visitor center is 6 miles west of Baker. From Ely travel 65 miles east on U.S. 50, then 5 miles south on Nevada 487. Travel 6 miles west on Nevada 488 to the park. From Salt Lake City, travel south on Interstate 15, then 165 miles west on U.S. 6, which joins with U.S. 50 at Delta, Utah. Turn south on Nevada 487 and drive 5 miles into Baker before heading west on Nevada 488 for 6 miles to the park.

We will make three stops, all marked on the map, before proceeding to the trailhead. We will begin at the visitor center then stop at Lower Lehman Creek Campground and Mather Overlook on our way up the mountain.

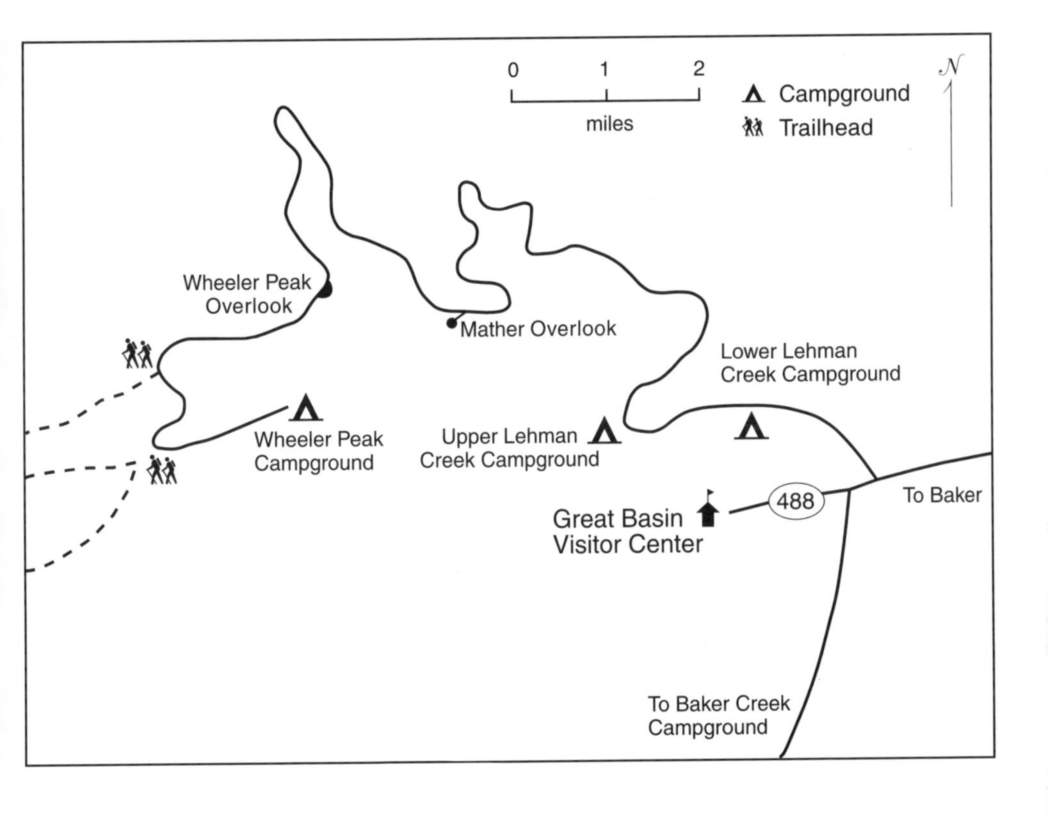

Just inside the park boundary, a well-marked road heads north to the Wheeler Peak trailhead. It passes Lower Lehman Creek and Upper Lehman Creek Campgrounds and several overlooks before reaching the trailhead at the entrance to Wheeler Peak Campground. The road gains 3,400 feet in elevation over 12 miles. A number of overlooks along the way provide stunning vistas above and below. The trail to the peak is a demanding 4.3 miles long with an elevation gain of 2,900 feet. The last mile or so is steep with difficult footing as the trail winds over blocky quartzite rubble. The decreased oxygen at Wheeler Peak's high altitudes may make breathing and walking difficult. Proceed at a slow and steady pace, rest as often as you need to, and turn back at the first sign of a headache or nausea. Those who do not wish to hike the trail may observe many of the features described in this vignette from Mather Overlook with a good pair of binoculars.

Wheeler Peak is named after George Montague Wheeler, a member of the U.S. Army Corps of Engineers who led a series of expeditions into the Great Basin in the middle to late 1800s. Since Wheeler's time, miners and ranchers have made eastern Nevada their home. Congress established Lehman Caves National Monument in 1922 and in 1986 expanded the national monument to include 120 square miles of the southern Snake Range, creating Great Basin National Park. Located far from population centers, Great Basin National Park is one of the least visited parks in our national park system. A visit here remains something of an adventure—it is easy to imagine oneself as a member of Lieutenant Wheeler's intrepid band, enduring scorching heat, lack of water, and other hazards while mapping an unexplored wilderness.

As we journey up Wheeler Peak, we'll examine the layers of sedimentary rocks that make up the underlying structure of the mountain. We'll also note how elevation influences the climate. We begin at Great Basin National Park visitor center, 6 miles west of Baker, for a look at the displays there. Several photos of the high peaks of the

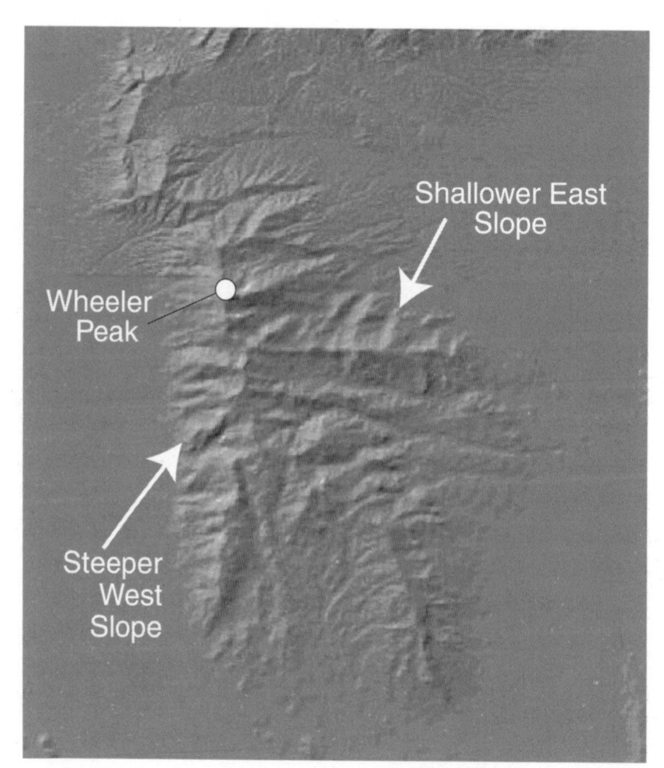

Tilting of the Snake Range along faults has resulted in a steeper western slope and a shallower eastern slope.

Snake Range grace these walls. The visitor center rests on limestone bedrock. The blocky, gray Pole Canyon limestone crops out behind the visitor center. Another good place to observe this limestone is the gray cliffs area on the way to Baker Creek Campground. There, Fremont-style rock art decorates the Pole Canyon limestone. The limestone formed in an ocean that covered Nevada during the middle part of Cambrian time, about 530 million years ago. In deep oceans, shells and skeletons of dead microscopic animals accumulate as a calcareous ooze on the bottom of the sea. The ooze later hardens into limestone. Limestone is made of the mineral calcite, which water can dissolve. This dissolution commonly leads to the formation of caves, such as Lehman Caves (vignette 21). While inspecting the rock faces here, keep an eye out for fossils of larger shelled animals.

Now drive to Lower Lehman Creek Campground; head downhill, then turn north onto the Wheeler Peak road. Pull into the campground and look south at the outcrop of Pole Canyon limestone. Sediment, such as that making up this limestone, may be deposited in a variety of terrestrial and aquatic environments, but gravity dictates that it is laid down in horizontal sheets. Once deposited, the sediment compacts and cements together in a process called lithification,

Dipping beds of Pole Canyon limestone at Lower Lehman Creek Campground

and becomes sedimentary rock. The limestone here clearly is not horizontal; it dips steeply to the east. Something has disrupted its initial orientation.

Let's continue our exploration of Wheeler Peak by driving to the Mather Overlook, a well-marked turnoff that offers a beautiful view of the peak. Wheeler Peak Overlook sits a little farther up the road, but Mather Overlook offers a more panoramic view. You can see Jeff Davis Peak to the south and Wheeler Peak to the southwest. You can also clearly see the elevation that marks the upper limit of tree growth and the barren, rocky landscape above it. If you choose not to take the peak hike, this overlook is a good place to view the upper elevations of the mountain as we continue to discuss processes here.

You are standing at the edge of a cliff that leads down into the Lehman Creek valley. Steep slopes abound. Below the upper rock-covered zone on Jeff Davis Peak, a tongue of debris extends into the forested zone. This is an avalanche chute, a particularly steep section of the mountain flank that channels unstable snow downhill. This chute and others like it see heavy avalanche activity during the winter months. Cascading snow uproots vegetation, creating the scar in

View from Mather Overlook of treeline, tundra, and avalanche chute below Jeff Davis Peak

Drunken trees, like these aspen, grow toward the sun as soil creeps downslope.

front of you. Vegetation stabilizes slopes as roots bind together and anchor soil and rock in place. Once avalanches have stripped away the vegetation, rock and soil move downslope more freely, so snow avalanche chutes feature rockslides as well.

Steep slopes also produce a type of very slow downhill motion called creep. As you explore Great Basin National Park, including the aspen- and pine-covered slopes near Mather Point, keep an eye out for trees with curved lower trunks and straight upper trunks. These so-called drunken trees demonstrate that the slope upon which they are growing is moving downhill. Trees grow toward the sun, so they reach straight skyward. But in mountainous regions, slopes are unstable. During much of the year, water percolates into the soil in daytime, then freezes when the sun sets. Because water expands when it freezes, the soil expands at night and contracts when the ice melts during the day. Consider a sand grain on the soil surface. Each expansion-contraction cycle forces the grain, and every other soil par-

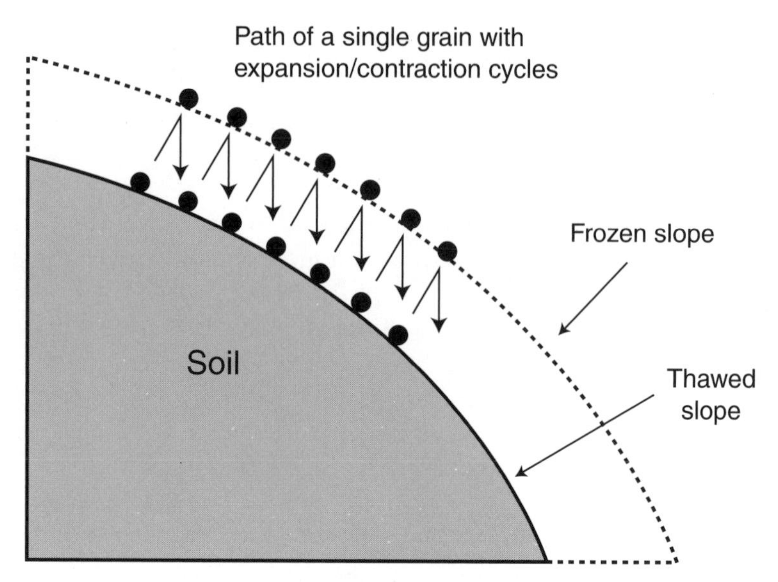

A single soil grain moves downslope as a result of expansion and contraction cycles.

ticle, to move incrementally downhill, resulting in movement of the entire slope. Movement is slow, on the order of fractions of an inch per year, but persistent. And the many drunken trees that grow on steep hillsides record that slow creep.

As you ascend the road all the way to the trailhead, you gain 3,400 feet of elevation in 12 miles. Geologic processes create elevation; elevation, in turn, strongly controls temperature and precipitation. You may notice, most likely in your ears, the atmospheric pressure decreasing with altitude as you drive toward the trailhead. Air at the earth's surface responds to changes in pressure by expanding and contracting. When pressure increases, the air at the surface contracts, and molecules in the air move closer together. At lower pressures, air expands, and molecules move farther apart. These processes of expansion and contraction influence the temperature of the air. When molecules move closer together, the temperature of the air rises, and when molecules move apart, the temperature drops. The net result is that temperatures are higher at lower elevations where atmospheric pressure is greater, and lower at higher elevations where atmospheric pressure is lower. The temperature changes about 10 degrees Fahrenheit per mile change in elevation. You may have noticed that the air felt cooler at Mather Overlook than it did at the visitor center.

Not only does temperature change with elevation but also precipitation changes. The temperature of an air mass controls the amount of moisture that air holds. Warm air can hold much more water than cool air can. As air rises over a mountain range, it cools, thus losing much of its ability to hold moisture. The moisture falls to the earth as rain or snow. The air then sinks into the adjoining desert valley and warms, once again able to hold large amounts of moisture. The Snake and Spring Valleys, east and west of the Snake Range, respectively, average about 4 inches of precipitation per year, but the higher peaks of the Snake Range may receive more than 30 inches of precipitation, mostly in the form of snow.

Mountains in the Great Basin, then, rise as climatic islands above the surrounding desert valleys. The mountain climates differ from those of the adjacent valleys in the long-term averages of temperature, precipitation, atmospheric pressure, wind, and solar radiation. Of these, temperature and precipitation most strongly influence local variations in plant and animal communities.

Great Basin National Park encompasses six distinct climatic zones, each supporting a unique community of plants and animals. As you drive the scenic road toward the Wheeler Peak trailhead, you pass through a series of climate zones, each reflected in the prevailing plant community. A sea of sagebrush surrounds Baker, just outside the park boundary, indicating the very dry conditions there. The park boundary itself lies in a juniper-piñon woodland, slightly cooler and moister than the low-lying valley floor. You pass from juniper and piñon pine into manzanita and mountain mahogany, then into aspen and Douglas fir as conditions cool and moisten even further. At the road's end, you enter a thick forest of Engelmann spruce and pine. And farther on lies the alpine tundra where trees and bushes cannot grow.

You'll explore further by foot, so pack a sandwich, a raincoat (weather here changes quickly with little warning), and lots of water. The trail to Wheeler Peak winds up through thick stands of spruce and pine to the treeline, the upper limit of tree growth. Just below treeline, the trail winds across alpine meadows. In the meadows, look for outcrops of grayish green Pioche shale, which dates to early Cambrian time, 550 million years ago. Shale is essentially clay that was deposited in still water, in this case a Paleozoic ocean, and later compacted. But shale perhaps is not the correct name for the rock we see here. Notice the glossy sheen and small crinkles, called crenulations, on the surfaces of outcrops in these meadows. These are characteristics of a rock called phyllite, a metamorphic form of shale. Metamorphic rock forms when heat and pressure alter pre-existing rock, in this case shale. New minerals grow and align themselves to minimize pressure; the aligned crystals give the rock its sheen. The Pioche shale contains pockets of metamorphic rock such as the scattered outcrops in this meadow.

Warped and stunted trees, called krummholz vegetation, greet us at treeline. While pines stood 50 feet tall at lower elevations, they barely reach heights of 6 feet here, and their contorted shapes reflect the extreme climate here. Low temperatures and very high winds prevent trees from growing at elevations above about 11,000 feet in the Snake Range, and the wind and weather batter any trees the take root at treeline. It is perhaps a non sequitur that the oldest trees in the world, bristlecone pines, live at treeline, the most extreme climate at which trees can grow.

As we continue upward, we enter the alpine tundra, where only hardy, low-lying plants grow. It is noticeably cooler and windier here, and patches of snow remain in protected depressions even in the warmest summer months. Large, blocky chunks of light-colored rock with dark blotches cover the ground here. The blotches are algal-fungal growths called lichens. The alga and the fungus of the lichen coexist in a mutually beneficial, or symbiotic, relationship. The fungus removes nutrients directly from rock, so the lichen does not need to grow in soil. The algal partner then uses those nutrients, plus water

Nearing 12,000 feet above sea level—and breathing hard—on the trail to Wheeler Peak. Angular boulders and cobbles protect scattered, low-lying alpine plants. —Dave Futey photo

Lichens feed on solid rock. This growth is about 6 inches across.

vapor from the air, to produce food through photosynthesis, which it shares with the fungus. This relationship allows lichen to grow in extreme conditions where other plants cannot survive. In extreme climates like this one, a lichen may take hundreds of years to grow even an inch.

The rock hosting these lichens is Prospect Mountain quartzite, originally a massive and widespread sandstone during earliest Cambrian time, about 570 million years ago. The Prospect Mountain quartzite is the metamorphosed version of that sandstone. Streams carried sediment that they dumped in the ocean. These streams deposited sand on a continental shelf, a gently sloping submarine platform like that in the Atlantic Ocean along the eastern coast of the United States today. If you look closely at the rock here, you will also see layers of rounded pebbles within the quartzite. These are conglomerate lenses that streams deposited as they meandered across the sand when sea level fell. Sea level eventually rose, and other sediment that ultimately became the Pioche shale and Pole Canyon limestone buried the shelf. The weight of overlying deposits compressed the sediments. This compaction and later cementation eventually changed them into sedimentary rock. Horizontal and vertical pressures from interactions between the earth's plates, particularly the convergence of giant plates, further altered this rock during late Paleozoic and early Mesozoic time, particularly about 225 million years ago, when the Rocky Mountains lifted to the east. Block faulting 5 million years ago—the same faulting that produced Nevada's alternating mountains and valleys—brought the Prospect Mountain

These ripple marks in the Prospect Mountain quartzite were preserved in sandstone, which later metamorphosed into much harder quartzite.
—Dave Futey photo

A lens of rounded stream pebbles in the Prospect Mountain quartzite. Metamorphosis has altered the pre-existing sandstone, but not enough to destroy this clue to the stream origin of the original sedimentary rock.

quartzite and other metamorphic rocks to the surface throughout the southwestern United States. These faults bound the Snake Range on the east and west. (For a general overview of plate tectonics, see this book's introduction, "Geology of Central Nevada.")

The Prospect Mountain quartzite, composed almost purely of quartz, strongly resists weathering, the chemical and physical alteration of rock at the earth's surface. The footing becomes difficult as you approach the summit because, while frost wedging has broken

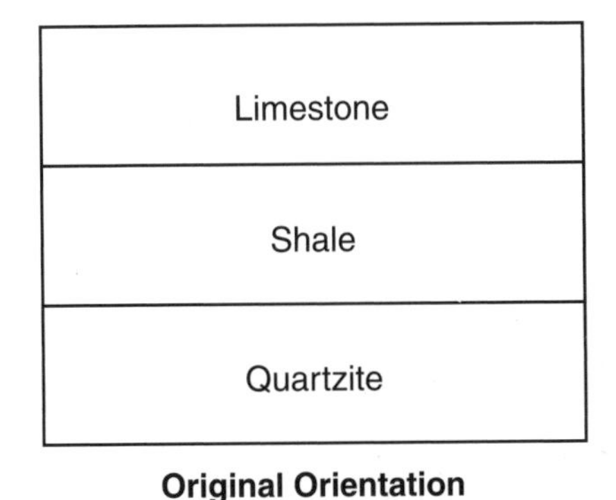

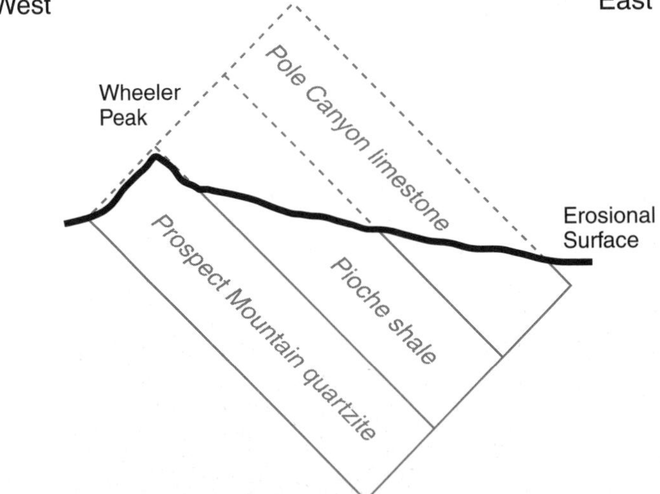

Tilting of the bedrock that makes up Wheeler Peak allows the older quartzite to crop out at higher elevations than the younger shale and limestone.

the quartzite into large pieces, the rock resists the chemical weathering that might otherwise weaken it and break it down.

In an undisturbed sequence of sedimentary rock, the oldest rock lies at the bottom of the stack and the layers get progressively younger above it. Our trip up Wheeler Peak has led us through Pole Canyon limestone, Pioche shale, and Prospect Mountain quartzite. Within this sequence, we find the youngest rock, Pole Canyon limestone, at the lowest elevations and the oldest rock, Prospect Mountain quartzite, at the highest elevations. This is contrary to our expectations, so something must have disturbed this stratigraphic column. Metamorphism altered the sedimentary layers, and faulting changed their orientation. The uplift that began 5 million years ago did not simply lift this section of rock. It tilted the whole stack to the east as it rose. Recall the outcrop of limestone at Lower Lehman Creek Campground—its layers dipped down to the east. This tilt, plus subsequent erosion, produced the situation we see at Wheeler Peak, wherein older rock crops out at higher elevations than does younger rock. Weathering and erosion affect different types of rock differently. Quartzite resists these processes better than either limestone or shale do. Once tilted and lifted, the two younger units—the limestone and shale—quickly wore away, while the sturdy quartzite stood tall to form Wheeler Peak.

As the highest point on this tilted stack of sedimentary rocks, Wheeler Peak offers an astounding vista that itself provides a lesson

Frost-shattered quartzite at the summit of 13,063-foot Wheeler Peak

in Great Basin morphology. To the east, you can see the Snake Valley, and beyond that the mountains and valleys of western Utah. To the west lie Spring Valley, the Schell Creek Range, Steptoe Valley, the Egan Range, White River Valley, the White Pine Range, and a continuing series of valleys and mountains across Nevada. Here is the Basin and Range in living color, its alternating green mountain ranges and brown desert valleys reflecting not only the region's geologic history but also the geologic controls on climate.

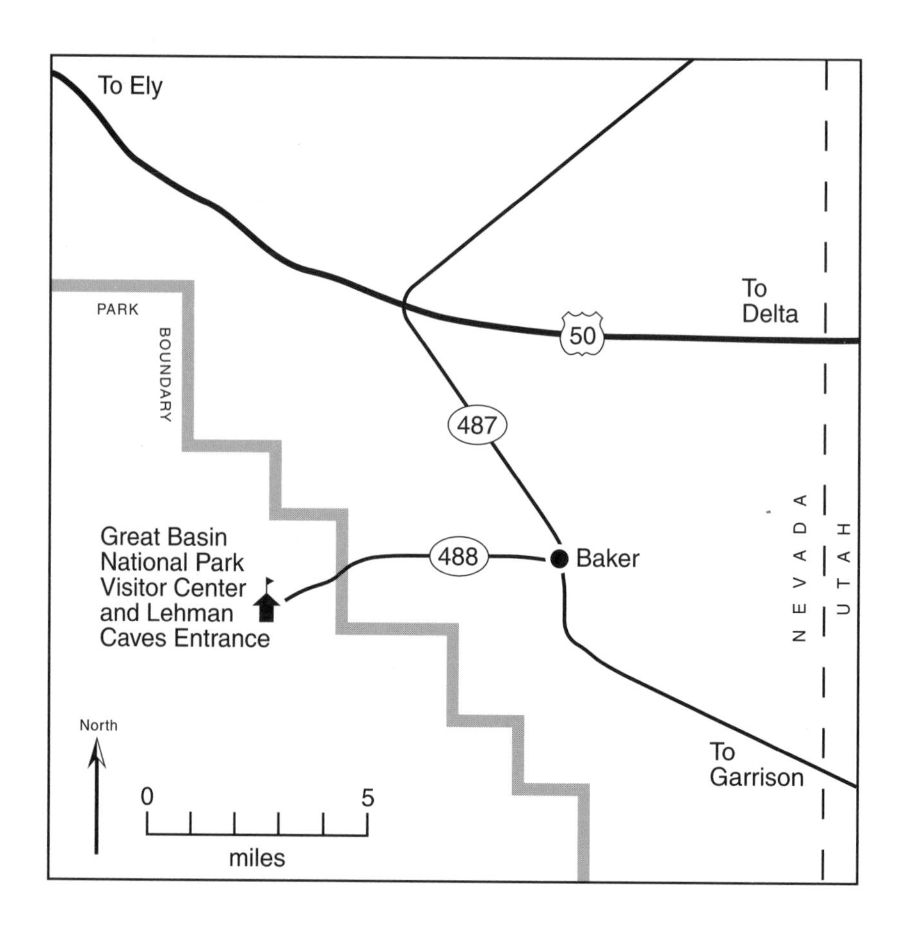

▶ GETTING THERE

Great Basin National Park lies in eastern Nevada near the town of Baker. From Las Vegas, drive 20 miles north on Interstate 15, then 234 miles north on U.S. 93. Turn east on U.S. 50 and continue 37 miles before turning south on Nevada 487. Travel 5 miles into Baker, then go west on Nevada 488. The park visitor center is 6 miles west of Baker. From Ely travel 65 miles east on U.S. 50, then 5 miles south on Nevada 487. Travel 6 miles west on Nevada 488 to the park. From Salt Lake City, travel south on Interstate 15, then 165 miles west on U.S. 6, which joins with U.S. 50 at Delta, Utah. Turn south on Nevada 487 and drive 5 miles into Baker before turning west on Nevada 488 for 6 miles to the visitor center.

Rangers lead cave tours every day of the week, and the caves are open year-round. You can make tour reservations at the main desk in the visitor center. Dress warmly as temperatures beneath the ground remain cool, near 50 degrees Fahrenheit, even during the warmest summer months. Flash photography is permitted within the cave, and we recommend high-speed film (ASA 400). As you tour the cave, be careful not to touch walls or features, as oils in your skin can damage the travertine. Let's protect this jewel for posterity.

21

Beauty Beneath the Surface

LEHMAN CAVES

Absalom S. Lehman came to the Snake Range as a miner in 1867 but astutely changed professions. He cleared land near a stream east of Wheeler Peak and began farming, supplying meat and produce to mining camps. In 1885, he chanced upon an opening in the ground. With rope and lantern, Lehman lowered himself into the earth and discovered a series of beautifully decorated, interconnected passageways. Lehman publicized his discovery in the local newspaper and began guiding visitors through the cave. This cave system, as well as

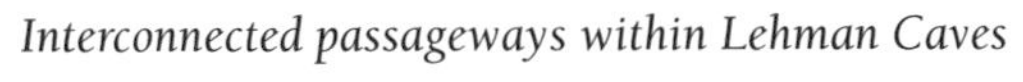

Interconnected passageways within Lehman Caves

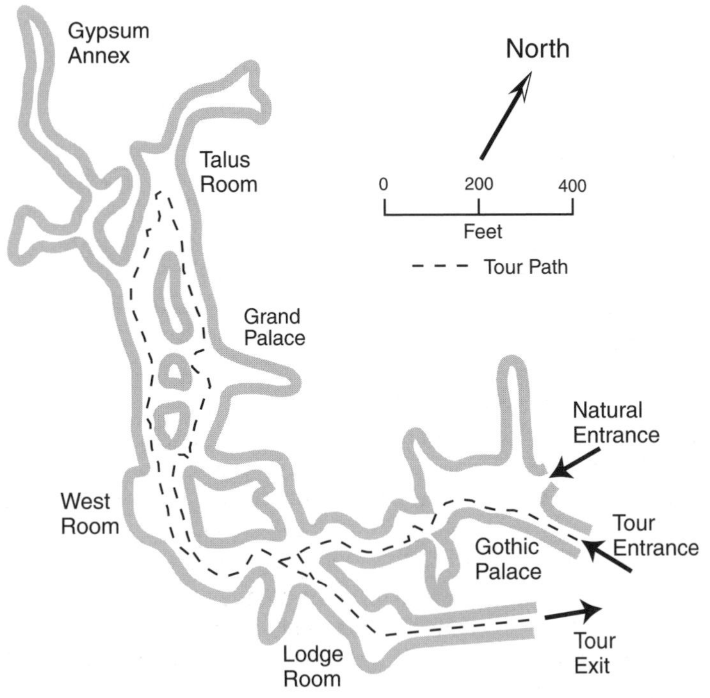

the land above it, became Lehman Caves National Monument in 1922. The monument was subsequently enlarged and established as Great Basin National Park in 1986. Eight fruit trees that once belonged to Lehman's orchard remain today, standing near the park's visitor center. The entrance to Lehman Caves—really only one cave, despite the name—lies immediately behind the visitor center. Access is restricted to ranger-led tours that enter the cave at regular intervals.

The Lehman Caves system developed within the Pole Canyon limestone, a rock unit composed almost entirely of calcium carbonate in the form of the mineral calcite. Approximately 550 million years ago, during Cambrian time, a warm ocean covered much of what is now Nevada. During the ensuing 400 million years, remains of sea creatures with shells made of calcium carbonate contributed to an ever-thickening accumulation of calcareous ooze on the sea floor. This ooze later hardened into the sedimentary rock limestone.

Before leaving on the tour, take a stroll behind the visitor center to look at the exposed outcrops of this blocky rock, the Pole Canyon limestone. It is light gray to white and the surface is pitted. Notice also that layers within this rock tilt down to the east. This limestone has undergone low-grade metamorphism, the alteration of rock by heat and pressure, which has made the rock both denser and whiter

Mineral-rich water has dripped from ceiling straws onto a sloping surface to form this flowstone tableau. —Dave Futey photo

The Lehman Caves tour winds through passageways like this one.
—Dave Futey photo

than it was originally. Metamorphism destroyed most of the fossils that limestone commonly contains and also recrystallized the calcium carbonate grains, giving this formation the characteristic crystalline texture of marble.

The tour guide, a park ranger, will present a short orientation before leading you underground through a locked door. This is not the natural entrance—entering through the natural opening requires a great deal more effort than entering through this walkway. The beauty of the cave and the variety of features decorating the walls will immediately strike you. Let's consider cave evolution.

Limestone and marble, both carbonate rocks, are especially susceptible to dissolution by water, especially water that has interacted with the atmosphere and soil. As rain falls through the atmosphere it reacts with atmospheric carbon dioxide, the fourth most common gas of our atmosphere, to form a relatively weak solution of carbonic acid:

$$CO_2 + H_2O \rightarrow H_2CO_3$$

Water becomes more acidic as it infiltrates through soil, reacting with additional carbon dioxide that the decay of plants and animals contributes to it. If one were to apply a weak acid solution, roughly equivalent to the acidity of fluids in the human stomach, to limestone or marble, the rock would fizz vigorously as it dissolved. During dissolution, calcium carbonate breaks into positive and negative ions that remain in solution. The entire chemical reaction is as follows:

$$H_2O + 2CO_2 + CaCO_3 \rightarrow Ca^{+2} + 2HCO_3^-$$

H_2O +	$2CO_2$ +	$CaCO_3$ →	Ca^{+2} +	$2HCO_3^-$
water	carbon dioxide	calcium carbonate	dissolved calcium carbonate	dissolved bicarbonate

Applying a weak acid solution to other types of rock does not produce this reaction, because rocks containing mainly silica resist dissolution much more than limestone or marble do. Infiltrating water in the natural environment is less acidic than the acid solution we apply in a laboratory setting to determine whether an unknown rock sample is limestone, but it is strong enough to profoundly influence the features that develop in a landscape dominated by carbonate bedrock. The pits that you saw on the limestone surface near the visitor center are dissolution features, as is this entire cave system.

Geologists use the term *karst,* a German adaptation of a Slavic word meaning "bleak, waterless place," to describe the landscape that forms on soluble rock as well as for the unique patterns of drainage that develop there. In addition to rock type, karst development relies on porosity and permeability of the soluble rock. Porosity is the total volume of empty spaces within the rock—its pore spaces. Porosity includes voids between grains that date from the formation of the rock as well as large, continuous openings along bedding planes and fractures that increase permeability as well as porosity. Permeability, the ability of a rock to transmit water, is vital to the development of karst because water must flow into and through rock before it can dissolve it. Permeability strongly depends on the interconnectedness of pore spaces. Recrystallization during metamorphism drastically reduces the spaces between grains, so bedding planes and fractures probably played a major role in the formation of Lehman Caves.

Most of the dissolution that created the connected caverns and passageways of Lehman Caves took place during Pleistocene time, or the Ice Age, from about 2 million years ago to 10,000 years ago. Changes in surface heating patterns on the earth resulted in alternating glacial and interglacial stages. The wet and cold glacial stages produced a great deal of precipitation and run-off, much of which

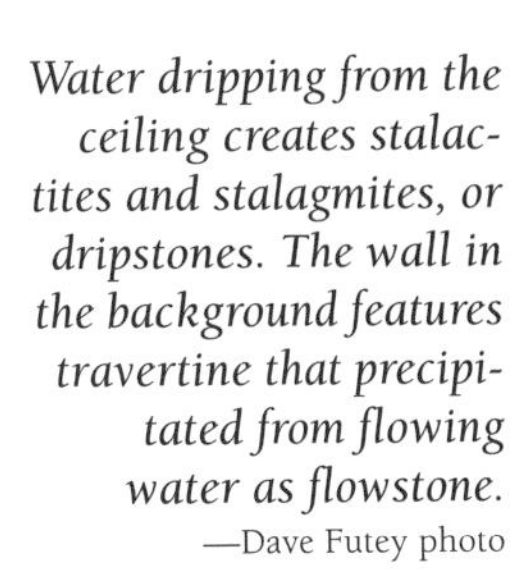

Water dripping from the ceiling creates stalactites and stalagmites, or dripstones. The wall in the background features travertine that precipitated from flowing water as flowstone.
—Dave Futey photo

infiltrated into porous and permeable bedrock. Most carbonate dissolution happens just beneath the water table, where water completely fills pore spaces. Here, flowing water dissolved portions of the Pole Canyon limestone, at times completely filling the caverns.

We noticed earlier that the limestone outcrops near the visitor center tilt down to the east. An important clue to the timing of cave formation lies in the orientation of cave passages relative to the orientation of those layers within the Pole Canyon limestone. In an undisturbed sequence of sedimentary rocks, strata accumulate as horizontal layers. The water table is more or less horizontal, too. As limestone dissolves, caverns develop with passageways parallel to the water table. Layers within the Pole Canyon limestone now tilt to the east, reflecting the asymmetric block faulting that began about 5 million years ago. Had dissolution of the cave taken place prior to faulting, passageways would originally have been parallel to both the water table and layering within the rock, and then tilted. But the cave passages are horizontal, so dissolution took place after faulting. Moving water would have initially smoothed and rounded the walls, roof, and floor

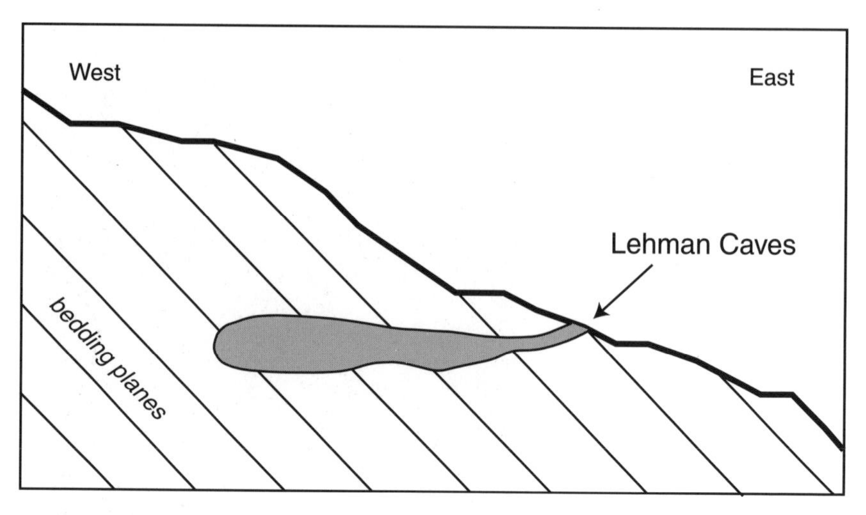

Side view of horizontal Lehman Caves and tilted bedding planes of the bedrock

of the cave system, but smooth walls are not what we encounter in Lehman Caves.

Many caves claim particularly impressive and unique characteristics. Kentucky's Mammoth Cave, with more than 350 miles of interconnected passageways, is the world's longest known cave system. Carlsbad Caverns in New Mexico boasts the Big Room, broad enough to accommodate fourteen football fields and tall enough to hold the United States Capitol. Although tiny compared to those massive systems, Lehman Caves bases its claim to fame on beauty, not size. We probably don't need to tell you this—so many stunning features surround you that you may be having trouble deciding where to look first. Lehman Caves is one of the most finely decorated caves in the world, which leads us to the topic of speleothems, or dripstone features. Like the passageways themselves, speleothems represent the work of flowing water. After the last glacial stage, the water table dropped below the cave floor. Water continued to infiltrate, though at a much slower rate. Small amounts of water percolated through the upper layers of the Pole Canyon limestone, then reached the cave. Upon entering the cave, flowing water released some of its dissolved carbon dioxide to the air in the cavern, which caused calcite to separate out of solution as the mineral travertine, calcium carbonate that commonly crystallizes as fine layers in caves. Look back at the dissolution reaction. Carbon dioxide (CO_2) is on the left side of the equa-

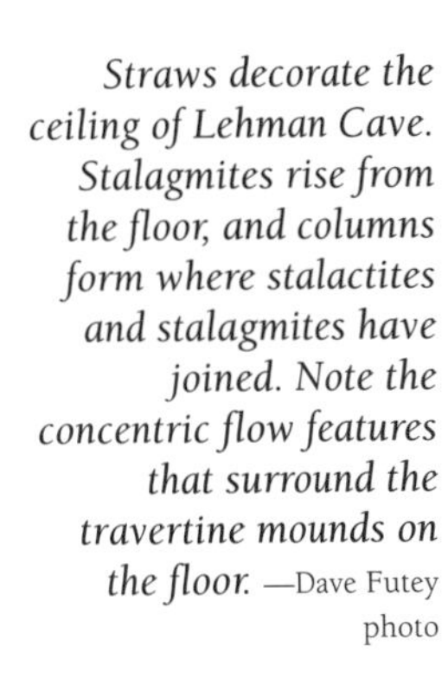

Straws decorate the ceiling of Lehman Cave. Stalagmites rise from the floor, and columns form where stalactites and stalagmites have joined. Note the concentric flow features that surround the travertine mounds on the floor. —Dave Futey photo

tion. If the system loses carbon dioxide, the reaction is driven to the left, and calcium carbonate crystallizes. Over an extended period of time, dripping water deposited, forming a variety of features whose shapes and forms strongly reflect the style of bedrock permeability.

Stalactites and stalagmites are perhaps the best known examples of speleothems. Look at the conical growths that hang from the ceiling and point up from the floor. When water drips from a single opening on the cave roof, calcite precipitates as a ring of travertine. With each ensuing drop, the initial ring builds downward into a thin tube known as a soda straw. Many of these hang from the ceiling here. The straw commonly clogs, and water begins to move along the outside of the tube, depositing calcite in thin layers as an elongated cone, the familiar stalactite. Over time, stalactites widen and lengthen as they reach for the cave floor. If you could slice a stalactite off—not good cave behavior—you would see on the sliced surface hundreds or perhaps thousands of concentric rings representing the history of travertine deposition. These concentric rings surround a very narrow, hollow center in a mature stalactite. Stalagmites grow simultaneously with stalactites. Water that drips from the roof-bound features deposits calcium carbonate on the floor, building an upward-reaching cone of travertine there. Stalagmites lack a hollow center because they don't evolve from straws, and they are usually broader and rounder than their associated stalactites. Often stalactites and stalagmites grow to-

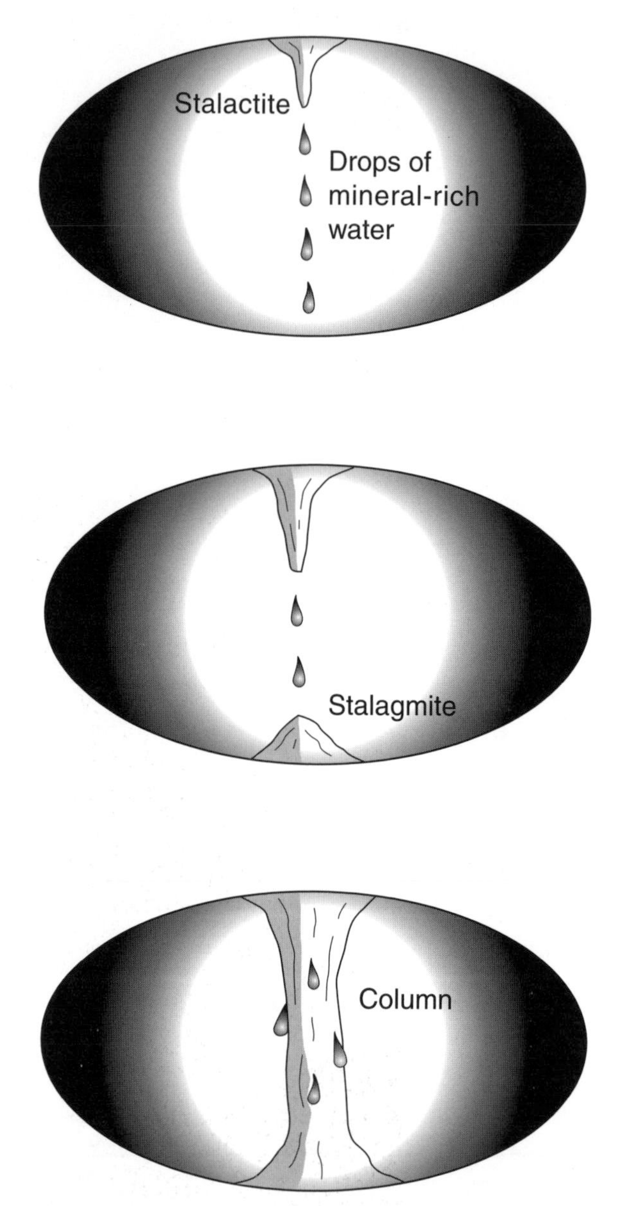

Evolution of stalactites, stalagmites, and columns within a cave

gether into a single column that broadens with continuing deposition. Columns adorn virtually every room in Lehman Caves.

Lehman Caves contains other types of speleothems as well. Water in Lehman Caves seeps directly out of tiny pores in the cave walls and deposits knobby protrusions called cave popcorn. Cave popcorn can be isolated or joined into great clusters. Where water runs down the cave's walls, we find travertine deposits that look like rounded waves or draperies. Helictites, which look like curvaceous twigs, spiral out of walls. Boxwork structures form where intersecting fractures in the roof or walls control the flow of water. Water seeps out of fractures and builds ridges of travertine that mirror the geometry of the linear features. Travertine deposits called rimstone bound ponded water within the cave. You may even see cave pearls within these ponds. Pearls are spheres of travertine that have precipitated onto sand grains. If dripping or flowing water constantly agitates the grains, the pearls cannot adhere to another surface and instead move freely within the pond. As they roll around, the grains accumulate travertine snowball-style, growing ever larger over time. Perhaps the most

Cave popcorn covers part of the wall in the lower left and center. Its nubby texture contrasts with the smoother flowstone on the vertical walls elsewhere. —Dave Futey photo

A shield, one of the more unusual speleothems, protrudes from the wall of the cave. Straws and stalactites hang from the bottom of the shield, while flowstone forms where water drips over the leading edge. —Dave Futey photo

unique speleothems in Lehman Caves are shields, also called parachutes. Shields look like tilted clamshells that have many small stalactites and columns protruding from the bottom side. How these features form remains a mystery.

The northernmost room on the Lehman Caves tour is the Talus Room. Talus is rock debris resulting from mass wasting, including free fall from vertical cliffs, rock and debris slides, slumping, and creep. Within the Talus Room we find a chamber whose floor holds a great deal of rock fragments that have fallen from the cave roof and walls. In fact, the entire roof for almost 300 feet north of the tour route has collapsed. Nevada is tectonically active—the state experiences relatively frequent earthquakes, reflecting the ongoing process of block faulting that has produced the Basin and Range province. A series of earthquakes 30,000 to 10,000 years ago probably was responsible for the damage seen here. As earthquake activity continues, we can expect more mass wasting within this room and the cave as a whole.

As you wander through Lehman Caves, you see and hear water dripping and flowing. And water is the key. Under cooler and wetter

Talus from rockfalls covers the floor of the Talus Room, the northern-most part of Lehman Caves. A series of earthquakes 30,000 to 10,000 years ago triggered the rockfalls. —Dave Futey photo

climatic conditions, groundwater dissolved limestone to produce featureless passages. Later, the climate warmed and dried, and the water table dropped. Rooms that water once drowned now held air, and the second stage of cave development began. Water percolated through soil and rock, entering the cave with dissolved calcium carbonate that then precipitated on the ceiling and walls to form the marvelous decorations that make this cave unique. Speleothems continue to grow in the interconnected rooms of Lehman Caves. You can even hear them—drip, drip, drip.

This bristlecone pine grows on glacial moraine containing mainly large pieces of resistant quartzite. —Dave Futey photo

22

Gnarled Elders of the Snake Range

BRISTLECONE PINES

Nestled among the high peaks of the southern Snake Range lie several groves of bristlecone pines. These gnarled trees live at the transition between forest and alpine tundra, and they occupy the harshest environment in which trees can grow. Stunted trunks and warped limbs attest to the wind and cold that shape life at this high elevation. Trees growing in these demanding conditions conserve energy and utilize available resources to their fullest. A bristlecone pine tree, for example, replaces its needles only once every 20 to 40 years, while other pines replace needles every few years.

The trail to the bristlecone pine grove heads southwest from the trailhead before turning to the east. It climbs gently—although the elevation may make it seem steeper than it actually is—through a tall pine forest, then turns north. Huge blocks of quartzite deposited by glaciers cover the ground here, and vegetation thins noticeably. Temperatures are cold year-round, winds blow with great gusto, and what little soil there is offers minimal nutrients. The only trees growing in this stressful environment are bristlecone pines. As you look upslope from the pine grove you will see that no trees grow at higher elevations. Even hardy bristlecone pines cannot tolerate colder temperatures and stronger winds than those found here.

Once you reach the pine grove, stand among these ancient trees and take a look around. Many bristlecone pines look dead, with only a thin strip of bark snaking up a mostly bare trunk. But looks can be deceiving. How many of the trees that initially looked dead to you actually have some living bark and green needles? These trees are not only alive, but they have been thriving for thousands of years. Many of the trees in the Wheeler Peak grove stood here when Napoleon met defeat at Waterloo, when Alexander the Great conquered the world, and when Egyptian kings and queens built great pyramids in the Nile Valley. Bristlecone pines do not display the great height of

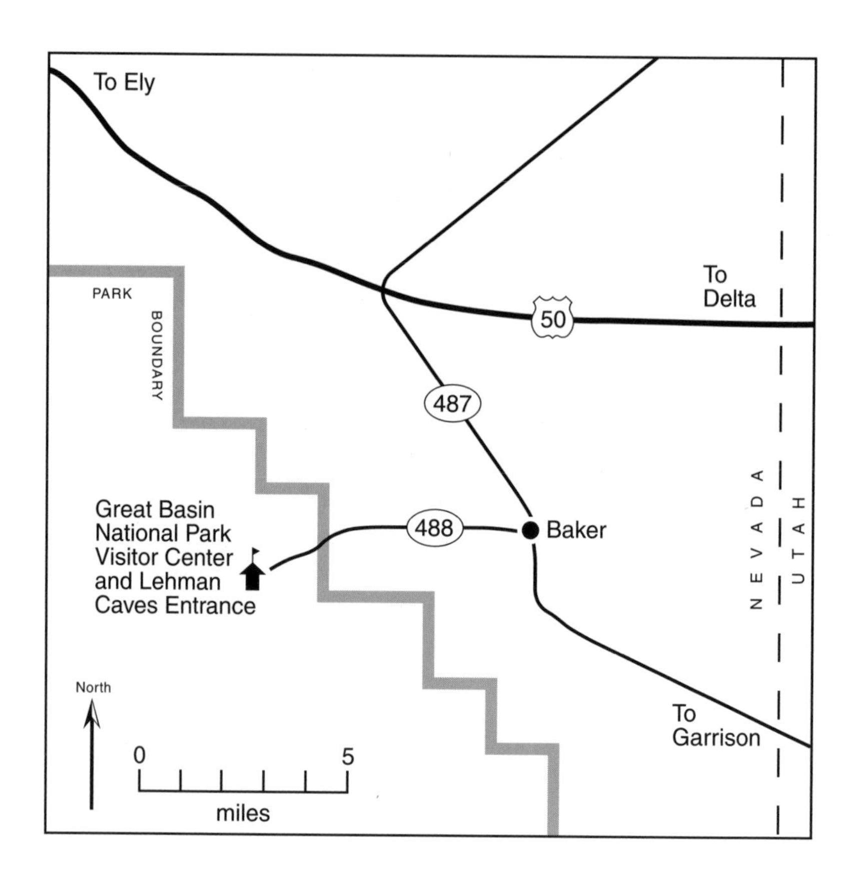

GETTING THERE

Great Basin National Park lies in eastern Nevada near the town of Baker. From Las Vegas, drive 20 miles north on Interstate 15, then 234 miles north on U.S. 93. Turn east on U.S. 50 and continue 37 miles before turning south on Nevada 487. Travel 5 miles into Baker, then go west on Nevada 488. The park visitor center is 6 miles west of Baker. From Ely travel 65 miles east on U.S. 50, then 5 miles south on Nevada 487. Travel 6 miles west on Nevada 488 to the park. From Salt Lake City, travel south on Interstate 15, then 165 miles west on U.S. 6, which joins with U.S. 50 at Delta, Utah. Turn south on Nevada 487 and drive 5 miles into Baker before turning west on Nevada 488 for 6 miles to the visitor center.

In the park visitor center, a sliced and polished section of a bristlecone pine shows the internal structure of the pine much more clearly than the

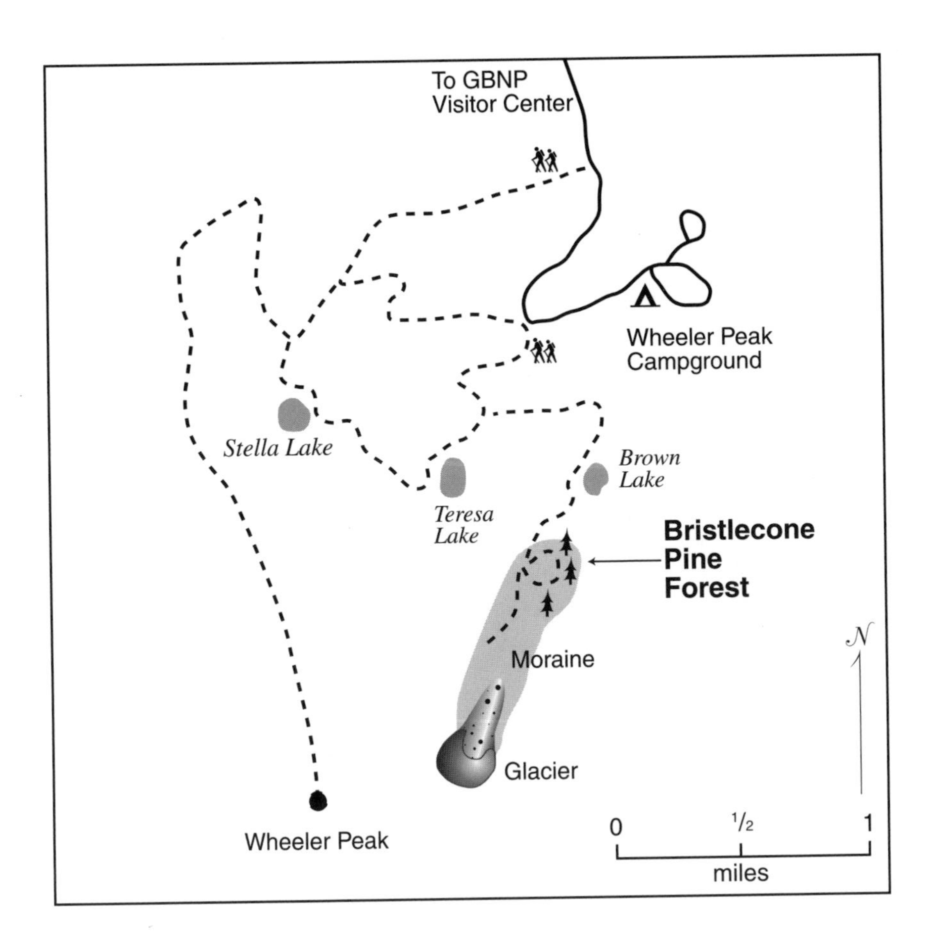

rough deadwood at the grove. Try to count the rings of the polished section. You will be amazed at how many there are.

Just inside the park boundary, a well-marked road heads north to the Wheeler Peak trailhead. It passes Lower Lehman Creek and Upper Lehman Creek Campgrounds. The trailhead sits at the entrance to Wheeler Peak Campground. The road gains 3,400 feet in elevation over 12 miles. There are a parking lot and restroom facilities at the trailhead. Follow the marked trail to the bristlecone pine grove. This is not a particularly steep hike—it gains only 600 feet in elevation—but the high elevation can make it strenuous. Go slowly and take plenty of water. Remember as you visit here that deadwood is as important to dendrochronology as live wood and that it is unlawful to remove or vandalize either.

Although much of this tree lacks both bark and needles, it is alive. Many of the ancient trees feature mostly bare wood with only thin strips of bark and a few needle-covered branches. Such trees may be older than 4,000 years. —Dave Futey photo

California's coastal redwoods nor the ponderous majesty of giant sequoia; they have another claim to fame. They are the oldest trees in the world, with some specimens approaching 5,000 years. Because of their great age, they act as storehouses of climatic information that those skilled in the field of dendrochronology—the use of tree rings for assigning dates—can glean.

In 1964, in a true affront to persistent longevity, the oldest known living tree was cut down. This ancient organism dwelt in the upper reaches of what would later become Great Basin National Park. Scientists removed it and sliced it into sections so they could investigate its growth rings—4,900 in all.

Leonardo da Vinci described seasonal growth rings in trees in the late fifteenth century. The relationship between the size of tree rings and climate was first published in Europe after the severe winter of 1708–1709. The narrowness of growth rings from that year drew scientists' attention as a record of extreme climate. A. E. Douglass, an astronomer, noted in the early 1900s that variations in the width of rings from trees in the arid southwestern United States might provide a record of climatic variation there. Douglass studied living trees and deadwood from archeological sites to establish the foundations of dendrochronology as well as dendroclimatology, the use of tree rings as indicators of past climate. Many scientists have since added to this pioneering effort, and one large step was the recognition of the importance of *Pinus longaeva*, the bristlecone pine.

In 1953, Dr. Edward Schulman, who studied tree rings at both the University of Arizona and California Institute of Technology, heard tales of very old trees in the White Mountains, on the border of Nevada and California. He was interested in establishing climate records from growth rings and had previously discovered ancient limber pines *(Pinus flexilis)* at treeline in Idaho. Schulman took core samples from

Close-up of needles on a bristlecone pine branch. To conserve energy, bristlecone pines only replace needles once every 20 to 40 years.
—Dave Futey photo

a number of trees, including bristlecone pines. He used an incremental borer to extract cores, removing a narrow dowel of wood extending from the bark to the center of the trunk. Removal of this material doesn't harm the tree, and it allows scientists to study growth ring patterns. One of Schulman's cores, from a bristlecone pine tree, revealed an incredible record of longevity. He had discovered the world's first 4,000-year-old tree. He named it Pine Alpha, signifying its importance to the field of dendrochronology. In subsequent years he found yet older pines, and in 1957 he cored a 4,700-year-old tree that he named Methuselah. Schulman died in 1958—just before *National Geographic* magazine published his research—leaving behind an important scientific legacy.

A sliced cross section of a tree trunk shows alternating light and dark rings. These are growth rings. Each ring is one year's growth of tissue in the cambium, the section of tree immediately under the bark. Look for growth rings on downed wood in the pine grove. A close look at a single ring reveals large, thin-walled cells on the inside giving way to ever smaller, thick-walled cells on the outside of the ring. The larger cells grow early in the growing season and

This schematic cross section of a tree shows growth rings.

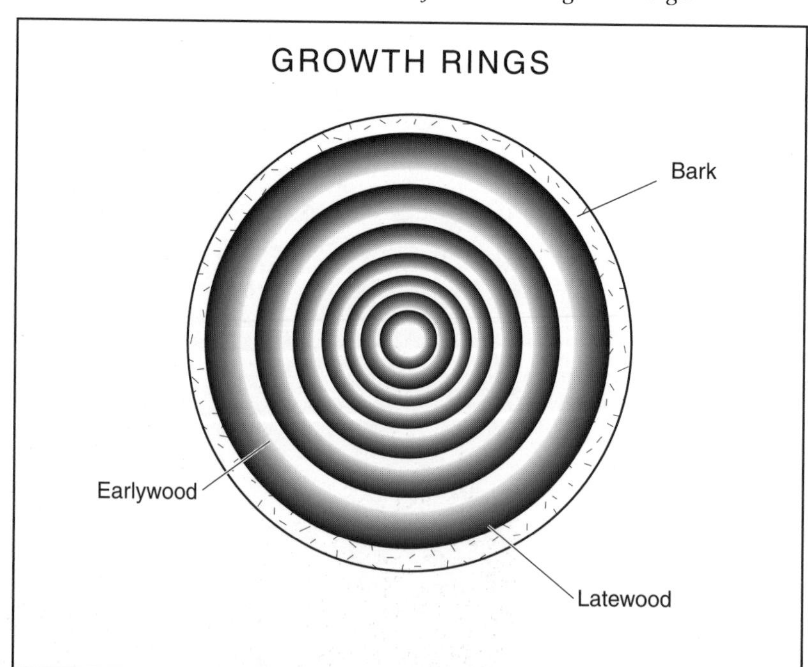

are called earlywood. The smaller cells, or latewood, grow at the end of the growing season. Latewood is darker because it contains more small cells with thick walls. The combination of earlywood and latewood makes up trunk growth for a given year. The average thickness of each ring depends on a great number of factors including soil type, climate, and species and age of tree. Dendroclimatologists therefore face the problem of distinguishing a tree's responses to many different factors and isolating changes in temperature or precipitation.

While early dendroclimatologists studied only ring width, many scientists now look also at ring density and chemical variations in rings. Both density and chemistry change as a function of precipitation, but also with changes in nutrient levels in the soil. Varying weathering rates and even wind direction can affect the nutrient levels and, consequently, the tree's growth.

Selection of tree species and location is an important factor in dendroclimatology. Long-lived trees that grow in areas of high stress, such as the trees growing near treeline in Great Basin National Park,

Close-up of wood grain on a bristlecone pine tree. Because of its high resin content and dense tissue, the wood does not decay when the tree dies, which makes it a good candidate for dendrochronology.
—Dave Futey photo

present the best climatic data. Growth-ring researchers divide trees into two broad categories: complacent and sensitive. Complacent trees grow in an area that provides abundant water and nutrients. Small changes in climate may reduce the quantity of available water or the length of the growing season, but not enough to affect the width of growth rings. Sensitive trees, on the other hand, grow near the limit of their stress tolerance. A tree living at the upper end of its growth zone cannot tolerate much cooler temperatures, so small changes in climate strongly influence its growth. We are essentially at treeline in the Snake Range—these pines are highly sensitive to subtle changes in temperature or precipitation.

The two main limiting factors for sensitive tree growth are temperature and moisture. Cooler or drier conditions produce narrower rings, while warmer, moister conditions produce wider rings. Alternating narrow and wide rings in sensitive trees therefore attest to variations in climate. The availability of water limits trees in arid and semiarid regions of the southwestern United States. These trees are thus sensitive to changes in precipitation associated with climate change. Temperatures decrease both with increasing elevation and increasing latitude, so trees growing near the upper limits of elevation or latitude in their life zone are very sensitive to changes in temperature. Trees like Great Basin National Park's bristlecone pines, which grow at treeline in a semiarid climate, are sensitive to both moisture and temperature changes. Their growth rings reflect a wide variety of climatic change.

The climate record that tree rings produce extends nearly 10,000 years into the past, yet the oldest known living tree—after the death of the Snake Range's 4,900-year-old pine—is 4,700 years old. Both living and dead bristlecone pines have contributed to this record through a process known as cross dating. Assigning ages to rings from living trees is simple. One counts back from the outermost ring to the innermost, subtracting one year from the present year for each ring. Around you lie many dead trees, some of which have lain here for thousands of years. Dead trees present a more challenging problem. We can count the rings, and we can look at changes in ring width to assess climate variability, but determining the absolute age of each ring requires correlation with living trees. To correlate ring sequences, we need an overlapping record of growth.

Consider the case of trees 1 and 2. Tree 1 is living, with a 2,000-year record of growth rings. Tree 2, on the other hand, has fallen down and is greatly weathered, indicating that it died a long time ago. Growth rings from tree 1 show distinct patterns of wide and

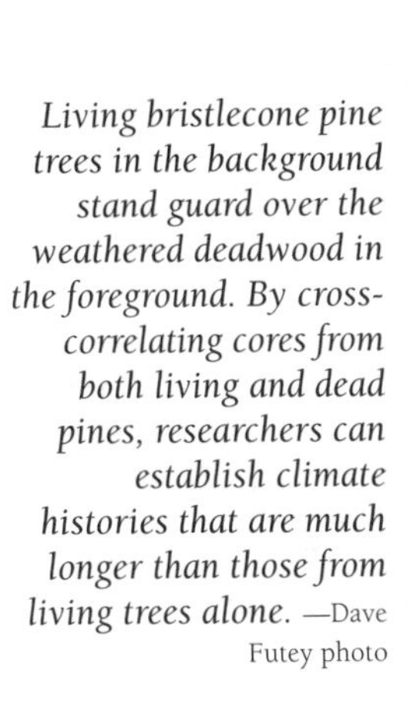

Living bristlecone pine trees in the background stand guard over the weathered deadwood in the foreground. By cross-correlating cores from both living and dead pines, researchers can establish climate histories that are much longer than those from living trees alone. —Dave Futey photo

narrow banding. For instance, for years 101 to 110 we observe the following sequence: 5 very narrow bands, 2 broad bands, 2 narrow bands, and 1 very broad band. Nowhere else do we see this pattern. Upon close investigation, we notice a similar pattern in rings from tree 2, this time for years 1,501 to 1,510 of its life. Careful inspection shows that rings before and after this sequence also match. So we can state that a correlation exists and that trees 1 and 2 coexisted for a period of time. Cross dating of trees 1 and 2 has created a ring record that extends 3,400 years into the past, 1,400 years more than that supplied by the living tree alone. Examining many living and dead trees in a given location makes this process more accurate and extends the record ever farther into the past.

The tree ring record from bristlecone pine trees has contributed more to science than simply a sequence of climatic variations. It has also acted as an important control in dating events in earth's history by helping assign ages to objects and processes. Deadwood,

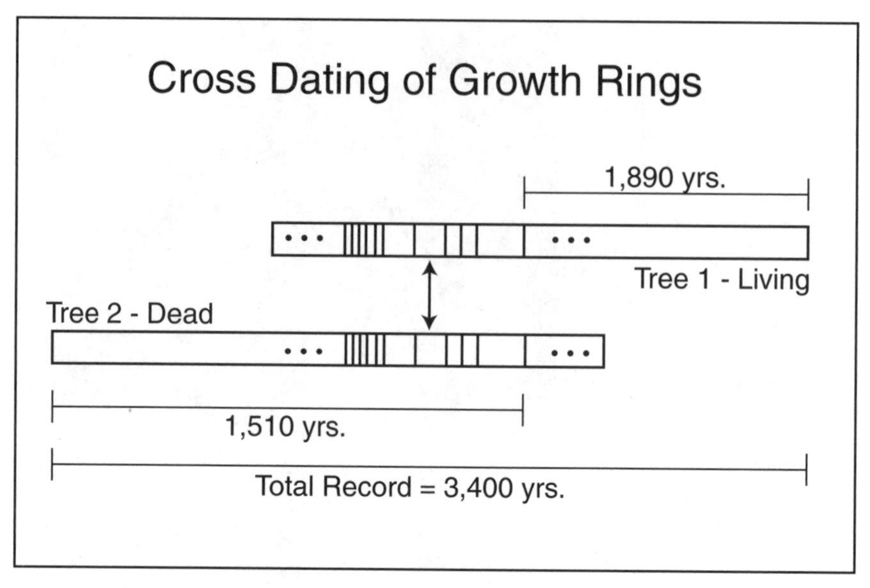

Cross dating of bristlecone pine tree rings by correlating overlapping growth rings from a live and a dead tree

a once-living material, is a good candidate for carbon-14 dating. All that is required is that this material once interacted with the atmosphere, taking up and releasing gases, including carbon dioxide.

Scientists use radiocarbon dating to assign ages to objects by comparing the amounts of radioactive carbon-14 and stable carbon-12. Carbon-14 forms when solar radiation bombards carbon in the earth's atmosphere. Like all radioactive elements, carbon-14 decays at a constant rate to produce a stable daughter product, in this case nitrogen-14, along with a beta particle and a neutrino. Because the sun's rays continually interact with the earth's atmosphere, the level of carbon-14 in the atmosphere remains constant. All living organisms take in carbon constantly, so the level of carbon-14 in plants and animals remains the same as that in the atmosphere. As soon as an organism dies, however, it no longer absorbs radioactive carbon, and the carbon-14 it contains decays at a steady rate. The ratio of carbon-14 to carbon-12 therefore helps pinpoint the date the organism died.

Understanding how ratios correlate to ages is integral to the use of radiocarbon dating. Bristlecone pine trees have proved to be an important control on that use. The record of bristlecone growth rings provided an accurate chronology that scientists used to test radiocarbon dating. Researchers dated wood in growth rings of known age

Windblown ice and dust sculpted and smoothed this bristlecone limb. —Dave Futey photo

using carbon ratios and discovered errors in the radiocarbon method. Carbon-14 dating was producing ages younger than the true age of these rings. The dendrochronologic record provided by *Pinus longaeva* allowed scientists to recalibrate the radiocarbon method. For this reason, some people refer to bristlecone pines as the trees that rewrote history.

The Wheeler Peak grove of bristlecone pine trees is growing on glacial moraine, debris dumped by glacial ice. This moraine consists predominantly of quartzite boulders. Two other groves of bristlecones grow in Great Basin National Park. The largest grove sits on Mt. Washington in the west-central section of the park. These trees grow on limestone bedrock and are accessible only by cross-country travel, which is exceptionally difficult in this rugged terrain. Another grove, also fairly inaccessible, grows on the linear ridge that separates the Baker Creek and Snake Creek watersheds. Bedrock here is limestone as well. The one constant among the three stands of trees is the harsh environment and its effect on the trees themselves. Dead trees and limbs lie scattered about the landscape. Because of its high resin content and narrow growth rings, bristlecone pine wood does not decay. Instead, windblown dust and ice sculpt it like rock. Bristlecones in Great Basin National Park stand as gnarled elders in the modern world, bent and twisted but persevering nonetheless.

Nevada's only modern glacier sits at the base of the cirque headwall below Wheeler Peak. The glacier feeds into an ice-cored rock glacier, the rocky mound in the center of the cirque. Foreground deposits are moraine. —Dave Futey photo

23

Nevada's Only Glacier

GLACIAL FEATURES IN GREAT BASIN NATIONAL PARK

The Snake Range of east-central Nevada once hosted many alpine glaciers that resembled rivers of flowing ice. Just a single one remains today, nestled within the deep valley below Wheeler Peak. Although Nevada boasts many tall mountains, Great Basin National Park is home to Nevada's—and indeed the interior Great Basin's—only glacier. What remains as well, here and on many neighboring peaks, is an alpine landscape shaped by moving ice, a testament to a very different climate that existed only a short time ago, geologically speaking. The beginning of this hike winds through a forest of tall pines, then takes you to the side of a ridge. As you turn to the north, you approach treeline, and vegetation thins noticeably. The trail continues over a distinctive mound of blocky quartzite deposited by the last glacier in this part of the valley. This is a good place to rest and consider the nature of glaciers.

The Pleistocene epoch, commonly called the Ice Age, extended from about 1.8 million years ago to 10,000 years ago. During Pleistocene time, the climate over much of the earth periodically changed from cooler temperatures to warmer temperatures, then back again. In cooler, wetter periods, called glacial stages, continental ice sheets and alpine glaciers grew and covered huge areas. The growth of ice sheets lowered global sea level—when continental ice sheets held water as ice, less water was available to fill ocean basins. During warmer periods, called interglacial stages, the ice retreated through melting, and sea level rose once again. Were today's ice sheets in Greenland and Antarctica to melt, sea level would rise disastrously and inundate all of the world's coastal cities.

A number of factors drove the dramatic swings in temperature during Pleistocene time. These factors included changes in the earth's tilt and orbital pattern. The position and orientation of the earth as it orbits the sun control the amount of incoming solar radiation, and

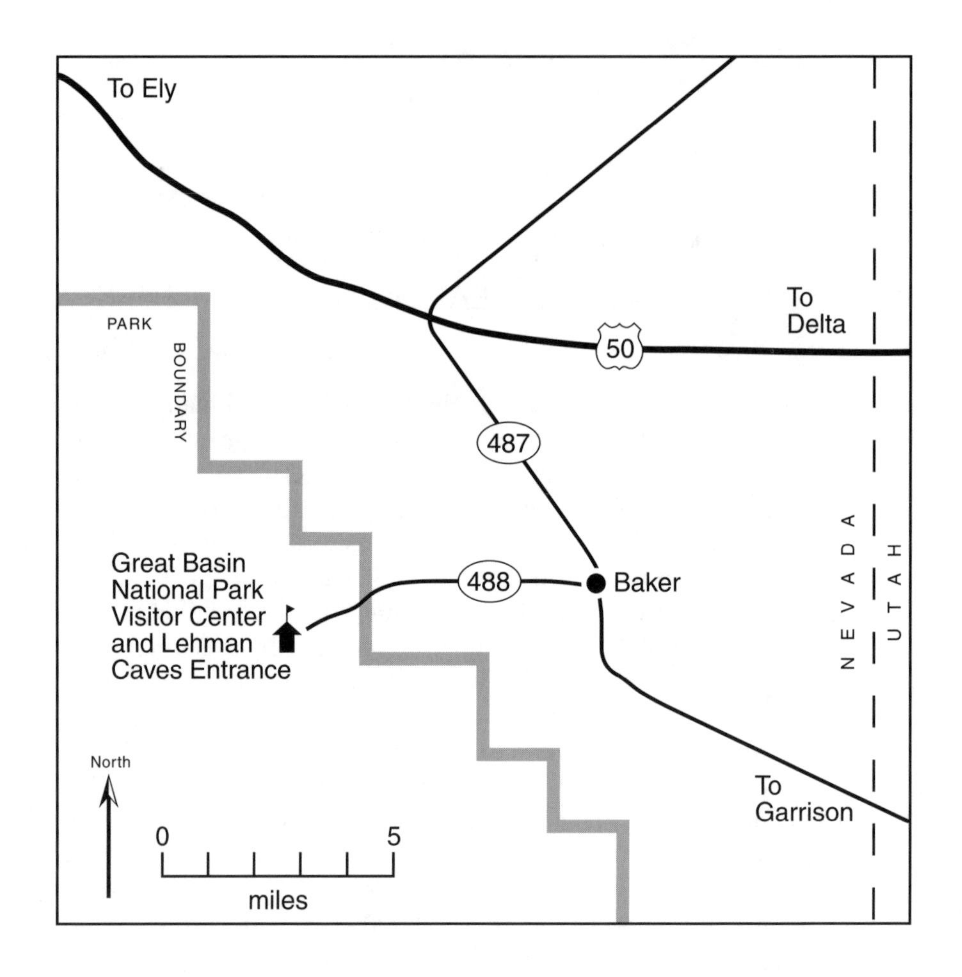

▶GETTING THERE

Great Basin National Park lies in eastern Nevada near the town of Baker. From Las Vegas, drive 20 miles north on Interstate 15, then 234 miles north on U.S. 93. Turn east on U.S. 50 and continue 37 miles before turning south on Nevada 487. Travel 5 miles into Baker, then go west on Nevada 488. The park visitor center is 6 miles west of Baker. From Ely travel 65 miles east on U.S. 50, then 5 miles south on Nevada 487. Travel 6 miles west on Nevada 488 to the park. From Salt Lake City, travel south on Interstate 15, then 165 miles west on U.S. 6, which joins with U.S. 50 at Delta, Utah. Turn south on Nevada 487 and drive

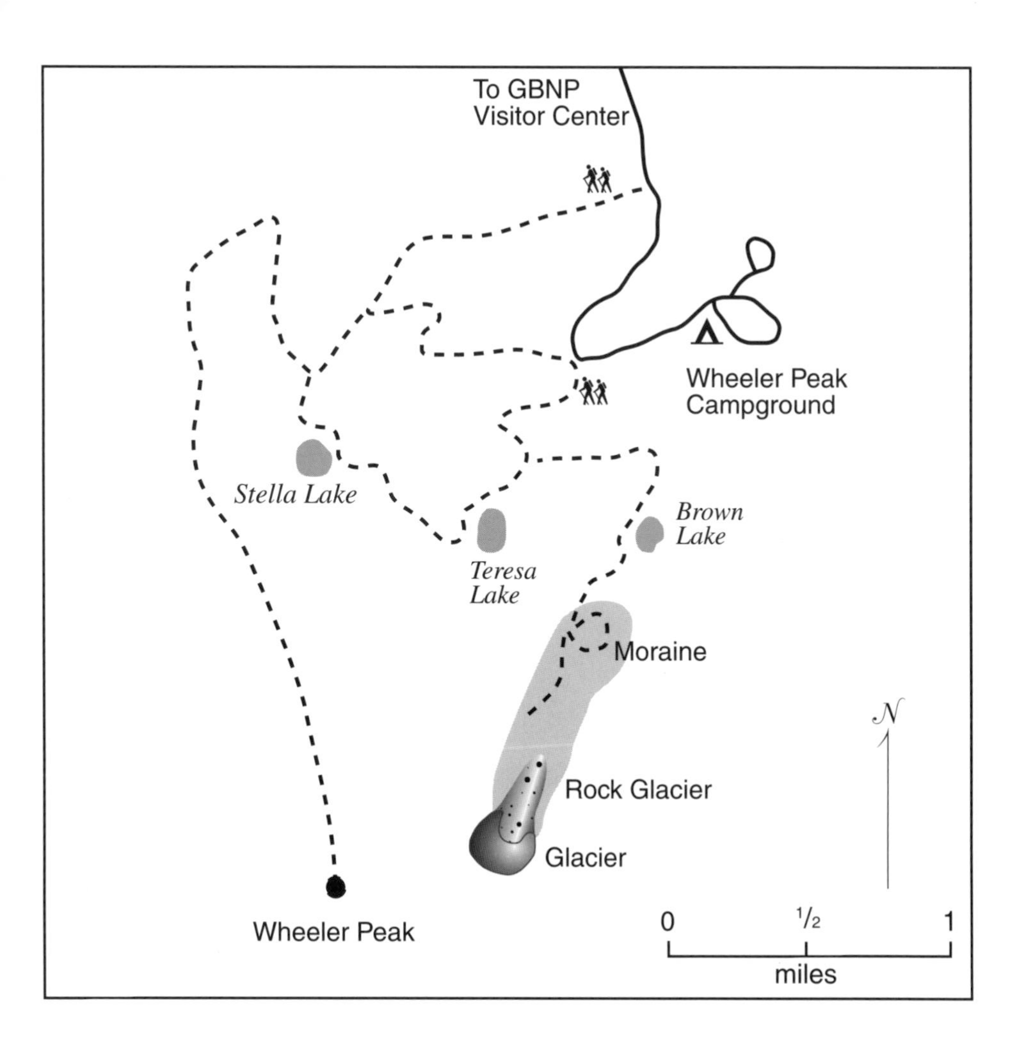

5 miles into Baker before turning west on Nevada 488 for 6 miles to the visitor center.

Just inside the park boundary, a well-marked road heads north to the Wheeler Peak trailhead. It passes Lower Lehman Creek and Upper Lehman Creek Campgrounds before reaching the trailhead at the entrance to Wheeler Peak Campground. The road gains 3,400 feet in elevation over 12 miles. There are a parking lot and restroom facilities at the trailhead. Follow the marked trail to the glacier. The trail is 2.8 miles long with an elevation gain of 1,100 feet. Take plenty of water, and go slowly as you acclimate to the high elevation.

thus control global heating patterns. Interestingly, the present configuration of orbital parameters resembles that which during the Pleistocene led directly into a glacial stage. Human activity on earth, such as the release of greenhouse gases and interaction with the hydrologic cycle, however, may prevent or alter this natural climatic progression.

Scientists divide glaciers into two broad categories: continental ice sheets and alpine glaciers. Continental ice sheets, such as those in Antarctica and Greenland, are huge masses of ice that accumulate in a central highland and flow outward to cover large areas. The Antarctic ice sheet, for example, measures up to 14,000 feet deep—well over 2 miles—and covers 5 million square miles. Similar ice sheets covered much of North America and Europe during Pleistocene glacial stages. Alpine glaciers, also called valley glaciers, originate in isolated highlands and usually move downslope in confined valleys. During the Ice Age, alpine glaciers formed in the Snake Range and other mountains of the Southwest while continental ice sheets developed to the north.

A valley glacier contains a zone of accumulation and a zone of ablation, which are separated by the snowline. Within the zone of accumulation, more snow falls in a given year than melts. New snow layers compress a given year's surplus, which eventually recrystallizes into a coarse, granular ice called firn. Increasing pressure transforms firn to glacial ice, which can flow. Ice within the zone of accumulation flows and slides downslope into the zone of ablation, where melting exceeds snowfall. The snowline literally marks the elevation above which we see fresh snow year-round on the glacier's surface.

Glaciers are agents of erosion. Like wind and water, glaciers can pick up, transport, and deposit sediment. Glaciers flow and slide downhill in response to gravity. Ice is a plastic material, meaning that it has characteristics of both a solid and a liquid. That's why ice flows. Meltwater flows along the base of the glacier, lubricating it and allowing it to slide more easily. Glaciers pick up sediment by a combination of plucking and abrading. Meltwater at the base of a glacier frequently refreezes, essentially gluing the rock on the valley floor to the glacier, in much the way that your tongue will stick to a very cold spoon. As the glacier moves, it tears up chunks of rock in a process called plucking. These pieces of rock embed into the ice at the base of the glacier, where they act like sandpaper grit, grinding away at other rock and removing more fragments that then incorporate into the ice mass. The glacier transports this debris to the zone of ablation, and finally

An ancient glacier carved the broad, roughly U-shaped valley that leads toward Nevada's only modern glacier. Ground moraine from the last ice age covers the valley floor. —Dave Futey photo

to the glacier's terminus, the lowest elevation at which we find glacial ice. There the glacier dumps its load of rocks and sediment in mounds as the ice melts. The glacier's meltwater picks up much of the sand, silt, and clay and carries it into streams. Wind, too, can remove fine sediment from glacial deposits and transport it over vast distances. Wind-borne dust of glacial origin, or loess, is common throughout the world.

Alpine glaciers erode bedrock into distinct features that mark a landscape. Whereas alpine streams cut valleys with a roughly V-shaped cross section, glaciers carve broad, roughly U-shaped valleys. As you continue along the trail, look up and down the valley. You can clearly see the valley's roughly U-shaped cross section, evidence that glaciers once filled and eroded this valley. You are now walking toward the head of the valley where the glacial ice first accumulated. At the head of such valleys, we find amphitheater-like depressions called cirques. Typically, several glaciers carve cirques and shape valleys on different sides of a mountain at once, leaving behind a faceted peak called a horn. Knife-edged ridges separate one glacial valley from another. As neighboring glaciers erode the bedrock, they deepen their respective valleys, leaving behind these narrow ridges of uneroded rock called aretes. When the glacier melts it exposes depressions in its valley that

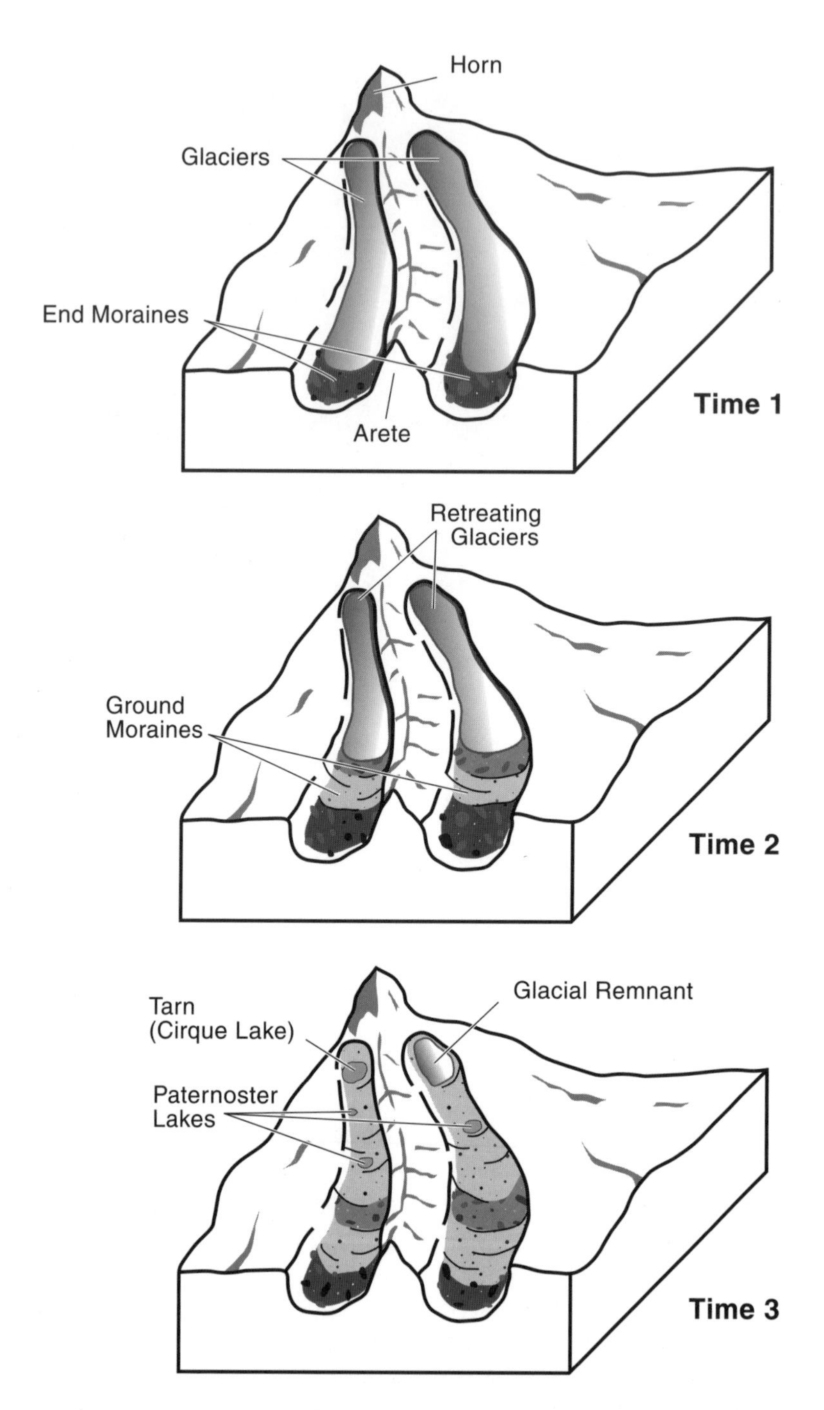

Development of moraines and lakes as alpine glaciers retreat

Wheeler Peak, Nevada's second highest mountain, is a horn, a prominent peak left behind as glaciers eroded deep valleys into the mountain's flanks. —Dave Futey photo

fill with water. Lakes that form in cirques are called tarns. Chains of small, circular lakes that fill bedrock depressions down the valley are called paternoster lakes. The name comes from the two Latin words for "our father," which reflects early glaciologists' impression that the strings of lakes looked like rosary beads.

Alpine glaciers also leave behind depositional features. Glaciers deposit their sediment in ridgelike accumulations called moraines. End moraines pile up in arcuate mounds of debris at the toe, or lowest extent, of a glacier. End moraines form when the glacier is stable, that is, when the glacier is neither advancing or retreating—which does not mean there is no movement; the glacier continues to flow and slide, but its rate of melting balances its forward movement. As climate warms and a glacier retreats, the ice may deposit a series of end moraines, each marking a period of relative climatic stability. A lateral moraine, on the other hand, contains sediment carried along the upper surface of the ice where the glacier met the valley walls. When a glacier melts, this debris accumulates as linear mounds along the valley sides. A glacier will also leave a ground moraine, a horizontal layer of rubble that it deposits directly onto the valley floor during retreat.

The mounds of rocky debris you are walking through are part of a moraine complex that includes end, lateral, and ground moraines. Two well-exposed end moraines punctuate this valley, and a ground moraine covers the valley floor. The upper end moraine, centered at about 9,000 feet above sea level, probably accumulated 20,000 years ago, during the last Pleistocene glacial episode. An earlier glacial advance left an end moraine at an elevation of 8,200 feet above sea level. These features have not been well studied, so we can only guess at their ages by comparing the moraines with those in other mountain ranges of the Southwest. Look at the huge, angular blocks of quartzite that the glacier plucked from the head of the valley and dumped here. Many rocks within moraines have ground against other rocks during transport and deposition. This contact scratched and chipped them. Can you see these scratches in the quartzite? Quartzite is a very hard rock, so the scratches are more difficult to find here than in most moraines.

Let's move on to the base of the linear mound of rock below the glacier. Glacial features surround you now. Notice the distinctive shape

Snow-covered ice, the true glacier, at the base of the headwall feeds into the rock glacier in the middle ground of this photograph. Steep slopes shed the talus after the glaciers receded.

of the valley that this glacier carved. The peaks that tower over us now, Jeff Davis Peak to the southeast and Wheeler Peak to the southwest, are horns. Ice that once filled the valleys surrounding these peaks clawed away at their flanks, eventually sculpting the prominent features you see today. Brown Lake, a paternoster lake filling an ice-carved depression, sits just downslope of us within the ground moraine. Teresa and Stella Lakes, both tarns, lie along the Alpine Lakes Trail that connects with our trail. Aretes separate our valley from the neighboring valleys of Lehman Creek to the northwest and the North Fork of Baker Creek to the south.

As you continue along the trail, you move into the natural amphitheater of the cirque, with a headwall that rises 2,000 feet to Wheeler Peak. In front of you, notice the mound of rock, elongated in the downslope direction. Lying immediately above that mound and partially hidden by it is Nevada's only glacier. Look for part of the glacier extending up into a small but deep and narrow gully, or couloir, to the southwest.

The tightly enclosed valley faces north, which explains why this small glacier can exist today. North of the Tropic of Cancer, which

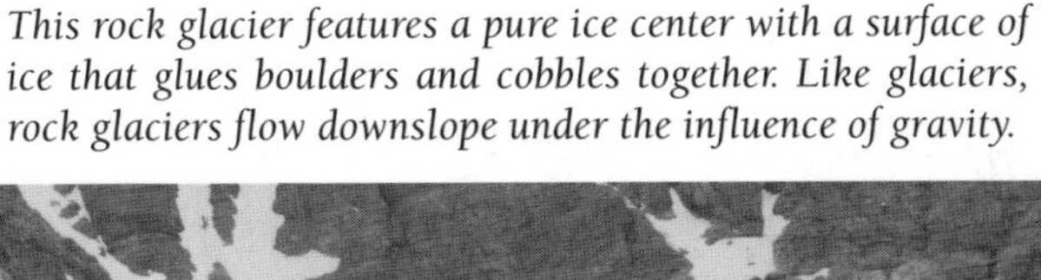

This rock glacier features a pure ice center with a surface of ice that glues boulders and cobbles together. Like glaciers, rock glaciers flow downslope under the influence of gravity.

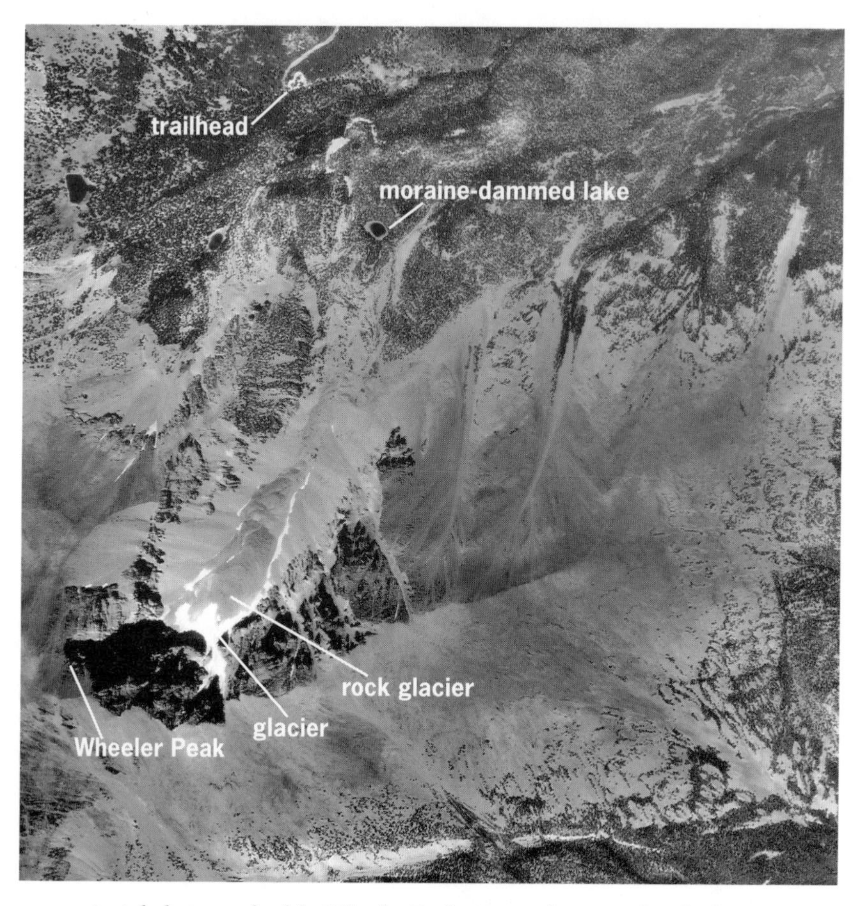

Aerial photograph of the Wheeler Peak cirque, glacier, and rock glacier. Flow lobes on the surface of the rock glacier indicate that it is moving downslope as snow and ice accumulate at the cirque headwall.

lies at 23.5 degrees north latitude, the sun always shines from the south. Steep, north-facing slopes like this one never feel the sun. The lack of solar radiation produces a microclimate, a local variation in regional climate, that remains much colder than surrounding areas. The microclimate of this valley has allowed the glacier to survive today. The mound of quartzite rubble that partially obscures the glacier is an interesting feature in and of itself. It is an ice-cored rock glacier, a moving mass of rock that ice consolidates. Loose rockfall debris called talus covers the upper part of the rock glacier, but ice holds together the middle and lower sections. The upper end of the rock glacier joins the terminus of the glacier above, which provides it with ice. Quartzite debris comes from the valley walls—listen for the sound of falling rock. The rock glacier is moving downslope and will continue to do so as long as ice is available to bind the rock together into a coherent whole.

Moving ice shaped this valley and those to the north and south. The Ice Age, roughly the last 2 million years of earth's history, saw alternating glacial and interglacial stages. The Snake Range holds a pristine record of the last glacial stage; the sculpted valleys, prominent horns, and moraines bear witness to the power of ice in altering a landscape. Climate is always changing, and as it changes, different processes dominate. Ice once ruled here, and if climate cools again, it will rule once more. As you visit Great Basin National Park, try to imagine how it once was—and may be again.

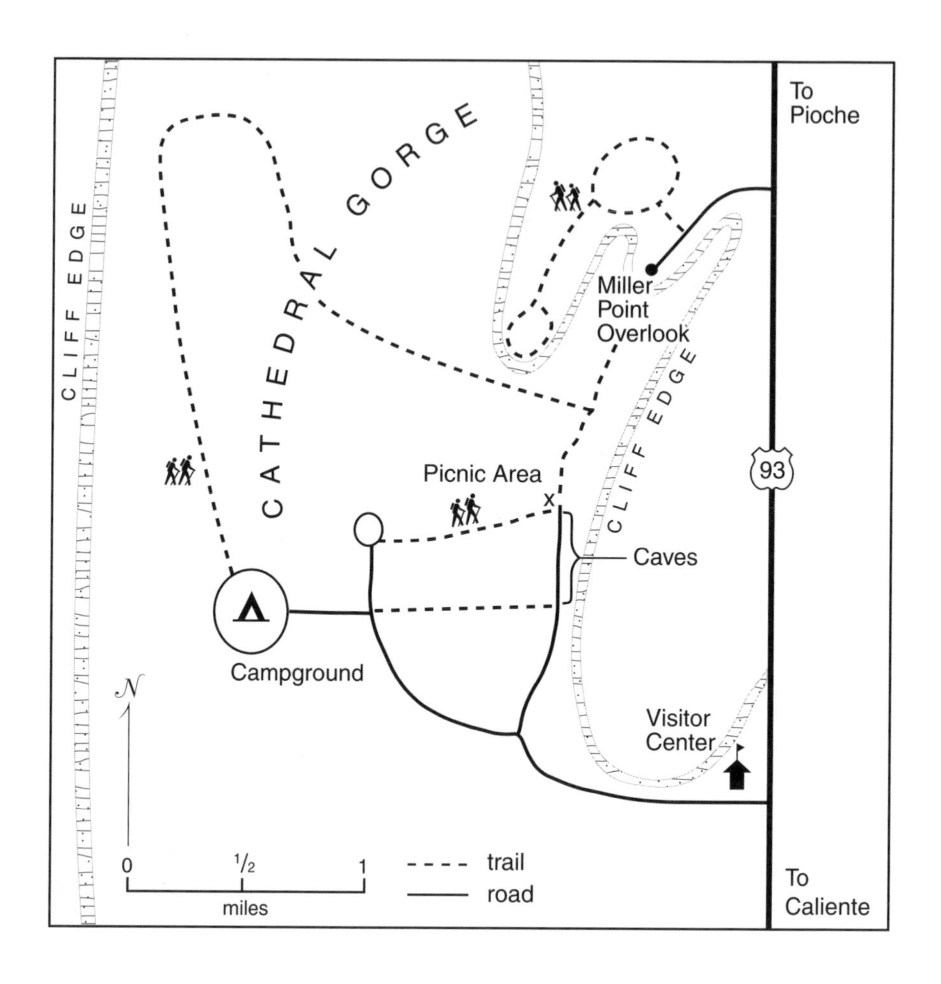

▶ GETTING THERE

Cathedral Gorge State Park lies 24 miles north of Caliente and 10 miles south of Pioche on U.S. 93. From Las Vegas, drive 20 miles north on Interstate 15, then 152 miles north on U.S. 93. Turn west (left) onto a clearly marked state park road. The visitor center is immediately to the right and within sight of U.S. 93. Continue past the visitor center for about 1.5 miles to the picnic area at the base of the eastern cliff. You will pass a road on the left that leads to the park campground. At the cliffs you will find several parking lots and picnic sites, as well as numerous trails that explore Cathedral Gorge. These trails are well marked and maintained. Take care when hiking on steep slopes, as footing on badlands such as these can be quite tricky. We will look at features near the picnic area first, then hike the trail to the Miller Point Overlook, a distance of 1 mile.

24

Hoodoos and Badlands

CATHEDRAL GORGE

What you probably notice first at the Cathedral Gorge picnic area are the steep, light cliffs to the east and the prominent spires and mounds rising from their base. Cathedral Gorge is a brilliant example of a landscape produced by a sequence of varied geologic processes. Modern erosion shaped the badlands of sculpted cliffs, spires, and deeply incised canyons that enclose the valley today. But the geologic story began much earlier and encompasses volcanism, structural upheaval, and water as both a depositional and erosional agent.

Walk to the cliff and examine the sediment held there. Very loosely cemented, it crumbles easily. Not very impressive stuff, really, but it certainly has an eventful history. During middle Cenozoic time, beginning about 40 million years ago, the Great Basin literally exploded into action. A wave of violent volcanism spread across what is now Nevada, each eruption laying down a layer of ash hundreds of feet thick and covering thousands of square miles. These were not lava flows, but explosions of glowing hot ash and hot gas. For tens of millions of years, one explosive eruption followed the other, burying the landscape beneath fused ash, or ignimbrites. The source of the volcanic ash in this part of Nevada was the Caliente Caldera Complex, which lies just south of Cathedral Gorge.

About 17 million years ago, 4 or 5 million years after Nevada's major volcanism ceased, block faulting broke Nevada's surface into the alternating mountains and valleys that characterize the Basin and Range province. One of the depressions this faulting created was the valley you are in right now, Meadow Valley. Mountains bound the valley to the east, west, and south.

Meadow Valley became a closed depression about 5.3 million years ago. It began to fill with water and soon held a freshwater lake. Rain that fell on the surrounding areas began to whittle away the ignimbrites that had spewed from the magma chamber to the

Foreground sediment eroded from the gorge walls in the background. Later rainfall events incised small rills into the deposited sediment. Valley walls are eroding too rapidly for vegetation to gain a foothold.
—Dave Futey photo

south. Streams carried the eroded sediment downslope and deposited it into the newly formed lake. The light tan cliffs in front of you are remnants of that lake. They are mostly silt and clay, the altered remains of volcanic ash.

If you look closely at the outcrops in the picnic area, both on the cliffs and the free-standing spires, you will see some variation in layers of rock and sediment. There are thin layers, about an inch thick, that are clearly more resistant to erosion than the other layers. These are lenses of limestone, a carbonate rock common in the deeper sections of a freshwater lake. You may also see some very interesting crescent-shaped accumulations of gravel exposed on the cliff walls. Most of these are several feet across and maybe 1 foot thick. The water level of this lake periodically changed during Pliocene time. As the level fell, streams flowed over lake deposits, carving channels and depositing coarse sand and gravel. The gravel crescents are cross sections of stream channels—imagine the stream flowing out of the wall toward you. A stream deposited these small rocks and sand as it

A cross section of a small stream channel within the Panaca formation
—Dave Futey photo

flowed on top of existing lake deposits when the lake level dropped. When the level rose again, the lake deposited silt and clay on the channel, essentially encasing the stream gravel. These channels at various elevations within the cliffs record sequences of change in lake level. This recurring cycle produced interlayered stream channel and lake deposits concentrated at the outer edges of the formation.

Block faulting continued through the life of our lake, as did broad doming of the Great Basin. The northern Great Basin domed more than the south, resulting in a north-south gradient that probably helped water breach the southern wall of Meadow Valley. Erosion to the south incised a channel into the Clover Mountains, through which the modern Carpenter River flows. Meadow Valley, once a closed basin, became instead a valley with an outlet at its southern end, through which water could now drain. Deposition within the lake ended and erosion began.

Scan the badlands topography at Cathedral Gorge and notice the fine drainage networks and steep, unvegetated slopes. Badlands regions like this one experience the highest rates of erosion anywhere. The underlying rock at Cathedral Gorge, the Panaca formation, is obviously poorly cemented and thus unable to resist the forces of wind and water. The Panaca formation contains much montmorillo-

Mud cracks in the surface of the Panaca formation in Cathedral Gorge State Park —Dave Futey photo

nite clay, a mineral that derives from the alteration of volcanic ash. This type of clay mineral absorbs moisture from rainfall and swells. Upon drying, the clay surface crumbles and shrinks, leaving characteristic mud cracks behind. These cracks are natural pathways for water, and they further enhance erosion.

Notice that the valley floor is relatively well vegetated and stands in stark contrast to the completely unvegetated walls. Cathedral Gorge receives little rain, so plants grow very slowly here—so slowly that growth cannot keep up with erosion. Normally, vegetation effectively stabilizes slopes, but plants cannot establish a foothold here. Erosion takes place during infrequent but heavy rainfall, which can cause entire slopes to liquefy and flow.

Thousands of channels, ranging from very large to very small, dissect the cliffs. Run-off from heavy rains concentrates in these channels and carves away the soft rock. Take a walk into one of the "caves" near the picnic area—there are some here and others at marked locations a few hundred feet south. While they certainly are dark and cool, these aren't really caves. They are deeply incised channels, tens of feet deep but only several feet wide. These canyons serve as conduits for large quantities of sediment during storms. The upper end of each of these channels is a vertical wall, a cliff face over which water pours. At the base of each canyon's headwall lies a plunge pool, a bowl that the falling water excavated.

The streams that intermittently flow within these deeply incised channels often disappear into or reappear from holes in the floor or wall. This is a process called piping. Walk along the cliff and look for holes, several inches to several feet in diameter, through which water has flowed. Water percolating through the Panaca formation follows natural lines of weakness and erodes sediment, carving small tunnels. Flowing water then enlarges the passageways, undermining portions of the rock. Eventually piping causes the wall to collapse, further exacerbating the already rapid erosion.

Look for lobes of debris at the mouths of the slot canyons and many of the lesser channels. Water enters the channels at the upstream end and immediately picks up and transports sediment. Within the confined drainages in the badlands of Cathedral Gorge, the water remains deep, but this changes when the water leaves the channel mouth and enters the valley floor. Here water spreads in a flat sheet, and the flow slows. This decrease in velocity brings a corresponding

Two mounds of eroded sediment from neighboring gullies have coalesced into a single deposit.

decrease in the ability to carry sediment, so deposition occurs. The lobes of debris contain sediment eroded from the channel walls and floor, then deposited at the mouth. Later flows typically then incise "badlands" gullies into these lobes. Water that exits the mouth doesn't deposit all of its load, however. If you were to visit during a storm, you would see that water flowing from these channels is white with suspended sediment. Even as it spreads onto the valley floor, the water can carry very small clay minerals. Each clay particle carries an electrical charge that allows it to repel other particles, and so it avoids aggregation and settling.

Perhaps the most impressive features within Cathedral Gorge are the prominent spires, also called monuments or hoodoos. Check out the variety of forms here. Some are tall with steep sides; others are smaller and more rounded. Look at the top of each hoodoo, and see if you can detect any difference in the rock there compared with the rock within the body of each spire. You should see resistant caprock that shields underlying layers to some degree from erosion. A layer of more resistant rock, such as limestone, might shield the poorly

A hoodoo, or erosional remnant, stands alone on the valley floor. An especially resistant layer at the top caps this landform and prevents it from eroding as rapidly as the surrounding rock, preserving it as a solitary tower. Thin white horizontal bands represent more resistant layers of rock within the spire. —Dave Futey photo

cemented sedimentary layers beneath it and allow the column to stand alone as the rock around it wastes away. A single cobble may be enough of a cap to initiate a hoodoo. Look around at the many spires visible within the gorge. Some are newly formed and still connected to the wall. Others stand alone, marking the former position of the wall before erosion removed it. Still others are no more than small mounds, remnants of tall towers that once stood here. Hoodoos waste away rapidly once the caprock erodes. The small mounds are dwindling remnants of once-tall spires.

Lake deposits commonly contain fossils, remains of living organisms. Rainwater and small streams wash bones and teeth into larger streams that carry the remains into a lake. There, sediment covers the remains before they can decay. The Panaca formation contains a wide variety of fossils of Cenozoic mammals such as the early horse *Pliohippus*, the early rhinoceros *Teleoceras*, the early muskrat *Pliopotamys*, and many rodents. Your chances of coming across the remains of a large Pliocene mammal are slim, but you can find abundant fossils of smaller animals. Look for concentrations of rodent bones within the walls of Cathedral Gorge. These bones are very small, typically less than an inch long. Identifying individual bones requires a sharp eye, but the careful observer may see leg bones, jaws with tiny teeth, and even entire skulls. Paleontologists, scientists who study ancient life, have identified new species of rodents within the Panaca

These rodent bones eroded from the walls of one of the incised canyons. While this formation has yielded bones of larger mammals, smaller bones like these are more common. —Dave Futey photo

formation. Researchers identified one such rodent, *Mimomys panacaensis,* whose name obviously derives from the formation, based on the examination of a set of jaws and teeth found near Cathedral Gorge. You may not collect fossils in the state park, but you certainly can take photographs of them.

Miller Point Overlook is a great place to get a panoramic view of Cathedral Gorge. The trail is 1 mile long and has very little elevation gain. Stairs at the end lead to the overlook, which is also accessible by car from U.S. 93 (see map). The trail takes you north along the base of the cliff, where you pick out features like those at the picnic area. The trail then moves into an incised canyon, much larger than the "caves" at the picnic area, before climbing to the overlook.

Cathedral Gorge opens to the south and west. Badlands rim the valley where erosion eats away the poorly cemented Panaca lake deposits too rapidly for stabilizing vegetation to gain a toehold. Adjacent to the cliffs stand tall hoodoos and smaller mounds that are remnants of older spires. Cathedral Gorge records a history of dynamic events—volcanism, freshwater lakes and channels, and now, rapid weathering and erosion. This is a good place to consider the dynamic nature of geologic processes and the cycles of change inherent in earth's history.

View from Miller's Point of the Cathedral Gorge badlands

Glossary

aa. Viscous lava that forms blocky, angular flow surfaces.

acid-sulfate alteration. A type of hydrothermal alteration in which hydrogen sulfide from hydrothermal fluid combines with oxygen—from either the atmosphere or another mineral such as pyrite—to form sulfuric acid, which chemically alters the rock.

alluvium. Unconsolidated sediments that running water deposited.

almandine. A dark red variety of the mineral garnet.

alteration. Changes in the chemical or mineralogical composition of a rock, usually as a result of weathering or the reaction with hydrothermal fluids.

amalgam. A mixture of mercury and gold resulting from the separation of mineral and waste.

amplitude. The height of a wave.

andesite. A fine-grained, usually grayish, extrusive igneous rock with a silica content intermediate between basalt and rhyolite.

arete. A knife-edge ridge that separates two glacial valleys.

ash. Fine-grained volcanic material erupted into the atmosphere.

ash flow. A hot, rapidly flowing mixture of ash, volcanic gases, and pumice, created during explosive volcanic eruptions.

ash flow tuff. A volcanic rock composed of the material from an ash flow.

assay. A test to determine economic viability of a particular ore.

asthenosphere. The part of the upper mantle that lies immediately below the lithosphere. Convective motion of the viscous material within the asthenosphere drives plate motion above.

badlands. A semiarid landscape with an abundance of small channels on poorly cemented bedrock.

basalt. A fine-grained, dark, extrusive igneous rock containing calcic plagioclase, pyroxene, and olivine, with little silica.

basement. Old rocks, generally igneous and/or metamorphic, that lie beneath the younger rocks and sediments in an area.

Basin and Range. A physiographic province within the western United States with north-south trending mountain ranges and desert valleys. This region has experienced broad doming and crustal thinning, resulting in the development of parallel fault-bounded ranges and valleys.

batholith. A large body of intrusive igneous rock generally having more than 40 square miles of surface exposure and no known floor.

bedding. The layered structure of sedimentary rocks.

bedrock. Solid rock, either exposed or lying beneath a mantle of unconsolidated material.

biotite. A common rock-forming mineral of the mica group, generally black or dark brown, with perfect basal cleavage that causes it to peel into thin sheets.

breccia. A rock composed of angular rock fragments in a fine-grained matrix.

caldera. A huge, bowl-shaped depression formed by the collapse of a volcanic cone following a large eruption.

carbonate rock. A family of rock that includes limestone, dolomite, and marble. Caves form in carbonate rock by dissolution.

chemical weathering. The breakdown of rocks and minerals into new chemical combinations at or near the earth's surface by chemical means.

chlorite. A group of greenish, platy minerals that are similar to micas.

cinders. Coarse, angular particles of hardened lava.

cirque. A bowl-shaped, glaically carved depression at the head of a mountain valley.

clast. An individual fragment of rock found in sediment or a sedimentary rock, formed by the physical disintegration of a larger rock mass.

clay. Extremely fine-grained sedimentary particles.

clay minerals. A group of water-bearing aluminum-silicate minerals with platy or sheetlike cleavage.

columnar jointing. Prismatic columns, polygonal in cross section, that form in some igneous rocks as a result of contraction during cooling.

crustal extension. The pulling apart of the earth's crust due to forces in the lower crust and mantle, such as the upwelling of magma.

debris flow. A large mass of unsorted rocks, sediment, vegetation, and water flowing downslope.

dendrochronology. Establishing a climatic history using tree ring cores from living and dead trees.

desert pavement. A tightly interlocking fabric of gravel and cobbles that forms on the desert floor.

desert varnish. A patina of iron, manganese, and clay that builds up on rock surfaces exposed for a long time in an arid environment.

dessication. Drying out, especially of a body of water.

diatom. Microscopic, single-celled aquatic plants that secrete a bivalved, siliceous cell wall. Diatoms grow in both fresh and salt water.

dike. A tabular intrusion of igneous rock that cuts across the rock into which it has intruded.

diorite. A coarse-grained, intrusive igneous rock of intermediate composition consisting of sodium plagioclase, hornblende, biotite, and pyroxene, with little quartz.

dip. The angle that any planar geologic surface, such as a bedding plane or fault plane, makes with the horizontal; measured perpendicular to strike.

disseminated deposit. An ore deposit in which a valuable mineral exists within microscopic pore spaces in the host rock.

dune. A hill-like accumulation of sand in a desert or coastal environment.

earthquake. Sudden vibrations within the earth's crust resulting from the abrupt release of accumulated strain.

epicenter. The point on the earth's surface directly above the point of rupture, or focus, of an earthquake.

erosion. The wearing away and transportation of rock and soil by natural processes such as water, wind, and ice.

evaporites. Minerals that precipitate during the evaporation of a standing body of water.

exfoliation. The peeling off of concentric sheets of rock from a large rock mass by physical or chemical means. Usually the result of differential stresses.

extrusive. Igneous rocks that have erupted onto the earth's surface.

fault. A fracture or fracture zone in rock or sediment along which one side has moved relative to the other.

fault block. A block of the earth's crust that faults bound on at least two sides and that behaves as a unit during tectonic activity.

fault breccia. Angular fragments of rock resulting from crushing and shearing during movement along a fault.

fault gouge. Pulverized rock that results from the grinding of two sides of a fault during movement.

fault scarp. A steep slope or cliff formed by the vertical movement along a fault.

fault trace. The intersection of a fault plane with the surface of the earth; the surface expression of a fault.

feldspar. The most widespread group of rock-forming minerals, containing primarily silica, aluminum, and oxygen combined with calcium, sodium, and/or potassium.

firn. Coarse, granular ice that represents a midpoint between new snow and glacial ice and that can flow under its own weight.

fissure. A fracture in rock, usually with some separation.

focus. The point within the earth where an earthquake originates; the actual point where the rocks rupture.

footwall. The lower or underlying side of an inclined fault plane. The footwall is the fault plane you could stand on if the overlying fault block were removed.

fossil. Remains of a once-living organism preserved in rock.

frost wedging. The mechanical disintegration of a rock due to the pressure that water within cracks or joints exerts as it freezes.

fumarole. A volcanic vent that emits only gas.

geochronology. Establishing a sequence of dated events in earth history.

geomorphic province. A region of the earth's surface identified by its characteristic landscape.

geothermal. Pertaining to heat within the earth's interior.

geyser. A specific type of hot spring that intermittently erupts hot water and steam.

glacier. A large mass of natural ice that flows slowly over the land surface.

graben. An elongate block of the earth's crust bounded by faults on its long sides that has dropped down relative to the surrounding blocks.

granite. A coarse-grained, intrusive igneous rock composed primarily of quartz and alkali feldspar, with lesser amounts of plagioclase and mica.

granitic. A general term referring to granite and related rocks of a similar makeup, such as granodiorite and quartz monzonite.

granodiorite. A coarse-grained, intrusive igneous rock with a composition midway between granite and diorite. Relatively equal parts of light and dark minerals give it a salt-and-pepper look.

gravity graben. A depression at the base of a fault scarp due to settling.

Great Basin. A region of the southwestern United States that drains inward. It is bounded by the Sierra Nevada Mountains to the west, the Rocky Mountains and Colorado Plateau to the east, and less obvious highlands to the north and south.

groundmass. The finer-grained material found between the larger crystals in a porphyritic rock. It contains microscopic crystals that grew as the magma cooled quickly while it was flowing onto the surface.

groundwater. Subsurface water contained within the pores and spaces in rocks and sediments.

half-life. The time required for one-half of a radioactive isotope to decay into a daughter product.

hanging wall. The upper or overhanging side of an inclined fault plane.

hoodoos. Prominent spires in badlands that form due to variations in resistance to erosion.

horn. A prominent mountain peak that glaciers carved as they eroded rock from valleys below.

hornblende. A common, dark, rock-forming silicate mineral in the amphibole group that often forms prismatic crystals.

hummocky. Terrain of small hills and depressions.

hydrothermal. Refers to processes, and the effects of those processes, related to hot subsurface water.

hydrothermal alteration. Chemical alteration of rocks brought about by the interaction with hot subsurface water.

hydrothermal deposit. Minerals precipitated from hydrothermal fluids.

ichthyosaur. Extinct group of marine reptiles such as those found within Ichthyosaur State Park, Nevada.

igneous rock. Rock that forms from the cooling and crystallization of molten rock (lava or magma).

ignimbrite. The rock formed by the consolidation of hot ash flows from explosive volcanic eruptions. Ignimbrites contain primarily pumice fragments and may or may not be welded.

infiltration. Downward movement of water through pore spaces in soil or rock.

intensity (earthquake). A measure of the amount of ground shaking during an earthquake based on human experience and damage to structures.

intrusive rock. A rock body that has emplaced as magma beneath the earth's surface.

joint. A fracture in a rock with no displacement.

joint set. A series of parallel fractures in a rock body with no displacement.

karst topography. Characteristic landforms that develop from dissolution of carbonate bedrock.

krummholz vegetation. Stunted and gnarled trees that live at treeline, the upper limit of tree growth in mountains.

lacustrine. Derived from or related to lakes.

lahar. A landslide or mudflow down the side of a volcano consisting mainly of material ejected from the volcano.

landslide. A general term for gravity-driven movement of rock and soil downslope.

lava. Molten rock material that volcanism has released onto the earth's surface.

leaching. Dissolution and removal of soluble components of rock and soil by percolating water.

left-lateral motion. A fault on which the movement is in the horizontal plane and the side opposite the observer appears to have moved to the left.

lichens. Algal and fungal organisms that coexist and take nourishment directly from rock.

limestone. A sedimentary rock composed of calcium carbonate.

lithification. The process of compaction and cementation that transforms sediment into sedimentary rock.

lithosphere. The brittle upper portion of the earth consisting of the crust and uppermost mantle.

loess. Windblown silt from glacially deposited sediment.

maar. A volcanic crater that forms due to the explosive interaction between groundwater and magma.

magma. Molten rock within the mantle or crust.

magma chamber. A reservoir of molten rock within the mantle or crust.

magnitude (earthquake). A quantitative measure of the energy an earthquake releases as determined from seismographs. Magnitude is measured on a logarithmic scale, with each unit of magnitude representing a tenfold increase in the amplitude of the seismic waves.

Mercalli scale. A scale of earthquake intensity that Guiseppe Mercalli developed in 1902. This qualitative scale measures the effects of an earthquake based on human experience and damage to structures, and ranges from I (detected by instruments, felt by few) to XII (almost total destruction).

metamorphic rock. A rock that heat and pressure change sufficiently in texture or composition to consider it distinct from the parent rock.

microclimate. A local variation in regional average climate.

midden. A deposit of primarily organic material; derived from the Danish word *mogdynge*, meaning "muck heap."

mineralization. The process by which minerals are introduced into an existing rock; usually refers to valuable minerals concentrated into an ore deposit. Also, the process of replacing organic materials with inorganic materials to form fossils.

moraine. An accumulation of unsorted, unstratified material that a glacier deposited.

mudflow. A flowing mass of fine-grained sediment and water that often is highly fluid.

normal fault. An inclined fault in which the hanging wall appears to have moved downward relative to the footwall, usually a result of extension of the crust.

obsidian. Dark volcanic glass that forms when lava cools so rapidly that no crystals form.

offset. The distance that two points on opposite sides of a fault have moved relative to one another. The amount of movement along a fault.

ooze. Fine-grained sediment deposited in the deeper part of a lake or ocean basin, composed primarily of the skeletal remains of microscopic organisms such as radiolarians or diatoms.

opaline. Having the minerology and characteristics of opal—that is, an amorphous, hydrated silica.

ore. An economically viable mineral deposit. This term most commonly describes deposits of metallic minerals such as gold, silver, and copper.

outwash. Sand and gravel that meltwater streams beyond the perimeter of a glacier deposited.

oxidation. The process of combining an element or compound with oxygen.

pahoehoe. Low-viscosity lava that cools with a ropelike surface.

paternoster lakes. A string of small lakes filling depressions within a glacial valley.

permeability. The ability of rock or sediment to transmit a fluid. Permeability is a function of the interconnectedness of pore spaces.

petroglyph. A drawing chipped or carved into a rock surface.

phenocryst. A large, isolated crystal that is surrounded by a finer-grained groundmass in a porphyritic rock.

physiographic. Pertaining to the earth's surface features and landscapes; also geomorphic.

plagioclase. Common rock-forming minerals of the feldspar group containing sodium and/or calcium.

playa. Evaporite minerals and fine-grained sediment representing material left behind during the evaporation of a pluvial lake.

pluton. A distinct body of igneous rock intruded deep beneath the earth's surface from a single body of magma.

plutonic. Rocks that form deep within the earth from a body of magma.

pluvial lake. A lake that varies in size as a function of climate change.

porosity. Percent of the total volume of rock or sediment that is made up of voids or pore spaces.

porphyritic. A texture in some igneous rocks in which larger crystals that grew early in the cooling process and at depth sit within a matrix of much finer crystals that grew as the magma extruded and cooled rapidly.

pumice. Lightweight, volcanic glass containing many bubblelike holes from gas that escaped from the lava.

pyroclastic. Fragmented rock material ejected during an explosive volcanic eruption.

quartz. A common, pale, rock-forming mineral composed of silica and oxygen.

radiocarbon dating. Using carbon-14 to establish an age for dead organic material. This method relies on the steady decay of carbon-14 to nitrogen-14.

recessional moraine. An accumulation of unsorted, unstratified debris deposited by a receding glacier.

resurgent caldera. A very large caldera in which the floor initially collapses, but later begins to rise as more magma moves into the magma chamber. Resurgent calderas are large depressions of volcanic origin with a central raised area.

reverse fault. A fault in which the hanging wall has moved up relative to the footwall.

rhyolite. A pale, silica-rich, extrusive igneous rock of granitic composition. Generally fine-grained or porphyritic.

right-lateral motion. Horizontal movement along a fault such that the side opposite the observer appears to have moved to the right.

roche moutonnée. An asymmetric, glacially sculpted knob of bedrock with the long axis parallel to the direction of ice flow.

rockfall. The more or less free fall of a freshly detached rock mass from a cliff.

rockfall avalanche. The downward movement of a huge mass of rock, debris, and dust resulting from the collapse of a cliff or mountainside. Rockfall avalanches can move rapidly over great distances.

rockslide. Downward, gravity-driven movement of a freshly detached bedrock mass that usually breaks into many smaller units.

saltation. Movement of sediment particles by bouncing along the desert floor or within a stream channel.

sedimentary rock. Rock made up of weathered remains of pre-existing rock.

seismic. Pertaining to earthquakes.

seismograph. An instrument used to record earthquake waves.

shear. Deformation resulting from forces that cause two rock bodies to slide past one another along a plane or zone.

silica. Common term for the compound silicon dioxide.

siliceous. Rich in silica.

sinter. A chemical deposit, usually siliceous, that precipitates from hydrothermal waters.

slickensides. A polished, grooved surface on a rock resulting from rock grinding against rock during movement along a fault plane.

speleothems. Features created by precipitation of dissolved calcite within caves. The most widely known speleothems are stalagmites, which grow on the floor, and stalactites, which hang from the ceiling.

spheroidal weathering. A type of weathering in which water enters the fractures in a mass of rock and chemically and physically disintegrates it. This process usually begins around the edges and works inward, resulting in rounded, fresh rock surfaces.

stamp mill. A facility for crushing ore prior to the separation of a valuable mineral from waste.

stope. A large ore pocket in a subsurface mine.

strain. The deformation of rocks caused by stress.

strata. Individual layers of sedimentary rock that are visually distinct from the layers above and below.

stratified. Layered.

stratigraphy. The study of physical and temporal relationships between rock units, as well as the environments of formation.

stress. Force per unit area acting on a rock.

striations. Parallel scratches or grooves that movement of a glacier or movement along a fault plane cut into rock.

strike. The direction of a line formed by the intersection of an inclined planar geologic surface, such as a bedding plane, with the horizontal.

strike-slip fault. A fault on which the movement is in the horizontal plane.

subduction. The process by which an oceanic plate descends beneath another plate.

Tahoe stage. Local term used in the Sierra Nevada, considered equivalent to the Early Wisconsin stage of continental glaciation.

talus. An accumulation of coarse, angular rock fragments at the base of a cliff or steep slope that shed them.

tarn. A small lake occupying a cirque, the amphitheater-like depression at the head of a glacial valley.

tectonic. Pertaining to forces involved in the deformation of the earth's crust.

terrace. A relatively flat surface breaking the continuity of a slope.

till. Unsorted, unstratified sediment that a glacier has deposited and water has not reworked.

Tioga stage. Local term used in the Sierra Nevada, considered equivalent to the youngest phase of Wisconsin glaciation.

travertine. A deposit of calcium carbonate that precipitates from either surface water or groundwater.

truncated. A geologic structure that has been cut off.

tufa. A deposit of calcium carbonate that forms in brackish, mineral-rich water.

tuff. A general term used to describe rocks formed of consolidated fragments of volcanic material, especially ash.

unloading. The removal of great quantities of rock, soil, and/or ice from above a rock unit by erosion or melting of the ice.

unsorted. Said of a sediment deposit that contains clasts of many different sizes.

vent. The opening through which volcanic materials reach the earth's surface.

ventifact. A streamlined rock that wind-carried sediment has abraded and polished.

viscosity. A measure of a fluid's resistance to flow. The higher the viscosity, the slower the flow.

viscous. Said of a fluid with a high resistance to flow. A fluid that flows sluggishly.

volcanic neck. An igneous intrusion composed of the solidified material filling a volcanic vent or pipe.

weathering. The physical disintegration and chemical decomposition of rocks at or near the earth's surface due to exposure to the atmosphere.

welded tuff. A glassy rock that forms from hot ash flows in which the pumice clasts weld together from the heat retained by the particles and flatten under the weight of the overlying material.

xenoliths. Inclusions of other rock found within volcanic rock at the earth's surface.

SOURCES OF MORE INFORMATION

General

Fiero, B. 1986. *Geology of the Great Basin.* Reno: University of Nevada Press.

Stewart, J. H. 1980. *Geology of Nevada.* Nevada Bureau of Mines and Geology Special Publication 4.

Tingley, Joseph V., and Kris Ann Pizarro. 2000. *Traveling America's loneliest road, a geologic and natural history tour through Nevada along U.S. Highway 50.* Reno: Nevada Bureau of Mines and Geology Special Publication 26.

1. Sand Harbor, 2. Cave Rock, and 3. Emerald Bay

Burnett, J. L. 1971. Geology of the Lake Tahoe Basin. *California Geology* 24(7):119–27.

Burnett, J. L., and R. A. Matthews. 1971. Geological look at Lake Tahoe. *California Geology* 24(7):128–30.

Purkey, B. W., and L. J. Garside. 1995. *Geologic and Natural History Tours in the Reno Area* (guidebook). Nevada Bureau of Mines and Geology Special Publication 19.

4. Genoa Fault Scarp

Harden, D. R. 1998. *California Geology,* 141–193. Upper Saddle River, N.J.: Prentice Hall.

Thompson, G. A., and D. E. White. 1964. *Regional Geology of the Steamboat Springs Area, Washoe County, Nevada.* U.S. Geological Survey Professional Paper 458-A.

5. Slide Mountain

Tabor, R. W., S. E. Ellen, M. M. Clark, P. A. Glancy, and T. L. Katzer. 1983. *Geology, geophysics, geologic hazards and engineering and geologic character of earth materials in the Washoe Lake area* (text to accompany map of the Environmental Series, Washoe City Quadrangle, Nevada). Nevada Bureau of Mines and Geology Open-file Report 83–7.

Watters, R. J. 1984. Engineering geology of the Slide Mountain rockslide and water flood-debris flow. In *Western Geological Excursions Guidebook*, vol. 2, field trip 12, 88–95. Geological Society of America Annual Meeting, November 1984, Reno, Nev.

6. Steamboat Hot Springs

White, D. E. 1985. Summary of the Steamboat Springs Geothermal Area, Nevada, with Attached Roadlog Commentary. In Tooker, E. W., ed., *Geologic Characteristics and Sediment- and Volcanic-Hosted Disseminated Gold Deposits—Search for an Occurrence Model.* U.S. Geological Survey Bulletin 1646:79–88.

White, D. E., G. A. Thompson, and C. H. Sandberg. 1964. *Rocks, Structure and Geologic History of Steamboat Springs Thermal Area, Washoe County, Nevada.* U.S. Geological Survey Professional Paper 458-B.

7. Mount Rose

Bonham, H. F. 1969. *Geology and Mineral Deposits of Washoe and Storey Counties, Nevada.* Nevada Bureau of Mines and Geology Bulletin 70.

Harden, D. R. 1998. *California Geology.* Parasmus, NJ: Prentice Hall, 141–93.

Thompson, G. A., and D. E. White. 1964. *Regional Geology of the Steamboat Springs Area, Washoe County, Nevada.* U.S. Geological Survey Professional Paper 458-A.

8. Lake Lahontan

Mifflin, M. D., and M. M. Wheat. 1979. *Pluvial lakes and estimated pluvial climates of Nevada.* Nevada Bureau of Mines and Geology Bulletin 94.

Purkey, B. W., and L. J. Gartside. 1995. *Geologic and Natural History Tours of the Reno Area.* Nevada Bureau of Mines and Geology Special Publication 19.

9. Grimes Point

Bard, J. C., C. I. Busby, and J. M. Findlay. 1981. *A Cultural Resources Overview of the Carson and Humboldt Sinks, Nevada.* Cultural Resources Series #2, U.S. Department of the Interior, Bureau of Land Management.

Heizer, R. F., and M. A. Baumhoff. 1984. *Prehistoric Rock Art of Nevada and Eastern California.* Berkeley: University of California Press.

Pendleton, L. S. A., A. R. McLane, and D. H. Thomas. 1982. *Cultural Resource Overview, Carson City District, West Central Nevada.* Cultural Resources Series #5, part 1, U.S. Department of the Interior, Bureau of Land Management.

10. Hidden Cave

Begley, S., and A. Murr. 1999. The first Americans. *Newsweek* (April 26):50–57.

Hurst, D. 1985. Archaeology of Hidden Cave, Nevada. *Anthropological papers of the American Museum of Natural History,* vol. 61, part 1. New York: American Museum of Natural History.

U.S. Department of the Interior. Bureau of Land Management. 1989. *Hidden Cave Interpretive Trail*. Bureau of Land Management, Washington, D.C.

11. Sand Mountain

Livingstone, I., and A. Warren. 1996. *Aeolian Geomorphology, an Introduction.* Essex, England: Addison Wesley Longman.

12. Fairview Peak

Caskey, S. G., P. Z. Wesnousky, and D. B. Slemmons. 1996. Surface faulting of the 1954 Fairview Peak (M 7.2) and Dixie Valley (M 6.8) earthquakes, Central Nevada. *Bulletin of the Seismological Society of America* 86:761–87.

Slemmons, D. B. 1957. Geological effects of the Dixie Valley-Fairview Peak, Nevada, earthquakes of December 16, 1954. *Bulletin of the Seismological Society of America* 47:353–75.

13. Berlin Ghost Town

Pomeroy, H. R. 1972. *Where and How to Search for Gold and Silver Mines* (Pomeroy's Mining Manual). Fort Davis, Tex.: Frontier Book Company.

Tingley, J. V., R. C. Horton, and F. C. Lincoln. 1993. *Outline of Nevada Mining History.* Nevada Bureau of Mines and Geology Special Publication 15, Reno.

Young, Otis E., Jr. 1970. *Western Mining.* Norman: University of Oklahoma Press.

14. Nevada's Ichthyosaurs

Allison, P. A., C. R. Smith, H. Kukert, J. W. Deming, and B. A. Bennett. 1991. Deep-water taphonomy of vertebrate carcasses: a whale skeleton in the bathyal Santa Catalina Basin. *Paleobiology* 17(1):78–89.

Camp, C. L. 1981. *Child of the Rocks: The Story of Berlin-Ichthyosaur State Park.* Nevada Bureau of Mines and Geology Special Publication 5.

Dupras, D. 1988. Ichthyosaurs of California, Nevada, and Oregon. *California Geology* 41:99–107.

Geraci, J. R., D. M. Anderson, R. J. Timperi, D. J. St. Aubin, G. A. Early, J. H. Prescott, and C. A. Mayo. 1989. Humpback whales (*Megaptera novaeangliae*) fatally poisoned by dinoflagellate toxin. *Canadian Journal of Fisheries and Aquatic Sciences* 46(11):1895–98.

Hogler, J. A. 1992. Taphonomy and paleoecology of *Shonisaurus popularis* (Reptilia: Ichthyosauria). *Palaios* 7(1):108–17.

Kosch, B. F. 1990. A revision of the skeletal reconstruction of *Shonisaurus popularis* (Reptilia: Ichthyosauria). *Journal of Vertebrate Paleontology* 10(4):512–14.

Massare, J. A. 1988. Swimming capabilities of Mesozoic marine reptiles: implications for method of predation. *Paleobiology* 14(2):187–205.

McGowan, C. 1991. *Dinosaurs, Spitfires, and Sea Dragons.* Cambridge, Mass.: Harvard University Press.

15. Hickison Summit

Francis, P. 1998. *Volcanoes, a Planetary Perspective.* Oxford: Oxford University Press.

Stewart, J. H., E. H. McKee, and H. K. Stager. 1977. *Geology and mineral deposits of Lander County, Nevada.* Nevada Bureau of Mines and Geology Bulletin 88.

16. Lunar Crater Volcanic Field and 17. Black Rock Lava Flow

Dohrenwend, J. C., A. D. Abrahams, and B. D. Turrin. 1987. Drainage development on basaltic lava flows, Cima volcanic field, southeast California, and Lunar Crater volcanic field, south-central Nevada. *Bulletin of the Geological Society of America* 99(3):405–13.

Foland, K. A., J. S. Kargel, C. L. Lum, and S. C. Bergman. 1987. Time-spatial-compositional relationships among alkali basalts in the vicinity of the Lunar Crater, south central Nevada. *Geological Society of America, Abstracts with Programs* 19(7):666.

Snyder, R. P., E. B. Ekren, and G. L. Dixon. 1972. *Geologic map of the Lunar Crater Quadrangle, Nye County, Nevada.* U.S. Geological Survey Miscellaneous Geologic Investigations Map I-700.

Turrin, B. D., and J. C. Dohrenwend. 1984. K-Ar ages of basaltic volcanism in the Lunar Crater volcanic field, northern Nye County, Nevada: implications for quaternary tectonism in the central Great Basin. *Geological Society of America, Abstracts with Programs* 16(6):679.

U.S. Department of the Interior. Bureau of Land Management. 1979. *Lunar Crater Volcanic Field.* 1979. Bureau of Land Management, Battle Mountain District, publication number 6.

18. Garnet Hill

Hollabaugh, C. L., B. D. Robertson, and V. L. Purcell. 1989. Petrology and vapor phase mineralogy of rhyolite and tuffs from Garnet Hill, White Pine County, Nevada. *Northwest Science* 63(5):201–10.

U.S. Department of the Interior. Bureau of Land Management. 1995. Garnet Fields Rockhound Area, Bureau of Land Management Fact Sheet. Washington, D.C.: Bureau of Land Management.

19. Robinson Mining District

James, L. P. 1976. Zoned alteration in limestone at porphyry copper deposits, Ely, Nevada. *Economic Geology* 71:488–512.

McDowell, F. W., and J. L. Kulp. 1967. Age of intrusion and ore deposition in the Robinson Mining District of Nevada. *Economic Geology* 62:905–09.

20. Wheeler Peak

Ellwood, B. B. 1996. *Geology and America's National Parks.* Upper Saddle River, N.J.: Prentice-Hall.

Harris, A. G., E. Tuttle, and S. D. Tuttle. 1997. *Geology of National Parks.* Dubuque, Iowa: Kendall/Hunt Publishing.

Unrau, H. D. 1990. *Basin and Range: A History of Great Basin National Park.* Denver, Colo.: U.S. Dept. of the Interior, National Park Service.

21. Lehman Caves

Halladay, O. J., and V. L. Peacock. 1972. *Lehman Caves Story.* Baker, Nev.: Lehman Caves Natural History Association.

Harris, A. G., E. Tuttle, and S. D. Tuttle. 1997. *Geology of National Parks.* Dubuque, Iowa: Kendall/Hunt Publishing.

22. Bristlecone Pines

Cohen, M. P. 1998. *Garden of Bristlecones: Tales of Change in the Great Basin.* Reno: University of Nevada Press.

Johnson, R., and A. Johnson. 1978. *Ancient Bristlecone Pine Forest: Living Then, Living Now.* Bishop, Calif.: Chalfant Press.

U.S. Dept. of Agriculture. Forest Service, Intermountain Region. 1994. *Bristlecone Pine: Nature's Oldest Living Thing.* Ogden, Utah.

23. Glacial Features in Great Basin National Park

Harris, A. G., E. Tuttle, and S. D. Tuttle. 1997. *Geology of National Parks.* Dubuque, Iowa: Kendall/Hunt Publishing.

24. Cathedral Gorge

Johnson, W. G., and B. A. Holz. 1998. *Archaeological, geomorphological, and paleontological investigations of a bison bone deposit at Cathedral Gorge State Park, Nevada.* Las Vegas: Quaternary Sciences Center, Desert Research Institute.

Index

aa, 188–9
abrasion, wind, 103
alluvium, 44
archeology, 114, 118–9, 245
aretes, 257–8, 261
artifacts, 114–5, 119
assay process. *See* gold assay
asthenosphere, 4–5

badlands, 265, 267, 269–70, 272
Baker, 216, 221
Basin and Range, evolution of, 3–6
Bates Mountain tuff, 167–71, 175
batholith, 13, 63, 71
beach bars, 87
Berlin, 145–53, 158
Berlin–Ichthyosaur State Park, 145–66
Black Rock lava flow, 181, 183, 185–92
breccia, volcanic, 75–6
bristlecone pines, 241–51
Burnt Cave, 118–9

calderas, 172–5, 265
Caliente Caldera complex, 265
California Emigrant Trail, 41
cambium. *See* growth rings
carbon–14. *See* radiocarbon dating
Carson Range, 11, 23; and faulting, 39, 45, 47; glaciation of, 31
Carson Sink, 94, 96
Cathedral Gorge, 265–72
cave pearls, 237
cave popcorn, 237
Cave Rock, 23–4, 26–7
Central Nevada Seismic Belt, 140
Churchill County Museum, 111
cinder cone volcanoes, 180
cirques, 258–9
Comstock Mines, 28, 41, 68, 145, 150
copper mines, 203–12
core, earth's, 4
Crater Lake, 117, 174
creep, 219–20
cross dating, 248–50
crossbeds, 126
crust, earth's, 4
Crypt Cave, 120

debris flows, 49, 52, 55–7
dendrochronology, 244, 247, 251
dendroclimatology, 245, 247
desert pavement, 125
desert varnish, 100–1, 104–6, 169
Diana Mine, 150
dissolution, limestone, 232–3
Dixie Valley, 131, 140–1
drunken trees, 219–20
dunes, 121–30

Eagle Falls, 37
earthquakes, 133–143; focus, 134; intensity, 137–9; magnitude, 139
Easy Chair Crater, 178–81, 192
Eetza Mountain, 107
Eight Mile Flat, 94, 96
Emerald Bay, 29–37
epicenter, 134
eutrophic lakes, 19–20
eutrophication, 20
evaporite deposits, 95–6
exfoliation. *See* unloading

Fairview Peak, 133–43
Fairview Valley, 128
Fallon, 99, 111, 113
Fannette Island, 34–5
faults, 25, 39; evidence for, 42–5; and hot springs, 66; movement along, 45; type, 139–40; and volcanism, 178
firn, 256
Fish Cave, 106, 113, 119–20
fissures, 64–5, 67
floods, 33
focus, earthquake, 134
Four Mile Flat, 95, 121
freeze/thaw cycles, 219–20

garnets, 195–6
Garnet Fields Rockhound Area, 193
Garnet Hill, 193–202, 212
gemstones, 195
Genoa, 39–47
geologic time, 1–2
geothermal energy, 61
geysers, 65
glacial deposits. *See* moraines
glacial erosion, 257–9
glaciation, 29, 33, 253–63
gold mining, 145–153
gold assay, 150–2
Great Basin, 6–7
Great Basin National Park, 213–63
Grimes Burial Cave, 119
Grimes Point, 93, 97–106, 112
Grimes Point Archeological Area, 99
growth rings, 246

Hickison Summit, 167–75
Hidden Cave, 106–7, 111–7,119–20
Hobo Hot Springs, 40
hoodoos, 265, 270–2
horns, 257–9, 261
hot springs, 40, 61–70
hummocky topography, 50, 60
hydrothermal alteration, 205, 209–10
hydrothermal fluids, 64; and mineralization, 68, 146–7, 209–10

Ice Age, 232, 253, 263
ichthyosaurs, 155–66
incised channels, 269
intensity, earthquake, 137–9
intrusions, igneous: sill, 209; stock, 209

Jacks Valley, 39, 41, 47
Jeff Davis Peak, 218, 261
Jeffrey pines, 69
jetstream, 93
joints, 15, 17, 53–4, 65

krummholz vegetation, 232

lahars, 26, 71, 77–8
Lahontan Mountains, 94, 97–9, 106
Lake Lahontan, 83–96; beach sand from, 129–30; deposits from, 110, 113–4
Lake Tahoe, 11, 25, 74, 81; bedrock, 13; degradation of, 19–21; and glaciation, 32
landslides, 31, 49–60
Lehman Caves, 216–7, 229–39
Liberty Pit, 203–7, 212
lichens, 222–3
lithosphere, 4–5
Long Valley caldera, 172
Lunar Crater, 181–2, 192
Lunar Crater Volcanic Field, 177–92
Luning formation, 155, 162

maar, 182
magnitude, earthquake, 139
mantle, earth's, 4
Mather Overlook, 218–20
Meadow Valley 265, 267
mercury. *See* gold assay
Methuselah pine, 246
middens, 118
mill, gold, 151–3

Miller Point Overlook, 272
mixotrophic bacteria, 101–2
Mono Craters, 118
moraines, 251, 258–60
Mount Mazama, 117, 174
Mount Rose, 71–81
mud cracks, 268
mudflows, 116

Nevada State Museum, 113

oligotrophic lakes, 19–20
oolitic tufa, 116
Ophir Creek, 49–50, 59
Ophir Mill, 58

pahoehoe, 188–90
paleogeography, 160–3
paleontology, 160–6
Paleo-Indians, 97, 104, 106–7
Pancake Range, 177, 191
permeability, 232
petroglyphs, 97, 110, 167, 172, 217; and desert varnish, 102–6; types of, 104–6
phreatic eruption, 182
Picnic Cave, 110–2
pictographs, 118–9
Pine Alpha, 246
piping, 269
placer gold mining, 147–9
plate tectonic theory, 3–6
playas, 94–6, 103, 183
plucking, glacial, 36, 256
plug, volcanic, 27
plutons. *See* intrusions
porosity, 232
pressure ridges, 189
Price Lake, 49
pumice, 171–2
pyroclastic flows, 172–3

radiocarbon dating, 3, 89–90, 114–5, 250–1
rainshadow effect, 31
red tides, 165–6
Relay Ridge, 80
Reveille Range, 192
rhyolite domes, 64
Richter scale, 139
rift valleys, 4
ripplemarks, 125–6, 224
Robinson Mining District, 203–12
roche moutonnée, 34–5
rock cycle, 7–8
rock glacier, 262–3
rockslides, 51, 55–6, 219
Ruth, 202–4

salt deposits. *See* evaporite deposits
Salt Wells Basin, 128
saltation, 123–5
sand dunes, types, 126–8
Sand Harbor, 11, 13
Sand Mountain, 121–130
Sand Springs Desert Study Area, 123
seismology, 137–9
shields, 238
Shoshone Mountains, 155, 160
Simpson Pass, 130
sinter, 63–4, 67, 70
Sinter Hill, 70
slickensides, 43, 45–6, 137
Slide Mountain, 49–60
Snake Range, 213, 216–7; climate of, 221; bristlecone pines of, 241, 248; glaciation of, 253, 263
snow avalanches, 218–9
snowline, 256
speleothems, 234, 237–9
spheroidal weathering, 16–7
Spirit Cave, 113, 119
springs. *See* hot springs
Spring Valley, 213, 221
stalactites, 233, 235–6
stalagmites, 233, 235–6
Steamboat Hot Springs, 61–70
Stillwater Mountains, 128, 130–1
stratigraphy, 165; of Hidden Cave, 114, 116–8, 120
striations, glacial, 37

strike and dip, 44
superposition, 206–7

Tahoe stage glaciation, 33
talus, 24, 238–9, 262
till, glacial, 79
Tioga stage glaciation, 33
Toquima Range, 175
travertine, 64, 234–5
triangulation, 134–5
Truckee River, 28, 33
tufa, 87–9, 110–3, 116. *See also* oolitic tufa
tuff, volcanic, 76, 183, 211; Bates Mountain, 167–75; formation of, 167–70; mineralization of, 198; welding of, 170–2
tundra, alpine, 222–3, 241
Twenty Mile Beach, 87

Union Mining District, 146, 153, 158–9
unloading, 13–4, 53–4

Van Sickle Station, 47
ventifacts, 124
vesicular basalt, 187
volcanic ash. *See* tuff

Walleys Hot Springs, 46
Walker Lake, 83–93
Walker Lake Recreation Area, 86
Wassuk Range, 87
wave–cut terraces, 86, 94
weathering, 201, 211, 225–6; spheroidal, 16–7
Wheeler Peak, 213–27, 253
White Mountains, 245

Richard L. Orndorff, Robert W. Wieder, and Harry F. Filkorn (left to right) *on top of Wheeler Peak*

About the Authors

Richard L. Orndorff, assistant professor of geology at the University of Nevada, Las Vegas, studies glacial geology in Nevada's mountains and the effects of climate change on lake systems in Nevada and California. Paleontologist and biologist **Robert Wieder** studies insects that harm agriculture for the California Department of Agriculture. **Harry Filkorn** is a paleontologist at Kent State University whose research includes mapping the geology of the Southwest. All three authors enjoy exploring the geology of Nevada's wide-open landscapes.

Nadi Jyotisha and *Graha Yuddha* which delve into some inconsistencies that she has discovered in her studies of the classics. She has thoroughly researched these subjects and the results contained herein are most illuminating.

This material has been the subject of mutual interest, and the focus of conversations between us since the mid-1990s. Edith used those conversations as the impetus for her subsequent research, and, given the implications for chart analysis, I believe the astrology community at large will significantly benefit from her findings. Specifically, she assesses Jupiter/Saturn conjunctions and their dominance within any particular element, (fire, earth, air and water), as they apply to charts of particular individuals, as well as understanding the charts of larger cycles.

The process of her investigation itself will be instructive to the novice looking to better understand how Vedic Astrology works, and the experienced astrologer will doubtless find her insights provocative if not directly helpful in their own work.

There is much for the reader to absorb in this book. Edith does an excellent job of defining the key components of Vedic Astrology, and describing the philosophy and principles of Vedic India. Additionally, in her historical survey she demonstrates her advanced skills in the practice of mundane astrology (the astrology of world affairs), as well as natal astrology, (the study of individual horoscopes). Each of these segments is engaging and enlightening, and reflects the scholarship involved - which is impressive.

I have long encouraged Edith to "always go back to the *Rashi Lagna*" as a pivotal essential in chart analysis. She has clearly taken this suggestion to heart. Her book offers the reader a study of Vedic Astrology specifically relating to the natal ascendant, which is deeply respectful of the knowledge and traditions of Vedic India, but is infused with the spirit of thoughtful inquiry. Her observations are original, her approach inventive, and her research is scrupulously executed.

Edith has masterfully chronicled her observations in this book *In Search of Destiny* which I am sure you will enjoy reading as much as I did.

Chakrapani Ullal
Vedic astrologer
Los Angeles, California, USA
November 2011

Preface

In Search of Destiny explores our common humanity in a unique way. When Vedic astrology – the astrology of India – is combined with biography, history, politics, the arts, and culture, a key missing component is provided to all of the above, being more attentive to time and cycles. This factor permeates the study of history, making one more conscious about time. But an awareness of history itself "has been progressively squeezed out," notes the late historian Tony Judt in a 2008 interview.[1] He attributes some of this unprecedented trend to "the sheer speed of contemporary change [and to]... a perverse contemporary insistence on *not* understanding the context of our present dilemmas, at home and abroad... on seeking actively to *forget* rather than to remember, to deny continuity and to proclaim novelty on every possible occasion." Without an awareness of history, Judt says, citizens operate without "the dimension of time. ... We have lost touch with much of what we need to recall from our own recent past.... We *think* we know – which makes it much more dangerous, because we invoke it all the time."

The awareness of time dominates astrology. Because of that, astrology can look at the events of history and biography in a unique way, informed at every turn by planetary patterns and signatures. The study of astrology exists without the context of history and vice versa, but a broader, deeper understanding of people and events is restored when astrology is combined with it.

The astrology of the ancient Vedic civilization began as a timing tool for the sacred rituals (often including fire ceremonies and chanting) and the planting of crops. The Seers or wise men among them understood that to conduct these rituals at the most auspicious time would greatly assist the intended outcome – to preserve and protect their civilization. For this, the planetary configurations were studied. Their system of astrology employs time units from the very small (4 seconds) to the very large (4.32 billion years times 100s of 1000s). The Vedic perspective is from the smallest time unit of the human breath – the *prana* – assigned to 4 seconds, up to the largest one – the *Mahakalpa*, a lifetime of Brahma, at 100 of his years, given that one *Kalpa*, or one day

---

[1] Tony Judt, 1948-2010, British author, historian, and university professor, in a television interview about his book ***Reappraisals: Reflections on the Forgotten Twentieth Century** (2008)*, on *Charlie Rose* (Public Television), June 6, 2008. Most quotes are from that interview. Second quote is from Judt's 2008 book, p. 2.

of Brahma = 4.32 billion years. One *Mahayuga* (or 4 *yugas*) = 4.32 million years. Our present *Kali Yuga* = 432,000 years, and we are about 5000 years into it. The Precession cycle contains from 24,000 to 25,900 years.[2]

Among these components there are smaller cycles that enable us to observe human events, and those most often involve Jupiter and Saturn. Their conjunctions occur every 20 years and are event markers for the collective that also radically affect many of the individuals included in this book. The larger cycles of the Jupiter-Saturn conjunctions run several hundred years each, and are characterized by being in one of the four elements that dominate for most of that time – fire, earth, air, or water. Mutation periods occur at the beginning and end of cycles, when the element gradually changes from one to the next. Chapter 1: ***Historical Context and Collective Destiny*** covers this topic and orients us to a larger time frame, so that when we encounter each of the individuals in this book, we understand how their personal destinies were affected by the larger historical context, and the collective destiny imprinted on anyone born or living at that time.

We discuss what it means to live in an EARTH dominant period, compared to one dominated by FIRE, AIR, or WATER. From this we learn why a FIRE dominant period is much more advantageous for national aspirations, whereas a WATER period favors the control of either royalty or religious authorities, and more often a combination of the two.

Major advancements in weaponry were certainly favored in the Mutation period from FIRE to EARTH, which ran from 1901 to 1961, and witnessed the start of the Atomic Age and the unleashing of the first atomic bombs along with the threat of their use. Along too came peaceful use of nuclear energy, though uranium is innately volatile and would bring far greater volatility in global interrelations. This was followed by the EARTH dominant period, which is more technologically capable, but generally less ethical than all of the others, and less interested in education and true knowledge. Among other factors, with Jupiter debilitated in Capricorn and overpowered by Saturn very strong in its own sign, special-interest money has increasingly distorted the democratic process, diminishing humanitarian concerns along the way and requiring much more effort on the part of individual citizens to counter these more dominant trends.

The **JU-SA conjunction in Capricorn on Feb. 18, 1961** marked the start of the EARTH dominant cycle, after a 60-year Mutation cycle from FIRE to EARTH. This 1961 conjunction impacts the next 238 years, up to 2199, and was the first Jupiter-Saturn conjunction predominating in sidereal earth signs since 1285 A.D., setting a whole new tone for the era. On the macro level, corporate conglomerates (Saturn) would gain dominance, with greater consolidation among governments, banks, and corporations, while each nation-state (Jupiter) would increasingly face issues of survival and well-being due to the rise of unbridled corporate power. Saturn (business and material power), typically voiced by the Conservative Right, often triumphs over ethics, humanitarian concerns, and social justice (Jupiter), typically voiced by the Liberal Left. This is how Jupiter and Saturn act for the collective. Also, Jupiter is signified as expansive and optimistic, while Saturn rules over fear and our individual sense of material security. Framed in their own moral terms and overtly expressed in post-1961 America, Conservative principles aim to reduce reproductive rights, regulation, taxation, unions, worker rights and the amount of damages in tort cases. Conservatives also want to exert more control over race, language, religion,

[2] See Glossary for Sanskrit and Vedic astrological terms.

and education (i.e., control over what is taught). They see this as their way of offering structure and order.

Scholars and historians associate Conservatives with **Thomas Hobbes** (1588-1679), a British philosopher and ardent monarchist, whose materialistic philosophy was more fear-oriented (as in his major work, *Leviathan*, 1651). He focused on man's innate evil or brutishness. Meanwhile, British philosopher, **John Locke** (1632-1704) focused in his works on the innate goodness and cooperative spirit in mankind. He espoused a more anti-authoritarian point of view typically associated with Liberals. Even so, they both agreed that Social contracts exist to preserve the common good and to protect private rights. But Hobbes was less trusting in the human being to look out for others, and more willing to let a government have absolute authority over its subjects and its institutions. Both of them wrote during the more royalist WATER period (1425-1723), but Locke strongly influenced the more nationalist FIRE period (1723-1921), especially the American and French revolutions, based on government by popular consent. With survival needs more prominent in the EARTH period, and at times the seeking of security at whatever cost to the individual, the society or the environment, the views of Thomas Hobbes may dominate in many public circles over those of John Locke, from 1961.

Jupiter is knowledge and education, and with Saturn dominating over Jupiter, real knowledge can be marginalized by those in power, with many aspects of life gradually commercialized, including science and academia. Coincidentally, historian Tony Judt remarked that "history was quite prominent as recently as the 1960s" – a decade also marking the last significant gains and clear (at times strident) voice of the liberal Left in the USA and elsewhere. The achievements of American Civil Rights legislation passed in the mid-1960s, described here in several biographies, reveal the strength of a message that was soon to diminish. In retrospect, and with greater understanding of the significance of the Jupiter-Saturn conjunction in early Capricorn in Feb. 1961, we can better recognize a clear trend for the collective destiny starting from that time.

The individual destiny can also be seen with greater clarity against this backdrop. In the process of discussing the larger picture – or Mundane astrology – we see that the assigned effects of planets when dealing with the collective destiny are slightly different from those in Natal astrology, dealing with the destiny of the individual. For instance, Jupiter has the meaning of royalty for the collective, but for the individual it has some other definitions. These are contained in Chapter 2: ***Introduction to Vedic Astrology and Philosophy***.

Among **personal event markers**, the Vedic *Dasha* system(s) – or planetary cycles – are extraordinary, and depend on the individual birth chart for accurate timing. The most important one (the *Vimshottari Dasha*) is featured here. As an example, most biographies on Albert Einstein cannot fathom why the last 30 years of his life produced none of the genius of his first 30 years. Einstein even tried to reject or dispute some of his earlier findings in that later period. According to Vedic astrology, the most productive period of Einstein's life would occur during his Venus *Dasha*. This 20-year cycle occurs once in 120 years, and would be excellent for him, running from Dec. 1895 through Dec. 1915. The subsequent 6-year Sun *Dasha* would bring him rewards for those achievements, including a Nobel Prize in 1921.

Vedic astrology examines the destiny from the vantage point of *Dharma* (right action and right work), and *Karma* (destiny that must be lived). The probable timing of when the fruits of the *Karmas* come due can be read through the *Dasha* cycles – for which Vedic astrology is famous

– and also through the transits of planets, especially the cycles of Jupiter, planet of *Dharma*, and Saturn, planet of *Karma*. In addition to *karmic* timing, the astrology of Jupiter and Saturn provide a deeper explanation of why luck (Jupiter) cannot operate with complete success without discipline, especially self-discipline (Saturn). Western minds often try to calculate which of these is superior, but in Vedic terms they are as inseparable as the two hands on the clock. Together, they also help to describe why collective forces of destiny come together at certain times on the earth, and how they are embodied in an individual – whose birth chart, in turn, shows how much he or she may be a major force in coalescing with history.

Vedic astrology places great emphasis on analyzing an astrological chart from the Ascendant, also from the Moon – unlike Western astrology's emphasis on Sun signs. If you are not familiar with Vedic astrology or principles of Vedic culture and philosophy from ancient India, you will benefit by reading my general Introduction (Chapter 2) before proceeding to the biographies, especially the sections with astrological details. The Introduction conveys how the traditions of the great Vedic culture remain a living force. A part of that spiritual inspiration can be seen and experienced through Vedic astrology. There is much condensed layering of information in this general Introduction, providing a backdrop as you proceed through the chapters on the different Ascendant signs.

The Chapter Introductions are also important to read as you start each chapter, depending on your knowledge of Vedic astrology and philosophy. A non-astrological biographical account precedes the astrological analysis of each biographical subject, told thematically, rather than completely chronologically. If the figure is important historically, I provide relevant historical background, and in all cases precise dates of various events in the life, when available. These are absolutely critical to any astrological studies.

The biographical subjects have been carefully chosen to show the power of the sign on the Ascendant as an indicator of the life destiny. Therefore, subjects include only those with accurate birth data so the correct Ascendant sign is unlikely to be in dispute.[3] Most of them are well-known figures internationally, but a few are less well known. The sequence is of great importance. For instance, the Aries Ascendant chapter begins with two pioneers in their own ways, two individuals who were forced to start new enterprises for various compelling reasons: Martin Luther King, Jr. and Robert F. Kennedy. Since Mars rules the sign of Aries and is the Ascendant lord, governing the chart in major ways, this can introduce an element of danger, depending on the position of natal Mars.

The last biographical subject in the Aries Ascendant section is Jacqueline Du Pré, the British cellist who developed multiple sclerosis and was forced to retire from a meteoric career at age 28. The very next biographical subject has a birth chart very closely parallel to that of Du Pré, except the Ascendant sign is Taurus rather than Aries. This is a significant factor in saving that person from the same fate. He has had good health and stamina into his late 60s, as well as a long-term marriage, while Du Pré suffered many chaotic problems in her marriage and relationships. This is a subject of some controversy, especially after the publication of her siblings' book, *A Genius*

[3] I have researched each biographical subject for accurate birth data, using the categories established by American astrologer Lois Rodden (1928-2003), renowned for her data research and collections. I have used mostly Class AA or A (birth record or memory), and only rarely Class B (biography), when it was clear that the time range was well established from comments from parents or others closely attending the birth.

in the Family (1997) and the film version of it, *Hilary and Jackie* (1999). Astrology provides some answers on this matter as well.

Some astrological biographies exist in print form and others are increasingly present on the Internet due to increased public interest in applying astrological knowledge to mainstream subjects such as biography. But with a few notable exceptions, they do not often provide longer, more in-depth studies with researched and nuanced biographical material that is both astrological and non-astrological, as I have tried to do in this book. Relatively few use Vedic astrology, though it is becoming more known in the West.

Furthermore, rarely does astrology enter the public domain of biography and history. On a larger topic, I have revisited a mainstream academic debate on Adolf Hitler. I provide further background from his astrological chart to contribute to a major historical debate. This surfaced in the 1950s between British historians Allan Bullock and Hugh Trevor-Roper and their counterparts: namely, was Hitler an opportunistic adventurer totally lacking in ethics or beliefs? (Alan Bullock) Or was he a man who did have beliefs – although repulsive ones – and those motivated his actions? (Hugh Trevor-Roper) Hitler's astrological chart gives much more credence to Hugh Trevor-Roper's general historical position on Hitler. With three planets in Sagittarius, including the Moon – indicating key mental tendencies – it is clear that Hitler had strong beliefs and they fuelled all his actions throughout his life. His political passions can be documented from his late teens. He was a monomaniacal person who tolerated little opposition, but he also had a major gift in magnetizing the people to him by tuning in to the same grievances shared by many German people at the conclusion of World War I.

The destinies of nations and their leaders are conjoined, with at times odd combinations of historical and astrological events. For example, Mahatma Gandhi and Adolf Hitler share the same Ascendant degree (within 12 minutes of arc) – in Libra, ruled by Venus. The outcome depends as always on the whole chart and how the scales (Libra) are tipped.

> If Mars is the planet of war, there is no one planet of peace, although Venus and Libra are often assigned that role due to their association with equilibrium, and with a vision of harmony or a way of harmonious living. Venus and Libra rule over love and contractual relationships such as marriage. But Venus is also one's desire nature, and may tear down everything in its way in order to achieve the goal.
>
> Chapter 9: Libra Ascendant, Chart #29, by the author.

I invite the reader to drink in some of the infinite richness of Vedic astrology, as it shines its light on some details from biography, history, and culture. In the process – as each of us goes "In Search of Destiny" – may it deepen our awareness of the dimension of time.

Edith Hathaway
La Jolla, California, USA
October 2011

CHAPTER 1

Historical Context and Collective Destiny

The Vedic view of time and of history is a very long-range and cyclical view, much closer to the movement of natural history, with its long expanded periods dealing with mega events in space, planets and the earth, and marked by periodic cataclysms. Astronomical events are disguised in the myths of the Puranic literature of India, but they reflect the idea of a World Age and of endlessly repeating cycles in which the universe continues to expand and contract. Other advanced ancient civilizations (the Egyptian, Mayan and Aztec) shared similar but not identical concepts of long cycles of human civilization that were destroyed periodically. In the Vedic view, the universe is again and again created, preserved, and then destroyed, echoing the functions of the three primary Hindu deities: Brahma, the Creator, Vishnu, the Preserver, and Shiva, the destroyer. In this philosophical system the soul is viewed as eternal and indestructible, living in a timeless reality on the one hand, but in a time-bound reality on the other, when incarnated in a physical body. Then issues of human awareness, Divine law, and concepts of *Dharma* and *Karma* arise.[4]

These very large time cycles seem more related to cosmic history than to earthly historical events. Even so, they are supposed to describe the rise and fall of *Dharma*. The time from creation to destruction is considered a day in the life of Brahma, the Creator. One day of Brahma is also called a *Kalpa,* and is 4,320,000,000 years (or 4.32 billion earth years), a unit of time given in the *Rig Veda* and in the *Bhagavad Gita,* Chapter 8, Verse 17. The Universe expands and contracts for a cycle equaling 100 years of Brahma, or one *Mahakalpa* (a large *yuga*, also called a divine *yuga*), equaling one lifetime of Brahma.

The *Surya Siddhanta* is probably the most important ancient Vedic astronomical text discussing time and its measurements. It describes units of time from the smallest microsecond to large expanses of time, such as a *Mahakalpa*. One *Kalpa* (day of Brahma) = 4.32 billion years, or 1000 *Mahayugas.* One *Mahayuga* = 4 *Yugas* = 4.32 million years, sometimes also called The Great Age. We are currently

[4] See Glossary for all Sanskrit and Vedic astrological terms, including how it regards a planetary conjunction.

in the *Kali Yuga*, lasting 432,000 years. It is the 4th and last of the cycle of four *yugas*.[5] Opinions differ about when the *Kali Yuga* began, but apparently we are only about 5000 years into it, as many consider it started in 3102 B.C. There are various sub-periods within the *Kali Yuga*, and according to *Manu Samhita*, we are in an Ascending phase, called *Dwapara Yuga*, from 1485 to 3885 A.D.

The larger cycle of the four *yugas* begins with themes of spiritual purity and virtue before deteriorating in the *Kali Yuga* towards their absence, when darkness predominates along with the lower qualities of human beings. The positive news is that spiritual knowledge and development shines even more brightly through the darkness. Since honesty and kindness are more uncommon, their light can shine all the more in the midst of corruption and violence. The Winter Solstice is currently close to the Galactic Center, and this helps to bring a new influx of spiritual energy.

The *Yuga* is a World age, and is based on the number 432, in multiples of 100s or 1000s. The *Kali Yuga* is 432 x 1000, or 432,000 years. *Yugas* are traditionally linked to the Precession of the equinoxes in Hindu thought. With 72 years for 1 degree of the zodiac, *Kali Yuga* is made up of 200 *Rashis* (zodiacal signs) starting in the Fire element and ending in Water (200 *Rashis* x 30 degrees each x 72 degrees = 432,000 years). Classical texts, such as *Manu Samhita*, link the *Yugas* to the Precession of the equinoxes. In his book, *The Holy Science*, the sage and astrologer Shri Yukteshwar cites *Manu Samhita*, though Yukteshwar assigns a Precession cycle of 24,000 years, whereas it is probably much closer to 25,900 years, the figure assigned by modern astronomers. The **Jupiter-Saturn conjunctions** occur approximately every 20 years, and they recur 59.575 times within one *Yuga*. If you multiply 432 x 59.575 the result is 25,736 years, or one Precession cycle. (The precession rate is variable, from 24,000 to 25,900 years.)

The **Precession of the Equinoxes** is a phenomenon in which the earth slips back in its relationship to the Fixed Stars, approximately one degree every 72 years, making a complete cycle of the zodiac. (In ca. 285 AD., the vernal equinox and the transiting Sun entering the sign of Aries coincided, using *Lahiri ayanamsha*.) This concept is presented metaphorically in the *Bhagavata Puranam* (Book IV, Chapter 9) when the young boy Dhruva prays to Lord Vishnu, who tells him:

> "You will not reach me now. You have to wait for a time of 26,000 years, which is a period when the stars will complete a revolution around the pole star. The lasting abode around which the stars and the great sages go round in circumambulation is permanent and it survives the dissolution of the three worlds, Earth, Heaven and …the intermediate space."[6]

The Sanskrit word *Dhruva* means "fixed, not moving" and also indicates the Pole star. The Precession was well known to Hindu astronomers, and called *Ayana Chalanam.* Varahamihira lived ca. 500 AD and noted that the precession (called *Ayanamsha*) was zero at that time. That is, there was no difference between the tropical and sidereal, or constellational zodiacs.

Jupiter and Saturn cycles: For earthly historical events, smaller time units provide an advantage over mega time units such as the *yugas*, even with their view of cosmic history. Jupiter and Saturn

---

[5] In Earth years, the four *yugas* in the Hindu/Vedic system are: 1) *Satya Yuga* – 1,728,000 Earth years; 2) *Treta Yuga* – 1,296,000 years; 3) *Dwapara Yuga* – 864,000 years and *Kali Yuga* – 432,000 years. Their combined total: 4,320,000 years.

[6] Quoted by Satguru Shri Sivananda Murty, in his article "Our Future – Projections, Prophesies and Predictions," ***The World Tomorrow: What They Say*** (A collection of Articles presented in the World Conference on Mundane Astrology [Oct. 2010]), Sanathana Dharma Charitable Trust, 2010, p. 268. (The original *Bhagavata* texts give "36,000 years," but this has been changed by later Indian scholars to coincide with the precessional cycle, which is closer to 26,800. In the translation above, the number has been rounded to 26,000.)

have cycles that are within our human grasp to observe and contemplate, especially their 20-year conjunction cycles, as well as their individual orbital cycles. Their conjunction cycles (at times combined with their individual orbital cycles) provide important milestones when the history of mankind can be assessed. Saturn is related to the structure of the physical body and its progress over Time, while Jupiter is related to the human consciousness and its aspirations. These two planets return to their same place in the zodiac approximately every 60 years. Jupiter's orbital cycle is exactly 11.9 years, and Saturn's is 29.6 years. Thus, five orbits of Jupiter, and two orbits of Saturn come close to 60 years, and in the lives of individuals or the collective their cycles provide pivotal life turning points between the ages of 58 and 60. For assessing worldly events, the Vedic system also uses a 60-year Jupiter cycle, called the Jovian years.[7]

There are many mathematical correlations between larger and smaller cycles in Vedic astrology. While the mega cycles employ the number 432 in multiples of 100s and 1000s, the smaller cycles use the numbers 12, 24, 30, and 60.[8] **Jupiter**, planet of *Dharma* and **Saturn**, planet of *Karma* are the hands of the clock, and in the larger scheme of things, they give us clues about socio-political-economic matters. Saturn's orbital cycle of 29.6 years is parallel numerically to the Moon's orbital cycle, exact at 29.6 days. Both of them can be used in segments of 28 years and 28 days, respectively. They return to the same sign position approximately every 28 years (Saturn), or every 28 days (Moon). Jupiter's orbital cycle can be used in 12-year segments for the same reason.

The Jupiter-Saturn conjunctions occur every 20 years. Within the 20-year JU-SA conjunction cycles, the most important event markers are the **JU-SA oppositions** that arrive 10 to 11 years after the conjunctions. The oppositions are defined by events coming to a crisis point, at which the issues of the era become very obvious, and more demanding of solutions. An opposition, like a Full Moon, helps us to see things more clearly, even if it is a confrontation of planets and of people. The time is auspicious to solve problems of the 20-year cycle or at the very least – identify the problems more clearly.

In turn, this 20-year cycle can be seen against a backdrop of the larger era in which the JU-SA conjunctions prevail in one of the four elements. These run for several hundred years. Also called the "triplicities," they refer to the three zodiacal signs in each element. **FIRE:** Aries, Leo, Sagittarius; **EARTH:** Taurus, Virgo, Capricorn; **AIR:** Gemini, Libra, Aquarius; and **WATER:** Cancer, Scorpio, Pisces.

Before moving more consistently into one element, the JU-SA conjunctions can occur in the previous element at various random intervals. These are **Mutation periods** and can last anywhere from 60 years up to 139 years, as shown in this upcoming set of JU-SA conjunction cycles, covering almost 1000 years, from 1405 A.D. through 2398 A.D. The individual astrologer can assign what constitutes a Mutation period. I define a Mutation period as ended when the next element has repeated itself in the JU-SA conjunctions at least twice in a row with no break. One anomaly in this particular sequence is the WATER to FIRE Mutation period. It has two sequential Fire conjunctions before returning to water once more. In an earlier WATER to FIRE Mutation

---

[7] Each Jovian year is assigned a Sanskrit name evocative of the theme for that year, starting at the Vedic New Year, which arrives annually at the last New Moon in Pisces prior to transiting Sun entering sidereal Aries. (Further exploration of the 60 Jovian years is not included here.)

[8] One *Vighati (or Pala)* = 24 seconds (1/60th of a *Ghati*); one *Ghati* = 24 minutes; 60 *Ghatis* = 24 hours (one day); one lunar month = 30 lunar days, from one New Moon to the next; One sidereal month = 30 sunrises; One solar month = the interval between the entry of the tr. Sun into one sign of the zodiac and when it enters the next sign. One year = 12 months. (These are only some of the Vedic time units given in the ***Surya Siddhanta,*** Chapter 1.)

period in the 9th century A.D., FIRE dominated very quickly as it entered the series, and was an anomaly to the other Mutation periods.

The triplicities: The JU-SA conjunctions occur in one element for an average of 243 years in this cycle, though these periods can vary between 198 and 298 years, using *Lahiri ayanamsha.*[9] The larger periods when the conjunctions predominate in a certain element (or triplicity) have significance for the world due to these factors: **1)** their element, and **2)** the sign and *nakshatra* in which the conjunction occurs.

Further, there is added significance or intensity when the JU-SA conjunction: **a)** occurs three times (triple conjunctions), **b)** is *Vargottama*, especially in a sign owned by Jupiter or Saturn, and **c)** coincides with Winter or Summer Solstice and/or Fall or Spring Equinox, or those events have transiting Sun close to the degree of the conjunction.

Triple conjunctions occur rarely enough to be noteworthy, and though they have not been clearly defined by Vedic classics, they should draw our attention to that particular year, the zodiacal sign and element, and the subsequent 20-year cycle. For instance, in 7 B.C.E., during a WATER dominant period, there were three JU-SA conjunctions in Pisces (at 25:06, 21:33 and 19:42 Pisces, all in *Revati nakshatra*). This was supposed to announce the coming of the King of the Jews. Thus, the Magi (the astrologers) went looking for him. Pisces is ruled by Jupiter, and is the most religious of the WATER signs, all of which bring in religious themes. Cancer is a water sign more associated with families and royal power, often aligned in this era with religious power. The **triple JU-SA conjunctions in Cancer in 1682-1683** were the last set of triple conjunctions prior to 1940-1941. They marked a time when the excesses of religious and royal power came to a peak, symbolized by the Portuguese, Spanish and Roman Inquisitions, which – though not ended until the mid-1800s – had decreasing authority once the WATER to FIRE Mutation period began in 1702, and even less influence from 1782 when the FIRE-dominant period began.

The first **JU-SA conjunction in Aries (in 1821)** marked a period of major growth and influence of the British Empire and other European empires, enhanced also by the new railroad travel. It was punctuated at the end by the **triple JU-SA conjunctions in Aries in 1940-1941**. By then, the British Empire was finished financially and territorially from the events of World War II. Nationalistic themes came to a grand finale, exploding as world conflict in the FIRE to EARTH mutation period. The EARTH dominant period began in earnest in Feb. 1961 and moved into full throttle with the **triple JU-SA conjunctions in Virgo in 1980-1981**. The 20th century could be predicted to have events of accelerated intensity since it contains two sets of triple JU-SA conjunctions within only forty years of each other. The 23rd century also has two sets of triple conjunctions, both in AIR signs, 41 years apart: in 2238 and 2279. Except for these two anomalies during a 1000-year period, the span between triple conjunctions is 257 years.

The Jupiter-Saturn conjunction is also a *Graha Yuddha*, or a Planetary War. This occurs between two planets within one degree. According to rules I have established using classical principles and

---

[9]My colleague, astrologer Raphael Gil-Brand and I each presented different aspects of the Jupiter-Saturn cycles in our lectures at the World Conference on Mundane Astrology, held in Vishakapatnam, India, Oct. 2010. Using *Bhasin ayanamsha*, Gil-Brand's periods differ, with his fire triplicity starting in Oct. 1682, some 41 years earlier. The average duration of the great conjunction cycles is around 900 years, this sequence totaling 973 years. My preference is for the *Lahiri ayanamsha*, which is also more widely used. Gil-Brand cites Persian astrology, which has the opinion that JU-SA conjunctions at the start of Aries foretell long term changes, and New Law of some kind.

years of research, Saturn is always the winner over Jupiter. [10] The winning planet can crush the losing planet even more so under various circumstances which can be specified. Here are significations for Jupiter and Saturn, especially as applied to the collective and its destiny, in worldly terms:

JUPITER: humanitarian and ethical concerns, law and lawyers, religion and its leaders, education, especially higher education, institutions of learning, teachers, journalism and journalists, publishing industry, urge for political and individual freedom, thus patriotism and aspirations for national identity (very strong in Sagittarius); the nation-state; the wealthy; royalty and ruling monarchs.

SATURN: business, status, consolidation of material resources and powers (very strong in Capricorn); material security, iron, lead, zinc, black metals, coal and fuel of all kinds, especially oil (black liquids), gasoline, chemicals, inferior grains, leather goods, weapons of defense; democracy, or looking like a democracy (weakened, especially with the JU-SA conjunction in Capricorn, when election results can be purchased); the transnational corporation.

Although Jupiter loses the *Graha Yuddha* to Saturn in any sign, its defeat is less devastating in FIRE and WATER. Jupiter owns one fire sign (Sagittarius) and one water sign (Pisces), and is exalted in Cancer. Therefore, when the JU-SA conjunction occurs in FIRE or WATER, Jupiter has a greater chance of enhancing its areas of significance, and brings nationalistic and religious themes to the forefront.

Saturn owns one earth sign (Capricorn) and one air sign (Aquarius), and is exalted in Libra. Therefore, Saturn's victory over Jupiter can be more definitive when the JU-SA conjunction occurs in either AIR or EARTH, but especially EARTH and especially Capricorn. This is because Jupiter is debilitated in Capricorn, and Saturn can crush the more ethical and humanitarian concerns of Jupiter under the weight of Saturn's tendency toward consolidation of material powers and resources in Capricorn. Saturn is weaker in both WATER and FIRE, but especially FIRE, and in Aries, where it is debilitated, and in Leo, ruled by Saturn's planetary enemy, the Sun. (Saturn can do somewhat better in Sagittarius, among all three fire signs.)

Nakshatras give finer details, but in this chapter we will pay most attention to the elements and events, following the strength of Jupiter and Saturn in each period, as correlated with some selected (though by no means exhaustive) historical and cultural events. Most of them concern Europe, the Americas, and European empires, notably the British Empire, and the disbursement of its colonies and national interests. We will scan a **1000-year period** from 1405 through 2398 A.D.

JUPTER-SATURN CONJUNCTIONS: 1405 to 2398

Jan. 16, 1405	**JU-SA conjunction at 8:12 Aquarius, in *Shatabisha nakshatra*. MUTATION PERIOD: AIR TO WATER (1405 to 1544), within WATER PERIOD.**	**AIR**
Feb. 14, 1425	**JU-SA conjunction at 1:28 Scorpio, in *Vishakha nakshatra*. 1st CONJUNCTION IN WATER series. Triple conjunctions in 1425. (Two more: March 18, 1425 at 0:42 Scorpio and Aug. 26, 1425 at 26:50 Libra. Next triple conjunctions in 257 years, in 1682-1683 in Cancer.)**	*****WATER**

[10]See Chapter 2: Introduction, under *Graha Yuddha*, with full description. See also Glossary.

July 14, 1444	**JU-SA conjunction at 22:52 Gemini, in *Punarvasu nakshatra.***	**AIR**
April 8, 1464	**JU-SA conjunction at 18:12 Aquarius, in *Shatabisha nakshatra.***	**AIR**
Nov. 18, 1484	**JU-SA conjunction at 6:30 Scorpio, in *Anuradha nakshatra.***	**WATER**
May 25, 1504	**JU-SA conjunction at 29:29 Gemini, in *Punarvasu nakshatra.***	**AIR**
Jan. 31, 1524	**JU-SA conjunction at 22:01 Aquarius, in *Purva Bhadra nakshatra.***	**AIR**
Sept. 8, 1544	**JU-SA conjunction at 10:35 Scorpio, in *Anuradha nakshatra.***	**WATER**
Aug. 25, 1563	**JU-SA conjunction at 11:25 Cancer, in *Pushya nakshatra.***	**WATER**
May 3, 1583	**JU-SA conjunction at 2:08 Pisces, in *Purva Bhadra nakshatra.***	**WATER**
Dec. 18, 1603	**JU-SA conjunction at 19:58 Scorpio, in *Jyestha nakshatra.***	**WATER**
July 16, 1623	**JU-SA conjunction at 18:00 Cancer, in *Ashlesha nakshatra.***	**WATER**
Feb. 24, 1643	**JU-SA conjunction at 6:13 Pisces, in *Uttara Bhadra nakshatra.* SPRING EQUINOX is at 11:08 Pisces.**	**WATER**
Oct. 17, 1663	**JU-SA conjunction at 23:48 Scorpio, in *Jyestha nakshatra.***	**WATER**
Oct. 24, 1682	**JU-SA conjunction at 29:43 Cancer, in *Ashlesha nakshatra.* Triple conjunctions in 1682-1683. (Two more: Feb. 9, 1683 at 27:16 Cancer and May 18, 1683 at 25:05 Cancer, both in *Ashlesha nakshatra.*** Next triple conjunctions in 257 years in 1940-1941 in Aries.)	*****WATER**
May 21, 1702	**JU-SA conjunction at 16:54 Pisces, in *Uttara Bhadra nakshatra.* START OF MUTATION: WATER TO FIRE.**	**WATER**
Jan. 5, 1723	**JU-SA conjunction at 3:20 Sagittarius, in *Mula nakshatra.* 1st CONJUNCTION IN FIRE.**	**FIRE**
Aug. 30, 1742	**JU-SA conjunction at 6:52 Leo, in *Magha nakshatra.***	**FIRE**
Mar. 18, 1762	**JU-SA conjunction at 21:48 Pisces, in *Revati nakshatra.***	**WATER**
Nov. 5, 1782	**JU-SA conjunction at 7:18 Sagittarius, on the Galactic Center. WINTER SOLSTICE is at 9:11 Sagittarius. START OF FIRE PERIOD. END OF MUTATION: WATER TO FIRE.**	**FIRE**
July 17, 1802	**JU-SA conjunction at 14:01 Leo, in *Purva Phalguni nakshatra.* (*Vargottama* status adds strength.)**	**FIRE**
June 19, 1821	**JU-SA conjunction at 3:17 Aries, in *Krittika nakshatra.* (*Vargottama* status adds strength.) 1st CONJUNCTION IN ARIES.**	**FIRE**
Jan. 26, 1841	**JU-SA conjunction at 17:15 Sagittarius, in *Purva Ashadha nakshatra.***	**FIRE**
Oct. 21, 1861	**JU-SA conjunction at 26:26 Leo, in *Purva Phalguni nakshatra.***	**FIRE**

April 18, 1881	JU-SA conjunction at 9:24 Aries, in *Krittika nakshatra.*	FIRE
Nov. 28, 1901	JU-SA conjunction at 21:30 Sagittarius, in *Purva Ashadha nakshatra.* START OF MUTATION: FIRE TO EARTH	FIRE
Sept. 9, 1921	JU-SA conjunction at 3:49 Virgo, in *Uttara Phalguni nakshatra.* 1st CONJUNCTION IN EARTH.	EARTH
Aug. 7, 1940	JU-SA conjunction at 21:25 Aries, in *Bharani nakshatra.* Triple conjunctions in 1940-1941. (Two more: Oct. 19, 1940 at 19:26 Aries, and Feb. 14, 1941 at 16:05 Aries, both in *Bharani nakshatra.* Next triple conjunctions in 1980-1981 in Virgo.)	***FIRE
Feb. 18, 1961	JU-SA conjunction at 1:52 Capricorn, in *Uttara Ashadha nakshatra.* (*Vargottama* status adds strength.) START OF EARTH PERIOD. END OF MUTATION: FIRE TO EARTH.	EARTH
Dec. 31, 1980	JU-SA conjunction at 15:54 Virgo, in *Hasta nakshatra.* Triple conjunctions in 1980-1981. (Two more: March 4, 1981 at 14:29 Virgo and July 23, 1981 at 11:20 Virgo, both in *Hasta nakshatra.* Next triple conjunctions in 257 yrs., in 2238-2239 with JU-SA in Gemini.)	***EARTH
May 28, 2000	JU-SA conjunction at 28:52 Aries, in *Krittika nakshatra.*	FIRE
Dec. 21, 2020	JU-SA conjunction at 6:20 Capricorn, in *Uttara Ashadha nakshatra.* WINTER SOLSTICE.	EARTH
Oct. 31, 2040	JU-SA conjunction at 23:40 Virgo, in *Chitra nakshatra.*	EARTH
April 7, 2060	JU-SA conjunction at 6:04 Taurus, in *Krittika nakshatra.*	EARTH
Mar. 14, 2080	JU-SA conjunction at 16:54 Capricorn, in *Shravana nakshatra.* START OF MUTATION: EARTH TO AIR.	EARTH
Sept. 18, 2100	JU-SA conjunction at 0:16 Libra, in *Chitra nakshatra.* (*Vargottama* status adds strength.) FIRST CONJUNCTION IN AIR.	AIR
July 15, 2119	JU-SA conjunction at 19:18 Taurus, in *Rohini nakshatra.*	EARTH
Jan. 14, 2140	JU-SA conjunction at 21:16 Capricorn, in *Shravana nakshatra.*	EARTH
Dec. 21, 2159	JU-SA conjunction at 11:53 Libra, in *Swati nakshatra.* WINTER SOLSTICE.	AIR
May 28, 2179	JU-SA conjunction at 26:41 Taurus, in *Mrigashira nakshatra.* LAST CONJUNCTION IN EARTH.	EARTH
April 7, 2199	JU-SA conjunction at 1:41 Aquarius in *Dhanishta nakshatra.* START OF AIR PERIOD. END OF MUTATION: EARTH TO AIR.	AIR
Oct. 31, 2219	JU-SA conjunction at 17:45 Libra in *Swati nakshatra.*	AIR

Sept. 7, 2238	**JU-SA conjunction at 9:25 Gemini, in *Ardra nakshatra*. (The first triple conjunction in 257 years, following those in 1980-1981 in Virgo. Triple conjunctions in 2238-2239. Next two: Jan. 13, 2239 at 6:14 Gemini, and March 22, 2239 at 4:28 Gemini, both in *Mrigashira nakshatra*.**	*****AIR**
Feb. 3, 2259	**JU-SA conjunction at 5:47 Aquarius, in *Dhanishta nakshatra*.**	**AIR**
Feb. 6, 2279	**JU-SA conjunction at 28:12 Libra, in *Vishakha nakshatra*. Triple conjunctions in 2279. (Two more: May 7, 2279 at 26:04 Libra and Aug. 31, 2279 at 23:21 Libra, both in *Vishakha nakshatra*.)**	*****AIR**
July 12, 2298	**JU-SA conjunction at 16:42 Gemini, in *Ardra nakshatra*.**	**AIR**
April 27, 2318	**JU-SA conjunction at 15:56 Aquarius, in *Shatabisha nakshatra*. (*Vargottama* status adds strength.) START OF MUTATION AIR TO WATER: 2318 to 2398.**	**AIR**
Dec. 2, 2338:	**JU-SA conjunction at 3:26 Scorpio, in *Anuradha nakshatra*. 1st CONJUNCTION IN WATER.**	**WATER**
May 23, 2358	**JU-SA conjunction at 23:41 Gemini, in *Punarvasu nakshatra*.**	**AIR**
Feb. 18, 2378	**JU-SA conjunction at 19:48 Aquarius, in *Shatabisha nakshatra*.**	**AIR**
Oct. 2, 2398	**JU-SA conjunction at 7:58 Scorpio, in *Anuradha nakshatra*.) START OF WATER dominance.**	**WATER**

SNAP SHOTS OF EACH TRIPLICITY & MUTATION PERIOD:

AIR TO WATER (1405 to 1544; 139 years)

Trends: Trade and commerce (AIR) are furthered and expanded, especially via water routes. Prevailing themes involve royalty, especially royal blood lines (both WATER), and prolonged conflicts over who could or should be crowned, and over what territories they should rule.

Events: New trade routes were discovered from sea explorations by intrepid and skilled sea voyagers, most notably from the Portuguese empire, followed by those from the Spanish Empire. They had now replaced the Vikings in this capacity.

WATER (1425 to 1723; 298 years)

Trends: WATER periods can bring the strength and growth of religion. Empires and monarchies are defined by their religious associations, and are closely intertwined. Even royal houses are intertwined through blood lines, but religion must also unite them. In the WATER period, there is no hesitation in subjugating other peoples and races or even destroying other cultures for the religion and/or the monarchy. The gestation process in many areas of life may take relatively longer in WATER periods.

Events: Sea voyages were undertaken for the glory of the monarchy, but also in most cases to expand Christendom or Islam.

WATER TO FIRE (1702 to 1782: 80 years)

Trends: Religion and monarchies (WATER) cling to their joint power and authority in the face of burgeoning national aspirations from groups with other ethnic, tribal or language affiliations (FIRE). But the populace too may be divided, not all of them revolutionaries, as they in turn cling to their long-time connections and obligations to royalty and to Empire. Having provided security and continuity in the past, it may feel safer than the chaos of revolution.

Events: In a previous WATER TO FIRE Mutation period, the long reign of Charlemagne came to an end, with its expansion of art, religion, and culture through the Catholic Church. His death in 814 A.D. occurred just as the **Nordic Vikings** were coming into prominence (800 to 1000 A.D.) through a combination of their ship-building, navigational skills, and ability to explore by sea where no one else dared to go. The Vikings' superiority as sailors and their fearless invasion of new territories meant that no other nation could oppose them from the sea, and often on land as well. Seeking their share of the massive wealth of the Christian Church, the Vikings were not afraid to go after it, and were known for their brutalities. They often took their captives as slaves, and had a lucrative slave trade. They were vengeful against Christian invaders who had ruthlessly forced Christian conversion on pagans, among other non-Christians. Vikings were pagans, and had their own set of Norse Gods. Viking men were required to own weapons, and their status depended on it. Tales of their plunder and atrocities, notably against Christian churches and strongholds began from 793 A.D. onward, with their first attack on British soil. Though fierce fighters, Vikings were also traders, colonists and mercenaries in the Byzantine Empire, profiting from international trade in many locations. Also in the 9th century A.D., the Chinese invented gunpowder. This occurred as Chinese alchemists searched for an elixir for immortality. It is the only known chemical explosive until the 19th century, coinciding with the FIRE period.

FIRE (1723 to 1921; 198 years)

Trends: While also religious, FIRE influence promotes more overtly nationalistic aspirations. This period is very patriotic, and lights the fires of national pride, pitting nationalistic interests and aspirations against each other, and demanding personal freedoms – especially from religious persecution – and equality among people, at least in theory, if not finally by law. New territories and trading realms are conquered by way of militarism. FIRE is associated with *Kshatriyas,* the Warriors and political leaders in Hindu/Vedic culture, who also had royal status. Thus, the fire signs have royal associations, but these long-term ties can finally be broken if royalty is experienced as tyrannical.

Events: Many nation-states achieved independence from colonial masters, especially in the 1800s, and notably in South America. Fire is associated with swift destruction. World War I (1914-1918) was highly destructive, with more than nine million combatants killed. Then In 1918-1919 – the last few years prior to the first EARTH conjunction (Virgo) in 1921 – a major pandemic swept through the world, attacking the immune systems of healthy young adults. Called the **Spanish flu,** it was estimated to have killed up to fifty million people worldwide, and was perhaps worse than the Black Death in its devastation. It moved rapidly, killing its victims sometimes within hours. Its origin was never known, but it may have started in China.

FIRE TO EARTH (1901 to 1961; 60 years)

Trends: The drive for personal freedoms and equal rights of all citizens (FIRE), including eventually those of slaves, minorities, and women are soon replaced by the drive for material security and profits (EARTH). Even if greater security is promised, at times the results are capable of inflicting enormous damage to the individual and to the environment. Territories and trading realms continue to be conquered by militarism. Weapons technologies advance, notably after 1942. Greater protection for the wealthiest members of society, especially corporations, tends to gain more traction and to replace the inspirational fire of justice and the rule of law (FIRE).

Events: With a greater chance to achieve national sovereignty during the FIRE dominant period, and up to Feb. 1961, some nations were still smoldering in 1961 under the yoke of colonialism. Among them, Vietnam became tragically involved in the crossfire of new political realities in which Superpowers fought for control over weak nations. This Mutation period saw the greatest rise of the lower and middle classes in U.S. history, as education and income opportunities expanded, in part through financial support from government for higher education for WWII veterans. As of 1970 this trend started to reverse, along with the level of educational achievement and/or job opportunities for graduates. Also by 1970, effective public transportation in most American cities was eliminated or dramatically reduced, and the automobile and trucking industries were already in dominance. The American example was copied in many countries, especially as American movies and culture spread globally from 1921 onward.

Though two World Wars marked this period, causing enormous damage and loss of life in Europe and Asia, the Second World War lifted the U.S. out of the economic Depression of the 1930s and by 1945 bestowed on it unquestioned global superpower status. The first controlled nuclear reaction occurred Dec. 2, 1942 in Chicago, and the U.S. bombed Japan in summer 1945, revealing to the world the first human use of nuclear energy. Atomic energy changed world political realities forever, paving the way for nuclear power and its attendant possibilities and dangers. Advertised as a means to security, **nuclear energy and nuclear weapons** provide one of the deepest ironies of the EARTH period, since – for instance – no one can insure nuclear power companies against catastrophic accident and the dangers of concentrations of radioactivity. Thus, there IS no insurance anywhere for it.

EARTH (1921 to 2199; 278 years)

Trends: EARTH promotes technology and its trade and exchange, providing solutions to problems of many kinds, especially at the crisis points coming at the half-way mark in each 20-year cycle, when even accidents or acts of nature can force needed reforms. Though technology thrives as never before, it is increasingly used for military purposes. Agriculture and mineral resources are the focus both for survival and as mere businesses, at times pitting profit motive (or state control) against ethical concerns for human safety and well-being. There is danger to the food supply from inferior grains (Saturn), and businesses profiting from their distribution. Slavery or slave labor during the EARTH dominant period is primarily driven by the profit motive.

Especially after 1961, nationalistic and religious themes subside, as more focus goes to secular matters such as business concerns and economic stability, with aims to improve living standards around the world, though at times achieving the opposite result. Material and/or financial security (EARTH) is sought, sometimes heedless of the results. Material advantage often goes first to those already possessing wealth and influence, while middle and lower classes may take too many risks attempting to attain greater financial security.

The **JU-SA conjunction in Capricorn in Feb. 1961** favors business to such a degree that corporate conglomerates increasingly dominate over nation-states, especially in the arena of weapons and pharmaceuticals, but also oil and nuclear industries – all of which often own telecommunications companies as well.[11] By owning the latter, they can control the message. Pro-business is equated increasingly with patriotism, which helps the military industries. The EARTH period also helps oligarchies and/or dictators to thrive, especially in oil-rich nations. (Saturn dominates more than ever in Capricorn and rules over fossil fuels.) Many enterprises are turned into mega-businesses, such as sports, entertainment and the arts. Formerly fringe businesses such as pornography and prostitution become larger via the Internet from 1990 and benefit from less regulation of almost all corporations.

Events: As of the early 1960s mass production was so successful that it was finally possible to have a mass-produced uniform culture. Debilitated Jupiter can lower standards of ethics, knowledge, and education while Saturn emphasizes business. Many arenas of life shifted gears: A new economic model came into being by the early 1980s in which the nation-state (Jupiter) was no longer the center of everything. Though it had previously assumed an important role in developing its own economic strength and helping its own citizens, nation-states became increasingly vulnerable to the rise of corporate power (Saturn), notably those where special interests could dominate legislatures. After World War II the escalating crisis of debtor nations set up a new susceptibility to the International Monetary Fund (IMF) and the World Bank – organizations established in late 1944 by the Bretton Woods agreement to create greater economic stability worldwide. Joined by foreign investors, mainly transnational companies, by the early 1980s they were direct agents of **globalization** – a term first used in the early 1960s by economists and social scientists, and reaching regular use in the mainstream media by the late 1980s. Its pros and cons continued to be debated, but as IMF and World Bank policies evolved they often resulted in destabilizing many of the debtor nations receiving assistance, requiring them to hand over sovereignty de facto when they were unable to meet payments. The IMF and World Bank then began privatizing and deregulating state assets, such as utilities and airlines. They also cut state spending on education, healthcare and infrastructure, resulting in a massive rise in poverty and widening economic and social inequality. These effects were not confined to developing nations. With their overwhelming legal, political and financial strength, corporate conglomerates began to control political elites around the world.[12]

In the EARTH-dominant era serious investigative journalism (Jupiter) also began to suffer. Called "muckraking" in the pre-WWI era, the groundwork for this was laid in the 1920s with

---

[11]See my article: "Corporate Conglomerates vs. Nation-States: Which Nations will Survive and Thrive?" (Oct. 2010). It features the Feb. 1961 JU-SA conjunction in Capricorn, and how it works as a seminal global influence ever since that time: Published in ***The World Tomorrow: What They Say*** (A collection of Articles presented in the World Conference on Mundane Astrology [Oct. 2010]), Sanathana Dharma Charitable Trust, 2010, pp. 30-58. Also available at: **www.edithhathaway.com.**

[12] On Jan. 21, 2010 the "Citizens United" U.S. Supreme Court ruling overturned a century of laws and practices that had limited corporate financing for political campaigns, and now enabled national or transnational corporations to influence U.S. elections. The decision says that corporations have the rights of 'personhood' but are exempted from the responsibilities of citizenship.

the burgeoning new Public Relations industry pioneered by Edward L. Bernays (b. Nov. 22, 1891, d. March 9, 1995). Bernays believed the manipulation of public opinion was necessary, especially in a democracy. He created "false realities" that became "news events," a model often copied in the EARTH era.[13] Journalists' coverage of news and politics became increasingly like the sports and entertainment mega-businesses, a trend more notable from the 1980s onward and marked by the triple JU-SA conjunctions in Virgo (1980-1981). With more advertising and public relations consultants protecting corporate interests and less vigilance from journalists investigating fraudulent practices (in part due to fewer of them being hired), more fake testing prevailed in numerous areas, including military products, food and drugs. Militarism continued, especially as a lucrative enterprise, often prolonging wars and/or starting them on false premises. A new and more ruthless capitalist model emulated from auto industrialist Henry Ford. It held that the competition is the enemy who should be crushed. For this tactic to succeed, regulations to limit corporate power were reduced as much as possible, while also reducing the power of the workers to bargain.

Coincidentally, the first non-family member to be appointed President of Ford Company was Robert S. McNamara, on Nov. 9, 1960, the day after John F. Kennedy was elected U.S. President. Five weeks later, McNamara accepted Kennedy's offer to become his new Secretary of Defense, a position he held from Jan. 21, 1961 through Feb. 29, 1968, presiding over the U.S. entry and escalation in the Vietnam War.[14] He was President of the World Bank from April 1, 1968 up to June 30, 1981.

As the EARTH period evolved, the fire of justice was more and more just lip service, and by the late 1960s the substance of the progressive or "liberal" message was largely exhausted. Increasing numbers of anti-war protests could not seem to end the Vietnam War, which dragged on until 1975. Two powerful opponents of that war were both murdered in 1968 – Dr. Martin Luther King, Jr. and Robert Kennedy, who was campaigning for the U.S. Presidency. By 2003, McNamara admitted the Vietnam War was wrong, and that the U.S. mistakenly saw a Communist threat instead of a civil war in which the U.S. should have not intervened.

There were some major gains for minorities (especially African-Americans) and for women, with the passage in the U.S. of the **Equal Pay Act of 1963** and the **Civil Rights Act of 1964,** requiring equal pay for women and men and prohibiting discrimination against women by any company with more than twenty-four employees. But within thirty years some major companies violated these laws and were hiring cheaper labor in foreign countries. Citizens once protected from deceptive banking practices by the U.S. Banking Act of 1933 ("the Glass-Steagall Act") found those economic safeguards disappearing with deregulation from 1993 to 2009. Glass Steagall was finally repealed Nov. 12, 1999, paving the way for predatory lending practices that continued to protect big banks and assure them of government bailouts (from 1999 onward), making them richer while defrauding the public. International banking practices were affected but never fully corrected – even after the major banking crisis in fall 2008, coupled with the largest bank bailout of all time. This crisis arrived a full Saturn cycle (28-29 years) after the start of the new economic model, which can also be dated from the triple JU-SA conjunctions in Virgo in 1980-1981.

[13] Edward L. Bernays worked on behalf of governments and corporations to promote their products, campaigns, or wars (e.g., U.S. entry into WWI). He opposed "muckrakers" (reform-oriented journalists) as unpatriotic, and his work contributed greatly towards eliminating them after WWI. He re-branded the slaughter of the First World War as "necessary and noble." His book ***Propaganda*** was published in 1928.

[14] McNamara was a Bomb Damage efficiency expert during WWII and a top advisor to Air Force Colonel Curtis Le May, commander of B-24 and B-29 bombers. For McNamara's statements on Vietnam, see **Youtube.com**.

From 1961, the number of billionaires increased, while the income gap increased between the wealthiest 1% and the rest of the U.S. population. Between 1961 and 2011, a 60% reduction in taxes on U.S. corporations and the wealthiest sectors of citizens was supposed to lead to re-investing in business (the "trickle-down" theory), but in reality led to the collapse of the middle and lower classes and loss of workers' power to unionize and engage in collective bargaining. This situation was challenged in 2010-2011, when worker and student protests exploded around the world. However, the larger trend still favored the corporations.

The increase of financial inequality since 1961 is a worldwide phenomenon, and in turn creates social and health problems. With the widest income gap of all nations in 2011, the U.S. has more exaggerated social and health issues. In 2011, the top 1% of the U.S. population held over 35.6% of the private wealth. Globally, the percentages were even more extreme.[15]

In spite of great gains for civil rights in the mid-1960s in the U.S., this focus was gradually replaced by more materialistic concerns from 1961 onwards. The Conservative movement in the U.S. suffered many humiliating defeats up through 1964, and then started gaining ground through building organizations and think tanks, and educating lobbyists to spread their influence very broadly. Amply funded by wealthy advocates seeking more unfettered power for U.S. corporations, their aim was to shield corporations from regulation and taxes. This resulted in vastly increased momentum for the Conservative movement in the U.S. from 1970 onward, firmly entrenched by the effects of the triple JU-SA conjunctions of 1980-1981, especially for the next 28 years. Dangers from products polluting humans, soil, air, and/or water were sometimes caught, but often minimized and their critics labeled as "extremists" or "alarmists." Environmentalists have often been targeted in this way, and struggled to build their credibility in a world in which facts are not always the best defense. When scientific results did not look good for company profits, fabrication of information has been common.

This is especially troubling in the arena of **climate change,** which was known about since at least 1988 when the United Nations established the **IPCC (Intergovernmental Panel on Climate Change)**. The technological capacity to implement alternative energy beyond either nuclear or fossil fuels is clearly possible in the EARTH period, but resistance to climate reality has been strong in the first fifty years of the EARTH period (since 1961). Saturn's strength in the EARTH period allows fossil fuels to persist alongside corporate giants, whose steamroller effect impacts all sectors of life. Thus, it may take an event of catastrophic proportions to convince corporate polluters to reassess their position and consider wiser use of fossil fuels as well as alternative technologies.

Predictions: The **EARTH conjunctions in Capricorn** continue to emphasize Saturn as king of business, with Jupiter marginalized. When *Vargottama* (first 3:20 of Capricorn, Aries, Cancer, or Libra; see Glossary), the emphasis is confirmed, as in 1961. In 2020, the JU-SA conjunction is again in *Uttara Ashadha*, "the later victor," giving even more power to the corporate conglomerates, gained increasingly from Feb. 1961 onward. It has further weight by occurring at the Winter Solstice. The JU-SA conjunction in 2080 in *Shravana nakshatra* is about "good listening," and may soften the corporate steamroller and bestow greater responsiveness on the corporate elite towards the people and their real needs.

---

[15] See **http://inequality.org/wealth-inequality/**Statistics from the U.S. Institute for Policy Studies, Washington, D.C.

The **EARTH conjunctions in Virgo** favor 20-year periods of growth for information technologies, notably in *Hasta nakshatra* in 1980-1981. When in *Chitra nakshatra*, ruled by Mars, as in 2040, the emphasis is on information technology for more militaristic uses. (This occurs again in 2100, when the JU-SA conjunction is in *Chitra nakshatra*, though at 0:16 Libra, an AIR sign, and part of the EARTH TO AIR Mutation period.)

With **EARTH conjunctions in Taurus**, especially *Rohini nakshatra*, as in 2119, the next 20-year period is less favored economically, as Taurus is an area of the zodiac associated with prosperity and growth, and Saturn tends to contract growth when it transits there. This is also true when Saturn is transiting through *Mrigashira nakshatra*, as in 2179.

EARTH TO AIR (2080 to 2199)

Trends: Practical concerns are such that private citizens are less willing to submit to corporate control over their lives. There is a desire for a broader cultural exchange between people without the impediment of government, corporate or religious surveillance or influence. Individuals and groups want the use of advanced technology without being limited in the free communication of ideas at every level of inquiry, whether personal, social, political, literary, or scientific. This also relates to travel and interest in the exchange of ideas. During this transition, there is opportunity for greater perspective on just how far technology should be involved in the lives of human beings and the environment, especially if it is not an enhancement.

Scientists from the **IPCC (Intergovernmental Panel on Climate Change)** have predicted a sea level rise of ca. half a meter (19.69 inches) and up to 88cm. (2.89 feet) between 2000 and 2100. Their estimates are based on studies of the rate of the warming of the oceans, the melting of mountain glaciers and the combined melt of Greenland and Antarctica. IPCC projects air temperature rises of from 3 to 7°F (2 to 4°C) over the 21st century – a more radical shift than has occurred in over 10,000 years. IPCC also predicts that the outcome of climate change may happen suddenly over several decades. If so, a turning point may well be at the first JU-SA conjunction in AIR on Sept. 18, 2100 at 0:16 Libra, in *Chitra nakshatra,* when environmental crises previously denied or ignored will be addressed more objectively.

Events: In a previous EARTH TO AIR period, Marco Polo (1254? -1324) travelled from Venice to central Asia and China, the first European to do so. He was both a merchant and an explorer, well educated, spoke four languages, and met the Kublai Khan, the great Khan of the Mogul Empire, 1260 to 1294, and founder of the Yuan Dynasty in East Asia (1271). The Kublai Khan had never met Europeans, and had many questions about the Church of Rome and the European political and legal systems. He requested Marco Polo deliver a message to the Pope, requesting to meet 100 Christians who were familiar with "the Seven Arts" (grammar, rhetoric, logic, geometry, arithmetic, music and astronomy).

AIR (2199 to 2398)

Trends: During the AIR period, trade and commerce are featured, especially the desire to establish communications and cultural exchanges across new boundaries. Language, literature, the arts, and the exchange of scientific ideas and rational thought can flourish. As long as dogmatism remains minimal, cultural reciprocity across territorial boundaries diminishes political, ethnic, tribal or religious barriers. With the intense desire to connect, learn and cross-pollinate with other cultures, rather than to destroy or subjugate them, the AIR period and its Mutation periods can spread cultures better and faster than most others. All results are dependent on the level of *Dharma* (sense of duty and integrity) of the people. Though *Kali Yuga* still reigns, enlightened individuals can still have an influence. The negatives during an AIR dominant period involve the momentary orientation generated by AIR signs. Thoughts, actions, and enterprises can lack depth if done hastily and without discriminating all the many layers of the subject at hand.

AIR to WATER MUTATION PERIOD: Jan. 1405 to Sept. 1544.	**139 years**
WATER: Dominant from Feb. 1425 to Jan. 1723.	**298 years**
WATER to FIRE MUTATION PERIOD: May 1702 to Nov. 1782.	**60 years**
FIRE: Dominant from Jan. 5, 1723 to Sept. 9, 1921.	**198 years**
FIRE TO EARTH MUTATION PERIOD: Nov. 1901 to Feb. 1961.	**60 years**
EARTH: Dominant from Sept. 9, 1921 to April 7, 2199.	**278 years**
(Previous EARTH dominant period ran from 1007 to 1305 A.D., 298 years.)	
EARTH TO AIR MUTATION PERIOD: March 2080 to April 2199.	**119 years**
AIR: Dominant from April 7, 2199 to Oct. 2398.	**199 years**
AIR to WATER MUTATION PERIOD: 2318 to 2398.	**80 years**

NOTE: the extra-long (139-year) Mutation period from AIR to WATER, from 1405 to 1544 A.D. Triple conjunctions occurred in 1425 (two of them in Scorpio, the last one in Libra). Sixty years elapsed between the first JU-SA conjunction in WATER in 1425 and the next one in 1484 (also Scorpio); then another 60 years between 1484 and the next JU-SA in WATER (again Scorpio) in 1544, starting a solid sequence of JU-SA conjunctions in WATER from 1544.

AIR TO WATER Mutation period (1405 to 1544)

WATER Dominant period (1425 to 1723)

Since WATER is by far the most expansive of the four elements, we note the expansion of trade specifically due to explorations by sea voyage. This in turn brought the expansion of the empires of Great Britain, Spain, France, and Portugal, led by Portugal, which between 1415 and 1542 was among the greatest seafaring nations of the era. Portugal's empire was enhanced by its discoveries of new trading routes and by establishing the first direct maritime trade with China and Japan. After Vasco da Gama's discovery of a sea route to India in 1499, Portugal held territories in India from 1505 up through 1752 and established headquarters in Goa in 1510. Portugal's claims to Goa extended through Dec. 1961.

Royalty is closely connected with religion all through the WATER dominant period. With issues of royal and religious authority at a premium, royal blood lines sometimes needed Papal Bulls to establish legitimacy, as in the case of **Isabella and Ferdinand of Spain**, who married in 1469. (Their grandfathers were brothers.) Following the Wars of the Roses, the 1486 marriage of **Henry Tudor and** his queen, **Elizabeth of York** restored stability to the British throne. The Papacy played a pivotal role in both cases.

The **Hundred Years' War** (1337–1453) was a series of conflicts fought between England and France in a dynastic struggle to establish which royal line would have control of their countries, as well as other surrounding territories. The House of Plantaganet, backed by the Holy Roman Empire, claimed to rule both England and France, which the House of Valois contested. Fought mainly in France and in the Low Countries (Belgium, Netherlands, etc.), the Plantaganets lost, and with it – their claim to Normandy. Soon after came the **Wars of the Roses** (1455–1485), a series of dynastic conflicts fought in England to determine which royal house would rule over England, the House of York (the white rose) or the House of Lancaster (the red rose).

The expansion of the **Spanish Empire** into the New World in 1492 was also an expansion of Christendom, which had just overcome the last of Moorish (Muslim) rule over the Iberian Peninsula and in Europe, since 711 A.D. Having united Spain through their marriage in 1469, Queen Isabella of Castile and King Ferdinand of Aragon were known as Catholic Monarchs. They were successful in greatly expanding the Spanish Empire and the cause of the Catholic Church, especially after its major loss of Constantinople (its Eastern capital) to the Ottomans in 1453. In 1492, Queen Isabella finally succumbed to the funding of Columbus's first sea voyage (after two years of his petitioning), and this became one of her triumphs. She found ways of truly conquering for Spain lands already occupied by Indian natives with other religions and loyalties that had to be banished. During her reign (1474-1504), she sent 250,000 young and single Spanish men to mate and propagate with the indigenous peoples in Mesoamerica.

Prior to this, in 1478, Isabella and Ferdinand requested powers from the Papacy called "**the Spanish Inquisition**." Infamous for its many excesses, it was under the direct control of the Spanish monarchs, who sought to police false or relapsed converts. Starting in 1492, many Jews and Muslims had raced to convert to Christianity to avoid being expelled from Spain. The main mission of the Spanish Inquisition was to combat heresy, and it was not finally abolished until 1834. By the mid-1700s, the start of the FIRE period and the Age of Enlightenment, it became more difficult to enforce the Inquisition and to hold back the spread of printed matter, which came by the end of the 1700s. The aspect of religion associated with WATER is the overwhelming sense that the world is emotionally connected through shared beliefs, and that those who do not share those beliefs need to be coerced or sacrificed to this larger cause. This deep sense, once learned or taught, is not easily eliminated.

Some of the same kind of excesses occurred in England in the mid-1600s, when a Catholic monarch (King James II), sought to bring Absolute power and Catholicism to the British throne. This led to **the Glorious Revolution in 1688,** which installed a Dutch Protestant prince (William III) on the British throne, prevented any future British monarch from having Absolute power, and gave it instead to Parliament. The events of 1688 insured that Catholicism would not be a part of the Church of England or its monarchy, thus removing any control by the Pope over England. Royal bloodlines often crossed through Europe, and were useful for religious or territorial reasons.

Along with the flourishing expansion of trade and cross-pollination of cultures and the new-found mobility came some new negatives, including **slavery** and **the spread of diseases.** The dawn of the 15th century saw Europe recovering from the loss of up to 60% of its population due to the bubonic plague, or Black Death, during the 1300s. It is believed to have started in China with oriental rat fleas carried to Europe on merchant vessels. The Black Death returned to Europe in waves: London (1603), Italy (1629-1631), Seville, Spain (1647-1652), London (1665-1666), and Vienna (1679). These pandemics were not confined to the WATER period, however. Eastern Europe and Russia were felled by plagues in the 18th century, and more plagues affected Europe through to the early 20th century (both FIRE periods), with the Spanish Flu pandemic of 1918-1919. The Catholic Church was especially hard hit by the Black Death of the 14th century, and this may have contributed to their zeal to conquer new territories. It also led to widespread persecution of minorities such as the Jews, foreigners, beggars, and lepers, whom they perceived could have spread the disease.

The European invasion and occupation of the Americas caused Native American populations to die in large numbers due to lack of immunity to diseases such as smallpox, influenza, bubonic plague and pneumonia. New sea explorations also led to the expansion of the slave trade, initially justified on the grounds that the conquerors were converting their captives to Christianity. Fanatic religiosity led to their "civilizing" those natives, while often brutalizing them or treating them as slaves. This process probably went more unchecked in the WATER dominant period.

Though there is no specifically Vedic reference to Portugal and Pisces, early Western astrologers assigned this watery zodiacal sign to describe Portugal, the early leading seafaring nation of the WATER period. Coincidentally, **the slave trade** began with the Portuguese Empire. In 1441, the Portuguese explorer, Antão Gonçalves, was the first European to purchase Africans as slaves from black slave traders, having gone to the continent to purchase gold and spices. He was sent by the Portuguese prince known as "Henry the Navigator (1394-1460), an important figure in the early development of European exploration and maritime trade with other continents. Henry the Navigator supplied maps and navigation instruments of the highest quality, and they in turn came from the Order of Christ, founded in 1318, also known as the Military Order of Christ, or the Knights Templar in Portugal.

When Portuguese subjects were enslaved by pirate attacks on Portuguese ships or villages, the Portuguese fought back, and started taking African slaves. In addition to martial prowess, the desire for naval superiority catalyzed the Portuguese to make War on Islam, and they saw this as their Christian duty. The Moors had dominated sections of Europe up through 1492, when Spain became victorious. Under Spain, the African slave trade began in 1517 in Haiti, later becoming the richest of France's colonies. Great Britain joined the lucrative Slave trade in the New World in 1562.

In the 1500s and 1600s, the ancient Mayan civilization in Mesoamerica was almost entirely wiped out by the Spanish conquerors, whose policy of mass destruction and forceful coercion to the Christian religion left only a few surviving records of the Mayans. The second bishop of the Yucatán ordered a mass destruction of Mayan books in 1562. Only three survived. Some of the Mayan downfall has been attributed to civil war, but the Spanish conquerors also brought infectious diseases, and possessed more sophisticated weapons. The Mayans finally succumbed in the last battle in 1697.

The technology of the **Steam Engine** arrived during the WATER period. The first steam engine came in 1679, and was based on the pressure cooker. Its first successful commercial version arrived in 1712, but James Watts' improved version (1763-1775) further accelerated the Industrial Revolution, as it could operate with 75% less coal than the previous model, and was much cheaper to run. (This invention came during the WATER to FIRE mutation period, 1702-1783.)

Culturally, the **visual arts and architecture** in Europe and elsewhere featured biblical or sacred themes. Many cathedrals and mosques were built during the WATER period, though not confined to this period. Built in Agra, India from 1632 to 1653, the Taj Mahal epitomized the Muslim architectural style for tombs, prevalent during the Mogul dynasty in India, 1526 through 1858, but in rapid decline after 1725.

The Bible was **the first printed book** in the Western world reproduced on a movable type press, around 1455. Though similar printing existed in China (from 1040 A.D.) and in Korea (ca. 1230

A.D.) during a previous EARTH dominant period 1007 to1305 A.D, it was unknown in Europe up to the time of **Johannes Gutenberg.**[16] Gutenberg was a German goldsmith credited with this invention ca. 1440, using an alloy of lead, tin, and antimony. He also printed the first Bible. (This followed the JU-SA conjunction in Gemini, promoting education, as ruled by Mercury, but within the AIR TO WATER Mutation period.)

Printing presses were very slowly available over the next few centuries, and printed matter featured mainly Bibles and religious treatises. The first printing press reached Asia (India) in 1556 via the Portuguese, who intended to print brochures to promote Roman Catholicism. The first printing press reached the American colonies in 1638. So strong was the hold of organized religion over all areas of life, the spread of scientific ideas often suffered the opposition of the Church if they countered Christian theology. Astronomer/astrologer **Galileo Galilei** (1564-1642) was forced to recant his views on the planets orbiting the Sun, and he barely escaped the wrath of the Inquisition.

The WATER period also saw the growth of sacred music, especially Christian music, as most of the music was commissioned for Church events or settings, as were the visual arts and architecture. **Johann Sebastian Bach (1685-1750)** dedicated his music to the glory of his Christian God, as did many others of his period, especially in Europe. His extraordinarily prolific outpouring of music occurred during the WATER TO FIRE Mutation period. **Wolfgang Amadeus Mozart (1756-1791)** lived in the latter part of this transitional period. In 1777, he left his post after six years as Concertmaster to the Archbishop of Salzburg, as he found too many limitations. His prodigious output of music spanned almost every genre, but was more often played at court and commissioned for secular occasions.

WATER to FIRE Mutation period (1702 to 1782)

FIRE dominant period (1723 to 1921)

Ludwig van Beethoven (1770-1827) was a key figure in the so-called Age of Enlightenment, also called the **Age of Reason**, which spanned across artistic, cultural, and national borders. It proposed that reason and logic should dominate as the primary source of legitimacy and authority. There is disagreement on when the Age of Reason began, but it is often assigned to the early or mid-1700s, thus coinciding with the WATER to FIRE Mutation period. With this period we see the yearning for a rationale not totally dominated by religious beliefs. Meanwhile, the FIRE dominant period, in full gear by 1783, would witness the Romantic period, known for its sweeping emotional displays in music, art and literature, especially including nationalistic aspirations. But in social terms the Victorian culture typically did not encourage such open displays. This gives us a small glimpse into the dichotomies between art and life.

More nationally oriented music came in this FIRE period with the great Russian and German composers, among them **Peter Tchaikovsky (1840-1893)** and **Richard Wagner (1813-1883),**

---

[16] China is credited with the inventions of paper, printing, compass and gunpowder. During the Song Dynasty (960-1270 A.D.), scientist and statesman Shen Kuo wrote that ceramic moveable type printing was invented by Bi Sheng, a common artisan (990-1051 A.D.). This was during a prior EARTH period, 1007 to1305 A.D. Wooden and metal moveable type was invented in China in the early 1300s and mid-late 1400s, respectively. Also invented in 11th century China was the partial de-carbonization method for iron production. It preceded the Bessemer steel process of 1855, enabling mass production of steel.

respectively, whose music was very evocative of national pride. Tchaikovsky was under tremendous pressure to produce a specifically Russian kind of art music. He was downgraded and severely criticized if he was overly imitative of his European counterparts. In Bohemia, Wagner and Franz Liszt were strong influences on **Bedrich Smetana (1824-1884)**, a Czech composer based in Prague, who was pressured to be distinctively Czech in his music. The country was then known as Bohemia, part of the Habsburg Empire, also known as the Austrian Empire (1804-1867), then Austria-Hungary (1867-1918). Smetana's musical style and output became closely identified with his country's aspirations for independent statehood, through his operas, and especially his symphonic cycle "Má Vlast" ("My Fatherland"), most famously the 12 minute segment called "Vitava." Better known by its German name "Die Moldau," it evoked the Moldau River, which flowed toward Prague. The leading melody later formed the basis for the Israeli national anthem, "Hatikvah."

Frederick Chopin (1810-1849) was born of French and Polish parents, and his music is very evocative of Polish tradition. He composed and performed many polonaises and mazurkas reflecting Polish nationalism. The Polish Uprising in Nov. 1830 against the Russian suppression was followed by a major flood of emigration of political elites out of Poland, between 1831 and 1870, many of them to France. Having grown up in Warsaw, and trained at the Warsaw Conservatory, Chopin settled in Paris in 1831. Though devoted to Polish culture and artists, he never returned to his homeland. His last concert was a benefit performance for Polish refugees; and at his Paris funeral, Polish soil was spread on his grave.

The quest for national aspiration was more successful elsewhere in the FIRE period, especially in the early 1800s. In the Americas alone, **16 nations** in Central and South America gained their **first sovereignty** between 1810 and 1825, most of them from the Spanish Empire, though Brazil was a territory of Portugal, and Mexico belonged for a time to France. Canada became a Dominion of the British Empire, gaining its first sovereignty in 1867, and fully so by 1931. Australia became a Dominion in 1907, with full sovereignty in 1942, and India and Pakistan gained independence in 1947.

FIRE describes both **royalty and patriotism.** FIRE dominant periods are marked not only by nationalistic uprisings, but by fierce loyalty to monarchs. Thus, huge colonial empires headed by monarchs still flourished during this time. The dates of the FIRE dominant period almost exactly coincide with **the Russian Empire**, which began in 1721 when Tsar Peter I was declared Emperor of All Russia, following the Treaty of Nystad (1721), and ended with the Russian Revolution in Nov. 1917. In the early 1800s, Russia was the largest country in the world, with the third largest population, after China and the British Empire.

Such loyalty to monarchs also helps to explain why the **American Revolutionary War** was also a virtual Civil War within the American colonies, though that has been much less known. Loyalists comprised around half a million men, women, and children, or 20% of the population at that time.[17] Those loyal to the British crown and the British Empire called themselves "Loyalists."

---

[17] Source: Maya Jasanoff, ***Liberty's Exiles: American Loyalists in the Revolutionary World***, 2011. Jasanoff confirms in historical terms the astrological strength of the 1780s, with the JU-SA conjunction in 1782 at the Galactic Center. Jasanoff says: "The 1780s stand out as the most eventful single decade in British imperial history up to the 1940s." She considers the "spirit of 1783" had more worldwide significance than the "spirit of 1776." She also gives enormous credit to the British for their handling of Loyalists, including repatriation of up to 100,000 Loyalists, who emigrated out of the former American colonies after the war.

Their enemies called them "Tories," even if they shared many of the same attitudes about popular representation in government, land distribution, and racial animosities. Loyalists opposed the break from Great Britain in 1776, and were often terrorized by the American revolutionaries, whom they fought against, joined by many African slaves (who were promised their freedom) and also by some Native American Indians. Some 20,000 black slaves left their masters and fought for the British. Among them were about 24 slaves owned by Thomas Jefferson. There were 500,000 slaves in the 13 American colonies in 1775.

The **American Revolutionary War** was fought between 1775 and 1783, and was not formally ended until the Paris Peace Treaty, signed Sept. 3, 1783. The JU-SA conjunction preceding the revolution was in Pisces in 1762, but a FIRE conjunction was necessary to bring the new nation to fruition, and to wrest these 13 colonies completely from their deepest royal bond (WATER). We can see the WATER influence right up through 1782, and the FIRE necessary for complete independence, which came with the 1782 conjunction in Sagittarius, strongly on the Galactic Center and in the sign that conjoins religion AND patriotism. (Many who immigrated to America sought freedom from religious persecution in other parts of the world.) In addition, there would be 139 years of uninterrupted JU-SA conjunctions in FIRE from 1782 through 1921. It would be the signal for the real start of the new nation. **George Washington** was inaugurated as first U.S. President on April 30, 1789 (through March 4, 1797). A few months later came the **French Revolution of 1789**, which is still regarded as the first truly modern revolution, as the people overthrew the reigning monarch after centuries of absolute monarchy. After ten years of struggle and conflict, they replaced it with a republic guaranteeing "inalienable rights" to its citizens and establishing the basis of a liberal democracy in the modern era. King Louis XVI was executed in 1793.

Though economics and social class structure were closely involved, **slavery and slave labor** in the FIRE period were also driven by the interests of the nation or Empire. In North America, Southern plantations thrived specifically due to the black African slaves. New Orleans had the nation's wealthiest plantations, and **the largest slave revolt in the U.S.** occurred there as well in **1811.**

George Washington (1732-1799) and **Thomas Jefferson** (1743-1826), were both from Virginia and both owned slaves, Jefferson over 1000 of them. These are little known facts in U.S. history as it is usually recounted. Each of them fought for independence from tyranny, and was a Founding Father and President of a newly liberated nation. But in the drive for equal rights and freedom for all, contemporary beliefs did not tend to include African slaves, minorities such as American Native Indians, or women. Slavery was not outlawed in the U.S. until 1865, at the end of the Civil War, with the passage of the **13th Amendment** to the Constitution.

The **U.S. Declaration of Independence** was written by free men, and mostly by Thomas Jefferson. It was intended to promote and protect "**certain unalienable Rights**,...among these are Life, Liberty, and the pursuit of Happiness." An earlier draft, written and adopted by the Virginia State Convention in June 1776, included this phrase: "the enjoyment of life and liberty, with the means of acquiring and possessing property, and pursuing and obtaining happiness and safety." Thomas Jefferson and Benjamin Franklin deleted the phrase on property ownership, creating the famous document that remains. But still it gave no rights to slaves or to women. A single or unmarried woman had no legal rights, and even a married woman had few rights at that time. She had no legal responsibility for her children and was unable to control her own property, own slaves, buy or sell land, or even obtain an ordinary license. She had no legal existence separate from her husband. This did not change radically until 1920, when women were able to vote for the first time in the U.S. Thus, even with greater public attention to human rights during FIRE periods, it would take the coming of the EARTH period in 1921 to make manifest some 144 years of agitation by American women.

Meanwhile, British women made gains some four decades sooner (1869) – a credit to their liberal constitutional empire that stood up well to the democratic republics being formed during the 19th century. There were a few educated American colonial women. Among them was Abigail Adams (1744-1818), wife of John Adams (2nd U.S. President, 1797-1801). She tried to influence the language of the Declaration of Independence and other early political documents, but she was overruled. (John and Abigail Adams of Massachusetts both believed slavery was evil, and a threat to the fledgling democracy.) Discrimination against women and minorities did not change or improve much in the U.S., even with the **14th Amendment**, adopted in July 1868, protecting citizens' civil and political rights, giving former slaves the rights and privileges of U.S. citizenship, and overruling the **Dred Scott decision** (1857), ruling that blacks could not be U.S. citizens.

The ***Plessy v. Ferguson*** **case in 1896** effectively institutionalized segregation in the U.S. Meanwhile, corporations were making great progress legally and financially, starting with the important legal precedent of ***Santa Clara County v. Southern Pacific Railroad Company***, May 10, 1886, which held that corporations are entitled to protection under the 14th Amendment to the Constitution, and could be considered "persons." Women and former slaves did not yet have that privilege, as it required high legal expenses to achieve. Not until the **Civil Rights Acts and Voting Rights Acts of 1964 and 1965** would these goals be reached, largely thanks to the efforts of Martin Luther King, Jr. He galvanized the Civil Rights movements from Dec. 1955 onward, inspired in turn by the rhetoric of U.S. President John F. Kennedy, 1961-1963.

The Women's Suffrage Movement was quiet in the U.S. until after the Civil War ended in 1865, and then became more vocal regarding married women's rights to own property, and the right of all women to vote. Several Western states gave voting rights to women from 1910 onward, and before that some Western states allowed women to hold public office. But it was not until **Aug. 26, 1920** that a Constitutional Amendment was passed granting **full woman suffrage across all states.** Tennessee was the last to ratify it. In the previous two decades, women's right to vote had been achieved in numerous European countries and parts of the Commonwealth (Australia and New Zealand), the process accelerating with WWI and the Russian Revolution in Nov. 1917. The earliest women's suffrage was in England in **1869**, in which unmarried women householders were allowed to vote in local elections. In **1894**, this was extended to married women, but still only in local, and not national elections. The education of women was still minimal in most countries, especially at the university level, but higher education for men was expanding in more advanced nations and territories.

The FIRE signs are expansive in a way that often uses legal arguments and "rights" to justify the takeover of lands and/or property. Initially, they are used in establishing the rights of a democratic republic, but could be used in justifying any form of government. An important harbinger for this trend was the **JU-SA conjunction in Nov. 1782 in Sagittarius,** on the Galactic Center, and within 2 degrees of the 1782 Winter Solstice. Ruled by Jupiter, Sagittarius is an innately optimistic and expansive sign, which adds to the sense of great self-confidence and even entitlement. Within a few years, on **Sept. 17, 1787, the U.S. Constitution** was adopted by the Constitutional Convention in Philadelphia, PA, replacing the previous Articles of Confederation (1777), and ratified by each of the 13 individual states (former colonies) from Dec. 1787 through May 1790. This established the Supreme Law of the United States, as well as the form of government. George Washington was elected President of the Convention, 1787-1789, prior to his Inauguration as first U.S. President April 30, 1789.

U.S. President James Monroe (1758-1831) presented **The Monroe Doctrine (1823)**, a policy stating that no further colonization of American states should be permitted, and would be regarded as acts of aggression requiring U.S. intervention. Though there was no credible American military to enforce it, the mighty British Empire was behind the new policy, as it served them in opposing Spain's efforts to further colonize Latin America. The Monroe Doctrine was tested many times, including in 1836, when the U.S. government objected to Britain's alliance with the newly created Republic of Texas. Closely related to The Monroe Doctrine, and enforcing it, U.S. President James Polk enunciated what he called **America's "Manifest Destiny" (1845).** It stated that the U.S. had a duty and a right to extend its influence and territory throughout North America.

America's westward expansion coincided with the invention and development of **the railroads.** For the first 100 years of the FIRE period, travel by land was the Age of the Horse and Carriage. For the next 100 years, railroads expanded trade and travel, notably after 1825-1830. The **first steam-driven self-propelled carriage** was invented in **1789** in Paris by Nicholas Joseph Cugnot, a French engineer. Its speed was 3 miles per hour (5 km. per hour). In early **1804** came the **first high pressure steam engine and the world's first working steam locomotive**, invented by Richard Tevithick, a British inventor and mining engineer, and the first to invent a heavier haulage for mining. The first railroads were used to haul minerals for mining purposes. Mining use of railroads increased, but passenger use took another few decades to arrive. By 1825, England had the first steam-powered freight and passenger train. And by 1826, the U.S. had a train for mining only, imported from Britain. The carriage was too heavy for most American tracks, and had to be rebuilt in the States. Travel by waterways and canals had dominated up to 1830, when U.S. railroad traffic finally surpassed that of the canals.

From **1836 to 1896**, the **Red Flag act** was in force in the U.S. and Britain, requiring a man to walk in front of any self-propelled vehicle, carrying a red flag by day and a red lantern by night. The maximum speed allowed was 4 miles per hour (6.4 km per hour). The Red Flag act and other legislation prevented growth of the auto industry in the 19th century, and since the self-propelled rail carriage COULD run on tracks, there was development of the railroad instead, instigated by needs for the mining industry. Only in the last decade prior to 1921 did the automobile become more prevalent, though mainly for the privileged classes. From the 1921 JU-SA conjunction in Virgo onward, trends moved away from public transportation and towards the privately owned automobile, greatly influenced by the long-range goals implemented by American car and truck makers.

Between 1851 and 1855, Bessemer steel was invented – the first inexpensive industrial process to mass-produce steel from pig iron. It was named after Englishman **Henry Bessemer,** who was awarded a patent on it in 1855, though American **William Kelley** had discovered it independently in 1851. The industrial revolution exploded with this invention, as the process had previously required large amounts of coal, with far greater cost and labor. Bessemer conducted his work between Jan. and Oct. 1855, and was spurred by interest in improving the military technology of artillery.

Communications: Prior to the telephone, the telegraph reigned, as did the handwritten letter. The **telegraph** was first in more regular use by the 1860s, with the first experiments in France in 1746, and important implementations in **1832** by Baron von Schilling von Canstatt in Russia. Germany had the first telegraph network in 1835-1836, and the U.S. by 1836. American scientist David Alter proposed the first electric telegraph one year before **Samuel Morse**, the American who independently developed and patented it. By 1837, the British were using the telegraph for police work. By 1846, the U.S. was sending its first telegraph messages, and during the 1850s the military was using the telegraph in Bulgaria and elsewhere. The first transcontinental telegraph system was established in the U.S. **Oct. 24, 1861**. The **telephone** was invented by both Elisha Gray and **Alexander Graham Bell** in Feb. **1876**, but Bell was awarded the patent.[18] The first long-distance telephone call was made Aug. 10, 1876. The locations were six miles apart. In **1915**, the first transcontinental phone call was made in the U.S., and the first transatlantic phone call was made **March 7, 1926.**

FIRE TO EARTH Mutation period (1901 to 1961)

The **first internal combustion engine** was invented in **1885** in Germany by **Karl Benz**, and an improved version was built in the same year by the German **Gottlieb Daimler.** But greater mobility via the **automobile** would not be possible until mass production made it more generally accessible. That occurred after the first EARTH conjunction in 1921. The invention of the hydraulic brake in 1918 facilitated this process to some extent, and prior to that, **Henry Ford** and other carmakers were hampered legally by a patent held by George Selden since 1895 for a two cylinder engine car. This was settled by a high court in 1911, clearing the way for cars with four-cylinder engines. Early carmakers such as Henry Ford made good profits for vehicles used in the First World War, though Ford had initially been a staunch isolationist. The Stanley brothers of Massachusetts manufactured steam engine self-propelled vehicles, known colloquially as "the Stanley Steamers." They were in operation from 1902 to 1924.

The same era witnessed the first airplanes, from the **first sustained flight of a manned glider in 1901,** and that of **an airplane Dec. 17, 1903** (12 seconds, 120 ft), both by the Wright brothers in North Carolina. They applied for a patent for a "Flying Machine" March 23, 1903; it was awarded May 22, 1906. By Oct. 1911, the military first used aircraft in the Italo-Turkish War, and throughout WWI (1914-1918), Britain, France, and Germany built and used military aircraft that became increasingly more sophisticated after 1915, when Germans had air superiority and maintained it through mid-1917, both with observation and early bombing capacity. The Germans began to lose this advantage a few months after the U.S. entered the war. In general, the use of airplanes in WWI increased their general positive impact on the public.

FIRE promotes **weapons and militarism.** The First World War was one of the deadliest in history due to advances in both fire power and mobility, with 9 million combatants killed. At the time, it was called the Great War, also "the War for Justice," and "the War to Preserve Civilization" (from the French and German languages). Woodrow Wilson, U.S. President, 1913-1921, said this war "would make the world safe for democracy." But it did not. There were **two World Wars** fought in the FIRE TO EARTH Mutation period, evoking the combination of forceful territorial expansion, increasingly powerful weapons or methods to eradicate the enemy (both FIRE), alongside diminishing ethical considerations (EARTH, with Jupiter's inability to wield much influence). In

[18] Alexander Graham Bell (1847-1922) was born in Edinburgh, Scotland. His family immigrated to Canada in 1870. Bell considered it an "intrusion" to have a telephone in his study, where he was working.

this one period, between them Hitler and Stalin's murderous regimes killed some 17 million people in their competing utopias based on class or race war. The figures rise to 21 million, 12 million from the Nazis, 9 million from the Soviets, if we include deaths from deportation, hunger, and sentencing to concentration camps.[19] Under Hitler, Germany is reported to have deliberately killed 11 million noncombatants, either through mass starvation, labor camps, or shooting, gassing, or hanging the victims. This included some 5½ million Jews, and some 3½ million Soviet prisoners in German captivity. Stalin's mass killing peaked in 1930-1933 with the forced starvation of over 5 million prosperous peasants ("kulaks"), an action Stalin engineered to enable the state to control agriculture. The lands suffering most were those occupied by the Soviets in 1939, the Germans in 1941, and the Soviets again in 1944. These lay between Berlin and Moscow, with Poland targeted repeatedly by both Nazi and Soviet atrocities. As this Mutation period came to a close, even larger numbers died during China's mass starvation under Mao Tse Tung, 1958 to 1961. Originally a 5-year economic plan, it was officially abandoned in 1961. Intended as agrarian reform to radically transform China into a modern industrial state, and called "The Great Leap Forward," this disaster deliberately starved between 36 and 45 million people, a fact largely obfuscated from the rest of the world until years later.[20]

National fervor burned hot all during the FIRE dominant period and included many **national movements.** In addition to Pan Germanism was the Zionist movement, which sought to find a homeland for the Jews and to bring them back to what they considered "their rightful homeland." The politics and momentum of this movement radically affected both World Wars, notably World War I. But it was **the assassination of royalty** in Sarajevo on June 28, 1914 that was the initial catalyst for World War I. A Serbian activist killed Archduke Franz Ferdinand of Austria-Hungary, and his wife Sophie, Duchess of Hohenberg. The Archduke was heir apparent to the Austro-Hungarian throne. Hostilities were declared by the first parties on July 28, 1914, followed by others in the next few days. Germany entered the war on Aug. 1, 1914, and a young eager recruit from Austria (Adolf Hitler) enlisted two days later, having avoided the draft for some years in his native Austria, and asking special dispensation to be allowed to fight for Germany, thus also for Pan Germanism.

Before World War I, almost all of Europe was ruled by monarchs related to King George V. But after 1914, as with Russia, the monarchies of Austria, Germany, Greece, and Spain all fell to revolution and war. The JU-SA conjunction in Virgo in 1921 was the first departure from the FIRE conjunctions, toward the end of the FIRE dominant period. Thus the 1901-1921 period, preceded by the JU-SA conjunction in Sagittarius, was a pivotal transitional period for nationalism and especially dangerous for monarchies, notably those identified with large empires or countries lit by the fire of revolution.

Saga of three kings: In addition to Duke Ferdinand and Duchess Sophie of Austria-Hungary, three kings should be cited here: **Kaiser Wilhelm II** of Germany and Prussia, **Czar Nicholas II** of Russia; and **King George V** of England. All three of them lost power within three years: from 1915 to 1918, and they were all first cousins. King George V fell off his horse Oct. 28, 1915

---

[19] These statistics appear in historian Timothy Snyder's book, ***Bloodlands: Europe between Hitler and Stalin***, 2010. Though World War II statistics are always subject to revision, Snyder bases his on information from Eastern European archives opened in the 1990s and previously inaccessible.

[20] Frank Dikötter, ***Mao's Great Famine: The History of China's Most Devastating Catastrophe, 1958-62***, 2010, p. 333. Dikötter differs from previous accounts, in saying the famine continued through late 1962.

while conducting a review of troops in France, and was badly injured. Together with his heavy smoking, this accident made him prone to pleurisy and chronic obstructive pulmonary disease. His decline due to poor physical health adversely affected the remaining two decades of his reign (May 6, 1910 to Jan. 20, 1936). Note that **the horse and the elephant** are animals associated with **royalty,** and **horses** especially with **Sagittarius.** The last JU-SA conjunction in Sagittarius occurred Nov. 28, 1901. The 64-year reign of Queen Victoria ended with her death on Jan. 22, 1901. King George V was King-Emperor of the British Empire, and in Dec. 1911 he and his Queen were presented to dignitaries in India as the Emperor and Empress of India. His empire expanded as all the other empires collapsed after WW I, though it was an increasing struggle to keep control over India, and his personal powers were depleted from 1915 onward due to the fall from his horse.

Kaiser Wilhelm II officially abdicated on Nov. 18, 1918. Because he was forced to leave Germany, he immigrated to Holland. He died there in 1941, just as Nazi Germany appeared to be winning WWII. He was not allowed burial in his native country. **Czar Nicholas II** of Russia was overthrown in the Russian Revolution of Nov. 1917. The British government wanted to extend asylum to the embattled Russian royals, but there was disagreement about whether this would be a wise decision. According to letters of his private secretary, Lord Stamfordham, King George V himself opposed the rescue of Czar Nicholas and his family, against the advice of his government. Therefore they remained in Russia, and on July 17, 1918 the Czar was shot along with his wife Czarina Alexandra and their five children. Even royal blood lines could no longer afford protection, and England was forced to turn against Germany as well. Britain declared war on Germany Aug. 4, 1914, when it did not withdraw from Belgium. Later in the war, to appease British nationalist feelings, **King George V** issued a royal proclamation on July 17, 1917, changing the name of the British Royal House of Saxe-Coburg and Gotha to the House of Windsor. This was a major turn of events for the royal houses of Europe, as the British royal dynasty had been predominantly German. Queen Victoria had little British blood in her, and her husband had none.

Kaiser Wilhelm II was **Emperor of Germany,** and **King of Prussia** (June 5, 1888 to Nov. 18, 1918). He was also the last German Emperor and King of Prussia. He was first grandchild of Queen Victoria of England, his mother being **Victoria, Princess Royal, of England.** His father, Frederick III, was far more liberal in his outlook, and strongly influenced by his temperate British wife. But he died only three months into his reign. Wilhelm II ruled over a mighty empire, but was crushed by Germany's defeat at the end of World War I that ended the monarchy in Germany. This was in part attributable to his impetuous militarism that ignored most civilian and diplomatic advice.

Manufacturing consent: The upcoming EARTH influence would spell a different kind of danger for democracies as well as empires. This could be seen just prior to the first JU-SA conjunction in EARTH in 1921. In part through more advanced technology, EARTH dominant periods would have a greater capacity to make more items manifest and accessible, and to influence the opinions of citizens. This would have both desirable and undesirable results, depending on the goals and the momentum of the initiators. Regarding World War I, various European powers and some within the U.S. were eager for the U.S. to enter the war. To this end, President Woodrow Wilson established a committee through Executive Order, called the **Committee on Public Information (CPI).** Under the guidance of the Secretaries of State, War, and Navy, this "independent agency" was headed by journalist George Creel. Its sole purpose was to "manufacture consent" in the U.S. to enter WWI, since large numbers of German-born immigrants to the U.S. were pro-Germany, and American public opinion throughout the war and in its aftermath was mostly non-interventionist.[21]

This can be seen from the timing, as the U.S. entered the war on April 7, 1917, and the CPI was formed six days later. The Creel Committee operated for just over 28 months, from April 13, 1917 to August 21, 1919. The Armistice signed Nov. 11, 1918 formally ended combat in WWI, though the Paris Peace Conference dragged on for six months, through late June 1919. George Creel defined CPI aims as: "not propaganda as the Germans defined it, but propaganda in the true sense of the word, meaning the 'propagation of faith.'"[22]

Soon the cornerstone of American advertising, its successful methods (even if often heavy-handed, with gross exaggerations) were long admired and copied by advertising and public relations mavericks as well as dictators. They learned that public opinion could be managed, with replicable outcomes. A big industry was duly born, hired as often by nations as by businesses, at times restoring complete confidence and credibility when major damage was done by a nation or corporation. The CPI campaign was successful well prior to the broader use of the telephone, automobile or even air travel. Gaining cooperation from the public would become surprisingly easier, even in democracies, through consolidation of the media. This increased dramatically in the EARTH period, even more so after the rapid acceleration of information technology from 1980-1981, easing the way towards control of the media and the message.

Following the triple conjunctions in Aries of 1940-1941, the U.S. unleashed the most powerful weapons ever devised. The first **atom bomb test occurred July 16, 1945** over the New Mexico desert. **On Aug. 6 and Aug. 9, 1945** U.S. military planes dropped atomic bombs over Hiroshima and Nagasaki, Japan, respectively. One of the seminal scientific and political figures in this period of history was **Albert Einstein,** whose career peaked in 1921 with his Nobel Prize in Physics for his discovery of the Photoelectric Effect. Having fled to the U.S. from Nazi Germany in 1933, he worried that Hitler might develop a bomb, and in fall 1939 informed **President Franklin Roosevelt** (U.S. President, 1933-1945) about scientific experiments that could lead to a new level of bomb. This led directly to the American involvement in nuclear weapons and nuclear energy.

Einstein was later haunted by his participation in this process and spent most of the last decade of his life (up to his death in 1955) involved in pacificism and in Zionism. He sought to establish organizations that would preserve international peace and abolish weapons of war, especially atomic weapons. His opinion on these matters was considered dangerous by the U.S. government. So too was that of former First Lady Eleanor Roosevelt, who shared Einstein's belief in world institutions

---

[21] The usual cause cited for raising U.S. sympathy for entry into WWI is an incident two years earlier – when on May 7, 1915, a German submarine sank the ship Lusitania (flying a British flag) off the Irish coast, causing the death of 1,195 passengers, including 128 U.S. citizens. Germany claimed it had warned Britain well in advance regarding ship traffic on those waters, but erred in ignoring American public opinion and not realizing the full extent to which it could be manipulated in favor of the US entering the war.

[22] Edward L. Bernays worked for the CPI in its Latin American division. With CPI's great success, Bernays (age 27) was invited by President Wilson to attend the prestigious Paris Peace conference in 1919. He soon became famous as "the father of public relations" in the U.S. He promoted cigarette-smoking for women in the 1920s as lighting up "Torches of Freedom." He also advised on many successive ad campaigns for the tobacco industry, especially to sow seeds of doubt that smoking could be harmful. Nephew of Sigmund Freud, Bernays believed that crowd mentality was irrational and needed to be controlled. His 1947 essay "The Engineering of Consent" defines this process as "the art of manipulating people... [and] the very essence of the democratic process, the freedom to persuade and suggest." See also ***Propaganda for War: The Campaign Against American Neutrality, 1914-1917***, c1939, 1968, 1997, by Horace Cornelius Peterson.

for preserving international peace. The **League of Nations** was established for that purpose and as a direct result of the Paris Peace Treaty (also called the **Treaty of Versailles**), signed June 28, 1919. Its first meeting was on Jan. 16, 1920. The League of Nations was later replaced by the **United Nations**, established Oct. 24, 1945. Though providing an important forum for nations to engage with each other, neither organization has been given any real legal clout to enforce their resolutions. Without such authority, there is minimal effect in limiting wars between nations or preventing one nation from invading another. This was tested early on with the Six-Day War in Israel in June 1967.[23]

At least two major world powers, **Israel** and **India**, achieved statehood in this Mutation period, and their final drive towards that end was accomplished within this time frame as well. Important astrological implications arise when combining **28-29 year Saturn cycles** with major turning points such as the **Treaty of Versailles**, which ended World War I and in many ways set up the conditions for World War II. The Treaty of Versailles contained many items of heated controversy, and was never ratified by the U.S. Senate. The last vote taken was on March 19, 1920, refusing to ratify, though the U.S. later signed a separate peace treaty with Germany. The U.S. disliked the League of Nations, among other items, and did not want to be engaged in more world conflicts. It saw the League as laying the groundwork for that. A major defeat for President Wilson, the League of Nations was the 14th of his famous 14 points. The terms of the treaty were especially harsh on Germany, and designed to cripple their economy, to make Germany solely responsible for the war, and to spread the spoils among the victors. Germany was already disavowing the treaty by 1933, giving Hitler early momentum in German national politics.

With this amount of trauma associated with the Treaty of Versailles, and no party fully satisfied, some major difficulties could be expected for entities born as a result of it. This would become clear at the conclusion of a Saturn cycle (28-29 years), and/or the return of both Jupiter and Saturn in 59-60 years. **The League of Nations** did not make it past its first Saturn return. It was intended to prevent future world conflicts, and was established as a direct result of the Treaty of Versailles, though without U.S. support. But it was dissolved April 20, 1947, and so endured just 28 years – one Saturn cycle.

Nationhood for Pakistan and India came Aug. 14 and 15, 1947, respectively, both also a full Saturn cycle (28-29 years) after the **Treaty of Versailles**, signed June 28, 1919. Israel declared its statehood May 14, 1948, some 29 years later. The Treaty supported the premises for a Jewish homeland in Palestine even more strongly than the Balfour declaration of Nov. 2, 1917, since the Zionists had major influence at the Paris Peace Conference. Meanwhile, India had been thwarted in its hopes to achieve independence from Britain following World War I, having sent 1.3 million Indian soldiers and laborers to fight for the British in WWI. This resulted in further resistance to British rule, with more and more demonstrations for agrarian and land reforms. The turning point for galvanizing Indian nationalism was the **Amritsar massacre of April 13, 1919**, in which British militia shot and killed over 1500 unarmed Indians at a religious festival gathering. The timing of this incident, within two months of the fateful Treaty of Versailles, did not bode well for the status quo in India, but neither did it instantly bring independence. We will explore some of the astrological reasons contributing to the historical events.

---

[23] The United Nations chart is one of 25 examples discussed in greater detail in my article "*Graha Yuddha:* Testing the Parameters of Astrology and Astronomy," 2003, revised & expanded 2010. Available at my website: **www.edithhathaway.com.**

National independence can be dated from pivotal events which occur: **1)** close to a JU-SA conjunction; and/or **2)** a full Saturn cycle (28-29 years) earlier. In addition to the likelihood of this 28-29 year delay for Israeli and Indian statehood, a JU-SA conjunction in a FIRE sign would make these long delayed events far more probable. This did not occur again for forty years after the 1901 conjunction in Sagittarius. The 1921 conjunction in EARTH (Virgo) delayed the process, in my opinion, and the 1940-1941 conjunctions in Aries accelerated it, with further emphasis from triple conjunctions bringing the FIRE period to a close. Unfortunately, it also emphasized the religious residue of FIRE, with still unresolved problems created by a Jewish state on territory also sacred to Muslims and Christians, and occupied for centuries by Palestinian Arabs.[24] For India, there was a national/territorial split along Hindu/Muslim lines that was never intended by most of the former Freedom Fighters, including Mohandas (later Mahatma) Gandhi. Much bloodshed followed close on the heels of both India's and Israel's declaration as independent states.

South American nations achieving first sovereignty or declaring independence from **1810** onward reached this goal 28-29 years after the important **JU-SA conjunction at 7:18 Sagittarius** on the Galactic Center in **Nov. 1782.** Though all of them did not achieve independence in 1810, the momentum for 16 states becoming sovereign in Central and South America began in 1810 and 1811 – with Argentina, Colombia, Mexico, Paraguay, and Venezuela. Most of the states declaring independence in 1810 were fully independent from Spain or Portugal by 1821, within close proximity to the **JU-SA conjunction in June 1821 at 3:17 Aries.**

Haiti had been in revolt from France from 1781 through 1804, when it became the first-ever republic ruled by people of African ancestry. However, the timing for **Haiti's first sovereignty in 1804** was not auspicious for the future of the new nation, though it was monumental in what it represented – former slaves declaring themselves free from their French colonial masters.[25] The **largest slave revolt in the United States** occurred in **1811** outside New Orleans, and was suppressed. (500 slaves were involved.) Though it occurred 29 years after the 1782 JU-SA conjunction, it was not the intention of the U.S. Constitution of 1783 to provide freedom or citizenship for black slaves. However, this non-provision was tested at the Saturn return of the U.S. Constitution. (This was the largest among 250 other such slave uprisings in the American South.)

---

[24] In 1947, British Colonial Secretary Arthur Creech Jones announced that "His Majesty's government [has decided] to lay down the Mandate, under which they have sought for 25 years to discharge their obligations to fulfill the growth of the Jewish national home and to protect the interests of the Arab population." Britain turned over its mandate to the United Nations, who then partitioned the area into two states, one Arab, one Jewish. This was rejected by the Palestinians and by neighboring Arab states. Soon after the final withdrawal of British troops from Palestine in May 1948, Israel proclaimed itself a state. Neighboring Arab states then promptly declared war on Israel. Over 700,000 Palestinians were forced from their lands by Zionist armed groups in 1948, and this remains a source of Arab-Israeli dispute ever since. It was preceded by the Palestinian Arabs' revolt in 1936-1939 against the British mandate and against the mass immigration of Jews to Palestine. Losses in that revolt and regional economic weakness left them vulnerable in 1948. Historically, religious or tribal loyalty among Palestinian Arabs superseded nationalistic sentiment up to 1834, when they revolted (unsuccessfully) against Ottoman rule. Further Arab revolts occurred in 1916-1918 but were compromised by Western powers vying for territorial control in the region. (A.C. Jones quote from a 1947 newsreel on ***Masters of their Own Destiny***, Episode 1 of a 6-part series, produced by Al Jazeera. First airing: July 13, 2009.)

[25] The national chart of Haiti is discussed on pp. 9-10 of my article: "Corporate Conglomerates vs. Nation-States: Which Nations will Survive and Thrive?" (Sept. 2010). Available at: **www.edithhathaway.com**.

EARTH Dominant period (1921 to 2199)[26]

Almost all African nations were formed after 1961, and thus without the influence of the FIRE dominance, and the strength it gives to nation-building. This explains a major weakness in their existence as nation-states, which need FIRE to bring patriotism and a sense of loyalty shared by a larger number of peoples towards a single political entity, whether a republic, a democracy, or even a dictatorship. Nation-states, with a few rare exceptions (such as the United Arab Emirates and Singapore), when founded after Feb. 1961 and in the EARTH dominant period, are more likely to be overrun by multinational corporations or dictators acting like they have the same authority. They appear to be beyond the reach of laws, whether local, state, or international. Further, there is accelerated movement of people, money, and goods outside the control of individual nation-states or their citizens. With dictators, normal rules of law may not be observed, but this can differ little from the potential lawlessness of multinational corporations and their frequent disregard for the environment or the people producing or using their goods.

Though its **focus is on material security**, the EARTH cycle often starts out with excesses of personal and collective spending that typically depletes the middle and lower classes, while enriching the very wealthy sector. The 1921 JU-SA conjunction in Virgo was followed by the madcap 1920s, various high profile financial scandals (the **Teapot Dome scandal** in the U.S., among others), and then the **stock market collapse in Oct. 1929.** The first JU-SA oppositions arrived in 1930-1931 demanding solutions to this dire situation, and from the early 1930s came the drive towards repairing ruined economies and establishing new regulations for financial trading. An EARTH theme that is a positive response to financial excesses is that of the repair and expansion of infrastructure. In the U.S., the Works Progress Administration was established April 8, 1935 and it built roads, highways, bridges, sewers, etc. via various economic projects within the New Deal set in motion by President Franklin Roosevelt from 1933 onward. The major halt to its momentum was American entry into World War II in Dec. 1941. Very little was done over the next 70 or more years to improve the American infrastructure, and signs of decay announced themselves with the collapse of bridges, among other items.

The EARTH period is more secular, and though religions do not disappear, they are not the major reasons for conflicts or power plays. EARTH favors nations joining together on economic policies, which may or may not favor the individual nation or region over the long run, especially land, water, and mineral rights. **The European Economic Community** (EEC), also called the Common Market (in English-Speaking nations) is more favored than most such organizations to do well within the EARTH cycle. Formed Jan. 1, 1958, the EEC entered its first major crisis in 1962 when member nations had to agree on financing the Common Agricultural Policy. Some member nations, France in particular, had concern about infringement of their national sovereignty, and being overpowered by other nations. By Jan. 1, 1999 the Euro was launched as the common currency of most EEC nations.

The EARTH period also favors technology, and the greater accessibility of **more advanced technology**, making products more available, including electronic equipment and **cars** (mass produced in greater quantities after hydraulic brakes were invented in 1918). The first passenger

[26] The previous EARTH dominant period occurred for 298 years, from 1007 to1305 A.D. During that time the Mongolian Empire was the largest empire, from 1206 to 1368 A.D., initially under the leadership of Genghis Khan. Known as "Moguls," they had power through financial wealth, as well as military strength. Their land covered 22% of the earth's surface, and they ruled over a population of 100 million people. The Mongols spread and exchanged various technologies, commodities and ideologies across Eurasia.

car equipped with four-wheel hydraulic brakes was in 1921, with the Model A Duesenberg. Though carmakers were not quick to use hydraulics, the basic braking system in cars was in place by 1921, with further refinements over the decades. Henry Ford started making luxury cars after 1921, though cars were not more prevalent until post WWII, when there was more prosperity and more roads and highways were built. These events coincided with the JU-SA conjunction in Virgo in 1921, the first in the EARTH element. After 1921 the American car industry was only briefly diverted from producing vehicles for private use. Shortly after entering WWII, and for three years (1942-1944), wartime President Roosevelt ordered all U.S. automakers to manufacture only military vehicles. All these developments greatly favored the growth of oil and gas industries.

As of 1920, 90% of Americans still used the electric trams and interurban railway systems, but by the early 1960s those had been mostly dismantled. The systematic dismantling of American public transportation is still regarded as controversial, though it is undisputed that carmakers and allied industries got away with it, with minimal attention from the government. It is also undisputed that in 1922 the head of General Motors, Alfred P. Sloan, established a special unit within the corporation whose task was to replace America's railways with cars, trucks and buses. Other companies (including Firestone Tires) joined them in this vast corporate project. They succeeded in reaching their goal, greatly eased by the 1921 and 1961 JU-SA conjunctions in EARTH. Indictments followed in 1949, but by the 1950s, most large American cities were already stripped of their electric trolleys and interurban railways. Other cities around the world were not immune from copying the American model.

Charles Lindbergh was a national and world hero after his **first solo transatlantic flight in 1927**. By the late 1930s, business men on expense accounts could afford to fly often, but **airplane travel** was still mostly confined to the wealthiest by the early 1960s, even with the invention of the first jetliner in 1959, and faster transatlantic flights as of 1965. More people could afford to fly only when President Jimmy Carter deregulated the airline industry in the U.S. in 1978.

Radio did not come into common use until **the 1920s** and later, though the first experiments in radio technology were in 1820 in Denmark. The first patent for radio was awarded in 1896 to the Italian inventor **Guglielmo Marconi** (1874-1937). Jagdish Chandra Bose, an Indian physicist in Calcutta, had invented it in 1894, but did not seek a patent for it. The first radio manufacturing plant was opened in England in 1898. (With Italian-Irish parents, Marconi sought and received more initial interest in England.) A Radio patent was awarded in 1897 in the U.S. to Serbian-born **Nikola Tesla** (1856-1943), who also did other important scientific work in electricity and cosmic rays, and unsuccessfully contested Marconi's rights to the Radio patent. In **1916**, the first regular radio broadcasts were in Morse code, and gave weather forecasts in Wisconsin. The first-ever **regular radio broadcasts** of news and/or entertainment began in **1920** in Argentina and in the United States, first in Detroit, Michigan, and shortly after in Pittsburgh, PA.

Within a few years after the first conjunction in EARTH in 1921, **transatlantic phone calls** were available (**March 1926).** In **1955,** the first **transatlantic cables** were laid. By **the 1960s** Bell Labs in the U.S. developed the electronics for the first cellular phones. Cell phones were invented in the 1960s, but not in regular use until the 1980s, with the first mobile phone circuitry between Sweden and Norway in 1981, coinciding with the triple JU-SA conjunctions in Virgo.

There were major developments with the **telephone**, which superseded the telegraph, then **radio** and **talking films.** The first short movies with sound technology (**"the talkies"**) made their debut in New York City. The first-ever feature film with sound was released in **Oct. 1927**. Called *The Jazz Singer*, it was a major success, leading to the global phenomenon of the talkies in the

1930s and to the cultural and commercial phenomenon of Hollywood, USA as a world center for filmmaking, and the dominant one prior to Bollywood – India's film counterpart from the 1960s, coinciding with the JU-SA conjunction in Capricorn in Feb. 1961.

Television developed from 1923 onward, but more quickly in the 1930s and 1940s. The Soviet Union initiated the first commercially manufactured television sets in 1932. The **first live radio broadcasts** were in a suburb of Washington D.C. in July 1928, and **first live television broadcasts** also in Aug. 1928 from New York City. By 1936, there were 200 TV sets in the world. The Germans broadcasted the 1936 Olympic Games on live radio – the first time ever for such events. They also experimented in television, and used an image iconoscope to broadcast the 1936 Olympic Games. This German device became the industrial television standard from 1936 to 1960. On June 2, 1953, the first most watched event on television was the coronation of Queen Elizabeth II, who wisely understood the power of television, even if monarchs had been crowned for 900 years in Westminster Abbey without it. The first televised U.S. Presidential debates in fall 1960 were groundbreaking events, as visual images of John F. Kennedy favored him over Richard Nixon, who should have won through greater political experience. Kennedy was elected by a narrow margin in Nov. 1960 and went on to be the **"first television president"** – just as Roosevelt had been the first radio president in the 1940s, and Winston Churchill's voice was pivotal on the radio, especially in WWII II. The Vietnam War, expanded by the U.S. in 1961, would be the **"first television war."**

Two months after the seminal JU-SA conjunction in Capricorn in Feb. 1961, Soviet cosmonaut **Yuri Gagarin** was launched into orbit around the Earth on Vostok I, the first human to **travel into space** and the first to orbit the earth (**April 12, 1961**). On **May 5, 1961** Naval aviator and astronaut **Alan Shepard** became the second human and the **first American in space** on the Mercury-Redstone3. At a Joint Session of Congress on **May 25, 1961** President Kennedy announced he intended to put a man on the Moon:

> "[The U.S.] … should commit itself to achieving the goal before this decade is out, of landing a man on the Moon and returning him safely to the earth. No single space project in this period will be more impressive to mankind, or more important for the long-range exploration of space; and none will be so difficult or expensive to accomplish."

The **first manned spacecraft landing on the Moon** did occur **July 20, 1969**, a testimony to the rapid acceleration of technology in the EARTH period, harnessed by the ability of business interests and greater American prosperity to finance such ventures. President Kennedy also wanted to win the Space Race with Russia, which Russia was winning as of 1961. To do so, Kennedy was prepared to spend whatever was necessary, not because he was interested in space exploration, but to win a contest and diminish Russia's perceived power in the world. The paradox between Kennedy's inspiring rhetoric in public on this subject and the more sardonic words spoken behind closed doors (such as at a Cabinet meeting on Nov. 21, 1962) echoes the JU-SA conjunction in Capricorn: It gives lip service to higher principles of knowledge and education (Jupiter), but in most cases the political or business angle (Saturn) wins out. In 1961 Russia and Communism were still viewed as major enemies of the state, but by Sept. 1963, in a speech before the United Nations General Assembly, Kennedy proposed that the U.S. and the Soviet Union join forces in their efforts to reach the moon. This was later rejected by Lyndon B. Johnson (U.S. President, 1963-1969).

A battle between capitalism and communism fuelled the **Space Race**, which started just before WWII ended in 1945, when the U.S. and Russia competed for the most brilliant German scientists, whose knowledge of rocketry surpassed that of other nations at the time. German rocket scientist

Wernher Von Braun (1912-1977) and his team worked for Nazi Germany developing V-2 rockets through WWII, then afterwards for the U.S.[27] His first series of rockets for the U.S. Army provided them with their first operational medium-range ballistic missiles.The Redstone rocket would later launch the first American satellite and piloted space missions and became the basis for both the Jupiter and Saturn family of rockets. The Jupiter rocket was renamed "Saturn" by Feb. 1959, with the **first Saturn rocket launched Oct. 27, 1961**. (Even in the rocket program Saturn wins over Jupiter!) Von Braun led the development of the Saturn V booster rocket that helped land the first men on the Moon in July 1969. This was considered his greatest achievement in the field of rocket science, but prior to that, it was he who saved public embarrassment and enabled the U.S. to compete with the Soviet Union's unexpected launch of **Sputnik I** on **Oct. 4, 1957**. Because of his expertise, the U.S. launched the Explorer I on Feb. 1, 1958.

Thus the stage was set for rapid technological advancement in the EARTH period as soon as German rocket scientists went to work for either the U.S. or the Soviet Union. The first results came in the late 1950s, but with greater projects underway in the early 1960s. In the U.S. this was under the auspices of the **National Aeronautics and Space Administration (NASA),** created by an act of Congress, and signed by President Eisenhower on July 29, 1958. Despite many setbacks, the Space Race eventually set the stage for rapid advances in technology, notably the telecommunications industry. The first photos of earth from space also showed the damage the earth was suffering due to increased greenhouse gases. The environmentalist movement was soon born, and its opponents started arming themselves to deny the significance of these photos.

In a similar way that national movements had a better chance of realizing sovereignty once the FIRE series started in an unbroken sequence from 1782, the EARTH series from 1961 was more decisive for the manifestation of many technologies, especially **information technologies.** Though the 1921 JU-SA conjunction in Virgo heralded a 20-year period of major steps forward for radio and television, it would be **the triple JU-SA conjunctions in Virgo in 1980-1981** that would fully revolutionize the Information Technology industry. Virgo is ruled by Mercury, the communicator (as is Gemini, an AIR sign), but Mercury is exalted in Virgo and can reach unprecedented technological modalities when in an EARTH sign. With the triple conjunctions in 1980-1981, the last for 257 years, and the first in Virgo with the more solid EARTH sequence (one last aberration in FIRE in 2000), they were also in the *nakshatra* of *Hasta* ("the hand," reflecting that which can be put in the hands of the user, giving "hands-on" use).

This confirms the major innovations that would occur in this arena between 1980 and 2000, along with culturally shared events made possible through telecommunications. Telegraphing the importance of what he called the new "electronic interdependence," in 1961 the Canadian philosopher and educator **Marshall McLuhan** wrote his seminal book *The Gutenberg Galaxy: The Making of Typographic Man*, published in 1962. In it he predicted the World Wide Web, and introduced concepts of "surfing" for information through multidirectional means, and also "The Global Village." His ideas were well known in the 1960s and 70s, especially after the publication of his 1964 book *Understanding Media: The Extensions of Man.* He introduced key media theory such as "The medium is the message," i.e., not so much the content but the medium itself is of critical importance, in his view, and the personal and social consequences of any new medium,

[27] In 1945, prior to the Allied victory in WWII, von Braun negotiated for himself and 500 of his top German rocket scientists to work for the U.S., prior to the capture of the V-2 rocket factory in Mittelwerk, Germany.

creating a change in attitude or action on the part of the audience. For instance, with scenes of war and violence commonplace in the average home or office, does this create a tolerance and even appetite for war and violence? Or the desire for more diplomacy and peaceful methods of resolving conflicts? The invention of the Internet in 1990 brought renewed interest in McLuhan.[28]

Here are some pertinent events in late 1980-1981: **1) Dec. 12, 1980**: First public offering (IPO) of Apple Computer, Inc., the largest IPO since Ford Motor Company IPO in 1956. All 4.6 million shares sell within minutes. Over 40 of Apple's 1000 employees became instant millionaires from stock options. Co-founder Steve Jobs' 7.5 million shares bring him $217 million in one day. (In summer 2011 just prior to the death of Steve Jobs (1955-2011), Apple Computer had $75.9 billion in cash reserves - $2.1 billion more than the U.S. Treasury Dept.); **2) Dec. 31, 1980:** Death of Marshall McLuhan (b. 1911), the same day as the first JU-SA conjunction in Virgo; **3) Jan. 20, 1981:** Inauguration of U.S. President Ronald Reagan, dubbed "The Great Communicator," and formerly an actor in films and General Electric spokesperson; **4) Feb. 13, 1981:** Purchase of London's *The Times* and *The Sunday Times* by Australian-born media mogul Rupert Murdoch and his News International (later News Corp.), expanding his media empire across the world.[29] (The second JU-SA conjunction in Virgo was March 4, 1981.); **5) June 25, 1981:** Incorporation of Microsoft by Bill Gates, software mogul.[30] **6) July 29, 1981:** Televised wedding of Prince Charles and Princess Diana, watched by a global audience of over 750 million people. Princess Diana became "the People's Princess" via the media. The last of the triple JU-SA conjunctions in Virgo was on July 23, 1981. (Diana was born July 1, 1961, 4½ months after the 1961 JU-SA conjunction in Capricorn.) **7) Aug. 12, 1981:** First-ever personal computer was released by IBM, called the IBM PC. Prior to this, in July 1980, IBM first met with Bill Gates and Paul Allen, hiring them to create an operating system for the new IBM PC, for which Gates and Allen retained the rights.

Between 1980 and 2000 **the personal computer** truly arrived, the first examples coming in 1975 with the Altair 8800, designed by Americans Ed Roberts and Forest Mims III, and sold as kits by mail order. This was soon followed by the cell phone in the early 1980s, the Internet (1990), and fiber optics in a wide range of uses. In **1982**, Japan's Sony Corp. sold the world's **first compact discs (CDs),** read by a laser beam and replacing long-playing record (LP) sales in Japan by 1987. Sony executive **Norio Ohga** (1930-2011) was impressed by demonstrations of music CDs in 1979 by Phillips (a multinational Dutch corporation) and pushed for its development by Sony. He is personally credited with the development and commercial production of the CD. He wanted a disc that was 4.8 inches (12cm) in diameter, with sufficient capacity at 75 minutes to store all of Beethoven's Ninth Symphony. Ohga became Sony CEO and President, 1982-1995. Working as a team, Philips and Sony paved the way towards this new digital recording technology. Philips pressed the first-ever CD at a factory in Germany on Aug. 17, 1982.

After the initial euphoria of greater freedom and choice for citizens, EARTH dominance placed more emphasis on consumerism at the expense of the flowering of individual cultures. Citizens

---

[28] In his ***Four Reasons for Eliminating Television*** (1978), former advertising executive Jerry Mander says television is not a neutral or benign instrument, that those controlling it will sell their goods and slogans, and in spite of its usefulness, it cannot easily be reformed from promoting autocratic control by a very few.

[29] By 1985 Rupert Murdoch owned so many U.S. media properties, he had to become a U.S. citizen. Among others, he founded Fox News Television in 1986, and purchased *The Wall Street Journal* in 2007.

[30] Bill Gates was working on computer software from the early 1970s, and left Harvard to do so in 1975. Microsoft was first traded on the New York Stock Exchange March 13, 1986.

become consumers of products and services in a culture of global mass-production, in which greater cross-cultural understanding can happen but marketing and censorship also affect results. For further entertainment and interest there are social networking websites, with nearly 800 million users on Facebook as of Jan. 2012. Launched in Feb. 2004, Facebook connects people socially and unites them in their social, cultural or political efforts. It also creates far more distractions, accelerates the natural daily rhythms and removes privacy – initially unacknowledged factors.

By re-inventing our world through all these technologies, greater economic and personal independence for the individual simultaneously facilitates the ability of governments and corporations to control information and to conduct surveillance, testing the boundaries for any democracy, where freedom of speech and voter participation is necessary to protect the democratic process. As of 2011, U.S. rulings allow the government to secretly amass private information related to individuals' Internet communications, feasible from 2000 or earlier but not fully understood by the public. Technology has no morality per se, and can lead to greater problems, especially if one does not have an awareness of them. Marshall McLuhan correctly predicted this in the early 1960s:

> "Instead of tending towards a vast Alexandrian library the world has become a computer, an electronic brain, exactly as an infantile piece of science fiction. And as our senses have gone outside us, Big Brother goes inside. So, unless aware of this dynamic, we shall at once move into a phase of panic terrors, exactly befitting a small world of tribal drums, total interdependence, and superimposed co-existence. [...] Terror is the normal state of any oral society, for in it everything affects everything all the time. [...] In our long striving to recover for the Western world a unity of sensibility and of thought and feeling we have no more been prepared to accept the tribal consequences of such unity than we were ready for the fragmentation of the human psyche by print culture."
>
> **Marshall McLuhan**, *The Gutenberg Galaxy: The Making of Typographic Man*, 1962, p. 32.

McLuhan refers to what astrology regards as Mercury's domain, as Mercury rules over the nervous system of the individual and the collective. With the new and more advanced technology comes the possibility of overtaxing the nervous system, on the way to potentially enlightening it. At some point the receiver of the information can tune out, no longer paying attention to what is important. McLuhan may be wrong that the viewer continues to notice content. More information does not guarantee that one is more discriminating and knowledgeable, especially if sources and facts are not examined. One of the ironies of the EARTH period is that the technology makes more research available to the public, but it does not guarantee that the public pays attention to the facts. With more laxness about accuracy of information, and less clout for solid investigative journalism, the new technology can both catch errors and criminality *and* spread false and/or biased information parading as the truth.

Marshall McLuhan advises people to observe the effects of a medium so they can stay ahead of it and remove or minimize the deleterious results. This is an excellent idea, but an ongoing challenge, as unfortunately astrological trends through 2199 tend to favor autocrats over citizens. They could change, but only if advocates for citizens – i.e., actual legislators – are as numerous and influential in government as corporate advocates. (Jupiter's weakness in the JU-SA conjunction keeps revealing itself, especially in Capricorn.)

From 1981 onward, there is the increasing occurrence of **Corporate-produced fake news** – commonly known as VNRs (Video News Releases), which are aired as actual news pieces, without stating the source (and obvious agenda). With a reduction of journalists in all sectors, especially investigative journalists, this is an ongoing danger to democracy, though a boon to the corporate elite, and not the only one. They have the power to own media empires (Rupert Murdoch's News Corporation, among others). They can buy elections, portray fiction as facts, and influence Congress (at least in the U.S.) to eliminate most opposition, such as public media, whose funding by the U.S. government has been under assault by Conservatives since 1980.

Climate change has also suffered from manipulation of information. On June 23, 1988 NASA scientist James Hansen testified to U.S. Congress that the planet was heating up to historically unprecedented degrees due primarily to human activities - use of fossil fuels and deforestation.[31] Hansen and other scientists from around the world predicted new extremes of weather patterns, along with melting of the polar caps. **Carbon dioxide in the atmosphere** has increased from 315.98 parts per million (ppm) in 1959 up to 391.7 ppm released in Feb. 2011. The upper safety limit for CO2 is 350 ppm, which the earth reached in 1985. The level has risen dramatically since 1958, when first measured at Mauna Loa Observatory, Hawaii, and it has remained above the upper safety limit since 1988, the year James Hansen first testified. The warmest year ever recorded (since 1880) was 2010, tied with 2005, and it produced extreme flooding both in 2010 and early 2011 in several nations.

Industry leaders have acted swiftly to protect their commercial interests, systematically funding efforts to manipulate public opinion on this subject, and to fight legislation related to climate change.[32] Called the "kingpin of **climate science denial**" by its detractors, Koch Industries has opposed environmental regulations of its oil refineries and pipelines since the mid-1960s. Other companies (ExxonMobil, Peabody Industries, et. al.) have acted in a similar fashion. As of 2011, they have successfully blocked any serious national or global initiatives to take bold action for reduction of carbon dioxide emissions. Climate scientists say these cause global warming, but their opponents say there is doubt as to the actual cause, and that global warming is an economic boon for various nations. The U.S. coal and oil lobby has embraced climate crisis denial, while nuclear power companies and lobbyists gain ground, not because they are completely safe in terms of *not* emitting radioactive particles, but because they have financial and corporate power, and promote their *not* emitting carbon dioxide. **Renewable energy** such as solar and wind power lost ground for decades in the U.S. when Ronald Reagan (U.S. President, 1981-1989), removed the solar panels at the White House installed in 1979 by his predecessor Jimmy Carter (U.S. President, 1977-1981), and allowed Carter's Solar energy tax credits and research subsidies to expire. In fall 2010 Barack Obama (U.S. President, 2009-) announced plans to re-install solar panels on the roof of the White House in spring 2011.

Pharmaceuticals and military weapons are particularly favored from 1961 onward, and in danger of massive fraud due to less and less regulation. In the EARTH period, there is far more likelihood of both these categories reaping huge markets. At the same time, inferior and unsafe drugs (both legal and illegal) are bought and sold, as are military weapons (both legal and illegal). In both cases, corporate profits drive these industries at the expense of public health and well-being.

---

[31] In 2008 James Hansen targeted oil company lobbyists and executives, calling for them to be put on trial for spreading disinformation, just as tobacco companies had done for decades, declining to link cancer with smoking. **http://www.guardian.co.uk/environment/2008/jun/23/fossilfuels.climatechange.**

[32] Jane Mayer, "Covert Operations," *The New Yorker magazine*, Aug. 30, 2010. Mayer's important investigative piece reveals to what magnitude the Koch brothers (David and Charles) have funded opposition to any environmental regulation of their industries and any reports supporting climate science/change. Annual revenues for Koch conglomerates were more than $100 billion per year, as of 2010.

Illegal drugs have swept into America ever more strongly through its wars from 1961 onward, especially from Vietnam and Afghanistan, the Opium capital of the world.

There are important associations timing-wise between **the Vietnam War** and the **JU-SA conjunction in Capricorn Feb. 1961**. Though military conflict in Vietnam, Cambodia, and Laos had begun from Nov. 1, 1955, it was relatively low key until 1961, when the United States tripled the number of its troops in Vietnam, and tripled them once again in 1962. In 1965, U.S. combat units were deployed and heavy bombing peaked in 1968. The Paris Peace Accords of Jan. 1973 were ineffective, as conflict continued until April 30, 1975 with the fall of Saigon. Thus, the U.S. deeper engagement in Vietnam lasted 14 years (half a Saturn cycle). The military strategy was one of containment of Communist takeover of South Vietnam and was regarded as part of the U.S. Cold War policy. North Vietnam regarded the war as a colonial war, and South Vietnam as a U.S. puppet state. But other factors played a part in expanding the war. **President Dwight Eisenhower** had warned of them in his Farewell address to the nation Jan. 17, 1961:

> "...Until the latest of our world conflicts, the United States had no armaments industry.... This conjunction of an immense military establishment and a large arms industry is new in the American experience. The total influence – economic, political, even spiritual – is felt in every city, every State house, every office of the Federal government. We recognize the imperative need for this development. Yet we must not fail to comprehend its grave implications. Our toil, resources and livelihood are all involved. So is the very structure of our society. In the councils of government, we must guard against the acquisition of unwarranted influence, whether sought or unsought, by the military-industrial complex. The potential for the disastrous rise of misplaced power exists and will persist."

With **Robert S. McNamara** (1916-2009) as U.S. Secretary of Defense, 1961-1968, there was a new business approach to Defense, which combined with the force of the corporate military establishment, quickly became a big budget, war-driven machine. After retiring from the Presidency of the World Bank in June 1981, and in the last decades of his life, McNamara came to regret U.S. escalation in the Vietnam War, and his key role in that war. He later rejected the popular "domino theory," made efforts to ban nuclear weapons on an international level, and was critical of George W. Bush's invasion of Iraq in 2003.[33] By 1968 McNamara was already critical of the expense and viability of **Anti-ballistic missile systems.** ABMs received renewed attention in 1981 with incoming U.S. President Ronald Reagan, and his deep belief in his Strategic Defense Initiative (SDI), which has since proved a needless expense to American taxpayers ($25 billion on research by 1993), and was derided then and later as unrealistic, and even unscientific. SDI was in defiance of the 1972 ABM Treaty with the Soviet Union, a treaty in force until 2002, when the U.S. unilaterally withdrew from the Treaty under the more aggressive military policies of President George W. Bush. The Vietnam War was the nation's first highly wasteful war in the EARTH era. Over five million Vietnamese civilians and military perished in that war, according to 1995 Vietnam government statements, while U.S. military casualties were 58,000. Beyond that, the American defoliant Agent Orange poisoned large areas of Vietnamese soil and groundwater and caused cancer and genetic abnormalities in large numbers of civilians and soldiers (and their offspring) in the decades following the war.

---

[33] In 2003, Errol Morris's documentary film was released: ***The Fog of War: Eleven Lessons from the Life of Robert S. McNamara***. It consists largely of interviews with McNamara, along with archival footage.

The fall of the Soviet Union, between Jan. 1990 and Dec. 1991, did not slow down American military spending, although Russia was "the evil Empire" to be fought – in the words of President **Ronald Reagan**. Russia, for its part, saw the United States as an "imperialist superpower" aiming to conquer the world. The Soviet government was crippled by heavy military expenditures in its decade-long war in Afghanistan in the 1980s, abetted by the U.S. proxy war against the Soviets, arming and training Jihadists in Afghanistan who were very effective in fighting the Soviets. The same Jihadists later turned against the United States, reportedly forming the nucleus of Al-Qaeda.

Over 50 years into the EARTH dominant period, the U.S. defense budget remains greater than the total spent on defense by all other nations combined. This trend has been seemingly irreversible since 1961. The **Weapons** industry benefits so greatly from wars that the EARTH period is fraught with wars manufactured and started under false pretenses. The Gulf of Tonkin incident in 1964 justified major escalation of the Vietnam War. Yet it was fabricated from the outset, and admitted as such in 2003 by McNamara, and in declassified reports from the National Security Agency, 2005. In his book *Prophets of War* (2011), William Hartung documents how tests on weapons are often faked, as with the multi-billion dollar Strategic Defense Initiative system. Hartung says this propels high profits for the manufacturers, regardless of whether the weapons are safe or even effective. The **true extent of militarism**, at least in the U.S., is largely disguised from the public, as are the known flaws in certain military weapons and vehicles, which, aside from the devastation of the wars and the loss of lives, cost the public billions of dollars through overpricing and no-bid contracts.[34]

The FIRE dominant period generated wars fought for freedom, independence, and fulfillment of national aspirations. In the EARTH dominant period, a random FIRE conjunction in May 2000 joins this energy together with several more factors, making war and war-like situations increasingly dangerous, and showing **these trends for war** during the EARTH period: **1)** the obfuscation of real reasons for war, which (aside from gains by the military industry) may also be generated by commercial interests of corporate conglomerates seeking rights to oil, minerals, water, or land without necessarily benefiting, and often minimizing local or national interests;[35] **2)** lack of resolution over the use of nuclear weapons and nuclear waste, causing continued buildup, and increased dangers from accidents or terrorism; **3)** advancement of weapons technology (including new laser technology) useable on the battlefield and also by policemen or criminals on unarmed, law-abiding citizens; **4)** lack of ethics and/or of precision in using advanced technology, some of it containing uranium and plutonium (including "bunker buster bombs"), thus radioactive material with dire long-term effects; or remote-controlled drones which drop bombs on targets, often harming or killing innocent civilians; or soldiers killing indiscriminately amidst the local civilian population, seeing war as sport, creating "staged killings," etc.; **5)** the use of mind-altering drugs on soldiers, keeping them permanently battle-ready, along with physical and psychological torture

---

[34] Hartung is a journalist and Director of the Arms and Security Initiative at the New America Foundation. His book is titled: ***Prophets of War: Lockheed Martin and the Making of the Military-Industrial Complex***, 2011. The author examines the history of one of America's major defense contractors, and its accumulated power and influence. Government and industry spokesmen have been quick to minimize the vast scope of its revelations and the accuracy of its documentation.

[35] See ***Fuel on the Fire: Oil and Politics in Occupied Iraq***, 2011, by British investigative journalist Greg Muttit. After a five-year struggle, Muttit obtained over 1000 official documents through the Freedom of Information Act. These reveal what British and U.S. governments firmly denied: that the agenda of leading oil companies had anything to do with the March 2003 US-UK led invasion and occupation of Iraq. (Iraq's oil reserves rank as the second largest in the world.) These and similar themes are also discussed in Canadian author Naomi Klein's book, ***The Shock Doctrine: The Rise of Disaster Capitalism***, 2007.

techniques used on "enemy combatants" or disloyal soldiers;[36] **6)** the vast increase in the use of mercenary soldiers, such as Blackwater (a private military company, renamed Xe Services), who perform tasks involving military intelligence, surveillance, and offensives, and often cannot be held accountable, as they can claim being outside the jurisdiction of any national or international law (similar to transnational corporations);[37] **7)** the huge increase in surveillance, rendering any citizen vulnerable to being misunderstood, and wrongly accused or categorized; and lastly **8)** the American commitment to the prison as an industry, with a population of 2.5 million in prison in the U.S. as of 2011, over 70% of them black or brown races; also, the continuation of virtual "black hole prisons," such as Guantanamo, as well as "extraordinary rendition," removing the right of *habeas corpus.*

Such advances in technologies are potentially of enormous advantage to all nations, but with Jupiter's capacity to bestow ethics so reduced in the EARTH dominant periods, it is far easier for war to become big business, and therefore – ongoing.[38] As **Jupiter** represents **journalists and journalism**, its relative weakness during EARTH dominant periods greatly reduces the power of journalists to document wars and conflicts and keep the balance of power between citizens and their government. Journalists, for their part, are increasingly swayed by access to power rather than by holding accountable those in positions of power. The results are potentially far more lethal for the entire world, compared to wars in previous eras. Citizens have to be increasingly diligent about their sources of information. Otherwise, they are pawns in the hands of the most moneyed and powerful. They have to learn how to make the new technology increase their individual rights, and this is a challenge in the EARTH period. Citizens have no real privacy from governments and corporations, who in turn do not hesitate to punish those who try to make *their* secrets known, as with the example of WikiLeaks.

The next JU-SA conjunction in Capricorn occurs Dec. 21, 2020 (6:20 Capricorn, in *Uttara Ashadha nakshatra*). It is especially potent, coming at the Winter Solstice, but the meaning of the *nakshatra* "unstoppable victory" is troubling for democracies and for the individual citizen, if it continues the corporate-dominant policies solidified from 1961. (See below for the significance of **the U.S. Presidential candidate** elected in **2020**. Any political or corporate appointee entering office in late 2020 has greater opportunity, but the American Presidential Election cycle makes it especially likely that party will remain in the White House for the next 28 years, with one exception, at the 12-year mark, or one complete Jupiter orbit.)

The Dec. 2020 conjunction is followed by 80 years of unbroken EARTH conjunctions until 2100, when the first intermittent AIR sign conjunction occurs. Two more **JU-SA conjunctions** occur **in Capricorn**: on March 14, 2080 (16:54 Capricorn, in *Shravana nakshatra*); and on Jan. 14, 2140 (21:16 Capricorn, also in *Shravana nakshatra*). These *nakshatras* hold greater hope for the corporatocracy listening to the voice of the people, and being more adaptable to the real needs of the times, rather than creating unnecessary needs. Throughout the EARTH dominant period,

---

[36] **http://www.armytimes.com/news/2010/03/military_pharmaceuticals_webb_032410w/**Psychiatric drug use by U.S. soldiers has risen 76% since the start of U.S. wars in Iraq and Afghanistan, from 2002.

[37] See Jeremy Scahill's ***Blackwater: The Rise of the World's Most Powerful Mercenary Army,*** *2007.*

[38] Not optimistic about the pattern of Permanent War being reversed, Andrew Bacevich writes on this topic in his book, ***Washington Rules: American's Path to Permanent War*** 2010. Bacevich retired as U.S. Army Colonel after 23 years, and is Professor of History and International Relations at Boston University.

but especially during these 20-year periods ushered in by a Capricorn conjunction, reinforcing the seminal Feb. 18, 1961 JU-SA conjunction, the **stakes are the highest for human survival** and for finding solutions to problems of the earth's environment, including overpopulation, climate extremes, nuclear energy, fossil fuel energy, and the safety and availability of food and water.

The decolonization process following World War II gave independence to some nations desiring it, but also bred a new level of local conflicts, based on differences in language, race, and religion – differences that had been held together before under the yoke of colonialism. One major difference in these fierce conflicts, at times breeding genocide, is that they are increasingly focused around the sharing of often dwindling resources, such as food and water. In favoring the corporations, the EARTH period has also favored the hiring of workers in other countries whose services can be purchased far more cheaply. Thus, workers are more beholden to corporate employers, often working abroad, with their rights diminished, and fewer jobs for educated young people to enter. The EARTH period can see the most complaints regarding financial inequities and workers' rights. As long as democracies can be bought by corporations, there is less likelihood these situations will be corrected, and they are among the major dilemmas of the EARTH dominant period, reaching peaks of chaos and rebellion at the opposition of Jupiter and Saturn, every 10-11 years following the conjunctions.[39]

[39] The **Virgo/Pisces axis** is associated with workers, work and health conditions, and sacrifices made in both arenas. It is also about nations, races, or segments of the population who have been victimized. Pisces can involve spiritual leaders and themes of salvation, or perceived as salvation. If a national chart has Sagittarius Ascendant (such as the U.S. or Japan), the Virgo/Pisces axis radically impacts its infrastructure, especially the oceans or water supply, with Pisces in the 4th house. The **first exact JU-SA opposition in Pisces/Virgo occurred Aug. 16, 2010.** The Deepwater Horizon Oil explosion occurred April 20, 2010, resulting in the largest-ever **oil spill in the Gulf of Mexico**, with 11 workers killed, vast damage to the ecosystem, fishing and tourism industries, and health of workers and local residents. A government commission found BP (British Petroleum), Halliburton, and Transocean lacking in safety and risk assessment. Federal regulation was also found to be very weak. On **March 28, 2011 the second (and last) exact JU-SA opposition in Pisces/Virgo** occurred. Numerous relevant events can be cited: **1)** From Jan. 2011 **unprecedented protests by citizens around the world** focused on inequities for workers and students created by autocrats and/or autocratic behavior favoring the very rich at the expense of the broader population of citizens. Many unarmed citizens were killed by government forces in several nations operating under long-term Emergency powers to protect their unlimited authority. **2)** On March 11, 2011, a **massive 9.0 earth quake and tsunami hit NE Japan,** killing around 24,000 people, with thousands more missing, and de-stabilizing the entire Japanese economy. Already in need of repair, six nuclear reactors in Fukushima, Japan were severely damaged, and evolved into the world's worst known catastrophe up to that point involving radioactive leaks from a nuclear plant. **3)** On March 14, 2011, **His Holiness the Dalai Lama XIV retired** from his position as Head of the Tibetan government in exile (in India), one he had held for 61 years, from March 17, 1950. On May 29, 2011 he delegated his administrative and political authorities to the democratically elected leaders of the Central Tibetan Administration. China has pressured other nations not to recognize him or his government since March 29, 1959, when he fled Tibet during its uprising against China. **4)** On March 29, 2011, the U.S. Supreme Court heard its first oral arguments in the case of **Dukes v. Wal-Mart Stores**, Inc., the largest class action suit (1.6 million women) against the world's largest private employer. On June 19, 2001, employee Betty Dukes first filed a class action sex discrimination lawsuit against Wal-Mart Stores, charging that Wal-Mart discriminated against women in promotions, pay, and job assignments in violation of Title VII of the Civil Rights Act of 1964. The case began in 2000, coinciding with the **JU-SA conjunction.** On June 20, 2011, the U.S. Supreme Court ruled against the 1.6 million women plaintiffs, saying that they may not seek a class action lawsuit as a group, but can do so individually. This overturned a lower appeals court ruling. **5)** On April 1, 2011, esteemed African-American scholar **Manning Marable** died, shortly before the release on April 4, 2011 of his life's work, ***Malcolm X: A Life of Reinvention***, a book revealing new facts about his subject, including three chapters of his famous *Autobiography* never published, and held back by co-author Alex Haley. Marable takes a deeper look into the history of black American slaves and slavery.

The U.S. Presidential election cycles: Within a **160-year time frame**, from **fall 1940 through fall 2100**, the **JU-SA conjunctions** coincide closely with **the U.S. Presidential elections** (occurring every four years). This makes them an excellent predictive model of a historical event combined with recurring 20-year JU-SA conjunctions, 28-year Saturn cycles, and 12-year Jupiter cycles. In this case, the historical event is either the U.S. Presidential Election or the Presidential Inauguration.[40] The JU-SA conjunctions prior to 1940 and after 2100 do not coincide closely enough to either the inauguration or the election. When they are one year or more distant from each other, the parameters do not meet the criteria I have established for these events to take place.

The JU-SA conjunction should fall at most within several months of either **1)** the Presidential Election – Tuesday following the first Monday in November) or **2)** the Presidential Inauguration – Jan. 20 every four years since 1937. (From 1793 to 1933 the Inauguration occurred on March 4.) The closer the proximity of either of these events to the JU-SA conjunction, the more effective in giving authority (real or mythical) to that Presidency. If both events are close, all is confirmed.

The triple **JU-SA conjunctions of 1940-1941** coincided closely with both the election *and* inauguration of a Democratic candidate (**Franklin Delano Roosevelt**). He was elected on Nov. 6, 1940 to his third term as president. The first of three conjunctions in Aries was on Aug. 7, 1940, followed by one on Oct. 19, 1940, and a third one on Feb. 14, 1941, just under four weeks after his Inauguration. This initiated a 28-year run (one Saturn cycle) of Democratic Party presidents, with a vulnerable point at the 12-year mark (one Jupiter cycle) in Nov. 1952, when Eisenhower and his Republican administration was elected for two successive terms. From the 1930s, Progressives in the U.S. were united into the **New Deal Coalition**, which dominated American elections up through 1968, in terms of winning the Presidency for the Democratic Party. By 1968, at the end of the Saturn cycle from 1940, Democrats were vulnerable once again.

Democrat **John F. Kennedy** was inaugurated one month before the **JU-SA conjunction** Feb. 18, 1961, and Republican **Ronald Reagan** was both elected *and* inaugurated within two months of the first of the **JU-SA triple conjunctions of 1980-1981**, this one on Dec. 31, 1980. Reagan's election coincided with the start of Republican Party dominance in the Presidency for the subsequent 28 years, with the exception of 8 years of Bill Clinton, a Democrat. This too occurred at the 12-year mark, with Clinton's election in Nov. 1992, and also in Nov. 1996. As the Virgo conjunction occurs in the 10th house of status of the USA chart (8:59 Sagittarius Ascendant), Reagan was assured of being deified, despite many of the hard facts of his presidency.[41]

Since this cycle is reliable and accurate within these parameters as of 1940, and since there is a JU-SA conjunction in Capricorn on Dec. 21, 2020 we can predict that the political party whose candidate is elected to the U.S. Presidency in **November 2020** is likely to stay in the White House and keep dominance for the next 28 years (a Saturn cycle), with a hiatus only at the 12-year mark (a Jupiter cycle) in 2032, as that is a point of vulnerability in the cycle. Due to characteristics of the EARTH period, that candidate and party will have to curry favor with the corporate elite, but that has been true for most American elections since 1961. The U.S. President elected in Nov.

---

[40] In part based on these principles, I successfully predicted the outcomes of several U.S. Presidential Elections, including that of Barack Obama, in early Feb. 2008. See link to this coverage on the Home page of my website: **www.edithhatha-way.com.** All other articles of mine cited here are also at that website.

[41] USA chart used is the James Kelleher chart: July 4, 1776, 6:30 PM LMT, Philadelphia, PA.

2020 is again likely to gain stature in historical terms, whether deserved or undeserved. The U.S. remains a dominant national power throughout the EARTH period due to the strength of its national chart and due to the status of the JU-SA conjunction chart set for Washington Feb. 1961.[42]

AIR dominant period (2199 to 2398)

This period seems to run fewer years than average for the triplicities, at around 120 years. At least this is true of the previous one and the one upcoming, though an earlier one ran for 239 years. The most recent AIR dominant period (1305 thru 1425 A.D.) coincided with the Age of the Troubadour, evoking romantic sentiments, in synch with the desire nature (*Kama*), expressed in all three air signs. With an orientation towards cultural exchange, AIR sign dominance brings the impetus to interact across territorial boundaries not governed by politics or religion.

In the AIR period from 332 thru 571 A.D., the Sassanid Empire reigned and its culture peaked during this time. This was the last pre-Islamic Persian Empire (224 to 651 A.D.), and is considered among the most important in Persian history. The cultural influence of the Sassanids extended far beyond the empire's territorial borders to Western Europe, Africa, China, and India. European and Asiatic medieval art was profoundly affected by the art and culture of the Sassanids. In addition to its language and its culture, the Sassanid Empire was known at its peak for its rational system of taxation, based on a survey of landed possessions. This AIR period was succeeded by a WATER period, which brought religious dominance. The larger Persian armies were finally overcome by smaller Muslim armies in 637 A.D., in the first of a series of battles that brought down this Empire.

Continuity of literary and scientific life is enhanced through great libraries. They tend to flourish during AIR periods, while WATER periods may bring the destruction of books due to dogmatism of one fanatical group or another. The great library of Alexandria was destroyed in a WATER period. In 48 B.C. during the Alexandrian War, Julius Caesar accidentally set fire to the building, after having ordered ships in the harbor to be set on fire.[43]

ANNOTATED LIST OF JU-SA CONJUNCTIONS: 1405 to 2020

ASTROLOGICAL PARAMETERS PLUS HISTORICAL EVENTS & DATES

Jan. 16, 1405	**JU-SA conjunction at 8:12 Aquarius, in *Shatabisha nakshatra.***	**AIR**
Feb. 14, 1425	**JU-SA conjunction at 1:28 Scorpio, in *Vishakha nakshatra.* First in WATER conjunction series. Triple conjunctions in 1425. (Two more: March 18, 1425 at 0:42 Scorpio and on Aug. 26, 1425 at 26:50 Libra.** Next triple comnunctions in 257 years in 1682-1683 in Cancer).	*****WATER**

---

[42] This is a method I have devised for predicting national viability over the long term, and is outlined in my 2010 article "*Corporate Conglomerates vs. Nation States: Which Nations will Survive and Thrive?*" Sept. 2010. (See earlier footnotes.)

[43] Source: ***Parallel Lives,*** by Plutarch (46 to 120 A.D.), written about 50 or 60 years after the event. Other versions of the burning of the Alexandrian library surfaced in later centuries, usually blaming Christians or Muslims. British historian Edward Gibbons was known to be anti-Christian and blamed Christians, and Bishop Gregory was anti-Muslim, and blamed Muslims. Consequently, both of these accounts may be more biased.

1440	The **printing press is invented by Johannes Gutenberg** (ca. 1400-1468), first time in the West, though first invented in China 1040 A.D. and Korea 1230 A.D.	
1441	**The Slave Trade begins** when a Portuguese explorer/trader buys African slaves from a black slave trader in Africa.	
July 14, 1444	**JU-SA conjunction at 22:52 Gemini, in *Punarvasu nakshatra*.**	**AIR**
May 29, 1453	**Final fall of Constantinople to the Turks** after a 53-day siege, ending over 1000 years of rule by the Byzantine Empire – a massive blow for Christendom and a great victory for the Ottoman Empire.	
1453	**End of the Hundred Years' War,** a series of conflicts between England and France and other nations, from 1337.	
1454 or 1455	The **first "Gutenberg Bibles"** are completed, begun in 1450 – the first printed books in the West.	
1455-1485	**The Wars of the Roses**, a series of dynastic conflicts fought in England between the Houses of York and Lancaster to determine who would rule over England.	
April 8, 1464	**JU-SA conj. at 18:12 Aquarius, in *Shatabisha nakshatra*.**	**AIR**
Oct. 19, 1469	**The marriage of Isabella of Castile** (1451-1504) **and Ferdinand of Aragon** (1452-1516) unites two Catholic kingdoms of Spain. Their reign lasts from 1474 to 1504, and is considered a great victory for the spread of Christendom.	
1478	**The Spanish Inquisition** is established by a Catholic tribunal ordered through Queen Isabella and King Ferdinand, strongly enforced up to the late 1700s, diminishing from mid-1700s.	
Nov. 18, 1484	**JU-SA conj. at 6:30 Scorpio, in *Anuradha nakshatra*.**	**WATER**
1492	**Columbus "discovers" America** – his voyage funded by Queen Isabella of Spain.	
May 1504	**JU-SA conj. at 29:29 Gemini in *Punarvasu nakshatra*.**	**AIR**
Oct. 31, 1517	German priest **Martin Luther posts his 95 theses,** disputing the theology and practices of the Roman Catholic Church, and saying salvation is by faith alone. Highly controversial, this is the virtual start of the Protestant movement within the Christian religion.	
1519	**Spanish conquerors enter Central America,** aiming to conquer new territories for Spain, and spread Christianity. Finding indigenous peoples, their aim is to eradicate Mayan culture and religion, replace it with Catholicism.	

Jan. 1524	**JU-SA conj. at 22:01 Aquarius, in *Purva Bhadra nakshatra.***	**AIR**
Sept. 8, 1544	**JU-SA conj. at 10:35 Scorpio, in *Anuradha nakshatra.***	**WATER**
1562	**Great Britain joins the lucrative Slave Trade in the New World.** In the same year, the second bishop of the Yucatán orders a **mass destruction of Mayan books.**	
Aug. 25, 1563	**JU-SA conj. at 11:25 Cancer, in *Pushya nakshatra.***	**WATER**
1564-1616	**William Shakespeare:** Many of his most important plays document British royal history, while other plays are on other eras of history or mythical subjects.	
Oct. 7, 1571	**The Battle of Lepanto.** The Ottoman fleet (though superior in numbers) is defeated by the Holy League galleys, involving ca. 150,000 personnel. This last major naval battle in the Western world is fought almost entirely between rowing vessels. Losing more than 200 vessels, the Turks have at least 20,000 casualties.	
May 3, 1583	**JU-SA conj. at 2:08 Pisces in *Purva Bhadra nakshatra.***	**WATER**
Aug. 8, 1588	**Defeat of the Spanish Armada:** British fleets thwart the attempt by Spanish fleets to overthrow Queen Elizabeth I of England, and to stop English dominance of new trade routes.	
1601-1947	**British rule over India** unofficially begins when the British-owned East India Company first enters India to compete with the Dutch Empire in India. The British government officially assumes rulership in 1858, following the rebellions of 1857. British Empire rules over India:1858-1947, in addition to its other holdings in Asia.	
Dec. 18, 1603	**JU-SA conj. at 19:58 Scorpio, in *Jyestha nakshatra.***	**WATER**
1610-1632	Astronomer/astrologer **Galileo Galilei (1564-1642) publicly supports the heliocentric view of the universe** first promoted by Copernicus in 1530 declaring the planets move in elliptical orbits around the Sun. In 1632, Galileo writes his *Dialogue on the Great World Systems,* banned by the Church: 1633-1835.	
1618-1648	The **Thirty Years' War** is fought between Protestants and Catholics in the Holy Roman Empire. It leads to the decentralization of the Holy Roman Empire, and loss of power and influence of the Catholic Church.	
July 16, 1623	**JU-SA conjunction at 18:00 Cancer, in *Ashlesha nakshatra.***	**WATER**
1633	**Galileo's ideas are denounced as heretical,** at a trial by the Inquisition. To avoid torture or imprisonment he agrees to recant his views on heliocentric theory. He is under house arrest and close surveillance until his death in 1642.	

Feb. 24, 1643	**JU-SA conjunction at 6:13 Pisces, in *Uttara Bhadra nakshatra.*** SPRING EQUINOX is at 11:08 Pisces.	**WATER**
Oct. 17, 1663	**JU-SA conjunction at 23:48 Scorpio, in *Jyestha nakshatra.***	**WATER**
1679	**First steam engine is invented,** based on a pressure cooker. **First commercial version: 1712.**	
Oct. 24, 1682	**JU-SA conjunction at 29:43 Cancer, in *Aslesha nakshatra.* (Triple conjunctions in 1682-1683. Two more: Feb. 9, 1683 at 27:16 Cancer and May 18, 1683 at 25:05 Cancer.** Next triple conjunctions in 257 yrs., in 1940-1941 in Aries.)	*****WATER**
1687	***Principia Mathematica...* by Isaac Newton (1642-1727).** He presents the Law of gravity. Many consider this is the start of a new era in scientific investigation.	
1688	**The Glorious Revolution in England** puts a permanent end to any possibility of re-establishing Catholicism as the Church of England. Papal control over England is now removed. William III, a Dutch Protestant prince, is crowned British king. Britain becomes a Constitutional monarchy, with no future monarch allowed Absolute Power, as more of it is ceded to Parliament.	
May 21, 1702	**JU-SA conjunction at 16:54 Pisces, in *Uttara Bhadra nakshatra.* START OF MUTATION: WATER TO FIRE.**	**WATER**
Jan. 5, 1723	**JU-SA conjunction at 3:20 Sagittarius, in *Mula nakshatra.* 1st CONJUNCTION IN FIRE.**	**FIRE**
Aug. 30, 1742	**JU-SA conjunction at 6:52 Leo, in *Magha nakshatra.***	**FIRE**
March 18, 1762	**JU-SA conjunction at 21:48 Pisces, in *Revati nakshatra.***	**WATER**
April 1775 to Feb. 1783	**The American Revolutionary War.** End of conflicts by Feb. 1783.	
July 4, 1776	**The Declaration of Independence of the 13 American colonies,** written primarily by Thomas Jefferson (1743-1826), American statesman, later the third U.S. President, 1801-1809.	
Nov. 15, 1777	The **Articles of Confederation** are adopted, the first Constitution of the United States, later replaced (Sept. 1787). Though the new government can print money, it *has* no money and cannot repay the foreign debt incurred.	
Nov. 5, 1782	**JU-SA conjunction at 7:18 Sagittarius, in *Mula nakshatra.* On the Galactic Center. WINTER SOLSTICE is at 9:11 Sagittarius. START OF FIRE PERIOD. END OF MUTATION: WATER TO FIRE.**	**FIRE**

Feb. 1783	**King George III announces Cessation of Hostilities in the American Revolutionary War.**	
Sept. 3, 1783	**Paris Peace Treaty is signed** – formal end of the American Revolutionary War. It gives **formal recognition to the 13 colonies as "free, sovereign, and independent states."** Britain cedes most of the land east of the Mississippi River.	
Sept. 17, 1787	The **U.S. Constitution** is adopted by the Constitutional Convention, Philadelphia, PA. (George Washington is elected President of the Convention.)	
1789	**French Revolution. George Washington is inaugurated as first U.S. President, April 30, 1789.**	
1789	French engineer **Nicholas Joseph Cugnot invents the first steam-driven self-propelled carriage in Paris.** Its speed is 3 miles per hour (5 km. per hour).	
1781-1804	**Haitian Revolution.** Black slaves revolt against their French masters, declaring themselves free. This is the first-ever republic ruled by people of African ancestry, with independence declared Jan. 1, 1804.	
July 17, 1802	**JU-SA conjunction at 14:01 Leo, in *Purva Phalguni nakshatra*. (*Vargottama* status adds strength.)**	**FIRE**
1804-1815	The **Napoleonic Wars** are fought, with the expansion and collapse of the French Empire, notably after its unsuccessful invasion of Russia in the War of 1812. Spain's Empire also starts to collapse, as does the Holy Roman Empire (962 to 1806).	
1804	British inventor and mining engineer **Richard Tevithick invents the first high pressure steam engine and world's first working steam locomotive** – the first to invent a heavier haulage for mining. First railroads are used to haul minerals for mining purposes. **Feb. 21, 1804:** world's first locomotive-hauled railway journey.	
March 25, 1807	**Britain outlaws its slave trade, with the Slave Trade Act of 1807**. Slave trade is abolished on moral grounds, but not slavery itself – still legal in British Empire thru 1833. Pressured by Britain, **the U.S. abolishes its African slave trade in 1807,** but not its internal slave trade.	
Jan. 8, 1811	**Largest slave revolt occurs in the United States,** in New Orleans, seat of the nation's wealthiest plantations. (500 slaves involved)	

1815-1914	The **British Empire dominates**, and the *Pax Britannica* (British Peace). Relative peace reigns throughout Europe, as Britain controls most maritime trade worldwide, including with Chinese markets. Britain's sea power is unchallenged.	
June 19, 1821	**JU-SA conjunction at 3:17 Aries, in *Krittika nakshatra*. (*Vargottama* status adds strength. FIRST conjunction in ARIES.)**	**FIRE**
Dec. 2, 1823	**U.S. President James Monroe presents the Monroe Doctrine**, a policy stating that no further colonization of American states should be permitted.	
1825	**England has the first steam-powered freight and passenger service train.**	
1826	**U.S.A. has the first locomotive train, for mining use only.**	
1832	The first of the **British Reform Acts** is passed, reforming corruption and flagrant abuses of patronage in the House of Lords, and reorganizing the entire representative basis of the House of Commons. Constitutional monarchy continues, as from 1688.	
1833	**Slavery Abolition Act of 1833 passes in Great Britain**, abolishing slavery throughout most of British Empire, with a few exceptions. The East India Company is exempt in the islands of Ceylon (now Sri Lanka) and Saint Helena (South Atlantic Ocean). The largest of slave-trading nations, Britain is the first to abolish slavery.	
1834	**English education is introduced into India as part of the British Empire**, breaking the legacy and tradition of learning Sanskrit, the ancient and traditional language of the Hindu/Vedic classics. Loss of this tradition is resurrected gradually after independence.	
June 20, 1837	**Queen Victoria ascends the British throne**, until her death Jan. 22, 1901. A Constitutional monarch, she also uses the title of Empress of India (from May 1, 1876).	
Jan. 26, 1841	**JU-SA conjunction at 17:15 Sagittarius, in *Purva Ashadha nakshatra*.**	**FIRE**
Dec. 2, 1845	**U.S. President James Polk** announces his policy of "**Manifest Destiny**" to Congress. In strictly enforcing the Monroe Doctrine, he says the U.S. should agressively expand into the Western part of the continent.	

1851 & 1855	**Bessemer steel is invented – the first inexpensive industrial process to mass-produce steel from pig iron.** Englishman **Henry Bessemer** is awarded a patent on it in 1855. Independent discovery occurs in 1851 by American William Kelley. (Predecessor to this process is invented in China in the 11th century).	
March 6, 1857	**Dred Scott decision: The U.S. Supreme Court determines that the Constitution does not protect black slaves** (those of African descent previously imported to the U.S.).	
May 10, 1857	**Indian Rebellion of 1857 (thru 1859).** The first War of Independence by East Indians against the British Empire.	
1858-1947	The **British Empire officially rules over India and other territories in Asia,** taking over control from the East India Company.	
Nov. 22, 1859	British naturalist **Charles Darwin's book *Origin of Species*** goes on sale, introducing his scientific theories of evolution by natural selection. His previous work is published anonymously in 1844: *Vestiges of the Natural History of Creation,* speculative natural history, initially causing a controversy, notably in theological circles.	
Oct. 21, 1861	**JU-SA conjunction at 26:26 Leo, in *Purva Phalguni nakshatra.***	**FIRE**
April 12, 1861 to April 9, 1865	**The American Civil War:** 11 Southern States (the Confederacy) declare their secession from the United States. They fight against the Union (led by the U.S. federal government, with 25 mostly Northern states, 20 of which had abolished slavery, but not 5 Border states. Heavy casualties are suffered. In 1865, the Confederacy loses, the union is restored, with a stronger federal government, and slavery is outlawed in all states.	
Sept. 22, 1862	**The Emancipation Proclamation** is issued by **U.S. President Abraham** Lincoln, making a new and central goal of the American Civil War the ending of slavery in the South. He proclaims the freedom of 3.1 million of the nation's 4 million slaves. (Southern states go against this federal ruling, effective Jan. 1, 1863).	
Dec. 18, 1865	**Thirteenth Amendment** to the Constitution ends slavery in the United States. Though slavery had been abolished as a war act (1862), it is now illegal.	
Jan. 18, 1871	**German Empire is created** after unification, becoming one of the most powerful industrial economies in the world. Its first ruler is Kaiser Wilhelm I, 1871-1888. This empire endures up to Nov. 1918.	

April 18, 1881	**JU-SA conjunction at 9:24 Aries, in *Krittika nakshatra.***	**FIRE**
1885	**First self-propelled vehicle with an internal combustion engine.** Three wheeled car built by German engineer **Karl Benz,** Germany. Improved version of same vehicle built in Germany same year by **Gottlieb Daimler.** Introduced in France in 1894.	
June 5, 1888 to Nov. 18, 1918	**Kaiser Wilhelm II is Emperor of Germany, King of Prussia,** following the death of his father, Frederick III, after a 3-month reign. Wilhelm II is the last German Emperor and King of Prussia. He is first grandchild of Queen Victoria of England. His mother is **Victoria, Princess Royal, of England.**	
1897-1920	**Steam-driven autos known as the "Stanley Steamers"** are produced by the Stanley brothers in Massachusetts, biggest manufacturers of this type of car.	
Jan. 22, 1901	**Death of Queen Victoria, after an unprecedented reign of 64 years. King Edward VII (her eldest son) ascends the throne.**	
Nov. 9, 1901	**Eldest son of King Edward VII is created Prince of Wales and Earl of Chester.**	
Nov. 28, 1901	**JU-SA conjunction at 21:30 Sagittarius. MUTATION: FIRE TO EARTH (1901-1961).**	**FIRE**
Dec. 17, 1903	**First successful flight of a flying machine by the Wright brothers,** Kitty Hawk, North Carolina, lasting 12 seconds.	
1908	**The Model T Ford car is manufactured by Henry Ford,** 15 million of them up to 1928, when they are discontinued.	
May 6, 1910	**Death of King Edward VII. His eldest son, King George V ascends the throne. Coronation June 22, 1911.** (His reign lasts until his death in 1936.)	
1914-1918	**World War I begins July 28, 1914,** sparked by the assassination of Archduke Franz Ferdinand of Austria-Hungary and his wife Sophie, Duchess of Hohenberg, in Sarajevo one month earlier, by a Serbian activist.	
July 17, 1917	**British Royal House of Saxe-Coburg and Gotha becomes the House of Windsor** – name change from longtime German ancestry by royal proclamation from **King George V.**	
Nov. 2, 1917	**The Balfour Declaration** is signed in Britain, indicating Britain's intention to support the creation of a national homeland for the Jews in Palestine.	

Nov. 19, 1918	**Kaiser Wilhelm II** officially abdicates the throne (unofficial announcement Nov. 9, 1918) and is forced to leave Germany. He immigrates to Holland, where he resides until his death in 1941. He is not allowed burial in Germany.	
1918	**Hydraulic brakes are invented by Malcolm Lougheed in California (later Lockheed, of Lockheed Aircraft Corp. in 1926,** originally founded 1912 under another name.) Hydraulic brakes not applied to cars until 1921, clearing the way for greater mass production of cars.	
June 28, 1919	**The Treaty of Versailles is signed,** after over six months of negotiations and conferences, concluding World War I. The treaty is disputed by many of its parties, and is not ratified by the U.S. Senate.	
Nov. 7, 1919	**Physicist Albert Einstein becomes an overnight celebrity** when *The Times* (London) headlines reports the results of May 1919 expeditions to observe a Total solar eclipse, successfully proving his Theory of Relativity (1905), declaring "A Revolution in Science." This was announced the previous evening in London.	
Sept. 9, 1921	**JU-SA conjunction at 3:49 Virgo, in *Uttara Phalguni nakshatra.*** FIRST CONJUNCTION IN EARTH.	**EARTH**
1921	**Ford Motor Co.** (after their financial crisis in 1921) **starts producing high-priced motor cars,** along with other vehicles. Hydraulic brakes are used in some cars. Public transport is still used by 90% of Americans residing in cities and interurban areas.	
May 20, 1927	**Aviator Charles Lindbergh flies the first successful solo transatlantic flight** from New York to Paris, in 33½ hours. Famous overnight, his flight is the harbinger of commercial airline travel.	
October 1929	**New York Stock Market Crash** leading to a 12-year Economic Depression and the greatest economic crisis of the 20th century. After the speculative boom in the 1920s, the stock market lost over $30 billion in two days at the end of Oct. 1929.	
Jan. 30, 1933	**Adolf Hitler is appointed Chancellor of Germany,** soon consolidating absolute power of his Nazi party and its radical racist and fascist policies in Germany.	
June 16, 1933	**Banking Act of 1933 ("Glass-Steagall Act") enacted,** separating commercial banks from investment banks and intended to protect consumers from fraudulent lending and speculation by banks.	

1939-1945	**World War II, from Sept. 1, 1939 through Aug. 1945.** Germany invades Poland Sept. 1, 1939. France and Britain declare war on Germany Sept. 3, 1939.	
May 10, 1940	**Winston Churchill is appointed British Prime Minister,** after 10 to 11 years of virtual political exile. On the same day Germany begins its assault on Western Europe, starting with the Low Countries (Netherlands, Belgium and Luxembourg). **June 22, 1940,** Germany occupies the northern half of France.	
Aug. 7, 1940	**JU-SA conjunction at 21:25 Aries, in *Bharani nakshatra.* (Triple conjunctions in 1940-1941. Two more: Oct. 19, 1940 at 19:26 Aries, and Feb. 14, 1941 at 16:05 Aries.** Next triple conjunctions in 40 yrs, 1980-1981 in Virgo.)	*****FIRE**
Nov. 6, 1940	**Democrat Franklin D. Roosevelt is elected to his 3rd term as U.S. President. Inaugurated Jan. 20, 1941.**	
Dec. 7, 1941	**Japan attacks Pearl Harbor. The U.S. officially enters WWII.**	
Dec. 2, 1942	**The first controlled nuclear chain reaction occurs on Dec. 2, 1942 in Chicago,** a major breakthrough in signaling U.S. leadership in the nuclear arms race and eventual dominance during WWII.	
July 1944	Allied nations (all 44 of them), with 730 delegates meet at **Bretton Woods,** NH, USA to set up a **system of monetary management** intended to rebuild the international economic system. All participating nations are obligated to tie their currencies to the US dollar, the only gold-backed currency and the one with the most purchasing power. **Bretton Woods creates: 1) the International Monetary Fund** (IMF) in late 1945, starting operations March 1, 1947 and **2) the World Bank,** initially named the International Bank for Reconstruction and Development (IBRD). Both are set up to bridge temporary imbalances of payments from debtor nations.	
July 16, 1945	**First nuclear weapons test of an atomic bomb** is conducted in New Mexico by the U.S. Army, effectively the start of the Atomic Age. British and Americans conduct feasibility studies from 1939. The project comes under U.S. control from 1942, with the first important breakthroughs in Dec. 1942.	
Aug. 6 & 9, 1945	**Atomic bombs on Hiroshima and on Nagasaki,** respectively, are dropped by U.S. military planes. On **Aug. 15, 1945** Emperor Hirohito announces the surrender of Japan, ending WWII in the Pacific. European theater ends in **May 1945.**	

Oct. 24, 1945	**The United Nations is established,** successor to the League of Nations.	
Aug. 14 & 15, 1947	**Statehood is declared for Pakistan and India,** respectively, splitting the new nation into Muslim and Hindu sectors not intended by many of the original Freedom Fighters, including Mahatma Gandhi.	
May 14, 1948	**Israel declares its statehood,** followed by war with neighboring Arab states, which are defeated by Israel.	
June 2, 1953	**Queen Elizabeth II coronation** occurs in London, the first-ever to be televised.	
Nov. 1, 1955 to April 30, 1975	The **Vietnam War.** U.S. military involvement is from the early 1960s, escalating in summer 1964, officially with the Gulf of Tonkin Resolution.	
Oct. 4, 1957	The **Soviet Union launches Sputnik I,** the first-ever vehicle to orbit the earth. This unexpected event creates shock waves, especially with the U.S. public, comparing the advanced status of the Soviet space program to that of U.S. education and achievements. Perceived as reflecting badly on the American space program, etc.	
Feb. 1, 1958	The U.S. responds to Sputnik with its **launch of Explorer I,** only possible through the efforts of German scientist **Wernher von Braun,** working for the U.S. space program (& U.S. Army) since 1945 with his mostly German team of rocket scientists.	
July 29, 1958	**The National Aeronautics and Space Administration (NASA) is created** through an act of U.S. Congress. President Eisenhower signs this Act into law.	
Jan. 20, 1961	**Democrat John F. Kennedy is inaugurated as U.S. President. Elected Nov. 8, 1960.**	
Feb. 18, 1961	**JU-SA conjunction at 1:52 Capricorn, in *Uttara Ashadha nakshatra.* (*Vargottama status* adds strength.)**	**EARTH**
April 12, 1961	Soviet cosmonaut **Yuri Gagarin is launched into orbit** around the Earth on Vostok I, the first human to travel into space and the first to orbit the earth.	
May 5, 1961	Astronaut **Alan Shepard becomes the second human and the first American in space on** the Mercury-Redstone3. Originally scheduled for Oct. 1960, and postponed several times, this was a ballistic trajectory sub-orbital flight.	

Aug. 12-13, 1961	**The Berlin Wall is erected by the Soviet Union,** preventing citizens from the Eastern Europe Bloc from emigrating to the West. A major Cold War symbol, it divides East and West Berlin until its demise Nov. 9, 1989.
Sept. 18, 1961	Called "the greatest statesman of the 20th century" by JFK, **Dag Hammarskjold is killed** in a suspicious plane crash over the African Congo on his way to a meeting in the Congo. **Secretary General of the United Nations (1953-1961),** he is the last major activist in this U.N. position, seeking to protect small nations and to help newly independent African nations to have access to their own resources. Swedish-born Hammaraskjold is winner of the Nobel Peace Prize in Dec.1961, the only person to receive it posthumously.
1964 -1965	**U.S. Civil Rights Acts and Voting Rights Acts,** major goals reached. Civil Rights movement is led from Dec. 1955 to April 1968 by **Martin Luther King, Jr.**
June 5-10, 1967	**Six-day War:** Arab armies fight Israel on three fronts (Egypt, Syria, Jordan) and are swiftly defeated. Israel gains control of most of the Golan Heights, West Bank, Sinai Peninsula (to 1982), and Gaza Strip (to 2005) though Gaza airspace is still controlled by Israel. (East Jerusalem was annexed in 1980). Israel's occupation of the Palestinian territories from June 1967 is a subject of ongoing dispute and periodic wars. The United Nations passes 131 resolutions between 1967 and 1989 regarding the ongoing Israeli-Palestinian conflict and/or Israeli conflict with neighboring states, which began May 14, 1948 (its statehood).
July 20, 1969	**U.S. lands a man on the Moon, first promised by President John F. Kennedy on May 25, 1961.**
Aug. 15, 1971	**Gold-dollar link is severed by U.S. President Richard Nixon,** formalizing the complete removal of the U.S. dollar from the gold standard (a process in the making from March 1968). This creates a free-floating and speculative currency market, and breaks the promise of Bretton Woods (July 1944). The USD is now a full "fiat currency" backed only by a promise from the federal govt., which is unable to maintain the gold-peg to the USD due to U.S. war expenses and the rise of the EEC & Japan. Domestic economic interests are favored over preserving an international monetary regime, lifting sanctions on international currencies and banking practices, easing the globalization process as well as greater potential for money laundering.

Dec. 3-16, 1971	**Pakistan-India War:** India takes a swift offensive after the Pakistani Air Force hits Indian airfields in northern India on Dec. 3. War results in **nationhood for Bangladesh** (formerly East Pakistan) and a decisive victory for India. This is regarded as a high point for Indira Gandhi as Prime Minister.	
1975	The **first personal computers** appear on the market, built with microprocessor chips. They come as a kit to be assembled. The **Altair 8800** is usually conceded to be the first personal computer.	
June 1, 1980	**The world's first 24-hour television news network is launched** with CNN (Cable News Network), in the U.S.	
July 1980	**IBM executives meet for the first time with Bill Gates and Paul Allen of Microsoft,** hiring them to create an operating system for their first-ever personal computer - the IBM PC - released on Aug. 12, 1981.	
Dec. 12-18, 1980	**First public offering (IPO) of Apple Computer Inc.** - the largest IPO since Ford Motor Company went public in 1956. All 4.6 million shares sell within minutes. More than 40 Apple employees become instant millionaires from their stock options. Co-founder Steve Jobs makes $217 million in one day.	
Dec. 31, 1980	**JU-SA conjunction at 15:54 Virgo, in *Hasta nakshatra.* (Triple conjunctions in 1980-1981. Two more: March 4, 1981 at 14:29 Virgo and July 23, 1981 at 11:20 Virgo.** Next triple conjunctions occur in 257 years, in 2238-2239, with JU-SA in Gemini.)	***EARTH
Jan. 20, 1981	**Republican Ronald Reagan is inaugurated U.S. President. Elected Nov. 4, 1980.**	
Aug. 12, 1981	**IBM releases its first-ever personal computer** - called **the IBM PC** computer.	
Aug. 17, 1982	**Philips Electronics presses the first CD (compact disc)** at a factory in Germany, having worked with Japan's Sony Corp. on its development, from 1979. **Sony sells the world's first CDs in 1982,** replacing LP (long-playing records) sales in Japan by 1987.	
March 24, 1989	**Exxon-Valdez Oil spill - largest oil spill in U.S. history, prior** to the explosion of the Deepwater Horizon oil rig in the Gulf of Mexico April 20, 2010, both disasters are in U.S. coastal waters. Congress passes the **Federal Oil Pollution Act Aug. 19. 1990,** intended to regulate commercial oil operations, though it does little to improve safety measures or regulate the oil industry, as of 2011.	

April 15 to June 4, 1989	**Tiananmen Square protest (Beijing) by Chinese students and intellectuals leads to a massacre, when government militia open fire, killing up to 800, and probably more** (statistics unavailable). The protesters use mainly non-violent means of civil resistance to seek economic reform and political liberalization in China. Further government crackdowns follow, with mass arrests and imprisonments, and foreign press banned.	
Nov. 9. 1989	**Fall of the Berlin Wall, End of the Cold War, followed by the breakup of the Soviet Union, from June 12, 1990.** The reunification of Germany takes place on Oct. 3, 1990.	
Dec. 25, 1990	The **Internet is invented by British scientist Tim Berners-Lee,** with his first successful communication between an HTTP client and server via the World Wide Web.	
1994-1995	**The era of massive bank bailouts** begins in Mexico with the Mexican peso crisis.	
Summer 1999	**Unprecedented level of bailout of Mexican banks and their lenders (American banks) by the U.S. Exchange Stabilization Fund ($20 billion),** sent without U.S. Congressional approval; and $18 billion from the IMF (International Monetary Fund). The bailout reaches a total of over $100 billion, a model repeated and amplified in fall 2008. The summer 1999 bailout was designed to save loans made by American banks. Europeans worry about the new IMF trend and "crony capitalism" replacing free market capitalism.	
Nov. 12, 1999	**Repeal of the Glass-Steagall Act (U.S. Banking Act of 1933),** paving the way for rampant speculation by financial institutions worldwide, using their clients' money and risking average taxpayer's funds.	
May 28, 2000	**JU-SA conjunction at 28:52 Aries, in *Krittika nakshatra*.**	**FIRE**
Nov. 7, 2000	**Republican George W. Bush is elected U.S. President. Inaugurated Jan. 20, 2001.** A disputed election result is decided by the U.S. Supreme Court on Dec. 12, 2000, an historically unprecedented move by this court, overturning a Florida Supreme Court decision from Dec. 8, 2000 to allow a statewide recount of Florida votes.	
Sept. 11, 2001	**The World Trade Towers are attacked in New York City.** The George W. Bush administration says it is caused by 19 terrorists using hijacked airplanes to destroy the towers and kill nearly 3000 Americans. The 19 hijackers are reportedly from: Egypt (1), Saudi Arabia (15), United Arab Emirates (2), and Lebanon (1). U.S. and British investigations identify Osama bin Laden and Al-Qaeda as key perpetrators.	

Oct. 7, 2001	**U.S.-U.K. forces invade Afghanistan in** retaliation for Sept. 11 attack on the U.S., with the goal of attacking Al-Qaeda, since some of their operations are in Afghanistan. The war and occupation are ongoing, as of early 2012.	
March 20, 2003	**U.S.-U.K.-led forces invade Iraq.** The stated goal is to depose Saddam Hussein, and remove his "Weapons of Mass Destruction." These WMD are never found, but war and violence continues. American troops leave Iraq in Dec. 2011, handing over 505 large U.S.-built military bases, though the U.S. retains a large (104 acre) embassy compound in Baghdad.	
Sept. 15, 2008	Start of the **Wall Street collapse of 2008-2009.** Failure of massive financial institutions in the U.S. and worldwide due to "toxic assets" (mortgage-backed securities based on untenable subprime loans), predatory lending practices and minimal regulations that increased from the 1990s, with risky investment practices escalating from the early 1980s.	
Sept. 19, 2008	**Huge bank bailouts:** Pres. George W. Bush announces the urgent need for U.S. Congress to give $700 billion (in taxpayer money) to rescue the banks that had overspeculated, using clients' funds. Major financial firms and industrial corporations in the U.S. and elsewhere are rescued by the U.S. govt. and U.S. Federal Reserve, thereby consolidating and increasing their profits. A first-ever Federal Reserve audit report in 2011 reveals some $16 trillion in unreported bailout funds (2007-2010) transferred to major financial firms and without Congressional oversight. Taxpayers lose millions of jobs, pensions and homes, suffering the largest-ever rate of home foreclosures from 2008 onward.	
Nov. 4, 2008	**Democrat Barack Obama is elected U.S. President. Inaugurated Jan. 20, 2009.** The first U.S. president of black African descent, his ancestry is Anglo and African (from Kenya).	
Jan. 21, 2010	**U.S. Supreme Court decision: Citizens United v. Federal Election Commission** grants corporations the right to spend unlimited amounts on political campaigns, giving national or foreign companies & wealthy individuals a decisive role in U.S. elections, thereby also affecting ongoing policy decisions.	
July 21, 2010	**Dodd–Frank Wall Street Reform and Consumer Protection Act** is passed to create greater financial stability and protect consumers, but many loopholes remain for banks to continue reckless practices.	

Jan. 25, 2011	**Pro-democracy protests erupt in Egypt, spreading across the Middle East.** Preceded by riots in Tunisia in Dec. 2010 and causing the ouster of longtime Presidents of Tunisia and Egypt (Jan. & Feb).	
July 28, 2011	**Apple Computer, Inc is (briefly) richer than the U.S. federal govt.** during the debt crisis. Apple's cash reserves at $75.9 billion are $2.1 billion more than the U.S. Treasury Dept., at $73.8 billion. The wild success of iPhone and iPad throughout the world, though vaunted, is underwritten by a production model that ships most jobs to China, South Korea, Taiwan and elsewhere where cheaper labor is available - often with slave labor conditions. This model is increasingly used by other large corporations.	
Dec. 21, 2020	**JU-SA conjunction at 6:20 Capricorn, in *Uttara Ashadha nakshatra*. WINTER SOLSTICE.**	**EARTH**
Oct. 31, 2040	**JU-SA conjunction at 23:40 Virgo, in *Chitra nakshatra.***	**EARTH**
April 7, 2060	**JU-SA conjunction at 6:04 Taurus, in *Krittika nakshatra.***	**EARTH**
Mar. 14, 2080	**JU-SA conjunction at 16:54 Capricorn, in *Shravana nakshatra.***	**EARTH**
Sept. 18, 2100	**JU-SA conjunction at 0:16 Libra, in *Chitra nakshatra.* (*Vargottama* status adds strength.) FIRST CONJUNCTION IN AIR.**	**AIR**
July 15, 2119	**JU-SA conjunction at 19:18 Taurus, in *Rohini nakshatra.***	**EARTH**
Jan. 14, 2140	**JU-SA conjunction at 21:16 Capricorn, in *Shravana nakshatra.***	**EARTH**
Dec. 21, 2159	**JU-SA conjunction at 11:53 Libra, in *Swati nakshatra.***	**AIR**
	WINTER SOLSTICE.	
May 28, 2179	**JU-SA conjunction at 26:41 Taurus, in *Mrigashira nakshatra.* LAST CONJUNCTION IN EARTH.**	**EARTH**

CHAPTER 2

Introduction to Vedic Astrology and Philosophy

Once created, the gods fell into the moving ocean. He brought them hunger and thirst.
They prayed: 'Find for us an abode where we may be well and eat.'
He brought them a bull. They said: 'This is not sufficient.'
He offered them a horse. They said: 'This is not enough for us'. He offered them a man.
They said: 'Well done, for a man, verily, is a thing well done.'
He said: 'Enter into your abodes.'
Fire became speech and entered the mouth. Wind became breath and entered the nostrils. The sun became sight, and entered the eyes. The directions of space became hearing, and entered the ears. Plants and herbs became hairs, and entered the skin. The moon became mind, and entered the heart. Death became the outward-breath (apana), and entered the navel. Waters became semen, and entered the virile member.
Hunger and thirst said to him: 'Find us an abode.' He said: 'I shall give you a place among the gods. I make you partakers among them.'
Hence, whenever oblation is made, hunger and thirst share it.

Aitareya Upanishad 1.2. v.1-5, "The Cosmic Powers in the Human Person,"
Gita Press, ed.; 4.2, Adyar ed.

Of all the systems of Indian philosophy, the *Sankhya* is the most basic one. From it we learn there are two basic principles: the principle of pure consciousness, or the seer (*Purusha*), and the principle of that which is seen (*Prakriti*). *Prakriti* is both matter and the stuff of experience. *Kala Purusha* is Time Personified. It extends across the signs of the zodiac, and shows God, or the Divine force entering human life. God is inextricable from *Kala Purusha*.

In the *Upanishads* of ancient India, the jewel of consciousness is seen to manifest itself in "all that shines: the sun, moon, fire, and speech." The Sanskrit word *Jyotish* means "science of light." Known in the West as Vedic astrology, it is considered sacred knowledge of the cosmos. From the beginning, the Sanskrit language was used to describe and teach it orally, deepening the power and clarity of the astrology with one of the very few sacred and mantric languages in existence. Using

Jyotish to identify the most precise and auspicious timing for the performance of Vedic rituals and for the planting of crops, the sages were able to preserve the heart of this ancient civilization of India. As *Jyotish* evolved, it gradually expanded to include the birth times of human beings, especially those of leaders, to study their destinies from the configuration of planets at birth. The astrological birth chart thus gives a snapshot of when consciousness enters matter. This moment is defined as when *prana* (the breath) enters the body. Being the primal life force, *prana* is vital for life and even has a unit of time assigned to it.[44] At the moment *prana* enters the body the Law of *Karma* begins to operate. Without physical manifestation it cannot operate. The birth moment as described in the heavens is the key to the *dharmic* and *karmic* destiny.

> "Time is the force behind the Creation of the universal elements. It is also the force that brings about the Destruction of all these. Even when the Universe is Asleep, Time remains in a state of Awakening. Time, which is thus ever alert, is an Inviolable Force. ... Time bakes the various living beings along with their souls."
>
> From "The Concept of Time," ***Hora Ratnam***, Chapter 2, *Shlokas* 327-328, p. 309. Transl. by R. Santhanam.

In general, *Dharma* describes the harmony of the universe. On a personal level *Dharma* is one's true nature, one's rightful purpose in life, and the duties and responsibilities therein. If one is expressing one's *Dharma*, it is easier to deal with *Karma* – our conscious actions on an ongoing basis, as well as the cumulative effects of our actions in this life as well as in previous lives. Vedic astrology enables us to identify the *dharmic* and *karmic* destiny of the individual, group, or nation. The chart of a nation's leader reveals group *Karma* and destiny, and is considered to be intimately interwoven with the destiny of the nation in question. To learn more deeply about what is revealed in a particular astrological chart – comparing that to the relevant biographical and historical details of an individual's life – is to understand how *karmic* destiny is inextricably linked to the expression of one's true nature or *Dharma*. It is always possible to overcome negative tendencies and grow with the strengths. When in the West we speak of "character," in Vedic terms the equivalent is how well one understands and accepts one's *dharmic* and *karmic* destiny, and how gracefully one evolves with it over time.

Because "Time bakes ... living beings along with their souls," it is not always possible to avoid difficult *Karma*, though one may learn how to soften the effects. Specific time periods indicate when negative or positive *Karmas* will come due. By studying the lives of others, we can examine how much these individuals have been at one with their destinies, how successfully they persevered through the less auspicious periods, how much were they prone to excess, brought down by life's inevitable defeats, raised up by triumphs, and how well they have used their innate talents, especially within the most auspicious time periods – the ones giving maximum benefit.

The eye of the *Vedas*: The *Vedas* are sacred scriptures of India, and considered Divine knowledge. The Sanskrit word comes from the root *vid*, "to know or to see." *Jyotish* is one of six *Vedangas*, or

---

[44] *Prana* = 4 seconds, or 10 long Sanskrit syllables. One *vinadi* = 24 seconds, or 6 *pranas*. 60 *vinadis* = 24 minutes (1 *nadi*). 60 *nadis* = 1 day. "That which begins with respirations (*prana*) is called real [or embodied]; that which begins with atoms (*truti*) is called unreal [or unembodied]." ***Surya Siddhanta***, Chapter 1, v. 11-12. Smaller time units are given in the *Vedas*, even if not of practical use for most Vedic astrology. For instance, a *Paramáńu* = ca.16.8 microseconds; a *truti* = ca. 1/3290th of a second; a *vedha* = 100 *trutis*; a *lava* = 3 *vedhas*; *a nimesha* = *3 lavas*, or the blink of an eye = 0.2112 seconds. (See *Prana* in Glossary).

limbs of the *Vedas,* and the one considered "the eye of the *Vedas.*" Vedic astrology uses only the sidereal zodiac, and classically **nine planets**: Sun, Moon, Mars, Mercury, Jupiter, Venus, Saturn, Rahu and Ketu. (Rahu and Ketu are eclipse points, or shadowy planets treated as planets. They are also the north and south nodes of the Moon.) Rahu is Saturn-like, and Ketu, Mars-like.) The planets represent **concentrations of life force.** They rule over the 12 signs of the zodiac and the 27 *nakshatras,* sometimes also called "lunar mansions." The *nakshatras* were referred to in the *Rig Veda*, considered the world's oldest book, long before there was reference to 12 signs. But from the earliest Vedic texts there were various references to spokes of the 360 degree wheel. The planets were revered by the ancients as gods, perhaps due to the reverence they had for the cosmos. They were called *grahas,* meaning "that which seizes you."

In Vedic astrology, the most important *graha* is generally considered **the Moon,** as we experience the world through the Moon, and our level of acceptance of life events depends on that Moon. The Moon is also the intelligence motivating our actions and it holds the memory. We always assess the impact of the Moon, along with the Ascendant lord, and thirdly - the Sun. The Ascendant lord is the planet ruling the sign on the Ascendant. We will assess most matters from the vantage point of the Ascendant, where the *karmic* destiny begins, and to some extent from the Moon.

In the West, astrology - and culture - is largely solar. **The Sun** or Sun-sign is seen as the most important factor. In Vedic astrology, while the Sun is crucial to physical vitality, will-power, and self-confidence, the Moon as the mind, heart and memory, is considered more critical in directing the life and to finding inner peace and happiness, no matter whether the physical body is weak or strong. Thus, the East is traditionally a lunar culture, giving prominence to the life of the spirit and the level of consciousness. Actions are important, but always have *karmic* consequences. The West is more of a solar culture, imposing its will on the material world through physical dominance and expedience, seeking happiness primarily in the material rather than the spiritual realm. To the average Westerner, *Dharma* and *Karma* are mostly foreign concepts.

The two zodiacs: Aries is a fire sign, and as the first sign it is the sign of beginnings and initiations. In the tropical zodiac, used in most Western tropical astrology, Aries is associated with the spring season in the Northern hemisphere only, based on the earth's orientation to the Sun. In the sidereal zodiac, used in Vedic astrology, the signs of the zodiac are wed to constellations, as the sidereal zodiac measures the position of the planets against a backdrop of the fixed stars. This is the observable zodiac, what we actually see in the heavens. The two zodiacs coincided ca. 285 A.D., and will do so again in about 25,900 years from that date, using Lahiri *ayanamsha* which increases by 83 minutes or 1.4 degrees every 100 years. The difference between tropical and sidereal zodiacs is calculated using an *ayanamsha*, the official *ayanamsha* of the government of India being Lahiri. The Precession of the equinoxes is due to a slight wobble of the earth as it rotates on its axis. The earth moves backwards very slightly about one degree of arc every 72 years against the backdrop of the fixed stars. As of Jan. 1, 2010 the two zodiacs were exactly 24 degrees apart.[45]

Thus the sign of Aries is not necessarily tied to the spring season in Vedic astrology, but it is tied to the idea of qualities. **The three qualities** identified in Vedic astrology are the same as the three states of matter. Qualities can be: **1)** ***Chara*** (moveable, active, or in a state of motion),

[45] For all technical terms, see the Glossary of Sanskrit and Astrological Terms.

2) ***Sthira*** (fixed), or 3) ***Dwiswabhava*** (dual, or oscillating). The first three signs of the zodiac run in this pattern (i.e., Aries – *Chara*; Taurus – fixed; Gemini – dual), and the following nine signs repeat this pattern; that is Cancer, the 4th sign then becomes the next *Chara* sign – though now in water. ***Chara* signs** are: Aries, Cancer, Libra, and Capricorn (Western astrology calls them "Cardinal signs"). **Fixed signs** are: Taurus, Leo, Scorpio, and Aquarius. **Dual signs** are: Gemini, Virgo, Sagittarius, and Pisces (Western astrology calls them "Mutable signs"). The classics assign dual signs with the greatest strength, perhaps through their adaptability. But for confirmation the whole chart must be examined.

> "An immovable *Rashi* is stronger than a movable one. A dual [sign] is much stronger than an immovable one."
>
> **Kalidasa,** *Uttara Kalamrita,* Section 4 (Planets in their several *bhavas* and their effects), *Shloka 31.*

In Western tropical astrology, the four elements (fire, earth, air, water) are pivotal to describing the 12 signs of the zodiac. In Vedic sidereal astrology, these elements certainly play a part, but in addition the ethers – or etheric, unseeable realm – play a role as 5th element. In the most ancient texts on Vedic astrology, many of which are still awaiting translation from Sanskrit, the elements are not mentioned as much as the four castes of people from classical Hindu society. We should keep an open mind about how this could help us to understand these signs, as the symbolism in the Vedic culture is always far broader than it first appears. Thus the fire signs (Aries, Leo, Sagittarius) are associated with the ***Kshatriyas*** (the warriors, and/or political leaders); the earth signs (Taurus, Virgo, Capricorn) with ***Shudras*** (the servants and the laborers who get things accomplished in the material world); the air signs (Gemini, Libra, Aquarius) with ***Vaishyas*** (the mercantile or business class, who desire to make contact with other people and to make profits); and lastly, the water signs (Cancer, Scorpio, Pisces) with ***Brahmins*** (the priests, spiritual teachers, those seeking higher consciousness). For artists, especially musicians, more planets may occupy water, or houses 4, 8, and 12. This may confer an emotional capacity to deliver a transcendent message to the public.

The 9 classical *grahas* (planets) in Vedic astrology – keywords: This sacred order of the planets reflects the order of the days of the week, Sunday being ruled by the Sun (from sunrise Sunday through sunrise Monday), Monday by the Moon, Tuesday by Mars, and so on. Rahu and Ketu are regarded as the 8th and 9th planets, and do not rule over any weekdays. Saturn should be propitiated on Saturdays, Venus on Fridays, etc. It can also be helpful to remember the planets as members of the "planetary cabinet."

The planetary cabinet is as follows: **Sun and Moon,** king and queen; **Mars,** military general; **Mercury,** first prince; **Jupiter and Venus,** counselors; **Saturn,** servant. **Rahu and Ketu,** foreigners, or outcasts. "To the *Kalapurusha* [Time personified], the Sun is the soul, the Moon is the mind, Mars is strength, Mercury is speech, Jupiter is knowledge and health, Venus is desire, and Saturn is sorrow." (***Brihat Jataka***, Ch. 1, *shloka* 1)

SUN (*Surya*): physical vitality, will power, dominance, status, father, soul, ego (rules Leo).

MOON (*Chandra*): the mind, happiness of the heart, emotions, capacity of feeling, memory, mother, females, public (rules Cancer).

MARS (*Mangala*): auspiciousness, courage, passion, war, sex, brothers, technical ability, work, real estate (rules Aries and Scorpio).

MERCURY (*Buddhi*): rational intelligence, discrimination, intellect, skin, nervous system, communications, research, publishing, writing, astrology (rules Gemini and Virgo).

JUPITER (*Brihaspati*): knowledge, religion, the grace of God, wisdom, philosophy, wealth, abundance, children (rules Sagittarius and Pisces).[46]

VENUS (*Shukra*): happiness in love matters, harmony, beauty, charm, charisma, art and artistic ventures, music (rules Taurus and Libra).

SATURN (*Shani*): truth, reality, time, *Karma*, duty, longevity, death, adversity, sorrow, separation, endurance, persistence (rules Capricorn and Aquarius).

RAHU: foreigners, exotic, manipulative behavior or appearance, desire for worldly benefits, our greatest desires.

KETU: foreigners, exotic or unpredictable behavior, intrigues, hidden or elusive qualities or resources, that which may be difficult to access, ascetic, non-attachment to worldly benefits; spiritual initiation and growth.

Natural, functional, and cyclical qualities of planets: The essential qualities of each planet are listed above. These are significators and known as *Karakas*. There is also the functional or temporary meaning of a planet – called *Adhipatya*. This can vary depending on how the planet is situated in an astrological chart by house position, and what house it rules. Even so, a planet that is a friend to the Ascendant lord will always do well for the chart, more so if well placed and unafflicted. *Lagnadhipati* is usually always favorable, as it is Ascendant lord. Cyclically, the planet will wield its greatest influence during is respective *Dasha*, whether in the major or minor (or shorter) time period. The planet's influence will depend on both its natural and functional qualities. All of these qualities of a planet must be assessed for the astrologer to make correct predictions. The condition and house position of a planet are crucial and will determine the degree of success of the affairs over which that planet rules.

The 12 Classical signs and houses, or fields of planetary action – keywords: Signs and houses are not identical. The Sanskrit word for sign is *rashi*, meaning literally "heap." It represents constellations and areas of the heavens, whereas the houses are fields of planetary action, called *Bhava* in Sanskrit, meaning literally "mood" or "attitude" – as if one has entered a different room of a house, a whole different environment where different fields of endeavor take place. The first house is always associated with Aries, though it may not be the sign of Aries in the chart. Nevertheless, there is always an overlay, in that issues and life themes related to Aries will always apply to House 1. Whole sign houses are the general rule in Vedic astrology, though there are other ways we also look at houses. The Sanskrit word for each sign follows the name of the sign, in italics. Each house and sign also rules over a part of the physical body, given first. This list applies to natal astrology in particular, though there are some parallels with mundane astrology (for the collective).

House 1 (**Aries** – *Mesha*): the head, brain, physical body in general and its vitality; sense of self, ego, appearance.

House 2 (**Taurus** – *Vrishabha*): face, eyes, teeth, neck, tongue; the speech; intake of food, financial income, *kutumba* (happiness from the family of origin).

[46] In Vedic culture, knowledge was considered wealth.

House 3 (**Gemini** – *Mithuna*): ears, nose, vocal cords, voice, throat, nervous system, lungs, shoulders, arms; courage, personal efforts, writing, short trips, foot travels, siblings.

House 4 (**Cancer** – *Karka*, or *Kataka*): chest (including lungs and heart), breasts; mind, happiness of the heart, mother, relatives, physical residence, educational degrees, education (South India), vehicles for land or water.

House 5 (**Leo** – *Simha*): heart, upper abdomen, stomach, liver, gall bladder, the womb; *poorva punya* (past life credit), intellect, creative intelligence, education (North India), fertility, children, devotional practices.

House 6 (**Virgo** – *Kanya*): nervous system, lower abdomen, large intestine, small intestine; health matters, disease in general, legal concerns, theft, open enemies, conflict, wounds, accidents.

House 7 (**Libra** –*Tula*): lower urinary tract, kidneys, uterus, ovaries, testes, prostate gland, semen, groin, anus; marriage or business partnership, sex drive, opponents, battlefield, victory over enemies, foreign residence.

House 8 (**Scorpio** – *Vrishikha*): external sex organs, pubic region; longevity, death, major transitions in life, research, obstacles, alienation, chronic or terminal illness, inheritance, partner's wealth

House 9 (**Sagittarius** – *Dhanu*): hips, thighs; *Dharma*, good fortune, grace of God, father, guru or teacher, religion, philosophy, long journeys, foreign travel, the future.

House 10 (**Capricorn** – *Makara*): knees, spine; *Karma*, status, visibility, profession, success.

House 11 (**Aquarius** – *Kumbha*): legs, calves, ankles; financial gains, goals, aspirations, networks of friends.

House 12 (**Pisces** – *Meena*): feet, sleep disorders; financial losses or expenditures, sacrifice, martyrdom, income from abroad, residence abroad, pleasures of the bed, private settings, liberation, the past.

The 27 *nakshatras*: *Nakshatra* means "that which does not decay." It refers to the 27 lunar constellations, whose movements are infinitesimal over long periods of time, whereas the planets are constantly moving. The average speed of the Moon per day is 13 degrees 20 minutes, the space occupied in the heavens by each *nakshatra*. In the Vedic myths of the *Puranas*, Soma – the Moon (as a man) – visited each of his 27 wives, the constellations, within a lunar month. Soma stays too long with *Rohini*, who is the most beautiful of his wives, causing the other wives to be jealous and necessitating the intervention of the Gods. For indulging in this worldly gratification, Soma receives the curse of decreasing light for two weeks, along with the boon of increasing light for the next two weeks. Thus *Rohini nakshatra* is associated with both earthly desires and artistic expression, but also jealousy or intense competition. A key planet here may indicate a *karmic* destiny to endure the jealousy of others, especially from other women but not confined to women. The key is the Moon's position in the chart, as planetary owner of *Rohini*. Each *nakshatra* has its own *Shakti* – which is the vibrational energy that infuses and empowers. It is that which inspires or motivates people. In Sanskrit, *shakti* means "power."

1. ***Ashwini*** *(owning horses)*: 00:00 – 13:20 Aries. Planetary ruler – Ketu; Deity – Ashwini Kumars; Symbol – horse's head; *Shakti* – speed, quick results.
2. ***Bharani*** *(she who carries)*: 13:20 – 26:40 Aries. Planetary ruler - Venus; Deity – Yama; Symbol – yoni (female sex organ); *Shakti* – power to carry things away.
3. ***Krittika*** *(cutters)*: 26:40 Aries – 10:00 Taurus. Planetary ruler –Sun; Deity – Agni; Symbol – razor, sharp cutting blade, flame; *Shakti* – power to burn or purify.
4. ***Rohini*** *(red, or growing)*: 10:00 – 23:20 Taurus. Planetary ruler – Moon; Deity – Brahma; Symbol – an ox cart; *Shakti* – the power of growth.
5. ***Mrigashira*** *(deer's head)*: 23:20 Taurus –6:40 Gemini. Planetary ruler – Mars; Deity – Soma; Symbol – deer's head; *Shakti* – power to give fulfillment.
6. ***Ardra*** *(moist, or moist one)*: 6:40 – 20:00 Gemini. Planetary ruler – Rahu; Deity – Rudra; Symbol – teardrop; *Shakti* – the power of effort.
7. ***Punarvasu*** *(good, or prosperous again)*: 20:00 Gemini – 3:20 Cancer. Planetary ruler – Jupiter; Deity – Aditi; Symbol – quiver of arrows; *Shakti* – prosperity.
8. ***Pushya*** *(nourishing)*: 3:20 – 16:40 Cancer. Planetary ruler – Saturn; Deity – Brihaspati; Symbol – cow's udder; *Shakti* – spiritual energy.
9. ***Ashlesha*** *(clinging, entwined)*: 16:40 – 30:00 Cancer. Planetary ruler – Mercury; Deity – Nagas. Symbol – coiled snakes; *Shakti* – poison.
10. ***Magha*** *(great one)*: 00:00 – 13:20 Leo. Planetary ruler – Ketu; Deity – Pitris; Symbol – throne room; *Shakti* – to leave the body.
11. ***Purva Phalguni*** *(former red one)*: 13:20 – 26:40 Leo. Planetary ruler – Venus; Deity – Bhaga. Symbol – conjugal bed; *Shakti* – procreation.
12. ***Uttara Phalguni*** *(latter red one)*: 26:40 Leo – 10:00 Virgo. Planetary ruler – Sun; Deity – Aryaman; Symbol – healing cot; *Shakti* – wealth through marriage.
13. ***Hasta*** *(hand)*: 10:00 – 23:40 Virgo. Planetary ruler – Moon; Deity – Savitr; Symbol – hand or fist; *Shakti* – to gain what we are seeking, placing it in our hand.
14. ***Chitra*** *(bright, multicolored)*: 23:40 Virgo – 6:40 Libra. Planetary ruler – Mars; Deity – Tvastr; Symbol – shining jewel; *Shakti* – accumulation of good *Karma.*
15. ***Swati*** *(independent one)*: 6:40 – 20:00 Libra. Planetary ruler – Rahu; Deity – Vayu; Symbol – young sprout; *Shakti* – power to scatter like the wind.
16. ***Vishakha*** *(forked one, or Radha – delightful one)*: 20:00 Libra – 3:20 Scorpio; Planetary ruler – Jupiter; Deity – Indragni; Symbol – archway, covered with leaves; *Shakti* – power to achieve many goals.
17. ***Anuradha*** *(another Radha)*: 3:20 – 16:40 Scorpio; Planetary ruler – Saturn; Deity – Mitra; Symbol – archway, covered with leaves; *Shakti* – devotion.
18. ***Jyeshta*** *(the eldest):* 16:40 – 30:00 Scorpio; Planetary ruler – Mercury; Deity – Indra; Symbol – circular talisman; *Shakti* – to conquer.

19. ***Mula*** *(root)*: 00:00 – 13:20 Sagittarius; Planetary ruler – Ketu; Deity – Niritti; Symbol – roots; *Shakti* – destruction, power to destroy or damage.
20. ***Purva Ashadha*** *(early victory)*: 13:20 – 26:40 Sagittarius; Planetary ruler – Venus; Deity – Apa; Symbol – winnowing basket; *Shakti* – power to energize.
21. ***Uttara Ashadha*** *(later victory)*: 26:40 Sagittarius – 10:00 Capricorn; Planetary ruler – Sun; Deity – Ten Vishvadevas; Symbol –tusk of elephant; *Shakti* – unstoppable victory.
22. ***Shravana*** *(ear)*: 10:00 – 23:40 Capricorn; Planetary ruler – Moon; Deity – Vishnu; Symbol – three footprints; *Shakti* – to connect diverse elements.
23. ***Dhanishta*** *(wealthiest)*: 23:40 – Capricorn – 6:40 Aquarius; Planetary ruler – Mars; Deity – Eight Vasus; Symbol – musical drum; *Shakti* – wealth and fame.
24. ***Shatabhisha*** *(100 physicians)*: 6:40 – 20:00 Aquarius; Planetary ruler – Rahu; Deity – Varuna; Symbol – circle, empty or invisible; *Shakti* – power of healing.
25. ***Purva Bhadra*** *(front lucky feet)*: 20:00 Aquarius – 3:20 Pisces; Planetary ruler – Jupiter; Deity – Aja Ekapada; Symbol –front of a funeral cot; *Shakti* – spiritual fire.
26. ***Uttara Bhadra*** *(rear lucky feet)*: 3:20 – 16:40 Pisces; Planetary ruler – Saturn; Deity – Ahir Budhnya; Symbol – back of a funeral cot; *Shakti* – power to bring rain.
27. ***Revati*** *(wealthy)*: 16:40 – 30:00 Pisces; Planetary ruler – Mercury; Deity – Pushan; Symbol – drum; *Shakti* – power of nourishment.

Nakshatras **and consciousness:** Since each of the 12 signs of the zodiac contain 2 ¼ *nakshatras*, this further qualifies the portions of each sign and further differentiates Vedic astrology from its Western tropical counterpart. *Nakshatras* work only with the sidereal or constellational zodiac and are Vedic in origin. They arise out of the ancient Vedic culture of India, which observed the heavens directly and followed lunar cycles closely and in great detail. This culture has much the same philosophy as modern Hinduism but is not characterized as a religion per se. Since each of the 27 *nakshatras* is 13 degrees 20 minutes in length, or the average speed of the daily transiting Moon, we can see the prominence Vedic astrology gives to the Moon as the foundation of this system. Although each planet is considered an aspect of consciousness, the Moon is the intelligence motivating our actions and thus fundamental to our being incarnated. **Moon** is the mind and heart, and holds the memory; it is the seat of the emotions. **Mercury** is the intellect, the decision-making faculty, the power to communicate and to organize information. Moon and Mercury, in that order, are the most receptive planets in the chart. Therefore the influences on them from other planets will show the type of mind and intellect the person has, and how open they are to being educated.

Just as in the heavens the Moon is a receiver of the Sun's light, for a human being the consciousness – symbolized by the Moon at birth – is the receiver of divine light. In this capacity the Moon is Mind. Ancient scriptures such as the *Upanishads* teach us that consciousness has to be reflected through something like the Mind, also called *Manas*. The intelligent aspect of *Manas* is part of the higher, more divinely connected part of the individual, and the emotional aspect of *Manas*, part of the lower individual. Emotional aspects of *Manas* connect us to survival, earthly concerns, likes and dislikes, and human life. Emotions can also distract us from an awareness of our divine connection, or our higher mind. Mercury symbolizes the ability of the intelligence to discriminate and to communicate. But primarily it is the Moon and the Moon's *nakshatra*, also called one's Birth Star (especially in India), that indicate the mental and emotional traits of the individual. Where the Moon is placed in the birth and *Navamsha* charts tells us how the person is likely to accept the circumstances of life.

As the Moon's *nakshatra* tells us how the mind will tend to operate, the most important timing of the *karmic* destiny is usually set from the Moon's *nakshatra* in the birth chart. One such destiny-timing scheme of prime importance is the ***Vimshottari Dasha.*** This is a sequence of 120 years, always in the same planetary sequence, and completely dependent on the planet's position in the individual birth chart and ***Navamsha* chart** for the results. Other *Varga* (divisional or harmonic) charts can also be examined, but in South Indian tradition, we concentrate here on the birth chart and the most important harmonic chart – the *Navamsha* chart (9th harmonic chart). The *Navamsha* chart is derived from the precise mathematical portion of the sign and *nakshatra* where each birth planet is placed, measured in nine portions. **The *Navamsha* chart *is* the connection between the *rashis* (the 12 signs of the zodiac, Aries through Pisces) and the 27 *nakshatras*.**

> ***Time is contained in the ascendant.*** *Based on Time, the natal ascendant ... along with auspicious and inauspicious things should be understood.... [The nine planets from the Sun onwards are also noted, especially in their positions in the 12 houses starting from the Ascendant, or House 1.]... The Kundali (the horoscopic diagram) is a Power. This Power contains in it the planets ... moving in the twelve-segmented zodiac. The 12 houses commencing from the ascendant should be erected and effects, good and bad, should be declared accordingly.*
>
> From "The Ascendant," ***Hora Ratnam***, Chapter 2, *Shlokas* 335-336. Transl. by R. Santhanam.

The sign and *nakshatra* on the Ascendant give the *Dharmic* thrust of the life, while the Moon shows how we are likely to deal with that *Dharmic* and *Karmic* destiny. Our focus is on understanding the charts, ascendant by ascendant, through the 12 signs of the zodiac. The life of the individual or of anything begins at the Ascendant – the juncture of the eastern horizon and the ecliptic, where the Sun – or any planet – appears to rise over the earth. In Vedic philosophy, the cosmic principle enters the incarnation at the ascendant. Time is considered God, and God is Time. Therefore God or divine energy enters the destiny and inhabits each of us from this moment, or from the start of anything important. For the birth of an individual, this is the specific point in material reality when the consciousness becomes embodied. Called the *Ahamkara*, it is neither entirely ego nor personality – but encompasses the ego and is the notion of individual existence. The *Ahamkara* enters from the *Param Atma*, or the pure consciousness permeating the entire universe and into which the *Ahamkara* goes at death. Since the *Param Atma* is BEFORE LANGUAGE it cannot be described. But once there is the moment of physical manifestation at the eastern horizon, language has the capacity to describe the life of the person or entity just born. **The individual *dharmic* and *karmic* destiny is thus inextricable from this moment of birth.** Astronomically, the Ascendant is also the fastest moving point in the Vedic astrological chart.

Vedic astrology contains five main branches or systems: **1) Parashari; 2) Jaimini; 3) Nadi; 4) Tajika**; and **5) Tantric**. The first two are named after the sage who is credited with codifying those particular *shlokas* (or verses) giving the astrological rules. Each of them was preceded by other rishis who transmitted the verses orally. In this book we will use mainly the systems of Parashara and *Nadi Jyotisha*, with some Jaimini components.

Nadi Jyotisha is one of two Nadi techniques that work more deeply with *nakshatra* analysis. The other is *Nadiamsha*, which works with 150 *amshas* or divisions for each *rashi* (sign) in the chart. *Nadi Jyotisha* is generally used only in natal astrology and examines lords of the *nakshatras* to see when the *yogas* will come to fruition. (See Definition of *Yogas* – upcoming) We will use only *Nadi Jyotisha*. This system within Vedic astrology examines the *nakshatra* lords of each planet, especially concentrating on the *nakshatra* of the Moon, the Ascendant lord, the Ascendant, the house lords and *Dasha* lords. *Nadi Jyotisha* considers **the *nakshatra* lord** as the control planet for the affairs of that planet, more so than the sign lord. This is a totally different way of understanding the strength of all planets. When we examine a planet we note in what *nakshatra* it is located. This is because the subtle level is considered more influential. *Nadis* deal with subtle levels of vibrations, *nadis* being channel systems or streams through which energy flows. Each *nakshatra* vibrates at certain frequencies, and when *prana* is passing through *nadis*, that *prana* vibrates according to the *nakshatra* that is influencing us in the chart.

Following are **the four aims in life recognized by Vedic science.** This enables us to properly describe the qualities and the elements in a way that is true to Vedic astrology.

***Dharma*:** Houses 1, 5 and 9; also associated with the fire signs – Aries, Leo, and Sagittarius. That which you must do; your proper work in the world – based on your true nature. Therefore *Dharma* also describes your true nature, and by understanding it and living according to it – you will fulfill your *karmic* destiny in this particular incarnation.

***Artha*:** Houses 2, 6 and 10; also associated with the earth signs Taurus, Virgo, and Capricorn. Material security, taking care of your physical and financial needs. Thus it is about eating and nourishment on the physical level – the survival needs, so that you can achieve status and/or recognition in the community and in the world.

***Kama*:** Houses 3, 7 and 11; also associated with the air signs Gemini, Libra, and Aquarius. The desire nature, especially the desire to be in relation with others; with siblings and neighbors, those who are immediately accessible (Gemini), partners – marriage or business (Libra), and friends, colleagues, and larger networks of people (Aquarius).

***Moksha*:** Houses 4, 8 and 12; also associated with the water signs Cancer, Scorpio, and Pisces. Liberation of the soul; the spiritual journey. The 4th house (Cancer) describes the foundation of the mind and heart, and thus the beginning and ending of all suffering. The 8th and 12 houses (Scorpio and Pisces) respectively, describe the suffering and loss that is often necessary before we start to realize we have a spirit housed in a physical body. The purpose of *Moksha* is often to show us this difference and this larger perspective. It is considered the most important of the four Hindu aims in life and is consistent with Eastern philosophy, which seeks happiness in the spirit rather than in material things.

Examining any Ascendant chart from this perspective, we realize several things immediately: that **the primary *Dharma* is always described by the sign on the Ascendant**, as it shows the basic thrust of the life. If it is the first or last degree of a particular sign, there may be some blending of the energies of the sign before or after. Even so, there is a dominance of that particular zodiacal sign as *Dharma* indicator. (See Peter Jennings, Chart #14. His Ascendant is 29:20 Gemini.)

Gunas*:** All states of mind where there is an "I" or "mine" are the results of various combinations or permutations of the three *gunas*, Sanskrit for "rope" or "strand." A *guna* indicates one of three natures that human beings or human activities will possess, to varying degrees. The three *gunas* are as follows: **1)** ***Sattva is light, flowing and truthful, with divine energy; **2)** ***Rajas*** is fiery and has ambition to achieve goals; **3)** ***Tamas*** is heavy, fixed and inert. It is satisfied with material pleasures and comforts. Vedic thought considers Sattvic energy the highest form of spiritual energy, but one has to have a combination of these *gunas* in order to survive most earthly destinies. During a more materialistic age, such as the *Kali Yuga* (the dark or Iron Age), *Rajas* and *Tamas* are necessary. *Sattva* by itself may not be competitive and aggressive enough. *Gunas* will not be emphasized in this book, but it is good to recognize that generally ***Sattvic*** **planets** are Mercury and Jupiter, ***Rajasic*** **planets:** Sun, Moon, and Venus; and ***Tamasic*** **planets:** Mars, Saturn, Rahu, and Ketu.

Any of these can be altered somewhat by the influences they receive by sign, *nakshatra*, sign lord, *nakshatra* lord, planetary aspects and house position. We notice especially when a *guna* is heightened or diminished, as in the case of Adolf Hitler, with *rajasic* Venus overwhelming the chart. Venus rules over politics and in this case, Venus in Aries becomes very militaristic. When in its own *nakshatra* Venus is always in a fiery sign and exacerbates Venus's already *rajasic* or passionate nature. (This condition especially affects Venus as Ascendant lord, sign or *nakshatra* lord, or placed in the Ascendant or a *Kendra*). It offsets the notion that a planet in its own *nakshatra* is always beneficial, and helps to understand the wide range of factors that must be weighed for the astrologer to delineate a chart.

The Ascendant Lord: The condition of the planet ruling the Ascendant sign is crucial to understanding that chart and that individual, including their physical health and vitality. For example, Mars rules the Aries Ascendant chart and indicates how that person is managing the heat and the action generated by fiery Aries as a motivating factor. There are three principle indicators of physical vitality: the Ascendant lord, Mars and Sun. With Aries Ascendant, Mars is both Ascendant lord and *karaka* (significator) of physical vitality. Mars has double importance in this role for Aries Ascendant. But for any Ascendant we assess the strength of the Ascendant lord in various ways:

1. its house position
2. its zodiacal sign
3. its *nakshatra* location
4. aspects it gives and receives, affecting other planets and houses
5. position and condition in the *Navamsha* chart
6. strength of the *kendras* & *kendra* lords (the angular houses: 1, 4, 7, 10)
7. strength of the *trikonas* & *trikona* lords (trinal houses: 1, 5, 9)

House position: Power and visibility can be bestowed on planets placed in *kendra* houses (the angular houses: 1, 4, 7, 10), whereas planets in trinal houses (1, 5, 9) can receive *karmic* and *dharmic* blessings and good fortune. The 9th house is deemed the best trinal house, bringing the most good fortune and above all the grace of God, no matter how we define our individual belief system. The Ascendant is both a *kendra* and a trinal house, both considered the most beneficial kinds of houses in terms of bestowing the person with blessings, gifts, and abundance. This will depend on the planets in the *kendras* and trinal houses. But a well-placed planet in the Ascendant or even a well aspected Ascendant with no planets in it can already bestow much positive energy to that individual or entity.

Digbala: Critical among the six factors in *Shadbala*, a complex Vedic technique for measuring the various strengths of a planet, *Digbala* measures the directional strength of a planet, assigning power to its house placement in a chart. (Rahu & Ketu are not included in *Shadbala*, and thus have no *Digbala* rankings.) If the planet is also in its own sign and/or exalted it is magnified in its ability to benefit the chart, depending on whether that planet is a friend to the Ascendant chart. Best angular house placements are as follows: East (House 1): Mercury and Jupiter; North (House 4): Venus and Moon; West (House 7): Saturn; and South (House 10): Sun and Mars. No single factor can be taken in isolation, and the chart must always be assessed as a whole, especially in relation to the other *Varga* charts, most prominently the *Navamsha* chart.

Aspects: The Sanskrit word for aspect is *drishti*, from the word root "to see." If the planets can see each other or one planet can see another in a one-way aspect, that affects its behavior and its relationship to other planets, as does the strength or weakness of the planet(s) making the aspect. Planets posited in the same sign (known as a **conjunction**) are not considered aspects in Vedic astrology, but they are associated with each other and influence each other, especially the closer they are by degree. For planetary conjunctions within one degree in the same sign or in adjacent signs, the ***Graha Yuddha*** should be considered. This excludes Sun, Moon, Rahu and Ketu. (See upcoming section; also Glossary) All planets can aspect planets in the opposite sign or House 7 from itself. Then, going forward in the zodiac, always in clockwise motion in the South Indian chart (counter- clockwise in the North Indian), certain planets have special aspects according to Sage Parashara. The basic Parashari aspects (and the ones used in this book) are as follows:

Mars : aspects houses 4, 7, and 8 from itself
Jupiter : aspects houses 5, 7, and 9 from itself
Saturn : aspects houses 3, 7, and 10 from itself

Planetary friends: Two separate groups: 1) Sun, Moon, Mars, Jupiter; 2) Mercury, Venus, Saturn, Rahu. (Traditionally, Rahu and Ketu are not included in Planetary friendships, but since Rahu is Saturn-like and Ketu is Mars-like, they are sometimes substituted as such, more often Rahu as Saturn.) Generally, planets in group #1 are enemies of those in group #2, with some exceptions: Sun and Mercury never travel more than 28 degrees apart. For Mercury – the Sun is a friend, but for the Sun – Mercury is a neutral planet. Jupiter and Saturn are neutral to each other, as are Venus and Mars. To Saturn, Mars is an enemy, whereas to Mars, Saturn is neutral. Mercury's enemy is Moon, but Moon has no enemies. To review: Rahu and Ketu are eclipse points, or shadowy planets treated as planets. Since they cause eclipses, they are enemies of the Sun and Moon. Rahu and Ketu do not rule over signs of the zodiac or houses, though they each own three *nakshatras* and are impacted by their house position, planetary associations, and planets owning their sign and *nakshatra*. (See Glossary: Planetary Friendships for the table of friendships)

In the astrological chart, planetary enemies rule at least one *Dusthana* house (6, 8, or 12), considered the most difficult houses. A planet ruling a *Dusthana* house can cause problems, though if also a great friend to the Ascendant lord its power to have a positive influence usually outweighs its status as *Dusthana* lord.

Definitions: In Vedic astrology, **a benefic planet** is a planet that has mostly positive effects, and eases the way in life. While **a malefic planet** can be very useful during the materialistic age of *Kali Yuga*, acting like a fierce watchdog, it can also bring more obstacles, delays and/or separations, especially during its *Dasha* period. Results are totally dependent on whether the benefic or malefic planet (classically and/or temporally) is a friend to the Ascendant chart, acting more harmoniously with it – in spite of being a malefic. For example, Saturn and Rahu are great friends to Venus and in the same planetary group of friends; therefore in Venus-ruled Ascendant charts Saturn and Rahu do less harm in general, even as major classical malefics. They become temporal benefics in this instance, bestowing protection and resilience. However, for best results, Saturn and Rahu should also be in favorable house placements from the Ascendant.

- **Classical benefics**: Moon, Mercury, Jupiter, and Venus, in ascending order; major classical benefics – Jupiter and Venus.[47] (A waxing Moon becomes benefic, though a bright waning Moon also has strength. Mercury and Moon are both conditional natural benefics. They absorb the qualities of planets influencing them through contact or aspect.)
- **Classical malefics:** Sun, Mars, Saturn, Rahu and Ketu: major classical malefics: Saturn and Rahu. (Waning Moon can act as a malefic, especially when weak or afflicted.)
- **Special rules for *Kendra* lords (*Kendradhipatya dosha*):**
  1. Planets ruling *kendras* become neutral, though they do not totally lose their essential nature, whether classically malefic or benefic. This is generally more damaging to classical benefic planets, especially during their *Dashas*. The ruler of a *kendra* which is also a classic malefic planet will benefit, even more so if it also owns a trinal house. The Ascendant lord is always an exception to this rule, being innately auspicious as ruler of a *kendra* and a trinal house.
  2. Ruler of *a kendra* which is also a classical benefic (notably Jupiter and Venus) can be harmful, especially during its *Dasha*, mainly because its beneficence is reduced or neutralized. This applies especially for *a Maraka* (7th house lord). A classical benefic that also owns a trine (e.g., Venus for Capricorn Asc.) or forming a *Mahapurusha yoga* can overcome many difficulties and be protective to the native. Strength of sign and *nakshatra* lords can play a role. Note also this exception: When a classic benefic planet rules *a kendra* and is not a planetary friend of the Ascendant lord, it can do more damage (such as Jupiter for Gemini and Virgo Ascendant charts), whereas Moon as 4th house lord for Aries Ascendant is less problematic. For Libra Ascendant, Moon as the 10th lord brings benefits as *Nadi yogakaraka.*
  3. House Lordship is always important to consider: Venus exalted in Pisces Ascendant chart is not advantageous as 8th house Lord in the Ascendant. Venus is also a planetary enemy of Pisces ruler Jupiter.
- **Temporal malefics:** Those planets ruling *Dusthana* houses (6, 8, and 12) that may also be classical benefics. (Examples: For Aries Ascendant, classical benefics Mercury and Jupiter rule the 6th and 12th houses, respectively, though Jupiter's strength as 9th lord – the highest trinal house – overrides its malefic status as 12th house lord. And in general, Jupiter is a friend to

[47] Some classic texts may vary in assigning the highest benefic role to Jupiter rather than to Venus. Either way, both planets in the same house may bring extraordinary (and also excessive) results. See Chart examples #2 and #25 (Robert Kennedy and John F. Kennedy, respectively): Each of them has Venus and Jupiter in the 9th house of father, and each benefited from the financial wealth and connections of their father, Joe Kennedy, Sr.

Mars, whereas Mercury is not. (See the list of Planetary friendships above and table in Glossary.) At times the Ascendant lord itself will rule over a *Dusthana* house. This applies only to Mars or Venus-ruled Ascendants: Aries, Scorpio, Taurus, and Libra. Most problematic is when the Ascendant lord also owns House 6, as with Taurus and Scorpio Ascendants. In this case the Ascendant lord can attract enemies as 6th lord, but can be victorious. In matters related to the 1st house the planet still gives good results, and in general the Ascendant lord is auspicious and that role takes precedence.

- **Retrograde planets:** When in Retrograde motion, a planet is closer to the earth. Its natural qualities intensify and become stronger. The astrologer has to make a judgment based on the whole chart.

Vantage point: Look from 1) the Ascendant itself; 2) from the Moon as Ascendant; and 3) from the *Dasha* lord as Ascendant. The first two are of primary importance, Moon even more so if the birth time is uncertain. Observe the planet owning each house (the house lord), planet(s) occupying each house and aspects to each house. As we understand more about the meaning of each planet and the layers of underlying meanings, we can see patterns the planets form in relation to each other, especially from these vantage points.

More on the Houses (*Bhavas*):

Houses 1, 4, 7, 10: *Kendra* houses; have power to do good in the life, to bring success and visibility, depending on the factors above.

Houses 1, 5, 9: Trinal houses: considered the very best houses, especially the 9th house – bringing the "grace of God," or ***bhagya***, fortune. Temporal and even classical benefic planets can be very effective in one of these placements.

Houses 3, 6, 10, 11: *Upachaya* houses: the affairs of these houses can improve over time, depending on the status of their respective house lords. The 11th house is one of increasing gain/improvement, and can deliver significant financial gain if well aspected, and if the house lord is well situated in a *Kendra* or trine – giving it more power.

Houses 6, 8, 12: *Dusthana* houses: these are considered the most difficult of all the houses, especially the 8th and the 12th. Protection from some of the worst effects of *Dusthana* house lords can come with one or more of the ***Viparita Raja yogas*** (reverse *Raja yogas*). They enhance material prospects but do not necessarily protect the physical body. This occurs when one or more of the *Dusthana* lords, the more the better, are situated in either their own house or another *Dusthana* house, and even more effective if aspecting its own house. This is as if two negatives make a positive. (Bill Gates, Chart #13, has all three *Viparita Raja yogas* reading from the Moon as Ascendant, shielding him from adversity and loss from competition, including lawsuits. Among the many hundreds of *yogas*, or planetary combinations, these are among the most protective, though not regarding physical health.)

Houses 2 and 7: *Maraka* houses: Lords of these houses can cause difficulties to the physical health, and even death – depending on the general indications for longevity, the *Dasha* running at the time, and a number of cumulative factors that need to be examined. , This is because Houses 3 and 8 indicate longevity, and Houses 2 and 7 are the 12th house (of loss) from the respective Houses of longevity. (See Derived houses)

Derived houses: The use of Derived houses enables one to examine the affairs of any given house. For instance, the 2nd house of finance from the 7th house of partnership will describe the partner's finances. Perhaps the most important Derived houses are called ***Bhavat Bhavam***. They are found by counting as many houses away (clockwise in the South Indian chart, counter-clockwise in North Indian chart) as the House number itself. Thus, we observe the 1st from the 1st (House 1 always being the Ascendant), 2nd from the 2nd, 3rd from the 3rd, etc. The 1st from the 1st would be House 1, 2nd from the 2nd is House 3 from the Ascendant, 4th from the 4th is House 7 from the Ascendant, and the 9th from 9th is House 5 from the Ascendant. *Bhavat Bhavam* tells a lot about the original house, and explains why House 3 is a House of longevity in addition to House 8, since House 3 from the Ascendant is also the 8th house from the 8th. If the lord of a house is in the *Bhavat Bhavam*, it is generally strengthening to the affairs of the initial house, except for Houses 6, 8, and 12, and the 6th, 8th or 12th houses from their respective houses or any house.

Combustion occurs when a planet is so close to the Sun it cannot be seen with the naked eye. It applies only to five planets: Mercury, Venus, Mars, Jupiter, and Saturn. Combust planets are also furthest from the earth in the case of Mars, Jupiter and Saturn. Inner planets Mercury and Venus may be combust on either side of the Sun. The effectiveness of the planet to operate may be greatly curtailed by being burnt by the Sun's rays. This affects the planet as House lord, ruling over the affairs of a particular house, and as *karaka* (significator). For example, Venus is *karaka* for artistic ability, fashion sense, love matters, and eyesight, among others. Mars is *karaka* for physical vitality and real estate, among others. If a planet is combust or diminished, those matters over which it is a significator or house lord can be adversely affected. If the planet is within 5 degrees of the Sun and at lower celestial longitude than the Sun, the effect is more severe. The widest range of orbs used on either side of the Sun for planets to be combust are these: Mars: 17 degrees; Mercury: 13 degrees; Jupiter: 11 degrees; Venus: 9 degrees; and Saturn: 15 degrees.[48]

Graha yuddha is Planetary War. It occurs when two planets are within one degree of each other in the same sign or in adjacent signs, and applies only to Mercury, Venus, Mars, Jupiter, and Saturn. The winner takes on the energy of the losing planet, while overpowering it with its own energy and agenda. Unless other factors soften the outcome, the affairs of the loser, as *karaka* (significator) and house lord are diminished considerably, especially during its major or minor *Dasha* (planetary period). The victorious planet is chosen in this descending order: **1)** Size and influence of the planet (Saturn and its rings over Jupiter); **2)** Effulgence, or brightness; **3)** Speed of orbital motion; **4)** If the victorious planet is also north of the defeated planet by either celestial longitude or declination, especially the latter, then it is also usually the decisive winner. But the winner may also be to the south; **5)** Mars is a special exception, being generally the loser in *Graha yuddha*, even when situated further north, or within one degree of Mercury – the one exception in which the losing planet is also the larger planet. But in general, the other four planets are all bigger, brighter, or faster than Mars, which as loser creates particular havoc.[49]

---

[48] The Introduction to Chapter 7 (Leo Ascendant) discusses combustion in more detail.

[49] These principles are based on both the Vedic classics (Varahamihira's ***Brihat Samhita***, and Kalidasa's ***Uttara Kalamrita***) and on 15 years of research, initially with astrologer Chakrapani Ullal, from the mid-1990s. It was first presented in my conference lecture Nov. 2002 in Calicut, India. See my article: "*Graha Yuddha:* Testing the Parameters of Astrology and Astronomy," 2003, revised & expanded version 2010 (40pp.) with 25 example charts. Published in three astrological journals: 2003 thru 2011. Also available at: **www.edithhathaway.com.**

***Yogas*:** *Yoga* means union, and astrologically, *yogas* yield specific results according to planetary combinations. Often there are obstructions to these *yogas* causing them to be minimized, so it is important to assess them carefully. Hundreds of such *yogas* exist and they impart their qualities throughout the life, but they are especially in evidence when one of the planets is *Dasha* lord. *Dashas* and their sub-periods show when favorable or unfavorable *yogas* are likely to manifest. Conscious awareness and spiritual practices help in minimizing or maximizing the yogas' effects, but cannot prevent them from occurring.

Parashari *yogas*: Credited to sage Parashara (who codified them), these *yogas* are the most dominant and numerous. They occur as a result of many factors: how planets are positioned vis-à-vis each other and their house relationships. *Yogas* may form due to a certain combination of house lords. For most *yogas*, the house position of planets are assessed: 1) from the birth Ascendant ; 2) from the Moon in the birth chart; 3) from the Sun in the birth chart; 4) from the *Navamsha* Ascendant. *Yogas* may depend on where the planets are concentrated or whether the majority of them are in certain signs, whether moveable, fixed or dual. *Yogas* may be favorable or unfavorable.

Jaimini *yogas*: Credited to sage Jaimini, but probably in part based on sage Parashara and other rishis, these *yogas* tend more to relate one group of signs to another group of signs. (Jaimini astrology is also included in the classic work *Brihat Parashara Hora Shastra.*) Also, degree rank is considered in assessing the capacity of a planet, especially during its *Dasha* or sub-*Dasha*. Highest degree planet is ***Atma Karaka (AK),*** and can be very favorable in the life, as it relates to one's soul (*atma*); next highest is ***Amatya Karaka (AmK)***, excellent for networking, and on down. Fifth highest relates to having children, 4th highest to motherhood. Lowest degree rank is ***Dara Karaka (DK),*** spouse indicator. During the *Dasha* of the *Dara Karaka* one is oriented towards the mate and may marry if not already married. The overall chart has to be examined for likelihood of this destiny. (Jaimini methods often include Rahu among *karakas*, while Parashari methods do not.)

Nadi yogas* – Background:** The planet owning the *nakshatra* where a house lord resides is considered the soul, spirit, and intention of that house. Its Sanskrit name is ***Jiva life, or life-giver; living being). In Vedic thought, intelligence as a motivating factor is key. We have already described how the *nakshatras* pick up the subtlest level of vibrations and the ones that have the most profound effects. (See under *Nadi Jyotisha*) *Nakshatra* lords are always affected by their house placement from the Ascendant, and greatly affected by their respective *gunas*. You can go further by examining the body, activity and action that derives from that *Jiva* by looking at the planet owning the *nakshatra* where the *Jiva* planet is placed. This planet is called the ***Sharira***. A planet in its own *nakshatra* becomes both *Jiva* and *Sharira*, doubling its influence on all planets in *nakshatras* it owns. In this book we focus only on *Jiva* and sign lord (dispositor).

***Nadi yogas* – Definition:** These *yogas* are caused by the *nakshatras* and their lords. For example, when a planet is located in the *nakshatra* of a *yogakaraka* planet, as judged in the *Nadi* or *Parashari* system, this creates a *Raja yoga*. It gives excellent results during the *Dasha* of that planet. Conversely, when a planet is located in the *nakshatra* of a planetary enemy to that Ascendant, results are less auspicious during the *Dasha* of that planet, as its *Jiva* is more problematic for the chart and thus for that individual destiny. A primary example of a *Nadi yoga* is a mutual exchange of *nakshatra* lords. These *yogas* occur even when the planets are not in mutual aspect. They are considered by some to be even more powerful than the hundreds of traditional *yogas* named in the *Parashari* and *Jaimini* systems, especially if they combine with them. If the *Nadi yoga* contains two planetary friends that are also planetary friends of that Ascendant chart (ideally even a *Nadi yogakaraka*), and they are well placed in *Kendra* or trinal houses, the resulting *Nadi yoga* can cause a major elevation in life starting in the *Maha Dasha* (major planetary period) of one of these planets. (See Mia Farrow, Chart #4, Aries Ascendant)

Nadi yogakarakas:[50] These are considered best planets for each Ascendant chart, in descending order. Mars is not included, either as Ascendant lord or as trinal house lord (another major difference from the Parashari system):

Aries : Sun, Jupiter, Moon
Taurus : Mercury, Saturn, Rahu
Gemini : Venus
Cancer : Jupiter
Leo : Jupiter
Virgo : Venus
Libra : Saturn, Rahu, Mercury, Moon
Scorpio : Sun, Moon, Jupiter
Sagittarius : Sun
Capricorn : Venus, Mercury
Aquarius : Mercury, Venus
Pisces : Jupiter, Moon

Three types of *Karmas*: "After inventing the Law of *Karma*, God was able to retire." (attributed to Mahatma Gandhi). *Karma* means action, and each action has effects over this and other lifetimes. In Vedic philosophy, there is no ultimate evil, only ignorance. The *Karmas* have both fixed and non-fixed qualities. These *Karmas* can be read to some extent in the astrological charts, especially through the Moon. For timing of *karmic* destiny in a given chart, we examine both ***Dashas* and transits** – including all of Saturn's transits, especially to the Moon and during ***Sade Sati*** (description below). **Saturn** is the chief dispenser of *Karma*, and **Moon** its chief receiver. Giving further emphasis to this connection we note Saturn's cycle averages 29 years and the Moon's cycle 29 days.

The motivation for our actions, as well as the results, plays a key role in the distribution of *Karma*. And as the Moon contains the intelligence, habitual thought patterns, and motivation with which our actions are imbued, the goal of most Vedic spiritual practice is to purify the contents of the mind, since thought precedes and informs all action. The *Bhagavad Gita* teaches that action is ideally performed without attachment to results or *Karmaphala* (fruits of our actions). The focus is on *Dharma*, or right action and intention.

---

[50] For material on *Nadi Yogakarakas*, see ***Nadi Jyotisha***, by R. Gopal Krishna Rao (also known as ***"Meena's Nadi"***), Chapter 2. See Selected Bibliography: Meena (pen name of R. Gopalkrishna Rao), ***Nadi Jyotisha or The Stellar System of Astrology***, 1st edition 1945, 2nd edition 1951, 3rd edition 1954.

1. *Sanchita Karma* – the totality of results emerging from our actions in all our previous lifetimes, pre-destined to some extent
2. *Prarabdha Karma*– the fruits of actions from previous lives that one is destined to experience in the present life
3. *Agami Karma* – the result of actions we perform in our present existence

Timing of the *karmic* destiny: The Vedic ***Dasha* systems**, or planetary periods, are numerous and sophisticated, at least 55 of them known. But the most important one to consider is the ***Vimshottari Dasha*** system, as given earlier under *Nakshatras and Consciousness.* "*Vimshottari*" means 120 and the *Dasha* runs for 120 years. It is based on the natal Moon's exact *nakshatra* position in the birth chart and starts there. Sage Parashara lists this *Dasha* system first in his *Brihat Parashara Hora Shastra*, and calls it "the most appropriate for the general populace." *Dashas* can only materialize what resides in the birth chart and the other *Varga* charts, primarily the *Navamsha* chart. The will of the individual and the level of consciousness can often, but not always, affect the *karmic* destiny. But all planetary configurations, especially the various *yogas* come to bear in the *Dashas.* Many but not all *yogas* and *Dashas* are inextricably linked. In addition, we look at **the transiting planets**, especially Jupiter, Saturn, Rahu and Ketu – through the houses and signs in question, especially when they go Stationary Retrograde or Stationary Direct. We observe the degrees of the Solar and Lunar eclipses, and for Jupiter and Saturn – their return to certain natal and stationary degree positions.

As planets of *Dharma* and *Karma*, respectively, the positions of **Jupiter and Saturn** in the birth chart and in the heavens are critical. We especially note the conjunctions of the two planets, occurring every 20 years. Next in importance are the oppositions of Jupiter and Saturn (a type of Full Moon in their mutual cycle), and their mutual trines.

Sade sati: In Sanskrit, *Sade sati* is 7 and a half. It refers to the on-average 7½ year period that Saturn takes to transit to Houses 12, 1, and 2 from the natal Moon. It occurs for all of us every 22½ years, from one *Sade Sati* to the next. Saturn's exact orbit around the Sun is 29.46 tropical years. *Sade Sati* is considered to push the parameters of one's life to an unprecedented degree – the extent of difficulty depending on the condition of the Moon in the chart. *Sade sati* often brings more responsibilities in life, depending on the houses involved. It brings a *karmic* test to the workings of the mind and heart, represented by the Moon. Have they gone on automatic, as mental patterns tend to do? Or are they conscious and awake?

Chart formats: Vedic astrology has traditionally been practiced using North or South Indian charts. Students often learn to use both of them but soon find they have a preference for one of them and gravitate towards that chart format, using it exclusively. Outside of India, the choice may depend on the teacher, books, or greater exposure to one of them. Within India, the astrologer in North India will most likely use the North Indian chart format, and in South India, the South Indian format. Mean node will be used throughout this book for Rahu and Ketu (the lunar nodes). They are always in retrograde motion, while the Sun and the Moon are never retrograde. It will be indicated when other planets are retrograde.

How to read a North Indian chart

North Indian Chart

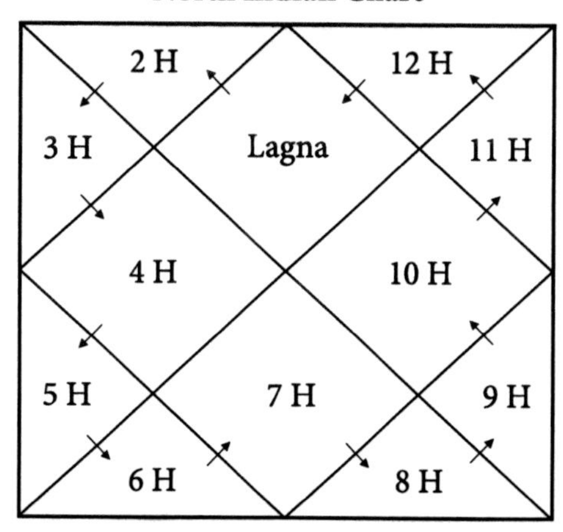

This chart is in the shape of a *yantra* (a mystical diagram), with diamond-shaped houses. The Ascendant (known as the *Lagna* in Sanskrit) is always located at the top central position of the chart. The houses are read counter-clockwise from the Ascendant. In this diagram, 2 H indicates the 2nd house, 3H the 3rd house, and so on. The arrows indicate the direction in which the houses are read. Since the *Lagna* is always set for this format, the typical North Indian chart will have numbers in each house to show what sign of the zodiac resides in that house. The zodiacal sign that is rising is indicated by a number from 1 to 12: Number 1 is Aries, 2 is Taurus, 3 is Gemini, etc. through each of the 12 houses. Thus, if number 3 is in the *Lagna*, Gemini is rising and is assigned to House 1, Cancer to House 2, Leo to House 3. All North Indian charts are read like this, and the birth chart is typically shown to the left or above the *Navamsha*, the 9th harmonic chart.

How to read a South Indian chart

South Indian Chart

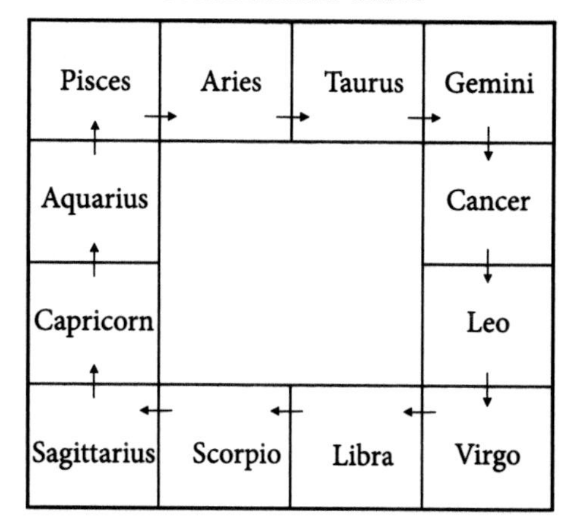

In this square chart the 12 zodiacal signs (known as *Rashis* in Sanskrit) are always in the same squares. The Dual signs are in the four corners. Pisces is always in the upper left hand corner, followed directly by Aries, then Taurus, and Gemini in the upper right hand corner. Virgo is in the lower right hand corner, Sagittarius in the lower left. The signs and houses are read clockwise around the zodiac. The Ascendant can be in any one of these 12 signs, and the planets distributed according to where they are in the heavens. All South Indian charts are read like this, and the birth chart is typically shown to the left or above the *Navamsha* (the 9th harmonic chart.)

CHAPTER 3

Aries Ascendant

Introduction

The first three signs of the zodiac contain both the element and the quality that reinforce each other's energies. This gives a primordial quality to Aries as an active fire sign, to Taurus as a fixed earth sign, and to Gemini as a dual air sign. Fire has a deeply active quality in Aries. In Vedic astrology, all fire signs are associated with the ***Kshatriya***, or warrior class, which includes political leaders. *Kshatriyas* are considered to have royal status. In addition, the Sun and Moon are the two royal planets, giving royal status to the respective signs over which they rule: Leo (fire) and Cancer (water). The Sanskrit word *Chara* (moving, moveable) well describes the force of the four *Chara* signs (Aries, Cancer, Libra and Capricorn). Western tropical astrologers use the word "Cardinal" to describe these four signs. Cardinal implies prominence in terms of the beginnings of the four seasons, identified by the spring and fall equinoxes, the summer and winter solstices. Cardinal implies something that is pivotal, and while it fits with the more seasonally oriented tropical zodiac, the term gives no sense of the attribute of movement in these signs, as used in the sidereal zodiac with Vedic astrology. *Chara* also better describes the visceral quality of initiating action, as it is designated in each of the four elements. Therefore, we will use the word *Chara* exclusively.[51]

The Sanskrit word for Aries is ***Mesha***, meaning literally a ram. "The abode of *Mesha* or the Ram is the surface of the earth containing precious stones and yielding minerals."[52] When we

---

[51] See Glossary, including for pronunciation of Sanskrit words. Sanskrit names for planets and signs are still used more exclusively in books and software programs published or produced in India. Those published or produced in the West usually provide Sanskrit language options. These names and terms are still taught to *Jyotish* students in the West, and while some students pursue Sanskrit language studies in greater depth, all learn the importance of the Sanskrit language to *Jyotish*.

[52] From ***Jataka Parijata***, Vol. 1, *Adhyaya* [Chapter] 1, *Shloka* [verse] 10. Translation by V. Subramanya Sastri.

combine the fire power of **Aries**, the initiatory properties of a *Chara* sign, its association with the *Kshatriya*, or warrior class, its position as the first sign of the zodiac and first *Dharma* sign – our attention is drawn to essential Aries life themes: to be independent, not deterred by the opinions of others, not afraid of being first, having the courage to do what few others have done, to go where they have not, and to do it alone, if necessary. Such actions may involve physical strength, but also the breaking of conventional social, political, or economic barriers – often because what needs to be done is long overdue. The Aries person has the indispensable quality of initiative for such action. But since the rest of the society may not yet be ready for the new steps or projects being offered, they may not accept them with open arms. Thus, popularity may remain elusive for a time for the Aries person, who may be more respected and honored later in life, if not posthumously.

The Aries Ascendant lord is the classic malefic planet **Mars**, a planet of action and of discontent. Many Sanskrit words are used for the Sun and the Moon (up to 108 of them), and Mars has several of its own, the major ones being *Kuja, Angaraka,* and *Mangala.*

Kuja means "born of the earth," while ***Angaraka*** is "fiery, or like coals," and ***Mangala*** is "auspiciousness or auspicious action." Mars as *Angaraka* reinforces the fiery nature of Aries, over which it rules. Mars is lord of auspicious action, courage, passion, war, sex, brothers or siblings, technical ability (especially when conjoined with Saturn), work, and real estate. At times any of these themes become dominant in the life. Whatever the nature of the courage and on whatever frontier of life it is needed, it is shown by numerous factors in the chart, including the condition of Mars itself as well as the entire chart configuration. At the very least – vitality, youthfulness and courage are emphasized. Aries is a youthful sign, and this is a characteristic of Mars in the First house or when aspecting the Ascendant (i.e., in the 6th, 7th, or 10th houses from Aries Ascendant).

Mars in the Ascendant in any sign can also give these features, though Mars is most dominant in the 10th house, where it receives *Digbala* (best directional strength). Mars is exalted in Capricorn, with maximum or "deepest degree" of exaltation at 28:00 Capricorn. The impact of Mars on or in the Ascendant can be personal attractiveness through sheer physical vibrancy. In addition, if other factors coincide, the person can have a revolutionary effect on his or her environment or field of endeavor. This can also occur with Mars contacts or aspects to the 10th house or 10th lord (Saturn for Aries Ascendant) and Saturn as *karaka* for the 10th house. Mars-Saturn combinations can also produce this outcome, more so when unafflicted.

In the Planetary Cabinet, **Mars** is the **Military General**. Vedic mythical stories surrounding the planets give important clues to their meanings in Vedic astrology. Thus, in Vedic, Epic, and Puranic literature Mars appears again and again as ***Skanda***, whose purpose above all is to have immense strength and be victorious in battle. No woman is involved in Skanda's birth, as he is born from the 'spilling" of Shiva's seed that is spread by the gods of Fire and Wind. He is depicted as forever young, vital, and unscathed by the battles and competitions of life. While Venus is the planet of sexual desire, Mars as Skanda is the most admired when he *contains* his sexual potency. Thus in this myth he is associated with chastity, and his only wife is the Army of the gods. His passion is most closely associated with taking action in battle. And though as a planet Mars is associated with sexuality, in his role as Skanda the emphasis is more on his ability to destroy the external enemies or the internal vices of his devotees. The teaching myth implies that when sexual energy goes untamed, it can distract the warrior from his central purpose. Further, we observe that when there is close interaction between desirous Venus and warrior Mars in the chart (through mutual aspect or contact), there is often a yearning to achieve worldly power either through marriage or victory in battle. This can lead to corruption or at the very least diversion from one's higher *Dharmic* path.[53]

Kuja means "born of the earth." In Vedic mythology, there is a long history of names of Mars relative to the son of the earth. In Vedic astrology, the connection then developed between Mars and land or real estate, with Mars as *karaka* of real estate. At times, any of these themes become dominant in the life. Many of the Vedic mythic texts include this myth, among them the *Vishnu Purana*: In this story the Earth goddess is saved from the demon Hiranyaksha, who has thrown the earth to the bottom of the ocean. The earth's rescuer is Vishnu, who takes the form of a boar. This Boar avatar, called Varaha, lifts the earth up deep out of the ocean. "He then rescued the earth and re-established it floating over the ocean like a large ship."[54] Varaha has to fight off the demon in the process. The Earth goddess is so grateful to Varaha that she approaches him and they make love for one year. Out of that union Mars is born, and Lord Shiva later gives Mars the status of a planet.

As the first of the three fire *Dharma* signs in the Ascendant, the Aries influence gives added impetus to the desire to act, to take the initiative, and to the basic thrust of life which demands we examine our true nature through each action. There may be a tendency to want to take action no matter what, but this will be modified by other factors, such as the condition of the Ascendant itself, Ascendant lord, Moon and Mercury, and their respective configurations with Jupiter and Saturn, classic lords of *Dharma* and *Karma*.

Kuja Dosha occurs when Mars (*Kuja*) is situated in House 1, 2, 4, 7 or 8 from the Ascendant or Moon.[55] Vedic astrology classics from Kerala include Mars in House 12, but not in House 1. (In North and West India, *Kuja Dosha* is called "Mangalik.") Such is the power of Mars that its heat is considered in Vedic astrology to be capable of ruining or undermining marital harmony when *Kuja Dosha* occurs in the chart. In classical Vedic or Indian culture, marriage is one of the societal institutions to be preserved above all. Originally *Kuja Dosha* meant the likely early death of the spouse; but it evolved to signify that the person would encounter aggressive energy in the marriage, either from the partner or from oneself, depending on the house position of Mars. If a woman possesses *Kuja Dosha*, it is considered especially bad for marriage, for if she is too assertive she is not as likely to submit to the will of her husband and her husband's family. (It is the custom in India for a young woman to live with the husband's family, along with the husband). We must consider the cultural nature of this delineation and the fact that in some, but not all Western and Eastern cultures, assertiveness in women has become more acceptable within the family and the society.

But in general, *Kuja Dosha* – without any cancellation *(Bhanga)* or compensating factors in the chart – can account for difficulties in marriage or one-on-one relationships. One is likely to assert oneself strongly in many relationship situations, though it can be masked at times by charm. *Kuja Dosha* is even stronger if it occurs in both birth chart and *Navamsha* chart. If it does not recur

---

[53] See the description of this interaction between Venus and Mars in Chapter 9 (Libra Ascendant): Introduction.

[54] ***Vishnu Purana***, 1.4.45-50

[55] This condition occurs in most of the charts in this book and in many charts in general. Since *Kuja Dosha* occurs so often, we will not place emphasis on it in this book. But we will always note the condition of Mars.

in the *Navamsha*, the condition may be eased over the lifetime and reduced through conscious awareness. Other corrections occur when a person with *Kuja Dosha* marries someone who also has *Kuja Dosha*; or when Mars is in its own sign or exalted; or when natal Mars appears in the same sign as natal Moon, creating a *Chandra Mangala yoga*. Other cancellations less generally accepted by most scholars are these: 1) *Kuja Dosha* is cancelled when marriage occurs after age 28, when Mars is considered to reach its maturity; or 2) *Kuja Dosha* is not cancelled until the 7-year Mars *Dasha* has passed. (The latter excludes many people who will not have Mars *Dasha* in their lifetimes.)

In the traditional Hindu wedding, Indian women wear *mangala sutras* (marriage necklaces) for extra good luck to the marriage. The *mangala sutra* is given to the bride instead of a wedding ring. Astrologers are often consulted in India for the compatibility between bride and groom as well as for an auspicious date and time for a wedding.

As ***Mangala*** signifies "auspicious action," we see from a spiritual perspective that to deal positively with taking action in life, auspiciousness in the Vedic sense relates to a pure motive, doing the action for what is needed, not for the expected personal rewards. *The Bhagavad Gita* contains much guidance on this subject, including an examination of who is the doer of actions, and how we should properly regard the mind, senses and body. In this short section (Chapter III, verses 6-9), Krishna instructs the warrior Arjuna:

> "He who outwardly restraining the organs of sense and action, sits mentally dwelling on the objects of senses, that man of deluded intellect is called a hypocrite.
>
> On the other hand, he who controlling the organs of sense and action by the power of the will, and remaining unattached, undertakes the Yoga of Action through those organs, Arjuna, he excels.
>
> Therefore do you perform your allotted duty; for action is superior to inaction. Desisting from action, you cannot even maintain your body.
>
> Man is bound by his own action except when it is performed for the sake of sacrifice. Therefore, Arjuna, do you efficiently perform your duty, free from attachment, for the sake of sacrifice alone."
>
> ***Bhagavad Gita***

Krishna's message to Arjuna is that if you are a *Kshatriya* (warrior), with a warrior's destiny, in certain circumstances you must fight. In the larger sense, Krishna is saying that conflict is inevitable and that in the midst of the battle of life the spirit should remain peaceful and detached as a witness. Though we think we are the DOER and ENJOYER of our actions, we learn we are neither one. Thus, one examines an Aries Ascendant chart carefully for the ability to act wisely and decisively, and for the power conferred to one's siblings and enemies, as one or more of them may bring extra responsibilities or unwanted notice. Since Mars will want to act – ideally the action should be auspicious and not headlong, selfish, or ill considered. As the planet of war and of discontent, Mars can be dangerous and bring accidents. It can be overly aggressive, courting competitive situations and often attracting them. Mars is the warrior, the scout, the adventurer, the innovator, often required to step beyond the previous parameters established by society or one's peers.

Mars is red and contains heat. When in distress Mars is feverish. The Aries Ascendant person may direct emotional heat outwards or have it directed at them. They have to carefully channel that heat, fire, and sharpness that is Aries. As it rules the period of youth, as well as youthful vitality, it is important to guide the young Aries Ascendant person so that the qualities of impetuousness and hot-headedness do not dominate and/or even cause premature death. Hot-headedness is very real here as the First house (and Aries) relates not only to the physical body but to the head. With Mars as ruler and a fire sign on the Ascendant – we look to see if the desire for action is tempered by other more moderate planets and by cycles indicating greater longevity.

Nakshatras: Aries contains the following three *nakshatras*: ***Ashwini*** 0 – 13:20 degrees Aries, (ruled by Ketu), ***Bharani*** 13:20 – 26:40 Aries (ruled by Venus), and ***Krittika*** 26:40 – 30:00 Aries (ruled by the Sun). The Aries Ascendant, or any planet situated in one of these three *nakshatras* is colored by being 1) in Aries, ruled by Mars and 2) in one of these three *nakshatras*, ruled by Ketu, Venus, or Sun, respectively; and 3) by aspects to the Ascendant and Ascendant lord Mars.

Sun, Moon and Jupiter are all beneficial planets for Aries Ascendant chart and considered *Nadi yogakaras*. (See Glossary) The Sun is exalted in Aries, maximum exaltation at 10:00 Aries. Saturn is debilitated in Aries, maximum debilitation at 20:00 Aries.[56]

Chart 1: Martin Luther King, Jr.

Birth data: Tuesday, Jan. 15, 1929, 12:00 PM CST, Atlanta, Georgia, USA, Long. 84W23 17, Lat. 33N44 56, *Lahiri ayanamsha* -22:51:44. Class A data (Mother said:"high noon.") Ascendant: 20:56 Aries. *(Time zone: Please note that Atlanta, GA operated in the Central Time Zone between Jan. 1, 1918 and March 21, 1941. Daylight savings time was used in spring to fall.)*

Biographical summary: Dr. Martin Luther King is the greatest hero of the American Civil Rights movement and its first major leader and activist working for social justice in the non-violent tradition of Mahatma Gandhi.[57] His leadership led to the passage of the Civil Rights Act of 1964 (signed July 2, 1964), outlawing segregation in the U.S., as well as the Voting Rights Act of 1965 (signed Aug. 6, 1965) outlawing discrimination in voting in many states. It enabled American ethnic minorities to vote for the first time in many states after a long history of disenfranchisement, especially among African-Americans. King was adamant about voting rights, and this law became the most effective civil rights legislation of the 20th century, even if it took years to enforce fully. For leading non-violent resistance to end racial prejudice in the United States, Dr. King won the Nobel Peace Prize in 1964. In his acceptance speech on Dec. 10, 1964, he said:

---

[56] Saturn in Sanskrit is *Shanishchara*, meaning "slow-moving," or *Shani*, meaning "slow." While Aries and its planetary lord Mars (youth) are associated with speed, Saturn (the elder) is not comfortable having to move faster. Saturn's influence often causes delays or hard focus. Its worst house placement is the first house, which classically has an overlay of Aries themes. Mars receives *Digbala* (best house placement) in the 10th house, and worst in the 7th house, where it can attract enemies. Best house placements for Mars are the *Upachaya* houses: 3, 6, 10 and 11, the best of these being House 10.

[57] Others such as Malcolm X (1925-1965) advocated a separate nation for African-Americans. The Muslim faith appealed to some African-Americans seeking social justice and equality in the U.S. (among them Mohammad Ali). Other civil rights activists wanted to meet violence with violence, and King opposed that.

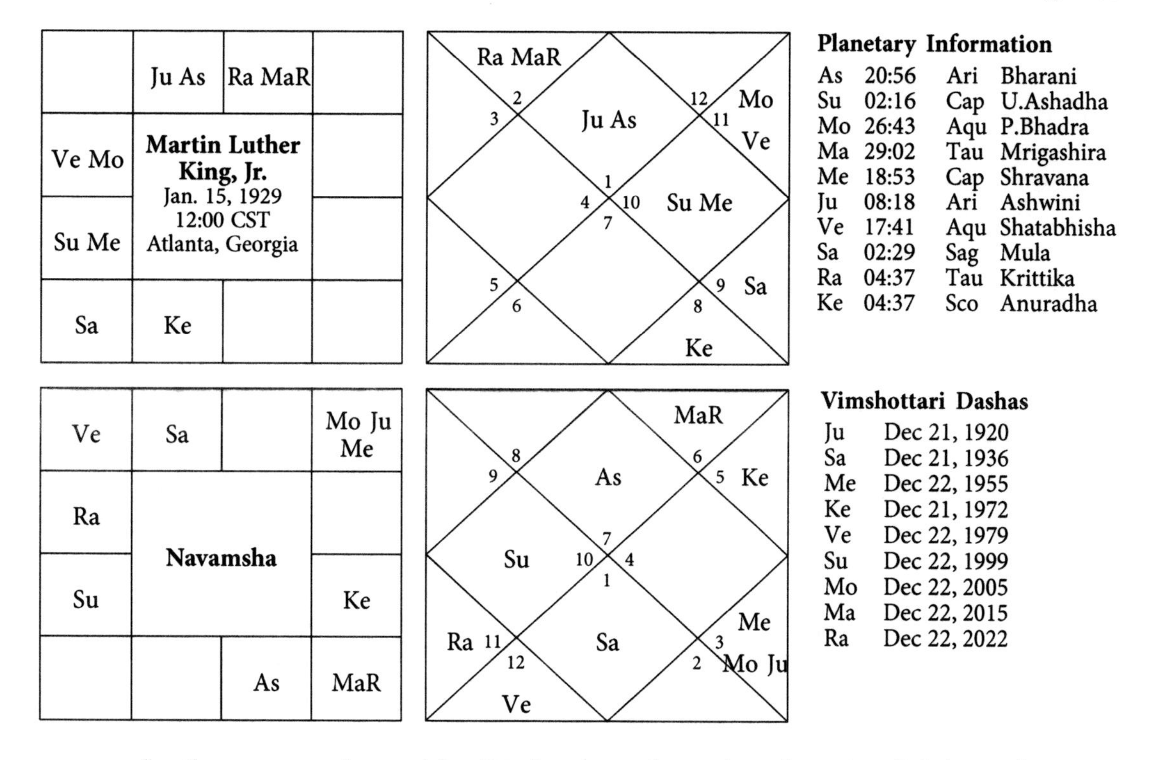

Planetary Information

As	20:56	Ari	Bharani
Su	02:16	Cap	U.Ashadha
Mo	26:43	Aqu	P.Bhadra
Ma	29:02	Tau	Mrigashira
Me	18:53	Cap	Shravana
Ju	08:18	Ari	Ashwini
Ve	17:41	Aqu	Shatabhisha
Sa	02:29	Sag	Mula
Ra	04:37	Tau	Krittika
Ke	04:37	Sco	Anuradha

Vimshottari Dashas

Ju	Dec 21, 1920
Sa	Dec 21, 1936
Me	Dec 22, 1955
Ke	Dec 21, 1972
Ve	Dec 22, 1979
Su	Dec 22, 1999
Mo	Dec 22, 2005
Ma	Dec 22, 2015
Ra	Dec 22, 2022

> "... the answer to the crucial political and moral question of our time [is] the need for man to overcome oppression and violence without resorting to violence and oppression."

He was the first major American public figure to speak out about civil rights of black people and against the Vietnam War.[58] In the second cause, he stood virtually alone. But he refused to listen to those of his advisers and supporters who cautioned against it. He understood a larger perspective when he said:

> "The arc of the moral universe is long, but it bends towards justice."

King grew up in a prosperous African-American minister's family, with an older sister and younger brother. He lived in a pious, proud and progressive community in segregated Atlanta, Georgia. His father was the minister at the Ebeneezer Baptist Church, where Martin was ordained in Feb. 1948 at age 19. Though he could have continued on at this church he chose to become pastor at Dexter Avenue Baptist Church in Montgomery, Alabama as of Oct. 31, 1954. He returned to Atlanta to be co-pastor with his father at Ebeneezer Baptist Church, from 1960 to 1968. On June 18, 1953, King married Coretta Scott (b. April 27, 1927, 4:00 PM, Marion, AL; d. Jan. 30, 2006). They had four children.

A gifted speaker, King won oratorical contests from the age of 14. He skipped the 9th and 12th grades, and due to his high college entrance exam scores, entered Morehouse College at age 15.

[58] Marginalized as a radical by the mainstream, Malcolm X did speak out publicly against the Vietnam War long prior to King, on Aug.10, 1963, in 1964 and up to his assassination Feb. 1965. For a serious re-evaluation of his life, see Manning Marable's biography: ***Malcolm X: A Life of Reinvention***, 2011.

He was one of 11 African-Americans in a class of 90, and graduated from Morehouse in 1948, having been ordained earlier that year. He went on to Crozer Theological Seminary, where he was student body president and class valedictorian. He then earned a Bachelor of Divinity in 1951 and a Ph.D. in Systematic Theology from Boston University June 5, 1955. King was awarded numerous honorary degrees, some posthumously, and he received the Nobel Peace Prize in 1964. At age 35, he was the youngest man, the second American, and the third black man to be so honored.

From the start, King's leadership was crucial to the success of the Montgomery Bus Boycott, which was sparked by Rosa Parks' arrest on Dec. 1, 1955 for refusing to give up her bus seat to a white person, as the whites-only section was full.[59] The following Sunday, King began his sermon to his packed church on Dexter Avenue in Montgomery:

> "You know my friends, there comes a time when people get tired of being trampled over by the iron feet of oppression."

Soon the agitated congregation was on its feet joining in cheers that became a resounding din. King had instantly become the delegated spokesman for the Montgomery Bus Boycott, which began the next day (Dec. 5, 1955) and lasted 381 days. It ended in bus de-segregation in the city and marked the first major event and first success of the American Civil Rights movement. It also led directly to King's founding of the Southern Christian Leadership Conference – an organization that backed numerous sit-ins, marches, and peaceful protests. King was its president from 1957 to 1968. His philosophy of passive resistance in the style of Mahatma Gandhi led to his frequent arrests and tours through the Bible belt, where he became known as "Alabama's Modern Moses." A powerful orator wherever he went, King is probably best known for his speech during the [Civil Rights] March on Washington Aug. 28, 1963. There he delivered his famous "I Have a Dream" speech to a crowd of 200,000 people.[60]

According to friends, family and colleagues, King was wiretapped at home and in every hotel he visited for many years up to the time of his death. His speech on Aug. 28, 1963 created intensified FBI interest in King. The rationale for their ongoing and rigorous surveillance campaign on Dr. King, officially from 1963 and probably earlier, was their fear of Communist infiltration of the Civil Rights movement. When no such evidence could be found, the FBI fabricated evidence to incriminate King and tried to blackmail him in 1963-1964 with the marital infidelities they did discover on the tapes. FBI surveillance of Coretta Scott King continued for some years after 1968 due to concern she might "tie the anti-Vietnam War movement to the Civil rights movement," as her husband had done.

The Gulf of Tonkin Resolution of Aug. 7, 1964 justified the rapid U.S. expansion of military involvement in Vietnam, though it was later admitted to be a false pretext for war. National dissension increased on this issue and Anti-war protests grew in size from April 1965 onward, often led by college student activists or veterans, but as of yet no one of major national clout.

---

[59] The legal precedent for Rosa Parks was Brown v. Board of Education, decided May 17, 1954 in the U.S. Supreme Court. This landmark case demolished the legal basis for segregation in the U.S., ruling that "separate educational facilities are inherently unequal." It was won by Thurgood Marshall (1908-1993), a civil rights lawyer who himself became the first African-American Supreme Court Justice. He was appointed by President Lyndon Johnson on June 13, 1967 and in office from June 17, 1967 through June 28, 1991.

[60] In this speech, King said: "I have a dream that one day this nation will rise up and live out the true meaning of the creed: 'We hold these truths to be self-evident, that all men are created equal.'"

On Feb. 23, 1967 Noam Chomsky published his influential Anti-war article: "The Responsibility of Intellectuals" (*New York Review of Books*). King's first participation came soon after, on March 25, 1967, when he led a march of 5,000 in Chicago. On April 4, 1967 he delivered his speech "Beyond Vietnam: A Time to Break Silence."[61] Three weeks later, 100,000 protesters marched in San Francisco and 500,000 in New York City, from Central Park to the United Nations building, where Dr. King was a featured speaker, along with Dr. Benjamin Spock and Vietnam veteran Jan Barry Crumb.[62]

King's wife and family remained loyal despite the marital infidelities and the many threats and dangers. King himself continued undeterred from his social and political causes, which took him away from home 25-27 days a month. This caused him to be the target of physical attacks and/or bomb scares on numerous occasions. Starting with the Montgomery Bus Boycott in 1955, King was arrested, his home was bombed and he was subjected to personal abuse. On April 3, 1968, his supporters gathered at the Mason Street Temple in Memphis, TN, where King delivered his legendary "I have been to the mountaintop" speech. His arrival had been delayed by a bomb threat to his airplane.

> "And then I got to Memphis. And some began to say the threats... or talk about the threats that were out. What would happen to me from some of our sick white brothers?
>
> Well, I don't know what will happen now. We've got some difficult days ahead. But it doesn't matter with me now. Because I've been to the mountaintop. [applause] And I don't mind. Like anybody, I would like to live a long life. Longevity has its place. But I'm not concerned about that now. I just want to do God's will. And He's allowed me to go up to the mountain. And I've looked over. And I've *seen* the promised land. I may not get there with you. But I want you to know tonight, that we, as a people, will get to the promised land! [applause] And so I'm happy, tonight. I'm not worried about anything. I'm not fearing any man. Mine eyes have seen the glory of the coming of the Lord."

The next evening, April 4, 1968 at 6:01 PM EST, he was shot in the neck by a single rifle bullet while standing on his hotel balcony in Memphis. The alleged lone assassin was James Earl Ray. King was rushed to the hospital and pronounced dead at 7:05 PM EST. Though urban riots persisted across the U.S. from 1964 through 1970, the assassination of King provoked the largest spate of them throughout April 1968 – in over 100 cities – and the worst racial riots since the Civil War.

On Dec. 8, 1999 in a civil trial spearheaded by his widow Coretta Scott King, the jury – after examining extensive evidence – concluded that government agencies, including the city of Memphis,

---

[61] King spoke to Concerned Clergy and Laity at the Riverside Church in New York City. He described why it was necessary for him to "bring Vietnam into the field of [his] moral vision." The entire speech is available at this website. **http://www.ssc.msu.edu/~sw/mlk/brkslnc.htm.**

[62] The Vietnam War lasted from Nov. 1, 1955 to April 30, 1975, with U.S. involvement escalating from the early 1960s. Not until late 1968 were a majority of Americans opposed to this war. Though 1967 was a turning point in broad American support for the war, it was still a dangerous time to oppose it. World Heavyweight boxing champion Mohammad Ali (Cassius Clay, b. Jan. 17, 1942) was arrested on April 28, 1967 for refusing to be drafted into the Armed Forces. Claiming "Conscientious objector" status, he opposed the Vietnam War from several vantage points, including his Muslim religion (from 1964). Found guilty in a trial on June 20, 1967, he was sentenced to five years in prison and barred from professional boxing for almost four years. He did not go to prison while his case was on appeal and finally reversed by the U.S. Supreme Court on June 28, 1971.

the state of Tennessee and the federal government were party to the conspiracy to assassinate Martin Luther King.[63] Meanwhile King's stature has only grown. Since 1986 a national holiday is observed annually to honor his memory; and on Oct. 16, 2011 a thirty foot high statue and monument on the National Mall in Washington, D.C. was formally unveiled. It was originally intended to mark the 48th anniversary of his famous "I Have a Dream" speech. The final design was approved in Sept. 2008, just prior to the election of the first (half) African-American president of the United States, Barack Obama.

Books by Dr. King include: *Stride toward Freedom: The Montgomery Story* (1958), *Why We Can't Wait* (1964), and *Where Do We Go from Here: Chaos or Community?* (1967).

Ascendant lord Mars and the activist preacher: Natal Mars is located in the 2nd house of speech and *Kutumba* (happiness from the family of origin). This shows where the energy of this life will go and that it follows the family lineage in some way – especially due to 9th lord Jupiter placed in the Aries Ascendant. Jupiter is the priest or spiritual leader and is considered a traditional influence. Jupiter's sign lord is Mars, which again points back to the voice. Mars is retrograde at 29:02 Taurus, within one degree of its Stationary Direct degree at 28:07 Taurus (12 days after he was born). This gives added strength to Mars. A retrograde planet is pushed even more into the activity of that planet, whether positive or negative, depending on its inherent nature. With Mars also strong in its own *nakshatra* and retrograde near its stationary degree, this added to the danger in his life. Since the nature of Mars is active and assertive, Dr. King's message and his goals were seen as a major threat to the status quo – even if his goals were long overdue and even if his approach was through persuasion and not coercion.

> "The ultimate measure of a man is not where he stands in moments of comfort and convenience, but where he stands in moments of challenge and controversy."
>
> **Martin Luther King, Jr.**

Another factor lending extra weight to Mars is its *Atma Karaka* status – a planet at the highest degree of longitude in any sign in the chart. Natal Mars is also well placed in Taurus in Mars-ruled *Mrigashira nakshatra* (symbolized by the head of a deer, or a sacrificial or wild animal.) As a planet of speed in a fixed earth sign, Mars is forced to focus and organize itself more than usual, especially at the Stationary Direct degree. This has implications for his entire life, and means that his voice and his message are likely to be taken very seriously. Though Mars is more dangerous when combined with Rahu or Ketu – as in this case – at least both Rahu and Mars are in the *nakshatras* of planetary friends to the Ascendant lord: Sun and Mars, respectively. This helps him to accomplish a lot of his goals and quickly. That King's path would differ in some way from his family roots is shown by Rahu (the foreigner) with Mars in the 2nd house.

Rahu gives a more raw and rough edge to the energy of Mars. It can and did give it a sense of urgency: Dr. King was known for the power of his voice and his electrifying oratorical skills. It would drive him to be courageous in extraordinary times, but it would also be dangerous for his personal health and safety. The condition of Mars in both his birth chart and *Navamsha* indicates

---

[63] A transcript of the trial is available at **www.thekingcenter.org.**

that he is likely to speak out on controversial issues and that his physical strength is not good for longevity, though Sun at the top of the sky gives him better stamina. Autopsy reports showed his heart was as worn out as someone twice his age – such was the grueling intensity of his work schedule. He was committed to what he often called **"the fierce urgency of now."**

The sign lord of Mars is sensuous Venus, idealistically placed with Moon in Aquarius in the 11th house. This describes the large number of supporters who revered him, including women friends and lovers. (Moon and Venus are the only two female planets.) Both birth Mars and *Navamsha* Mars are in earth signs, indicating the pragmatic bent to the life: Since this is the situation I am in, what can I say or do that will yield the most effective result? Birth Mars in Taurus in the house of speech, plus *Navamsha* Mars in Mercury's sign of Virgo identified from birth what would be the key vehicle of his dharmic destiny.

Navamsha Mars is opposite exalted Venus in Pisces: This shows how King could be devoted to a much higher cause than his own personal status or even personal safety. There is a surrendering to a larger benevolent force (Pisces) for the greater good (Venus in Pisces). King saw this as his calling: **social justice** that would eliminate inequalities between the races and financial inequalities between all peoples. He was killed shortly before his scheduled Poor People's March on Washington, DC, an event delayed by his participation in a strike in Memphis, TN by Sanitation workers who received far lower wages and under worse terms than white workers doing the same job.

Navamsha Mars is in the 12th house in Virgo – the sign of conflict. The 12th house is also secret or foreign places, and includes pleasures of the bed. Mutual Venus and Mars contacts indicate sexual passion, which can cause controversy when involving houses 6 and 12. In this case, the FBI tried to use his extramarital affairs as blackmail.

Venus in the 6th house is in its worst house position. *Navamsha* Venus exalted in Pisces in the 6th house confirms two other factors: 1) Sacrifices in his love life, and his wife could suffer; and; 2) Personal sacrifice in the cause of the greater good. Since Venus is also *Navamsha* Ascendant lord, this weakens King's physical health, while putting him in a position involving conflict.

On the fast track to education and honors: Winning honors and oratorical contests from age 14, King made a rapid sprint through secondary school, college (from age 15) and graduate school, with two graduate degrees by age 26 and a Nobel Peace Prize at age 35, the youngest man ever to receive the award and the third black man. Such velocity and focus in a short, action-packed life is attributable in part to the extraordinary power of Ascendant lord Mars, supported and amplified by Jupiter in the Ascendant. Both Jupiter and the Ascendant are in fire signs – which tend to initiate action. From a timing perspective, the action-oriented life is confirmed by the sequence of *Dashas* (planetary periods) up to Dec. 1955: Jupiter *Dasha* (birth to Dec. 21, 1936), then Saturn *Dasha* (Dec. 21, 1936 to Dec. 22, 1955). Both planets are in fire signs and located in either angular or trinal houses from the Ascendant in birth and *Navamsha* charts, giving added success.

Another factor supporting the fast track is that both **Jupiter and Saturn** (*Dharma* and *Karma* lord, respectively) are in synch, with both of them in dharmic fire signs, and Jupiter aspecting Saturn in the sign of Sagittarius, which it owns. From an early age there is no doubt about what motivates him to act. His only hesitation was in the traditional Baptist influence that he inherited. An Aries Ascendant person generally needs some independence, but in this case – he did not disappoint his paternal lineage, though he did forge entirely new ground within that context. This was fitting for an Aries Ascendant destiny, with traditional Jupiter on the Ascendant.

The 4th house of education and higher educational degrees is read both from the Aries Ascendant and from the Moon. The Moon rules the 4th house from the Ascendant, itself strongly placed in the 11th house of the birth chart and in the 9th house of the *Navamsha* chart. His Moon is aspected by Saturn in both the birth chart and *Navamsha.* This causes affliction but also makes him work hard and take on many responsibilities. In the birth chart, the 4th house is aspected by 5th lord Sun, considered excellent – as 5th house rules intelligence. Mercury is lord of the 6th house of conflict; thus its aspect to the 4th house gives some controversy, perhaps due to an African-American achieving doctoral status in a still segregated America and accomplishing it so speedily. *Navamsha* 4th lord Saturn confirms a rise from some significant disadvantage, as it is *Digbala* (best possible angle) yet debilitated in Aries.

From the Moon in Aquarius, the 4th house is Taurus, whose lord is Venus. Venus is with the Moon – a positive influence. Also from the Moon, Mars and Rahu are situated in the 4th house of education. Mars gives speed and movement, agitated by its contact to Rahu. This had the effect of speeding up the education rather than cutting it off – which can also happen. In a certain way, his secondary education was shortened so he could quickly move on to a lengthier higher education.

The waxing Moon is a classical benefic, and overall is very positive for this chart, especially as a great good friend of Ascendant lord Mars and located in the *nakshatra* of Jupiter, another planetary friend to Mars. This mostly overrides Moon's ownership of a *Kendra* for the Aries Ascendant chart, which falls under the rules of *Kendradhhipatya Dosha,* rendering a benefic more neutral. (See Glossary) The Moon and 4th house gave King an excellent education and a good family foundation in life. But also, when a planet is situated in the 8th house from the house it owns, as with the Moon, it could and did cause harm to domestic matters. He travelled so much in later years he was rarely at home with his wife and children. And after he became more politically active in Dec. 1955, the safety of his home and family was increasingly threatened. His home was also under constant surveillance.

Leadership qualities: The Sun at the top of the sky is excellent for leadership, but does not by itself guarantee it. However, it is a good planet for the Aries Ascendant chart and is in its best possible angle of the chart – *Digbala* – where it gains strength and dignity and lends visibility to the person. With the power of his moral authority, Reverend King's leadership in opposing the Vietnam War was perceived as a major threat by the U.S. government. Mainstream media coverage of his 1967 "Beyond Vietnam" speech was mostly negative, but in later years it began to be quoted or replayed with reverence, especially on his birthday.

> "We still have a choice today: a non-violent coexistence or violent co-annihilation. We must move past indecision to action. We must find new ways to speak for peace in Vietnam and justice throughout the developing world – a world that borders on our doors. If we do not act we shall surely be dragged down the long, dark and shameful corridors of time, reserved for those who possess power without compassion, might without morality, and strength without sight. Now let us begin. Now let us rededicate ourselves to the long and bitter – but beautiful – struggle for a new world."
>
> **Martin Luther King, Jr.,** from his "Beyond Vietnam" speech,
> Riverside Church, New York City, April 4, 1967.

The Sun can signify royalty as well as physical vitality. In the Vedic planetary scheme, the Sun and Moon represent King and Queen, respectively. His mother said he was born at "high noon."

Sun was exactly culminating at 11:47 AM CST. This gave him superior vitality and leadership, especially as Sun is *Vargottama* and in its own *nakshatra* of *Uttara Ashadha* ("the later victor"). *Vargottama* is the repeat of a planet in the same sign in the *Navamsha* chart (Capricorn, in this case). Sign lord and *nakshatra* lord are well placed in either trinal or *Kendra* houses in both birth chart and *Navamsha*. In addition, the Sun is in an excellent yoga – the *Ubhayachari yoga*. This occurs when planets other than Moon, Rahu and Ketu occupy houses 2 and 12 from the Sun, and confers on the native a large network of friends and supporters.[64] The 10th House lord (Saturn) is also well placed in the 9th house. All these factors put him in a position of leadership and helped him to continue to lead despite daunting circumstances that continually threatened is own personal health and safety.

,iter and his destiny within the tradition of the Baptist ministry: Although it was expected he might follow in the footsteps of his father and grandfather, King was wary of the emotionalism of southern Baptist churches and questioned whether a life in this church could be "intellectually respectable as well as emotionally satisfying." Even so, he was pastor of only Baptist churches.

Jupiter is a traditional planet, especially when on the Ascendant of the birth chart, and in this position it gives the impetus for Dr. King staying within the family tradition of the ministry, even if he briefly considered a profession in law or medicine. Jupiter is excellent in the Aries Ascendant; thus the best choice for him was to stay within the family tradition. After choosing a ministry in Montgomery, Alabama (1954-1960), where his involvement in the Bus boycott brought him to national attention as of Dec. 5, 1955, he returned to Atlanta to co-pastor the Ebeneezer Baptist Church with his father (1960-1968).

Jupiter has *Digbala* in the Ascendant (best possible angle) and as a great planetary friend to Ascendant lord Mars its transits and major or minor *Dashas* will have a potent effect in the life, especially as King was born in Jupiter *Dasha*. Jupiter is the priest or minister, the advisor to kings. It rules over philosophy, religion, law, and education.

Jupiter's strong placement in the Ascendant can bring any one of these professions into the life, and King contemplated being a lawyer or a doctor before deciding on the ministry. If Saturn sets boundaries, Jupiter expands the realm of tolerance and forgiveness. Jupiter's prominence in King's chart reminds us of Jupiter's capacity to bless and protect. Jupiter is the spiritual leader capable of imparting ethical teachings.

King's Jupiter would also give him an excellent education. Only Jupiter is allowed to cast a trinal aspect on to other houses, in classical Vedic astrology. And from the Ascendant, Jupiter also aspects its own house – the 9th house, generally considered the most fortunate house in the chart. From these factors alone, we can see his good fortune in being brought up in a prosperous minister's family during a time of racial segregation.

Dharma **– Fire signs:** The fire signs are all considered Dharmic signs; thus, a fire sign on the Ascendant, as well as both Jupiter and Saturn placed in fire signs accentuates the *Dharma*-driven nature of this destiny, or the urgent need to identify and act in harmony with his rightful purpose in life. Aries is a *Chara* sign, thus very active, and Mars can be a revolutionary. Thus, the Aries Ascendant person has to be willing to go first, to do or say what needs to be said when no one else has yet done or said it. Although Mars signifies muscularity as well as vitality, as Ascendant

---

[64] The *Ubhayachari yoga* also occurs in Chart #2 (Robert F. Kennedy).

lord for Aries ascendant, Mars will not be about muscularity primarily. Emphasis is on the courage necessary to take initiatives in life, and at times when they could be dangerous for whatever reason, often because they threaten the status quo. On a philosophical level, this fiery planetary combination led by Jupiter enabled King to speak with complete conviction. Saturn in Sagittarius helped give the sound ethical basis, in service to the greater good. This quote is from his renowned "I Have a Dream speech," delivered Aug. 28, 1963:

> "I have a dream that my four little children will one day live in a nation where they will not be judged by the color of their skin but by the content of their character. I have a dream today."

Fortune, prizes and discipline - *Dharma* and *Karma* lords: The positions of Jupiter and Saturn in any astrological chart are extremely important in establishing the expansion of good fortune and the boundaries of commitment and self-discipline, respectively. But in the Aries Ascendant chart **Jupiter and Saturn** have a double role: Jupiter not only rules the 9th house of *Dharma* (Sagittarius), but it is the classic *Dharma* lord. Likewise, Saturn rules the 10th house of *Karma* (Capricorn) and is also the classic *Karma* lord. Generally, the most powerful placements for these planets are in *kendras* (angles) of the chart or in trinal houses. Dr. King's Jupiter is placed very strongly in the Aries Ascendant - best *kendra* for Jupiter, and also a trinal house; Saturn is in the 9th house (a trinal house) in Sagittarius. Saturn is in perhaps its most ethical sign placement in Sagittarius. It is very well placed for leadership due to a *Raja yoga:* 10th house lord is placed in the 9th house, and the 9th lord Jupiter is aspecting Saturn.

Fortune and discipline are dependent on stable and strong mental planets (Moon and Mercury), as well as reasonably strong physical planets (Mars and Sun, in this case). Then there is an opportunity to be of great service in an endeavor of philosophical or ethical importance, whether or not one sees the results or is rewarded for them. Normally also, Jupiter as friend to the Ascendant lord in the Ascendant protects the physical body, but if the Ascendant lord or Sun are not well placed in either birth chart or *Navamsha* chart, there is just so much that Jupiter can do to protect. In this case, as we noted earlier, Mars has definite weakness in its configuration - which did not give good longevity. The 8th house of longevity also has Ketu placed in it and the 8th lord is afflicted by Rahu.

Tr. Jupiter-Saturn conjunction at 1:52 Capricorn (Feb. 18, 1961): From this date, Dr. King was likely to have an even more important public role. This conjunction occurred very close to his natal Sun at 2:16 Capricorn in his 10th house of career, status, and maximum visibility. Though the Montgomery Bus Boycott brought him into national prominence from Dec. 1955 onward, his expanding role as Civil Rights leader in the 1960s brought him even more responsibilities and notoriety.

Tr. Saturn entered Capricorn on Feb. 2, 1961, also marking the start of King's ***Sade Sati*** (7 ½ year period when Saturn transits houses 12, 1, and 2 from natal Moon). Considered one of life's biggest test periods, it brings greater responsibilities. *Sade Sati* for King ran from Feb. 2, 1961 thru June 17, 1968. **Tr. Saturn's passage through Aquarius** (Jan. 18, 1964 thru April 9, 1966) was the strongest house and sign position for King. During this period King witnessed the passage of both the Civil Rights Act of 1964 and the Voting Rights Act of 1965 - goals he and his coalition had worked hard to achieve. Saturn can and did bring truth, reality, and manifestation, after long delays.

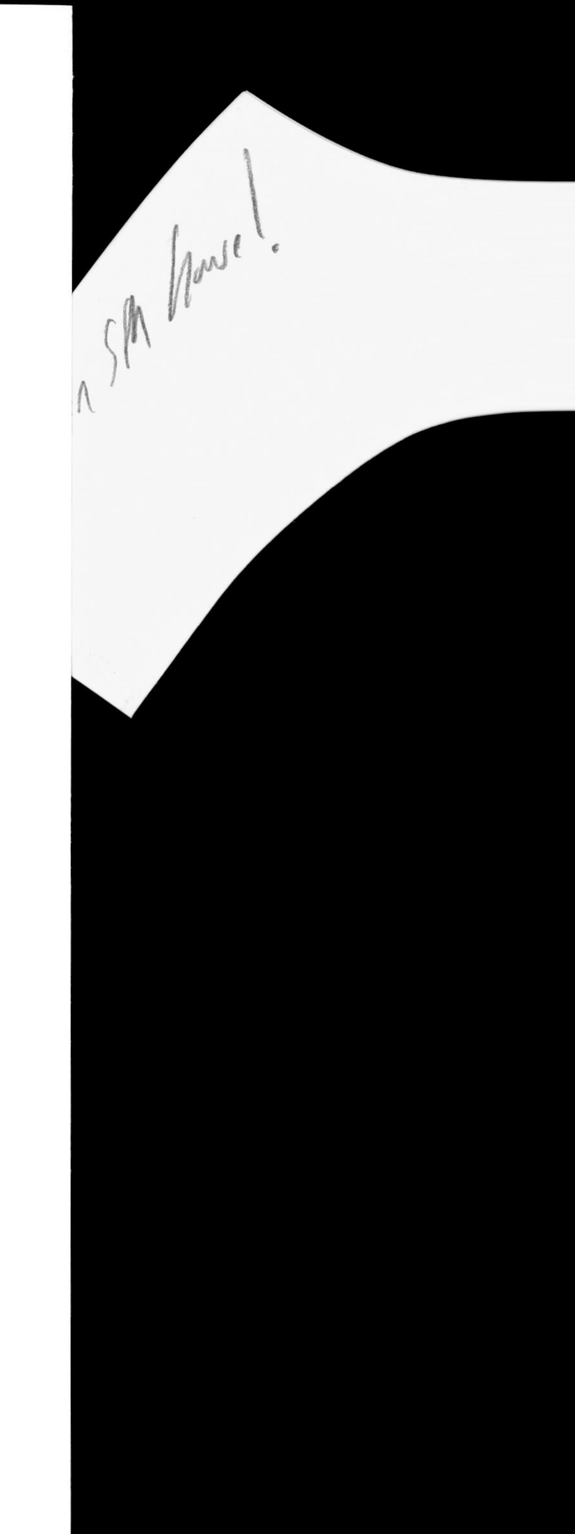

Dr. King's natal **Jupiter** forms the most important *Raja yoga* in his chart; thus the transits of Jupiter should indicate when he is honored. When King accepted the **Nobel Peace Prize** on Dec. 10, 1964, **tr. Jupiter** retrograde was at 24:20 Aries, near its Stationary Direct degree of 22:44 Aries, close to his Ascendant at 20:56 Aries. Five months earlier, on July 2, 1964, the **Civil Rights Act of 1964** was signed into law. Tr. Jupiter was then at 24:55 Aries, near the exact same degree as when he received the Nobel Prize. King's "I Have a Dream" speech at the March on Washington (Aug. 28, 1963) was widely regarded as paving the way for the Civil Rights Act. Even so, Jupiter's honors also made King more threatening to the status quo. The FBI, for one, stepped up its surveillance and harassment of King after this date. When he was killed, he was greatly mourned and honored: tr. Jupiter was in Leo aspecting his Ascendant, Jupiter, Moon, Venus, and Saturn. He was also in a Jupiter sub-period of Mercury *Dasha,* as of Jan. 6, 1968. (See below)

Mercury – and the master of communication: Natal Mercury is in Moon-ruled *Shravana nakshatra*, connecting Mercury to Moon. In the *Navamsha* chart, Moon, Mercury, and Jupiter are all situated in the 9th house in Mercury's sign of Gemini. This gave Dr. King the power to speak voluminously (Mercury-Jupiter) and in a way that was most accessible (Gemini) to the largest number of people (Moon). Moon's interaction with Mercury in both charts gave emotional potency to his speech, but without the grounding of Jupiter and Saturn this could have been excessive and unfocused. Even so, Mercury can bring controversy for Aries Ascendant, especially if connecting with the 6th house of conflict in the *Navamsha*, as in this case. Jupiter rules the 6th house in the *Navamsha* chart, increasing conflict in the Jupiter sub-period of Mercury *Dasha*: Jan. 6, 1968 thru April 13, 1970.

Regarded strictly as *karaka* (significator) for communication, Mercury is well placed in both the birth chart and *Navamsha* chart. At the top of the birth chart in the 10th house, the most visible placement, and in Capricorn, this gives him a sense of obligation about what he says in public as well as a duty to speak out. His natal **Mercury** is placed **in *Shravana nakshatra*** (meaning "the ear" or "to hear"), and whose distinction is to be a good listener and to connect diverse elements. *Shravana* is ruled by the Moon and gives a strongly emotional approach. This is confirmed by Mercury's placement in the *Navamsha* 9th house in Gemini along with *Navamsha* Moon and Jupiter. Fortunately, Moon is an excellent planet for Aries Ascendant, and its good house placement in the birth and *Navamsha* charts lends more emotional power to Mercury. For a public figure, Moon also symbolizes the public. King's legendary oratorical skills would carry enormous weight apropos of least two major socio-political events: the Civil Rights movement and the Anti-Vietnam War movement.

Natal Moon in Aquarius is in *Purva Bhadra nakshatra,* ruled by Jupiter. This gives a fiery temperament and an aggressive, active nature, amplifying the action orientation of the Aries Ascendant person. With Jupiter as *jiva* (*nakshatra* lord) so strongly placed in the Ascendant, the mind and spirit are expansive, optimistic and self-confident. Moon in Aquarius further defines the individual as very idealistic about improving human relations, especially among larger groups of people (11th house Moon). Together with Venus in the 11th house, King was idealistic about manifesting more equality in the largely segregated society he knew as a child and entered as an adult. Mercury and Moon, respectively, are the two planets showing the cast of the mind and heart. Both are in signs owned by Saturn: Capricorn is more responsible in a pragmatic way, while Aquarius is motivated by philosophical ideals. Natal Saturn aspects natal Moon, doubling the impact of a very conscientious person who is likely to work hard. Jupiter's influence gives joy as *jiva.*

Danger from one's enemies: Situated in the 10th house, Mercury's prominence and visibility is assured. The Sun joins Mercury in the 10th house in its most dominant placement. Mercury gains even more strength by being situated in the *Navamsha* chart in its own sign of Gemini in the 9th house. In the birth chart, Mercury's sign lord Saturn is well placed in the 9th house and its *nakshatra* lord Moon is also well placed in the 11th house. However, Mercury is a planetary enemy of Mars, especially here as lord of the 6th house of conflict. Thus a strongly positioned 6th lord can put one often in the midst of conflict, especially when the voice and message are both powerful in their effect and personally dangerous to him. With **Mercury *Dasha*,** King's life took on new purpose and also became suddenly more dangerous for him. The most definitive events of his life as an activist preacher began from the start of his Mercury *Dasha* (**from Dec. 22, 1955** until his death). This *Dasha* began 17 days after the start of the Bus Boycott on Dec. 5, 1955.

In the last year of his 19-year Saturn *Dasha*, King moved to Montgomery, Alabama to become pastor of the Dexter Avenue Baptist Church: on Oct. 31, 1954. Then on Dec. 1, 1955, also in Montgomery, black seamstress Rosa Parks was arrested. Her single, courageous act was to have far-reaching effects in segregated America. But first the people needed to be united behind this common cause. The Bus Boycott was organized to provide that purpose and King showed his ability to unite his people. Not quite 27 years old, he led the **Montgomery Bus Boycott**, starting **from Dec. 5, 1955 (and lasting 381 days).** It resulted in the triumphant end of segregation on buses and led the way towards outlawing most forms of racial segregation in the U.S. – although this came later with the Civil Rights Act of 1964 (July 2, 1964), and the Voting Rights Act of 1965 (Aug. 7, 1965). The Bus Boycott also made King a target for his enemies. Among others, the FBI (Federal Bureau of Investigation) considered Dr. King dangerous for his oratorical skills and for providing the Civil Rights movement with such strong leadership.

King had many successes during *Mercury Dasha*, but without **the power of his Moon and Saturn**, he would not have done as well during this *Dasha*. He may not have survived for as long as he did, had he not had Jupiter's generally protective powers in the Ascendant. Also, Mercury's *nakshatra* and sign lords (Moon and Saturn, respectively) are well placed. Even so, for upholding his allegedly radical views he was stoned, physically attacked and his house bombed – starting from his Mercury *Dasha* and the Montgomery Bus Boycott.

MAJOR LIFE EVENTS & DATES:

1. **Moves family to Montgomery, Alabama (Oct. 31, 1954):** When he took over the pastorship of the Dexter Ave. Baptist Church, King was in the last sub-period of Saturn *Dasha* (Saturn-Jupiter, from June 9, 1953 to Dec. 22, 1955). Tr. Saturn was at 18:20 Libra, approaching its maximum exaltation at 20 Libra. It was in its strongest *kendra* in King's chart, the 7th house, pushing him to take on new responsibilities, including marriage (from June 18, 1953). Five transiting planets were in *Chara* (active) signs, galvanizing his birth chart: Mars in Capricorn, Jupiter in Cancer, and Saturn, Sun, Mercury in Libra. Three planets were exalted (Mars, Jupiter, and Saturn, and one debilitated – tr. Sun in Libra). An endeavor begun with so many planets in exalted (or debilitated) signs tends to bring extreme situations. Further, Saturn is within four degrees of tr. Sun, creating a combust Saturn, similar to when King was killed, though on that date tr. Saturn was within 1/60th of a degree of the Sun in Pisces. (See under Assassination) Sun-Saturn shows extreme focus, also someone challenging the status quo. Tr.

Mars opposite Jupiter pushes the level of religious or philosophical idealism, especially with both planets exalted.

2. **Ph.D. received from Boston University (June 5, 1955):** King was awarded his PhD. in Systematic Theology during the last sub-period of Saturn *Dasha*, **Saturn-Jupiter period.** In his 4th house of education was tr. Jupiter at 5:21 Cancer (close to its maximum exaltation degree of 5:00 Cancer). Saturn was transiting at 22:42 Libra – also exalted – aspecting his 4th house of education and higher degrees as well as his Ascendant at 20:56 Aries. Tr. Saturn was close to its Stationary Direct degree (21:16 Libra), adding power, especially as Saturn's maximum degree of exaltation is 20:00 Libra. Combined with his birth chart, and with Jupiter or Saturn *Dasha* (he had both), these factors point to the extraordinary level of his personal academic achievement, especially for an African-American in 1955. The completion of his academic education during the last few months of Saturn *Dasha* is fitting: Saturn is 10th house lord of worldly status, situated in the 9th house of higher education. Though these were all excellent omens, the day also marked a **Full Moon eclipse** at 20:54 Scorpio, exactly on the cusp of his 8th house, portending intrigues and turmoil but also spiritual progress. Two weeks later there was a **Total Solar eclipse** at 4:50 Gemini, opposite King's then-*Dasha* lord Saturn, natally at 2:29 Sagittarius. This was the last in a series of three eclipses, intensifying the effects.

3. **Montgomery Bus Boycott begins (Dec. 5, 1955):** Tr. Mercury was at 20:00 Scorpio (close to tr. Sun and Rahu), implying danger to him from his speech, as it was in his 8th house and repeating the eclipse degree of June 5, 1955. (See above) The Boycott lasted 381 days, and ended on Dec. 21, 1956 with the successful de-segregation of public transport. King's 17-year **Mercury *Dasha*** began on Dec. 22, 1955 and was already having an influence in early Dec. 1955. Both prominence and controversy are apparent from the position of Mercury in his birth and *Navamsha* charts. This boycott catapulted King into the national spotlight, and into leadership of the burgeoning Civil Rights movement. His safety and that of his family were suddenly in much danger due to entrenched opposition to their goals. On Dec. 5, 1955, four planets were transiting his 8th house in Scorpio: Saturn, Sun, Mercury, and Rahu. Outgoing *Dasha* lord Saturn and incoming *Dasha* lord Mercury were both in his 8th house of loss and radical changes. The events of early Dec. 1955 were also surrounded by two eclipses: On Nov. 29, 1955 there was a Lunar eclipse at 13:26 Taurus in his 2nd house of voice and family happiness. A **Solar eclipse** occurred on Dec. 14, 1955 at 28:16 Scorpio in his 8th house, opposite his natal Mars 29:02 Taurus. Though not a Total eclipse, as an Annular eclipse it had the distinction of lasting 12 minutes 9 seconds, the longest duration of all eclipses in the 351-year period from 1700 to 2050.[65] At the *gandanta* degree of Scorpio, it was also in an extremely treacherous degree area, fraught with danger. Just opposite his Ascendant lord Mars, it reflected his hazardous path from 1955 onward, already shown by the start of Mercury *Dasha* within 8 days of the eclipse. On Jan. 30, 1956 King's house was bombed, but there were no injuries.

[65] Source: Neil Michelsen's ***Tables of Planetary Phenomena***, 1990, p. 38. Annular eclipses last longer than Total eclipses, whose maximum duration is 7½ minutes, but are considered generally more intense.

4. **[Civil Rights] March on Washington (Aug. 28, 1963):** He was in Mercury-Moon *Dasha/ Bhukti* from Jan. 21, 1963 to June 21, 1964. Natal Mercury in Capricorn is in Moon-ruled *Shravana nakshatra*, increasing the emotional power of his speech, especially during this period. Tr. Mercury was exalted in early Virgo and tr. Moon in late Scorpio most of the day. King's "I Have a Dream" speech was the high point of this historic gathering at the nation's capitol, with 200,000 peaceful demonstrators. One of the stated purposes of the march was to promote the Civil Rights Act, though many considered it too little, too late. Tr. Mars was at 27 Virgo, close to its position on Aug. 7, 1965 when the Voting Rights Act was signed. (See below) On July 20, 1963 there was a **Total Solar eclipse** at 4:03 Cancer opposite King's natal Sun at 2:16 Capricorn. The FBI's interest in King and his peaceful movement intensified from this march, trying to prove a Communist link. Four transiting outer planets hovered near King's house cusps: Rahu and Ketu at 25 Gemini and Sagittarius, Saturn retrograde at 25 Capricorn, and Jupiter retrograde at 25 Pisces. Jupiter and Saturn were strong in their own signs, and tr. Jupiter opposite tr. Mars provided the philosophical zeal. Many feared a violent outcome to this day, but the malefic planets (Sun, Mars, Rahu, Ketu, and Saturn) were not linked with each other, and the condition of Jupiter and Saturn elevated the proceedings. Tr. Jupiter helped to tame tr. Mars and gave optimism to the Moon. Tr. Venus was tightly combust tr. Sun in Leo (and in *Magha nakshatra*), avoiding a more hedonistic event.

5. **Civil Rights Act of 1964 signed (July 2, 1964):** He was in Mercury-Mars *Dasha/Bhukti* from June 21, 1964 to June 19, 1965. On this day, tr. Jupiter was at 24:55 Aries, near his Ascendant and on the exact same degree as when he received the Nobel Prize. (See below) King received much credit for the passage of this legislation, though he would not rest until he pressed President Johnson for a Voting Rights Act. On June 24, 1964 a **Total Lunar eclipse** occurred at 10:08 Sagittarius. (The Ascendant of the USA chart (Kelleher) is at 8:59 Sagittarius.)

6. **Nobel Peace Prize received (Dec. 10, 1964):** The danger quotient increased for King with fuller recognition from each of his achievements and awards, especially as he sought to challenge the status quo. One week prior to his accepting the Nobel Peace Prize there was a **Full Moon eclipse** at 18:34 Scorpio (Dec. 3, 1964), again in his 8th house. At that time, tr. Rahu and Ketu were at 29:36 Taurus and Scorpio, contacting King's natal Mars at 29:02 Taurus in the 2nd house of speech and amplifying the personal volatility created by his message. Transiting Mercury (and *Dasha* lord) was retrograde at 11:19 Sagittarius, close to its Stationary Retrograde degree at 11:31 Sagittarius, and in his 9th house of highest *dharma* – true life purpose. For him a reliable signal of rewards and triumphs, tr. Jupiter was at 24:18 Aries, near its Stationary Direct degree and on his Ascendant (its strongest position). Tr. Saturn was at 6:15 Aquarius, near its Stationary Direct degree of 4:58 Aquarius. King was in the midst of *Sade Sati* (a pivotal life passage of 7 ½ years. (See Glossary) It would not end until Saturn finished its passage through Pisces on June 17, 1968. Saturn yields its strongest results toward the end of a sign, as does the Moon. As if to draw attention to Saturn, on Dec. 18, 1964 there was a **Total Solar eclipse** at 3:52 Gemini, closely opposite his natal Saturn, lord of the 10th house of status and awards. From March 1965 his sermons and speeches began to express doubt about America's involvement in the Vietnam War.

7. **Voting Rights Act of 1965 signed (Aug. 7, 1965):** King was in Mercury-Rahu period (June 19, 1965 to Jan. 6, 1968). Tr. Mercury was retrograde at 6 Leo, tr. Venus at 22 Leo, both opposing tr. Saturn retrograde (22:34 Aquarius), King's natal Moon at 26:43 Aquarius and his natal Venus at 17:41 Aquarius. Tr. Saturn turned Stationary Retrograde at 23:50 Aquarius (June 1965), and Stationary Direct at 17:07 Aquarius (Nov. 1965), impacting both his Moon and Venus. Tr. Saturn's passage through Aquarius (Jan. 28, 1964 to April 9, 1966) was the cornerstone of his *Sade Sati,* and his strongest ranked house – in the sign of philosophical idealism. Tr. Rahu and Ketu that day were at 17:01 Taurus-Scorpio, close to his 8th house cusp. Tr. Mars was in a warrior position, at 28:40 Virgo, *Vargottama* in Mercury's sign of exaltation – the sign of conflict, and in Mars-owned *Chitra nakshatra.*

8. **First participation as leader in Anti-Vietnam War protests (March 25 to April 15, 1967):** Preceded by his statements from March 1965 onward questioning the Vietnam War, his first major public involvement was on March 25, 1967 with a march of 5,000 people in Chicago. On April 15, 1967 the march was much larger, with 500,000 people in New York City. He spoke at the U.N. building at the end of the march. His "Beyond Vietnam" speech on April 4, 1967 at Riverside Church in Manhattan is less well known, but is one of the seminal speeches of the era, on how "silence" [in not speaking out about the war] was "a betrayal." His stature as Civil Rights leader gave the Anti-War movement more legitimacy, as prior to 1967 it lacked significant national leaders. King was in his Mercury-Rahu *Dasha/Bhukti,* and from March to early May 1967 there were several markers of transiting planets: **a) March 8, 1967: Tr. Mars was Stationary Retrograde** at 9:48 Libra – opposite King's Jupiter at 8:18 Aries. Mars and Jupiter goad each other to speak out, especially in fire and air signs. Tr. Mars was in *Swati nakshatra,* ruled by Rahu, giving a more raw-edged and urgent influence echoing that of King's natal Mars-Rahu; **b) March 15, 1967: Mercury was Stationary Direct** at 11:10 Aquarius. Mercury's idealism is strong in the sign of Aquarius and imbued with a desire to heal, in *Shatabhisha nakshatra,* also ruled by Rahu; **c) March 21, 1967: Jupiter was Stationary Direct** at 1:02 Cancer – opposite King's Sun at 2:16 Capricorn. Near its maximum exaltation at 3:00 Cancer, Jupiter is very expansive, and widened King's leadership role while impacting his natal Sun, planet of leadership; **d) April 24, 1967: Total Lunar eclipse** at 10:13 Libra (repeat of Mars SR on March 8, 1967, giving it more weight); **e) May 9, 1967: Solar eclipse** at 24:53 Aries, four degrees from King's Ascendant.

9. **Assassination (April 4, 1968):** In an irony often buried in history, the most stirring political speech of Bobby Kennedy's career was the one in which he announced Dr. King's death in Indiana.

> "... Martin Luther King dedicated his life to love and to justice between fellow human beings. He died in the cause of that effort."
>
> **Robert F. Kennedy**, April 4, 1968, Indianapolis, IN[66]

---

[66] Four days earlier, on March 31, 1968, President Lyndon Baines Johnson gave a lengthy speech on the Vietnam War, at the end of which he shocked the nation when he said: "I shall not seek, and I will not accept, the nomination of my party for another term as your President." Forced to bow to fierce opposition to the Vietnam War within the U.S. and within the Democratic Party, the capitulation of such a domineering leader was equally astonishing. The tumultuous events of spring 1968 were punctuated by eclipses: a solar eclipse at 14:54 Pisces on March 28, 1968, and a Total lunar eclipse at 29:54 Virgo on April 12, 1968 (EDT).

As Attorney General, Robert Kennedy gave the order in Oct. 1963 to have King wiretapped at all times so that his every movement would be known. FBI Director J. Edgar Hoover was vehemently against King and angered by his rising reputation. Though there was no law enforcement justification for it, he conducted this surveillance both before and after its authorization. Documents reveal the FBI sought for years to destroy or discredit Dr. King and thus weaken the Civil Rights movement. Through harassment they hoped to break his spirit. Then on April 4, 1968 at 6:01 PM EST King was hit by a single bullet to the right side of his neck. It smashed his throat and travelled down his spinal cord before lodging in his shoulder. King was pronounced dead at 7:05 PM EST in Memphis, TN, and in the wake of his death, riots broke out in over 100 cities. An alleged lone assassin was soon named.

King was in **Mercury-Jupiter** *Dasha/Bhukti* from Jan. 6, 1968 to April 13, 1970. Mercury and Jupiter in the *Navamsha* chart show both the fame and the conflict produced by his speeches. (*Navamsha* Jupiter rules the 6th house of conflict, and is situated with Mercury and Moon in the *Navamsha* 9th house.) At the time of his death, 17:54 Virgo was rising in Memphis, almost King's exact 6th house cusp. Ketu was nearby in Virgo and five planets were in Pisces (sign of surrender and martyrdom) in his 12th house of loss: in this order, Venus, Mercury, Sun, Saturn, and Rahu. Sun and Saturn were within five minutes of each other, Sun at 21:51 Pisces, Saturn at 21:56 Pisces. Mutual contacts of Sun and Saturn signify the iconoclast challenging the status quo. Saturn combust the Sun means Saturn cannot be seen with the naked eye, and it is temporarily hard to see any logic, truth, or sense of order in the world. Tr. Mars in Aries was near King's Ascendant, tr. Jupiter in Leo aspected his Ascendant (but unable to protect him), and tr. Moon was alone at 9:00 Gemini in *Ardra nakshatra*. Ardra's symbol is the teardrop or the human head, and its *shakti* (energy) is effort. Its deity is Rudra, ruler of storms, and a fierce form of Shiva. *Ardra nakshatra* is closely associated with suffering. King accepted his own suffering, but not that of racial intolerance. His vision of social justice and equality would long outlive him.[67]

Chart 2: Robert F. Kennedy

Birth data: Friday, Nov. 20, 1925, 15:11 EST, Brookline, Mass., USA, Long. 71W 07 18, Lat. 42N19 54. *Lahiri ayanamsha* -22:49:03. Class A birth data, from his office. Ascendant: 13:03 Aries.

Biographical summary: An American politician born into a famous and wealthy family, Robert (Bobby) Kennedy was the seventh out of nine children of Joseph P. Kennedy, Sr. (b. Sept. 6, 1888, d. Nov. 8, 1969) and Rose Kennedy (b. July 22, 1890, d. Jan. 22, 1995). President John F. Kennedy was his older brother.[68] Though he did not see active duty during WWII, Bobby enlisted in Oct. 1943 in a Naval reserve program and continued through college with a V-12 Naval College

---

[67] At the moment of King's assassination, transiting Moon at 9:00 Gemini was within 1/60th of a degree to the Descendant of the USA chart (Kelleher chart: July 4, 1776, 6:30 PM LMT, Philadelphia, PA). That Ascendant is 8:59 Sagittarius.

[68] John F. Kennedy's life and astrological indicators are discussed in Chapter 8: Virgo Ascendant, Chart #25.

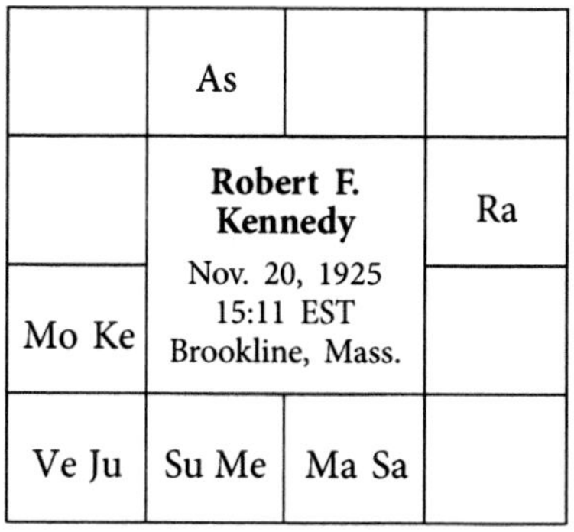

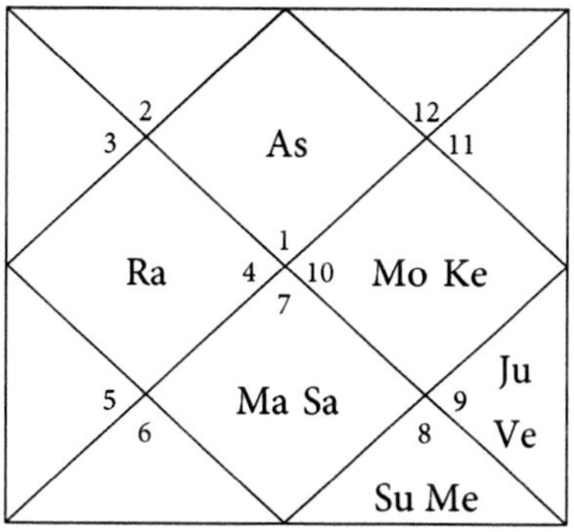

	As		
	Robert F. Kennedy		Ra
Mo Ke	Nov. 20, 1925 15:11 EST Brookline, Mass.		
Ve Ju	Su Me	Ma Sa	

2 3 As 12 11
Ra 4 1 10 7 Mo Ke
5 6 Ma Sa 9 8 Ju Ve
Su Me

Planetary Information

As	13:03	Ari	Ashwini
Su	05:11	Sco	Anuradha
Mo	05:37	Cap	U.Ashadha
Ma	12:01	Lib	Swati
Me	27:00	Sco	Jyeshta
Ju	27:20	Sag	U.Ashadha
Ve	22:14	Sag	P.Ashadha
Sa	25:29	Lib	Vishakha
Ra	05:39	Can	Pushya
Ke	05:39	Cap	U.Ashadha

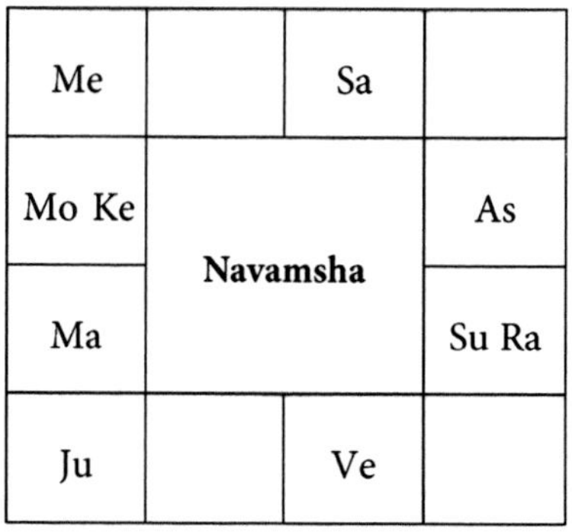

Me		Sa	
Mo Ke	**Navamsha**		As
Ma			Su Ra
Ju		Ve	

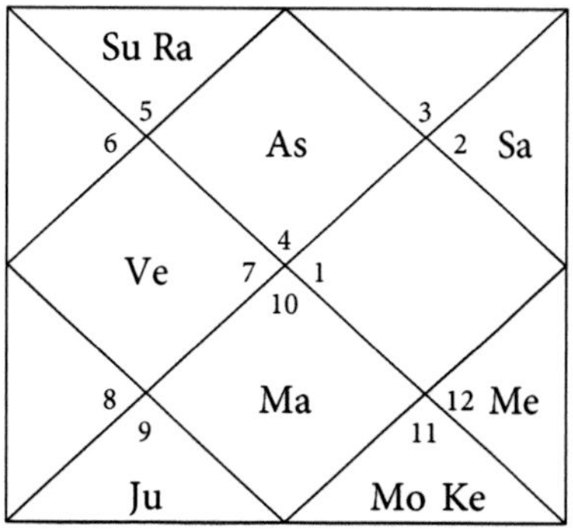

Su Ra
5 6 As 3 2 Sa
Ve 7 4 1 10
8 9 Ma 12 Me 11
Ju Mo Ke

Vimshottari Dashas

Su	Nov 9, 1921
Mo	Nov 9, 1927
Ma	Nov 9, 1937
Ra	Nov 9, 1944
Ju	Nov 9, 1962
Sa	Nov 9, 1978
Me	Nov 9, 1997
Ke	Nov 9, 2014
Ve	Nov 9, 2021

training program. After the war he completed his undergraduate degree at Harvard (a B.A. in government) in March 1948, followed by a law degree from University of Virginia in June 1951. In the 1950s he was legal counsel to a U.S. Senate committee investigating labor unions, leading to his well-known feud with Teamsters union leader Jimmy Hoffa. He also managed three successful political campaigns for his brother John F. Kennedy: for the U.S. Senate in 1952 and 1958, and for the U.S. Presidency in 1960. Much at the insistence of his father Joe Kennedy, Bobby became Attorney General in the Kennedy administration. He had not yet argued a case in a courtroom – and only 35 at the time he was sworn into office – he was one of the youngest Attorney Generals in American history.

Closest advisor to Jack Kennedy from spring 1961, Bobby was U.S. Attorney General from Jan. 20, 1961 to Sept. 3, 1964. In turn, Joe Kennedy was the pivotal advisor and supporter to his sons. When he suffered a stroke on Dec. 19, 1961, it left him mentally intact but unable to speak or walk up until his death in Nov. 1969. After Jack's assassination Nov. 22, 1963, Bobby continued on as Attorney General under President Lyndon Johnson, though their relationship was notoriously difficult. Resigning from that post in Sept. 1964, he ran for U.S. Senator from New York, an office he held from Jan. 3, 1965 thru June 6, 1968.

In several of his government posts, especially as Attorney General, Robert Kennedy was in a very powerful position to fight crime. This also made him a more potent target to his enemies. Ironically, it is well documented that his father Joe Kennedy himself had courted and used organized crime. Bobby was a target again when as a freshman senator in Feb. 1965 he took the risky step of opposing the still popular Vietnam War. He was largely rebuffed at the time and accused of lacking patriotism. President Johnson would go on to escalate from 16,000 American troops in Vietnam in Jan. 1964 to 360,000 troops by Dec. 1967. When Johnson made it clear by Feb. 1968 he would not change course in Vietnam, Kennedy declared himself a Democratic Presidential candidate, on March 16, 1968. For this move he was accused of opportunism, though as his Anti-war stance became more vehement, many feared for his life from the start of his presidential campaign.

On a campaign stop in Indiana on April 4, 1968, he learned that Martin Luther King had just been killed. He delivered this news to a crowd gathered in Indianapolis, who moaned and screamed upon hearing it. The impromptu speech he gave at that moment is still considered one of the finest and most impassioned speeches in American politics. It was entirely Kennedy's speech from his own hastily prepared notes, turning down those of his speechwriting staff. Kennedy risked his own personal safety at a time when local police could not guarantee it. And in spite of riots in over 60 cities that night, Indianapolis was quiet, perhaps in large measure due to the powerful effect of Bobby Kennedy's words:

> ".... We have to make an effort in the United States. We have to make an effort to understand, to get beyond, or go beyond these rather difficult times.
>
> My favorite poem, my — my favorite poet was [the Greek] Aeschylus. And he once wrote:
>
> 'Even in our sleep, pain which cannot forget falls drop by drop upon the heart, until, in our own despair, against our will, comes wisdom through the awful grace of God.'
>
> What we need in the United States is not division; what we need in the United States is not hatred; what we need in the United States is not violence and lawlessness, but is love, and wisdom, and compassion toward one another, and a feeling of justice toward those who still suffer within our country, whether they be white or whether they be black.
>
> So I ask you tonight to return home, to say a prayer for the family of Martin Luther King — yeah, it's true — but more importantly to say a prayer for our own country, which all of us love — a prayer for understanding and that compassion of which I spoke.
>
> ...The vast majority of white people and the vast majority of black people in this country want to live together, want to improve the quality of our life, and want justice for all human beings that abide in our land.
>
> And let's dedicate ourselves to what the Greeks wrote so many years ago: to tame the savageness of man and make gentle the life of this world. Let us dedicate ourselves to that, and say a prayer for our country and for our people."

Two months later Kennedy won the California Democratic primary, making him the strongest likely presidential candidate for his party. He had just delivered a victory speech at his campaign headquarters in Los Angeles, CA when he was shot on June 5, 1968. He died 26 hours later. Kennedy's declared assassin was Sirhan Sirhan. Together with Dr. Martin Luther King's assassination just two months earlier, 1968 has been called the year in which "the Dream died," and one of the most cataclysmic years in American politics.

Kennedy was author of *The Enemy Within* (1960) and *The Pursuit of Justice* (1964). A movie version of *The Enemy Within* was stalled by many factors apparently related to Mafia power in Hollywood. Close to the start of filming, the project died in July 1962 along with its producer Jerry Wald, then 49. Bobby's third book was published posthumously in 1969: *Thirteen Days: A Memoir of the Cuban Missile Crisis.* On *The Enemy Within:*

> "It was not about communism, but about organized corruption, spreading from the underworld into labor, business, and politics–expressing the moral sickness of a greedy society."
>
> **Arthur Schlesinger Jr.**, author of *Robert Kennedy and His Times,* 1978.

Schlesinger praised Robert Kennedy as the most politically creative man of his time, but acknowledged he had played a larger role in trying to overthrow Cuba's Fidel Castro than the author had previously stated in his earlier book on John F. Kennedy, *A Thousand Days* (1966). Other paradoxes included Bobby's support of Martin Luther King, Jr. in 1968, after ordering federal wiretapping on him from 1963; and his opposition to the Vietnam War from 1965, after participating in escalating that war as Attorney General in the Kennedy administration. But these also reflect the evolution in Kennedy's position on civil rights and the Vietnam War.[69]

However, it was his brother Jack's death that caused Bobby to grow in many ways, eventually transforming his profound grief and torment into a major redemptive effort on behalf of the poor, the minorities and the disenfranchised. In those last years of his life he took major risks in not only putting his political weight behind the powerless and the oppressed, but by opposing the Vietnam War. In doing so, he took a dangerous course – rather than waiting his turn for the White House.

On June 17, 1950 Kennedy married Ethel Skakel (b. April 11, 1928), and together they had 11 children, from July 1951 up to Dec. 1968. A daughter (Rory) was born six months after her father's death. He and Ethel both strongly embraced the Roman Catholic faith and Bobby was regarded as a "Radical Catholic," in the sense of his faith informing his politics.

> "If the single man plant himself indomitably on his instincts, and there abide, the huge world will come round to him."
>
> **Ralph Waldo Emerson**, one of Bobby Kennedy's favorites.

Ability to initiate effectively: This Aries Ascendant is in the first *nakshatra* of *Ashwini* – giving a great deal of initiative and an undaunted spirit, especially with Ascendant lord Mars aspecting its own house very closely, so much so (almost one degree) that Mars repeats its position on the Descendant in many of the important *Varga* (or harmonic) charts, including the all-important *Navamsha* chart, where Mars is exalted in Capricorn. From these factors we see a person of above-average courage who will test the waters and push life's parameters – a common theme in the Aries Ascendant charts. Normally if the Ascendant lord aspects its own house it is good for longevity, especially along with the fortunate Jupiter aspect. But Mars in the 7th house can also diminish longevity, turns *maraka* and attracts injuries or accidents.

All the physical planets are placed in a continuous sequence between the 7th and 10th houses (a *Shakti yoga*), showing a person with an extroverted life, as these planets are all above the horizon. However, Sun in the 8th house in Scorpio modifies this pattern somewhat, giving the tendency for introspection.

Natal Rahu is very powerful for several reasons. Rahu and Ketu are Stationary Direct in the chart. Though no Stationary positions occur using the Mean node of Rahu and Ketu, it is worth

---

[69] In 1963 Bobby Kennedy gave written approval of limited wiretapping of Dr. King, which the FBI expanded under director J. Edgar Hoover and without Kennedy's knowledge. Bobby's relationship with King evolved gradually from mutual suspicion to shared aspirations. His position on Vietnam also evolved, causing many to accuse him of opportunism when he jumped into the campaign after Eugene McCarthy showed good results as a relatively unknown but rising Democratic Party candidate with an Anti-Vietnam War platform.

noting when the True node is Stationary Direct or Stationary Retrograde: it gives greater intensity to Rahu and Ketu, especially during their *Dashas.*[70] Moreover, Rahu is fortified by sign and *nakshatra* lords that are both angular in the birth chart: Moon and Saturn, respectively. All this gives tremendous potency to **the 18-year Rahu *Dasha*** (Nov. 9, 1944 to Nov. 9, 1962), making it the most beneficial *Dasha* of his lifetime. Five planets are in *kendras* of the birth chart. In addition to the Moon at the top of the sky, there are four strong classical malefics: **Mars, Saturn, Rahu, and Ketu**. This gives an ability to deal with challenges in the material world and a tendency to attract them, especially with Mars and Saturn in the 10th house from natal Moon, and exalted Saturn *Digbala* from the Aries Ascendant.

> "He was sort of a hard, passionate parish priest, the avenger type – all faith and justice… I always thought of Jack [his brother], on the other hand, like an English lord."
>
> **Haynes Johnson** (American journalist & author) on Robert Kennedy, quoted in *Brothers: The Hidden History of the Kennedy Years*, by David Talbot, 2007, p.12.

The courage factor – Ascendant lord Mars: The natural assertiveness of the Aries Ascendant destiny is further defined by the house position of Mars, planet of courage and valor in battle. Placed in the 7th house, it can bring challenges or dangers from one's enemies. With Mars very close (within one degree) of the Descendant and exalted in the *Navamsha* chart, the challenges and dangers increase exponentially. Bobby's first book, ***The Enemy Within* (1960)** gives an important clue about our subject and his life's preoccupations. A movie version of the book was aborted in July 1962 during Bobby's **Rahu-Mars** *Dasha/Bhukti* (Oct. 22, 1961 to Nov. 9, 1962). This was a perilous sub-period for him: Mars is featured as sub-lord, elevating the power of his enemies and the tendency to attract enemies. Mafia power in Hollywood seemed determined to block the movie version of his book. Coincidentally, Marilyn Monroe died on Aug. 5, 1962. Rumors of her close connection to Robert Kennedy that summer ran wild but were never confirmed. On Dec. 19, 1961, two months into his Rahu-Mars *Dasha/Bhukti*, his father suffered a massive stroke. He survived, but his capacity to act as family protector suddenly vanished.

Classically, the last sub-period of Rahu *Dasha* is fraught. Worldly gains, honors, happiness that come in the beginning of the *Dasha* are liable to be lost at the end of it. And though the karmic fruit of a planet comes due throughout its *Dasha*, there is special significance for the end of any *Dasha*, and at the change in *Dashas*. Rahu *Dasha* would bring him many worldly benefits during those 18 years (from Nov. 1944), while the subsequent **16-year Jupiter *Dasha*** (from Nov. 1962) would prove far more precarious for him, no matter how much courage Bobby possessed. This is indicated from Jupiter's condition in the chart(s).

Courage in action is apparent on several levels here: **1)** speaking out when one's viewpoints are not popular and when it is necessary to change a position; and **2)** being unafraid to admit that change or having made a mistake. Bobby shifted from having supported the Vietnam War, even participating in its expansion during his brother's administration, to opposing it as a freshman senator in Feb. 1965. (From March 1965 Martin Luther King, Jr. also began to question the

---

[70] For some examples of charts with Ascendant lord at SD or SR degree, see Chapter 8, Chart #25 (John F. Kennedy) and Chart #26 (Barbara Walters). Classical reference for planetary motional strength (*Chesta Bala)* is given. See also Chart #1 (Martin Luther King), with Ascendant lord Mars within one degree of SD degree.

war in his public speeches and sermons.) Meeting with stiff resistance and accusations of being both antipatriotic and against the military, Kennedy was forced to retreat on this issue until his 1968 presidential campaign, which began with his first campaign speech on March 17, 1968. Astrologically, the period was even more dangerous for him, as he entered his worst sub-period within Jupiter *Dasha*, that of Mercury – enemy of the Aries Ascendant. **Jupiter-Mercury** *Dasha/ Bhukti* ran from July 11, 1967 to Oct. 16, 1969. The Vietnam War was *the* pivotal campaign issue of 1968 and a red hot issue. It proved dangerous for others such as Dr. King, who shared some similar astrological components such as Aries Ascendant and Mercury major period – rather than Mercury sub-period.

This particular Aries Ascendant chart shows a native who will be focused on his enemies and will go after them, his pursuit in this case being morally and passionately reasoned by what he perceives as injustices to society or to his family.

Whirlwind of energy – more Mars effects: When Mars is the Ascendant lord and placed in a *Kendra*, the person is often called upon to be a warrior, a competitor and/or an innovator. Where Mars is situated in terms of sign, house, and *nakshatra* gives us many clues, as well as aspects to Mars. Aspecting the Ascendant from a *Kendra*, Mars gives youthful looks and good vitality, both of which Robert Kennedy possessed. For the most part he enjoyed excellent health and was considered extremely energetic. He started coming into his own during the **7-year Mars *Dasha*** (from Nov. 1937), ages 12 to 19, and out of his more timid phase. He was the younger brother and of shorter stature.

This Mars is very powerful through its aspects to the Ascendant, protecting the physical body to some extent – even if in a shortened life – and giving him added physical energy. Mars aspects to the 10th house galvanize career aims and ambitions, and give energy to his status. With a busy family and social life, he had many opportunities for advancement through his father, his brother, and numerous others. In a short period of time – from 1961 to 1968 – Kennedy was the U.S. Attorney General, a Senator from New York and a U.S. Presidential candidate for the Democratic Party. He won the California primary shortly before he was killed in June 1968. By aspecting the 2nd house, Mars energizes the speech, as well as heating it up and sharpening it. By aspecting the Moon, Mars speeds up the mind, making it overactive, less content and difficult to slow down. Mars also heats up the house of *Kutumba* (family happiness, for the family of origin). His words or actions can stir up the family, activate it, politicize it or anger it.

There are other *yogas* adding to this whirlwind of energy: **1)** a *Rajju yoga*: Ascendant is in a *Chara* (active) sign as well as five planets in *kendras* in *Chara* signs, giving a great deal of action: not only the physical body is active, but also the mind, with a potential for overactivity and for moving about excessively. **2)** *Shakti yoga*: All planets (excluding Rahu and Ketu) are in four houses: houses 7 through 10, giving *shakti*, or energy and a capacity to accomplish a lot careerwise, and perhaps neglecting the marriage.[71] Thoroughness and discipline came through Saturn's aspect to the Ascendant and its contact to Ascendant lord Mars. Moon is in Saturn's sign in the birth and *Navamsha* charts, and *Navamsha* Saturn aspects the *Navamsha* Ascendant and Moon, further confirming Saturn's strong influence.

[71] A *Shakti yoga* is one of the 32 *Nabhasa yogas* in the classical texts. They have to do with the arrangement of the seven classic planets in the heavens. The non-physical planets Rahu and Ketu are excluded.

Many geographical moves and changes: The *Dasha* sequence of his early years (ages 2 to 19) shows the timing and the effects of the *Rajju yoga*. Bobby entered his **Moon *Dasha*** at age two (for ten years) and into **Mars *Dasha*** for the next seven years. At age two, his family moved houses and two years later they moved again. After two different public schools from kindergarten through 5th grade, he attended numerous boarding schools, changing schools every one to two years, starting from 5th grade through 12th grade. (The Moon *Dasha* would be difficult emotionally, especially for a young person, due to Moon on the Rahu-Ketu axis and *Navamsha* Moon in the 8th house, again with Ketu.) He entered Mars *Dasha* Nov. 9, 1937 and during his father's ambassadorship spent 1938-1940 in England. He enlisted in the Naval reserve in October 1943 and attended several colleges, graduating from Harvard after many interruptions for the V-12 Naval College training program. When his Mars *Dasha* ended in early Nov. 1944 he had completed his V-12 training at Harvard and avoided active military service in WWII due to the war ending by spring and summer 1945 when he was still in college. With his Mars *Dasha* ending shortly before his 19th birthday, Bobby was in less danger and more likely to move into a worldly period, ready to build the foundation for a successful career. As Mars signifies brothers, Bobby's Mars *Dasha* was a perilous period for any of them: brother Jack was badly injured on Aug. 2, 1943 in his PT boat in the Pacific, and Joe, Jr. was killed on Aug. 12, 1944.

Powerful enemies: The 7th house contains marriage and/or business partners, as well as one's opponents. The latter is also read from the 6th house of enemies and conflict, ruled by Mercury here. As Mercury is 6th house lord placed in the 8th house, this combination tends to protect his enemies and give them cover.[72] The hidden quality comes from natal Mercury in the hidden sign of Scorpio; and the treacherous way they could achieve that cover is evident from Mercury at 27:00 Scorpio in the *gandanta* sector of Scorpio, considered the most volatile of the three water-sign *gandanta* sectors. Mercury is placed in its own *nakshatra*, fortifying the power of the enemy and in the sign of Mars, which is exceedingly strong here as an adversary in the 7th house.

When Mars and Saturn are placed together in a *Kendra* or trinal house, the classic Vedic treatise *Saravali* indicates that the native can be quarrelsome, clever in speech, versed in the use of weapons, and in danger from poison or injury.[73] Ascendant lord Mars placed in the 7th house is going to seek out his partners *and* his enemies. Both are highly placed in society and with intrinsic power due to Saturn exalted in the 7th house in the birth chart and Mars exalted in the 7th house in the *Navamsha* chart. Formidable partners *and* enemies are strongly indicated, even though he himself (with Mars as Ascendant lord) was also a formidable opponent. This is especially due to *Navamsha* Mars exalted in Capricorn in the 7th house – a theme confirmed by the closeness of Mars to the Descendant degree, thus repeating in many *Varga* (harmonic) charts.

The most powerful bureaucrat of the era, J. Edgar Hoover had a long list of enemies, and the Kennedy brothers were on it. For 37 years (1935-1972), Hoover was the much feared and unscrupulous Director of the Federal Bureau of Investigation (FBI). Every president deferred to him, as did Bobby Kennedy as Attorney General. And though he was credited with being the first Attorney General to gain control of the FBI, pushing it to act on organized crime and on civil rights, in doing so he also ceded more power to Hoover. In turn, Hoover's unchecked anti-

---

[72] This is a *Viparita Harsha Raja yoga*, which in this case may only protect Kennedy in debate or argument.

[73] This reference is from the ***Saravali*** of Kalyana Varma, Vol. 1, Chapter 15, Verse 16.

Communist obsession gave him the rationale to oppose Martin Luther King, Jr. and any other influential voices opposed to the Vietnam War. When Bobby became more vocal in his anti-Vietnam War stance, he also became more and more a target of Hoover's FBI. The Mafia was also perennially powerful and angered by Bobby's declared war on organized crime, especially after doing so many favors for the Kennedy family at Joe Kennedy's behest.

Partnership theme: Astrological indicators have shown us Bobby's partners would be highly placed in society: His brother Jack was a close collaborator for over a decade. Bobby ran Jack's U.S. Senate campaigns in both 1952 and 1956 as well as his Presidential campaign in 1960 before becoming Attorney General in his administration. His wife Ethel was born into a very wealthy family.[74]

> "... [Ethel] was almost as complicated as he was: recklessly frank yet guarded, canny and guileless, brash and sensitive, an observant Catholic who threw wild parties and hobnobbed with celebrities.... She understood him better than anyone, believed in him more, was convinced he would be a great president, and knew he would never forgive himself if he sat out the [1968 presidential] race."
>
> Excerpt from **Thurston Clarke**'s *The Last Campaign: Robert F. Kennedy and 82 Days That Inspired America*, 2008, from *Vanity Fair* magazine, June 2008, "The Last Good Campaign," by Thurston Clarke.

Saturn is placed in the 7th house of partnerships, and has significant strength in its best angular position *(Digbala)* and its sign of exaltation. This confers a special *yoga* on Saturn – the *Mahapurusha Shasha yoga*, though with this yoga many times Saturn can act out of fear and a desire to remove the conditions that induce that fear. Saturn is also conjoined with Asc. lord Mars. Because of this we know: **1)** powerful partnerships are likely in the life and; **2)** Saturn major or minor periods in the life will bear results that are more extraordinarily fruitful in this regard. Since the major period of Saturn did not occur in his lifetime, we examine the last sub-period of Saturn: **Rahu-Saturn** *Dasha/Bhukti* ran from Dec. 15, 1949 thru Oct. 21, 1952. Bobby's closer friendship and partnership with his brother Jack began from Oct. 1951, with a 7-week Asia trip they shared, along with sister Patricia; and his first job managing his brother's U.S. Senate campaign started from **June 6, 1952**. His marriage to Ethel Skakel occurred on **June 17, 1950**. Together they had 11 children, from July 1951 through Dec. 1968. Having met during the 1945-1946 academic year, their marriage would not be auspicious until his Saturn sub-period. Marriage gave Bobby greater self-confidence, as the seventh of nine children within a large and competitive family.

Partnership themes dominate with Ascendant lord Mars in Libra – a sign of sociability. Ethel was somewhat of a madcap, more vivacious and outgoing than her husband, especially in the early years, active in numerous ways and a playful counterpoint to Bobby's more thoughtful and serious nature. With a house full of children and always active athletically (touch football, golf, swimming, skiing, etc.), she sponsored or supported many charitable events both during and after Bobby's death.[75] Her husband's equal in his fierce and unforgiving loyalty to the family, Ethel

---

[74] Ethel's father George Skakel was a self-made millionaire and founder of the Great Lakes Carbon Corporation. Her family was not yet on the Social Register in 1938, but Ethel's 1950 wedding to Bobby Kennedy at the Skakel estate in Greenwich, CT provided the definitive social leap, if one was needed.

[75] Into her 80s, Ethel Skakel Kennedy has continued her active support of the Robert F. Kennedy Memorial, funding humanitarian projects and promoting human rights around the world. She helps orphans and homeless or abandoned children, and supports organizations such as the Earth Conservation Corps, which sponsors environmental cleanup programs. She is also steadfastly devoted to her children and to the memory of her husband.

continues to be the keeper of the flame for the family. Saturn exalted in the 7th house shows Ethel as the extremely responsible mate and one who believed in him so strongly she urged him to run for President in 1968 when many other key advisors (including Ted Kennedy and Arthur Schlesinger, Jr.) were against it and thought it would endanger him and his family in a highly polarized era. Fortified by a strong Catholic faith, she attends Mass daily. She is focused on family and entertaining, but has never liked cooking and does not cook.

The rambunctious nature of the wife is also described by the condition of natal Venus in Bobby's chart, as Venus is the lord of the 7th house of marriage, and together in mutual contact Venus and Jupiter can be excessive. The sign of Sagittarius is both athletic and religious-minded. Before marriage, Ethel had seriously considered becoming a nun. She was religious and stoic, as well as enthusiastic, cheerful, mischievous and competitive.

Venus-Jupiter in Sagittarius (in the 9th house of father) also hints at the voracious appetite for women that Joe Sr. espoused and encouraged in his sons. His dictum was to marry, stay married, have lots of children, and enjoy as many women as desired. Rumors circulated of some dalliances by Robert Kennedy; but given the strong **Saturn in his 7th house of marriage**, Bobby was probably less egregious in his infidelities than his father and brothers, and more constrained to be loyal to his wife.

Marrying in a Saturn sub-period was beneficial to his marriage: he was in Rahu-Saturn *Dasha/Bhukti* on his wedding date June 17, 1950. This brought karmic results from both natal and transiting Rahu and Saturn. Natal Rahu is situated in *Pushya nakshatra*, owned by a strong natal Saturn in the 7th house. Transiting Rahu at 10:30 Pisces was in *Uttara Bhadra nakshatra*, also Saturn-owned. Tr. Jupiter and Saturn acquired extra potency close to their Stationary degrees, and both planets aspected the 7th house of marriage (tr. Saturn at 20 Leo and tr. Jupiter at 14 Aquarius). Tr. Venus in Aries also aspected the 7th house of marriage. These combinations tend to promote auspiciousness for marriage. Venus and Jupiter are the great benefics, Venus the planet of love, and Saturn ruling over commitment and marriage in general, especially as an institution for social continuity.

Partnership with brothers: Mars in Libra exactly on the Descendant also describes the physical/athletic activities for which the Kennedys were known. They were taught to be highly competitive with each other intellectually and physically and to use these skills in the larger world. Given every advantage of status, education, and wealth, they were still asked to prove and improve themselves constantly, politics and history being a common interest.

Mars is the *karaka* for brothers (and siblings), and for Aries Ascendant this can be an important theme. Robert Kennedy had three brothers and five sisters. Educated and trained as a lawyer, he devoted many years to helping advance his brother John's career until he was killed in Nov. 1963. Following his death, Bobby zealously pursued his personal investigation of the assassination. His total devotion to his brother Jack and his suffering on account of his death are major factors in his life. He rarely if ever spoke to his aides about his brother's death, but it can be felt in several of his speeches, including in Indianapolis on April 4, 1968 after King's death. At the Democratic Party National Convention, Atlantic City, NJ, August 27, 1964, he was asked to introduce a film about his brother. Greeted with tumultuous applause for 22 minutes, and fighting back the tears, he delivered these lines about his slain brother, suggested to him by Jackie Kennedy:

> "When I think of President Kennedy, I think of what Shakespeare said in *Romeo and Juliet* [Act. III, Scene II]:
> ... When he shall die
> Take him and cut him out in little stars,
> And he will make the face of heaven so fine
> That all the world will be in love with night
> And pay no worship to the garish Sun."

Other ways to examine **the condition of brothers in the chart** is through the 3rd house, its lord and *karaka* (Mars), also the 3rd house from the *karaka* itself. Natal Venus and Jupiter in Sagittarius aspect the 3rd house from the Ascendant. They are also situated in the 3rd house from Mars, indicating an abundance of brothers and happiness from them – further confirmed by *Navamsha* Mars exalted in Capricorn, again in a *Kendra*. This can make the native himself personally strong and heroic. However, 3rd lord Mercury is posited in the natal 8th house of loss. Even in the *Navamsha* chart, 3rd lord Mercury is weak in Pisces in its sign of debilitation and owned by Jupiter. *Navamsha* Jupiter in turn is situated in the 6th house of conflict and enmities. In the birth chart, natal Mercury's sign lord Mars is also dangerously positioned. This hints at why two brothers were killed prior to Robert Kennedy's own assassination at age 42. The only other younger brother, Ted Kennedy, lived to age 77, with a long (47-year)and respected career in the U.S. Senate. In spite of substantial legislative achievements, his much-anticipated presidential chances were diminished.[76] Jack, Bobby and Ted are buried near each other at Arlington National Cemetery, a tribute to their closeness and to their national service.

As seen through Bobby's chart, **the omen for the JFK presidency** was not auspicious. His election and inauguration occurred in Bobby's **Rahu-Moon** *Dasha/Bhukti*. These two planets are very closely opposite each other (within 2 minutes, or 2/60ths of a degree) on the 10th house axis, destabilizing the Moon – his most visible planet. *Navamsha* Moon is not well placed in the 8th house of loss and political and sexual intrigue, while natal Moon is in the *nakshatra* of the Sun, located in turn in the 8th house. Natal Sun and Rahu are both in Saturn's *nakshatra*, bestowing a strong Saturn influence on the Sun, a mutual planetary enemy. (For his election and inauguration Jack Kennedy was in his own **Rahu-Mars** *Dasha/Bhukti*, the last sub-period in the Rahu *Dasha* and the most dangerous.) These factors did not bode well for a long-lived administration or for one completing his allotted term in office. Even Bobby's next *Dasha* – the Jupiter *Dasha* – continues the theme of conflict. Bobby's Rahu *Dasha* was generally his most powerful lifetime *Dasha*, though it too courted danger.

Personal wealth and a wide network of friends & relatives: When the natal Sun is hemmed in on either side (i.e., houses directly before and after) by planets other than Moon, Rahu, or Ketu, this is a special yoga called an *Ubhayachari yoga*, a powerful combination for personal

---

[76] The condition of Mars in Bobby's chart or his brothers' charts shows the destiny of his brothers to some extent, especially relative to their innate talents and life circumstances. Ted Kennedy was born Feb. 22,1932, 3:58 AM EST, Dorchester, MA; died Aug. 25, 2009. Ted's natal Mars is combust and loses a Planetary War, weakening it and making him fight harder in order to regain the expected competitive advantage. As 5th lord, it also causes problems for his children. His 7-year Mars *Dasha* (from Nov. 18, 1967) was the worst of his life. It included Bobby's assassination June 6, 1968, their father's death Nov. 8, 1969, his son Ted Jr's loss of his right leg to bone cancer in Nov. 1973, and Ted's own accident on July 18,1969 at Chappaquiddick. Driving his car off a bridge, his passenger (Mary Jo Kopechne) drowned, as did virtually all his presidential chances. His negligence and delay in reporting the accident were negative factors to his reputation for many years.

networking.[77] Even for someone in less extraordinary circumstances, there is ongoing access to people who can be helpful in advancing career or other life aspirations. Furthermore, Bobby Kennedy has several *yogas* for financial wealth: he was born into a very wealthy family and married into one. He has the *Lakshmi yoga*: Ascendant lord is strong, while 9th lord (of *bhagya*, fortune) occupies a *Kendra* or trine in own sign, exalted or *mulatrikona*. He also has *Dhana yogas*: Ascendant Lord is with the 11th lord in a *Kendra*; also 2nd and 9th lords are in the 9th house – a merging of wealth factors in houses associated with wealth or good fortune.

Personal power for leadership: Long accustomed to being in a support position to his brother Jack, Bobby Kennedy was less comfortable asking for favors for himself. Even so, there are various astrological clues that show us he would be able to step into the primary leadership role, even if briefly. Leadership is shown by the strength of the **10th house lord, Sun, Moon, and Ascendant lord**. Tenth lord Saturn is exalted and has *Digbala* (best possible angle) in the 7th house, giving high status to his career and conjoined with Ascendant lord Mars. Moon is also well placed for political leadership in the 10th house, though with definite disadvantage for maintaining stability so close to Ketu – a point confirmed by the condition of *nakshatra* lord *(Jiva)* for planets situated in the 10th house (Moon and Ketu here). They are both placed in *Uttara Ashadha nakshatra*, ruled by the Sun. As noted earlier, natal Sun is placed in the 8th house. This can bring loss, along with abrupt or radical changes in career. With or without this connection, 8th house Sun can indicate a good investigator, researcher, or psychologist, someone who likes to penetrate below the surface, and can be a trusted authority in business or finance.

As **Venus** is **the planet of politics**, we examine its condition in the chart to show the ability to magnetize people to one's self. Bobby did not fully reach that point until the last few months of his life during his presidential campaign. People were so eager to see him they ripped buttons off his jackets and shirts and other items of clothing. The condition of his Venus indicates he himself could be overly zealous in politics in general. This is due to Venus in a fire sign and in its own *nakshatra*. Since both are fiery, *rajasic* conditions, they tend to reinforce the zealous nature of the political drive. As a **"Radical Catholic,"** his politics were strongly influenced by his Catholicism, far more so than his brothers. Bobby was motivated by large causes, fighting organized crime, and later evolving his position on both civil rights and the Vietnam War. Since Venus is in the sign of its planetary enemy – Jupiter – this causes some excess in embracing large and overarching themes. Jupiter also closely contacts Venus in the same sign.

Fortunately, Jupiter is a great good friend to the Aries Ascendant chart, and well placed in Sagittarius in the 9th house. But while Jupiter's influence does not lessen the level of zeal, it may give more of an ethical foundation to the efforts that are undertaken with such passion and dedication. Even that foundation is interpretable, however, as Sagittarius can justify its efforts by joining together religion and patriotism, increasing its fervor through self-righteousness and the desire to eliminate the perceived injustice(s).[78]

---

[77] His sister Jean introduced him to her roommate Ethel Skakel while they were at Manhattanville College during their freshman year, 1945-1946. Other important associations came through his father or other siblings.

[78] For Robert Kennedy, Venus is his one planet in a Venus-ruled *nakshatra*. It is situated in Sagittarius in the 9th house with Jupiter. Amplifying some of the same fiery components, the chart of Adolf Hitler (Chart #28) contains five planets in Venus-ruled *nakshatras*. Hitler's Venus is in a fire sign in its own *nakshatra*, with Mercury, Sun, Mars, and Venus all in Aries in the 7th house, and three planets in Sagittarius in the 3rd house. With 7 out of 9 planets in fire signs, the passion-driven motivation is operating in high gear.

Personal timing: For this we examine the *Dashas* and transiting planets, especially Jupiter, Saturn, and the *Dasha* lord. A sudden rise in status was well indicated by the **Jupiter-Saturn conjunction on Feb. 18, 1961** in his 10th house of career and status near his natal Moon. But maintaining that status would be jeopardized by natal Moon so close to the Rahu-Ketu axis. In mid-April 1961, three months into serving as Attorney General in the new Kennedy administration, the Bay of Pigs fiasco occurred. From that time onward he became his brother's most trusted advisor, almost unofficially "Prime Minister." But that was cut short in late Nov. 1963 with his brother's assassination. He stayed on as Attorney General under President Johnson, but given the deep enmity between the two men, it is all the more extraordinary he did stay. On Sept. 3, 1964 Bobby resigned as Attorney General to run for the U.S. Senate.

Bobby entered the 16-year **Jupiter *Dasha*** Nov. 9, 1962. Though well placed in its own sign of Sagittarius in the 9th house of the birth chart, and *Vargottama* in the *Navamsha* chart, Jupiter is placed in the *Navamsha* 6th house, continuing the theme of conflict and unease, something always to be fought for or against. And with *Dasha* lord in Sagittarius, justice and ethics are major themes. His later run for the Presidency in 1968 was during **Jupiter-Mercury** *Dasha/Bhukti*, the most dangerous sub-period. A few months later it was ended with his assassination.

SUN - physical strength, self-confidence, and father: Bobby gained much overall from his wealthy and influential father, but in his childhood he was often the target of his domineering ways. Natal Rahu (the foreigner) in the 4th house echoes this theme. From Aries Ascendant, the Sun is not well placed for visibility in the 8th house. Situated in the sign of Scorpio, owned by Mars – its great good friend – Sun is diminished by its placement in *Anuradha nakshatra*, owned by Saturn, planetary enemy of the Sun. The Sun-Saturn interplay shows some antagonism with the father, and a tendency for Bobby to confront the status quo, especially as Saturn is so strong in this chart.

One of the ironies here is that pre-eminence is given to the enemies. Also, the circuitry of the planetary connections keeps linking back to **the father**: The *karaka* (significator) for father is **the Sun**, as well as **9th house** and **9th house lord** – Jupiter here. Tracking these planets between birth and *Navamsha* charts, 9th lord Jupiter is in the *Navamsha* 6th house of enmities and conflict. The strongest planet in his birth chart is Saturn – exalted in Libra, whose sign lord is Venus, placed in the 9th house. Natal Saturn is in *Vishakha nakshatra*, ruled by Jupiter, also placed in the 9th house, strong in its own sign and in *Uttara Ashadha nakshatra*, ruled by the Sun. Sun is placed in the 8th house of death, loss, and financial inheritance. Along with his eight siblings, Bobby benefited enormously from his father's financial largesse and his determination to give his sons political and social prominence. But Bobby went after many public enemies, especially organized crime. Because his father courted and dealt with organized crime figures on a regular basis, especially prior to Jack's election, this was bound to be a dangerous course for Bobby. With Sun in Scorpio in the 8th house of secrets and intrigue, this fact about the father is not generally known, though well documented. Both father and son were keepers of many secrets.

Scorpio is a sign of money, suffering and death. **Sun in Scorpio** as 5th lord placed in the 8th house can indicate a higher level of intelligence and intuition. Sun, Moon, or Ascendant lord in the 8th house can bring a fondness for political and/or sexual intrigue.[79] With a different orientation,

---

[79] His brother John F. Kennedy had his Ascendant Lord in the 8th house – not good for longevity, but confirming a taste for political and/or sexual intrigue, or having a secret part to his life.

one might aspire to a deeper level of spirituality. The 8th house is a *Moksha* house and can push one beyond the veil of everyday life in those ways. One can keep looking for more subtle clues in life, or potentially guarding many secrets.

Joe Kennedy, Sr. suffered a massive stroke on Tuesday, Dec. 19, 1961. It was a Mars-ruled day and two days prior to the Winter solstice and a Full Moon at 6:36 Gemini in *Mrigashira nakshatra*, ruled by Mars. The solstices can be very demanding on the heart, physically or metaphysically; likewise the Full Moon – this one opposing Bobby's 9th house of father. Bobby was in his **Rahu-Mars** *Dasha/Bhukti* (Oct. 22, 1961 to Nov. 9, 1962), during the last sub-period of Rahu *Dasha*, known to bring losses and difficulties. Transiting Rahu-Ketu entered the Cancer-Capricorn axis Nov. 9, 1961 through May 1, 1963, impacting his *Dasha* lord and family of origin by return to the 4th house. On Dec. 19, 1961, tr. Saturn at 5:00 Capricorn closely contacted Bobby's natal Moon at 5:37 Capricorn and Ketu at 5:39 Capricorn. Tr. Mars was closely combust tr. Sun, both in early Sagittarius in his 9th house of father.

Joe Kennedy's stroke marked the end of his unquestioned dominance as family patriarch and protector.[80] Bobby's **Rahu-Mars** *Dasha/Bhukti* contained other sudden deaths of people and/or projects: **1)** In **July 1962,** the planned film version of Bobby's book ***The Enemy Within*** (documenting the rise of organized crime in the USA) was abruptly halted near the start of filming, never to be resumed, due to the death of its producer, age 49; **2)** On **Aug. 5, 1962** Marilyn Monroe died, age 36. The Kennedy brothers were widely rumored to be having affairs with Monroe. She intended to hold a press conference two days after her still mysterious and controversial death, initially called a suicide but later ruled a homicide by the coroner assigned to the case and the prosecutors investigating it.[81]

MOON – mental and emotional strength: Natal Moon is placed in *Uttara Ashadha nakshatra* ("the later victor") in the sign of Capricorn. It is practical and dutiful towards family and country in the sign of Capricorn, and driven to succeed in the 10th house. The person is responsible and hard working, with the mind more oriented toward civic issues and larger societal obligations. In spite of his privileged background, Bobby had a keen sense of life's injustices. Moon in the 10th house is an excellent position for a politician, as it gives a good instinct for the public pulse, puts one in better touch with the public, and gives greater visibility in the world in general – whether they choose it or not. Moon in the 10th house also focuses the mind on 10th house issues such as career and status.

However, emotional balance is not easy to maintain with Rahu or Ketu so close to the Moon, in this case it is within two minutes of arc (2/60ths of a degree), repeating in the *Navamsha* in the 8th house – a house that, for the Moon at least – can involve much emotional suffering that is difficult to avoid.[82] Known to be timid as a child, this was likely behavior during his **10-year Moon *Dasha***, from ages 2 to 12. As he matured, he would deal better with the influence of Mars, Saturn and the Sun, planets impacting the Moon through aspect or ownership of sign (Saturn)

---

[80] After Jack had won the Presidency, with Bobby appointed Attorney General, prior to his stroke Joe Kennedy insisted Ted Kennedy run for Jack's vacant Senate seat in Mass. as soon as he was eligible (age 30) in 1962. He spearheaded efforts to set up this legal reality, with a "care-taker Senator" in place.

[81] Marilyn Monroe's life story is covered in Chapter 6: Cancer Ascendant, Chart #17.

[82] Prince Charles of Wales (Chapter 6, Chart #19) also has Moon in the 10th house closely conjunct Rahu, though in an improved position in his *Navamsha* chart.

or *nakshatra* (Sun). Bobby took his role as a public servant very seriously, but was also known for his fierce attacks on anyone who criticized or made trouble for his brother Jack or for the extended Kennedy family.

Moon-Ketu in Capricorn pushes the force of obligation initially inwards to find its source of inspiration and renewal, and occurs in the *Navamsha* chart in the idealistic sign of Aquarius. In his last years Kennedy showed a more heightened humanitarian vision, perhaps coming out of the depths of grief for his brother's death, but also because now he was no longer in the accustomed support position. Indeed, his greatest moments seemed to come *after* his brother's assassination – reminding us of the essence of Moon in *Uttara Ashadha nakshatra* ("the later victory"). With *Navamsha* 8th house Moon, personal suffering can always be mitigated by rigorous transformative work – either spiritual work, or some kind of humanitarian effort that can transform a broader spectrum of society.

SATURN – self discipline and career success: We have discussed natal Saturn's strong position in its sign of exaltation and in a *Kendra* in its best possible angle *(Digbala)*. This creates a *Shasha Mahapurusha yoga* (special status to the person through Saturn's role). As Saturn is also the 10th house lord, this benefits the career:

> "...One will always gain through royal patronage and in business."
>
> **Parashara**[83]

Saturn is also the planet of the masses, it accounts for the huge obligation Bobby felt to his public and how that sentiment was returned, especially by African-Americans and other minorities. Saturn aspects to the Ascendant, Ascendant lord and/or Moon can produce a conscientious person willing to work hard for what they believe in. We can see this as well in the *Navamsha* chart with Saturn aspects to *Navamsha* Ascendant and *Navamsha* Moon in Aquarius.

Reading from the natal Moon, Mars and Saturn are in the 10th house, very powerfully situated, and Sun is well placed in the 11th. The Mars-Saturn combination in *kendra*s from both Aries Ascendant and the Moon (as sub-Ascendant) is a potent factor, though also challenging. They are in separate planetary camps, and Mars is a planetary enemy of Saturn, though Mars is exalted in Saturn's sign. By the rules of *Kendradhipatya Dosha* (See Glossary), the malefic nature of Saturn is neutralized, especially during its *Dasha,* but this occurred after his death. (Mars is less affected by KD, being Ascendant lord.) But together they can show technical abilities, ability to organize, and sometimes the capacity to oversee revolutionary change. Bobby's organizational gifts were much praised by his brother Jack.

Mars-Saturn mutual contacts can also be very positive for personal growth, and can give strong persistence in the face of seemingly insurmountable obstacles. But their placement in the 7th house from the Aries Ascendant also gives enormous powers to Bobby's enemies, especially with such a strong Saturn in the 7th house.

When he became Attorney General Jan. 20, 1961 Bobby was in the period of **Rahu-Moon-Saturn.** His brother's presidential election occurred in the previous period of **Rahu-Moon-Jupiter.** Both Jupiter and Saturn link back to the 9th house of father, perhaps alluding to the power of

---

[83] ***Brihat Parashara Hora Shastra***, Volume 1, Chapter 21, Verse 4. Parashara goes on to say that 10th lord's connections with a malefic will cause grief. Thus, despite Saturn's strength, mixed results occur from Saturn inhabiting the sign of a benefic, while also contacting a malefic planet. (Both effects are present, especially as in this case Mars is also Ascendant lord and as such must bring some beneficence, unless severely afflicted.)

their father in his ability to help get his son elected (Jupiter in the 9th house of father in its own sign). This was a plan in the making for several decades, though eldest son Joe Jr. was to have been the candidate until his death Aug. 12, 1944. On Inauguration day 1961 Jupiter and Saturn were transiting Bobby's 9th house in Sagittarius, more confirmation of good fortune in general and through the father. He insisted on Bobby as Attorney General despite his age and relative lack of experience. The **Jupiter-Saturn conjunction Feb. 18, 1961** in Bobby's 10th house also indicated his sudden rise, with all other astrological factors brought into the equation.

The passionate communicator – Mercury in Scorpio: Robert Kennedy was passionate about many political and social issues, but none more so in his last few years of life than the U.S. engagement in the Vietnam War. He regarded the war as one of the nation's "misguided policies" – a still very controversial position in 1968. And though increasing numbers of the American public questioned the wisdom of the Vietnam War, there was a backlash from those who believed the anti-war position was anti-patriotic and unsupportive of American soldiers fighting the war. In short, this was a deeply polarized environment for any political candidate.

> "I am...glad to come to the home state of ... [a] great Kansan, who wrote, 'If our colleges and universities do not breed men who riot, who rebel, who attack life with all their youthful vision and vigor then there is something wrong with our colleges. The more riots that come on college campuses, the better the world for tomorrow.' ... The man who wrote these words was that notorious man William Allen White—the late editor of *The Emporia Gazette.*"[84] ...[White] is an honored man today; but when he lived and when he wrote, he was often reviled as an extremist—or worse—on your campus and across this nation. For he spoke as he believed. He did not conceal his concern in comforting words; he did not delude his readers or himself with false hope or with illusion. It is in this spirit that I wish to talk to you today."
>
> **Bobby Kennedy**, first speech of his Presidential campaign, March 17, 1968, Kansas State University, Manhattan, Kansas.

> "Our country is in danger: not just from foreign enemies; but above all, from our own misguided policies—and what they can do to the nation that Thomas Jefferson once said was 'the last, great hope of mankind.' There is a contest on, not for the rule of America but for the heart of America. In these next eight months we are going to decide what this country will stand for—and what kind of men we are." *Ibid.*

Bobby would inspire many Americans, especially in the last few years of his life, notably in his last few months. But the location of his natal Mercury – by sign, by degree, and by house position – is not good for a politician. Mercury's natural functioning as a planet of speech and communication does not do as well in the element of water. What should be direct, reasoned, and adept is more emotional in the water signs, and potentially careless verbally. This is because the element of water rules over the emotions, and in the sign of Scorpio Mercury exhibits the heat and passion of its planetary ruler, Mars. There is danger of speaking too much on the "wrong" subject – that is, any topic likely to bring trouble or disfavor.

However, Scorpio is a hidden sign and Bobby was careful in what he was willing to reveal, and famously unwilling to engage in small talk. Further, natal Mercury is at 27:00 degrees of Scorpio,

---

[84] William Allen White was also a close friend of former Kansas governor Alf Landon, and an icon to Kansas Republicans. If this quote had been nationally broadcast, it could have been lethal for Bobby's campaign; but it was prescient in describing the rest of 1968 – the most turbulent year on American college campuses.

in the most treacherous area of *gandanta*.[85] This created a certain dichotomy in his speech: natal Mercury is secretive and very private in Scorpio, but also very emotional and Mars-driven in the *gandanta* degree area of Scorpio. It was rare for Bobby to so much as mention his brother's assassination, even to his closest aides. But when he did so at the 1964 Democratic Party National Convention, he could barely contain his tears.

If *Navamsha* Mercury had improved, it might lessen the severity of natal Mercury's condition and effects. Instead, it is debilitated in Pisces, owned by Jupiter. *Navamsha* Jupiter in turn is in the 6th house of conflict in its own sign of Sagittarius. This compounds the indicators for difficult life events during the major or minor planetary period of Mercury. His **Jupiter-Mercury** *Dasha/Bhukti* (July 11, 1967 to Oct. 16, 1969) gives the timing for when the fruits of this Mercury *karma* would come due. It spanned the time period when the strongest yet anti-Vietnam War sentiment was rising in the nation, and Kennedy was deliberating whether to enter the presidential race. This depended for him on whether President Johnson would continue to broaden the war. And in Feb. 1968 he did just that.

It was extremely courageous for Bobby to speak in Indianapolis on the day Martin Luther King was killed.[86] Police expected riots in response to Dr. King's death. Instead, Bobby's speech showed a remarkable ability to heal while educating and informing. Five planets were transiting through the watery sign of Pisces, including Mercury, which is debilitated in Pisces. Bobby's *Dasha* lord Jupiter was transiting in Leo, elevating his stature, and sub-*Dasha* lord Mercury was in the 8th house from Leo. Already watery in birth and *Navamsha* charts, tr. Mercury's position amplified the danger of the situation, the emotional impact of his brief speech and the emotional backdrop of his life that he unveiled briefly and fittingly in referencing his own brother's assassination. Pisces is a sign of surrender to the greater will and the greater good. Mercury does less well there precisely because it can reveal too many feelings that have not been processed. But Bobby understood this moment very well. He was a conduit for the strong feelings all around him, while unleashing a new level of courage and directness as a leader. His note of truth electrified the crowds, while the government had concern that his speeches were mirroring those of Martin Luther King, Jr.

His assassination:[87] As noted earlier, the biggest timing clue for personal danger to Bobby Kennedy comes from Mercury's sub-period within Jupiter *Dasha*. **Jupiter-Mercury** *Dasha/Bhukti* ran from July 11, 1967 to Oct. 16, 1969: Bobby Kennedy announced his presidential candidacy March 16, 1968 and was fatally wounded minutes after declaring victory in the California Primary June 5, 1968. He was shot at about 12:15 AM PDT in Los Angeles and died 26 hours later as a result of the three bullet wounds.[88] Full Secret Service protection was not yet provided for U.S. presidential

---

[85] See Glossary. The Scorpio-Sagittarius juncture is the most troublesome of the three *gandanta* areas, and the one most likely to bring some form of treachery.

[86] A major excerpt from this speech is given in the Biographical summary on Robert F. Kennedy.

[87] This topic is also discussed under Chart #31, Jacqueline Kennedy Onassis, Events and Dates (Item #11).

[88] Forensic scientists Robert Joling and Phillip van Praag studied the topic, Joling for 40 yrs., co-authoring a book: ***Open and Shut Case***, 2008. In the Foreword Paul Schrade (Kennedy aide, witness, and victim of the shootings) says: "...The Los Angeles Police Department never really opened the case. It quickly shut down its limited investigation, and ignored or destroyed key evidence." Joling and van Praag used more recent computer analysis of acoustical evidence. Their results contradict the single gunman theory, and further support Coroner Thomas Noguchi's statement that Bobby Kennedy died from bullet wounds entering the back of his head from a point blank range of no more than 3 inches. According to witnesses, gunman Sirhan Sirhan was facing Kennedy – within 3 to 6 feet. Further, Sirhan's gun held only 8 bullets, whereas acoustical analysis shows 13 to 14 shots were fired, some too close together to come from the same gun.

candidates. At that moment, tr. Mercury was at 8:34 Gemini, about to turn Stationary Retrograde the next day at 8:37 Gemini, in *Ardra nakshatra.* (Tr. Moon at King's death was at 9:00 Gemini.) *Ardra* means "teardrop" and is associated with storms and suffering. Tr. Mars was at 25:48 Taurus, closely aspecting both natal Jupiter and Mercury, his *Dasha* and Sub-*Dasha* lords, respectively. Tr. Saturn was at 28:59 Pisces, in the treacherous *gandanta* area of Pisces and on the eclipse axis closely opposite the previous eclipse (Total lunar) on April 12, 1968 at 29:54 Virgo. Saturn tends to produce its full effects at the end of any given sign, and it would exit Pisces June 17, 1968. The full combination was tr. Saturn-Rahu in his 12th house opposite tr. Moon-Ketu in his 6th house of conflict. Both tr. Jupiter in Leo and tr. Saturn in Pisces aspected *Dasha* lord Jupiter in Sagittarius, tr. Saturn within two degrees. Jupiter, planet of *Dharma*, and Saturn, planet of *Karma,* are the two hands on the clock.

Bobby Kennedy courted danger, as one might expect from an Aries Ascendant person with Mars in the 7th house. There is also a confluence of dangerous elements in his birth chart, and this must be weighed when assessing his karmic timing. Along with the theme of personal and physical courage in the face of danger, Mercury shows a major threat to his personal safety. We discussed how Mercury provides some protection, as it is one of the *Viparita Raja yogas* (a planet in one of the *Dusthana* houses – 6, 8, or 12 – is placed in another *Dusthana* house). But that protection is largely in terms of enhancing material prospects and does not offer protection to the physical body. As planetary enemy of the Aries Ascendant, Mercury's position at 27:00 Scorpio already gives his enemies exceptional power and very little exposure.[89]

Danger is clearly indicated during his **Jupiter-Mercury** *Dasha/Bhukti* due to this natal Mercury, as well as Mercury's vulnerability in the *Navamsha* chart. *Navamsha* Mercury in the 9th house brings public speech or writing but suffers through its debilitation in Pisces. Its sign lord Jupiter, poorly placed in the 6th house of conflict, sets up a problematic dynamic when the *Dasha* periods of Jupiter and Mercury are combined. Pisces is a sign of compassion and surrender, and in his last years Kennedy surrendered his life on behalf of the poor and the oppressed of the world, not just in the United States, much as did Martin Luther King, Jr. who was killed two months earlier. (Also with Aries Ascendant, King died in **Mercury-Jupiter** *Dasha/Bhukti.*) Each man was considered a threat to the governing powers and various other factions. Each man received a huge outpouring of grief in the nation. But while Kennedy's Mercury was not practical for a politician in the long run, it made him fearless and inimitable. Bobby's personal magnetism in the years after Jack Kennedy's death is still considered a benchmark for genuine political passion.

Chart 3: Henry Miller

Birth data: Sat., Dec. 26, 1891, 12:30 PM EST, Brooklyn, NY, USA. Long. 73W56 00, Lat. 40N38 00. *Lahiri ayanamsha*: -22:20:41 Class A data, from memory: From a letter from his father, "You were born on a Saturday – mid-day – between 12:30 and 12:45, can't remember exact time." [EH:

[89] This point is discussed earlier in this section on Chart #2 (Robert Kennedy), under *Powerful Enemies.*

	As	Ra	
Ju	**Henry Miller** Dec. 26, 1891 12:30 EST Brooklyn, NY		
Ve			
Su MeR	Ke	Ma Mo	Sa

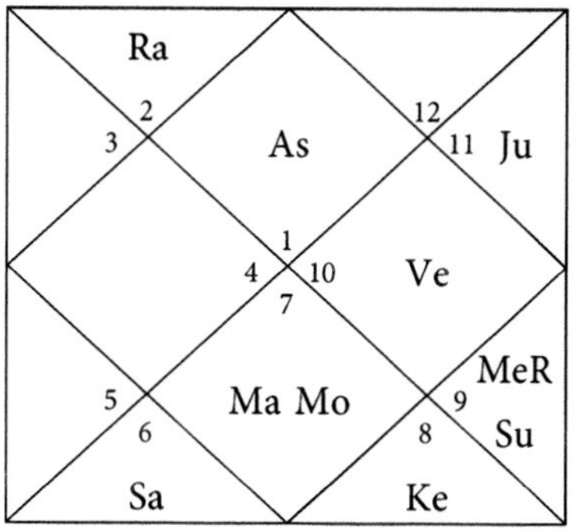

Ra 2 3 As 12 11 Ju 1 4 10 Ve 7 MeR 5 Ma Mo 9 6 8 Su Sa Ke

Planetary Information

As	00:37	Ari	Ashwini
Su	12:21	Sag	Mula
Mo	20:26	Lib	Vishakha
Ma	19:20	Lib	Swati
Me	17:29	Sag	P.Ashadha
Ju	20:16	Aqu	P.Bhadra
Ve	06:48	Cap	U.Ashadha
Sa	07:37	Vir	U.Phalguni
Ra	01:49	Tau	Krittika
Ke	01:49	Sco	Vishakha

Ve Sa Ma	Ju Mo As		
	Navamsha		Ke Su
Ra			
			MeR

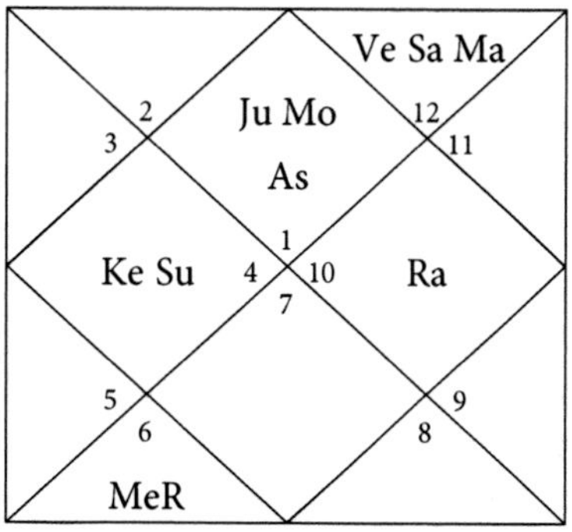

Ve Sa Ma 2 Ju Mo 12 3 11 As 1 Ke Su 4 10 Ra 7 5 9 6 8 MeR

Vimshottari Dashas

Ju	Jun 14, 1891
Sa	Jun 15, 1907
Me	Jun 15, 1926
Ke	Jun 16, 1943
Ve	Jun 16, 1950
Su	Jun 16, 1970
Mo	Jun 15, 1976
Ma	Jun 16, 1986
Ra	Jun 16, 1993

At 12:37 PM the *Navamsha* Ascendant shifts from Aries to Taurus, with *Navamsha* Venus in the 11th house and *Navamsha* Mercury in the 5th house. If this were true, it is not clear why Miller did not get published more widely during Mercury *Dasha* (1926-1943), nor does it properly describe his ongoing vitality despite periods of poverty and starvation. Many biographical factors support Aries *Vargottama* Ascendant.] Ascendant: 00:37 Aries.

Biographical summary: Born and raised in Brooklyn, New York, Henry Miller was a writer whose earliest books were banned as obscene in the United States and other countries. For many years they were smuggled into the country by tourists. He was first published at age 42 while living in Paris, France. The Obelisk Press in Paris initially published his most important books. French law at the time had a loophole for publishing controversial books as long as they were published in English.

Miller was married five times, with numerous liaisons to women who fascinated him and whom he wrote about, most especially his second wife June Smith (also called June Mansfield), and writer Anaïs Nin, both of whom gave him financial support to write and publish. The following are considered among his most well known novels and perhaps his best. They are largely preoccupied with sex and sexual relations: *Tropic of Cancer* written 1931-1932, published on Sept. 1, 1934; *Black Spring* in 1936; and *Tropic of Capricorn* in 1939. Miller was a prolific writer whose style was garrulous, lusty, earthy and eccentric. Here are some commentaries about him from some famous writers:

> "[*Tropic of Cancer* is] the most important book of the mid-1930s." [Miller is extremely important as an] "imaginative prose-writer…among the English-speaking races."
>
> **George Orwell**
>
> "There aren't many people like Miller in all literature."
>
> **Kenneth Rexroth**

> "Miller is a compendium of American sexual neuroses, and his value lies not in freeing us from such afflictions, but in having had the honesty to express and dramatize them."
>
> **Kate Millet**, *Sexual Politics*, 1970, p. 295.

When Grove Press published *Tropic of Cancer* for the first time in the U.S. (June 24, 1961) it led to an obscenity trial that tested U.S. pornography laws in the 1960s. The U.S. Supreme Court decided the case on June 22, 1964, declaring the book could no longer be banned.

In addition to his earlier works focused on sex and sexuality, he is also known for his social and political commentary, and his travelogues – his favorite being *The Colossus of Maroussi* (1941), a travel book of modern Greece. He was a maverick in his field, breaking new ground and breaking the rules. He defied convention in many ways. His 1945 book *The Air-Conditioned Nightmare* and its sequel *Remember to Remember*, 1947 – were biting commentary on the USA he found upon returning after years spent in Europe (from 1928) – mostly France. He resettled in the USA (California) in 1942 at age 51. He died there in Pacific Palisades, California on June 7, 1980 at age 88.

Miller disliked dealing with banks or any machinery, even simple machinery like dial telephones. Special interests included bicycling, painting abstract expressionist pictures and astrology. An extrovert who liked to socialize and travel, one of his biographies is titled: *Always Merry and Bright*. He had many close and fruitful friendships: His correspondence with Lawrence Durrell was published in 1963 and his letters to Anaïs Nin in 1965. Another close friendship was with Norman Mailer, who wrote *Genius and Lust: A Journey Through the Major Writings of Henry Miller* (1976).

Good longevity and vitality – clues from the Ascendant: The birth chart is generally strengthened, with a few exceptions, when the Ascendant is in the first *navamsha* (00:00 to 3:20 degrees) of a *Chara* sign, and thus *Vargottama*.[90] A fire sign Ascendant that is *Vargottama* greatly benefits the physical health and vitality. It was a major factor in Miller's longevity and helped to sustain him through years of poverty and near-starvation before he became more successful as a writer.

Bobby Kennedy also has an Aries Ascendant with Mars in the 7th house. Yet Kennedy was killed at age 42 and Miller lived to be 88 years old. The difference is that Mars here is 19 degrees from the exact Descendant and is considered fortunate when conjoining the Moon, especially for Aries Ascendant, as Moon and Mars are planetary friends. In Miller's *Navamsha* chart, Mars is not ideally located in the 12th house. But its mutual exchange with Jupiter in Aries is critical to improving its status in Pisces. Located in the 12th house, *Navamsha* Mars is less openly confrontational, thus less dangerous for his personal safety. Also, *Navamsha* Mars benefits from its contacts to exalted classic benefic planets. Though with Saturn, *Navamsha* Mars is also with exalted Venus opposite exalted Mercury. The 12th house can be residence abroad, and concerns private pursuits, such as the pleasures of the bed. He lived in Paris for many years, and the 12th

[90] *Vargottama* planets or Ascendants are repeated in the same sign in the *Navamsha* chart. A fixed sign planet or Ascendant repeats when in the 5th *navamsha* (13:20 to 16:40 degrees); a dual sign repeats when in the 9th *navamsha* (26:40 to 30:00 degrees). With natal Mars at 3:20 Capricorn, Mia Farrow (Chart #4) just misses *Vargottama* Mars by one minute of arc.

house placement of *Navamsha* Mars (and Venus) confirms that foreign residence could be very beneficial for him.

Examining his Ascendant, Ascendant lord Mars, and Sun, here is a summary of the **good indications for longevity**:

1. ***Vargottama* Ascendant**, especially in a **fire sign**.
2. No aspects by malefic planets to the Ascendant. (Mars is absolved as Asc. lord.)
3. Ascendant lord aspecting its own Ascendant, thus protecting it. (Exception: Mars exactly on the Descendant can be dangerous, as with Robert Kennedy.)
4. Jupiter aspect to the Ascendant lord of the birth chart. (Though Bobby Kennedy also has a Jupiter aspect to the birth Ascendant, it is diminished by its close connection to Venus. In Miller's case Jupiter is within one degree of an exact aspect to Asc. lord Mars. Jupiter is also strong in its own *nakshatra*, receiving no malefic aspects.)
5. Though hemmed in by malefics (a *Papa Kartari yoga*) in the 7th house of the birth chart, this Mars receives no aspects from malefics and only contacts from Moon and Jupiter, both good planetary friends and classic benefic planets.
6. ***Maha Bhagya yoga***: For a male, born in daytime with Sun, Moon, and *Lagna* (Asc.) in odd (masculine) signs, this *yoga* is achieved.[91] It promotes masculine qualities, gives good longevity, and can bring fame and prosperity. *Maha bhagya* means great good fortune. B.V. Raman says: "The *Lagna*, the Sun and the Moon form the tripod of life ruling as they do the body, the soul and the mind respectively...."[92]
7. Moon and Jupiter (two classical *and* temporal benefics) in the *Navamsha* Ascendant.
8. *Parivartana yoga* (mutual exchange of signs and energies) between Mars and Jupiter in the *Navamsha* chart. This has a protective effect on the physical body, especially as it involves the Ascendant lord. It is good for success in publishing (9th house) abroad (12th house), and for success in sexual matters – his main topic. *Navamsha* Mercury is exalted in the 6th house of competition: a *Viparita Raja yoga* protecting him and helping him to win his major legal battles.
9. The 12th house placement of *Navamsha* Mars could be detrimental to the physical health but is corrected by the Mars-Jupiter exchange and the good placements of planetary friends of Mars in the *kendras* of the *Navamsha* chart. Also, Pisces is the sign of a natural benefic and Jupiter is placed in its best angle, thus having *Digbala* in the *Navamsha* Ascendant, giving this Jupiter more power to assist Mars.
10. Natal Sun is well placed in the 9th house, with sign lord Jupiter well situated in the 11th house, *nakshatra* lord Ketu well placed in the sign of Scorpio. Any negativity from Ketu's influence is corrected by *Navamsha* Sun in Cancer in a *Kendra*, though not an ideal *Kendra* (4th house) for the Sun. Sign lord Moon is in the *Navamsha* Asc. with Jupiter, helping to correct other indications of a hidden or restricted Sun. Women (Moon) often came to his rescue financially.

---

[91] For a woman, *Maha Bhagya yoga* comes with night time birth, and Asc., Sun, and Moon in even signs.

[92] From B.V. Raman, ***Three Hundred Important Combinations***, 1983, 9th ed., p. 55, comments on this yoga.

Independent persona with divine discontent: As with Robert Kennedy, natal **Mars** is **in *Swati nakshatra***, ruled by planet Rahu (the foreigner) and by the deity Vayu, god of the wind. Being the constellation of the star Arcturus, placed well off the ecliptic – it refers to those who may go far afield from the prescribed course in order to find what they are looking for. It gives lightness, speed, adaptability, and independence to the planet situated here, especially Moon – describing the mind, but any planet here, with emphasis on Ascendant lord (Mars in this case), Sun, and Moon. As the planet of discontent, Mars placed in *Swati nakshatra* has even more of the usual qualities associated with divine discontent. Its placement here keeps him roaming, and in his case – trying to escape from what he perceives to be unnecessary restrictions in his life.

Even so, Miller was unfailingly optimistic, due to his well placed and unafflicted **Moon**, closely contacting Jupiter, planet of expansion and optimism. He was self-disciplined on his own terms, and this kept him writing through periods of poverty and starvation. His strongly placed **Saturn** in the 6th house helped him to maintain his competitiveness and perseverance. Saturn also interacts well with the rest of the chart, causing less difficulty here because of its placement in the 6th house of the birth chart, not aspecting the Ascendant, Ascendant lord, Sun, or Moon. Some constricting effect may have come from being born at the beginning of a *Sade Sati* (natal Saturn in the 12th house from Moon), but it may also have made restrictions feel familiar rather than formidable. Thus he works well within the restricted circumstances of his life and we know there will be restrictions due to the *Papa Kartari yoga* to Mars. (See item #5 above) The ***Viparita Raja yogas*** protect him from utter physical and financial ruin. All three of the powerful and protective *Viparita Raja yogas* are present in the *Navamsha* chart. It is more rare to have all three of them, and occurs from the *Navamsha* Aries Ascendant, where lords of *Dusthana* houses (6, 8,12) are in their own or another *Dusthana* house.[93] The 6th lord is in the 6th (exalted here), 8th lord is in the 12th house, and 12th lord is virtually in the 12th due to its *Parivartana yoga* (mutual exchange of sign) with Mars.

Natal Mercury is **Retrograde**, giving him the tendency to review what he says, to keep going back over it and return to some of the same subject matter repeatedly – as he did, and also in his case – to have to wait for years to see his most famous works published more widely. This coincided with greater public acceptance of his works and eventual changes in the censorship laws. Mercury can do damage and create conflicts, as enemy of the Aries Ascendant lord. However, it is still the *karaka* of communication in any chart. For Miller, it is well enough placed for him to eventually triumph over his enemies or detractors. Situated in the *nakshatra* of Venus, and opposite exalted Venus in the *Navamsha* chart, this Mercury would do more for his visibility as a writer during the 20-year Venus *Dasha* (June 1950 to June 1970), though he was ceaselessly productive for many years prior to 1950.

> "I have no money, no resources, no hopes. I am the happiest man alive.... Confusion is a word we have invented for an order which is not understood.... Every man has his own destiny: the only imperative is to follow it, to accept it, no matter where it leads him."
>
> **Henry Miller**

[93] *Viparita Raja yogas* can be read from the Ascendant or Moon. Bill Gates, founder of Microsoft (Chart #13) also has all three of these *yogas*, in his case from natal Moon. Further, Gates has the very beneficial *Adhi yoga*, bestowing wealth and high status: classic benefic planets in Houses 6, 7, and/or 8 from Moon or Ascendant.

The passionate, prolific bohemian author: With Moon and Ascendant lord Mars in the 7th house of partnership the mental/emotional/physical attention is on partnerships – all aspects of them, including sexual. (For Aries Ascendant, of course, Mars is also 8th lord, and 8th house can focus on sexuality, depending on the rest of the chart.) Mars aspects Venus in the birth chart, galvanizing Venus as both an artistic planet and as a planet pertaining to the love life. What adds to the passion of Venus and Mars here are the mutual contacts between Mars (planet of the sexual urge), Moon, and Jupiter.[94] Venus and Mars share mutual contacts in the *Navamsha* chart, in addition to Mercury, Jupiter and Saturn contacts, so there is much planetary action to describe the preoccupation with 12th house activities. How can we tell this would be sexual rather than spiritual, and therefore more ascetic? After all, 12th house is a house of *Moksha* (spiritual liberation). We can tell from the ongoing participation of Mars, Venus, Moon and Mercury in both the birth chart and *Navamsha* chart. This tends to keep him more secular.

Mercury-Jupiter contacts tend to give one a lot to say, and prolific output can come with Saturn's discipline. Four planets in watery signs in the *Navamsha* chart add the emotional propensity. Breaking from conventional patterns in life can come from 6th lord interacting with the 9th house, or vice versa. It assures there will be conflict with what is expected from the family of origin. Coming from a German Catholic heritage and the first generation of immigrants – he was completely pagan.

> "[Miller was]...free from the sense of guilt or sin that comes from a Christian or Jewish conscience. This offends a lot of people."
>
> **Kenneth Rexroth**

This also partially accounted for how he managed to preserve an innocence of the practice of literature while transcending conventional literature. It fits that he would be innocent, yet not unsophisticated in his writing and content. He would also be raw and immediate. This would stem from beginnings he was not afraid to leave, and a higher education he preferred to do on his own. Two months at City College of New York was enough. He was extremely well read, but his lack of academic preparation perhaps freed him to dismiss literary conventions, allowing him to write about many subjects, especially sexuality, in a revolutionary way. His iconoclasm spread to many areas of life, including his refusal to deal with banks.

A Flamboyant original – pushing the parameters of censorship laws: When writer Anaïs Nin met Henry Miller in Paris in late 1931 she wrote about it in her diary:

> "I saw a man I liked. In his writing he is flamboyant, virile, animal, magnificent. He's a man whom life makes drunk, I thought. He is like me."

Anaïs Nin would be his sometime lover and steadfast friend until her death in 1972. It was her money, lent to her by psychoanalyst Otto Rank, that enabled Miller to publish *Tropic of Cancer* (on Sept. 1, 1934). For the Aries Ascendant chart, **Jupiter** is pivotal for publishing success as 9th house lord (of publishing): Jupiter is placed in the 11th house of friends. **Saturn** is important as 10th lord (of career, status, and visibility) and is well placed in the 6th house of competition.

[94] See material in Chapter 9 (Libra Ascendant) Introduction about Venus interacting with Mars, and about the Puranic myths of Lakshmi (Venus) and Skanda (Mars). When in addition, Venus or Mars also own the Ascendant, one of the overriding themes can be sexual conquest. For Henry Miller, it was a major theme of his writing and his daily life for decades.

Mercury is *karaka* (significator) of the writer and of communications: he was in Mercury-Moon *Dasha/Bhukti* on publication day of his first book.

His **Mercury *Dasha*** running at that time would be a key astrological factor describing the controversial nature of this book, and the subsequent lengthy legal battle to allow it to be published in the USA. It had also been banned in England. Miller's 6th house lord (of conflict) is Mercury, placed in the 9th house (of publishing), and Retrograde, causing review and delays. The 9th house can be a good placement for Mercury, but for Aries Ascendant it brings some controversy into the area of the 9th house due to its 6th house connection, especially during Mercury *Dasha*. The 9th house includes the realms of education, law, religion, and publishing. Further confirmation of this long-term strife comes from the placement of *Navamsha* Mercury in the 6th house of conflict. However, due to Mercury's exaltation in Virgo, there would be eventual good results.

A propos **Mercury-Moon** *Dasha/Bhukti* on publication date, *Navamsha* Mercury and Moon are in a difficult 6 - 8 house relationship. (Mercury is 6th house from Moon in Aries, and Moon is 8th house from Mercury.) *Navamsha* positions of *Dasha* lords are crucial to confirm the effects of sub-periods and in fact all planetary configurations in the birth chart. The subsequent 7-year Ketu *Dasha* would also not be advantageous for wider publication, though Miller continued to write and publish other books through these years. Fortunately, he was unlikely to suffer personal torment or depression over these delays. The condition of his **natal Moon** describes his tendency to remain buoyantly optimistic about his future, due to close Moon-Jupiter contacts in both birth and *Navamsha* charts, and no other afflictions to the Moon other than dealing with some restrictions – shown by the *Papa Kartari yoga* to the Moon (and Ascendant lord).

On publication date (Sept. 1, 1934), tr. Jupiter was at 29 degrees Virgo, and transiting Saturn was retrograde at 1 degree Aquarius. Tr. Jupiter was not well placed in Miller's 6th house of conflict. Just one week later, tr. Jupiter's entry into Libra would have been much more beneficial for this event, aspecting the Aries Ascendant, his natal Moon and Mars in Libra, and both natal *and* transiting Saturn in Aquarius. Instead, the publication date more closely matched the destiny that was due to unfold, according to the Vedic chart, delaying his more widespread acceptance until after 1961, well into his Venus *Dasha*.

The joining of Jupiter *(Dharma)* and Saturn *(Karma)* is significant. Therefore we note when transiting and/or natal Jupiter and Saturn meet by sign. On publication date, tr. Saturn contacted natal Jupiter in Aquarius, and tr. Jupiter contacted natal Saturn in Virgo. We know that fulfillment of *Dharma* and *Karma* in career matters will not be a smooth journey due to natal Jupiter and natal Saturn in a mutual 6 - 8 house relationship. Jupiter is 6th house from Saturn in Virgo and Saturn is 8th house from Jupiter. Called *Shashtashtaka,* it can bring sudden changes– favorable or unfavorable depending on all factors, including their *Navamsha* positions. Usually it occurs in the *Dasha/bhukti* of the two planets, but the mutual contacts of tr. Jupiter and Saturn with their natal counterparts have a similar impact.

The agony and the ecstasy: With ***Navamsha* Venus** exalted in the 12th house in Pisces (the sign of sacrifice and surrender), much had to be sacrificed to earn the fruit of his efforts. A good portion of his early books relate to his personal experiences, including emotional suffering over the events he experienced with his second wife June, though written in fictionalized accounts. It would be the focus again in the *Rosy Crucifixion* trilogy, *Sexus* (1949), *Plexus* (1953) and *Nexus*

(1960). Early on he was inspired and supported by June's belief that he could be a writer and by Anaïs Nin, who gave him financial support, as did June in the early years. These themes are hinted at by the power of Venus and Moon in the chart, both angular in the birth chart. Venus is a planet of desire and sexuality. Venus and Mars interactions emphasize this theme, more so depending on their placement and contacts to other planets. Miller struggled for some years in self-imposed financial poverty due to his desire to be a writer. Both help and loss come through Venus, as 2nd lord of income (and 7th lord of partnerships) placed in the *Navamsha* 12th house of financial loss and gains from foreign income.

Landmark 1964 decision: Miller's book *Tropic of Cancer* spawned a 30-year censorship debate. Grove Press published the book in the USA in late June 1961, thus pushing the legal parameters even further. It was favored by several astrological factors: his change in *Dasha* and the transiting **Jupiter-Saturn conjunction** in Capricorn in **Feb. 1961**. The latter would be more lenient towards a wider range of literature, including pornographic, as long as it was profitable. The case was finally decided in the U.S. Supreme Court on June 22, 1964 in Miller's favor. They ruled that Miller's work could not be considered obscene as it had artistic merit. On this date in 1964, Saturn - planet of *Karma* - had returned again to Aquarius, the same sign it was in on the original publication date (Sept. 1, 1934). Tr. Jupiter was at 23 degrees Aries on his Ascendant, closely aspecting his Moon and Mars in the 7th house of partners or opponents and his Sun and Mercury in the 9th house of publishing. This was positive and expansive for personal and career matters.

Another decisive factor yielding good results for Miller was the change in *Dashas* - the planetary periods that deliver the karmic fruit. **Venus *Dasha* (June 16, 1950 to June 16, 1970)** was much more favorable for him than Mercury *Dasha*. In particular, **Venus-Jupiter** *Dasha/Bhukti* **(Aug. 12, 1960 to April 13, 1963)** had the power to activate exalted *Navamsha* Venus and its exaltation lord Jupiter as no other sub-period prior to it. When an exalted planet becomes *Dasha* lord, there is a chance for some extraordinary results - whether the planet be in the birth or *Navamsha* chart, but even stronger in the *Navamsha* chart, as it shows future trends and confirms or improves birth chart planets, if that planet is unafflicted by other factors. For instance, Venus is the only planet that does not suffer unduly in the 12th house. In this case, it delivered gains from a foreign source, making him more exotic, if not also controversial.

We can also credit the strong **Venus** in Miller's 10th house of career and status in the birth chart. It is in a *Shubha Kartari yoga*, hemmed in by classic benefic planets Mercury and Jupiter, giving it protection in matters related to Venus and/or the 10th house. Venus is in the *Uttara Ashadha nakshatra*, owned by the Sun, a great good friend to Mars and the Aries Ascendant. In turn, the Sun is well placed in the 9th house of publishing. All this further elevates Venus, enabling Miller to elevate his status during Venus *Dasha* or *Bhukti*, especially if Venus is interacting with many other planets, including Ascendant lord. In this case, Venus receives aspects from four other planets in the *Navamsha* chart. (Jupiter is included through its mutual exchange with Mars.) Thus after some initial suffering and surrender, greater success was due, especially if he wrote on 12th house matters.[95]

---

[95] Spiritual concerns are also included in the 12th house, but if one is not oriented in that direction, it can also apply to foreign residence, residence in pastoral places and/or to the pleasures of the bed.

Grove press: The first American edition of *Tropic of Cancer* was published by Grove Press on June 24, 1961. Given the promising features of Venus, we might wonder why it took 11 years into the 20-year Venus *Dasha* for Miller to get this book published in the USA. But ***Navamsha* Venus** exacts the price of major sacrifice when not only exalted in Pisces but in the 12th house of loss, even if rewards come later. We have this hint from natal Venus in *Uttara Ashadha nakshatra* ("the later victor"). This also applies to events within the span of Venus *Dasha*, some sub-periods being more favorable than others. Further, **the Jupiter-Saturn conjunction** on Feb. 18, 1961 would bring a higher level of status and visibility for the Aries Ascendant person, if all else was ready and favorable, including the *Dasha* sequence. This is particularly true in the case of Henry Miller, as the conjunction occurred at 1:52 Capricorn in the 10th house is exactly at the top of his chart, almost to the degree, and in the same *nakshatra* as his Venus. This enabled him to publish *Tropic of Cancer* (as well as his other most famous books from the 1930s) for the first time in the USA in 1961. His 1964 Supreme Court case enabled him to publish more widely, with his books no longer banned in the U.S. and elsewhere. His last book focused on sexual relations was *Nexus,* published in 1960. Having reached a goal of sorts, he was to spend the last twenty years of his life focusing more on his watercolors and his books about painting. The Jupiter-Saturn conjunction of 1961 finally gave him more literary stature, but his most serious literary efforts were over. He lived to age 88.

Chart 4: Mia Farrow

Birth data: Friday, Feb. 9, 1945, 11:27 AM PWT, Los Angeles, CA, USA, Long.118W14 34, Lat. 34N03 08. *Lahiri ayanamsha* -23:05:10. Class AA data, from birth certificate. Ascendant: 17:50 Aries.

Biographical summary: Reflecting her strong Catholic heritage, American-born actress Mia Farrow was named Maria de Lourdes Villiers Farrow at birth. She was third of seven children (three brothers and three sisters) born to celebrity parents: writer-director John Farrow (b. Feb. 10, 1904, d. Jan. 27, 1963) and actress Maureen O'Sullivan (b. May 17, 1911, d. June 22, 1998). She made her film debut in 1947 in a short subject with her mother: celebrity family members modeling the latest fashions. At the age of nine she contracted polio, from which she recovered, though it was a life-transforming experience to be exiled as a child to a Los Angeles hospital ward among seriously ill and dying patients. Her brother Michael (1939–1958) died four years later in a plane crash, and her parents' marriage deteriorated around the same time.[96] Mia's father tried to dissuade her from an acting career, in part by sending her to all Catholic schools. She was nearly 18 and attending a Convent school in England when he died of a sudden heart attack. Soon after, she won her first off-Broadway role as Cicely in Oscar Wilde's comedy *The Importance of Being Earnest*, opening July 2, 1963.

[96] Her brother Patrick Farrow (b. Nov. 27, 1942, d. June 16, 2009) was a sculptor for 35 years. He and his wife owned an art gallery in Vermont. He died of a self-inflicted gunshot wound

Ve	As		SaR Ra
	Mia Farrow Feb. 9, 1945 11:27 PWT Los Angeles, CA		
Ma Me Su			
Mo Ke			JuR

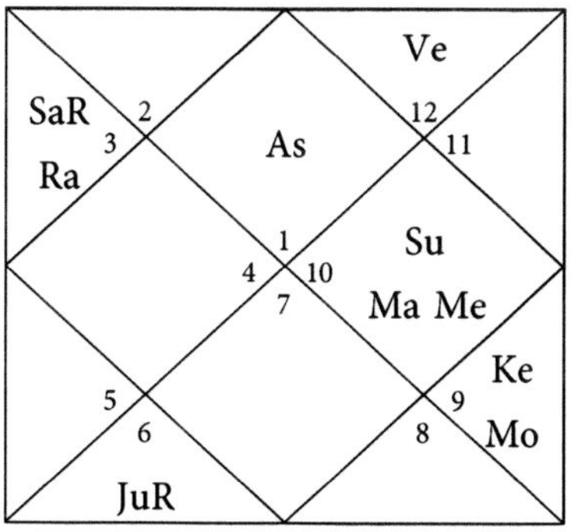

Ve
SaR 2 3 Ra
As
12 11
1 4 10 7
Su
Ma Me
Ke
5 6
9 8
Mo
JuR

Planetary Information

As	17:50	Ari	Bharani
Su	27:34	Cap	Dhanishta
Mo	18:35	Sag	P.Ashadha
Ma	03:20	Cap	U.Ashadha
Me	13:45	Cap	Shravana
Ju	03:08	Vir	U.Phalguni
Ve	14:15	Pis	U.Bhadra
Sa	11:16	Gem	Ardra
Ra	23:36	Gem	Punarvasu
Ke	23:36	Sag	P.Ashadha

		Ra Me	
Ma	**Navamsha**		
SaR JuR			
	Ke Ve		Su As Mo

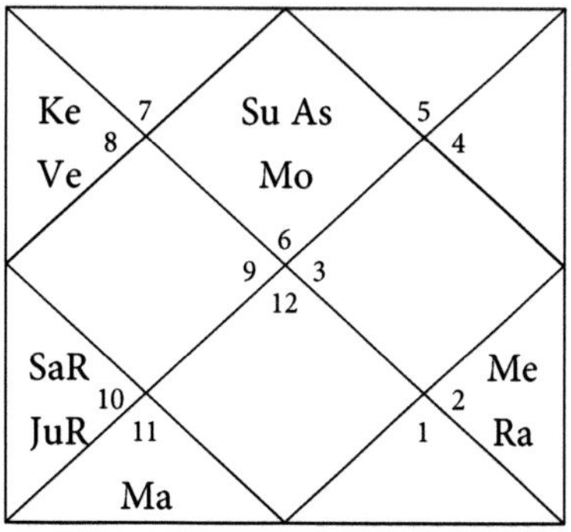

Ke 7 8 Ve
Su As
Mo
5 4
6 9 3 12
SaR 10 JuR 11
Ma
Me 2 1 Ra

Vimshottari Dashas

Ve	Mar 22, 1937
Su	Mar 22, 1957
Mo	Mar 22, 1963
Ma	Mar 22, 1973
Ra	Mar 22, 1980
Ju	Mar 22, 1998
Sa	Mar 23, 2014
Me	Mar 22, 2033
Ke	Mar 23, 2050

From age 14, she played very small roles in films and began her training in drama in New York City in early 1963. Then with only the one off-Broadway play as professional experience she won the role of the lead teenager (Allison MacKenzie) in a new television series, *Peyton Place* (1964-1966). From this she rose to instant stardom at age 19. Her third film, *Rosemary's Baby* (1968) was a critical success and leading roles in numerous feature films followed. In the mid-1970s, she appeared on the London stage, and was the first American actress to be cast in various productions at the Royal Shakespeare Company.

In Oct. 1964, Mia began dating Frank Sinatra (b. Dec. 12, 1915, d. May 14, 1998). She was 20 and he, 49. They were married briefly from July 19, 1966 to Aug. 1968. Her subsequent liaison with conductor-composer André Previn led to a nine-year marriage (Sept. 10, 1970 to Feb. 1979), three biological children, and three adopted children, two from Vietnam and one from South Korea (Soon-Yi Previn, birth date given as Oct. 8, 1970 though her precise age was in question; adopted 1977). By 1997, Farrow had a total of 15 children, 11 of them adopted - many of them with disabilities. She had one biological child (a son) with Woody Allen (b. Dec. 1, 1935), her partner from spring 1980 to Aug. 1992 - when their split was official. It came to a close in a blistering court case, in which Farrow accused Allen of molesting 7-year old adopted daughter Dylan, current name Malone Farrow. That charge was later dropped but began after she found evidence in mid-Jan. 1992 of Allen's sexual relationship with her adopted daughter Soon-Yi Previn, then age 21 or possibly 19. (She became Allen's third wife in Dec. 1997.) In a much publicized celebrity case, there were bitter public accusations from both sides. The judge called Allen's behavior with some of their children "grossly inappropriate," though Allen vehemently denied the charges. Mia Farrow was ultimately awarded custody of all the children, including her only biological child with Allen, Satchel (b. Dec. 19, 1987), later renamed Ronan Farrow.[97]

Before their breakup, Farrow and Allen enjoyed a fruitful artistic partnership.[98] She appeared in 13 of his films over a 12-year period. Her acting career in films slowed down over the next ten years, with only six films, two of them for television. As of 2005, her children ranged in age from 11 to 35 years old. The same year she returned to the New York stage to critical acclaim, with more film and theatre projects in subsequent years. Between 2004 and 2009 she appeared in six films, one for television. She has become active in various humanitarian projects with organizations such as UNICEF (United Nations Children's Fund), and honored for her work, especially in Africa. Her memoir *What Falls Away* was published in 1997. Its major focus was on her years with Woody Allen. From 1995 she has been linked with author Philip Roth.

Youthful vitality and beauty: Ascendant lord Mars has extraordinary power here in the 10th house of the birth chart, *Digbala* (best directional strength, i.e., best house position), in the sign of its exaltation, just one minute of arc past *Vargottama*. This is a *Ruchaka Mahapurusha yoga*. The Sanskrit root "*rucha*" means bright and radiant, giving sharpness to the mental focus. The person is daring, energetic and hard working. Exalted Mars confers unusually high physical vitality. (Farrow has said she rarely needs more than four hours of sleep per night, for example.) It can also confer youthful exuberance and beauty, and Mars in the 10th house is excellent for career visibility. Further enhancing these qualities is Jupiter's exact aspect to exalted Mars in a *Kendra*, Jupiter being a classic and temporal benefic. Known as a *Chamara yoga*, this shows strength of character with philosophical tendencies, eloquence, and sometimes being born into a royal family or honor through a king.[99] Ascendant lord aspecting the Ascendant generally protects the physical body from harm. However, when Mars rules the Ascendant, it is more advantageous when it aspects the Ascendant from this superior position. (There was danger for Bobby Kennedy with Mars exactly on the Descendant.)

In the *Navamsha* chart, Mars is located again in Saturn's sign, but in Aquarius. It is well placed in the 6th house from Virgo Ascendant, though this indicates she might have to deal with disease and/or controversy in her life and would fight back well. In the birth chart, Mars is in a ***Nadi yoga*** exchange with the Sun: Mars is in a *nakshatra* owned by the Sun, and Sun is in a *nakshatra* owned by Mars. Sun's *nakshatra* is *Uttara Ashadha* ("the later victor"), and that of Mars – *Dhanishta* ("the wealthiest"). This inter-relationship of two great planetary friends, both fiery masculine planets and

---

[97] Since 1992, Ronan Seamus Farrow (birth name Satchel Ronan O'Sullivan Farrow) has been estranged from his father, Woody Allen. Ronan is a lawyer, human rights activist and freelance journalist. A child prodigy, he graduated from college at age 15 and was accepted to Yale Law School at age 16, graduating in 2009. He deferred his entry until 2006 in order to work for Richard Holbrooke, former United Nations ambassador, and to do special projects for UNICEF, notably in Africa. A UNICEF Spokesperson for Youth in Nigeria, Angola and Sudan (2001-2009) and Special Advisor to the State Dept. on Humanitarian and NGO Affairs (from 2009), he received the prestigious Rhodes Scholarship to study at Oxford University, England starting from fall 2012.

[98] Woody Allen is discussed in Chapter 7: Leo Ascendant, Chart #22 (also more on Soon-Yi Previn).

[99] *Chamara* = the tail of a yak, insignia of royalty. Its *yoga* is formed when the Ascendant lord is exalted, in an angular house and aspected by Jupiter.

both *Digbala* in the 10th house, not only confirms the physical vitality but indicates public success and visibility. It could also give some hot-headedness. Ascendant lord Mars aspects the 4th and 5th houses, the latter from its special aspect to the 8th house from itself. This speeds up the mind and the intelligence, and often aborts the higher education in terms of academic degrees. Not yet 18, she left convent school to study acting in New York. Her father's death from a sudden heart attack (Jan. 27, 1963) accelerated this process, as he had opposed her acting career.

Childhood bout with polio: When the 6th house lord contacts the Ascendant Lord, there can be physical problems for the person, in this case neuro-muscular problems. Mercury (6th lord) is in the 10th house with Mars in Capricorn. This can also give conflicts of a very public nature (10th house). When Farrow suffered from polio at age nine, she was still in **Venus *Dasha***. Natal Venus is exalted in the 12th house of the birth chart. Locations associated with 12th house are foreign residences, hospitals, prisons, or ashrams – hidden away from the normal public life. Exalted Venus may require a big sacrifice from which much is learned. As polio was thought to be contagious at the time, she was placed in an isolation ward at the hospital and all her personal belongings were burned. Other than her hospital experience, she spent most of her 20-year Venus *Dasha* in Convent schools, for 13 years up to age 18, or in foreign residences – mostly in Europe where her parents were filming. Fortunately, her 12th lord Jupiter aspects its own house, minimizing her losses, both physically and financially. Natal Jupiter also aspects Sun and Mars, two planets signifying physical strength, including muscular strength (Mars). These factors, along with her very strong Ascendant lord and the ***Nadi yoga*** **between Sun and Mars** (See above) account for her apparent complete recovery from polio.

Farrow contracted polio during her **Venus-Mercury** *Dasha/Bhukti* (March 22, 1953 to Jan. 21, 1956), ushering in for her both the problematic nature of Mercury, as well as that of Saturn: **Mercury and Saturn** are in ***Parivartana yoga*** (mutual exchange). These two planets are also in a difficult 6 – 8 house relationship: Saturn is 6th house from Mercury, and Mercury is 8th house from Saturn. In addition, transiting Saturn was at its Stationary Retrograde degree closely opposite her Ascendant February 17, 1954 and closely aspecting her natal Moon as well. It hovered within close range for another nine months, and in contact to transiting Mars, amplifying the effect, for 6 ½ months up to mid-October 1954. As a planetary enemy of Mars, the effects of this long Saturn transit were to bring a major illness, restructure her life, add a new seriousness and deepen her understanding of life around her. Jupiter is a great good friend to the Aries Ascendant and is the only classical and temporal benefic planet that can act protectively. Thus, tr. Jupiter's 13-month transit in Cancer (from Sept. 10, 1954) opposite Ascendant lord Mars in Capricorn would finally come to her aid in speeding up the recovery process.

Luminous Venus and the seemingly fragile persona: There are some special rules from a classic text on Vedic astrology, the *Bhavartha Ratnakara,* which apply to this chart and this situation, and served to protect Farrow during a potentially perilous **Venus *Dasha*** (from birth to age 12: up to March 22, 1957). Venus is lord of both 2nd and 7th houses, situated in her 12th house in Pisces. It is *maraka* planet for Aries Ascendant, and thus could bring death during its *Dasha*.

> "**For one born in Mesha [Aries], if the lord of the 2nd is in the 12th**, he becomes good. For those born in other *Lagnas* the 2nd lord does not become good if he is placed in the 12th."
>
> ***Bhavartha Ratnakara***, Chapter I: on *Mesha Lagna*, Stanza 7 (B.V. Raman translation).

> "The person will be fortunate in respect of that *Bhava* whose *karaka* is situated in the 12th from Ascendant." *Ibid.*, Ch.VIII, Stanza 6.

We make an important distinction between house lord and *karaka* (significator) in the 12th house from the Ascendant. If a house lord is placed in the 12th house, the affairs of the house(s) over which it rules are adversely affected. If a *karaka* is placed there – the matters over which it governs in general are protected and blessed. In this case, 7th lord placed in the 12th house is not beneficial for partnership matters. However, since Venus is also *karaka* for love affairs and the arts – Venus exalted in Pisces in her 12th house ultimately protects these areas of her life from the losses that would happen for other charts with a 7th lord other than Venus situated in the 12th house.

This **Venus in Pisces** also contributed to a kind of luminous quality Mia Farrow had from an early age. Its greater influence could stem from the Ascendant degree in *Bharani nakshatra*, ruled by Venus exalted in the 12th house, and Venus *Dasha* occurring in the first 12 years of her life. These were formative years and included some of her 13 years of Convent education. A devout Catholic, Farrow is also a very unconventional person. She would go on to teach her children about many other religions and cultures. We see this from **Moon-Ketu** in her birth chart and **Venus in Scorpio** in her *Navamsha* chart on the Rahu-Ketu axis. Both influence the 9th house of foreign travel, philosophy and religion, among other things, and the Rahu-Ketu influence on Moon and Venus give an unusual twist to her mind and her love life. Through her long career, marriages, love affairs, and children – most of it was on her own terms. The independent nature of this Aries Ascendant person would bridle under too many restrictions. She continued to work, even though her first two husbands preferred she either not work or work only on certain projects.[100] Woody Allen was not accustomed to being around so many children (she had nine children when they met). Their solution was to maintain separate living quarters in Manhattan during their 12-year period together.

Unconventional partnerships with older men: Reading this chart from the Moon as Ascendant, we see many very fortunate *yogas* (planetary combinations), some even more apparent than from the Aries Ascendant. This adds to why **Moon *Dasha*** was so fruitful. Natal Saturn opposite natal Moon made her a hard worker and brought her marriages or partnerships with men considerably older than herself, from 9 to 29 years older (Frank Sinatra, André Previn, Woody Allen, in that order). Since they started dating in 1995, her name has been linked with her neighbor, friend and author Philip Roth (b. March 19, 1933). Older partners or a big age difference with partners can occur with Saturn in the 7th house of partnerships from Moon or from the Ascendant. The fact that each partner to date was famous in his own right comes from exalted Venus as lord of the 7th house of partnerships. Being part of a famous couple is indicated by Sun and Moon both in *Kendra* houses in the birth chart or *Navamsha* chart. The King (Sun) and Queen (Moon) show some degree of leadership, especially during *Dashas* of Sun and/or Moon, depending on the chronological age.

At age 18 Farrow entered the 10-year **Moon *Dasha*** (March 22, 1963) and acquired her first starring role on television within six months. Prior to that, her father and mother (also Sun and Moon in any chart) were considered a glamorous Hollywood couple for many years. Her celebrity parents gave Mia Farrow name recognition, early visibility, and social and career opportunities.

[100] Frank Sinatra wanted her to appear in his film, ***The Detective***. When Mia refused, due to a scheduling conflict with ***Rosemary's Baby*** (a Roman Polanski film), Sinatra served Farrow divorce papers on the set where she was working, in front of cast and crew. Normally cool in public, Farrow broke down in tears.

As natal Saturn is in the 7th house from *Dasha* lord Moon, this period brought the first of several older partners: Frank Sinatra entered her life in Oct. 1964.

Success as an actress: Reading from natal Moon in Sagittarius, Venus is exalted in the 4th house and Jupiter is in the 10th house (*Amala yoga*). This yielded extremely favorable results during **Moon *Dasha***, though since Moon is aspected by three malefics - Saturn, Rahu, and Ketu - this could also make success difficult to manage in some ways due to increased responsibilities and emotional turbulence. Even so, there are several *Raja yogas* from the Moon: five other planets in angles from the Moon, all in dual signs, including exalted Venus; Mars-Mercury-Sun in the 2nd house from Moon - creating a *Dhana yoga* of wealth. When together in a financial house, the 5th, 9th and 10th lords from the Moon are conducive to bringing wealth and success. Her ability to generate her own income is shown from the *Dhana yoga* in the 2nd house from the Moon (or Ascendant).

In her birth chart, the first excellent indication for high visibility is key personal planets both in the angles of the chart, especially above the horizon, and *Digbala*. Natal Sun and Ascendant lord Mars each have *Digbala* at the top of the sky - the same combination that gives her excellent physical vitality. Success in the area of theatre or film is hinted at from interactions between lords of the 3rd and 5th houses, or planets located in these houses. There is a *Parivartana yoga* (mutual exchange) between Mercury and Saturn. (Saturn in Mercury-ruled Gemini in the 3rd house exchanges energies with Mercury in Saturn-ruled Capricorn in the 10th house.) Fifth lord Sun is also in the 10th house. In addition, 10th lord Saturn is strongly placed in the *Navamsha* 5th house in its own sign of Capricorn alongside Jupiter, creating a *Raja yoga* capable of bringing one visibility through 5th house activities. (The 5th house is artistically and/or biologically creative - through the arts, education, or children). Mia Farrow is well known for her many roles as an actress and was the first American actress ever invited to perform at the Royal Shakespeare Theatre in London.

The artistic and personal fruitfulness of her partnership with Allen lasted 12 years - a full Jupiter cycle - beginning and ending with transiting Jupiter in Leo in her 5th house. Their last of 13 films together - *Husbands and Wives* (1992) - was about bickering spouses.

Major success from an early age: Best planetary friends to this chart are Sun, Moon, and Jupiter. They are all in important combinations in the birth chart, and we see this accelerating in the *Navamsha* chart. Sun and Moon are powerful in the Ascendant of the *Navamsha* chart, while Jupiter and Saturn are also powerful together in a trinal house, the 5th house.[101] *Navamsha* Jupiter aspects the Sun and Moon. *Navamsha* Mars in the 6th house aspects the Ascendant, from its special 8th house aspect. A *Jaimini yoga* again joins the Sun and Moon in the *Navamsha*, giving them *Jaimini Raja yoga* status. (The highest degree planet is *Atma karaka*, and the next highest degree planet is *Amatya Karaka.)* This is a *yoga* of power, strength, and leadership, especially when it involves the Sun and Moon. When Farrow's **Sun and Moon *Dashas*** occur, this important *yoga* can first become manifest, though numerous factors point to the Moon *Dasha* as the real start.

[101] *Jyotish* computer programs now contain the option of numerical degrees in the *Navamsha* chart, but the planetary degrees in the birth chart are the ones that have predominance. When assessing birth planets in the *Navamsha* chart, Jaimini *karakas* are transferred to the *Navamsha* chart as they occur in the birth chart, and should not be confused with the degrees given in the *Navamsha*, which provide minute detail, but are not used as degrees per se other than to show how close a planet is to an angle or to another *Navamsha* planet. Primarily the *Navamsha* house and sign positions should be observed, as well as *yogas* and aspects.

Sun *Dasha* came when she was 12 to 18 years old: March 22, 1957 to March 22, 1963. Her father John Farrow (also the Sun) died on Jan. 27, 1963 of a heart attack. Sun is lord of the *Navamsha* 12th house of loss, and spiritual or foreign experience. **Moon *Dasha*** (March 22, 1963 to March 22, 1973) covers the years when she blazed her way into the public eye as a young actress with relatively little professional experience. **Moon in *Purva Ashadha*** ("the early victor") gives some indication of this result. Its closeness to Ketu-Rahu axis (7 degrees away) shows the emotional turbulence of the life and the many unconventional decisions. But Moon and Mercury contacts to Saturn help to modify that flaw, as they lend discipline and a willingness to work hard. Saturn's role hints also at the help she received from among others, her mother's friend, actress Vivien Leigh. Her preeminence enabled her to draw Farrow's New York stage debut to the attention of major agents and directors.

Tr. Jupiter-Saturn conjunction in Feb. 1961 near her natal Mars in the 10th house was another indicator that she could gain early and sudden public status in the 20-year period starting Feb. 1961. Her Moon *Dasha* from March 1963 brought these events into sharp focus. After her father's death in late Jan. 1963, Mia was freer to pursue an acting career, as he was resistant to her becoming an actress. She made her off-Broadway debut in *The Importance of Being Earnest*, which opened July 2, 1963. This resulted in her meeting with an agent looking for a leading young actress for a new television series, *Peyton Place*. She made her television debut in this series in September 1964.

Two eclipses intensified the month of her career debut – one she originally hoped would lead to an acting career on the New York stage. A **Lunar eclipse on July 6, 1963** at 20:45 Sagittarius was closely conjunct her natal Moon (and *Dasha* lord) at 18:35 Sagittarius. A **Total Solar eclipse on July 20, 1963** at 4:03 Cancer was closely opposite her Ascendant lord Mars. Sudden or exotic changes in her career direction would come through 10th lord Saturn conjunct Rahu (the foreigner) in the birth chart. Some of Farrow's most famous roles throughout her career came to be associated either with sexual scandal (the *Peyton Place* series), or with a demonic twist (horror movies such as *Rosemary's Baby* (1968) and *The Omen* (2006 remake)). Farrow has always maintained that

> "the capacity for terrible things... is a human component within all of us."

The worst years: Sexual scandal in her private life became public in late 1991 and into 1992, when Farrow endured the betrayal of her life partner and her oldest daughter (Soon-Yi Previn), and the subsequent loss of those relationships as well as a valued artistic partnership. We have previously noted the astrological significance of the split with Woody Allen coming **12 years** after the start of their union in 1980 – a **Jupiter cycle**, Jupiter being the planet of *Dharma*, or right action. It also came **29 years** after her 1963 career debut and sudden rise to stardom – a **Saturn cycle**, Saturn being the planet of truth, reality, and *Karma*. From these factors alone, we know there would be some kind of major shift for her in 1992. The *Dasha* and its sub-period would determine the nature of that shift, depending on their positions in the birth chart.

These events occurred during her **Rahu-Venus** *Dasha/Bhukti* (Oct. 10, 1991 to Jan. 9, 1994). She was born in **Venus-Rahu** *Dasha/Bhukti*, the reverse of the above and an important karmic indicator. The two planets together show unconventional, unpredictable trends in matters of love and/or the arts. Rahu and Ketu are shadowy planets that cause eclipses and magnify events. They are foreign influences that take us away from expected or traditional patterns. Their contacts to the Moon can upset the mind and heart. With Venus, they can destabilize the love life and/or one's life in the arts, or bring an exotic component to it. Between the birth chart and *Navamsha* chart she has both combinations. This upset reached its karmic peak for Farrow during Rahu-Venus *Dasha/Bhukti.*

This situation in 1991-1992 was further exacerbated for her by the **Saturn transit in Capricorn** (Dec. 14, 1990 to March 5, 1993): Tr. Saturn in her 10th house in Capricorn gave Farrow maximum visibility and at the same time put maximum pressure on 7th house (partnership) matters.[102] Most of the damage would be done from late 1991 when tr. Saturn was passing back and forth over her 10th house cusp. Tr. Saturn at the top of the chart can bring great visibility and/or publicity, and it can also bring a fall from power. After her split from Woody Allen she made far fewer films in the ensuing years.

Jupiter causes some havoc for her with children during Rahu and Jupiter *Dashas*. This is due to Jupiter's role as *nakshatra* lord of *Punarvasu nakshatra,* where Rahu is located; also due to Jupiter's position in the 6th house of conflict. From the start of the 16-year **Jupiter *Dasha*** (March 22, 1998), Farrow suffered some career disadvantages due to being: 1) no longer in close alignment with a very prolific and gifted film director, and to: 2) her decision to adopt five more children on her own after 1992, many of them with disabilities.

Sade sati: (See Glossary) We know that for Mia Farrow this 7½ year period would be more difficult than for some others due to the condition of her natal Moon. We have already described the emotional turbulence that is likely with Moon on the Rahu-Ketu axis, either with Rahu or Ketu, the closer by degree the more excruciating, especially if it repeats in the *Navamsha* chart - which it does not here. Farrow's *Sade Sati* began Sept. 17, 1985 when tr. Saturn entered Scorpio (with no further retrogrades into Libra). It ended March 5, 1993 when tr. Saturn exited Capricorn.

Another negative factor was **Rahu's transit in Sagittarius, 1991-1992**: This occurred from mid-April 1991 to mid-Oct. 1992. Rahu's transit was even more significant as she was then in **Rahu *Dasha*.** Since tr. Rahu was passing over **natal Moon-Ketu opposite Saturn-Rahu** it was a very rough combination. Even though Moon *Dasha* brought spectacular success, this natal configuration is **the most difficult configuration in her chart.** It also describes the unconventional nature of the Farrow-Allen relationship, starting in her **Rahu *Dasha*.** (Rahu *Dasha* began March 22, 1980. Her first solo lunch date with Allen was April 17, 1980.) Their overriding joint artistic successes can be attributed to the redeeming effect of her other remaining five natal planets and the overall strength of Sun, Moon, Jupiter, Saturn and Mars in the *Navamsha* chart. This era in her life coincided with **three Jupiter-Saturn conjunctions in Virgo (Dec. 1980 to July 1981),** all on her natal Jupiter in the 6th house of conflict, but in the 10th house of publicity from the Moon.

Large number of children: As with Robert Kennedy, who had 11 children (all his own biological children), Mia Farrow had 15 children, 11 of them adopted. For this we look to Jupiter as *karaka* of children, and to the 5th house from the Ascendant and from the Moon. Fifth lord from the Ascendant is the Sun, receiving directional strength in the 10th house, as noted. Fifth house from Moon is Aries, which receives an aspect from its own house lord Mars, whose extraordinary strength and vitality we have noted. However, Mars and Saturn are both classical malefics and therefore separative planets and they both aspect the 5th house from the Ascendant. This indicates the conflict issues regarding children, with the threat of separation from them in her legal custody battle with Woody Allen. Farrow's likely success in any custody dispute is shown from the strong

---

[102] Saturn's special aspects are always forward in the zodiac, to the 3rd, 7th and 10th houses from itself.

Navamsha 5th house and 5th house lord, with no malefic aspects to the *Navamsha* 5th house. In mid-June 1993 she gained custody of all the younger children but lost contact with her adopted daughter Soon-Yi, Woody Allen's next partner, later wife. Allen was exonerated from Farrow's claims of child molestation.

Jupiter is well placed in Sun-owned *Uttara Phalguni nakshatra*, but in the 6th house of conflict in the birth chart. In the *Navamsha* chart Jupiter is debilitated in Capricorn in the 5th house of children and artistic endeavors.[103] These factors show the likelihood of some extreme situations in these areas, starting with some disadvantages: one of her children (Ronan) is a child prodigy whose relationship with his father was severed by his parents' dispute and their 1992 court case. Major strife about children ended both their personal and artistic partnership.[104] Their 12-year relationship lasted a full **Jupiter cycle**, with tr. Jupiter returning to her 5th house in Leo in 1992. Farrow went on to adopt several other children who had disabilities and she helped many other children collectively through UNICEF.

Chart 5: Jacqueline Du Pré

Birth data: Friday, Jan. 26, 1945, 11:30 AM GDT, Oxford, England, Long. 01W15 00, Lat. 51N45 00, *Lahiri ayanamsha*: -23:05:08. Class A data, from memory. Ascendant: 3:59 Aries.

Biographical summary: A musical prodigy who began her cello studies at age five, Jacqueline Du Pré first came to serious public notice at age ten, at the same time her older sister Hilary, also a prodigy as a flutist was making her debut. Their concert pianist mother taught them both initially, but Jacqueline's major cello teacher was William Pleeth. She became a world-class cellist and musical superstar while still in her teens, making her London debut in March 1961 at age 16. She played cello concertos with major orchestras, and was best known for her interpretation of the Elgar Cello Concerto in E minor. She also played other cello concertos by Elgar, Dvorak, and Schumann.

> "...Du Pré almost single handedly opened up the predominantly male field of cello playing.... [She is] an exuberant, musical dynamo known for powerful, evocative and provocative playing ...[as well as her] unorthodox personal life."
>
> ***Kirkus Reviews***

> "Few cellists have penetrated the [Elgar Cello] concerto's inner recesses so deeply, or produced a performance of such burning intensity."
>
> ***The Rough Guide to Classical Music***

---

[103] A cancellation of debilitated Jupiter is provided by Saturn in Capricorn in the same *Navamsha* house. Known as a *Neecha Bhanga Raja yoga*, the affairs over which the planet rules – in this case Jupiter (ruling over children and artistic endeavors) – gain improvement after some initial disadvantages and difficulties.

[104] Tr. Jupiter's return to Leo marked Allen's personal renewal, with his own Ascendant in Leo (Chart #22), and his Moon and Saturn in his 7th house. Tr. Jupiter in the 7th house from Moon can bring new partnerships.

	As		SaR Mo Ra
Ve	**Jacqueline Du Pré** Jan. 26, 1945 11:30 GDT Oxford, England		
Su			
Me Ma Ke			JuR

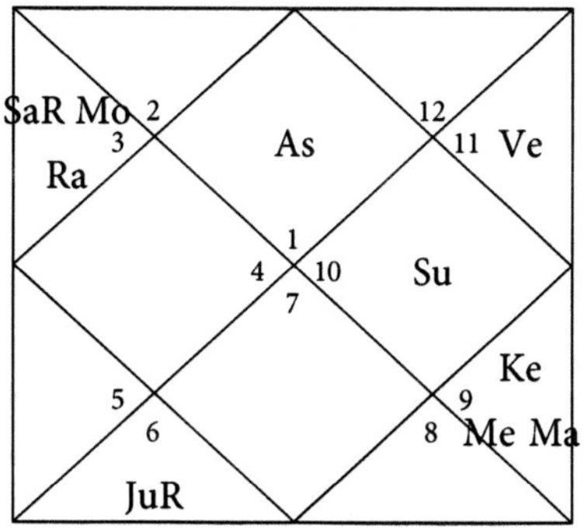

SaR Mo 2
3
Ra
As
12
11 Ve
1
4 10 Su
7
Ke
5
9
6
8 Me Ma
JuR

Planetary Information

As	03:59	Ari	Ashwini
Su	13:02	Cap	Shravana
Mo	21:20	Gem	Punarvasu
Ma	22:25	Sag	P.Ashadha
Me	22:20	Sag	P.Ashadha
Ju	04:05	Vir	U.Phalguni
Ve	29:44	Aqu	P.Bhadra
Sa	12:03	Gem	Ardra
Ra	24:22	Gem	Punarvasu
Ke	24:22	Sag	P.Ashadha

	Mo Su	As Ra	Ve
JuR	**Navamsha**		
SaR			
	Ke	Me Ma	

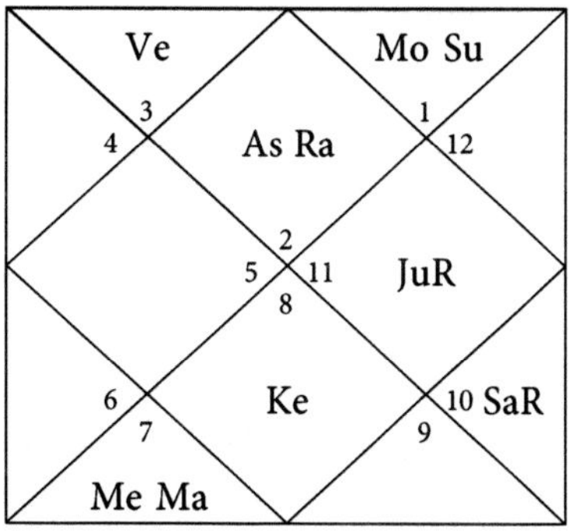

Ve
Mo Su
3
4
1
As Ra
12
2
5 11 JuR
8
6
Ke
10 SaR
7
9
Me Ma

Vimshottari Dashas

Ju	Jun 17, 1943
Sa	Jun 18, 1959
Me	Jun 17, 1978
Ke	Jun 18, 1995
Ve	Jun 18, 2002
Su	Jun 18, 2022
Mo	Jun 17, 2028
Ma	Jun 18, 2038
Ra	Jun 18, 2045

A famous chamber music trio was formed in 1967 when, at age 22, Du Pré joined two Israeli musicians – violinist Pinchas Zukerman and conductor-pianist Daniel Barenboim.[105] She married Barenboim in Jerusalem on June 15, 1967, after a whirlwind courtship from their first meeting on New Year's Eve 1966. They flew to Jerusalem shortly after the Six-day War, and Du Pré converted to Judaism in one day.[106] They were a celebrated couple, dazzling the music world with their charisma, passion and musicianship. The years from 1967 to 1970 were a personal and artistic triumph for them both.

Prior to that, from 1963 to 1965, Du Pré suffered from self-doubts about whether to continue to play the cello. By 1966, while studying with the great Mstislav Rostropovich in Moscow, she received great encouragement that temporarily lessened her doubts about her career. At the end of her study with him, Rostropovich (1927-2007) declared that she was:

> "the only cellist of the younger generation that could equal and overtake [his] own achievement."[107]

---

[105] Daniel Barenboim was born Nov. 15, 1942, 11:35 AM, Buenos Aires, Argentina. Barenboim's parents were Russian-Jewish, and the family emigrated from Argentina to Israel in 1952.

[106] The Six-Day War occurred June 5-10, 1967, and was fought between Israel and its neighboring states - Egypt, Syria, and Jordan. At the end of the war, Israel gained control of the Sinai Peninsula, Gaza Strip, West Bank, East Jerusalem, and the Golan Heights. Intensified sentiments for and against Israel prevailed in 1967, and the result of this war continues to have a radical effect on geopolitics over 40 years later. Though converting to Judaism overnight is considered highly unusual, special arrangements were made due to the couple's celebrity status, the pressures of the war era, and of their concert schedules, which they cancelled suddenly to make the wedding trip to Israel. While there, they performed various benefit concerts.

[107] Elizabeth Wilson, "Jacqueline du Pré: A 60th year Anniversary celebration," *BBC Music Magazine*, Feb. 2005, pp. 22-26. (Also a cellist, Elizabeth Wilson was a student of Du Pré.) **http://www.jacquelinedupre.net/memorabilia/bbc-mag_200502/source/bbc_mag_jackie_5.htm.**

In the midst of the same period – in 1964 – she first noticed some periodic numbness in her hands. By the spring of 1971 she took time off from her hectic concert career to recoup from her physical and mental exhaustion. To family members she was clearly suffering a nervous breakdown through 1971-1972. Though her symptoms were early warnings of multiple sclerosis, they were repeatedly misdiagnosed as "psychosomatic stress" related to the challenges of her concert career. Du Pré returned to a full international performance schedule by early 1972, though increasingly in 1972-1973 she was experiencing more numbness in her hands as well as dizzy spells and a serious loss of mental and muscular control. After several inconclusive medical tests from February 1973 onward, her condition was at last correctly diagnosed in October 1973 as multiple sclerosis. Soon after that she retired abruptly from the concert stage, as the disease was progressing rapidly. She was almost 29 years old. Her last studio recording session was in December 1971, though no one knew it at the time.

After leaving the concert stage she gave master classes over the next few years. But her withdrawal from public life soon became complete. She slipped into further physical and emotional decline, and after many years in seclusion, died Oct. 19, 1987. She suffered greatly from her disease and from extreme emotional swings and profound imbalances in her personality. Jacqueline had been almost deified as a musician, and together with her husband, treated as musical royalty.

In 1997 her two siblings Hilary and Piers published a book called *A Genius in the Family*, recounting her life as they remembered it. The book was soon made into the movie *Hilary and Jackie* (1999). They created a storm of controversy in breaking the myth around the woman and artist. It made much of the sibling rivalry between Jacqueline and her older sister, Hilary, a flutist. It also delved into Jacqueline's emotional and mental turbulence and the effect of her demanding career on her family and loved ones.

Other biographies were published, as well as numerous articles, most of them by friends of the cellist, the first being Carol Easton's *Jacqueline du Pré: A Biography*, 1990. Another was Elizabeth Wilson's *Jacqueline du Pré: Her Life, Her Music, Her Legend*, 1999. Extensive in its coverage and intent to be the definitive Du Pré biography, it was written with the full support of Jacqueline's husband Daniel Barenboim.[108] But in her attempt to counter the siblings' book and the film based on it, Wilson's biography lacks objectivity in overly defending its subject, giving her mythical stature in the process. Wilson represents those who feel Du Pré should be remembered only as a great musician, ignoring other details.

An astrological analysis of her birth chart describes the controversy surrounding the reputation of Jacqueline Du Pré. It supports the siblings' story of their sister's mental and emotional imbalance, perhaps also accounting for some of the emotional extravagance in her performances, in spite of her musical greatness. Their point of view comes closer to acknowledging the intense complexity of their sister, as they grew up in a household revolving for many years around the cellist and her career. Their mother Iris both inspired and goaded the sisters to compete with each other, and

[108] Upset about the Du Pré siblings' book and subsequent film, Barenboim said: "Couldn't they have waited until I was dead?" Barenboim travelled often, and was music director of the Orchestre de Paris from 1975 to 1989. He lived in Paris from 1974, eventually with Russian pianist Elena Bashkirova, whom he married in 1988. Their two sons were born in 1983 and 1985. Though he essentially abandoned Du Pré in London, Barenboim arranged for all her care as an invalid for the last 14 years of her life up to the time of her death Oct. 19, 1987. The Du Pré family learned of his "other life" in Paris several years after Jackie's retirement – a fact he tried to keep hidden from his wife. He may have succeeded, as her mental clarity deteriorated.

she tended to ignore the loser – including Jacqueline herself when she lagged behind her sister musically and again after she became ill. The accountant father was distant, and brother Piers eventually opted out of music altogether to become an airline pilot.

Jacqueline Du Pré was made Officer of the Order of the British Empire in 1976, and left behind a rich legacy of her recorded performances.[109]

Fierce and passionate warrior-musician: The Ascendant contains *Ashwini nakshatra*, giving the initiative and the undaunted spirit also reflected by natal Mars. Mars is the *Kshatriya* (warrior), its fiery and competitive passion increased by being in a fire sign and a *rajasic nakshatra* (*Purva Ashadha*, owned by Venus). Mars is in an excellent house (9th house) and sign (Sagittarius) for the Aries Ascendant chart. However, Mars is also within one degree of Mercury, creating a *Graha yuddha* (Planetary War).[110] Since Mars loses to Mercury – its planetary enemy – Mars is fighting for dominance from a weakened position. It does not give up easily and often behaves badly in the process. Since Mars is the weakest planet in the chart, and is both Ascendant lord and *karaka* for siblings, several things are described immediately: **1)** The tendency for the person not to feel seen, and to fight to be seen in their realm of life, whether personally or professionally; **2)** The tendency to suffer problems in physical health, especially related to the nervous system or the muscular system, realms of Mercury and Mars, respectively: **3)** The tendency for the siblings to feel disempowered or overpowered by their sister – in this case by her forceful personality and her career success; **4)** The tendency to focus on 3rd and 9th house matters (3rd house – music, and 9th house – long distance travel) excluding all else.

Woman merges with cello: The child who asked for a cello at age four soon became wed to her instrument, as musicians often are. In her birth chart, the same concentration of planetary energy that describes **her musical gifts** also describes the **many layers of controversy** surrounding her, including the mental and physical devastation that would eventually pull her down from the fullness of her life as a concert cellist. Du Pré possessed a maniacal combination of planetary energies: Saturn-Moon-Rahu in Gemini opposite Mercury-Mars-Ketu in Sagittarius. All nine planets in the chart are linked to these six planets through either house lord or *nakshatra* lord. Further, music is assigned to the 3rd house and the sign of Gemini. It is strong as its house lord Mercury aspects its own house.

The problem for the Aries Ascendant person is that Mercury here is far stronger than Mars, and Mercury is the planetary enemy of Mars, the warrior who fights to the end. Three planets are in the 3rd house, including Moon; but the indications are that suffering would be immense, along with her ebullient passion for life. Receiving aspects from four classic malefic planets,

---

[109] Some of these can be seen in Christopher Nupen's film ***Remembering Jacqueline Du Pré***, 1994, 56 min. Nupen holds 95% of all film archive of Du Pré in performance. In 2001 he said: "It is a sad fact that ever since illness touched her, at the age of 28, the public image of her personality has been progressively distorted, particularly by the fiction film ***Hilary and Jackie***."

[110] *Graha yuddha* is defined in Chapter 2 and in the Glossary. Another Mercury-Mars example is Chart #26 (Barbara Walters). Mercury wins over Mars, but as Mercury is Ascendant lord for the Virgo Ascendant chart, her physical and mental health were not impaired. She has maintained an active career past the age of 80.

Mercury and Moon show emotional and mental imbalances. Nor are they as stable when situated in a *kendra* from each other in dual signs. Together Saturn, Moon and Rahu demand extreme discipline and hard work, and for the same reason can produce heightened emotional impact – especially opposite Ketu-Mars-Mercury. This is due to the inherent disequilibrium of Saturn's rigidity exerting too much control over Moon (mind), which needs to be freer. Saturn presses and focuses, while Mars pushes back; thus the pendulum swings to the extremes. Even so, the interplay of Mars and Saturn can also produce fine technical proficiency in some area.

Ebullience is described by the strong interconnection between **Moon, Mercury and Jupiter**. Much of the circuitry between Moon and Mercury can be traced back to Jupiter in Virgo in the 6th house of conflict. Moon is situated in *Punarvasu nakshatra*, owned by Jupiter. Moon is also in Gemini, the sign owned by Mercury; Mercury in turn is in Sagittarius, owned by Jupiter in Virgo – the sign owned by Mercury. A *Parivartana yoga* (mutual sign exchange) exists between Mercury and Jupiter, linking their energies and functions. Saturn provides the element of discipline with its strong involvement on the Gemini-Sagittarius axis. But just as strong is the expansionist and publicity-prone component coming from Jupiter and the planets in Sagittarius, setting up a contradictory desire to escape restrictions imposed by the demands of being a world class cellist, and before that a family background in which "… emotional censorship [was] a way of life."[111] This was at the root of the emotional extravagance, self-indulgence and lack of control reported in some of her performances. It also accounted for the attractiveness of her buoyant personality and the publicity she received before and after her untimely demise.

Layers of intensity and controversy within her life can be seen through the close concentration of planets between the 3rd and 9th houses. Emphasis on either Sagittarius or the 9th house can bring religious and/or ideological zeal into the life. When there is 6th house involvement with the 9th house, a person often leaves the religion of the family of origin, creating potential conflict in this arena of life. In this case, Jupiter (9th house lord) is in the 6th house of conflict in the sign of Mercury. As winner of the *Graha yuddha* with Mars, Mercury can bring extremes. Her **overnight conversion to Judaism** in June 1967 was undertaken in dramatic circumstances in Israel right after the Six-Day War, and shortly before her marriage to Barenboim. Though proud of her conversion her entire life, it was not well received by her parents. She described her mother as "an anti-semitic Christian."

The sibling rivalry with older sister Hilary appears to have been a compelling factor throughout her life, though they were close. In the early years of childhood, flutist Hilary was considered the superior musician and performer. Jacqueline eventually won this rivalry. Later she yearned for the stability she saw in her sister's marriage and family life, and set about competing in that arena as well. **Saturn and Rahu** are the two malefic planets that form a *Papa Kartari yoga* to her Moon, hemming it in on either side. One can feel confined by the circumstances of one's environment, including with siblings. These same planets also aspect her Ascendant lord Mars. Saturn is lord of 11th house of elder sibling and Mars is *karaka* for siblings.

While separated from her husband in 1971-1972, Jacqueline carried on **a 16-month affair with her brother-in-law** Christopher (Kiffer) Finzi, a musical conductor who was radical in his opposition to the life imposed on a professional musician. This also suited his hippie-like orientation, in which

[111] Carol Easton, ***Jacqueline du Pré: A Biography***, 1990, p. 126.

he and Hilary retreated to their pastoral family setting, with their four children and a chicken farm. Their **unorthodox attempt to help** the struggling cellist regain her balance was conducted with wife Hilary's acquiescence. It probably did not help Jacqueline achieve that end or heal the rift with husband Daniel, not only due to her extreme emotional vulnerability, but due to Kiffer's fierce opposition to the trappings of her professional concert career, which included extensive musical collaboration with her husband. Along with the terrifying and mystifying symptoms not yet diagnosed as multiple sclerosis, this affair had to bring some added confusion.[112]

Jacqueline's **ardent focus on music** was always intense: From age eleven she began a grueling practice schedule, departing from both a regular school life and thus a normal childhood. This coincided with the **first Mars sub-period** of her life, in this case Jupiter-Mars *Dasha/Bhukti*.[113] Due to her weakened Mars she was compelled to work much harder whenever Mars was featured in a planetary cycle. Beyond the world of music, for most of her life she knew little else, her general education and culture being otherwise limited. She preferred a more narrow music repertory that played to her emotional style. She also preferred to perform herself and listen to her own performances rather than discover new music or musicians. Her husband Daniel Barenboim greatly broadened her perspective in this regard.

Another way of seeing this is that Moon (lord of 4th house of higher education) is in the 12th house from the one it owns, giving the potential for loss or destruction to 4th house matters, including family life along with higher education. Separative planets Mars, Saturn, Rahu and Ketu also aspect this Moon, separating the person from further education and confirming problems in the life of the family of origin – all 4th house matters. Since the Moon also symbolizes the mother, this naturally creates difficulties for the mother and sets up the emotional separation between them that evolved when Jacqueline was forced to abandon her concert career. The birth chart shows the extent to which Jacqueline and her sister (also Mars – sibling) embodied their mother's ambitions. She was also their first teacher and inspired them towards music and musical careers.

Music and the love life: Music and string-playing are seen from the 3rd house and from Venus as *karaka* of music (and love matters). **Venus** also describes the marriage partner, as lord of the 7th house in the birth chart. The **3rd house** in the birth chart is ruled by Mercury aspecting its own house, giving strength to that house but also contention, as it is in a Planetary War with Ascendant lord Mars. This puts tremendous pressure on the entire physical, emotional, and mental equilibrium. Venus is strong here as *Atmakaraka* (planet in the highest degree of a sign) in the sign of a friend, giving the power to bring benefits from Venus-ruled activities, especially financial rewards since it provides a *Dhana yoga* of wealth. This benefit is confirmed in the *Navamsha* chart, with Venus in Gemini in the 2nd house of income. But since Venus is also in a *Parivartana yoga* (mutual sign exchange) with *Navamsha* Mercury in Libra the 6th house, there is some inner and outer conflict about her artistic endeavors and love life (both Venus) and some turbulence in the family that could arise from her personal and professional life. The 6th house also concerns health issues, and both *Navamsha* Mercury and Mars are placed there.

---

[112] Hilary's daughter Clare Finzi later disputed her mother's account of the affair, claiming her father made a habit of seducing other women openly in their home, and that her mother Hilary was alternately passive and furious about the situation. She said her parents took advantage of her aunt when she was very vulnerable.

[113] Jupiter-Mars *Dasha/Bhukti* ran from Feb. 16, 1956 to Jan. 22, 1957. The *Maha Dasha* (major period) of Mars did not occur in her lifetime.

Even so, Venus was bound to bring benefits during its major or minor period. During the **Venus sub-period** of Saturn *Dasha* (April 8, 1966 to June 8, 1969), Du Pré experienced her first meeting, courtship, and subsequent marriage to husband Daniel Barenboim, followed by some of their best and most successful years together personally and professionally. Barenboim himself was a musical child prodigy, with his first public piano concert at age 7 in Buenos Aires, followed by his debut as a pianist in Vienna and Rome in 1952, in Paris in 1955, and in London in 1956. He has gone on to have a long and distinguished career in music, remaining prominent in the field. His versatility as pianist, conductor, and educator are shown here by *Navamsha* Venus in Gemini, along with the conflict his marriage with Du Pré would bring, especially in regard to her deteriorating health and the mysterious and disturbing conditions that preceded her eventual diagnosis. Venus escapes the difficulties of most other planets here, just as Barenboim was able to escape to Paris in the end. But it was Du Pré who initiated their first separation in 1971-1972.

Mental and emotional affliction: Rahu can eclipse the Moon, so when placed close to the Moon or anywhere in the same sign – the closer the more conclusive – mental and emotional stability are not easy to achieve. With two malefics hemming in her Moon in a *Papa Kartari yoga*, the person may experience both the heavy discipline of Saturn and the emotional extremes of Rahu, including a sense of confinement. Nor is the *Navamsha* Moon well situated in Aries in the 12th house with aspects from Sun, Mercury, and Mars. Especially when afflicted, Moon in the 12th house can manifest as fearful and full of worries. This is less appropriate for a public career, as with this Moon position one needs more privacy and time for introspection. The exception is if one is living and working abroad, or in a 12th house setting such as a pastoral venue, a prison or a hospital. (Her last 14 years were spent in virtual confinement, ignored by her mother.) When in the 3rd house, natal Moon often brings major issues revolving around courage, and confusion about when to take action and when not to take action. There is a lot of sexual energy in Gemini, but in this case it can create wild diversions and/or abnormal repression due to extreme career demands.

Further, there is an adverse effect on mental and emotional equilibrium due to Moon and Mercury both in dual signs in mutual *kendras*, receiving aspects from four classic malefic planets. This potential for greater emotional turbulence can adversely affect partnerships – whether romantic, artistic, or business.

The mother (Moon) was a musician and early teacher to the siblings. Her own fierce musical ambition was embodied in her daughters. She both inspired and goaded them to compete with each other, Hilary, the flutist, being the early winner. When Jacqueline was forced to retire from her concert career in late 1973, her mother found this new reality unbearable, and soon distanced herself completely from her daughter's life.[114]

Comparison to Mia Farrow chart: Born exactly two weeks prior to Mia Farrow, Du Pré shares several of the same astrological factors: Both are born mid-day on a Friday (Venus-ruled day) with natal Sun at the top of the sky in Capricorn, and Aries rising. The **Jupiter-Saturn conjunction**

---

[114] Guyanese-born Ruth Ann Cannings was hired as Du Pré's nurse for the last 11 years of her life, the last 8 years as live-in nurse. She was a puritanical evangelical Christian whom Jacqueline secretly called "Purity," and struggled under her dominance while increasingly losing her powers. Ruth Ann believed her patient was being punished for converting to Judaism, and did not hesitate to say so. Jackie's parents had never been comfortable with her religious conversion. And her mother saw her daughter's disease as divine punishment for her sexual promiscuity. (Source: Carol Easton, ***Jacqueline du Pré: A Biography,***1990.)

in the 10th house **Feb. 18, 1961** would trigger major career success for each of them. Du Pré's London debut at Wigmore Hall was in March 1961. Farrow's New York theatre debut was within two years of that and was delayed until her more powerful Moon *Dasha* began March 22, 1963 – according to her *Dasha* scheme.

While Farrow overcame childhood polio, with strong physical stamina well into her 60s, and a long career spanning 40-plus years, Du Pré suffered deteriorating health and loss of emotional/mental/physical equilibrium by her late 20s. Self-doubts plagued her for at least ten years before she was forced to retire from the concert stage. With the exception of Moon, Mercury, Venus, and Mars, all their planets are in the same signs and houses. Differences in the condition of Ascendant lord Mars are critical, as are their respective Moon positions. Due to the natal Moon's *nakshatra* position, Du Pré's *Dasha* sequence **(karmic timing)** is less favorable, as most of her adult life is spent in *Dashas* ruled by planets unfriendly to the Ascendant lord. Though Jupiter is an excellent planet for Aries Ascendant, and is the first *Dasha* up to June 18, 1959 (birth to age 14½), it will also bring conflict and/or health issues of some kind due to its 6th house position in the birth chart.

Moon and Mercury determine emotional and mental strength**.** Though Farrow's Moon suffers some affliction due to aspects from Saturn, Rahu and Ketu, it is better situated in her *Navamsha* chart. Likewise, Du Pré's natal Moon is located on the Rahu-Ketu axis, but 180 degrees opposite Farrow's Moon, and hemmed in between malefics Saturn and Rahu in Gemini, thus much more afflicted. Du Pré's *Navamsha* Moon in the 12th house is also not well situated for a public career or for emotional stability, unless one is more spiritually inclined.

Because it rules over the 6th house in the Aries Ascendant chart, Mercury can cause health problems and conflicts for Mars, including with siblings. This is especially true with natal Mercury-Mars contacts, called an ***Arishta yoga*** (the mutual association of Ascendant Lord with *Dusthana* lord). When it occurs, the situation can become worse during **Mercury *Dasha* or *Bhukti*.** In Saturn-Mercury *Dasha/Bhukti* (June 20, 1962 to Feb. 27, 1965) Du Pré first experienced numbness in her hands, and problems with lack of confidence. Farrow also has this same *Arishta yoga*, but it is corrected by: **1)** exalted Mars *Digbala* in the 10th house (best possible house placement); **2)** a close aspect of Jupiter to Mars; and 3) by a wider orb between Mercury and Mars (10 degrees apart), placing them in different *nakshatras*, and in different houses in the *Navamsha* chart. Farrow's Mercury *Dasha* also does not come until Feb. 2033 – at age 88, while for Du Pré, it came in June 1978.

Time of reckoning: Saturn is the planet of *Karma*, and its return by sign is one of life's moments of reckonings, every 28-29 years for all of us. On June 11, 1973 tr. Saturn entered Gemini, and during this year both Farrow and Du Pré experienced the Saturn return to natal position in very different ways. Farrow had the advantage of higher than average physical strength and stamina due to the superior condition of her Ascendant lord Mars. She also began her **Mars *Dasha*** in 1973, with Mars truly bearing its fruit. (She was three years into her second marriage, with several children and an active career.)

Du Pré was in **Saturn *Dasha* from June 18, 1959 to June 17, 1978**. This 19-year *Dasha* encompassed the entire rise of her extraordinary concert career, from early 1961 up to her enforced retirement in late 1973. (This also coincided with a full 12-year Jupiter orbit around the Sun.) But as *Dasha* lord, Saturn had more capacity to cause affliction to her physical and mental vehicle than in Farrow's case, largely because natal Saturn aspects natal Mars in the chart of Du

Pré, afflicting her Moon, Mercury and Mars – all planets of emotional/ mental/physical health and well-being.

> "[Jacqueline Du Pré is like]...the lightning passage of a comet which, with remarkable intensity — but all too briefly — illuminated our lives."
>
> **Zubin Mehta**, Conductor

Multiple sclerosis: With multiple sclerosis, Du Pré suffered serious afflictions to her mental and emotional stability along with the known severe affliction to the nervous system. With this disease, the myelin sheaths protecting the central nervous system are damaged or destroyed. It is characterized by speech defects and loss of muscular coordination. "Sclerosis" is an abnormal hardening of body tissues or parts, especially of the nervous system as in this case. **Saturn** has the effect of contracting, tightening or hardening, and is located in the 3rd house associated with the nervous system. Saturn aspects or contacts five other planets, including Mercury and Mars, as described earlier. The years encompassing Du Pré's debut, stellar rise and precipitous fall are all contained within **Saturn *Dasha***. Saturn moves slowly, taking time to reveal itself. We have also noted how Saturn presses and focuses, and Mars pushes back. With multiple sclerosis, experts describe even a type of "clinical euphoria." This could contribute to wide mood swings in attempting to deal with reality as opposed to expectations. Moon and Mercury's close circuitry with Jupiter also has the effect of encouraging optimism.

Ascendant lord Mars is in the 9th house in Sagittarius, joining the number of malefic planets already afflicting both Moon and Mercury, and is itself afflicted by Saturn, Rahu, and Ketu. As described above, natal Mars and Mercury form an ***Arishta yoga*** (weakening the physical health unless modified), and also – a ***Graha yuddha*** – a Planetary War in which Mercury wins over Mars.[115] When Mars loses and is also Ascendant lord, it is especially challenging for the physical health and self-esteem. This is a situation shared by the famous opera singer **Maria Callas.**[116]

Reviewing ***Graha yuddha***, the nervous system (Mercury) is being attacked by Mars (the muscular system), and vice versa, so in the process both of them are weakened, even if Mercury wins. *Karakas* for the nervous system include chiefly Mercury, but also Ketu, 3rd house, and Gemini, the natural 3rd house. Mercury and Jupiter are in *Parivartana yoga*, further reinforcing Mars, but from the 6th house of disease and controversy. The major and minor *Dasha* periods of **Mercury** are bound to be problematic. Her **Mercury *Bhukti*** in Saturn *Dasha* ran from June 20, 1962 to Feb. 27, 1965. In 1964 she experienced her **first symptoms** of numbness in the hands as well as serious problems with lack of confidence.

---

[115] Although Mars is slightly higher in celestial longitude in this instance, Mercury wins the Planetary War as it is 1) brighter in the heavens, 2) moving faster, and, 3) further north by declination and latitude. Mars is the one exception in *Graha yuddha* where the losing planet is also the larger planet, as with Mercury. In an odd paradox for the warrior planet – it always loses in Planetary War and does not give up any battle easily. Mars often behaves badly as the loser, especially as Ascendant lord. (See also Glossary and Chapter 2)

[116] Both charts have Mars as Ascendant lord losing the Planetary War. (The natal Mars of Callas loses to Saturn in the 12th house.) Compensating for Du Pré is Sun at the top of the sky, and for Callas – Sun exactly rising in Scorpio. This allowed each of them to rise to prominence career-wise, but they dealt with the ongoing feeling of not really being seen, despite much public adoration and obvious indications to the contrary. It made it difficult for each of them to sustain the full-blown career visibility due to the inner and ultimately outer fragility.

For several years there was ongoing **misdiagnosis** of the disease. Du Pré was told she was suffering from nervous exhaustion and with symptoms of a nervous breakdown in 1971-1972. Doctors continued to dismiss her symptoms as psychosomatic, and though personable, she was dramatic and idiosyncratic. **Ketu** can indicate wrong diagnosis or confusion with diagnosis, notably when in contact with the Ascendant lord, Sun or Mars. In this case, Mars and Ketu are just three degrees apart and in *Purva Ashadha nakshatra.*

Her enforced retirement came in late 1973 in **Saturn-Rahu** *Dasha/Bhukti*. She lived another fourteen years, in a steady downward spiral mentally and physically. Her mental and emotional equilibrium decreased dramatically with the years, especially during the eleven tortuous years of her **Mercury *Dasha***. Du Pré had a quick wit (Mercury-Mars), and for many years a strong sense of self-discipline and musical technique (Mars-Saturn), but the Mercury-Mars *Graha yuddha* indicates that the 17-year Mercury *Dasha* would bring further rapid deterioration of her health. Additional emotional restrictions were due to her virtual isolation, and with a nurse who, although devoted, was very judgmental of her patient. Mercury *Dasha* began June 18, 1978. Especially challenging for her survival would be **Mercury-Mars** period from Dec. 17, 1986 to Dec. 14, 1987. She died Oct. 19, 1987, just eight days before the end of **Mercury-Mars-Venus** period: Aug. 28, 1987 to Oct. 27, 1987. Her natal Mercury and Mars are placed in the Venus-owned *nakshatra* – ***Purva Ashadha*, "the earlier victor."**

Below is a list outlining the astrological timing factors that would adversely affect Du Pré's already troubled Ascendant lord Mars, starting with tr. Mars in 1971, as tr. Mars resonates strongly for a Mars-ruled chart:

1. **Tr. Mars in Sagittarius**, March 2, 1971 to April 22, 1971. Returning to its natal position, tr. Mars heats up an already overactive natal Gemini-Sagittarius axis. **Tr. Mars in Capricorn,** April 22, 1971 to Oct. 25, 1971 (six months), with Mars retrograde July 11, 1971 to Sept. 9, 1971. Mars exalted in Capricorn as well as retrograde acts as if in debilitation. This combination would exacerbate Du Pré's tendency to be hyperactive, while needing to rest from her exhaustion and veritable nervous breakdown. Tr. Mars in Capricorn coincides with tr. Rahu-Ketu in Capricorn-Cancer, lending Mars even more raw and uncontrollable energy.[117]

2. **Saturn-Mars** *Dasha/Bhukti*, Dec. 20, 1971 to Jan. 28, 1973. The major or sub-period of the Ascendant lord places attention on the person's physical health and well-being. During 1972 Du Pré experienced the first serious indications of multiple sclerosis. Her last recording studio session took place in late Dec. 1971.

3. **Saturn-Rahu *Dasha/ Bhukti***, Jan. 28, 1973 to Dec. 5, 1975. Continued focus on the afflicted Gemini-Sagittarius axis in the natal chart, including on Asc. Lord Mars, and on Mars-Mercury. Multiple sclerosis was finally diagnosed in Oct. 1973. This coincided closely with tr. Saturn SR Oct. 17, 1973. (See #10 below)

4. **Tr. Saturn in Gemini**, June 11, 1973 to July 24, 1975. Tr. Saturn is on the Gemini- Sagittarius axis for the first time since the Jupiter-Saturn conjunction Feb. 18, 1961, marking the

[117] See also Chart #11: Albert Einstein (Chapter 5: Gemini Ascendant). His natal Mars and Rahu are in Capricorn. Normally a Mars return is beneficial for a Mars-ruled Ascendant chart, except when Mars is weakened by *Graha yuddha* or other severe afflictions, causing Mars to overreact.

start of her meteoric 12-year career. Tr. Saturn in Gemini indicates the 2nd third of her ***Sade Sati*** (tr. Saturn in houses 12, 1, and 2 from natal Moon), thus very serious pressures on her at all levels, especially emotionally and mentally. Her *Sade Sati* began April 28, 1971 with tr. Saturn in Taurus.

5. **Tr. Jupiter in Capricorn,** Jan. 25, 1973 to Feb. 10, 1974. Jupiter returns to Du Pré's 10th house, where it was at the start of her career. Unfortunately, due to its debilitation in Capricorn – without an aspect or an exchange from Saturn, it is powerless as greater benefic to be helpful, especially to protect her physical and mental status. Nor is it aspecting her Ascendant, Asc. lord Mars or Moon.
6. **Tr. Rahu-Ketu in Sagittarius-Gemini,** Sept. 27, 1972 to March 19, 1974. Through emphasis, this shows further affliction to the natal Gemini-Sagittarius axis, true of all items on this list, except #5, which cannot soften these effects.
7. **Total Solar eclipse** on natal Mercury-Mars-Rahu-Ketu at 25:08 Gemini, July 10, 1972.
8. **Solar eclipse** on natal Mercury-Mars-Rahu-Ketu at 20:40 Sag., Jan. 4, 1973.
9. **Total Solar eclipse** on natal Saturn at 15:02 Gemini, June 30, 1973.
10. **Tr. Saturn Stationary Retrograde** at 11:16 Gemini, closely conjunct natal Saturn (also *Dasha* lord), Oct. 17, 1973, with natal Moon and Rahu also in Gemini. The transiting planet that is currently *Dasha* lord is more important during its *Dasha*. She was already in the midst of the difficult ***Sade Sati***. (See #3 above) With earliest symptoms of multiple sclerosis in 1964 and in greater severity from 1971, she was consistently misdiagnosed, including in Feb. 1973 tests. The correct diagnosis finally came in Oct. 1973, coinciding with tr. Saturn Stationary Retrograde. Du Pré retired soon afterwards from her concert career.

REVIEW OF ARIES ASCENDANT CHARTS

Give the astrological reasoning in each case.

1. What factors in Martin Luther King's chart show his capacity to be compassionate as well as daring? Why would he be more likely to choose the ministry over the law?
2. What about King's Mars position indicates he would have an unusual and powerful way of speaking? How could both his style and content of speech bring him trouble?
3. What single factor in Bobby Kennedy's chart shows that he would be likely to attract enemies and open conflicts with them? Hint: Check *Navamsha* chart for confirmation.
4. Why was the major or minor (sub-period) of Mercury *Dasha* more dangerous for both King and Kennedy? (King was killed in Mercury-Jupiter period, Kennedy in Jupiter-Mercury period. Why is Bobby Kennedy's speech potentially dangerous for him? Why would that tendency increase during his Jupiter *Dasha*?

5. Why would Henry Miller's natal Mars be comparatively less dangerous for his physical safety? (He attracted controversy, but it did not kill him or end his career.)
6. What astrological factors indicate Henry Miller would be more recognized professionally once he entered Venus *Dasha*? How did that help him win a landmark censorship case in the United States Supreme Court? (June 22, 1964)
7. What factors in Mia Farrow's chart indicate both physical vitality as well as professional visibility? Why did the Jupiter-Saturn conjunction of Feb. 1961 mark a major turning point in her career – as well as for all five biographical subjects in this chapter?
8. What in Farrow's chart shows her propensity to have many children, including adopted children, some with afflictions, and to endure a legal case at some point over them? (Court case against Woody Allen in 1992.) Hint: Study Jupiter, the *karaka* for children, as well as 5th house from Moon and/or Ascendant. Same applies to Bobby Kennedy.
9. What caused the physical body to be weaker in the case of Jacqueline du Pré?
10. How is it that Saturn's condition in her chart is such that – during its *Dasha* – it brought her both a wider audience for her musical talents *and* the effects of multiple sclerosis?
11. How does Mia Farrow's *Navamsha* chart indicate she would find greater emotional and mental equilibrium than Du Pré? That Henry Miller (and to some extent Martin Luther King) would be the most exuberantly optimistic?
12. What factors link all five Aries biographical figures in their breaking new ground in some way?

CHAPTER 4

Taurus Ascendant

Introduction

The Sanskrit name for Taurus is *Vrishabha*, which is literally "a bull." The second sign of the zodiac, Taurus is the first of three ***Artha*** houses, concerned with personal and material security at different levels. The ***Shudra*** class is associated with the earth element; thus those who serve by their labor to produce something tangible in physical terms. As a fixed earth sign, with the *nakshatras* therein concerned with growing and flourishing in various stages – we see the principle of acquisition and manifestation of results in its incipient stages. This forms the essence of the dharmic theme for the Taurus Ascendant chart.

For a **Taurus Ascendant** person, **Venus** is the dominant ruling planet – even though any planet in turn can be dominated by another planet that is more forceful by house position, sign and/or *nakshatra*, i.e., overall strength. Venus often lends a more gentle tone to the chart, as it is a classical benefic planet.[118] Basic questions for a Taurus Ascendant person are: What do I consider of lasting value? What can I create that lives on long after I die? What am I doing to manifest physical and social security for my loved ones and myself? How can I create something concrete that will be of ultimate significance in the world? And how am I contributing towards that with what I communicate? The latter is due to the 2nd house rulership over the face, the mouth, and the tongue, thus also speech and eating, or that which goes out of the mouth and that which comes into it. With a strong Taurus emphasis or Ascendant one can be very attracted to the comforts and luxuries in life and wanting to acquire them, especially when they are assigned high value and closely associated with prosperity and personal success.

---

[118] Exceptions occur when Venus is aspected by malefics or in a fire sign in its own *nakshatra*, emphasizing its *rajasic* (passionate) nature, or when Venus receives contacts or aspects from other *rajasic* planets – Sun and Moon. Much will also depend on the condition of the mind (Moon) in the chart.

Such a person often creates both artistic *and* commercial success, so strong is the aspiration to manifest in material reality. Taurus and early Gemini and the *nakshatras* therein of burgeoning growth are associated with ***Lakshmi*** – the goddess of wealth and abundance, especially in India, but also associated with fertility and the sensual pleasures. She is considered an extremely important aspect of Venus. Physical beauty is important to Taurus, and *Rohini nakshatra* – in the middle of Taurus – is a testimony to that.[119]

Material and/or social security in some form is a major motivational force for Taurus. In addition, the second sign of Taurus as well as the second house in any chart is also the house of *Kutumba*, Sanskrit for happiness in the family of origin. Such matters may be emphasized in the life, along with the pleasures of Venus. Venus is the planet of desire and rules over matters of love, sensual pleasures and the arts. How well one fares in relationships, love or friendship, and in diplomacy in general depends on the condition of Venus in the chart, though the Moon's condition is always pivotal in assessing one's emotional and mental balance. Being a fixed earth sign, Taurus may be good for longevity and staying power. If many factors coincide, it can contribute to a long life and the establishment of something lasting. The Sanskrit word for the fixed sign is *Sthira.*

Venus rules over both the Taurus Ascendant and the 6th house (Libra). We have said that as a classic benefic planet Venus is a relatively gentle planet. But it has its more fierce side, especially in its desire nature, and also as the attractor of enemies when lord of the 6th house. As planetary lord of Taurus Ascendant, Venus's dual role is comparable to that of Mars for Scorpio Ascendant, since it too rules over both Ascendant and the 6th house (Aries). The Ascendant lord is always a benefic planet when ruling over the affairs of the first house (the physical body, and general well-being and good fortune); but as lord of the 6th house, Venus can attract enemies, as Mars does for the Scorpio Ascendant. However, Mars is by nature aggressive and competitive and gravitates to matters of the 6th house, whereas unafflicted Venus by nature is more peaceful and always suffers when associated with the 6th house or 6th sign of Virgo, where it is debilitated. Thus Venus suffers some innate antagonism or sacrifice as 6th house lord.[120]

The Sanskrit word for Venus is ***Shukra***, signifying "brilliant light," "heat," also the male reproductive fluid. *Shukra* is invoked for graciousness, refinement and good taste in life, for the art of loving and for all the arts – music, painting, poetry, drama, and dance. *Shukra* in Hindu mythology was also a man and at times a demon guru, not a human being, who had immense powers of persuasion, the ability to foresee the future and to revive a person who has died. His abilities as a seer were so strong that in one myth in the *Puranas*, he was blinded in one eye as a reprimand. Because of this, both Venus and the 2nd house are associated with eyesight, along with the two luminaries: Sun and Moon. The 2nd house also refers specifically to the right eye, whereas the left eye is the 12th house.

---

[119] See Chapter 2 under *Nakshatras*. The Puranic myth of *Soma* and *Rohini* describes why key planets in *Rohini nakshatra* may bring themes of jealousy from others.

[120] Houses 6, 8, and 12 are *Dusthanas*, and considered the most difficult of all the houses. When a *Dusthana* lord is in its own house or that of another *Dusthana*, this paradoxically creates one of the three *Viparita Raja yogas* of protection and power. But in spite of this powerful *yoga*, when 6th lord is in the 6th house, and that planet is also Ascendant lord – as for Taurus and Scorpio Ascendants – there can be some problematic 6th house issues for the individual. They can attract enemies and conflicts, but they also have the capacity to fight them and win. This is due to the planet's ultimate significance as Ascendant lord. With Libra Ascendant, Venus rules over Houses 1 and 8; and for Aries Ascendant, Mars rules over Houses 1 and 8. As 8th house relates to longevity, this does not diminish the condition of the Ascendant lord when situated in its own sign in the 8th house.

Shukra, as the son of the seer Bhrigu, agreed to instruct the *Asuras* (known also as the demons or titans in these mythical stories). They were often at war against the *Devas* (gods), who were advised by Brihaspati (Jupiter), towards whom Shukra had enmity. Shukra possessed a high level of powers, especially as sage and advisor to the *Asuras.* He understood that to rule or advise most effectively one needed to use charm and delicacy, and to know the hidden workings of the mind. In this context, we can see how *Shukra* was considered the inventor of political science. He was also the author of the *Shukra-niti,* a famous code of behavior and politics. On a lower level of Venus, we see glamour as a source of distraction, and further – the lure of sex and violence to hypnotize the masses. We understand how this works when we see how sex scandals or crimes can instantly command the attention of the masses, far beyond more serious or important events. At a higher level, Venus (or *Shukra*) can transform and inspire the ordinary mind to use artistic imagination for personal and collective upliftment. At the highest level Venus is able to transmute sexual power into spiritual power.

> "In contrast to the gods, the antigods are the inclinations of the senses which, by their nature, belong to the obscuring tendency, and which delight (*ra*) in life (*asu*), that is, in the activities of the life energies in all the fields of sensation."
>
> Sankaracharya, commentary on ***Chandogya Upanishad*** 1.2.1 [204].

In the Planetary Cabinet, Venus and Jupiter are both counselors. **Venus** was advisor to the *Asuras,* while Jupiter was advisor to the *Devas,* as described earlier. They share Brahmin status as the priestly class of planets, and both are the major classical benefic planets. But they are in different planetary groups, and while Jupiter is neutral to Venus, Venus is a planetary enemy to Jupiter. From sage Parashara we learn this about Venus:

> "Venus is charming, has a splendorous physique, is excellent or great in disposition, has charming eyes, is a poet, is phlegmatic and windy and has curly hair."

The phlegmatic and windy nature refers to the ayurvedic nature of Venus, corresponding to *Kappha* (water) and *Vata* (air), thus potentially both emotionally and mentally bright. With such love of appearance, style, and fashion, Venus can be a seductress in life, or – she can raise us up to new levels. Venus's placement in the birth chart is crucial to determine how her energy will be used – in which sign and *nakshatra,* in which house, and in what contact to other planets. Position in the *Navamsha* chart will also be critical in confirming the above.

Nakshatras: Taurus contains the following three *nakshatras*: ***Krittika*** (starting at 26:40 Aries, in the previous sign) 00:00 Taurus – 10:00 Taurus (ruled by the Sun), ***Rohini*** 10:00 – 23:20 Taurus (ruled by the Moon), and ***Mrigashira*** 23:20 – 30:00 Taurus (ruled by Mars). The Taurus Ascendant, or any planet situated in one of these three *nakshatras* is colored by being 1) in Taurus, ruled by Venus and 2) in one of these three *nakshatras,* ruled by Sun, Moon, or Mars, respectively; and 3) by aspects to the Ascendant and Ascendant lord Venus.

Mercury, Saturn and Rahu are all considered *Nadi yogakarakas,* and beneficial planets for Taurus Ascendant chart. Saturn is the best *Raja yogakaraka* for this Ascendant, as it rules both 9th and 10th houses from the Ascendant. The Moon is exalted in Taurus, maximum exaltation at 3:00 Taurus.

Chart 6 : Male Client

Birth data: Sunday, Jan. 7, 1945, 15:02 PST, San Diego, CA., USA, Longitude: 117W09 23, Latitude: 32N42 55. *Lahiri Ayanamsha*: -23:05:05. Class AA data, from birth certificate, reported to the author. Ascendant: 13:50 Taurus.

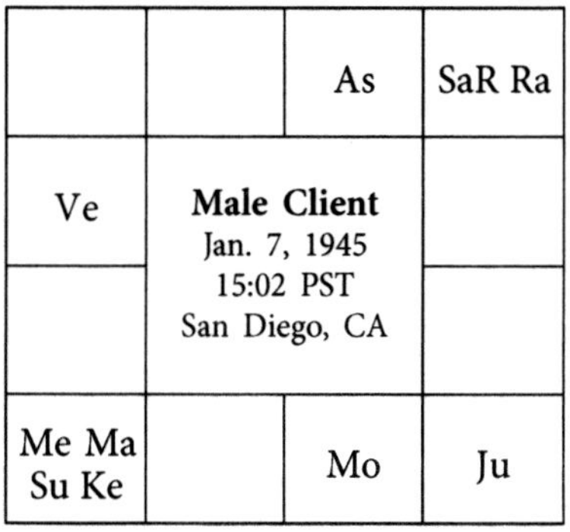

		As	SaR Ra
Ve	**Male Client** Jan. 7, 1945 15:02 PST San Diego, CA		
Me Ma Su Ke		Mo	Ju

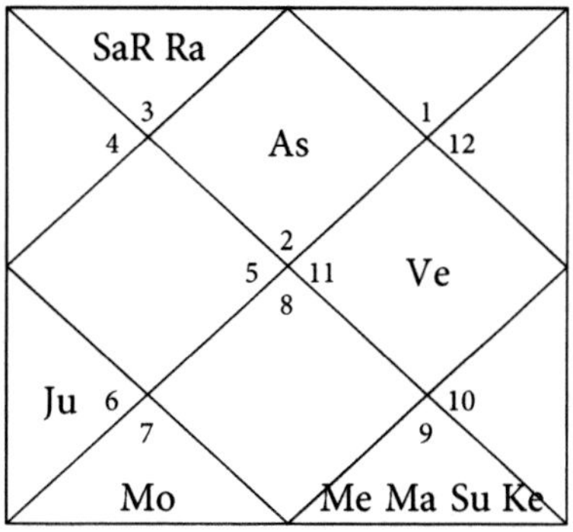

SaR Ra
3
4
As
1
12
2
5
11
Ve
8
Ju 6
10
7
9
Mo
Me Ma Su Ke

Planetary Information

As	13:50	Tau	Rohini
Su	24:11	Sag	P.Ashadha
Mo	09:19	Lib	Swati
Ma	08:29	Sag	Mula
Me	01:28	Sag	Mula
Ju	04:22	Vir	U.Phalguni
Ve	09:19	Aqu	Shatabhisha
Sa	13:25	Gem	Ardra
Ra	25:21	Gem	Punarvasu
Ke	25:21	Sag	P.Ashadha

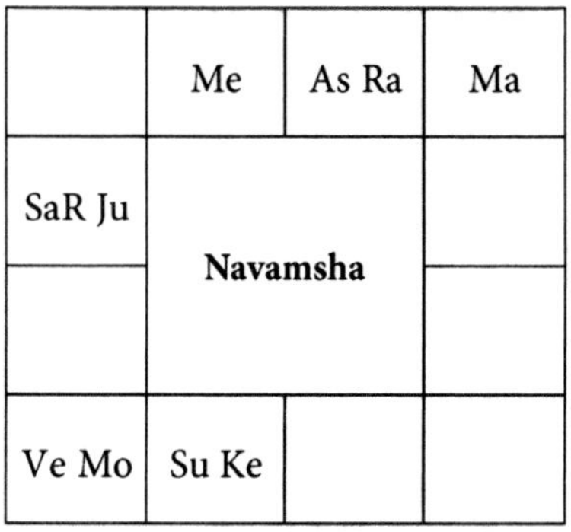

	Me	As Ra	Ma
SaR Ju	**Navamsha**		
Ve Mo	Su Ke		

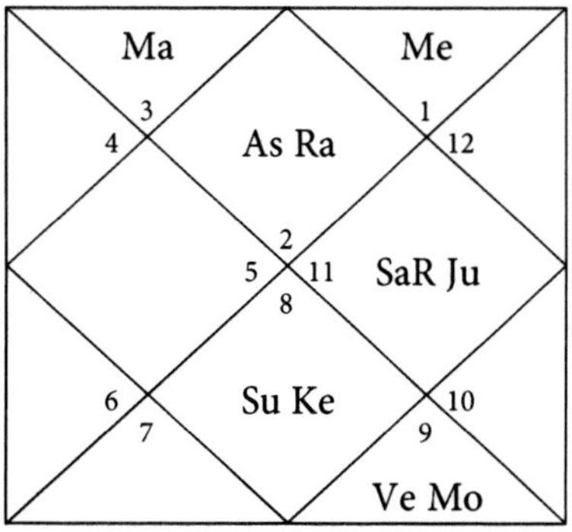

Ma
Me
3
4
As Ra
1
12
2
5
11
SaR Ju
8
6
Su Ke
10
7
9
Ve Mo

Vimshottari Dashas

Ra	Jun 4, 1941
Ju	Jun 5, 1959
Sa	Jun 5, 1975
Me	Jun 5, 1994
Ke	Jun 5, 2011
Ve	Jun 5, 2018
Su	Jun 5, 2038
Mo	Jun 5, 2044
Ma	Jun 5, 2054

Biographical summary: This man's chart is of particular interest because of the comparison to the two previous charts.[121] He is not a celebrity or someone in the public eye. He enjoys excellent health, physically and mentally, and has done so for much of his life. He spent five years in the Marine Corps (from 1963 to 1967) and served in the Vietnam War. Later he worked on ships of various kinds before settling down to marry and raise a family. He then worked for 33 years as a millwright – designing, building and installing machinery – not just in mills, but in various industrial and technological settings. This has been his major source of livelihood for most of his life, and involves heavy physical work requiring a high level of physical strength. It also appeals to his sense of fulfillment working with metals and materials in a very tactile way.

In his mid to late 50s, he had some joint and muscular issues due to his physically active life. Partly because of this, in 2002 he seriously considered a career change to a less physically demanding occupation and one that attracted him – career counseling of young people or a child counselor. He had already shown some talent for it. But since it would have involved more education

[121] The chart belongs to a male client of mine. He has given his permission for his data and life events to be used, as long as he remains anonymous.

and a sacrifice in retirement benefits, and was also against the wishes of his wife, who had already retired from her full-time job and career, he decided to remain working as a millwright. He was able to continue in this strenuous work up to the age of 62½. Since he was then eligible for retirement benefits he retired at that time. He was fortunate to be in good health, and this continued well into his retirement years, allowing opportunities for travel for himself and his wife.

His family life is stable, with one long marriage and two grown children. He has had a primary meditation and spiritual practice for over forty years, and his wife participates in that. In his retirement, he devotes much of his time and attention to spiritual practice, activities involving and supporting his place of worship, along with some foreign travel, often to a destination that has spiritual significance for him.

Comparison to Farrow and Du Pré charts: This chart is the third in a series of three charts (Chart examples #4, #5, and #6) that share many of the same planetary configurations due to their birth dates occurring within the same five-week period: Jan. 7 to Feb. 9, 1945. Saturn and Rahu are in Gemini, Ketu is in Sagittarius – with more planets joining this already demanding axis in each chart. This difficult planetary configuration on the Gemini-Sagittarius axis, shared also with Farrow and Du Pré, has been mitigated to a large extent by the change from Aries to Taurus Ascendant. Astrologically this makes the pivotal difference, as well as the natal Moon not being involved with the Gemini-Sagittarius axis, which contains five classic malefic planets in addition to Mercury. Nor is his Ascendant lord Venus involved, or receiving aspects or contacts from malefic planets. Therefore Venus is considered unafflicted.

Since both his Ascendant lord Venus and the Moon (in Venus's sign of Libra) are untouched by aspects from malefic planets, either temporal or classic, it brings a physical and mental advantage. Farrow is also blessed to have her Ascendant lord (Mars in her case) free from contacts of the planets on the Gemini-Sagittarius axis, though the placement of her natal Moon is not as fortunate. For Du Pré, both Ascendant lord and Moon were severely afflicted, impacting both her mental and physical health, respectively.

From the Taurus Ascendant, most of his planets are in the 2nd and 8th houses, or in the 3rd and 9th houses from the Moon. By comparison to Du Pré's chart, **Mars is not in a *Graha yuddha*** (Planetary War). This frees Mercury *and* Mars to be of greater benefit in the life. Mars is seven degrees ahead of Mercury in the same sign of Sagittarius. Though they are in the same *nakshatra*, they are in different *pada*s and thus situated in different houses in the *Navamsha* chart, also known as different *Navamshas.*[122]

Saturn and Rahu rendered more harmless: Saturn is an excellent planet for a Venus-ruled Ascendant chart and lends the steadiness, discipline and reliability that Taurus values so highly. Saturn rules both 9th and 10th houses from the Ascendant, giving Saturn *yogakaraka* status. As Rahu is considered Saturn-like, there are two major classical malefics that become temporal benefics. By natal position and by transit they do not tend to do as much harm as in charts where they are both classical and temporal malefics, as with Aries Ascendant.

---

[122] A *pada* is a "foot," or one quarter of a constellation, equaling 3 degrees 20 minutes of arc.

The same principle applies to Saturn *Dasha*: For this man, it was much easier on many levels than for Du Pré, especially regarding physical health. Her first major symptoms of multiple sclerosis became evident in Saturn-Mars *Dasha/Bhukti*; and she announced her retirement from her career as a performing musician during her Saturn-Rahu *Dasha/Bhukti* at age 28. For this man (Chart #6), it was not until six or seven years into his Mercury *Dasha* at around age 55 that he felt the need to ease up somewhat on his more physically strenuous work. This was due to some joint and muscular issues, including his back; but he continued the same work regimen for another 7½ years. The difference in the manifestation of the Saturn *Dasha* has everything to do with the different Ascendant chart, as both transiting Saturn and Saturn *Dasha* will be likely to yield better results in the case of Taurus Ascendant.

Tremendous physical strength: We expect to see a strong Ascendant, given this man's long-term physical strength and stamina, which includes serving and surviving several years in the Vietnam War. The condition of the Ascendant is indeed excellent, being *Vargottama*. (The sign of the birth chart Ascendant repeats in the *Navamsha* chart.) This is a very fine indication for the health and longevity of the person, if other factors do not diminish from this advantage. Natal Jupiter aspects the Ascendant, which is favorable as a classic benefic, a role that Jupiter enjoys here despite being 8th house lord for Taurus Ascendant. Fixity of purpose comes from a fixed sign on the Ascendant combined with Asc. Lord in a fixed sign and five planets in fixed signs in the *Navamsha* chart: thus, the long career in one area, one long marriage, and two (now grown) children. Seven natal planets in dual signs could override such fixity, but the *Navamsha* chart has a dominant effect, as well as the position of the Ascendant lord and the Moon, showing the predominant mental traits.

Loyalty, steadiness, fixity: This is a very good example of a Taurus Ascendant chart supported and amplified in its fixity by numerous factors. Ascendant Lord Venus in Aquarius and five *Navamsha* planets are in fixed signs. This together with the *Vargottama* Taurus Ascendant emphasizes components of: 1) *Artha* – focus on material security, and; 2) earth element – focus on material results that are practical and if possible, aesthetically pleasing. These astrological factors support one long career, one long marriage, a stable family and a continued spiritual life. Though they are on the problematic Gemini-Sagittarius axis, Mercury and Saturn are excellent friends to the Taurus Ascendant. Their *Dashas* (planetary periods) during the prime of life benefit this individual, especially financially.

The **Saturn *Dasha*** ran for 19 years from early June 1975, thus ending the less propitious Jupiter *Dasha* (a time that included some drug use), and marking the beginning of his career as a millwright. **Mercury *Dasha*** runs from early June 1994 for 16 years up to 2011. One might have expected a career change in Mercury *Dasha* to something more exclusively mental, but it did not happen. Natal Mercury is in the fire sign of Sagittarius and *Navamsha* Mercury is in another fiery and active sign (Aries) in the 12th house, explaining this man's satisfaction with behind-the-scenes physical work. The combination of the physical and mental work of a millwright kept him strongly engaged.

In Jan. 2002 he considered a career change to something less physically taxing. Some impetus for change was coming from his Mercury-Moon *Dasha/Bhukti* and from tr. Saturn turning Stationary Direct at 14:09 Taurus exactly on his Ascendant, aspecting his 10th house of status and career. This Saturn effect would cause him to take a serious look at whether his life and career were appropriate and fulfilling. The physical burdens of his heavy work would be more challenging at that time. The emotional strain would be intense as well, due to tr. Saturn in the 8th house from his natal Moon in Libra (June 6, 2000 to April 3, 2003). This condition is called *Ashtama Shani*. (See Glossary)

Though he considered changing professions in early 2002, in the end he did not do it. His reasons for not making the change, as of summer 2006, describe his fixed nature and the Taurean need for material security inherent in the destiny: **1)** A career change required further education and training; **2)** His physical health and strength were still excellent; **3)** He would receive more retirement and pension benefits by staying with his career as a millwright; and **4)** His wife had already retired by 2004 and was not in favor of him beginning another career when they would finally have time to travel together.

The tendency of the fixed Ascendant person with dominant fixed planet energy (confirmed in the *Navamsha*) would be to keep on with the current career and work routine, especially if the physical strength and health supported it. Curtailing his workload would be more likely when tr. Saturn moved to his 4th house in Leo (July 16, 2007 thru Sept. 9, 2009), as this placement of transiting Saturn tends to bring lulls in one's career, or even retirement. At that time, the career tends to be at a low, and one has less visibility. In July 2006, his plan was to retire in another year, also yielding maximum financial benefits. He carried through with his plan, retiring Sept. 25, 2007, exactly as tr. Saturn was giving the same message. It would have been challenging to start a new work project at that time. Tr. Saturn was closely opposite his Ascendant lord Venus, within one degree – a fitting symbol for retirement. Tr. Saturn had entered Leo almost eleven months earlier.

Financial security and financial benefits: There is a *Parivartana yoga* (mutual exchange of signs) between Mercury in Sagittarius and Jupiter in Virgo in the birth chart. This occurs between the 5th and 8th house, both financial houses, and is beneficial for financial matters and for the condition of each planet. Mercury has the energy of being in Virgo – its sign of exaltation; and Jupiter acts as if in Sagittarius, its own sign and also *mulatrikona* (root trine, where it functions very well – similar to being in exaltation). It could have caused difficulties for his children (5th house – children), but so far it has not. It has given him good intuition as well as financial benefits due to the combination of 2nd, 5th and 11th house lords, creating a wealth-producing *Dhana yoga*. Strengthening the wealth factor, Mercury aspects its own house – the 2nd house of financial income; and Jupiter aspects its own house, the 11th house of financial gains. This can protect matters concerning both of those houses.

The **Mercury-Jupiter** *Dasha/Bhukti* from June 2006 to mid-Sept. 2008 would yield maximum financial benefits in his career as a millwright, and this was the time period during which he chose to retire. It was also a time during which he took a major foreign trip and went on a spiritual pilgrimage. He departed on Nov. 23, 2006 and returned on Jan. 8, 2007. When he departed, tr. Sun, Jupiter and Venus were all in his 7th house of foreign residence. (He stayed at a temple in a foreign land.) On Oct. 27, 2006 tr. Jupiter had just entered Scorpio (his 7th house), and on Nov. 1, 2006 tr. Saturn entered Leo (his 4th house), indicating the time was right to turn inwards and slow down.

Venus in action: Ascendant lord Venus is very strong in the 10th house of the birth chart in Aquarius – the sign of great good friend Saturn. A *Kendra* house position is excellent to give some power of action in the life and the ability for achievements on the outer plane. (From the Moon, Venus is in a trinal house, the 5th house – also excellent.) Venus in Aquarius is very idealistic in love matters and in all matters ruled by Venus. The Ascendant lord in a fixed sign gives the individual some staying power and potentially some stubbornness, and in Aquarius in the 10th house is very idealistic about choice of career.

In addition to Venus's sign lord being Saturn, its *nakshatra* lord is Rahu, which is Saturn-like – and both are together in the 2nd house from the Taurus Ascendant. Both these planets have some control over Venus, so the fact that they are considered good friends to Venus and participate in some *Raja yogas* with the planets in the 8th house has a generally beneficial effect. Only the presence of Rahu and Ketu along that axis gives some potentially volatile results. But in his case it kept him working with the raw energy of metals and at the same time kept him devoted to his spiritual practice and active in his local temple. Venus in the 8th house of the *Navamsha* chart shows the spiritual orientation.

Spiritual practice serves as a balance: In 1973, in the last sub-period of his Jupiter *Dasha* (**Jupiter-Rahu** *Dasha/Bhukti),* he found his spiritual path for this lifetime. Though any *Bhukti* involving Rahu or Mars can create upheavals, especially at the end of a *Dasha*, this one was helpful to him. Tr. Jupiter in Capricorn was then in his 9th house of religion, aspecting the Ascendant, and tr. Saturn returned to its natal position of Gemini by early June 1973 in the 9th house from natal Moon in Libra. The first Saturn return at ages 28-29 is usually a pivotal one, bringing an awareness of the deeper meaning of life.

Moksha is spiritual liberation from the confines of earthly life. Ketu is *Moksha* significator in the 8th house of *Moksha*, along with two other *Moksha* house lords, 4th house lord - Sun, and 12th house lord - Mars.) This orientation would become even stronger in the 7-year Ketu *Dasha*, starting June 5, 2011. Four natal planets in Sagittarius in the 8th house show a tendency to be intuitively open and move towards some spiritual path. This is confirmed with the placement of Moon and Venus in Sagittarius in the *Navamsha* 8th house, which can also be dangerous for the physical body in terms of accidents or disease, and could show an obsessed, unbalanced mind. But with a strong spiritual practice, other levels of the 8th house become more dominant. As a *Moksha* house, the 8th house can bring either suffering or spiritual liberation. When Moon is in the 8th house, the person may have intuitive gifts and empathic abilities.

As noted earlier, Saturn and Rahu are less harmful to the Taurus Ascendant chart because they are friends of the Ascendant Lord Venus. Thus the Saturn-Rahu placement in the 2nd house of the birth chart is not as difficult for the Taurus Ascendant chart as for the Aries Ascendant chart. Also, Saturn and Rahu move to two different *Kendra* positions in the *Navamsha* chart, benefiting him career-wise. The Mercury-Mars *Parivartana yoga* in the *Navamsha* chart could be difficult, especially for financial losses, as it involves 2nd house of income and 12th house of losses. But he survived it well, and his regular spiritual practices (also 12th house) may have minimized this karmic indication. It may also be why he did not take early retirement or depart from a financially reliable profession – even with all its heavy physical demands.

When classical malefics are in the 2nd house, there may be bad eating habits, harsh or coarse speech, or difficulties deriving from the family of origin, as 2nd house is also house of *Kutumba* (happiness from the family of origin). There was some drug use earlier on in his life (in Jupiter *Dasha*), but due to his long-term spiritual practice he has been able to surmount some of these more difficult influences and calm the mind and heart.

Chart 7: Theodore Prostakoff

Birth data: Monday, July 20, 1953, 01:25 AM EDT, Brighton, MA, USA, Long. 71W09 25, Lat. 42N21 00, *Lahiri ayanamsha*: -23:12:45, Class A data, from memory. Ascendant: 06:11 Taurus.

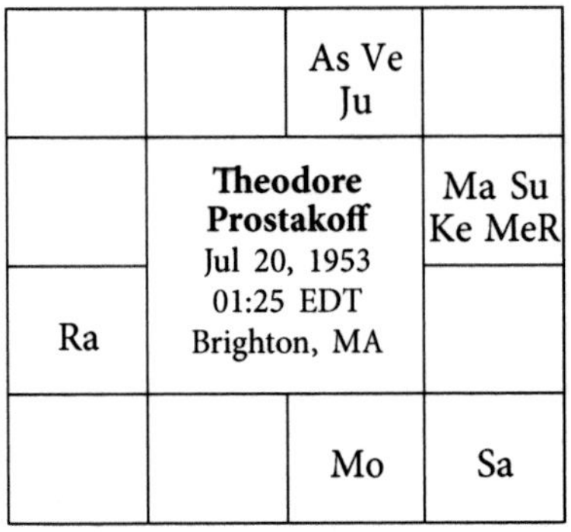

		As Ve Ju	
	Theodore Prostakoff Jul 20, 1953 01:25 EDT Brighton, MA		Ma Su Ke MeR
Ra			
		Mo	Sa

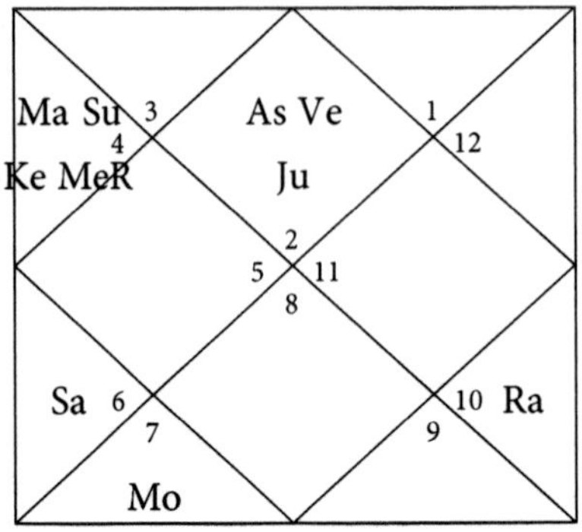

Ma Su 3 4 Ke MeR
As Ve Ju
1 12
2 5 11 8
Sa 6 7 Mo
10 Ra 9

Planetary Information

As	06:11	Tau	Krittika
Su	03:59	Can	Pushya
Mo	15:31	Lib	Swati
Ma	00:34	Can	Punarvasu
Me	12:25	Can	Pushya
Ju	22:58	Tau	Rohini
Ve	20:23	Tau	Rohini
Sa	27:54	Vir	Chitra
Ra	10:15	Cap	Shravana
Ke	10:15	Can	Pushya

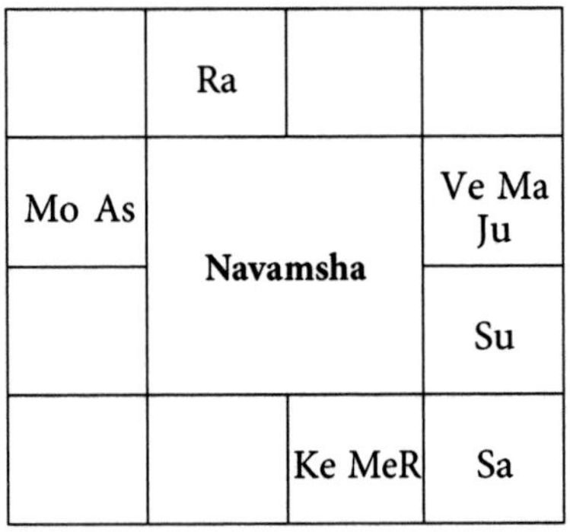

	Ra		
Mo As	**Navamsha**		Ve Ma Ju
			Su
		Ke MeR	Sa

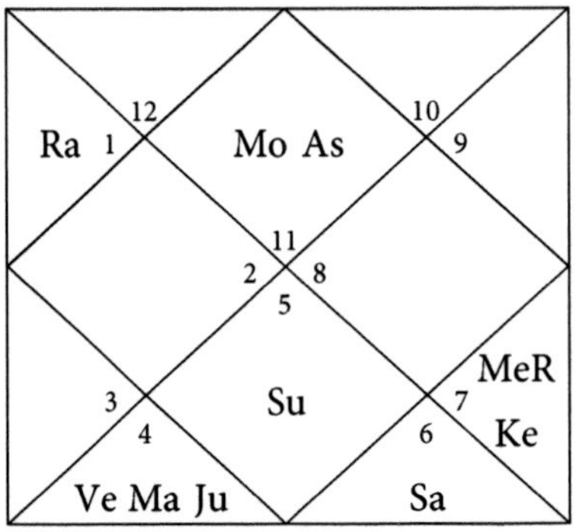

12 Ra 1
Mo As
10 9
11 2 8 5
3 4 Ve Ma Ju
Su
MeR 7 6 Ke
Sa

Vimshottari Dashas

Ra	Aug 4, 1941
Ju	Aug 5, 1959
Sa	Aug 5, 1975
Me	Aug 5, 1994
Ke	Aug 5, 2011
Ve	Aug 5, 2018
Su	Aug 5, 2038
Mo	Aug 4, 2044
Ma	Aug 5, 2054

Biographical summary: A musical prodigy, Prostakoff began playing the piano at age three, auditioned for conductor Arthur Fiedler at age five and made his professional debut with the Boston Philharmonic at age seven. By the age of 16 he had played with most major orchestras in the world, earning $10,000 per concert. By 1969-1970, at ages 16 and 17, a whirlwind tour of over 250 concerts led to a complete physical and emotional breakdown, resulting in his retirement from the concert stage. Upon recovery he did only benefit concerts, and those included performances at the Vatican, at Buckingham Palace, and at the White House for three presidents.

His education was unusual, as he never attended regular school, though he studied at Juilliard and was a student of the great pianist Vladimir Horowitz for over a year. He lived alone in New York from the age of ten.

Venus – lord of music: Ascendant lord Venus is situated in the Ascendant in its own sign. This is excellent, as it can control the affairs of its own house and can give the person self-confidence and generally good health under most normal circumstances. In addition, Venus in a *Kendra* in its own sign (or exalted) gains the status of *Malavya Mahapurusha yoga*, bestowing a graceful appearance and a handsome body, along with learned qualities, power, and financial wealth. It can also give charm and popularity.

As it rules over music, Venus is well situated in the Moon's *nakshatra Rohini*, the most favored of the *nakshatras* in Vedic mythology. Part of the karmic destiny is to deal with being so favored, and the jealousy that can engender in others. Ideally, the individual recognizes this destiny and aims to modify any ill effects through an ongoing attitude of humility, being detached from the fruits of one's labors.

The position of Venus in the *Navamsha* 6th house in Cancer (with Mars and Jupiter in this case) does not favor a long-term public career and portends health problems from some manner of burnout, from the time of birth. *Navamsha* Venus is somewhat modified by Venus in the 5th house of the *Dashamsha* chart (10th harmonic chart, dealing with career matters), but does not override the all-important *Navamsha* chart. The excellent *Lagna* position of Venus in the birth chart is not confirmed or amplified in the *Navamsha* chart. The 6th house is among the least auspicious house positions for Venus, implying its sign of debilitation (Virgo) along with the issues of the 6th house, including conflict and ill health.

Musical talent: In addition to Venus, other indicators of music in the chart are **the 3rd house**, as well as the sign of **Libra**, where Moon is placed. Four planets in the 3rd house can show a strong focus on music (Du Pré had four planets in her 3rd house). In this case it was combined with a powerful Venus in Taurus and Moon in Venus-ruled Libra. The Moon's 6th house position indicated some form of conflict in carrying out his musical career. The 5th house is also involved, with a strong *Vargottama* Saturn in Virgo in the sign of a planetary friend. A concentration of planets in water can show the emotional intensity that often leads to artistic creativity, especially when other planetary patterns indicate that tendency. Add to this two great benefics in the Ascendant, Venus and Jupiter, one of them – Venus – ruling over music and powerful in its own sign. There are potential pitfalls with this combination (we discuss them below), but they can also bring many opportunities for success. The climax of his musical output at age 16 occurred in Jupiter-Venus *Dasha-bhukti*.

So many planets in the third house emphasize the issue of courage and of *Kama* (the desire nature). He wants to do what he wants to do, especially with a combust, debilitated Mars, and on the Rahu-Ketu axis. Add to this a Moon in *Swati nakshatra*, which likes to be independent and values movement. (*Swati* is ruled by Vayu, god of the wind.) Six planets in *Chara* (action) signs in both the birth chart and the *Navamsha* chart give a tremendous amount of activity, but the fixed Ascendant in the birth and *Navamsha* chart show that there will be tendency to stay with one career or life theme.

Potential for excess (*and* wealth) – Venus-Jupiter effects: Venus is within 2½ degrees of 8th house lord, Jupiter. This can be challenging for the well being of the individual, especially in terms of giving a sense of balance and proportion. Since the two great benefics are planetary enemies from Jupiter's perspective (neutral from Venus's perspective), when located together in the same house this can create excesses. As 8th house lord Jupiter is lord of *a Dusthana* house. When combined with the Ascendant lord this creates an *Arishta yoga*, one that can be physically damaging to the health. The plus side is that as ruler of the 11th house of financial gains Jupiter's presence in the Ascendant also creates a *Dhana yoga* of financial wealth. Further confirming financial wealth in the life, Venus and Jupiter are in the 8th house from natal Moon, the *Chandra-Adhi yoga*, another *Dhana yoga*.

The Taurus Ascendant person is always concerned with material security, and very motivated throughout their life to see what they can earn in financial benefits, and how much security it can provide for themselves and any family members. But while this *Dhana yoga* is excellent for financial well being, the chart indicates the destiny for accumulating financial wealth through some form of excessive activity. This is due specifically to the Venus-Jupiter combination in the Ascendant.

Prostakoff entered **Jupiter *Dasha*** in early August 1959, at the age of six The first year of Jupiter *Dasha* would coincide with his debut at age seven as a musical prodigy, performing with the Boston Philharmonic. When he entered the sub-period of Venus in the Jupiter *Dasha*, this karmic destiny for excess was most likely to manifest, especially as Jupiter was the *Dasha* lord and had dominance over Ascendant lord Venus for 16 years starting from July 1959. The **Jupiter-Venus** *Dasha/Bhukti* ran from June 17, 1967 to Feb. 15, 1970, and encompasses the period when his world concert tours were running at their most frantic pace, leaving him in a state of physical and emotional collapse by 1970. Though they were extremely lucrative years, they exacted a heavy price on his health and on any long-term professional career as a performing concert pianist.

Venus and Jupiter recur together in the *Navamsha* 6th house of health issues and competition, this time in Cancer, alongside *Navamsha* Mars. Mars is debilitated in Cancer and Jupiter is exalted in the same sign, so we have a *Neecha Bhanga Raja yoga* in the *Navamsha* chart confirming the Venus-Jupiter excesses of the birth chart, but also creating the remedy for them through the necessity of tending to health issues, which in this case led eventually to philanthropic service. Thus a disadvantage is already foretold through excesses (Jupiter-Venus, fuelled by Mars) that would be corrected in time through a life of service and humanitarian efforts.

This 6th house theme is repeated by the fact that natal Venus and Jupiter are situated in the *nakshatra Rohini*, owned by the Moon, which in turn is placed in Libra in the 6th house of the birth chart. *Navamsha* 6th lord Moon is also in the *Navamsha* Ascendant. Thus, health and competition will definitely enter the life as major issues to triumph over emotionally, mentally, and physically. Sixth house themes show where and when there is a conflict for the individual, and if Moon is present in the 6th house, there is a lot of personal mental and emotional focus on this issue throughout life.

Successful physical and mental recovery after 1970: Once the Jupiter-Venus *Dasha/Bukhti* was completed Feb. 15, 1970, the worst would be over. By Aug. 5, 1975 his troublesome 16-year Jupiter *Dasha* was over, bringing to an end Jupiter's power as *Dasha* lord. The Jupiter-Saturn opposition in the heavens March 8, 1970 also indicates a new set of influences and/or solutions, especially occurring at 12:05 Aries-Libra (Jupiter in Libra, Saturn in Aries), closely opposite his natal Moon at 15:31 Libra. Further, here are several innate factors in his birth and *Navamsha* charts that account for Prostakoff's likely physical and mental recovery from his debacle in 1970:

1. **Venus** strongly placed in the Ascendant of the birth chart, though poorly placed in the *Navamsha* 6th house.
2. The *Neecha Bhanga Raja yoga* in the *Navamsha* chart – **Mars-Jupiter-Venus** in Cancer in the *Navamsha* 6th house – as described directly above, creating the potential correction for any excesses that would be likely to occur.
3. **Sun** angular in its own sign in the *Navamsha*, though 6 degrees from Ketu in the birth chart (a Solar eclipse effect), and in Saturn-owned *Pushya nakshatra*. (Saturn is planetary enemy of the Sun, so its influence on the Sun can cause it to work too hard or deplete its physical

vitality in some way.) Natal Mars and Sun in the 10th house from the Moon are excellent for his general physical strength and vitality.

4. **Mars** *Vargottama* in Cancer, though debilitated in the sign of Cancer, receives cancellation in a *Neecha Bhanga Raja yoga* through multiple factors that bring improvement in the end. Both Sun and Mars are *karakas* (significators) of physical health and vitality. Mars is severely combust the Sun, and as both are fiery planets in a water sign – the potential for physical excess is great here too. Combust Mars often does not understand its physical earthly limits. (A combust planet is not visible to the naked eye; thus a person with a very combust planet may not be able to see what is happening in the arena of life over which that planet rules.) Luckily, however, Mars is not Ascendant lord, and a combust and debilitated planet will have more adverse effects over the houses it rules, the 7th and the 12th in this case. This curtails his financial losses, but weakens his partners, and either way does not favor foreign residence, as 7th and 12 houses both rule over foreign residence.
5. **Moon** is in the urban *nakshatra* of *Swati*, which is ruled by the deity Vayu, god of the wind. Like the wind, Swati likes to move in many directions and does not like to be confined. It prefers living in cities and being among other human beings. Its karmic purpose is to spread itself in such a way to disseminate what it has to give to the world. Placed in the 6th house of health matters, this almost led to his demise in 1969-1970 at age 16. However, Moon is better situated in the *Navamsha* chart in the Ascendant in the fixed sign of Aquarius opposite a Sun in Leo, receiving no bad planetary aspects except for an 8th house aspect from debilitated Mars.
6. **Mercury** improves in the *Navamsha* from a natal 6th house watery position to the 9th house in its friend's sign of Libra. Mercury is Retrograde, so has a tendency to go back over things in life, and close to Ketu, may search for spiritual answers in an intense way. As both Moon and Mercury improve in the *Navamsha*, this is a good indication his mental/emotional state deepened and grew stronger over the years. Mercury can give quick wit when closely associated with either Mars or Ketu. He has both, and is known to have a good sense of humor.

No loss of reputation: After his recovery, Prostakoff resumed his life as a classical pianist, though in his own time and on his own terms. Removing himself from the commercial scene also appeared to restore his energy and confidence, and probably added many years to his life. He was protected from ultimate losses by the strength of his Ascendant and Ascendant lord. The Jupiter factor was destructive, especially during the Jupiter *Dasha*, but there are no other bad aspects to the Ascendant or Ascendant lord. Moreover, two powerful *Viparita Raja yogas* from the Moon help to protect him from ultimate ruin during his lifetime. This *yoga* occurs when the lords of the *Dusthana* houses (6, 8, 12) from the Ascendant or the Moon contain planets ruling any of the *Dusthana* houses. In this case, Venus and Jupiter in the 8th house from the Moon provide this situation (while also bringing wealth: the *Chandra-Adhi yoga*.) They protect his reputation, though birth chart 10th lord Saturn in the 8th house of loss in the *Navamsha* shows that the original career plans will be altered in some way.

A sure sense of destiny – no matter what the rewards: Saturn is classically the lord of *Karma*, and for Taurus Ascendant rules both 9th and 10th houses of ***Dharma*** **and** ***Karma*** respectively. Its position in the natal chart is pivotal for the success of that person, as well as the transits of Saturn.

Secondarily, we note the transits of the classical *Dharma* lord Jupiter. Tr. Jupiter and Saturn were in mutual opposition during the climactic 1969-1970 time period, tr. Jupiter in Libra in his 6th house with natal Moon, and tr. Saturn in Aries (Saturn's sign of debilitation) in the 12th house from his Ascendant. After three months in Aries in 1968, tr. Saturn re-entered Aries March 7, 1969. The house location of this particular Saturn transit would not modify the Jupiter-Venus *Dasha/Bhukti.* Instead, it would tend to push him to work even harder and take on too many responsibilities prior to the exact Jupiter-Saturn opposition March 8, 1970.

Another important factor in offsetting this crisis and providing a long-term steadying influence is the trine of natal Jupiter and Saturn, **classical *Dharma* and *Karma* lords**. They are in the earth houses of Taurus and Virgo, respectively. This further solidifies the desire of the person to do well on a practical, financial level. There is support for this in the life and there is no conflict over this motivation, another amplification of the Taurus Ascendant, giving further reason to do whatever necessary to accept opportunities for profit when they first arose – when he was no doubt too young to know better and perhaps did not receive advice on moderating his schedule. From the Moon, Jupiter is in a *Dusthana* house (6, 8, or 12) in both the birth chart and *Navamsha* chart. Known as a *Shakata yoga*, this combination tends to bring strong fluctuations of fortune; but as Jupiter is in a *Kendra* from the Ascendant, the *Shakata yoga* is cancelled. Jupiter is strong, having *Digbala* in the Ascendant. He lived alone in New York City from the age of 10.

Review of this Taurean *Dharma*: With the Taurus Ascendant, the life of this musical prodigy was not about breaking some new boundary in the realm of piano performance or piano literature. But in commercial terms, at the age of 16 he was breaking records for his earnings per concert and per concert season. The blitz of a public concert career starting at age 7 and ending at age 17 comes especially from the effects of **Jupiter as *Dasha* lord** – pushing him ever outward and bringing too much too soon. This originated from his love of music and performing music (Venus-Jupiter). For career success, we assess the strength and position of the 10th lord in the birth chart and follow it in the *Navamsha* chart for confirmation.[123] The birth chart shows 10th lord Saturn well placed in the 5th house, but this is not confirmed in either the *Navamsha* or *Dashamsha* charts. Saturn is in the 8th house in the *Navamsha* and in the 12th house in *Dashamsha*. Therefore, losses or a change in career direction are indicated, starting any time but especially likely during Saturn *Dasha:* from August 1975 to August 1994, ages 22 to 41.

Saturn should also give some benefits and bring him into the deeper aspects of himself, as Saturn is *Atma karaka* – planet at the highest numerical degree of celestial longitude in the birth chart. It is also *Vargottama* in the *Navamsha* in a friend's sign– giving strength to the planet – and in its own sign of Aquarius in the *Dashamsha* chart. This is very good, especially as Saturn is *Yogakaraka* planet for Taurus Ascendant, ruler of both the excellent 9th and 10th houses, and a good indication his destiny might proceed more smoothly after a jet-propelled beginning in life. The focus of this karmic destiny is on how this person will handle the **Venus-Jupiter** combination, especially during its planetary period in 1969-1970 and in its aftermath.

[123] The *Dashamsha* chart (for 10th house/career matters) can be checked for these matters, though *Dashamshas* are generally not provided in this book. If indications are not present in the *Navamsha* chart, they cannot be overridden by the *Dashamsha* chart.

Chart 8: Martha Graham

Birth data: Friday, May 11, 1894, 6:00 AM EDT, Pittsburgh, PA, USA, Long. 79W59 46, Lat. 40N26 26, *Lahiri ayanamsha*: -22:22:51, Class A data, from memory. Ascendant: 13:05 Taurus.

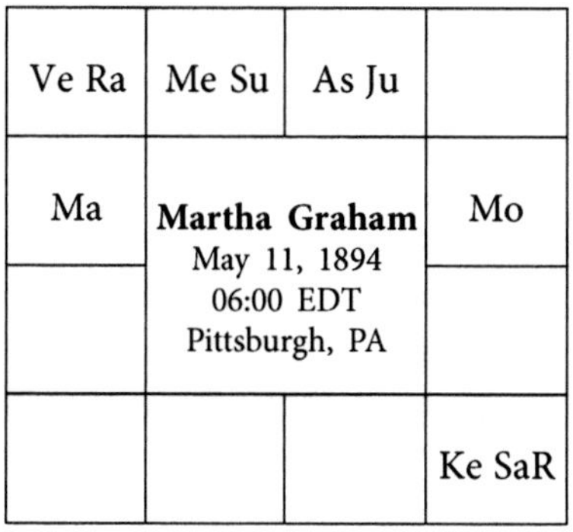

Ve Ra	Me Su	As Ju	
Ma	**Martha Graham** May 11, 1894 06:00 EDT Pittsburgh, PA		Mo
			Ke SaR

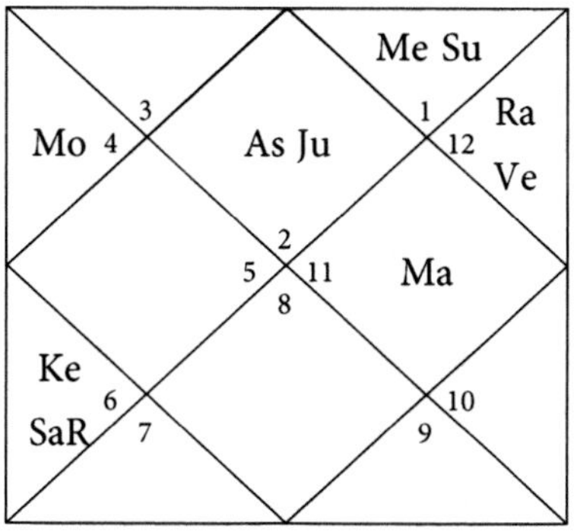

Me Su
3 1 Ra
Mo 4 As Ju 12 Ve
2
5 11 Ma
8
Ke
6 10
SaR 7 9

Planetary Information

As	13:05	Tau	Rohini
Su	28:19	Ari	Krittika
Mo	17:45	Can	Ashlesha
Ma	08:39	Aqu	Shatabhisha
Me	17:40	Ari	Bharani
Ju	15:55	Tau	Rohini
Ve	12:46	Pis	U.Bhadra
Sa	27:20	Vir	Chitra
Ra	15:53	Pis	U.Bhadra
Ke	15:53	Vir	Hasta

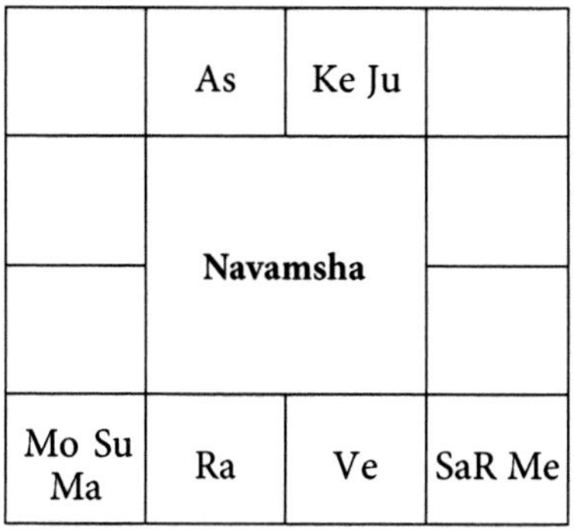

	As	Ke Ju	
	Navamsha		
Mo Su Ma	Ra	Ve	SaR Me

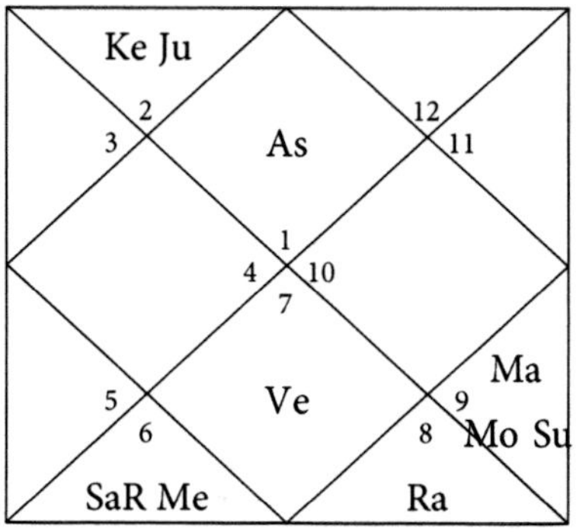

Ke Ju
2 12
3 As 11
1
4 10
7
Ma
5 Ve 9
6 8 Mo Su
SaR Me Ra

Vimshottari Dashas

Me	Dec 21, 1892
Ke	Dec 22, 1909
Ve	Dec 22, 1916
Su	Dec 22, 1936
Mo	Dec 23, 1942
Ma	Dec 22, 1952
Ra	Dec 23, 1959
Ju	Dec 23, 1977
Sa	Dec 23, 1993

Biographical summary: Sometimes referred to as "the mother of American dance," Graham was a legendary modern dancer, choreographer and teacher whose active career spanned 75 years. She devised a new vocabulary in modern dance, one that she intended to "increase the emotional activity of the dancer's body." This was at a time when classical ballet was still the norm in the world of dance.

Graham was almost 17 years old when she first saw a performance of Denishawn, a dance troupe led by Ruth St. Denis and her husband Ted Shawn. Revolutionary for the era, the performance was in late April 1911 in Los Angeles. Graham was totally overwhelmed by the experience, and five years later she enrolled in their dance school. She had been fascinated with dance and movement since the age of ten, but her father saw it as frivolous and was unsupportive. He was a medical doctor with a thriving general practice and a special interest in psychology. Because of this, he studied people's movements, and was known to say: "Movement never lies."

Martha Graham was the oldest of three daughters in an upper-middle class family. She received an education in the arts, history, languages and science, both by private tutoring and public schools. In 1908 Dr. George Graham moved his family to Santa Barbara, California from the coal mining area of Pittsburgh, Pennsylvania – due to his second daughter Mary's increasing problems with asthma. He insisted on a good traditional education for his daughters and wanted his oldest daughter to attend an academic college such as Vassar. But Martha was strong-willed, and insisted on a junior college in Los Angeles that was more experimental, and oriented toward arts and performance – the Cumnock School of Expression.[124] In the fall of 1914 while in her second year at Cumnock, her father died suddenly of a heart attack. Though his unexpected death left the family in financially reduced circumstances, it freed Martha Graham to pursue her dream of becoming a performing artist, notably a dancer. Some 15 years younger than Dr. Graham, her mother was excited by her daughter's aim to become a dancer and encouraged Martha to pursue it. She continued at Cumnock, graduating in June 1916. She developed a strong, lean body, and though not a beauty, she inherited her mother's large deep-set eyes and shiny black hair, often with a severe part down the middle. She had a fierce but quiet determination, combined with a fiery temper if she did not get her way. She was petite, at 5 feet 2 inches tall.

Martha Graham began her professional dance training at age 22, dangerously late for a dancer. But even so, she had found her destiny. Though not the first modern dancer to break from classical ballet, she established her own unique style that helped to legitimize the whole world of modern dance and give it dramatic content. Many of her methods are still taught in dance schools today; and her choreographic style of fierce pelvic contractions and intense floor work became the cornerstone of postwar modern dance. Some criticized her style and stories as being too angular and angst-ridden, but she remained undeterred. For decades she was the high priestess of a dance company, and she presided over it with hurricane-force strength.

> She said she wanted to "chart the graph of the heart" through movement. She also said: "That driving force of God that plunges through me is what I live for." "One becomes in some area an athlete of God."[125]

She formed her own dance company in 1926 and proceeded to attain much artistic attention and power right up to the late 1960s, with her biggest breakthroughs in the 1940s and 1950s. In 1935 she established a school of modern dance at Bennington College. Some argue that she did not produce any new choreography of lasting interest after 1950 (age 56), but she continued to receive much critical attention from her adoring press. She also continued to perform occasionally up to the age of 74 in 1968, and was reluctant to give it up even then. She received the Presidential Medal of Freedom in 1976 – the first ever awarded to a dancer. Meanwhile, she maintained an active work schedule until the last two months of her life. She died on April 1, 1991 at age 96 from complications from pneumonia she contracted while on tour with her company in the Far East.

In 1930 (at age 36) she formed a long professional and romantic relationship with Louis Horst, whom she met in 1922 when he was music director with Denishawn. And in 1948 (at age 54), she married her fellow dancer and protégé Eric Hawkins, at his urging, after eight years of living together. The union fell apart within the next two years.

She choreographed 181 works over her lifetime, including "Letter to the World" (1940), "Appalachian Spring" (1944) and "Cave of the Heart" (1946). For decades she collaborated with numerous artists, including composer Aaron Copeland and sculptors Isamu Noguchi and Alexander

---

[124] In her 1991 autobiography, ***Blood Memory***, Graham calls it the "University of Cumnoch," though the school's full name at the time she attended it was Cumnock School of Art and Expression.

[125] The latter quote appears in Edward R. Murrow's ***This I Believe***, Volume 2, 1954, p. 58.

Calder. This long-term mutual association with leading artists in their respective fields raised the level of her creative work. Her dance company lives on and remains America's oldest dance company and among the most renowned. It celebrated its 80th anniversary in 2006. Graham's autobiography *Blood Memory* was published in 1991.

> "No artist is ahead of his time. He *is* his time; it is just that others are behind the time."
>
> **Martha Graham**

Qualities of the pioneer: Martha Graham has many of the same qualities of the Aries Ascendant person. With her courage, originality and vision Martha Graham was considered a pioneer in her field. This is confirmed astrologically by the close aspect of Mars to the Ascendant. Mars has *Digbala* (best possible angular house) in the 10th house. It gives some of those qualities as well as excellent physical strength and stamina. If in addition to Mars the Ascendant lord and Sun are strong in both birth and *Navamsha* charts, as in this case, then physical strength and stamina are even stronger. The *Navamsha* Ascendant is located in Aries, ruled by Mars.

As *karaka* of physical vitality, the Sun is not damaged by its placement in the 12th house of loss, according to the Vedic classic *Uttara Kalamrita.* As her natal Sun is exalted in the sign of Aries and located in Venus's *nakshatra*, it gave good results during her 20-year Venus *Dasha* (Dec. 1916 to Dec. 1936) and her 6-year Sun *Dasha* (Dec. 1936 to Dec. 1942). However, as the Sun is 4th house lord (of family foundation and foundation of the mind) placed in the 12th house, it can give mental and emotional turbulence and a sense of separation from the family of origin. This is amplified by the condition of the natal Moon.

Venus - lord of the dance: Ascendant lord Venus - exalted in Pisces and well situated in the 11th house of the birth chart - is excellent for the destiny of this individual. Further accent on Venus comes from her being born after sunrise on a Friday, the weekday ruled by Venus. The Taurus Ascendant with its fixed earth modality already bodes well for longevity; thus the excellent condition of the Ascendant lord could also account for her longevity (nearly 97 years). In turn the planets ruling over the sign and *nakshatra* where Venus resides (Jupiter and Saturn, respectively) are also well placed. Nor does Venus receive bad aspects, except from being on the Rahu-Ketu axis. (Rahu is easier than Ketu in this case, since Rahu is Saturn-like, and Saturn is a great friend to Venus.) The positive power of Ascendant lord Venus is likely to manifest at some extraordinary level due to being strong in both birth chart and *Navamsha* chart. Venus is situated in a *Kendra* position in the 7th house of the *Navamsha* chart in its own sign of Libra, providing a distinguished *Mahapurusha Malavya yoga*. Venus in Libra is sophisticated, urban, intellectual and musical. Such is the overlay of *Navamsha* Venus.

Exalted Venus in Pisces gives everything to its love and/or its chosen art form. That she gave so much to her students and dancers and the world of dance in general is a testimony to the 11th house placement of Venus in the birth chart. Over the decades, she had a large group of dedicated students and dancers, as well as actors she trained in movement styles. The classical 12th sign of Pisces involves the greatest surrender and it is the highest level of love for which Venus is known and revered. Venus in Pisces is devotional in quality, and worships at the feet of its love. Normal earthly relationships may not last under its influence, depending of course on other factors. (We will discuss the Venus-Jupiter exchange shortly.)

The major or minor period of Venus would be bound to bring benefits, and it was in her **Ketu-Venus** *Dasha/Bhukti* from May 20, 1910 to July 20, 1911 when Martha Graham attended her first-ever professional dance performance in late April 1911: Ruth St. Denis and the Denishawn company. Graham's 7-year Ketu *Dasha* was not yet strong enough to bring her into the full-time dance training she desired, and it was not until she completed her junior college education in June 1916 that she was freer to do that. This was five months prior to the start of her Venus *Dasha* on Dec. 22, 1916. Meanwhile, her artistic and academic training leading up to Venus *Dasha* provided a foundation for her later career.

Delayed start – influence of the father: The father's influence can be seen from the 9th lord Saturn as well as the Sun, *karaka* for father. With Saturn opposite Venus on the Rahu-Ketu axis, this already describes some opposition from the father. He would have dominance over her until she came into her own. But given the strength of her Ascendant lord Venus, her *Vimshottari Dasha* sequence indicates the father would be unlikely to hold her back or prevail over her once her Venus *Dasha* began. When her father died unexpectedly of a heart attack in the fall of 1914, she was in Ketu-Jupiter *Dasha/Bhukti,* and in her second year at the Cumnock School of Expression. Jupiter is lord of the 8th house from her Taurus Ascendant. The 8th house rules over matters of death and longevity. Jupiter is also located in the 9th house (of father) from *Dasha* lord Ketu. (Sub-charts using the Ascendant of the *Dasha* lord can be very informative for timing during that particular *Dasha.*)

As 8th house lord for Taurus Ascendant, Jupiter and its major or minor periods would not be as favorable in the life. Nevertheless, Jupiter gives some benefits as a classical benefic planet that is: 1) V*argottama* (repeats in the *Navamsha* in the same sign) and 2) Placed in the Ascendant, where it is *Digbala* (best possible angle for Jupiter in the chart). Since Jupiter exchanges signs with exalted Venus, the Ascendant lord, this Jupiter can give better results than might be expected. Despite her father's death, Graham was able to continue at Cumnock School due to her mother's support. She took courses in what was then called "interpretive dance," and as she became more serious about it she stopped playing basketball to save her legs for dancing. She graduated in June 1916.

During this same period, Graham was in the midst of ***Sade-Sati*** (a 7½ year period with tr. Saturn in houses 12, 1, and 2 from natal Moon). This can bring irrevocable changes in the life; and in her case, it had bearing on her life path and a positive outcome in enabling her to pursue her goal of a career in dance. Martha's *Sade Sati* began June 20, 1914 – just a few months prior to Dr. Graham's death – when transiting Saturn entered Gemini. Two years later, tr. Saturn entered Cancer Aug. 20, 1916 to June 2, 1919), aspecting her natal Sun and Mercury in Aries and contacting her Moon in Cancer. Saturn contacts to both Sun and Moon put extra pressures and responsibilities on her parents. The last third of *Sade Sati* was from June 2, 1919 to Nov. 17, 1920. By this time she was starting to perform with Denishawn. It was a very successful residential dance school with its own touring company. In 1920, marking the last year of her *Sade Sati,* she made her debut in Ted Shawn's *Xochitl,* which was created for her. Tr. Saturn in the 4th house can destroy old foundations and create new ones.

A dance visionary is unleashed: It is clear this exalted Venus must give above-average results, especially if Venus *Dasha* occurs during the prime of life. The *Karma phala* (fruits of the *Karma*) would come due during Venus *Dasha* and indeed they did. In Vedic astrology, the planetary positions and *yogas* are inextricably linked to the *Dashas*, and thus the karmic timing. We know that Graham's artistic and love life would have a good chance of flowering during Venus *Dasha* and beyond. It is precisely then – at the start of Venus *Dasha* – that she was finally able to pursue her passion for dance and movement. This *Dasha* also brought her a long professional and romantic association with Louis Horst, officially from 1930, though they met at least eight years earlier at Denishawn.

Venus in a mutual sign exchange with Jupiter (a *Parivartana yoga*) acts as if it is also in the Ascendant in its own sign. This further confirms Graham's physical strength and personal dominance in her field. Venus is located in *Uttara Bhadra nakshatra*, ruled by Saturn, the best planetary friend to this Ascendant chart. Saturn also aspects Venus, and these two factors can bring the delaying side of Saturn's influence, as Graham was relatively old to be starting a dance career at age 22. However, Rahu is considered Saturn-like, and in the same house with Venus, within 5 degrees, is less damaging than in other charts. Even so, Venus on the Rahu-Ketu axis does give a rough edge to the chosen art form, the persona, and to the love life, all matters ruled over by Venus. There is a more exotic component that dominates over the security and predictability that a Taurus Ascendant person prefers. Her unconventional style and her terse, dramatic ways, however inventive and original, would bring her both accolades and criticism.

If people criticized her for being ahead of her times, she remained supremely confident in her chosen form of artistic expression. From an astrologer's perspective, this shows a good understanding of Saturn, lord of Time and excellent planet for the Taurus Ascendant chart as *Yogakaraka* and lord of 9th and 10th houses. Situated in the 5th house from the Taurus Ascendant, Saturn's placement is excellent for her artistic career, but not favorable for having children. This is confirmed by four classic malefic planets (Mars, Saturn, Rahu and Ketu) aspecting the 5th house, and 5th house ruler (Mercury) in the 12th house of loss, a *Dusthana* house.

Venus-Jupiter effects – Boom or bust: In the previous chart of Theodore Prostakoff, we see how an intimate connection between Venus and Jupiter can bring creativity – but to the level of a blowout, especially with his Jupiter-Venus period coming at such a young age. However, this planetary combination in a Taurus Ascendant chart does not have to lead to a sudden or early loss of a public career. Martha Graham had an exceptionally long and illustrious career as a dancer, choreographer, and teacher. Here are some differences between Graham and Prostakoff:

1. Jupiter's influence on Venus is less overpowering in Graham's case because the two planets do not coexist in the same *Navamsha*, as with Prostakoff. Though Jupiter is in the Taurus Ascendant, it is in a *Parivartana yoga* (mutual exchange of signs) with Venus in Pisces, giving the same *Dhana* (wealth-producing) *yoga* due to contacts between 1st and 11th house lords. Prostakoff's 20-year Venus *Dasha* arrives later in his life (at age 65), and is not as favored as Graham's Venus *Dasha* (from age 22) due to several factors, mostly related to the condition of his *Navamsha* Venus.

2. The 16-year **Jupiter *Dasha*** came early for Prostakoff: ages 6 to 22, whereas Graham's Jupiter *Dasha* spanned ages 83 to 99. She died in Jupiter-Mars period at almost 97. She was fortunate her Jupiter *Dasha* came much later in her life. The interplay of Venus and Jupiter in her chart would still lead to some excesses. Throughout her career she was considered spectacular but arrogant in her oddly angular style of choreography. But in the end, her style was accepted, studied and copied.

Saturn's modifying influence probably played a part in the acceptance of Graham's artistry and techniques, as her natal Venus receives the aspect of Saturn (along with Rahu and Ketu), giving her excellent discipline, even in her tendency to go too far at times with her creative projects. Like Prostakoff, she had both **natal Jupiter and Saturn in earth signs** in a beneficial trine, in Houses 1 and 5 in both cases. But hers was even stronger, as both Jupiter and Saturn are *Vargottama*, recurring in the same signs in the *Navamsha* chart. This gives a very concrete and practical orientation to one's life goals. For an artist, there is the desire for some tangible results from one's artistic pursuits: performances, and a record of the body of work, among others.

During his **Jupiter-Venus** *Dasha/Bhukti*, Prostakoff was ages 16 and 17 and came to his greatest period of excess, at the end of which he put the full brakes on an exploding concert career. Graham's **Venus-Jupiter** *Dasha/Bhukti* occurred Feb. 2, 1927 to Oct. 22, 1929. At that time she was in her early 30s, starting her own dance company, studying many different forms of dance and teaching at a dance adjunct of the Eastman School of Music in Rochester, New York. Her series of concerts at the school starting from April 18, 1926 began her role as mentor to thousands of dancers and students and a visionary to millions. In fact, she considered 1926 her real beginning. Because Venus and Rahu are situated in the natal 11th house of friends, networking, and financial gains, Venus-Rahu *Dasha/Bhukti* would be very likely to bring her most important networks of friends, associates, and collaborators. Those dates were: Feb. 22, 1924 to Feb. 21, 1927.

If Venus-Rahu period created the foundation, **Venus-Jupiter** *Dasha/Bhukti* was her career launch pad, and she profited because it was during the *Maha Dasha* of her well-placed Ascendant lord. This factor gave her a very different destiny from Prostakoff. The strength of the lords of subsequent *Dashas* gave good longevity to her dance career right up to the Jupiter *Dasha*. Though she would live for another 14 years into the Jupiter *Dasha*, it was not the period for which her choreography was most noted. In fact, one could study her **Jupiter-Venus** period to see what possible excesses occurred: Nov. 4, 1985 to July 5, 1988. We know that in Oct. 1987, at age 93, she premiered her final dance, "Persephone," choreographed to Stravinsky's Symphony in C.

A woman apart – "charting the graph of the heart" through movement: Her natal Moon is placed in the 3rd house from the Ascendant. The 3rd house is dance, drama, and music – central to her life – also courage and bravery. In this case, 3rd house lord Moon is situated in its own house, receiving no bad aspects. Moon in the 3rd house is always concerned with what initiative it should or should not take. Being a *Chara* (active) Moon, this promotes a life of action. Reading from the natal Moon in Cancer, there is an excellent placement of planets. Sun in the 10th house from the Moon is excellent for visibility. Venus exalted in the 9th house from the Moon is also excellent. The *Parivartana yoga* between Venus and Jupiter occurs between 9th and 11th houses, favoring finances – always a challenge for a dancer or a dancer-choreographer seeking funding for their own company. It also favors a wide network of friends and associates. Saturn is well placed in the 3rd house from the Moon, and Jupiter is good for finances in the 11th house from the Moon. It also lends expansive energy to the 5th house of education and creative projects.

Only the Moon itself, with no planets in houses either side of it (*Kema Druma yoga*), would indicate some loneliness and some sense of separateness from other people. There is some correction for this from the Sun and Mercury being in *Kendra* houses from natal Moon. It is in its own emotional sign of Cancer, made more complicated by its location in *Ashlesha nakshatra*, ruled by

Mercury. Further psychological complexity comes from *Navamsha* Moon in Sagittarius surrounded by two fiery planets, Sun and Mars. This describes someone with a very fiery mind and a strong, almost fundamentalist belief system about everything she did. It is a combination that brought some extremes of emotional trauma that permeated her life and her work. She wanted to evoke strong emotions through dance. *Navamsha* 9th house Moon shows that her desire to educate and teach, as well as to choreograph and perform would create her lasting influence – something for which a Taurus Ascendant person yearns in terms of establishing something concrete. She was revered for all of these, but her own preference was to be remembered above all as a dancer. It was difficult for her to retire from performing as a dancer even at the age of 74, already an advanced age for such activities.

Her deep interest in the psychological aspect of her art form may be attributed to her *Ashlesha* Moon. A person with **Moon in *Ashlesha nakshatra*** is never content with the meaning of something unless they have plumbed the depths of its contents. During her 10-year **Moon *Dasha*** (from Dec. 23, 1942), Graham choreographed some of her most notable works, including *Appalachian Spring* (1944) and *Cave of the Heart* (1946). Her most fruitful *Dashas* professionally were those of Venus, Sun, and Moon, spanning 36 years, though she would persist in her art form and in her considerable influence well beyond those years.

Long, sustained career of a High priestess of dance: We examine first the condition of the ***karaka* for the arts,** especially dance, and that is **Venus**. It must be strong for a long, sustained career in the arts. As Venus is also Ascendant lord, its condition doubles for the physical well being of the person, and we have already noted its excellent status as exalted planet in the birth chart, going to Libra in a *Kendra* position in the *Navamsha* chart, and again exalted in the *Dashamsha* in the 11th house of friends and associations. Graham had many artistic collaborations that were successful and long-term, including with sculptor Isamu Noguchi and composer Aaron Copeland.

Once we have established the condition of the Ascendant lord and the *karakas* for health and vitality and found them supporting a long and successful career, next we examine the **10th house lord in the birth chart** and track its position and aspects in the *Navamsha* and *Dashamsha* charts. In the birth chart, Saturn rules the 10th house and is very well placed in the 5th house of the arts and education. It is also in the sign of its friend Mercury, though Mercury itself is not favorably placed in the 12th house from Taurus Ascendant. The good position of the *nakshatra* lord makes the difference. Saturn is in *Chitra nakshatra*, ruled by Mars. Since Mars is well positioned in the 10th house of the birth chart, this gives dominance in one's personal and professional life, even if Mars and Venus are neutral to each other.

In addition, Saturn is *Vargottama*, recurring in the same sign in the *Navamsha* chart, along with sign lord Mercury exalted in Virgo, both receiving an aspect from *Vargottama* Jupiter in Taurus. Placed in the *Navamsha* 6th house, this is favorable for Saturn, as a malefic can do well in the 6th house to face the obstacles and conflicts that will come in life. Saturn in the *Dashamsha* chart is well situated in the 10th house in its own sign of Aquarius. *Dashamsha* Venus recurs exalted in Pisces in the 11th house opposite friend Mercury exalted in Virgo, confirming other very positive indications for the career, artistically and intellectually.

The combination of the fixity of the Taurus Ascendant and the strong Venus and Saturn show the strong sense of discipline enabling her to withstand the volcanic forward propulsion of her career, and to establish something more lasting in one of the most ephemeral art forms. Five planets in dual signs in the *Navamsha* chart show the immense versatility and multi facetedness Martha Graham possessed. She often designed her own costumes and lighting and had an interest in every aspect of her art form, digging deep into the culture in which she found herself: the indigenous cultures of the North American continent as well as many of the Greek myths.

> "I am absorbed in the magic of movement and light. Movement never lies. It is the magic of what I call the outer space of the imagination. There is a great deal of outer space, distant from our daily lives, where I feel our imagination wanders sometimes. It will find a planet or it will not find a planet, and that is what a dancer does."
>
> **Martha Graham**

Chart 9: Julia Child

Birth data: Thursday, Aug. 15, 1912, 23:30 PST, Pasadena, CA, USA Long. 118W08 37, Lat. 34N08 52, *Lahiri ayanamsha*: -22:38:09, Class AA data, from birth certificate. Ascendant: 13:16 Taurus.

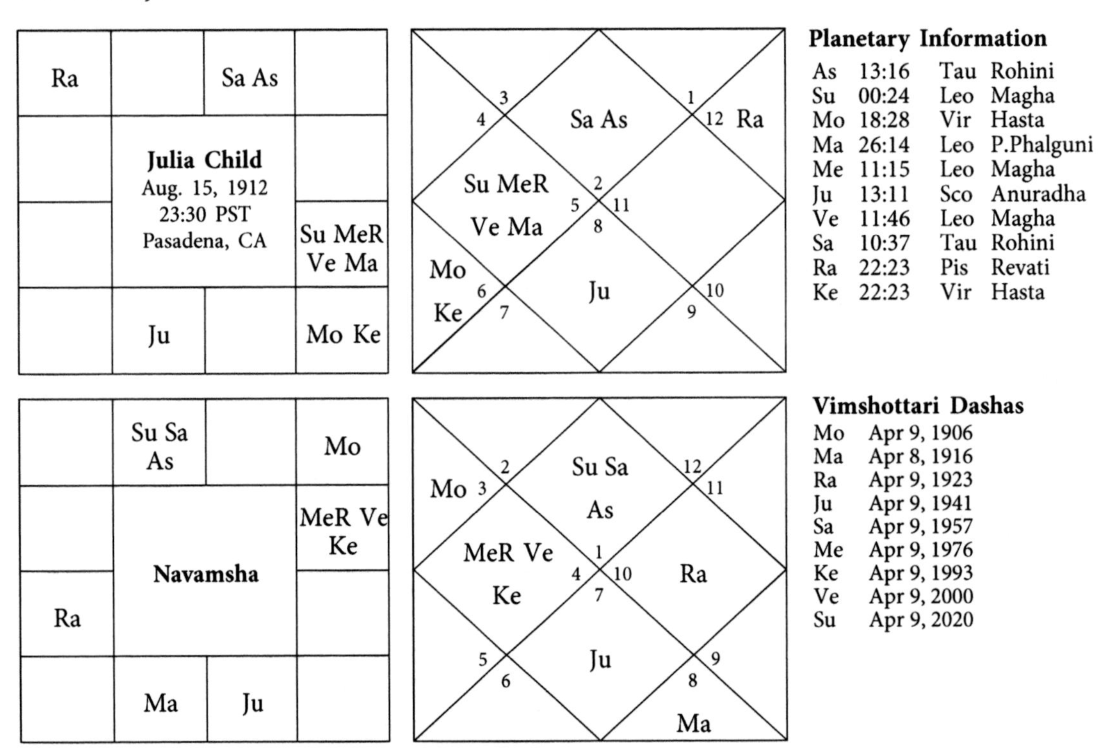

Planetary Information

As	13:16	Tau	Rohini
Su	00:24	Leo	Magha
Mo	18:28	Vir	Hasta
Ma	26:14	Leo	P.Phalguni
Me	11:15	Leo	Magha
Ju	13:11	Sco	Anuradha
Ve	11:46	Leo	Magha
Sa	10:37	Tau	Rohini
Ra	22:23	Pis	Revati
Ke	22:23	Vir	Hasta

Vimshottari Dashas

Mo	Apr 9, 1906
Ma	Apr 8, 1916
Ra	Apr 9, 1923
Ju	Apr 9, 1941
Sa	Apr 9, 1957
Me	Apr 9, 1976
Ke	Apr 9, 1993
Ve	Apr 9, 2000
Su	Apr 9, 2020

Biographical summary: Born Julia Carolyn McWilliams into a wealthy and distinguished southern California family, she had two younger silblings, a brother and a sister. With a privileged education and upbringing and surrounded by servants, her future as anything but a happy-go-lucky socialite looked unlikely, especially up to 1942. But she became a Master chef, writer or co-author of

17 cookbooks, entrepreneur, and television personality. All this originated from her most important and influential cookbook: *Mastering the Art of French Cooking* (1961). It was co-authored with two French women she met in Paris in 1949: Simone Beck and Louisette Bertholle. Simone Beck (b. July 7, 1904, d. Dec. 20, 1991) became Julia's close friend and her most important French collaborator.

Active as a culinary expert and spokesperson right up to her retirement in Nov. 2001 at the age of 89, Julia Child died on Aug. 13, 2004, two days before turning 92. Much beloved for her charm and expertise as well as her personal foibles, she reigned as America's celebrity chef for over forty years. A natural exuberance and unpretentious knack for spontaneous humor enlivened her culinary talents. In a 1966 cover article in *Time* magazine, she was dubbed "Our Lady of the Ladle."

As a teenager she had vague notions of becoming a novelist or a basketball player, especially with her 6 foot 2 inch frame. She received an Ivy League college education, though she was never a distinguished student and was known more for her pranks than for her academic prowess. She called herself "an adolescent until I was thirty." It may not have helped that her father was a staunchly conservative Republican Presbyterian who tended to equate intellectuals with Communists. Her mother was more of a free spirit, though not aggressively so. She graduated from Smith College and Julia followed her there, graduating in 1934 with a major in History. She did various jobs, including copywriting, and turned down a marriage proposal in summer 1941 from Harrison Chandler, scion of the Los Angeles-based publishing dynasty and Times Mirror printing executive, 1938 to 1968.

When the U.S. entered WWII she was eager to participate, but was rejected from entering military service due to being "too tall." Then in 1942 her desire to become a spy for the Secret Service led to a job in the newly formed Office of Strategic Services (OSS), a precursor to the Central Intelligence Agency (CIA). Her posts were in Washington, D.C., Kandy, Sri Lanka (then Ceylon), and briefly in Kunming and Chungking, China. Arriving in Ceylon in March 1944, Julia was put in charge of the OSS Registry, containing all classified documentation for the China-Burma-India Theatre. This included military plans and operations and top secret documents between U.S. government officials and intelligence officers. Her organizational skills quickly became obvious. She had the highest security clearance and was considered a very able and effective intelligence officer. By April 1945 the OSS was considering her as "spy material," but the war came to an end before that materialized. Prior to that she was also assigned to an experimental project – the Emergency Sea Rescue Equipment Section – in which she helped to develop a shark repellent. This was a critical tool during WWII, as it was coated on explosives targeting German U-boats.[126]

In July 1944 in Ceylon she also met her future husband, Paul Cushing Child (b. Jan. 15, 1902, d. May 12, 1994). A member of the diplomatic service and an OSS officer ten years her senior and some four inches shorter, he was in charge of the OSS Visual Display units (various war rooms with maps, drawings, and photographs), and he worked closely with the generals. She called him a "one-man factory." He was also an exceptionally cultivated man, fluent in several languages, an artist, cartographer, photographer, poet, violinist, and a 4th degree black belt in judo. She was immediately smitten but he needed some time to be won over to this enthusiastic but still relatively unsophisticated woman. Their courtship developed later in China in 1945 and in the States in 1946. To match his considerable gourmet tastes Julia started paying attention to the culinary arts, and in general made

[126] On Aug. 14, 2008, the National Archives released 750,000 pp. of previously classified OSS files, including 130 pages of Personnel file on Julia Child's OSS career, showing that she worked more extensively for them than as a file clerk. See also Jennet Conant's 2011 book, ***A Covert Affair: Julia Child and Paul Child in the OSS.***

every effort to be someone who would interest Paul Child. At the end of WWII, they both returned to the U.S. and after some further courtship were married Sept. 1, 1946 in Lumberville, PA.

In 1948 Paul was reassigned to the U.S. Information Service to run the Visual Presentation Dept. at the American Embassy in Paris. Julia was uncertain how

> "… a six-foot-two-inch, thirty-six year old, rather loud and unserious Californian [would fare in France]."[127]

The couple arrived aboard the S.S. America in Le Havre on Nov. 3, 1948. This was her first day ever in France and also the day she discovered French cuisine. On the road trip to Paris she and Paul had a memorable lunch in Rouen, one she would later call "the most exciting meal of my life."[128] They lived in France (five years in Paris, and one year in Marseilles) until late Oct. 1954, when Paul was transferred to other parts of Europe: first to Bonn, Germany, and later Brussels, Belgium and Oslo, Norway.

> "Those early years in France were among the best of my life."
>
> **Julia Child**[129]

In early April 1955 Paul was asked to return to Washington, D.C. and to his surprise underwent a full loyalty investigation, part of "the McCarthy era" when U.S. fears of Communist infiltration ran hot. Paul was targeted because of his friendships with other OSS staff members, notably Jane Foster, later found to be a Soviet spy. A full year of tension ensued for the couple who, though Left-leaning politically, were otherwise innocent of the accusations. Julia's wealthy Right-wing family connections helped to put an end to this chapter, even if Paul's government career advancement may have been affected. He retired from the Foreign Service in summer 1961 and soon became Julia's road manager, agent, general factotum and designer of the kitchen in their Cambridge, Mass. home that appeared on many television shows. Rarely apart, they had a very spirited and harmonious marriage. The couple was married for 48 years, and had no children. Paul suffered several strokes in 1989 and lived in a nursing home his last five years, up to age 92 (1994).

When not travelling they lived in Cambridge, Mass. from summer 1961, spending winters in Santa Barbara, CA from 1980. In 1963 they built a country home in Plascassier, near Cannes, France on property owned by Simone (Simca) Beck and her husband, Jean Fischbacher. Having enjoyed it for 29 years, Julia gave up the house in France in 1992, and the Cambridge, Mass. home in 2001, donating the kitchen to the Smithsonian National Museum of American History in Washington, DC, and her office and house to Smith College, who later sold it.[130] She donated her papers (1925-1993) to the Schlesinger Library, Radcliffe Institute at Harvard University.

---

[127] Julia Child & Alex Prud'homme, ***Bon Appétit: My Life with Paul in France***, 2006, p. 13.

[128] *Ibid.*, p.19. Julia recalled her first meal in France as an "epiphany"— half-dozen oysters on the half shell, sole meunière, green salad, cheese, a bottle of Pouilly Fumé wine and coffee. They dined at La Couronne (the crown), a local restaurant in business since 1345, and the oldest inn in all of France.

[129] *Ibid.*, p. 3.

[130] In her arrangement with Smith College, Julia formally donated her home to the college in 1990 but lived in it as long as she wanted. Purchased for $48,000 in Jan.1959, and renovated in 1961, the house sold for $2.35 million in early 2002. Funds went towards the building of a new 60,000 sq. foot Campus Center at Smith College – opened in Sept. 2003 – where students, faculty, staff, and the larger community could meet, and the values of "hospitality, welcome and exchange" could be cultivated.

Her interest in cooking first developed in 1945 while in China with the OSS, but it was while living in France that the spark of inspiration was ignited. She took intensive classes at the Cordon Bleu and also with the best chefs while polishing up daily on her college French. She went about her studies with whirlwind energy and dedication, encouraged by her husband, himself an enthusiastic gourmet and a lover of fine wines. His influence was central to her new career.

Her cookbook, *Mastering the Art of French Cooking*, quickly became famous and very successful. It was ten years in the making, from 1949 through 1958, and grew out of Julia's intense desire to make French cooking accessible to American women and to a larger American public. Her co-authors Simone Beck and Louisette Bertholle had already begun this project and were looking for an American collaborator. In 1951 the three women opened a cooking school in Julia's Paris kitchen: L'École des Trois Gourmandes (School of the Three Happy Eaters). Its insignia, designed by Paul Child, appeared on *Mastering the Art of French Cooking* and throughout Julia's Public Television show *The French Chef*. The book was released in Aug. 1961, with an official publication date of Oct. 16, 1961. The show made its on-air debut Feb. 11, 1963 and ran for 206 episodes. It became the longest-running program in the history of public television. Other of her television series followed, up through the 1990s. She was preceded in this field only by James Beard and Dione Lucas, distinguished teaching chefs with much shorter runs of cooking shows on American television (1946-1948). Julia Child was the first to make a television career of it, at least in the USA, and became a celebrity in her own right. Her presentation of French cooking was well received, and she was recognized with numerous awards in France as well as in the USA. Throughout her long career, she considered her most important role to the public was as a teacher, and she scrupulously avoided any commercial endorsements.

Personal interest in Julia Child continued to run strong as of her television debut in 1963. Numerous television, radio and film projects and skits have used her as a subject, sometimes as parody. Her personal favorite was Dan Aykroyd's 1978 sketch on *Saturday Night Live*. In the skit, he continues the cooking show even though his cut finger is bleeding profusely. The movie *Julie and Julia* opened Aug. 7, 2009, starring Meryl Streep as Julia Child and Amy Adams as Julie Powell. Based on a screenplay by Nora Ephron, one half of the story focuses on Julia Child's life in post WWII France, where she entered the field of cooking in a more serious way. It is based on both her cookbook and her memoir *Bon Appétit: My Life with Paul in France*, published posthumously in 2006 and co-authored with Alex Prud'homme, her great-nephew. In Ephron's screenplay, Julia Child's story is linked with that of blogger Julie Powell, who in 2002 cooked her way through all 524 recipes in Julia Child's first cookbook and wrote a lively account of it. Major themes include each woman finding her courage, passion and center of inspiration through cooking.[131]

"You have to want to learn, and you have to love to eat."

Julia Child in October 2001, talking to students at Smith College, while visiting as a famous Alumna-in-residence.

---

[131] Julie Powell's blog, *The Julie/Julia Project*, became her 2005 memoir ***Julie & Julia: 365 Days, 524 Recipes, 1 Tiny Apartment Kitchen***. (The paperback version was retitled ***Julie & Julia: My Year of Cooking Dangerously.***) Julia Child told a reporter she did not approve of the project, calling it "a mere stunt." Powell took this in stride, saying that Child's feisty, independent quality is what she most admired about her.

Haute cuisine with a unique voice: The classical 2nd house and sign Taurus is concerned with the mouth – both eating and speaking, or that which comes out of the mouth as well as that which goes into it. We can learn a lot by studying the chart of a Taurus Ascendant person who was well known both for her cuisine and talking about how to do it. **Mercury** is the planet of speech. In this chart it loses a *Graha yuddha* (Planetary War) with Venus. We will examine this shortly, especially in terms of her strange voice, as Mercury suffers some diminishment through losing the *Graha Yuddha.* Child had a high-pitched and warbling speaking voice that ended up being a part of her iconic personality. Charming, funny, knowledgeable and gracious, she could also be clumsy person, and her voice was part of that equation.

Side effect of weak Mercury – infertility: Since Julia Child's natal Mercury is also lord of the 5th house of children, its weakened condition contributed to her inability to have children. The couple's infertility is well documented. Astrologically, this can be seen through the diminished Mercury in addition to aspects of malefic planets Rahu and Ketu to the 5th house. This would also be true if 5th lord or *karaka* for fertility (Jupiter) contacts Rahu and Ketu. (Rahu and Ketu are foreign and can be separative to the process of bearing or raising children.) Natal 5th lord Mercury is situated in the *Navamsha* chart with Venus in Cancer, both of them also on the Rahu-Ketu axis. Further confirming the infertility issue, Sun is 5th house lord in the *Navamsha* chart and is afflicted by an aspect from classic malefic planet Saturn, both of which are mutual planetary enemies.[132]

Big breakthrough and long-term success: For this to happen, rulers of both the Ascendant and the 10th house must be well placed. In this case, Venus and Saturn, respectively, are the planets in question. Both of them reside in *Kendra* positions in the birth chart and *Navamsha* charts, Venus in its best *Kendra* position in the 4th house, giving it *Digbala.* Some malefic aspects to Venus and Mercury indicate Julia's eccentric style, and one that may come from a foreign source. Venus and Mercury reside in *Magha nakshatra*, owned by Ketu. Saturn is in *Rohini nakshatra*, owned by the Moon; and Moon is close to Ketu in the 5th house. This brought a flurry of creativity and an unusual career but denied children. Her first book was immediately recognized by major culinary experts and chefs:

> "Probably the most comprehensive, laudable, and monumental work on [French cuisine] was published this week, and it will probably remain as the definitive work for nonprofessionals . . . [It is] a masterpiece."
>
> **Craig Claiborne's** review in *The New York Times* when ***Mastering the Art of French Cooking*** was first published on October 16, 1961.

Master timing for a Master chef and entrepreneur: Saturn's placement in the Ascendant is excellent for the Taurus Ascendant person, as it is the best planet for the chart. At the same time its presence in the Ascendant is very demanding. It requires much hard work, but also creates several highly beneficial *Raja* and *Dhana yogas.* In addition, the close mutual opposition of Jupiter and Saturn in *kendras* indicates very good entrepreneurial skills. Part of the great karmic-dharmic blessing of this chart is also the timing of her 19-year **Saturn *Dasha*,** from April 9, 1957 to April 9, 1976, and also the **Jupiter-Saturn conjunction** Feb. 18, 1961. Because Saturn is so favorable in the chart, Saturn *Dasha* was bound to be excellent for her personally. But the outward manifestation of her

[132] For a woman, this is a difficult combination for getting pregnant. In a man's chart, he might have children, but have infrequent access to them due to life circumstances, including official or unofficial loss of custody.

success would not be apparent until the Jupiter-Saturn conjunction in Feb. 1961 in her 9th house of higher education and publication, and communication on a broader scale. (Her first cookbook was published eight months later.) At this time, the American public was perhaps more open to French cooking due to the recent arrival of John and Jacqueline Kennedy at the White House, along with their French chef and the First Lady's personal style of French chic. The Jupiter-Saturn conjunction occurred one month after Kennedy's Inauguration.

Delays and early setbacks are seen in numerous ways astrologically: A prominent Saturn can delay the events of one's life, even if it is a favorable planet for the chart. Saturn's name in Sanskrit is *Shani* (slow, or slow-moving one), as it is the planet the furthest out that can be seen with the naked eye and the last one classically considered in Vedic astrology. *Navamsha* Saturn is debilitated in Aries, but since some cancellation is provided in several ways, this shows that in Saturn *Dasha* she will start out with some apparent weakness that gradually turns into an advantage.[133] The corrections are as follows: **1)** *Navamsha* Saturn is with *Navamsha* Sun exalted in Aries; **2)** Venus and Mars are in *kendras* of the birth chart; and **3)** Venus is in a *kendra* in the *Navamsha* chart. The condition of *Navamsha* Saturn also shows both her slower start in life and the fact that she is more likely to rise up out of some disadvantaged or delayed situation and make a personal success out of it. Her extreme height probably delayed finding a mate. In the end it set her apart and contributed to making her an icon. She was also more likely to find a mate while in a foreign country. This is largely due to Jupiter situated in the 7th house of marriage and foreign residence in both birth and *Navamsha* charts. (More details on this will be discussed in an upcoming section).

Julia seldom dated in her 20s and rejected a marriage proposal in summer 1941. She was already 32 when she met her future husband in Ceylon in July 1944 ("late" for the era). Their courtship began in China in 1945 and they married in Sept. 1946 when she was 34. Regarding her first cookbook, it was only after rejection of their manuscript and the breaking of their first contract that she and her French collaborators found another publisher. A new young publisher (Alfred A. Knopf) and a 25-year old editor (Judith Jones) were able to see the potential of the manuscript, much to the later chagrin of those who had turned it down, notably Houghton Mifflin. They had the original contract but considered the project too encyclopedic: the published book contained 734 pages and weighed three pounds. Julia was 49 when the book was finally published in fall 1961, 13 years after the manuscript was begun. She was 50 when she began her PBS television series in Feb. 1963.

Though she considered herself too tall to look chic in the typical French fashions geared for smaller women, she made up for it with her cheerful, enthusiastic manner. Julia's personal style became her hallmark. A woman of class, *and* classy, while known for her down-to-earth ways, she was much adored by her widespread audience of viewers and readers. Brought up in a wealthy home with a hired cook, Julia Child refers in her first cookbook to the "servant-less home." This earthy Taurus showed her public how to prepare French-style cuisine, and **Moon in *Hasta nakshatra* (the hand)** in the earth sign of Virgo confirms the hands-on practical approach. Meanwhile, natal Saturn close to the Ascendant is the servant *(Shudra)* in the Planetary Cabinet. Through her example she showed the American people how to be of service in the kitchen and how to do it oneself. Her years of effort preparing French recipes for an American audience paid off.

---

[133] *Navamsha* Saturn qualifies as a *Neecha Bhanga Raja yoga*. The cancellation (*Bhanga*) of the debilitation *(Neecha)* qualifies the yoga as a *Raja yoga*, after some initial struggles. (See Glossary).

The Kitchen: Astrologically the kitchen is assigned to the affairs of the **4th house**, the heart of the home and Julia's domain. She has **four planets in the 4th house**: Sun, Mercury retrograde, Venus and Mars in Leo, ruled by the Sun – which resides in its own sign, and thus also owns the house. Mars is also associated with cooking, as it cuts and heats and its fiery energy is necessary both to cook food and digest it. The artistry and love of cooking relates to Venus, and communicating about anything relates to Mercury. Each one of these planets is in the 4th house, assigned to the kitchen. Mercury and Venus recur in the 4th house of the *Navamsha* chart, along with Ketu, this time in Cancer. (As the classical 4th house sign, Cancer concerns food and cooking.)[134] The *Navamsha* 4th house receives aspects from several malefic planets, as noted above, so it is surprising that her home (also 4th house) was not more emotionally turbulent. What these planets seem to be reflecting instead is that Paul and Julia Child's kitchen in Cambridge, Mass. was the hub of an enormous amount of activity. It was extensively remodeled, and served as the set for three of her television shows over many years. It was the 17th kitchen they designed together, and a home they owned for 43 years. The fixed sign emphasis contributed to the durability. Four planets in Leo contributed to the drama in the kitchen – Julia's cooking shows.

Venus and Mercury are within one degree of each other in the same sign, and thus in ***Graha yuddha*** **(Planetary War)**. Normally, in a *Graha yuddha*, the condition of the losing planet is destroyed, especially its capacity to rule over the houses in the chart that it owns. Venus is the winning planet here and Mercury is the loser. Fortunately in this case, there are some modifications that do not cause as much devastation to Mercury as might otherwise occur Otherwise, she could have had a major problem with her voice and finances. These modifications are as follows:

1. The *Graha yuddha* is between planetary friends, especially friends of the Ascendant lord.
2. The *Graha yuddha* is placed in a *Kendra* or a trine, though not in the sign of a friend to Venus.
3. The *Graha yuddha* is well placed by sign and house in both *Rashi* and *Navamsha* charts.
4. The sign lord (Sun) of the *Graha yuddha* is well placed and in good condition.
5. The *nakshatra* lord (Ketu) of the *Graha yuddha* is well placed and in good condition.
6. The losing planet participates in a *Raja* and *Dhana yoga* with the winning planet.
7. The *Graha yuddha* is *not Gandanta*, especially not in Scorpio.

If we examine item #6 above, we see that this Mercury-Venus duo forms a powerful *Raja yoga* with the Sun and Mars, as lords of the 1st, 4th, 5th, and 7th houses from the Ascendant. This combination includes two trinal houses from the Taurus Ascendant, and two *Kendra* houses. When Lords of both these houses are interacting strongly in a *Kendra* or trikona house and not afflicted by classical or temporal malefic planets, they enable the person to combine the qualities of personal talents with good timing and connections with people.

Her first cookbook was published Oct. 16, 1961 during **Saturn-Mercury** *Dasha/Bhukti* (April 12, 1960 to Dec. 21, 1962). We have discussed the astrological reasons why her initial disadvantages would be overcome as she moves further into Saturn *Dasha*. It took over six years to find a publisher and even then, the publisher (Knopf) was considered very daring to take on such a project. Julia and her co-authors were completely unknown and untried, but they had worked long and hard on their

[134] Even the address of Paul and Julia Child's Cambridge home reflects the number four: 103 Irving Street.

book. Saturn-Mercury period would reflect the weaknesses of both *Dasha* lord Saturn (debilitated in Aries in the *Navamsha* chart) and *Bhukti* lord Mercury (loser of the *Graha Yuddha* in the birth chart), though each planet is granted an uplift, especially Saturn. Saturn improves due to the *Neecha Bhanga Raja yoga* and Mercury's condition is mitigated by various factors listed above.

Saturn-Mercury period brought both Paul's retirement from the Foreign Service in 1961 and his immersion in Julia's new career. Jupiter is in the 7th house of partnership in both birth and *Navamsha* charts, and Mercury is in the 4th house in both charts, and thus the 10th house (of career and status) from Jupiter. Paul had hoped for greater advancement in his Foreign Service career, but his career with Julia actually became more personally rewarding and lucrative. Mercury's close contact with Venus, planet of love, becomes another defining factor, even if loser to Venus in the *Graha Yuddha*. Paul never sought to be in the spotlight and was uncomfortable there. But he was always close by attending to details for Julia and insuring that all would go smoothly.

Julia's 17-year **Mercury *Dasha*** (April 9, 1976 to April 9, 1993) was successful for her career, confirming that early difficulties brought by Mercury would be overcome, and in fact Mercury brought even more wealth during this period with the success of more cookbooks and television projects. However, it would not necessarily bode well for her husband during **Mercury-Jupiter** *Dasha/Bhukti* (April 24, 1988 to July 31, 1990), as he suffered several strokes in 1989, and never recovered. He lived for the next five years in a nursing home. Mercury sub-periods from 1944 onward brought big events concerning her mate. (They met in Jupiter-Saturn *Dasha/Bhukti* and their courtship and marriage took place in **Jupiter-Mercury** *Dasha/Bhukti*. With Mercury as 5th lord and Jupiter placed in the 7th house of marriage, these two planets (both angular) with their mutual contact through *Dasha-bhukti* would bring the joining of creative, romantic and business interests. Paul and Julia enjoyed all of it together and their individual talents enhanced their joint projects.[135]

As ruler of the 2nd and 5th houses, Mercury is *Dhana yogakaraka*, a planet of wealth for this chart. Julia Child grew up in a wealthy upper middle class family, and then moved into a marriage that became more financially comfortable, followed by great success in her own independent career. Therefore, Mercury was clearly not ruined as a bringer of wealth, though it did not give children (as ruler of the 5th house) and gave an unusual quality to her communication style, including an odd, warbling voice. Though her voice was flawed for a public speaking career, nonetheless it was part of her outspoken and unmistakable style. Her easy and cheerful manner worked well on her television shows. If something did not go as planned, she could turn it instantly into an amusing experience, especially in the earliest PBS shows when there was no budget for rehearsals. This endeared her to her audiences, and immediately put her on the television map. With natal **Retrograde Mercury** there is a tendency to review information far more than the average person. Despite appearances to the contrary, Julia was a perfectionist and often prepared 19 hours for a 30-minute television segment. In the 10-year period while writing her first cookbook, she repeatedly tested all the recipes, often with feedback from a close circle of family and friends.

---

[135] When important *Dasha-Bhuktis* appear in reverse, we can often expect the beginning and ending of an important chapter in life or an echo effect. Jacqueline Kennedy Onassis watched her first husband get shot in the head in Dallas, TX in her Moon-Rahu period, and was herself diagnosed with a terminal brain tumor in her Rahu-Moon period. Her natal Moon in Aries – in the 7th house of partnership – gives clues pertaining to the head. (See Chapter 9, Chart #31.) Marilyn Monroe was born in Mars-Jupiter period and died in Jupiter-Mars period. This volatile and vulnerable planetary combination was in her 8th house. (See Chapter 6, Chart #17.)

The Educator inspired by foreign experience: Mercury forms a *Raja yoga* (as 5th house lord placed in the 4th house) that favors the education and capacity to educate others. Waxing Moon in the 5th house can give high intelligence and a strong ability to educate and be educated. (In the 1930s, only 5% of American women received a college education, and a third of Julia's class of 645 dropped out before graduating. Julia did graduate, as her mother before her.) But as **Moon** is with **Ketu**, she is less likely to be inspired by the usual cultural setting for her era. Natal Moon contacting Ketu explains her offbeat style as well as her desire for foreign travel and experience. Rahu or Ketu gives an odd or foreign twist to the education. She became galvanized to investigate cooking styles when she traveled to the Far East, especially China, then later to France. Prior to that, her first recipes for the OSS were for shark repellent, as sharks could set off underwater explosive devices intended for enemy submarines. This is the influence of Ketu. Later on, with the success of her television shows, for over twenty years she and Paul resided half the year in their second home in France, built in 1963.

Success emanating from foreign residence is shown by several factors: **1)** Jupiter has a powerful role in the 7th house of foreign residence from Taurus Ascendant; **2)** Jupiter is also in the 7th house in the *Navamsha* chart, even more powerful as 9th house lord; **3)** Saturn in Taurus is well placed in the 9th house of foreign travel from natal Moon in Virgo; **4)** Four planets are in the 12th house (of foreign residence) from natal Moon; and **5)** Mars and Saturn in Aries are in the 12th house in the *Dashamsha* chart (10th harmonic, career matters). All of her knowledge of fine cuisine and fine wines came out of France, and began in the Far East, Chinese cuisine being her second favorite to that of the French. A dual sign emphasis would have turned her focus to both styles. Instead, her fixed sign emphasis gave her the tendency to go in one direction and to be known for one major endeavor in its various facets. From this, she revolutionized the way Americans regard French cooking and French cuisine. She also championed fresh ingredients in an era when American housewives were dependent on frozen and canned foods. However, it was before the era of organic produce, and when this movement became more predominant, she did not embrace or welcome it.

The revolutionary quality evoking an Aries Ascendant is shown in this case as natal Mars in the 4th house contacts Ascendant lord Venus in the same house and sign; and natal Mars aspects the 10th house by opposition aspect. In addition, this is someone who does what she loves (Venus) in collaboration with her husband (Mars, and also Jupiter), and has a wonderful time doing it. Supporting this outcome is the position of Saturn in the Taurus Ascendant, and in the Moon's *nakshatra*. Moon in turn is in its own *nakshatra* of *Hasta* (the hand) – showing the importance of her working with her hands. Natal Moon is well placed in the 5th house of education. Saturn in an earth sign works extremely well for Taurus Ascendant for business and financial success, especially in this case: Saturn forms a *Dhana yoga* of wealth with Jupiter (9th and 11th house lords combined). Thus Saturn is a great blessing in this chart, as is Jupiter. Their mutual opposition in both birth and *Navamsha* charts in *kendra* houses confirms the likelihood she would flourish as an entrepreneur.

Success in marriage and business partnership: When the birth chart contains a fixed ascendant and six planets in fixed signs, there is generally a steadiness and a desire for continuity. For Taurus Ascendant, Saturn is a stabilizing factor in marriage when it aspects the 7th house of marriage and partnership in both birth and *Navamsha* charts, if other afflictions to Venus, Mars or Moon do not interfere. This can also indicate a significant age difference from one's mate. (It was over 10½ years in this case.) Connection between the 1st and 7th lords, Venus and Mars, respectively, is made here, as each is situated in the same house in the birth and *Navamsha* charts. All these factors indicate the destiny for a harmonious relationship. Mars placed in the *nakshatra* of Ascendant lord Venus bodes well for the mate in her destiny. Some unusual, exotic, or foreign element in love matters occurs when Venus is on the Rahu-Ketu axis, especially if located in one of the relationship houses (3, 7, or 11). In this case, Rahu and Ketu contact Venus only in the *Navamsha* chart, and then on the 4-10 house axis. This describes some turbulence or extra activity in the home or career. Julia Child's public career was based out of her kitchen in Cambridge, MA. That kitchen is now on exhibit at the Smithsonian Institute in Washington, D.C. – another confirmation of her status as a cultural icon. Paul was uniquely equipped to help her in all aspects of her new career, including with his training as a Visual display artist for the OSS and later the U.S. Information Service.

Jupiter is exactly on the Descendant of this chart, and would bring many opportunities for travel and partnerships, especially from the start of her 16-year **Jupiter *Dasha*** April 9, 1941. As we have noted, Jupiter provides a *Dhana* (wealth-producing) *yoga* by its position in the *Kendra* of the chart. While 8th lord in the 7th house is not normally good for the longevity of the partner or of the marriage, sometimes it seems to work in the West if there is some distance or independence between the partners, or some foreign connection. The theme is repeated in the *Navamsha* with 12th lord Jupiter in the 7th house. However, *Navamsha* Jupiter is also 9th lord, and so is an excellent planet for the Aries *Navamsha* Ascendant chart. Ninth and 12th houses are associated with foreign travel and residence, respectively, and she met her husband while working in Ceylon (now Sri Lanka) in 1944. **(Jupiter is also the *karaka* for knowledge and for foreign travel.)**

Jupiter's potentially damaging effects on her marriage are minimized by the excellent house positions of Jupiter's sign and *nakshatra* lords, Mars and Saturn respectively. Located in Saturn-ruled *Anuradha nakshatra*, Jupiter operated more favorably in the life and brought good results during its *Dasha*, though best career results would not occur until Saturn *Dasha*. Jupiter served to broaden her horizons at many levels, and did not perform as excessively as in some of the other Taurus Ascendant charts we have seen, again in part due to Saturn's aspect and its influence as lord of the *nakshatra* where Jupiter resides. This would also temper natal Jupiter's effect in the sign of Mars, aspected by Mars in Leo. This shows Paul's tempering effect in her life, along with his energizing, expansive influence. As Jupiter rules the 11th house of friends, it shows her many friends. Among them was her lifelong editor at Knopf, Judith Jones (from 1960) and Avis DeVoto (from 1952), wife of Bernard DeVoto (1897-1955), the American historian and critic. Avis championed Julia's first cookbook and acted as unofficial agent and editor from 1953, steering Julia to potential American publishers.[136]

Here is a summary and review of the astrological factors behind the late start to her marriage and career:

1. **Saturn** in the Ascendant aspecting 7th house can cause delays with marriage, though it is an excellent planet for Taurus Ascendant, providing *Dhana* and *Raja yogas* by its presence in the Ascendant. Even so, it can also cause delays because Saturn's inherent nature is slow and deliberate. In addition, *Navamsha* Saturn is debilitated in Aries, and though *Neecha Bhanga* (corrected) it brings some early disadvantages and/or delays in Saturn *Dasha*. Further, Saturn's transit plays a role.

[136] See ***As Always, Julia: The Letters of Julia Child & Avis DeVoto: Food, Friendship, and the Making of a Masterpiece,*** edited by Joan Reardon, 2010. This close and lively friendship began in March 1952, with Julia's fan letter to Bernard DeVoto at *Harper's magazine*, and lasted until Avis DeVoto's death in 1989.

2. **Jupiter *Dasha* (April 9, 1941 to April 9, 1957)** was most likely to bring marriage and partnership, as *Dasha* lord Jupiter is placed in the 7th house of marriage and partnership in both birth and *Navamsha* charts. (The previous 18-year Rahu *Dasha* was too volatile to bring commitments, with Rahu contacts to natal Moon in the birth chart and to Venus-Mercury in the *Navamsha* chart. She was jilted by her first beau in 1936, during her Rahu-Venus period.) Summer 1941 immediately brought a marriage proposal from the wealthy Harrison Chandler; but Julia turned it down as she felt lukewarm about it. Dating seldom in her 20s and early 30s, Julia met her future husband in July 1944 during Jupiter-Saturn *Dasha/Bhukti*. The Saturn sub-period (May 28, 1943 to Dec. 8, 1945) was the pivotal period during the *Dasha* for foreign experience and for meeting a future mate in a foreign setting. This theme is further emphasized as the two planets are on the marriage and partnership axis (Houses 1-7) in both birth and *Navamsha* charts. Marriage was in Jupiter-Mercury.

3. **Jupiter-Rahu *Dasha/Bhukti* (Nov. 14, 1954 to April 9, 1957)** was the last sub-period of Jupiter *Dasha*. Rahu as sub-lord can take away what was given at the start of the *Dasha*. On Oct. 23, 1954, Paul and Julia arrived in Germany, a bureaucratic USIS decision re-posting them from their beloved France and not totally logical, as Paul spoke no German. It was also far less convenient with Julia no longer residing in France, as she and her French co-authors continued working on their cookbook. Negotiations with Houghton Mifflin publishers in Boston began in late 1953 and a contract was signed June 1, 1954. Julia received one third of the $750 advance. (She was still in Jupiter-Mars *Dasha/Bhukti.)* Then on April 7, 1955 a telegram came summoning Paul back to Washington, D.C. – a message Julia was hopeful meant a promotion for Paul. But it was a month-long ordeal of anti-Communist loyalty investigations and the start of a year-long period of anxiety and suspension as to where this would lead for Paul's career and reputation and how it would impact both their lives.

4. **Saturn *Dasha* (April 9, 1957 to April 9, 1976)** is most likely to bring career success in Julia's case. Saturn is lord of the natal 10th house of career and placed in Venus's sign of Taurus. Venus in turn is well placed in *a Kendra* of the char*t, Digbala* – or bes*t Kendra* location for Venus. Saturn is located *in Rohini nakshatra,* owned by the Moon, which is well placed in the 5th house, though with Ketu. Projects initiated after the start of her Saturn *Dasha* in April 1957 would have good fortune and longevity, but after some initial delays or setbacks in the first few years of the *Dasha*. This occurred in several areas of her life and are shown by*:* **a)** natal Saturn on the Ascendant; **b)** Saturn debilitated on th*e Navamsha* Ascendant; and **c)** 10th lord Saturn transiting Sagittarius in Julia's natal 8th house (of loss), called *Ashtama Shani*: Nov. 8, 1958 to Feb. 1, 1961, then again briefly Sept. 18 to Oct. 8, 1961. The book's publication date was Oct. 16, 1961. (During Saturn's next transit in Sagittarius Dec. 1987 to Dec. 1990, Paul suffered several debilitating strokes and Julia's close friend Avis DeVoto died, both events in 1989. Sagittarius is a sign of long ascension, so transits there last longer.)

   House in Cambridge, MA: The couple bought their house in Cambridge, MA in Jan. 1959 (in Saturn-Saturn *Dasha/Bhukti*), just prior to departing for Oslo, Norway – their last foreign post, by mutual decision. While in Oslo Jan. 1959 to July 1961, they rented out the Cambridge house. Upon their return to the States in July 1961 they lived in it for forty years, using the kitchen for many of Julia's television shows. In Nov. 2001, Julia retired to California, donating the house to Smith College and the kitchen to the Smithsonian Institute.

Book contract with Houghton Mifflin in Boston: After six years of negotiating with Houghton Mifflin and 5½ years after signing a contract, Julia received Houghton Mifflin's final formal rejection Nov. 6, 1959: "...too expensive to print, no prospects of a mass audience."[137] At noon that day in Oslo, tr. Mercury and Jupiter were in a Planetary War exactly opposite her Ascendant: tr. Mercury at 13:06 Scorpio, tr. Jupiter at 13:04 Scorpio. Literary efforts (Mercury) lose to Jupiter (larger forces, including the size issue). This planetary combination exactly describes the publisher's problem with the length of the cookbook manuscript: The authors had promised to condense it but had not done so and were reluctant to give up their original concept. Though tr. Saturn near her 8th house cusp was an indicator of endings, tr. Jupiter's return to its natal position at 13:11 Scorpio indicated some silver lining to this bad news. By Dec. 1959, Julia's friend Avis DeVoto had sent the same manuscript to editors at Alfred A. Knopf publishers, and by spring 1960 negotiations began, with Knopf insisting the title be changed from *French Recipes for American Cooks.* In Aug. 1961, over four years into her Saturn *Dasha, Mastering the Art of French Cooking* was first printed by Knopf. The official publication date was Monday, Oct. 16, 1961 – fortuitous, as tr. Jupiter and Saturn had both turned Stationary Direct in late Sept. On Oct. 8, 1961 tr. Saturn exited Julia's 8th house and entered her 9th house of publishing, joining tr. Jupiter, both in the *nakshatra* of *Uttara Ashadha*, "the later victor." It was a waxing Moon on a Monday, a day ruled by the Moon – in turn, ruler of food and kitchens. Tr. Moon was in *Purva Ashadha*, "the earlier victor." After so many years of effort, this seems a tongue-in-cheek message. But at age 49, Julia's long public career in cuisine was just beginning.

5. **Jupiter-Saturn conjunction Feb. 18, 1961** was exquisitely timed to trigger the publication of her book (the culmination of a 10-year project), and the start of a business partnership that was already a very successful 14-year marriage. Paul retired from the Foreign Service in summer 1961, six months earlier than planned, and soon moved into full partnership with Julia in her budding new career, both in television and books. Born Jan. 15, 1902, Montclair, NJ, time unknown, Paul was born just seven weeks after the **JU-SA conjunction Nov. 28, 1901** at 21:30 Sagittarius. At age 59, he experienced a major turning point with the return of both his *Dharma* and *Karma* planets: Jupiter and Saturn, respectively. His natal chart has four planets in Capricorn, including a very combust Jupiter in early Capricorn and natal Saturn in late Sagittarius, both strengthened by being *Vargottama* and in *Parivartana yoga* (mutual sign exchange). The 1961 JU-SA conjunction and the year leading up to Julia's publication date were all the more intense for him because of its close resonance with his own natal Jupiter and Saturn. His natal Sun and Jupiter were within one degree of the 1961 JU-SA conjunction, and his natal Saturn within five degrees, all of them in the same *nakshatra* of *Uttara Ashadha*. Though he had planned to retire to do his art, so much planetary energy in Capricorn would instead promote the expansion of business responsibilities, especially as of Feb. 18, 1961.

---

[137] *Ibid.*, front flap.

Physical health: Julia Child had mostly good health and vitality throughout her long life, and was known for her excellent physical endurance and hearty appetite. Sun and Mars are both well placed, but Ascendant lord Venus can be tempted by social habits, especially on the Rahu-Ketu axis in the *Navamsha* chart. She was a heavy smoker up to the time she was diagnosed with breast cancer in Feb. 1968 and soon after underwent a radical mastectomy of her left breast. There was a second cancer scare two years later, and she slowed down her schedule for a time, but that was the end of it. Astrologically, the **4th house** and the **Moon** are both associated with the breasts. At the time of the diagnosis and treatment, she was in the **Saturn-Sun** *Dasha/Bhukti*, with **Saturn-Moon** *Dasha/Bhukti* starting March 12, 1968. We can also track these periods from the *Navamsha* chart, as both *Dasha* lord Saturn and *Bhukti* lords Sun and Moon have some affliction in the *Navamsha* chart. *Navamsha* Saturn is debilitated in Aries, aspecting both Sun in Aries and Moon in Gemini. *Navamsha* Mars in Scorpio also aspects *Navamsha* Moon (with its 8th house aspect). Thus, Moon as 4th lord in the *Navamsha* chart is aspected by two classic malefic planets, as is the *Navamsha* 4th house, being aspected by Rahu and Ketu. Fortunately, *Navamsha* Sun and Mars are strong as classic significators of physical vitality. Sun is exalted and Mars is in its own sign in the 8th house of longevity.

At the time of her breast cancer, tr. Saturn in Pisces was opposite her natal Moon in Virgo, from April 9, 1966 to March 6, 1969. Since the Moon is also *Navamsha* 4th lord, we have Saturn aiming its focus on several symbols of hearth, home, nurturing, and breasts. Tr. Saturn opposite the Moon can be stressful on the emotional and physical health, compounded by the subsequent Saturn transit in the 8th house from natal Moon, from 1969-1971. Though Saturn is normally not as difficult for Venus-ruled Ascendant charts during its transits or *Dashas,* transiting Saturn did have an even larger capacity to affect her life while it was *Dasha* lord from 1957-1976. And in its capacity as a first class malefic and as *Dasha* lord, Saturn can still make an impact as it aspects the emotional planet Moon. In this case, the 4th house of the birth chart contains four planets. As three of these planets in Leo are in *Magha nakshatra,* ruled by Ketu, we pay special attention to Ketu's close proximity to the Moon in Virgo (2 degrees apart). It is not easy on the nerves or – in the 5th house – on one's confidence that one knows everything one should know.

With planets in Virgo, perfectionism can reign. This situation was no doubt exacerbated by the aspect of tr. Saturn in 1966, and probably coincided with a great deal of overwork. She was in the throes of launching her television shows and additional book projects. Saturn's transits to the Moon can bring more responsibilities and more work. However, natal Moon is in an excellent trinal house in its own *nakshatra,* even if on the eclipse axis. Since 10th lord Saturn is also located in Moon-ruled *Rohini nakshatra,* this points to the education that would feature her books and her television programs as her creative legacy. All royalties from her book sales go to the Julia Child Foundation for Gastronomy and the Culinary Arts, an organization promoting culinary education. Founded by Julia in 1995, the Foundation awarded its first grants in December 2006.

Retirement and death: Julia preferred to stay active throughout her life and was wary of retiring. She kept up a dynamic work and social schedule to the end, increasingly centered around her charming one bedroom apartment in Montecito, near Santa Barbara, California – her retirement home after leaving Cambridge November 9, 2001 at the age of 89. In late May 2000 the JU-SA conjunction in late Aries was soon followed within days by tr. Jupiter and Saturn in her Ascendant. This marked the third Saturn return to natal Taurus, and the Jupiter-Saturn combination would give her a last spurt of activity. This planetary combination was beneficial for her and resonated with her own *Raja yoga* formed by natal Jupiter and Saturn. For the last 7 to 8 months of her life she worked with her great-nephew Alex Prud'homme on the memoirs that were published posthumously in 2006. After her death, her co-author had the benefit of many family letters, diaries, and photographs. Paul Child had suggested this project in 1969, but it got delayed until early 2004.

Julia entered the 20-year **Venus *Dasha*** April 9, 2000, and the Venus-Moon sub-period on August 9, 2004. She died of kidney failure four days later – on Aug. 13, 2004, at 2:50 AM in Montecito, CA, just two days short of her 92nd birthday. Venus was just rising in Gemini, with Moon and Saturn in late Gemini in *Punarvasu nakshatra*, signifying "return of the light." Venus rules the kidneys, and any vulnerability in that area would be more featured in Venus *Dasha*. But Venus also rules over matters of love and the sweetness and pleasures of life, while Moon rules over cooking and hospitality. These were her lasting hallmarks.

> "I had always been content to live a butterfly life of fun, with hardly a care in the world. But at the Cordon Bleu, and in the markets and restaurants of Paris I suddenly discovered that cooking was a rich and layered and endlessly fascinating subject. The best way to describe it is to say that I fell in love with French food—the tastes, the processes, the history, the endless variations, the rigorous discipline, the creativity, the wonderful people, the equipment, the rituals. I had never taken anything so seriously in my life – husband and cat excepted – and I could hardly bear to be away from the kitchen."
>
> **Julia Child**, quoted in *Bon Appétit: My Life in France with Paul,*
> by Julia Child with Alex Prud'homme, 2006, p. 63.

REVIEW OF TAURUS ASCENDANT CHARTS:

Give the astrological reasoning in each case.

1. What two key factors in Chart #6 (the Male client) and Chart #5 (Jacqueline Du Pré) account for the difference in their longevity as well as physical and mental health?
2. What factors in the Taurus Ascendant chart show the tendency to want material security in life? How could this become exaggerated or even a source of conflict or illness with a Venus-Jupiter contact in the chart? Hint: Study whether their house and sign position improves in the *Navamsha* chart, and what is the condition of their *nakshatra* lord(s).
3. Why did the Venus-Jupiter contact generally work better for Martha Graham than for Theodore Prostakoff? (Hint: Note also the timing of their respective *Dashas.*)
4. What combination of astrological factors shows that Theodore Prostakoff would be unlikely to resume his stellar concert career after age 17? Hint: See both his birth and *Navamsha* charts. Note that 10th lord Saturn aspects no other planets.
5. What in Martha Graham's chart and timing assured that: a) she would not begin her dance career until age 22 (a very late start for a dancer); and b) even so, she would be highly successful?

6. Why does the combination of Taurus Ascendant, along with numerous planets in fixed signs tend to promote one long career, one long marriage, and a stable family life? Hint: Check birth *and Navamsha* charts for Charts #6 and #9 (Male client and Julia Child).
7. What in Martha Graham's chart accounts for her exotic and unusual artistic output, but at the same time proved unreliable for marriage and her love life?
8. What single planet accounts both for the delays in Julia Child's career – in terms of getting established (she was age 49) – as well as its sustained success?
9. Why was it highly unlikely that Julia Child's first cookbook would be published prior to 1961? What astrological indicators show her entrepreneurial talents?
10. What single astrological indicator shows Julia Child's strange and unique voice?
11. Why would Julia Child's Jupiter *Dasha* tend to be beneficial for foreign travel and marriage? Why would marriage not be favored until Jupiter *Dasha* (1941-1957)?
12. Although the Taurus Ascendant chart tends to show persistence, stubbornness, and the tendency to stay with one activity – what could indicate lack of continuity in marriage and partnership matters? Hint: Check the condition of Venus in all charts, both in birth chart and *Navamsha*; also, the relationship of 1st and 7th lords. Venus in a 6th house placement can indicate sacrifice in matters of art and love.

CHAPTER 5

Gemini Ascendant

Introduction

The third sign of the zodiac is Gemini. The ***Vaishya*** class – merchant or trading class – is equated with all three air signs in ancient Vedic literature. Gemini is a **dual air** sign with the Vedic aim of ***Kama***, or the desire to interact with other human beings. At times this can be motivated by the desire to do business or by sexual desire, as Mercury (its ruler) frequently has its mind focused on sexual union. We may not think of Gemini as sexual, especially as Mercury itself is a neutral and mental planet signifying the intellect, logic and discrimination. But it is colored by whatever is near to it, specifically the qualities of the planets that are either in the same sign or aspecting it.

The intertwined couple is a symbol for Gemini, though it is more generally known for the twins, which is the literal translation of the Sanskrit word for Gemini: ***Mithuna***. The myth of Soma and Tara helps in understanding this sexual orientation.[138] In the Puranic myth, Mercury is born to Jupiter's wife Tara, but is not his own child – though he is later forced to accept him as his own. Mercury comes from the union of Tara and Soma – the Moon in an incarnation as a man in this story. Mercury is finally accepted by Jupiter in large part to quell dissent among the gods, and also because the child Mercury is so witty and clever; thus the association of Mercury with youthfulness, mischief, and sexual desire, sometimes bringing messy relations due to the inability to settle for one choice and the discomfort with an abundance of confusing choices.

Another understanding from this teaching myth is Mercury's relationship with both Jupiter and the Moon (Soma). Though Mercury is heavily dependent on Jupiter for higher knowledge, and Jupiter's aspect to Mercury is considered the blessing of a guru, even so there is an uneasy relationship between them. For Jupiter, Mercury is the surrogate child. The same is true to some

[138] The Introduction to Chapter 6 (Cancer Ascendant) also discusses this myth.

extent between Mercury and Moon, in spite of Moon's acceptance of everyone. (The Moon has no planetary enemies, whereas for Mercury – Moon is its enemy. Soma is the interloper in the myth.)

With such mental focus on human interaction in Gemini, and at times sexual interaction, the Gemini Ascendant person may face more challenges deciding on a mate or a career, and when confronted with going only one route with any endeavor. There is a natural versatility which in turn often brings a variety of choices. Career-wise it is best to seek at least two directions or approaches. Saturn's influence in the chart shows how well self-discipline works in dealing with such a range of choices.

Vedic scriptures emphasize the need for constant awareness of the contents of the mind. Likewise in Vedic astrology we pay close attention to the Moon and Mercury to examine the emotional and intellectual aspects of the mind, respectively. As the ruler of Gemini Ascendant, Mercury represents not only the physical body but the workings of the rational mind and the architecture of the brain; also, the nervous system and its capacity to make connections throughout the body. As the planet of speech and communications, Mercury rules over the lungs. Mercury is associated with neurotransmitters in the brain, which are so vital to keep healthy and viable in order to have an empowered brain. Thus if Mercury is diminished in an astrological chart, there is potential for the person to suffer damage to the nervous system, and – if that connection adversely affects the Moon (memory) – there can be short or long-term memory loss. Therefore, just as we count on our computers to hold vast quantities of memory and respond to our touch pad commands, we count on a strong Mercury to give us access to our nerve-command systems. All of these functions are ruled by Mercury and are related to the sign of Gemini.

> **"Gemini constitute[s] a pair of human beings bearing a harp and a mace."**
>
> ***Jataka Parijata,*** Chapter 1, *Shloka* 8.

> **"Gemini Described: The sign Gemini ... represents a male and a female holding a mace and lute."**
>
> ***Brihat Parashara Hora Shastra,*** Chapter 4, *Shloka* 9-9.5.

These images of Gemini from Vedic astrology classics remind us of the contradictions inherent in any given Ascendant chart, ones that become especially notable in the dual sign Ascendants – Gemini, Virgo, Sagittarius, and Pisces. Perhaps the most iconic example of this phenomenon of "bearing a harp and a mace" is Albert Einstein (Chart #11), whose scientific discoveries led to the making and use of the atom bomb, but whose later years were spent as a passionate peace activist – so much so that he was on the F.B.I. "Watch List." This obvious paradox is also a tribute to the intellectual curiosity of Mercury and its desire to make its discoveries accessible. But with the application of Jupiter's higher knowledge, Mercury may discover how its mundane activities in the realm of science or commerce may have much larger unintended consequences – in this case, the unleashing of man-made weapons of mass destruction.

Especially because it is ruled by Mercury and is an air sign, Gemini is concerned with the rational mind in its most expansive mode. (In its exalted and ownership sign of Virgo, Mercury works in its most practical mode.) Planets in air signs and houses like to interact with other humans, and if the natal Moon is in an air sign, the person often prefers a more urban environment, close to the social, cultural, political and economic interplay with other human beings. But while Moon can thrive in Mercury's sign, the reverse is not true, as Moon is a planetary enemy of Mercury, and water signs in general are difficult for Mercury, adding an emotional (even if intuitive) component where neutrality and logic are needed. The same holds true if Moon contacts Mercury, or is in a mutual *Kendra*. Being the first of the air signs, Gemini is especially focused on developing mental and communication abilities. Gemini is concerned with language itself, literacy and numeracy (efficiency using numbers), and thus mathematics and music, as well as astrology. Each has a complex manipulation of a vast array of numbers and/or symbols. Not surprisingly, both the 3rd house in the astrological chart and the 3rd sign of Gemini are associated with music and the dramatic arts.

Gemini is focused on knowledge of the abstract and education in general, or the cultivation of the mind and intellect. Being colored by whatever it is near – Mercury-ruled Gemini is like the process of education itself. Sometimes the planet acts like a benefic and sometimes like a malefic, depending on the planets with which it associates. The Sanskrit word for Mercury is ***Buddhi.*** Lacking the higher powers of Jupiter, it is dependent on Jupiter for its higher knowledge and greater breadth of learning. **Gemini can be obsessed with making knowledge accessible.** Whether it takes that tendency too far, thus lowering the cultural level, will depend on various factors – including whether classic benefic planets Venus and Jupiter are well aspected and well situated. Have Venus and Jupiter added refinement to this picture, or subtracted it?

Being also a sign of commerce, especially at a local level, a Gemini Ascendant or a planetary emphasis in Gemini can make one a very successful marketer of one's ideas, whether selling Coca-Cola or a set of Sanskrit learning disks for a larger public.[139] A frequent theme of the Gemini Ascendant chart is that the person often does well without the traditional or formal education. There may be an obsession to learn particular subjects and an unwillingness to study other more conventional subjects, or ones expected of them. We see this to some extent with each of the Gemini example charts. Bill Gates and Peter Jennings chose to cut short their formal higher education in favor of commercial experience or research in the fields that interested them the most. With an auspicious configuration around the Gemini Ascendant, as well as around Mercury and the Moon (both planets of the mind), there is often a unique ability to focus completely on a specific area of learning that fascinates the individual. This can be accompanied by a dislike for formal schooling and the necessity to master a wide range of other subjects that are not compelling. Often this is compensated for by self-study, self-mastery of a subject – opening the way for a unique perspective.

The Gemini Ascendant person is often extremely curious about many subjects. With the dual nature of the sign they adapt easily to many modes of being, including different cultures and/or languages. Even if they have just one career, they will have at least two strong activities within any given field. Otherwise, boredom sets in. Dual signs can bifurcate, which is their strength and their weakness. With a zest for a high influx of information, they can easily handle the increasingly omnipresent information technology. A person with Gemini Ascendant or planetary emphasis tends to embrace the benefits of constant marketing and accessibility of information. Whatever its flaws, Gemini can usually sell anything, as it will find a way to make it compelling and interesting. To compensate, they also need to learn how not to be scattered,

---

[139] The USA chart has four planets in Gemini in the 7th house: Mars, Venus, Jupiter and Sun. This chart is discussed under Chart #2 (Robert F. Kennedy) and under Chart #25 (John F. Kennedy). See also Appendix.

since everything interests them at some level. Exhaustion arises from too much enthusiasm and non-stop mental activity. The mind becomes overloaded and stressed with too much information.

Add all these components to the portion of the zodiac associated with **Lakshmi, the goddess of prosperity, beauty, and auspiciousness (both material and spiritual).** Her association with abundance extends also to fertility, and chances are there is a fertile imagination in advertising and in the dramatic arts, and in stretching the parameters of what the public thinks of as possible. The capacity of Mercury to focus its energies successfully will be further described by how Mercury is configured in the chart: its house and *nakshatra* position, and aspects to it as well as to the Gemini Ascendant. By comparison to the fixed earth sign Taurus, the **oscillating nature of Gemini**, a **dual sign**, with its airy curiosity and mental dexterity, we now have an orientation that may seem less stable, but is capable of adjusting quickly to new circumstances and making its influence more widespread. The Vedic classics have even declared that the dual signs are stronger than the fixed signs, and the fixed stronger than the *chara* signs. The power of a dual sign Ascendant chart is enhanced if it is well aspected and if the planets are well placed from the Ascendant and from each other. Since Fire and Air signs are naturally complementary, fire and air can work well together. However, information (air) spread like wildfire can bring very negative results if the contents of the information and the motivation for spreading it are not dharmically pure.

Basic questions for a Gemini Ascendant person are: What fascinates me the most? How can I learn about it and go deeper into it? How can I share that information with others? In fact, how can I share it with the maximum number of people? Who can I meet partner-wise who will be interesting both sexually and mentally? Since I want to have both, will I need more than one partner? How can I avoid boredom in my relationships and career? How can I be in command of all the many interests I have?

The best planet for Gemini Ascendant is Venus, which is also *Nadi Yogakaraka*. The Sun is also considered friendly to Gemini Ascendant. Since Saturn rules the 8th house, its benefic role as 9th lord is significantly reduced. Thus, classically, Saturn is considered neutral to Gemini Ascendant, along with Mars and Jupiter, although Mars and Jupiter can be problematic to Gemini Ascendant. **Jupiter, in particular, can cause difficulties for the Gemini Ascendant person,** as it owns two angular houses (*Kendras*), and due to rules of *Kendradhipatya Dosha* (See Glossary), Jupiter's auspiciousness as a classic benefic is removed and it is rendered neutral. And as lord of the 7th house and thus a *Maraka* planet, Jupiter can inflict damage. There is also unease between Mercury and Jupiter, reflecting the myth of Tara and Soma. But as always, Jupiter's house position, sign and *nakshatra* need to be examined, along with aspects to it.

As with Aries Ascendant, Jupiter and Saturn are said classically not to be able to produce full *Raja yoga* as 9th and/or 10th lords.[140] These planets are also neutral to each other. But several notable examples in this chapter (Albert Einstein and Peter Jennings) show us that when Jupiter and Saturn are placed in the 9th and 10th houses from the Ascendant, especially in a mutual exchange of signs (*Parivartana yoga*), some extraordinary results can occur, if there are not a lot of afflictions to the chart. This is helped further by the *Raja yoga* created by Jupiter and Saturn from natal Moon in

---

[140] Parashara, ***Laghu Parasari*** *(Jataka Chandrika)*, Translation and notes by O.P. Verma, Ranjan publications, 2002, Chapter 5, "Auspicious and Inauspicious Planets," pp. 133, 140.

each case.[141] From Gemini Ascendant, both men have natal Jupiter in Aquarius in the 9th house and natal Saturn in Pisces in the 10th house in a *Parivartana yoga*. Jupiter is the classic planet of *Dharma* (9th house), just as Saturn is the planet of *Karma* (10th house). The 9th house is also the most powerful *Dharma* house while the 10th house is the most powerful *Karma* house, so this too adds to why planetary exchange between these houses brought great achievement and visibility in their respective fields.

Nakshatras contained in Gemini are: ***Mrigashira*** (starting from 23:20 Taurus, in the previous sign) 0 – 6:40 Gemini (ruled by Mars), ***Ardra*** 6:40 – 20:00 Gemini (ruled by Rahu), ***Punarvasu*** 20:00 – 30:00 Gemini (ruled by Jupiter). The Gemini Ascendant, or any planet situated in one of these three *nakshatras* is colored by being: 1) in Gemini, ruled by Mercury; 2) in one of these three *nakshatras*, ruled by Mars, Rahu, or Jupiter, respectively; and 3) by aspects to the Ascendant and Ascendant lord Mercury.

Again, Venus is the best planet for Gemini Ascendant and is *Nadi yogakaraka*.

Chart 10: Srinivasa Ramanujan

Birth data: Thursday, Dec. 22, 1887, 18:00 LMT, Erode, India, Long. 77E44 00, Lat. 11N21 00, *Lahiri ayanamsha*: -22:17:20, Class B Data "soon after sunset." [Sunset: 5:48:49 PM LMT] from the biography: *The Man Who Knew Infinity*, 1999, p. 11, by Robert Kanigel. Gemini Ascendant from 5:14 PM to 7:24 PM LMT, thus Gemini Ascendant known. Birth time rectified by the author. Ascendant: 10:40 Gemini.

Biographical summary: His full name was Srinivasa Ramanujan Iyengar, but he was commonly known as Ramanujan (Rah-MAN-ooh-jan). He was a genius at mathematics and though of Brahmin class – he came from a poor family and had many gaps in his education. In addition, he often suffered from malnutrition or starvation. Thus it was even more remarkable that after years of working alone in obscurity, he became arguably the most creative and original mathematician in the history of the world, and an inspiration in India and worldwide. Unfortunately, his early struggles and nutritional deficiencies probably cut short the unfolding of his brilliance, as mathematical genius usually flowers in the early years of life, and he died young at age 32.

As a student, Ramanujan was fascinated with mathematics, especially prime numbers, equations, patterns and sequences; but he was not motivated to study the other required subjects. This made it difficult for him to complete his schooling and attain the required credentials. At age ten he entered the local town school at Kumbakonam, and probably first came into contact with higher mathematics at that time, though his mother was gifted at math and may have been an early influence. By age twelve he had mastered trigonometry, and by age fourteen his brilliance was becoming apparent.

---

[141] Einstein's natal Jupiter and Saturn are placed in the 4th and 5th houses from natal Moon in Scorpio, creating a *Raja yoga* there as well as from the Gemini Ascendant. With natal Moon in Leo, Peter Jennings has Jupiter and Saturn forming a *Raja Yoga* from the Moon, being both 5th and 7th lords from the Moon.

Mo			As
	Srinivasa Ramanujan Dec. 22, 1887		SaR Ra
Ke	18:00 LMT Erode, India		
Su	Ju Me	Ve	Ma

SaR Ra
4 2
5 As 1
3
Ma 6 12 Mo
9
Ve 7 Su 11
8 10
Ju Me Ke

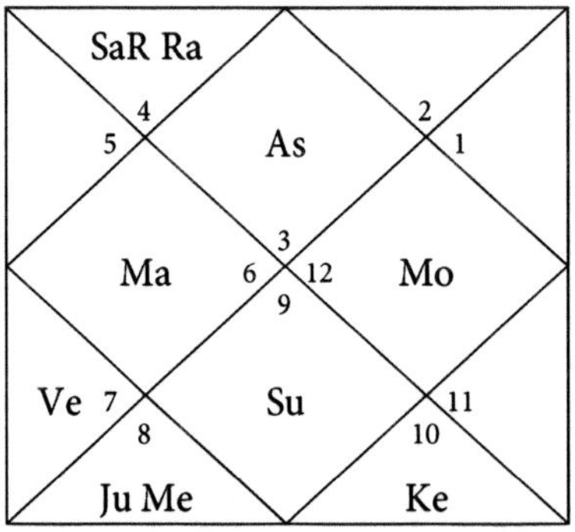

Planetary Information

As	10:40	Gem	Ardra
Su	08:07	Sag	Mula
Mo	10:40	Pis	U.Bhadra
Ma	15:54	Vir	Hasta
Me	23:01	Sco	Jyeshta
Ju	03:23	Sco	Anuradha
Ve	22:28	Lib	Vishakha
Sa	13:10	Can	Pushya
Ra	19:27	Can	Ashlesha
Ke	19:27	Cap	Shravana

	Ve	Ma	Su Ke
	Navamsha		
As Me			Ju
Ra		Mo SaR	

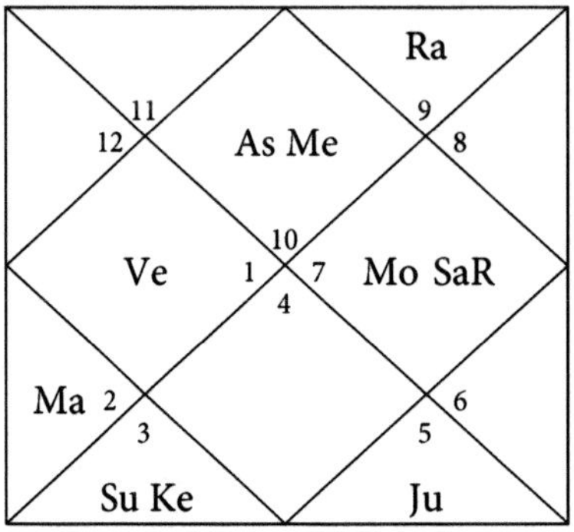

Ra
11 9
12 As Me 8
10
Ve 1 7 Mo SaR
4
Ma 2 6
3 5
Su Ke Ju

Vimshottari Dashas

Sa	Jul 5, 1877
Me	Jul 5, 1896
Ke	Jul 7, 1913
Ve	Jul 6, 1920
Su	Jul 7, 1940
Mo	Jul 7, 1946
Ma	Jul 7, 1956
Ra	Jul 7, 1963
Ju	Jul 7, 1981

But it would be another twelve years before there would be teachers, professors, or sponsors who fully recognized this brilliance. In the meantime Ramanujan suffered intermittent ill health, undergoing surgery in April 1909. He married July 14, 1909 and soon after began searching for work as a cleric or accountant. He was hired March 1, 1912 by the Madras Port Trust as an accounts clerk, and worked there for two years while continuing to pursue his mathematical ideas. He also received several scholarships from the University at Madras. Then in late 1912 and early 1913 he wrote to several academics in India and in England. Only G.H. Hardy, a top mathematician of his day and professor at Trinity College, Cambridge recognized his genius and invited him in the spring of 1913 to take a degree and study with a team of mathematics professors there. It was not an easy decision for Ramanujan to go, as orthodox Brahmins considered they lost their caste by traveling abroad. The issue was apparently resolved when Ramanujan's mother had a dream in which the family goddess Namagiri told her not to forbid her son's travel. Thus he went to England, leaving his family and young wife behind. He spent the years 1914-1919 there, leaving India by boat March 17, 1914 and arriving in London April 14, 1914. On March 16, 1916, he graduated from Cambridge with a Bachelor of Science by research, called a Ph.D. from 1920. (Except for the first four months, his stay in England coincided with the entire length of World War I. England declared war on Germany Aug. 4, 1914 and hostilities continued through Nov. 11, 1918, with the Treaty of Versailles signed June 28, 1919.)

There were many difficulties for him living outside of India, which greatly compromised his strict vegetarian diet and spiritual practices as a Brahmin. Food shortages in England during the war no doubt contributed to his general malnutrition and the rapid deterioration of his health. He became ill in England in 1917 and tried to commit suicide by throwing himself under a train in Jan. 1918. After numerous struggles, he was able to return home to India, leaving England Feb. 27, 1919, and arriving in India March 13, 1919. Upon his return to India in spring 1919, having received numerous high honors, such as Fellow of the Royal Society and other prestigious British societies, it was both an intellectual triumph and a religious dishonor for a Brahmin who left his country. Suffering from tuberculosis, he died April 26, 1920 at the age of 32. As his legacy, Ramanujan left behind numerous collected papers and notebooks which continue to inspire generations of mathematicians. His widow, who was nine years old when they married, lived for many years outside Madras, and died in 1994. She was devoted to him in their eleven years of marriage, enduring his absence for five years. By local custom, they did not live together until she was twelve.

> "An equation for me has no meaning, unless it represents a thought of God."
>
> **Ramanujan**

Ramanujan credited his acumen to his family goddess Namagiri and looked to her for inspiration in his work. Namagiri was a consort of the Lion God Narashima, Lord Vishnu's 4th incarnation. Because his British patron G.H. Hardy was an atheist, Ramanujan was unlikely to have shared with him his devotion to Namagiri and his general spiritual orientation to life. But though emotionally aloof, mathematician G.H. Hardy appeared to be Ramanujan's greatest champion during his short lifetime. In an anecdote about his brilliant protégé, he says:

> "I remember once going to see Ramanujan when he was lying ill at Putney. I had ridden in taxi cab number 1729 and remarked that the number seemed to me rather a dull one, and that I hoped it was not an unfavorable omen. 'No,' he replied, 'it is a very interesting number; it is the smallest number expressible as the sum of two cubes in two different ways.'" [The number 1729 turns out to be of huge interest in other mathematical ways, including being divisible by the sum of its digits.]

Workings of the mind of a math genius: To examine mental qualities we study the *karakas* Mercury and Moon, as well as the 4th and 5th houses and their lords. For this we will shortly discuss the powerful, but also problematic *Parivartana yoga* between Mercury and Mars, 4th and 6th lords respectively. We also note the *Sarva Ashtakavarga* bindus in those houses. (See Glossary) Higher than average numbers of bindus (30 and above) may show mental brilliance. In Ramanujan's case, the 4th house is ranked 37 and the 5th house 29. Thus, innate mental qualities are high, as are the possibility of higher educational degrees. The 5th house contains Venus in its own sign in Libra – excellent for Gemini Ascendant, and the best planet for this chart according to Nadi principles. But Venus would be much stronger here if not in a Jupiter-owned *nakshatra*. Jupiter can cause problems for the Gemini Ascendant chart, for reasons stated in the chapter Introduction. In this case, Jupiter is also located in the 6th house of conflict and health issues.

Moon in a watery dual sign (Pisces) and **Mercury** in a watery fixed sign (Scorpio) give good intuitive abilities. By his own admission this is the source of Ramanujan's originality. Moon in the 10th house aspected by Jupiter allowed him to receive some public attention for his extraordinary abilities, and in **Ketu *Dasha*** (July 7, 1913 to July 7, 1920). Ketu is placed in Moon-owned *Shravana nakshatra*. Its symbol is "the ear" and it gives a special capacity to listen.

Mercury and its attributes become a dominant theme for the Gemini Ascendant. We ask: How is the mind working? What is the nature of the mental capacity? What is the nature of the education? In this case, his education is incomplete and unconventional, lacking in substance for the most part. Yet somehow Ramanujan had enough extraordinary brainpower to come up with unique advanced ideas well beyond those of the leading mathematicians of his time.

Mercury is the planet most associated with mathematics, and in this case his intense obsession with mathematics is shown by Ascendant lord Mercury in Scorpio (a fixed water sign and the most obsessive sign), which, since it is also in the 6th house of conflict, accounted for his problems in neglecting other areas of his education and his difficulty finding colleagues who could understand his genius. The latter is also impacted by the isolation of natal Moon in *Kemdruma yoga*, with no planets in houses adjacent to the Moon, though this yoga is largely corrected by two factors (Jupiter's aspect to the Moon and Sun in a *Kendra* from the Moon). The inspirational effect of the family goddess on his intellectual mind, her direct whisperings in his ear of entire equations, he said, can be attributed to the five planets in water signs, especially Ascendant lord Mercury, Moon, 9th lord Saturn (divine powers are in the 9th house), and Jupiter (*Dharma* planet and classical lord of the 9th house). The fact that these whisperings would materialize into something more tangible, even in posterity in this case, is shown by Mercury in its own *nakshatra* of *Jyeshta*, situated in the *Navamsha Lagna* in its friend's sign of earthy Capricorn and receiving no malefic aspects. *Navamsha* Mercury is well placed (*Digbala*) and provides some much needed grounding for Mercury in Scorpio in the birth chart. It allows for the retention of numbers in the memory, as numeracy in this case was combined with the ability to memorize and recognize instantly a wide variety of numbers and combinations of numbers. In addition, we know the karmic fruit of Mercury would come due during **Mercury *Dasha*** (July 5, 1896 to July 7, 1913). Indeed, this was the most pivotal formative period for his mathematical genius.

This Mercury is also in a ***Parivartana Yoga*** with 6th lord Mars in the 4th house in Virgo. It makes Mercury's energy very intertwined with Mars as well as with Jupiter in the same house. Mercury's exchange with Mars enables Mercury to act as if it is in Virgo, the sign of its exaltation, and Mars acts as if in Scorpio, a sign it owns. Mercury in its exalted sign is far superior to its functioning in Scorpio. Mercury in Scorpio is in the water element, and while good for intuition, it is not a good element for Mercury in terms of retaining information and having a solid factual grounding in the chosen field. In the upcoming case of Albert Einstein, Chart #11, with his Mercury in Pisces, the facts will always be questioned. In Ramanujan's case, his general education was so incomplete that he was ignorant of other areas of math and culture in general, and thus had little context for most of his findings. This in part accounted for his innovations and in part made it difficult for him to have dialogues with the most advanced mathematical minds of his time. He was beyond them and yet had to communicate with them what he could perceive.

The influence of Mars can create some difficulty as 6th house lord, but Mars can also bring some benefit in its influence on Mercury. Its flaws involve overheating and overexciting the nervous system and the communications. (For **Mars-Mercury problems**, please review Chart #5: Jacqueline Du Pré. With her Mars-ruled Aries Ascendant, Mercury is a planetary enemy.) The benefits involve the incisiveness that Mars can lend to Mercury by sharpening its powers of observation and speeding up the mind to make it more adventurous and original. (See also Chart #13: Bill Gates, with Mars-Mercury in the 4th house, and Chart #25: John F. Kennedy, with Mars-Mercury in the 8th house.)

Health problems: It is generally not a favorable house position for the Ascendant lord to be in the 6th house of the birth chart, especially for the longevity of the person and their ability to deal well with disease. In addition, he earned very little when he did make a living - and for one reason or another he was often suffering from starvation throughout his life. The latter could arise from the afflicted 2nd house of income and of eating, as three malefics affect this house (Saturn, Rahu and Ketu), though Jupiter as classical benefic provides some relief by its trinal aspect to the 2nd house as well as to 2nd house lord Moon in Pisces. This provides some alleviation and brings some optimism through challenging years. We have discussed how Jupiter can cause particular havoc for the Gemini Ascendant person. We know it will affect his health, with both Mercury and Jupiter in the 6th house of health maters, even if Jupiter gives benefits to career visibility as 10th house lord contacting the Ascendant lord.

We have spoken of **Mercury *Dasha*** as being fortunate for the intellectual flowering of Ramanujan. But it was also very hard on his health and we know that the condition of Mercury in the chart would bear that out. Fortunately, Mercury is in its own *nakshatra* of *Jyestha* and improves in the *Navamsha* chart. This would enable him to survive through Mercury *Dasha*. He also made it nearly to the end of the 7-year **Ketu *Dasha***, which is even more demanding health-wise, as Ketu is located in the 8th house in the birth chart and in the 6th house with Sun in the *Navamsha* chart. This shows a potential for upheaval through major life transitions. At least Ketu (the foreigner) is helped by being in Moon-owned *Shravana nakshatra*. His five years in England (1914-1919), all within Ketu *Dasha*, coincided with war time and war-time food shortages. Ketu, Rahu and Saturn afflict the 2nd house of eating.

Congenital physical strength is also shown through the *Dasha/bhukti* at birth. Ramanujan was born in **Saturn-Sun** *Dasha/Bhukti* (because of natal Moon located in the third *pada* of Saturn's *nakshatra*, *Uttara Bhadrapada*). Sun is *karaka* of the physical vitality and Saturn is its enemy. This would not bode well for having a strong constitution throughout life and also hints at the difficulties he would have dealing with authority figures in his life.

Lack of formal education: With very little formal education and training, Ramanujan came up with ideas that were considered brilliant in his field. He solved many problems that no one else had ever solved. But it was difficult for him to properly convey the arguments for his math solutions and he had challenges to convey this to both his Indian and British colleagues. Education, especially degrees in higher education, is shown by the strength of the 4th house and by the disposition of both Mercury and Jupiter. The 4th house receives aspects from two malefics, Mars and Saturn, indicating his education would be afflicted in some way. There were delays during **Saturn *Dasha***, up to July 6, 1896. But later, Saturn – as 9th house lord – was beneficial in bringing him an unusual way to enter a university setting, being conjunct Rahu. This signifies anything or anyone foreign. Fiery Mars in the 4th house can show impatience with conventional education; but as 11th house lord (friendships and networks of helpful people) it brought him patrons and sponsors – enough so to continue his work. Venus in its own sign of Libra in the 5th house of intellect has some very beneficial effects on the level of his intelligence and the educational process, but is less stable being in *Vishakha*, a *nakshatra* owned by Jupiter – a problematic planet for the chart situated in the difficult 6th house. Nevertheless Jupiter is also the 10th house lord of career.

Mercury's exchange with Mars, as discussed above, is critical in transforming the early condition of obscurity into the highest form of national and international recognition in his field, though far more recognition would come posthumously. That whole process began during **Mercury *Dasha***.

The ***Navamsha* chart** created by my rectified time of 6:00 PM LMT gives a Capricorn Ascendant and provides explanation for his educational reprieve and for his ability to jump from obscurity in India to Trinity College, Cambridge. This is shown astrologically by *Navamsha* Venus and Mars in *Parivartana yoga* in the 4th and 5th houses. These houses are pivotal for a successful education and for the development of a first class intellect. This further confirms Venus in its own sign of Libra in the 5th house of the birth chart, giving enormous strength to his intellect and empowering his education, despite the problematic position of Mercury, which is guaranteed to cause health problems and potentially also psychological problems – being over-agitated by the Mars influence. *Navamsha* Ascendant lord Saturn is exalted in the 10th house conjunct *Navamsha* Moon, the lord of the 7th house of one's associates in the sign of Libra showing his career benefits through relationships.

Solutions for major cultural dissonance: The fact that ***Dharma* and *Karma* lords Jupiter** and **Saturn** are both in water signs shows that there would be a solution to his social, economic, and cultural problems – especially with Ascendant lord Mercury interacting with both of them and in Saturn's sign in the *Navamsha*. This can give determination and disciplined persistence. There would be some unusual solution here, enabling him to study abroad and to be among his peers in the field – even if there were racial tensions for a poorly educated Tamil Nadu Brahmin arriving in England and upstaging his colonial masters in academia. His departure from India coincided with the tail end of Mercury *Dasha*. It was in the **Mercury-Saturn-Jupiter** period.

Radical changes in the life: The 7-year Ketu *Dasha* began July 7, 1913. Eight months later he was on a ship travelling to England. Ketu is exotic or foreign: He spent most of Ketu *Dasha* in England, where he resided for five years, receiving higher education and honors there in mathematics. Since Ketu is located in the 8th house of the birth chart opposite Saturn-Rahu, this is already an indication of a radical change to his life that might be overwhelming. That his health suffered greatly from his years in England is also apparent from Ketu's position in the 8th house of the birth chart. This included the upheaval of being separated from his wife and family and from his culture of choice. Among other factors, Ketu's location in a Moon-owned *nakshatra* enabled him to survive most of Ketu *Dasha*. He died on April 26, 1920 during **Ketu-Mercury-Jupiter** period, just ten weeks prior to the start of his 20-year Venus *Dasha*.

Travel abroad: Natal Moon is in *Uttara Bhadrapada nakshatra*. It signifies "the latter one with lucky feet." It is also known as "the warrior star." Moon in this *nakshatra* usually denotes a person with a love of travel to faraway places. This 10th house Moon shows the strong influence of his mother, who made it possible for him to leave India. She also apparently had a gift for mathematics, though her formal education was minimal.

Reading the chart from the Moon as Ascendant (*Chandra Lagna*), there is excellent indication for foreign travel and study. Ninth house of foreign travel contains Mercury-Jupiter in Scorpio, with Mercury in a *Parivartana yoga* with Mars, interaction here between the favorable 7th and 9th house lords. Lords of the 4th and 5th houses from Moon (Mercury and Moon) are in the Ascendant and 9th house respectively, with a strong motivation to travel for educational purposes. Lords of the 4th, 5th and 9th houses aspect each other and are in *kendra*s or trinal houses from the Moon. Further, by studying these planets both from Moon and *Dasha* lord Mercury as Sub-*Lagna*s (applying to the period of Mercury *Dasha* only, 1896-1913), you can see a strong 9th house pattern involving either natal Saturn or natal Jupiter, respectively. Saturn on the Rahu-Ketu axis brought much hardship during his foreign experience. The same process can be used with ***Dasha* lord Ketu as sub-*Lagna*,** (applying to Ketu *Dasha* only, 1913 to 1920). From Ketu in Capricorn, Mars in Virgo falls in the 9th house. Since the *Parivartana Yoga* occurs between the 9th and 11th houses, that helped win the necessary support for his patronage at Trinity College.

Chart 11: Albert Einstein

Birth data: Friday, March 14, 1879, 11:30 AM LMT, Ulm Baden-Württemberg, Germany, Long. 10E00 00, Lat. 48N24 00, *Lahiri ayanamsha*: -22:10:27, Class AA data, from birth certificate. Ascendant: 19:28 Gemini.

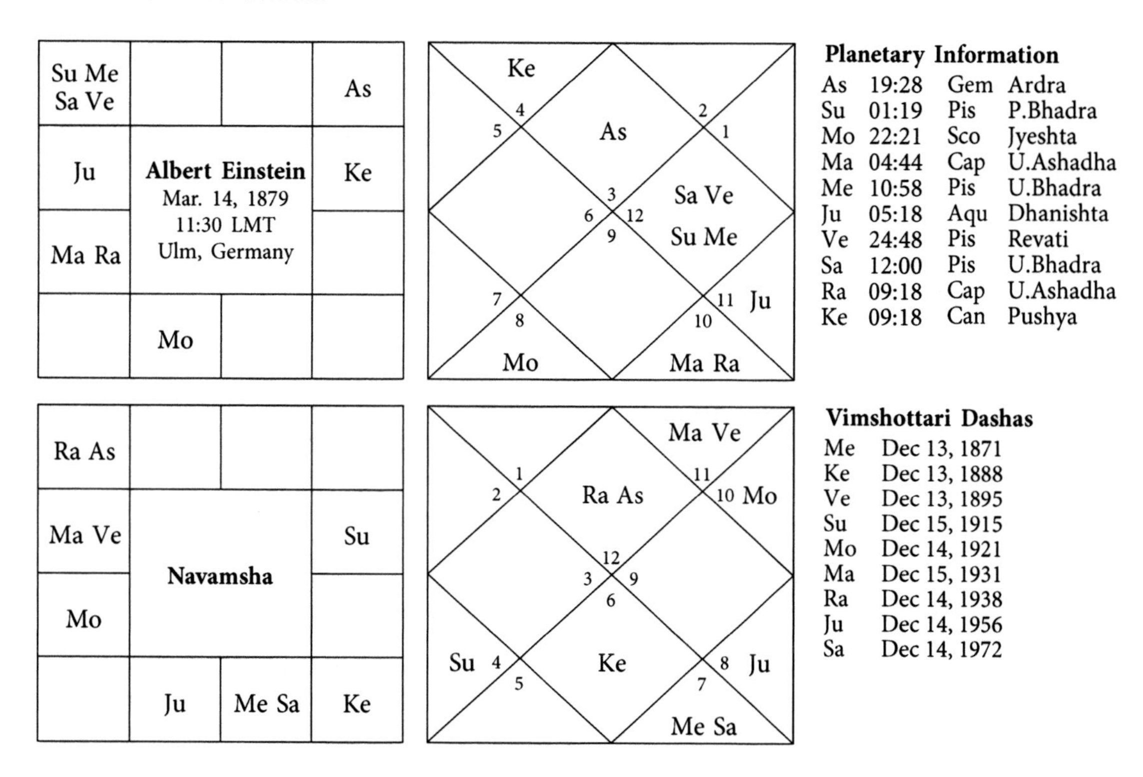

Biographical summary: A German-Swiss-American theoretical physicist who published more than 300 scientific works and more than 150 non-scientific works, Einstein is considered one of the greatest physicists of all time, and known for his Theory of Relativity. From 1914 to 1933 he was Professor and Director of Physics at the Kaiser Wilhelm Institute in Berlin, Germany. Einstein was lecturing in the U.S. when Hitler and the Nazi Party took control in Germany on Jan. 30, 1933. Hitler gained emergency powers one month later. On his return to Germany in mid-March 1933, Einstein (being Jewish) suddenly found his life in danger and his residential properties confiscated by the Nazis.

Forced to resign from the Kaiser Wilhelm Institute, he aborted his plans and never again entered the country. Instead, he immigrated to the U.S. in Oct. 1933. He became an American citizen in Oct. 1940, though retaining his Swiss citizenship, acquired in 1901. (In 1896 at age 17, he had renounced his German citizenship to avoid military service – with his father's approval – and only reinstated it in 1914 as a requirement to accept his post at the Kaiser Wilhelm Institute.) In March 1933 he renounced it again and in 1945, rejecting Nazi Germany's role in WWII, he said:

"I am not a German but a Jew by nationality."[142]

From 1933 to 1955 he resided in the U.S., holding a position at the Institute for Advanced Study in Princeton, New Jersey, devoting himself to research, with no administrative or lecture duties. By April 1955, he saw himself more as "a human being" rather than as a Jew or an American citizen. But he had always seen himself as a citizen of the whole world.

Although a pacifist for much of his life, he was concerned about Hitler developing an atomic bomb, and he advised President Franklin Roosevelt to begin investigating the new scientific experiments using fission and high-grade uranium to make a totally new level of bomb. Other scientists with a similar view did not have access to the President, while Einstein's world stature assured that result. Persuaded to convey this information to FDR, he did so in a fateful letter dated Aug. 2, 1939, though not delivered to Roosevelt until Oct. 11, 1939. Einstein sent another letter to the President in March 1940 on the urgency to act on this venture, which was started but not accelerated until after Oct. 9, 1941, when FDR approved a crash program to develop an atomic bomb. Einstein did a small amount of scientific work on it himself in Dec. 1941. The Manhattan Project followed in early 1942 (to 1946), and with it, America's long-term nuclear weapons program. After Aug. 1945 Einstein changed his views, and saw the bomb as a force so destructive that it had to be contained and stopped. He spent much of the next ten years of his life working on behalf of Zionism and pacifism. He was courted by the Zionists and even offered the Israeli presidency in 1952. To him, pacifism meant establishing world institutions that would preserve international peace by abolishing war and the use of weapons and by retaining control over military power. For this orientation he was considered dangerous by the U.S. government and put on FBI surveillance, along with internationalists such as former First Lady Eleanor Roosevelt.

> "Einstein's pacificism, never doctrinaire, seems to have first developed from his great revulsion to World War I, and the chauvinistic attitudes that had provoked it. Thus, in March 1915 he wrote to Romain Rolland [the great French writer and pacifist, who resided in Switzerland. Einstein sought Rolland's support in his efforts to bring the war to an end]: 'When posterity recounts the achievements of Europe, shall we let men say that three centuries of painstaking cultural effort carried us no further than from religious fanaticism to the insanity of nationalism? In both camps today even scholars behave as though eight months ago they lost their heads.' ... But by the war's end, he was convinced that it was Germany that had to change [through a process of 'swift and radical democratization']."
>
> **Bernard T. Feld,** "Einstein and the politics of nuclear weapons," *Bulletin of the Atomic Scientists*, March 1979, Volume 35, No. 3, p. 10. [Letter dated March 22, 1915 quoted from *Einstein on Peace*, edited by Otto Nathan & Heinz Norden, 1960, from the Estate of Albert Einstein, p. 13.]

[142] Quoted in Walter Isaacson, ***Einstein: His Life and Universe***, 2007, p. 506. Note #52, p. 634. Letter to Clara Jacobson, May 7, 1945, AEA (Albert Einstein Archives) File #56-900.

Einstein's early years did not indicate a promising future, especially not as an intellectual. He had language difficulties and was unable (or unwilling) to form complete sentences in his native German until the age of nine. At times he had a violent temper, was fiercely independent, disliked organized sports, and had atypical likes and dislikes. For instance, while his peers were excited by the sight of marching soldiers, he was disturbed by them. His mother taught him to play the violin and piano from the age of five, the same age his father showed him a pocket compass – a seminal event in his life, as he wondered what made it work. From the age of ten, he thrived under the tutelage of an impoverished Polish medical student (Max Talmud, later Max Talmey) who came to dinner every Thursday night for six years and taught him key texts in science, mathematics and philosophy.

His father wanted him to become an electrical engineer and work in his company, but young Albert resisted these expectations along with most conventional education except for what interested him. And since nothing other than physics really interested him, he worked mostly on his own. He had trouble getting an advanced degree and when he could not get a teaching post, he worked in a Patent office for several years. He did not meet many other physicists until he was about 30 years old (1909). In the fall of 1909, after much struggle, he finally obtained a university lecture position in Zurich. An outsider of outsiders, especially prior to 1905, Einstein was considered scientifically "provincial" – just as Ramanujan was seen as mathematically provincial. However, this may have also facilitated greater independence and originality of ideas. Einstein's non-conformity freed him to look at space and time in daring, new ways. In 1905 he published four groundbreaking works, all while working at the Patent office in Bern, Switzerland. This was later called his "Annus Mirabilis."

Einstein established some of the major foundations for modern physics and the essential structure of the cosmos, challenging the Newtonian theory of mechanics. One of his major concerns was to understand the nature of electromagnetic radiation. His Special Theory of Relativity (1907) was expressed in the equation E=mc2 (energy=mass times the speed of light squared). Energy and matter are seen as aspects of a single phenomenon. A large amount of energy could be released from a small amount of matter, a principle later illustrated by the atom bomb. In 1905 he developed his Theory of Relativity, in which space and time are no longer viewed as separate, and later also his General Theory of Relativity. He finished work on it in 1915 and published a 46-page paper in 1916. It expanded on his 1907 Special Theory to include the proposal that matter causes space to curve. Gravity is seen not as an exterior force but as a property of space and time, or spacetime. In his 1916 paper Einstein says:

> "Matter tells space how to bend; space tells matter how to move."

According to Einstein's theory, it could be known exactly to what extent a light beam would be bent when it passes near the sun – a theory that could be tested by observing a total solar eclipse. Because of the Great War (WWI), this project was delayed for several years until May 1919, when two expeditions set out – both organized by British astronomers Arthur Stanley Eddington and Frank Watson Dyson: one to an island in the Gulf of Guinea (led by Eddington), and one to a town in Brazil (Sobral). Studies of the May 29th total solar eclipse did prove Einstein's theories correct, and Eddington reported the findings of both expeditions at a joint meeting of the Royal Society and the Royal Astronomical Society in London. The next day, on Nov. 7, 1919, the front page of *The Times* newspaper called it "A Revolution in Science," and newspaper headlines around the world echoed this message. Einstein became an instant celebrity. And though the theory continued to be disputed, experiments in the late 1960s finally provided full validation of the Relativity Theory, showing that the amount of deflection was the full value predicted by General Relativity.

Einstein was awarded the Nobel Prize in Physics in 1921 for his work in Theoretical Physics, notably on the Photoelectric Effect. The expected proceeds from the prize were crucial in his lengthy divorce negotiations from 1916 to 1919, and should have gone to his first wife Mileva Marić and their two sons. But Einstein's family letters (released in 2006) reveal that most of the prize money was invested in the U.S. and lost during the Depression. (Mileva was resistant to divorce, and there is no proof to the rumors that she was responsible for his theories.) After the early 1920s, his scientific work was considered generally less important than what came earlier. He married twice, in 1903 and 1919, and had three children and two stepdaughters. There were many liaisons with women before, during and after both marriages. He died of an aortic aneurysm on April 18, 1955 at about 1:15 AM in Princeton, NJ, at age 76. At his memorial, physicist J. Robert Oppenheimer said of Einstein:[143]

> "He was almost wholly without sophistication and wholly without worldliness . . . There was always with him a wonderful purity at once childlike and profoundly stubborn."

Extraordinary destiny: An outstanding life – and on the world stage – is shown here by six out of nine planets either exalted, debilitated (with cancellation of debilitation), or virtually in their own signs due to mutual sign exchange (*Parivartana yoga*) between the best houses: 9 and 10. In addition, a 7th planet – the Sun – has *Digbala* (its best angular house) in the 10th house of maximum visibility, another form of *Raja yoga*. The *Navamsha* does not repeat the extent of *Raja yogas* in the birth chart, though Moon is improved in the *Navamsha* and there are several *Raja yogas* from the Full Moon in Capricorn, owned by an exalted Saturn. The *Parivartana yoga* between Venus and Saturn creates a protective *Viparita Raja yoga*.

Jupiter is again in the philosophical 9th house aspecting a Pisces *Navamsha* Ascendant. Jupiter aspects to both chart Ascendants protect him physically, giving a longer life, though not extra-long, as Ascendant lord Mercury is in the *Navamsha* 8th house. Even with a life full of controversies and conflicts, his ongoing reputation is protected by Jupiter's excellent placement as 10th lord. As

---

[143] J. Robert Oppenheimer (b. April 22, 1904, d. Feb. 18, 1967), American theoretical physicist, also spent his last years at the Institute of Advanced Study in Princeton, NJ, 1947-1967. As technical director of the Manhattan Project (1942-1946), his leadership was essential to its success. He had a key position with the Atomic Energy Commission, Jan.1947 to June 1954, when he lost his security clearance and was betrayed and destroyed for some years by his political enemies, notably the scientist Edward Teller. They claimed he had "substantial defects of character." But what they really objected to was his belief (shared with Einstein) that the use of the bomb as a weapon of genocide was "morally wrong." Scientists with this attitude were on a collision course with presidents and generals who wanted to ignore this issue and saw the building of nuclear weapons as the next logical stage in American military policy. Oppenheimer's reputation has been resurrected in recent years, when more details have come out on his demise in the McCarthy years, systematic efforts to "ruin" him, and accusations of spying for the Soviets disproved. At the first bomb explosion July 16, 1945 in Alamogordo, NM he recalled a verse from the *Bhagavad Gita*, which he read in the original Sanskrit: "I am become Death, the destroyer of worlds."

technology advances, many of his predictions – most of them from 1905 – continue to be proven correct, such as the bending of light in a gravitational field, gravitational waves, gravitational lensing, and black holes. Rahu in the *Navamsha* Ascendant shows him as ever the foreigner or outsider in his immediate environment.

The sweep of his life is governed by the sequence of the *Vimshottari Dashas*, and since seven out of nine planets have *Raja yoga* status (either through *Neecha Bhanga* or through other means), the *Dasha* sequence is especially relevant in pinpointing when extraordinary events will happen This revolves around Venus and the Sun. Venus is the best planet for Gemini Ascendant and is also the planet providing correction for the two debilitated planets (Moon and Mercury). These two *Dashas* were by far his best: Venus from 1896 to 1916; Sun from 1916 to 1921. For Einstein, his most important achievements as a scientist were prior to 1922. His wider fame came on Nov. 7, 1919, when London newspapers announced the validation of his Theory of Relativity. In 1921 he was awarded the most prestigious international award in his field, the Nobel Prize in Physics for the year 1921, though a technicality pushed the ceremonies into 1922. His *Vimshottari Dasha* sequence supports this result, again showing the karmic fruits – what is likely to be due and when.

The mind and body of a genius in physics: In the birth chart Ascendant lord Mercury is located in the 10th house in Pisces, its sign of debilitation, and in the *nakshatra* of *Uttara Bhadrapada*, ruled by Saturn.[144] Even in the house of maximum visibility, natal Mercury is isolated and hemmed in by malefics Sun and Saturn on either side of it. *Navamsha* Mercury is situated in the 8th house in Libra, a much more hidden placement, which can be good for intuition and for research but may also account for some chronic health problems, which he had for much of his life from 1917 onward, starting in Sun-Rahu *Dasha*.

"I rarely think in words at all." **Albert Einstein**

Natal Mercury in the sign of its debilitation shows he would start life with some distinct cognitive disadvantages and would always be balancing a sense of emotional and intellectual disequilibrium. Fortunately, this Mercury is in a *Neecha Bhanga Raja yoga*, a planetary combination in which the debilitation is corrected over time. In addition, he was born in the *Dasha-bhukti* of **Mercury-Moon**, the two mental planets. Considered a planetary enemy of Mercury, debilitated Moon in Scorpio also has *Neecha Bhanga Raja yoga* status – again confirming a start in life with some problems and disadvantages that he gradually overcomes. (Even the family life was disrupted with a move to Munich six weeks after Albert was born.) Given all these factors and that Einstein was in Mercury *Dasha* from birth to nearly 10 years old, we know the weaknesses of Mercury will reveal themselves in those specific years and that this condition is likely to be congenital, showing itself in infancy and childhood. He was a non-conformist and had obsessive behavior at times, often unable to deal properly with many social relationships or conventions, and focusing on an issue until it was solved to his satisfaction. Both of these traits could be seen as **signs of genius**. Others would assign it to a disease such as Asperger's Syndrome, a mild form of autism not uncommon among gifted people.

---

[144] This is the same *nakshatra* where Ramanujan's natal Moon is located, within minutes of arc, and in the same *Navamsha* of Libra. Ramanujan (Chart #10) is the previous chart in this chapter.

Ascendant lord Mercury is placed in Jupiter's sign, but within 1:03 degrees of Saturn, both planets in Saturn's *nakshatra*, *Uttara Bhadrapada*. This gives **Saturn** a lot of power in this chart as ***Jiva*, or karmic control planet**. *Navamsha* Mercury recurs with Saturn in Libra, wedding the two planets still further. In the birth chart, **Jupiter and Saturn** form a powerful *Parivartana yoga*, uniting both classical planets of *Dharma* and *Karma* from the Gemini Ascendant, as well as – in this case – the 4th and 5th houses from the Moon (*Chandra Lagna*). The exchange between Jupiter and Saturn allows Jupiter (acting as if in Pisces) to have more direct influence on Mercury, bringing the fluency of speech and volume of writing after an initial period of seeming stagnation or stoppage. He was in Mercury-Jupiter *Dasha* at age five when the violin and the compass entered his life. As a combination, natal Jupiter and Saturn also indicate that despite difficult beginnings with his education and his apparent cognitive faculties, Einstein would overcome all of it and rise to great heights during his lifetime, even if he never lived through his Jupiter or Saturn major *Dashas*. (The 16-year Jupiter *Dasha* began in Jan. 1957, less than two years after his death.)

Superior intellect: Intelligence and education are revealed from both Moon and Mercury, as shown, and also from the lords and occupants of the 4th and 5th houses from the Ascendant and the Moon. In each case there are excellent connections between 4th and 5th lords: From the Moon, we have already noted Jupiter in *Parivartana yoga* with Saturn. Natal Venus aspects the 4th house of education and higher degrees (Virgo), along with the *Neecha Bhanga* Mercury, which aspects its own house – thus protecting it from damage from early setbacks. From the Ascendant, Mercury (4th lord) contacts Venus in Pisces (5th lord). Because it is exalted and in a *Kendra*, Venus gives *Malavya Yoga*, one of the five *Mahapurusha* (great person) *yogas*, which distinguishes his Venus in this lifetime. In his case it also demands that his mental capacities be given in broader service to mankind.

Intellectual strength is enhanced by exalted 5th lord Venus within 2 degrees of its maximum exaltation in Pisces. (The 5th house governs intellect, creative intelligence, and education.) Placed in the 10th house of status, exalted Venus also brought Einstein many friends in high places who usually sought him out. The 5th house is aspected by classic benefic Jupiter, though it too would bring challenges due to the rules of *Kendradhipatya Dosha*. (See Glossary and Chapter Introduction) A superior mind and education can develop, but with early and periodic struggles. Mercury as 4th lord in the *Navamsha* 8th house of obstruction indicates he might abandon his original family, family home, and/or country of origin and that he might depart from some previous academic foundations he himself established.

Capacity for genius/ Problems with childhood language skills: We see that sometimes genius is born out of seeming disadvantage. Ketu in the 2nd house of speech can cause speech problems.[145] Even if speech (Mercury) is only an outward indication of a child's growing language skills, we note that young Albert Einstein was unable (or unwilling) to speak through most of his early childhood. He began to utter some words at age three but was not reasonably fluent in his mother tongue (German) until the age of nine. This process may have accelerated from age seven, when he entered the **Mercury-Saturn** period (April 4, 1886 – Dec. 13, 1888), forcing

[145] Ketu in the 2nd house also occurs with Winston Churchill and Barbara Walters – both noted public speakers with speech issues, Churchill most famously. (See Chapter 8: Charts #24 and #26, respectively)

him to be more concrete in his verbal expression. Saturn gives structure and focus to a subject, even when it is not easy to grasp it. But too much influence of Saturn on Mercury causes Mercury to self-edit, to be self-critical, perhaps dismissing or correcting ideas or thoughts before they have taken shape. Saturn is *Shani*, the slow-moving planet: it can and did delay language development. As 9th house lord from Gemini Ascendant Saturn is beneficial, but as 8th house lord it can be destructive. Even at age nine he would hesitate a long time before answering questions.

But from the age of ten his educational life was blessed by the arrival of Max Talmud (later Max Talmey), a Polish medical student who dined with the family every Thursday night for six years and gave lessons. (Jupiter is the planet of learning and rules over Thursday. Its full orbit around the Sun is twelve years, with a half-cycle of six years.) These informal lessons continued until the family was forced to move from Munich for economic reasons, and coincided with most of Einstein's **7-year Ketu *Dasha*, from Dec. 13, 1888**. Ketu is the foreigner or the outcast: Max was impoverished, Polish and Jewish. But he was also brilliant and his teaching captivated his young student, whom he introduced to science, mathematics and philosophy texts. Among the most influential texts were Immanuel Kant's *Critique of Pure Reason* and *Euclid's Elements*, which Einstein called his "holy little geometry book." He was enthralled by analytic geometry and calculus and studied these subjects mostly on his own. After reading Charles Darwin, among others, he moved away from religion, which was another required subject in that era. He did not attend his own Bar Mitzvah in 1892, declaring himself "a free-thinker." (His parents were non-observant Jews.)

As the *Moksha* planet, Ketu necessitates detachment during its planetary period and tends to bring losses or separations, but can also increase divine knowledge, philosophy, and intuitive perception. Situated in the 2nd House of verbal knowledge and expression, its benefits were clear at this time. The 2nd house is also the House of *Kutumba* (happiness from the family of origin). Ketu here could and did cause problems in this arena, mainly for their finances, which only began to improve after Albert's Ketu *Dasha* ended. The family moved to Milan, Italy in 1894 after repeated failures of the family business - a small shop that manufactured and sold electric machinery.[146] In 1895 Albert went on to finish high school in Aarau, Switzerland, graduating in 1896.

The **water element** plays an important role here in assessing genius, especially when we observe that his *Dasha* lords from birth to Dec. 1931 - nearly 53 years- were all planets in water signs (**Mercury, Ketu, Venus, Sun, and Moon**). After Dec. 1931 his *Dasha* lords (in the birth chart, at least) are in earth signs - demanding a kind of practicality and concrete logic that water signs tend to resist. We see the challenges in Ketu and Moon *Dashas* in particular: Natal Ketu is in Cancer in *Pushya nakshatra*, with sign lord Moon debilitated in Scorpio in the 6th house of conflict, and *nakshatra* lord Saturn in Pisces in the 10th house, all three planets in the emotional water element. On the positive side, the extensive water influence up through 1931 (especially prior to Moon *Dasha*) helped him develop his intuition and he continued playing the violin and piano, begun at age five. His early and life-long receptivity to music can be seen from the concentration of six planets in water signs, as music is closely associated with water. On the negative side, he had a violent temper as a child and his parents were concerned about his

[146] Albert's father had invested in equipment using direct current of electricity, and alternating current won out.

rebellious behavior. Most conventional education bored him and he disliked most of the schools to which his parents sent him. When his father asked one headmaster what profession young Albert might adopt, he said:

> "It doesn't matter. He'll never make a success of anything."

The rebellious behavior would continue through adulthood in his non-conformist and bohemian ways, and was part of his tendency to see things differently from the mainstream. He would be known for his voluminous and eloquent writing on a variety of subjects, in articles, speeches, and letters. The condition of Mercury in the chart accounts for both these early disabilities as well as his later genius. Having both his cognitive planets Moon and Mercury in water signs (same as Ramanujan), Einstein came upon much of his most important scientific discoveries intuitively, mostly in his case in images or dreams during sleep or daydreams. Even his description of how a scientist works shows the intensive watery nature of his birth chart – combined with the Gemini drive to share it.

> "Each [scientist, like a painter, poet or philosopher] makes this cosmos and its construction the pivot of his emotional life, in order to find in this way the peace and security which he cannot find in the narrow whirlpool of personal experience."
>
> **Albert Einsten,** in his keynote speech at a 60th birthday celebration for Max Planck in Berlin, April 1918, quoted in Dennis Overbye, *Einstein in Love: A Scientific Romance,* 2000, p. 337.

But later in life, following his Nobel Prize in 1921, Einstein started to question the validity of some of the ideas for which he was famous, especially after 1925. His 10-year Moon *Dasha* (from Dec. 14, 1921) was bound to be challenging, since his Moon is: 1) debilitated (though corrected); 2) in a difficult house (the 6th); and 3) relatively isolated. There are no planets in houses either side of it (a *Kemadruma yoga*), though it receives some cancellation from Jupiter in *Kendra* to the Moon. (Corrections to debilitated natal Moon and Mercury show his ultimate triumph.) The conflicts of the Moon *Dasha* became more pronounced when it coincided with 7 ½ years of *Sade Sati* – also extremely demanding years: Oct. 15, 1923 to Dec. 25, 1931. His financial losses may have occurred in fall 1929 with the N.Y. stockmarket crash, and in 1930 his younger son was diagnosed with schizophrenia.

The relative ungroundedness of Einstein's natal Mercury, both in the birth chart and in the *Navamsha* chart, shows even more reason for self doubt and/or a tendency to question ideas that came intuitively or during sleep. Saturn's heavy influence on Mercury plays a part in this. Libra is a good sign for *Navamsha* Mercury, but its placement in the 8th or 12th house keeps Mercury in the classically watery intuitive realm. This in itself is not negative, but Saturn's ongoing contact to *Navamsha* Mercury (repeated from birth chart) shows that he would not easily accept this type of extraordinary knowing. *Navamsha* Saturn exalted in Libra is a hard taskmaster and demands concrete evidence. In addition, *Navamsha* Mercury and Saturn are in a money house (house 8), aspecting another money house (house 2), adding further difficulty to his **ability to manage money** well or to hang on to it, no matter how many worldly honors he eventually received, including the Nobel Prize in Physics, most of which was invested in the United States and "lost during the Depression," according to letters released in July 2006 from the Albert Einstein Archives. (Given his astrological timing, a more likely scenario was that his losses came from the 1929 New York stockmarket crash.) Moon *Dasha* generates more controversy for him financially: how he disposed of his Nobel Prize money (part of a 1919 divorce settlement), and why he did not deliver his acceptance speech in Sweden until July 11, 1923 (his Moon-Rahu period). This caused a delay in payment of seven months, and his divorce agreement was predicated on it. By custom the Nobel prize money was not transferred until after the acceptance speech.

Chronic health issues: Three planets concern physical health and vitality in this chart: Ascendant lord Mercury, Sun and Mars. Each of them has some affliction in the *Navamsha* chart. *Navamsha* Mars is in the 12th house and *Navamsha* Sun (*Navamsha* 6th lord) is aspected by Saturn. Einstein was deathly ill throughout 1917, and though he was reported to have recovered fully by 1918, these health issues actually lingered for the remainder of his life – unknown to most of his public – and attributable to the hiddenness of the 8th house *Navamsha* Mercury. These have been listed as liver ailments, stomach ulcer, inflammation of the gall bladder, jaundice and intestinal pains. (Einstein's longtime collaborators Phillipp Frank and Banesh Hoffmann and his secretary Helen Dukas confirmed these details he was not eager to disclose.) Moon rules the stomach, and Einstein's debilitated Moon in the 6th house is not favorable for digestion, while Ketu in the 2nd house can cause inexplicable dietary problems, with the Mars aspect to Ketu compounding the issue. (Nor did his regular cigar-smoking help matters.) Jupiter rules the liver and gallbladder and is lord of the 10th house, where four planets reside, including Sun and Ascendant lord Mercury. Jupiter can create more extreme results, as it occupies Mars-owned *Dhanishta nakshatra*, and Mars is in the destructive 8th house of the birth chart. In addition, the 5th house rules the heart, upper abdomen, stomach, liver, and gall bladder.

Severe illness in 1917: In early 1917, Einstein collapsed and nearly died. A liver ailment and a stomach ulcer coincided with the war years in Europe, as well as the global Spanish flu pandemic causing more fatalities in Germany and Austria than in any other part of Europe. Food shortages were common, and he lost 50 pounds in body weight that year. Prior to his illness there were 2½ years of transiting Saturn in Gemini (to Aug. 2, 1916), contacting his Ascendant, aspecting his four natal planets in Pisces, and adversely affecting his health (and marriage) through aspects to Sun and Mercury as well as 7th house. Adding to this confluence was ***Ashtama Shani***, when Saturn transits the 8th house from natal Ascendant or natal Moon (Moon, in this case), one of the most difficult Saturn transits. Further, tr. Ketu was in Gemini from Nov. 18, 1916 through June 7, 1918, again on his Ascendant, destabilizing health and marital issues. Throughout 1917 and up to Sept. 19, 1918, tr. Saturn was in Cancer in his 2nd house, restricting intake of food and/or income.

For most of 1917 he was in **Sun- Rahu**, period and in 1928 – when he was forced to recuperate the entire year from heart problems – he was in **Moon-Mercury** *Dasha/Bhukti.* (He was born in Mercury-Moon *Dasha/Bhukti.*) In 1928 Helen Dukas was first hired to nurse him through that year and stayed on as a fiercely loyal secretary and housekeeper, virtually a part of the family, up to her own death in 1982.[147] The Sun rules the heart, and *Navamsha* Sun is in the 5th house, with afflictions as stated earlier. Since the 5th house lord of the birth chart is exalted, this allowed him to recover. But even if exalted, planets in Pisces tend to require acts of devotion and surrender. In summer 1917 his cousin Elsa Löwenthal got him an apartment in her building in Berlin and

[147] Helen Dukas (1896-1982) was devoted to the Einsteins, and immigrated with them to the U.S. in 1933. She lived at their home in Princeton, NJ until her death. Albert Einstein named her as one of two trustees of his estate (with Dr. Otto Nathan): executors of his literary heritage. She also co-authored a biography on Einstein.

nursed him back to health, becoming his second wife in 1919. This was after a delay of seven years from when they began their affair and after much struggle for Einstein to obtain a divorce, a process he joked about to his first wife:

"I am curious what will last longer: the World War or our divorce."[148]

Married life and children: Venus rules over marriage and love matters, and since Venus is exalted near the maximum degree in the 10th house of career, this demands great sacrifice and devotion to one's profession. All other matters related to Venus, including women and children in this case, are surrendered on the altar of career and public responsibilities. Women, including his wives and lovers, would often serve him with great devotion, most especially his long-suffering second wife Elsa, his longtime secretary, Helen Dukas, his sister Maja, and his stepdaughter Margot. Venus is lord of the 5th house of children (also intelligence and learning, which benefited greatly). But except for his stepdaughters his children did not benefit for the most part. He had three children from his first marriage and two stepdaughters from his second marriage. He probably never saw his first child (a daughter, born in 1902 and probably put out for adoption that year in Hungary). As she was born out of wedlock to the woman he eventually married, there would have been barriers to career and respectability in that era. Their younger son was fragile mentally and physically, and he rarely saw either of the two sons after his divorce in Feb. 1919 up to 1933, though he provided for them financially. Nor did he see them much before that due to work or travel, but he saw more of Hans Albert after his older son immigrated to the U.S. in 1938.

Men generally found Einstein to be more detached and complained about it, despite his sociability, kindness, and love of humanity, whereas women saw him differently, perhaps because he showed them more warmth. Thus, he was closer to women and closer to his two stepdaughters, Ilse and Margot.[149] After he had finally obtained a divorce from Mileva in Feb. 1919, his dilemma was whether to marry Elsa or Ilse, as he had fallen in love with the more beautiful daughter, Ilse (1897-1934). And since he could not decide which one to marry, he left it up to them. Elsa won out as Ilse saw him more as a father. When Albert was travelling he wrote daily to Elsa and often to Margot. He gave many details of his love affairs along with other topics. Elsa revered him, was devoted to him for 24 years and liked the benefits of having a famous husband. But even with her large capacity for tolerance she found life with him "exhausting and complicated," especially due to the numerous extramarital affairs he justified, claiming "men and women were not naturally monogamous."

The sign lord of Venus (Jupiter) and *nakshatra* lord (Saturn) are also involved with the 10th house in the birth chart, again bringing matters of love and children always to the altar of career,

---

[148] Einstein's letter to Mileva, April 1918, quoted in Dennis Overbye, ***Einstein in Love: A Scientific Romance***, 2000, pp. 340-341.

[149] Margot Löwenthal Einstein (1899-1986) was Elsa's younger daughter, a sculptor, extremely shy, and close to her step-father. She separated from her Russian husband in 1934, divorcing him in 1937. In 1934 she immigrated to the U.S. and moved into the house in Princeton, NJ. In 1938 Albert's sister Maja also left her husband and moved to Princeton. Along with Helen Dukas, both of them lived at the house at 112 Mercer St. until their respective deaths. (Maja died in 1951.) A fierce defender of Einstein's legacy, Margot stipulated that none of the ca. 1400 letters (3500 pp.) written 1912-1955 be released until 20 years after her own death. The Hebrew University in Jerusalem, holder of the Albert Einstein Archives, duly released the letters July 8, 2006.

status and public duties. Unpredictability in marriage or partnerships, or marriage to a foreigner is indicated by Rahu-Ketu on the *Navamsha* 1-7 axis. Also, likely problems in partnership are indicated by his ***Kuja Dosha*** ("the blemish of Mars") in both birth and *Navamsha* charts. Varied opportunities for partnership come from the Dual sign ascendant, the number of planets in dual signs, and the emotional nature (six planets in watery signs) which can take the form of spirituality, the arts and intuitive gifts, or sexuality, in descending order. The collection of Einstein's letters confirms his many female admirers and the numerous opportunities he took for womanizing. *Navamsha* Venus-Mars in the 12th house confirms this further astrologically, especially as the 12th house is the pleasures of the bed.

Einstein's first-ever romance began in late 1895/early 1896 at the start of his Venus *Dasha* when he was enamored of a daughter of the Winteler family, with whom he and his sister boarded in Aarau, Switzerland. He married the Hungarian Mileva Marić on Jan. 6, 1903 in his **Venus-Mars** period.[150] They met as students in fall 1896 at the Zurich Polytechnic Institute. She was a mathematician, aspired to be a physicist and was his intellectual equal. Both their families opposed the union, which lasted until their separation late June 1914 and their divorce Feb. 14, 1919. They had a daughter (Lieserl) born in 1902 and two sons: Hans Albert, 1904–1973, and Eduard, 1910–1965. Hans became a civil engineer, and Eduard was studying medicine and planned to become a psychiatrist when he had a nervous breakdown in 1930 and was diagnosed with schizophrenia. From 1932 onward he spent many years in and out of a Zurich mental institution and died there at age 55. He was gifted intellectually and musically and his illness was very painful for his father to accept.

Einstein's second marriage was to his first cousin Elsa Löwenthal on June 2, 1919 during his **Sun-Saturn** *Dasha/Bhukti*.[151] They became reacquainted at Easter 1912 during his **Venus-Mercury** *Dasha/Bhukti*, and this created difficulties for his first marriage from that time onward, though Mileva was reluctant to divorce. Elsa died in Dec. 1936, during his **Mars-Venus** *Dasha/Bhukti*. Thus, two marriages over 33 years began with Venus-Mars period and ended with Mars-Venus period. Planetary "bookends" can appear with major life chapters. Also, his first marriage took place three months after the death of his father at age 55 on Oct. 10, 1902. His second marriage took place eight months before the death of his mother at age 62 on Feb. 8, 1920. Both parents strongly opposed the first marriage.

---

[150] Mileva Marić was born December 19, 1875 in present-day Vojvodina, Serbia and died August 4, 1948 in Zurich, Switzerland. She was one of the first women in Europe to study mathematics and physics. Upon marrying Einstein she gave up any career of her own, and her academic career was first interrupted when she became pregnant in 1901 with their daughter Lieserl. It is unknown whether the child died as an infant or was put out for adoption. The couple resided in Zurich when in spring 1914 Albert accepted a post in Berlin. His affair with Elsa Löwenthal (who lived in Berlin), his new stronger positions on politics and Mileva's unease being in Germany at the onset of WWI played a part in their separation. He also laid out a new set of harsh rules in their relationship that she found intolerable. Negotiations for their divorce were ongoing from 1916 through 1919, creating bitterness especially for Mileva. This continued until her death, as he did not deposit the entire sum from the Nobel Prize in a Swiss bank account, as promised in the divorce agreement. She would have withdrawn interest payments, but had to keep asking him for more money, including for the care of Eduard (hospitalized for several decades) and for her own medical treatment for tuberculosis. Over the years, however, Einstein did pay her more than he received from the 1921 prize: $28,000 (now worth 10 times that).

[151] Elsa Löwenthal was born Jan.18, 1876 in Hechingen, Germany, and died Dec. 20, 1936 in Princeton, NJ. Her first marriage to Max Löwenthal ended in divorce in May 1908. She and Albert began their affair in spring 1912. They lived in Berlin from 1919, together with her two young daughters, Ilse and Margot.

Preference for solitude: Married or not, Einstein seemed to require intellectual isolation. Birth chart 10th house placement of his Mercury presented numerous opportunities for visibility and public stature through his career as a scientist, while *Navamsha* Mercury does not confirm the sustained personal visibility, showing instead his propensity for privacy and solitude to do his work. Despite so many planets above the horizon of this birth chart, usually creating an extrovert, there is a *Papa Kartari yoga* to the Ascendant lord (malefic planets on either side of Mercury) as well as an isolated Moon in *Kemadruma yoga*, though Jupiter provides some correction being in a *kendra* from Moon. These *yogas* provide the key astrological explanation for Einstein's relative isolation in life, starting with the early years of not speaking. He was a somewhat solitary person for most of his life, despite his public fame. He had great devotion to his work, and generally wanted to be left alone to do the work that he considered helped him know "the mind of God." Einstein said:

> "I live in that solitude which is painful in youth, but delicious in the years of maturity."

The birth chart has four planets in the 10th house of visibility, while only the shadowy planets Rahu and Ketu are in *kendras* in the *Navamsha*. The absence of physical planets in angles of the *Navamsha* chart makes him less accessible to his public, at least on an emotional level, but also physically. He spent his last 22 years of life at the Institute of Advanced Study in Princeton, estranged from the mainstream of contemporary physics, working on a unified field theory and not at ease with his earlier triumphs or the content of those ideas. He found solace in playing the violin and the piano throughout his life, especially works by Bach and Mozart, which he played most often at the end of his life. Even with a sense of isolation, he received a steady stream of visitors to his house in Princeton, NJ the last ten years of life, 1945-1955. This mainly revolved around his activities supporting pacifism and Zionism and getting other scientists and intellectuals to support pacifism in particular.

Highest achievements as scientist prior to 1922: The first two *Dashas* of his life (Mercury and Ketu) were unlikely to bring high achievements or prominence, up to Dec. 13, 1895 (age 16 yrs. 9 months). However, when the 20-year Venus *Dasha* began on Dec. 13, 1895, it freed him from military service (at age 17), brought his first marriage and family and his greatest work in physics, while the 6-year Sun *Dasha*, from Dec. 15, 1915 brought the rewards for work done mostly during Venus *Dasha*. Sun has *Digbala*, or best house placement in the 10th house. **Sun-Venus** *Dasha/Bhukti* ran for one year: from Dec. 14, 1920 to Dec. 14, 1921. It brought the announcement of his 1921 Nobel Prize in Physics (received in 1922) as well as his first visit to the U.S. as a celebrity (described below) – with some of the wild enthusiasm Americans seem to reserve for celebrities. Spring 1905 was the most concentrated creative era of his entire life, and 1905 was called Einstein's "Annus Mirabilis," or Miracle Year. It was his **Venus-Rahu-Venus** period. (Natal Rahu is in the 11th house from exalted Venus and closely contacting exalted Mars. This brought excellent results, though not immediately. Tr. Saturn in Aquarius was contacting his natal Jupiter in the 9th house and aspecting tr. Jupiter in Aries, triggering his natal Jupiter-Saturn exchange.)

In the early 1920s Einstein started moving in a very different direction. He remained unconvinced that quantum mechanics could be correct. This cut him off from the latest developments in physics, as he believed it was only a provisional step on the way to finding the right theory of atomic physics. He also lacked the same spark and instincts he had previously as a scientist, and according to many – relied too heavily on mathematics to provide answers to the great questions in physics:

> "According to Banesh Hoffman, one of his assistants, 'The search was not so much a search as a groping in the gloom of a mathematical jungle inadequately lit by physical intuition.'"
>
> **Lee Smolin,** "The Other Einstein," *The New York Review of Books*, June 14, 2007, p. 80.

Einstein's abrupt change and apparent decline within the scientific realm after 1921 is described astrologically by the Vimshottari *Dasha* sequence, which becomes much less favorable for creative work after the Venus and Sun *Dashas*, lasting 26 years in all. The Dual sign Ascendant person also needs to do at least two different things, and the strength of his humanitarian interests drew him increasingly towards his efforts to establish world peace. This gives more weight to his political life, which some biographers have tried to minimize. By 1933, he had already parted ways with most of his colleagues in science, though most of modern physics was based on his early work up to 1915 when his Venus *Dasha* ended. He remained honored but unheeded by scientists, especially once he lost his Directorship of the prestigious Kaiser Wilhelm Institute in March 1933. The Institute of Advanced Study in Princeton could not replace it. But he continued to be glorified by a larger public who did not understand his work. In that role he reached unprecedented superstar status as a scientist.

Venus was the most powerful *Dasha*, followed by the Sun *Dasha*. The Nobel Prize in Physics for 1921 should have been presented to Einstein on Dec. 10, 1921, just four days prior to the end of his Sun *Dasha*. It was for work he did in 1905 on the Photoelectric Effect. Due to a technicality, however, the prize was delivered on Dec. 10, 1922 (in his Moon-Mars *Dasha*), and the German ambassador gave a speech, since Einstein could not be present. His own acceptance speech in Sweden occurred July 11, 1923. Natal Venus is just over two degrees from maximum exaltation in the 10th house of the birth chart, so it must give extraordinary results, especially as Nadi *yogakaraka* and best planet for Gemini Ascendant. Venus is less well placed in the *Navamsha* 12th house in Aquarius, though it can work better when one lives or works outside the country of origin. (He lived in Switzerland for most of Venus *Dasha* up to April 1914, when he returned to Germany and the maelstrom of WWI.)

Navamsha Venus also shows his ongoing tendency to be a loner, working in isolation from (or excluded by) other scientists until at least 1909, in **Venus-Saturn** *Dasha/Bhukti*. (There is some modification and protection with *Navamsha* Venus in *Parivartana yoga* – mutual exchange – with *Navamsha* Saturn in Libra in the 8th house, also qualifying as a *Viparita Raja yoga*, protecting against losses and competitors and in his case anti-semitism, which took various forms – including opposition to his Theory of Relativity. His politics was also used by some to attack his theories.) Einstein's natal Sun is in the 10th house of birth chart and 5th house of *Navamsha*, well placed in both charts for personal recognition and leadership. During **Venus-Mercury** *Dasha/Bhukti* (spring 1914) he accepted the prestigious offer as Professor and Director of Physics at the Kaiser Wilhelm Institute in Berlin. The year of his Nobel Prize was 1921– his **Sun-Venus** *Dasha/Bhukti*. Even in sub-cycles Venus has to bring excellent results.

Instant fame: In addition to natal Venus and Sun, his fame is shown by factors such as the favorable 9th house position of Jupiter, 10th lord – of worldly status and position – both in birth chart and *Navamsha* chart. This helps to maintain a good reputation over a long period, including posthumously. His instant celebrity came on Nov. 7, 1919 (a Solar eclipse at 21:46 Aries), just five weeks into his **Sun-Mercury** *Dasha/Bhukti:* from Oct. 2, 1919 to Aug. 8, 1920. Sun is a *karaka* for fame, and his Asc. Lord Mercury personalizes it, with both planets situated in the 10th house of status and maximum visibility in Jupiter's sign of Pisces. *Neecha Bhanga Raja yoga* to Mercury shows he is elevated out of obscurity or a disadvantaged position. (In 1905 his work was still largely ignored.) Following the publication of his 1916 paper on Relativity, there were several years when fact-finding expeditions to validate his theories could conceivably have happened. But they were delayed by the events of World War I, and results of the May 1919 expeditions to observe the solar eclipse were not announced in London for another five months. These results were also more ambiguous than the ones reported. British astronomer Eddington already believed in the Theory and was looking for an event to prove its correctness, which was not definitively proved until the late 1960s. Nevertheless, Einstein achieved star status from Nov. 7, 1919 onward, during his Sun-Mercury period.

Because of his sudden rise to fame in Nov. 1919, Einstein was courted by many, including the Zionists, who sought funding for their planned Hebrew University in Jerusalem and their dream of a Jewish homeland in Palestine. Their leader, Dr. Chaim Weizmann, enlisted Einstein to do a lecture tour of the U.S. (April 2 to May 30, 1921).[152] It was Einstein's first trip to the U.S. and a triumphant but exhausting tour. (Weizmann did not raise as much money as hoped, and Einstein would not become a Zionist or go to live in Palestine, though he was supportive of the Hebrew University for many years and of Israel, from 1947. Offered the Presidency of Israel in 1952, he turned it down.) Einstein's reception in the U.S. in 1921 was wildly enthusiastic; such was his perceived celebrity and perhaps a welcome relief from the war. This was during his **Sun-Venus** period: Dec. 14, 1920 to Dec. 14, 1921.

Birth of a new arms race: Nuclear fission was discovered at the Kaiser Wilhelm Institute in Berlin in Dec. 1938. On Aug. 2, 1939 Einstein sent his first letter to President Roosevelt regarding the urgent need for the U.S. to research nuclear fission. As progress was still slow with this work in the U.S., in March 1940 Einstein (along with fellow scientists Leo Szillard and Spencer Weart) wrote again to FDR, urging greater speed with the nuclear weapons program, as they believed the Germans were making progress with their uranium research. In this way Einstein had a significant impact on the speed at which the U.S. undertook its nuclear weapons program, and many mistakenly assumed he was responsible for it, calling him the "Father of the Atom Bomb." He entered his 18-year **Rahu *Dasha*** Dec. 14, 1938, and was in the seminal Rahu-Rahu period for 2 yrs 8.5 months up to Aug. 27, 1941. The triple Jupiter-Saturn conjunctions of 1940-1941 in Aries would also favor weapons programs, the first of the three conjunctions occurring Aug. 7, 1940 at 21:26 Aries. All three of the conjunctions were in *Bharani nakshatra*, ruled by Yama, god of death and destruction.

As Mars is the planet of war and weapons, we examine Einstein's exalted Mars in Capricorn in the 8th house of the birth chart. (The 8th house is the house of death and suffering.) There is

---

[152] Born Nov. 27, 1874, Motal, Russia (now Belarus), Chaim Weizmann was the president of the World Zionist Organization, and later the first president of Israel, up to his death Nov. 9, 1952. A chemist by profession, a British resident (1904-1948) and president of the British Zionist Federation from 1917, he worked with British political leaders David Lloyd George and Arthur Balfour to promote the cause of Zionism. This led to the Balfour Declaration in Nov. 1917. He was leader of the Zionist delegation at the Paris Peace Conference (1919), pushing for the British mandate over Palestine at the League of Nations, also in 1919. A major force in the Zionist movement from 1901 through 1952, he met with U.S. President Truman in March 1948.

a very high *Sarva Ashtakavarga* ranking of the 8th house (41 bindus), and a low ranking of the 9th house at 20 bindus. This shows how Einstein is much more commonly associated in posterity with the bomb (8th house) than with pacificism (9th house, as a house of philosophy). *Navamsha* Mars goes to the 12th house in Aquarius. (The 8th and 12th houses are considered hidden houses, their contents not easily seen by others, or by the individual with this position in their chart.) There is always the sense of hidden power, or the ability to develop something the full effect of which is initially unknown. Even so, it is a myth that Einstein was primarily responsible for the atom bomb. His ideas of mass and energy made the concept more possible, but he did not execute them. His two letters to FDR were in fact his most important contribution to the bomb-making project, though undeniably timely.

Rahu aspects or contacts to Mars can bring very raw, even ruthless physical energy, as Rahu makes Mars less steady or predictable. In this case, the power of Mars is even more exaggerated due to being in its sign of exaltation in Capricorn. **Mars-Rahu** aspects his 2nd house of speech, initially cutting it off, then later enabling the power of his ideas or speech to unleash hugely destructive powers in the world – which he understood fully only after they were unleashed. After Aug. 1945 Einstein resumed the pacificism he had kept hidden during the years leading up to and during World War II. Roosevelt died April 12, 1945, and Einstein did not believe he would have dropped the bombs on Japan and said so in his front page article in *The New York Times*, Aug. 19, 1946. The headline was: "Einstein Deplores Use of Atom Bomb." Here are some other relevant quotes from him on the subject:

> "When the war is over, then there will be in all countries a pursuit of secret war preparations with technological means which will lead inevitably to preventative wars and to destruction even more terrible than the present destruction of life."
>
> **Albert Einstein,** in a letter to physicist Niels Bohr, Dec. 1944.

> "As long as there are sovereign nations possessing great power, war is inevitable."
>
> from his article "Einstein on the Atom Bomb", *The Atlantic Monthly*, Nov. 1945.

> "Technological progress is like an axe in the hands of a pathological criminal."
>
> **Albert Einstein**

> "Shall we put an end to the human race; or shall mankind renounce war? People will not face this alternative because it is so difficult to abolish war."
>
> **Albert Einstein**

Though he had celebrity status as a scientist, Einstein's pacifism assured his loss of political power leading up to the EARTH dominant era from Feb. 1961. This era would view nuclear weaponry as part of a logistical and lucrative enterprise, and by June 1954 it also brought the political demise of J. Robert Oppenheimer. His deep questioning of the moral issues of genocidal weapons was systematically derailed, sending a strong message to future generations of scientists and/or physicians knowledgeable about the actual destruction caused by such weapons.

Life in danger: From the Vedic perspective, we know the karmic fruit of this powerful **Mars-Rahu** combination will come due specifically during the *Dashas* of Mars and/or Rahu, the last two *dashas* of his lifetime. The 7-year **Mars *Dasha*** ran Dec. 13, 1931 to Dec. 14, 1938, and the 18-year **Rahu *Dasha*** from Dec. 14, 1938 to Dec. 14, 1956. These covered the years of the rise and fall of Nazi Germany, and specifically when he was about to return to Germany in mid-March

1933. His **Mars-Rahu** *Dasha/Bhukti* ran from May 12, 1932 to May 30, 1933, and encompasses the time he could have been killed had he returned to Germany as planned. Mars-Rahu period also coincides with tr. Saturn in Capricorn (Dec. 25, 1931 to March 16, 1934), thus for Einstein his ***Ashtama Shani*** – Saturn's most difficult transit, in this case following 7½ years of *Sade Sati*, also extremely demanding years. Tr. Saturn was exactly on his 8th house cusp in mid-March 1933, his planned return to the new Nazi Germany. But after becoming more fully informed of the perilous situation that awaited him, he resigned from duties in Berlin as Professor and Director of Physics. He travelled to Belgium and England before returning to the U.S. on a permanent basis, setting sail from England Oct. 7, 1933 on a 10-day voyage, never to return to Europe. Mars-Rahu period was pivotal in saving his own life and giving him first-hand experience of the real dangers of Hitler's Germany. Mars in the *Navamsha* 12th house favored foreign residence, and in his case made it mandatory. He obtained U.S. citizenship in Rahu *Dasha*, which again favors the foreigner (or outsider) in a foreign land, with Rahu prominent in the *Navamsha* Ascendant, aspected by benefic Jupiter. He became a citizen Oct. 1, 1940 in Rahu-Rahu-Venus period.

On Aug. 2, 1939, in his Rahu-Rahu-Jupiter period, and with tr. Mars close to an exact return to natal Mars, Einstein sent the letter to President Roosevelt informing him about the current scientific capacity to produce a bomb, along with his concerns that Hitler's Germany might produce one first. Roosevelt received the letter Oct. 11, 1939, when Einstein was in his Rahu-Rahu-Saturn period. This promoted the physical manifestation of his ideas, though he himself was not involved in the actual making of the bombs. In Einstein's Rahu-Saturn-Sun period the bombs were dropped on Japan, after Germany and Italy had already surrendered. Saturn-Sun can show conflict with authorities and reflects the inner conflict he would later feel about sharing information with FDR that led to the Manhattan Project.

Mars is also *karaka* for siblings and as we know is very strong. In 1938 his sister Maria "Maja" Einstein left her husband in Italy to come live with him in Princeton, NJ. She stayed from 1938 until her own death in 1951, coinciding with the last year of his Mars *Dasha* and first 13 years of his Rahu *Dasha*. Natal Rahu is closely conjunct exalted Mars. Maja was Albert's only sibling and a close confidante over many years. She was born Nov. 18, 1881 in Munich and died June 25, 1951, Princeton, NJ in the *dasha* of Albert's Rahu-Venus-Mars.

Caught between pro-war and anti-war currents: The Moon's location in the chart shows where the mind is focused in the life. In the birth chart Moon is in Scorpio in the 6th house of conflict and of enemies, while *Navamsha* Moon is situated in Capricorn in the 11th house of friendships, opposite *Navamsha* 6th lord Sun. Thus, conflict was unavoidable. His 10-year **Moon *Dasha*** identified the period when his political beliefs would become even more of an issue, bringing conflict where he sought unity. This *Dasha* ran from Dec. 15, 1921 to Dec. 14, 1931. As this Moon is opposite Sun in the *Navamsha* chart, it would necessarily involve the previous 6-year Sun *Dasha* as well (Dec. 15, 1915 to Dec. 15, 1921). His political activism began in summer 1914 with the outbreak of war in Europe. He was against World War I and in Oct. 1914 signed an anti-war "Manifesto to Europeans." (He was only one of four German scientists to do so and considered very daring. It was in response to the pro-war Manifesto signed by 93 German scientists, intellectuals and artists.) For this he was applauded by the pacifists and scorned by the German nationals, who were fuelled by the growing pan-Germanism. Since science and politics were so interwoven, he encountered periodic opposition as a German-born Jewish pacifist and as a scientist espousing his particular scientific theories, including his Theory of Relativity. This reached such a fever pitch that he considered leaving Germany in the aftermath of World War I. In spring 1933 the Nazis burned his books (among many other "un-German books") and denounced him, calling his Theory of Relativity "Jewish Physics."[153] Most of the book burnings occurred in Einstein's Mars-Rahu period.

His Moon *Dasha* even brought conflict concerning the Nobel Prize in Physics, which came later than expected, during his Moon-Mars period. It was hotly debated prior to the Nobel Prize ceremonies in Dec. 1922 as to whether he was primarily a German or a Swiss citizen. The German Ambassador eventually prevailed, delivering the speech in German, as Einstein was travelling in Japan. For technical reasons the prize was given in 1922 rather than in 1921, the year for which it was announced, and for his contributions to Theoretical Physics – specifically for his discovery of the law of the Photoelectric Effect. Einstein gave his Nobel acceptance speech in July 1923 (during his Moon-Rahu period), surprising all by its exclusive focus on his Theory of Relativity, which was by then his major passion in physics, though not the subject for which the prize was given. Similarly, having promoted the bomb in the years 1939-1941, Einstein made another about face from 1945 onward.

> "In November 1954, five months before his death, Einstein summarized his feelings about his role in the creation of the atomic bomb: 'I made one great mistake in my life... when I signed the letter to President Roosevelt recommending that atom bombs be made; but there was some justification – the danger that the Germans would make them.' In fact the mistake was a double one. The danger from the Germans never materialized; but in America an almost superhuman technological effort gave the lie to [physicist Niels] Bohr's [claim it would be] 'impracticable.' "
>
> **Ronald W. Clark,** *Einstein: The Life and Times*, 1971, 1984, p. 672.

The extent of his pacifism – even if largely unknown to his public – is shown by exalted Venus in the 10th house. This Venus is hidden in the 12th house of the *Navamsha* chart in the sign of idealistic Aquarius, along with Mars, planet of war. Thus is described the true legacy of Einstein: his making possible an unprecedented new level of warfare with the atom bomb, a legacy he would spend the rest of his life trying to destroy. But his personal destiny intersected with a collective destiny that favored major advancement of weaponry, since he lived most of his life in the **Fire to Earth Mutation period (1901-1961)**. This was followed by the EARTH dominant period, which would further commercialize these developments.

Though he was primarily known as a scientist, Einstein believed fervently in the necessity of peaceful cooperation between nations and in the power of international organizations to solve problems between nations. The League of Nations was established to do just that, but his association with the organization lasted only three years, 1920-1923, as conflict arose with his French colleagues due to his being born a German Jew, though his family had immigrated to Switzerland and he had been against World War I being fought at all. He was criticized by pacifists for remaining silent in the pre-war years 1933-1939 and for later promoting nuclear bomb research by the Americans after he came to consider World War II unavoidable. Others distrusted his pacifism that re-surfaced after 1945. In his *Handbook on Pacifism*, Einstein said:

---

[153] Mass book burnings in Germany occurred from April 6 through June 21, 1933, with thousands of citizens participating. At one public book burning on May 10, 1933 in Berlin, 40,000 citizens gathered and Joseph Goebbels, Nazi Minister of Propaganda, declared an end to "the era of extreme Jewish intellectualism."

"A human being who considers spiritual values as supreme must be a pacifist."

In the last ten years of his life he strove to educate the world in how NOT to prepare for war, including persuading scientists NOT to participate in nuclear research. Astrologically, with six out of nine planets in water signs (Sun, Moon, Mercury, Venus, Saturn, and Ketu), we observe how what may begin as an emotional reaction to events can evolve into a deeper spiritual perspective. The fact that four of them were in dual signs, along with his Gemini Ascendant, brings more complexity to his views, making him someone not easily categorized and therefore seemingly inconsistent. He promoted Zionism, yet he was not a Zionist; he was a pacifist, yet he was a warrior-pacifist, and did not think Hitler should be the only one with nuclear weapons. He rejected religion at age 17, but reinstated his Jewishness in 1914, along with the birth of his political activism. He did not believe in a personal God, yet he was not an atheist and could accept a God of beauty and harmony (Spinoza). He was against militant nationalism, but actively supported the idea of uniting Jewish people in a homeland. He was admired and celebrated by world leaders, but put under surveillance by the U.S. government for his anti-military stance. His passion for the mysteries of science equaled the depth of his enduring concern for humanity. He believed scientists had a moral responsibility and he acted on that tirelessly from 1914 onward, impacting both World Wars.

Chart 12: Steven Spielberg

Birth data: Wed. Dec. 18, 1946, 18:16 EST, Cincinnati, Ohio, USA, Long. 84W27 25, Lat. 39N09 43, *Lahiri ayanamsha*: -23:06:42, Class A from birth certificate, as quoted by Joan Negus. Ascendant: 17:40 Gemini.

Biographical summary: Steven Spielberg is a major American filmmaker and one of the most successful filmmakers of all time. His commercial success and prodigious output are unquestioned, though the consistency of his artistic level is debated by film critics, perhaps because there is such a wide range of material, and dazzling pyrotechnics can obscure less worthy material. His first major breakthrough film was *Jaws,* a shark thriller. In the summer of 1975 it surpassed all records for previous films and set new standards for the Hollywood blockbuster, whetting studio appetites to do more of them.[154] A cultural phenomenon, *Jaws* was nominated for an Academy Award for Best Picture in 1975, though Spielberg was not nominated for Best Director. That recognition

[154] Pauline Kael (b. June 19, 1919, 2:00 AM, Petaluma, CA; d. Sept. 3, 2001) as film critic 1951-1991, saw the new commercialism as "the death of the movies." In her article "On the Future of the Movies," (*The New Yorker*, Aug. 5, 1974, p. 43.) she warned young directors and writers to work independently of major studios, or the cultural level of movies would descend further. A benchmark article, it is often cited as an accurate prediction of cinematic history, though her favorite decade of movies was to be the 1970s. In any case, Spielberg's rise to major financial success coincides with the astrological EARTH dominant period (from Feb. 1961) in which commercial factors would take precedence over most others. After the Jupiter-Saturn conjunctions in Virgo in 1980-1981, films would benefit even more from technology and the new Information Age. (See Chapter 1)

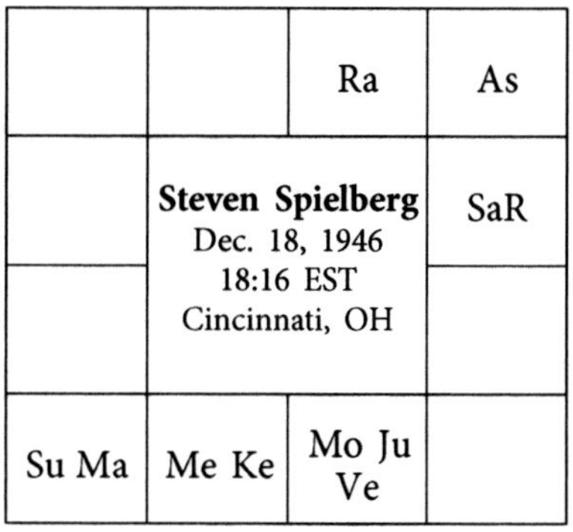

		Ra	As
	Steven Spielberg Dec. 18, 1946 18:16 EST Cincinnati, OH		SaR
Su Ma	Me Ke	Mo Ju Ve	

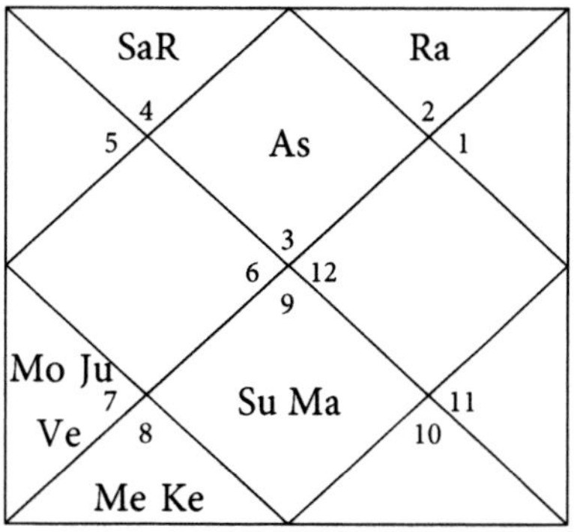

SaR 4 5 As Ra 2 1 3 6 12 9 Mo Ju 7 Ve 8 Su Ma 11 10 Me Ke

Planetary Information

As	17:40	Gem	Ardra
Su	03:20	Sag	Mula
Mo	13:44	Lib	Swati
Ma	08:01	Sag	Mula
Me	14:41	Sco	Anuradha
Ju	24:49	Lib	Vishakha
Ve	26:08	Lib	Vishakha
Sa	15:02	Can	Pushya
Ra	17:43	Tau	Rohini
Ke	17:43	Sco	Jyeshta

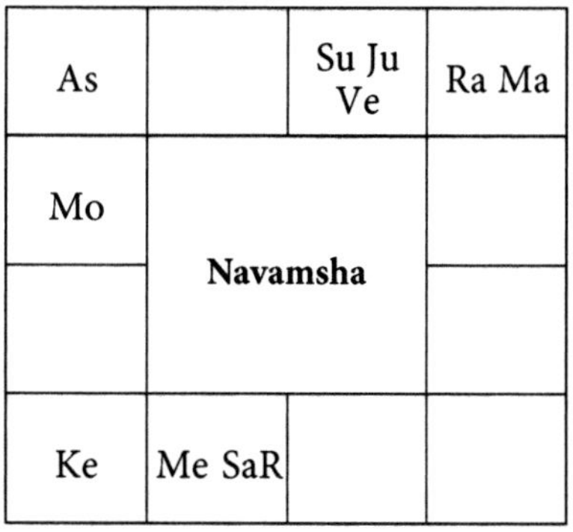

As		Su Ju Ve	Ra Ma
Mo	**Navamsha**		
Ke	Me SaR		

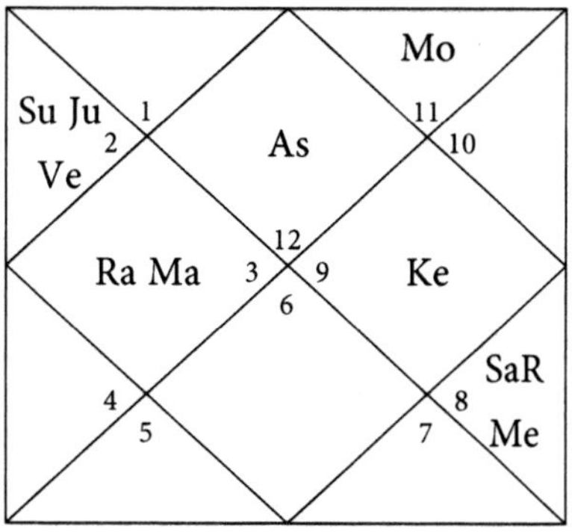

Mo Su Ju 1 2 Ve As 11 10 Ra Ma 3 12 9 6 Ke SaR 4 8 5 7 Me

Vimshottari Dashas

Ra	Jun 3, 1937
Ju	Jun 3, 1955
Sa	Jun 3, 1971
Me	Jun 3, 1990
Ke	Jun 4, 2007
Ve	Jun 4, 2014
Su	Jun 4, 2034
Mo	Jun 3, 2040
Ma	Jun 4, 2050

finally came in 1993 and 1998. But his arrival in 1974 as a director of feature films was noted by the influential film critic Pauline Kael, who admired "his technical assurance" and regarded *The Sugarland Express* as "one of the most phenomenal debut films in the history of movies." (*The New Yorker*, March 18, 1974):

> "*The Sugarland Express* is like some of the entertaining studio-factory films of the past (it's as commercial and shallow and impersonal) yet it has so much eagerness and flash and talent that it just about transforms its scrubby ingredients. The director, Steven Spielberg is twenty six, I can't tell if he has any mind, or even a strong personality... but he has a sense of composition and movement that almost any director might envy ...[155] He could be that rarity among directors—a born entertainer.... If there is such a thing as movie sense - and I think there is (I know fruit vendors and cabdrivers who have it and some movie critics who don't) - Spielberg really has it. But he may be so full of it that he doesn't have much else. There's no sign of the emergence of a new film artist (such as Martin Scorsese) in *The Sugarland Express*, but it marks the debut of a new-style, new-generation Hollywood hand."

Spielberg started making home movies at age 8, and grew up with an unabated fascination for the movie industry. His 24 minute movie *Amblin'* (1968) reaped him a 7-year contract with a major Hollywood studio (Universal), the first to be awarded to someone his age - then 22 years old. Between 1971 and 2010 he made 26 feature films, including the 1971 thriller *Duel*, originally made for television. Between 1964 and 2010 he directed, wrote, produced, and/or acted

[155] Spielberg was then age 27, but often took a year off his age, perhaps to promote his "wunderkind" status.

in a total of 94 films. Ten film projects are scheduled from 2011 through 2013, with Spielberg as director of four of them and producer of seven of them. In 1994, he co-founded the movie studio Dreamworks. Prior to that, from 1980 onward, he was involved increasingly as producer of films and often as director, but less often as writer – which he had done more from the 1970s up to 1983. Up through 2011, he appeared as an actor in 13 of his films, usually playing himself. In 1987 he received the Irving G. Thalberg Memorial Award for his work as a creative producer, and has since received several Lifetime Achievement Awards and frequent nominations for his films for either Best Picture or Best Director. He has received a long list of honors and awards or award nominations from Hollywood and around the world.

Spielberg credits much of his success to his happy second marriage to actress Kate Capshaw (b. Nov. 3, 1953, Fort Worth, TX) and the family life they share with their seven children. The couple married on Oct. 12, 1991. His first marriage Nov. 27, 1985 was to Amy Irving (b. Sept. 10, 1953, Palo Alto, CA), with whom he had a son Max Samuel (b. June 13, 1985). The marriage ended in 1989 and was marked by a controversial divorce settlement of over $100 million to Irving, after the judge dismissed a Pre-nuptial agreement written on a napkin. Though it should have protected Spielberg financially, the judge rejected it because Spielberg was already involved with Kate Capshaw, who became his second wife. The third costliest divorce on record, it was finalized April 24, 1989. After the divorce they shared joint custody of their son.

Criticized by some for too much pure sentimentality, too much orientation to children's themes and inability to produce a truly adult film, Spielberg finally answered his detractors in 1985 when he tackled themes of slavery in his film version of Alice Walker's novel *The Color Purple*. It was also the year of his first marriage and first child. *The Color Purple* received numerous awards, though not yet Best Director for Spielberg, who was then age 39. His first Oscar for Best Director came at age 46 for *Schindler's List* (1993) – for which he also received Best Producer. More prizes came internationally in 1994 for this film. His second Oscar award for Best Director came in 1998 for *Saving Private Ryan (1998)*. For Spielberg, 1993 was another watershed year, as he directed the science fiction adventure film *Jurassic Park*, which grossed $914.7 million. This outpaced his 1982 film *E.T. the Extra Terrestrial*, which grossed $782.9 million. (Both figures are from July 2011.) Most of his films have been commercially successful, some of them spectacularly so – starting in 1975. Film critic Peter Biskind (*Seeing Is Believing*, 1983, 2001), comments:

> "[Spielberg's] 'aesthetic of awe' helped to reduce an entire culture to childishness. ... To infantilize the audience, to reconstitute the spectator as child, [George] Lucas and [Steven] Spielberg had to obliterate years of sophisticated, adult movie-going habits.... The trend toward interesting Hollywood films of the 1970s was destroyed by Spielberg's jackpots, and the art is just now beginning to recover from...the catastrophe of success."

Whatever the case, his name has become synonymous with the Hollywood blockbuster, and no doubt Spielberg's true artistic merit will be assessed over the years alongside the sheer technical mastery for which he is known and respected. He continues to be one of the most powerful and influential figures in the entire film industry and at the close of the 20th century, *Time magazine* cited him as one of the "100 Most Important People of the Century," while *Life magazine* named him "the most influential person of his generation."

Gemini Ascendant – uniting the familiar and the exotic: Gemini reveres accessibility of knowledge and information, and Spielberg's major commercial success in the field of motion pictures is unprecedented. Whatever his critics say, this occurred because he has used a successful combination of both the familiar and the exotic. He has family themes featuring parent-child relationships, the child feeling the distance from the parent, and/or parent or child trying to get home or to find each other. Then there are all the exotic creatures from outer space, wild animals, or a fiendish trailer truck that appears to be driving itself (*Duel*), since we never see the driver. The narratives are often very simple and aim to please children and young people. As if to prove this point, Spielberg has the largest known private collection of Norman Rockwell paintings as well as the largest private collection of meteorites.

God of storms and wild animals: The Ascendant is located in the *nakshatra* of *Ardra*, ruled by the deity Rudra, god of storms and of wild animals. Many of Spielberg's films feature wild animals (*Jaws*, *Jurassic Park*) or exotic creatures (*E.T.*). *Ardra* in Sanskrit means "moist" and is symbolized by a teardrop. The teardrop alludes to suffering, either personal suffering or tears shed for the suffering of others. As a child, Spielberg heard many stories about the Holocaust from his grandmother. His first film on the subject (*Schindler's List*) was the one that brought him his first Oscar award as Best Director in March 1993, after numerous previous nominations and following stupendous commercial success for his earlier movies. Such was the overwhelming response to *Schindler's List* that in 1994 Spielberg formed the Survivors of the Shoah Visual History Foundation, a lengthy filmed testimony of Holocaust survivors.

This Ascendant is aspected by classical benefic Jupiter and by two malefic planets, Sun and Mars. The Mars aspect is considered especially inauspicious for a Gemini Ascendant, as it is 6th lord, and its aspect on the Ascendant can bring ill health and or conflict and competition. At the very least, it hampers marriage matters, as the planet ruling over conflict is situated in the house of marriage. This is also *Kuja Dosha*, a blemish with which Mars delivers difficulties in marriage. (His first marriage lasted four years, while the second marriage has endured over 20 years.) In addition, the Ascendant is hemmed in by malefics Rahu and Saturn, a *Papa Kartari yoga*. This *yoga* can hamper the affairs of the house affected by the malefics. It can also cause the individual to work extra hard to escape from the circumstances of their early environment. Spielberg calls his high school years "the worst years of his life" and "Hell on earth." Others have said such claims were much exaggerated. In any case, as a young person he was known as a loner and otherworldly. His parents' unhappy marriage also wore on him. He saw his concert pianist mother as close and supportive and his electrical engineer father as emotionally distant and unavailable. They divorced when he was 19, and he remained close to his mother and three younger sisters.

Physical health and competitive edge: Ascendant lord Mercury is not well placed in the 6th house from the Gemini Ascendant. This is an *Arishta yoga* and not good for the physical health, though it is offset to some extent by Mercury *Vargottama* (repeats in the same sign in the *Navamsha* chart), and in a more beneficial house – the 9th house. Further, *Navamsha* Mercury in Scorpio is in a *Parivartana yoga* with *Navamsha* Mars in Gemini in the 4th house.

Technical expertise can come with Mars-Saturn combinations, and this occurs with Mars "virtually" in Scorpio through the exchange. Since *Navamsha* Mercury is "virtually" in Gemini through the exchange, *Navmasha* Mercury-Rahu gives a unique or exotic outlook, as does *Navamsha* Mars-Rahu and Mercury-Ketu in the birth chart. All of these tend to give an aborted education (with 4th house involvement), but also a very active mind and major success: A *Raja yoga* is formed by trinal and angular lords in combination. Spielberg's dynamic and consistent output would demand high-level vitality. Proving the effectiveness of the *Parivartana Yoga* and of his "technical assurance," Spielberg's first major blockbuster *Jaws* (1975) was released during Saturn-Mercury period (June 6, 1974 to Feb. 13, 1977).

Mercury is strengthened by its *Vargottama* status in several other *Varga* charts, including the *Dashamsha* (career chart). The most important *Vargottama* position, however, occurs in the *Navamsha*, where Mercury improves by moving from a *Dusthana* house (6, 8, or 12) to a trinal house (1,5, or 9). Located in *Anuradha nakshatra*, owned by planetary friend Saturn, Mercury gains some strength and discipline from this factor and some originality from Ketu so close to it in the same sign. This also gives some unpredictability and some introversion to the intellect. Ketu in the 6th house helps him to be very competitive, as a classical malefic does well in the house of competition and conflict.

Natal Mercury is better placed from the vantage point of the Moon, but a strongly competitive quality would push Spielberg towards his goals and towards fulfilling his creative ambitions over the decades. It would also create some degree of controversy around him, perhaps more pronounced in Mercury and Ketu *Dashas*. Spielberg has the same Mercury position as Ramanujan, not good for health and prone to obsession in Scorpio. Ascendant lord in the 6th house can also experience some conflict in how they are seen and how they want to be seen. But as with Ramanujan, some of Spielberg's most important accomplishments (to date) occurred during Mercury *Dasha*, a testimony to Mercury's improved condition in the *Navamsha* chart. Ketu could also be prominent in the destiny, being in Mercury's *nakshatra* and placed in the 10th house of the *Navamsha* chart.

Born in **Rahu-Mercury** *Dasha/Bhukti*, he came into **Mercury-Rahu** *Dasha/Bhukti* Nov. 30, 1999 to June 19, 2002. (The reverse of the birth *Dasha/Bhukti* is often a pivotal time period.) In the *Navamsha*, these two planets are in a 6 - 8 house relationship; that is, Rahu is 8 houses away from Mercury and Mercury is in the 6th house from *Navamsha* Rahu. This is not ideal for the physical health, though modified somewhat by being in *Kendra* and trinal *Navamsha* houses. On Feb. 7, 2000, two months into the **Mercury-Rahu** period, Spielberg's publicist announced that "an irregularity had been spotted on Spielberg's kidney, and the kidney had been removed." It was diagnosed as renal cell carcinoma, a form of kidney cancer. He received no further treatment, and by all reports has done well since then. With his Ascendant lord in the hidden sign of Scorpio, Spielberg preferred little to no mention of this in the media.

Aborted education: Some accounts suggest that the young Spielberg may have suffered from Asperger Syndrome – a mild form of autism that leads to obsessional interests – often with very positive results. There are no delays in language or intelligence. If so, Spielberg would join an already illustrious group of Gemini Ascendant individuals who share some similar characteristics. In addition to their obsessive interests, they often do not go the conventional educational route. In Spielberg's case, after graduating from high school, he tried several times to get into film school. Failing entry, he entered university as an English major and dropped out after several years – leaving to take an unpaid job at a Hollywood studio. This enabled him to make his first short film *Amblin'* (24 minutes). *Amblin'* won numerous awards and landed him his first major Hollywood contract.

Astrologically, the 4th house tells about higher educational degrees. Malefic aspects or contacts to this house may cut short the education or cause problems in some way. In his case, 4th lord Mercury is located in the 6th house of conflict, aspected by malefics Rahu and Ketu. So there would be some problems with 4th house matters, though Spielberg finished his college degree after a hiatus of 34 years. Through special projects from spring 2001 up to spring 2002, he attained a Bachelor of Arts degree in Film Production and Electronic Arts, conferred in ceremonies held May 31, 2002. This occurred in his **Mercury-Rahu** period. Due to the *Parivartana yoga* in the *Navamsha* chart between Mercury and Mars, as discussed earlier, a good resolution to career and educational matters can occur in **Mercury-Rahu** period (Nov. 30, 1999 to June 19, 2002).

Gift for movie-making: Venus is the best planet for Gemini Ascendant and is well placed here in the birth chart 5th house in its own sign of Libra, and in the *Navamsha* 3rd house in its own sign of Taurus. Venus is also the primary significator for the arts. Fifth house is strongly associated with the visual arts, and 3rd house with music and dramatic arts, so when both 3rd and 5th houses are strong, from Ascendant and/or Moon, there is a possibility for success in a film career. Another hint is that the 10th lord of career (Jupiter) is placed in the 5th house along with Moon and Venus. Thus three classical benefic planets are in this pivotal house for a filmmaker, and because of this rich combination in an excellent house for a filmmaker we could predict excellent career results during **Mercury-Venus** period. (See below under Major Career Awards).

Also, just prior to this, **Mercury-Ketu** period (Oct. 30, 1992 to Oct. 27, 1993) supplied the needed component for him to achieve serious critical attention in terms of subject matter not dependent on the technical wizardry for which he had become famous. Ketu is a spiritual significator in the chart, and is closely contacting Mercury in the emotional sign of Scorpio, where Ketu is considered by many to be exalted, and at 20:00 Scorpio. Spielberg's Ketu is close to the maximum exalted degree.[156] Ketu performed well for Spielberg and elevated him to a level he had been seeking for several decades, even if he had long surpassed most financial and worldly criteria for success in his field. Further, Ketu is in *Jyestha nakshatra*, owned by Ascendant lord Mercury and reflects Mercury's influence in Scorpio as well as that of Mars (lord of Scorpio) located in Sagittarius, where it is close to the Galactic Center. *Navamsha Ketu* is in the 10th house in Sagittarius and as 10th house planet would give career visibility. The major success during Ketu's sub-period would also bode well for the 7-year *Ketu Dasha* from June 4, 2007 to June 4, 2014.

Even during the earlier **19-year Saturn *Dasha*** (1971-1990) we could predict success, based on the fact that Saturn is *Yogakaraka planet* from the Moon. Reading from the Moon in Libra, Saturn rules over both Houses 4 and 5 and is located in a *Kendra* house (10th house) in the sign of Cancer. Saturn is also well placed in the *Navamsha* chart in the 9th house with Mercury.

Major career awards: To experience worldly success from innate gifts and talents is by no means guaranteed. But when the Ascendant lord is strong enough, the physical and emotional stability good enough, along with other supporting factors, the unfolding of the *Dashas* together with the timing of Jupiter and Saturn transits will give us the answers. In this case, 10th lord Jupiter

[156] The exaltation of Ketu in Scorpio and Rahu in Taurus are not universally accepted. The sign and *nakshatra* lords will be very directive as well. The Vedic classic ***Brihat Parashara Hora Shastra*** only discusses exaltation for the seven physical planets, Rahu and Ketu being the "shadowy" planets. Opinions vary on what signs constitute exaltation and/or debilitation for Rahu and Ketu. Some postulate that they can always do well, or at least be less harmful, in signs owned by either Mercury or Jupiter. However, that can be adversely affected if classic malefic planets are contacting Rahu and Ketu. In his book, ***Panchadhyayee: A Compendium of Predictive Astrology***, 2005 edition, Ch. 2, verse on Rahu and Ketu, Dr. Suresh Chandra Mishra says: "Generally Rahu and Ketu placed alone in any house assume the effects of the house lord."

is contacting the Moon and Venus. And though Venus and Jupiter are not planetary friends and can cause excess when together, in this case they rule over some beneficial houses: Jupiter the 10th house of career (a *Kendra* house); Venus the 5th house of artistic creativity (a trinal house); and the Moon, the 2nd house of personal financial income, as well as the personal voice. These three planets in turn are placed in a trinal house (5th house) from the Gemini Ascendant. Getting his first big break at age 20, during **Jupiter-Sun** or **Jupiter-Moon** period shows us that from Jupiter *Dasha* onward (age 9), Steven Spielberg would forge enthusiastically onward with his creative ideas, in this case for movie making.

As Jupiter is a planet unfriendly to Gemini Ascendant, we see some hindrances during its 16-year *Dasha* (1955-1971), compensated to some extent by Jupiter being well situated in a trinal house and in its own *nakshatra* (*Vishakha*). Jupiter also forms a ***Gaja-Kesari yoga*** with the Moon in both birth chart and *Navamsha* charts (Moon and Jupiter are in *kendra*s from each other). This *yoga* can confer competitive success on the individual, if there is no affliction to either the Moon, Jupiter or the Ascendant lord in the chart.

On March 29, 1993, with tr. Venus retrograde near the top of his chart (at around 19:19 Pisces), Spielberg received the coveted Oscar award for Best Director, an award that had eluded him throughout several decades of outstanding as well as commercially record-breaking filmmaking. (Tr. Mercury, his *Dasha* lord, was close to the cusp of his 9th house of good fortune.) His winning film was *Schindler's List*, a film about the suffering of the Jews during the Holocaust in WWII. In 1998, he again received Best Director Oscar for *Saving Private Ryan*, a film set during the Invasion of Normandy in World War II. *Schindler's List* was made during his **Mercury-Ketu** period (Oct. 1992 to Oct. 1993) and rewards were received within the same period and extending into the **Mercury-Venus** period (Oct. 27, 1993 to Aug. 27, 1996). This same *Dasha/Bhukti* coincided with the production and release of *Jurassic Park* (1993), his largest-ever grossing film, while *Schindler's List* gave him the long-sought directorial prestige. (He was nominated previously: in 1977 for *Close Encounters of a Third Kind;* 1981 for *Raiders of the Lost Ark;* and in 1982 for *E.T. The Extra Terrestrial.*) Spielberg would also benefit from the **triple JU-SA conjunctions in Virgo, 1980-1981.** They set the stage for new technological breakthroughs affecting film and film-making, which in turn engendered a greater appetite for techno-thrills on the part of film-going audiences.

We could predict that Spielberg would garner more awards during Mercury *Dasha* (from June 1990) in part due to Mercury's strong position in the *Navamsha* chart. ***Navamsha* Mercury** is *Vargottama* (same sign as in birth chart) and in the 9th house, in exchange with Mars in Gemini. Since the exchange involves both a trinal house and an angular house it creates a *Raja yoga* and thus very successful results, as mentioned earlier. However, since *Navamsha* Saturn is with Mercury – rewards were somewhat delayed. Mercury is in the Ascendant (*Digbala* – best possible angle) in the *Dashamsha* chart of career, again in Scorpio. This keeps compounding the possible good results for Mercury during its *Dasha*. Also, Saturn, Mercury and Ketu mutual contacts in the birth chart and/or *Navamsha* charts can bring three successive *Dashas* producing continuously good results.

Dhana yogas* for wealth:** In March 2010 Forbes Billionaire List gives Spielberg's net worth as $3.0 billion. There are *Dhana yogas* from both the Gemini Ascendant and from the Moon. Saturn owns the 8th and 9th houses and is placed in the 2nd house, aspecting its own house, the 8th house (Capricorn), which is also the income of the spouse or business partner, being the 2nd house from the 7th. Spielberg's expensive divorce from his first wife came at the end of Saturn *Dasha*. With a malefic in the 2nd house of income, aspecting and affecting three of the financial houses (2, 8, and 11), we see how a classical malefic has the power to separate you from the contents of those houses. Reading from ***Chandra Lagna, (Moon as Ascendant), we also have several more *Dhana yogas*. From Moon in Libra, 9th lord Mercury is placed in the 2nd house in Scorpio - though with unpredictable Ketu. In his adult years Spielberg has been a quiet philanthropist, contributing generously to his favorite projects. The strength of the 3rd house in the *Navamsha*, and excellent placement of most planets from the Ascendant and/or from the Moon, along with favorable *Dasha* sequence - further confirm the major talent combined with major financial success.

Relationship and marriage factors: For a Gemini Ascendant there is likelihood of at least two major relationships or marriages, and of some difficulty making the choice. We have already noted that the 6th lord of conflict (Mars) occupies the 7th house of marriage in the birth chart. There is also *Kuja Dosha* caused by natal Mars in the 7th house. All these factors are bound to cause conflict around marriage - more so depending on the auspiciousness of the timing of that marriage. Spielberg's marriage to actress Amy Irving lasted from Nov. 27, 1985 to April 24, 1989, with a very costly and somewhat controversial divorce settlement. (Their son Max was born on June 13, 1985.) As the first marriage was during Spielberg's **Saturn-Rahu** period it was not ideal. At the time of marriage, tr. Saturn and Rahu, both in Libra, were in Spielberg's 5th house of children containing natal Moon, Jupiter and Venus, thus also impacting the 7th house of marriage from natal Moon. While we expect to see tr. Saturn aspecting the relevant house, its contact with Rahu is damaging. The rupture of the marriage occurred while tr. Saturn in Sagittarius was in his 7th house from the Gemini Ascendant.

Three classical benefic planets in the 5th house, including house lord Venus gives many children (especially with the potentially excessive Venus-Jupiter), though Jupiter in the 5th house is not a positive indicator in this regard in a woman's chart. Also, in Spielberg's case, Jupiter is not a planetary friend to Gemini Ascendant. Thus no biological children came into his life during the 16-year Jupiter *Dasha* (ending June 3, 1971, at age 24).

Vedic astrology defines the Mercury *Dasha* as the period most likely to bring children into his life, if there are not too many impediments indicated.[157] His 17-year Mercury *Dasha* began June 3, 1990. His relationship with actress Kate Capshaw began in 1989 (or before) and they married on Oct. 12, 1991. Their biological daughter, actress Sasha Spielberg was born May 14, 1990, three weeks prior to the start of his Mercury *Dasha*. Together he and Capshaw have seven children, including Max, three biological children between them, and two adopted, plus a stepdaughter (actress Jessica Capshaw, b. August 9, 1976, from Kate Capshaw's first marriage). Their youngest child was born Dec. 1, 1996 during Spielberg's Mercury-Sun *Dasha/Bhukti*. All three of their biological children were born during his Mercury *Dasha*, with one adopted together also born in 1996, highlighting the fruitfulness of Mercury *Dasha* for him for expanding his family with children.

[157] Natal planets with the 4th and/or 5th highest degree rank (Jaimini *karakas*, or significators) can bring children into the life, especially during their *Dashas*, if other factors in the chart do not rule against fertility. (Rahu and Ketu are not generally included.) The 4th highest degree planet is ***Matrikaraka*** or **MK**: Mercury is MK planet in Spielberg's chart. After his first child was born in June 1985, the next of seven children was born May 14, 1990. Mercury *Dasha* began June 3, 1990. The 5th highest degree planet is *Putrakaraka*, or PK. The Moon is PK and its 10-year *Dasha* starts when Spielberg is age 93 (June 3, 2040).

Chart 13: Bill Gates

Birth data: Friday, Oct. 28, 1955, 21:15 PST, Seattle, WA, USA, Long. 122W19 51, Lat. 47N36 23, *Lahiri ayanamsha*: -23:14:40, Class B data, as given in the biography: "shortly after 9 PM" in the biography, *Hard Drive: Bill Gates and the Making of the Microsoft Empire*, by James Wallace and Jim Erickson, 1992. The author prefers this time (rectified to 21:15 PM) to a later time of 22:00, or 10 PM given informally by Bill Gates at a Microsoft luncheon talk. (See below) Ascendant: 24:58 Gemini.

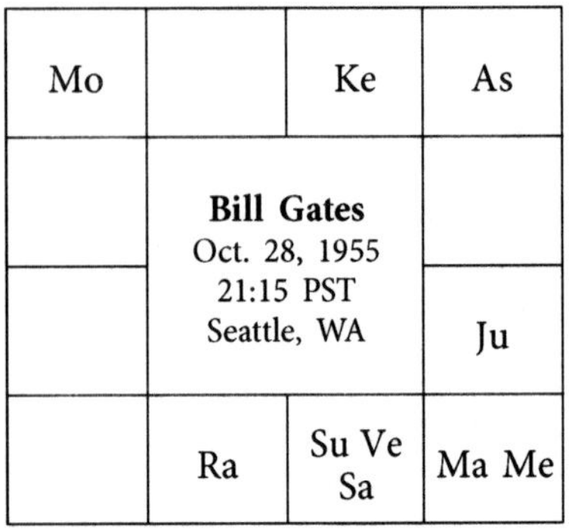

Mo		Ke	As
	Bill Gates Oct. 28, 1955 21:15 PST Seattle, WA		
			Ju
	Ra	Su Ve Sa	Ma Me

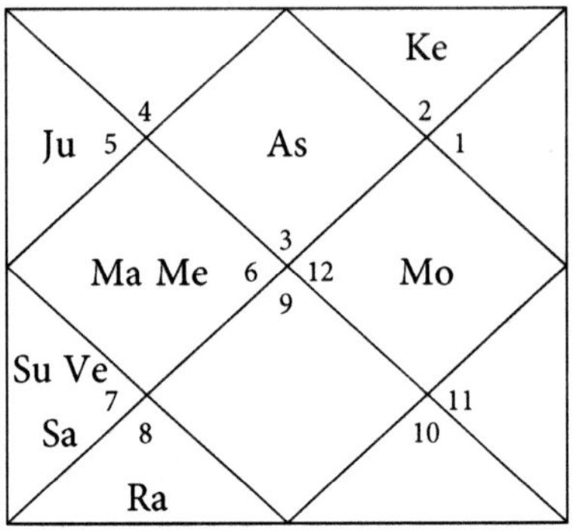

Ke 2 1
Ju 5 4 As
Ma Me 6 3 12 9 Mo
Su Ve 7
Sa 8 10 11
Ra

Planetary Information

As	24:58	Gem	Punarvasu
Su	11:45	Lib	Swati
Mo	14:33	Pis	U.Bhadra
Ma	16:51	Vir	Hasta
Me	23:19	Vir	Hasta
Ju	04:32	Leo	Magha
Ve	26:56	Lib	Vishakha
Sa	28:20	Lib	Vishakha
Ra	26:13	Sco	Jyeshta
Ke	26:13	Tau	Mrigashira

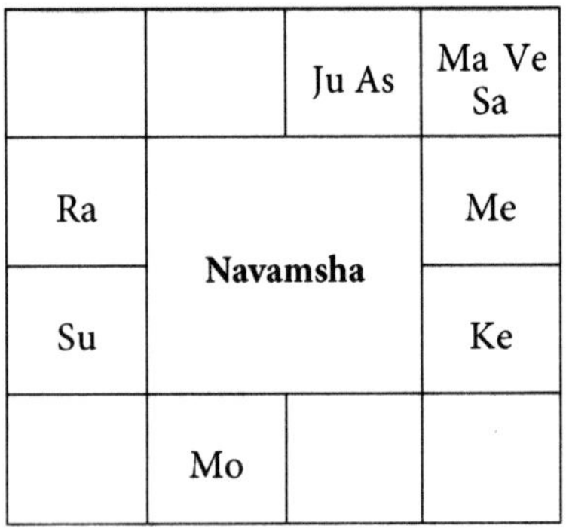

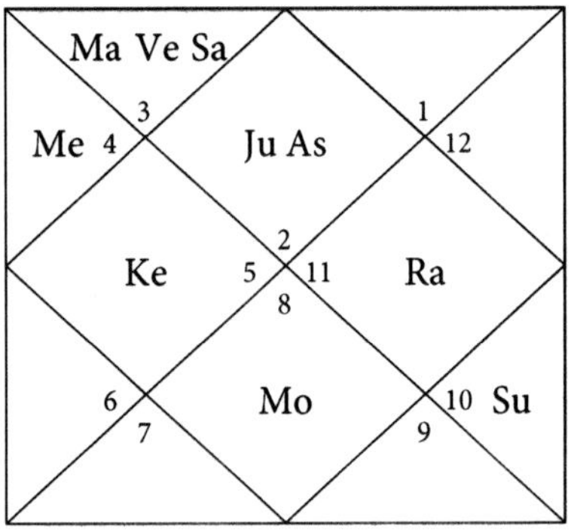

		Ju As	Ma Ve Sa
Ra	**Navamsha**		Me
Su			Ke
	Mo		

Ma Ve Sa
Me 4 3 Ju As 1 12
Ke 5 2 11 8 Ra
6 7 Mo 10 9 Su

Vimshottari Dashas

Sa	Oct 27, 1939
Me	Oct 27, 1958
Ke	Oct 27, 1975
Ve	Oct 27, 1982
Su	Oct 27, 2002
Mo	Oct 27, 2008
Ma	Oct 27, 2018
Ra	Oct 27, 2025
Ju	Oct 28, 2043

Debate over birth time: At a Microsoft function in recent years, when asked about his birth time Gates said he was quite sure it was 10 PM. There was no reference to his birth certificate. Previously his birth time had been given as "shortly after 9 PM" in the 1992 biography of Bill Gates by Wallace and Erikson, as stated above. This later birth time pushes the birth ascendant from Gemini to Cancer. This does not work for the Vedic chart, in my opinion, and does not describe many key themes in his life. As a public person of such monumental stature commercially and financially, his chart energies are bound to be magnified. This applies also to the Ascendant. The intellectual qualities of Gemini, the zeal for information and for technological advances and the commercial nature associated with Gemini gives prominence to Mercury as Ascendant lord, and to Moon as 2nd house lord (of personal income and voice). With a Cancer Ascendant, starting at 9:40:36 PM PST, we would not see the planetary configurations for a large fortune and protecting that fortune, nor for his creative and restless mind or his aborted higher education. Also missing would be the emphasis on the New Information Age, which is more classically symbolized by Mercury's sign than by that of the Moon. Whatever your vantage point, Bill Gates and Microsoft will remain towering icons as we assess the Age of the Personal Computer, especially in the West.

Biographical summary: The richest man in the world from 1993 through 2007 (*Forbes 400*), the personal wealth of American Microsoft billionaire Bill Gates briefly surpassed $101 billion in 1999. In June 2008 at age 52 he shifted his focus to his philanthropies. However, he was again no. 1 in 2009 and no. 2 in 2010 and 2011, with a fortune of $56 billion in 2011. Thus for 19 years he remained either number 1, 2, or 3 richest person in the world, and number 1 for 15 of those years, even with widespread financial losses suffered in 2008.[158]

Gates became a billionaire in 1987 at the age of 32, and the world's youngest self-made billionaire (actually $1.25 billion) up to that time.[159] This came at a young age through a fiercely competitive spirit (encouraged by his lawyer father and educator mother), intense dedication, hard work, and a fine-tuned intellect with an eye to the future of the personal computer. He entered Harvard as a freshman in fall 1973 and left in his junior year in fall 1975 after he and Paul Allen (b. Jan. 21, 1953) successfully programmed software for the first personal computer. (The pair met in 1968 in high school in Seattle and became close friends through their mutual fascination with computers. Allen was working as a programmer for Honeywell in Boston, having dropped out of Washington State University to pursue that interest.) Both of them had feverishly pursued computer programming since the late 1960s - Gates since age 13. Both of them correctly saw the rapid changes ahead for computer use by a larger public. In the world of personal computers, 1975 was a critical turning point: the Altair 8800 was announced on the cover of *Popular Electronics* in Jan. 1975 and became available in April 1975, though it came in a kit requiring more than 1000 solder connections. Only the most avid computer hobbyists could handle it.

From Dec. 1974 the future entrepreneurs worked on a marathon schedule to create the first computer program for Altair 8800. They gave the first demonstration of their BASIC program in March 1975 and a month later licensed and sold it to MITS (also owned by Altair) in Albuquerque, NM. They named themselves Micro-soft (Allen's idea), dropping the hyphen within a year. Microsoft was first established as a legal entity April 4, 1975 in Albuquerque, NM, with Gates as CEO and taking a leave of absence from Harvard by Nov. 1975. They both worked for MITS in Albuquerque, Allen from spring 1975, and they separated Microsoft from MITS by late 1976. Microsoft annual sales exceeded $1 million in 1978 and in Jan. 1979 its headquarters was moved to Bellevue, WA. The next big break came in July 1980. IBM executives, having failed in prior negotiations with Digital Research, hired Allen and Gates to create an operating system for the new personal computer (the IBM PC) to be released Aug. 1981. IBM granted Microsoft the right to sell its PC operating system to other computer hardware firms, a strategic advantage that Allen and Gates understood and acted on quickly. IBM paid them a fee of $50,000 and Microsoft kept the copyright for MS-DOS. By providing the standard operating system they also kept the software industry independent of the hardware. This was the start of their fortune, as software had

[158] In 2008 Warren Buffett (b. Aug. 30, 1930) of the investment company Berkshire Hathaway was the number 1 richest person at $62 billion; Mexican telecom mogul Carlos Slim Helú (b. Jan. 28, 1940) was no. 2 at $60 billion; and Gates was no. 3 at $58 billion. These three remained at the top of the Forbes list through 2011. In March 2011 Gates moved to the no. 2 spot at $56 billion ($59 billion by Sept. 2011) behind Helú at $74 billion, with Warren Buffett no. 3 at $50 billion. They all recouped their losses from their 2009 lows, when Gates was at $40 billion, Warren Buffett at $37 billion and Helú at $35 billion, after which Helú increased his lead.

[159] Becoming a billionaire at age 26 in 2010, Facebook founder Mark Zuckerburg (b. May 14, 1984) has in common with Gates that he launched Facebook also in his sophomore year at Harvard University (Feb. 4, 2004, from his dormitory). He too dropped out after the sophomore year, intending to return.

not previously come with a price tag but had been open source (and still is in certain countries in South America). Co-founder Allen became ill in 1982 and left the company in 1983 with 36% of the shares. Gates negotiated for 64% since he had to drop out of Harvard. Microsoft was incorporated June 25, 1981 and re-incorporated Sept. 23, 1993, Olympia, WA. It was first traded on the New York Stock Exchange March 13, 1986 (9:30 AM EST),

In his 2011 autobiography *Idea Man*, Paul Allen describes Gates as a "cold-eyed pragmatist" and a "tough taskmaster" known for his aggressive and at times harsh management style. His disdain for the rules was well-known from childhood. Allen also notes Gates' "rare gift for programming," and his ability to focus on a problem and sort out the less worthwhile solutions. Allen saw himself as Microsoft's "visionary," while Gates ran with the ideas and made the company highly competitive and successful. Some feel that the company lost its visionary quality after Allen's departure in Feb. 1983, but most agree that Gates has been the dominant force in the success and expansion of the company.

On Jan. 1, 1994 Gates married Microsoft executive Melinda French (b. Aug. 15, 1964, Dallas, TX.) They met at a company dinner four months after Melinda joined Microsoft in early 1987. They started dating that fall and became engaged in early April 1993. Their three children were born 1996, 1999, and 2002. In mid-Sept. 1997 they moved into their 4-storey mansion in Medina, WA built into a hillside (much of it underground) overlooking Lake Washington. Construction began in 1990 on the 66,000 square feet building. The house occupies 50,000 sq. ft., and garages and outbuildings another 16,000 sq. ft. Gates bought the land (5.12 acres) in 1988 for $2 million. After their wedding, Melinda insisted on changing the architectural design to be more family-oriented. By 1997 the entire property was worth $97 million and by 2011 over $200 million, with annual property taxes over $1 million. The house is notable for its futuristic design and its many technological innovations, including an estate-wide server system and over 40 miles (64 km) of optic fiber. The couple owns additional homes in New York City and in Nice, France.

Bill Gates and his wife Melinda have focused their energies increasingly on their philanthropic work. The Bill and Melinda Gates Foundation was started in Jan. 2000, though it was formerly known as the William H. Gates Foundation (1994-1999).[160] Bill Gates announced June 15, 2006 that he would remain CEO but intended to step down as President of Microsoft in July 2008 to become full-time Chairman of their foundation. This occurred June 27, 2008. Gates' decision coincided with an unprecedented gift of $31 billion from Warren Buffett to the Bill and Melinda Gates Foundation, which was already worth $30.6 billion. The announcement was first made on Sunday, June 25, 2006, and the papers signed the next day. This was the largest single donation in the history of private philanthropies and may well become a much larger amount – a testimony to the close friendship between Bill Gates and Warren Buffett, the two richest men in the world, respectively, in spring 2007. Knowledgeable critics of this merger have said this is also an excellent way for both men to preserve their capital from taxes, without really changing the widening disparity between the rich and the poor. But the stated focus of the Gates Foundation is on global health, education, libraries, and the Pacific Northwest. Strategies and technologies are aimed to improve reproductive health and to combat infectious diseases of the world's poorest people,

[160] The largest privately owned foundation in the world as of 2006, it began in Dec. 1994 as the William H. Gates Foundation, named after Bill's father, William H. Gates, Sr. (b. Nov. 30, 1925). In Jan. 2000 it merged with the Gates Learning Foundation and was renamed the Bill and Melinda Gates Foundation. By Dec. 31, 2010 its assets totaled $37.3 billion, an increase of $2 billion in 13 months. Source: **http://www.gatesfoundation.org/about/Pages/foundation-timeline.aspx.**

notably HIV/AIDS, malaria, and tuberculosis. It also aims to reform U.S. high school education, which Bill Gates considers "obsolete."

An active software developer in the early years of the company, the role of Bill Gates became more that of executive and management. As such he defended Microsoft against widespread charges of questionable and monopolistic business practices, insisting the company's market dominance is due to being innovators. Its market dominance gained the attention of the Federal Trade Commission from 1991, and a settlement was reached July 15, 1994. But Gates ran into more legal troubles from the late 1990s, when his company was accused of breaking antitrust laws and abusing its monopoly on the PC operating system market. The first major antitrust case against Microsoft was filed on May 18, 1998 by the U.S. Dept. of Justice and twenty U.S. states. On April 3, 2000, Federal Judge Thomas Jackson made a ruling against Microsoft for breaking antitrust laws. Appeals went on for over a year until – with the ascendancy of the George Bush presidency in Jan. 2001 – court rulings became far more lenient on Microsoft, reducing the more severe penalties of the previous Federal Judge. On Sept. 6, 2001, the Dept. of Justice announced it was no longer seeking to break up Microsoft and would impose lighter penalties on the company. A settlement was reached Nov. 2, 2001 that was favorable for all corporate conglomerates.[161] Microsoft continued for some years to pay out penalties for its monopolistic practices both inside and outside the USA, including some large fines to the European Union. But it appears that Bill Gates will be as fortunate in his battles to maintain his fortune as was J.P. Morgan (1837-1913), the American financier and philanthropist.

Accessibility par excellence: A pivotal person in the history of personal computers and computer software, Gates has enabled millions more people to have regular use of personal computers and later to connect to the Internet. Thus we observe the quality of Gemini that is both information and communication networks; also the desire to share information and to market it. Further, Gemini is a commercial sign, being associated with the *Vaishya*, or merchant and trading class. No one had sold software before Gates. And despite ongoing monopolistic issues, through his foundation Gates intends to increase access to technology in public schools and libraries in the United States and around the world. The plan is to assist residents in low-income and disadvantaged communities through partnerships with public libraries nationally and globally. With Gemini-Sagittarius as the axis of the nervous system of the world, this indeed describes the Internet, and the products of Bill Gates are closely associated with it. Three *Navamsha* planets (Mars, Venus, and Saturn) are located in Gemini and are surrounded by two classic benefics, a *Shubha Kartari yoga,* protecting the Gemini arena. *Navamsha* Mars and Saturn together in any sign can denote technical expertise.

[161] In her *New York Times* column on Sept. 9, 2001 Maureen Dowd railed against the legal unleashing of Microsoft on Sept. 6, 2001, "signaling media conglomerates that they can conglomerate away, accelerating the centralization of American power into the hands of a very few very rich people." She quoted writer Paddy Chayefsky, calling him "eerily prescient." The quote is from the 1976 movie ***Network,*** about the takeover of television networks by corporate conglomerates: "There is no America. There is no democracy. There is only IBM and ITT and AT & T and DuPont, Dow, Union Carbide and Exxon. Those are the nations of the world today. The world is a business ... and there is not a single law on the books to stop them!" (See Chapter 1. regarding the EARTH dominant period, 1921 to 2199, strongest from Feb. 1961. It promotes technology and its trade and exchange and favors corporate conglomerates and their increasing power over nation-states.)

Information technology is ruled by Mercury, ruler of Gemini and Virgo. *Navamsha* Mercury is in the 3rd house, repeating the Gemini theme as Gemini is the classic 3rd sign.

Zeal for technological innovation: In the birth chart Ascendant lord Mercury, planet of the intellect is placed in its own sign and exalted in Virgo in the 4th house. There are mutual contacts to Mars and Moon, its *nakshatra* lord – giving a lot of passion and drive both physically and mentally. Most of the energy is mental, however. There is seemingly no separation between what feeds his emotions and what feeds his intellect, as both planets are in close contact by opposition. As 6th lord, Mars is not an easy planet for Gemini Ascendant, but in this case the tempering effects of other planets as well as *Navamsha* positions give a lot of heat and drive to the intellect. And though some imbalance can come from that obsessiveness, it has not damaged his ability to stay focused and healthy.

The strength of the 5th house (the fruit of the mind and intellect) and of the *Navamsha* chart enable this Moon-Mercury-Mars to work well. All three planets plus the Ascendant in dual signs shows how dual signs can be advantageous in their ability to adapt to new circumstances. This is certainly a hallmark of Bill Gates, who has spent most of his life leaping ahead of the expected technological changes, first as a computer software engineer and later as the head of a global business empire. From there he moved on to the new "philanthrocapitalism," heading the world's largest private foundation. Though **Mars** is **the planet of discontent**, it can be a positive force for creative change. (*Mangala*, one of the major Sanskrit names for Mars signifies "auspiciousness.") With a strong and well-placed Saturn (the planet of discipline), the excesses of Mars and Mercury together can be channeled constructively. Natal Jupiter and Saturn (the business and economic planets) are placed harmoniously from each other in a 3-11 house relationship, which gives him no major economic or dharmic issues. That is, he will not hesitate about what he wants to do in life, and will not lack for money or career opportunities throughout his life.

The Mars contact to Mercury in the 4th house can create insomnia and mental turbulence, though the very same mental restlessness can bring a brilliant and innovative mind, always solving problems – even for recreation. Gates worked ceaselessly, taking no vacation for 13 years in the early era of Microsoft, and even after that he often spoke of vacations as "think weeks." Saturn keeps the discipline while the "thinking man" never seems to take a break. Malefic planets in contact with the Ascendant lord or Mercury (one and the same here) can also give technical expertise and the ability to work with advanced technology in some way. His mathematical abilities were clear at a young age: he scored 1590 out of 1600 on the math SAT test (American college entrance exams).

Aborted college education & early success: The astrological factors that aborted his Harvard education are the same ones that give a restless, energized mind, uneasy at complete rest and potentially unstable. The 4th house is the *Bhava* of higher educational degrees. The separative, heating, cutting influence of Mars created some conflict for him whether to finish his Harvard education. Placed in the 4th house or aspecting the 4th house, Mars can abort the higher education. With this tendency already in place, the *Dasha* sequence tells the rest. Gates entered Harvard University in fall 1973 with the intention of becoming a lawyer. It was the last sub-period of his Mercury *Dasha*, featuring two well-placed and exalted planets: Mercury and Saturn. During Mercury-Saturn *Dasha/Bhukti* (Feb. 16, 1973 to Oct. 27, 1975) he and his friend Paul Allen successfully produced the first software for a personal computer – in March 1975, after three months of non-stop work. They founded Microsoft on April 4, 1975.

Entering university in the last sub-period of a *Dasha* is likely to bring a change of direction from the original plan of study. But whatever the change, this *Dasha-Bhukti* of Mercury-Saturn has to bring excellent results. The planetary transits support the indications in the natal chart: On April 4, 1975, tr. Sun, Mercury, and Jupiter were all in Pisces in his 10th house of status, and tr. Saturn in Gemini was in his Ascendant aspecting his 10th house. It is auspicious for career when both *Dharma* and *Karma* lords (natal and/or tr. Jupiter and Saturn) influence the 10th house as well as Ascendant lord Mercury. (On Feb. 20, 1975 tr. Jupiter entered Pisces for twelve months, while tr. Saturn continued in Gemini up to July 24, 1975.) On April 4, 1975 there was also a *Neecha Bhanga Raja yoga:* Mercury's debilitation is cancelled by its Jupiter contact, indicating a sudden rise from a disadvantaged or murky position (a college student using university computer labs for independent commercial projects). His father's influence and legal advice throughout his career no doubt minimized or averted certain crises. Father (9th house lord) is exalted Saturn in the fortunate 5th house.

Gates entered the **7-year Ketu *Dasha*** on Oct. 27, 1975 and left Harvard in Nov. 1975 to work on computer programming projects with Paul Allen and their newly founded company. Gates always considered he was just "on leave" from Harvard, but the start of Ketu *Dasha* suggested a new chapter altogether. Some continuity was present in that most of his time at Harvard was spent buried in the computer labs: a setting similar to Ketu in the 12th house.

Fantastic wealth & the protection to win monolithic legal battles: There are numerous *yoga*s for financial wealth in the birth chart of Bill Gates, including those created by contacts between Moon, Mercury, and Mars (lords of houses 1, 2, and 11). Most notably there is a full-fledged ***Adhi yoga***, when classic benefics (Venus, Jupiter, and Mercury - preferably not combust, i.e., too close to the Sun) occupy the 6th, 7th, and/or 8th houses from the natal Moon and/or Ascendant. Gates has all three of these from his natal Moon. When malefic planets participate in the *Adhi yogas*, as we have here, with Mars, Sun and Saturn involved, it is said in the classics that this can spoil the *yoga*. However, it does appear in modern times that malefic planets signify technical education and scientific professions, as mentioned above. These planets can also bring more power, ruthlessness, and aggressive energy, and on the whole they bring a more materialistic orientation to the mind.

His ***Chandra Mangala yoga*** is also excellent for wealth. This occurs when Moon and Mars are situated in the same house, or aspect each other. In this case they are opposite each other in mutual *kendra*s, power positions in the chart. Furthermore, all the planets ruling over the houses of wealth, Houses 2, 5, 9, and 11, are well placed from the Gemini Ascendant. Except for Venus (5th lord) and Saturn (9th lord), they are all in mutual aspect with Ascendant lord Mercury.

Benefits derived from ***Dhana yogas*** for wealth are strengthened when ***Viparita Raja yogas*** are also present. These are the most important *yoga*s to have for protection against personal and/or financial loss, and can occur from the Moon or the Ascendant. In the case of Bill Gates, all three of these *yoga*s occur - which is more rare, and they occur only from his Moon in Pisces. This is when the lords of the 6th, 8th and/or 12th houses (from either the Ascendant or the Moon) are placed either in their own house or one of the other *Dusthanas*. Thus, it is as if two negatives make a positive. Ideally the *yoga* should contain no combust planets or Planetary Wars to be fully beneficial. If there is no damage to the *yoga*, the person enjoys status, fame and rise in wealth during the *Dasha* period of the planet(s) involved in the *yoga*. When all three ***Viparita Raja yogas (Harsha, Sarala, Vimala)*** are present, the person is protected from their enemies and from misfortune in general. The person can ride through life's difficult periods relatively unscathed, keeping their wealth and fending off their enemies extremely well. These *Viparita Raja yogas* describe wealth and continued progress, and though there may be challenges, success always follows. It can also bring high status and government position.

The planetary combination in Libra is further vaunted because not only is Venus an excellent planet placed in a trinal house from the Gemini Ascendant, considered fortunate in itself especially for the fructification of the intellect, but Saturn is exalted in Libra, and Sun – though debilitated in Libra – gets cancellation by being placed with Venus and Saturn, a ***Neecha Bhanga Raja yoga***. This is in addition to Sun's status in the *Harsha Viparita Raja yoga* from the Moon. From disadvantage can come major advantage. We might have expected a big drop after Venus *Dasha* (Oct. 27, 2002), but it did not occur due to the Sun's protective position in the *Viparita Raja yoga*. As a student both at high school and at Harvard, Gates often broke the rules, notably regarding computer lab use. This too could have been far more problematic. The issues he encountered in later years could have had crushing legal repercussions, especially the Dept. of Justice hearings, during which Gates' image became tarnished and Microsoft nearly broken apart as of spring 2000. But all was resurrected in the Sun *Dasha* by his new philanthropic orientation and the court decisions of 2001, in which government regulation of Microsoft was radically reduced. Helpful too in deflecting damage was the Jupiter-Saturn conjunction in Aries in his 11th house of finances: on May 28, 2000.

Gift of long-term & powerful friendships: In addition to being a house of wealth, the 11th house is also the house of friendships. (Oddly, for Gemini and Sagittarius Ascendant charts, the 6th lord and the 11th lord are identical. Thus, it may be difficult at times to distinguish friends (11th lord) from enemies (6th lord), especially if this house lord is poorly placed.) Bill Gates' childhood friend **Paul Allen** (two years older) persuaded him to drop out of Harvard in 1975 to found Microsoft. IBM executives hired them in July 1980 to create an operating system for the new personal computer (the IBM PC), released in Aug. 1981. Allen remained with Microsoft until 1983, when he had to resign for medical reasons. He remained on the board until 2000. **Steve Ballmer** (b. March 24, 1956) was a Harvard classmate who lived down the hall from Gates. Ballmer became Microsoft's first business manager when he joined the company on June 11, 1980. For some years, Gates retained control of the company's technological vision and Ballmer handled the finances, becoming Chief executive officer in Jan. 2000. He has been with the company continuously since June 1980. Introduced to **Warren Buffett** in 1991, Gates was astonished to find they not only had much to talk about, they had the substance for a long-term and powerful friendship that has resulted in uniting the fortunes of America's two richest men through philanthropy – in June 2006. Buffett (25 years older) tutored Gates in his own belief to give back to society rather than to foster dynastic wealth.[162]

Friendship is assessed through the **11th house** from both the Ascendant and the Moon. This is also the house of financial gains, especially as a result of one's career, being the 2nd house from the 10th house. A strong 11th house can enhance and magnify those gains. In the chart of Bill Gates, the 11th house is Aries, ruled by Mars in Virgo, situated in the *nakshatra* of *Hasta* ("the hand")

[162] Warren Buffett was born Aug. 30, 1930, Omaha, Nebraska. His birth time, given "from memory" as 15:00, shows 2:35 Sagittarius Ascendant and five planets in the versatile and eclectic dual signs. Buffet's birth time probably has more reliability than Gates' birth time "from memory." There are many *Dhana yogas*, in any case.

– in turn ruled by the Moon. Moon is an excellent financial planet for the Gemini Ascendant person, ruling the 2nd house of personal income. In opposition to Mars, it forms one of the *Dhana yogas* for wealth: the *Chandra Mangala yoga*, as described earlier. With Mercury and Mars both in *Hasta nakshatra*, Gates has always been a "hands-on" person at Microsoft, extremely involved in the product design and research end of the company, initially working closely with Paul Allen. From July 2008 onward he stated his intention to continue with a very hands-on role while also managing his foundation, which as of 2006 was the largest in the world. Under the conditions of Buffett's gifts to the Gates Foundation, either Bill or Melinda Gates must be alive and in charge of the foundation in order to receive the payments. Such is the immense trust Buffett has in the capabilities of the Gates couple to make a real difference with their foundation – and in their stated purposes.

The Blessing of good timing: Both the *Dasha* sequence and the timing and placement of the Jupiter-Saturn conjunctions every 20 years are crucial for the intersection of personal and collective destinies. Fortunately for Bill Gates, his ***Vimshottari Dasha* sequence** is formidable for worldly success. It reflects the many planets involved in his *Dhana yogas* for wealth. All planets except for Rahu and Ketu are involved in either the *Dhana yogas or the Viparita Raja yogas*; thus the *Dasha* sequence in itself is extremely beneficial. Reviewing the *Maha Dashas* from birth onward, Mercury, Venus and Jupiter are the three *Dasha* lords involved in the *Adhi Yoga* from the Moon, and the Moon *Dasha* reflects the *Chandra Mangala yoga*. As we shall see, his Moon *Dasha* is particularly suited to philanthropy.

His first major rise in wealth was in his **20-year Venus *Dasha***: Oct. 27, 1982 to Oct. 27, 2002. This was by far the largest gain in his personal wealth and it came at a chronological age when he could use it to advantage. Some changes in direction came during his Sun *Dasha* (Oct. 27, 2002 to Oct. 27, 2008), mainly a shift of focus towards his foundation. He announced in June 2006 that he would become full-time Chairman of the Gates Foundation July 2008. Buffett's gift was given in June 2006.

Prior to that, the **triple Jupiter-Saturn conjunctions of 1980-1981** occurred in *Hasta nakshatra* in Virgo. This 20-year social-political-economic indicator signaled a leap forward in the technological Information Age, with Mercury ruling Virgo and Moon ruling *Hasta nakshatra* ("the hand") – indicating that the larger public might benefit through a hands-on approach to the computer. This conjunction that was followed in late Oct. 1982 by his highly favorable Venus *Dasha* sent Bill Gates galloping through this 20-year period, fast becoming the richest man in the world at age (32) in 1987. The following year he bought land in Washington and built his high-tech mansion – also 4th house – between 1990 and 1997. This resonates strongly with his Mars-Mercury in Virgo in his 4th house. (The source of his wealth came from an interest first generated as a teenager in his Mercury-Mars period.)

The Microsoft phenomenon brought technological innovation and through its software provided the public with greater access to the personal computer, especially in the second half of that Jupiter-Saturn cycle up to spring 2000. Gates was at the center of this revolution, which included the birth of the Internet in 1990. With the culmination of the Jupiter-Saturn cycle in 2000 would come his highest-ever income ($101 billion in 1999), as well as his toughest legal battles with numerous states, other companies (domestic and foreign), and the U.S. Dept. of Justice. Virgo is also a sign of conflict and dispute, and his Ascendant lord Mercury is in Virgo along with Mars, the 6th lord. Thus, part of his *Dharma* has involved defending himself and his products at great length and at great expense. A softening effect (remedy or *Upaya)* came through philanthropic endeavors of the Gates Foundation – its intention, vision, and actual projects – whether or not they all succeed. This has also helped to soften the more negative images of Bill Gates that surfaced during his DOJ hearings.

Venus shines – best planet for the chart: As with Steven Spielberg, Venus is extremely well placed in the 5th house from the Ascendant in its own sign. Venus not only participates in various *Dhana yogas* for wealth but brings personal happiness and powerful intellectual gifts, especially during Venus ***Dasha*** (Oct. 27, 1982 – Oct. 27, 2002). **Ketu** *Dasha* (1975-1982) would not yield nearly the same level of success, in part due to Ketu's position in the hidden 12th house and its *nakshatra* lord being 6th lord Mars. Its sign lord Venus, however, would protect Ketu *Dasha*. Important and life-changing results would also be generated during the preceding Mercury *Dasha*, especially during: **1) Mercury-Mars** (April 27, 1967 to April 24, 1968), when he first discovered computers and was magnetized to working with them; and **2) Mercury-Saturn** (Feb. 16, 1973 to Oct. 27, 1975), when he and Paul Allen produced the first software for a personal computer and founded Microsoft.

Although Gates began work on his software programs in 1975, it was not until Venus *Dasha* that his major wealth came. On March 13, 1986, during **Venus-Sun** period Microsoft first traded on the New York Stock Exchange. By 1987 he was officially a billionaire, and the same year Melinda French (his future wife) came to work for Microsoft. He met her in Venus-Moon period, and married her in Venus-Jupiter (Jan. 1, 1994). For Gemini Ascendant, Venus in the 5th house shows both **personal creativity and emphasis on education** as a lifetime interest. Children are also described in the 5th house: All three Gates children were born during Venus *Dasha,* from 1996 through 2002. Their first child, Jennifer Katherine was born April 26, 1996, during his Venus-Saturn period. (Both planets are in his 5th house.) Son Rory John was born May 23, 1999 – in Bill Gates' Venus-Mercury period, and daughter Phoebe Adele was born in 2002 (no date available) in his Venus-Ketu period.

In the latter half of his Venus *Dasha*, a series of **major legal battles** also raged. They required huge legal fees that would probably extinguish a lesser company. The most demanding of these cases was filed May 18, 1998 (in his Venus-Saturn period), leading up to April 3, 2000 (Venus-Mercury) when a Federal Judge ruled that Microsoft was in violation of antitrust laws. Its breakup looked imminent, except for several astrological factors. Natally, Ascendant Lord Mercury exalted and in close contact to Mars gives Gates' a relentless combative quality, emphasized through being in a *Kendra* from both Ascendant and Moon. His unyielding belief in his position and his fierce (some would say arrogant or petulant) ability to keep pressing his case brought numerous appeals and delays that enabled him to hold out until a more favorable era for corporations: the George W. Bush presidency. Timing-wise, the Jupiter-Saturn conjunction May 28, 2000 in Aries turned national and government attention away from Microsoft and on to military matters, to include increased surveillance at home and abroad. (Microsoft and its technology may have been useful to the government in this regard.) Venus *Dasha* was his best one and the subsequent Sun *Dasha* also protected him, since Venus, Sun, and Saturn all participate in the *Viparita Raja yoga*. Mutual Sun-Saturn contacts often show conflicts with authority figures, and both Sun and Saturn surround this natal Venus. If each planet (as *Dasha* lord) was not as fortified on so many levels and the Ascendant lord were not as strong, Gates would have lost his case.

Philanthropy and marriage delayed: Bill Gates was criticized for years for doing too little philanthropically relative to the financial wealth he possessed. The classical *karaka* **(significator) of wealth** *and* generosity is **Jupiter**, which is also a potentially challenging planet for the Gemini Ascendant chart. Thus, Jupiter can cause problems in his life, especially during its major or minor periods. Jupiter is lord of the 10th house of career and of 7th house of marriage and business partnerships. However, this Jupiter is in a friend's sign and aspects its own house (the 7th house), bringing strong and self-reliant partners, whether marriage or business. Further, as the planet at the lowest degree of longitude, Jupiter also qualifies as *Darakaraka*, or spouse significator. Though Melinda French joined Microsoft in early 1987 and they began dating that fall, Bill Gates did not marry her until Jan. 1, 1994, when he was age 38 and in **Venus-Jupiter** *Dasha/Bhukti* (from Dec. 26, 1992 to Aug. 27, 1995). This was auspicious since Jupiter was sub-lord. On their wedding day, tr. Jupiter in Libra and tr. Venus in Sagittarius were in a mutual exchange ***(Parivartana Yoga)*** compounding the good omens for the couple. On that day tr. Jupiter was at 16 degrees Libra. It made another complete orbit around the Sun by the year 2006, when Warren Buffett announced his historic gift to the Bill and Melinda Gates Foundation. This occurred on June 25, 2006, when tr. Jupiter was retrograde at 15:12 Libra. Ten days earlier transiting Jupiter was nearly 16:00 Libra when Bill Gates announced his intention to work more closely with his wife at the Foundation from July 2008.

In Jan. 2000, the William H. Gates Foundation (1994-1999) was merged with the Gates Learning Foundation and renamed the **Bill and Melinda Gates Foundation.** This was during Bill's Venus-Mercury *Dasha/Bhukti*, the same year his personal wealth reached its peak at $101 billion (and legal battles threatened to break up Microsoft). Initially funded in December 1994 from $94 million of Microsoft stock and worth $30.6 billion by 2006, the Foundation received its first outside gift – about $31 billion from Warren Buffett. On the same day of the announcement of Buffett's gift – June 25, 2006 – a New Moon occurred at 10:01 Gemini, in *Ardra nakshatra*, with a benefic aspect from transiting Jupiter. This unprecedented donation was signed and sealed the next day. This New Moon was on Gates' Ascendant. His *Dasha* period was **Sun-Saturn**, with both tr. Jupiter in Libra and tr. Saturn in Cancer aspecting his 11th house of financial gains and also, of course – friendship. It was "happenstance," Buffett said, that Gates announced ten days earlier he would be taking over as full-time Chairman of the Gates Foundation from July 2008. This coincided closely with the start of his 10-year Moon *Dasha* on Oct. 27, 2008.

The Moon *Dasha* was far more likely than previous d*ashas* to involve Gates in global philanthropy. As owner of the 2nd house of income and speech, natal Moon is placed in the 10th house of status and visibility. A major *Adhi Yoga* (benefics in houses 6, 7, *and* 8 from Moon or Ascendant) emphasizes wealth. Moon is in the sign of Pisces – a sign whose best qualities are associated with surrender to the greater collective good. Gates' mother (also Moon) devoted herself for many years to philanthropic work. She died in June 1994 and the Gates Foundation was established in Dec. 1994.)[163] Natal Moon is in a *nakshatra* owned by Saturn, exalted in the excellent

[163] Mary Maxwell Gates (b. July 5, 1929, d. June 10, 1994) died of breast cancer. She died during Bill Gates' Venus-Jupiter period, five months after his wedding. Though fortunate for his finances and his love life, it was not a fortunate period for his mother's health, as the *Dasha* lord (Venus) and sub-lord (Jupiter) are located in the 8th and 6th (*Dusthana*) houses from natal Moon. Very challenging for the mother physically or for Bill Gates emotionally, Venus-Jupiter would still bring major financial gains for him: Dec. 26, 1992 to Aug. 27, 1995. In early 1993 Gates became the richest person in the world, through 2007, and has remained very near the top.

5th house. Expansive Jupiter is Moon's sign lord and has *Digbala* (best angle) in the *Navamsha* Ascendant. *Navamsha* Moon is in Scorpio in the 7th house. Its angular position in the *Navamsha* chart continues the visibility and personal force of Bill Gates in the world, at least through fall 2018. Debilitated *Navamsha* Moon gains benefit and correction through Jupiter's aspect. All of this brings a strong Moon *Dasha*.

From 1994 to June 2011 the Gates Foundation disbursed $25.4 billion, far more than the Rockefeller Foundation had distributed since its founding in 1913, even adjusting for inflation. The Gates Foundation aims to alleviate extreme poverty and disease around the world and to improve education nationally and globally, especially through computer technology. (Focus on health and technology is seen in Gates' key planets in Virgo, while five natal planets reside in the 4th and 5th houses of education.) The Gates Foundation insists on transparency and seeks to avoid the pitfalls typical of foundations – results that are seldom measured and failures that are often hidden, with administration costs unjustifiably high. They have set high standards for accountability in a field that has seen many failures along these lines. Another major challenge is how to stay "socially responsible" in the new era of "philanthrocapitalism," in which philanthropy becomes more like for-profit capital markets. As always with Bill Gates, his critics will follow him and question how this Foundation is run, how socially responsible it is, and who receives funding. Others have a lot of confidence in how the Gates couple operates. Jimmy Carter (U.S. President, 1977-1981) gives the Gates Foundation high praise, calling it "the most important organization in the world.... They know what they're doing."[164]

Chart 14: Peter Jennings

Birth data: Friday, July 29, 1938, 5:00 AM EDT, Toronto, Canada, Long. 79W25 00, Lat. 43N40 00, *Lahiri ayanamsha*: -23:00:12, Class A data, from memory. Ascendant: 29:20 Gemini.

Biographical summary: One of the most successful television journalists of the modern era, Peter Jennings was first on the air in radio at age nine, on CBC (Canadian Broadcasting Corporation), hosting his own radio show, called *Peter's Place*. His father Charles Jennings (1908-1973) was a prominent figure at CBC, and in later years Peter sought to please him and emulate him. The family environment was sophisticated and cultured, but also warm and loving. His mother was lively and supportive, and he had one sibling, a younger sister (Sarah Jennings). Peter left high school early, and after a brief stint as a bank teller he pursued his interest in broadcasting, working initially in Canadian radio and television. In 1964 he did his first stint in American television with a short segment for ABC (American Broadcasting Corporation) News. He then joined ABC News Aug. 3, 1964.

[164] **http://nonprofit.about.com/od/profiles/p/gatesfound.htm.** Carter is quoted here, along with *Forbes magazine*, which cites Bill Gates as the billionaire who gave the most to charity in 2011.

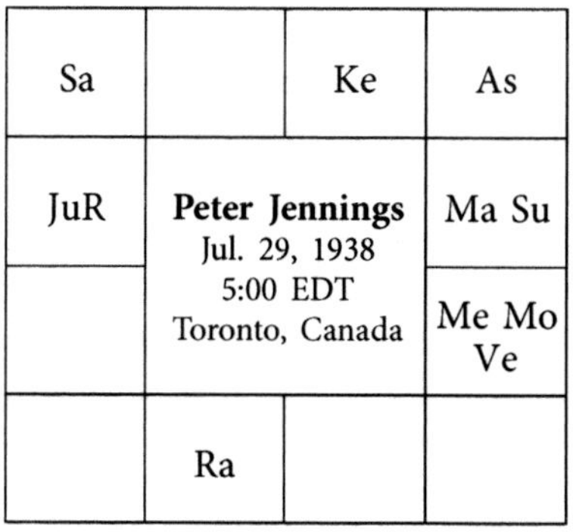

Sa		Ke	As
JuR	**Peter Jennings** Jul. 29, 1938 5:00 EDT Toronto, Canada		Ma Su
			Me Mo Ve
	Ra		

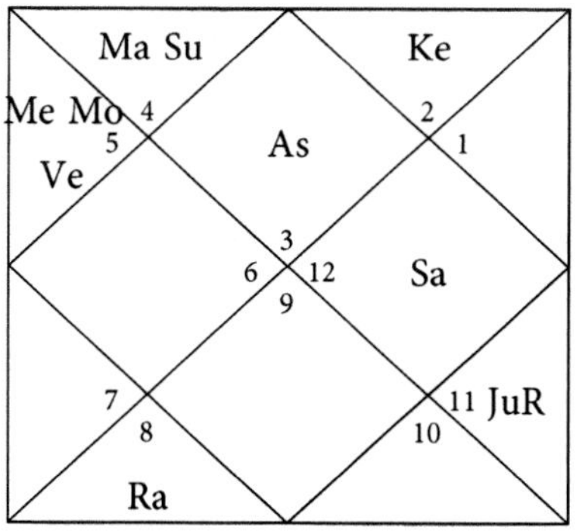

Ma Su 4 Ke 2
Me Mo 5 As 1
Ve
3
6 12 Sa
9
7 11 JuR
8 10
Ra

Planetary Information

As	29:21	Gem	Punarvasu
Su	12:34	Can	Pushya
Mo	13:33	Leo	P.Phalguni
Ma	11:08	Can	Pushya
Me	09:41	Leo	Magha
Ju	07:04	Aqu	Shatabhisha
Ve	24:20	Leo	P.Phalguni
Sa	25:02	Pis	Revati
Ra	00:07	Sco	Vishakha
Ke	00:07	Tau	Krittika

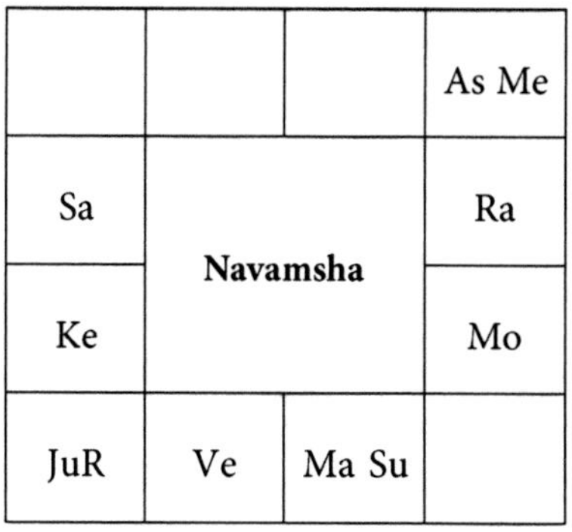

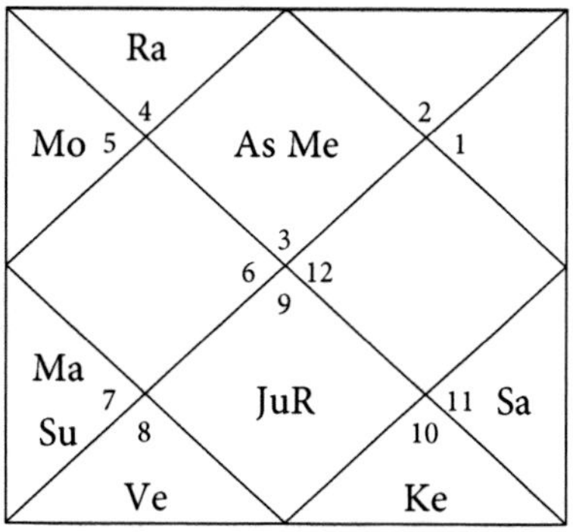

			As Me
Sa	**Navamsha**		Ra
Ke			Mo
JuR	Ve	Ma Su	

Ra 4 2
Mo 5 As Me 1
3
6 12
9
Ma 7 JuR 11 Sa
Su 8 10
Ve Ke

Vimshottari Dashas

Ve	Mar 29, 1938
Su	Mar 29, 1958
Mo	Mar 29, 1964
Ma	Mar 29, 1974
Ra	Mar 29, 1981
Ju	Mar 30, 1999
Sa	Mar 30, 2015
Me	Mar 30, 2034
Ke	Mar 30, 2051
Ve	Mar 30, 2058

Becoming an anchorman for ABC Evening News on Feb. 1, 1965, Jennings was the youngest person ever to fill that post, at age 26. He held the position from 1965 through 1968, when – due to poor ratings – he left to establish the first U.S. television news bureau in the Arab world as ABC bureau chief in Beirut, Lebanon, 1968-1974. Returning to the U.S. 1975-1976, Jennings was News anchor for A.M. America, and then during 1978-1983 was Chief Foreign Correspondent for ABC News and foreign desk anchor for *World News Tonight*. He returned to New York to become Anchor and Senior editor of ABC's *World News Tonight*, starting Aug. 9, 1983. He held this post for 22 years. His last television appearance on the program was April 5, 2005, announcing he had been diagnosed with lung cancer. He died four months later on Aug. 7, 2005 at around 11:30 PM EDT in Manhattan, NY.

Perhaps the last of a very refined breed in his field and always passionate about broadcasting, Jennings considered himself a reporter first. His departure from ABC marked a four-month period in which three well known anchormen retired: Dan Rather, Tom Brokaw, and Jennings. Brokaw, a close friend and colleague, describes Jennings as "unrelenting and unremitting" in his dedication to his work.

> "... I think what he did as a broadcast journalist, especially in that network chair, was compel America to look beyond its borders constantly. He never lost his sense of internationalism, especially in the Middle East and ... in Europe. And it is to his credit that *ABC News* distinguished itself especially in Bosnia."
>
> **Tom Brokaw,** on PBS *Evening News with Jim Lehrer*, Aug. 8, 2005.

As a leader in his field, Jennings was truly a witness to history, being present for some of the milestone events from the early 1960s right up to his death in 2005. He covered the Civil Rights movement in the American South, the Voting Rights Act of 1965, the construction of the Berlin Wall in mid-Aug. 1961 and its fall 28 years later in early Nov.1989. He was in the Soviet Union and Eastern Europe to report the fall of communism; was one of the first newsmen to go to Vietnam in the 1960s, and saw firsthand the South African struggle against apartheid. On Sept. 11, 2001 and all that week he was ABC news anchor.

Jennings received numerous awards and accolades in journalism, the first most notable one being the Peabody Award in 1974 for his documentary on Anwar al-Sadat. This established his excellence as a foreign correspondent, an experience that gave him greater maturity as a journalist. He also received 16 Emmy awards, as well as two Edward R. Murrow Awards, one of them in 2004. In 1995 *The Boston Globe* noted that Murrow had passed on the torch to Peter Jennings. In turn, his father Charles Jennings was considered "the Edward R. Murrow of Canada."[165] (Jennings was married four times, his fourth marriage to Kayce Freed, Dec. 6, 1997. He had two children: Elizabeth, b. 1980 and Christopher, b. 1982.)

Spreader of information – newsman par excellence: The dual sign Ascendant gives versatility and *Vargottama* Ascendant emphasizes the themes of Gemini destiny, including here the differences Jennings felt being a Canadian living and working in the USA, also living and working in Europe and the Middle East. Gemini can reflect the internationalism of its opposite sign, Sagittarius. (His U.S. citizenship came in 2003.) Since Gemini is the third sign of the zodiac, the third house in any chart also has an underlying Gemini feel to it and is concerned with the same themes. Jennings' Ascendant lord Mercury is placed in his third house – already a kind of double indication of Gemini influence, even if the sign is Leo. *Navamsha* Mercury is in Gemini, as we shall discuss shortly. News and information was much more compelling for him than a standard education. Music also interested him, especially jazz; and in his teens and early 20s he performed in musical theatre. He always had in mind to pursue a career in journalism. When he did, a whole new level of self-discipline and sense of responsibility entered his life that had not been present before. A well-placed Saturn can supply this quotient whereas an educational degree may not.

[165] A pioneer in American broadcast journalism, Edward R. Murrow (April 25, 1908– April 27, 1965) was highly regarded for his honesty and integrity in combination with his hard-hitting approach to investigative journalism, notably his live radio coverage from Europe during WWII, and later his television show *See it Now*. It played a major role in bringing down Sen. Joseph McCarthy and his Communist witch hunts in the mid-1950s. Murrow was critical of the increasingly commercial and entertainment orientation of television, and greater subservience to corporate sponsors regarding content. He clashed with CBS bosses following his last episode of *See it Now* in July 1958. His last major TV milestone was on the *CBS Reports* installment "Harvest of Shame," on the plight of migrant farm workers in the U.S. It was aired in late November 1960 just after Thanksgiving. Though CBS's most respected journalist and an American icon, his time at CBS came to an end. In Jan. 1961 the newly elected Pres. Kennedy offered him a position as head of the U.S. Information Agency. But Murrow's health was weakened from spring 1961 onward from indications of lung cancer, reducing his role more often from leader to advisor. His brief trajectory in news broadcasting from 1938 to 1960 (all of it at CBS) can be seen against the backdrop of the EARTH period dominating from Feb. 1961. The early steamroller effects of the EARTH period are evident within the realm of television broadcasting leading up to early 1961. Murrow is often cited as a standard bearer whose legacy should be upheld, though later journalists faced the same issues he did with CBS.

Personally, Peter Jennings was tall, handsome, and considered by all who met him a gracious, thoughtful, and urbane man – also very cosmopolitan, amiable, with a good sense of humor. He was very persuasive, though he tried to understand and not to judge. He was known for his loyalty to his many friends. His **intense curiosity about life and the world** can be seen by the many planets with contacts to his natal Mercury: Moon, Venus, Jupiter and Saturn (the latter two planets through their mutual exchange).

Mercury is a neutral planet colored by the planets associated with it, and in this case two of those planets are female planets: Moon and Venus. That gave his strong emotions, his love of women and of the more sensual side of life. It also enabled him to connect at a deep emotional level with many who knew him personally, in addition to his larger public. He was known to cry easily. He was equally known for his remarkable composure under stress and a restrained sense of theatre. This reflected the Saturn influence, which also gave him self-discipline and the ability to edit, even if Saturn competed with the more expansive influences of Jupiter and Venus. Jupiter in close opposition to Mercury gave him the ability to speak voluminously without effort, while Moon in Leo can bring dramatic appeal as well as elegance, since Leo is a royal sign – amplified by contacts from Mercury, Venus and Jupiter. He is bound to be noticed (three planets in Leo opposite Jupiter, including Moon opposite Jupiter, creating the ***Gaja Kesari yoga*** and giving him further prominence.

Mercury has *Digbala* (best possible angle of the chart) in the *Navamsha* Ascendant and is also in a *Bhadra Mahapurusha Raja yoga*. This *yoga* arises when Mercury is in its own sign or exalted (Gemini or Virgo, respectively) and in an angle of the chart. It confers a commanding stature, physically and mentally, with emphasis on learning, intelligence, and virtue. The native earns wealth through his own endeavors. *Navamsha* Mercury is opposite *Navamsha* Jupiter in its own sign of Sagittarius, repeating the **Mercury-Jupiter** opposition from the birth chart. Both planets are strong in their own zodiacal signs in angular houses, thus further supporting the ability to speak or deliver a lot of information well. Neither planet receives malefic aspects. (His death in **Jupiter-Mercury** *Dasha/Bhukti* brought an outpouring of deep respect and affection from people around the world.)

Physical vitality and self-confidence: The *Navamsha* chart confirms the strength or weakness of any planet in the birth chart. When a planet is in its own sign in the Ascendant, as is ***Navamsha* Mercury** – this factor alone is positive for establishing good health, vitality, and a great sense of self-confidence, which Jennings had from a very young age. His sister Sarah describes him as a very "high-spirited child." This vitality continued for most of his life, his health deteriorating only in the last year of life. The Ascendant and Moon are both *Vargottama*, repeating in the same sign in the *Navamsha* chart. A *Vargottama* Ascendant generally gives good health and vitality. It protected Jennings through a very active life, one in which he traveled to dangerous areas of the world. *Navamsha* Sun debilitated in Libra is weaker than the birth chart Sun, even though corrected by Mars. He was a chain smoker for many years, starting to smoke at age 11 and only dropping the habit in 1988, after 39 years. He took it up again in fall 2001, reportedly due to the emotional strain of the Sept. 11, 2001 attacks in Manhattan and the extra responsibilities it entailed for him. When malefic planets tenant the 2nd house (of oral habits, including food and drink intake) poor habits can occur.

Aborted education & departure from country of origin: Jennings attended high school as well as three colleges, though he never graduated from high school or college. Describing himself as "bone lazy" while at school and a bad student, bored by most subjects, he liked ideas but far preferred sports – football and hockey – to most academics. (Sailing and tennis came in adulthood.) He also liked girls. He said he dropped out of high school "out of pure boredom."[166] Jupiter, the planet of learning and education, is in a *Nadi yoga* with Rahu, the foreigner, the exotic, the unpredictable. This is one of the astrological indicators that a traditional academic education (no matter how privileged) would not work well for him and was a waste of time and money. He left at 16 and took a job as a bank teller, plotting how he would get into broadcasting. While formal education did not bring self-discipline, working for a paycheck had that immediate effect. He started to educate himself unstintingly.

The 4th house of educational degrees receives only one aspect – from malefic Saturn. Although 9th house lord (a good influence), Saturn is also 8th house lord and as a classical malefic aspecting a house – can separate us from matters pertaining to that house. Moreover, 4th lord Mercury is situated in the 3rd house, which is the 12th house (of loss and/or foreign residence) from the 4th. This creates losses or separation for him in 4th house matters by leaving Canada, his country of origin. In addition, Mercury is situated in *Magha nakshatra*, whose ruling deities are the Pitris (the departed forefathers). This gives a strong sense of inheriting something from the paternal lineage. His father Charles Jennings, a famous Canadian broadcaster and CBC's first news anchor, was an enormous influence on his only son, who sought to emulate him. *Magha* is ruled by Ketu, which in turn is placed in the 12th house of foreign residence from the Gemini Ascendant. Jennings left Canada in 1964 and was a constant traveler for some years before settling in New York City in 1983. He maintained his Canadian citizenship but also obtained U.S. citizenship on May 30, 2003.

Though Venus is the best planet for the Gemini Ascendant chart, it must be well-placed and aspected to give its best results. The pleasure-seeking component dominated during **Venus *Dasha*,** from birth up to March 1958 – a few months prior to his 20th birthday. This Venus would not give the best results for his education or career, as it catered too much to the enjoyment of social activities and sports to the exclusion of academic work, which was a struggle for him. In the *Navamsha* chart Venus is not well-placed in the 6th house in an exchange with Mars. As we will discuss later, themes of marital conflict arise from excess of passion and do not evaporate once Venus *Dasha* is over in 1958. In the birth chart, Venus is better situated in the 3rd house, but in its own *nakshatra* of *Purva Phalguni* it is more *rajasic* (passionate), giving a strong desire nature and a love of beauty and art.

Venus and Moon are both in Leo in the same *nakshatra* contacting Mercury also in Leo. This indicates a colorful personality, elegant in appearance and manner. Jennings was noticed early on by both Canadian and American producers. Partly due to his father's exposure he was offered his own radio show at age 9 in 1947. But it was not until **Sun *Dasha*** (March 29, 1958 to March 29, 1964) that he began his broadcasting career in radio: 1959 to 1961, television from 1961. In 1964 at the start of **Moon *Dasha*** (March 29, 1964 to March 29, 1974) he left Canada to join ABC News in New York City. Foreign residence would be more likely in Moon *Dasha* due to the mutual exchange (*Parivartana yoga*) of Sun and Moon, with Sun being in the 12th house (of foreign residence) from natal Moon. Also, the 6-year Sun *Dasha* would not bring the rise that came in the 10-year Moon *Dasha*.

---

[166] ***Peter Jennings: A Reporter's Life***, edited by Kate Darnton, Kayce Freed Jennings, & Lynn Sherr, 2007, p. 6.

Royal status in the news business: Sun and Moon are royal planets and the fire signs are also royal, especially Leo, which is closely associated with the Lion and the pride of rulership. The Sun- Moon exchange in the birth chart further strengthens the royal component. Moon *Vargottama* in Leo repeated in the 3rd house of the *Navamsha* chart emphasizes the extent to which Jennings loved news and information (3rd house) and went after it with both passion and a sense of authority that eventually give him royal status in the profession. He had self-confidence, though never arrogance. Ascendant Lord Mercury in Leo resides in *Magha nakshatra* in the natal chart. *Magha* means "the great one," and is symbolized by the throne room. In addition, several well-placed planets in Leo have the potential to give one personal charisma. He was placed on the anchorman's throne a bit too early, but in the end he gained from the experience. He would also be greatly helped by the timing of the Jupiter-Saturn conjunctions of 1980-1981 in Virgo, ushering in the Information Age at a whole new level. Television was a part of this and he would remain at the center of the broadcasting industry.

Self-discipline and the ability to give his fullest effort to his journalistic projects come from the influence of Saturn, mostly from its contact to the Moon – as the most important mental planet. In the *Navamsha* chart, Moon in Leo is opposite Saturn in Aquarius, while also receiving a trinal aspect from *Navamsha* Jupiter in Sagittarius. In the birth chart, the *Parivartana yoga* (mutual exchange) between Jupiter and Saturn gives Saturn surrogate action from Aquarius, as if in opposition to the natal Moon in Leo. The *Navamsha* Moon-Saturn opposition is between 3rd and 9th houses: 9th house is foreign travel and information. It was during **Moon-Saturn** period in 1968 that Jennings left ABC to take a less senior position abroad as foreign correspondent in the Middle East. Saturn is the servant, and may involve sacrifice and a pulling inwards in order to face greater responsibilities. He was willing to work harder and do what was necessary not only to advance, but to help others.

> I grew up in a household where to be a journalist… was seen as an opportunity to be a public servant. I think of myself as a broadcaster who has the privilege of access to the public airwaves."
> *Ibid.*, p. 282.

Rare sense of purpose and destiny: With Gemini Ascendant, the intermingling of Jupiter and Saturn as both classical planets of *Dharma* and *Karma* (and natural 9th and 10th house lords) makes this a powerful combination for career, status, and recognition for one's professional achievements. In addition to exchanging signs in the birth chart, Jupiter and Saturn are strengthened in the *Navamsha* by being in their own signs and in either angular or trinal houses, confirming the power of the ***Parivartana yoga*** and ***Raja yoga*** in the birth chart. Major *yoga*s in the 9th and 10th houses can catapult one to career visibility, even if the *Dasha* periods of these planets do not occur during the lifetime, as with Einstein (who had the same configuration), or as with Jennings, only in the last six years of his life. Jupiter is also a problematic planet for Gemini Ascendant, being *maraka*, among other factors. Its major or minor periods could bring challenges. For Jennings, the **Jupiter-Saturn** *Dasha/Bhukti* ran May 17, 2001 to Nov. 28, 2003 – tumultuous times in American and world history, with the terrorist attacks on the World Trade Towers Sept. 11, 2001 in Manhattan, NY. As anchor for ABC News, Jennings was again at the center of world events and was tested as never before. On the day of the attacks he anchored for 17 hours straight and logged 60 hours that week. *TV Guide* called him "**the center of gravity**." *The Washington Post* praised his "Herculean job of coverage."

Ideally the Ascendant lord shares in the *Raja yogas* with Jupiter and Saturn, as happens with both Peter Jennings and Albert Einstein. Also, for Einstein the exchange occurred in the 4th and 5th houses from the Moon (*Chandra Lagna*), doubling the career and mental impact. For Jennings, the exchange of 7th and 8th lords from the Moon contributed to marital turbulence and to his having four marriages and numerous affairs. However, the overriding force of these *yogas* would also lend to Jennings a very conscientious and energetic desire to fulfill a very public destiny. It was said he attacked each project "like a tiger." He never wavered from a career in broadcasting and took great pride in handling many aspects of it, including in his last 22 years when he was both anchorman at ABC News and also Senior Editor of *The Nightly News*. He loved the unpredictability and complexity of the news business. His ability to be kind and yet piercing with his interview subjects is still praised. But he was critical of the tone of the "the national conversation" [especially starting around 1988] and how it was too often "a shouting match… much of it infected with venom."[167]

> "Our modern communications system and other technologies have increased the volume of public argument to an unholy racket. Our national conversation sometimes feels impoverished as a result….Civility doesn't just promote decency; it also leads to a fairer exchange of ideas and a greater chance of finding workable solutions to the kinds of problems we face."

Extra responsibilities came for Jennings with Saturn transits for nine consecutive years from April 1998 thru July 2007, also putting extra pressure on his health. This coincided with his most difficult *Dasha*, the 16-year Jupiter *Dasha*, from March 30, 1999. The combination of Jupiter *Dasha* and Saturn transits to the Ascendant and to key planets representing the physical body (Ascendant lord Mercury, Sun and Mars) was likely to be hard on the physical body, especially if there was already an area of weakness. Gemini and the 3rd house rule over the lungs and can place emphasis there – like a magnet of sorts. Jennings announced on April 5, 2005 he had lung cancer and died of the disease Aug. 7, 2005. Here is a brief look at those **Saturn transits, 1998 to 2007:**

1. Tr. Saturn in Aries : April 18, 1998 to June 7, 2000 : aspect to Gemini Ascendant.
2. Tr. Saturn in Taurus : June 7, 2000 to April 8, 2003: aspect to natal Sun and Mars.
3. Tr. Saturn in Gemini : April 8, 2003 to May 26, 2005 : contacts Asc., aspects natal Mercury, Ascendant lord.
4. Tr. Saturn in Cancer : May 26, 2005 to July 16, 2007 : contacts natal Sun and Mars. Start of *Sade Sati* (Saturn transits houses 12, 1, and 2 from natal Moon for 7½ years.)

Saturn's impact from April 18, 1998 onward brought added responsibilities to both his personal and/or professional life, starting 4½ months after his 4th marriage. For the Millennium Eve celebrations Dec. 31, 1999 Jennings was the only anchor to appear live for 25 consecutive hours – an almost superhuman task, viewed by over 175 million Americans.

Destiny outside of country of origin: We have discussed how foreign residence was likely in Moon *Dasha* due to the mutual exchange of Sun and Moon, with Sun in the 12th house (of foreign residence) from natal Moon. He entered the 10-year **Moon *Dasha*** on March 29, 1964. Natal Moon

---

[167] *Ibid.*, p. 124.

is *Vargottama* in Leo, along with the *Vargottama* Ascendant in Gemini – thus repeating themes of kingly power or aspirations (Leo) for broadcasting (Gemini). The strong support he enjoyed from his family and family connections is apparent from the planets around the Sun and Moon, including Venus – best friend to Gemini Ascendant. Additional protection to this Moon comes from the ***Shubha Kartari yoga***, with benefic planets on either side of the Moon – Mercury and Venus, in this case.

Natal Sun-Moon exchange also indicated good finances during Sun *Dasha*, but even better opportunities – especially abroad – during Moon *Dasha*. This well describes his success abroad, as it was in 1964 at the start of his **Moon *Dasha*** that Jennings first moved his base of operations outside of Canada, his country of origin. Not only is the Moon *Dasha* stronger than the Sun *Dasha*, but it brought the possibility of foreign residence. In fact, had he not gone abroad (or to a 12th house setting) at this time, it would not have been as favorable for him. (Canada did not allow nepotism in its broadcasting system, so he could not have worked for them as long as his father was also there, up to his retirement in 1971. He died in 1973.) But since the Sun and Moon represent the King and the Queen, it means that this particular exchange will be favorable. He received a kingly offer as ABC Anchorman. It would come with some drawbacks, due to the nature of Mars in this chart. Therefore, Mars major or minor *Dashas* will have difficulties followed by gains.

He started as **ABC anchorman Feb. 1, 1965** during his Moon-Mars *Dasha/Bhukti*. It was unprecedented to have such a young anchor, but ABC's ratings were down and they were desperate to regain them.

> "They were willing to try anything, and to demonstrate the point," Jennings said, "they tried me."[168]

As a 26-year old he had a distinct disadvantage up against very seasoned journalists on competing networks: Walter Cronkite, Chet Huntley and David Brinkley. Jennings lasted three years until poor ratings forced a change.

Disempowered Mars: For a Gemini Ascendant chart, Mars is the 6th house lord and as such can cause problems for the person if it is overly strong. In this case, Mars is not only debilitated in the sign of Cancer, but combust the Sun. Fortunately for Jennings, this disempowers his enemies or competitors and enables Jennings to establish himself very well against the competition. Paradoxically, the person with debilitated Mars is often a far more competitive person, as they know they have some sort of disadvantage (such as lack of a high school diploma, etc.) and are eager to make up for it. As 11th lord, it had the effect of giving him many close and loyal friends to whom Peter seemed to walk on water. He was gifted in social interactions and in his ability to make all those in his presence feel heard and appreciated, including children and young people.

On another level, as Mars is significator of physical vitality, this would tend to give Jennings a strong, virile quality (the close Sun-Mars combination), but also the tendency to burn himself out physically and be easily emotional (Sun-Mars in a water sign). Though his style of delivery was considered impeccable and straightforward, his staff knew they might have to do several takes when there was a particularly emotional subject.

---

[168] *Ibid.*, p. 19.

Mars *Dasha* – turning a disadvantage into an advantage: Debilitated Mars in Cancer represents life situations that start out as disadvantages, even when Mars is *Neecha* (cancelled), as in this case. This would occur during major or minor periods of Mars, but more emphasized during the 7-year Mars *Dasha*. For Jennings, **Mars *Dasha*** ran from March 29, 1974 to March 29, 1981. It looked to the world like a step down when he left ABC News in 1968 after three years as the youngest Anchorman ever. He eschewed the many offers to anchor at other local TV stations. Instead, he sought more experience as a reporter that would give him the depth and maturity he needed. Choosing to spend those years mostly in the Middle East later gave him the advantage of knowing more in depth about the area and its politics, society and culture than most other Western news broadcasters. Jennings entered Mars *Dasha* March 29, 1974 – the same year he received the coveted Peabody award for his special report on Anwar Sadat.

Four marriages: Both Moon and Venus are located in ***Purva Phalguni nakshatra***, owned by Venus and symbolizing the conjugal bed. The deity ruling over this *nakshatra* is *Bhaga*, God of good fortune and prosperity. It protects marital happiness and bestows family inheritance. It also seeks the pleasures of the bed, and in the case of Peter Jennings, he had four marriages and two children. There were also numerous romantic connections both during and between marriages, except apparently for the fourth and last one. Though it is not unusual In the Western world to have several major relationships or marriages, the *Vargottama* Ascendant in a dual sign (or even numerous planets in dual signs) can bring out this tendency still further. More auspicious timing for the marriage(s) could have helped somewhat, but this was also his destiny. We have already discussed how an emotional nature comes with Jennings' natal Mercury-Moon-Venus placement in Leo opposite Jupiter. Venus-Jupiter in itself can be excessive, and Moon and Mercury being the key emotional and mental planets, respectively, are colored by the planets aspecting them. The mind and heart can fluctuate between self-indulgence and sensuality vs. conscientiousness and strong discipline. This is shown by the shared influences of Jupiter and Saturn and the effect of their mutual exchange. The strong desire nature has also been described. Further, Moon and Venus together in the same house can bring too many (or an extra) woman into the picture in a competing role; there is fluctuation (Moon) in matters of love or art (Venus).

The *Navamsha* chart confirms this changeability in partnerships and/or unpredictability of the partner. At the very least there is an enthusiastic desire to be in partnership and some difficulty deciding with whom. With Gemini *Vargottama* Ascendant, and *Navamsha* Jupiter in the 7th house of marriage, this **Mercury-Jupiter** opposition also emphasizes the abundant choice in partnerships, as well as a potentially strong or distinguished partner, as *Navamsha* Jupiter is strong in its own sign. (His 3rd and 4th wives are well known professionally and his 2nd wife was a professional photographer in Lebanon.) Another factor intensifying emotional and/or sexual relationships is the **Venus-Mars** mutual exchange in the *Navamsha* chart: Venus in Scorpio, Mars in Libra. Also, Venus is not favorable in the *Navamsha* 6th house, indicating there are sacrifices and/or conflicts with love matters. This certainly was the case up to the 4th and last marriage in Dec. 1997. And though far more stable, this marriage endured even greater responsibilities for the couple, in terms of the workload from 1998 onward and the health issues of Peter Jennings that surfaced publicly in 2005.

Jennings' third marriage to journalist and author Kati Marton was by all accounts the most turbulent one. They met in 1977 in the midst of his 7-year Mars *Dasha* and married in

May 1979 during his **Mars-Venus** period, with an intense desire to have children together. This would bring to a peak the karmic destiny of emotional marital conflicts, reflecting the problematic *Navamsha* Venus-Mars exchange and its heightened passions, as discussed above. During the several months they were separated in 1987 their private lives were splashed across the newspapers on a daily basis – a low point for Jennings, by his own admission. Then on Aug. 13, 1993 a public announcement was made of their final separation. It was during his **Rahu-Venus** period (Oct. 16, 1992 to Oct. 17, 1995). They divorced in April 1994. They had two children: daughter Elizabeth (b. 1980) and son Christopher (b. 1982). Jennings remained close to both children. During his final months of life, in **Jupiter-Mercury** period, there was said to be a major reconciliation between Jennings and Marton. (This final *Dasha-Bhukti* features *Maraka* planet Jupiter in a *Maraka* house in the *Navamsha* chart. The symbolism resonates on many levels. *Bhukti* lord is Ascendant lord Mercury in opposition aspect to Jupiter in both birth chart and *Navamsha* chart. The Ascendant or Ascendant lord always pertains to one's physical health.)

Jennings fourth marriage was much happier. On Dec. 6, 1997 he wed ABC producer Kayce Freed, during his **Rahu-Moon** period. Rahu is in the *nakshatra* of Jupiter, 7th lord of marriage and Rahu *Dasha* often brings marriage, if other factors do not prevent it. Natal Moon is in the 10th house from Rahu, showing Freed's visibility and status in the same world Jennings moved in for many years. Rahu is also considered well-placed in the 6th house.

Personal financial wealth: Peter Jennings left an estate valued at $50 million. He was a generous contributor to various charities, having already given away millions. He personally worked with the homeless, including in soup kitchens serving homeless and needy people. *Dhana yoga*s for wealth are evident through contacts between lords of houses 1, 2, 5 and 9 (9th lord Saturn through the *Parivartana yoga* with Jupiter). House lords 1, 2 and 5 are situated in the 3rd house of communications opposite 10th lord Jupiter. This favors wealth and success through some endeavor related to communications.

Chart 15: Francis Poulenc

Birth data: Saturday, Jan. 7, 1899, 16:00 LMT, Paris, France, 2E20 00, 48N52 00, *Lahiri ayanamsha,* -22:27:06, Class AA, from birth certificate. Ascendant: 21:53 Gemini.

Biographical summary: One of the most well known and beloved French composers of the 20th century, Poulenc was also an accomplished pianist. Over his lifetime he composed three operas, two ballets, four concertos, five pieces of chamber music, five major choral works, over 100 songs and numerous piano compositions. Symphonies interested him much less. Before the mid-1930s, his music was lighter in tone, more jocular, and always influenced by popular music. One of his operas, *The Dialogues of the Carmélites* (1955) is considered part of the major opera repertoire of the 20th century and noted for its strong religious and political themes. However, even in 1944 he produced *Les Mamelles de Tiresias* – an opera with more farcical themes and in the secular mode.

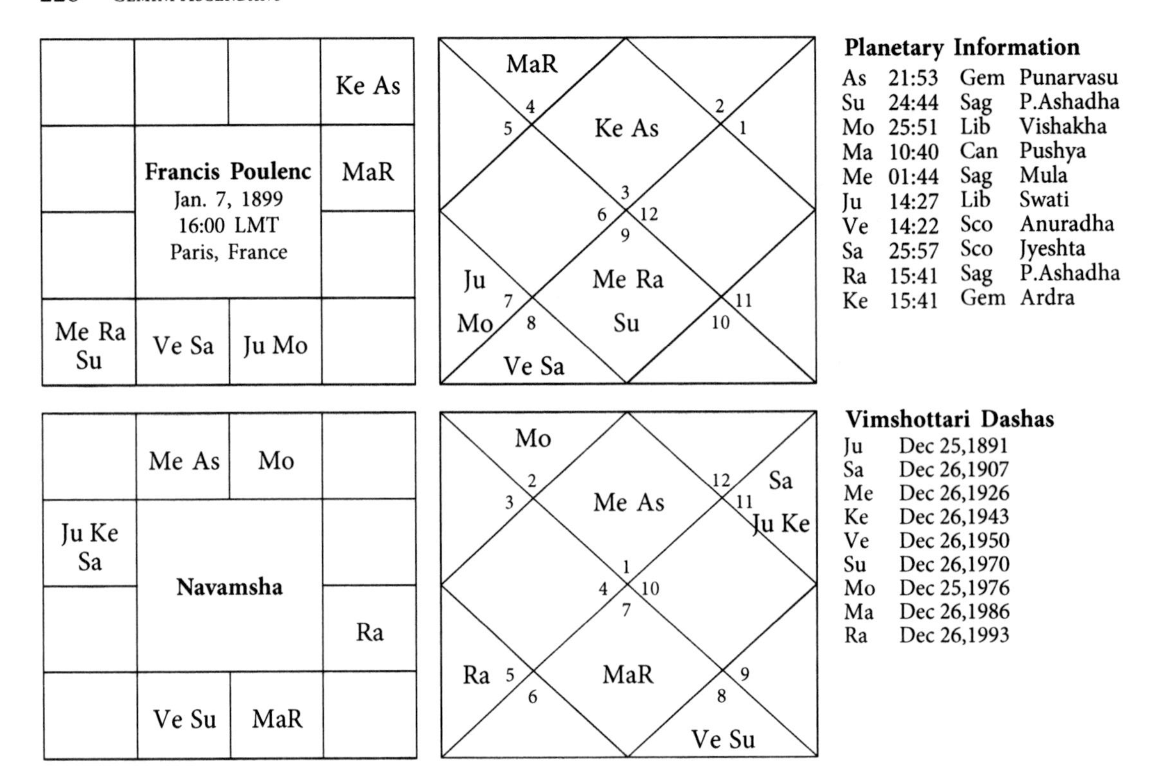

Planetary Information

As	21:53	Gem	Punarvasu
Su	24:44	Sag	P.Ashadha
Mo	25:51	Lib	Vishakha
Ma	10:40	Can	Pushya
Me	01:44	Sag	Mula
Ju	14:27	Lib	Swati
Ve	14:22	Sco	Anuradha
Sa	25:57	Sco	Jyeshta
Ra	15:41	Sag	P.Ashadha
Ke	15:41	Gem	Ardra

Vimshottari Dashas

Ju	Dec 25,1891
Sa	Dec 26,1907
Me	Dec 26,1926
Ke	Dec 26,1943
Ve	Dec 26,1950
Su	Dec 26,1970
Mo	Dec 25,1976
Ma	Dec 26,1986
Ra	Dec 26,1993

His music is notable for its lyrical melodic lines, either soaring or interspersed with surprising flippant touches.

Poulenc was known as a sophisticated eccentric and he loved to shock the bourgeoisie. In his early years, even preceding his formal training from 1921, he was one of a group of composers called "Les Six." Irreverent mockery and self-mockery was one of their ways to shock, but it caused him problems with the music establishment. After showing his music at age 18, Poulenc was contemptuously rejected by the director of the Paris Conservatory. But his mother supported and encouraged his musical tastes – including his admiration for then avant-garde composers Erik Satie and Igor Stravinsky – though Poulenc's musical style would become more complex than Satie and warmer and less intellectual than Stravinsky. He loved the music hall and the vaudeville popular in that era and this greatly influenced the eclectic nature of his music, alongside the more sacred themes.

Poulenc was born into a very wealthy industrialist family, whose wealth came from the Rhône-Poulenc chemical and pharmaceutical company. Enabled to work and study independently, his musical gifts are distinguished by being out of the mainstream (though fundamentally tonal), in part because he was largely self-taught as a composer. (His mother taught him piano.) His formal training in music composition began in 1921 with composer Charles Koechlin (1867-1950), who in turn was a student of Gabriel Fauré (1845-1924).

Poulenc was born and raised a Roman Catholic, and this contributed to his inner struggle as probably one of the first "openly gay" (also bisexual) composers. He was reported to have fathered a daughter, Marie-Ange, but never acknowledged it. In 1936 he had a major religious experience that brought him back to his Catholic faith and to compose many more sacred works. But in July

1950, French music critic Claude Rostand (*Paris-Presse*) dubbed Poulenc "le moine et le voyou," or half-monk, half-guttershipe – a label that stuck with him for the rest of his career. He died of heart failure at age 64, on Jan. 30, 1963.

Musical eccentric: Mercury can be playful. Rahu's contact to Ascendant lord Mercury adds some eccentricity and volatility to this playful character and mind. Rahu and Ketu are unconventional and they are prominently featured, close to the Ascendant axis. Natal Sun is also in Sagittarius, but more than 20 degrees apart from Mercury, forming a *Budha Aditya yoga*. This bestows mental cleverness and an extra level of skill, intelligence, productivity and practicality. But Rahu-Ketu so close to the Ascendant degree and on the relationship axis (Houses 1-7) brings some notable turbulence in personal relationships, as does Venus in the 6th house of conflict, the least advantageous house position for Venus. (Peter Jennings had *Navamsha* Venus in the 6th house, also in Scorpio.) Jupiter, planet of abundance, is 7th lord of partnerships. Placed in the 5th house, it gives him many lovers. Problems arise in this arena with sign lord Venus in the 6th house and *nakshatra* lord Rahu in the 7th house. Mercury is in *Mula nakshatra* ruled by Niritti, goddess of death and destruction, tending to disrupt connections and associations to the planet.

Poulenc's unique and original style of music came in part from being largely self-taught as a composer. His mother, an amateur pianist herself, taught him the piano and encouraged his independence. He became an accomplished pianist, later performing and recording some of his own works. At age 18 he was rejected by the Paris Conservatory, as his musical style was not in the conventional mode. (This was in his Saturn-Venus *Dasha/Bhukti*, with both planets in his 6th house of conflict. In the *Navamsha* chart Saturn aspects Venus and Sun in the 8th house, also not ideal for acceptance by the orthodoxy.) It was another four years before he found his private teacher, Charles Koechlin – in 1921, during his Saturn-Rahu *Dasha/Bhukti* and near the start of his *Sade Sati*, which began when tr. Saturn entered Virgo Aug. 8, 1921. The 7½ year period of *Sade Sati* would bring more discipline into his life, in this case from the first outside teacher. On Sept. 9, 1921 the Jupiter-Saturn conjunction in Virgo in his 4th house (of higher education) was another major marker for when it would be auspicious to begin his (first and only) formal education in music. This confluence of fortuitous timing helped him grow as a composer and musician without breaking his unconventional spirit.

For physical vitality and longevity Ascendant lord Mercury is well placed in the 7th house aspecting its own house. Classic benefic Jupiter aspecting the Ascendant gives physical protection. If there are no other afflictions to the Ascendant or its lord, these factors combine to give a reasonably long life.[169] Other factors weaken the longevity, however. Even with Mercury well-placed in the *Navamsha* Ascendant (*Digbala*), it is aspected by two classic malefics: Saturn and 8th lord Mars. Natal Rahu contacts Mercury in the same sign (within 14 degrees), causing affliction to the physical health, mostly strain on the nervous system as he was prone to emotional extremes. But he survived two stints in the military service during WWI, the first one starting in Jan. 1918, during his Saturn-Sun period. He did not abide by the strict rules there either. (Saturn-Sun shows confrontation with authority.)

[169] Classic Vedic astrology texts assign a short life as from birth to age 32, medium life span from 32 to 64, and a long life is anything beyond 64. Poulenc lived to 64. We study various planetary factors in combination.

Financial wealth: Born into a wealthy family, Poulenc was independently wealthy throughout his entire life. Since there was no need to make a living from his music, he enjoyed greater artistic freedom and independence. This is shown astrologically by numerous *Dhana yogas* for wealth in both birth and *Navamsha* charts. From the Gemini Ascendant, 11th lord Mars is in the 2nd house and 2nd lord Moon is in the 5th house with Jupiter. Classic benefic Jupiter is better placed than Venus in this chart, assuring him of dependable financial wealth throughout life. Jupiter helps his career as well, being 10th lord in the 5th house. From natal Moon in Libra, there are further *Dhana yogas*: House Lords 1 and 5 in the 2nd house from the Moon, and Lord of 2nd house (Mars) aspects the Moon. *Navamsha* Moon is exalted in the 2nd house, aspected by *Navamsha* Venus, Sun and Mars. *Navamsha* 11th house contains both 9th and 11th house lords, along with Ketu.

So many wealth indicators gave him the possibility of more autonomy to establish his own musical identity and standards. Though freed from financial worries, he was a serious craftsman. Saturn's location in the 6th house from the Gemini Ascendant and in the 2nd house from the Moon gave him the necessary discipline to manage such freedom. Saturn also gives one a competitive edge when in the 6th house. Staying power is confirmed with seven out of nine planets in fixed signs in the *Navamsha* chart, while *Navamsha* Mercury is *Digbala* (best possible angle) in the Ascendant.

Overcoming obstacles and protecting wealth: Special *yoga*s that help to overcome obstacles, expenditures, and enmities are called ***Viparita Raja yogas.*** (See Bill Gates, Chart #13.) These occur when the lords of the most difficult houses (6, 8, and 12) are situated in one of these same houses from the Ascendant or the Moon. In this case they come only from the Ascendant, where there are two out of three of the *Viparita Raja yogas.* This protects him from his detractors and enables him to deal well with his competition. There were certainly detractors who thought Poulenc should adhere to a more recognizable mold of music. His sexual proclivities and love relations were also less conventional. Natal Mercury-Rahu in the same sign aspecting the Ascendant contributed to this. It also gave him some of his eccentricity and his unusual charm and wit. Ketu in the Ascendant made him more mysterious and less predictable. This tends to make people more eager to identify you. Thus in July 1950, late in his 7-year Ketu *Dasha* (in Ketu-Mercury period), a prominent French music critic described him as "le moine et le voyou," translated by some as "part monk, part bad boy." Though this label remained with him for the rest of his career, Poulenc defied his critics and was highly productive. He worked hard at his music composition on a daily basis and at night he would typically go out to the music halls, jazz clubs or the circus. He loved all of it and all of it became components of his music. With natal Moon in Libra showing preference for urban life, Paris was the center of his universe.

Constant theme of duality: The strong Gemini influence contributes to this theme, along with Ascendant lord Mercury and three other planets in dual signs in the birth chart. Poulenc composed both sacred and secular music. His secular music often has a particular air of insouciance, while his sacred works are far more somber in tone, with beautiful melodic lines permeating whatever he wrote. He had an important re-conversion experience to Catholicism shortly after the sudden death of a friend on Aug. 17, 1936.[170]

---

[170] Pierre Octave-Ferroud (b. Jan. 6, 1900, d. Aug. 17, 1936) was a rival French composer who was killed (decapitated) in a car accident in Hungary. Poulenc suffered deeply from the deaths of several close friends. He was warmly regarded by his colleagues and very affected by them. Ferroud's death triggered a pilgrimage.

After this he concentrated much more on sacred music, and began to produce many more works devoted to sacred Christian themes. However, **the themes of sacred and profane** continued to move back and forth throughout his life. **Poulenc's life-altering religious experience was at the Shrine of Our Lady of Rocamadour** in France, and brought him back to his Catholic faith. Soon afterward he wrote his ethereal *Litanies à la Vierge noire* (Litanies to the Black Virgin, 1936) for three-part choir of children's or women's voices and organ. It was composed in one week and was in honor of the conversion experience, which occurred during his **Mercury *Dasha***, soon after mid-Aug. 1936: during **Mercury-Rahu** (June 22, 1936 to January 10, 1939). The volatility of his Mercury-Rahu-Ketu connections would always be reflected in their *Dasha-Bhuktis*. Further, in Aug.-Sept. 1936 tr. Saturn was in Aquarius in his 9th house of religion and tr. Jupiter was in Scorpio, contacting his natal Saturn (9th lord). Both *Dharma* and *Karma* lords are placing focus on 9th house matters.

We have already spoken of the turbulence in his personal relationships, which were bisexual but mostly gay. In part this was due to his inner turmoil from his Catholic upbringing. (Saturn, 9th lord of religion, is placed in the 6th house of conflict, along with Venus, planet of love.) This is one of many astrological factors indicating that relationships might not go smoothly. It was so upsetting for him when it did not go well in this arena of his life that he would be unable to produce or create for up to a year or more. His first serious relationship was in his late 20s with the painter Richard Chanlaire. Poulenc was ecstatic, dedicating his harpsichord concerto to him: *Concerto Champêtre*. His 19-year Saturn *Dasha* was less likely to bring key relationships than Mercury *Dasha*, from Dec. 26, 1926 (age 28): Mercury contacts or aspects the 7th house of partnerships in both birth and *Navamsha* charts, and is *Darakaraka* (spouse indicator), as the planet at the lowest degree of celestial longitude.

Venus rules over music, and his 20-year **Venus *Dasha***, beginning Dec. 26, 1950 set the stage for the creation of some of Poulenc's finest works, especially since Venus is the best planet for Gemini Ascendant chart and is well placed from the natal Moon. But it was not good for his health or his longevity placed in the 6th house from the natal Ascendant and in the 8th house in the *Navamsha* chart. Any planet has more influence during its *Dasha*, when it is also *Dasha* lord. The 8th house gives more profundity, since this is the house that brings immense changes in life and can bring greater spiritual awareness, along with the greater suffering. Though it caused turmoil in his love life, the Venus-Saturn combination in the end reunited his music with deeply sacred themes. Composed entirely in Venus *Dasha* and completed five years into the *Dasha*, his opera *Dialogues of the Carmélites* was based on a story by George Bernanos, and set during the French Revolution. The emotional power of his re-conversion to Catholicism in 1936 influenced much of his subsequent work, infusing this opera as well. The central character (Blanche) struggles between earthly and heavenly concerns, believing that if she becomes a nun she can escape all her earthly suffering. Her quest for inner peace has surprise results. Poulenc died 13 years into his Venus *Dasha*, in Venus-Jupiter period. Jupiter is a *maraka* planet.

> "Too much introspection has been gnawing away at me for months. My consuming love for Lucien, which far from abating only seems to grow more intense, made me fall into a blind panic.... My work is the only thing that will pull me out of this, and in that respect I have not lost my touch. In an extraordinary moment of emotion and turmoil the final moments of *Les Carmélites* came to me, Blanche's arrival and her march to the scaffold.

> Looking at this music coldly, I honestly believe that it is overwhelming in its simplicity, in its resignation, and ... in its peace."
>
> **Francis Poulenc** in a letter to Pierre Bernac, responding to his Nov. 1954 letter chiding him for "an inclination to let himself go completely while... suffering a broken heart." Wilfred Mellers, *Francis Poulenc* [Oxford Studies of Composers], 1993, p. 181.

The 3rd house is the house of music in the Vedic astrological chart. In the case of Poulenc, the 3rd house from the Gemini Ascendant receives a very high numerical ranking of 39 in the *Ashtakavarga* rating system. A ranking of 30 bindus starts to be above average, so 39 is very high. When he got into the emotional realm (that of the Moon), the 3rd house from the Moon receives a numerical ranking well below average – at 21. This house contains Mercury, Rahu and Sun in Sagittarius. Rahu destabilizes the whole combination and can pull it down, especially in emotional terms, challenging his ability to produce when in emotional distress or regain his equilibrium. Natal Mars is in the emotional, watery sign of the Moon (where it is debilitated), and also aspects the Moon, heating up and often irritating the emotional nature. In the *Navamsha* chart, Mars aspects both Mercury and Moon, further confirming the tendency for extreme volatility of his mind and heart. On a positive note, however, from natal Moon this Mars is well placed in the 10th house – giving good career visibility.

Review of Poulenc's Mercury: Mercury is always affected by planets situated near it in the same sign, being a neutral planet that picks up the influences of the planets contacting or aspecting it. With Poulenc, Mercury's influence in the birth chart comes from Rahu, Ketu and Sun. Rahu's influence gives the non-conformity. Sagittarian influence gives the breadth, the eclecticism, the versatility to his character, and later the religious re-conversion during Mercury *Dasha*. But Ketu also plays a part in that. Mercury is in Ketu-owned *Mula nakshatra*. Niritti, goddess of destruction rules over *Mula* and has the power to break things apart so they can arise anew. Ketu is a significator for *Moksha*, or spiritual liberation, and pushes the energy inwards. With Ketu in the Gemini Ascendant, the dapper Poulenc, frequenter of music halls becomes much more mysterious, more unknowable, someone we cannot easily categorize because he is too elusive for that.[171] Thus, unknowable Ketu on the Gemini Ascendant creates a real paradox: Though Gemini is eager to make knowledge and information of many types accessible to all peoples, Ketu changes the dynamic. Picture a man speaking with a bullhorn. But he is hidden discreetly behind a large black curtain, close to a church altar or maybe in a jazz club. It depends on his state of mind.

REVIEW OF GEMINI ASCENDANT CHARTS

Give the astrological reasoning in each case.

1. Why is the nervous system potentially more vulnerable for Gemini Ascendant? How did that shorten the life of Ramanujan? Hint: Check Mercury's position in his birth chart and in his *Dasha* sequence.

---

[171] For another biographical study with Ketu in the Ascendant, see Jacqueline Kennedy Onassis (Chart #31), Chapter 9, Libra Ascendant charts. Queen Elizabeth II (Sagittarius Ascendant) also has Ketu exactly conjunct the Ascendant, thus repeating in the *Navamsha* chart.

2. With so much Gemini (dual sign) versatility, what factors tend to lessen the attendant confusion and difficulty in decision-making? Explain Peter Jennings' singular focus from childhood to become a broadcast journalist. Also, Ramanujan's obsession with mathematics.
3. What astrological factors account for Albert Einstein's superior intelligence? What also made him appear lacking in intelligence in the early years of his childhood? Hint: This same factor accounts for the abnormally slow start in his language skills.
4. What astrological factors describe the largely intuitive source for Ramanujan's mathematical brilliance? For Einstein's scientific genius? And for Spielberg's creative and commercial genius?
5. Why was Venus *Dasha* the most productive period for Albert Einstein? Why is Venus *Dasha* generally the best one for the Gemini Ascendant chart? Hint: Note the early start it gave to Peter Jennings, and the crowning glory to the careers of Albert Einstein, Bill Gates, and Francis Poulenc. Ramanujan was less fortunate. He died at age 32, ten weeks prior to the start of his Venus *Dasha*.
6. How can Mars work against Mercury in the Gemini Ascendant chart? How can Mercury and Mars also be an effective alliance? Hint: See the chart of Bill Gates. Note also that Einstein's Mars *Dasha* was the most dangerous for him, especially Mars-Rahu, and provided the basis for his temporary departure from pacificism – notably his decision to contact President Roosevelt in Aug. 1939.
7. How could Einstein be so well known and yet so personally isolated?
8. Why does the Mercury-Mars combination make someone less likely to want a more traditional education? Hint: Review the educational paths of all six biographical subjects with Gemini Ascendant.
9. Why is an air sign Ascendant a good candidate for large wealth, judging by the 2nd house? What planetary configurations could erase that advantage? Hint: Study the chart of Einstein, who was unable to deal well with money or keep it.
10. How could Microsoft founder Bill Gates keep overcoming his competition for so many years? What enabled him to stave off the breakup of his company in the major court case against Microsoft in the late 1990s, concluding in Sept. 2001?
11. What astrological indicators show that Bill Gates would have a tendency to attract legal troubles, along with the ability to remain dominant in spite of ongoing court battles? (in the realm of anti-trust and unfair trade practices)
12. Why is the *Raja yoga* between Jupiter and Saturn curtailed somewhat for Gemini Ascendant chart? Even for Albert Einstein and Peter Jennings? Hint: This shows that career and life paths would be elevated but there would be ups and downs.

CHAPTER 6

Cancer Ascendant

Introduction

The Sanskrit name for Cancer is ***Karka***, or ***Kataka***, meaning literally "the crab." Its domain is in the midst of water or a sandbank near water. Cancer is a ***Chara*** sign, the second *Chara* sign after Aries. It is moveable and active in nature. Its element is water and its Vedic noble aim or goal in life is *Moksha,* or spiritual liberation. The 4th sign and the classical 4th house both rule over the foundation of the mind, and thus have much to do with its outcome – the intelligence motivating our actions, and the ability to be educated. Cancer is concerned with the mind as consciousness. Its lord is the Moon, and as the receiver of the Sun's light it is considered the most important planet in Vedic astrology. The Moon shows how we are likely to receive the experiences of this life. It is like the giant satellite dish of our lives. Thus primarily how we receive and accept the experiences of this life accounts for our happiness and is associated with our sense of security, emotionally and physically.

The Sun and the Moon are the only two **royal planets**, the King and the Queen, respectively, bestowing royal status on Sun-ruled Leo and Moon-ruled Cancer. (All three fire signs also have royal associations, due to being ***Kshatriyas*** (warriors and/or political leaders), while the water signs are ***Brahmins***, the priestly caste, often also associated with artists, especially musicians, and those who interpret large cultural moods and themes, with an emotional capacity to deliver a transcendent message to the public. There are 108 names for the Moon in Vedic myth, 108 names also for the Sun. *Chandra* is the most commonly used name for the Moon, as is *Soma* – which means immortal nectar. Sun, Moon and Venus are all associated with vision and eyesight, being the three brightest objects in the sky.

The Moon and the 4th house/sign concern the happiness of the heart, the repository of the memory, the mother, mother's relatives, and relatives in general. The physical home and real estate matters are included within Cancer, including the first abode – the mother's womb. Not surprisingly, the memory contains experiences within the mother's womb and can be confused at times with feelings and intuitions not belonging to us. Rather they belong to the mother. Such is the natural watery nature of the Moon, the intuitive or feeling part of the brain. It is innately habitual in its likes and dislikes and in its emotional, mental patterns. As a planet of fluctuating light, the Moon shows our tendency to be changeable in our mental tendencies, mistaking at times our transient likes and dislikes for our deepest feelings or an invincible truth. Likewise, the *Bhava* (house) where Moon is located in the chart shows where we put our mental and emotional focus and experience many changes.

As the water signs are more emotional, the water sign Ascendant person will be more obsessed with what makes them happy or what *could* make them happy in the future. This is more obsessive in Scorpio, the Moon's sign of debilitation, but it is a theme identifying all water sign Ascendants or clusters of planets in water. For the Cancer Ascendant person, happiness and security are linked with **family or a sense of family**, especially the mother, as described above, as well as maternal relatives and relatives in general. Cancer and the 4th house describe both the roots of the life and the end of life; the family and home, the foundation of the mind and of the life – thus the importance of the mother, the memory, and the learning process. We learn about these from the condition of the Cancer Ascendant and its lord (the Moon), also the 4th house lord, planets in the 4th house, and aspects to all of these, especially to the Moon – as natural *karaka* for 4th house matters as well as Ascendant lord. The birth chart with Cancer Ascendant identifies someone who wants to maintain the status of the home (physically and otherwise) and who may go to great lengths to find it, keep it and defend it, especially being so tied to family and country. A very tribal sign, Cancer has instinctual emotional bonds, and the mother's role may be one of mythic proportions. She might have been extraordinary or even absent, so her presence looms large due to her absence.

As with Mercury, the Moon is strongly affected by other planetary contacts or aspects, but even more so as its watery nature makes it more porous. Since the Moon IS the mind, it tells us more about a person than any other planet in the chart, notably the mental and emotional nature of the person.

> "Water takes on the color of the vessel in which it is contained."
>
> **Rumi** (1207-1273), Sufi poet and mystic.
>
> "The materials used by the senses are a doorway to the mind.... First, the education of the senses, then the education of the intellect."
>
> **Maria Montessori** (1870-1952), Educator (Chart #16).

For the Cancer Ascendant person, the emotions may have a huge impact on what makes the person act. As with the Gemini Ascendant – there is an extra challenge and some extra pressure on the mental and emotional being, since the planet ruling the Ascendant (representing the physical body) also rules over the mind, increasing the impact of the mind's ability to affect the physical health.

The contents of the mind are also regarded as sound (i.e., vibrations stimulating the auditory nerves), and thus are very affected by sound of any kind. That is why *mantra* is considered so important in its capacity to purify the mind. When we consider the watery nature of Cancer and the nature of sound and mind as auditory vibrations, we note that sound moves more quickly through a denser medium such as water. It moves the fastest through solids – such as beryllium, but it moves faster through liquids than through air, or gases, and the fastest through ocean water. That is because it has molecules weighing it down. Sound cannot travel through a vacuum. The parallel to the astrology is that air and gas relate to the more mental aspects of the mind, while water relates to its more emotional, intuitive aspects. The watery emotional nature moves far more rapidly than purely intellectual components. It can connect people more quickly and more deeply, either positively or negatively. Water is closely related to the emotional drive to create sound, such as music. It, in turn, can be an instant memory trigger, along with the sound of the human voice.

The Moon is not considered to have any enemies, as it is supposed to be a friend to all planets. However, Saturn as 7th and 8th house lord can be very difficult for Cancer Ascendant. When aspecting the Ascendant or Moon, it can bring even more obligations to family or nation than the Cancer Ascendant person already feels. Or it increases the sense of loss if the family is not present. Mercury as 12th house lord for Cancer Ascendant can also be problematic if poorly placed in the chart. And for Mercury, Moon is a planetary enemy. **The myth of Soma and Tara** is helpful to review in this context, and from some other angles.[172] In this story from the *Puranas*, Soma (the Moon) is a man and is courting Jupiter's wife Tara. Jupiter is busy with all the requirements of his ritual ceremonies as a priest – so much so that he is perhaps neglecting his wife. Thus far they have failed to produce an offspring. When Tara becomes pregnant from her tryst with Soma, Jupiter is jealous and offended, and much havoc ensues. But in the end, the child (Mercury) is so witty, clever, and charming that Jupiter is persuaded to keep him.

This teaching myth shows the Moon's ability to connect with all things and all peoples. Moon as Soma could impregnate Tara, later accepting Mercury as his true child. In spite of his high status as a priest, Jupiter may not always be able to connect at the human level. But as the planet of tolerance and forgiveness, Jupiter eventually forgives everyone involved, even if there is some residual discomfort in his relationship with Mercury. Meanwhile, Mercury is in awe of Jupiter's wisdom and knowledge, and somewhat resentful of Soma (the Moon) as interloper, even if they come from the same blood. Very intelligent and educable, Mercury is enhanced by both Jupiter (higher knowledge) and the Moon (the mind, especially its emotional, intuitive intelligence). Thus Cancer Ascendant, as well as the 4th sign and 4th house, is concerned with the educational process and the formal education, including the higher educational degrees likely to be attained.

Nakshatras: Cancer contains the following three *nakshatras*: ***Punarvasu*** (starting from 20:00 Gemini, in the previous sign) 0 – 3:20 Cancer (ruled by Jupiter), ***Pushya*** 3:20 – 16:40 Cancer (ruled by Saturn), and ***Ashlesha*** 16:40 – 30:00 Cancer (ruled by Mercury). The Cancer Ascendant, or any planet situated in one of these three *nakshatras* is colored by being: 1) in Cancer, ruled by the Moon; 2) in one of these three *nakshatras*, ruled by Jupiter, Saturn, or Mercury, respectively; and 3) by aspects to the Ascendant and Ascendant lord Moon.

Jupiter is the best planet for Cancer Ascendant, and is *Nadi yogakaraka*. Jupiter is also exalted in Cancer, maximum exaltation at 5:00 Cancer. Mars is debilitated in Cancer, maximum debilitation at 28:00 Cancer. Note how this plays out in the chart of Maria Montessori (Chart #16), especially as there is some correction of Mars by its *Nadi yoga* with Jupiter, creating more extraordinary results in the life due to rising up from the initial disadvantage that comes from debilitated planets.

---

[172] The Introduction to Chapter 5 (Gemini Ascendant) also discusses the myth of Soma and Tara.

***Gandanta* degrees:** They occur at junctions and are usually adverse if a natal planet or Ascendant is located there or if an auspicious action is performed with planets or Ascendant at that location. A natal planet in *Gandanta* degrees shows weakness and problems in the life that need grounding and strengthening. (This pertains to the house lordship and significations of the *Gandanta* planet.) The Sanskrit word *Ganda* is a knot; *Anta* is the end. As a cyclical reference it is the end of one grouping and the start of another. Energy cannot flow smoothly through the obstruction of the knot. (Of the three types of *Gandanta,* we will not cover *Tithi Gandanta.*) *Rashi* (sign) and *Nakshatra Gandantas* refer to the last 3 degrees 20 minutes of a water sign and first 3 degrees 20 minutes of a fire sign. These junctures are considered dangerous and critical because both the zodiacal sign and the *nakshatra* come to an end. There is more vulnerability at the end of water, especially at the end of Scorpio, which is considered the most treacherous of the three junctures. But the first 3 degrees 20 minutes of fire signs (also a *nakshatra pada)* are not immune.

Ashlesha nakshatra is ruled by the Vedic deity, *Sarpa* – the Serpent. When *Gandanta* occurs in *Ashlesha*, results may not coincide with appearances. *Ashlesha* energies must be used carefully, as the Serpent (or any area ruled by *Sarpa*) is both perceptive and seductive, and carries poison that can either kill or cure. *Ashlesha nakshatra* is also within a *Sarpa drekkana.*

***Sarpa drekkanas*:** These occur only in water signs. The 2nd and 3rd *drekkanas* (divisions of 10 degrees) of Cancer are considered *sarpa*, or snake-like in quality: In Cancer, this is 10:00 to 20:00 and 20:00 to 30:00 Cancer; in Scorpio, the first two *drekkanas*: 00:00 to 10:00 and 10:00 to 20:00 Scorpio; and in Pisces – the 3rd *drekkana*: 20:00 to 30:00 Pisces. A planet in *Sarpa drekkana* must be handled with care, as greater volatility is associated with planets in these areas.

Chart 16: Maria Montessori

Birth data: Wednesday, Aug. 31, 1870, 3:23 AM LMT, Chiaravalle, Italy, Long.13 E 19, Lat. 43N 36, *Lahiri ayanamsha* -22:02:47. Ascendant: 22:30 Cancer. Class AA data, though two options are given: Birth certificate quoted from Rodden's *Astrodatabank* gives 2:40 AM GMT, with 14:15 Cancer Ascendant. Rodden also cites Gauquelin, Vol. 2, #1509 with 3:30 AM Rome time, and 22:28 Cancer Ascendant. The discrepancies in Rodden's data arise from her using GMT instead of LMT, the latter being correct. At www.dominantstar.com Rob Couteau gives 3:33:20 AM LMT, Chiaravalle, Italy, with a 24:28 Cancer Ascendant, and Aquarius *Navamsha* Ascendant. (The author prefers Capricorn to Aquarius *Navamsha* Ascendant, rectifying to 3:23 AM LMT, Chiaravalle, Italy, a time corresponding exactly with the Gauquelin data at 3:30 AM LMT Rome time, for Rome, Italy. Capricorn *Navamsha* Ascendant gives five planets in *kendra*s, including four classic benefic planets. This *Navamsha* chart more accurately reflects Montessori's life.) Ascendant: 22:30 Cancer.

Biographical summary: Although she is known largely as an educator, Maria Montessori was first a doctor of medicine, and has the distinction of being the first woman in Italy to receive a medical degree: in 1896. Special permission to allow her entry into medical school came from the Pope, among others. She was trained as a psychiatrist and an anthropologist and became a pioneer in

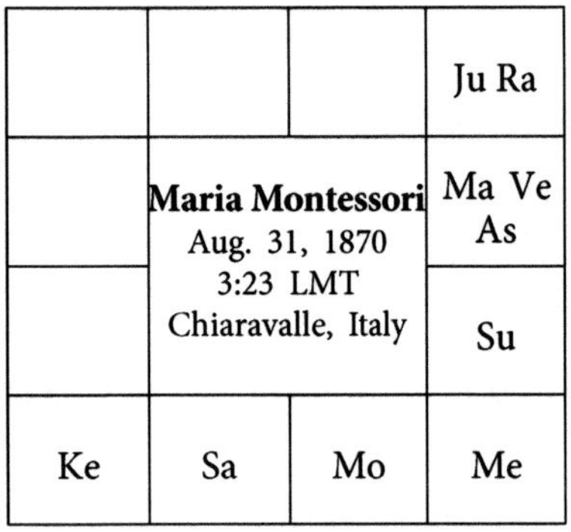

			Ju Ra
	Maria Montessori Aug. 31, 1870 3:23 LMT Chiaravalle, Italy		Ma Ve As
			Su
Ke	Sa	Mo	Me

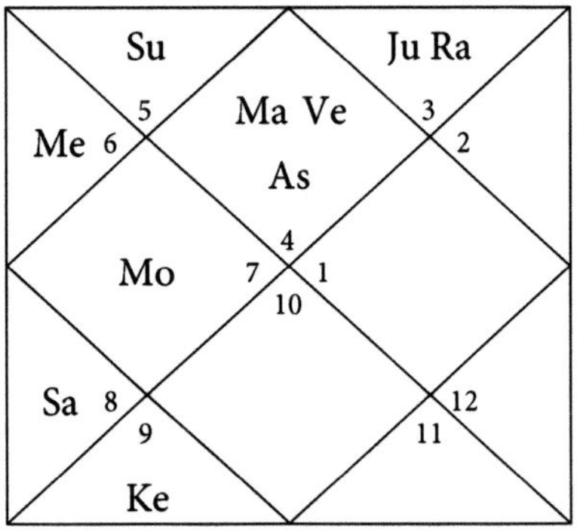

Su
Ju Ra
5
Ma Ve
3
Me 6
2
As
4
Mo 7
1
10
Sa 8
12
9
11
Ke

Planetary Information

As	22:30	Can	Ashlesha
Su	15:24	Leo	P.Phalguni
Mo	12:26	Lib	Swati
Ma	02:28	Can	Punarvasu
Me	10:58	Vir	Hasta
Ju	01:17	Gem	Mrigashira
Ve	20:23	Can	Ashlesha
Sa	29:53	Sco	Jyeshta
Ra	24:30	Gem	Punarvasu
Ke	24:30	Sag	P.Ashadha

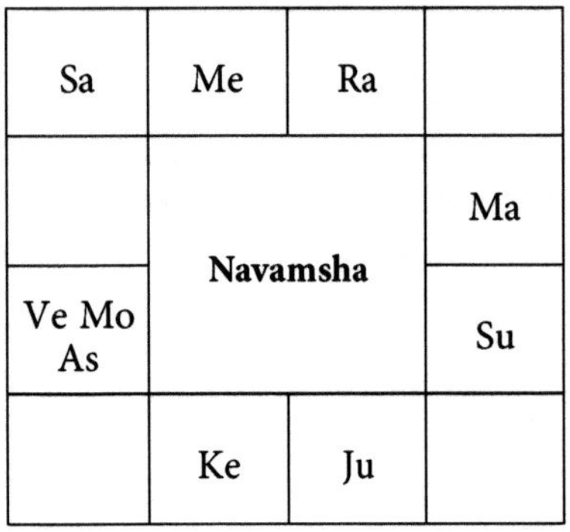

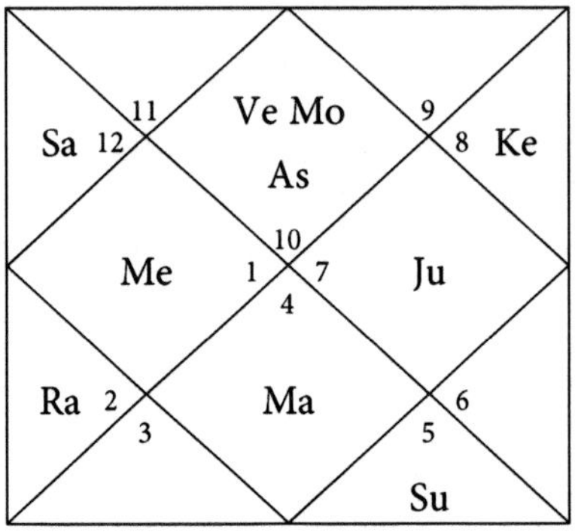

Sa	Me	Ra	
	Navamsha		Ma
Ve Mo As			Su
	Ke	Ju	

11
Ve Mo
9
Sa 12
8 Ke
As
10
Me 1
7 Ju
4
Ra 2
Ma
6
3
5
Su

Vimshottari Dashas

Ra	Nov 14, 1862
Ju	Nov 14, 1880
Sa	Nov 14, 1896
Me	Nov 16, 1915
Ke	Nov 15, 1932
Ve	Nov 16, 1939
Su	Nov 16, 1959
Mo	Nov 16, 1965
Ma	Nov 16, 1975

the field of education due initially to her interest in retarded and disadvantaged children. What she began to see very early on in her psychiatric work, especially with retarded children, was that these children did not need medical treatment. What they needed was a proper education. So she set about trying to find ways to do that, using her ideas from medicine, education, and anthropology. The methods she used were experimental and yielded some miraculous results. With an improved and more stimulating environment, nurturing each individual child – they soon passed tests no retarded child should have been able to pass. Montessori said:

> "The essential thing is for the task to allow such an interest that it engages the child's personality.... When you have solved the problem of controlling the attention of the child, you have solved the entire problem of education."

On January 6, 1907, she opened her first school, called Casa dei Bambini, in the poor San Lorenzo district of Rome. From 1907 to the mid-1930s Dr. Montessori spent much of her time developing schools throughout Europe and North America. She made two trips to the USA in 1913 and 1915. Among her strongest American supporters were Thomas Edison, Alexander Graham Bell and Helen Keller. Up through 1947 she traveled to India and Ceylon (renamed Sri Lanka after 1972). There she trained thousands of teachers in the Montessori curriculum and methodology. Subsequent Montessori schools were established in Russia, China, Japan, Chile, Argentina and elsewhere. She wrote over a dozen books and founded numerous schools and training centers. Among her most influential books were *The Montessori Method* (first edition in Italian, 1909; first English translation, 1912) and *The Absorbent Mind* (1949). She was nominated for the Nobel Prize in 1949, 1950, and 1951. She died of a cerebral hemorrhage May 6, 1952 at age 81.

Montessori never married, but had a son (Mario), who was born in March 1898. The child was raised in a family in the countryside outside Rome, and she visited him whenever possible, though there were long separations. The child's existence was kept secret while she began to build her career. But eventually they reunited when he was about 15 years old and went to live with her for the first time. They then worked together for decades, and he carried on his mother's work up to the time he died in 1982. Maria's granddaughter in turn has also been involved in carrying on the Montessori legacy.

Mario's biological father, Dr. Montesano, was a colleague with whom Maria worked at the Orthophrenic School in Rome. She promised not to reveal their relationship or his identity as the father, and part of their mutual agreement was that neither of them would ever marry. While still working together on a daily basis, Dr. Montesano broke his promise by marrying someone else, and later they had four children. Because of the broken promise, Montessori left the Orthophrenic School; the child was sent to a wet nurse and later to boarding school.

Along with life themes that were iconoclastic and revolutionary for her times, Maria Montessori was a devout Catholic and a woman of much compassion and understanding. She was later regarded as a kind of mystic. Her educational philosophy comprises some of the following principles: 1) The teacher should pay attention to the child, rather than the reverse. 2) The child should be free to proceed at his or her own pace in a controlled learning environment. 3) Imaginative teaching materials are pivotal to the success of the process. 4) Each part of the system is self-correcting, the goal being to develop each child to his or her fullest potential.

Benito Mussolini came to power in Italy in 1922. He established complete power, a police state, strict censorship and state propaganda by 1926. Although an early supporter of the Montessori schools, his fascistic ideals were antithetical to the Montessori methods and their intended results. Thus she was forced to close all her schools in Italy and leave the country by 1934. By 1936 Hitler and Mussolini created "the Rome-Berlin Axis". She and son Mario fled initially to Spain for two years, but had to leave there too in 1936 due to the Spanish Civil War. They were rescued by a British cruiser and went on to Holland, which became their new home base though they continued to travel worldwide. While working in India in 1940, after India entered WWII, she and son Mario were interned as enemy aliens – though still permitted to conduct their training courses. The Association Montessori Internationale (AMI) was founded in Amsterdam in 1929, and still carries on the life work of Maria Montessori.

Revolutionary woman in Victorian times: At a time when most women had very little legal or social independence apart from their husbands or families, Maria Montessori lived a relatively independent life, surviving through war and political turmoil. It was rare for a woman to be well educated in that era – let alone a scientist, a doctor, and a world-traveler who commanded respect in many countries. All of this demands a strong astrological chart at any time, but much more so for a woman born in 1870. First of all the mind, body and spirit have to be strong, focused, and able to overcome big obstacles. The timing also has to be favorable for her. For a Cancer Ascendant person, the natal Moon has to be excellent. In this case, natal Moon has *Digbala* placement (best angular house) in the 4th house. Thus, the Moon is considered very well placed in terms of both physical and mental health, also for self-confidence. Maria was close to both parents, but enjoyed the strongest support for her advanced education from her mother, who was very literate in an era when most women were lucky to be able to sign their names. These factors are amplified when we see the *Parivartana yoga* between Venus and Moon, Venus in the Ascendant in Cancer in mutual exchange with Moon in Libra. Venus is lord of the 4th and 11th houses. As lord of the 11th house, it can bring her good finances and a powerful network of friends and colleagues. There is also a *Raja yoga* in the Ascendant between Venus and Mars. As planetary owners of angular and trinal houses, they bestow power on the individual.

For the most part, Montessori's planets are very well placed to help her succeed in life. Jupiter in the 12th house is the major challenge, as well as Saturn at 29 degrees Scorpio – a *gandanta* degree. Natal Jupiter and Saturn are almost in opposition by sign, and just miss by 7 minutes of arc, or 7/60 of a degree. A Jupiter-Saturn conjunction or opposition is regarded as one of the major signatures of an entrepreneur. In this case, the positions of Jupiter and Saturn also tell us something potent about the 5th house (house of the intelligence, educational development, thus mental and also biological creativity in the form of children). It shows strong likelihood of a partner, a child, and the potential loss of that child (born in 1898), and that this would occur near the start of Saturn *Dasha*, beginning Nov. 14, 1896. Saturn is lord of the 7th house of partnerships, placed in the 5th house of education children, and love affairs. Jupiter is *karaka* of children. A *karaka* is not adversely affected by being in the 12th house, but its capacity as house lord and/or *nakshatra* lord is adversely affected. (The *Jiva* is *nakshatra* lord of a house lord. It becomes a controlling factor for a house: Jupiter is *Jiva* for Houses 5 and 10 and owns Houses 6 and 9.)

When we assess Jupiter, planet of learning and knowledge from Montessori's natal Moon, we see that it is in the 9th house – the best house – and a house of higher education. We expect to see something unusual about her philosophy and her education because Rahu (the unconventional) is with Jupiter (the keeper of traditional values). While not ideal in Gemini, Jupiter does hint at making information and the educational process more accessible to the student – in this case, very young students and many of them deprived. (Accessibility is a key for Gemini). Also, 9th house is one's religion or philosophy, and she is the devout Catholic living an unconventional life for a Catholic woman of the era.

Though her own biological child was normal, her educational philosophy and aims all centered upon the disadvantaged or retarded children that she first worked with in the poor San Lorenzo district of Rome. This is where she began and this is what is shown in the astrological chart. The *Dasha* sequence is a blessing in that each planet brings some benefit, as each is strongly placed, if not in the birth chart – then in the *Navamsha* chart.

The Moon is in *Swati nakshatra.* Vayu, god of the wind is the deity of *Swati*, and shows she would be an independent thinker for her times, and would need a certain amount of freedom, independence and movement. She would have the desire to spread whatever information was dear to her, just as the wind disseminates the seeds. This enhanced her abilities as a revolutionary educator and scientist. Ruled by Rahu (the outsider or foreigner), *Swati* can breed a restless and adventurous mind, especially if Moon is placed there. Saraswati, goddess of learning is also associated with *Swati.*

The Education of a gifted educator: Maria Montessori scored extremely high on all her medical entrance exams, and eventually persuaded the powers that be (even the Pope) to allow her entry into medical school, as women had previously been denied entry. This was not the first social barrier she broke. For such an historic achievement, not only must the Ascendant lord be strong but the 4th house of the chart, as well as the 4th house lord, aspects to it, and planets occupying the house. **Education** is featured here with the 4th sign in the Ascendant and the 4th house lord almost exactly on the Ascendant in mutual exchange with the Ascendant lord, and Mars also in the Ascendant. (Mars is favorable as lord of both an angular and a trinal house, 10th and 5th, thus a Parashari *yogakaraka.*) Having 4th and 5th lords together is a kingly combination for education, and for the cultivated mind, especially if both are in the Ascendant or aspecting the Ascendant lord – both of which are true here. The Moon-Venus exchange adds power and versatility to the individual, as a Mars-Moon combination is excellent for Cancer Ascendant, and Moon can act as if in the Ascendant due to the *Parivartana yoga* with Venus. Venus in turn can act as if in its own sign in the 4th house, with Mars aspecting the 4th house as well as Moon and Venus.

The nature of the mind: As the mind is the 4th house, the 5th house is what it is capable of. Thus, the 4th and 5th houses tell about the nature of the mind. Just as education means "to lead one from," the 2nd house from the 4th house tells us the likely outgrowth of this mind. Any positive linking between these two houses favors all of the above, and can enable that person to be educated and perhaps to educate others according to what ignites or inspires them. This is even more favorable if the Ascendant lord is involved as well as other benefic planets. We have all of the above.

The 4th house: Montessori focused on **"the educator as the keeper of the environment."** In Vedic astrology, the 4th house is the domestic environment and is connected to the education as a whole, the educational degrees, but also that first environment of not just the womb, but also the educational environment. So this domestic environment where we begin is already considered educational. This is where the child's spirit is formed and the human potential educated, and this is exactly Montessori's focus. She even called her very first school *Casa dei Bambini* (1907). In Italian *casa* is house, and she intended to give children the safe and caring environment of a home, not an institution. This is where the child would find his or her place in what she called "the cosmic plan." Cancer emphasis has a strong desire to maintain the home.

The water signs are all *Moksha* signs, and Houses 4, 8, and 12 relate to *Moksha* as well. *Moksha* is spiritual liberation. So on the highest level – the Cancer Ascendant person, or one with 4th house focus (we have both here) will be very concerned with both education and spiritual pursuits, endeavoring to raise up the human spirit at all levels. Montessori was a scientist and a very pragmatic one, but also highly spiritual in her pursuit of truth. Not that all charts with Cancer Ascendant or 4th house emphasis guarantee spiritual lives. But as the first water sign, the 4th sign, and the first *Moksha* sign, it is the first beginnings of opening up to an awareness of the spiritual pursuit, and the happiness of the heart. As she writes:

> "Our care of the child should be governed, not by the desire to make him learn things, but by the endeavor always to keep burning within him that light which is called intelligence."

The 5th house: The house lord is **Mars**, located in the watery Cancer Ascendant in *Punarvasu nakshatra*, owned by Jupiter. **Jupiter** in turn is located in *Mrigashira nakshatra*, owned by Mars. This exchange of *nakshatra* lords is a ***Nadi yoga*** of tremendous power for the Cancer Ascendant chart, as Jupiter is *Nadi yogakaraka* (best planet in the chart, and 9th house lord), while Mars is Parashari *yogakaraka,* as lord of both 5th and 10th houses. As owner of both a *Kendra* and a trinal (*Kona*) house here, Mars has a lot of authority in this chart. As Mars is debilitated, though with correction (a *Neecha Bhanga Raja yoga*), it indicates there will be a struggle from some position of disadvantage. She started out at an all-boys technical school (usually preparation for engineering), which was already very unusual for a girl. Then she switched to medicine - which was also unprecedented for a woman in Italy. She was not initially accepted into medical school, nor did her father initially approve. Montessori was also at a disadvantage as a woman doctor starting out in the late 1890s. Among other factors, it forced her to separate from her own child for some fifteen years. This theme reappears from another perspective in her ongoing commitment to improve the lot of disadvantaged and mentally retarded children through education. Her courage as a true pioneer of the era is shown through natal Mars, though it would not necessarily favor marriage. Mars in ***Kuja Dosha*** (See Glossary) in both birth and *Navamsha* charts indicates difficulties and/or impediments to marriage; but it also gives her the necessary assertiveness to pursue her education and subsequent career.

The *Navamsha* chart is strong, with four classic benefics in the *kendras*. This tends to upgrade the condition of most of her natal planets, thus benefiting the *Dashas*, each and every one as they occur. Mars and Jupiter are in 2-12 relation to each other in the birth chart, which is not ideal, but the situation improves in the *Navamsha* chart. *Navamsha* Mars is *Vargottama*, repeating in Cancer, and *Navamsha* Jupiter is in Libra - both in *Kendra* houses, thus improving their status, especially for Jupiter. In addition, Jupiter is *Jiva* for the 5th house. It is situated in the 12th house of loss in the birth chart, though in the *Navamsha* 10th house, a much better placement. Thus, from her early losses (especially not being able to bring up her own child), she gained spiritual understanding and transcendence. In her books and in her teaching there is spiritual wisdom. She sees the intelligence within the child as "a light" that must be lit and kept lit - ideally growing brighter all the time. This is a common concept in ancient wisdom teachings, including in the Vedic spiritual tradition out of which Vedic astrology arises.

As lord of the 9th house of *bhagya* (fortune), the father, knowledge and wisdom, Jupiter is located with Rahu (the outsider) in the 12th house of loss and of foreign residence. Her father was overtly opposed to her pursuit of medicine and on the verge of withholding his support for it, but her mother was ardently behind her and very modern in her views. In choosing first a technical school and then medical school, Maria departed sharply from her father's wishes. She would also later travel and reside outside of Italy. When she received her medical degree in 1896, Maria was in **Jupiter-Rahu** *Dasha/Bhukti*, in the last months of the 16-year Jupiter *Dasha*. She was the first Italian woman to accomplish this and as such still a relative outsider (Rahu).

Hidden child - the metaphor and the reality: Montessori's only child, son Mario was born in March 1898, two years into her Saturn *Dasha*. We expect to see both the biggest heartache and the biggest advance in her personal growth during **Saturn *Dasha*.** This is because Saturn as configured here is both advantageous and a big problem. As the planet at the highest degree, at 29:53 Scorpio, Saturn is *Atma karaka*, and thus bound to bring her into her own greater powers. Her most innovative educational ideas came during the 19-year Saturn *Dasha*, which began Nov. 14, 1896. On the down side, Saturn is placed in the hidden sign of Scorpio in the sign of its enemy Mars, and in a *gandanta* degree, being in the last 3 degrees 20 minutes of a water sign. Any planet residing in a *gandanta* degree represents problems in the life, Scorpio *gandanta* being more extreme. It would be a very emotional period, and would bring some unreliable people into her life - with the potential for a marriage partner *and* for a child. Saturn is lord of the 7th house of marriage and partnerships, situated in the 5th house of education, lovers, and also - children. A watery 5th house is the most fertile of all, unless aspected by malefics. The ongoing theme is that whatever Saturn brings into her life is not how it first appears: The father of her child became a faithless lover; the child that was hidden from the public for years and lived apart from his mother became central to her life; and the children that at first appeared retarded revealed themselves to be otherwise when exposed to Montessori's educational techniques and environment.

As their child was born in March 1898, the love affair with Dr. Montesano can be dated to early in Montessori's Saturn *Dasha*, in **Saturn-Saturn** *Dasha/Bhukti*. The separation from her son lasted around 15 years and was due to the conservative social and economic realities of the early 20th century.[173] She lived apart from son Mario for most of her Saturn *Dasha* and they seldom saw each other. But they were reunited by the end of it, during Saturn-Jupiter *Dasha/Bhukti*, and had an important working partnership for the next several decades through to her death in 1952. He continued this work until his own death in 1982. Following in his footsteps was Montessori's granddaughter, Marilenna Henny-Montessori, who says this about her father, Mario Montessori:

> "...All his many loves were nothing compared to his love for his mother and her work – an all encompassing love which dominated his whole existence. His dedication to her was a conscious and free choice..."

In an era when children were largely seen and not heard, Maria Montessori was able to perceive **the child's innate ability to absorb culture.** Here are some astrological factors that would show these traits: Mars is the fiery planet of revolution. To have a revolutionary effect, Mars should ideally aspect or contact the Ascendant, Ascendant lord, 10th house, 10th house lord, and/or Saturn – a *karaka* for 10th house matters. For Cancer Ascendant Mars is already 10th lord, and is in the Ascendant aspecting Ascendant lord Moon in the 4th house. Mars also contacts 5th lord Venus, and is *Vargottama*, repeating in the *Navamsha* in Cancer, again in a *Kendra*, aspecting Jupiter, Venus and Moon. A competitive advantage would also come from at least a couple of *Viparita Raja yogas*, enabling her to deal well with competition and prejudice. In this case, from the Ascendant, 6th lord Jupiter is placed in the 12th house, aspecting its own house. Also, from Moon, 12th lord Mercury is situated in its own house. So there is some good protection from losses. In this case, the loss of the son's father, and the years away from the son eventually turned out in a very positive way. She also spent those 15 years of **Saturn *Dasha*** as well as the rest of her life devoted to all children, especially those she saw who were poor or with mental or emotional problems that were not getting solved through ordinary means. The fact that all this would turn out to everyone's great advantage is indicated by several astrological factors:

1. **Saturn** is placed in Mercury's *nakshatra*, *Jyestha*. This makes **Mercury** the *Jiva* (life force) for Saturn. Fortunately, Mercury is placed in its own sign of Virgo, exalted. So there would be some superior results, after some initial despair. Luckily too, Mercury *Dasha* follows after the 19-year Saturn *Dasha*. Mercury *Dasha* would run 17 years from Nov. 1915 to Nov. 1932, and would be extremely prolific years as well.
2. Saturn's sign lord, **Mars**, is strongly placed in Cancer, *Vargottama*, and in a *Kendra* in both birth chart and *Navamsha*. Mars rules excellent houses in the birth chart, and is in a *Neecha*

[173] As discussed in Chapter 5, Albert Einstein's first child (Lieserl) was born in 1902, out-of-wedlock. He and his future wife Mileva Marić feared the same consequences of scandal in the same era. There was the same hidden component, and it remains unknown whether Lieserl died as an infant or was put out for adoption.

Bhanga Raja yoga, cancelling its debilitated condition. After some early setbacks, especially as a woman in that era, 10th house matters (status and career) and 5th house matters (education and children) would go well.

3. As 9th lord, **Jupiter** is the best planet for Cancer Ascendant, even though as 6th house lord (and on the Rahu-Ketu axis here) it can bring 6th house problems, especially during Jupiter and/or Ketu *Dasha*: debt, disease, competition, and legal troubles. **Ketu *Dasha* (Nov. 1932 to Nov. 1939)** brought turbulence, as the Montessoris fled a series of countries in war torn Europe and established a new home base in Holland in 1936, while continuing to travel worldwide. They even faced trouble in India in 1940 as Italian nationals during World War II. (Jupiter in a *Viparita Harsha yoga* helped in this regard, rescuing them each time from political chaos and physical danger.) Jupiter is *karaka* for children and for learning and knowledge. As *Jiva* for the 5th house of children, placed in the 12th house of loss, and foreign, hidden locations, there would always be some degree of loss in this area that could turn around once Jupiter and Saturn *Dashas* came to a close. By around 1913 Maria was reunited with her son, and from then on until her death in 1952 they worked together as a close and productive team.

Chart 17: Marilyn Monroe

Birth data: Tuesday, June 1, 1926, 9:30 AM PST, Los Angeles, CA. USA, Long. 118W14 34, Lat. 34N03 08, *Lahiri ayanamsha* –22:49:29, Class AA data. Birth certificate in hand from Bob Garner. Photo of birth certificate in Robert F. Slatzer's book *The Curious Death of Marilyn Monroe*, 1975. Ascendant: 20:15 Cancer.

Biographical summary: Marilyn Monroe was one of the most famous female sex symbols of the 20th century. Born Norma Jean Mortenson, she took the name Marilyn Monroe in 1946 and dyed her brunette hair blond. The film industry wanted her to have a more glamorous first name, and they approved her choice of her mother's maiden name as surname. Starting as a model, she moved to an acting career in films. She signed her first movie contract with 20th Century Fox on July 23, 1946, and made 29 films in her 16-year career, 24 of them in the first eight years of her career, between 1946 and 1954. Her first serious acting jobs were bit parts in two 1950 films: *Asphalt Jungle* and *All about Eve*. Other films for which she is remembered, often as a comedienne, are: *Some Like it Hot* (1959), *Bus Stop* (1956), *Seven Year Itch* (1955), *How to Marry a Millionaire* (1953), and *Gentlemen Prefer Blondes* (1953). After the latter two films, *Photoplay* magazine voted her "Best New Actress of 1953."

> "She had a luminous quality, a combination of wistfulness, radiance, and yearning that set her apart and made everyone wish to be part of it – to share in the childish naiveté which was at once so shy and yet so vibrant."
>
> **Lee Strasberg**, speaking at Monroe's funeral in Aug. 1962.

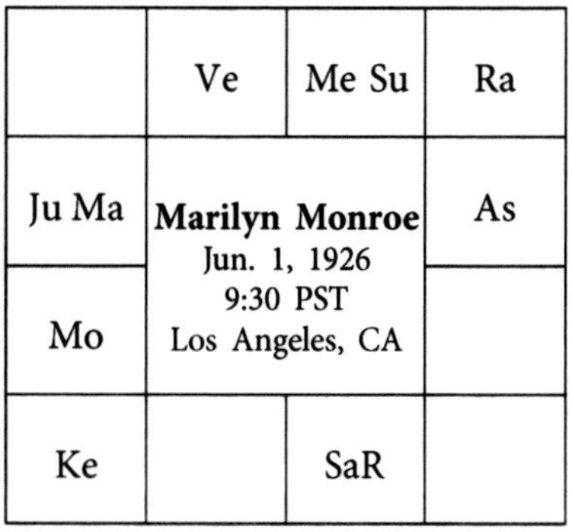

	Ve	Me Su	Ra
Ju Ma	**Marilyn Monroe** Jun. 1, 1926 9:30 PST Los Angeles, CA		As
Mo			
Ke		SaR	

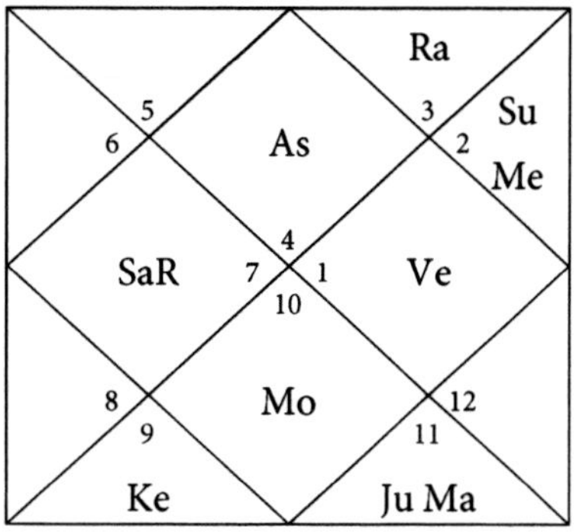

Ra
5 3 Su
6 As 2 Me
4
SaR 7 1 Ve
10
8 Mo 12
9 11
Ke Ju Ma

Planetary Information

As	20:15	Can	Ashlesha
Su	17:37	Tau	Rohini
Mo	26:16	Cap	Dhanishta
Ma	27:54	Aqu	P.Bhadra
Me	13:57	Tau	Rohini
Ju	04:00	Aqu	Dhanishta
Ve	05:55	Ari	Ashwini
Sa	28:37	Lib	Vishakha
Ra	25:26	Gem	Punarvasu
Ke	25:26	Sag	P.Ashadha

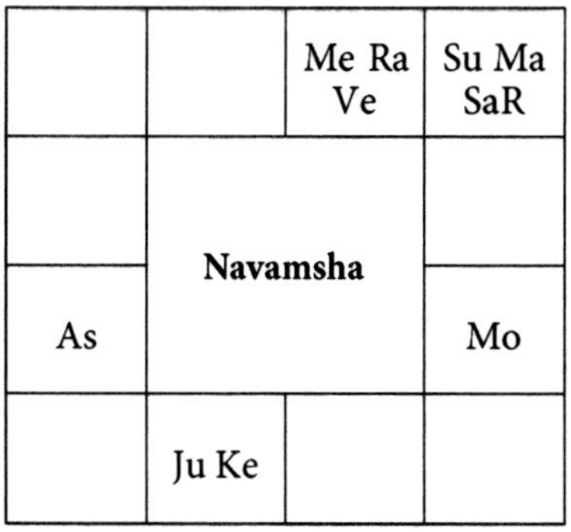

		Me Ra Ve	Su Ma SaR
	Navamsha		
As			Mo
	Ju Ke		

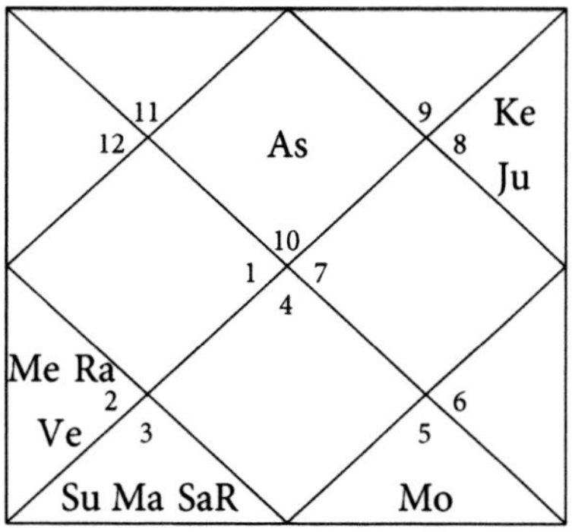

11 9 Ke
12 As 8 Ju
10
1 7
4
Me Ra
2 6
Ve 3 5
Su Ma SaR Mo

Vimshottari Dashas

Ma	Nov 14, 1924
Ra	Nov 15, 1931
Ju	Nov 14, 1949
Sa	Nov 15, 1965
Me	Nov 14, 1984
Ke	Nov 15, 2001
Ve	Nov 15, 2008
Su	Nov 15, 2028
Mo	Nov 15, 2034

Desiring to become a more serious actress, she took a temporary hiatus from her film studio contract in 1955 to study acting with Lee Strasberg in New York City. With her own production company (from Dec. 31, 1955), she made one film, *The Prince and the Showgirl* (1957) with Sir Laurence Olivier. Her last completed film was *The Misfits* (1961), with a script written for her by her then-husband, playwright Arthur Miller. She won a Golden Globe award in 1960 as Best Motion Picture Actress in a Musical/Comedy for her role in *Some Like it Hot*, which still enjoys commercial and artistic success.

Both her mother and maternal grandmother were mentally unstable and unable to support her. Though Marilyn never knew her biological father, he is believed to be Martin Edward Mortensen, a Norwegian who immigrated to the USA and left Los Angeles before his daughter was born. She spent most of her childhood in different foster homes, at times in an orphanage, and very little time with her mother Gladys. From this lack of stability and the inherited mental history, Monroe's mind was also disturbed. By her own admission, she lacked confidence in large part due to her difficult childhood and being unsure of her family or lineage.

> When astrologer Richard Ideman asked her: "Is New York as lonely a town as they say?"
> Monroe replied: "Any town is lonely when you don't know who you are."

Monroe suffered for most of her adult life from depression, insomnia, some alcohol abuse and drug-induced paranoia. She sought ongoing psychiatric treatment, which in those years included regular prescriptions of barbiturates and tranquillizers. These in turn probably impacted her chronic lateness and problems remembering her lines. In spite of her reputation for unreliability on the set, director Joshua Logan (*Bus Stop*) commented:

> "I found Marilyn to be one of the great talents of all time.... She struck me as being a much brighter person than I had ever imagined, and I think that was the first time I learned that intelligence and, yes brilliance have nothing to do with education."

She had three marriages and no children. During her third marriage to Arthur Miller, she suffered several miscarriages due to endometriosis. When she was 16, her foster parents at the time gave her the choice of returning to an orphanage or marrying her neighbor, James Dougherty, then 21. She chose marriage over the orphanage, and it took place June 19, 1942. Dougherty was away in the Merchant Marines the first few years and they divorced in 1946. A second marriage to the baseball legend Joe DiMaggio occurred Jan. 14, 1954 and lasted only nine months. But they remained friends and planned to remarry Aug. 8, 1962. Her third marriage to the famous playwright Arthur Miller took place June 29, 1956 and soon after she converted to Judaism. Her marriage to Miller lasted 4 ½ years up to their divorce in January 1961. Each husband was older than her: Dougherty 5 years older, DiMaggio 12 years older, and Miller 11 years older.[174]

Marilyn Monroe was photographed from the time she was 16 and working in a parachute factory. At 23 she posed nude for a calendar – "Miss Golden Dreams" – a venture that would haunt her some years later as she sought to become a serious actress. But for a larger public, her nude image as Miss Golden Dreams has perhaps outlived any memories of her in an acting role. She posed for this photo on May 27, 1949 and it became the famous centerfold for the Dec. 1953 inauguration issue of *Playboy magazine*, also featuring Monroe on the cover. (Its founder Hugh Hefner was unsure of the future of the magazine, so it was undated.) Another of her iconic images is from the movie *Seven Year Itch* (1955). She stands on a Manhattan (New York) subway grate, caught suddenly with her skirts blowing high above her while a man looks on shyly. The director originally conceived this scene as a publicity stunt, filming it on Sept. 14, 1954.

At her death Marilyn Monroe left total control of 75% of her estate (then $1.6 million) to Lee Strasberg, her acting coach from 1955 and director of the Actor's Studio in New York City. She left the remaining 25% to Dr. Marianne Kris, her psychoanalyst. The licensing of Marilyn's name and likeness continues to generate more and more each year for the Monroe estate, up to around $4 million per year by 2000, according to *Forbes magazine*. But since her death in 1962 there have also been legal battles over "the right to publicity" of her images. Finally a New York court ruling on Sept. 4, 2008 confirmed a California ruling from March 17, 2008, saying that Monroe's estate did not have exclusive rights to profit from her image.

Marilyn Monroe's death was and is still controversial. It was listed as a probable suicide, but many details did not add up. Her diary and personal notes were never found. She died Sun. Aug. 5, 1962, officially of an overdose of barbiturates, although later reports from Coroner Dr. Thomas Noguchi and other prosecutors on the case indicate it was a homicide. Many who knew her did not accept the cause of death as suicide, especially as her life was taking some positive

---

[174] Miller's play ***After the Fall***, opened in New York Jan. 23, 1964, and outraged some due to the deeply personal view of his marriage to Monroe. The lead was a Monroe-like character called Maggie. Arthur Miller was born Oct. 17, 1915, Manhattan, N.Y. (His birth time of 5:12 AM has a Rodden data rating of C, to be used with caution.) Miller remarried in 1962, to the photographer Inge Morath. Their marriage lasted 40 years, until her death in 2002. Miller died at age 89, on Feb. 10, 2005. Joe DiMaggio was born Nov. 25, 1914 in Martinez, CA. (His birth time of 7:00 AM is dubious: also a "C" Rodden rating.) DiMaggio was devastated by Monroe's sudden death in Aug. 1962, and took charge of the funeral arrangements. For the next 20 years he had six red roses delivered three times a week to her crypt. He never remarried and died at age 84, on March 8, 1999.

new turns. She was planning a press conference for Monday, Aug. 6, 1962 and a remarriage to Joe DiMaggio Aug. 8, 1962.

Source of her magnetism: The Ascendant lord **Moon** is just setting in the chart, opposite the Ascendant – thus positioned powerfully in a *Kendra* and aspecting its own house. This gives enormous emotional presence and an immediate potential for emotionality, if not vulnerability – as one's feelings, and the happiness of one's heart will be immediately known. A planet aspecting its own Ascendant bestows innate physical strength. Venus, planet of beauty and sensuality is also prominent in the 10th house. However, this Moon loses power in the *Navmasha* chart due to its location in the *Navamsha* 8th house, the most difficult house in the chart, especially for the Moon. An 8th house Moon can experience more suffering, at times through betrayal – real or imagined. Strong spiritual discipline is needed to handle this house position. The churning of the ocean happens here, so the emotional suffering can be more extreme, whether Moon is in the birth chart or *Navamsha* – though worse in the *Navamsha.* This is exacerbated by the fact that the *Navamsha* Moon is isolated, with no planets in adjacent houses (*Kemadruma yoga*). However, there is some correction, with five planets in angular houses to the *Navamsha* Moon.

There is a quality of inaccessibility about the *Moksha* houses (4, 8, and 12), especially the 8th and 12th houses, as the ego must surrender in these houses. The secret of her magnetism may lie in that very inaccessibility for most other people *and* for herself. Her sense of mystery and sexual allure (8th house planetary concentration) also coincide with her sense of vulnerability, as in the "dumb blond" roles she perfected as a comedienne. The *Navamsha* Moon in Leo in the 8th house is not a good indication for her mental/emotional stability, or – for that matter – for her physical health, as Moon is also her Ascendant lord in the birth chart. Her physical health was considerably compromised by her dependence on tranquillizers and barbiturates. Due to her self-preoccupation motivated by trying to find herself and her roots (*Navamsha* Moon in Leo), there is exaggerated and unconscious behavior during which she suffers a lot and may do much psychological exploration before finding any relief (again, Moon in the 8th house, especially in Leo). If she had a stronger and more supported Moon in the *Navamsha* chart and more indications for a conventional career, she herself might have become a psychologist. Instead, she was the person endlessly projected on to by her adoring fans, especially male fans. Anything can be projected on to the Moon; and another level of the Moon's symbolism is the public, the people. Thus perhaps a major factor in her sex appeal was that she was looking to her public at all times for the acceptance, approval and love she never found on a primal level from any immediate family or partnership. She may have understood this very well when she said:

> "I knew I belonged to the public and to the world, not because I was talented or even beautiful, but because I had never belonged to anything or anyone else."

Absent mother, father and family: Moon is also the symbol for mother and maternal relatives. Neither her mother nor her maternal grandmother was able to take care of her. Her mother was constantly in and out of mental institutions, and unable to provide a home for herself and her daughter. The young Norma Jean Mortenson lived in a series of foster homes and an orphanage, carrying with her the sense she was not wanted. She said as much many times. While married to Arthur Miller, she tried to have her own children and miscarried several times due to her endometriosis. Astrologically, the 5th house (of children) contains afflictions from several perspectives. From Cancer Ascendant, the 5th house lord of Scorpio is Mars – located in the 8th house of loss and major changes. This 5th house is aspected by the Sun (too much heat from a malefic) and in a *Papa Kartari yoga* (surrounded on either side by malefics Saturn and Ketu), causing constraint to matters of this house. We can find similar afflictions reading the 5th house from natal Moon in Capricorn and also from the *Navamsha* Ascendant. In the *Navamsha* 5th house, Rahu-Ketu creates too much confusion for a successful pregnancy, though it brought an exotic component to her artistic output and to her love life. (The 5th house is also the house of lovers and love affairs.)

In this Cancer Ascendant chart, there is a yearning for family and tribe due to its almost complete absence and unreliability for so many years. Thus began the long search for someone she could depend on, often including the professionals she trusted, such as her acting coach, psychiatrist, and psychoanalyst. For the **father** to be totally gone, the 9th house lord and/or the Sun should be beleaguered. In Monroe's case, the 9th house lord Jupiter is in the 8th house of loss, suffering and/or *Moksha* (spiritual liberation). The 9th house lord located in a *Dusthana* (house 6, 8, or 12) creates a *Nirbhagya yoga*. This is a loss of fortune (*bhagya*) which may be due to the removal of the father or guru (teacher/advisor). Because of this *yoga*, it may also be difficult for her to get good advice – an unfortunate combination with her vulnerable Moon.

Comparison with the Maria Montessori chart: As Montessori's Moon was not afflicted, she was able to find equilibrium, even in the midst of great personal or political turbulence. She also had the ongoing support of her mother, and eventually both parents. Even the Pope provided support in giving her permission (as the first woman) to enter medical school in Italy. As a doctor and a scientist she was able to create an educational system out of her research efforts, and could use the contents of her mind (Moon) in a more constructive way than Monroe, whose *Navamsha* Moon position is very difficult and isolated. There are also no planets in angular houses in Monroe's *Navamsha* chart, adding to the isolation. Monroe's formal education ended in the early years of high school, as she married and went to work in a factory by age 16.

Montessori and Monroe each have a *Nadi yoga* involving a **Mars-Jupiter** exchange. Mars and Jupiter are both considered good friends to Ascendant lord Moon. In Monroe's case Mars and Jupiter are both poorly placed in the 8th house in Aquarius, and they are even more pivotal to the entire chart. Montessori's Mars is well placed in the Ascendant of the birth chart (though not ideal for marriage), and Jupiter is in the 12th house – again not ideal, but not as damaging as Jupiter in the 8th house. With Jupiter in the 12th house, Montessori could use some of that energy for spiritual purposes or to go abroad – which became necessary when she had to leave Fascist Italy in 1934. She ran into more political trouble in Spain and India.

Since Jupiter is considered the best planet for Cancer Ascendant, we would like to see it located somewhere in the chart where it can be beneficial, and not pulled down by a difficult placement or aspects, either in the birth or *Navamsha* chart. Monroe's Mars-Jupiter combination in the 8th house would indeed cause her trouble, though it also gives her some protection for a time due to Jupiter's position as 6th lord in the 8th house – a *Viparita Raja* yoga. But as 9th house lord Jupiter is weakened by its location in the 8th house (of loss) – a *Nirbhagya yoga*, as mentioned earlier, removing the good fortune from the 9th house.

Miss Golden Dreams: The planet of beauty, sensuality, love and the arts is Venus. Prominent in the 10th house of the birth chart, **Venus** is in Aries in *Ashwini nakshatra*, owned by Ketu. **Ketu** in turn is in *Purva Ashadha nakshatra* owned by Venus. This is one of two powerful *Nadi yogas* she possesses in this chart, the second being Mars-Jupiter in the 8th house. Venus and Ketu closely linked through this Nadi *yoga* transport her beauty, her artistry and love life into another realm: exotic, unknowable, and unpredictable. Ketu is the foreigner. Symbolized by the flag, it can draw attention to any planet associated with it, even if Ketu's highest purpose is a spiritual one. Ketu is unconventional and can bring fears and phobias, especially of not being seen. Third husband Arthur Miller said of her:

> "She was endlessly fascinating, full of original observations... There wasn't a conventional bone in her body."

He looked on her as an ideal and was shocked to discover she was a human being. Similarly, her second husband Joe DiMaggio hoped to have a nice housewife and was startled to find people mobbing her every public appearance.

Navamsha Venus is in Taurus, aspected by Mercury, Rahu, Jupiter, and Ketu. Though Venus rules Taurus, and is in a creative house placement, it is pulled in too many directions by aspects from 6th lord Mercury, 12th lord Jupiter, and malefics Rahu and Ketu. Venus-Jupiter aspects already bring in potential excesses. With six planets in fixed signs in the *Navamsha* chart, it is a big challenge for her to make major changes, and to recover from life's disappointments. She herself is loyal but has few indications of loyalty in return, with an exotic though turbulent love life and artistic career. Due to her excesses, she was in constant conflict with her film studios. But they continued to renew her contracts and agree to her terms as she became an ever bigger box office draw.

Turbulent love life: Moon in *Dhanishta nakshatra* ("the wealthiest") brings significant financial wealth, though often unhappiness in marriage and love affairs. (Princess Diana of Wales also had Moon in *Dhanishta*.) This *nakshatra* is owned by Mars, whose placement in the 8th house brings further instability in marriage. It becomes more dangerous *and* glamorous, especially as *Jiva* (life-giver) to the Ascendant lord Moon. Amplify that several times when you consider that either Mars or Jupiter (both in the 8th house) play the role of *Jiva* to five planets: Rahu (which cannot be a house lord), Moon, Mars, Jupiter, and Saturn. The affairs of seven houses are thus affected by some loss or turbulence: houses 1, 5, 6, 7, 8, 9, and 10. Venus in contact with Rahu or Ketu can give a foreign element to the love life, as well as turmoil and unpredictability. In addition to the *Nadi yoga* between natal Venus and Ketu, *Navamsha* Venus contacts Rahu and is opposite Jupiter and Ketu.

In general, the Cancer Ascendant person has some challenges in love matters, as Saturn is the lord of the 7th house of marriage and partnership matters, and Saturn and Moon do not have an easy relationship. Moon wants to feel secure and loved while Saturn imposes discipline, delays, heavy responsibilities and/or older people (often partners) who form an important part of the karmic destiny. Monroe's three husbands were all older than herself, by 5 to 12 years. Moon's placement in the 7th house put her mind constantly on the partner or on her public. Moon in Capricorn as well as exalted Saturn's aspect to the Ascendant gave her some seriousness and sense of discipline, though it would be hard to carry through on it for several reasons: 1) *Navamsha* Moon is more problematic in the 8th house, and; 2) natal Saturn resides in the *nakshatra* of

Jupiter, whose placement in the 8th house of loss alongside Mars is not favorable, reducing exalted Saturn's ability to give its full benefits. Its retrograde status is discussed below, as that too can reduce Saturn's benefits.

Famous husbands and lovers: Monroe's husbands were famous, except for the first one, James Dougherty, whom she married at 16 – her only option to avoid returning to an orphanage. Her subsequent husbands were baseball legend Joe DiMaggio, and playwright Arthur Miller. DiMaggio was set to remarry Monroe on Aug. 8, 1962, but she died three days earlier. Among her other lovers were well-known figures such as actor Marlon Brando, singer and actor Frank Sinatra, actor Yves Montand and director Elia Kazan. In the last few years of her life she was linked with President John F. Kennedy and his brother Robert Kennedy, in-laws of actor Peter Lawford.

As lord of the 7th house of husbands and partnerships, Saturn is placed in a *Kendra* in its sign of exaltation. This can give more extreme situations for the affairs of the houses ruled by Saturn (Houses 7 & 8). It gave her strong partners, or at least those of high social status. Unfortunately, this exalted Saturn is also retrograde, which can reverse the benefits of the exaltation, making Saturn act as if debilitated.[175] Though two of the husbands were famous, their fame did not shelter her. Instead, her fame overwhelmed them. In the 4th house of family foundation and mental and emotional happiness, retrograde exalted Saturn did not help her in this area either. Saturn – as 7th lord of partnership – aspects the Ascendant, ruled by the Moon; but the Moon does not aspect Saturn There is also a one-way aspect from Saturn to Moon in the *Navamsha* chart, showing the possibility of partners who can exert superior power over her, perhaps in their effort to compete with her fame. Exalted in a *Kendra*, natal Saturn forms a *Shasha Mahapurusha yoga*: It shows she experienced fears from a very early age and tried to accumulate wealth as a means of overcoming these fears. This is in addition to seeking relationships to fill the gap of the missing family.

Hidden indicators behind her legend: The core energy of Monroe's life seems to resonate with her sexual magnetism, her insecurity regarding who she was and where she came from, and the sexual and political intrigue that engulfed her at the end of her life and may well have killed her. **Mars and Jupiter in the 8th house** describe many of these themes. Mars is the planet of physical vitality, and Jupiter has an expansive effect on it. Placed in the 8th house of mostly hidden matters, those that thrive behind-the-scenes, the energies of Mars and Jupiter together provide a potentially explosive effect. (Albert Einstein had Mars-Rahu in the 8th house.) The combination describes what Monroe is probably best remembered for: her mysterious sexual allure and mysterious death, themes that linger around the legend of Marilyn Monroe more than any others, especially her controversial death. It may account for why she transcended her Superstar status as "Blonde bombshell" to become a longstanding icon, not because she was the best actress or model of the era, but because she evoked such a strong response in people and died with many pieces of the puzzle still unsolved.

The 8th house is a *Moksha* house but often plays out with themes of sex, death, and intrigue. Though Mars and Jupiter provide the underlying motivating factors, it may not be evident at first glance. This is because Ascendant lord Moon is located in *Dhanishta nakshatra*, owned by Mars; Mars in turn is in *Purva Bhadra nakshatra* owned by Jupiter. By this exchange of *nakshatra*

---

[175] This reversal does not apply in certain cases, as with American financier J. P. Morgan (1837-1913). His exalted natal Saturn retrograde in Libra was in *Vishakha nakshatra*, owned by exalted Jupiter in Cancer.

lords, a powerful ***Nadi yoga*** is created, especially as Mars is *Jiva* (or life force) for the Moon, and Jupiter is in turn the *nakshatra* lord of Mars. Placed in the most difficult house - the 8th house, and in the most idealistic sign of Aquarius, **Mars-Jupiter**, especially Mars - denotes idealism about sexuality and sexual relations. Combined with her already extraordinary sexual magnetism, it gives a level of expectation that would be hard to meet on a personal level. But on a collective level, Marilyn Monroe became the Pin-up girl to countless millions. Her image as unattainable sex goddess has endured.

For Cancer Ascendant, **Jupiter and Mars** are also lords of *Dharma* and *Karma* houses, respectively: Houses 9 and 10. In Vedic astrology we examine the sense of purpose in life through the ***Dharma* lord** (classically Jupiter and the 9th house lord: same planet in this case). The karmic destiny is examined through the ***Karma* lord** (classically Saturn and 10th house lord: Mars in this instance). In Monroe's birth chart, Mars and Jupiter are together in the 8th house of loss and suffering from the Ascendant. The 8th house is also a house of sexual and/or political intrigue (one of the lower levels of *Moksha*, spiritual liberation). This shows either that she would depart from an initial career path or destiny, or she would suffer in the process of pursuing that destiny. Whatever her career, it would change or be cut short due to Mars (10th lord of career) in the 8th house of loss and major changes.

Psychology and psychological research - plumbing the depths - is also an 8th house theme. Monroe is on record as saying she felt she could advance more rapidly as an actress if she could have both her acting coach and her psychoanalyst at her side on a daily basis; hence, her estate provisions, leaving 75% of her estate to her acting coach and 25% to her psychoanalyst.

Marilyn's image as a marketable icon: Another 8th house factor is inherited money and money passed on as one's legacy. When key planets are in the 8th house, they can bring inherited money during the lifetime and leave behind legacies after death. Monroe's natal 8th house has the power not only to generate money during the lifetime, but to bestow a significant legacy in financial terms. This is due to 8th lord Saturn exalted in a *Kendra*, and its *nakshatra* lord Jupiter situated in *Dhanishta nakshatra* ("the wealthiest"). From her death in 1962 up to 2008 the licensing of Marilyn's name and likeness brought high profits but also legal battles. Both political and financial intrigues arise with *Raja yoga* planets Jupiter (9th lord) and Mars (10th lord) situated in the 8th house. Legal battles are usually seen from the 6th house, whose lord Jupiter is placed in the 8th house (normally protective in a *Viparita Harsha yoga*). However, 8th lord Saturn is also aspecting the 6th house. As stated earlier, Saturn's exalted state can suffer from its retrograde status.

The **JU-SA conjunction of Feb. 1961** marked the last short chapter of Marilyn's life. It did not benefit her level of personal happiness, but it did increase her legend and her marketability long after death. With Marilyn's self-obsession, her strong desire to please her public and the public's obsession with her as sex icon, her mystifying death only increased that obsession. Coinciding with tr. Saturn exiting Sagittarius, her Jan. 1961 divorce from her third husband Arthur Miller occurred one month prior to the JU-SA conjunction in Capricorn. Sagittarius was the strongest sign placement of tr. Saturn during her *Sade Sati* (Saturn's passage through Houses 12, 1 & 2 from her natal Moon in Capricorn). The difficulties and complications she suffered with partners would only increase during Saturn's transit through Capricorn. This reflected the challenges from her own natal Saturn also aspecting natal Moon and Cancer Ascendant, either in the birth or *Navamsha* chart, and in a *Kendra* from her Moon in the birth chart. The three phases of her *Sade Sati* were as follows:

1. Tr. Saturn in Sagittarius: Nov. 8, 1958 thru Feb. 2, 1961
2. Tr. Saturn in Capricorn: Feb. 2, 1961 thru Jan. 28, 1964
3. Tr. Saturn in Aquarius: Jan. 28, 1964 thru April 9, 1966

Controversy over cause of death: Much of what we know about Monroe's life and her life's purpose is contained in the Mars-Jupiter energy in the 8th house, so we will be focused on the *Dashas* of these two planets. We know there is a dramatic change to her life that will happen in the *Dasha/ Bhukti* of **Mars-Jupiter** or **Jupiter-Mars**. She would be especially vulnerable at birth, as she was born in Mars-Jupiter *Dasha/Bhukti*. This caused danger to her even in childbirth, as her mother and grandmother were both mentally unstable and their social and financial circumstances were also unstable. Later on, there would be another peak crisis period during Monroe's **Jupiter-Mars** *Dasha/Bhukti* (July 16, 1962 to June 22, 1963). She died, or was murdered a few short weeks into this dangerous period. Planets in the *Moksha* houses, especially Houses 8 and 12 are meant to diminish the ego in the process of developing *Moksha*, or spiritual liberation. But this is far more difficult if most of the childhood occurs in *Dashas* whose planetary lords are placed in *Moksha* houses (especially Houses 8 and 12), where self-identity is diminished unless the person lives in a foreign country or in a 12th house setting, such as an ashram. In Monroe's case, she spent her entire life span of 36 years in such *Dashas*, i.e., each *Dasha* lord up to her death was situated in either House 8 or 12. Though it can catalyze spiritual development, it is harder for the person to build a basic sense of self-confidence in life, as the ego may never be well established. While this could be an advantage in a spiritual setting, it is an obvious disadvantage in Hollywood, dream-capital of the world.

In the last few months of her life, her involvement with the Kennedy brothers, then the President and Attorney General of the U.S., respectively, had reached a point of no return. She had been told not to contact them any more, especially not at the White House. Soon afterward, she scheduled a press conference, to be held on the following Monday after she died. The question was never answered as to whether she intended to reveal her intimate connections with the Kennedys. Her diary and personal notes were removed before they could be examined. She was also scheduled to remarry Joe DiMaggio in a few days.

The sub-period of **Jupiter-Mars** started July 12, 1962 and ran for the next 11 months. Using a three-level, more detailed look at the sub-periods, **Jupiter-Mars-Mars** ran from July 16, 1962 to Aug. 5, 1962, compounding the Mars energy and the hidden dangers. Monroe was found dead early on the morning of Aug. 5, 1962. On that date the *Dasha* planets switched to **Jupiter-Mars-Rahu**; then on Sept. 25, 1962 up to Nov. 9, 1962 it was **Jupiter-Mars-Jupiter.** Since natal Jupiter and Mars are in the 8th house and natal Rahu in the 12th house from her Cancer Ascendant, it is much easier to disguise the circumstances of her death in this three-month time frame. Rahu is the foreigner and is also associated with poison. All three planets are in houses of *Moksha*, or spiritual liberation. While demanding the suspension of the ego, these positions can also be difficult for the health and safety of the individual during their periods, as chronic or hidden factors reside in these houses.

Jupiter in general represents teachers or advisors. Among her many advisors was her psychiatrist, who was on constant call that summer and prescribed a heavy combination of tranquillizers and barbiturates – customary for psychiatrists of that era. Assessing *Dasha* lords Jupiter and Mars in the *Navamsha* chart we note that *Navamsha* Jupiter is in Scorpio in the 11th house of friends (on the Rahu-Ketu axis opposite Mercury-Venus), implying an exotic and expanded team of advisors and friends. But *Navamsha* 11th house lord (Mars) is placed in the 6th house of enemies, along with malefics Sun and Saturn. *Navamsha* Mars and Jupiter are in 6 – 8 house relationship to each other, showing a likely rapid change of events in this 11-month sub-period and some difficulty for her to distinguish her friends from her enemies. We know some crucial evidence was tampered with and removed from the scene of death. Though the case was reopened several times, the cause of her death remains unsolved.

Chart 18: Indira Gandhi

Birth data: Monday, Nov. 19, 1917, 23:11 IST, Allahabad, India, Long. 81E51, Lat. 25N27, *Lahiri ayanamsha* –22:42:53, Class A data, from memory. (Robert Jansky quotes her private secretary, "41 *Ghatis* 52 *Phalas* and 23 *Viphas* = 11:11:14 PM IST." Ascendant: 27:22 Cancer.

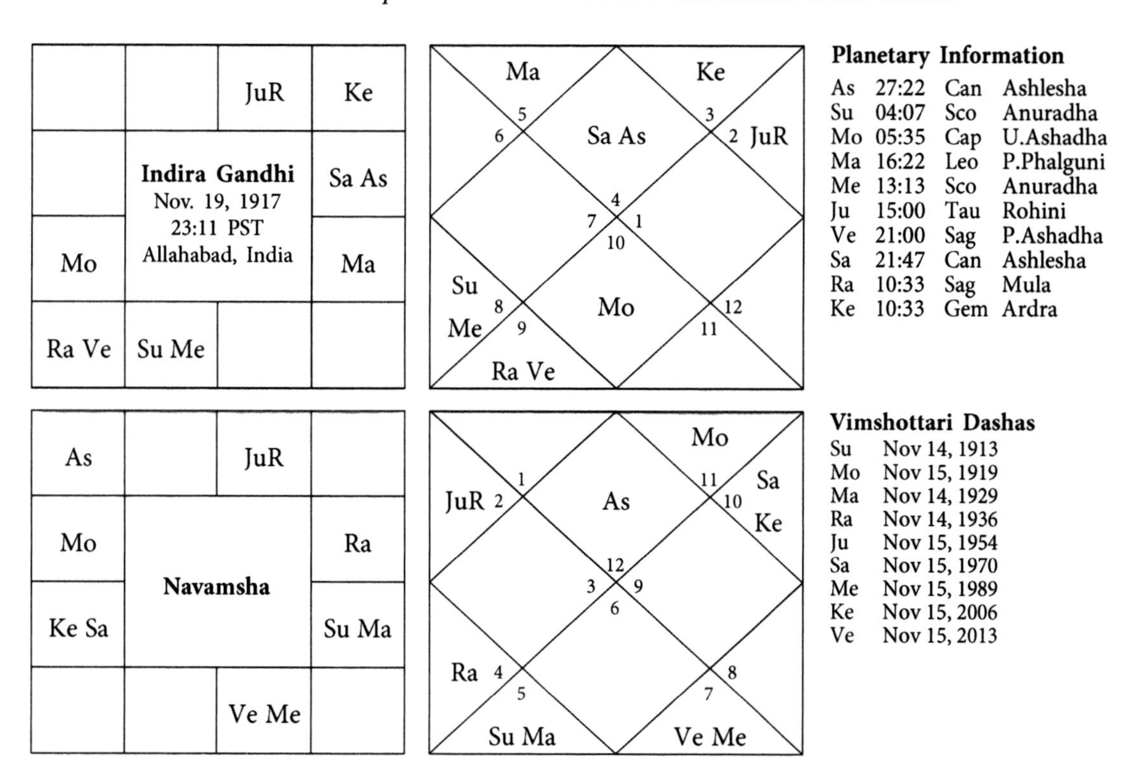

Biographical summary: Indira Gandhi was the first woman ever to be elected to a high leadership position in a democracy. She was Prime Minister of India from Jan. 19, 1966 to March 24, 1977 and again from Jan. 14, 1980 to her death on Oct. 31, 1984. Born into a prosperous and powerful family, her paternal grandfather Motilal Nehru was a famous lawyer, well known public figure and major financial supporter of Mohandas (later Mahatma) Gandhi. She was the only child of Kamala Kaul Nehru (b. Aug. 1, 1899, d. Feb. 28, 1936) and Jawaharlal Nehru (b. Nov. 14, 1889, d. May 27, 1964). Her father was India's first Prime Minister from the independence on Aug. 15, 1947. He was in office until his death May 27, 1964.[176] Along with her father and her eldest son Rajiv, Indira Gandhi is considered part of the "Nehru dynasty." Rajiv Gandhi was Prime Minister from Oct. 1984 to Oct. 1989. The Nehru dynasty ruled India for 37 years, from 1947 through 1989, with a few breaks in between. Indira was considered to cast a magic spell over India through her own personal charisma. However, intense controversy also surrounded some of her years as Prime Minister, especially during the mid-1970s when for 19 months she declared a National Emergency, and again in June 1984 with the infamous Operation Blue Star. Thus her legend remains mixed. For some, she will always be "Mother India," and for others a malevolent despot. Meanwhile, the Gandhi family continues to be a powerful force in Indian politics into the 21st century through Indira's daughter-in-law Sonia Gandhi (b. 1946), widow of Rajiv Gandhi, and her two children: Rahul Gandhi (b. 1970) and Priyanka Gandhi (b. 1972). Sonia Gandhi has been the longest serving President of the Indian National Congress Party, re-elected in Sept. 2010 for the 4th time since 1998.[177] Also active in politics are Sanjay's widow Maneka Gandhi (b. 1956) and her son Varun Gandhi (b. 1980).

Indira Gandhi grew up surrounded by politics, political leaders, and the heat of agitation for Indian independence. From the age of 12 she was actively involved in politics when she joined the "Monkey Brigade," whose purpose was to help end British control in India. In early 1930 she met her future husband. Due to their activities in the "Quit India" movement (1942-1947), Indira and husband Feroze were both imprisoned by the British on charges of subversion, Indira for 8 months, Sept. 11, 1942 – May 13, 1943, Feroze longer. She married journalist Feroze Gandhi on March 26, 1942, at age 24. (Feroze: b. Aug. 12, 1912, d. Sept. 8, 1960. No relation to Mahatma Gandhi.) The marriage created family disapproval and a national outrage due to Gandhi's Parsi status. Parsis emigrated originally from Persia, and follow the religion of Zoroastrianism. They comprise a very small minority of the Indian population.

Indira's childhood was a lonely one, and this was to become a theme throughout her life. Even when she became Prime Minister she felt lonely and isolated. She was an only child, her parents were frequently in prison due to their efforts for Indian Independence, and her mother Kamala was ill with tuberculosis for the last ten years of her life. Along with her paternal grandfather Motilal, Kamala was Indira's strongest support within the family. Motilal Nehru died in 1931 and Kamala Nehru in early 1936. Her mother was also a major spiritual influence on Indira and helped to alleviate the largely hostile relations with her paternal aunt and great aunt from the time she was a child. They never fully accepted Kamala into the family and continually belittled both Indira and her mother. This created discord in the family home and strained relations with her cousins. Out of this Indira developed a strong inferiority complex regarding her

---

[176] The national chart of India is based on the moment India achieved independence from Britain, Aug. 15, 1947, at 00:00 hours in Delhi. (See Appendix) It is discussed later on in this section on Indira Gandhi.

[177] Born Dec. 9, 1946, 21:30 local time, Lusiana, Italy, Sonia Gandhi has 27:11 Cancer rising with natal Saturn in the Ascendant at 15:26 Cancer. This repeats Indira's *Gandanta* Cancer Ascendant and Saturn in Cancer. Sonia's Moon is at 9:50 Gemini in the 12th house of foreign residence. She met her husband in England in 1964, married him in 1968 and became an Indian citizen in 1983. She has lived in India since 1968. Initially her non-Indian birth caused disapproval in the family, including from Indira. But later they became very close.

looks and her abilities. She was considered plain compared to all the other beautiful women in the family.

After her mother's death on Feb. 28, 1936, with the exception of the early years of her marriage, Indira lived for many years with her children at her father's residence. She had two sons: Rajiv (b. Aug. 20, 1944, d. May 21, 1991); and Sanjay (b. Dec. 14, 1946, d. June 23, 1980). She often acted as her father's hostess and confidante, especially after he became Prime Minister in 1947. Indira lived apart from her husband for much of their 18-year marriage after their second son Sanjay was born in Dec. 1946. In her last 20 years, Indira was increasingly dependent on her two sons, and prey to son Sanjay's opportunistic and troublesome influence prior to his death in a plane crash in 1980.

There were two events pivotal in freeing Indira to pursue a higher political position: the sudden death of her husband Feroze from a heart attack in Sept. 1960, and the death of her father in May 1964. It took her another several years to experience her grief and gather more self confidence. Then she was able to come into her own as a skillful politician who was also enormously popular with the masses. Prior to becoming Prime Minister she was elected to the Indian National Congress in 1959, and in 1964 appointed Minister of Broadcasting and Information, a fairly insignificant position at the time. Her political competitors did not consider her a threat in the Cabinet, as few could see her potential leadership qualities and political acumen. As Minister of Broadcasting and Information, she promoted inexpensive radios and their distribution to address the problem of illiteracy in India. She also started a family planning program.

After her father's death on May 27, 1964, ending his 17-year run as Prime Minister, Lal Bahadur Shastri became Prime Minister. Five days after Shastri's sudden death Jan. 14, 1966, Indira succeeded him as P.M., in a battle with former Finance Minister Morarji Desai. She was later elected and served several successive terms up through March 1977, followed by a last term Jan. 14, 1980 to her death on Oct. 31, 1984. Indira Gandhi had an obsession with national sovereignty and desired to improve the condition of the poor of India through the socio-economic guidelines of her *20 point program.* She spoke the rhetoric of socialism, while consolidating a lot of personal power. Even so, she led India to become one of the fastest growing economies. She improved relations with Russia and China, and she promoted science and technology, including India's first satellite into space (1971) and its first nuclear test explosion (1974) – for "peaceful purposes," its assigned codename "Smiling Buddha." Major problems she encountered were overpopulation, low food production, financial difficulties, and poor foreign relations. Indira's handling of the Pakistan-India War in Dec. 1971 was regarded as one of her finest hours as Prime Minister.

Her most controversial act in her last term as Prime Minister was ordering the Indian Army to attack the Golden Temple at Amritsar, where Sikh separatists were hiding amidst the worshippers. This occurred June 4-6, 1984, and so angered the Sikh community in India, that Indira was killed in reprisal by two of her Sikh bodyguards Oct. 31, 1984. In response, some 4000 Sikhs were killed in India in the days following her assassination. Upon Indira's death, her son Rajiv Gandhi was appointed Prime Minister.

Major obligations to the parents – Powerful political family: Saturn's message is written very large in this chart, as it is the most prominent planet. It has just risen in the birth chart, within

6 degrees of arc.[178] Saturn is in a *Parivartana yoga* (mutual sign exchange) with Ascendant lord Moon in Capricorn, and the two planets are also in a mutual opposition aspect. This combination tends to produce a hard-working, conscientious person and forces them to confront heavy responsibilities in life from a very young age. In addition to the themes we have come to expect from Cancer Ascendant – yearning for a sense of family and nation, especially for deep nurturing from the mother or a mother figure – we see that Saturn's influence here requires a great deal more sacrifice and personal obligation to family or national duties, especially during the *Maha Dasha* (major period) of Saturn. Indira's 19-year Saturn *Dasha* began Nov. 15, 1970.

Saturn is also *Atmakaraka*, planet at the highest degree of celestial longitude in the chart, and so must bring some good fortune for the individual, within the context of their karmic destiny. When *Dasha* lord is also *Atmakaraka*, during that *Dasha* the person will come into their own. In this case, Saturn-Moon *Dasha/Bhukti* (Nov. 5, 1980 to May 20, 1983) promises personal fulfillment if important obligations are met, but also isolation and loneliness. Indira's political misjudgments in later years, some of them very large, cannot be explained away. But in seeking some relief for the weight of the load she carried (Saturn is heavy), she may have leaned too much on the wrong advice. For instance, some people thought Indira was guided too much by her younger son Sanjay.

Saturn is lord of the 7th and 8th houses of this chart. As 7th lord of marriage, it could have brought her a much older marriage partner. (Feroze was only 5 years older.) This is because Saturn rules over elders and those with greater authority or experience. Saturn is *Shanishchara*, Sanskrit for "slow-moving one." In a certain way this description was virtually fulfilled, as Indira served as a partner of sorts to her father for many years after her mother's death Feb. 28, 1936. Her ongoing obligations to her father did not end until his death May 27, 1964. When he became Prime Minister in 1947 he needed her more than ever. She traveled with him on numerous international trips and assisted him at a close personal level. Soon after the birth of her second son in Dec. 1946 she returned with her children to live in her father's house, and without her husband Feroze, though they did not divorce.

Just over 28 years elapsed between the death of each of Indira's parents, or a complete Saturn cycle. (Saturn's orbit around the Sun averages 29 years). With Saturn as lord of the 8th house of death and major changes, we see that precisely at the death of each parent, she became elevated both within the family and within the country. Each time, tr. Saturn in Aquarius was in the 8th house from her Ascendant. Saturn's passage through the 8th house from either Ascendant or Moon typically brings some level of unexpected loss and suffering. In this case, she endured the loss, but was simultaneously handed a greater role within the family and within the nation. Similarly, at her own sudden death, her son Rajiv became Prime Minister at age 40.

The level of sacrifice was likely to be a heavy burden for her, and ultimately she was killed for some of her decisions as Prime Minister. With the strong exchange between natal Moon and Saturn, it is not easy for the Moon to receive Saturn's energies. It can cause seriousness to the point of depression, but it does give self-discipline and a conscientious quality, not afraid of hard work. The natural expression of the emotions is curtailed and rendered cautious with so much Saturn influence. The person receives a very powerful influence from the mother, first of all – as

---

[178] Saturn in the 1st house (Ascendant) is its least favorable house position, especially if it is a difficult planet for the chart – as in this case, as 8th lord. When in Libra (exalted), or Capricorn or Aquarius (its own sign), Saturn in any *Kendra* produces a *Shasha Mahapurusha Raja yoga*, or "great soul." See Ch. 9, Chart #30 (Ammachi).

Moon symbolizes the mother, and secondly, influence from the mate or partner, as Saturn rules the 7th house of partnership. A partner in life is likely through marriage or business; and with Saturn's tendency to endure, any such partnership could have some longevity. An overriding sense of personal duty dominates, but in this case it would be to the family and to the nation rather than primarily to the husband. Moon is also the public, the people, the tribe or nation – in the collective sense. Just as with the individual mind, anything can be projected on to the collective mind, or mass consciousness.

When the Moon is in the 12th house of loss and surrender, as in Indira's *Navamsha* chart, it is a difficult placement for a public person unless they reside out of their country of origin or in some other 12th house setting. Moon in the 12th house can denote a strong need for privacy and/or a tendency for depression, more so with Saturn associations or contacts: Indira's *Navamsha* Moon is in Aquarius, birth Moon in Capricorn, both Saturn-ruled. It highlights the huge loss created by her mother's frequent absence and her early death.

Difficult marriage: How do we know the marriage to Feroze would be so difficult? We know because Venus (planet of love and marriage) is in the 6th house of conflict and contacts Rahu – which can create problems due to unorthodoxy or an unorthodox choice. Because he was a Parsi, a minority group in India, Feroze was not approved by Indira's family for years in advance. He was also a Freedom Fighter, a journalist, and later parliamentarian. We know from the condition of **Venus** in Indira's chart that a marriage in Mars, Rahu, or Jupiter *Dasha* could bring her trouble, as Venus is the sign lord of Jupiter and the *nakshatra* lord of Mars and Rahu. Thus Venus has some control over the results of the *Dashas* of Mars, Rahu and Jupiter: This occurs for 41 successive years from Nov. 1929 to Nov. 1970, ages 12 to 52. *Navamsha* Venus is in the 8th house, which does not improve the condition of Venus from the birth chart, though it is in its own sign of Libra, and thus gives some protection from losses as a *Viparita Raja Yoga* (*Sarala*).

Four months into the impetuous Mars *Dasha*, in March 1930, Indira and her mother Kamala met Feroze. He became one of Kamala's adoring assistants and disciples through her final six years of struggle with tuberculosis up to her death in early 1936. Indira and Feroze were close for some years, and finally Indira defied her family and married Feroze on March 26, 1942, during her **Rahu-Saturn** period. It was 12 years (one full Jupiter cycle) after they first met. Though the marriage produced two children, it was ultimately unhappy.

Venus-Rahu indicates an unpredictable or exotic element in love matters. Feroze was known to be a womanizer and too dependent on Indira's father for income. Though they never divorced, they mostly lived apart after Dec. 1946, when Indira returned to live in her father's house, taking the two children with her. She and Feroze reunited briefly in 1958 when he suffered his first heart attack. His sudden and fatal heart attack occurred Sept. 8, 1960. Born Aug. 12, 1912, he was just 48. Their legal marriage lasted 18 years, and their friendship for 30 years, just past one complete Saturn cycle. Even in the presence of unconventional or destabilizing factors, natal Saturn on Indira's birth Ascendant denotes endurance or loyalty. (His natal Sun and Moon in Cancer were on her Ascendant, creating another deep bond.) Other indications of conflict in marriage are shown by ***Kuja Dosha*** in the birth chart, by *Navamsha* Saturn and Mars aspecting the *Navamsha* Ascendant, and by *Navamsha* Mercury and Jupiter in 6 – 8 house relationship.[179]

[179] The 6 – 8 house relationship between planets has tension and unease; called *Shashtashtaka*. See Glossary.

(Jupiter rules House 1, and Mercury, House 7 of partnership. Feroze died in Indira's Jupiter-Mercury *Dasha/Bhukti*.) Only five months after Feroze's death, in mid-Feb. 1961, the pivotal **Jupiter-Saturn conjunction** at 1:52 Capricorn occurred. It was close to Indira's natal Moon in the 7th house of partnership. This helped to elevate her status. And now, without the scandal of divorce, she was liberated from the political liability of her marriage to Feroze – always controversial in the public mind.

Childhood and mother's influence: Indira's mother Kamala Nehru was a Kashmiri Brahmin middle class woman who was not westernized at the time of her marriage. Her family ran a flourmill in Delhi. Though she was considered a great beauty, she would not have been the first choice of bride for Jawaharlal Nehru except that her horoscope indicated she would give birth to a very distinguished lineage. By Hindu custom, the parents pre-arranged the marriage, and astrologers examined horoscopes of potential wives and husbands for propitiousness and compatibility. (Kamala Nehru was born Aug. 1, 1899 in Kashmir: birth time and precise location unknown. Jawaharlal Nehru: Nov. 14, 1889, 11:05 PM IST, Allahabad, India. This time is as rectified by astrologer K.N. Rao, 1993.)[180] In this case, Nehru's father Motilal and Kamala's aunt arranged the marriage. But Nehru's aunt, sisters and nieces considered it a mistake, and they had a major influence on his household. This led to an atmosphere of friction, with several lifelong effects on Indira. She and her mother were both belittled and berated by the Nehru women, and in a household of beautiful women, Indira was considered plain and unaccomplished. An only child, her strongest support came from her grandfather Motilal and her mother. They died when she was 13 and 18, respectively.

Kamala Nehru was healthy at the time of her marriage on Feb. 8, 1916. But after Indira's birth in 1917 there were three other pregnancies that led to miscarriages or premature birth. Physically weakened, she struggled with tuberculosis for the last ten years of her life before her death in 1936 at age 36. She had a youthful appearance, though was sickly in later years. Even so, she continued agitating for Indian independence and lived a deeply devoted and spiritual life. Many considered her the jewel of the family. Three years after her death, Mahatma Gandhi wrote:

> "I have not known a truer and braver and more god-fearing woman... a true devotee of her country and a woman of great spiritual beauty."
>
> Quoted from his weekly *The Harijan*, Nov. 25,1938.

From Indira's chart, her mother's shorter life and misfortune is shown by: **1)** the 12th house Moon in the *Navamsha* chart, as mentioned earlier; **2)** natal Venus, lord of the 4th house (the mother, family of origin) in the 6th house of conflict and health problems; and **3)** from the natal Moon, 4th lord Mars in the 8th house of chronic illness and suffering. Since the 8th house is also a spiritual house, we note Kamala's influence on Indira was also spiritual. The *Navamsha* chart confirms this, as 4th house lord Mercury is in the 8th house of *moksha* (spiritual gain, worldly

---

[180] K.N. Rao presents this rectified time (*Lahiri ayanamsha*) in his book ***The Nehru Dynasty: Astro-political portraits of Nehru, Indira, Sanjay & Rajiv***, 1993, p. 61. In his book, ***Notable Horoscopes*** (6th edition, 1991, p. 317), B.V. Raman, rectified it to a time of 11:33 PM IST, as Nehru's birth time was unknown. Raman first presented this time in the Oct. 1942 issue of *The Astrological Magazine*, using his own *ayanamsha*. K.N. Rao's rectified time gives 24:47 Cancer Ascendant with natal Moon at 17:56 Cancer, Venus *(Digbala)* and Mercury in Libra in the 4th. Nehru's 17-year run as Prime Minister began auspiciously in his Moon-Venus *Dasha/Bhukti*.

loss), along with 8th lord Venus. Kamala died during Indira's **Mars-Moon** *Dasha/Bhukti*. These two planets are in 6 – 8 house relationship (strain, tension, changes) in the birth chart, and straddle *Navamsha* houses 6 – 12, respectively.

While Kamala's marriage to Jawaharlal Nehru was not as inharmonious as Indira's marriage to Feroze, it was full of turbulence due to the political upheavals of the era. The couple was often in and out of jail as Freedom Fighters, and Kamala was often seeking treatment for her tuberculosis. Nehru was ten years older than Kamala, which is common for Moon-Saturn interactions, especially in opposition. She also endured much physical suffering. For Indira, her mother Kamala's death was truly the end of an era. It marked the end of her mother's support – a rare bulwark in the family – the end of her formal education (4th house is also education and higher degrees), and the beginning of her importance as a behind-the-scenes partner for her father, from whom she gained much political knowledge and experience.

Though steeped in politics since early childhood, the sadness and loneliness of Indira's childhood left her reluctant to enter politics. If she was not dealing with difficulties in the household, she was away at various schools. Then from ages 8 to 18, Indira devoted much time caring for her tubercular mother. Such is the powerful emphasis and influence of the **Moon-Saturn** exchange across the Ascendant axis. With their planetary energies so intertwined, and Moon being her Ascendant lord, her duties (Saturn) would continue unabated for most of her life, only increasing during **Saturn *Dasha***, from age 53.

Non-British education and aborted higher education: The 4th house not only describes the mother and the foundation of the conscious mind, but one's higher educational degrees. We will see some similarities in the difficulties for education, as for other matters we have explored regarding Venus, lord of the 4th house, and the challenge of 4th lord in a *Dusthana* (House 6, 8, or 12). Indira was well educated, though in many different schools – over seven of them, in Switzerland, England, and India – and often separated from her family. In protest of the British occupation, her parents sent her to mostly non-British schools in India and Europe. Venus, 4th house lord, is with Rahu (the foreigner) in the 6th house of conflict and/or health issues. Indira suffered from health issues, and though her own tubercular condition was not as serious as her mother's, it was chronic until fully cured in 1957 by new medicine available. She passed no examination except her secondary education in Pune, India. She was also forced to leave her higher education behind when her father needed her assistance after her mother's death in 1936. In the *Navamsha* chart, 4th lord Mercury is in the 8th house of loss or major change. Rahu is in the 5th house (also education) and brings what is foreign or unconventional in the context of the person's life. Born into a Brahmin, yet westernized family, *not* attending British schools was an anomaly. One outcome of that was Indira's complete fluency in French throughout her life.

Complexity of Indira's character: Following are some planetary links that would assure ongoing strength of character and purpose, no matter what hostile environment surrounded her. Ascendant lord Moon is aspected by both Jupiter and Saturn – one of the many powerful *Raja yogas* in this chart: Jupiter is 9th lord (trinal house), and Saturn is 7th lord (*Kendra* house). As 6th and 8th lords, they can also do her harm. Further, there are three *Parivartana yogas* (mutual sign exchanges): **1) Moon-Saturn; 2) Jupiter-Venus**; and **3) Sun-Mars**. Lastly, **Mercury** and **Saturn** are in a *Nadi yoga* (*nakshatra* lord exchange). These dynamic exchanges create strong interrelationships in her life, whatever the obstacles. The Jupiter-Venus exchange is more problematic, as we will see. Saturn often worked in her favor as *Atmakaraka*, but Saturn or any classic malefic in the Ascendant is demanding on the physical health. Exacerbating the situation are malefics on either side of the Ascendant (12th and 2nd houses) – a *Papa Kartari yoga*. One can be hemmed in by difficult life circumstances and seeking continually to extricate oneself from those constraints.

In addition, Cancer Ascendant or any key planet in a *gandanta* degree (the last 3:20 of a water sign) can bring difficulties to the individual. *Gandanta* is considered generally treacherous and unknowable. (See Glossary) Planet Mercury rules *Ashlesha nakshatra*, as does the deity *Sarpa*, the snake. Its wily, labyrinthine ways cannot be perceived at first glance. We may not know what it wants, though there is often great emotional force and will power behind it. By her own accounts Indira never fully recovered from the harsh judgment she experienced from her paternal aunt and great-aunt during her childhood. Thus an emotional vulnerability comes with this *gandanta*, especially apropos her status within the family (Cancer). Saturn's presence nearby hides it from the world for the most part and contributes to the chilly emotional façade – even if opposite the Moon, planet of emotions.

> "In power, she seemed a woman of supreme self-assurance, exuding a haughty *froideur*; but in private, she spoke of self-doubt and diffidence – 'I was so sure I had nothing in me to be admired,' she confided to one of her close friends days before her death."
>
> **Sunil Khilnani** on Indira Gandhi.

Saturn in the Ascendant gives the air of responsibility and trustworthiness. The person is likely to take on many responsibilities – however they are able to handle them. Further, Saturn in Cancer Ascendant tends to place the mantle of the family expectations on the shoulders of this individual. Though the Moon-Saturn combination is likely to show itself as contained emotionally, at least in most public situations, it already accounts for potential melancholy or seriousness along with a zeal for hard work. This can be seen by everyone around her, as these planets are both in *kendras* of the birth chart, therefore more immediate. Her Saturn has just risen and so is visible in the heavens – as in life. Indira's sense of personal isolation can be seen from Saturn's direct aspect to the Ascendant lord Moon, along with 12th house *Navamsha* Moon, with its yearning for privacy and thus a tendency to separate from others. It also describes her at times authoritarian style of leadership. A very demanding ruler, she was not above conniving to keep her power, reflecting the influence of Rahu-Ketu on Saturn in the *Navamsha* chart. It was at the onset of her **Saturn *Dasha*** in Nov. 1970 that Indira felt more confident with the great power she wielded, so much so that Saturn's more authoritarian side began to show in her leadership style.

Supporting the *gandanta* theme is Mercury (planet of speech and communication) in Scorpio, an emotional placement, contacted or aspected by Sun, Mars and Jupiter. Sun and Mars in particular are heating up the speech, and giving it sharpness, speed, or incisiveness. Natal Mercury is lively and intelligent in the 5th house, but in Scorpio it is not always straightforward about what it says it wants. Mercury's sign ruler is Mars, giving it emotional heat; its *nakshatra* ruler is Saturn, accounting for the overriding care and certainty with which she learned to speak publicly, especially in later years. Saturn, as *Jiva* (or life-giver) planet for Mercury, wins in the end. There is also a *Nadi yoga* (exchange of *nakshatra* lords) between **Mercury and Saturn**. (Saturn edits or aborts the speech and youthfulness of Mercury.) Further, in the *Navamsha* chart, Saturn in Capricorn aspects the Pisces Ascendant as well as Mercury in the 8th house. These factors tend to supersede the Mars-Sun heating effect on Mercury, as well as Jupiter's more expansive aspect on natal Moon. As an example, the voluminous exchange of letters between Indira and her father became especially heated in the years prior to her marriage to Feroze in 1942. Much was said and revealed, whereas in person together she and her father were far more stiff and reticent.

Mercury also rules over **astrology.** Mercury in Scorpio may hide its intentions, but with Saturn's influence can be both traditional and calculating. Indira and her father had both been largely Europeanized and wanted to appear modern. Though they regularly consulted with astrologers in private, they condemned astrology in public. By her second term in office, however, Indira never condemned the subject but caused damage to its credibility by hiring corruptible astrologers to write predictions favorable to her political goals. Such goals included restoring her son Sanjay's reputation.

Sanjay and Rajiv - continuation of the dynasty: The theme of the *Nadi yoga* between **Mercury and Saturn** continues when we examine children in Indira's chart. Mercury in the 5th house of children indicates controversy and problems for Indira's children, due to Mercury's ownership of the 12th house of loss. We have discussed Saturn's aspects on Mercury in the *Navamsha* chart. In addition, Saturn, Rahu and Ketu aspect the *Navamsha* 5th house, confirming these effects: Her children have responsibilities placed on them by their family or country. Rajiv's 5-year term as Prime Minister ended in October 1989, coinciding with the end of Indira's impactful 19-year **Saturn *Dasha*** on Nov. 15, 1989. This is fitting, even if posthumous.

The prominence of Indira's children is shown by her 5th lord **Mars**, which is strong as a *Vargottama* planet in a friend's sign, **in exchange with sign lord (Sun),** and aspecting its own house. Reading from the 5th house as Ascendant, the planets are well situated for leadership, with Mars in the 10th house and Sun rising. Birth chart 5th lord (Mars) in the *Navamsha* is a strong competitor in the 6th house of conflict. But potential loss of children comes from *Navamsha* 5th lord Moon situated in the 12th house of loss. Further, natal Mars is located in the *nakshatra* of *Purva Phalguni*, ruled by Venus. **Venus** in turn is located in its own *nakshatra* of *Purva Ashadha*, giving a strong desire nature to achieve perceived goals. Venus has a strong role controlling the affairs of Venus and Mars, lords of Houses 4, 5, 10, and 11. As *Jiva* (life force), Venus is not well placed in the 6th house of conflict and competition on the Rahu-Ketu axis. Both sons experienced controversy around them, most especially Sanjay, the younger son. (Sanjay was born Dec. 14, 1946, 9:27 AM Delhi, India: Capricorn Ascendant, with Saturn in Cancer, Sun in Scorpio, and Moon in Leo.)

Lenient in dealing with both her sons, Indira showed exceptional weakness with Sanjay, due in part to his Saturn close to her own natal Saturn in Cancer. She was grooming him to be Prime Minister, an unfortunate and unsuitable choice, as he proved to be an unethical person whose excesses and influence on his mother became harmful to the very well-being of her Prime Ministership. Indira won re-election as Prime Minister Jan. 14, 1980. Some say her major motivation for returning to power after the 1977 defeat was to rid the many charges against her son Sanjay. He was one of her closest advisors for 10 to 12 years before he died in an airplane crash on June 23, 1980.

After Sanjay's death, Indira urged her elder son **Rajiv** to enter politics. Though he had no prior political ambitions, his character and destiny were well suited to leadership. (Rajiv was born Aug. 20, 1944, 8:11 AM (war time), Bombay, India: 14:36 Leo Ascendant, with five planets in Leo, including Sun, Moon, Mercury, Venus, and Jupiter.) He won his brother's seat in Uttar Pradesh in Feb. 1981, and many believed his mother was now grooming him to be Prime Minister. At his mother's death, he was pressed to take on the Prime Ministership, and led India from Oct. 31, 1984 through Dec. 2, 1989. He was head of the Congress Party and running for re-election when he was killed on May 21, 1991 by a female suicide bomber, a supporter of the Sri Lankan separatist Tamil Tigers. Rajiv's death was considered to mark the end of the "Nehru dynasty."

Success later in life - after fulfillment of responsibilities to both parents: Though she was a member of a prominent political family and a recognized public figure prior to becoming Prime Minister in Jan. 1966, Indira did not yet have a political power base or a real political vision of her own. She was a poor speaker and at a disadvantage as a woman in a man's world. Yet she had an unwavering sense of obligation to her family and country. Saturn is a planet of perseverance and of delays, maturing later - as would befit the "slow moving one." Indira seemed to have no particular appetite for power until she herself became Prime Minister. Then her life took on new meaning, and her self-confidence suddenly blossomed. We see this confirmed with Ascendant lord Moon placed in *Uttara Ashadha nakshatra* ("the later victor"). She was a young child during **Moon *Dasha*** (1919-1929), and too young to benefit from the **Moon-Saturn exchange** in terms of her own worldly success. Instead it brought many obligations towards her parents or longing for them during their absence as India Freedom Fighters. (Her **Saturn-Moon** *Dasha/Bhukti* came in Oct. 1981.) After her husband's death in 1960 and her father's death in 1964, she had fulfilled her largest family responsibilities and was freer to come into her own politically. In addition, she would be exceedingly fortunate in life in terms of abundance and success. This is seen from the *Maha Bhagya yoga* (a *yoga* of major good fortune.) It is formed when, for a woman, she is born during the night, with the Ascendant, Sun and Moon in even signs.

Indira's first years as P.M. coincided with the last years of her Jupiter *Dasha*. But the real experience of personal power came soon after the start of her 19-year **Saturn *Dasha***, from Nov. 15, 1970. Indira was riding a wave of popularity in the early 1970s. She was re-elected in March 1971, running on a campaign slogan of "eliminate poverty," and won in a landslide election. A **high point** for Indira as Prime Minister was in Dec. 1971, with her decisive victory in the **Pakistan-India War**, Dec. 3-16, 1971. Under her leadership, India took a swift offensive after the Pakistani Air Force hit Indian airfields in northern India on Dec. 3rd. The war resulted in nationhood for Bangladesh (formerly East Pakistan). In addition to her favorable **Saturn-Saturn** *Dasha/Bhukti* at the time, tr. Jupiter in Scorpio aspected natal Sun, Mercury, Jupiter, Saturn and Ascendant, opposite tr. Saturn in Taurus. A tr. Jupiter-Saturn opposition brings out into the open that which has been seeded at the previous Jupiter-Saturn conjunction (Feb. 18, 1961).

Destiny for leadership with controversy: For leadership we examine the 10th house from the Ascendant, Moon and Sun. We check planets occupying the house, significators, aspects, and its major *karaka* (significator): the Sun. From birth Moon in Capricorn: 10th house lord of Libra is Venus, located with Rahu in the 12th house of loss. From Sun, Mars is very strong in the 10th house, in a *Parivartana yoga* (mutual exchange) with the Sun. From the Ascendant, Mars in Leo is 10th lord, *Vargottama,* in exchange with its sign lord, the Sun. (This theme is echoed by Sun and Mars in Leo in the *Navamsha* 6th house.) Natal Mars is located in the Venus-ruled *nakshatra* of *Purva Ashadha* ("the earlier victor"). Venus in turn is in its own *nakshatra*. Thus, Venus's passionate influence is strong as *Jiva* (life-giver) to the matters of that house. With Venus situated in the 6th house of conflict in the birth chart, and in the 8th house in the *Navamsha*, all things coming through Venus (including love matters and marriage) are fraught with difficulty and require sacrifice. They also have a certain unpredictable quality due to Venus also in Sagittarius on the Rahu-Ketu axis.

The condition of Venus is protected somewhat by the *Parivartana yoga* (mutual exchange) with Jupiter in Taurus. But they are also planetary enemies, in exchange between the house of enemies and competitors (6th house) and house of friends (11th house). This **Jupiter-Venus exchange** brings both powerful friends and powerful enemies and makes it more difficult for Indira to distinguish friend from foe, or to receive the best possible advice. However, it does give her the ability to face fierce competition. The two planets are in 6 – 8 house relationship to each other in both birth chart and *Navamsha*. In the end, it was those she considered her friends and bodyguards who killed her.

Prologue to becoming Prime Minister: Prior to his death May 27, 1964, Nehru was trying to set the stage for Indira to succeed him very soon, if not immediately. His Cabinet Minister Lal Bahadur Shastri became P.M. in May 1964, but ruled only briefly until his own sudden death while travelling in Russia in Jan. 1966. Indira succeeded Shastri in a bitter contest with Morarji Desai until she was elected to the office. She won elections in 1967 and 1971. This lead-up period occurred during her **Jupiter-Venus** *Dasha/Bhukti* (Sept. 27, 1962 to May 28, 1965). The *Dasha* of Jupiter-Venus shows how this planetary exchange plays out in Indira's karmic destiny. A competitive period that paved the way for her ascent to greater power, it also indicated death or major change for the father. Father is 9th lord Jupiter, and Venus is in the 8th house from Jupiter. Venus and Jupiter repeat their 6 – 8 house relationship in the *Navamsha* chart. Other factors contribute to this picture, including tr. Saturn in her 8th house in Aquarius.

Since this is far larger than a personal matter we can also study the **India national chart** (See Appendix). The India Independence chart is set for 00:00 IST, Aug. 15, 1947, Delhi, India. The nation started a new *Dasha* (Mercury) in Sept. 1965, after a 19-year Saturn *Dasha* – during most of which Jawaharlal Nehru was Prime Minister. Mercury as *Dasha* lord hints at younger leadership. Indira's chief opposition, Morarji Desai, was 21 years older, born Feb. 1896.

Indira Gandhi became Prime Minister on Jan. 19, 1966, when she was in **Jupiter-Sun** *Dasha/Bhukti*. It was a more auspicious period for her political rise than Jupiter-Venus, as Jupiter and Sun are planetary friends and in mutual aspect in her birth chart. In addition, tr. Jupiter, her 9th house lord (*Dharma* lord) was in Taurus, its natal sign position, even if retrograde and late in the sign. Tr. Sun, Venus, and Mars (the latter exalted) in Capricorn in her 7th house favored her gaining power. Together with Jupiter and Rahu in Taurus, five transiting planets were in earth signs when she set out the first time as P.M. She was Prime Minister a total of 15 years, though her initial successive terms ran just over 11 years, coinciding with a Jupiter cycle. Only tr. Saturn in Aquarius in her 8th house (from Jan. 1964 to April 1966) shows how death brings her in and out of office.

National Emergency: 1975-1977: Indira's greatest popularity as Prime Minister was in the early 1970s. By 1975 her policies and programs proved to be highly controversial, and some of them were initiated and executed by her son Sanjay. On June 12, 1975 the Allahabad High Court declared her recent election null and void due to her corrupt election practices. The verdict effectively ordered her removal from her Parliamentary seat and banned her running for election for another six years. When news of the decision spread, it caused escalating riots and so much civil unrest that rather than step down, she imposed a National Emergency on June 26, 1975. This period is still considered one of the darkest and most controversial episodes in post-Independence India. She suspended civil liberties, jailed her opponents, and ordered the arrest of the main opposition leaders and censorship of the press. Many said she played castes, religions and political groups against one another, contrary to her father's ideals. Indira was accused of being paranoid and ruthless. She was increasingly isolated, lacking in confidence, and taking too much advice from her son Sanjay. The National Emergency continued for 19 months through March 24, 1977 when an overwhelming Anti-Indira Coalition voted her out of office. By late March 1977, tr. Jupiter returned by sign to early Taurus. This should have favored her, especially as Saturn was also in its second return. (Both returns coincide only once at ca. age 59). But this also coincided with the first Saturn return to the India national chart, and such a transit would tend to promote a change of leadership or direction for the country, especially if the current leader was out of synch with the Constitution.

The **India national chart** has Taurus Ascendant and five planets in Cancer – Sun, Moon, Mercury, Venus, and Saturn. Natal Sun is at 27:59 Cancer, Saturn at 20:28 Cancer. Indira Gandhi's natal Saturn is at 21:47 Cancer, her Ascendant at 27:22 Cancer, both in *Ashlesha nakshatra* ruled by *Sarpa*, the snake. India's national chart birth Moon at 3:59 Cancer is just opposite Indira's Moon at 5:35 Capricorn. Such important planetary contacts to the India national chart can bring strong links between an individual destiny and a national destiny.[181] People would tend to identify her with India, especially Indians. At her best, she was their "Mother India." But both her Saturn and Ascendant are in *Sarpa drekkanas*, as are India's national chart Mercury, Venus, Saturn, and Sun, the latter three also in Sarpa-ruled *Ashlesha nakshatra,* opposing the 9th house of religion. (*Sarpa drekkanas* can bring troubles through treachery and deceit, notably: the *Gandanta* degrees – the last 3:20 of a water sign.) The context of the tribal nature of Cancer adds to the emotional turbulence generated while trying to unite the many religious factions under one harmonious roof and in one secular state.

Return to power 1980: After her resounding defeat in the election March 24, 1977, Indira Gandhi's return to power as Prime Minister Jan. 14, 1980 was considered astonishing. But **Saturn** is strong as *Atmakaraka,* and in exchange with **Moon,** Saturn acts as if in Capricorn, in its own sign in the 7th house – where it has *Digbala* – best possible angle of the chart for Saturn. This spells endurance and perseverance. In addition, the **Jupiter-Saturn conjunction of 1980-1981** in mid-Virgo was in the 9th house (best house) from her natal Moon, though only 3rd house from the Ascendant. The conjunction was in *Hasta nakshatra*, ruled by the Moon. These conjunctions started a new 20-year socio-political-economic cycle throughout the world. (See Chapter 1)

With **Saturn-Moon** pivotal in Indira's chart, **Saturn-Moon** *Dasha/Bhukti* would be important in her life and probably successful. It ran from Oct. 18, 1981 to May 20, 1983. But it was during **Saturn-Mars-Moon** period (May 25, 1984 to June 27, 1984) that she ordered the fateful attack on the Sikh Golden Temple (June 4-6, 1984), an action that set up the revenge killing by her two Sikh bodyguards. She was assassinated while in **Saturn-Rahu-Rahu** period (June 27, 1984 to Dec.

[181] Using K. N. Rao's rectified time, Jawaharlal Nehru's Ascendant at 24:47 Cancer and natal Moon at 17:56 Cancer are also closely tied to the India national chart, with its five planets in Cancer. Nehru wanted to keep the Prime Ministership in the family (Cancer), with his daughter succeeding him.

1, 1984). With natal Saturn in *Sarpa drekkana* and Rahu capable of delivering poison, this would be a dangerous period, especially with *Navamsha* Rahu opposite Saturn-Ketu in the 11th house of friends and natal Saturn and Rahu in 6 – 8 house relationship in her birth chart, a dynamic repeated all of 1984 by transiting Saturn and Rahu, also in 6 – 8 house relationship.

On a Monday morning (Oct. 31, 1984) in her tranquil Delhi garden, Indira walked towards her office with five security guards just behind her. Two Sikh guards stood at attention ahead of her more than half way down the path. One was new and the other was a favorite she had known for ten years. At 9:08 AM IST they fired 33 bullets into her at close range, killing her instantly. Doctors tried for hours to revive her, refusing to believe she was gone. Her death was not announced until 1:45 PM. (The Ascendant was 26:26 Capricorn, just opposite that of Indira and the natal Sun of the India national independence chart at 27:59 Cancer.)

At the moment of her assassination, 16:34 Scorpio was rising, with tr. Venus and Ketu both in Scorpio opposite tr. Rahu (the foreigner) in early Taurus in her 11th house of friends. Tr. Saturn and Mars – both malefic planets – closely aspected her Cancer Ascendant. Tr. Mars and Jupiter in Sagittarius were in her 6th house of enemies, and tr. Moon in Capricorn in *Uttara Ashada*, her birth *nakshatra*. Tr. Saturn, Sun, and Mercury were all in Libra in her 4th house – the roots of the life and its ending, as well as the domestic environment, most significant for the Cancer Ascendant, being the 4th sign. Saturn's transit through the 4th house is generally a career low point in its cycle, notably near the 4th house cusp. As current *Dasha* lord, Saturn's transit carried extra weight. It brought her to her greatest heights and as 8th house lord, it was also likely to bring her down. The night before her death she spoke to a large, enthusiastic crowd in Bhubaneshwar, Orissa's capital city:

> "I am not interested in a long life. I am not afraid of these things. I don't mind if my life goes in the service of this nation. If I die today, every drop of my blood will invigorate the nation."

Chart 19: Prince Charles of Wales

Birth data: Sunday, Nov. 14, 1948, 21:14 GMT, London, UK, Long. 00W10 00, Lat. 51N30 00, *Lahiri ayanamsha* –23:08:24, Class A data. (News report) Ascendant: 12:14 Cancer.

Biographical summary: His Royal Highness Prince Charles of Wales is the 21st Prince of Wales, a title that has been invested since 1301 but only to the Heir Apparent to the throne, and thus a male who cannot be displaced in the succession to the throne by any future birth. The Heir Apparent is the oldest son of a monarch. The current duties of Prince of Wales are publicly stated as follows:

> "working as a charitable entrepreneur, undertaking royal duties in support of The Queen, and promoting and protecting nationalization, virtues, and excellence."[182]

---

[182] Queen Elizabeth II was born April 21, 1926, 2:40 AM, London, UK., Ascendant at 28:33 Sagittarius. Prince Philip, the royal consort, was born June 10, 1921, 21:46, Kerkira, Corfu, Greece. Ascendant 20:48 Sagittarius.

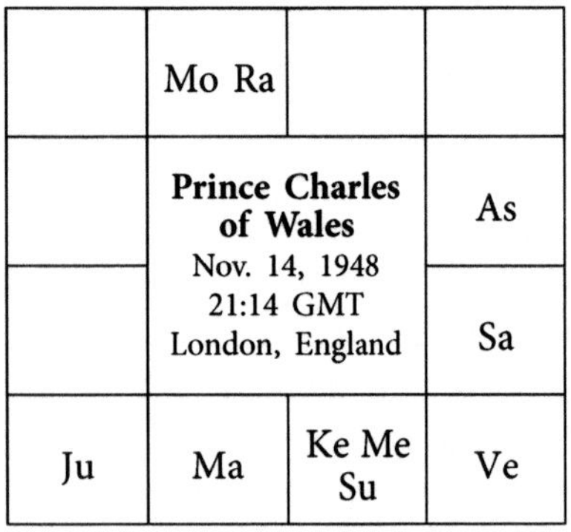

	Mo Ra		
	Prince Charles of Wales Nov. 14, 1948 21:14 GMT London, England		As
			Sa
Ju	Ma	Ke Me Su	Ve

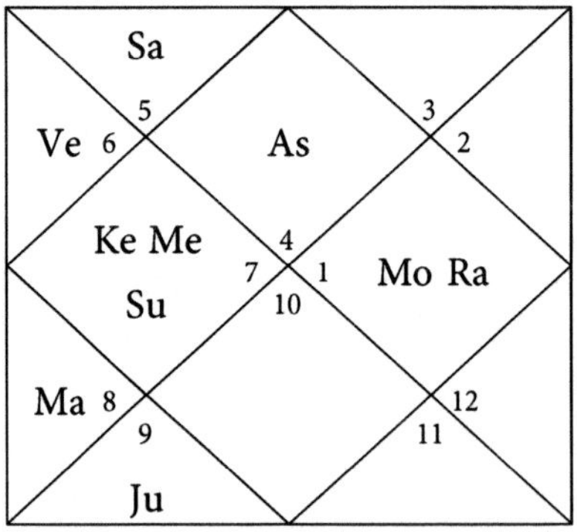

Sa
5 3
Ve 6 As 2
Ke Me 4
7 1 Mo Ra
Su 10
Ma 8 12
9 11
Ju

Planetary Information

As	12:14	Can	Pushya
Su	29:16	Lib	Vishakha
Mo	07:17	Ari	Ashwini
Ma	27:48	Sco	Jyeshta
Me	13:49	Lib	Swati
Ju	06:44	Sag	Mula
Ve	23:14	Vir	Hasta
Sa	12:07	Leo	Magha
Ra	10:47	Ari	Ashwini
Ke	10:47	Lib	Swati

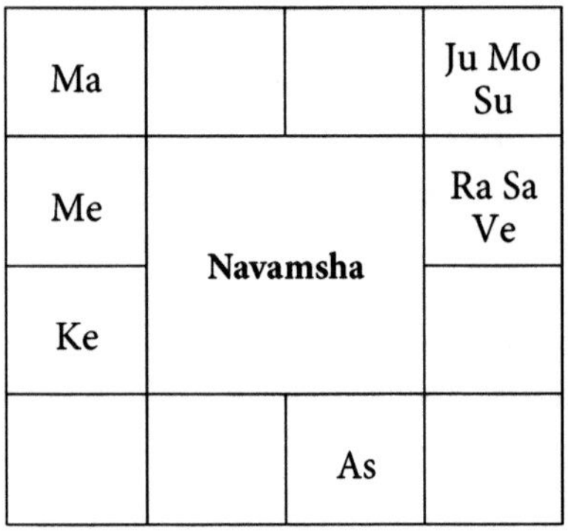

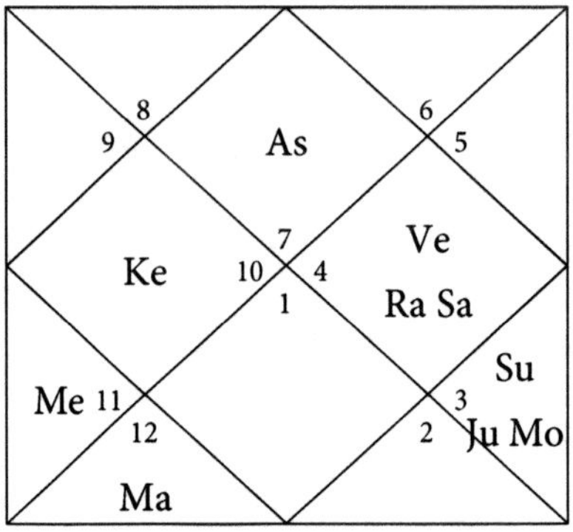

Ma			Ju Mo Su
Me	**Navamsha**		Ra Sa Ve
Ke			
		As	

8 6
9 As 5
7 Ve
Ke 10 4
1 Ra Sa
Su
Me 11 3
12 2 Ju Mo
Ma

Vimshottari Dashas

Ke	Jan 16, 1945
Ve	Jan 17, 1952
Su	Jan 17, 1972
Mo	Jan 17, 1978
Ma	Jan 17, 1988
Ra	Jan 17, 1995
Ju	Jan 17, 2013
Sa	Jan 17, 2029
Me	Jan 18, 2048

His position has not allowed him to become involved in politics. Instead, his major task as Heir Apparent to the throne of England has been to prepare himself to take on this role, and to attend to his duties, as given.

The Prince's childhood was somewhat restrained by the typically formal relations in this British royal family, especially with his father, whom he idolized. Prince Charles was the first British royal to attend a public school, and did so from the age of 8. The following year he went to boarding school, and from the ages of 13 to 18 attended a boarding school in Scotland. His private schooling was challenging for him, as he was not easily accepted and the emphasis was on a kind of macho behavior that was foreign to his sensitivities. In 1967, at age 19, he attended Trinity College, Cambridge where he studied history, government, archaeology and geography. He also learned his civic duties and the Welsh language. By British royal tradition, at age 21 he entered military training as a sailor, airman and soldier. He flew helicopters and spent five years at sea, commanding his own minor vessel. His favorite sport was polo, which he played extremely well. He was known increasingly for his dry wit, intelligence, and offbeat sense of humor. His many friends regard him as warm, kind, loyal and trustworthy. He has always loved horses and the country life and developed his own organic farm. In 1976 he set up a Small Business Trust to help Englishmen with their entrepreneurial ventures. Beyond fulfilling his royal duties, the Prince's interests have included painting, the occult, philosophy and ecology – a combination sometimes perceived as eccentric.

As Heir Apparent, Prince Charles was expected to marry and produce his own heirs. His marriage partner had to be of impeccable aristocratic breeding as well as a virgin. This prospect put much family and social pressure on Prince Charles from an early age. He was considered a highly eligible bachelor, and had a very active social life, dating various women briefly – including Camilla Shand. They met at a polo match in 1970, but she was not considered aristocratic enough and was not a virgin. But she remained a close friend and more. (She was born July 17, 1947, 7:00 AM Zone -2, London, UK. She married Andrew Parker-Bowles July 4, 1973, had two children with him in the mid-1970s, and was divorced March 3, 1995.) By 1980 the Queen put more pressure on Charles to choose a bride, as he was already 32. The press followed him closely from age 18; they tended to describe him as cold and stodgy. Thus began an uneasy long-term relationship with the media that has been at times hostile.

In early Feb. 1981, after a six-month courtship, Prince Charles proposed to Lady Diana Spencer, age 19, daughter of the 8th Earl of Spencer and of excellent lineage dating back to the 15th century. (They first met at a garden party in 1977 when Charles was dating her sister Sarah.) She accepted immediately and the engagement was announced officially on Feb. 24th. The fairy-tale wedding with all its pageantry took place July 29, 1981 in London. Lady Di became Diana, Princess of Wales. (b. July 1, 1961, 19:45, Sandringham, UK.)[183] Two sons were born within the next few years: Prince William on June 21, 1982, 21:03 GDT, Paddington, UK; and Prince Henry (Harry) on Sept. 15, 1984, 16:20 GDT, Paddington, UK.

The subsequent demise of this royal marriage was due in part to their incompatibility and in part to Princess Diana's insecurities, which were exacerbated by bulimia. She was naïve, impulsive, highly emotional, and afraid of abandonment. She needed a lot of reassurance from Charles that he was disinclined to give. Much blame has been placed squarely on him due to his previous emotional commitment to Camilla Parker-Bowles. On her wedding day Princess Diana became more aware of Charles's deep devotion to Camilla. But it took several more years for the royal marriage to come apart. They led increasingly separate lives, with the strain between them openly apparent by 1987. Princess Diana's popularity rose exponentially, eclipsing Prince Charles from the start, much to his annoyance. Her image in public was impeccable: she was beautiful, charming, and caring, while in private she was frequently in tears. The turmoil of their married life soon spilled out and became world news. After years of accusations, counter-accusations, revelations in tabloids, television interviews and scandalous biographies, their official separation announced on Dec. 10, 1992 came as a welcome relief. The Divorce decree was declared absolute on Aug. 28, 1996 at 10:27 AM, London. In the midst of starting a new affair, Princess Diana was killed in a car crash in Paris Aug. 31, 1997. Her death brought a huge outpouring of emotion from the public for "the People's Princess" and a new crisis for the Queen.

Though there were rumors of an engagement to Camilla as early as 1997, Prince Charles was required to wait several more years after Princess Diana's death before he could introduce Camilla into royal circles. Their joint public appearances were delayed until 1999, and not until 2000 did they attend the same function with the Queen, her official acknowledgement of their relationship. Due to their long-term liaison outside their respective marriages, Camilla's perceived role in destroying the royal marriage brought strong prejudice against her from a public that adored Diana. But Prince Charles was adamant that his relationship with Camilla was "non-negotiable," and in April 2002 the Church of England changed its rules so Prince Charles could marry a divorcee whose husband is still alive. This finally cleared the way for Charles to marry Camilla.

[183] Princess Diana's birth time is listed officially as 19:45 GMT. Source: "from her memory." But at the time of her engagement her mother said she was born at 14:00, "just before the start of play at Wimbledon."

On Feb. 10, 2005 they announced their engagement and their plans to marry on April 8, 2005. Six days prior to the event, the ailing Pope died. The wedding had to be postponed one day, as the funeral of Pope John Paul II was held on April 8th in Rome and Prince Charles was required to attend. Camilla became Duchess of Cornwall, and retains that title, regardless of whether Prince Charles becomes King.

The Moon dominates: The Moon is far stronger in the birth chart than the Sun. Fortunately Moon is the Ascendant lord, and at the top of the sky it keeps Prince Charles on the world stage. It put him there strongly during his **Moon *Dasha*** (1978–1988). Born two days before the exact Full Moon, Prince Charles has a bright Moon in Aries. It receives aspects from five planets, including Jupiter, which in Jaimini astrology is considered to elevate the person in status, recognition, and success. Moon is in the first *nakshatra* of *Ashwini,* meaning also "owning horses." Moon in *Ashwini nakshatra* prominent in the chart likes to act, to keep moving, and to be innovative. It likes change, and needs much room to roam around, as with the animal that rules the sign of Aries, the Ram. In addition to the (active) Ascendant sign of Cancer, five planets in *Chara* signs – Sun, Moon, Mercury, Rahu and Ketu – define a very active person, one who is also very interested in natural healing modalities. *Ashwini* is ruled by the Ashwini Kumaras, the twin horsemen who have the heads of horses and are the physicians to the gods. There is a natural connection to herbs, gardening, agriculture, and horses. Moon in *Ashwini* also prefers pastoral settings. Prince Charles's ongoing commitment to organic farming and environmental issues is notable, and he has been an avid horseman and polo player from a young age. He could be described as headstrong in his approach to life – despite – or in part because of – his destiny as King-in-Waiting.

Ascendant lord Moon in the 10th house is very good for lifelong public visibility or a political career, except for being on the Rahu-Ketu axis, which brings mental and emotional turbulence. A Total Solar eclipse 13 days prior to birth was at 15:35 Libra, close to the degree of Charles's Mercury and Ketu and opposite Moon at 7:17 Aries. Eclipses necessarily involve Rahu and Ketu in the chart, and set up sensitive degree areas. Moon on the Rahu-Ketu axis is already overly sensitized and this becomes even more exaggerated when Moon is also opposite Mercury. Planet of the feelings (Moon) opposite planet of the rational mind (Mercury) can waver between the irrational and the rational. As this involves the 10th house of the birth chart, such wavering and/or indiscretion becomes public knowledge – not a happy combination if you are British royalty and wanting to preserve your reputation and privacy. The Full Moon in the 10th house is a modifying influence to the weakened Sun; but Moon with Rahu in the 10th house tends to air its dirty laundry in public – a propensity that is magnified in the electronic age of microphones, scanners and phone hackers, causing great embarrassment. Several of his private phone conversations with Camilla Parker-Bowles were recorded, transcribed and distributed. One especially passionate exchange on Dec. 18, 1989 – dubbed "Camillagate" – was distributed through private channels and then published in Jan. 1993 in two British newspapers. Some of Diana's phone calls were also recorded and publicized.

The **Moon** represents all the important women in his life, including his mother, Queen Elizabeth II. As reigning monarch since Feb. 6, 1952 (20 days into his Venus *Dasha*), the Queen has eclipsed her eldest son for obvious reasons. Princess Diana also eclipsed Prince Charles in her lifetime.

This could easily happen with a marriage partner, especially if he marries during **Moon *Dasha*** (1978–1988). Rahu eclipses the Moon, and Rahu is with the Moon in the most visible house. There is a life-long issue about being eclipsed by certain women in his life. This same influence can also make one sensitive to psychic impressions: thus his strong interest in the occult.

King-in-Waiting: The **Sun** is *karaka* for leadership in any chart. The condition of the Sun in this chart is not favorable for leadership. It shines more brightly in private circles rather than in the glare of the public eye. The Sun is at the bottom of the sky, debilitated (*Neecha*) in Libra, without receiving any correction for this debilitation (least favorable sign position for the planet). Libra is owned by Venus, which in turn is debilitated in Virgo. Further weakness for this sign and house is caused by the location of Venus in the 12th house (of loss) from the 4th – indicating loss in 4th house matters. Though carefully groomed for leadership, Prince Charles's personal sense of authority and self-confidence is diminished by the Sun and other factors, which give him more confidence and enjoyment away from the public and the media. This affliction is especially clear during his **Sun *Dasha*** (1972-1978).

The Sun is situated in Jupiter's *nakshatra Vishakha*," the star of purpose." Unfortunately, Jupiter's placement in the 6th house from the Ascendant keeps placing the person at the nexus of some controversy, out of which they are likely to emerge victorious, but not without a lot of gut-wrenching verbiage. This is because Jupiter is in its own sign of Sagittarius, which forms several *yoga*s that both protect the person from losing face or reputation in conflict situations, and insure that the *dharmic* destiny will attract these very same situations to the individual. (The latter is due to 9th lord in the 6th house: *Nirbhagya yoga*; the former is due to a *Viparita Raja yoga*, 6th lord in the 6th house, lending protection in fiercely competitive situations.) This protection may not seem apparent at first, when in the throes of controversy,

Another source of protection is the Sun's role as *Atmakaraka*, planet at the highest degree of celestial longitude. Thus, the Sun has to bring some benefit to this destiny, and it has indeed brought much wealth and status, but a long wait for accession to the throne. Compare this with the *Atmakaraka* of Maria Montessori: Saturn at 29:53 Scorpio in her 5th house of children. As her *Atmakaraka*, it brought her much success after initial suffering. It brought her leadership in the field of children's education. For Prince Charles, the Sun is further protected by its improved house and sign position in the *Navamsha* chart, and by its *Ubhayachari yoga* in the birth chart, with planets in Houses 2 and 12 from the Sun, not including Moon, Rahu or Ketu. This combination insures being surrounded and supported by a strong network of friends and/or relatives throughout life. Maria Montessori shares this fortunate *yoga*.

The Sun is most dominant and fortunate (*Digbala*) at the top of the sky, where it is most visible. A late night birth, depending on the latitude and time of year, will bring 4th house placement and potential problems regarding visibility, especially if it is expected or wanted. The 4th house is one's home and family of origin. Since the Sun has a lot of fiery energy, when located in the 4th house it will feel it should have a dominant position in the family of origin. However, in this case Saturn's aspect to the Sun cools and delays that dominance, or the desire for it. Natal Saturn is placed in the royal sign of Leo in the *nakshatra* of *Magha*, whose symbol is the throne room, and whose desire is to flourish in the realm of the ancestors. *Magha* is ruled by the *Pitris*, the great fathers of humanity. Saturn is the planet of delays, and of events requiring patience and perseverance. Saturn in *Magha* aspecting the Sun is another symbol for delays in the development of royal status – *not* good news for a King-in-waiting.

The right choice at the right time: Due to the relatively weak condition of the natal Sun, the **Sun *Dasha*** is bound to be a weak period for leadership and marriage issues. Regardless of his Heir Apparent status, self-confidence and sense of direction in life would be at a low point for Prince Charles during the 6-year Sun *Dasha*, starting Jan.17, 1972. Such is the karmic destiny likely to come due precisely at this time in life – when he was expected to make a choice of marriage partner. During the first year of his Sun *Dasha* in 1972 Charles was still courting Camilla Shand, whom he met in 1970. She has said she did not believe he would ask her to marry him. Meanwhile, Charles went off on a naval tour of duty in 1973 and during this time Camilla married Andrew Parker-Bowles (July 3, 1973). During his weakest *Dasha* it was unlikely that Prince Charles would easily make a choice of marriage partner or of anything important. Tr. Saturn was in Taurus during 1972 (aspecting his Ascendant and Mars), with tr. Jupiter in its return sign of Sagittarius. This kept up the pressure on his sense of responsibilities but was not favorable for choosing a wife. Prince Charles instantly regretted losing Camilla, but he was not strong enough at that time to make an auspicious choice of marriage partner. The Vedic chart shows this as part of his destiny. Destiny is character plus timing.

Prior to the Sun *Dasha* is the 20-year **Venus *Dasha*.** Venus is the *karaka* for love and marriage. For Charles it ran from Jan. 1952 to Jan. 1972, when he was ages 3 to 23. Natal Venus is debilitated in Virgo, but does receive correction (called *Neecha Bhanga Raja yoga*), with Mercury in a *Kendra* from the Ascendant and from the Moon. (Lord of Virgo or Pisces must be so placed.) This means Venus's weaknesses can be overcome after some initial obstacles, and that he will have a love life and marriage(s). But a commitment for marriage is not likely during the *Dasha* of his weakest planet – the Sun. Marriage is also unlikely during the preceding Venus *Dasha:* Venus has its own weaknesses – being in the 6th house (of sacrifice and conflict) from the Moon, debilitated in Mercury's sign and in Moon's *nakshatra*. Moon and Mercury are both on the Rahu-Ketu and Full Moon axes across Houses 4 and 10. The message is that though there are plenty of choices for love and an active social life, the making of the decision is fraught with turbulence due to the intense expectations of the royal family, the inevitable public scrutiny, the number of choices and the inability to decide.

Marriage – the most crucial life issue: As we can see, marriage is by far the most crucial life issue for Prince Charles. His destiny required that he marry well and produce heirs, both of which are shown to be problematic in his chart, though it is clear he would have love interests and he would have a marriage as well as children. Princess Diana turned out to be an apparently bad choice for him, but she was the immediate darling of the public and the media. Cameras hovered on her irresistible image far more than that of Prince Charles. Since he did not choose a bride during Sun *Dasha* (1972-1978), this delay pushed him into the likelihood of marriage during Moon *Dasha* (1978-1988) and the destiny of marrying someone who would capture much public attention in a way that was unnerving for Prince Charles. Such is the **major shift between his Sun and Moon *Dashas***, reflecting the difference between Sun and Moon in his chart. SUN-MOON is also the marriage axis, if they are located in the same sign or opposite signs. (In addition to husband and wife, it describes the parents.) We examine this axis in both the birth and *Navamsha* charts, as well as 7th house lord Saturn, and Venus. In this case, the Sun and Moon and their respective *Dashas* are pivotal.

Emotional upsets in public: We have discussed how Moon with Rahu can produce mental and emotional turbulence; and when in the 10th house of greatest exposure, private emotional issues become very public. Since this Rahu is on the exact 10th house cusp from the Ascendant, it expands the publicity still further, and sign lord Mars has turmoil associated with it, as does *nakshatra* lord Ketu. Marrying Princess Diana in Moon *Dasha*, Charles would be likely to have a bride who outshines him and brings mental and emotional unrest – exacerbated by the major exposure they had as a couple, but featuring the woman. They married in his Moon-Jupiter *Dasha/Bhukti* (Dec. 18, 1980 to April 18, 1982), but their courtship began during his Moon-Rahu *Dasha/Bhukti*, a less favorable sub-period. Whatever happens during the **Moon-Rahu** period may well explain everything before and after it, as it indicates the emotional backdrop for the choice of mate. For Prince Charles, his Moon-Rahu *Dasha/Bhukti* ran for an 18-month period: June 18, 1979 to Dec. 17, 1980. On Aug. 27, 1979, Lord Louis Mountbatten was killed by an explosion on his boat. He was Charles's beloved uncle and a favorite relative. Charles was devastated by the loss. In his more vulnerable state he was under mounting pressure to find a bride.

Marriage proposal: In the summer of 1980 he became reacquainted with Lady Diana Spencer, whom he had met in 1977. A six-month courtship began, followed by his marriage proposal on Feb. 3, 1981 – a (Mars-ruled) Tuesday evening during a private dinner for two at Buckingham Palace. It was the night before a New Moon (solar) eclipse at 22:46 Capricorn, thus a very weak Moon. Four planets were in Capricorn in his 7th house of marriage: Sun, Moon, Venus, and destabilizing Ketu. This is a crowded 7th house, which would have been helped by an aspect from tr. Jupiter. But Jupiter was in a *Graha Yuddha* (planetary war) with Saturn in Virgo, in which dutiful Saturn won. Therefore, Jupiter's blessing and protection was greatly diminished. Moon was a key *nakshatra* lord, ruling over four planets. The excellent *Nadi yoga* between tr. Sun and Moon would have helped more, if only other factors supported it. The Jupiter-Saturn conjunctions in 1980-1981 were in the 6th house of conflict from the natal Moon of Prince Charles. Combined with the individual patterns of destiny, these were the prevailing collective influences that set the stage for what was to come. Some 14 years later, when asked if Camilla Parker-Bowles was a factor in the breakdown of her marriage, Diana said:

> "Well there were three of us in this marriage, so it was a bit crowded."
> (BBC 1 Panorama interview with Martin Bashir, broadcast in Nov. 1995.)

The official announcement of their engagement was again on a Tuesday: Feb. 24, 1981 at 11 AM, from Buckingham Palace. The royal wedding took place five months later at St. Paul's Cathedral, London on July 29, 1981, with vows completed at 11:17:30 AM local time. The first public signs of marital strain came in 1987, toward the end of his **Moon *Dasha*.** But the years of the couple's greatest public and private battles spanned his 7-year **Mars *Dasha*** (from Jan. 1988 to Jan. 1995). The official separation was announced Dec. 10, 1992, leading up to the divorce, Aug. 28, 1996.

Unafraid to defy royal tradition: In many ways, **Ketu** is the secret to the timing here. Ketu is the ***Jiva*** (life-giver) of this chart. This is because Ascendant lord Moon is in Ketu-ruled *nakshatra*, *Ashwini*. Several planets reside in Ketu-ruled *nakshatras*: Moon, Jupiter, Saturn, and Rahu, indicating that Ketu has ultimate control over the affairs of these four planets and the houses they rule (1, 6, 7, 8, 9). Prince Charles's love of pastoral settings comes from this, as well as 12th lord Mercury in the 4th house. Ketu is the foreigner and has an introspective, introverted energy which favors spiritual practices and study. Its condition in the birth chart and *Navamsha* chart will reveal in what way. Ketu is placed in the 4th house in both charts with numerous planetary influences; thus it remains committed to the family but is not afraid to defy royal tradition in order to experience his true heart and retain his privacy and relative freedom of choice. Venus, lord of the 4th house is in a mutual exchange with Mercury. Being in the 12th house from the 4th house and in exchange with the12th lord, the quiet or even concealed nature of the domestic life becomes clearer, further qualifying Ketu. Prince Charles runs the following *Vimshottari Dasha* sequence in his lifetime:

- the 10-year **Moon *Dasha***, from Jan. 1978
- the 7-year **Mars *Dasha*** from Jan. 1988
- the 18-year **Rahu *Dasha***, from Jan. 1995
- the 16-year **Jupiter *Dasha*** from Jan. 2013
- the 19-year **Saturn *Dasha*** from Jan. 2029

Except for Mars, each of these planets is located in a Ketu-ruled *nakshatra*. This is a total of 63 years from mid-Jan. 1978, and 53 consecutive years from Jan. 1995 with Ketu influence.

The start of his Jupiter *Dasha* in Jan. 2013 could bring a change in the British monarchy, but as always with Charles, there is conflict and resistance to doing what is expected. *Dasha* lord Jupiter in the birth chart is in the 6th house of conflict in *Mula nakshatra*, ruled by Niritti, goddess of destruction. Niritti's influence could contribute to breaking the line of succession.

The children: For children, we examine the lord of the 5th house (Mars), as well as Jupiter, *karaka* for children and abundance.[184] Since **Mars** is in the 5th house of children, the two young princes would be key factors in why the divorce was not likely to occur during **Mars *Dasha***, but only the marital separation. The condition of Mars in the chart sets the stage for some of the drama of the 7-year Mars *Dasha*: Jan. 17, 1988 to Jan. 17, 1995.

This Mars can cause a fair amount of conflict. It is *gandanta* at 27:48 Scorpio in the 5th house of children and of love affairs. In the *Navamsha* chart, Mars is 7th lord and situated in the 6th house of conflict, aspecting the Ascendant. Reading from Mars as sub-Ascendant in the birth chart – which is applicable during Mars *Dasha* – Moon and Rahu are in the 6th house. Though strong in its own sign, Mars is situated in its enemy Mercury's *nakshatra*: *Jyestha*. This is considered the most difficult *gandanta* at the end of Scorpio. (For Maria Montessori, Saturn was at 29:53 Scorpio in the 5th house, also problematic. She suffered the absence of her son for the first 15 years of his life, all of it during her Saturn *Dasha*. Something is hidden or not absolutely clear, though it turned out well for her, especially after Saturn *Dasha*.) For Prince Charles, his two children were perhaps the major factor holding the marriage together for most of its duration, especially after Sept. 1984 and the birth of their second and last child. Diana did not want to divorce, but Charles pressed for a separation and divorce. Even as duty runs strong with Saturn aspecting the natal Sun of Prince Charles, his Mars as 10th lord of status also chafes against the at times suffocating constraints of royal duty.

---

[184] For children and fertility, Saturn could also be examined in this chart as Jaimini *Putra karaka*, the 5th *karaka* in descending order of longitudinal degrees.

The royal separation: The increasingly public battles between Charles and Diana occurred during the Prince's **Mars *Dasha*,** from Jan. 1988. They often involved their sons and concerned who was the more fit, loving and caring parent. As Charles could be absent-minded and otherwise occupied at times, his good press rarely approached Diana's press, especially as the story of "the King's mistress" became public knowledge. Saturn is the planet of Truth, and can bring reconciliation. But tr. Saturn in Capricorn in Charles's 7th house of marriage unearthed truths too bitter to be reconciled. (This was where the joint agreement to marry had begun on Feb. 3, 1981, with four transiting planets in Capricorn: Sun, Moon, Venus and Ketu. And now Saturn was coming around to test it.) Tr. Saturn in Capricon ran from Dec. 15, 1990 to March 6, 1993, and coincided with the most openly contentious years between the royal couple. This was during Charles's **Mars-Mercury** *Dasha/Bhukti* (July 19, 1991 to July 15, 1992) and **Mars-Ketu** *Dasha/Bhukti* (July 15, 1992 to Dec. 11, 1992).[185] Natal Mars is located in the Mercury-owned *nakshatra Jyestha*, as mentioned. Mercury is a planetary enemy to Mars and they are placed in 12-2 houses to each other (an uneasy relationship) in both birth and *Navamsha* charts. As 12th house lord placed in the 4th house, it is more difficult to achieve peace of mind, especially in this case as Mercury has many emotional influences, being close to natal Ketu and Sun opposite Moon-Rahu.

The problematic nature of Mars in the *gandanta* degree of Scorpio symbolizes the likelihood of Charles' love life being in some way clandestine. On another level it gives an interest in the occult. As a planet of aggression and action, Mars at the tail end of watery Scorpio comes out of the deep to deliver its stinging attacks and/or revelations. These were made more public due to Moon-Rahu in the visible 10th house. Prince Charles has a reputation for great wit and also biting sarcasm and Diana became equally stinging. The biggest worldwide exposure came with Andrew Morton's book, *Diana, Her True Story*, published June 7, 1992, apparently with Diana's full cooperation. Morton was considered a credible and respected journalist. His revelations were mostly sympathetic to Diana and damning to Charles, in terms of his long-term devotion to Camilla throughout his marriage to Diana. The royal marriage was "irreparable," as Charles said, and some said the damage to his image was also "irreparable."

The official separation of the royal couple on Dec. 10, 1992 occurred within hours of a **Total Solar eclipse** at 11:40 PM London time on Dec. 9, 1992, at **24:24 Taurus,** opposite the Prince's natal Mars at 27:48 Scorpio in the 5th house of children, impacting them perhaps more than anyone. It also occurred on the last day of Prince Charles's **Mars-Ketu** period and one day prior to the start of his **Mars-Venus-Venus** period. The final divorce decree (Aug. 28, 1996) occurred in his **Rahu-Rahu-Ketu** period and Princess Diana's death occurred Aug. 31, 1997 in his **Rahu-Rahu-Mars** period. At these pivotal turning points either Venus (planet of love) enters as a sub-period, or Rahu and Ketu (the eclipse axis). When contacting his Moon, they are associated with the women in his life, both the Queen and the Princess. The Princess lost her royal title with the divorce, and the Queen was impatient to restore order in the monarchy after so many years of rumor and scandal surrounding the royal marriage.

---

[185] Their official separation was announced in Parliament on Dec. 10, 1992, causing upheaval for the monarchy, as there was no modern precedent for a divorced or separated monarch. "...The royal couple did not intend to divorce and would continue to carry on their royal and constitutional duties separately as the future King and Queen of England." William E. Schmidt, "Charles and Diana are separating 'amicably,'" *The New York Times*, Dec. 10, 1992.

Two weddings: Prince Charles was married both times close to an eclipse. The second time he narrowly missed being married on the actual eclipse day. Only due to the funeral of Pope John Paul II did he miss that fate. For his marriage to Princess Diana, we would like to assume no reputable astrologer was involved in selecting this date and time, as it is so unfavorable. The Moon was waning, and a **Total Solar eclipse** at 14:15 Cancer would occur shortly (**July 31, 1981**), within two degrees of Prince Charles's Ascendant.[186] In addition, Moon, Mars, and Mercury were all transiting Gemini, Moon and Mars in *Ardra nakshatra*, symbolized by the tear drop and suffering, and ruled by Rudra, god of storms. A very weak Moon is not recommended for the start of important ventures, such as marriage. Ideally a wedding Moon should not have contact to malefic planets, such as Mars. However, the fact that the wedding took place within six days of the **Jupiter-Saturn conjunction in Virgo** (the third and last of the sequence in Virgo, on July 23, 1981), this gave the event added power and symbolism in the new Information age. The conjunction occurred at 11:20 Virgo, in *Hasta nakshatra*. His Venus is at 23:14 Virgo and Diana's Ascendant is at 23:52 Virgo, using the 2:00 PM birth time given by her mother.

In the wedding chart (July 29, 1981) Mercury and Moon in the same house and sign, especially a dual sign, is not recommended. The heart is less expressive with the intellect trying to enter in. The Moon, in some ways the bride, is pushed around by too many influences to find steadiness in the marriage. There is also intense media coverage, as Diana proved so popular from the start. Some overly rational component is involved, perhaps in trying to rein in five planets in dual signs: Moon, Mars, Mercury, Jupiter, and Saturn. With Jupiter and Saturn close to their conjunction degree in the Virgo Ascendant, this looks more like a publicity event for a corporate merger. The day was a Wednesday, ruled by Mercury, planet of communications, and the wedding was watched by over 750 million people. The subsequent divorce from Princess Diana was announced Aug. 28, 1996, also a Wednesday. After her death a year later, Prince Charles became the single parent to the two young Princes and as such received some renewed public sympathy.

The second wedding to Camilla Parker-Bowles, later Duchess of Cornwall, occurred on April 9, 2005, first at a civil ceremony begun at 12:30 PM BST at Guildhall, followed by a prayer service at 2:30 PM at St. George's Chapel at Windsor Castle. It had been scheduled for the previous day, the day of a **Solar eclipse (at 25:10 Pisces)** opposite his natal Venus, but Prince Charles was required to attend the funeral of Pope John Paul II on April 8th in Rome.[187] (Pisces is a sign associated with religion and also with royalty, a water sign, amplified by Jupiter's rulership over Pisces.) The papal funeral was the largest gathering of statesmen in history.[188] Some of the dignitaries later attended

---

[186] The royal wedding took place at St. Paul's Cathedral, London on July 29, 1981, with vows completed at 11:17:30 AM local time. With a Solar eclipse, the Moon temporarily blocks the light of the Sun. As the Sun symbolizes the leader or the king (the future king in this case), we note that a total solar eclipse blocks his light near to the time of his wedding to Diana. Their relationship would cause his popularity to plummet, while vastly increasing hers, especially with her sudden death in late Aug. 1997.

[187] The eclipse on April 8, 2005 was Annular-Total. That is, the eclipse was total for part of its path and annular for the rest of it. Annular is not total as it is too far from the earth for the apex of its shadow to touch the Earth's surface. The two-tiered type of eclipse is more rare, and echoes the two events originally scheduled for the same day: one royal, one religious.

[188] In another Pisces pairing of royal and religious events, the next British royal wedding – that of Prince William (son of Prince Charles) to Kate Middleton, a commoner, took place Friday, April 29, 2011, the day following the exhuming of the coffin of Pope John Paul II prior to his official beatification ceremony in Rome (on the way to becoming a saint), held on Sunday, May 1, 2011. Five planets were transiting Pisces: Moon, Venus, Mercury, Mars, and Jupiter. For more material on historical perspectives for the intertwining of royal and religious themes, see Chapter 1.

the installation of Pope Benedict XVI on April 24, 2005, the day of another Solar eclipse at 10:23 Libra, this one closely opposite Prince Charles's natal Moon at 7:17 Aries and natal Rahu-Ketu at 11:49 Aries/Libra. Ideally one should avoid scheduling a wedding near to an eclipse, especially one impacting natal Moon or Ascendant. Destiny drew this twice for him. The saving grace for this wedding date is although it was very close to an eclipse, the eclipse was on *Chaitra Shukla Pratipada*, the Vedic New Year, and considered generally very fortunate. (This occurs every year with the New Moon in Pisces prior to the transiting Sun entering sidereal Aries.)

The royal wedding of Charles and Camilla took place on a Saturday, ruled by the planet of duty and delays. For the civil ceremony, Cancer was rising, with transiting Saturn at 29 degrees of Gemini aspecting tr. Sun, Mercury, Venus and Rahu in Pisces (with Sun at 26, Venus at 28 and Rahu at 29 degrees.). Seven out of nine planets were in dual signs. Venus, planet of love, was near the maximum degree of exaltation at 27 degrees Pisces, and so is extremely strong in this chart, though also combust the Sun and hemmed in between malefics Sun and Rahu. Indeed the couple's love for each other endured many trials and tribulations over several previous decades, once they discovered they married the wrong people. Four planets in Pisces were in the favorable 9th house (of religious ceremonies and legal matters). Pisces conjoins both religious and royal matters, and the Church of England had to change its rules in April 2002 so that Prince Charles could marry Camilla. Contrition, sacrifice, love, and duty – all were themes of the day, though the union of Charles and Camilla had no doubt destroyed their previous marriages. Tr. Moon at the top of the sky was in the *nakshatra* of *Ashwini*, "owning horses." About Camilla, always a very private person, one friend had said, "She doesn't want to be queen, she wants to go hunting."

REVIEW OF CANCER ASCENDANT CHARTS:

Give the astrological reasoning in each case.

1. What are the challenges for the Cancer Ascendant person in terms of emotional and mental stability? Why would they tend to look to the traditions of family (mother in particular) and community to establish stability and emotional support?
2. What explains Marilyn Monroe's tendency to lack those early connections? How could this cause emotional and mental turbulence? Hint: See her *Navamsha* Moon, and observe its relative isolation in the 8th house.
3. How do Monroe's 8th house positions also account for her charisma? Why was that likely also to change or cut short her career? Hint: Study both her 10th lord and her *Dasha* sequence.
4. Why would children be such a central issue for the Cancer Ascendant destiny? Explain Maria Montessori's dramatic story with her only son; Monroe's tendency to be unsuccessful in pregnancy; the powerful destinies of the children of Indira Gandhi and Prince Charles. Hint: Observe the position of Mars, lord of the 5th house of children, as well as Jupiter, *karaka* (significator) of children.

5. Why are national issues and a sense of patriotism involved with Cancer themes? Consider the charts of Indira Gandhi and Prince Charles, both irrevocably linked with their family ties and obligations. What planet can bestow a heavier family duty?
6. Why does the Cancer Ascendant tend to bring so much pressure and/or seriousness with marriage issues? Hint: Note the 7th house lord and its relationship with Moon, and how it created intense difficulties for Indira Gandhi.
7. Why is education or the lack of it such a big theme for Cancer Ascendant? What planets indicate this life theme for Maria Montessori? Hint: Note the interaction of 1st and 4th lords (Moon and Venus) in both birth and *Navamsha* charts.
8. How did Maria Montessori keep her private life more hidden from the public? Hint: Note how many planets are below the horizon in her chart. Even the 12th house, though above the horizon, is considered more hidden astrologically.
9. How does Prince Charles's Moon indicate he would be more visible, as would the key women in his life? How could these women overshadow him and be magnified in the public realm? Hint: Note how Rahu eclipses the Moon.
10. What factors in the charts of both Marilyn Monroe and Indira Gandhi brought them physical danger? How were their respective *Dasha* sequences pivotal in describing these eventualities?
11. Why would the Mars *Dasha* of Prince Charles prove so unfortunate for his marriage and children? Even so, why was divorce unlikely during his Mars *Dasha*? Hint: Note the degree and house position of Mars, in both birth and *Navamsha* charts.
12. Why would it be challenging for Prince Charles to make good decisions during his Sun *Dasha* (1972-1978)? Why would that also create adverse consequences in his love life over the next several decades? Hint: Note the condition of *Dasha* lord Sun, also that of its sign and *nakshatra* lord, Venus and Jupiter, respectively.

CHAPTER 7

Leo Ascendant

Introduction

Leo is ***Simha*** (the lion) in Sanskrit. After Aries, it is the second *Dharma* sign in the zodiac. Leo is fire and a fixed sign, leading towards consolidation. Vedic astrology associates all three fire signs (Aries, Leo, and Sagittarius) with the ***Kshatriya***, or warrior class, which includes political leaders. *Kshatriyas* have royal status, as do the Sun and the Moon, the two great luminaries and symbols of the King and the Queen, respectively. This includes the signs over which they rule: the Sun rules Leo and the Moon rules Cancer. Since each luminary rules only one sign, their respective Ascendants are important for that reason, among others. Of the three fire signs all associated with royalty, Leo is considered the most royal sign, especially with *Magha nakshatra* at the start of Leo. Its symbol is a throne room.

The Sun, the Moon and Venus are the three brightest objects in the sky. As such, all three planets are associated with vision and eyesight. When they are all strong and unafflicted in the chart, these planets give good eyesight. If all three are afflicted, there is poor eyesight or even blindness. The Sun symbolizes domination, authority (also the government), will power, leadership, self-confidence, physical vitality, and the physical heart. As a *karaka* (significator) for the Ascendant in general as well as lord of Leo Ascendant in particular, the Sun is also the *karaka* for *Ahamkara* – called the "ego-principle" by Yogananda. Others say it encompasses the ego, but is neither entirely ego nor personality. (See also Glossary)

As both Ascendant lord and *karaka* for physical vitality, its position in the chart is doubly important for determining physical health. As with the Moon, the Sun has 108 Sanskrit names. The most common one is *Surya* (the luminous); also *Aditya* (son of the primordial vastness; or the source) and *Savitr* (procreator or nourisher). The Sun is the immortal soul, or that which inspires us from a deep level. We look to the Sun for constancy on a personal level, and likewise the Leo persona prefers the personal touch and everything related back to the Self. Here lies

Leo's greatest strength *and* weakness. In seeking the Self, the ego easily becomes inflated. Just as the lion's image – and roar – is invoked to lead us with pride and self-confidence, the need for respect from others can also be excessive.

The Leo Ascendant person often has an innate sense of being special. They feel they have inherited a unique position in life, and thus may be driven to seek a leadership position or to seek recognition for their talents, which often lie in the dramatic or visual arts, music, politics or education. They are often associated with the performing arts, because they can be seen and appreciated. While Cancer shows the innate intelligence, Leo shows the creative intelligence. (This is because the outgrowth of the 4th house or sign of Cancer can be seen in the 2nd house from it – the 5th house or sign of Leo.) The Leo Ascendant person is also attracted to wealth, and may draw it to themselves – depending on the rest of the chart, especially on the strength of the Sun and Moon. Marriage may bring wealth, and may be very important in the life, even more so if key planets reside in either *Purva Phalguni* or *Uttara Phalguni nakshatras*, from 13:20 Leo to 30:00 Leo.

With a fire sign Ascendant, once again we have the natural sequence of the houses in terms of the four Vedic noble aims: *Dharma* – fire; *Artha* – earth; *Kama* – air; and *Moksha* – water. When the natural *Dharma* element is on the Ascendant, there is a strong yearning for a sense of purpose in the life. The fire Ascendant person burns with the intensity of this need for a purpose. With Leo Ascendant the personal story of the individual becomes dominant, even if the ultimate aim of the life is a spiritual one. The search for fulfillment must be both personally satisfying and engender respect and a good reputation. As the Sun rules over the father and spiritual authority, the Leo Ascendant person usually has a strong theme with his or her father, and at times a spiritual teacher or guru. The house placement of the natal Sun will indicate how this theme plays out, along with aspects to the Sun in the birth chart and in the *Navamsha* chart. And in general, the house placement of the Sun will show where the person puts a lot of his or her energy. The Sun's best house position is the 10th house, where it has *Digbala*. There it gives off the most light and most heat. From the 10th house the Sun can shine down on everything, and maintain the greatest amount of control. When in the 10th house from natal Moon, the Sun has a similar effect.

The Sun is also a classical malefic planet, as it can burn and cause separation due to the very power it possesses: its light, heat, and dominance. In human terms, this may be the heat of the Sun's power and ambition. When planets are too close to the Sun, they are said to be **combust**. This means their effectiveness to operate may be greatly curtailed by being burnt by the Sun's rays. In some cases, especially with Mercury or Venus, when they are within one degree of the Sun, the qualities of that planet suffer somewhat through burnout. But they may also receive extra brilliance, with their powers magnified. This is further described by the sign and house location of the combust planet and Sun. Note also the houses over which the combust planet rules and its role as *karaka*. Since Mercury always travels within 28 degrees of the Sun, its combustion is not considered as serious; but if it is at least 10 degrees away from the Sun in the same sign, this condition produces a *Budha Aditya yoga*, a *yoga* of cleverness. If Mercury is unafflicted and at least 20 degrees away from the Sun in the same sign, the condition is greatly enhanced. The **specific range of combustion** is related to how far away from the Sun the planet must be in order to be visible from the earth. Five degrees either side of the Sun is a very strong combustion, especially if the planet is 5 degrees behind the Sun in celestial longitude. Moon, Rahu and Ketu do not qualify for combustion. For planets to be combust, the widest range of orbs used on either side of the Sun is the following: Mars: 17 degrees; Mercury: 13 degrees; Jupiter: 11 degrees; Venus: 9 degrees; and Saturn: 15 degrees.

With Leo Ascendant, we are one step further along in the *Dharmic* sequence, once Aries has initiated the action. Being a fixed fire sign and the natural 5th house, Leo relates to the consolidation of the firepower begun in Aries. We have already discussed how Leo (and or the 5th house) relates to the fruitfulness of the mind and thus one's creative intelligence and acquired education. It is also one's past life credit, or *Poorvapunya*. If creative intelligence and artistic creativity are featured in the 5th house and sign, biological creativity is there as well, and thus – children. If malefics are placed in the 5th house or aspect the 5th house from the Ascendant (or from the Moon), one or more of the various levels of the 5th house may become afflicted. For instance, it will become difficult to either bear children or keep them – even if one becomes pregnant. This may be due to miscarriage, abortion, or stillbirth. Also, one may lose a child due to a custody battle, or some situation where the child is removed from the parent. If the 5th house is well disposed and aspected, the affairs of the 5th house go smoothly. Further confirmation resides in the *Navamsha* chart. If the *karaka* for children and fertility (Jupiter) resides in the 5th house, this does not bode well for pregnancy, especially for a woman, but it can also apply to a man. However, it is good for creative intelligence and the ability to learn and/or be a scholar or poet. Jupiter's aspects from its own house to the Ascendant and the 9th house are also beneficial to the native in terms of their own physical health and safety. This position also brings the *Bhagya yoga*, a yoga of great good fortune, when a strong benefic aspects both the Ascendant and the 9th house.

When there is a fixed sign Ascendant with at least several planets also in fixed signs, this is called a *Musala yoga*. (A stricter definition demands all planets be in fixed signs.) With a *Musala yoga* the nature of fixity is emphasized in the life. On the positive side, loyalty, continuity, and steadfastness of purpose are featured. On the negative side, the person can be stubborn, and fixity of purpose or of ideas will veer towards the judgmental and biased, not being able to see another way of doing things or of relating to people and events. With Leo, there is always the possibility the ego can become too attached to pride. Royalty cannot be snubbed, but when it feels fulfilled, there is great magnanimity and often an illustrious appearance.

Nakshatras: The *nakshatras* contained within Leo are as follows: ***Magha***: 00 – 13:20 Leo (ruled by Ketu); ***Purva Phalguni***: 13:20 – 26:40 Leo (ruled by Venus); and ***Uttara Phalguni***: 26:40 – 30:00 Leo (ruled by the Sun). (*Uttara Phalguni nakshatra* continues for 10 degrees into Virgo.) The Leo Ascendant, or any planet situated in one of these three *nakshatras* is colored by being: 1) in Leo, ruled by the Sun; 2) in one of these three *nakshatras*, ruled by Ketu, Venus, and Sun, respectively; and 3) by aspects to the Ascendant and Asc. lord Sun.

Jupiter is the best planet for Leo Ascendant and is the *Nadi yogakaraka*.

Gandanta: The first 3:20 of fire signs are *gandanta* and thus more vulnerable, though not as vulnerable and unpredictable as the last 3:20 of water signs. Ketu's house and sign position in the chart will be important to determine the outcome of any planets situated in *Magha nakshatra*.

Chart 20: Paramahansa Yogananda

Birth data: Thursday, Jan. 5, 1893, 20:38 LMT (Zone -05:21), Gorakhpur, India, Long. 83E 22, Lat. 26N45, *Lahiri ayanamsha* -22:21:38. Class A data, from memory. (*Mercury Hour*, July 1976 quotes his ashram, The Self-Realization Fellowship Institute.) Ascendant: 6:30 Leo. (Vedic astrologer K. N. Rao uses an Ascendant of 3:59 Leo, and gives a time of 20:14 LMT. His time zone is based on 5h 21m rather than the LMT I use: 5h 33m 28s.[189] Rao's source: the late astronomer-astrologer Nirmala Das (Calcutta), who was in close touch with the late N. C. Lahiri of the famous ayanamsha. K. N. Rao: "Nirmala Das was very particular in collecting his data, and invariably got it from the family sources. I trusted it always....")

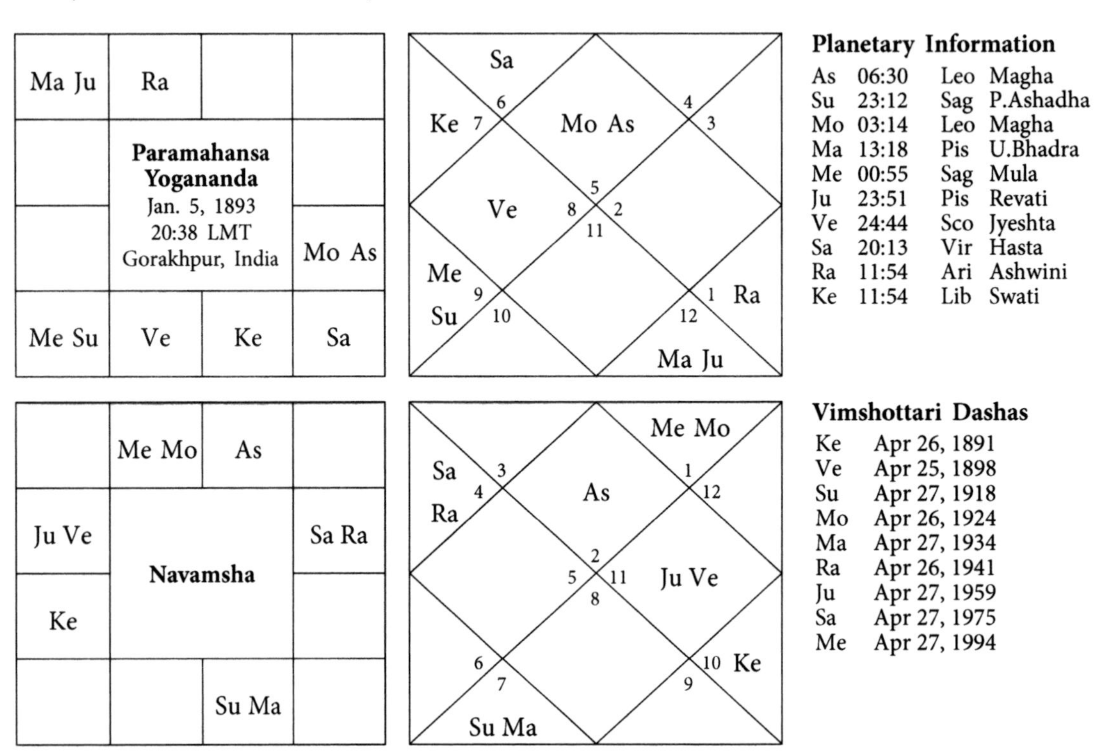

Planetary Information

As	06:30	Leo	Magha
Su	23:12	Sag	P.Ashadha
Mo	03:14	Leo	Magha
Ma	13:18	Pis	U.Bhadra
Me	00:55	Sag	Mula
Ju	23:51	Pis	Revati
Ve	24:44	Sco	Jyeshta
Sa	20:13	Vir	Hasta
Ra	11:54	Ari	Ashwini
Ke	11:54	Lib	Swati

Vimshottari Dashas

Ke	Apr 26, 1891
Ve	Apr 25, 1898
Su	Apr 27, 1918
Mo	Apr 26, 1924
Ma	Apr 27, 1934
Ra	Apr 26, 1941
Ju	Apr 27, 1959
Sa	Apr 27, 1975
Me	Apr 27, 1994

Biographical summary: Born Mukunda Lal Ghosh, he was the second son and fourth of eight children in a devout and prosperous Bengali family. His father, Bhagabati Charan Ghosh (1853-1942) was a high ranking officer in the Bengal Nagpur Railways. His mother, Gyana Prabha Ghosh (1868-1904), was a disciple of Lahiri Mahashaya, as was her husband. Throughout his childhood Mukunda demonstrated unusual intuitive gifts and deep spiritual yearnings. From the age of twelve he tried several times to run away to the Himalayas in search of the Divine. His mother encouraged him in these pursuits, and he was devastated by her death when he was only age eleven. After her death, she left him a written account of some extraordinary events surrounding his birth. Her guru Lahiri Mahashaya had observed the boy as a baby and in the womb. He told her:

[189] See Glossary for LMT, or Local Mean Time. The variable here depends not only on a difference in clock time, but on whether Standard time or LMT was used in the locale. My sources use LMT.

> "Little mother, thy son will be a yogi. As a spiritual engine, he will carry many souls to God's kingdom."

This and other revelatory messages were to be conveyed to young Mukunda one year after her death. His older brother Ananta, who was at their mother's deathbed, had delayed for 14 months. Then he finally relayed these prophetic messages, along with a powerful silver amulet for Mukunda, initially given to his mother. These events further catalyzed his journey towards seeking the Divine, an illumined master and a monastic life.

Having often envisioned his spiritual teacher in his mind, Mukunda finally met Shri Yukteshwar in 1910 at age 17 and studied with him for the next ten years. He learned *Kriya yoga* and meditation, and absorbed much about ancient Indian philosophy. In 1915, he took formal vows as a monk of India's venerable monastic Swami Order and was given the name Yogananda, meaning bliss (*ananda*) through divine union (*yoga*). Yogananda quotes his master on *Kriya Yoga*:

> "[It is] an instrument through which human evolution can be quickened.... The ancient yogis discovered that the secret of cosmic consciousness is intimately linked with breath mastery. This is India's unique and deathless contribution to the world's treasury of knowledge. The life force, which is ordinarily absorbed in maintaining heart action, must be freed for higher activities by a method of calming and stilling the ceaseless demands of the breath."
>
> **Paramahansa Yogananda**, *Autobiography of a Yogi*, pp. 278-279.

At Shri Yukteshwar's insistence, Yogananda finished his formal education at Calcutta University, while studying spiritual discipline in the Indian tradition with his teacher. Yogananda earned his B.A. degree from Calcutta University in June 1915. This was a surprise to many, as he had spent most of the intervening years in the presence of his master rather than in the classroom or studying. Nevertheless, through divine intervention and with his naturally bright and attentive mind he passed all his exams.

Various Swamis and yogis had predicted that Yogananda would travel to the West to spread the teachings of *Kriya yoga* and promote greater spiritual understanding between the East and the West. He himself had a detailed inner vision of lecturing before Westerners. Within two days of the vision he was invited to be India's delegate to an International Congress of Religious Liberals, to be held in Boston in early Oct. 1920 under the auspices of the American Unitarian Association. Yogananda prepared himself and was fortunate to receive a generous check from his father, enabling him to stay for several years. He set sail in August 1920 on a two-month ocean voyage to America. There were many obstacles in obtaining a passport at this time, and his passenger ship was the first to sail this route after World War I ended. The result was that not only was his maiden lecture well received, but it led to his extensive lecturing and teaching throughout the U.S. from 1920 for the next 32 years of his life.

Self Realization Fellowship was formed to spread his many teachings, which included the underlying unity of all the world's great religions. He introduced the yogic philosophy of India to the West, using sacred Vedic texts such as the *Bhagavad Gita,* the *Upanishads,* and others to raise the consciousness of humanity. He wanted to "present Spirit as a central Fact of man's existence..." (*Ibid.*, p. 206) The international headquarters of SRF was founded in Los Angeles in late 1925 and property in Encinitas, California was donated and built in 1935-1936. In late Dec. 1935, while traveling in India, Yogananda received from his master Shri Yukteshwar the monastic title Paramahansa. (In Sanskrit, *Parama* is highest, and *hansa* or *hamsa* is the swan, the highest manifestation of Jupiter, planet of knowledge and wisdom. The white swan is the vehicle of Brahma, the creator.) Later SRF texts sometimes refer to Yogananda's arrival date in the U.S. (Sept. 19, 1920) as the founding date of the organization, but in fact that was his arrival in the West.

Paramahansa Yogananda became a spiritual master who elevated the spiritual level of society. Through his book, *Autobiography of a Yogi* (1946), he was able to reach millions more people. This book has been translated into 25 languages, and millions of copies have been sold. It is considered a spiritual classic of the 20th century, and remains a perennial best seller. Included in the book are accounts of his childhood, his ten years with Shri Yukteshwar, his years in the West, and his meetings with Mahatma Gandhi, Rabindranath Tagore, Luther Burbank, and the Catholic stigmatist Therese Neumann, among others. Yogananda was the first Indian Swami ever to be an official guest of a U.S. President, when he was received at the White House by President Calvin Coolidge on Jan. 27, 1927.

Over his lifetime, Paramahansa Yogananda published 11 books, some of them containing his poetry. During the last 20 years of his life he also set about translating the *Bhagavad Gita*. Through the efforts of his devotees, this great project, with commentary was finally published in 1995, 43 years after his death. Yogananda's death on March 7, 1952 occurred soon after he gave a banquet speech in Los Angeles in honor of the Ambassador from India. Considered *Maha Samadhi*, the conscious exit of a yogi from his physical body, Yogananda's body remained perfectly preserved for the next 20 days, with no signs of physical deterioration. This extraordinary event was witnessed by many and included in a notarized statement by the local mortuary:

> "The absence of any visual signs of decay in the dead body of Paramahansa Yogananda offers the most extraordinary case in our experience.... No physical disintegration was visible in his body even twenty days after death."
>
> *Ibid.*, p. 575.

A Higher Being: Those with a Leo Ascendant or with a strong focus of planets in Leo tend to be performers – because they want to be seen, or at the very least respected. We start with a Higher Being whose birth chart has a Leo Ascendant, in part to illustrate that in the more evolved sense what wants to be seen is the Self expressed as the inner Light of the soul. Yogananda's naming of his organization "Self Realization Fellowship" is already a tribute to the more enlightened side of Leo and its central question: Who am I? Where did I come from? And where am I going? There is an interesting paradox when *Magha nakshatra* rises or is prominent in planetary positions, especially Sun, Moon, or Ascendant lord. The major *shakti* or energy of *Magha* is to be out of the body. That is perhaps because the Self understands it *is* the Spirit and *has* a body. The spiritual side of *Magha* is the desire to return to the world of the spirit and the deathless essence. The negative side is when the desire is simply to avoid the rigors of earthly life and the periodic discomforts of being in the physical body. *Magha* gives much intuitive energy and a strong desire to return to the world of the ancestors, as this *nakshatra* is ruled by the *Pitris* – the ancestors. Further, when *Magha* is prominent, especially Moon or Ascendant, those born under this star seek a high reputation for themselves, and this star can bestow fame that endures for generations. The symbol of *Magha* is the throne room, and literally Yogananda's living quarters at the Los Angeles "Mother Center" of SRF have been kept as a shrine since his death in 1952.

The Moon's position here in the Leo Ascendant gives an immediate *Moksha* influence due to its ownership of the 12th house. It also gave him his luminous Moon-like face. It is posited in the 12th house of the *Navamsha*, confirming a strong *Moksha* influence in the life. Even so, the person can be known for their personal beauty or personal life. A Leo Ascendant person can be more egotistical than many others, especially if the Sun is strong in the chart. But with *Magha* influence the native is more often challenged to surrender the ego on an ongoing basis. For a yogi, that process is accelerated from birth.

Yogananda has both Ascendant and Moon in *Magha nakshatra*. And though a great friend to the Sun, Moon's role as 12th lord – together with the *shakti* of *Magha* to leave the body – describes a person who yearns for a life path that focuses on a spiritual reality more than a physical one. When the spiritual task is done, his life in the physical body is over. This was clear from his birth and childhood onward and was predicted to his mother (also the Moon) when he was in the womb. Her keen perception of her son led him to understand the larger picture of his destiny. The Moon's *Moksha* influence in this case gives spiritual yearning that can be an ongoing emotional drive. It can bring early loss in the life and also residence abroad, since 12th house rules over both matters. Confirming all these points, *Navamsha* Moon is in the 12th house. Thus the Moon brings these desires and this kind of destiny, especially during the **Moon**'s major or minor *Dasha* periods or of those related to the 4th house. When his mother died (age 36), Yogananda was 11 years old and was either in his Venus-Moon *Dasha/Bhukti* (August 26, 1902 to April 26, 1904) or Venus-Mars *Dasha/Bhukti* (April 26, 1904 to June 26, 1905). His natal Venus is placed in the 4th house of the mother, and 4th lord Mars is in his 8th house of loss and suffering, the usual route to *Moksha*.

Path of the celibate: This path is discernible due to several astrological afflictions to the 7th house of partnership, to the 7th lord, and to Venus, *karaka* of love relations. We have all three of these here, as: **1)** Moon, 12th lord (of loss and of *Moksha*) aspects the 7th house; **2)** Saturn, lord of the 7th house is aspected by 8th lord Jupiter and malefic Mars; **3)** Venus in Scorpio is closely aspected by Saturn and its sign lord Mars is placed in the 8th house (of loss and of *Moksha*). No doubt due to the three obvious afflictions to the 7th house or its lord, the family astrologer had predicted three marriages, and being widowed twice. The family tried to get young Mukunda married, but he refused to go along with the customary plan and with the astrologer's predictions. He later wrote that he burned these prophecies, as he did not wish to be controlled by them. He intended to go the path of the celibate and the yogi, which required a far greater discipline.

Yogananda took his monastic vows in 1915 during his **Venus-Mercury** *Dasha/Bhukti*. Natal Venus is placed in *Jyeshta nakshatra* ruled by Mercury, an asexual planet in a fire sign in both birth chart and *Navamsha* charts. The lord of the 5th house of children (Jupiter) is situated in the 8th house, tending to deny children. In this case, he did not marry or have any children of his own, but he had many "surrogate children" through his role as teacher.

Where is the Sun shining?: From natal Moon and Leo Ascendant, the Sun is situated in Sagittarius in the 5th house of learning and creative intelligence. This is an excellent placement for the Ascendant lord, in a trinal house and in the sign of a great good friend – Jupiter, the *guru* or teacher. *(Guru* in Sanskrit means literally "the one who removes the darkness.") Jupiter in turn is placed in its own sign of Pisces, though not so easy for matters of the 5th house and occupying planets when the 5th lord goes to the 8th house. This is not ideal for a worldly life but the intuition and spiritual matters may well flourish. In fact, Yogananda was able to do a lot of fund-raising for his spirituality-based organization. (The 8th house is also a house of financial legacies.) He spent his entire life practicing what he taught, later called "Self Realization" techniques.

With the light and power of the Sun (as Ascendant lord) in the 5th house of the birth chart, there is much energy given to 5th house matters, which also include the practice of mantras. Yogananda introduced mantras and chants to the West and wrote books on the subject. Being the 2nd house from the 4th house of education, showing the fruits that come from the 4th house, the 5th house also describes the intellect and the results of one's education. This often relates to the visual arts, and some ancient texts refer to musical talents. It is also a financial house, and includes speculation and gambling. Yogananda's Sun and Mercury form a *Dhana yoga* of wealth in the 5th house of speculation (as Lords of houses 1, 2 and 11, respectively). This brought him ongoing financial wealth, initially from his father – who enabled him to make his first trip to the West in 1920. Later his philanthropic network made the Self Realization Fellowship a thriving physical and legal reality in the West, also financing his travels, lectures, and publications.

Education with divine guidance: In June 1915, during Venus-Mercury *Dasha/Bhukti* – the same period when he took his monastic vows – Yogananda finished his higher education, graduating from Calcutta University with a B.A. degree (Bachelor of Arts). ***Dasha* lord Venus** has *Digbala* (best angular house), located in the 4th house of education and higher educational degrees. Venus is in the *nakshatra* of Mercury, in turn well placed in the 5th house, showing the *nakshatra* lord's power to affect the outcome, especially during Venus-Mercury *Dasha*: Yogananda attended few classes and spent most of his time with his guru Shri Yukteshwar. Only through divine guidance and assistance was he able to pass all his university exams. A similar situation happened a few years earlier with his high school examinations, enabling him to pass those as well. The interweaving of spiritual practices and academic exams comes from 8th house contacts to the 4th and 5th houses and/or house lords. Mars and Jupiter own Houses 4 and 5, respectively, and are situated in the 8th house of *Moksha*. This can bring loss of worldly position but gain of spirituality or intuition. These factors also explain his seemingly instant fluency in the English language in 1920. At the time of his university graduation in June 1915, tr. Jupiter, planet of *Dharma* was in Pisces in his 8th house, and tr. Saturn, planet of *Karma*, was transiting Gemini, aspecting both his Leo Ascendant and natal Jupiter and Mars in Pisces in the 8th house. Therefore, the 8th house of spiritual education was more featured even at that moment of academic accomplishment when typically the major aspects are to Houses 4, 9 or 10.

Yogananda gave his maiden speech in Boston on Wed. Oct. 6, 1920 on "The Science of Religion." Prior to that, on the transatlantic journey, he was asked to give his very first informal lectures in English. He had no idea how he would accomplish this, with only a short amount of English language study. But according to witnesses of the event, after ten minutes of agonizing silence he gave an interesting talk for 45 minutes in flawless English. Inwardly, he had implored his guru to help him and he came through. Though his natal Mercury is well placed, it is in the 12th house in the *Navamsha*. There can be loss of speech unless in a foreign or spiritual (12th house) setting. Also, natal Saturn in the 2nd house of speech can bring some initial delays and setbacks. Tr. Jupiter and Saturn, classical *Dharma* and *Karma* lords, were on his Leo Ascendant the night of his maiden speech and transiting Sun in Virgo was in his 2nd house of speech, lending its vital energy to this arena.

Yogananda came into greater contact with astrology through Shri Yukteshwar, who was also a renowned *Jyotishi* (Vedic astrologer). Although Yogananda did not accept deterministic astrology, Shri Yukteshwar altered his view through his superior knowledge of the subject:

> "A child is born on that day and at that hour when the celestial rays are in mathematical harmony with the individual *Karma*. His horoscope is a challenging portrait, revealing his unalterable past and its probable future results. [Referring to the horoscope as a map, he continues:]... It is only when a traveler has reached his goal that he is justified in discarding his maps. During the journey, he takes advantage of any convenient shortcut.... There are certain mechanical features in the law of *Karma* that can be skillfully adjusted by the fingers of wisdom."
>
> *Ibid.*, pp. 188-189.

A well financed teacher and writer: Mercury rules over astrology and communications. Always transiting within 28 degrees of the Sun, it is often in the same sign. Several fortunate *yogas* are formed by natal Sun and Mercury together in the 5th house in Sagittarius. First, there is a *Budha Aditya yoga*, the *yoga* of cleverness – when Sun and Mercury are in the same sign but at least 10 degrees apart. When they are at least 20 or more degrees apart there is even more discrimination and independence that the Mercury has from the Sun. Thus it is freer to operate more efficiently, especially if there are no bad aspects and no affliction to sign and *nakshatra* lords. Though there are no aspects from malefic planets, we note that Mercury is in *Mula nakshatra* owned by Ketu (*Moksha* significator) and in Jupiter's sign. Jupiter is in the 8th house of *Moksha* and Ketu in the 3rd house of communication. Thus did Yogananda tirelessly lecture, teach, and write about Eastern philosophies in the West, bringing great skill to this process. *Navamsha* Mercury in Aries aspected by Sun and Mars gives him incisiveness in his speech. For those less evolved, it could easily be fiery, impulsive speech, though *Navamsha* Saturn's aspect lends caution and circumspection to *Navamsha* Mercury. Contacting *Navamsha* Moon, this gave him more of an emotional nature.

Mercury is also *Dhana yogakaraka* (wealth significator) for Leo Ascendant. As mentioned above, **Sun-Mercury** together form a *Dhana yoga*, since Sun rules the Ascendant and Mercury rules over two financial houses (2 and 11). Placed in the 5th house – one of the financial houses (speculation) – Sun and Mercury are very beneficial here for personal wealth. Yogananda was successful in raising substantial funds to support his spiritual organization, assuring its stability and continuation. However, no matter how much wealth and fund-raising are well indicated in this chart, the life purpose is primarily spiritual.

Some of the same factors apply as with the *Budha Aditya yoga* above. Ketu is a *Moksha* dispenser, and is lord of the *nakshatra* in which Mercury is placed. Sign lord Jupiter rules over the sign in which Mercury is placed and is itself placed in Mercury's *nakshatra*. In the 8th house in its own sign of Pisces, Jupiter gives several possibilities: inheritance, or moneys passed on for spiritual purposes, or because of the loss of a relative. The 8th house is closely related to fund-raising and/or inheritance because it has to do with the transfer of large amounts of money at death or by marriage, being the 2nd house of income from the 7th house of marriage. When the 8th lord is in its own house it serves to protect against losses. Called a *Viparita Sarala Raja yoga*, it also confers learning and prosperity.

The gifted educator in a foreign country: Yogananda was one of the first of the Indian masters to come to the West and make the spiritual techniques of India accessible to Westerners. To do this, he renounced his life in India for most of the last 32 years of his life.

> "The great masters of India who have shown keen interest in the West have well understood modern conditions. They know that, until there is better assimilation in all nations of the distinctive Eastern and Western virtues, world affairs cannot improve. Each hemisphere needs the best offerings of the other."
>
> *Ibid.*, p. 557.

Jupiter's improved position in the *Navamsha* chart – the 10th house, with *Navamsha* Ascendant lord Venus – enabled Yogananda to succeed in the outer world to a greater degree, combining spiritual teachings with enough financial stability to promote his teachings. *Navamsha* Jupiter also favorably aspected his 6th house Sun and Mars in Libra, four factors in the communications element of air.

Yogananda often said he renounced family life only to inherit many children in life who were his students or his charges, especially at his first school in India, which he started in 1917 but established in a more expanded form in 1918 at the start of his 6-year **Sun *Dasha*** (April 27, 1918 to April 26, 1924).[190] His Sun *Dasha* also brought the first opportunities to travel abroad. He began his first voyage to America in Sun-Jupiter period, and remained in Boston to lecture, teach and write for several years after his initial lecture in early Oct. 1920. Favoring such travel is his natal Moon in the 9th house (of long-distance travel) from natal Sun. The **Sun-Jupiter** period ran from May 14, 1920 to March 3, 1921. From early Aug. to early Oct. 1920 the *Dasha* lord Sun transited through Houses 12, 1, and 2 (Cancer, Leo, and Virgo). Throughout this same time span, tr. Jupiter was in early Leo in *Magha nakshatra*, on both his Ascendant and Moon, aspecting his natal Sun and Mercury in the educational 5th house. Transits of both the Sun and Jupiter were favorable, reflecting the *karmic* themes of *Dasha* and *Bhukti* lord.

Yogananda's 10-year **Moon *Dasha*** ran from April 26, 1924 through April 26, 1934. In 1924, near the start of his Moon *Dasha* he began a transcontinental tour of the U.S., lecturing in many of the major cities. (This is very fitting, with Moon as planet of the mind and of education.) By late 1925 he established an international headquarters for his Self Realization Fellowship in Los Angeles. Tracking *Dasha* lord Moon, his foreign residence during Moon *Dasha* could be predicted from his *Navamsha* Moon in the 12th house of foreign residence. *Navamsha* Mercury, Lord of the 2nd house of income in the *Navamsha* chart, is also situated in the 12th house, indicating either: 1) financial losses and/or expenses or 2) income from foreign sources. The latter was spectacularly the case, as Yogananda would move into beautiful and palatial surroundings in southern California, his home for the next 28 years. This began during his Moon *Dasha*. During the subsequent **Mars *Dasha*** a large estate in Encinitas was established.

Retreat by the ocean: His guru Shri Yukteshwar had told him many years earlier he would have "a retreat by the ocean." While Yogananda was traveling in India and Europe from June 1935 through Oct. 1936, a large headquarters for the Self Realization Fellowship was built overlooking the Pacific Ocean in Encinitas, California. James Lynn, one of his devotees, donated the land and paid for the buildings. Later known as Rajarsi Janankananda, he was a very successful businessman as well as a practitioner of *Kriya yoga*. Timing-wise, Yogananda was passing through the *Dasha-bhuktis* of **Mars-Rahu**, **Mars-Jupiter**, and **Mars-Saturn**. Mars is considered *karaka* of real estate,

---

[190] Through the generosity of the Maharaja of Kasimbazar, in 1918 Yogananda relocated and expanded his new school to the Kasimbazar Palace in Ranchi, 200 miles from Calcutta. Called the *Yogoda Satsanga Brahmacharya Vidyalaya*, its educational principles were based on those Yogananda discovered in 1916.

and is placed natally in Jupiter's sign of Pisces and in Saturn's *nakshatra, Uttara Bhadra*. The entire building project was kept secret from Yogananda until he arrived to see it in person on his return in Oct. 1936. For most of this time period transiting Jupiter in Scorpio was in his 4th house of real estate (from Oct. 1935 through Oct. 1936), while transiting Saturn in Aquarius was also aspecting his 4th house. When both Jupiter and Saturn are aspecting a particular house by transit, lasting about one year, the affairs of that house receive extra focus and plans have an excellent chance of materializing. Jupiter blesses, protects, and gives expansive energy, while Saturn makes it real. Tr. Mars was in Scorpio in his 4th house from March through Sept. 1935, a pivotal planning period for this major real estate project and headquarters for the burgeoning organization. Water signs are the most expansive and Scorpio, a hidden sign, in the 4th house is no exception. Venus has *Digbala* in the 4th house of the birth chart, benefiting real estate in general as well as education for the individual.

The Father and the Guru: For Leo Ascendant, themes of father and son or of father-like figures assume a role of enormous importance in the life. Yogananda spent eight years at the feet of his spiritual master, Shri Yukteshwar, from 1910 to 1917, and the next few years he visited Calcutta periodically from his school in Ranchi, 200 miles away. Jupiter is the planet symbolizing the guru, as is the 9th house lord – Mars in this case, located with Jupiter in the 8th house of *Moksha*. Father is also symbolized by Sun and/or 9th house or 9th lord. The 16-year Jupiter *Maha Dasha* occurred after his death, but in **each Jupiter sub-period** from 1910 onward there was a major event involving either the guru or international travel (also Jupiter and/or 9th house). Yogananda met his guru at age 17 in his Venus-Jupiter *Dasha/Bhukti*. Financed by his father, he left for America in Aug. 1920 in his Sun-Jupiter *Dasha/Bhukti*, residing there for most of the rest of his life. He received his title of "Paramahansa" from his guru in late Dec. 1935, a few months prior to the death of his guru, both events in Yogananda's Mars-Jupiter *Dasha/Bhukti*. The 10th house can also symbolize father as authority figure.[191] In the *Navamsha* chart Venus and Jupiter are both in the 10th house in Aquarius. Together they form an important Jaimini *Raja yoga* of power and prestige: Venus is AK and Jupiter is Amk. (AK is *Atmakaraka*, planet at the highest degree of celestial longitude, while Amk is *Amatyakaraka*, planet of next highest degree rank. Both are classic benefic planets.)

In the days just prior to his death in March 1936 at the age of 81, Shri Yukteshwar spoke the following words to his disciple of over 25 years, finally uttering what was unspoken for so many years:

> "During my married life I often yearned for a son, to train in the yogic path. But when you came into my life I was content; in you I have found my son.... Yogananda, I love you always."

Yogananda's own biological father, Bhagabati Charan Ghosh, died in Calcutta in 1942 at the age of 89. He had been an exemplary if not somewhat stern father, a widower for the last 38 years of his life and a disciple of Lahiri Mahasaya of Benares. Bhagabati wanted his son to take a position at the Bengal-Nagpur Railway, where he himself worked for years as an accountant. Only by 1918, when Mukunda (by then called Yogananda) had established his Yagoda Satsang school

---

[191] In North India, father is typically assigned to the 10th house, while in South India it is the 9th house. Either one can be used, but 10th house has the connotation of authority figure, while 9th lord is more the advisor/guru.

for boys in Ranchi did Bhagabati finally relent. He realized the path of the *sanyasin* (renunciant) and teacher more befitted his son than the railway job he had envisioned for him.

At the time of his father Bhagabati's death in 1942, Yogananda was at the beginning of **Rahu *Dasha***, with transiting Rahu on his Ascendant, and Ketu in the 7th house. These transits are echoed in Yogananda's birth chart, along with several related factors in his life: **1)** Natal Rahu (the foreigner) is in the 9th house of the father and the guru, indicating Yogananda would take a different path of learning or religion from that of his father; **2)** 9th lord Mars is situated in the 8th house of *Moksha* (His formal education would lose out to spiritual practices; even so, Yogananda passed all his exams - just barely, with the help of his guru); and **3)** Jupiter is in a *Dusthana* (house 6, 8, or 12) from the Moon: This is a *Shakata yoga*. Yogananda seems to have escaped most of the adverse effects of *Shakata yoga* due to his yogic powers. Also, the yoga is not repeated in the *Navamsha* chart and the same combination creates an *Adhi yoga* bestowing wealth. Usual descriptions of *Shakata yoga* indicate having a wealthy father, but nonetheless being ever ailing, miserable, and lacking comforts.

The death of Shri Yukteshwar occurred at 7:00 PM on March 9, 1936 in Puri, India. It was during Yogananda's **Mars-Jupiter** *Dasha/Bhukti.* Natal Mars and Jupiter reside in Yogananda's 8th house of death and transformation, a *Moksha* house. From birth, this combination would always indicate major changes in his life. The Mars-Jupiter period encompassed the period when Yogananda returned to India, initially at the inner summons of his guru, who informed him his death was near. They had not seen each other in 15 years, since Yogananda's departure for the West.

***Dharma* and *Karma*:** *Dharma* and *Karma* lords Jupiter and Saturn are in a close opposition on the Virgo-Pisces axis of Yogananda's birth chart, reflecting the hard work of Virgo combined with the sacrifices often demanded by Pisces. When this planetary combination is in mutual aspect, either the conjunction (same sign) or opposition, it can be very powerful in bestowing entrepreneurial abilities. Natal Jupiter and Saturn reside in two money houses (2 and 8), and form a *Raja yoga*, as lords of 5th and 7th houses aspecting each other. Even so, the 8th house has larger implications for *Moksha*, and for the passing on of larger amounts of money as one's legacy. (The sign of Pisces also has *Moksha* implications.) If there is personal strength and leadership already indicated, then this is all the more definitive.

A life may take pivotal turns at the time of a **Jupiter-Saturn conjunction**. When this occurs in a favorable house in the birth chart, then the destiny is galvanized to start new ventures or solidify projects already begun. Though not an angular or trinal house in the chart, the JU-SA conjunction in Virgo marked his Saturn return to its natal sign and on his JU-SA opposition axis. His first trip to the West in early fall 1920 was followed a year later by the **JU-SA conjunction in Virgo.** It was exact at 3:49 Virgo (*Uttara Phalguni*) on Sept. 9, 1921 and in the 2nd house from both his Ascendant and Moon. The Jupiter return had come six years earlier in June 1915 when he earned his B.A. degree from Calcutta University. From Sept. 1921, the subsequent 20 years between JU-SA conjunctions were especially important for Yogananda in terms of expanding his teachings and establishing his International headquarters in the West. It followed the 20-year imprint of the JU-SA conjunction in Virgo. The earth element of this pivotal conjunction helped in materializing and solidifying an organization that brought Eastern spiritual teachings to the West. When his *Autobiography of a Yogi* was first published in 1946, not long after the start of the Atomic Age - with the bombing of Hiroshima and Nagasaki in Aug. 1945 - it spread Yogananda's teachings still further.

> "The Western day is nearing when the inner science of self-control will be found as necessary as the outer conquest of Nature. The Atomic Age will see men's minds sobered and broadened by the now scientifically indisputable truth that matter is in reality a concentrate of energy. The human mind can and must liberate within itself energies greater than those within stones and metals, lest the material atomic giant, newly unleashed, turn on the world in mindless destruction. An indirect benefit of mankind's concern over atomic bombs may be an increased practical interest in the science of yoga [i.e., the Eastern philosophical concept], a 'bombproof shelter' truly."
>
> *Ibid*, p. 265.

The joint return of Jupiter and Saturn happens once in a lifetime at ages 59-60, and is considered extremely significant in Vedic terms. On the night of his own death March 7, 1952, transiting **Jupiter** and **Saturn** were almost exactly at their natal positions. The return of Jupiter-Saturn by transit need not spell one's end. But depending on their placements in the natal chart, combined with other factors, they can indicate that *Dharmic-Karmic* factors are either complete or about to shift into even higher gear. As Ketu, the *Moksha karaka*, had such a prominent role, his spiritual exit was more keenly probable. Ketu was transiting over his Ascendant at that time and he was in **Rahu-Ketu** *Dasha/Bhukti*. Natal Rahu is in Ketu's *nakshatra* (*Ashwini*) in the sign of Mars. Ketu is thus *Jiva*, or life force for Rahu, the *Dasha* lord. Natal Mars in turn is in Pisces in the 8th house of *Moksha*. His smile of benediction can be seen in a photo taken one hour prior to death. When an enlightened master dies, the passing of the soul from the body is called ***Maha Samadhi***. *Samadhi* is a yogi's conscious exit from the body. *Maha* is major, thus the final exit from this body.

Chart 21: Maya Angelou

Birth data: Wednesday, April 4, 1928, 14:10 CST, Saint Louis, MO, USA, Long. 90W11 52, Lat. 38N37 38, *Lahiri ayanamsha* -22:51:01, Class AA data (Birth certificate quoted in *Contemporary American Horoscope*.) Ascendant: 1:07 Leo.

Biographical summary: African-American poet, author, historian, actress, playwright, civil rights activist, producer, director, college professor of American studies – Maya Angelou was born Marguerite Annie Johnson. She has published over twelve books and countless magazine articles, and was the first black woman director in Hollywood. She has written, produced, directed, and starred in productions for stage, film, and television.

She wrote and produced several prize-winning documentaries, including *Afro-Americans in the Arts*, and was twice nominated for a Tony award for acting: first for her Broadway debut in *Look Away* (1973), and again for her performance in *Roots* (1977). She has been called "America's Renaissance Woman" and catapulted into international fame when she was invited to write and deliver a poem at President Bill Clinton's first Inauguration in Jan. 1993. The poem was called *On the pulse of morning*. Angelou has continued to teach, write, act, produce, and record. She won Grammy Awards for the spoken word for the years 1993, 1995, and 2002. In 1981 she accepted

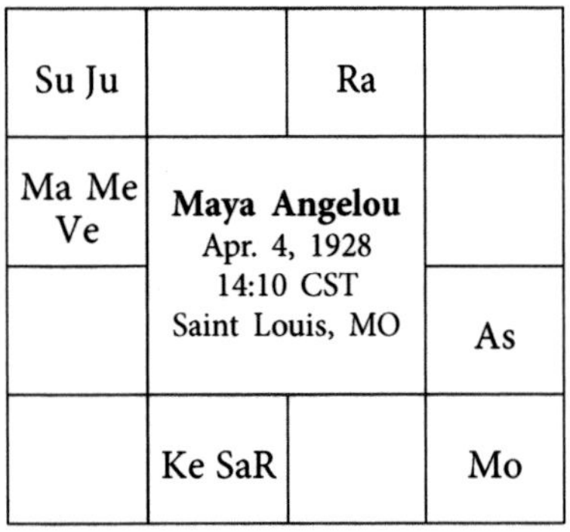

Su Ju		Ra	
Ma Me Ve	**Maya Angelou** Apr. 4, 1928 14:10 CST Saint Louis, MO		
			As
	Ke SaR		Mo

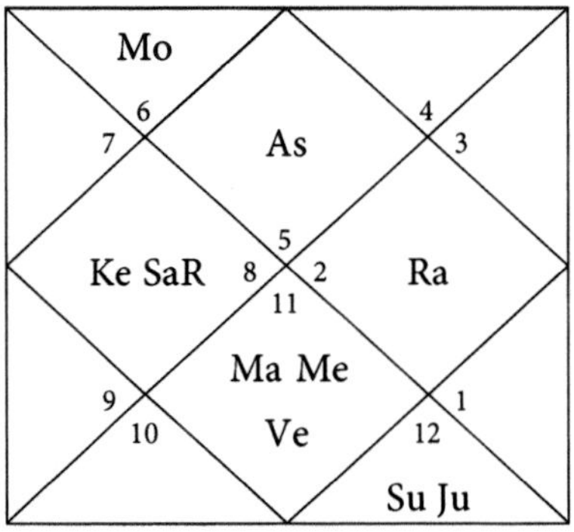

Mo
6 4
7 As 3
5
Ke SaR 8 2 Ra
11
Ma Me
9 1
10 Ve 12
Su Ju

Planetary Information

As	01:07	Leo	Magha
Su	21:57	Pis	Revati
Mo	18:30	Vir	Hasta
Ma	05:02	Aqu	Dhanishta
Me	27:34	Aqu	P.Bhadra
Ju	23:16	Pis	Revati
Ve	28:53	Aqu	P.Bhadra
Sa	26:14	Sco	Jyeshta
Ra	19:46	Tau	Rohini
Ke	19:46	Sco	Jyeshta

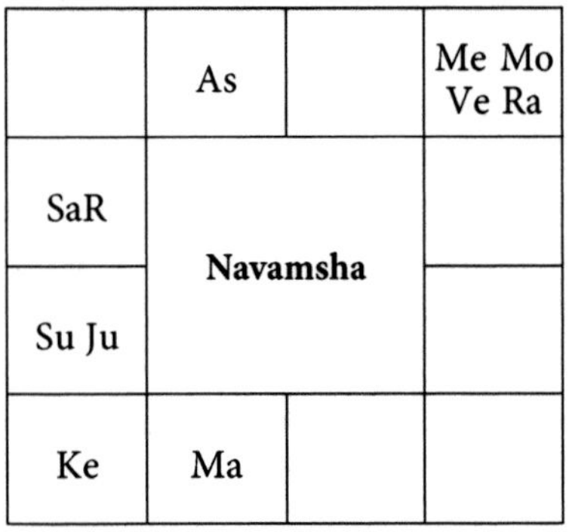

	As		Me Mo Ve Ra
SaR	**Navamsha**		
Su Ju			
Ke	Ma		

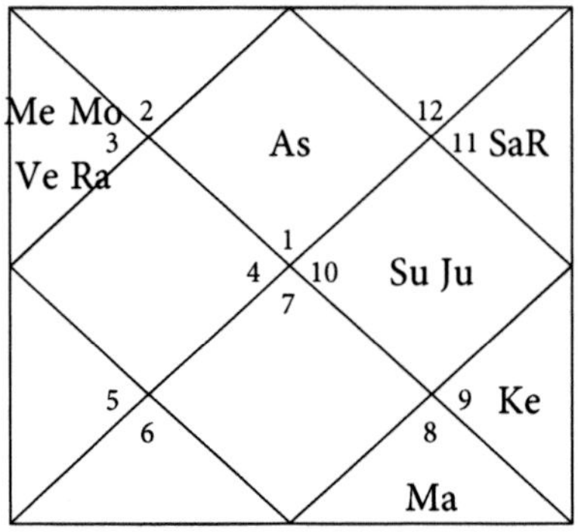

Me Mo 2 12
3 As 11 SaR
Ve Ra
1
4 10 Su Ju
7
5 9 Ke
6 8
Ma

Vimshottari Dashas

Mo	Nov 18, 1921
Ma	Nov 19, 1931
Ra	Nov 19, 1938
Ju	Nov 18, 1956
Sa	Nov 18, 1972
Me	Nov 19, 1991
Ke	Nov 19, 2008
Ve	Nov 19, 2015
Su	Nov 20, 2035

a lifetime appointment of Reynolds professorship of American Studies at Wake Forest University in Winston-Salem, NC. Although her formal education ended with high school, she has received over fifty honorary doctorates.

Angelou is both multi-talented and multi-lingual. She speaks six languages: English, French, Spanish, Italian, Arabic and West African Fanti. Her sensibilities towards racism and social justice were honed during her childhood in the still segregated Deep South. In 1959 Martin Luther King, Jr. asked her to become the northern coordinator for the Southern Christian Leadership Conference. In 1960 she married her second husband, Vusumzi Make, a South African activist and lawyer, and moved with him to Cairo, Egypt. Her interest in civil rights causes developed further and she began writing and working as an editor in Egypt and in Africa. After her divorce in 1963, she and her son moved to Ghana. While in Ghana, she met Malcolm X and returned to the U.S. to work with him. Soon after she arrived back in the U.S., however, Malcolm X was assassinated in late Feb. 1965. She then began to work again with Dr. Martin Luther King, Jr. whom she had known and worked with in the late 1950s. That too ended with Dr. King's assassination on her 40th birthday. She was so shattered by the event that only James Baldwin could coax her out of her depression in the late 1960s, urging her to write her first book. It described her turbulent childhood.

Her parents divorced when she was three years old, and she and her brother Bailey Johnson, Jr. were sent to live with their paternal grandmother and uncle in Stamps, Arkansas. Her childhood was spent largely shuttling between her grandmother in Stamps and her mother in St. Louis, more rarely to her father in California. On one visit to her mother and her boyfriend in the fall of 1935, the mother's boyfriend raped her. She remained mute for the next 5 1/2 years while living with her grandmother. Young Maya never expected to speak again, though she read voluminously, thanks to the efforts of a helpful neighbor, Mrs. Bertha Flowers. Her grandmother always urged her to speak, and said one day she would be a preacher and teach people about morals. Maya credits her grandmother and her extended family for a strong set of values that helped her to stay positive during a chaotic childhood. She credits Mrs. Flowers with instilling in her a love for language and literature and for coaxing her out of her silence. Her brother Bailey was also protective during that time.

She became pregnant at the age of sixteen, having decided she wanted to have her first sexual experience as a young adult, though unprepared for the possible consequences. Shortly after finishing high school she gave birth to her son Guy Johnson. She took a series of jobs to support them both and stay independent. In 1949 she began her first short-lived marriage to a Greek sailor whose last name Angelos formed the basis of her new name, along with her brother's childhood nickname for her: "Maya." (She used the name Maya Angelou from the start of a career as a nightclub singer and dancer in the early 1950s.) She and Angelos divorced in 1952. A second marriage to South African Freedom Fighter Vusumzi Make followed in 1960, ending in 1963. Married a third time in Dec. 1973 to Paul de Feu, former husband of Germaine Greer, they separated in 1978, divorcing in 1980.

If Angelou found her voice and her self-acceptance in autobiographical books, it made her one of the most powerful African-American voices in the United States. Her first book is still probably her most famous one: *I Know Why the Caged Bird Sings* (1969). It was nominated for the National Book Award in 1970 and remained on the *New York Times* paperback bestseller list for two years. It was the first in a six-volume series. Among the many books that followed, some had religious themes evident even in their titles. She also published numerous volumes of poetry, including *Just Give Me a Cool Drink of Water 'Fore I Diiie* (1971), which was nominated for the 1972 Pulitzer Prize.

> "Her genius as a writer is her ability to recapture the texture of the way of life in the texture of its idioms, its idiosyncratic vocabulary and especially in its process of image-making. The imagery holds the reality, giving it immediacy. That [the author] chooses to recreate the past in its own sounds suggests to the reader that she accepts the past and recognizes its beauty and its ugliness, its assets and its liabilities, its strengths and its weaknesses. Here we witness a return to the final acceptance of the past in the return to and full acceptance of its language, the language a symbolic construct of a way of life. Ultimately Maya Angelou's style testifies to her reaffirmation of self-acceptance, [which] she achieves within the pattern of the autobiography."
>
> **Sidonie Ann Smith**, in the *Southern Humanities Review,* Fall 1973.

Phenomenal woman – and her autobiographical statement: She grew up a descendant of slaves, an African-American woman fighting to find her personal identity in a turbulent social and political environment, seething with the injustices of racism and segregation. Her personal situation in childhood and later was often not secure or even safe. Yet astrologically, the constellation rising in her chart is *Magha*, meaning "great" or "mighty one," and symbolized by the throne room. A constellation entirely within Leo, *Magha* is near the lion's forehead and the royal star Regulus. Thus, the innate tendency is to yearn for greatness and to feel as if one has inherited royal status, no matter what humble circumstances one is born into. In her life as in her poem *Phenomenal* *woman*, six foot tall Angelou learned to break through all that. (Paramahansa Yogananda broke through other boundaries with *Magha* prominent in his chart, Chart #20.) When the Ascendant is within the first 3 degrees 20 minutes of a fire sign, it is considered in *Gandanta*, but is not as problematic as the last 3 degrees 20 minutes of a water sign. In fact, the first *pada* of a fire sign assures the *Navamsha* Ascendant will be Aries, thus doubling the impact of the fire sign Ascendant, as with Henry Miller (Chart #3). This adds to the physical vitality and personal initiative, although it also depends on where the *Navamsha* Ascendant Lord is located. Her *Navamsha* Mars is in the 8th house in Scorpio, its own sign. This repeats the message of some danger to her physical body during Mars *Dasha*, ages 3 to 10.

The Sun as Ascendant lord is not well placed in the 8th house for physical well-being, but gives excellent intuitive powers and the potential for powerful spiritual transformation – *or* total self-destruction. Fortunately the Sun is in the sign of its great good friend Jupiter, contacting Jupiter and opposite the Moon – a Full Moon chart, very close to exactly full within a few hours or about three degrees. The close Sun-Jupiter combination is in a less difficult house in the *Navamsha* chart. They recur together in Capricorn in the *Navamsha* 10th house, improving the health and vitality and giving her more status. Sun in the 10th house enjoys maximum visibility and influence, and the person with both Leo Ascendant and Full Moon has a double need to feel seen and recognized. However, *Navamsha* Jupiter is in its sign of debilitation (Capricorn), and thus will bring situations of disadvantage that she must keep rising above. She also has to be practical many times when she would prefer to follow her ideals. This theme will strongly affect her career and status, especially during her 16-year **Jupiter *Dasha***, from Nov. 18, 1956 to Nov. 18, 1972.

In 1969 she published her first book: *I Know Why the Caged Bird Sings*. It was the first of her series of autobiographies, and was nominated for the National Book award. In it she recounted her early childhood, almost as a salve to recover from her despair at the death of Martin Luther King. Although she had danced, sung and acted in various productions from Gershwin's *Porgy* and *Bess* to Brecht's *Mother Courage* and Genet's *The Blacks*, starting out in the early 1950s with a Caribbean accent as "Miss Calypso," she was irresistibly drawn as a poet and a writer to the autobiographical statement. The Leo Ascendant person will make a special effort to seek his or her own personal identity, and with Sun in the 8th house of *Moksha*, one's path of self-discovery may well lead to a far greater spiritual perspective.

Jupiter's close contact to the Sun can give a religious tone to the person, and does so in this case. First husband, Tosh Angelos (1949-1952) was said to be an atheist – causing her more than a basic ideological challenge, even in her early 20s. Natal Sun and Jupiter are both in Jupiter's watery sign of Pisces and in Mercury-ruled *Revati nakshatra*. The person's influence can be pervasive with the influence of water, which spreads quickly, and of Jupiter, the largest planet and best planet for Leo Ascendant. Her interests and concerns are universal, though they tend to focus around race. *Revati nakshatra* is associated with prosperity, talent in the arts and also love of travel. Maya Angelou traveled widely for much of her life, starting from childhood. Travel tends to bring benefits, as the deity Pushan rules over *Revati*. Pushan is the god of safe travel and is also the nourisher. Angelou has always loved to cook and entertain. In 2004 she published *The Welcome Table: A Lifetime of Memories with Recipes*.

From Mute to Full Moon Voice: In her book, *I Know Why the Caged Bird Sings*, Angelou wrote about the most traumatic experience in her childhood. She was seven years old in fall 1935 when she was raped by her mother's boy friend. He threatened her not to tell anyone, and when she did tell her brother a few days later, her uncles then kicked the man to death. Maya's guilt was severe over having spoken and thus in her own mind causing his murder. She went into a self-imposed mute state for the next 5 ½ years, later describing it as horror at the negative power of her own speech.

Her Full Moon in the 2nd house of speech makes all 2nd house matters more emotional. As lord of the 12th house of loss, Moon can bring a sense of loss initially, leading later to *Moksha*, or spiritual liberation – which is the more evolved meaning of the 12th house.

She was in **Moon *Dasha*** the first three years of her life. At the start of the 7-year **Mars *Dasha***, her parents split up, greatly destabilizing the family. The rape occurred during **Mars-Mercury** *Dasha/Bhukti*. (Together Mars-Mercury can mean strong speech, or violence through speech or communications.) Both planets reside in Aquarius in the 7th house of the birth chart, though 22 ½ degrees apart. Natal Mars is very close to setting in the chart, as in the case of Robert Kennedy (Chart #2), giving Mars a strong influence. And even though a friend to the Ascendant lord Sun, as it rules good houses (4 and 9), and brings closeness to brothers, Mars can still be dangerous as a violent and aggressive force. This is especially true during its own *Dasha*, as this Mars is strong in its own *nakshatra Dhanishta* and in its own sign of Scorpio in the *Navamsha* 8th house.

Transits of classic malefic planets such as Mars and Saturn to her Ascendant and/or Ascendant lord Sun will identify critical time periods within potentially troublesome *Dashas*, when extra precautions can be taken in advance. We should be able to see a confluence of such factors showing physical danger, confirming the impact of the Mars *Dasha* and its timing. It occurred Oct. 19, 1935 for the next seven weeks. Crucial to the timing of her mute period was the passage of **tr. Saturn in Aquarius** (Dec. 7, 1934 to Feb. 25, 1937); **in Pisces** (Feb. 26, 1937 to April 28, 1939); **and in Aries** (April 28, 1939 to June 18, 1941). Successively, these refer to tr. Saturn in her 7th house, aspecting her Ascendant and contacting three planets in Aquarius: Mars, Mercury, and Venus. (Mercury is a *karaka* for voice, contacted by *Dasha* lord Mars, which also aspects Moon in the 2nd house of speech.) Tr. Saturn in Pisces is doubly difficult as it is not only opposite the Moon and afflicting the Full Moon axis, but in the 8th house from the Leo Ascendant. (***Ashtama Shani*** is the most difficult, being 8th house from natal Ascendant or natal Moon.) Next is tr. Saturn in Aries in the 8th house from natal Moon in Virgo. Thus we have three successive Saturn transits, the latter two in particular – that are extremely challenging in an already demanding *Dasha*, and to a very emotional Moon.

She remained mute by choice for the remainder of the 7-year **Mars *Dasha***. Mars has the capacity to heat or cut. From the natal 7th house in Aquarius, natal Mars aspects the 2nd house of speech. (Mars aspects Houses 4, 7 and 8 from itself and impacts planets in the same house.) Mars contacts Mercury, planet of speech, and aspects Moon, planet of emotions placed in the 2nd house of speech and *kutumba* (family happiness). Thus, though her muteness was self-imposed, during its own *Dasha* Mars cut off her speech. This Mars has raw power and had a visceral effect on her speech, although at least she did not stop eating – another 2nd house activity. Mars can give incisiveness to the speech, galvanizing it, as it did in later years after this initial childhood trauma.

Fortunately, **Jupiter** provides some measure of physical protection, as natal Jupiter is closely conjunct natal Sun and in Mercury's *nakshatra*. Both of these are benefic planets to the chart, and Jupiter is best friend to the Sun. As *Nadi yogakaraka*, Jupiter is the best planet for Leo Ascendant chart. Jupiter gains by being in its own sign, but is weakened and also exaggerated by being **severely combust the Sun**, within 1:19 degrees. (This may have contributed to her concern for the fate of her rapist, at her own expense.) Thus, the two luminaries Sun and Moon are aspected by the best planet for the chart. Jupiter is also the 8th lord in its own house. The most difficult houses are 6, 8, and 12, and the 8th lord brings *Moksha* concerns directly to that individual, insuring that they will learn about the spirit through loss and suffering. Suffering is a special domain of the 8th house, and can come from Jupiter (as 8th lord) aspecting Sun and Moon. The safety net comes with the *Viparita Sarala Raja yoga*, since 8th lord Jupiter resides in its own house, and so protects the affairs of its own house. Whatever physically menacing situations she encounters, especially during Mars *Dasha*, her spiritual awareness is likely to be raised up to new levels.

We have established the benefits of the Full Moon in this chart. Even so, there is always some inherent danger or voltage attached to the quality or content of her speech, as Ascendant lord Sun is contacting both the 8th and the 12th lords, Jupiter and Moon, respectively. But the greatest difficulties came during **Mars *Dasha*** (Nov. 1931 – Nov. 1938). She started speaking again when she was almost 13, several years into **Rahu *Dasha***. It was Rahu-Rahu period (Nov. 19, 1938 to Aug. 1, 1941). As it is a worldly planet, Rahu tends to bring one out into the world. In Angelou's case, natal Rahu is also in the most visible house, the 10th house, in the sign of artistic and sociable Venus and in the *nakshatra* of *Rohini*, owned by the Moon. Moon in turn is located in the 2nd house of speech. As the *Jiva* (life force) for Rahu and the Moon itself, Moon is drawing our attention back to 2nd house matters, such as speech. The floodgates of her own verbal expression were about to open. Rahu *Dasha* brought her first flowering as an artist and as a woman. She had a child in 1944, married in 1949 – the first of three brief marriages, and started performing in nightclubs in the early 1950s.

Prodigious vitality and versatility: Jupiter is responsible for many noble attributes in this chart. As the planet of abundance, it also gives a large appetite for life when it aspects the Ascendant lord, and in this case both luminaries: Sun and Moon, which are both planetary friends of Jupiter. In addition, when many key planets are in **dual signs** in both birth chart and *Navamsha* chart, this yields tremendous range, as planets in dual signs give the greatest versatility. Mercury and Jupiter rule over the dual signs: Gemini and Virgo (Mercury), Sagittarius and Pisces (Jupiter). **Mercury and Jupiter** also connect through a powerful ***Nadi yoga*** (exchange of *nakshatra* lords) in this chart. Sun, Moon, and Jupiter are in dual signs in the birth chart, and five planets are in dual signs in the *Navamsha*, amplifying the versatility and changeability: Moon, Mercury, Venus, Rahu, and Ketu. *Navamsha* Moon is not easily placed in Gemini along with Mercury-Venus-Rahu opposite Ketu. Moon's influence is always fluctuating in any case, but even more so in a dual sign with contacts from five other planets, including Mars. Thus, versatility can come at an emotional price. The **Venus-Rahu-Ketu** combination has yielded a colorful if somewhat unstable love life, and an artistic style that is also unique and exotic (Rahu and Ketu influence.) So much Gemini impact in the *Navamsha* chart has given her the ease of being accessible. The 3rd house is associated with music, drama, and dance, but also with courage. Personal courage is another major theme of Maya Angelou, not just to overcome obstacles, but courage to become who you must be: an enduring Leo theme.

We know matters of speech and communication will be pivotal in the life. Not only is the Full Moon in the 2nd house of speech, strong in its own *nakshatra* of *Hasta*, and in Mercury's sign, but also Mercury, planet of communications, is the *Jiva* (or life force) for the Ascendant lord Sun. This is because the Sun – as well as Jupiter, Saturn and Ketu – resides in Mercury's *nakshatra Revati*. Thus, Mercury plays a key role in the life, including financially, as 2nd and 11th lords.

The close connection of **Mercury and Jupiter** to each other and to the Ascendant lord Sun tends to promote an outpouring of speech and language. With Mercury in air signs in both the birth chart and *Navamsha* chart, the intelligence is quickened, giving her an eagerness to learn. This applies also to learning other languages, as she speaks six languages. Even with a formal education that ended with high school, she has received over fifty honorary degrees and was offered several professorships. Not knowing what the limits of her learning were supposed to be, her appetite for learning knew no bounds. This is the Jupiterian emphasis, as Jupiter expands whatever it touches. Jupiter also aspects the 4th house of higher academic degrees in both the birth and *Navamsha* charts, overriding other contra-indicators.

What especially captures our attention is that after being mute for 5½ years in her childhood, Maya Angelou went on to be so stunningly prolific and in so many artistic arenas. Being prolific with words often comes when natal Jupiter, planet of abundance, aspects natal Mercury, planet of speech, or when they are interacting in some important way – as in this case, with the ***Nadi yoga*** between **Mercury and Jupiter.** We know the mute state could not have happened during Jupiter or Mercury *Dashas*, as their close association assures a vibrant profusion of communications. Several other factors helped to break her silence: the onset of her Rahu *Dasha*, and the Jupiter-Saturn conjunction Feb. 1941 in Aries in her 9th house (of *bhagya*, good fortune, and of higher learning).

The condition of natal Mercury is of course pivotal for a writer, actress, performer, or commentator. In her case, the prominence of her natal Mercury, Venus, and Mars in an angular house in Aquarius – the most idealistic of all the signs – gives her fluidity of the mind and the intellect. Natal Moon in Mercury's sign of Virgo benefits from Mercury's mental agility and cleverness. Also, Mercury has the chance of becoming stronger over the years, due to its importance in her lifetime *Dasha* sequence. Only 14 months into the 17-year **Mercury *Dasha*** Maya Angelou delivered her poem at President Bill Clinton's Inaugural Jan. 20, 1993. Though she was already known to many in the U.S., especially in the film, entertainment and civil rights worlds, this event put her solidly onto the world stage, where she has been every since. Her Mercury *Dasha* ran from Nov. 24, 1991 to Nov. 23, 2008.

Often referred to as "America's Renaissance Woman," she has done a lot both in the fields of education and the arts. She is a very international person (another Jupiter trait) – not only through her residence and friendships abroad – but through her multi-lingual capacities and her philosophical and ethical sensibilities. She was a spokesperson for two major Civil Rights leaders (both ministers) during her **Jupiter *Dasha*** (Nov. 1956 – Nov. 1972), and they were each assassinated: in 1965 and 1968, respectively. Since her *Navamsha* Jupiter is debilitated in Capricorn, we know that if and when she gains status, she gains it through or in spite of initial disadvantage(s). Her most significant literary and spoken output is bound to be from the onset of *Dashas* involving Mercury: either Mercury itself, or Mercury as *nakshatra* lord – which gives even more subtle power as *Jiva* than as sign lord. This began with the 16-year **Jupiter *Dasha*** in November 1956, and has continued unabated with the subsequent **Saturn *Dasha*** (1972 – 1991), **Mercury *Dasha*** (1991– 2008), and **Ketu *Dasha*** (2008 – 2015). Even **Venus *Dasha*** (2015 – 2035) is rich with Mercury's influence, as the two planets are only 1 degree 19 minutes apart in the same sign of Aquarius. They recur together in Gemini in the *Navamsha* chart.

The Preacher within - a combust Jupiter: Maya Angelou has always considered herself "a very religious person," aware of what she calls "the constant, unwavering, unrelenting presence of God." In her birth chart, Ascendant lord Sun is in Pisces, closely bound with the religious concerns of Jupiter. Sun and Jupiter are in the 8th house, where things tend to be hidden or below the surface. The planets are in a very tight conjunction and Jupiter is combust. When a planet is combust, we cannot see it in the heavens with the naked eye, as it is too close to the Sun. Therefore, as in life, we are not always sure it is there. We tend to emphasize and exaggerate it in various ways to make ourselves believe it is really there.

In such a way the themes represented by a combust planet can become almost as if scorched, they are so hot! This condition of Jupiter both enlivens all of Jupiter's characteristics and tends to burn them out periodically. There is an enormous sense of curiosity about the world and with fewer limits on the explorations, it can be exhausting. Jupiter close to her Ascendant lord Sun also gives size as well as largesse. Angelou is over six feet tall, with a towering personality and a full rich voice. Initially active as a dancer, singer and actress, she later became a poet, novelist and social commentator. Martin Luther King, Jr. strongly influenced the rhythm and content of her speech, as she comments below:

> "The music of the 'I Have a Dream' speech is a replication of the music which comes out of the mouths of the African American preacher, singer, blues singer, jazz singer, rap person. It is so catching, so hypnotic, so wonderful that, as a poet, I continue to try to catch it, to catch the music. If I can catch the music and have the content as well, then I have the ear of the public. And I know that is what Martin Luther King was able to do, not just in the 'I Have a Dream' speech — although that has become a kind of poem which is used around the world - but in everything he said there was the black Southern Baptist or Methodist preacher, singing his song, telling our story — not just black American story either, but telling the human story. And as a poet, if I can replicate that, I am okay, Jack."
>
> **Maya Angelou**

With both natal Jupiter (classic planet of *Dharma*) and natal Saturn (classic planet of *Karma*) in watery signs in the birth chart, there is greater likelihood that artistic expression would be her vehicle to both teach and entertain. They are inextricable in her case, and linked most often with themes of social justice, including civil rights. Water placements are more emotional and tend to drive the artistic process. Dr. King had natal Jupiter and Saturn in fire signs with Jupiter on the Ascendant, making religion or law inevitable for him. But even with her emphasis on civil rights, Maya Angelou is quick to move to the larger and more Jupiterian context. On race, she says the following:

> "When we talk about racism, we have to see that we are not just talking about acts against blacks, we are talking about vulgarities against any human being because of her - his - race. This is vulgar. That is what it is, whether it is anti-Asian, whether it is the use of racial prejudices about Jews, about Japanese, about Native Americans, about blacks, about Irish, it is stupid, because what it is really is it is poison. It poisons the spirit, the human spirit."
>
> **Maya Angelou**, interview from Museum of Living History,
> High Point, NC, Jan. 1997; revised Oct. 2006.

Chart 22: Woody Allen

Birth data: Sunday, Dec. 1, 1935, 22:55 EST, Bronx, NY, USA, Long. 73W54 00, Lat. 40N51 00, *Lahiri ayanamsha* –22:57:59, Class AA data. (Birth certificate in hand, Lois Rodden.) Ascendant: 9:20 Leo.

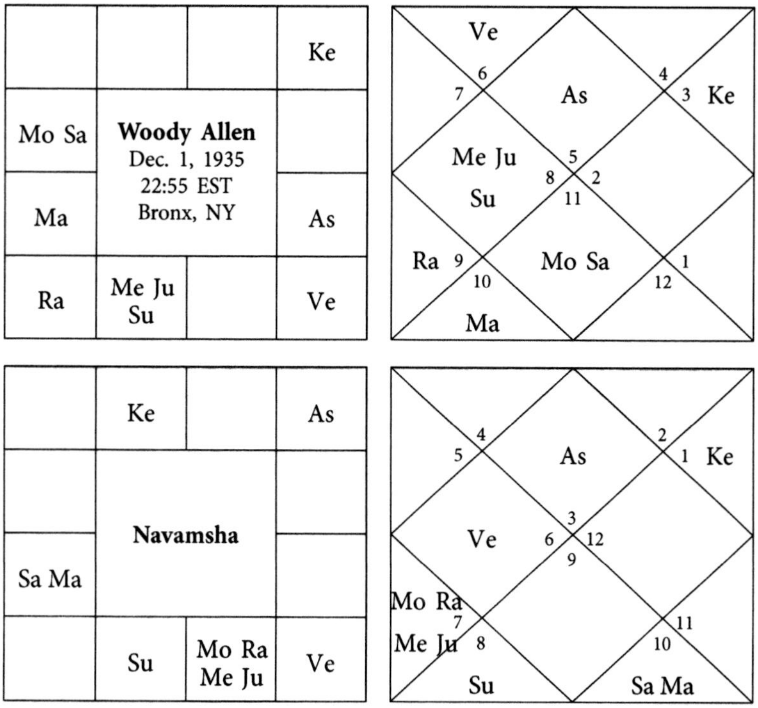

			Ke
Mo Sa	**Woody Allen** Dec. 1, 1935 22:55 EST Bronx, NY		
Ma			As
Ra	Me Ju Su		Ve

Ve
6
7 As 4
3 Ke
Me Ju Su 5
8 2
11
Ra 9 Mo Sa 1
10 12
Ma

Planetary Information

As	09:20	Leo	Magha
Su	16:04	Sco	Anuradha
Mo	01:08	Aqu	Dhanishta
Ma	03:08	Cap	U.Ashadha
Me	11:32	Sco	Anuradha
Ju	12:09	Sco	Anuradha
Ve	29:55	Vir	Chitra
Sa	11:03	Aqu	Shatabhisha
Ra	21:31	Sag	P.Ashadha
Ke	21:31	Gem	Punarvasu

	Ke		As
	Navamsha		
Sa Ma			
	Su	Mo Ra Me Ju	Ve

4 2
5 As 1 Ke
3
Ve 6 12
9
Mo Ra
7 11
Me Ju 8 10
Su Sa Ma

Vimshottari Dashas

Ma	Oct 28, 1931
Ra	Oct 27, 1938
Ju	Oct 27, 1956
Sa	Oct 27, 1972
Me	Oct 28, 1991
Ke	Oct 27, 2008
Ve	Oct 28, 2015
Su	Oct 28, 2035
Mo	Oct 28, 2041

Biographical summary: Born Allen Stewart Konigsberg to Jewish parents of Austrian and Russian ancestry, he grew up in New York City. His father Martin Konigsberg (b. Dec. 25, 1900, d. Jan. 13, 2001) was a jewelry engraver and a waiter. His mother Nettie (birth name Cherrie; b. Nov. 8, 1906, d. Jan. 27, 2002), was a bookkeeper at her family's delicatessen. His grandparents were immigrants who spoke Yiddish, Hebrew and German. Allen also spoke Yiddish in his early years. His one sibling, Letty Aronson (birth name Ellen Konigsberg; b. 1943) has produced a number of his films. One of the most prolific American filmmakers and directors, Allen began his career as a stand-up comedian and comedy writer. To make extra money he started to write comic material professionally when he was sixteen years old. At that same time he took the name Woody Allen. Though painfully shy and introverted as a child, he was known for his skill with magic and card tricks. After eight years in Hebrew school, his subsequent years in New York public schools showed him to be an uncooperative, unruly student with little interest in any classes except English. He then enrolled at New York University, studying film and communications. Considered a poor student, he failed the class in Motion Picture Production and was suspended from NYU.

By 2011, however, at age 75, he had written and/or produced 49 films, 11 plays, numerous articles and 7 books, which included collections of his short stories. He has directed and acted in most of his films and plays from 1965 onward, and has written or produced countless material for himself and others for television, starting with Sid Caesar and his *Your Show of Shows,* the

Ed Sullivan Show, Candid Camera, The Tonight Show and later *Saturday Night Live*. He was an actor in several other films with screenplays other than his own, and he released three comedy albums (1964 to 1966). In his first two feature films, *What's Up, Pussy Cat (1965)* and *What's Up, Tiger Lily?* (1966), he lost control over the execution of his original screenplays. From then on he insisted on very tight control over any film for which he wrote the original screenplay – over the script, directing, shooting, editing, and casting, often with an ensemble of many of the same actors through his films. This is far more rare in the world of filmmaking. His vaunted reputation became such that he was not even required to submit a script for studio approval.

> "There's nobody else in American film who comes anywhere near him in originality and interest.... One has to go back to Chaplin and Buster Keaton, people who were totally responsible for their own movies, to find anyone comparable to him."
>
> **Vincent Canby**, film critic, 1986.

Known for his signature dry wit and humor and for his one-liners, his particular brand of humor also capitalizes on self-deprecation and a certain morose view of life, one that is endlessly given to self-analysis. Allen himself entered psychoanalysis in 1959 and continued to see a psychiatrist for over thirty years – at least once a week and at times far more often. But soon after he began his intimate relationship with Soon-Yi Previn in fall 1991, Allen abruptly ended his life-long consultations with psychiatrists. He does not drink, smoke or take drugs, and though he played baseball and ran track as a youth, he is not athletic in adulthood. A persistent hypochondriac for much of his life, nevertheless he has enjoyed excellent health and vitality. Allen is of slight build and five feet six inches tall. An accomplished clarinetist, he has played for years in a jazz band that performed weekly in Manhattan, New York, and still performs around the United States.

In spring 2002 Allen received the prestigious *Palme des Palmes* award at the Cannes Film Festival for his lifetime achievement in films, an award only ever granted previously to Ingmar Bergman, one of his idols. The recipient of numerous national and international film awards, Allen has been the most frequently nominated of all Oscar nominees throughout his professional life, especially for Original Screenplay – 14 times. His most successful years artistically and commercially were from 1977 to 1987. In this period there were 11 films he wrote and directed (and acted in all but three of them). Among these are: *Annie Hall* (1977), *Manhattan* (1979), *The Purple Rose of Cairo* (1985), and *Hannah and Her Sisters* (1986). In the 1990s his films became more dark and bleak, as did his critical and financial successes. The tide turned, however, with *Match Point* (2005), the first of a series of movies he made in Europe, this one set in London. A tense murder mystery in an upper class British setting, as of 2007 Allen considered *Match Point* the best film he had ever made. It won various awards and also earned $23 million in two years, more than any of his films in the previous 20 years. Since then, his romantic comedy-fantasy film *Midnight in Paris* (2011) was very successful at the box office in the U.S. It pays homage to contemporary Paris and to Paris in the 1920s. Though in his material he has consistently used his own life story, philosophy, and fears (of water, heights, darkness, death, etc.), he has almost as consistently denied he was an auteur filmmaker or that his films are particularly autobiographical. His critics have argued to the contrary, saying many of his films were self-indulgent versions of his own life.

His first marriage was at age 19 to Harlene Rosen, then a 16-year old student. They were married from 1954 to 1959. His second marriage to actress Louise Lasser was on Feb. 2, 1966.They were divorced in 1969. Lasser starred in two of his films. His other major love relationships with actresses were with

Diane Keaton and Mia Farrow, who also starred in his films. Keaton met Allen in the stage production of his first play called *Play it Again, Sam*, and was featured in eight of his films. Allen was linked with Keaton from 1969 through 1977. Mia Farrow was in thirteen of his films over the 12-year span of their relationship, from 1980 through 1992. Their relationship blew up in a bitter legal dispute in summer 1992 once Farrow discovered nude pictures of her 21-year old adopted daughter Soon-Yi Previn in Allen's apartment in mid-Jan. 1992, uncovering a romantic relationship begun in the fall of 1991, though Allen claimed it started Dec. 1, 1991.[192] In a court case beginning Aug. 4, 1992 Farrow accused Allen of molesting their adopted daughter Dylan. This charge was later dropped in court. On Aug. 13, 1992, Allen then sued Farrow for custody of three of their jointly adopted children (Satchel, Moses, and Dylan) on the unprecedented grounds they were "separate but equal adoptive parents." (Satchel Farrow, later renamed Ronan Seamus Farrow, was their one biological child together. An intellectual prodigy, he was born Dec. 19, 1987.) Allen lost the custody battle in mid-June 1993 and was allotted only strict visitation rights to the children. Relations between the opposing parties were broken irretrievably. Allen married Soon-Yi Previn on Dec. 23, 1997 in Venice, Italy. They have two adopted daughters born in 1998 and 2000, Bechet Dumaine Allen and Manzie Tio Allen.

It's not about *me!* An indefatigable Film auteur, Woody Allen is averse to conceding autobiographical content beyond "a few true facts." This in itself has sparked an ongoing debate. From the mid-1960s he has written and directed his own films, maintaining tight artistic control over every aspect of the films from 1969 onward. Since he also often stars in his own movies, there is enormous audience interest and connection between the private Woody and his on-screen comic persona. He even uses some of the same filmmaking techniques used in autobiographical documentaries, such as stream of consciousness. But for his audiences and many of his fans, the private and public persona of Woody Allen is almost inextricable. His audiences recognize and connect with many things about him, including his philosophy of life, whether or not they are true in his private life.

Reading from the birth chart Ascendant, **Sun is in the midnight position**, in the 4th house where it is the least visible. Therefore it brings up all the issues around being seen. Does one feel properly seen and recognized? And if not, this can be a definite problem for the Leo Ascendant person, though Sun in Scorpio brings some inherent contradictions. Scorpio is considered a hidden sign of the zodiac and key planets in Scorpio indicate the desire to remain somewhat unknown and apart, even if sharing one's obsession with plumbing the depths of the psyche, and of sex and death in particular. This only increases the mystery and intrigue of the **Scorpio emphasis**, and is even more a major theme if: **1)** the Ascendant lord resides in Scorpio, and **2)** the Ascendant lord is also the Sun. Sun as the brightest and hottest planet wants its influence to be seen and felt by everyone. Its purpose is to illuminate, so it cannot help dominating and shedding light on the matters over the house where it resides. (Scorpio is research, psychology and the psychoanalytic process, while the 4th house is the interior mental and emotional life, as well as one's domestic environment.)

[192] In an Appendix of her 1997 memoir, ***What Falls Away***, Mia Farrow includes in entirety the New York State Supreme Court decision of the 1992-1993 Custody trial between herself and Allen. Among the pre-trial incidents reported, in fall 1991 Dylan witnessed an intimate relationship between Soon-Yi and Woody Allen.

Before orienting himself to Europe, Allen has even preferred filming in his own New York City neighborhoods. With natal Mercury and Jupiter in close contact and both of them close to the Sun, the person is longing to talk a lot – but just does not want it known in some way. Sun in the 4th house shares similar issues with Sun in the 8th or 12th houses – both of which are considered hidden placements for most planets. (Maya Angelou has natal Sun in the 8th house.) As with Maya Angelou and Paramahansa Yogananda, the Ascendant degree is in *Magha nakshatra*, with its symbolism of the throne adding to the royal theme. One feels born to greatness.

In astrological terms, Ascendant lord Sun placed in the 4th house in Scorpio is largely responsible for Allen's protestations that his films are not autobiographical. Even though it would appear that many of them track his life rather closely, he is adamant they are unrelated and dislikes being questioned about his private life. *Annie Hall* (1977), for instance, is a nostalgic look back at a romantic relationship that no longer exists. Featured are Woody Allen and actress Diane Keaton (birth name Diane Hall), whose real life romantic relationship together ran for eight years and ended the same year *Annie Hall* was released. It also reflects Allen's Sartrean existentialist view that romantic love is doomed to oblivion. The movie *Husbands and Wives* (1992) in turn reflects the end of his 12-year relationship with Mia Farrow. Allen and Farrow star as a married couple whose 10-year marriage is ending due to Allen's infatuation with a 20-year old student, a close parallel to his own real life situation, though apparently written well beforehand. In *Hollywood Ending* (2002), Allen stars as a washed-up Hollywood director. During that same time period, he filed a lawsuit against Sweetland Films, the producers of his previous eight films. (His long-term contracts with Tri-Star and Orion had ended amidst the scandals of 1992.) By 2004, Allen started making more movies in Europe in an environment he considered to have a more liberal and creative attitude than Hollywood at this time.

While still in production, Allen's film projects – especially his screenplays – are veiled in secrecy. Scripts are often not allowed out prior to production except in rare circumstances. If so, a courier may be waiting outside for the actor or actress to read the script in one sitting. Scorpio values privacy and secrecy, and Allen generally avoids publicity unless it is absolutely necessary. Scorpio relates to what may be concealed or camouflaged. He uses psychoanalytic theory in his films, with Freudian concepts of desire, repression, anxiety and sexuality. With over thirty years of psychoanalysis, Allen had much personal experience with this modality, even going on a daily basis for eight years. The aim of the process is to bring to the surface the inner contents of the subconscious mind. But soon after Mercury *Dasha* began, he ended his psychoanalysis abruptly. This coincided with the start of his intimate relationship with Soon-Yi in fall 1991. Mercury's loss in the Planetary War to Jupiter plays a part in this sudden change, to be discussed shortly.

What lends a public face to the private persona here is the whole chart as seen from the Moon (the mind, and also the public). To offset the Sun in the 4th house from the Ascendant, if we look from natal Moon in Aquarius – the Sun is situated in the 10th house. Moreover, Sun, Mercury, and Jupiter are all in the 10th house from the Moon. This to some extent corrects the deficit of having the Sun (especially as Ascendant lord) in the 4th house and enables the person to receive more credit for what they do in the world. It gives some limelight to a person who tends to shun it, and though correcting one problem, creates another in terms of bringing public attention to an essentially private person who cannot help sharing his thoughts and turning them into screenplays.

With Allen in a major role, or even if played by another actor, there is often a theme of self-deprecation. His weaknesses are exaggerated for effect, as he plays the Yiddish *schlemiel*: the neurotic, shy, and nervous figure that became famous in his long string of movies. An inveterate New Yorker and confirmed city person, Allen has set most of his films in contemporary Manhattan and – increasingly since 2004 – in European cities. The taste for urban life comes from Moon in air signs, and the zest and compulsion for work from Saturn. In addition to the Sun's placement, his shyness and disinclination to publicity come from Moon in tight contact to Saturn, both prominent in the chart. But the rigid daily work routine stands in stark contrast to his scatter-brained comic persona.

> "I make [movies] because if I don't work then I become depressed because I have time on my hands and I reflect and get into morbid introspection."
>
> **Woody Allen**, interview in *The Guardian (UK)*, Sept. 27, 2001.

> "The real Allen holds himself in reserve. He is, like all great funny men, inconsolable… [and his] antidote to anxiety is action."
>
> **John Lahr**, profile of Woody Allen, "The Imperfectionist," *The New Yorker*, Dec. 9, 1996.

The Inconsolable clown: The astrological signature of depression or its tendency is a strong Saturn influence in the chart, especially on the Moon. In this case Saturn closely contacts the Moon and also aspects Mercury, Sun, Jupiter and the Ascendant. Saturn doubly influences the three planets in Scorpio as they all reside in *Anuradha nakshatra*, owned by Saturn. Thus Saturn becomes the *Jiva*, or life force for Sun, Mercury, and Jupiter. They have extra power being in an angular house of the birth chart, and close to the house cusp, as are Moon and Saturn on the Descendant. Moon and Mercury are the emotional and mental planets, while Sun rules over the physical body.

Saturn is the strongest planet in this chart, residing in its own sign of Aquarius and in its best possible angle of the chart (*Digbala*). It remains in its own sign of Capricorn in the *Navamsha* chart. Such a powerful Saturn (and forming a *Shasha Mahapurusha yoga*) lends discipline to the character, though not necessarily contentment. Allen has always been a very hard worker and extremely productive, with at least one film per year. Though not an ascetic, he is a tee-totaler and modest in his dietary habits. Moon-Saturn prominent in the chart adds to his self-contained quality, at times bringing the sense of gloom and doom.

At the same time, **Jupiter's influence** on the Sun is also strong, being combust the Sun within four degrees. Moreover, Jupiter overwhelms Mercury, as they are within one degree of each other in a Planetary War (***Graha yuddha***) and Jupiter wins the war. While Jupiter's effect is expansive, Saturn's is contractive. Jupiter is pushing the energy outwards, but Saturn is simultaneously pushing it inwards – "the morbid introspection" Allen describes. Thus we see a real pressure-cooker effect on the person. The release comes through **Mars in Capricorn**, driving him to work and to compete, even though Allen has said "art shouldn't be competitive." Also, the Sun-Jupiter-Mercury combination resides in the emotional sign of Scorpio, ruled in turn by Mars, planet of action. Mars is exalted and *Vargottama* (repeats in the *Navamsha* in the same sign), and in the natal 6th house of competition. Mars thrives here, though it also attracts legal suits, especially from partners, with 6th lord Saturn in the 7th house of marriage and partnerships. Saturn rules over Capricorn, making Mars in Capricorn more effective in the world. With a prominent Saturn, one is ambitious, serious and always seeking to establish a strong material base. Allen found ways to make extra money from the age of 15 and earlier, auditioning for television shows and selling his comedy one-liners professionally by age 16, when he also changed his name.

For many years Allen was a self-professed hypochondriac. But since early 1992 and the developing relationship with Soon-Yi Previn, Allen shifted more to touting his vitality and excellent lineage for healthy longevity, as both his mother and father lived to ages 95 and 100, respectively, and in good health. But his parents' longevity may not coincide with his own, as shown by the astrological indications. In this case, however, longevity, good physical health and vitality look excellent. His natal Sun is *Vargottama* and in the sign of a good friend, Mars, which in turn is very strong as a *Vargottama* planet that is also exalted. These factors give excellent physical vitality and promote good longevity. Saturn's close aspect to the Ascendant and Sun affect the physical being and persona, giving a slim, wiry body, moderate to short height, and a serious demeanor. Saturn-Moon contacts bring the emotional contraction, perhaps for self-protection, along with rigorous work habits. Saturn is a *Maraka* planet, but its *Dasha* was highly successful. Venus is also a *Maraka* planet, located in the 2nd house. (*Maraka* houses are 2 and 7.) Thus, the 20-year Venus *Dasha* could leave him vulnerable, but Allen will be almost 80 years old by its start in late 2015.

In summary, there is a double pressure cooker situation in the chart of Woody Allen caused by Jupiter and Saturn. This is further both elevated and alleviated by exalted Mars and by the sheer power and influence of Jupiter, Saturn, and Mars on so many key factors in the chart. It results in a person who is ambitious, conscientious, and dedicated to his work – though he will not be conservative in his private life or in his art. The outrageous and zany content of his material comes from natal Mercury which is both in control and out of control – it is so overwhelmingly pushed into gear by Jupiter first of all and secondly by a combination of Sun, Mars *and* Saturn (hot *and* cold, fast *and* slow). The simultaneity of these disparate energies, combined with excellent cerebral and physical energy, creates an ongoing mental and emotional searchlight reaching in many directions at once, aided by the sharp-focusing lens of Saturn, the discriminator. Because of Saturn's ability to provide the necessary discipline and focus in this chart, it raises Allen from mere star status to that of a legend.

The ***Navamsha* chart** gives further confirmation of such characteristics. Mercury, Jupiter, and Moon reside in Libra in the 5th house (of creativity) on the Rahu-Ketu axis, aspected by Saturn in Capricorn. Sign lord Venus is *Vargottama*, repeating its debilitation in Virgo. This flaw is corrected in part by its 4th house location (*Digbala* for Venus), and the *Parivartana yoga* (mutual sign exchange) with Mercury – an excellent *Raja yoga* for high intelligence and the ability to speak sweetly or eloquently. The zany quality comes from the simultaneous influence of Rahu, Ketu, Jupiter and Saturn. Rahu and Ketu are the foreigners, so give the unexpected incongruities and the element of surprise. Jupiter expands Mercury, making it more talkative, even verbose. Saturn's influence is discriminative, pushing him to be more concise and practical within the rambling, whacky realms he creates. Simultaneous worldly influences from Mars, Saturn, Rahu, and Ketu can easily overwhelm a person, making it difficult to find the inner discipline to work and stay productive, as Allen has done year after year. These same worldly influences have pushed him away from religion and metaphysics and more towards the secular. But Jupiter's influence has given him enough optimism to stay interested in the unpredictable journeys of the human psyche. His observations are often astute and unexpected. The following scene from *Play it Again, Sam* (Paramount Pictures, 1972) is an example:

WOODY ALLEN: That's quite a lovely Jackson Pollock, isn't it?

GIRL IN MUSEUM: Yes it is.

WOODY ALLEN: What does it say to you?

GIRL IN MUSEUM: It restates the negativeness of the universe, the hideous lonely emptiness of existence, nothingness, the predicament of man forced to live in a barren, godless eternity, like a tiny flame flickering in an immense void, with nothing but waste, horror, and degradation, forming a useless bleak straightjacket in a black absurd cosmos.

WOODY ALLEN: What are you doing Saturday night?

GIRL IN MUSEUM: Committing suicide.

WOODY ALLEN: What about Friday night?

The *"I Hate School"* Success story: The Sun and Saturn are planetary enemies, and vie for authority with each other. With Saturn's aspect to his Sun there is a tendency to bridle under too much outside authority. This contributes to creating an iconoclast or a maverick. Allen is his own best taskmaster and his voluminous life's work is proof of that, despite his lack of success in formal schooling. His dislike for authority and for formal education is also shown through the influence of Mars. The 4th and 9th houses are related to education and each is ruled by exalted Mars placed in the 6th house of conflict. This has several effects: A classical malefic in the 6th house is excellent for dealing well with competition and gives excellent physical strength and stamina. But as 9th lord of *Dharma* situated in the 6th house, it brings some inner conflict or discontent regarding his path in life, even if he is successful in diverting from the expected path. The 5th house also relates to education: 5th lord Jupiter is combust the Sun and burning too hot. Rahu, the foreigner, the outcast is situated there.

All three planets in the 4th house are located in Saturn's *nakshatra*, *Anuradha*, making Saturn the *Jiva*, or life force, of the 4th house. Saturn's influence is the pressure to keep working and to avoid bad health and other problems it can bring as 6th lord (debt, disease, divorce). As a separative planet, Saturn's aspect to the 4th house of educational degrees can cut off the process and shorten the education. Mars as sign lord of the 4th house can shorten the formal education due to a restless and creative mind, combined with an ambitious nature. Saturn is 6th *and* 7th lord, and as lord of the 6th house of conflict, its aspect to the Ascendant and to his natal Sun shows a tendency to buck the authorities, starting with one's father and teachers. It makes him rebel against his formal education. Allen suffered through most of his schooling, as mentioned earlier. But he dazzled the authorities at opportune moments and started getting paid for his comic gifts from age 16. (At 15 he also took up the clarinet, favoring jazz and Dixieland.) Contributing to this is a *Saraswati yoga*. Saraswati is the Hindu goddess of learning. To qualify, one must have Mercury, Jupiter, and Venus in an angle, trine, or 2nd house of the birth chart. Jupiter, planet of knowledge, must be in its own sign, exalted, or in the sign of a planetary friend (Sun, Moon, or Mars). Thus we have the cerebral comic who may love to learn, but cannot bear the restraints of school and is far happier working independently. Moon in *Dhanishta nakshatra* gives a strong interest in music and often brings wealth and an interest in wealth, as it literally means "the wealthiest." The desire of *Dhanishta* is to revolve around the summit of the gods. Often the person is famous, if all else in the chart supports it.

From the age of three through age 21, he was in the 18-year Rahu *Dasha*. Allen remembers seeing his first movie at age three (*Snow White and the Seven Dwarfs*), and being so excited he rushed up to touch the screen. Rahu is a worldly planet, making us long to participate in the outer world. Strongly placed in the 5th house (of creativity) in both birth and *Navamsha* charts, Rahu resides in Jupiter's sign and Venus's *nakshatra Purva Ashadha* ("the earlier victor"). In **Rahu-Venus** *Dasha/Bhukti* (May 1950 – May 1953) he started playing the clarinet and earned his first moneys as a comic writer. Allen considers his life really began in 1952, when he took his new name.

All the astrological characteristics of a cerebral person are here, but far too unruly to make a scholar. Mercury is the trickster planet and the planet of communication. It can be elevated by Jupiter's influence, but in this case Mercury is overpowered and overwhelmed by Jupiter (losing to it in the *Graha yuddha*), and additionally by Saturn and the Sun, though Mercury is often close to the Sun. **Mercury-Jupiter** contacts give a profusion of speech. Exalted Mars (as sign lord of Sun-Mercury-Jupiter in Scorpio) gives speed and a competitive spirit to the speech, along with a savvy business sense. Jupiter is the planet of abundance and *Bhagya* (good fortune). Allen has often said, especially in later years, that he has been very lucky in his life. He was indeed lucky to have the 16-year **Jupiter *Dasha*** from age 21, with **Jupiter-Mercury** *Dasha/Bhukti* from June 27, 1961 to Oct. 03, 1963. In 1961 he made his debut as a standup comedian, with his first serious review in 1962. It was favorable. (He had already been writing for television.) None of this was associated with academic success, but was totally in synch with his inventive, more independent self, already married at age 19.

Love, Death, and When do we eat? As 7th lord of partnership, a prominent Saturn in its own house keeps Allen in almost constant partnerships – business, marriage or common law. Saturn here demands loyalty and constancy, and is strained when pitted against the wildness of Mercury-Jupiter-Sun in Scorpio, which are fairly popping out of the confines of societal norms. Moon and Saturn in the 7th house can be difficult for partnerships, but the best antidote to its negative side is either to have a working relationship with the romantic partner (as he did with actresses Louise Lasser, Diane Keaton, and Mia Farrow), and/or a big age difference between the partners, as with his third and most recent marriage to Soon-Yi Previn, who is 35 years younger. (He was also linked with a 17-year old actress in his 1979 film, *Manhattan*, played by Mariel Hemingway. According to actress Stacey Nelkin, it was based on a real-life relationship with her when she was a 17-year old high school student in NYC.) As planetary enemy of the Sun, Saturn's aspects on the Sun and Moon are troublesome for marital relations, which are already hampered by Moon in *Dhanishta nakshatra*, Moon and Saturn in the 7th house of marriage and partnerships, and debilitated Venus, even though it receives correction, to be discussed.

Venus in general rules over love matters, as does **Saturn** as **lord of the 7th house** of marriage and partnerships in this chart. The status of both planets assures a full and rich love life, though fraught at times and subject to frequent turbulence, including legal battles. This began in 1967 when Allen's first wife Harlene Rosen (1954-1959) sued him and NBC for "holding her up to scorn and ridicule." In 1967, tr. Saturn was in Pisces, the 8th house from Leo, and thus *Ashtama Shani*, the most difficult placement of Saturn from natal Ascendant or Moon. It was also opposite natal Venus. Reviewing Venus, we see it is *Vargottama* in Virgo, but is also *Atmakaraka* (planet at the highest degree of celestial longitude). Both of these factors are excellent for strength and status. However, Venus is debilitated in Virgo, and this brings some disadvantage in love matters: He is not conventionally handsome and has a tendency to over-analyze his relationships and to believe that romantic love is doomed. Also, his celebrity status made public any of his romances or court battles with partners. But due to the *Neecha Bhanga Raja yoga* (cancellation for debilitated Venus), drawbacks can be turned into advantages: 1) His actress-lovers have done well by him and vice versa; 2) They appreciate his cleverness; and 3) Existentialist *angst* about romantic love can make good comedy. The problem lies with Mercury's weakness as sign lord of Venus, overwhelmed by Jupiter – always bringing more choices of partners as part of the destiny. Mars as *Jiva* (life force) of Venus is powerful as an exalted *Vargottama* planet. It elevates his status as an attractive partner beyond what would be expected. It gives ongoing vitality and excitement to the love life as well as to the life in the arts. (Venus also rules over music and the arts).

The ***Navamsha* chart** also tells us about marriage and partnership issues. With a *Navamsha* Gemini Ascendant, the lords of the 1st and 7th houses are Mercury and Jupiter, respectively. These planets are together in Libra in the 5th house, showing a tendency for him to unite with partners due to creative projects and/or children. The 5th house concerns children as well as the fruitfulness of the mind – thus artistic creativity, and also love affairs. With contacts from Rahu, Ketu, Moon and Saturn, the influences on the 5th house are complicated. There are ongoing emotional and/or mental upheavals. With both Jupiter and Saturn (planets of *Dharma* and *Karma*) playing strong roles, Jupiter continues to provide new outlets and choices, while Saturn tames the wilder impulses. An intense curiosity about life (Mercury-Jupiter) keeps partnerships interesting.

Mercury and Jupiter at war: Mercury and Jupiter are within one degree of each other and in a ***Graha yuddha*** (Planetary War). Though Mercury is further north in latitude, Jupiter wins the war as it is: **1)** higher in celestial longitude, **2)** further north in declination, **3)** larger and **4)** brighter. The latter two factors are definitive. This normally means that the loser is diminished immeasurably, especially in its capacity as lord over the houses it rules, both financial houses in this case: houses 2 and 11. Since Mercury is *Dhana yogakaraka* (wealth significator), Allen could suffer financially due to Mercury's diminishment, especially during its major and minor *Dashas*. But there are some exceptions, as both its *nakshatra* lord (Saturn) and sign lord (Mars) are strong and well placed. Allen also has a special *yoga* for wealth here, as is noted in a classic work:

> "For a person born in Leo Ascendant, if the Sun, Jupiter and Mercury are combined together, good wealth is indicated."
>
> ***Bhavartha Ratnakara***, Ch. 1, *Simha Lagna* [Leo Ascendant], Stanza 2.

The *Graha yuddha* occurs in the 4th house (of domestic environment) in the natal chart, and in the 5th house of children and film arts in the *Navamsha* chart. The houses identify in what realm of his life the battle is likely to play out. The *Dasha* sequence tells us when: The major or minor periods of Mercury will be the most eventful for him, especially the first subperiod of **Mercury *Dasha***. The severity of the *Graha yuddha* is alleviated by the *Raja yoga* and *Parivartana yoga* (mutual exchange of signs) between *Navamsha* Venus and Mercury in the 4th and 5th houses in the *Navamsha* chart. This softens the financial and personal blow considerably, as do the *Dhana* and *Raja yoga*s formed by natal Sun, Mercury and Jupiter in a *Kendra*. Moreover, a strong Saturn enabled Allen to continue his prodigious filmmaking unabated, even in the midst of such personal upheaval.

The fateful Mercury *Dasha*: In late fall 1991 Allen's 19-year Saturn *Dasha* came to an end and his 17-year Mercury *Dasha* began. Saturn is the elder and Mercury is youth. Though influenced by both Jupiter and Saturn, Mercury's loss to Jupiter in the Planetary War gives Jupiter's optimism and leap of faith more sway during Mercury *Dasha*. Upon entering the *Maha Dasha* (major period) of a planet losing the *Graha yuddha*, its *karmic* consequences would now be due. This was in spite of the compensating good *yogas* and overall strength of his birth chart.

Allen's Mercury *Dasha* began Oct. 28, 1991. Exactly as Mercury *Dasha* began, Allen became more deeply involved with a woman 35 years younger, soon creating complete chaos in his inner and outer worlds. He would later say in court (in spring 1993) that he did not see the consequences of his actions. His love interest was Soon-Yi Previn, Mia Farrow's then 21-year old adopted daughter.[193] On Jan. 13, 1992, Farrow discovered the affair and this triggered the scandalous and very messy breakup of his 12-year personal and professional relationship with Mia Farrow, throughout 1992 –1993. Allen's press coverage was bad due to Farrow's accusations of child molestation combined with his new link with Soon-Yi. Although it was an unconventional family, it was also a close one in many ways. He was viewed as the custodial parent and she the stepdaughter. They both claimed it was never a father-daughter relationship and Allen saw no particular "moral dilemma." But this did not diminish the poor public perception. André Previn spoke out on Mia's behalf:

> "I am terribly shocked and saddened that he would choose to have a relationship with Soon-Yi. As a father, I don't think I have enough colorful vocabulary to tell you what I think. It is an unspeakable breach of trust which has caused a great deal of anguish in the family."
>
> **André Previn** quoted in *What Falls Away: A Memoir*, by Mia Farrow, 1997, p. 279.

Allen suffered his own personal turmoil as well as extra legal expense during Mercury *Dasha*, especially in its opening years but also later. Mercury-Mercury *Dasha-bhukti* (from Oct. 28, 1991 to March 26, 1994) included the breakup with Mia Farrow and their lengthy child custody battle for three of the younger children. (Two were jointly adopted, one of them their biological child.) In mid-June 1993 Allen eventually lost the custody battle, though charges of child molestation were dropped. Allen and Soon-Yi Previn married Dec. 23, 1997, during Mercury-Venus period. They later adopted two baby girls, born in 1998 and 2000. Their relationship with Farrow and her other children was effectively severed from Aug. 1992 onward. When questioned in 2001 about himself and Soon-Yi, Woody Allen said:

> "The heart wants what it wants. There's no logic to those things. You meet someone and you fall in love and that's that."
>
> Interview with Walter Isaacson, *Time magazine* June 24, 2001.

> "...It is ironic that my marriage to [Soon-Yi], which was seen by many as so irrational, to me is the one relationship in my life that worked."
>
> **Woody Allen** in David Kamp's review of the book *Conversations with Woody Allen* (2007, by Eric Lax), *The New York Times magazine*, Nov.18, 2007.

---

[193] Soon-Yi Previn was adopted in 1977 by Mia Farrow and then-husband, conductor André Previn, who were approved as adoptive parents in 1976 and one month later assigned to Soon-Yi. She was living at a State orphanage in South Korea, after being abandoned on the streets of Seoul. Though her birth date is listed as Oct. 8, 1970 in South Korea, the prospective parents were told in 1976 that "her precise age was not known but estimated to be around five; there was no other history." (Source: Mia Farrow, ***What Falls Away: A Memoir,*** 1997, p. 162.) In 1977 Farrow went to extraordinary lengths to bring Soon-Yi to the U.S., even launching an effort to lobby Congress to repeal a federal law then preventing U.S. families from adopting more than two foreign children. (They had already adopted two Vietnamese girls.) Once the federal law was changed, Soon-Yi joined the Previn family in 1977.

Mercury *Dasha* brought further personal and legal conflicts on and off, with the exception of Jupiter subperiod. In 1992, his longtime producers Orion Pictures went bankrupt. Tri-Star Pictures also appeared to drop Allen after his scandals of 1992, though he claimed his contract had simply come to an end. Meanwhile, to fill the gap, his longtime friend Jean Doumanian (formerly a producer for *Saturday Night Live)*, offered to finance and produce Allen's upcoming movies. She and her billionaire partner at Sweetland Films did so for the next eight films, most of which were not as successful, either artistically or commercially. Then, due to some major differences between them, in May 2001 Allen sued Sweetland Films for $12 million in moneys he said he was owed, and there were other areas of dispute. The court awarded Allen only $98,000 in legal expenses. Later the two parties reached an undisclosed settlement, but their 30-year friendship ended.

With the onset of Mercury *Dasha*, the attributes of Mercury, both good and bad, come to the forefront of the life. As the trickster planet, one might think this is the perfect planet to express Woody Allen, the clown. But Mercury is overwhelmed by Jupiter's potential for excess and loses to it in the *Graha yuddha*. Both Mercury and Jupiter are in turn combust the Sun, giving the burnout effect and more possibility of not being able to see the full ramifications of one's words and actions. Jupiter in itself has an expansive effect – especially in a watery sign, and Mercury rules over youth and youthfulness, as described. His birth chart shows many choices for partnership and the *Navamsha* chart confirms this. Thus, Allen may succumb to some outside influences at this time. Another immediate effect of Mercury *Dasha* was the abrupt end to over 30 years of psychoanalysis sessions. Jupiter is the advisor and wins the Planetary War with Mercury. Therefore Mercury's *Dasha* would either let the psychiatrist take over completely, or Allen takes over that entire role himself. He chose the latter option. His Sun is still the ruler of the chart and not *Dasha* lord Mercury.

Compounding the predictable volatility at the start of his Mercury *Dasha* is the heavy weight of his ***Sade Sati***, which occurs once every 22 years and lasts for 7½ years. (Saturn transits Houses 12, 1, and 2 from the natal Moon.) It generally brings greater responsibilities into the life and tests the mind and heart in major ways. It is considered pivotal in the life. This began for Allen Dec. 14, 1990 through April 18, 1998 and was his third *Sade Sati*. Since Allen's natal Moon is in the 7th house of partnerships, it shows the mental/emotional challenges would center upon partnerships – both personal and business. Saturn as both 6th and 7th lord shows the potential for legal conflicts with partners. His third marriage to Soon-Yi Previn (Dec. 23, 1997) occurred four months prior to the end of this *Sade Sati*. The matter was finally settled, legally in any case.

The **previous *Sade Sati*** began in early Feb. 1961, just prior to the Jupiter-Saturn conjunction in Capricorn very close to his exalted Mars, and coinciding with his **Jupiter-Mercury** *Dasha/Bhukti*, from June 27, 1961 to Oct. 3, 1963. With a strong confluence of factors, it triggered both the start of his new career as a standup comic in 1961 and his first professional review as a comic in 1962. The *karmic* destiny implied in the Mercury-Jupiter *Graha yuddha* was likely to come due at this time. At age 25, it laid the groundwork for him as musician, actor, playwright, screenwriter, and director, bringing both wealth and status. Even so, his early marriages were vulnerable. His first marriage ended in 1959 and there were ongoing love affairs until his second marriage Feb 2, 1966, simultaneous with his Saturn return to the 7th house of partnerships. The couple divorced in 1969.

With the converse combination, in **Mercury-Jupiter** *Dasha/Bhukti*, we can expect some important and life-changing events involving both career and partnerships. This period ran Nov. 12, 2003 to Feb. 17, 2006, some 13 years into Mercury *Dasha*. Financial adversities of some sort would be expected during Mercury *Dasha*, with Mercury as wealth significator (for Leo Ascendant) losing the *Graha yuddha* with Jupiter. But we also know **Jupiter sub-period** can reverse the downtrend of luck, as Jupiter is victorious in this Planetary War. Not only did Allen find new financial backers, he started making movies in Europe in 2004 – a fruitful decision, as it turned out. He wrote and directed his masterpiece *Match Point* (2005), set in London, UK, a film Allen considers his best to date. It also made $23 million in two years, far more than most of his previous films. He made a $7 million profit selling his Manhattan townhouse the same year and bought another on E. 70th Street. Further films followed, also set in Europe.

Saturn is the strongest planet in the chart, and has command over the three key planets in Scorpio. Thus its 19-year *Dasha* (Oct. 27, 1972 to Oct. 28, 1991) would be most likely to bring good results professionally and it did. The last sub-period, **Saturn-Jupiter** (April 16, 1989 to Oct. 28, 1991) also probably served to keep the continuity with Mia Farrow. His Mercury periods would be more eventful, but Saturn can tame Mercury's wildness. Another very successful sub-period artistically and financially was **Saturn-Mercury** *Dasha/Bhukti* (Oct. 31, 1975 to July 10, 1978). Allen's breakthrough movie *Annie Hall* (1977) received the most awards and pushed his artistry to a new level. This movie marked the start of his longest and best continuous period of artistic and financial success, 1977 – 1987. Many considered his films during this period of **"high cultural relevance."** Mercury's signature, even as loser in the *Graha yuddha* seems always to light up the Woody Allen legend.

Chart 23: James Taylor

Birth data: Friday, March 12, 1948, 17:06 EST, Boston, MA, USA Long. 71W03 37, Lat. 42N21 30, *Lahiri ayanamsha* –23:07:49, Class AA data (Birth certificate in hand from Frances McEvoy.) Ascendant: 21:57 Leo.

Biographical summary: One of the most popular and successful singer/songwriters of folk, rock and popular music in the modern era, James Taylor first broke through to national and international prominence with his 1970 album *Sweet Baby James*. It contained perhaps his best song, *Fire and Rain*, and the one for which he is best known. His introspective, bitter-sweet songs epitomized the country's transition from the radical 1960s into the less political, more inward-looking 1970s. He was tall (6 feet 3 inches), dark and handsome as he began his first concert tour in summer 1969 at age 21, and he has continued to perform around the world ever since. More public attention came with a *Time* magazine cover story in March 1971 and a celebrity marriage in Nov. 1972 to the gifted singer/songwriter Carly Simon (b. June 25, 1945, New York City.) But as his fame increased, so too did his problems with heroin, alcohol, and depression. The couple divorced in 1983.

Mo	Ve Ra		
Me Su	**James Taylor** Mar. 12, 1948 17:06 EST Boston, MA		SaR MaR
			As
Ju		Ke	

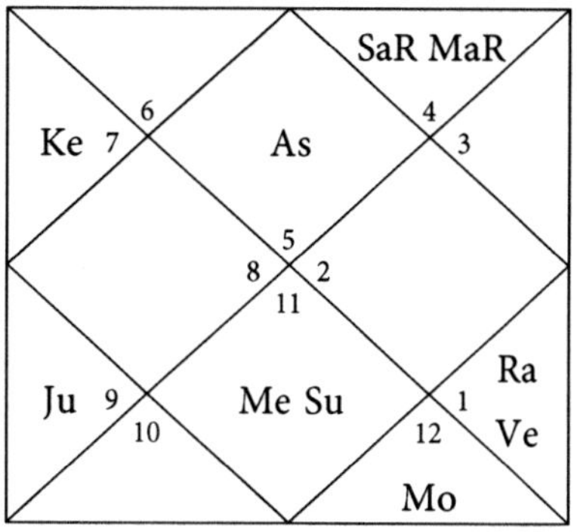

SaR MaR
Ke 7 6 As 4 3
8 5 2 11
Ju 9 10 Me Su 12 1 Ra Ve
Mo

Planetary Information

As	21:57	Leo	P.Phalguni
Su	29:07	Aqu	P.Bhadra
Mo	22:00	Pis	Revati
Ma	26:43	Can	Ashlesha
Me	02:02	Aqu	Dhanishta
Ju	04:06	Sag	Mula
Ve	12:09	Ari	Ashwini
Sa	23:42	Can	Ashlesha
Ra	23:53	Ari	Bharani
Ke	23:53	Lib	Vishakha

MaR		Ke Ju	Su
SaR	**Navamsha**		Ve
Mo			
	Ra	As Me	

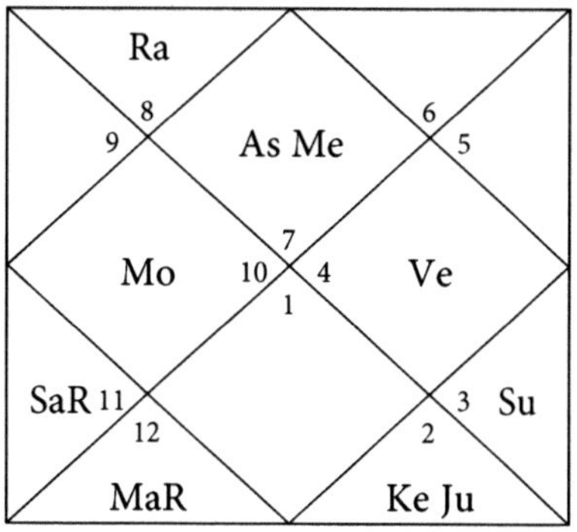

Ra
9 8 As Me 6 5
Mo 10 7 4 1 Ve
SaR 11 12 2 3 Su
MaR Ke Ju

Vimshottari Dashas

Me	May 20, 1941
Ke	May 20, 1958
Ve	May 20, 1965
Su	May 20, 1985
Mo	May 21, 1991
Ma	May 20, 2001
Ra	May 20, 2008
Ju	May 21, 2026
Sa	May 21, 2042

> "I've often thought that the drugs saved my life," says Taylor. "It's at a huge cost, of course. In the end, it's nothing but a waste of time. It's a lot like walking around dead to be permanently drug-addicted. It kills a lot of people. But generally speaking, over the course of my 18 years as a hard-drug user, I was self-medicating in order to function."
>
> **James Taylor** in an interview with Joan Anderman,
> *The Boston Globe magazine*, Oct. 1, 2006.

Much of the content of his song lyrics come from his personal experience and were initially inspired by hymns, carols and Woody Guthrie (1912-1967), singer/songwriter and folk musician. Taylor's best selling album, *Greatest Hits* (1976) sold over 11 million copies, reaching Diamond status. Other successful albums were released over the years: 15 solo studio albums of original music, and another 8 albums up to early 2007 consisting of revised and reissued releases. He produced his first 12 albums between 1968 and 1981, but considered retiring in the mid-1980s before his career was revitalized with a concert in Brazil in Jan. 1985. This in turn led to the release of his first album in four years: *That's Why I'm Here* (Oct. 1985). In a 42 year-period (1967-2009) he recorded and released 26 albums.

James Taylor was born into an affluent family, with four siblings, one older brother Alex (1947-1993), three younger siblings, Kate, Livingston and Hugh, and three step-siblings. His parents separated in 1971 and his father remarried. All his siblings are musical and Kate and Livingston have had careers as performing and recording artists. His mother Gertrude was a classically trained soprano and his father, Dr. Isaac Taylor, was a medical doctor, medical professor and Dean (1964-1971) of the medical school at University of North Carolina. In 1951 the family moved to North Carolina from the Boston area, though maintaining a summer home on Martha's Vineyard from 1953. The father's absence in 1954-1956 on a voluntary expedition to the South Pole and his subsequent descent into alcoholism had a profound effect on James and the entire family. His mother raised the five children virtually alone. Meanwhile, expectations were heavy that the siblings would follow the long Taylor lineage of doctors and lawyers.

James played cello as a child but switched to guitar in 1960. Sent to Milton Academy in Mass., he adjusted poorly there and left at age 16 to play in a band in New York City with his brother Alex. James suffered from mental depression and in 1965 he spent ten months at McLean Psychiatric Hospital in Belmont, MA. He finished high school there in 1966, played music and started songwriting. He formed his first professional group, *The Flying Machine* in 1966 and it was unsuccessful. The band consisted of Taylor on vocals and acoustic guitar, guitarist Danny "Kootch" Kortchmar, and drummer Joel O'Brien, who introduced him to a broader range of music and to hard drugs. From 1965 onward Taylor suffered from both depression and drug addiction (including heroin), and moved to London in 1968 to kick the habit. In London he signed with the Beatles' Apple label, and released his first album, *James Taylor,* Dec. 6, 1968 in the U.K. (Feb. 1969 in the U.S.). It was not a commercial success until after the release of his *Sweet Baby James* album in Feb. 1970 in the U.S. Nor was the heroin habit gone and he spent five more months in rehab in 1969. But his drug and alcohol addiction problems persisted. Along with his constant touring, they greatly contributed to the collapse of his first marriage in 1983. He returned to rehab once again that same year.

There were two more marriages: to actress Kathryn Walker (Dec.14, 1985 -1996), and on Feb. 18, 2001 to his girlfriend of six years, Caroline "Kim" Smedwig. When they met, she was the longtime Director of Public Relations and Marketing for the Boston Symphony Orchestra. Taylor has four children, two by his first marriage, both also singer/songwriters: Sally Taylor (b. Jan. 8, 1974) and Ben Taylor (b. Jan. 22, 1977); and two by his third marriage (their own biological children by a surrogate mother), twin boys Rufus and Henry, born early April 2001.

Taylor has been honored for his humanitarian efforts for the environment, arts education, and children's health programs. In addition to his most famous song *Fire and Rain* - a quintessential early Taylor work - others include: *Carolina on my Mind, Sweet Baby James, Country Road, Shower the People, Handy Man, Don't Let me Be Lonely Tonight,* and *Something in the Way She Moves.* He also sang songs by others, and is well known for performing Marvin Gaye's *How Sweet It is (to be Loved by you)* and Carole King's *You've Got a Friend.* He has received five Grammy awards, and numerous awards and honors from the recording industry as well as an honorary doctorate degree in 1995 from the Berkelee School of Music in Boston, MA. Berkelee says of its recipients:

> "Their contributions have the enduring qualities that define the musical era in which they played a leading role."

Intensely, intentionally personal: James Taylor is famous for putting autobiographical material into his songs, including his earlier stints in mental hospitals, or whatever is going on in his life. There is a self-reflective quality to the work of the Leo Ascendant person, as the Self is what they care about the most. It must be dealt with before anything else. This is especially applicable to writers, artists, and performers.

> "...Most of my work is, for better or worse, self-referred and autobiographical. I think everybody's writing music about themselves, essentially. But mine is admittedly so, and if it has value, it's that it's emotionally useful to people."
>
> **James Taylor**, interview with Joan Anderman, *The Boston Globe magazine*, Oct. 1, 2006.

Taylor also makes no secret of what he calls his "lust to perform." This is perfect for a Leo Ascendant person, as Leo is the primary royal sign and likes to be seen and recognized, and if various other factors are present – likes to perform in public. For instance, if the natal Sun is at an angle of the chart, especially 10th house or Ascendant, this is more likely. (This can also be examined from natal Moon.)

Ascendant lord Sun here aspects its own house from the 7th house. This gives some benefit to the strength and vitality of the physical body and to the self-confidence, though psychologically there is also some difficulty for the Leo Ascendant person who is born close to sunset, as the Sun – which likes to spread its light and influence – is about to disappear below the horizon. With the Sun in the 7th house, there is a focus on partnerships and/or a tendency to attract very strong partners, overwhelmingly so at times. This accounts for many songs about what he is feeling in relationship. Therefore, even though the Sun has strength in an angle and aspecting its own house, it is more problematic in House 4 or 7. (The Sun has strength in Houses 3, 6, 10 and 11, and receives *Digbala* in House 10.)

This Sun is also *Atmakaraka* (planet at highest degree), which fortifies the Ascendant lord, in spite of being in the sign of Aquarius – owned by Saturn, its planetary enemy. Saturn in turn is poorly located in its enemy's sign Cancer in the 12th house of loss and expenditure. It indicates the likelihood of internal conflicts and battles, including time of seclusion or time in 12th house locales – such as foreign residence, pastoral settings, hospitals or prisons, at times in the process of pursuing better health and well-being. The Sun's *Jiva* (life force) is Jupiter – the best planet for the Leo Ascendant chart, and *Nadi yogakaraka*.

Sources of protection and survival: In addition to the Sun's aspect to its own Ascendant, there are five other key astrological factors that serve to protect the physical and emotional body, and enabled Taylor to survive and somehow prosper through 18 years of alcohol and drug abuse: **1) Jupiter**'s aspect to the Ascendant in the birth chart, giving added protection to the physical body; **2) Sun**'s improved position in the *Navamsha* chart. Not only is it in the 9th house, but it is surrounded by benefic planets (a *Shubha Kartari yoga*), when classic benefic planets are in both adjacent houses; **3) Jupiter** is the Sun's *Jiva* (life force) planet, since natal Sun is situated in *Purva Bhadra nakshatra*, owned by Jupiter, fortunately a great good friend, situated in a favorable house – a trinal house; **4)** 6th house lord **Saturn** is placed in the 12th house, a *Viparita Raja yoga*, giving happiness, health, and fame. Saturn also aspects its own house (the 6th house of health matters), thus strengthening it. *****5)** The ***Navamsha* chart** is quite an improvement in many ways over the birth chart, with only temporal and classical benefics in the *kendras* (angles): Mercury, Moon, and Venus, with Moon and Mercury in their *Digbala* positions. With three benefics in the *Navamsha kendras*, this tends to give enough protection for the individual to survive physically and emotionally through severely turbulent periods, as long as *Navamsha* Moon is unafflicted, as it is here. Though in Saturn's sign of Capricorn, *Navamsha* Moon is aspected by benefics Jupiter and Venus. Moon in Capricorn also gives perseverance and good work ethics. And in general, the *Navamsha* chart indicates the life progression is generally in a more positive direction, in spite of whatever outward indications to the contrary, especially during Venus *Dasha*.

Marriage and partnerships: The Ascendant is in *Purva Phalguni nakshatra*, whose symbol is the conjugal bed and whose *shakti* (driving force) is procreation. Therefore both biological and artistic progeny will be pivotal in the life. It is ruled by planet Venus, and the deity Bhaga, god of wealth, especially wealth that is shared with the family or with the partner. There is usually a strong interest in finding a partner in life, as *Purva Phalguni* bestows much sexuality and passion and easily attracts partners. With natal Sun in the 7th house of partnerships, this puts tremendous fiery energy into this area of life and so can have a destructive effect. Also, the spouse can dominate overmuch, and at the very least is a very strong person. Mercury is a good planet for the Leo Ascendant chart. Its placement with the Sun is good for wealth (as the two planets form a *Dhana yoga* of wealth), and good for mental cleverness, as Mercury is in a *Budha Aditya yoga* with the Sun. However, Mercury in the 7th house can also bring a more mental or intellectual quality to the partnerships, and some mercurial qualities. Offsetting the intellectuality in partnerships is the 7th lord (Saturn) situated in the 12th house, which increases the sexuality with partners, as the 12th house rules over pleasures of the bed. However, the 12th house also signifies losses, expenditures or journeys to faraway places, and generally the dutiful Saturn does not relax into the 12th house arena.

The ***Navamsha* 7th house** of marriage and partnership is also afflicted, as 7th lord Mars is situated in the 6th house of sacrifice and conflict. But in general with this chart, there are bound to be ongoing sexual relationships, with partnerships and/or marriages going through some ups and downs. With a fixed sign on the Ascendant and two planets in fixed signs, including Ascendant lord Sun – there will also be a strong desire to maintain a committed relationship in theory – if not in practice. Taylor has been in marriage or partnership almost continuously since his early 20s. He has apparently found happiness and stability with his third wife, who is much more low-key and understated than he is, both in public and private.

Venus is *karaka* (significator) for love matters and for music and the arts. Situated in the 9th house in Aries on the Rahu-Ketu axis, Venus is 10 degrees from the Rahu-Ketu axis, but close enough for love matters to be destabilized by their influence. This gives some turbulence to both love life and career, as Venus rules 10th house of career. Natal Venus is aspected by Jupiter, Saturn, Rahu and Ketu – giving a lot of action in this area of life. Venus is both in Ketu's *nakshatra* (*Ashwini*), and aspected by Ketu, giving an exotic and otherworldly aspect to his aspirations in both love matters and his music. Ketu is the *Jiva* (life force) for Venus. Situated in the sign of Aries, Venus is also impacted by its sign lord Mars and the condition of Mars in the chart – which is weakened by its 12th house location. There it is hidden and has to work out a lot behind the scenes. Though in its debilitated sign of Cancer, Mars achieves some correction through its contact with natal Saturn (lord of Capricorn, where Mars is exalted). This means he can rise up out of some disadvantaged situation and succeed unexpectedly, so long as he can find some self-discipline. Although there is an initial sense of loss and direction in life, he battles it out to get results. This is echoed by the *Nirbhagya yoga*, where the 9th lord of *Dharma* is situated in a *Dusthana* house (6, 8, or 12), weakening the affairs of the 9th house. The *Raja yoga* of Mars and Saturn is afflicted by being in the 12th house of loss (and hospital settings), but he has also brought his personal emotions out into the public through his music. He has even gained from them financially, reflecting the *Dhana yoga* of wealth formed by Mars and Saturn from the Moon, ruling the 2nd and 11th houses from the Moon. Also, his first big break occurred when he moved to London in 1968 and recorded his first album abroad (12th house), released in Dec. 1968.

Tough enough to be sensitive: When a singer is remarkable within a given era, it means they have somehow touched a nerve within the public through their voice and their music. For decades audiences have connected in an intense way with the warm baritone of James Taylor, and with the evolving man behind the voice. According to friend Jimmy Buffet:

> "[We each started out as] a young angry folksinger [and progressed to] a middle-aged not-so-angry folksinger.... We had to either clean up or die."

Beyond that, Taylor had a way of sharing deep into his heart through his songs, and this has earned him a large audience and many devoted fans and friends.

> Hey mister, that's me up on the jukebox
> I'm the one that's singing this sad song
> Well, I'll cry every time that you slip in one more dime
> And let the boy sing the sad one, one more time.
>
> *Hey Mister, That's Me up on the Jukebox*, by **James Taylor**, 1971.

His gift for friendship is shown by a strong Mercury (ruler of the 11th house of friends), the same planet that rules over the 2nd house of voice. (Houses 2 and 11 are also houses of financial income). Fortunately it is extremely well situated in both birth chart and *Navamsha* chart, in angles of the chart and in air signs, and in *Dhanishta nakshatra* ("the wealthiest"). Mercury prospers in the air signs, as it communicates easily in the air element. It also gives financial benefit through Taylor's own efforts and through his communications – his songs. He is also multi-lingual, speaking English, French, and German.

Navamsha Mercury is in Libra on the Ascendant, where it is favored in the best angle of the chart (*Digbala*), ruled by Venus (music and the arts), placed in the *Navamsha* 10th house of status and career. *Navamsha* Mercury in the Ascendant creates an excellent *Raja yoga*, with 9th lord in the Ascendant, confirming the financial wealth indicated in the birth chart. Confirming the wide network of friendships and associations is his *Durudhara yoga*, in which planets (excluding Sun, Rahu, and Ketu) are situated on houses either side of the natal Moon. In this case, results are even better with two classical benefics Mercury and Venus. Coincidently, his only #1 hit single to date is his rendition of Carole King's *You've Got a Friend (*1971*)*.

Fire and Rain **– the addiction issues:** The second house has to do with what comes in and out of the mouth: eating (and drinking or drug) habits, speech and the voice. We examine it from the Ascendant and from the natal Moon. Everything that makes the voice unique and wonderful can also contribute to problems in some way. In this case, both Saturn and Moon aspect the 2nd house from the Ascendant, causing afflictions to this house. As lords of the 6th and 12th houses, respectively, both *Dusthana* houses, they can bring difficulties to this area of life. Moon is emotional, and imbibes what it will – driven by feelings. Saturn can cut off one's speech, make it more spare, and make one more careful in what one says. But as a planetary enemy of the chart, it can cause affliction to the 2nd house. So one might eat less and take some other substance instead. Nevertheless, the strength of 2nd house lord Mercury compensates greatly for these deficits – though it cannot entirely remove them from acting upon the destiny.

Reading from the Moon in Pisces as 1st house (and sub-Ascendant), Venus is 8th lord, thus a temporal malefic. Situated on the Rahu-Ketu axis, it lends unpredictability and potential turbulence.

Natal Moon is extra potent as an unwieldy emotional force, as it is exactly on the 8th house cusp from the Leo Ascendant. Further, natal Rahu (poison) is exactly on the 2nd house cusp from the Moon, closely aspected by Saturn. These are a lot of malefic aspects to the second house. Jupiter's aspect to the 2nd house from the Moon is protective. (Taylor is financially prosperous and still in good voice and good health at age 60 plus.) But Jupiter's aspect to Venus and Rahu brings both abundance and the possibility of excess, seeking after the exotic. During Venus *Dasha* he had a reputation for womanizing and hard drugs.

Mental health issues: Several planetary factors indicate problems to this area of the life. First is natal **Moon** exactly on the 8th house cusp from the Ascendant, indicating intense emotional suffering – which can also lead towards greater intuition and *Moksha* (spiritual liberation). Fortunately, it receives correction through its *Viparita Raja yoga* (12th lord in the 8th house) and is much better placed in the 4th house of the *Navamsha* chart – *Digbala* (best possible angle). Further, the 4th house is the foundation of the mind and the happiness of the heart. Thus we examine the condition of the 4th house and its lord. As there are no planets in the 4th house and no aspects to it, our attention goes to the *karakas* for the 4th house (**Moon** and **Mercury** – both discussed earlier), as well as 4th house lord **Mars.** Mars is debilitated in Cancer and located in the 12th house of loss and expenditures. Taylor's first major sense of loss was when his father went on a voluntary expedition to the South Pole, 1954-1956, when he was 6 years old. For the Leo Ascendant person, the father's presence or absence is of major importance. Also for Leo Ascendant, Mars rules over both mother (4th house) and father (9th house) and so relates to the experience of both parents. (We can also examine the *Karakas* for father and mother: the Sun and Moon, respectively.) When the 4th house lord is debilitated and situated in the 12th house of loss (as in the birth chart), one can experience an early loss within the family that destabilizes the mental and emotional foundation. All of this can be good for *Moksha*, of course, though the mother can suffer a lot as well. Natal **Mercury and Mars** are in a *Nadi yoga* (*nakshatra* exchange), exacerbated by Mars placed in the 12th house. Mars can sharpen or over-agitate the intelligence and/or nervous system (Mercury).

Fortunately, this **Mars** also enjoys *Neecha Bhanga Raja yoga*. Although in its debilitated sign, it receives correction through its conjunction with Saturn (its exaltation lord). This brings the likelihood of rising up out of a disadvantaged situation to achieve something much better. Such could definitely be said of James Taylor's years immediately following his time in various mental hospitals – including McLean Psychiatric Hospital in Mass. He capitalized on his experiences there and made it his trademark early on. One of his earliest songs in 1969 was *Knocking ' Round the Zoo*:

> "... There's bars on all the windows and they're counting up the spoons...."

McLean provided Taylor with a reassuring daily routine:

> "Above all, the day was planned for me there, and I began to have a sense of time and structure, like canals and railroad tracks."
>
> **James Taylor**, *Time* magazine, March 22, 1971.

His lyrics to *Fire and Rain* speak eloquently of his struggle with both depression and drugs, and of the loss of a friend (Suzanne) whom he had known at McLean hospital and who committed suicide. His friends kept it from him until he had completed his debut album in London in fall 1968:

Just yesterday morning, they let me know you were gone.
Suzanne, the plans they made put an end to you.
I walked out this mornin', and I wrote down this song;
I just can't remember who to send it to.

I've seen fire, and I've seen rain.
I've seen sunny days that I thought would never end,
I've seen lonely times when I could not find a friend,
But I always thought that I'd see you again.

Won't you look down upon me Jesus?
You gotta help me make a stand.
You just got to see me through another day.
My body's achin', and my time is at hand.
I won't make it any other way....

The agony and ecstasy of Venus: Taylor's addiction problems with hard drugs and alcohol lasted for 18 years: 1965 – 1983. This 18-year period was contained entirely within **his 20-year Venus *Dasha*.** It echoes the difficulties indicated above from the condition of Venus in the birth chart. Although natal 10th lord Venus (music) is placed in the 9th house, excellent for career, it is on the Rahu-Ketu axis, bringing turbulence throughout most of his Venus *Dasha*, both in his first marriage and with his addiction problems. (We have noted Venus-Rahu in the 2nd house from an afflicted Moon.)

The beginning of the 20-year **Venus *Dasha*** in May 1965 brought initial trouble for Taylor, but within a few years it also brought him fame and fortune through his music. Thus it could be said that Venus has brought him his greatest struggles and obstacles, along with his first greatest artistic creativity and successes, his first marriage and children. Themes of excess as well as desire for continuity ran simultaneously throughout most of his Venus *Dasha*. The worst of it subsided as of Dec. 21, 1984, when transiting Saturn exited the sign of Libra. (Saturn's 2.3 year passage through the 8th house from his natal Moon would bring intense emotional affliction, and/or spiritual progress – often inextricable). Exactly five months later the 20-year Venus *Dasha* was over, and Sun *Dasha* began. Indeed, Taylor was considering retiring from his musical career in 1983-84 in the midst of Saturn's fire test. But his career became revitalized in Jan. 1985 with his appearance at the Rock in Rio concert in Brazil. Later the same year there were two happy events: the release of his next album Oct. 1985 (the first in four years), *That's why I'm here*; and his second marriage to Kathryn Walker in mid-December 1985.

The root cause for Taylor's problems with drug and alcohol addiction would seem to relate to his depression and anger from feeling abandoned by his father. Dr. Isaac Taylor became more and more remote from the family, returning from two years at the South Pole with what James calls "the family demon" – alcoholism. His mother raised the five children virtually as a single parent. In addition, the expectation of following a long Taylor lineage of doctors and lawyers lay heavy on his shoulders and those of his siblings. All of them suffered in various ways, most of them also spending time at McLean Psychiatric hospital. Brother Alex died of heroin addiction at age 46 (1993). Their father died three years later. Ironically, Taylor would give special credit to his father for helping him to find stability through various stints in psychiatric institutions, though in that era they were not yet really equipped to handle serious drug addiction. In any case, for the Leo Ascendant person the father is a pivotal figure and can have exaggerated importance in the life.

To understand how Venus *Dasha* is likely to operate, we must understand the condition of ***Dasha* lord Venus** within the birth chart and *Navamsha* chart. In the birth chart, Venus rules houses 3 (music) and 10 (career and status), and is situated in the excellent 9th house. From the 9th house it aspects its own house – the 3rd house – strengthening it. As 10th lord, Venus is well situated in the 9th house and aspected by 5th lord Jupiter: thus a *Raja yoga).* But in general Venus is made more volatile by its contact to Rahu and Ketu and by its location in *Ashwini nakshatra*, giving Ketu power over it**. Venus** improves greatly in the *Navamsha* chart. Located in the 10th house in a watery emotional sign, it is opposite its sign lord **Moon** in Capricorn, perhaps indicating the sweetness *and* bitter-sweetness of his songs – many of which relate to the happiness of the heart (4th house), and to love matters with women (Moon and Venus). The improvement of Venus in the *Navamsha* chart shows the likelihood Taylor will move in a more positive direction during an emotionally fraught, but highly creative **Venus *Dasha*** (1965-1985). What saves this entire picture and enables Taylor to both survive his addictions and keep producing successful artistic works? It is the vast improvement in the entire *Navamsha* chart over the birth chart, especially in the excellent house positions of Venus and Moon, as well as Mercury. *Navamsha* Venus bodes very well for career as it links both *Navamsha* Ascendant lord (Venus) and *Navamsha* 10th lord (Moon), the latter aspecting its own house. Unafflicted, Venus is in a *kendra* from other classical benefics and well placed house-wise.

Venus-Moon turns the tide: Both birth and *Navamsha* charts indicate what he must undergo *karmic*ally and when, but the *Navamsha* is critical. By examining each *Dasha* lord in turn, we can ascertain how its *Dasha* is likely to bear fruit. As we shall see in the chronology below, it was in his **Venus-Moon** *Dasha/Bhukti* (Sept. 19, 1969 to May 21, 1971) that he experienced his first major breakthrough to international status as a singer/songwriter. It was the absolute turning point in Taylor's 20-year Venus *Dasha*, and would bring his first notable success, even though he had had some great opportunities during the previous period of **Venus-Sun** (Sept. 19, 1968 to Sept. 19, 1969), including live concert appearances in summer 1969 and the release of his first album out of London, *James Taylor*, in early Dec. 1968 on the Beatles' Apple label, with George Harrison and Paul McCartney on individual cuts. But its impact on the public was inconsequential at the time. Though the Apple album was not successful, Peter Asher – Taylor's producer at Apple – believed strongly in his musical talents. Asher left Apple and the UK and moved to Warner Brothers in the U.S., where he produced Taylor's next album the following year. Asher then became Taylor's producer for several decades. (Born June 22, 1944 in London, Peter Asher is a guitarist, singer, and record producer, very successful in his own right.)

We know that **Venus-Moon** *Dasha/Bhukti* would boost Taylor's career well beyond the previous sub-period due to the placements of Venus and Moon in the ***Navamsha* chart**: Venus is in the 10th house of career and status in the watery, emotional sign of Cancer, opposite Moon in Capricorn. Both *Dasha* lord and *Bhukti* lord are in *kendras*, and both rule *Kendra* houses. Furthermore, a third benefic planet – Mercury – is in the *Navamsha* Ascendant in its best possible angle, receiving *Digbala*. Transits of Saturn and Jupiter to the birth chart and *Navamsha* charts are also beneficial, with tr. Saturn approaching the top of the birth chart from the 9th house. There are many improved planetary placements in the *Navamsha*. Only marriage is still adversely affected, as 7th lord Mars is situated in the 6th house of sacrifice and conflict. But career will improve, and over the years he has developed a warm relationship with his audiences with his quiet, easy-going way. Initially he was more shy and removed.

The first few years of Venus *Dasha* prior to the Venus-Moon period would be likely to bring mixed results, including some big problems. This is because natal Venus is more afflicted than *Navamsha* Venus and therefore it must bring the *karmic* fruits of what is described in these two charts, moving in chronological order from birth chart to *Navamsha*. If the *Navamsha* chart were worse, it would be the reverse, starting more positively – with a tendency to descend. This is especially true if *Navamsha* Moon is afflicted by sign, house or aspect. Fortunately this *Navamsha* Moon improves radically from the birth chart, making a huge difference in the destiny. In the birth chart, Venus is conjoined with Rahu and aspected by Jupiter, Saturn, and Ketu. Since Saturn, Rahu and Ketu are malefic planets and enemies of Leo Ascendant, they have to bring some negative results. These will relate to: **1)** the affairs of life ruled by Venus – love, the arts, diplomacy in life; and **2)** the affairs of the houses over which Venus rules, houses 3 and 10: communication and siblings, and career and status, respectively.

Like a Rocket: Sudden escalation of status through professional success and/or marriage can be identified in Vedic astrology as to: **1)** whether it will occur at all, and if so, **2)** when it will occur. The fruits of the *Karma* can be timed best through the *Vimshottari Dasha* sequence, and we study which planet(s) are likely to produce more extraordinary results during their *Maha Dashas* and Sub *Dashas*. Secondly, we look at the transits of Saturn and Jupiter, Rahu and Ketu, especially Saturn, and its relationship to natal Ascendant and Moon, and to transiting Jupiter. We also watch the transiting planet that is the current *Dasha* lord. (When it is a much faster moving planet, its transits will be less important.) Reading from *Dasha* lord Venus as a Sub-Ascendant, Moon is in the 12th house, a place of some difficulty for the emotions. (This Sub-Ascendant is relevant only during the 20-year Venus *Dasha*.)

Below is a list of events with astrological corollaries. These help to define the most pivotal era of his early life in the establishment of his career with the debut album, his first marriage and family, in the years surrounding 1965-1972. Setting the stage for the most serious achievements in his life was the ***Sade Sati***. This is the approx. 7.5 year period when Saturn transits through Houses 12, 1, and 2 from natal Moon. It invariably brings more responsibilities in life and often more success, though mental and emotional anguish can accompany that if the natal Moon is afflicted or not well placed house-wise. Natal Moon in Houses 6, 8, or 12 causes the most trouble, the 6th house placement being the least difficult, and 8th house placement the most difficult, as 8th house is associated with radical transitions, suffering, loss and chronic illness. The Moon is especially difficult exactly on the 8th house cusp from the Ascendant (by degree), as in this case:

1. ***Sade Sati*** : Jan. 28, 1964 to April 29, 1971. **Tr. Saturn in Aquarius**: Jan. 28, 1964 to April 9, 1966; **Tr. Saturn in Pisces**: April 9, 1966 to June 18, 1968; Sept. 29, 1968 to March 8, 1969. **Tr. Saturn in Aries** June 18, 1968 to Sept. 29, 1968; March 8, 1969 to April 29, 1971. With natal Moon in the 8th house, the worst of the suffering was likely to occur while tr. Saturn was in both Aquarius and Pisces, especially in Pisces. In 1964 James Taylor dropped out of Milton Academy. In 1965 he moved to New York City. His first album, *James Taylor*, was released Dec. 6, 1968 in London. On Dec. 21, 1968 Saturn turned Stationary Direct at 25:16 Pisces, within 3 degrees of his natal Moon, a very intense and sobering marker. Tr. Jupiter was then opposite Saturn by sign. Also, Rahu-Ketu was on the Pisces-Virgo axis Jan. 14, 1968 to July 9, 1969. This is contacting his natal Moon in Pisces as well as the 2-8 house axis. Second

house is voice (and food or drug intake); 8th house is inherited money and also emotional suffering and loss.

2. ***Ashtama Shani*** **- Tr. Saturn in the 8th house from the Ascendant and/or natal Moon**: This is a time when people tend to suffer the most emotionally, especially if the Moon is afflicted or placed in the 8th or 12th houses in the birth chart, though spiritual practice can minimize suffering and take one deeper into *Moksha* (spiritual liberation). Occasional exceptions to this occur, especially if tr. Jupiter conjoins tr. Saturn in the 8th house. (See Winston Churchill, Chart #24.) If mental or emotional problems already exist, Saturn's transits through these houses relative to the birth chart will be the make or break period – when the person will have to find equilibrium amidst inner turmoil or suffer bitter, irretrievable consequences. In the case of James Taylor, transiting Saturn in Pisces coincided with the middle of his ***Sade Sati***, an already demanding period.

   a) **Tr. Saturn in Pisces** (in the 8th house from Leo Ascendant): April 9, 1966 to June 16, 1968. In this period James began his heroin use (1968) and immersed himself in the music world after ten months at McLean hospital where he was given the anti-psychotic drug Thorazine, but found solace in the regular routines.

   b) **Tr. Saturn in Libra** (in the 8th house from natal Moon in Pisces): Oct. 7, 1982 to Dec. 21, 1984; May 19, 1985 to Sept. 17, 1985. His drug habits worsened and wife Carly Simon gave him final ultimatums. He refused to curb his incessant touring or his drug use. They divorced in 1983 and he entered rehab for the third time since 1965. In 1983-1984 he considered retiring from his musical career.

   c) During the **preceding Tr. Saturn in Libra** (1953-1955), Taylor suffered the absence of his father, whose two years away from the family (1954-1956) and subsequent alcoholism and withdrawal from the family began at that time and was probably a central catalyst for James' mental depression.

3. His 20-year **Venus *Dasha*** began May 20, 1965. While in New York City, he suffered from mental depression and sought treatment at McLean's Psychiatric Hospital in Belmont, MA for ten months in 1965-1966. There he finished his high school education and played music. He wrote the song *Knocking 'Round the Zoo* in this era. The first sub-period, **Venus-Venus** ran May 20, 1965 to Sept. 19, 1968. He resided in London, UK during **Venus-Sun** period, Sept. 19, 1968 to Sept. 19, 1969, signing with the Beatles' Apple label, and releasing his debut album, *James Taylor*, on Dec. 6, 1968 in London; in Feb. 1969 in U.S. After five months in rehab he began his first live tours in July 1969, including the Newport Folk Festival, but got detoured soon after by a motorcycle accident, breaking both hands. He was unable to play guitar for several months.

4. **Venus-Moon** *Dasha/Bhukti*, Sept. 19, 1969 to May 21, 1971 was the most pivotal subperiod of the entire Venus *Dasha*. It encompassed the following events:

   a) Dec. 1969: *Sweet Baby James* album was recorded. Its release was in Feb. 1970 on Warner Bros. label, with a 10-member band, second vocalist – Carole King. It was immensely successful within one month, especially due to the hit single *Fire and Rain*. Interest began in Taylor's debut album (*James Taylor*, originally released Feb. 1969) on the Apple label, notably in the song *Carolina on my mind*. In 2003, *Sweet Baby James* was ranked #103 on *Rolling Stone* magazine's list of *The 500 Greatest Albums of All Time*, and #77 on greatest albums listed by TV Network VH1.

b) March 22, 1971: *Time* magazine cover story on James Taylor.

c) Spring 1971: third James Taylor album released, *Mud Slide Slim and the Blue Horizon*, going Platinum the first month (over 1 million copies sold).

d) Spring 1971: First Grammy award for Best Pop vocal performance.

e) Spring 1971: His parents separate.

f) Spring 1971: Meets Carly Simon. Their relationship begins that fall.

REVIEW OF LEO ASCENDANT CHARTS:

Give the astrological reasoning in each case.

1. Why does the Leo Ascendant person tend to focus on themes of self-identity and autobiography? What in his chart indicates Yogananda's more spiritually oriented destiny? Why might this also entail an extended residence abroad? Hint: Note the impact of 12th house planets (either 12th lord or occupant of the 12th house).
2. Why is a planet combust the Sun more pivotal for a Leo Ascendant person? Both Woody Allen and Maya Angelou have Jupiter combust the Sun. Why then is Angelou very religious in her personal outlook while Allen is distinctly secular?
3. How can the same combination describe Yogananda's talent for writing and publishing as well as his ability to be a good fund-raiser?
4. At age 7 Maya Angelou stopped speaking for 5 1/2 years. What astrological factors indicate she would be very unlikely to stay mute for the rest of her life?
5. What astrological factor does James Taylor share in common with Maya Angelou, making their 2nd house issues highly emotional? Hint: Study the 2nd house, its lord Mercury, Mercury-Mars contacts, as well as Moon's aspects to the 2nd house.
6. Why would Woody Allen be unlikely to succumb to drugs or alcohol? Hint: Note Saturn's strong influence on natal Sun, Moon, Ascendant, plus two other planets. This also accounts for his strong work ethic.

7. What in Woody Allen's chart makes him a reluctant celebrity? Hint: Note house and sign position of Ascendant lord Sun.
8. What in his chart describes the wild and unexpected humor of Woody Allen? What keeps him high spirited? What describes his ongoing existentialist angst?
9. What do Maya Angelou and Woody Allen have in common in their birth and *Navamsha* charts? Why would they both be so prolific? What other characteristics would they probably share? Hint: Check Mercury and Jupiter contacts.
10. What are the astrological factors that describe how James Taylor's toughest period is also his most prolific and successful one?
11. Why is James Taylor less secretive about his private life than Woody Allen?
12. What planetary combination explains the more chaotic situation within James Taylor's family of origin that Woody Allen did not experience? Why so? Hint: Study the condition of 4th lord Mars as well as planets in the 4th house.

CHAPTER 8

Virgo Ascendant

Introduction

The Sanskrit word for Virgo is ***Kanya***, meaning "the maiden" or "virgin." Its symbol is a young woman, often pictured in a boat, holding wheat (or corn) in one hand and fire in the other.[194] Given this symbol, we know Virgo is greatly concerned with purity – both in food and agriculture and in mental and verbal discrimination. And if not, the body strongly rebels and may well suffer ill health. Virgo is often asked to carry the fire of dispute and to play the role of arbiter. It is a dual sign – *Dwishvabhava*, in the earth element, with the Vedic noble aim of ***Artha*** – oriented towards material security. Ruled by Mercury, it shows the more practical, logical energy of Mercury. In the dual earth sign – there is a versatility that seeks practical avenues and goals. Virgo is associated with the ***Shudra*** (servant) class. It refers to those who serve with their labor to get things done on a practical level, often producing something tangible in physical terms. Mercury is a neutral planet highly influenced by its sign, house, and *nakshatra* position, as well as by aspects and contacts from other planets. Exalted in Virgo, Mercury in this sign is associated with high intellect and with an excellent level of discrimination, especially as a sign closely linked with work, service and the business community. Virgo can be elevated in these areas, though it often has to deal with tough competition, critics and condemnation in the process of achieving its goals.

All the meanings of the 6th house come into the 6th sign of Virgo: debt, disease, enemies, and competitors. Thus Virgo can bring struggle in various forms, including with health, conflict, and sacrifice. With Virgo, or 6th house emphasis, we are dealing with all these issues, some of them naturally difficult – as they bring discord and the collision of opinions. Virgo Ascendant or emphasis can be argumentative and overly critical, depending on the strength and interactions of Mercury and Mars, but also highly skilled in debate. Virgo emphasis often requires the destiny to deal with dissension of varying degrees, including wars, or correcting wrongs – through military

---

[194] "Virgo is a woman standing in a boat carrying with her corn and fire." ***Jataka Parijata***, Chapter 1, *Shloka* (verse) 8.

means or through legal or medical procedures. This includes arbitration or mediation. Similarly, if an important transit or *Dasha* focuses on this area in the chart, there may be a short-term destiny to deal with such matters. Virgo is associated with the working efficiency of the body, which puts emphasis on health, but also on work itself as an efficient process, especially the service aspect of work. Occupations often associated with Virgo or the 6th house are these: medicine or the healing arts, diagnostics, law, the military, accountancy, statistics, cleaning or dry cleaning.

As with Gemini Ascendant, also ruled by Mercury, the physical body may be overly sensitive – as the Ascendant lord rules over the nervous and digestive systems and the skin, making the person more vulnerable to input from the outside world, but often equally gifted for understanding what to do with that input. In the earth sign Virgo, results can be very practical, and with prominence of *Hasta nakshatra* it may produce a concrete body of work. Health and speech may be adversely affected if Ascendant lord Mercury is poorly placed and aspected, as well as the *karakas* for vitality: Sun and Mars. With Virgo Ascendant, Mars is well placed in its own sign in the 8th house, as is the Sun in the 12th house from the Ascendant. These are two *Viparita Raja yogas*, where lords of *Dusthana* houses reside in their own houses or in each other's houses (6, 8, or 12, clockwise from the Ascendant of a South Indian chart, counter-clockwise for North Indian chart). Fortifying the affairs of those *Dusthana* houses, these *yoga*s are considered excellent protection from losses and competition, especially when the Virgo-Pisces axis brings innate themes of sacrifice and hard work which may or may not result in victory.

"Mars is powerful, even in his enemy's house."
Kalidasa, *Uttara Kalamrita*, Section IV, *Shloka* 16.

We pay particular attention to the warrior planet Mars – to whom Mercury is a planetary enemy. Though Mars is technically neutral to Mercury, as 8th house lord it is a challenge to Virgo Ascendant. When it is located in the Virgo Ascendant, it can bring enemies to one's doorstep. However, as *karaka* of vitality, it can also bring youthful exuberance. The person may appear brimming with energy, especially if Mercury and Sun are strong and well situated. We look at three examples of Mars on the Virgo Ascendant, the most famous of which is Sir Winston Churchill – a man both praised and criticized for being a "war-monger." His Mars *Dasha* (Nov. 30, 1913 to Nov. 30, 1920) should have finished his career and reputation, but he survived it. (There were huge losses during the Dardanelles military campaign in World War I due to his bad decisions.)

Venus is the best planet for Virgo Ascendant and is *Nadi Yogakaraka*. It has *maraka* status as 2nd house lord, but escapes most of the difficult connotations of *maraka* due to its ownership of the excellent 9th house. There are special rules from the classic *Bhavartha Ratnakara, Kanya* [Virgo], Stanza 1:

> "For a person born in Virgo Ascendant, if the Sun is related to Venus or the Moon by mutual conjunction, aspect, etc., etc. there will be access to wealth in the course of Sun's *Dasha*." Then in Stanza 2: "The person becomes bereft of wealth in Venus *Dasha*. Mixed results will happen in the course of Moon's *Dasha*."

Considering the Sun rules the 12th house of loss and expenditures, this is an interesting turn of affairs and just opposite to what you might suppose. But all this depends on whether Venus contacts the Sun in the same sign, which is frequently the case, as they are never more than 48 degrees apart. And on the whole, Venus remains the best planet for Virgo Ascendant. But as with all planets, its *Navamsha* position will be absolutely crucial to the outcome of the birth chart, being the fruit of what is promised in the natal chart.

There are few other favorable planets for the Virgo Ascendant chart. Saturn – the other planetary friend of Mercury – is problematic since it rules both the 5th and 6th houses from the Ascendant. While the trinal (5th) house is favorable, the *Dusthana* (6th)) house is not. Benefic Jupiter is also problematic as a *Maraka* planet ruling the 7th house and as a benefic owning a *Kendra* – neutralizing its ability to bestow auspiciousness as a classical benefic. This condition is called *Kendradhipatya Dosha.* (See Glossary) But as noted before, such as with the Aries Ascendant, less harm is done if the planet owning a *kendra* is also a planetary friend to the Ascendant lord: Moon, in that case. Since Jupiter is not a planetary friend to Mercury, it can cause particular problems for the Virgo Ascendant person, especially in the realm of relationships, as it rules the 7th house of partnerships. Beyond those reasons, Mercury and Jupiter also have an uneasy relationship, as described in earlier chapters in the myth of Soma and Tara. Even so, Mercury can benefit from Jupiter's greater knowledge.

The triple Jupiter-Saturn conjunctions in Virgo in late 1980 and 1981, all in *Hasta nakshatra*, were the harbingers for the next twenty years and beyond in social/economic/ political terms, especially as the second set of triple JU-SA conjunctions to occur in the 20th century. (See Chapter 1) *Hasta nakshatra* is noted for its ability to gain what we are seeking and manifest it, particularly with the use of our hands – as *Hasta* means "the hand." A new Age of Electronic Information became possible, with the hand-held communication device as a key factor. Set in the EARTH-dominant era and magnified by occurring three times, the JU-SA conjunctions in Virgo would have to bring practical results, illuminating Virgo on a larger scale. They signaled the intense acceleration of events and unprecedented progress in telecommunications and the information technologies. We witnessed the practical side of Mercury that most concerns itself with serving a broad range of people and changing the way people work. The working efficiency of the developed and developing world changed profoundly, as did most technological forms of communications, with the proliferation of the personal computer (more and more portable), more accessible software, cell phones, and by the early 1990s – the Internet, perhaps the most radical development of all. This helps to understand Virgo's inherent nature and capacities, especially its versatility and adaptability as a dual earth sign, and its rulership over the nervous system of the individual and of the world.

Nakshatras: Virgo contains the following three *nakshatras*: ***Uttara Phalguni*** (starting at 26:40 Leo, in the previous sign) 00:00 Virgo – 10:00 Virgo (ruled by the Sun), ***Hasta*** 10:00 – 23:20 Taurus (ruled by the Moon), and ***Chitra*** 23:20 – 30:00 Virgo (ruled by Mars). (*Chitra* continues into Libra up to 6:40.) The Virgo Ascendant, or any planet situated in one of these three *nakshatras* is colored by being: 1) in Virgo, ruled by Mercury; 2) in one of these three *nakshatras*, ruled by Sun, Moon, or Mars, respectively; and 3) by aspects to the Ascendant and Ascendant lord Mercury.

Again, Venus is the best planet for Virgo Ascendant, and is *Nadi yogakaraka*. Ascendant Lord Mercury is exalted in Virgo, maximum exaltation at 15:00 Virgo. Venus is debilitated in Virgo, maximum debilitation at 27:00 Virgo.[195]

[195] It may seem confusing at first that Venus is debilitated in Virgo, when at the same time Venus is the most auspicious planet for a Virgo Ascendant chart. While Venus and Mercury are planetary friends, Mercury-ruled Virgo is the 6th sign and closely associated with the classical 6th house – considered the least advantageous house position for Venus. Furthermore, a planet is always debilitated in the sign opposite its exaltation sign. (Venus is exalted in the 6th house of the Libra Ascendant chart – both beneficial and problematic. See Ch. 9.)

Chart 24: Sir Winston Churchill

Birth data: Monday, Nov. 30, 1874, 1:43 AM GMT, Woodstock, England, USA, Long. 1W19 00, Lat. 51N51 00, *Lahiri ayanamsha*: -22.06.29. Class B data (biography). John Addey quotes father's letter 1:30 AM; T. Pat Davis quotes *Jennie*, by R. G. Martin. Virgo rises from 00:46 AM to 3:36 AM GMT in that location/day/year. Birth time rectified to 1:43 AM by this author to obtain Aries *Navamsha* Ascendant (1:43 AM to 2:01 AM LMT). Ascendant: 10:06 Virgo.

	Ra		
	Sir Winston Churchill Nov. 30, 1874 1:43 GMT Woodstock, England		
Sa			Mo
	Su VeR	Ju Ke Me	As Ma

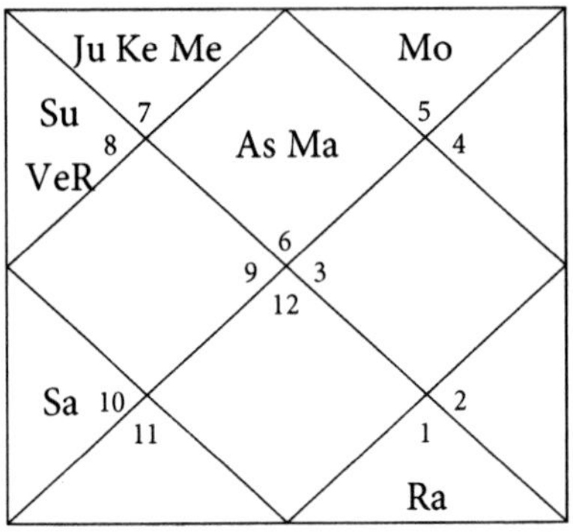

Ju Ke Me 7 · Mo 5 · Su 8 · VeR · As Ma · 4 · 6 · 9 · 3 · 12 · Sa 10 · 11 · 2 · 1 · Ra

Planetary Information

As	10:06	Vir	Hasta
Su	15:37	Sco	Anuradha
Mo	07:37	Leo	Magha
Ma	24:26	Vir	Chitra
Me	25:29	Lib	Vishakha
Ju	01:27	Lib	Chitra
Ve	29:54	Sco	Jyeshta
Sa	17:29	Cap	Shravana
Ra	02:15	Ari	Ashwini
Ke	02:15	Lib	Chitra

VeR	As Ra	Me	Sa Mo
	Navamsha		
			Ma
	Su	Ju Ke	

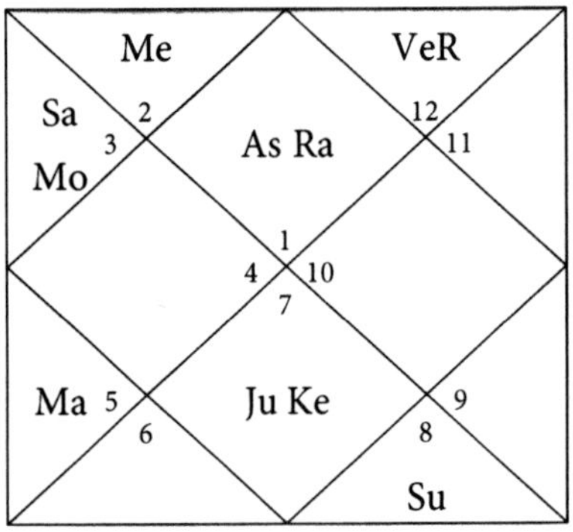

Me 2 · VeR 12 · Sa 3 · Mo · As Ra · 11 · 1 · 4 · 10 · 7 · Ma 5 · Ju Ke · 9 · 6 · 8 · Su

Vimshottari Dashas

Ke	Nov 29, 1870
Ve	Nov 29, 1877
Su	Nov 29, 1897
Mo	Nov 30, 1903
Ma	Nov 30, 1913
Ra	Nov 30, 1920
Ju	Nov 30, 1938
Sa	Nov 30, 1954
Me	Nov 30, 1973

Brief biographical summary: A towering figure on the world stage, Sir Winston Leonard Spencer Churchill was a soldier, statesman, politician, orator, and author. He served as British Prime Minister from 1940-1945 and again from 1951-1955. He is considered one of the most important statesmen in modern history, if not of the 20th century. With his indomitable spirit and energy he inspired England and the other Allied nations during World War II. The power of his oratory was considered crucial to Britain's very survival, especially throughout the Nazi bombing raids on London. He strengthened the nation's determination in his speeches on radio and in Parliament:

> "We shall not flag nor fail. We shall go on to the end. . . . We shall fight on the beaches . . . we shall fight in the fields and in the streets . . . we shall never surrender."
>
> **Winston Churchill**, his Speech on Dunkirk, House of Commons, June 4, 1940.

Also crucial were the inter-party alliances he formed, as well as those with the United States and Russia. His friendship with President Franklin D. Roosevelt and his efforts to enlist

American aid to Britain were among the decisive factors in the outcome of the war. Although his presence on the world scene was large for over fifty years, physically Winston Churchill stood only 5 feet 8 inches tall, with red hair that became sandy with the years. He was also known for his signature suit and bowtie, often a bowler hat and a cigar in his mouth.

FAMILY: A direct descendant of the Duke of Marlborough (1650-1722), he was born to Lord Randolph Spencer Churchill (1849-1895) and Jennie Jerome (1854-1921), an American from a wealthy New York family. Both families discouraged the marriage of this impetuous young couple. But it finally took place on April 15, 1874 at the chapel of the British Embassy in Paris, with the conspicuous absence of the Churchills. The following November Winston was born two months premature. His younger brother Jack was born in 1880. Many doubts circulated as to Jack's paternity, since his beautiful mother was known to have many lovers. Lord and Lady Randolph were aristocratic and self-absorbed, often ignoring and neglecting their children. Winston was closest to his nanny, Mrs. Elizabeth Everest, who was like a surrogate mother to the two brothers until her death in July 1895. From that time on, his mother – recently widowed – proved to be an important political mentor. She guided him in establishing his political career, first through military service, journalism, and then election as a Member of Parliament, like his father before him – though his father's promising but tumultuous career was cut short through ill health, intemperate habits and a lack of diplomacy. Winston inherited his fiery temper, his speech defect, his taste for politics and horses, and his love for cigars and whisky. He also assumed he would inherit his short lifespan, and was always in a hurry to experience and write about many events. Often lacking in diplomacy like his famous father, and with many enemies and critics, yet he was magnanimous in forgiving them and still managed to gain enduring respect from his colleagues and the public. In this way he surpassed his father in overcoming the more disagreeably headstrong side of his nature. Theirs was a distant relationship and Lord Randolph was often critical of his son. Though Winston was more devoted to his own children, the same pattern was repeated between Winston and his son Randolph.

On Sept. 12, 1908, Winston married Clementine Ogilvy Hozier (b. April 1, 1885, d. Dec. 12, 1977). She was won over by his "dominating charm and brilliancy." They had five children: four daughters and one son. One daughter died at age 3. The marriage was a great source of security and happiness for him, though as two strong personalities they had their quarrels. Winston refused a Life Peerage in his later years, but Clementine accepted one for her charitable work after Winston's death in Jan. 1965. Known as Baroness Clementine Ogilvy Spencer-Churchill, she died at age 92. Churchill College, Cambridge holds her papers as well as some 84 files of letters between Winston and Clementine during their long marriage, 1908-1965.

POLITICS: First elected in 1900, Winston Churchill was a Member of Parliament for over 60 years and a Cabinet member on and off for 25 of those years.[196] He held a sequence of Cabinet

[196] His Cabinet positions from 1905 through 1940 were as follows: 1905-1908 Colonial Undersecretary; 1908-1910 President of the Board of Trade; 1910-1911 Home Secretary; 1911-1915 First Lord of the Admiralty; 1917-1919 Minister of Munitions; 1919-1921 Secretary of State for War and Air; 1921-1922 Colonial Secretary; 1924-1929 Chancellor of the Exchequer; 1939-1940 First Lord of the Admiralty.

positions from 1905 through 1940, when he became Prime Minister (1940-1945, and again 1951-1955). From 1945 to 1951 he continued on as Conservative M.P. and Leader of the Opposition. In 1920 he remarked:

> "Politics are almost as exciting as war, and quite as dangerous. In war you can only be killed once, but in politics many times."

After graduating from Harrow, and then from the Military College at Sandhurst in 1895, he fought with the British Army in India and Sudan. He was then sent to South Africa as a War correspondent, where his dispatches from the Boer War first brought him to public prominence, especially after he escaped from a prison there. His wartime journalism and his heroism helped him win his first election in 1900 to the House of Commons.

Politically he was considered a maverick, as he was sometimes a Liberal and sometimes a Conservative. He rose quickly in government ranks, and took unusual and at times very controversial positions, claiming to place policy above political party. He entered the House of Commons as a Conservative in 1900. By 1904 he joined the Liberals, though after 1924 he was again a Conservative. His views on some matters were very forward-looking: For example, when he was First Lord of the Admiralty (1911-1915), he modernized the Navy, changed British naval ships from coal-burning to oil-fuelled, and encouraged more high-powered tanks and the immediate use of military aircraft. In fact he was among the first to grasp the military potential of aircraft. Though he was in and out of Liberal and Conservative governments, and formed a Coalition government during World War II, Churchill's overriding tendency was as a Conservative. He was an ardent monarchist who believed that Empire was the key to Britain's greatness and should be preserved at all costs. These views at times left him politically isolated.

His intense interest in both military and public affairs also led to imperial overreach at times as well as bad judgment militarily, when he should have left more decision-making to his generals. The disastrous British defeat at the Dardanelles in March 1915 cost him his cabinet position and nearly ruined his career and reputation. To ease the pain of this personal and professional disaster he started painting watercolors, and this quickly became his confirmed avocation. But he was not content being powerless to influence national war policy. He rejoined the British Army on the Western Front, where he commanded his own battalion: the Royal Scots Fusiliers. Churchill was never far from the battlefront, one way or another. By 1917, he was appointed Minister of Munitions and in 1919 Secretary of State for War and Air. After World War I, he advocated a new war on Bolshevism, which only increased the perception of him as war-hungry. This was not inaccurate, as he was perpetually obsessed with military matters.

After 1924 his political fortunes lay mostly with the Conservatives. But earlier on, as a member of the Liberal government he helped to lay the foundations of the welfare state. He imposed a limit on work hours, established a minimum wage, introduced old-age pensions, health care and unemployment insurance and labor exchanges. He also advocated various social reforms, specifically in British prisons. Liberals mistrusted him for using troops to break strikes when he was Home Secretary. And along with his visionary and humanitarian aspects – often defying strictly party policies – he was also anachronistic and racist in some of his views, especially in regard to India and Iraq. Indeed, some of his nation-building ideas and strategies proved to be disastrous in the long run. In the British colonialist tradition, he was against giving autonomy to India (even by the 1940s), disliked Mahatma Gandhi, and as Colonial Secretary created an artificial monarchy of Iraq after World War I, forcing together three unfriendly peoples under a single ruler, Emir Feisel, making him king over a land to which he had no connection.

Though the British and the French jointly carved up the Ottoman Empire after WWI, Winston Churchill was one of the chief architects of the current map of the Middle East. As British Colonial Secretary in 1921, what he created in the Middle East, especially in Iraq, has been called "Churchill's Folly" and has led to calamitous results and destabilization in the entire region. His nation-building mainly involved Iraq, Saudi Arabia and Palestine. An ardent Zionist for many years, he favored a permanent homeland for the Jews, with the idea of forcing his Arab allies to acquiesce. But he was not in favor of relinquishing the British mandate in most cases, and in 1942 he said:

> "[I have] not become the King's first minister in order to liquidate the British Empire."

He also said: "I will not preside over a dismemberment." But many historians feel that Churchill's policies even accelerated the empire's collapse.

Churchill's public career as a political leader appeared to be finished in 1929. He was out of the center of power from 1929 to 1939, his opinions in disfavor and his warnings about Hitler's rising power disregarded. All that suddenly changed when Hitler invaded Poland on Sept. 1, 1939. Neville Chamberlain reluctantly appointed Churchill First Lord of the Admiralty, as Churchill was "a war man" and Chamberlain was not. Churchill was then appointed Prime Minister on May 10, 1940, hours before the Nazis invaded France. It was an historic moment for Churchill and for the nation. In his first statement as Prime Minister in the House of Commons, May 13, 1940, he said:

> "I have nothing to offer but blood, toil, tears, and sweat."

He was to repeat this phrase often during the war. And in general, his oratory and his strategic and moral leadership during WWII are regarded as his crowning achievements.

POLITICS - POST 1945: Though he remained popular in July 1945 and was still revered for his wartime role, the public voted him out of office at the end of the war by a wide margin.They disliked his comparing Clement Atlee and his Liberal party to the Nazi Gestapo, and distrusted the Conservative Party to run a postwar England. But by late Oct. 1951 Churchill returned a second time and served as Prime Minister through April 5, 1955, when he was forced to retire due to ill health. His second term as P.M. was largely ineffective and overly prolonged. However, he backed a united Europe and a strong United Nations. But his strongest opinions in those years concerned the development of the hydrogen bomb. On this subject he differed radically from Dwight D. Eisenhower (U.S. President, 1953-1961). In private Churchill gave him a hostile appraisal as a president, notably for his strong support for the atomic bomb, though in later public statements he applauded the U.S. for using the atomic bomb as a deterrent. Churchill's Private Secretary J. R. Colville recorded these notes during the Bermuda conference, Dec. 6, 1953:

> "He [Eisenhower] said several things that were noteworthy. The fact was that whereas Winston looked on the atomic weapon as something entirely new and terrible, he looked upon it just as the latest improvement in military weapons. He implied that there was no distinction between conventional weapons and atomic weapons: all weapons in due

> course become conventional weapons.' This marked a big gulf. Churchill believed that an H-Bomb exchange meant effectively the end of the human race. Eisenhower, with his much more limited imagination, thought that it would be the battle of the Rhine crossings writ larger."[197]

Churchill's health had been deteriorating since the summer of 1954. He lived another ten years and died of a stroke (cerebral thrombosis) on Jan. 24, 1965, at age 90. He was given a state funeral and was the first and only non-royal ever to receive such an honor. He had served six British monarchs from Queen Victoria through Queen Elizabeth II. Since his death in 1965 the world continues to quote him, study him and praise him and argue hotly over his enormous legacy – full of inspiration *and* controversy, especially in the Middle East. Even so, in a Nov. 2002 British nationwide poll of more than a million votes, Churchill was named the greatest Briton of all time, receiving nearly half of all the votes.

WRITING: A prolific writer of history, politics, and biography, his collected works – including books, speeches and letters – fill some 72 volumes, and are still expanding as more private government papers become available. Among his larger works were: *The World Crisis 1911-1918*, 1923-31 (6 vols.); *The Second World War*, 1948-53 (6 vols.) and *A History of the English-Speaking Peoples*, 1956-58 (4 vols.). Churchill had once predicted that history would treat him kindly since he himself would write it! In 1953 he was knighted and in the same year awarded the Nobel Prize in Literature for his writing and his oratory, though what he coveted most was the Nobel Peace Prize. His Nobel Prize in Literature cited him for:

> "...his mastery of historical and biographical description as well as for brilliant oratory in defending exalted human values."

Circumstances around the birth: Winston Churchill was born two months premature (as was his brother John in 1880). Mars in the Ascendant indicates some element of hurry or potential danger in the circumstances around the birth. All the preparations had been made for his birth in London two months later. His parents went to the country to go shooting in the grounds of Blenheim Palace, the Churchill family seat in Oxfordshire. But while following the guns, the very pregnant Lady Randolph – then age 20 – took a nasty fall and also went on a very rough drive in a pony carriage. This brought on her labor pains and an agonizing 8-hour labor without chloroform. The mother is the **Moon** in the chart. She is American-born and thus in foreign residence in England and in a pastoral environment – both 12th house settings.

Mars prominent in the chart shows the haste, the guns, the upset due to sudden circumstances, a lot of it fuelled by youthful exuberance. Mother is also 4th house lord, which is Jupiter in this chart. Jupiter is a powerful planet here, in the same *nakshatra* as Mars, *Chitra* – ruled by Mars. Thus Mars is the *jiva* (life force) propelling Jupiter. Jupiter is also close to Ketu on the Rahu-Ketu axis. Also situated in *Chitra nakshatra,* Ketu is Mars-like and associated with what is foreign, exotic and unpredictable. Lady Randolph was all of those things – a noted beauty of the era, accustomed

---

[197] Roy Jenkins, ***Churchill: A Biography,*** 2001, p. 849. Jenkins quotes from J.R. Colville's ***Fringes of Power: Downing Street Diaries 1939-1955***, p. 685. By April 5, 1955, in his last speech as Prime Minister, Churchill publicly backed Britain's manufacture of the H-bomb, though he emphasized negotiation over military might.

to a very social and aristocratic life, having her way - with charm - and wildly impetuous. Risk to his personal physical safety as well as to that of his mother is further shown by **Moon in the 12th house** in *Magha nakshatra*, owned by Ketu. *Magha* is symbolized by the throne room and has royal connotations. He was born unexpectedly at the family seat at Blenheim Palace. Ancestral pride and power are featured with *Magha*, as it is greatly concerned with one's lineage and is ruled by the deities the *Pitris*. Later on it would manifest as his "born to rule" attitude.

As 8th lord in the Ascendant, Mars is potentially dangerous to the physical body early in life, and can cause ill health or accidents as well as many upheavals throughout the life. Countering the effect of Mars in the Virgo Ascendant and giving protection are the ***Shubha kartari yogas*** (classical benefics on houses either side of the Ascendant). These occur in both his birth Ascendant and *Navamsha* Ascendant. Fortunately a local doctor could be summoned and the baby was born healthy and sound, despite being two months premature. Winston was born in Ketu-Jupiter *Dasha/Bhukti*. Both planets are situated in the 2nd house in the sign of Venus - a great good friend to the chart. This overcomes its *maraka* status by being also 9th lord and diminishes the *maraka* effect of planets placed in the 2nd house.

Physical health & longevity: As a boy Winston was considered fragile. He had a serious bout of pneumonia at age eleven but grew stronger in his teens. Throughout life he had bouts of bronchitis, pneumonia and jaundice. Though he rarely succumbed to being sick in bed he was often "unwell" as an adult. In addition to the condition of the Virgo Ascendant - with accident-prone Mars in it - astrologically, the basic vitality of Winston Churchill should be assessed by the condition of Ascendant lord **Mercury**, and *karakas* for vitality and physical health: **Sun** and **Mars.** Mercury (ruling the lungs) has some volatility being on the Rahu-Ketu axis, but is well placed in both birth chart and *Navamsha* in Venus's sign: Mercury is in Libra in the birth chart and Taurus in the *Navamsha*. In both cases it is in the 2nd house of speech, with the influence of several planets - notably Mars, Venus and Jupiter. Mars has dominant placement both in birth and *Navamsha* charts - in a *Kendra* of the birth chart and a trinal house of the *Navamsha*. But Jupiter has the angular dominance in the *Navamsha* chart, not Mars. This helps longevity, along with the *Shubha Kartari yogas*, as described above.

Reinforcing the physical strength, Sun and Mars form a *Parivartana yoga* (exchange of signs) in the *Navamsha* chart between the 5th and 8th houses, and Mars is *Navamsha* Ascendant lord. Sun is strengthened by its *Vargottama* status in the sign of Mars-ruled Scorpio. However, Sun resides in *Anuradha nakshatra*, owned by its planetary enemy, Saturn. (Sun-Saturn associations engender a more confrontational nature, as one must face one's antagonists throughout life and deal with weakness of the teeth or bones. Churchill had dental problems from the time he was 16 and extending well into his 20s. Eventually he wore a special set of dentures that even accommodated his lisp.)

Not temperate in his life habits, he smoked cigars, drank whiskey, and copious amounts of alcohol with meals. He slept very little and worked and played very hard. He also suffered from insomnia, often taking "cat-naps" to compensate for lack of sleep, even in wartime. As a young man he was physically active and a skilled polo player, playing polo until 1927 (age 53). But in later years he took no regular exercise and was overweight. Natal **Mars in Virgo** puts more emphasis on aggressive thinking and mental strategies rather than on athletic or physical prowess per se. This is confirmed by *Navamsha* Mars in the 5th house of the intellect. Meanwhile the field of Virgo, while competitive and military-minded, is more geared for mental and business pursuits. Being the sign where Mercury is exalted, it tends to re-orient the muscular planet Mars in a mental direction. Virgo rules over the nervous system, so the Mars influence here can periodically bring major burnout to the nervous system. From the Ascendant, Mars also aspects Houses 4, 7, and 8. The 4th house is where the mind comes to rest, especially at night. Mars aspecting the 4th house can thus ignite the mind and bring creativity. It can also disturb peace of mind and cause insomnia, amplified by Jupiter in the 2nd house with its desire for sweets. (Excess sugar intake can cause insomnia.) With a warrior chart like this, one yearns for the speed and momentum of life and one is prone to burnout of the nervous system as well as accidents.

There were several serious accidents through his life: On Dec. 13, 1931 at around 10:45 PM local time in New York City, he was crossing Fifth Avenue as a pedestrian when he was hit by a car and dragged along the street. This was caused through his own error. He spent the next two weeks in the hospital. In addition to cuts and bruises, the more significant long-term effects were aftershock and depression. At the time, he described the injuries euphemistically as "slight." He was halfway into his **Rahu-Ketu** *Dasha/Bhukti*, classically an unwieldy period, being the eclipse axis. *Dasha* lord Rahu is situated in his natal 8th house of losses and impacts the 2nd house of income (and speech), where Mercury, Ketu and Jupiter are located. This coincided with tr. Saturn in Sagittarius – which marked a low point for Saturn transits through his chart.[198] Some of his largest financial losses were during this period as well, coinciding with the Wall Street crash Oct. 29, 1929.

He also nearly drowned while swimming at Lake Lausanne in 1895 at age 20. This short list of physical mishaps does not include the risk of being a soldier at numerous battlefronts 1895 to 1900, and again 1915 to 1917. Placing oneself in competitive and potentially dangerous situations is the ongoing influence of a prominent Mars. The same planets protecting him at birth protect him throughout life, though we know **Mars *Dasha*** (Nov. 30, 1913 to Nov. 30, 1920) could be especially hazardous: Mars is lord of the 8th house and is thus challenging to Ascendant lord Mercury. Though he was not physically harmed, his professional career and reputation were nearly ruined by the disastrous results of his military strategy for the Dardanelles campaign in March 1915, culminating on March 18, 1915 during **Mars-Rahu** *Dasha/Bhukti*, both planets in the potentially difficult 6 – 8 house relationship.[199]

He suffered several strokes from Aug. 1946 onward – kept secret from the general public. With this information withheld and with the "little green pills" (amphetamines) from his doctors, Churchill was able to prolong his productivity and his public career, even though many close to

---

[198] See also under Churchill: *Saturn at its low point in the chart.* Saturn transits in Sagittarius last three years, since it is a sign of long ascension. This one ended only 18 days after his accident: Dec. 31, 1931. Preceding and foreshadowing the event was a Total lunar eclipse (9:50 Pisces) opposite his Ascendant: Sept. 28, 1931 and a Partial solar eclipse (24:20 Virgo) on his natal Mars: Oct. 11, 1931. The accident was on a Sunday, with 16:36 Leo rising; Ascendant and Day Lord Sun in the 4th house at 28:16 Scorpio: its weakest angle and a difficult degree *(gandanta)* for any planet. Tr. Mars was at 10:08 Sagittarius at that moment, exactly on Churchill's natal 4th house cusp. Tr. Rahu (*Dasha* lord) was at 8:19 Pisces on his Descendant, Ketu on the Ascendant (within 1:49 orb).

[199] This was the largest loss suffered by the British Royal Navy since the Battle of Trafalgar in fall 1805. Some estimates list over 250,000 military dead among the British and French, with many more wounded or missing, and numerous British or French battleships and submarines sunk or destroyed. From Oct. 1914 the Ottoman Empire had closed the Dardanelles to Allied shipping. The entire Dardanelles campaign lasted from Feb. 19, 1915 to Jan. 9, 1916 and was a decisive victory for the Ottomans, who became allied with the Germans. (The Dardanelles is a narrow strait in north-western Turkey connecting the Aegean Sea with the Sea of Mamara.)

him questioned his ability to carry on effectively as Prime Minister for as long as he did. But once his **Saturn *Dasha*** arrived in late 1954 – the day he turned 80 – destiny was likely to slow him down finally, even with energetic Mars prominent on his Ascendant. Four months later he was forced to retire from his second Prime Ministership. Saturn is the one planet *not* influenced by Mars or Ketu: by sign, *nakshatra* or aspect – either in the birth chart or *Navamsha* chart. Therefore Saturn would also be the first *Dasha* lord in his lifetime not influenced by Ketu or the youthful (and accident-prone) energy of Mars. And though Saturn is strong in its own sign of Capricorn, its *nakshatra* lord is Moon – poorly placed in the 12th house unless one is in 12th house settings – in a foreign residence or travelling abroad. He spent much of his last ten years on long trips or cruises in foreign waters and in visits to the south of France. This protected his well-being and longevity in Saturn *Dasha*.

Mental health: Churchill was probably **manic-depressive** (later renamed "bi-polar"), with enormous energy followed by periods of severe depression throughout his life: He named it "the black dog on my back." Natal **Moon in the 12th house** would be the major cause of his periodic depression. Natal Moon in Houses 6, 8, or 12 can be problematic, and in the 8th or 12th can indicate a tendency for depression, deep brooding, or spiritual practices – if there is a proclivity for that. Churchill was not inclined in that direction. He was very concerned with public moral issues but he gave them no particular religious context. (Houses 4, 8, and 12 are houses of *Moksha*, or spiritual liberation.) Even being a very public figure, Moon in the 12th house shows the need for privacy to recoup energy.

Confirming the potential for depression is ***Navamsha* Moon with Saturn** in Gemini, as Saturn can weigh on the mind (Moon) and manifest as a sense of relentless responsibility. In the *Navamsha* 3rd house, the cause of the upset would be the inability to act, as 3rd house is the house of courage and initiative. It is also the house of writing and communication, and the Saturn influence can bring the needed self-discipline to get results. Since *Navamsha* Jupiter (*Vargottama*) also aspects this *Navamsha* Moon and Saturn, it provides some antidotal expansiveness to the heart and mind. Therefore the ongoing flow of written and spoken communication would help to overcome depression, as did his painting and his marriage and family life. Churchill was known for his great wit and humor, along with his ability to focus on the job at hand.

Courage with eloquence from an unlikely orator – Ascendant lord Mercury: As a young boy Churchill's voice was weak. He had a pronounced lisp from childhood and talked through his nose. All this was a source of anxiety and unhappiness, and not a situation that seemed likely to produce a great future orator. He worked hard to overcome a **speech impediment** throughout his life. The Churchill Centre in Washington, D.C. has concluded that he did not stutter, but had a lisp. They also concluded that Churchill's speech impediment was most likely cluttering (*tachyphemia)* – a hereditary speech disorder in which there is inattention to unimportant details, with the result that listeners cannot easily understand what is being said due to a rapid rate of speaking, an erratic rhythm, poor syntax or grammar, and words or phrases that appear to be unrelated. There is no lack of self-confidence or social skills, as is often the case with stuttering. With cluttering, there is also no problem putting thoughts into words, but those thoughts become disorganized during speaking. Churchill visited several speech specialists in his early 20s, and later had four sets of dentures fitted especially so that his impediment would not be altered; for even though he improved it over the years, it had become part of his trademark as a famous orator.

The astrological factors strongly confirm the Churchill Centre's case for **cluttering**, as opposed to **stuttering** – where there is a distinct hesitation in the speech. Hesitation would be caused more by a predominant Saturn influence, and though there is a Saturn aspect to Mercury and to the 2nd house (commonly denoting stammering) – there are other more numerous influences that galvanize the speech. Saturn strong in its own sign helps him to correct the problem and he does so in a unique way – with Ketu contacting Mercury. The 2nd house of speech contains three planets: Jupiter, Ketu, and Ascendant lord Mercury. The influence of Mars and Jupiter, speed and volume, respectively, both predominate. This is because Mercury is in *Vishakha nakshatra* (owned by Jupiter) and the sign lord of the house is Venus in Scorpio (owned by Mars). Jupiter and Ketu are located earlier in the sign of Libra in *Chitra nakshatra*, ruled by Mars. Ketu in the 2nd house can indicate a speech problem, as we saw with Albert Einstein, who did not speak fluently until he was almost nine years old. Furthermore, 2nd house lord Venus at 29:54 Scorpio, in a *gandanta* degree, indicates something problematic, most likely due to an outpouring of thoughts and emotions that have been restrained. Churchill's childhood was said to be very lonely.

In his first year at Harrow (1888-1889) he won an oratory contest when he recited 1300 lines of poetry from memory. This occurred during his **Venus *Dasha*,** in **Venus-Jupiter** period. Venus and Jupiter are both closely related to the voice due to their close association with Mercury and the 2nd house. Mercury is pivotal in the destiny as both Ascendant lord and planet of speech and communication. It is imbued with all the qualities of its sign lord (Venus) and *nakshatra* lord (Jupiter), as well as the planets that contact it and aspect it. Being more porous and neutral planets, this is true even more so for Mercury and the Moon.

Oratory and writing: A prolific writer as well as a military strategist and statesman, Churchill wrote often about his early years as a soldier and in later years about his experiences as either Cabinet Member or Prime Minister. He was fascinated with history, especially history he helped to make – but also that of his forefathers. His books, speeches, letters, and essays fill some 72 volumes, and with more archives becoming available even more of his writing will probably be published in addition to more being written about him.

> "Let us ... brace ourselves to our duties, and so bear ourselves that if the British Empire and its Commonwealth last for a thousand years, men will still say: 'This was their finest hour.'"
>
> **Winston Churchill** in a Speech in the House of Commons, June 18, 1940.

Mercury-Venus contacts or exchanges can bring eloquence. With natal Mercury in the sign of Venus, and Venus in the *nakshatra* of Mercury, the astrological factors for eloquence are in place. Mercury as *Jiva* planet for Venus adds refinement to his speech and writing. In addition, Mercury is in the 2nd house of speech in the birth chart *and Navamsha* chart. From the very start of his career in 1895, he went to combat zones and wrote dispatches for British newspapers – from Cuba, South Africa, and India. His grasp of military and political history and his fascination with the battle of life enabled him to write uncommonly well. Boldness in speech and writing manifests through a Mars association with Mercury and/or the 2nd house (as sign or *nakshatra* lord). Jupiter is located in the 2nd house and owns the *nakshatra* where Mercury is placed, making Jupiter the *Jiva* planet for Mercury. Jupiter's effect on Mercury can bring greater knowledge and increase the amount of writing – voluminous in this case! Also, there is the prevalence of planets in his 2nd and 3rd houses, houses of speech and writing, respectively. The lords of the 1st, 2nd and 3rd houses also interact: Ascendant lord Mercury is in the 2nd house, 2nd lord Venus is in the 3rd house, 3rd lord Mars is in the 1st house, and owns the sign where 2nd lord Venus is placed.

Churchill's ***Dasha* sequence** continually emphasizes the 2nd or 3rd house – either through the birth chart or though the *Navamsha* chart. And pervading all of this is the electrifying urgency of Mars. American journalist Edward R. Murrow commented on Churchill's oratory:

> "He mobilized the English language and sent it into battle."
>
> *I Can Hear it Now* [1933-1945], Vol. 2, 1948.

Furthermore, Ascendant lord Mercury in Libra continues to have the influence of Venus, Jupiter, Mars, Rahu and Ketu. This gives a widespread and very socialized influence, as the air signs yield the strongest human social influence. From the Moon in Leo, this theme continues: Mars is in the 2nd house from the Moon and is well placed in Virgo – as *yogakaraka* from Moon in Leo. (Mars rules both *Kendra* and trinal houses from Leo.) And though potentially challenging in speaking too harshly or abruptly, it gives financial wealth through speech and writing. Saturn in Capricorn in the 6th house from natal Moon is excellent for the competitive spirit and the ability to win amidst extraordinary circumstances.

The classic sweetness of Venus is dominated by three factors here: **1)** the Mars influence, **2)** being in the *Gandanta* degrees of Scorpio (last 3:20 of a water sign); and **3)** being in Retrograde motion, when a planet is closer to the earth. This intensifies the natural qualities of the retrograde planet: Venus becomes obsessively devoted to action or writing about it, preferably military action. With this more volcanic energy of Venus, any planets in the sign of Venus (Mercury, Jupiter, and Ketu in Libra) are inextricably linked to it. The more extreme nature of Venus and Jupiter in this case accounts for the early speech impediment as well as later speech habits and gifts. Churchill's thoughts often came in great torrents, and when he dictated to his secretaries they knew they dared not interrupt his flow. Even a specially muffled typewriter was used in order not to disturb his train of thought.

The Rahu-Ketu influence gives an unexpected, unconventional, and at times unwieldy quality to the speech. This would relate to his four special sets of dentures that preserved his lisp and his special way of speaking. Jupiter lends expansiveness and curiosity about the world, while the influence of Saturn in its own sign enables the person to work hard and effectively. Saturn is a positive antidote to a chart brimming over with enormous physical and mental energy. But its 19-year *Dasha* (from Nov. 30, 1954) brought a marked downturn in physical force and vitality. He resigned April 5, 1955 from his second term as Prime Minister and for the last four months he was not available to preside over most of the Cabinet meetings. Following his resignation as P.M., he was a Member of Parliament up to 1964. But for those last nine years he remained silent and never spoke again in Parliament.

Exceptional resilience: A Dual sign on the Ascendant gives oscillating tendencies and it may appear that someone is going in two different directions at the same time. That is the natural state of the dual signs and as such they can be the strongest of the modalities in a storm – just as the tree that can bend low in a strong wind will not be broken or ripped from its roots. There is a tenacious quality that comes from being adaptable, and Virgo is always concerned with practicality and survival.

> "Personally I'm always ready to learn, although I do not always like being taught."
>
> **Winston Churchill**

Viewed as a perpetual "war man" and obsessed by military matters, this thrust him into office as well as out of office. He coveted the Nobel Peace Prize but instead received a Nobel Prize in Literature. He was periodically disgraced or dismissed for his decisions, such as the Dardanelles military campaign in 1915. Out of higher office and out of favor for ten years, 1929-1939, he managed to survive politically – keeping a public presence through his writing and speaking. He knew instinctively how to shift gears and how to find another way to focus and operate that harnessed his prodigious energies, even when a sense of defeat and mental depression threatened to derail him. He had many failures *and* successes, winning and losing elections, making and losing large amounts of money – including in the 1929 New York stock market crash.

Churchill had enormous energy, though it was mostly mental – especially from 1927 onward when the amount of his physical exercise was minimal. He had an equal appetite for work as for play and self-indulgence, mainly food, fine wines and liquor, cigars and gambling.

> "I was ...increasingly struck by Churchill's extraordinary combination of an almost puritan work ethic with a great capacity for pleasure, even for self-indulgence."
>
> **Roy Jenkins**, from his lecture at the Guildhall, London, UK, Nov. 16, 2001 on his biography of Winston Churchill, 2001.

When Virgo is strong it can give great mental agility, as the orientation is realistic, responsive, and competitive. But its sign lord Mercury must be strong and well placed, as do Venus and Jupiter. If we compare the birth chart of Bill Gates, with his Mars-Mercury in Virgo (in the 4th house from Gemini Ascendant), we recall the intellectual power combined with the oscillating energy of the dual earth sign. This gives the basic versatility and adaptability of the individual, who must usually focus on at least two major endeavors in life. While it has the potential to scatter the energies, it can also harness the considerable mental capacities. These qualities combined with Mars strong in the Ascendant gave Churchill the tendency to seek out life's battles, and ones requiring unusual ideas and inventions in times of emergency or full-scale war. A political maverick, at times championing both Liberal and Conservative policies and parties, he had the capacity to form effective alliances and was magnanimous to his political foes. Many cite his tribute to Neville Chamberlain in the House of Commons Nov. 12, 1940.

Secular Churchill: Churchill had ambivalent feelings towards both religion and war. We examine his *Dharma* from the 9th house (Taurus), ruled by Venus. In this case the 10th lord (of *Karma*) is also the Ascendant lord, Mercury. Thus for a Virgo Ascendant person, any interactions between Mercury and Venus (9th lord) will fortify the ability to meld talent with destiny, also to harness communication skills in a more powerful way. For religious inclinations we study the condition of **Jupiter** as well as **9th house lord – Venus** in this case. Jupiter is in the sign of Venus, and in Mars-owned *Chitra nakshatra*. Both of these are secular planets, with a tendency to be full of worldly desire and ambition. In the *Navamsha* chart, Venus and Jupiter exchange signs, a *Parivartana yoga* between the 7th and 12th houses. Though 12th house is a *Moksha house*, Venus in that house – even if exalted in Pisces does not guarantee a spiritual path. It can also be pleasures of the bed and private pleasures. Churchill pursued his painting as much as possible in his spare time.

Jupiter is the classic *Dharma* planet and is close to the Rahu-Ketu axis, with Ketu in the same sign – *Vargottama* – both of them repeating in the same sign in the *Navamsha*. With this combination (i.e., Jupiter with Ketu or Rahu) the native often finds something untrustworthy about religion and those who lean on religion. He may not consider it a conduit to honesty and integrity but more often a cloak for extreme actions. The fact that 9th lord Venus is *Gandanta* at 29:54 Scorpio only adds to his suspicion. Not averse to large or small philosophical questions, his life is much more governed by Mars. And Mars is more likely to ask: What are the necessary and expedient actions that should be taken under the circumstances? Religion, especially Catholicism he viewed as "a delicious narcotic." Though a member of the Anglican Church, he considered that most religions could easily lead to fanaticism. Instead he subscribed to what he called "**the Religion of Healthy Mindedness**." It involves making your best effort to live an honorable life, doing your duty, being faithful to friends and kind to the weak and poor. If you did this, then:

> "...It did not matter what you believed or disbelieved. All would come out right."
>
> **Winston Churchill**

Inspiration *and* controversy: His legacy of both inspiration and controversy is inextricable from the many important roles he played, as well as the many volumes he wrote. The subject which galvanized him the most was the battle of life, for goodness and integrity, for moral rectitude against the face of evil, usually seen as the crushing power of totalitarianism: the Bolsheviks or the Nazis. This was not a religious question for him, but a moral and philosophical one. In this context not even democracy was sacrosanct. For him it was:

> "...the occasional necessity of deferring to the opinions of other people."

Mars in Virgo often generates controversy. Mars in the Ascendant or aspecting the Ascendant can be headstrong, but also courageous and capable of admitting mistakes in order to correct them. We see how courage and danger (both Mars) are so deeply linked. Mars likes the competition, likes to succeed, and in the process of going after his enemies may create further enemies as well as alliances. And there were some extraordinary alliances – such as with Joseph Stalin. Prior to WWI Churchill correctly foresaw the strategic importance of oil in the Middle East. He was always seeking ways to preserve the British Empire, and was against giving independence to India.

What he called "the biggest blunder in his life" was returning Britain to the gold standard. This occurred on April 3, 1925 while he was Chancellor of the Exchequer, 1924-1929. The infamous Dardanelles campaign of March 1915 has continued to be studied and re-evaluated. And though many view it as one of his worst mistakes, he defended it as an operation which should have brought WWI to a rapid close.[200] However, he strongly regretted authorizing the fire bombing of Dresden in Feb. 1945 as well as the use of gas against German citizens to retaliate for the Nazi bombings of Southeast England. But he did not write of it in his memoirs. In the earlier years, Churchill's biggest failure was over-confidence in the fledgling British Air Force, and underestimating the power of enemy ground forces. In that respect, his massive misjudgment of the Iraqi tribesmen in 1921 characterizes a typical arrogance of British imperialism or of imperialism in general. In 1920

---

[200] Though Churchill was dismissed in May 1915 as First Lord of the Admiralty for his role in the Dardanelles campaign, he strongly defended his concept of that operation in his book ***The World Crisis, Vol. II***, 1923, pp. 217-219. He vowed that it should have been the key to defeating the Ottomans and viewed his own strategical plan as technically sound but the execution as flawed, in part due to continual delays fatal to its success, notably the last one-month's delay from February to March 1915. However, later scholarship reveals material different from what Churchill wanted us to believe: that he viewed the Dardanelles as relatively low-risk and was not as focused on it as on more northern theatres.

the fledgling Royal Air Force dropped 97 tons of bombs to curb an uprising of over 100,000 armed tribesmen, killing 9000 Iraqis. When this plan still failed to end Arab and Kurdish resistance to British rule, Churchill suggested that chemical weapons should be used "against recalcitrant Arabs as an experiment." He added:

> "I am strongly in favour of using poisoned gas against uncivilised tribes [in Iraq]. The moral effect should be so good that the loss of life should be reduced to a minimum. It is not necessary to use only the most deadly gasses: gasses can be used which cause great inconvenience and would spread a lively terror and yet would leave no serious permanent effects on most of those affected."
>
> **Churchill papers**, May 12, 1919, War Office.

These are examples of how controversies deepen into legacies, and how a strong Mars in Virgo lives on at various levels, even if – for example – Churchill's role as architect of the beleaguered modern Iraq has faded from public memory. Instead he is mostly remembered for his valiant courage as a wartime leader, and this appears to be closely related to his own role as writer of the history he lived. As the planet of communications and sign lord of Virgo, Mercury is strong as lord of the 10th house of reputation. The errors and omissions he made in telling that history reveal a classical truth: Historical figures will try whenever possible to show themselves in a good light. About his six volumes on The Second World War he said:

> "This is not history, this is my case."

Military strategist and statesman: Mars is the red planet. Its prominent placement on the Ascendant gave Winston Churchill shorter than average stature for a man, at 5 feet 8 inches tall, red hair, a fiery temperament and a love for the military in his blood. He relished the danger and excitement of war, and his courage and zest for military matters was seemingly unlimited. Even when playing soldiers as a boy he insisted that his friends call him "General," as he preferred to be in charge of military maneuvers whenever possible. He had a large collection of toy soldiers, purchased between 1880 and 1900.

> "Nothing in life is so exhilarating as to be shot at without result."
>
> **Winston Churchill**, *The Malakand Field Force*, 1898.

As the most prominent planet in the natal chart of Churchill, Mars gave him his dominance, his unwavering belief in himself, and – since it is in the Virgo Ascendant – a tendency to attract bellicose circumstances in his life. This is an excellent example of an iconic figure who led his country at a time when the enemy was at its doorstep and bombing its cities. Often the chart of a national leader is synonymous with the fate of that nation at a given time. (Vedic philosophy would say this is in order to fulfill the *karmic* destiny that must be lived.) Though Mars in Virgo can be argumentative it can also bring firepower to the speech – through incisiveness and brilliance, as well as through anger. These combinations can produce a courageous, impulsive and unconventional leader who is unlikely to surrender.

Churchill was directly involved in the military, both as soldier and later as Cabinet Minister. At times he was labeled "a war man" and "a warmonger," though his unheeded warnings regarding Nazi Germany proved to be correct. As an advocate for action, Mars placed in the Ascendant in its own *nakshatra* can be full of courage, though sometimes reckless. Its *nakshatra* is *Chitra*, with its image as sharp and shining like a jewel. The presiding deity of *Chitra* is Tvastr, the divine architect.

In the *Rig Veda* Tvastr is the ideal artist and artisan who sharpens and carries the great iron axe and forges Indra's thunderbolts. This further strengthens the warrior planet Mars and galvanizes Jupiter, as it too is located in *Chitra nakshatra* though in the subsequent sign of Libra, where it can be more effective in this chart. And since Jupiter carries the influence of Mars, though on the unwieldy Rahu-Ketu axis, **Jupiter *Dasha*** becomes the ***Dasha* of maximum power** in the lifetime. This is when all of his powers as both military strategist and statesman are more likely to succeed. Jupiter's influence also means this is about a much larger picture, not just one battle.

> "... I felt as if I were walking with destiny, and that all my past life had been but a preparation for this hour and for this trial."
>
> **Winston Churchill**, May 10,1940.

Years of maximum power: It is generally agreed that Churchill had maximum power on a national and international stage when he was a national wartime leader. This occurred during his 16-year **Jupiter *Dasha*,** from Nov. 30, 1938 to Nov. 30, 1954**.** To review, Jupiter is strong because: **1)** it carries the power of Mars; **2)** it is strengthened by being in the sign of Venus (best planet for Virgo Ascendant); and **3)** it has *Vargottama* status (repeated in its same sign in the *Navamsha* and in many other harmonic or *Varga* charts). During his Jupiter *Dasha* Churchill served twice as Prime Minister of England: 1940-1945 and 1951-1955. (The first time he was appointed; the second time he won the General Election: on Oct. 26, 1951.) Prior to the first Prime Ministership he was First Lord of the Admiralty (Sept. 3, 1939 to May 10, 1940) and already lobbying for a very controversial alliance against Hitler: England, France, and Russia. With the collapse of Neville Chamberlain's government on May 10, 1940, King George VI invited Churchill to become Prime Minister and to form a Coalition government.

His ascent to power on May 10, 1940 was extraordinary – even if everything seemed, as he said, to be "but a preparation for this hour and for this trial." He was in **Jupiter-Jupiter** *Dasha/ Bhukti*, and on May 7, 1940 there was a **New Moon in Aries** – ruled by warrior planet Mars. At 23:44 Aries, the New Moon was in *Bharani nakshatra*, ruled by the deity Yama, the restrainer, and the lord of death, who guides the soul to the astral plane. As the first New Moon in which tr. Jupiter and Saturn came into contact in Aries prior to the **triple JU-SA conjunctions in Aries** between Aug. 7, 1940 and Feb. 14, 1941, it was the harbinger of a major new 20-year social, economic and political cycle, one that would accelerate the world into the Atomic Age. From an astrological perspective Churchill could not have risen to this height of power prior to these astrological events:

1. Start of his Jupiter *Dasha* from Nov. 30, 1938
2. Entry of tr. Jupiter (his *Dasha* lord) into the sign of Aries April 16, 1940
3. Entry of tr. Mars into Gemini on May 7, 1940
4. New Moon in Aries on May 7, 1940 (less pivotal than #1 through 3, but still important as a short-term timing indicator)

His sudden rise to power would have been highly unlikely during a Mars transit in Taurus (March 22, 1940 to May 6, 1940) – its absolute weakest position for Churchill. Late on May 6, 1940 transiting Mars exited the sign of Taurus and entered Gemini at the top of his chart, also bestowing *Digbala*: its best angular position. Tr. Mars was also in Taurus when Churchill lost his bid for re-election July 26, 1945 and when, after many delays, he finally resigned from this second Prime Ministership April 5, 1955.[201]

Prior to that, Saturn had begun its 2 1/3 year passage though Aries in his 8th house (April 27, 1939 to June 18, 1941) – not a generally favorable transit for him from his Virgo Ascendant, but favorable in the 9th house from his natal Moon in Leo. Tr. Jupiter also provided a modifying effect when it entered the royal sign of Aries in mid-April 1940, especially as it was now *Dasha* lord and brought the *karmic* fruits of Jupiter into play.

Also on May 7, 1940 at 1:07 PM London time a **New Moon in Aries** ignited the fiery passion of right action (*Dharma*). At 23:44 Aries, it was in close contact to the natal Ascendant lord of Adolf Hitler at 24:23 Aries, and in tight opposition to Churchill's Ascendant lord Mercury at 25:29 Libra. With 6:35 Leo rising in London – close to Churchill's natal Moon at 7:37 Leo, this New Moon was fitting for the start of a new administration that had to deal with war, even if victory would come at a great price after five more years. There were five planets in Aries: Sun, Moon, Mercury, Jupiter and Saturn. Tr. Venus and Mars were in Gemini, and Rahu-Ketu on the Virgo-Pisces axis were located within one degree of Churchill's natal Mars. Mercury and Mars in a *Parivartana yoga* (exchange of signs), indicate strong words or a war of words, as well as major work with military intelligence and code-breaking. (The US-British intelligence operation during WWII was vast, unknown to the public until the 1970s.) Tr. Mars had virtual interplay with the five planets in Aries through its exchange with Mercury, thereby fortifying the war themes. Though not as auspicious in the 8th house from Churchill's Virgo Ascendant, this New Moon and the upcoming Jupiter-Saturn conjunction were well placed from both his Moon in Leo and his *Dasha* lord Jupiter. Furthermore, the usual difficulties of **Saturn in the 8th house** from the Ascendant (*Ashtama Shani*) were definitely mitigated by the fact that Jupiter was also in Aries.

Bitter defeat turns into financial boon: On July 26, 1945 Churchill experienced the most bitter defeat of his entire career when he lost his bid for re-election as Prime Minister, although he remained Leader of the Opposition. But the public was tired of war and he was again perceived as a warmonger. His opponent Clement Attlee and the Labour Party won in a landslide election. Tr. Saturn in Gemini in his 10th house showed Churchill at the height of power and also a place from which one can take a big fall. In July 1945, tr. Rahu in Gemini was alongside tr. Saturn at the top of his chart, indicating there could be an unexpected twist to this "fall." Tr. Sun with this combination from June 15 to July 15, 1945 gave the signature of confrontation of the status quo. The General Election was held on July 5, 1945 but delayed until July 12 and July 19 in certain constituencies. All the votes were not counted and declared until July 26, 1945. Unfortunately for Churchill, tr. Mars was in Taurus July 13 to Aug. 27, 1945.

For best results in defining events, we look for a combination of *Dasha/bhuktis* and transits, especially including the *Dasha* lord – Jupiter, in this case. During the election tr. Jupiter was in his 12th house in Leo until July 18, 1945 when it moved on to his Virgo Ascendant and in the 2nd house from natal Moon. Then for most of 1946 tr. Jupiter was in its natal position in Libra in the 2nd house from the Ascendant. As *Dasha* lord and wealth significator with *Vargottama* status, this **Jupiter return** had the potential to yield excellent results financially.

---

[201] Though not emphasized in this book, *Ashtakavarga* is another of many thousands of valuable methods in Vedic astrology and accounts for why we know tr. Mars will be extremely weak in Taurus. For Churchill, it ranks zero out of a maximum of 8 bindus, the lowest possible ranking while in Taurus. (See Glossary for explanation of bindu number.)

His greatest moments of triumph as well as this re-election defeat occurred during **Jupiter-Mercury** *Dasha/Bhukti* (Aug. 1, 1943 to Nov. 6, 1945). The Ascendant lord involved as *Bhukti* or *Dasha* lord is part of the key. Mercury and Jupiter (large quantities of words) are both situated in the second house of income and are separated by planet Ketu, which as eclipse planet can push us inwards and bring unexpected events in this area of life. Thus his 1945 election defeat turned into a tremendous boon financially and helped him to parlay his formidable reputation into a lucrative project – his six-volume opus: *The Second World War*. It changed his personal financial history radically for the better. He made the equivalent of over $18 million USD in today's money, enabling him to keep his beloved Chartwell – the country estate in Kent he purchased in 1922. It had become unaffordable by 1945. Tr. Saturn in his 10th house aspects the 4th house of real estate and brings action regarding 4th house matters. In the process of negotiating the contracts, his literary agent Lord Camrose and sixteen other benefactors purchased Chartwell. The joint trust fund they set up in 1946 made it possible for Churchill to continue living there and he did so until his death in 1965.

Jupiter is likely to be his most lucrative *Dasha*. We can see this reading the chart from *Dasha* lord Jupiter in Libra (as a Sub-Ascendant: valid only during **Jupiter *Dasha***): Moon is well placed in the 11th house (of financial gains), with Venus and Sun in the 2nd house (of income) from Jupiter. They form *Dhana yogas* of wealth. From Jupiter, five planets in *Chara* (active) signs are in angular houses, including Saturn powerful in practical Capricorn. The unpredictable twists and turns in the 16-year Jupiter *Dasha* are largely due to Jupiter being on the Rahu-Ketu axis in both his birth chart and *Navamsha* chart. Eclipse planets in close contact with Jupiter also make it harder for him to consolidate power in the traditional ways.

Re-election: Churchill was re-elected Prime Minister Oct. 26, 1951 during his **Jupiter-Mars** *Dasha/Bhukti* (Aug. 1, 1951 thru July 7, 1952). Though nearly age 77, this would be a strong sub-period due to Jupiter's ongoing ability to carry the power of Mars and Venus (*Vargottama* in Venus-ruled Libra and in a Mars-ruled *nakshatra,* as discussed). Mars and Jupiter are good friends, well placed in trinal or *Kendra* houses in the *Navamsha*. At the time of the election, transiting Jupiter and Saturn were straddling his Ascendant axis, Jupiter in Pisces opposite Saturn in Virgo, spurring him on to further public endeavors and to confront current issues. In 1953 he was awarded the Nobel Prize in Literature for his oratory and writing, and in the same year he was knighted, becoming Sir Winston Churchill. This was during **Jupiter-Rahu** *Dasha/Bhukti* (July 7, 1952 to Nov. 30, 1954). Additional responsibilities would come due to the ***Sade Sati*** period (July 27, 1947 to Aug. 22, 1953): tr. Saturn passing through Houses 12, 1, and 2 from natal Moon. His health was increasingly weak, with a severe stroke in late June 1953 (again hidden from the public), but Churchill's presumed successor – Anthony Eden – was in even weaker health. These factors, combined with using his world stature and clever delaying tactics, prolonged Churchill's second term well beyond the expected one or two years. The astrology also confirms his timing, with both *Sade Sati* and *Jupiter Dasha*.

End of maximum power: His Jupiter *Dasha* ended Nov. 30, 1954 – also his 80th birthday. During **Saturn *Dasha*** it became more difficult to overcome the effects of the strokes he had suffered since 1946, and Churchill's increasingly bad health could not be disguised from others, including his deafness – made more inconvenient socially with his refusal to wear a hearing aid. This forced his long deferred resignation from his second Prime Ministership on April 5, 1955. He was just over four months into his **Saturn *Dasha***. As lord of his 5th and 6th houses, Saturn is not an easy planet for Virgo Ascendant. From Saturn, Jupiter, Ketu, and Mercury are in the 10th house, but Moon is in the 8th house. He continued on as a Member of Parliament up through 1964, though not giving any speeches; and he continued to produce more books: *A History of the English-Speaking Peoples*, 1956-1958 (four volumes).

Legendary persistence: Persistence is less likely without a strong **Saturn**, or without a well–balanced mind (**Moon**), which in Churchill's case was problematic, as we have described. He suffered from periodic deep depression, which can also result from a difficult Saturn and its relationship to the Moon. In fact, his natal Saturn is well placed in Capricorn (its own sign) in the 6th house of competition from natal Moon. It enables an already aggressive person to succeed in highly competitive situations such as elections, wars and political frays. But its *Jiva* is the Moon, as Saturn is located in *Shravana nakshatra*, owned by the Moon, which in turn is located in the 12th house – a difficult placement for the Moon. With Moon as *Jiva* for Saturn, the Moon and Saturn are linked in the birth chart, and in the *Navamsha* chart Saturn is contacting the Moon in Gemini, though alleviated by Jupiter's aspect. We discussed earlier how these factors can bring a tendency for depression, but also an appetite for hard work and the ability to take on many responsibilities.

We have discussed how spoken and written communications are emphasized by a cluster of natal planets in the 2nd and 3rd houses from Ascendant and Moon. Saturn's position vis-à-vis these planets is such that Saturn transits to Churchill's chart are always aspecting – thus impacting – his key personal planets as well as the Ascendant. Saturn (natal or transiting) aspects Houses 3, 7, and 10 from itself. Also a planet affects other planets it contacts in the same house and sign. Thus, he does not have any "off" years when Saturn transits are a bit easier, except possibly when in the natal sign of Capricorn, its return position. When that occurs it aspects all the financial houses, potentially causing money drains and making it necessary for him to earn more to pay off debts and expenses. His country house at Chartwell was referred to as "a money pit," and generally he lived the life of privileged aristocracy but without the requisite inheritance. For this reason he was constantly writing and publishing books and was well paid for his efforts. He also continued to work tirelessly on behalf of his country even when out of power and out of favor, 1929-1939. During this period he wrote and published some 16 books, including *While England Slept,* 1938.

In his book on Churchill, *The Last Lion (1989),* Vol. 2, William Manchester argues that Churchill was at his most powerful in the years 1932 to 1940, precisely because he refused to disappear and refused to stop haranguing Britain about the reality of Nazi Germany. This kind of persistence is not always recognized, though it had to play a part in why he was able to command so much loyalty from the British people in May 1940 and throughout the war years. Manchester's argument has merit, but the astrology confirms the majority opinion: Churchill was at his height of political power from May 10, 1940 to July 26, 1945.

"The price of greatness is responsibility." **Winston Churchill**

Strong sense of duty – prolonged Saturn effect: Generally a period of greater responsibilities for 7 1/2 years, ***Sade Sati*** occurs when Saturn transits through Houses 12, 1, and from the natal Moon in any chart. In Churchill's case, since natal Moon is in the 12th house from the Ascendant, *Sade Sati* occurs when Saturn transits through Houses 11, 12 and 1 the Ascendant. Prior to that,

Saturn passing through his 10th house creates maximum visibility and exposure, and one can be at the top of one's profession. One can also fall the furthest from there, especially with natal Saturn in the 10th house. (Adolf Hitler had this configuration, as did Richard Nixon.) Directly following the *Sade Sati*, tr. Saturn moves to Churchill's 2nd house in Libra, and then 3rd house in Scorpio – both houses containing a total of five planets, including Ascendant lord Mercury and the Sun, which can suffer physically from a direct Saturn transit. Continuing from there, all key planets and/or the Ascendant are constantly receiving an aspect from transiting Saturn. Saturn in the 10th house from the Ascendant is simultaneously in the 11th house from his natal Moon. This is considered a favorable position and is always a prelude to *Sade Sati*. This may have also helped to save a loss of reputation when tr. Saturn was in the 10th house from his Ascendant.

Saturn at its low point in the chart: The greatest amount of stress for Churchill would be likely to come during Saturn's transits in Sagittarius – in his 4th house – at the bottom of his chart. Transiting Saturn in the 4th house often demands great patience as it generally gives the least amount of power and/or visibility, so may give difficulty to find work. Tr. Saturn aspects the Ascendant from the 4th house, and can also bring health issues as the natural vitality may be more curtailed. For a man who is an advocate for action and prefers to be at the center of power this Saturn transit is surely the most excruciating. Saturn orbits around the Sun in 29.46 tropical years, and Saturn's passage in Sagittarius occurred three times in Churchill's lifetime. Since Sagittarius is a sign of long ascension, planets spend a longer time than usual in this sign of the zodiac. For Saturn, it is 3 years instead of the average 2.33 years:

1. **Saturn in Sagittarius:** Feb. 22, 1899 to Feb. 11, 1902
2. **Saturn in Sagittarius:** Dec. 25, 1928 to Dec. 31, 1931
3. **Saturn in Sagittarius:** Feb. 8, 1958 to Feb. 1, 1961

To this information, we must also weigh the *Dasha-bhukti* running at the time, as this will also give us more detail about the destiny. He was first elected to Parliament in Oct. 1900 after an unsuccessful bid in 1899. He was in Sun-Saturn *Dasha/Bhukti* in Oct. 1900 when he was first elected. (Conversely, he died in Saturn-Sun *Dasha/Bhukti* some 64 years later.) During the second lifetime passage of Saturn in Sagittarius, he was in Rahu *Dasha*, and by the third passage 1958-1961 he was in his Saturn *Dasha*. Tr. Saturn in Sagittarius never coincided with his Jupiter *Dasha*.

Saturn's second passage in Sagittarius had the greatest impact on Churchill. He was out of office again after the Conservative defeat May 30, 1929, and began what he called his "Wilderness years" – when he was out of the center of political power. This lasted for ten years. First he lost a large amount of personal money in the Oct. 29, 1929 NY stock market crash. Then there was the serious pedestrian accident on Dec. 13, 1931 in Manhattan, NY. (See under earlier section: *Physical Health & Longevity*.) Following this and from the early 1930s extending all through the decade Churchill observed Hitler's rising power and strongly warned Britain to re-arm itself. He was not taken seriously and his claims were largely ignored – until Sept. 1, 1939.

Saturn at its high point in the chart: On the two occasions Churchill considered his most bitter losses, tr. **Saturn** was **in Gemini** in his 10th house (top of the sky), a position of maximum power from which one can fall the furthest in terms of public and professional stature: **1) March 18, 1915** (defeat of the Dardanelles military campaign, followed by the near loss of his career and reputation). He was in **Mars-Rahu** *Dasha/Bhukti*. The *Dasha* period made this a potentially lethal fall, but he quickly recovered and re-entered the Cabinet in 1917 as Minister of Munitions. **2) July 26, 1945** (defeat in his bid for re-election as Prime Minister, won by his opponent in a landslide election.) This is discussed earlier under the section: *Bitter defeat turns into financial boon.*

Hidden Venus – "The Syndicate": Mercury's sign lord Venus is situated at 29:54 Scorpio, and thus becomes *Atmakaraka* – the planet at the highest degree in any sign in the chart – and auspicious in that regard. But the *gandanta* degree also brings unseen and potentially harmful influences to the affairs over which Venus rules, as well as the houses over which it rules: houses 2 and 9 here. In the late water signs, a *gandanta* planet will express a high degree of emotionality and aggressive ambition, further defined by its house and sign position and aspects to it. Churchill was considered a big spender with a gambler's temperament. Related to speech or writing, Venus *gandanta* in Scorpio can bring an influx of highly charged emotion, but also – due to the sign of Scorpio – something that remains hidden for a time. Being in the 3rd house of written communications – it shows the tireless, perhaps obsessive drive to write. In his case he wrote about many of the historical situations he experienced first-hand or of historical periods and themes that interested him.

This **Venus *gandanta*** also describes Churchill's zealous desire both to add to his income as needed and to restore his reputation by writing, using whatever means he could to do so. Though his energies were prodigious, he was still too busy with various engagements to give all his time to literary endeavors. In the 2005 book by David Reynolds, *In Command of History: Churchill Fighting and Writing the Second World War*, it became generally known for the first time ever that Churchill's six-volume work (5000 pages illustrated by maps and plans) was largely accomplished through a team of **ghost writers**. In private they were called **"The Syndicate"** and included two retired generals, a former naval officer and an Oxford historian. They were responsible for the research and wrote the first drafts. Churchill then supplemented this with his reminiscences of central events.

His means of being paid for these literary endeavors also fits into the snake-like arena of Venus at 29:54 Scorpio, a sign of secrets. Since it would have been impossible for him to avoid paying 97.5% tax on profits from his own books, given the new British tax structure under the new Labor government, Churchill worked with his lawyers to create a more palatable route to profits. In addition to the trust fund created to purchase Chartwell (the country estate in Kent), his chief literary agent Lord Camrose negotiated lucrative deals with publishers in 15 countries, and even more lucrative syndication deals with 50 newspapers and magazines in 40 countries. Churchill's chart shows both heavy expenses and income from foreign venues. This is due to *Navamsha* Venus exalted in the 12th house in exchange with *Navamsha* Jupiter in the 7th house (both foreign venues). Jupiter-Venus interactions can bring excessive and/or large-scale enterprises.

Hidden Venus – his painting: The condition of Venus in the chart also describes another artistic area that was little known by others for much of his life. Churchill began to paint sometime after the disastrous Dardanelles Campaign in March 1915 – during **Mars-Rahu** *Dasha/Bhukti*. Mars is the sign lord of Venus and helps him to take action of various sorts.

> "If it weren't for painting, I couldn't live; I couldn't bear the strain of things."
>
> **Winston Churchill**

Over his lifetime this became an enormous source of joy and relaxation. He produced a total of 570 paintings and two sculptures that are considered to have artistic merit and can be seen in various collections. His first exhibit was under an assumed name: Charles Morin. He had only a few exhibitions under his own name, and though his painting was largely unknown in his lifetime, a complete digital record of his paintings can now be viewed at the new Churchill Museum in London. He published his essay *Painting as a Pastime* in 1921 and as a book in 1948, with 34 pages of text and 16 pages of color plates. In 1948 he was recognized by the Royal Academy of Arts.

Family: Winston's mother was "like a shining star," he said, and someone he "admired from a distance." She was largely inaccessible through much of his childhood. This can be the case with Moon in the 12th house. His mother became more important after 1895, serving as an ally and an important political mentor. Winston's father (Lord Randolph Spencer Churchill) died Jan. 24, 1895, and his beloved nanny Mrs. Everest six months later. That summer Winston himself nearly drowned at Lake Lausanne, towards the end of his 20-year Venus *Dasha*. Venus is lord of the 9th house of father, and is not well indicated in the chart at 29:54 Scorpio. An extreme situation is described, and indeed Winston's father and mother were mostly involved with Lord Randolph's political career during this time, to the neglect of their children. His father's health was always uneven and went rapidly downhill in his last few years of life. He had periods of intense activity followed by depression. Lord Randolph died at age 46, during Winston's **Venus-Mercury** period. Rumors still circulate that Winston's father died of syphilis; but they are unfounded, as otherwise his wife and children would have been infected and they were not.

Winston entered the 10-year **Moon *Dasha*** Nov. 30, 1903. (Moon is mother and women.) He met his future wife in 1904 and married her Sept. 12, 1908, during **Moon-Saturn** *Dasha/Bhukti*. This Moon-Saturn combination appears together in the *Navamsha* 3rd house and is aspected by Jupiter. (The 3rd house rules over writing but also desire for relationship: *Kama*). Their marriage was affectionate and enduring, lasting 56 years. Clementine was his rock of support when he appeared to have none elsewhere. Saturn's influence can add responsibility but also endurance and loyalty. Jupiter's aspect to *Navamsha* Moon and Saturn helps to lighten the sometimes crushing weight of Saturn's duties. Jupiter is also lord of the natal 7th house of marriage and is placed in the same sign with Ascendant lord Mercury in the birth chart, though nearer to Ketu. Its strength in the chart brings a strong wife, whose merits would be known to all especially during his **Jupiter *Dasha***, 1939-1954.

Strong ancestral pride comes with **Moon in *Magha nakshatra***, whose deities are the Pitris, the fathers or ancestors. Sir Winston died at age 90 on Jan. 24, 1965 – exactly 70 years to the day from his own father's death at age 46. Though they had minimal contact, as his father preferred to keep him at a distance, Winston wrote a two-volume biography of his father, *Lord Randolph Churchill* (published 1906). From 1933 to 1938 he also wrote a four-volume biography of his ancestor John Churchill, the first Duke of Marlborough (1650-1722). *Magha nakshatra* has the image of the throne room and of royalty and greatness. At the end of his life Churchill refused a Life Peerage, mainly because his son Randolph wanted a career in the House of Commons and could not do so with a title which could not be returned or rescinded. However, his widow Clementine accepted a title after her husband's death. Churchill was the first and only non-royal to be honored with a state funeral. It was attended by leaders from all the major countries of the world. His body lay in state for three days while mourners paid their respects. Born and bred of British aristocracy, with a mother from the American nouveau riche, he was a confirmed monarchist. Clementine said he was probably the last modern to believe in "the divine right of kings."

Chart 25: John F. Kennedy

Birth data: Tuesday, May 29, 1917, 15:00 EST, Brookline, Mass., USA, Long. 71W07 18, Lat. 42N19 54, *Lahiri ayanamsha*: -22:42:29. Class A data: Garth Allen quotes JFK's mother in May 1960. (Same data from Doris Kearns Goodwin, *The Fitzgeralds and the Kennedys*, 1984, p. 274.) Astrologer Jim Lewis used a rectified time of 3:17 PM to put relocated Pluto at the Midheaven over Dallas, Texas and natal Saturn at the Midheaven over the birth place. This author prefers the 3:00 PM chart. Relocation-wise, natal Saturn is still only 3 degrees from birthplace Midheaven; Sun is at the top of the sky near Hollywood; Venus sets in Berlin, Germany – a scene of major personal triumph; and by zodiacal degree (1:16 orb) Pluto sets in Moscow, the home of his nemesis, Nikita Krushchev. Ascendant: 27:17 Virgo.

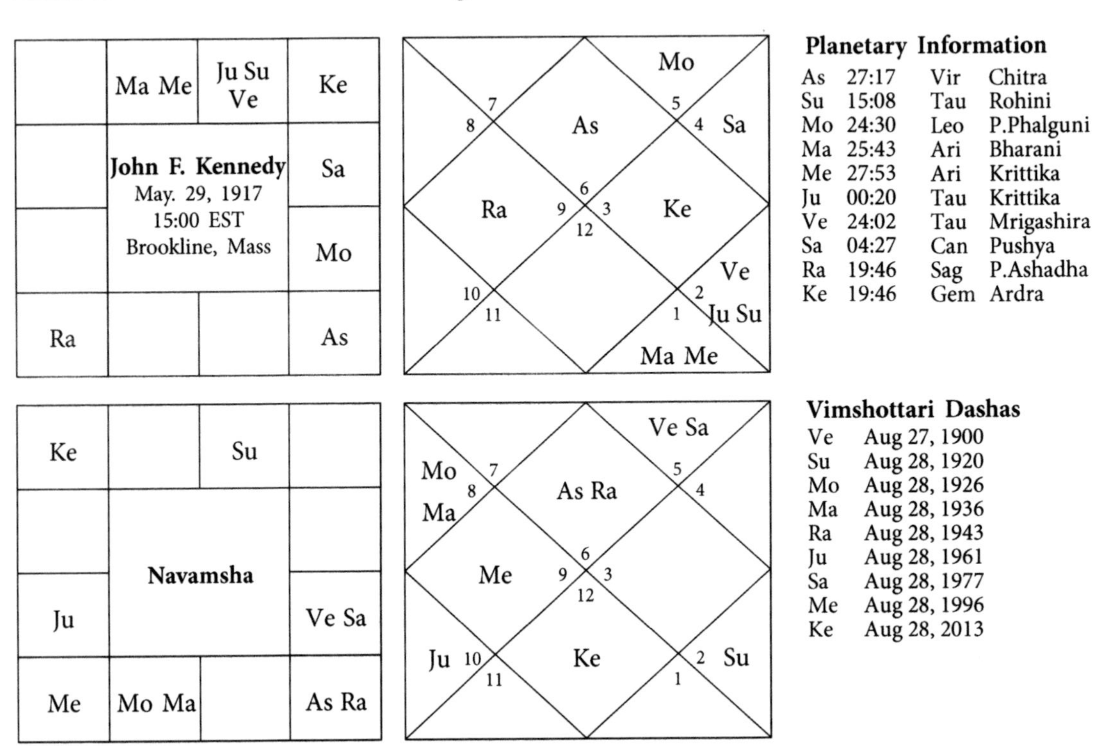

Ma Me
Ju Su Ve
Ke
John F. Kennedy
May. 29, 1917
15:00 EST
Brookline, Mass
Sa
Mo
Ra
As
Planetary Information
As 27:17 Vir Chitra
Su 15:08 Tau Rohini
Mo 24:30 Leo P.Phalguni
Ma 25:43 Ari Bharani
Me 27:53 Ari Krittika
Ju 00:20 Tau Krittika
Ve 24:02 Tau Mrigashira
Sa 04:27 Can Pushya
Ra 19:46 Sag P.Ashadha
Ke 19:46 Gem Ardra
Navamsha
Ke
Su
Ju
Ve Sa
Me
Mo Ma
As Ra
Vimshottari Dashas
Ve Aug 27, 1900
Su Aug 28, 1920
Mo Aug 28, 1926
Ma Aug 28, 1936
Ra Aug 28, 1943
Ju Aug 28, 1961
Sa Aug 28, 1977
Me Aug 28, 1996
Ke Aug 28, 2013

Brief biographical summary: The 35th U.S. President, John F. Kennedy was in office for just under three full years, 1961-1963. But in that time and in the time since he was assassinated Nov. 22, 1963 – he evoked some of the greatest outpouring of admiration and grief bestowed on any U.S. President. The myth and legend surrounding this "Camelot presidency" is very large, and television only magnified it, starting with the Kennedy-Nixon debates in Sept. – Oct.1960, the first such debates to be televised. The first one on Sept. 26, 1960 had the most impact and was watched by 70 million viewers, according to the Museum of Broadcast Communications. Kennedy also initiated live televised press conferences, and held one every 16 days throughout his term in office. The first one on Jan. 26, 1961 had 65 million viewers. At a time of relative innocence in American life, investigative reporters were easy on Kennedy, and most negative press was kept at bay by his family money, power, and influence, as well as a large circle of loyalists. (The Vietnam

War and Watergate were to change reporters' attitudes towards the presidency.) The camera loved Kennedy and his handsome good looks. His wit, intelligence and understated elegance also went over well. Men and women quickly fell in love with him; many journalists had great affection for him and he stirred the excitement and aspirations of the public in a profound way. Radiating self-assurance, and valuing courage above all, he promoted active citizenship. His Inaugural address is still regarded with reverence:

> "Let the word go forth from this time and place, to friend and foe alike, that the torch has been passed to a new generation of Americans..."

Watching him for some years, the public was fascinated by Kennedy, a man groomed and financed by his family for the presidency since 1946. When they saw him get shot in a Texas motorcade in Nov. 1963, most were ill-prepared for the gut-wrenching farewell. Enormous shockwaves went out across the country and the world, marking the end of an era and the death of perhaps America's most charismatic president, enhanced by television. Although he was the 4th president to be assassinated and the 8th to die in office, his assassination was the first to be witnessed by television audiences. The fact that he is still recognized and admired around the world is indisputable. Whether he deserves the extent of adulation and mystique surrounding him and his presidency is another matter. His public cannot easily separate the politician from the man of surpassing charm and movie star qualities.

Historians have more objective criteria, and most do not rank Kennedy as a great or even near-great president, as so much remained unfinished at his death. However, he was the first in many categories: first television president, first Roman Catholic president, and youngest president ever, elected and inaugurated at age 43. In 1960 he was still a neophyte statesman, a rich playboy candidate with ambition and a young American aristocrat whose elegance made mockery of the typical prejudice against Irish Catholics. Joseph Kennedy's unstoppable determination was responsible for his son's rapid rise to the presidency. Eldest son Joe Jr. (b. July 15, 1915) was originally to play this role but died in a plane crash in WW II Aug. 12, 1944. Jack was next in line to carry out the Kennedy dream. His rich father was U.S. Ambassador to Great Britain, 1937-1940, but by late Nov. 1940 he had sabotaged his own chances for higher office. Jack was the second of nine children, four boys and five girls. He soon developed plenty of his own drive and political ambition.

As with Churchill, the typical route to political power was first through a good education at established institutions. In Kennedy's case this meant the best Eastern Protestant schools (Choate and Harvard) in order to shed the Irish Catholic parochialism. Then there was a stint of military service – preferably heroic service in an important war – followed by some journalism. Kennedy was only briefly a journalist in 1945 after he was discharged from the Navy as a full Lieutenant. On Aug. 2, 1943 he sustained a serious back injury when a Japanese destroyer hit the PT boat he commanded in the Pacific. He swam six miles to safety, rescuing most of his crew. For this he received a Purple Heart in 1945.

BUILDING CAREER & FAMILY: His political career began in 1946 when he was elected to the U.S. Congress as a Representative from Massachusetts. He was re-elected in 1948 and 1950, then ran successfully for the U.S. Senate in 1952, unseating the well-established Henry Cabot Lodge. The following year, on Sept. 12, 1953, he married the glamorous Jacqueline Lee Bouvier (b. July 28, 1929, d. May 19, 1994), and together they had four children: Arabella (b. Aug. 23, 1956, stillborn), Caroline (b. Nov. 27, 1957), John Jr. (b. Nov. 25, 1960, d. July 16, 1999), and Patrick (b. Aug. 7,

1963, lived only 39 hours.) On Jan. 1, 1956 his book *Profiles in Courage* was published to lavish praise, bringing him a Pulitzer Prize in 1957. His one other book was based on an honors thesis he wrote during his senior year at Harvard (1936-1940). Published in 1940 and re-titled *Why England Slept*, it was modeled on *While England Slept* (1937) by Winston Churchill – the man he most admired and also his literary and political model. During the Democratic Convention in summer 1956, Kennedy narrowly missed being nominated as Democratic Vice Presidential candidate. He also delivered the nominating speech for Adlai Stevenson. Although his career as Congressman and Senator was unremarkable and he was often absent due to illness, Kennedy was re-elected to the Senate in 1958, and then to the U.S. Presidency on Nov. 8, 1960. His Inauguration was on Jan. 20, 1961, and his brother Robert soon became Attorney General as well as his closest advisor. Jack Kennedy would have liked to be his own Secretary of State – such was his preference for foreign policy over domestic policy.

ESTABLISHMENT MYSTIQUE: Kennedy believed in rational intelligence and pragmatism above all. He preferred to be at the center of power as much as possible, using fewer bureaucratic procedures than his predecessor. He believed many solutions for national and global problems would come through assigning the best possible experts as Cabinet Members, with whom he would work as a team. In 1960 there was a rare consensus on foreign affairs. It had been rigid and centrist: Communism was dangerous, containment was good, and foreign aid bills needed to be passed to keep the Third World safe from the expansion of Russian and Chinese influence, under the guise of Communism. Kennedy emerged among the new breed of Democrats who were hard-liners, committed to increased defense spending and not wanting to appear militarily weak or soft on Communism. His excellent rapport with the press helped him to express and promote his ideas and programs, also his infallible gift for using all forms of media to his advantage, especially television. Even so, his critics considered him so devoted to rationalism that he would never risk political defeat for a great moral issue.

His "New Frontier" approach and his first speeches brought a jolt of youthful energy to the country and to the world stage. Everything seemed possible: dealing with problems at home and abroad and setting newfound goals for the space program. He laid out the inspiration for this in his Inaugural address, still considered among his finest speeches. He wanted to distill American dreams and empower a new generation, to invoke both patriotism and a strong sense of the power of the individual. He succeeded, inspiring even civil rights advocates far more than he knew or perhaps intended. Towards the end came these lines:

> "And so, my fellow Americans, ask not what your country can do for you; ask what you can do for your country. My fellow citizens of the world, ask not what America will do for you, but what together we can do for the freedom of man."

1961-1963: In early 1961 Kennedy introduced the 10-year Alliance for Progress for aid to Latin America; also the Peace Corps, and programs to accelerate America's participation in the space program, bringing the first Americans into orbit. He suffered his first major defeat in mid-April 1961 with the failure of the Bay of Pigs invasion in Cuba, and this led to the entrenchment of the Communist-backed Castro regime. It also led to his trusting far fewer advisors, with the ongoing exception of his brother Bobby. CIA records show that both of them were insistent Fidel Castro had to be "eliminated" and ordered the CIA to do so. When the installation of Soviet offensive missiles was discovered in Cuba in 1962, Kennedy ordered a naval blockade around the island. A serious threat of nuclear war prevailed until Soviet premier Khrushchev ordered the removal of the missiles. Kennedy's handling of the Cuban Missile Crisis Oct. 15-29, 1962 is generally well regarded in terms of his avoiding World War III. Under constant pressures to go to war, he managed to stand up to the hard-liners within the National Security establishment by out-foxing them. Then on Aug. 6, 1963, a limited test-ban treaty was signed in Moscow. Having faced the brink of nuclear war, Kennedy came to a far more realistic position on nuclear weapons by summer 1963 and to a more conciliatory tone in general. Some consider this his greatest contribution as president.

However, this achievement is offset by his many serious miscalculations in his first year in office. Some scholars consider the Cuban Missile Crisis came precisely because of Kennedy's mistakes in 1961. Author Frederick Kempe calls 1961

> "...one of the worst performances in foreign policy of any U.S. President."[202]

This included the Bay of Pigs, the Vienna Summit with Khrushchev (June 1961) and the erection of the Berlin Wall by the Soviets (Aug. 12-13, 1961). Kennedy did not object to its being built as long as West Berlin was unaffected. He thought that allowing the Soviets to solve an emigration problem might avoid a war. But these events convinced Krushchev that Kennedy was a weak and indecisive president, as well as young and inexperienced. Kennedy himself referred to 1961 as "a string of disasters."

In domestic policies, Kennedy's proposals for medical care for the elderly and aid to education were defeated, but later he won important victories on minimum wage, trade legislation, and other measures. Forced to deal with civil disorders due to accelerating racial unrest, he proposed sweeping civil rights legislation which was pending, along with an $11 billion tax cut to bolster the economy – when he was killed in Dallas on Nov. 22, 1963.

ASSASSINATION: The Kennedy presidency will forever remain unfinished due to his assassination, which is still controversial in many ways: who perpetrated it and why, and how many killers were involved. Over 1000 books have been written on the subject to date. From spring 1961, the Kennedy brothers did not trust the FBI, the CIA or even their own Joint Chiefs of Staff, and were confounded by the CIA's independent actions, especially their secret war on Castro. The official explanation of the Warren Commission in 1964 was that there was a lone gunman, Lee Harvey Oswald. This has been discredited by numerous sources, including an 8 mm. color home movie by Abraham Zapruder, a Dallas resident. He filmed the entire assassination and showed Kennedy being hit from in front and from behind. Though some have disputed the authenticity of this film, in 1998 a Kodak engineer verified that the film was an "in camera original" and that any alleged alterations were not feasible. Only photo stills of the Zapruder film were allowed in public prior to its first-ever television broadcast in 1975 on *Good Night America*. This ignited widespread public outrage, and distrust in the Warren Commission findings. Scientific acoustical analysis of the film was done in 1978, indicating there were four shots, the third coming too rapidly after the second to be fired from the same gun.

1960 ELECTION: Other controversies about the Kennedy presidency include the election itself – which he won by the narrowest margin since 1916 – only 113,000 votes. The votes in Illinois

---

[202] Frederick Kempe: ***Berlin 1961: Kennedy, Krushchev, and the Most Dangerous Place on Earth***, 2011.

(especially Cook County) and Texas remain "officially" unresolved and irresolvable. He won 45.9 % of the popular vote to Richard Nixon's 45.7%. But long concealed reports from FBI electronic surveillance indicate Joe Kennedy bargained with Mafia and union bosses to deliver the vote for his son in several key states, including Illinois. In return, the powerful Chicago crime syndicate probably expected some level of immunity from federal prosecution. But they especially wanted Kennedy to eliminate Castro, as a new Cuban government could restore the lucrative casino, hotel, and prostitution businesses operated by the Mafia up through 1959. Kennedy did try repeatedly to have Castro killed, a plan previously promoted by Eisenhower. Meanwhile, the FBI reported to the Attorney General the election had been stolen in Illinois. Joe Kennedy had insisted Bobby Kennedy be appointed Attorney General. In this role he could and did stop any full-scale investigation into election fraud. FBI director J. Edgar Hoover knew the election had been stolen but could not act on what he knew due to the illegal nature of the wiretapping. He was assured he would stay on in his position in spite of the federal government mandatory retirement age of 70.[203]

VIETNAM: Whether Kennedy would have deepened America's involvement in Vietnam is another ongoing controversy of his presidency. Kennedy loyalists such as Arthur Schlesinger, Jr. and Ted Sorensen insisted he was about to begin disengagement in Vietnam. They point to an executive order he signed on Oct. 11, 1963 just prior to his death. They also say that despite massive military buildups, he was secretly searching for peace and for a world in which nations did not lead by intimidation or nuclear weapons. Several of his speeches attest to that, especially in 1963. But other historical records conflict with that view. For example, *The Pentagon Papers* (1971) make it clear how deeply involved Kennedy was in Vietnam. On April 29, 1961 "he approved a series of clandestine actions to escalate the war."[204] In June 1961 *New York Times* reporter James Reston recounted an off-the-record conversation with Kennedy just after his meeting with Khrushchev in Vienna. To Reston's astonishment, Kennedy said he intended to show Russia his firmness specifically by American actions in Vietnam. Further, his most trusted advisers were all hawks; military spending was increased by $3.25 billion, and Kennedy signed off on the murder of Ngo Dinh Diem in Vietnam (done Nov. 2, 1963) thereby assuring further American military involvement there. He did not want to be accused of losing Vietnam to the Communists, having accused his predecessors of losing China to the Communists. Even if in public he sounded conciliatory, key evidence indicates that Kennedy set everything in motion during his presidency to fight a more extended war in Vietnam.

SECRETS: Kennedy had many sexual liaisons right up to the end of his life, including with Marilyn Monroe. There was an ongoing parade of prostitutes, starlets, female staff members

---

[203] Source: Seymour Hersh, ***The Dark Side of Camelot***, 1997. Extensive documentation is given on these and other points. Born April 8, 1937, Hersh is one of America's top investigative journalists, notably on military and security matters. His 1970 Pulitzer Prize brought him initial recognition for his 1969 work exposing the My Lai massacre during the Vietnam War. Before accepting the commission to write a book on JFK, for several years he turned it down. His research unearthed much worse than what he expected: "Kennedy was much more corrupt than other post-war Presidents, by a major factor. Much more manipulative, though Nixon was a close second... [but] an amateur compared to Kennedy. Kennedy's beauty made him more corrupt. He was above the law; he didn't think anything could stop him." (Hersh interview Jan. 8, 1998 in *The Atlantic On-line.*)

[204] *Ibid.*, pp. 220-221. ***The Pentagon Papers*** (1971) are cited.

and assorted society women, notably Mary Pinchot Meyer.[205] Kennedy's addiction to sex was a persistent problem for his Secret Service and his closest associates – how to protect him on so many levels – including from blackmail and from imminent threat of scandal. One of his many lovers (Judith Campbell Exner) was also involved with a Mafia boss and carried messages for both men.)[206] His public was mostly shielded from the knowledge, though his wife was aware of it to some extent. But he refused to stop. Perhaps an outcome of his sickly childhood, he was determined never to be bored, frustrated or alone, and those near him had to remember those cardinal rules. It is clear he did not mind breaking the rules to get whatever he wanted, and part of that was being rich enough to be able to do so.

There is also the subject of his true state of health. Before 1970, medical records of patients remained confidential. Kennedy's medical records remained closed for years, then partially open in 1992, and fully open and available for the first time in Nov. 2002. It became clear he lied repeatedly about his true medical condition and did so initially to get into the Navy. Rejected from the Army, he got into the Navy only through his father's influence and with no physical examination. But with such serious medical conditions he should not have been in the military at all. And if the public had known the full facts about his health, his election to the Presidency would have been doubtful.

Kennedy was a walking medical marvel. Hospitalized at least 38 times between Feb. 1919 and Sept. 1957, he received the last rites at least four times. As President, he took over 12 serious medications per day, including chronic steroids to treat his Addison's disease (a total failure of the adrenal glands), which is terminal without daily cortisone treatments. Only in the late 1940s did synthetic cortisone become more widely available. At the same time doctors learned that Addison's disease could be treated successfully with cortisone. Kennedy was diagnosed in 1947, but probably had had it for years if not decades. By 1936, he still had a skinny and accident-prone frame. Since childhood he suffered from a degenerative back problem (a birth defect), along with intestinal ailments from age 13, and lack of weight gain. To treat colitis, for some years he had been on steroids, with the typical side effect of being constantly susceptible to infection, including in his case – ongoing venereal disease since 1940. Steroids can produce euphoria and over-optimism followed by severe depression. Kennedy loyalists either deny his Addison's disease (Arthur Schlesinger, Jr. in his *Journals, 1952 to 2000*, ©2007) or insist the hiding of his medical condition was defensible, even heroic, and that his private life was his own business. Independent historians with greater objectivity are starting to examine more closely the possible effects on his presidency of both his heavy drug use (prescribed and unprescribed, including LSD) and his rampant promiscuity.

---

[205] Mary Pinchot Meyer (1920-1964), a Washington D.C. socialite noted for her great beauty, was also an artist and married to C.I.A. official Cord Meyer from 1945 to 1958. They were close friends of the Washington elite, including John F. Kennedy, with whom Mary had apparently at least "30 trysts," starting from Oct. 1961. This was documented in Mary's diary that was read and burned after her death by her twin sister, Antoinette, known as Toni. (At that time Toni was married to Ben Bradlee, *Newsweek* journalist and later Executive editor of *The Washington Post*, 1968-1991.) The diary reportedly contained detailed information about Mary's relationship with President Kennedy, but not about his assassination. While walking in Georgetown on Oct. 12, 1964, Mary Meyer was shot by two bullets at point-blank range, two weeks prior to the release of the Warren Commission report. The murder has remained unsolved.

[206] See Ch. 9, Chart #31: Jacqueline Kennedy Onassis for more material on Judith Campbell Exner and ***The Church Report*** (Nov. 1975); also for more on the Kennedy marriage from Jacqueline Kennedy's perspective.

Though he exuded vitality in public, Jack Kennedy was unwell and in pain most of his life. His bad back was a congenital condition and did not originate from football or war injuries, as campaign briefs stated. He had scarlet fever, diphtheria, and whooping cough from age two, and later suffered from celiac disease, some deafness in the right ear and a baffling range of allergies and fevers. By 1934 he was diagnosed with colitis (severe inflammation of the large intestine), for which he was soon treated with steroids. Steroids first became available in the early 1930s. These and later daily cortisone treatments for his back would have further side effects – such as duodenal ulcers and osteoporosis. They may have also caused his Addison's disease, first diagnosed in 1947. He was then given one year to live. He weighed 95 lbs. at age 14 (1931) and was only 149 lbs. five years later at six feet tall, and 155 lbs in 1955. His Naval tour of duty in the Pacific brought malaria, various infections, a ruptured disk, and compression fractures in his low back, which was operated on in June 1944. On Oct. 21, 1954 and again in Feb. 1955 he underwent two back surgeries, initially a plate inserted into his spine to stabilize it and reduce pain. In Feb. 1955 the plate had to be removed due to infection. Doctors were reluctant to do the surgeries at all, as no one with Addison's disease had ever survived traumatic surgery. Kennedy gambled on the surgery as his back pain was so great. He remained hospitalized for nine months and had post-operative complications.

Physical health and longevity: The physical body is shown here by Ascendant lord Mercury, Sun and Mars (*karakas*), as well as the condition of the Ascendant itself. The Sun is *Vargottama* in the 9th house, and in a *Shubha Kartari yoga*, surrounded by classic benefics. But while Sun and Mars are strong and the Ascendant is *Vargottama* (normally giving good health), **Ascendant lord Mercury** is poorly placed in the 8th house of chronic disease. It receives contacts from two classical malefic planets, Mars and Saturn, and its mutual contact with Mars forms an *Arishta yoga* of ill health, while their joint 8th house placement shortens longevity. To add emphasis to this theme, Mercury is within half a degree of being exactly on the 8th house cusp. This is very dangerous for the health and safety of the physical body (unless the Ascendant lord is in its own sign). It can also cause mental imbalance through 8th house emphasis – including indiscriminate expression of sexual energy. The 8th house is also a hidden placement, thus the person is likely to keep some major personal factors concealed from others. Kennedy's afflicted Ascendant lord is mitigated somewhat by being at the exact Stationary Direct degree; and being Mercury, this also gives notable communication abilities.[207]

Raised in a highly competitive and athletic family, he spent much of his childhood alone in his sickbed reading books while his siblings played outside. **Moon in the 12th house** may place one alone a lot or seeking privacy. Settings are often pastoral, or away from others: bedrooms, hospitals or prisons, or residence abroad. Moon in Leo is more sociable and likes attention, so he prefers to have company in these settings. (The Moon's 12th house position also describes Kennedy's lack of physical closeness to his mother during childhood, same as Churchill and his mother.)

[207] ***Brihat Parashara Hora Shastra***, Vol. 1, Chapter 27, "Evaluation of Strengths," refers to motional strength (*Chesta bala*). Planets close to the Stationary degree possess the highest amount of *Chesta bala*. Planets are "*sama*" or stable and thus stronger if they are moving more slowly. Planets moving at the most rapid rate, and referred to as *bhita* (Sanskrit for "anxious") are weakened due to this deficit in motional strength.

His closest family connections were to his father (Venus, Sun), maternal grandfather (Ketu) and brothers (Mars), especially Bobby.

The Kennedy family emphasized physical toughness, not complaining about physical pain and making things look effortless. Thus Jack Kennedy learned very early to hide the physical pain he suffered most of his life. During his 1960 presidential campaign he even capitalized on his "youthful vigor" as an asset. This required a charade of monumental proportions and efforts to beat the pain through a daily battery of medicines – pills, pellets and injections – some on an hourly basis. One of his doctors, the trendy Dr. "Feelgood" (a.k.a. Dr. Max Jacobson) became part of the president's inner circle and even traveled with him. He injected Kennedy with a powerful mixture of painkillers and amphetamines, starting just before one of the televised debates with Nixon in fall 1960. Kennedy's reliance on these shots is questionable, as side effects were more dangerous: an exaggerated sense of power and capabilities with accompanying paranoia. By late April 1975 Jacobson lost his medical license but he quoted Jack Kennedy on the subject:

"I don't care if it's horse piss. It's the only thing that works."

Mars and Mercury interact very tightly in the 8th house in the sign of Aries. We have already described the *Arishta yoga* between Mercury and Mars. Mercury is weakened by its association with 8th lord Mars – at least in terms of physical health – though it helps his intellect and sharpens his wit and humor. When overly strong, Mars heats up and overextends the natural limits of the physical body. Kennedy suffered from constant infections and frequent high fevers. **Virgo rules the intestines as well as disease in general.** Kennedy suffered from colitis (severe inflammation of the large intestine) and celiac disease (faulty absorption of food in the intestines), both requiring a lifelong bland diet and anti-diarrhea medicine. Because of its close contact to Mercury in the 8th house, Mars is likely to bring very serious problems during the 7-year **Mars *Dasha*:** Aug. 28, 1936 to Aug. 28, 1943. On Aug. 2, 1943 his PT boat was hit by a Japanese destroyer. He escaped, but his injuries further weakened his already fragile condition and required back surgery in June 1944. His numerous medical problems along with his life habits perpetuated his physical vulnerability, radically draining his energy reserves and his immune system.

Mars also rules over the **adrenal glands.** Of all his many physical problems his Addison's disease was the most serious and potentially life threatening. With no functioning adrenal glands, the immune system is weakened and the energy level is greatly reduced, as well as the libido, which in his case was restored and amplified through drugs. Cortisone injections to treat his Addison's disease may have heightened his sex drive, and as steroids they can increase a sense of personal power and self-confidence. He also took daily doses of **testosterone** and was reported to have "a bull-like libido." The *Nadi yoga* (*nakshatra* lord exchange) between **Venus and Mars** involving the 8th and 9th houses may expand the sphere of sex and intrigue, unless those energies are focused in a more spiritual direction.

Though Addison's disease was always denied, the November 1963 autopsy report showed:

"no evidence of adrenal glands... and severe Addison's disease."

Richard Reeves, *President Kennedy: Profile of Power*, 1993, p. 668 notes.

As with most other Kennedy medical records that were suppressed, this confirmation was not available until 1992. However, it was easily detectable in medical journals in the mid-1950s. A still larger cache of his medical records became available November 2002. **The 8th house** is a hidden place, and Mercury rules over medicine as well as his physical body (as Ascendant lord). Therefore, just as in life Kennedy concealed his true state of health, there were decades of secrecy after his death regarding his medical records.[208] The Kennedy family, led by Robert Kennedy – then Attorney General – considered them "privileged communication."

"Youthful vigor": The 8th house has magnificent **public relations** capacities, as the manipulation of the public mind is kept secret or made less obvious. By 1960 Kennedy was identified with fitness, vigor, and preparedness, even as a leader in physical fitness, though he declined to take the famous 50-mile hike. In Feb. 1961 he established the President's Council on Youth Fitness, later Physical Fitness. His emphasis was timely, as Eisenhower had started a Council on Fitness but gave it little personal attention. For Kennedy it conveniently addressed a national issue and showed his youthfulness in a positive light, while diverting attention from his relative political inexperience and his innate bad health. The latter was vehemently denied, though his bad back could not be fully disguised and was attributed strictly to football and war injuries. Only a few factors protect the physical body in this case: **1)** Ascendant lord Mercury at the Stationary Direct degree; **2)** a strong **Sun** (*karaka* for physical vitality), *Vargottama*; and **3) Virgo Ascendant** also *Vargottama*, normally excellent for health, though best when in a fire sign. It receives an aspect from *Navamsha* Jupiter, though a wide orb (27 degrees) and not a positive factor due to Jupiter's status as *maraka* planet (potentially death-producing) for this chart.

Longevity would be adversely affected in the 16-year **Jupiter *Dasha*** (from Aug. 28, 1961) due to its *maraka* status and its weakness as a debilitated planet in the *Navamsha* chart. Even if *Neecha Bhanga* (See Glossary) the corrective planets for *Navamsha* Jupiter are themselves weakened by sign or house position from the *Navamsha* Ascendant. (This partly accounts for why Kennedy's Jupiter *Dasha* was far less favorable for him than it was for **Winston Churchill**, who shares the Virgo Ascendant.) However, Kennedy was somewhat protected during the three previous *Dashas*. This is because **Venus** is the *Jiva* (life force) for each *Dasha* lord, since it is *nakshatra* lord in each case.[209] It is well placed in the birth chart and though in the 12th house of the *Navamsha chart*, Venus is not as afflicted as most other planets when in the 12th house. Moon, Mars, and Rahu are the three *Dashas* preceding Jupiter *Dasha*. The 10-year Moon *Dasha* begins Aug. 28, 1926, at age 7 and takes him up to age 17. The 7-year Mars *Dasha* from Aug. 28, 1936 is very dangerous but has some protection from Mars well situated in the *Navamsha* chart, and Venus as *Jiva*. Though badly injured on Aug. 2, 1943 in a WWII incident, he swam to safety. *Mars* (*karaka* for brothers) was a strong period for his brothers. Joe Jr.'s death was not until Aug. 12, 1944. The 18-year Rahu *Dasha* from Aug. 28, 1943 is discussed more at length in the upcoming section: *Rise to Power*.

***Karmic* destiny:** The *Navamsha* chart shows whether the *karmic* destiny promised from the birth chart is likely to bear fruit and the *Dasha* sequence tells us when. Since the *Navamsha* chart is not

---

[208] The cycles of *Dharma* and *Karma* planets Jupiter and Saturn can be observed in many ways and have significance when timed from major events. A full 29-year Saturn cycle after his death – in 1992 – came the autopsy and medical reports long concealed by his family. A full 12-year Jupiter cycle later came ***The Church Report*** (Nov. 1975): the first major exposure of the lies and scandals of his presidency. In 1975 also came the first television airing of the Zapruder home movie taken of Kennedy's assassination. Prior to 1975, only stills were allowed. This film caused outrage, as it contradicted the findings of the Warren Commission (Oct. 1964).

[209] Each of Kennedy's lifetime *Dashas* had *Dasha* lords in Venus's sign or *nakshatra*. But the latter proved to be more protective, supporting the power conferred on the *nakshatra* lord as *Jiva* (life force) in *Nadi Jyotisha*.

nearly as strong as the birth chart, it is unlikely Jack Kennedy will be able to fulfill the promise of his innate talents, even though he achieves the Presidency. At the very least, we would say he is unable to achieve what he is capable of, even in the highest office. This may seem odd, but it is all relative to the *karmic* destiny promised in the birth chart. His state of health hampered him throughout his life, as did his life habits and demand for constant excitement. Especially in light of very strong enemies (Mars), these would be negative factors. Meanwhile, much was well hidden behind a cool and elegant façade. Afflicted Ascendant lord in the 8th house brings strong manipulative tendencies that are further supported by natal Mercury at the Stationary Direct degree. Natal 8th house and *Navamsha* Scorpio factors helped him to mask his vulnerability and to keep a lid on his scandals and on rabid pro-war forces in his administration, at least for a time.

Although the *Vargottama* Ascendant is a good indicator for health and vitality, Ascendant lord Mercury does not improve in the *Navamsha* chart. Though in a better house (the 4th), it resides in Jupiter's sign (Sagittarius) and *Navamsha* Jupiter's condition is weak, placed in its sign of debilitation in Capricorn (though *Neecha Bhanga*). *Navamsha* Jupiter is in the 5th house, which expresses the intellectual capacity and judgment. These factors can be diminished if the house lord and/or occupants are weak. *Navamsha* 5th lord Saturn is placed in the 12th house of loss and/or preoccupation with foreign affairs or love affairs - also 12th house matters. *Navamsha* Jupiter - already in its debilitated sign- suffers from this placement. Mercury's inherent strength is thus further weakened in the *Navamsha*, while 8th lord Mars is strengthened, being in its own sign of Scorpio in the *Navamsha* 3rd house. The outcome is that without taking measures to compensate for personal flaws - especially a tendency to overextend - innate judgment is weakened and enemies are stronger.

The timing of the *Karmic* destiny: This is seen primarily through the *Vimshottari Dasha* sequence, reflected in turn by the strength and weaknesses of each *Dasha* lord portrayed in the birth chart and *Navamsha* chart. Although his natal Ascendant lord is severely afflicted in the 8th house, fortunately Kennedy was born in **Venus *Dasha*** - highlighting the strength of Venus as 9th lord of good fortune and wealth and bringing some luck through modern medicine and the father's wealth to maintain and fortify the physical body. But there is an ongoing problem of trying to use more energy than is available in his body due to the *Nadi yoga* between Venus and Mars in Houses 8 and 9. This is reflected in his daily intake of numerous drugs and his sexual excesses; also, his father's insistence that he join the military and follow the most ambitious career path, making all of it possible through his vast wealth and connections. Venus's influence in the chart no doubt contributed to resurrecting him from the dead several times. In Vedic mythology, Venus was known under several names, Ushana and Bhrigu, and was bestowed with the knowledge of raising the dead.

Detrimental to his physical health - exacerbating the Mercury-Mars contact - this same ***Nadi yoga* between Venus and Mars** also fuels Kennedy's own **quest for power**. (See Chapter 9 Introduction for more on why Venus-Mars contacts, notably strong in a *Nadi yoga*, tend to bring a strong quest for unbridled power, some of it sexual power but also political supremacy.)[210] As the planet of desire, when Venus interacts with Mars the desire voltage increases markedly and may well distract from one's *Dharma* (life purpose and right action).

---

[210] See Chapter 9: Jackie Kennedy Onassis (Chart #31), who also has a *Nadi yoga* between Venus and Mars.

Add to this a weak Ascendant lord and a 9th house and its lord that do not retain their strength in the *Navamsha*, then this combination confirms the challenges to carrying out one's *Dharma*. The potency of the 9th house is reduced here through the *Nadi yoga* between natal Venus and Mars, especially as Mars is the strongest planet in the chart in its own house – the 8th house – and weakening Ascendant lord Mercury in an *Arishta yoga*. Further, the 8th house placement of Mars taints Venus and the affairs of the 9th house unless they are spiritualized. If not, it brings potential treachery through the father and/or other advisors and also to the *Dharma* or goals of Kennedy himself. In a hidden house, it would not be obvious to others how removing so many obstacles for Jack also brought him danger.

As the best planet for Virgo Ascendant, **Venus** is further weakened in the *Navamsha* chart, thus confirming its loss of benefits from the birth chart 9th house position. While the sex life is expanded, the *Dharma* is inhibited via ongoing distractions. *Navamsha* Venus is in the 12th house in the sign of its enemy the Sun. Compounding this result is an exchange of signs (*Parivartana yoga*) between *Navamsha* Sun and Venus, from 9th to 12th house. This brings some loss to all matters associated with Venus, Sun, and Jupiter. When in the same sign, Venus and Saturn emphasize sexual pleasures, while the 12th house indicates the pleasures of the bed as well as benefits from travel or experiences in foreign countries. Though Venus and Saturn are planetary friends, both suffer from being in the 12th house of loss in Leo – an enemy sign for both planets. In addition, Venus-Saturn contacts make love matters more business-like, diminishing the idea of romance and accentuating a strong sexual desire. Saturn is serious about sexual fulfillment when placed with Venus, while Venus placed in the 12th house benefits the sex life.

Afflicted Venus in the 12th house can bring problems due to lovers, including venereal disease. Kennedy was treated for venereal diseases throughout his adult life and no doubt passed it on to his numerous partners. Staving off blackmail was another ongoing issue and a reckless risk of his reputation. **Marriage and children** suffer, as 5th house also concerns children: *Navamsha* Jupiter is ruler of 7th house of marriage and partnerships. Jupiter is debilitated in Capricorn (though *Neecha Bhanga*) and *karaka* of children. It is not well placed in the house of children (5th house), though it benefits the intellect. The couple had two healthy children who grew to adulthood, but two other children were either stillborn or died soon after birth. His wife Jacqueline also suffered several miscarriages.

Four planets in the birth chart go to weaker sign or house positions in the *Navamsha* chart. Both *Navamsha* Moon and Jupiter are in their debilitated signs, though they both have *Neecha Bhanga*. Jupiter receives no aspects from benefics and its sign lord Saturn is poorly placed in the *Navamsha* 12th house, as discussed. This brings difficulties in Jupiter *Dasha* in particular, especially related to wife and children and also to father – with natal Jupiter in the 9th house. The father in this case is also THE major benefactor and advisor. The fact that the Kennedy presidency began at the tail end of his Rahu *Dasha* was not auspicious, especially during **Rahu-Mars** period. Further, Mars *Dasha* or sub-*Dasha* brings far greater risk of personal danger. When starting any major new endeavor, it is better to start when the energy is still growing, as with a waxing Moon. Otherwise, not enough fresh and renewable energy is available to complete the project. Kennedy's **Jupiter *Dasha*** began Aug. 28, 1961, seven months after his Inauguration. Not even four months into the new *Dasha* – on Dec. 19, 1961 – his father suffered a massive stroke, leaving him mostly paralyzed and bereft of speech until his death Nov. 8, 1969. Regarding children: on Aug. 9, 1963, he and Jackie lost their baby son Patrick, born Aug. 7, after 39 hours.

Inheritance: Venus is extremely well placed in the 9th house of the birth chart and in its own sign. It brings great benefits, especially through the father. Jack's father had extraordinary good fortune and wealth, even avoiding loss in the 1929 New York stock market crash by selling out two years in advance. As the best planet for Virgo Ascendant, Venus also rules over two houses of wealth. Jack's considerable inheritance began with a trust fund of $10 million at age 21.[211] On his 21st birthday Jack himself was too ill to celebrate, though he was to live on the principal of his trust fund for the rest of his life. An office was set up in New York so he would never have to pay his own bills. He rarely dealt with cash or with personal payments of any kind, requiring others to pay for him when in social situations.

When he was just over 19 years old, his 7-year **Mars *Dasha*** began: on Aug. 28, 1936. Though dangerous for him physically it was profitable financially. On his 21st birthday, Kennedy was in **Mars-Jupiter** *Dasha/Bhukti*. From the Virgo Ascendant Mars is in the 8th house of inheritance and is strong in its own sign. However, the financial houses (Houses 2, 5, 9, 11) are not as favorable from the Virgo Ascendant as they are from both the Moon and from *Dasha* lord Mars in the birth chart. Reading from his Mars in Aries as *Dasha* lord (and Sub-Ascendant, viable only during Mars *Dasha*), the 2nd house of income contains Jupiter, Sun, and Venus in its own sign of Taurus. This is a major *Dhana yoga* of wealth, joining the lords of the 2nd, 5th and 9th houses from Mars. The *Nadi yoga* (*nakshatra* exchange) between Venus and Mars links the 1st and 2nd houses in this sub-chart, good financially especially during Mars *Dasha*. During Jack's Mars *Dasha* his father sabotaged his own presidential ambitions, and began to re-direct them toward his sons' presidential chances.[212] His encouragement and assistance to them became legendary, grooming and financing first Joe, Jr. Then when he was killed in mid-August 1944, the same monumental attention was directed toward Jack Kennedy, now the eldest son and next in line.

Marriage: The wedding was on Sept. 12, 1953 in his **Rahu-Mercury** period. Natal Mercury placed in the 8th house of sexuality is less discriminating. Rahu is in the Ascendant of the *Navamsha* chart, and with Rahu and Ketu in Houses 1 and 7 (or vice versa) there are often unpredictable or exotic factors in marriage and relationship, and at times less ability to make a commitment. But it was necessary for Jack Kennedy to marry well to proceed on the road to the presidency. His father again played a pivotal role in pushing for his marriage and approving the choice of bride.

Jack Kennedy's marriage to the beautiful and refined Jacqueline Lee Bouvier gave him added pedigree and status, smoothing out his rough edges and placing him higher on the cultural barometer, if only through association. It also enhanced his celebrity status. Though he was by no means faithful to her, he benefited enormously from their marriage, as can be shown by **Moon** placed in ***Purva Phalguni nakshatra***. This *nakshatra* is about sensual pleasure and passion and

---

[211] On his 21st birthday there was a Total Solar eclipse at 14:31 Taurus – on his natal Sun (15:08 Taurus) in the 9th house, an adverse influence for Jack's physical health that year.

[212] Joe Kennedy's personal reputation declined after 1940 due to his isolationist views against U.S. assistance to Britain in WWII, his willingness to allow Nazi Germany to continue – as a useful foil against Communist Russia, and his abuse of his Ambassadorship to Britain by profiteering. By not being against Hitler's Germany, he was on the wrong side of the great issue of his times, forcing his resignation as Ambassador in late Nov. 1940, and official in early 1941. His former alliance with FDR was irretrievably broken and he never served in public office again. His fall from grace came during the Mars-Mercury period in Jack's chart. (From Virgo Ascendant, the 8th house containing Mercury and Mars is also the 12th house (of loss) from the 9th house of father.) Joe Kennedy's death on Nov. 8, 1969 was a full Saturn cycle following his forced resignation in late Nov. 1940.

is symbolized by the conjugal bed. It is ruled over by the deity *Bhaga*, God of good fortune and prosperity. If unafflicted, it gives good financial wealth, good fortune through marriage, and protection of marital happiness. However, natal Moon is not well placed in the 12th house, and goes to its debilitated sign in the *Navamsha*. There are also problems with Venus and Jupiter in the *Navamsha*. As 7th house lord, Jupiter is the planet symbolizing the wife or partner(s), and Venus is *karaka* for love matters. Venus-Saturn in the *Navamsha* 12th house damages marital fidelity. And debilitated *Navamsha* Jupiter in the 5th house indicates the wife has problematic issues with children: she suffered miscarriages and two children who died in infancy. Though she lends status to Kennedy (Jupiter aspects birth chart and *Navamsha* Ascendant), she has to accept two factors she dislikes: politics and promiscuity.

Rise to power: Within 26 days after the PT109 boat explosion, Mars *Dasha* was over, and the powerful 18-year **Rahu *Dasha*** came into effect: from Aug. 28, 1943 to Aug. 28, 1961. Physically unfit for the military and only there through his father's influence – Jack Kennedy risked his life and health to attain military status. He also created a political identity and gained hero status. His political rise occurred during the 18-year Rahu *Dasha*, a planet notable for giving worldly success during its *Dasha*, especially if Rahu is well placed, as well as its sign and *nakshatra* lords. Rahu is placed in a *Kendra* of both birth chart and *Navamsha* chart, giving it power – even better in the *Navamsha* – as Rahu is in the Ascendant, where it gives personal charisma and makes one sought after by others. Rahu is the foreigner and has exotic qualities. Placed in Sagittarius in the birth chart, the sign lord is Jupiter, well placed in the 9th house. Natal Rahu is in *Purva Ashadha nakshatra* ("the earlier victor"), ruled by Venus – well placed in its own sign in the 9th house. Thus in spite of Jack Kennedy's poor physical health, Rahu *Dasha* had to deliver increasing worldly power, with the help of the father.

Rahu-Ketu *Dasha/Bhukti* (Feb. 27, 1954 to March 17, 1955) is classically the most volatile sub-period within Rahu *Dasha*, joining both eclipse planets. During this period he endured two painful surgeries on his spine and nearly died from the first one Oct. 21, 1954. Coinciding with this difficult sub-period was tr. Saturn in Libra (Aug. 21, 1953 to Nov. 12, 1955). Tr. Saturn aspected his natal Mercury and Mars in Aries, putting further pressure on the physical body and slowing him down. He was hospitalized for nine months. And due to his long absence from the Senate that year, Kennedy avoided voting on the controversial issue of Sen. Joseph McCarthy's censure (Dec. 2, 1954). McCarthy was a close friend of the Kennedy family and commanded loyalty from them; but to support him in the Senate was dangerous politically. Thus, a major health crisis provided a political safety valve.

A fragile start: In the period **Feb. 8, 1958 to Feb. 1, 1961** tr. Saturn in Sagittarius was at the bottom of Kennedy's chart. For Virgo Ascendant, this was not an auspicious Saturn transit, but since tr. Jupiter conjoined it in the same sign (Jan. 26, 1960 to Feb. 11, 1961), a new foundation was being laid in his life and in the world. On Jan. 2, 1960 he announced his candidacy for President; on July 13, 1960 he received the Democratic nomination for President, and on Nov. 8, 1960 he was elected U.S. president. His Election and Inauguration occurred shortly before the **Jupiter-Saturn conjunction** in Capricorn on Feb. 18, 1961, marking the start of a new 20-year social/political/economic cycle. It occurred in his 5th house in both birth and *Navamsha* charts, signaling his contribution to education and the arts, also to inspiring young people. But these events also occurred during his **Rahu-Mars** *Dasha/Bhukti*, the last sub-period of Rahu *Dasha* (Aug.

10, 1960 to Aug. 28, 1961). By coincidence, his brother Bobby Kennedy was also in Rahu-Mars *Dasha/Bhukti* in the same period.

Classically, Rahu the snake tends to remove in the last sub-period what it has given at the beginning of its 18-year *Dasha*. Jack Kennedy's succession to power within his own family occurred in the first year of Rahu *Dasha*, when older brother Joe Jr. was killed in action in a WWII plane crash Aug. 12, 1944. At the end of Rahu *Dasha*, though Jack Kennedy was in the highest office – his presidency was far more fragile than it appeared. Along with his own recklessness, there were major crises with Cuba and Russia, and mutinous forces within his administration that were eager for war and enraged at Kennedy for not starting one. The public was unaware that Kennedy's meeting with Krushchev in Vienna in June 1961 was a total humiliation for him. Having insisted on meeting him one-on-one, Kennedy found Krushchev by far the master of the situation. The Soviets were also still ahead in the Space Race and had dominated that arena for several years starting in Oct. 1957 with the success of Sputnik, the first-ever vehicle to orbit the earth. Then on April 12, 1961, Soviet cosmonaut Yuri Gagarin was launched into orbit around the Earth. The U.S. met this challenge by launching Alan Shepard into a sub-orbital flight on May 5, 1961, making him the second man in space after Gagarin. Kennedy was not personally interested in space programs, but he wanted to win the competition with Russia. He expressed this in his statements in private meetings, including at a Cabinet meeting on Nov. 21, 1962.[213]

Leadership: With **Moon in Leo**, Kennedy's leadership style was dramatic and self-referenced. He preferred to be at the hub of the wheel and involved in all decision-making. Shortly after his Inauguration, he initiated live televised press conferences that occurred every 16 days throughout his administration. His personal and presidential command of the media even prompted some to resent his "news management." Nor did he mind asserting his power over others in personal ways, often holding meetings by the bathtub or swimming pool – where he was naked while others were fully dressed in business suits. Winston Churchill, with a similarly placed Moon, sometimes dictated books or speeches from the bathtub. The Moon rules over water and tides and is in the 12th house of private settings.

Reading the chart from the Moon as Ascendant (*Chandra Lagna*), Kennedy's planets in relation to the Moon in Leo are far better than from the Virgo Ascendant, showing his good emotional connection to the public and good political instincts. From the Moon in Leo, the 10th house (Taurus) contains excellent *Raja yogas* that are formed between Jupiter, Sun, and Venus. Secondly, the 9th house from Leo contains Mercury and Mars, forming a major *Dhana yoga* of wealth. They also form a *Jaimini Raja yoga* of leadership: *Atmakaraka* (planet at highest degree – Mercury) is in the same sign with the planet at next highest degree – Mars (*Amatyakaraka*). Further, the important *Nadi yoga* between Venus and Mars is far better between 9th and 10th lords, instead of between 8th and 9th lords, as from Virgo.

Virgo Ascendant gives versatility, and with **Mercury** ruling both Ascendant and 10th house of career and status, the condition of natal and *Navamsha* Mercury is pivotal in assessing the personal and professional success. In both cases it has fiery influence, and on the 8th house cusp excels at public relations: He is sensitive to the public pulse and to changes in the political currents, intuiting how he can deliver what the public wants. He knows how to perform in public and exalts "grace

[213] See Chapter 1. in the section on the EARTH Dominant Period (1921-2199), covering the Space Race.

under pressure" – his definition of courage. Both his critics and supporters see him as someone with "great potential." Even after Kennedy's death these assessments were made.

On reviewing the impact of planets from Virgo or Leo as Ascendant (*Chandra Lagna*), it is still clear that his career and status would undergo some type of radical change or loss, as in each chart the 10th house contains associations with the 8th house (of loss or radical change). From Virgo Ascendant, 10th lord Mercury is in the 8th house with 8th lord Mars. Mercury *Dasha* did not occur in his lifetime, but in Mars *Dasha* he entered the Navy, and though nearly killed when his PT boat exploded, he gained hero status. Venus as *Jiva* planet for Mars enabled him to survive Mars *Dasha*. Reading from the Moon in Leo, Jupiter is 5th and 8th lord placed in the 10th house. As 8th lord it brings changes in status during Jupiter *Dasha* due early in the 16-year *Dasha* as Jupiter is at an early degree of the sign it inhabits. In the first year of Jupiter *Dasha* Kennedy was in the midst of the Cuban Missile Crisis (Oct. 1962) and had already suffered an important defeat in the Bay of Pigs debacle mid-April 1961, late in Rahu-Mars, a period likely to be a rough ride, as discussed earlier.

Fortunately for him, in a pre-scheduled speech on April 21, 1961, he turned a major failure into a positive story by placing the emphasis on his own learning and responsibility. From then on, even his critics saw him as a "learner," someone who could learn the real messages of history, someone thoughtful and intelligent enough to keep learning. Though Kennedy glossed over the real U.S. involvement in the Bay of Pigs invasion, evaded acceptance of responsibility for it in certain speeches and took it on in others, he managed to achieve a public relations coup. His ratings went up 10 points directly after the Bay of Pigs fiasco with his public approval at 83%. **The strength of this Moon chart** (*Chandra Lagna*) indicates that his public (Moon) tends to see him favorably no matter what he does. In a single phrase he dispersed his critics and charmed his adoring fans:

"... Victory has a hundred fathers, and defeat is an orphan."

The **Sun** is a major *karaka* for leadership and this Sun is *Vargottama* in Taurus in the *Navamsha* 9th house. *Navamsha* Sun opposite Moon-Mars highlights the inspirational and educational role he would have, as the 3-9 house axis is about communications, written and spoken. The 9th house has more widespread, often international publicity. In the birth chart, the Sun is in the 10th house from natal Moon, thus in an excellent position to convey a sense of solid leadership, even if effective presentation is not always matched by actual deeds. There is a sense of a solid and accomplished leader, especially in **foreign affairs** – which interested Kennedy far more than any area of domestic politics, including civil rights. Matters pertaining to foreign countries relate to these houses in the chart: the 7th house (especially for foreign residence), the 9th house (foreign travel and education), and the 12th house (travel and/or residence, with income from foreign sources). Kennedy has key planets in the 9th and 12th houses, both in birth and *Navamsha* charts. Rahu and Ketu (the foreigner or exotic person) are across the *Navamsha* 1-7 house axis and are in *Kendras* in both birth and *Navamsha* charts. As the sickly one, he was the "foreigner" in his own family, which vaunted rough-and-ready athletic events and horseplay.

Though bad for his physical health, natal **Mercury-Mars** in Aries gave Kennedy a sharp and ready wit and a fine intelligence. However, at times he lied about his qualifications. His official biography said he attended The London School of Economics and studied there with Harold Laski. But he was too ill to do so, as he often was. This was to cost him dearly in his first meetings with Krushchev, who had a superior grasp of economics and out-maneuvered Kennedy on a number of points. Nor was he the speed-reader he claimed to be, though he was a quick study. As British Prime Minister Harold Macmillan noted in late 1961, Kennedy was "an extraordinarily quick and effective operator" when it came to specific issues. But on large issues such as nuclear war, the struggle between East and West, capitalism and communism, he seemed "rather lost."

Astrologically, the Moon and Mercury tell us about the workings of the mind. Kennedy's Moon and Mercury have fiery qualities in both birth and *Navamsha* charts, but the Moon is in fixed signs in both charts. Though the dual sign Ascendant gives versatility, Moon in fixed signs with fiery influence likes to focus on what needs immediate attention. What is the problem? And how can I fix it? Kennedy saw himself as "a managerial politician." Pragmatism ruled the day, and though his speeches presented grand and inspirational ideas, he did not tend to think for the long term. In late Dec.1961 Macmillan wrote the following in his diary, just after their day-long conference together in Bermuda:

> "I feel in my bones that President Kennedy is going to fail to produce any real leadership. The American press and public are beginning to feel the same."
>
> **Harold Macmillan**, British Prime Minister

By then Kennedy was four months into the less fortunate Jupiter *Dasha*. Macmillan also noted Kennedy's poor state of health, mostly due to severe back pain.

Managerial politician: The Kennedy administration took an activist stance from the beginning and wanted to flex its muscles on many issues. Kennedy often faced opposition from conservative southern Democrats, but he could have taken the lead in creating a freeze in the nuclear arms race. Instead he chose to embrace it. The strength of 8th house Mars in his chart indicates there may be a distinct difference between what he says in public and what he says and does in private. Tough talk in public was said to be balanced by compromises in private that were meant to prevent going to war. He campaigned on "a missile gap" with Russia he knew from classified information was untrue. And in general, his policy tended to support a resolute and firm approach, especially in confrontation with the Communists.

But although the Soviet and Communist influence looked to be gaining over that of the U.S. and capitalism in 1960, the **Jupiter-Saturn conjunction in Capricorn in Feb. 1961** indicated otherwise. It was a pivotal indicator of things to come on a global level (up through 2199) and was the start of the EARTH dominant period following the FIRE to EARTH transition period, 1901 to 1961.[214] Because Saturn is strong in its own sign of Capricorn, business and financial interests would grow to unprecedented proportions, affecting all nations whether capitalist or communist. At the same time, Jupiter debilitated in Capricorn is overwhelmed by Saturn's influence here. Thus, ethical and humanitarian concerns would be sacrificed increasingly to those of business and finance. Even civil liberties – formerly sacred touchstones of democracies – would be reduced in both democratic and communist countries. Corporations could now supersede governments in power and influence, even starting and prolonging wars, though governments would appear to be in charge.

---

[214] See Chapter 1 for more on the EARTH dominant period (1921-2199) and why it became stronger as of the JU-SA conjunction Feb.18,1961.

Kennedy was inextricably linked to these broader forces set in motion Feb. 18, 1961. With his managerial style, Kennedy set a new precedent for the conjoining of business and government. Elected on Nov. 8, 1960, five weeks later Kennedy appointed Robert McNamara as his Secretary of Defense. On Nov. 9, 1960 McNamara had become President of Ford Motor Company. McNamara's career sequence became a decisive new pattern: first a leader in business of a major corporation; then government – handling military defense budgets and huge corporate interests; then the World Bank (President, 1968-1981). The effect was economic domination, if it could not also be achieved militarily. Economic sanction of recalcitrant nations was one of many approaches, buttressed by military force.

In 1959 Fidel Castro ousted Fulgencio Batista, the U.S.-backed dictator in Cuba. The anti-imperialist Castro soon became Cuban Prime Minister. To counter the U.S. threat against Cuba (and himself), he forged an alliance with the Soviet Union that was alarming to the United States. Eisenhower wanted to remove Castro from power and Kennedy campaigned on a promise to overthrow Castro.[215] But his Bay of Pigs decision in mid-April 1961 defied common sense. Agreeing to a plan obviously doomed to failure, he had little real understanding of the situation in Cuba. Others close to the center of action suspected an even more fundamental issue.

If in the birth chart and/or *Navamsha* chart, the *Dharma* planets lack strength or connection to the other planets, there may be a lack of **moral compass**. For this we examine natal Jupiter (the planet of *Dharma*) and the 9th lord. Kennedy's 9th lord Venus suffers from its close connection with 8th lord Mars, and Jupiter is greatly weakened in the *Navamsha* chart, as discussed. Historians seem divided as to whether Kennedy was really responsible, as covert CIA actions undermined him. Some of his advisors knew the Cubans were well aware of the American invasion plans, but they did not want to bother Kennedy or give him information conflicting with his current plan. Loyalty to the Kennedy team was paramount, even at times when other voices should have been heard. Nor was there much questioning of the moral right to launch such an attack. Soon after the Bay of Pigs fiasco Chester Bowles, then Undersecretary of State, wrote in his diary (May 1961):

> "The question which concerns me most about this new Administration is whether it lacks a genuine sense of conviction about what is right and what is wrong. I realize in posing the question I am raising an extremely serious point.... The Cuban fiasco demonstrates how far astray a man as brilliant and well intentioned as Kennedy can go who lacks a basic moral reference point."
>
> **Chester Bowles** quoted in David Halberstam, *The Best and the Brightest*, 1972, p. 69.

The Bay of Pigs remained unfinished business for Kennedy, setting the stage for the Cuban Missile Crisis in Oct. 1962. Fuelled by vengeance against Castro, the man who defeated him so

[215] Fidel Castro's birth date of Aug. 13, 1926 is questionable and rated DD (Dirty Data). He survived numerous assassination attempts by the U.S. govt., including from 1961 to 1963, when President Kennedy ordered the CIA to "eliminate" Castro. (Source: *Alleged Assassination Plots Involving Foreign Leaders*, known also as "The Church Report," published Nov. 1975 by the Senate Select Committee to Study Governmental Operation with respect to Intelligence Activities; documented in Seymour Hersh, ***The Dark Side of Camelot*** 1997, pp. 3-5; 162; 220; 463, etc.) Castro was Cuban Prime Minister: 1959-1976; and President: 1976-2008. He retired in Feb. 2008 due to ill health, though he continued as close advisor to his brother Raúl Castro who succeeded him as President in 2008. In April 2011, Fidel stepped down as head of the Cuban Communist Party, which he headed from 1961, having declared Cuba a socialist state on May 1, 1961. Communist Cuba was established Jan. 2, 1959, 1:00 AM in Havana, Cuba, with a live broadcast speech by Fidel Castro. (Source: Nick Campion, ***The Book of World Horoscopes***, 2004, pp. 89-90.) Prior to the successful Cuban Revolution in Jan. 1959, Batista had cooperated closely with American multinational corporations and American Mafia.

badly in 1961, Kennedy's handling of the Cuban Missile Crisis was portrayed successfully as a sign of his growing presidential maturity and wisdom. But it was another public relations coup that masked no real victory and in fact far greater recklessness.

Oratory and writing: Just as Winston Churchill said history would treat him kindly since he himself would write it, Jack Kennedy intended to have his version of history written chiefly by Ted Sorensen and Arthur Schlesinger, Jr.[216] He had enough astrological indicators to do this successfully for some decades. Almost eight hours before Kennedy was born Mercury turned Stationary Direct in the heavens at 27:52 Aries. Kennedy's natal **Mercury** at 27:53 Aries still has the power of the **Stationary Direct** degree, bestowing on him enormous ability to communicate, all dependent on Mercury's placement in the birth chart by house, sign, and *nakshatra*, and what planetary aspects it receives. Mars in very tight contact has the effect of heating up Mercury and we have seen how this was very detrimental to his physical health. Mercury in the 8th house shows his love of intrigue and gossip and his many secrets. But much positive energy can go towards speech and the intellect. The 8th house placement also enabled Kennedy to be a master of public relations, especially in his instinctive command of the television medium and the press in general. He often avoided blame where it was due, while receiving credit where it was not due.

> "Television is a medium that lends itself to manipulation, exploitation, and gimmicks. Political campaigns can actually be taken over by the public relations experts, who tell a candidate not only how to use TV, what to say, what to stand for, and what kind of person to be."
>
> **John F. Kennedy**, 1959, in his article for *TV Guide* on the influence of television on politics.[217]

When the **Berlin Wall** started going up on Aug. 12-13, 1961, two weeks before the start of Kennedy's **Jupiter *Dasha***, what was not known - and what could have ruined his reputation - was that he allowed the wall to be built as a compromise solution. He perceived that Krushchev had a refugee problem, and the Berlin Wall served as a deterrent. Then in 1963, the year he was torn between Birmingham and Saigon, Kennedy went to Berlin on June 26, and gave his now famous *"Ich bin ein Berliner"* speech - shaking his fist - as did many other Western leaders, at this major symbol of the Cold War and epitome of the partitioning of Europe. In this way, he mastered the public relations issue and at the same time appeared to defuse the threat of nuclear war with Russia. As a master of irony himself, he must have found it ironic that one of the most overwhelming receptions he received anywhere in the world during his presidency was in Berlin on that day in late June 1963. He also delivered one of his most powerful speeches:

> "Two thousand years ago the proudest boast was 'Civis Romanus sum'. Today, in the world of freedom, the proudest boast is, 'Ich bin ein Berliner!' There are many people in the world who really don't understand, or say they don't, what is the great issue between the free world and the Communist world. Let them come to Berlin! There are some who say that Communism is the wave of the future. Let them come to Berlin!..."
>
> **President John F. Kennedy**, June 26, 1963.

In the *Navamsha* chart Mars in Scorpio adds galvanizing power to Moon opposite Sun on

---

[216] Ted Sorensen: b. May 8, 1928, d. Oct. 31, 2010; Arthur Schlesinger, Jr.: b. Oct. 15, 1917, d. Feb. 28, 2007.

[217] Quoted by Barry Levinson, film director, "Surviving the Media Circus Without Becoming a Clown," *Huffington Post*, Oct. 29, 2009.

the 3 - 9 house axis of communications. As noted earlier, this is excellent for written or spoken communications. With the help of speechwriters Kennedy excelled in all of these areas. He owed a lot to writer-journalist Arthur Krock for his first book, *Why England Slept*, and to Ted Sorenson for his second book, *Profiles in Courage*. He had a very close working relationship with Sorensen, his major speechwriter from 1953 onward. Throughout his presidency Kennedy put a very personal stamp on his speeches, avoiding what some have called "the ghostwritten presidencies" of later years, when presidents give speeches written by hired professionals, speeches with which they have very little real connection, knowledge or experience. Kennedy's personal participation in the process, his high intelligence and quick wit made even his impromptu speeches or remarks memorable.

***Profiles in Courage*:** Kennedy's first book *Why England Slept* (1940) was modeled on Winston Churchill's *While England Slept* (1937), and concerned British foreign policies in the 1930s. His second book was modeled on another Churchill book, *Great Contemporaries*, 1922, a study of world leaders at the start of the 20th century. It begins with this statement:

> "Courage is rightly esteemed the first of human qualities because it is the quality that guarantees all others."
>
> **Winston Churchill**

In his *Profiles in Courage* (1956) Kennedy starts by saying that courage is **"the most admirable human trait."** The planet **Mars** represents courage as well as dispute and war.

> "For without belittling the courage with which men have died, we should not forget those acts of courage with which men have *lived*.... A man does what he must – in spite of personal consequences, in spite of obstacles and dangers and pressures – and that is the basis of all human morality."
>
> **John F. Kennedy**, *Profiles in Courage* (©1956,1964 Memorial edition), p. 266.

Churchill's Mars in the 1st house was the most prominent planet in his chart, making it a major factor in his life destiny. Kennedy's 8th house placement of Mars with Mercury would turn courage and courageous acts into something far riskier for his own health and safety. In addition, though the impetus for *Profiles in Courage* was his, originally intended as a magazine article, Ted Sorensen did the bulk of the research, writing and re-writing. Kennedy's personal stamp was on it, though he received far more credit than he deserved – especially the Pulitzer Prize. Stories disputing Kennedy's authorship drew many counterattacks from the usual Kennedy team, who staunchly defended his reputation during his life and long after. Finally, with his 2008 memoirs, *Counselor: A Life at the Edge of History*, Ted Sorenson revealed that Kennedy paid him a large sum in 1957, the same year the Pulitzer prize was bestowed on Kennedy for work largely done by Sorenson. Controversy may surround Virgo Ascendant people both during and after the life.

Extraordinary charm & charisma: Astrologically, **Venus** is the planet of charm, diplomacy, and love matters. In its exalted state Venus is selfless, devotional, or universal love. In its worldly forms, its many aspects show the social graces, the ability to enjoy the company of others – in love, laughter, sociability – plus artistic sensibilities or talents, depending on where Venus is in the chart. At its worst, Venus is seductive and manipulative, corrupting through charm and flattery, and promising what is wanted – which is often more than can be delivered. This Venus, though strong in Taurus in the 9th house, is tainted by the influence of Mars and Saturn. One of the effects was that style became close to substance.

> "The reaction to the Kennedys in the White House was amazing, considering that more than half the country had voted against him a few months earlier. Once in the White House, Kennedy had to go on national television to ask people to stop sending telegrams and letters congratulating him and his wife."
>
> **Richard Reeves**, *President Kennedy: Profile of Power,* 1993, p. 73.

> "He made up for his sickliness with charm, good humor, and winning zest for life that kept him beloved by his peers, as it would throughout his life."
>
> **Seymour Hersh**, *The Dark Side of Camelot*, 1997, p. 15.

Kennedy was a cultural phenomenon, whose every taste in clothes and restaurants started trends and confirmed the success of a new style or product. As 9th lord, Venus also rules advisor(s) and father, whose vast wealth and political influence we have discussed. Joe Kennedy was the major catalyst for everything in his son's life. He remained his top advisor and role model until his stroke Dec. 19, 1961. In business, he could "fix" anything. In love matters, his example to his sons was to marry well, stay married, have lots of children, and enjoy as many women as desired. Marriage was viewed as no obstacle to maximum pleasure. Charm and money could be used to acquire whatever was wanted.

> "John Kennedy's life was sequential seduction and there were few complaints from the seduced. 'Social flattery,' it was called by [Arthur] Krock, who had given Kennedy his editorial skills and his valet before they became estranged when the columnist began criticizing the President in print. 'I have myself been infused with the warmth of good will engendered by this courtship of a suitor of such charm and unique distinction.'"
>
> **Richard Reeves**, *President Kennedy: Profile of Power,* 1993, p. 478.

The Kennedy mix of money, power and glamour made Jack an irresistible magnet. His chosen career of politics was already one that magnified charm and institutionalized seduction. This combined to make him perhaps the most charismatic U.S. president ever. It would also enable him to keep his secrets hidden for many years through his charm and money. Women who knew him, either as lovers or friends, spoke of his "overwhelming attractiveness." Many more around him would strive to please him, to win his approval, trust and friendship and to save him from boredom – as he was easily bored. These included members of the press: consequently, Kennedy's press coverage was mostly favorable during his years in the presidency and for the next decade, especially in the aftermath of his assassination, when America had lost its King of Camelot. In order to explain such magnetism on both a personal and public level, we expect to see some extraordinary qualities in Venus and in the Ascendant lord Mercury – as well as a *Dasha* sequence that serves to protect his reputation, even if it failed to protect him from being killed at age 46. We also expect to see astrological factors that might corrupt the original promise and/or cut it short.

Charm as power: Venus is not only the planet of charm but the planet of politics. In Vedic myth, Venus as *Shukra* is considered to be the inventor of political science and composed the *Shukra-niti* – the most famous code of behavior and politics in Vedic literature. We have discussed the positive and negative attributes of **Venus** in Kennedy's birth chart. (See the previous sub-section: *Karmic Destiny*.) In addition to being the best planet for Virgo Ascendant and in its own sign in the best house (9th house), reading from the Moon, Venus is in the 10th house, to the exact degree.

Sun in the 10th house from Moon is also excellent for a public figure, and Jupiter in an angle from the Moon is a *Gaja Kesari yoga,* often bestowing superior visibility in competitive situations.

For best results, Venus should not lose strength either in the birth chart or *Navamsha* chart, as in this case. Further, Kennedy's natal Venus is combust the Sun, though by its widest orb. Even so, a combust planet cannot be seen with the naked eye due to its closeness to the Sun. With combust Venus, often one is outwardly very charming but inwardly craves the things that Venus brings because either they cannot be seen or one is not fully aware of having them. Though in part influenced by his medications, which included cortisone and artificial testosterone, Kennedy was known for his relentless sexual appetite and craving for women. He compartmentalized his marriage and refused to deal with those who criticized him about anything public or private. As Kennedy's best planet, Venus enables him to radiate charm and self-confidence; but his focus is increasingly directed towards foreign affairs and sexual pleasure-seeking, to the detriment of his physical health and *Dharma*.

The public perception vs. the reality: In Vedic myth, Venus was the advisor to the *asuras*, and was famous for knowing the psychology of the human mind. Venus (*Shukra*) understood more than any other of the gods how mastering the human mind is the key to mastering both the individual and the collective.

> "[John F. Kennedy] was ... a master, as many people have become since, of using candor in lieu of truth. He was an extremely candid man, as many people are nowadays – telling most everything about things that used to be considered private. Others walk away thinking they've been told truth. But, in fact, they've really been told nothing of true importance. The small and candid moments set up the big lie."
>
> **Richard Reeves** on JFK in *Character Above All: Ten Presidents from FDR to George Bush*, ed. Robert A. Wilson, 1995, p. 95.

> "The manipulation [of the public perception] was extraordinary. The president was living a public lie as an attentive husband and hardworking chief executive, a speed reader who spent hours each night poring over bulky government files. But the Secret Service agents assigned to the White House presidential detail saw Jack Kennedy in a different light: as someone obsessed with sex, and willing to take enormous risks to gratify that obsession. They saw a president who came late many mornings to the Oval Office, and was not readily available for hours during the day to his immediate staff and his national security aides; a president, some thought, whose behavior was demeaning to the office. The mythmaking and media wooing began soon after Kennedy took office."
>
> **Seymour Hersh**, *The Dark Side of Camelot*, 1997, pp. 222-223.

With his Ascendant lord in the 8th house and on the 8th house cusp, Kennedy not only loved what was concealed – especially political and sexual intrigue – but he became a master of hiding his private self from his adoring public. This was useful politically, to deflect attention from intended actions, and helped him deal with his enemies as well as his supporters. The discrepancy between his public and private self showed in his sense of **irony** and **detachment.** He had long grown accustomed to masking physical pain, though he was not averse to taking massive amounts of medication to deal with it. Having received last rites several times, he was fatalistic about life.

To date no medical historian/detectives have tackled a serious investigation of Jack Kennedy's medical history and its effects on the presidency. But some medical experts have said that if a military person were taking even *one* of the many serious drugs Kennedy took daily in large doses for his various conditions, they would be temporarily exempt from carrying out their duties. Side effects included depression, paranoia, euphoria and over-optimism, along with compulsive sexual desire. At death, he had nongonorrheal urethritis and chlamydia.[218] Chlamydia is a serious venereal disease which can cause reproductive disorders and problems with infants, if a woman is infected while pregnant. For a man who hated "messy situations," he entered them all the time, politically and sexually. This was yet another contradiction in his nature and reflected in so many ways, including his compulsive need to shower up to five times daily, often with a complete change of clothes each time. In public he spoke of lofty humanitarian aspirations and in doing so he brought out the best in the American people. But in private his behavior was reckless and irresponsible, often using people for his own amusement in a race against boredom and death.

> "In the end, I come to the conclusion that perhaps what is most important is not the character of the President of the United States, but the character of the American people."
>
> **Richard Reeves** on JFK, *Character Above All: Ten Presidents from FDR to George Bush*, ed. Robert A. Wilson, 1995, p. 104.

Destiny of a nation: Astrology might agree with Richard Reeves – to the extent that a President or a national leader represents the heart and soul of a nation during his term in office, and a lot about the people on so many levels. However, this probably occurs with greater unease if the election is fraudulent. If Reeves is correct, and that is debatable – then Kennedy also represents **"the American dream"** we had for a brief time, a movie we loved to watch – even if much of it was based on false premises, starting with his physical vitality. This was no doubt a spectacular acting job. Americans wanted to believe in his leadership and in his poetic vision, but they also wanted to be charmed and impressed, to feel love for their president and his family. He made them believe they could achieve great things together. Then they watched him get shot. If he were just another movie star, that would have been hard enough, but as our movie star leader, our **"Superman in the Supermarket,"** as Norman Mailer called him, he succeeded in winning over the hearts and minds of the public for the most part, even when they might not agree with his policies.

With 8th house power, you can achieve anything for a time, even great fame – by magnetizing people to you, and making them believe you have that magic ingredient. Kennedy was successful at this and on a grand and global scale due to the following astrological factors, summarized again here:

1. Mercury exactly on 8th house cusp in the Stationary Direct degree
2. Strong Venus near the 9th house cusp
3. The *Nadi yoga* between Venus and Mars

The first factor is bad for his health. But all these factors elevate his ability to be seductive while at the same time reducing his chances of achieving his life's purpose. Add to this the dual sign Virgo Ascendant, and we see even more clearly the discrepancy between **what was promised** – which seemed powerful but was an illusion of sorts – and **what was lost forever**, also an illusion, perhaps even more so.

---

[218] The full autopsy report was not published at the request of the Kennedy family. Navy pathologists who conducted the autopsy on the night of Nov. 22, 1963 noted this condition. (***The Dark Side of Camelot***, p. 230.)

The USA chart (July 4, 1776, 6:30 PM LMT, Philadelphia, PA),[219] with 8:59 Sagittarius Ascendant, has Mercury and Rahu in the 8th house in Cancer, a perfect combination for influencing the public will through public relations and advertising. What used to be called "propaganda" or deceptive rhetoric becomes legitimized in order to put a more positive face on anything at all. Information easily becomes entertainment. **Mercury in the 8th house** can also focus on research or secret information; with Rahu, it becomes magnified and manipulative. In the USA chart, **Mercury-Rahu** *Dasha/Bhukti* (March 25, 1958 to Oct. 12, 1960) encompassed most of Jack Kennedy's presidential campaign and his first televised debates with Richard Nixon. Those who listened on radio thought Nixon had won, especially as he was far more experienced politically. But television viewers assumed Kennedy had won. Their policies were nearly identical, but Kennedy's television persona was likeable and impressive and Nixon's was not.

The entire 17-year **Mercury *Dasha*** (from Sept. 26, 1948) in U.S. history was a pivotal period in the development and use of television and public relations. Kennedy's Mercury in his natal 8th house has an important resonance with this national chart. As our first masterful television president and the first real American litmus test for the success of television *and* public relations on this scale, he enjoyed the good will of the public because he consistently won over American hearts and minds, whether or not the information delivered was truthful. Among his many inspirational speeches, it may be impossible to know what parts were truthful – due his instincts for public relations.

In another fortunate coincidence for Kennedy, soon after his Inauguration Jan. 20, 1961, the **Jupiter-Saturn conjunction** occurred **in Capricorn** (Feb. 18, 1961), giving him prominence in history. It also brought in financial prosperity at unprecedented levels, especially for American corporations, heralding a new era of corporatocracy well beyond government influence. Thus, whatever his original plans or intentions, Kennedy was subject to the inescapable momentum of this larger cycle. Whether he fought it or joined it is still hotly debated. But he presided over national expansion on many levels, including militarily. The forces of the military-industrial complex were stronger as of Feb. 1961 and he was weaker from his Jupiter *Dasha* starting in late Aug. 1961. A strong Mars in his birth chart says his innate judgment can grow weaker and his enemies stronger. His chart also indicates he was unable to overcome larger negative forces. But they existed within himself as well, and he had very little resistance to them.

The USA chart has Venus, Sun, Jupiter and Mars in the 7th house in Gemini. This describes a nation with an abundance of aggressive charm along with the persuasive (often deceptive) rhetoric of Mercury-Rahu in Cancer in the 8th house. The combination is good for business and national expansion, as Gemini has mercantile interests and features accessibility. Mars and Jupiter in the same house are more militaristic, while Sun and Venus can lead culturally.

Assassination: Relentlessly controversial, Kennedy's assassination raised questions that still persist about who killed him and why; and what he would have accomplished in his presidency had he

---

[219] This **USA chart** was rectified by Vedic astrologer and colleague James Kelleher in the late 1990s. I have used the chart continuously since Sept. 11, 2001, when I became satisfied with its accuracy. See Nicholas Campion's ***The Book of World Horoscopes*** (1995, 2nd edition, revised & enlarged, pp. 348-376) for a lengthy discussion of the various historical charts used for the national chart of the USA. This reflects the amount of ongoing research and varying opinions there are on the subject. However, many astrologers, both Western and Vedic, tend to agree on a Sagittarius Ascendant for the USA chart. The second most popularly assigned Ascendant is Gemini, where there are four planets, sidereally. Vedic sages have tended to give the sign of Gemini to the USA and the sign of Capricorn to India, as general qualities of each nation. On the basis of this, some Vedic astrologers have watched for important astronomical events in these zodiacal areas.

lived longer. Even his intentions are hotly debated. His Ascendant lord Mercury is in a charismatic but dangerous position on the 8th house cusp. He had poor health, lived life dangerously and courted danger. However, he got numerous reprieves – especially prior to Jupiter *Dasha*, which adversely affects his longevity. Jupiter is a *Maraka* planet that becomes much weaker in the *Navamsha* chart. A violent death can be shown by the overpowering strength of Mars in relation to the Ascendant lord. Further, he began his presidency in the last few months of Rahu *Dasha* in the Rahu-Mars sub-period.

Kennedy's political legitimacy was amplified by the proximity of his Inauguration to the major Jupiter-Saturn conjunction on Feb. 18, 1961, which also gave greater power to business interests. Coincidentally, he was in **Jupiter-Saturn-Saturn period** at the time of his assassination. When a planet is in the earliest degrees of a sign, results can come early in its *Dasha*. These results are likely to be less advantageous here due to his weakened *Navamsha* Jupiter. Also, Jupiter and Saturn are in 6 – 8 house relationship in the *Navamsha* chart, both with a problematic placement house and/or sign-wise. Therefore, a dramatic change of some sort comes during the *Dasha/Bukhti* involving both planets.

Assassination chart: At 12:30 PM CST Nov. 22, 1963, Dallas, TX, the Ascendant is 26:00 Capricorn, in *Dhanishta nakshatra* – meaning "the wealthiest." Moon and Saturn in Capricorn (public grief) are in the Ascendant, with four planets in hidden Scorpio in the 11th house of friends: Sun, Mercury, Venus, and Mars. It is difficult to distinguish between friend and enemy. There is also notable resonance with the Scorpionic nature of Kennedy's chart, with his two key planets in the 8th house, and with his Venus-Mars natal combination. Transiting Mars at that moment is at 27:11 Scorpio and tr. Venus is at 28:06 Scorpio, both of them in the *gandanta* area of Scorpio, the most treacherous and tumultuous area of the zodiac. In a Planetary War with each other, Mars loses to Venus and is far more troublesome as loser. With all of these factors placed in Mercury's *nakshatra*, false information is likely to be conveyed on what caused the event and why. An exact aspect of Jupiter (16:27 Pisces) to Mercury at (16:23 Scorpio) reflects the sheer volume of media coverage and books (over 1000 of them) generated from this one event. Many planets in Scorpio, including Mercury, echo how much information remains suppressed even decades after his death. This only serves to perpetuate ongoing debate and public dissatisfaction with the Warren Commission's "official" conclusions.

Chart 26: Barbara Walters

Birth data: Wed., Sept. 25, 1929, 6:50 EST, Boston, MA, USA, Long. 71W03 37, Lat. 42N21 30, *Lahiri ayanamsha*: -22:52:20. Class A data: Frances McEvoy quotes the date from the birth certificate, and her for the time, given in an interview. Ascendant: 11:11 Virgo.

Biographical summary: Barbara Walters' career in American broadcast journalism is remarkable for its sheer longevity. She has also remained at the top of her field for over thirty years. With her abilities and experience in research, writing, filming, and editing, she was news and public affairs

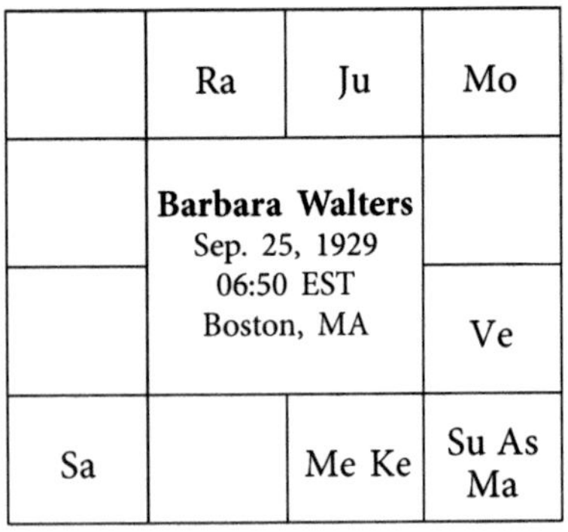

	Ra	Ju	Mo
	Barbara Walters Sep. 25, 1929 06:50 EST Boston, MA		
			Ve
Sa		Me Ke	Su As Ma

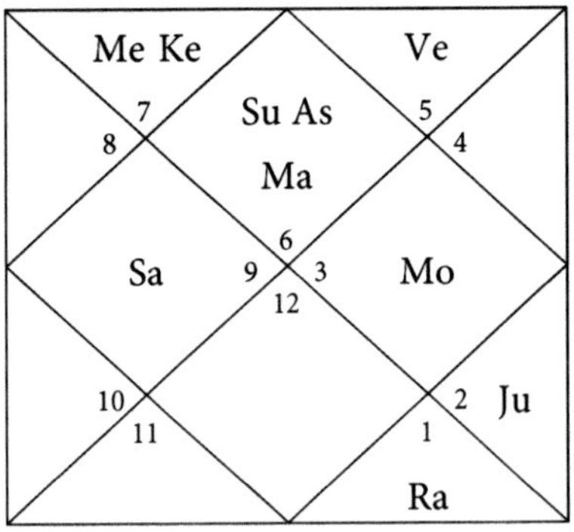

Me Ke 7 8 · Su As Ma 6 · Ve 5 4 · Sa 9 · 12 · 3 Mo · 10 11 · 2 Ju 1 Ra

Planetary Information

As	11:11	Vir	Hasta
Su	09:00	Vir	U.Phalguni
Mo	00:38	Gem	Mrigashira
Ma	29:40	Vir	Chitra
Me	00:29	Lib	Chitra
Ju	23:22	Tau	Mrigashira
Ve	06:51	Leo	Magha
Sa	01:37	Sag	Mula
Ra	21:14	Ari	Bharani
Ke	21:14	Lib	Vishakha

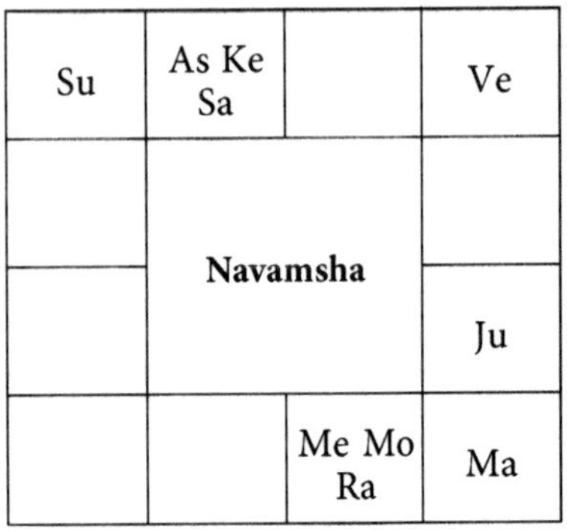

Su	As Ke Sa		Ve
	Navamsha		
			Ju
		Me Mo Ra	Ma

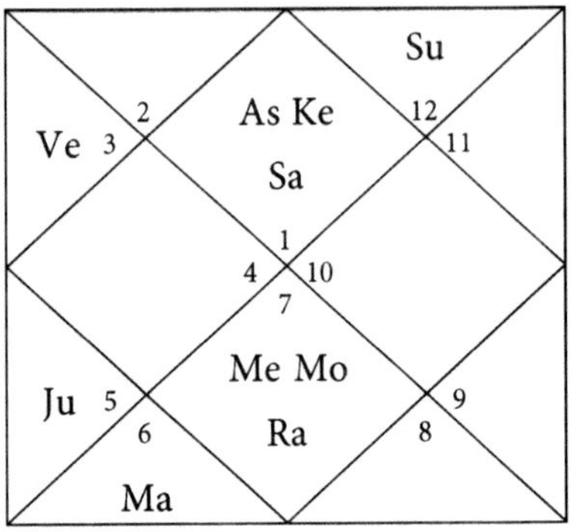

Ve 3 2 · As Ke Sa 1 · Su 12 11 · 4 · 10 · 7 · Ju 5 6 Ma · Me Mo Ra · 9 8

Vimshottari Dashas

Ma	Nov 23, 1925
Ra	Nov 23, 1932
Ju	Nov 23, 1950
Sa	Nov 23, 1966
Me	Nov 23, 1985
Ke	Nov 24, 2002
Ve	Nov 23, 2009
Su	Nov 23, 2029
Mo	Nov 24, 2035

producer for CBS before moving to NBC and later ABC television. Mastering the art of celebrity revelations, she has interviewed more statesmen and celebrities than any other journalist in history. This includes accidental celebrities such as Monica Lewinsky and people in closely followed legal cases, such as Martha Stewart and Claus Von Bulow. At times this involved extensive negotiations to secure the deal. She is credited with breaking new ground for her direct and enterprising style of interviewing as well as her sheer ability to land "plum" interviews, often exclusive, and at opportune times. She also broke new ground for women in her field, becoming the youngest producer at a New York City television station (where she began by writing news releases), and the first woman to co-anchor a news broadcast in 1974. Before Barbara Walters, women television journalists were relegated to "soft news," with many constraints on their discussing politics, economics, or other "hard news" topics.

Walters has been criticized for going for the emotional angle whenever possible and for becoming too personally involved with some subjects. Some even say her style has led to the blurring of lines between news and entertainment, and to the cult of personality in television journalism. For her part, Walters says she too is unhappy with the changing nature of television news and news magazines, and how they are increasingly driven by ratings. But over the years, starting with her first appearance on television in 1964, she is still best known for her exclusive interviews, claiming her mentor is *60 Minutes* correspondent Mike Wallace. She has interviewed every American president since Richard Nixon; but her highest rated interview was with Monica Lewinsky, who was involved in a sex scandal with President Clinton. It was on March 3, 1999, two hours' long, and watched by 74 million viewers. The other highest rated interviews of her career were with Michael Jackson in 1997, and with Fidel Castro in 1977. (She interviewed Castro a second time in 2002.) Also in 1977 she made journalism history by arranging the first joint interview with Egypt's President Anwar Sadat and Israel's Prime Minister Menachem Begin. She travelled to India with Jackie Kennedy and to China with Richard Nixon. In 1970 she published her book *How to Talk to Practically Anybody about Practically Anything* (written with ghostwriter June Callwood). The book has been unexpectedly successful, with eight printings by 2008 and translations into at least six other languages.

Having grown up surrounded by show business, she was not overly awed by celebrities. Her father Lou Walters was a nightclub owner and theatrical producer. He was also reckless with money and a compulsive gambler. He made and lost several fortunes, losing his business and both family homes in the end. He died in Aug. 1977. Barbara intended to get a Masters degree in Education and go on to teach after graduating from Sarah Lawrence College (B.A. in English). But directly after college – from the time she was age 22 or 23, she had to go to work to support her family: her parents and her older sister Jackie, who was very dominant in her life. Jackie was mentally challenged and later died of ovarian cancer in 1988. Her older brother Burton died of pneumonia in 1932. Barbara provided financial support for her family for the rest of their lives. She also had four marriages, the last two to the same husband: Her first marriage, 1955-1958, was to Robert Henry Katz, a baby bonnet manufacturer. Her second marriage, 1963-1976, was to Broadway producer Lee Guber. Their adopted daughter Jackie (born 1968 and adopted that year) was named after her sister. Her third marriage, 1981-1984 was to Merv Adelson, a television producer and Hollywood mogul. They re-married 1986-1992. Barbara Walters' television credits include:

1961-1976 *Today Show* (co-host from 1974-1976)

1974-1976 *Not For Women Only* (moderator)

1976-1978 *ABC Evening News* (co-anchor)

1976-present *The Barbara Walters Specials*

1979-2004 *20/20* (co-host; contains her most prominent interviews)

1993-present *The 10 Most Fascinating People* (year-end reviews of newsmakers)

1997-present *The View* (co-producer and moderator)

In 1976 Walters made headlines and was at the center of controversy for her $1 million per year/five-year contract to leave the *Today* show on NBC and go to *ABC Evening News* to co-anchor with Harry Reasoner. Half of her salary was as news co-anchor and half for her celebrity specials, four per year. As the first woman co-anchor in a male-dominated field, she also became the highest paid television journalist. (Top salaries for news anchors were then $400,000.) By 1996 she was earning $10 million per year. In 1997, she created a daytime news talk show called *The View*, which she co-owns with ABC. It was similar to a program in the 1970s she had hosted – *Not for Women Only*. She insisted on moderating the new show, though ABC President Roone Arledge warned against her being moderator *and* co-producer. *The View* had its debut Aug. 11, 1997 and became an instant hit. It also made her over $100 million within ten years. However, Arledge's predictions did prove correct: Her dual role as moderator and producer finally did create conflicts of interest for her. It started to tarnish her good reputation and her credibility, and was perhaps the first serious blow to her image in a long career. This issue first became more public in June 2006 with the on-air resignation of panel member Star Jones Reynolds, and soon exploded further out of control six months later with a dispute between panel member Roseanne Barr and Donald Trump. However, Walters has long had to deal with criticism as a serious news journalist.

Dealing with tawdry subjects has not helped her in this regard, nor have certain types of interview questions. This has caused Walters to be regarded with ambivalence from many quarters. But she says the business has required her to do more movie stars than heads of states and other more serious guests, as ratings radically improved with stars.

Comedian Gilda Radner was well known for her spoofs of Barbara Walters during her years at *Saturday Night Live* (1975-1980). Her "Baba WaWa" sketches were classic parodies of Walters' distinctive style, including her pronounced lisp. Some consider them timeless critiques of the cult of personality in television journalism. But Walters has prospered and survived in a very competitive business. She has remained a national celebrity herself, with many powerful and loyal friends. She has won numerous awards and honorary degrees over the years: In 1990, she was inducted into the Television Academy Arts and Sciences Hall of Fame. In 1996 she was honored by the Museum of Television and Radio for all her contributions to broadcast journalism. In 2003 she won Best Talk Show Daytime Emmy for producing *The View*, a frequent nominee. Her autobiography, *Audition: A Memoir* was published in 2008.

Broadcast journalist and national celebrity: Walters benefits from the accentuated power of Ascendant lord Mercury at a Stationary degree: tr. Mercury turned Stationary Retrograde 7½ hours after her birth.[220] This same phenomenon occurred with Jack Kennedy's Mercury (also his Ascendant lord) – which turned Stationary Direct 8 hours prior to birth. Its 8th house placement is much less fortunate for health and longevity. (See Chart #25) Due to its motional strength (*Chesta bala*), this Mercury brings advantages through the ability to communicate and bring financial wealth, as Walters' Mercury is situated in the 2nd house of speech and financial income. She was born in **Mars-Mercury** *Dasha/Bhukti*, giving a powerful signature in her life with this *Graha yuddha* (Planetary War). Mars in Virgo can denote a strong intellect, but in the Ascendant it can also bring a lot of enemies and/or competitors. However, because Mars loses the Planetary War, enemies and competitors are diminished. She has the ability both to speak well and be victorious over others.

In a male-dominated field, rife with competition and controversy, Barbara Walters has managed to stay in the eye of the storm for over forty years. And while in the spotlight interviewing controversial figures and being present at major news events, she has mostly avoided being controversial herself. The astrological configuration best describing this is her strong natal **Mercury** at its stationary degree winning in a ***Graha yuddha*** with natal **Mars**, dangerous as 8th lord. Though in adjacent signs, Mars and Mercury are within one degree of each other, and both of them are *Vargottama* – repeating in the same sign in the *Navamsha* chart. Mercury's strength accumulates as it is in a *Kendra* (angular) house position in the *Navamsha* chart. Mars serves her in the sign of the victorious planet, and in the *Navamsha* 6th house of conflict – where Mars thrives.

As a testimony to Mercury's relative mastery over Mars, no major disputes or career ordeals occurred during her **Mercury-Mars** *Dasha/Bhukti*, from May 25, 1994 for the next year. Her divorce from her third and last husband was in 1992. Some major battle may have occurred behind the scenes and she emerged victorious, as usual.

[220] Planets close to the Stationary degree possess the highest amount of motional strength (*Chesta bala*). Source: ***Brihat Parashara Hora Shastra***, Vol. 1, Chapter 27, "Evaluation of Strengths."

Thus the *Graha yuddha* is critical to this chart, and signifies that the very same components that could crush her in fact raise her up due to the victory of Mercury and how it is configured. Being located in adjacent signs gives Mercury and Mars some independence from each other, even though they are within one degree to qualify as a *Graha yuddha*. With persistence and hard work it enables her to receive the fruits of her talents in her lifetime, as it progresses. It gives her longevity and good luck, especially in a field related to communications, or something ruled by Mercury. Mercury in close contact with Mars brings a lively mind and a quick wit, as we note in the charts of Jack Kennedy and Bill Gates. Meanwhile, her **natal Moon opposite Saturn** keeps her working hard, especially as it repeats in the *Navamsha*, and in *Kendra* houses. Moon is in Gemini, an air sign owned by Mercury: air signs favor business and communications. Moon at the top of the chart keeps one attuned to the public pulse and able to sense what the public wants. Considered good for politicians, Walters has often used this gift in her interviews. It is also her emotional signature through her work. The Moon-Saturn opposition, as well as the Mercury-Mars configuration describes her straightforward style of questioning and her relentless, yet charming way of probing until she gets what she wants.

Pioneering spirit: The potency of the Mercury-Mars theme may also explain why even into her 80s and a very wealthy woman, Walters still remains active in her career. It also accounts for her pioneering spirit. Because women television journalists were so restricted when she first entered the field, she initially sought to do her interviews outside the studio in order to have more freedom and control of the interview process. This then developed into a very successful lifetime project. Dominant Mars on the Ascendant, as loser in the Planetary War, and *Jiva* (*nakshatra* lord) for four planets demands that she continue to secure her place in the world. Mars is a planet of survival, aggression, and discontent, and here it compensates constantly for its loss against Mercury. This is despite the victory over her competitors. A planet works through its destiny on both the internal and external planes. The uncertain nature of the world makes Mars all the more battle-ready. Having witnessed her father making and losing several fortunes, she took nothing for granted. Walters' income went immediately towards expenses – supporting her family of origin for many years. From Virgo, Venus rules both 9th house of father and second house of financial income. Venus is located in the 12th house of loss and expenditures, confirming her own expenses that derived from her father's losses. However, the 12th house is also private settings and pleasures of the bed, and Venus manages to survive rather well in the 12th house.

Health and longevity: In addition to a strong Ascendant lord, what keeps protecting her personally are several factors related to the Virgo Ascendant: Classical benefics Venus and Mercury surround the Ascendant, from the 12th and 2nd houses – a *Shubha Kartari yoga* to the Ascendant. In addition, since Sun is in the Ascendant, and benefics other than the Moon surround the Sun in adjacent houses – this is an especially beneficent *Ubhayachari yoga:* the Sun is considered extremely well fortified here and the Sun is a key planet of self-confidence, will power and physical vitality. The *Ubhayachari yoga*, especially with benefics, as in this case, bestows good looks, wealth, pleasures, a balanced outlook, and the capacity to take on many responsibilities. Her stylish good looks have continued even after four decades in the business. Further, classic benefic Jupiter aspects the Ascendant. Sun rising generally gives prominence in any chart and Sun is also in its own *nakshatra*, *Uttara Phalguni*.

However, for Virgo Ascendant, Sun is ruler of the 12th house of loss and Mars is lord of 8th house of suffering, both *Dusthanas*. Both of these planets in the Ascendant create an *Arishta yoga,* normally bestowing ill health and curtailing longevity if all the other factors above were not in place to reinforce good physical health and longevity. We have already discussed Mercury's victory over Mars, plus favorable *yogas*. Her *Dasha* sequence is also favorable, the 6-year Sun *Dasha* not arriving until she is 100. In spring 2010, about six months into her 20-year Venus *Dasha* she had open heart surgery to replace a faulty aortic heart valve. It had been an issue for some time though she had no symptoms. It was successful. Venus is located in the 12th house of loss or retirement. She continues to do television specials but in Jan. 2004 announced her semi-retirement from ABC's 20/20.

Speech defects or problems with speech can come with Ketu, the eclipse planet, located in the 2nd house of speech. We saw this with Winston Churchill, Albert Einstein and Maya Angelou. Walters has a pronounced lisp, though it seems to have lessened somewhat with the years. She credits it to her Boston accent, which typically drops the "r's". Hers is also not an ideal speaking voice. Natal Ketu in the 2nd house gives her voice a flawed, but distinctive quality. Sharing this astrological signature, Churchill went to some lengths to preserve his speech flaws - in order to stay unique as a speaker. Therefore, as a public speaker, uniqueness of speaking voice counts for a lot and sets her apart - even if it contributes to making her the subject of classic parodies.

Siblings: Mars is *karaka* of siblings. As loser of the *Graha yuddha* it gives weakness or ill luck for the siblings, especially during her **Mars *Dasha*.** Born in the 3rd quarter of the 7-year Mars *Dasha*, this would not be favorable for any of her siblings. Barbara was in Mars *Dasha* from birth up to Nov. 23, 1932, at age three. Older brother Burton died of pneumonia in 1932. Her older sister Jackie was mentally handicapped from birth and died of ovarian cancer in 1988. The most difficult *karmic* fruit came early and may suggest why Barbara remained so close to her family and continued to support them for many years. If this provided the initial motivation, the Moon-Saturn contacts assure she would work diligently.

Marriages: Walters has had numerous opportunities to marry a 5th time but stopped at four marriages. (The last two were to the same man.) Her lack of stability in this area comes from the volatility of the 7th house axis (Houses 1 and 7) in both birth chart and *Navamsha* chart. In the birth chart two fiery planets, Sun and Mars, aspect the 7th house of partnerships. This places many demands on marriage and partnerships of any kind. Extra responsibilities through marriage also come with natal Saturn in the 7th house from natal Moon and with *Navamsha* Saturn aspecting the Navamsha 7th house. Five planets In the *Navamsha* chart are contained in Houses 1 and/or 7: the relationship axis. Rahu and Ketu add changeability and a surprising choice of partners. *Navamsha* Saturn in the Ascendant in Aries makes her very conscientious, especially opposite Moon and Mercury. She tries to edit herself and perhaps others, but with Rahu and Ketu there she cannot always be in control of it. This may also account for her wide-ranging interview subjects and the inevitability of having to please a larger crowd. The 7th house is the other in intimate situations, including her interviewees.

Her first marriage was on June 20, 1955 in her **Jupiter-Saturn** *Dasha/Bhukti*, and was annulled in 1958. Her second marriage was on Dec. 8, 1963 in **Jupiter-Mars** *Dasha/Bhukti* and ended in divorce in 1976. They had one child: a daughter adopted in 1968: Jacqueline Dena Guber, born 1968, during Walters' **Saturn-Saturn** *Dasha/Bhukti.* Her third marriage (1981-1984) was in Saturn-Rahu *Dasha/Bhukti.* They re-married on May 10, 1986, in her Mercury-Mercury *Dasha/Bhukti.* The couple separated in 1990, divorcing in 1992. As the planet at the lowest degree of celestial longitude,

Mercury is *Darakaraka* (spouse indicator); thus its *Dasha* could be more auspicious for marriage. But Mercury is still fraught, as it is aspected by Rahu and Ketu and in the *Navamsha* receives aspects from Rahu, Ketu, Saturn and Moon. This shows many conflicting mental and emotional demands on any marriage, even if *Navamsha* Mercury and Venus are in exchange, lending much charm and social enjoyment. Therefore she might have considered marriage throughout Mercury *Dasha* (Nov. 24, 1985 to Nov. 24, 2002), but though she was linked with various partners, she said she would not marry again.

Slow building period: Walters would not be likely to reach any great heights in career, status or visibility in her **Jupiter *Dasha*** (Nov. 24, 1950 thru Nov. 24, 1966). This is in part because Jupiter's sign lord Venus is in the 12th house of loss and its *Jiva* (*nakshatra* lord) Mars loses the Planetary War. Though *Navamsha* Jupiter is in Leo in the favorable 5th house of education, it is pulled down due to its *Parivartana yoga* (sign exchange) with *Navamsha* Sun – in turn located in the 12th house of loss and expenditure. In **Jupiter *Dasha*** Walters had her heaviest expenses supporting her family. This is when her father lost his business and had to give up both family homes: the penthouse on Central Park West (New York City) and the house in Florida. The 9th house includes matters of the father, as well as her *Dharma*, or original life purpose – which had been initially to train as a teacher. *Dasha* lord Jupiter situated in the 9th house and 9th lord Venus in the 12th house (of loss) can cause major financial reverses for the father and extra expenses for his daughter. During Jupiter *Dasha* she began to support her family, also shifting from her original career intention of teaching. She began to work in television and did very well. The change of direction can be identified from Jupiter *Dasha* and could have been predicted as such.

Extraordinary career progress: Her biggest career breaks started to come in the 19-year **Saturn *Dasha*** (from Nov. 24, 1966). It would be bound to bring improvement over the 16-year Jupiter *Dasha*, as Saturn is more favorable in numerous ways. First, it is located in a *nakshatra* owned by Ketu, which though Mars-like does not lose a Planetary War, as Mars does with Mercury. Since Saturn is stronger, it brings stronger results in its *Dasha*. Saturn is angular in both birth and *Navamsha* charts, giving it power. From Saturn in Sagittarius in the birth chart, using this as a Sub-Ascendant for *Dasha* lord (during Saturn *Dasha* only), Sun and Mars in Virgo fall in the 10th house, an excellent position for career visibility and prominence. In the *Navamsha* chart Saturn is debilitated in Aries, showing how she starts the *Dasha* period at some disadvantage, but quickly makes gains due to the correction supplied by Venus and the *Parivartana yoga* (exchange of signs) between *Navamsha* Venus in Gemini and *Navamsha* Mercury in Libra. (Venus acts as if in the 7th house in Libra, sign of Saturn's exaltation.) Thus a disadvantage turns into a major advantage.

Her greatest professional advancement came in early Sept. 1976, in her **Saturn-Venus** *Dasha/Bhukti*, with a move from NBC to ABC television network, at a salary of $1 million per year and a five-year contract – unprecedented for any broadcast journalist. This was also the year her 2nd marriage ended. Transiting Jupiter (planet of *Dharma*) and transiting Saturn (planet of *Karma*) both aspected the Virgo Ascendant in early Sept. 1976, indicating an auspicious window for furthering her personal ambitions. Tr. Saturn was in Cancer, in the 11th house from her Ascendant – her strongest house, with highest *Ashtakavarga* count and well placed house lord – also considered an excellent house for Saturn transits. Tr. Jupiter was in her 9th house in Taurus, in the sign of its return, though near 7 degrees Taurus in *Krittika nakshatra*, ruled by the Sun, a better placement for her than in *Mrigashira nakshatra*. Tr. Saturn was near 17 degrees Cancer, in *Ashlesha nakshatra*, owned by Mercury.

Tr. Mercury was in turn around 12 to 13 degrees Virgo, in its sign of exaltation and rulership, and close to its degree of maximum exaltation at 15:00 degrees Virgo. Mercury was slowing down and would turn Stationary Retrograde at 14:17 Virgo on Sept. 8, 1976, magnifying its power to help her, just as it did near the time of her birth. Exalted Mercury by transit benefits her enormously as her Ascendant lord and in the best possible angle of the chart (*Digbala*). Further, the entire first week of Sept. 1976, tr. Mars, Mercury, and Venus were all in Virgo, Mercury and Mars within one degree of each other most of that week – a *Graha yuddha* in mid-Virgo that Mercury wins. This resonates with her natal *Graha yuddha* between Mercury and Mars, helping to overcome any opposition to her huge salary increase as a female broadcast journalist and first-time co-anchor on evening news. Her auspicious Saturn-Venus *Dasha/Bhukti* delivered its ripe *karmic* fruits.

Saturn's transits are especially important during the 19-year Saturn *Dasha*. With natal Moon opposite natal Saturn, there is a voracious appetite for work, especially in *Sade Sati* – when Saturn transits the 12th, 1st and 2nd house from her natal Moon in the 10th house of status and career. *Sade Sati* usually brings more responsibilities in life, especially related to the house where Moon is situated, as seen from the Ascendant. For Walters, this occurred whenever Saturn transited through Taurus, Gemini, and Cancer, as it has three times to date: **1)** March 4, 1942 to July 27, 1948; **2)** April 29, 1971 to Sept. 7, 1977; and **3)** June 6, 2000 thru July 16, 2007. Though each one would keep her working harder, the *Sade Sati* in the 1970s during her Saturn *Dasha* would have the most impact in her lifetime. Her **Saturn *Dasha*** ran from Nov. 23, 1966 to Nov. 24, 1985. And while Saturn's *Dasha* lasts 19 years, the orbit of Saturn around the Sun averages 29.46 years, and it can bring home some important events 28 to 30 years after the start of something major in the life, including the return to its own natal position.

As a planet of *Karma* and reality, Saturn will bring some important *karmic* messages when, for example, it returns to the sign of Cancer, especially near to the exact same degree of her major new contract with ABC in early Sept. 1976, near 17:00 Cancer. This occurs for the first time in winter 2005 and again summer 2006. Her next *Dasha* was more volatile – **Ketu *Dasha***, from Nov. 24, 2002 thru Nov. 24, 2009. Tr. Saturn was near 16 degrees Cancer on June 27, 2006, when Star Jones Reynolds, one of the panelists on Walters' daily news talk show *The View*, announced on-air her resignation from the show. This created great embarrassment for Walters and her co-producers, and she was forced to give a live explanation of this whole debacle soon after. More loss to her image and reputation occurred some six months later, when others quarreled on *The View*: Donald Trump and Roseanne Barr. As both moderator and co-producer, she was caught in a compromising position. By 2004 she retired from *20/20* after 25 years. She then concentrated on her specials, *The View*, and writing her autobiography, *Audition: A Memoir* (2008).

IMPORTANT DATES IN WALTER'S TELEVISION CAREER

1. **First week of Sept. 1976** – Walters accepts $1 million per year and a 5-year contract to co-anchor ABC evening news. She is in **Saturn-Venus** *Dasha/Bhukti* and tr. Saturn is near 17:00 Cancer. When tr. Saturn returns to the same degree in late June 2006, Walters suffers the first major breakdown of her public image and reputation: a panel member resigns on-air from

The View. She is in **Ketu-Rahu** *Dasha/Bhukti*, a classically volatile sub-period. Each time *Sade Sati* brings more career responsibilities and more pay, but a different *Dasha* alters the effect of the Saturn transit.

2. **Mon. Aug. 11, 1997** – launch date of *The View*, co-produced and moderated by Walters, earning her over $100 million within ten years. The day is ruled by the Moon, and she is in **Mercury-Rahu** *Dasha/Bhukti*. Both *Dasha* and *Bhukti* lords are in financial houses (Houses 2 and 8) from her natal Ascendant and from natal Moon (Houses 5 and 11), as well as in the 7th house of the *Navamsha* chart, strong in a *Kendra*. Tr. Jupiter in Capricorn and tr. Saturn in Pisces exchange signs, forming an excellent *Raja yoga* in her birth chart: between 5th and 7th houses from her Ascendant. This is less auspicious emotionally, being 8th and 10th houses from natal Moon.

3. **Wed. March 3, 1999**, 9 to 11 PM EST, New York City – the Monica Lewinsky interview on *20/20*, Walters' highest-ever rated interview, with 74 million viewers. On a Mercury-ruled day with a Mercury-ruled Ascendant, at 9:00 PM EST 27:51 Virgo is rising, strong in the *Vargottama* sector. Tr. Moon at 10:00 Virgo has just entered *Hasta nakshatra*, ruled by the Moon. *Hasta*'s symbol is "the hand" and its *shakti* is "to gain what we are seeking." Tr. Moon is near Walter's Ascendant degree of 11:11 Virgo and closely opposite tr. Mercury, Jupiter, and Venus in watery Pisces. Four classic benefics are in *kendras*, amplifying this event. Walters was in **Mercury-Jupiter** *Dasha/Bhukti* and both of these planets by transit are in her 7th house: Mercury debilitated in Pisces, corrected (*Neecha Bhanga*) by Jupiter in its own sign and exalted Venus. Tr. Venus and Jupiter both in her 7th house give her an extra large audience. The two women (Moon-Venus) discuss matters of sex, relationship and scandal, with Lewinsky several times in tears trying to explain herself as Bill Clinton's sex interest and a cause célèbre that threatened to bring down his Presidency.[221] (He was acquitted on Feb. 12, 1999.) With tr. Moon angular and three planets in a watery sign, emotions run strong and key planets in Pisces denote themes of martyrdom and accepting blame. Lewinsky herself has three natal planets in watery signs and an 8th house Moon. Her interview made broadcasting history as the highest-rated news program ever broadcast on a single network – a major credit to Barbara Walters and a predictable if not lamentable reflection on what subject matter draws the largest audiences.

[221] Monica Lewinsky was born July 23, 1973, 12:21 PM PDT, San Francisco, CA. Her Ascendant is 22:09 Virgo in *Hasta nakshatra*. Asc. lord Mercury and natal Sun are in Cancer. Natal Moon is at 15:30 Aries in the 8th house. Her White House Internship began in July 1995 and her affair with President Bill Clinton began in Nov. 1995 (to March 1997). She was transferred from the White House to the Pentagon in April 1996, as her superiors saw her spending too much time around Clinton. Her 10-year Moon *Dasha* began April 14, 1996. If not spiritualized, natal Moon in the 8th house can bring major problems during its *Dasha*, including sexual obsessions and lack of mental and emotional clarity. Natal Mars (8th lord) at 26:38 Pisces in the 7th house can bring danger from partners. Her affair became public in Jan. 1998 after a colleague secretly taped Lewinsky.

Chart 27: Françoise Gilot

Birth data: Sat. Nov. 26, 1921, 1:00 AM, Neuilly-sur-Seine, France, Long. 2E16, Lat. 48N53, *Lahiri ayanamsha*: -22:46:02. Class AA data: Birth certificate in hand from Edwin Steinbrecher. Ascendant: 1:10 Virgo.

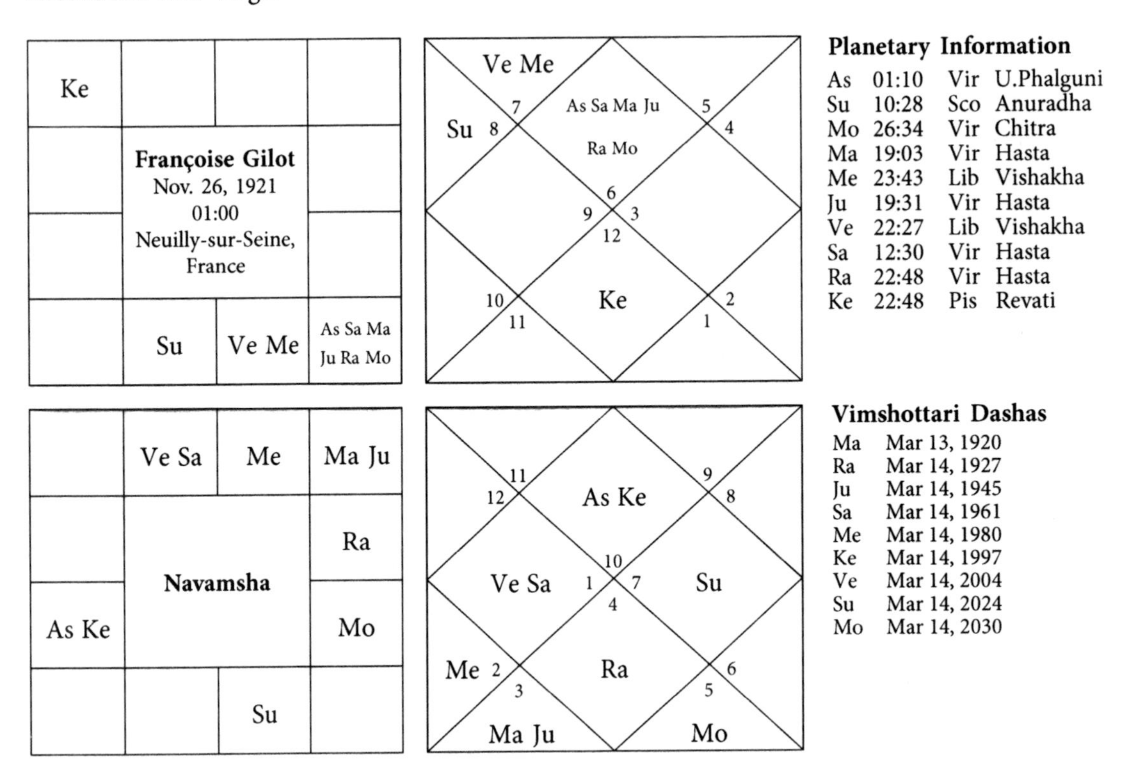

Planetary Information

As	01:10	Vir	U.Phalguni
Su	10:28	Sco	Anuradha
Mo	26:34	Vir	Chitra
Ma	19:03	Vir	Hasta
Me	23:43	Lib	Vishakha
Ju	19:31	Vir	Hasta
Ve	22:27	Lib	Vishakha
Sa	12:30	Vir	Hasta
Ra	22:48	Vir	Hasta
Ke	22:48	Pis	Revati

Vimshottari Dashas

Ma	Mar 13, 1920
Ra	Mar 14, 1927
Ju	Mar 14, 1945
Sa	Mar 14, 1961
Me	Mar 14, 1980
Ke	Mar 14, 1997
Ve	Mar 14, 2004
Su	Mar 14, 2024
Mo	Mar 14, 2030

Biographical Summary: Françoise Gilot is a visual artist, internationally recognized as a painter, printmaker, lecturer and author of seven books, including a biography of Pablo Picasso and an account of his friendship with Henri Matisse. She became known initially as Pablo Picasso's mistress, from 1944 to 1953. He was forty years her senior. Shortly before she met Picasso in May 1943 she began painting in a non-figurative manner and was later very influenced by the Cubists and her extensive contact with Picasso (b. Oct. 25, 1881, d. April 8, 1973). When she separated from Picasso in late Sept. 1953 he never forgave her for it or for writing a book on their years together: *Life with Picasso*, 1964, co-authored with Carlton Lake. Picasso initiated three lawsuits to block its publication in France, but lost all of them as of late fall 1965.

> "I was always my own self.... I think that is very important because I don't think you could be following in somebody's footsteps and become an original afterward."
>
> **Françoise Gilot**

Gilot believes it was this sense of independence that preserved her from the fate of Picasso's other partners: two of them went mad and another two committed suicide. She maintained a sense of gratitude for all she had learned and experienced in their years together.

Born of prosperous French lineage, her parents Madeleine Renoult-Gilot (1888-1985) and Émile Gilot (1889-1957) were married in Neuilly in Dec. 1920. Her mother was an accomplished watercolorist and her father an agronomical engineer by training who built up several thriving manufacturing businesses in chemicals, including perfumes. He was passionate about literature and had a large library of French literature, which Françoise devoured. Her mother gave her her first drawing lessons at age seven. She started to paint at age seventeen, but her father was strongly against her doing it full time and wanted her to become an international lawyer. He was a very willful and exacting man, very successful in business and accustomed to dominating anyone in his world – including his wife and daughter. As an only child, Françoise was treated more like a son than a daughter and was encouraged to be fearless in athletic activities and in her studies. She was fearful and shy up to about the age of eight, and then soon after became remarkably fearless, stoical, and strong-willed. Much of this had to do with intense training by her father, which eventually worked against his being able to impose his will on her. By encouraging her to master situations that frightened her in the least, it gave her the tendency to gravitate towards challenging people or situations throughout her life.

During the German occupation of Paris, it was a dangerous time to be a law student, and Françoise also got in trouble with the German authorities for joining in a student demonstration on Armistice Day, Nov. 11, 1940. This impeded her law studies and for a brief time she considered becoming a fashion designer. She was a student of literature and law at the Sorbonne when she met Pablo Picasso for the first time in a Paris restaurant on May 12, 1943. He invited her and her friend to visit his studio five days later. Gilot's first exhibition in Paris had just opened, running from May 8 to May 24, 1943. She had been cutting classes to paint secretly. But when she announced to her parents in Oct. 1943 that she wished to leave university and paint full-time, her father beat her in his rage at this unacceptable choice of profession. He severed all financial support and made it impossible for her to return home again. Leaving behind all her clothes and possessions, Françoise went to live with her grandmother for several years and painted in her attic studio. Though her grandmother's house was nearby the family home, Françoise was estranged from the Gilot family for years, reconciling with her father on a formal basis only, at her mother's request. This was after her grandmother's funeral in Aug. 1951. He died in Oct. 1957.

Most of her male friends were at least ten years older than herself, but though Pablo Picasso was forty years older, his age was not an issue for her. She found a miraculous affinity in being able to talk with someone at that level. He became her lover and companion from Feb. 1944 and they lived together from May 1946 until the day she left with their two children, Sept. 30, 1953. During their years together Gilot inspired Picasso to return to ancient Mediterranean sources for his art. Her ability to understand him and his art, and their mutual bond kept them together for close to ten years. Her eloquent descriptions of their life together and of his art and artistic process are contained in her book *Life with Picasso*. Many consider them among the most valuable and authentic commentaries of this kind on Picasso. An astute observer, she had excellent recall of many situations and lengthy conversations in great detail and on a wide variety of subjects – especially related to art history, philosophy and techniques. She did not keep diaries but credits her "fantastic memory." The same applies to her book *Matisse and Picasso: A Friendship in Art* (1990).

> "I don't write diaries and things like that," says [Gilot], "but I have a fantastic memory. I call that like a magic carpet. I can really concentrate and travel back in the past I don't know how many years from now and evoke that space if I wanted."
>
> Article by **Irene Lache**, *The Los Angeles Times*, March 6, 1991
> ("A Place of Her Own: Culture: Francoise Gilot, Picasso's former lover and Jonas Salk's wife, wants to be known not as the companion of great men, but as their equal.")

From Oct. 1953 through the next year after her departure, Picasso tried to persuade Gilot to return. Meanwhile there were his various infidelities, especially as of 1951. His bohemian ways made it acceptable for him to wander but not for her. The relationship became unbearable and confining for her, so she refused to return and later married Luc Simon, a French painter she had known since her teens. Shortly after her marriage to Simon in early July 1955, Picasso removed or destroyed all her possessions (including books, paintings, papers, and personal letters to her from Henri Matisse) all contained at the home they had shared together in Vallauris, France, and which she owned. Only a few items remained, unnoticed in the attic. Thus her ruptures with her father, Émile Gilot and Pablo Picasso – both titanic personalities – were very turbulent. Friends and colleagues she had known with Picasso shunned her for the rest of her life; such was his tyrannical hold on most of them, including directors of art galleries. Their circle included poets, philosophers, writers, and many legends of the art world, such as Georges Braque, Marc Chagall, Jean Cocteau, and Henri Matisse.

While with Picasso, Françoise gave birth to their two children: Claude (b. May 15, 1947) and Paloma (b. April 19, 1949). Her third child, Aurélia was born Oct. 19, 1956 during her marriage to Luc Simon. Never legally married to Picasso, Françoise had continued to push for their children's legitimacy. Then in Jan. 1961, Picasso secretly married Jacqueline Roque. There is conflicting information about whether the marriage was in Jan. or March 1961, why it was done secretly and whether the Picasso-Roque marriage was intended as an act of revenge against Gilot for leaving him. According to Picasso's supporters, in late Feb. 1961 he had persuaded Gilot to divorce Luc Simon and promised to marry her himself in order to secure her children's rights. At her website, Gilot says she secured the Ruiz-Picasso name in late 1960, separated from Simon in late 1961, divorcing in 1962.[222]

She began writing her book on Picasso in Feb. 1961, so it seems unlikely she would seek to marry him at this same time, if ever. She last saw him in 1954. Meanwhile Picasso was finally free to marry in 1955, with the death of his first wife Olga Khohklova. Separated since 1935 from Olga, he had been reluctant to divorce, as French law required him to share half his estate. Nor did he leave a will at his death in April 1973. It was contested for some years, and after a five-year legal battle, Paloma, Claude and their half-sister Maya inherited $90 million each in 1978 from an estate valued at $250 million. Claude Picasso is a photographer as well as the sole legal administrator of the Picasso estate, court-appointed to deal with over 70,000 works by his father as of his death. Paloma Picasso became a very successful designer of jewellery, cosmetics and fashion accessories, with her own business.

After divorcing Simon, Gilot left Paris in 1964 to live in London where she continued to paint and make art. She also became interested in the New York school of art, and began spending part of each year in a New York studio. While in Los Angeles for an exhibition in 1969 she traveled

[222] **http://www.Françoisegilot.com/bio50s.php©2010.**

to La Jolla, California where she met Dr. Jonas Salk, the immunologist famous for inventing a polio vaccine. After a brief courtship they were married in Paris on June 29, 1970, and enjoyed their life between California, New York and France for the next 25 years until Salk's death June 23, 1995. From that time Françoise kept studios only in New York City and in Paris. She has had numerous exhibitions over the years, much recognition for her vibrant and colorful art, and close contact with a wide network of people, especially her three children: Claude, Paloma, and Aurélia. An accomplished writer and poet, her books after *Life with Picasso* (1964) include: *The Fugitive Eye* (1976); *Interface: The Painter and the Mask* (1983); *Françoise Gilot: An Artist's Journey* (1987); and *Matisse and Picasso: A Friendship in Art* (1990). In 1990 she received the prestigious *Chevalier de la Legion d'Honneur* in France.

Independent spirit and imagination: With so much intensity from five planets in the Virgo Ascendant of the chart, it is crucial that Ascendant lord **Mercury** be well placed and reasonably independent from these many competing planetary forces. Fortunately, Mercury is very well positioned in the sign of Libra, bestowing charm and stylishness and the ability to discriminate in an esthetically pleasing manner. It also gives her language skills. (She is completely fluent in French and English.) Venus owns Libra and is a great good friend to Mercury. Mercury is in the 2nd house of voice, speech, and personal financial wealth alongside Venus. They are 1 degree 23 minutes apart, narrowly missing the *Graha yuddha* (Planetary War) – which must be within one degree of each other in celestial longitude. If in a *Graha yuddha*, Mercury would suffer from Venus's dominance. Instead, Mercury benefits from contact with its sign lord Venus, bestowing wealth, sweet speech, and the ability to communicate well and effectively – as is clear from her published works, including her own poetry. The 2nd house is also the face, and Gilot is known for her striking good looks. She is an elegant, thoughtful and highly intelligent woman in addition to being a gifted artist.

Mercury has an excellent position in the 5th house of fine arts and education in the ***Navamsha* chart**, again in the Venus-ruled sign of Taurus. Natal **Venus and Mercury** together in Venus-ruled Libra provide charm and greater stability, given the turbulence written into the concentration of planetary forces in the Virgo Ascendant: Saturn, Mars, Jupiter, Rahu and Moon. Venus and Mercury would also provide financial security every time it seemed it was about to disappear, and not just during Mercury or Venus major or minor *Dashas* – which would accentuate the financial abundance.

The artist survives and flourishes: The visual arts are ruled by Venus and the 5th house is their field of action. But success depends on perseverance (Saturn) and the ability to focus the mind (Moon) and intellect (Mercury and Jupiter). Saturn (5th lord) is dominant in the Virgo Ascendant. Saturn is emotionally linked through being in Moon's *nakshatra* (*Hasta*) and in the same house with natal Moon. Moon-Saturn contacts tend to pressure one to work hard and be very conscientious. (We see this in Barbara Walters' chart.) All five planets in Virgo also derive strength from Mercury's good placement. But there is danger of the mind having too many influences, especially with Rahu so close to the Moon. Nor is emotional stability helped by *Navamsha* Moon in the 8th house, though it keeps the person seeking for more profound sources of strength. Initially these can be sexual, and later artistic and/or spiritual.

However, the potential for ongoing emotional turmoil is offset somewhat by the presence of both Jupiter and Saturn in the Ascendant. She was born 2½ months after the **Jupiter-Saturn conjunction** at 3:49 Virgo and her Ascendant is close to the exact conjunction degree (within 1:39 orb). With both these planets placed in a *Kendra* they add power and intensity to the destiny along with entrepreneurial abilities. Jupiter's expansive influence and Saturn's disciplinary influence are both present. Meanwhile, versatility reigns with the Dual sign Ascendant and six planets in Dual signs. At least two major interests should be pursued simultaneously; and for Gilot, her talents in the visual arts equal those as a writer and lecturer. Her versatility is also reflected in her ambidexterity. She was born left-handed, but her father insisted that she learn to write with her right hand at an early age.

Physical health: Over all, Mercury's good placement and lack of afflictions in the birth and *Navamsha* charts bodes well for physical health and vitality. There is generally excellent health and stamina throughout her life, though a weak Mars can cause blood problems or loss of blood. Her first two pregnancies were generally smooth, though she had **hemorrhaging** after the birth of her second child Paloma in spring 1949.[223] Gilot needed surgery to correct the problem, and she was too thin and possibly anemic for two years following Paloma's birth. The third pregnancy and birth of daughter Aurélia (Oct. 19, 1956) was more difficult. The closest planet to the Ascendant is Saturn, which gives slimness and seriousness, though Ascendant lord Mercury contacting Venus in Libra would compensate for this somewhat. Gilot suffered from **insomnia** from childhood and read a lot in the night. The Moon has many planets contacting it and this can overstimulate the mind. Further, natal Mars aspects the 4th house, heating up the mental activity, at times leading to insomnia. Mercury's *Jiva* (*nakshatra* lord) is Jupiter. Though not the best planet for the Virgo Ascendant chart, at least Jupiter wins the *Graha yuddha* (Planetary War) over Mars, which is more problematic as 8th house lord. Jupiter in the Ascendant also has *Digbala* (best angular placement) and is generally beneficial for health when placed there. The amount of stimulus to the mind (Moon) may also account for Gilot's **"fantastic memory."** It has the intensity of all the planetary influences (except for the Sun), the balancing of Jupiter and Saturn, and the *Nadi yoga* between Moon and Mars, which can also bring financial wealth.

The drama of *All or nothing*: Gilot grew up amidst financial prosperity, but there were three points in her life when she had to give up her house and/or possessions because of turbulent situations. One event disrupted her studies. With all natal planets in three or less houses, extreme situations tend to occur. The *Dashas* and the transits give the timing:

1. The first event occurred in **spring 1940** during the early days of the German occupation of France, when a truck was bombed that was carrying the Gilot family heirlooms along with most of Françoise's early drawings and watercolors. All these were lost to the marauding German soldiers. Then at the student demonstrations in Paris on **Nov. 11, 1940** (the Armistice) she was arrested, requiring daily reports to the police station for many months, and making her attendance at law school more dangerous, as law students were targeted. (With Moon in her Virgo Ascendant, ***Ashtama Shani*** – tr. Saturn in the 8th house – has a double capacity

[223] Claude Ruiz-Picasso b. May 15, 1947; Paloma Ruiz-Picasso b. April 19, 1949, 19:45, Paris, France. Class B data (Biography). Gilot says Paloma was born "toward eight that evening." (***Life with Picasso,*** 1964, p. 222).

for affliction, as there is no compensation such as Churchill had with his Moon in a different house (Leo). Tr. Saturn was in Aries: April 28, 1939 to June 19, 1941; Dec. 1941 to March 4, 1942. Also, a Lunar eclipse at 9:59 Virgo on her Ascendant occurred March 23, 1940. A Total Solar eclipse at 15:08 Virgo on Oct. 1, 1940 preceded her arrest.)

2. In **Oct. 1943**, her father expelled her from the family home because she wished to be an artist and a painter. She was beaten and forced to leave behind all her possessions. Rahu eclipses the Moon (family, home). Thus a sub-period involving both planets (already in mutual contact in the birth chart) is bound to be emotionally turbulent. It also sets the tone for life with Picasso. (**Rahu-Moon** *Dasha/Bhukti* ran from Aug. 25, 1942 to Feb. 24, 1944.)

3. In late **August 1955**, returning from her honeymoon in Venice with Luc Simon, she discovered the house in southern France (Vallauris) she had shared with Picasso – and that she owned – had recently been vacated and emptied completely by Picasso, except for a few items he overlooked in the attic. He had removed all her books, drawings and paintings, those of Picasso's gifted to her, and other personal items, including numerous letters to her from Henri Matisse. (As the two classic benefics, Jupiter and Venus can bring excessive situations in their mutual sub-period. **Jupiter-Venus** *Dasha/Bhukti* ran from Jan. 24, 1953 to Sept. 25, 1955. This includes the time of her definitive split from Picasso on Sept. 30, 1953.)

Her two marriages occurred during a Venus sub-period – generally fortuitous for marriage, but less so when Venus is paired with a planet that is not a mutual planetary friend: Jupiter. Five planets in the Virgo Ascendant occur in this order by numerical degree: Saturn, Mars, Jupiter, Rahu and Moon. This kind of concentration of planetary energy gives enormous intensity to the life. It is considered much more difficult to maintain balance when the planets are in fewer than four houses in the chart, not including Rahu or Ketu. This particular planetary combination is called a ***Sankhya Shula yoga***: All planets, except Rahu and Ketu, are contained in three houses of the birth chart. *Shula* means "thorn," and is associated with struggling, and also "thorn in one's side." When the focus is in the Ascendant, the person has much complexity of character and – depending on the planets – a great deal of personal drive and potential for career success. Partnerships can be a major issue, as the majority of planets are also aspecting the 7th house of marriage and partnership. As two of these planets are **Jupiter and Saturn,** classical planets of *Dharma* and *Karma*, respectively, she has leadership qualities and a tendency to be very powerfully motivated towards her *dharmic* destiny. Even so, she will be severely tested in her choice of that destiny, and herein lie the key controversies in this Virgo Ascendant chart. The first major obstacle was her father's will; the second was that of Pablo Picasso.

When four or more planets are in one house and one sign, this is a ***Sanyasin yoga***. Usually this does not include Rahu or Ketu. *Sanyasin* means "renunciate," and though it is not necessarily that – it may describe someone who renounces almost everything in order to do one thing, or to live a certain kind of life. (To be a real ascetic, the 10th lord must be involved in the *Sanyasin yoga*. In this chart, 10th lord Mercury is situated in Libra.) By age 21, Gilot knew she did not want to continue studying at the Sorbonne and following the plans her father had for her as an international lawyer. She wrote him a letter from the south of France, where she was vacationing, and his reaction was swift. When they met in Paris he gave her half an hour to change her mind, and when she still refused – he beat her violently, threatening to put her into a mental institution.

He also demanded she undergo rigorous psychiatric testing. She underwent the testing numerous times and finally the ordeal was over. She broke with her father, but her maternal grandmother took her in to live with her. She took up her life as an artist, earning only sporadic money as an instructor of horseback riding, and painting much more of the time. She also started to frequent Picasso's studio again as of Nov. 1943.

Emotional turbulence: We have discussed why Rahu-Moon causes emotional turbulence. The first major ruptures in her family life occurred mostly during **Rahu-Moon** *Dasha/Bhukti* (Aug. 25, 1942 to Feb. 24, 1944): the break with her father and leaving home to live with her grandmother. This coincided with her first exhibition in Paris, May 8 - 24, 1943, and her first meeting with Pablo Picasso in a Paris restaurant May 12, 1943. They became lovers in Feb. 1944. Mars is also only 7 ½ degrees from the Moon; Rahu is even closer, giving some emotional turbulence and suffering throughout the life – especially with *Navamsha* Moon in Leo in the 8th house. When Rahu (the foreigner) is associated with Moon or 4th house, it can also give a tendency to live in foreign places, as she has done. In addition, **natal Moon** in a *Nadi yoga* (*nakshatra* exchange) with Mars gives emotional intensity, and is always compensating for something – as Mars loses its Planetary War with Jupiter. Moon in the Ascendant shows the importance of her maternal lineage in giving her support. Her maternal grandmother was more independent-minded than her mother, who was unable to oppose her husband and give Françoise her full support. But she eventually smuggled out some of her clothes, as Françoise left with only the clothes on her back.

Navamsha* Moon** in the 8th house gives radical life transitions with periodic emotional suffering and *Navamsha* Moon's relative isolation brings the tendency to have to keep facing life on one's own, regardless of the circumstances. She wrote of needing a lot of solitude, though at times finding it unbearable, especially in her last years with Picasso. *Navamsha* Moon receives no classic Parashari aspects and has no planets in adjacent houses. Without the corrections of at least one planet in a *Kendra* from *Navamsha* Moon or *Lagna*, a full ***Kemadruma yoga would be in force, one of the most difficult of the lunar yogas, as major difficulties arise from being or feeling unsupported. Fortunately for Gilot, *Navamsha* Mercury is in a *Kendra* from *Navamsha* Moon and three planets are in *kendras* from the *Navamsha Lagna*. With the latter correction of the *Kemadruma yoga*, there is a ***Kalpadruma yoga,*** bestowing many comforts on the individual, which she eventually had after each crisis.

Children: Having children was Pablo Picasso's idea. He was adamant about it in order to give her more femininity, more life experience and perhaps to keep her there – although he himself had a tendency to reject too much domesticity. Her three children, two with Picasso, became major positive factors in her life and great sources of stability, although she never abandoned her art on account of them. The 5th house is children, and Jupiter is *karaka* (significator) for abundance and for children, while Saturn is 5th house lord. Both planets are strong in the Ascendant in Mercury's sign and in Moon's *nakshatra* of *Hasta*, meaning the hand (and often giving talents with the hands). Further, Moon is in *Chitra nakshatra*. *Chitra* means bright shining jewel, and gives the desire to have wonderful children. Her three children were all born during **Jupiter *Dasha*.**

> "...I learned very early that no matter how fond you might be of Pablo, the only way to keep his respect was to be prepared for the worst and take action before he did."
>
> **Françoise Gilot**, *Life with Picasso*, 1964, p. 142.

Powerful and dominating men in her life: This topic can be described astrologically in several ways: first, by the strength of **Jupiter** in this chart, and by Mars and Jupiter in a ***Graha yuddha*** (Planetary War). Mars loses the war due to Jupiter's superior size and brightness. Jupiter is also higher in celestial longitude, though Mars is further north in both declination and latitude. Jupiter is lord of the 7th house of partnerships and signifies partner or spouse for Virgo Ascendant. As the 8th house lord Mars is a difficult planet for Virgo Ascendant, but since Mars loses, it eliminates her competitors. (Barbara Walters shares this combination.) However, with Jupiter victorious, it gives enormous power to her partners, especially during her 16-year **Jupiter *Dasha***, from March 14, 1945. Fortunately for her, Gilot's Ascendant lord Mercury is not involved in the *Graha yuddha*, giving her greater independence. Mercury also derives strength from being in *Vishakha*, the *nakshatra* of the winning planet – Jupiter.

The *Navamsha* chart also indicates that out of some initially disadvantageous circumstances (**debilitated Sun and debilated Saturn in opposition**) in which she faces strong, probably male authoritarian forces, she gains from these experiences and rises above them. This affects both her status and her career (10th house matters). *Navamsha* Saturn in Aries in the 4th house is in mutual aspect to Sun in Libra in the 10th house: a *Neecha Bhanga Raja yoga*. Both planets are debilitated but receive correction through Venus in the *Navamsha* 4th house with Saturn. An opposition aspect indicates confrontation. Her Mars-Jupiter *Graha yuddha* recurs in the 6th house in the *Navamsha* chart in the sign of Gemini, showing she might have some conflicts, legal issues or at the very least struggles to deal with a very dominant mate, especially during **Jupiter *Dasha***: Jupiter is the lord of the 7th house of partnership from Virgo Ascendant. The *karmic* fruits of Jupiter come due in its own *Dasha*.

Her ability to stand up to her father and also to Picasso, the only one of his women to do so, were major themes in her life – along with her ongoing artistic creativity. She has later said she was grateful to Picasso for cutting her off. Otherwise, she would not have been left as creatively free:

> "Pablo had told me, that first afternoon I visited him alone, in February 1944, that he felt our relationship would bring light into both our lives. My coming to him, he said, seemed like a window that was opening up and he wanted it to remain open. I did, too, as long as it let in the light. When it no longer did, I closed it, much against my own desire. From that moment on, he burned all the bridges that connected me to the past I had shared with him. But in doing so he forced me to discover myself and thus to survive. I shall never cease being grateful to him for that."
>
> *Ibid., p. 367.*

Challenges of childhood: Exhibiting more of the extremes of her chart, Gilot went from very fearful to very fearless early in life. She was born in the *Dasha* of **Mars**, the planet losing the *Graha yuddha* to Jupiter. Mars *Dasha* ran from birth up to March 14, 1927. This would make the early years of her childhood very difficult. Only the strength of her Ascendant lord Mercury, in Libra together with Venus and in the *nakshatra* of Jupiter – winner of the *Graha yuddha* – gave her enough personal will, physical strength and stamina to overcome the weak Mars, ruling over strength, courage, and blood. But it insures that in her early years she would be dominated by someone stronger than herself. Too young to have a partner or a husband, her father took this role. She describes his influence:

> "When I was very young I was afraid of everything, particularly of the sight of blood. If I had a cut that bled freely, I would faint. I remember also being afraid of the dark and of high places. My father reacted vigorously against all that. He used to make me climb up on to high rocks and then jump down.... I could protest for hours but in the end I had to do it. And as soon as I had accomplished one thing, he forced me to do something else, even harder. I felt powerless in the face of his will.... He wanted me to learn to swim, but I was afraid of the water. He forced me to learn and once I had learned, he made me swim faster and faster and always for greater distances than the week before. By the time I was eight, I was afraid of nothing; in fact, my nature had changed so that I sought out difficulty and danger. I had become another person, really. He had made me fearless and stoical... Later on, that psychology worked against me, too. As I was growing up, whenever anything frightened me in any degree, it fascinated me at the same time. I felt the need of going too far simply to prove to myself that I was capable of it. And when I met Pablo, I knew that here was something larger than life, something to match myself against. The prospect sometimes seemed overpowering, but fear itself can be a delicious sensation. And so I had the feeling that even though the struggle between us was so disproportionate that I ran the risk of a resounding failure, it was a challenge I could not turn down."
>
> *Ibid.*, pp. 108-109.

Any major or minor period of Jupiter would help her find her own strength, but would be likely to pit her against a partner or a more dominating force, probably male, as Jupiter is associated with male energy. She entered **Rahu-Jupiter** *Dasha/Bhukti* Nov. 24, 1929, two days prior to her 8th birthday – the age at which she says she became "fearless and stoical." Her 18-year **Rahu *Dasha*** (March 14, 1927 to March 1, 1945) was followed by the 16-year **Jupiter *Dasha*** (March 15, 1945 to March 14, 1961). Her most important years with Pablo Picasso occurred during the Jupiter *Dasha* and included the birth of her three children.

Gaining emotional stability: Françoise met Picasso in May 1943 during **Rahu-Moon** *Dasha/Bhukti* – an omen of mental, emotional turbulence – as these planets are close together in her Ascendant. These were also years of the German occupation of France. She and Picasso became lovers in Feb. 1944 during Rahu-Moon or Rahu-Mars *Dasha/Bhukti*, the latter from Feb. 24, 1944. Moon and Mars are in a *Nadi yoga* (*nakshatra* exchange), so they have a strong energy exchange. Her relationship with Picasso went on for several years before he was able to persuade her to leave her grandmother's house and come live with him. But her grandmother was so loving and supportive and had rescued her from her father's tyranny. Gilot was unwilling to leave there and apprehensive of living with Picasso.

Astrologically, he would be unlikely to persuade her prior to the 16-year **Jupiter *Dasha*** (from March 14, 1945), as Jupiter is her spouse significator, and brings her much more strength as well, including artistically and creatively. Their relationship might have blown up immediately if she had gone to live with him any earlier than her Jupiter *Dasha*. But by Jupiter *Dasha*, she is more self-confident, just as in Rahu-Jupiter, at age 8. Picasso finally convinced her he needed her more than her grandmother did, and that he loved her – though that was always an ambiguous issue. She moved into his Paris apartment and they began to live together in late May 1946, during **Jupiter-Jupiter-Venus** period. Even so, it was hard for her to find the courage to leave her grandmother, and it was Picasso who drafted the letters to her grandmother and her mother. She left with no warning, not even saying where she intended to go. For the next month she never left Picasso's apartment.

Pivotal life period – Jupiter *Dasha*: Many factors would contribute to making her Jupiter *Dasha* a period when some of the richest *and* most difficult *karmic* fructification would occur. Jupiter is a pivotal planet in this chart, as lord of the 4th house of foundations and 7th house of partnerships. Victorious in Planetary War over Mars, Jupiter is situated in the Ascendant, where it has *Digbala* – best possible angle of the chart. Jupiter is in *Hasta nakshatra*, owned by the Moon. Moon in turn is in a *Nadi yoga* (*nakshatra* exchange) with Mars, loser of the Planetary War with Jupiter. All of this spells a great deal of emotional intensity, as well as intellectual curiosity, as Jupiter in both birth chart and *Navamsha* chart is in a sign owned by Mercury, planet of mental agility and communications. We have discussed the potential for legal and health issues during Jupiter *Dasha*, with *Navamsha* Jupiter in the 6th house. Jupiter and Mars are both beholden to Mercury and the Moon: Mercury owns Virgo and Moon rules *Hasta nakshatra*. This makes Moon the *Jiva*, or true control planet for Mars, Jupiter, and Saturn. Emphasis goes to the 7th house of partnerships, since Jupiter rules the 7th house and all three planets aspect the 7th house.

Almost the entire **Jupiter *Dasha*** could be said to involve Picasso in Gilot's life, as she was in close relationship with him when it began March 14, 1945, and went to live with him in May 1946, for 7 1/2 years. During **Jupiter-Venus** *Dasha/Bhukti*, on Sept. 30, 1953, she separated from Picasso – for which he never forgave her, as no woman had ever left him. She took their two children with her. She married Luc Simon in early July 1955, also in Jupiter-Venus *Dasha/Bhukti* – not exactly auspicious, even with a Venus sub-period. By the late 1950s her artistic career took off, and she began to develop her own style, breaking away from Picasso's influence. By the end of 1960, she secured the legal name of Ruiz-Picasso for Claude and Paloma. And in Feb. 1961, she began writing *Life with Picasso* with Carleton Lake. In early 1961 Picasso secretly married Jacqueline Roque. Some controversy continues as to whether Picasso had persuaded Gilot to divorce Simon and marry him – to secure their children's legal rights – or whether this was merely a vengeful rumor. All of this takes us up to the end of Gilot's Jupiter *Dasha*, ending March 14, 1961.

The **Jupiter-Saturn conjunction of Feb. 18, 1961** marked the beginning of a new period personally and collectively. This occurred at 1:52 Capricorn, very close to the exact 5th house cusp of her chart. The 5th house symbolizes one's creative work, both biologically and artistically – thus her children and her art. In that same year she was irrevocably split from Picasso, though she has always claimed that gave her strength and the ability to carry on as an artist and a creative and independent person. Most others who lived close in Picasso's shadow did not fare as well. The 1996 movie *Surviving Picasso* was largely based on Gilot's book *Life with Picasso* (1964). As she was born just eleven weeks after the Jupiter-Saturn conjunction at 3:49 Virgo, and with her own natal Jupiter and Saturn together in Virgo, Gilot would resonate with each new 20-year Jupiter-Saturn cycle, especially the one in Virgo in 1980-1981. The Jupiter-Saturn conjunction in Feb. 1961 also gave her renewed ability to carry on in a practical way. Located in *Uttara Ashadha* ("the later victor") in an earth sign in her 5th house, it was a good omen for her own artistic and financial productivity as well as that of her children. In late Feb. 1961 Gilot also made her first trip to the United States, and a few years into Saturn *Dasha*, she spent more and more time in foreign countries: first England, then the USA, returning regularly to France, and traveling even more widely after her marriage to Dr. Jonas Salk in late June 1970.

Her **Saturn-Jupiter** *Dasha/Bhukti* (Aug. 31, 1977 to March 14, 1980) was also likely to be fruitful for her career as an artist and for her children. In 1978, after a five-year legal battle, her two children with Picasso were finally successful in winning $90 million each from his

$250 million estate. He left no will at the time of his death April 8, 1973. His widow (Jacqueline Roque) barred Claude and Paloma from attending the funeral, a reflection of Picasso's antagonism with his children the last few years of his life. He refused to see them in the end, due to their insistence he include them in his will, which he refused. (Jacqueline committed suicide on Oct. 15, 1986.)

Laying the foundations – Saturn *Dasha*: From Virgo Ascendant, Saturn rules the 5th house of children, creativity and education. It also rules the 6th house of conflict, legal battles, and health issues. Thus Saturn brings mixed results for Virgo Ascendant charts, depending on how it is configured in both the birth chart and *Navamsha* chart. As 6th house lord in the Ascendant of the birth chart, it brings 6th house matters into sharp focus in the life. Fortunately Ascendant lord Mercury is not contacting Saturn in either chart. Gilot would gain strength in Saturn *Dasha*, as Saturn is in angular houses in both birth and *Navamsha* charts, even if debilitated in Aries in the *Navamsha* chart. This would bring initial struggles and disadvantages early in Saturn *Dasha*, though she could rise up significantly from them – especially in the Venus sub-period. (Her 2nd marriage occurred at that time.) Venus is the planet that provides the correction (*Bhanga*) to the debilitated Saturn.

Her 19-year Saturn *Dasha* began March 14, 1961. One month earlier Gilot started writing *Life with Picasso* in collaboration with Carlton Lake. It was finished in late 1963, after she hired lawyers to review the text for libel. Published in Nov. 1964 in the United States, the book was very successful and immediately established Gilot with a larger public. It was soon translated into several other languages, and one million copies were sold within the first year. Picasso and his supporters considered the book an outrage, and over the next year he initiated three separate lawsuits to try to stop publication of the book in France. Gilot won all of them as of late fall 1965. Her victory was likely as: **1)** she was in **Saturn-Mercury** *Dasha/Bhukti* (March 17, 1964 to Nov. 25, 1966); **2)** her natal Saturn and Mercury are each strong; and **3)** tr. Saturn was in her 6th house in Aquarius (Jan. 28, 1964 to April 9, 1966), one of its best and most competitive placements from Ascendant or Moon, both in Virgo. Tr. Saturn was in Aquarius during the entire period of the book's publication and subsequent legal battles with Picasso.

Picasso had less ability to defeat her during her Saturn *Dasha*. His dominance in her life was more firmly established in her Jupiter *Dasha*. Jupiter's orbital cycle is twelve years, and the final legal battles coincided with a complete Jupiter cycle from the time she left Picasso Sept. 30, 1953. Though tr. Jupiter was again in early Gemini at the top of her chart, it was no longer her *Dasha* lord, and lacked the same *karmic* power. By late autumn 1965, admitting defeat, Picasso phoned Gilot to congratulate her on her victory. Not liking what she had done, he still acknowledged she had won the battle. It was the last time they spoke.

Marriage: Venus and Saturn are mutually great planetary friends. Therefore, their mutual sub-period brings far better results than with Venus and Jupiter, as they are in different planetary camps and can have a volatile effect when conjoined. This is usually due to excess of some kind and the greater number of possibilities that create both an opening and more confusion. Gilot started up with Picasso in Rahu-Moon *Dasha/Bhukti*, and moved in with him in **Jupiter-Jupiter-Venus** period. Her first marriage (to Luc Simon) was in **Jupiter-Venus** *Dasha/Bhukti* and lasted just over six years, from July 1955 to1962. She and Simon separated amicably in late 1961 and divorced in 1962 during her **Saturn-Saturn** period.

Saturn-Venus *Dasha-Bhukti* ran from Jan. 3, 1968 to March 5, 1971. Her second marriage to Dr. Jonas Salk (June 29, 1970) lasted almost 25 years until his death. They met in Oct. 1969, he proposed the next Easter (March 29, 1970) and the wedding was exactly three months later. **Saturn-Venus** *Dasha/Bhukti* provided much more stable planetary energy than **Rahu-Moon** or **Jupiter-Venus**, when she started with Pablo Picasso and Luc Simon, respectively. Gilot married Salk during **Saturn-Venus-Saturn** period. Venus is not only a *karaka* of love and marriage, it is the best planet for Virgo Ascendant chart. Venus and Saturn are together in Aries in the 4th house of foundations in her *Navamsha* chart. During Saturn-Venus *Dasha/Bhukti* she established a home base in the USA and her marriage to Salk further solidified her artistic career and reputation. The Venus-Saturn combination can bring business-like themes, so she establishes herself even more effectively on a commercial and artistic basis during this time.

As 6th house lord, Saturn can bring legal conflict or health issues, or – as in this case – more contact with people in the medical field, such as her famous second husband, Dr. Jonas Salk. Natal Saturn at 12:30 Virgo also shows that some real stabilizing of her life could start to occur about 2/5ths of the way into the 19-year *Dasha*, or 7.6 years: fall 1968. By 1968 she was spending more time in the United States and she met Jonas Salk in Oct. 1969. (Her son Claude met him in March 1969 on a photography shoot of the Salk Institute.) With many planets aspecting the 7th house, there is likelihood of many partnerships, business and/or marriage. Several of her books were collaborations, and Carlton Lake was her most important collaborator, for *Life with Picasso* – the book that put her on the world stage and on her own terms.

REVIEW OF VIRGO ASCENDANT CHARTS:

Give the astrological reasoning in each case.

1. Why does the Virgo Ascendant person tend to draw controversy? Can it be avoided or minimized? Hint: Study the Ascendant and its lord, also the effects of the 6th sign and house.
2. What factors describe the natural versatility of the Virgo Ascendant person?
3. How could Mercury help in making the Virgo Ascendant person an excellent speaker or writer? Why does Mercury also give a tendency for health problems?
4. Mars in Virgo characterizes Winston Churchill more than any other planet. Why? Why would this have significance for the nation over which he ruled for a time?
5. What issue regarding authorship & prize-winning books unites Churchill & Kennedy?
6. As lord of the 8th house of longevity, how did the timing of Mars *Dasha* adversely affect Winston Churchill and John F. Kennedy? What challenges did it present for Françoise Gilot and Barbara Walters?
7. How do Gilot and Walters both benefit by Mars in Virgo in the Ascendant, with Mars losing in a *Graha yuddha* (Planetary War) in each case?

8. Why do both Gilot and Walters tend *not* to suffer from depression, despite Saturn-Moon contacts, whereas Kennedy and Churchill both dealt with serious depression?
9. Why was Kennedy's *Dasha* timing unfavorable for the longevity of his presidency?
10. What planets best describe Kennedy's poor health? His proneness to accident and/or violent death? How can these same planets also describe his long-held secrets?
11. Why was Kennedy's father a financial success – establishing a family dynasty, while Barbara Walters' father gained and lost several fortunes, ending in loss? What astrological indicator points to great difficulties for Churchill's paternal legacy? Hint: Read each chart from the Virgo Ascendant as well as from natal Moon. Study the 9th lord's condition and the house position of the Sun (father) from the Moon as Ascendant.
12. Why is Jupiter *Dasha* the time period for Françoise Gilot when Pablo Picasso had the most power over her – even after she separated from him? Why also was Jupiter *Dasha* a period when relations with her family of origin were dramatically severed?

CHAPTER 9

Libra Ascendant

Introduction

Libra is an active or *Chara* sign, the third *Chara* sign - after Aries and Cancer - and the only one to occur in the air element. As an air sign it is associated with the ***Vaishyas***, the merchant or trading class. ***Tula*** is the Sanskrit word for Libra; it signifies balance and is often symbolized by the balancing scales. Libra expresses ***Kama*** - the desire nature, especially the desire to be in relationship and to create a more harmonious environment. At times the intensity of this desire can manifest in a very unbalanced way: Ascendant lord Venus is itself the planet of desire. Thus we observe closely the condition of Venus in the chart and as always, the natal Moon - ruling over the mind. In the process of seeking harmony and balance the Libra person may well create just the opposite. As Indian astrologer K.N. Rao is fond of saying: Libra and 7th house matters are either "marital" or "martial."

Libra is the second sign to be ruled by Venus, called ***Shukra*** in Sanskrit, and signifying "brilliant light," "heat," and the reproductive fluid. But now we move into the more moveable and active aspects of Venus, as opposed to Venus in fixed earth (Taurus).[224] In the more urban air sign, a sign of human interaction, we deal now with Venus in terms of the qualities of refinement and its more mental aspects in constructing harmonious relationships through marriage, and arranging a more harmonious environment through aesthetics. As ruler of Libra, Venus rules over everything that is refined or sensual. This includes the fine arts, ornaments, luxury, vehicles, and physical beauty; also, the humanitarian qualities of love, sympathy and compassion. Venus rules over sex, sexual relations, and the seminal fluids. In the realm of Libra, we are now dealing with civilized behavior in one-on-one relations, so we are concerned with what rules apply to support such civilized behavior. Saturn, the planet of rules and structures is in its exaltation in the sign

[224] There is important material on Venus in the introductory pages to Chapter 4: Taurus Ascendant, also Venus-ruled.

of Libra – as Saturn sets the boundaries for how civilized relationships should go. It also rules over Time.

Venus is the *karaka* (significator) of marriage. While Venus rules over Taurus and Libra, it is exalted in Pisces – where it expresses the highest level of love. In Vedic spiritual terms this is devotion, or *bhakti*. When you are completely devoted to a person, you give without reservation. True devotion is selfless love: what can I give you that you need, without thought of how this benefits me? In Libra, relationship is more contractual, considering what is needed by each person. It is more human, and less godly. Nevertheless, bearing in mind human frailties and capabilities, Libra is the arena where human beings strive to work out their one-on-one relationships, notably in marriage and committed relationships. Though Vedic classic texts refer to the more traditional marriage between a man and a woman, in modern times Libra and the 7th house need not be confined to that concept.

Since all air signs care about business and profits, being the *Vaishya* (merchant class), Libra cares about business relationships. The Libra-Aries house axis, as well as the classical houses 1 and 7 will always express what we hope to achieve not only with marriage or romantic partners, but also with our clients and business partners. If there are several planets in Libra or in the 7th house – in the birth chart or *Navamsha* chart – we will probably deal with a lot of partners or clients. The details of this will be described by the nature of the planets. For instance, the separative planets may create unions that quickly separate. Separative planets are the more tamasic planets: Saturn, Mars, Rahu, and Ketu. We have seen how Venus in association with Rahu and Ketu creates turbulence and unpredictability in relationships. Unlike Mercury or Moon, Venus is not as porous or as easily colored by the planets near it, but Venus is influenced to some extent by planets in contact with it – as with Rahu and Ketu above – or planets aspecting it. Venus is also affected by which sign it is in and in which house – from the Ascendant and secondarily from the Moon, both in the birth chart and the *Navamsha* chart. Further, how Venus operates is strongly related to the condition of its sign lord and its *nakshatra* lord. If in her own sign (*swakshetra*) of Taurus or Libra, Venus is very comfortable and confident. Whether friend or foe to Venus, much depends on whether the sign ruler is doing well in the chart and whether it is well situated by house, sign, and *nakshatra*. The *nakshatra* lord as *Jiva* (life force) will also be critical and generally performs better if it is a planetary friend to that Ascendant.

When Venus resides in the sign of a planetary friend it can do well, though there are exceptions: Venus is debilitated in the sign of Virgo, ruled by its friend Mercury. Venus is exalted in the opposite sign Pisces, ruled by 6th lord Jupiter. To Venus, Jupiter is technically neutral, but to Jupiter, Venus is planetary enemy. Though both belong to the priestly class, there is unease between Venus and Jupiter for many reasons, one of which is described by the stories in the Puranas, in which Jupiter is advisor to the *devas* while Venus advises the opposite camp, the *asuras*. Though also in different planetary camps, Venus and Jupiter in mutual contact can bring great munificence. Even so, their enemy status to each other (notably Venus to Jupiter) is testimony to the likely possibility of excess, especially in their joint *Dasha-Bhukti* periods, whether Venus-Jupiter or Jupiter-Venus.[225]

The danger of excess is also present when Venus is in its own *nakshatra*. In *Nadi Jyotisha*, Venus is not considered to bring good results when located in its own *nakshatra*. Venus belongs to the active and passionate *Rajasic guna*, and since it rules three *nakshatras* occurring in the three fiery

[225] See Chapter 4, Taurus Ascendant, Charts #7 (Theodore Prostokoff) and #8 (Martha Graham).

signs, the already *rajasic* condition of Venus is exacerbated.[226] It brings too much attachment to action and passion, especially if Venus is the Ascendant lord placed in the Ascendant or a *Kendra*, as in the chart of Adolf Hitler. And for a national leader there are larger implications, as the 7th house becomes foreign relations and/or disputes; also, the battlefield. (Hitler's Venus resides in the 7th house, along with three other planets. He saw Germany as "the world's battlefield" where the future of the globe would be decided.)

When the Ascendant lord of any chart is in the 6th house (or 6th lord is in the Ascendant), there is an *Arishta Yoga,* problematic for the physical health, more so if the Sun, Mars, and/or Ascendant are afflicted. Venus exalted in Pisces is the 6th house position for Libra Ascendant, but is still not protected from health problems. As the *karaka* for love matters, Venus in the 6th house (even if exalted) is the least beneficial house placement, even if finances are protected, especially inheritance or income from a partner. For Libra Ascendant, this is due to the *Viparita Raja yoga*, with one *Dusthana* house lord in another *Dusthana* house – 8th lord in the 6th house. Since Virgo is the classic 6th sign of the zodiac, to some extent Venus in the 6th house will tend to act like a debilitated Venus – an irony for the Libra Ascendant chart. And while sacrifice in and of itself is not bad for Venus, it brings some level of competition, conflict or sacrifice that has not been chosen voluntarily. The partner may be unavailable, ill, absent, or has other partner(s) he or she prefers.

The exaltation degree of Venus is at 27:00 Pisces in *Revati nakshatra*. If natal Venus is in *Revati nakshatra* there is often a major focus on service to those in need and/or to animals, as *Revati*'s deity Pushan is the Nourisher and Protector of animals (including cattle) and financial wealth. Pushan is also caretaker of souls in transition from the land of the living to the land of the dead. Thus Venus in *Revati* may lead to a deep interest in helping the dying. Generally, exalted planets bring exaggerated life situations, and with Pisces there may be more possibilities for *Moksha* (spiritual liberation), depending on various other factors.

Both Taurus and Libra Ascendants have the advantage that classic malefic planets Saturn and Rahu are good friends to Ascendant lord Venus and in the same planetary group. Saturn also has excellent status as *Yogakaraka* planet: it is both angular and trinal lord, owning Houses 4 and 5 from Libra Ascendant. ***Nadi yogakarakas***, or best planets for Libra Ascendant in *Nadi Jyotisha* are: Saturn, Rahu, Mercury and Moon, in that order. Since Venus rules the 8th house as well as the 1st house, it does not achieve *Nadi yogakaraka* status.

Yogakarakas are planets that can deliver some good results to any Ascendant, especially in their *Dasha* or sub-*Dasha* (*Bhukti*) period. For Libra Ascendant, Moon is an odd friend to Venus here, as the only other outright female *graha,* or planet and one that is in direct competition with it. In fact, though the Moon has no planetary enemies, both Sun and Moon are traditionally planetary enemies from the perspective of Venus in the Parashari system. However, Moon as *Nadi yogakaraka* to Libra Ascendant is also an exception to the rules of *Kendradhipatya dosha*, in which *Kendra* lords become neutral.

---

[226] See Glossary and Chapter 2 on *Gunas*. Mercury and Jupiter are classical benefics assigned to *Sattvic gunas*. (Thus Mercury's *Sattvic guna* also lessens Venus's sting when debilitated in Virgo.) As such, Mercury and Jupiter are generally less troublesome and operate more smoothly and peacefully unless they are afflicted by difficult configurations. This is also generally true of zodiacal signs ruled by classical benefics, both in the birth chart or the *Navamsha* chart. But Venus in its own *nakshatras* departs from this rule, as does any planet or combination of planets that are more afflicted.

Venus is both **the planet of charm and the planet of politics**, as we see in the charts of politicians. Vedic mythology presents Venus as *Shukra*, the inventor of political science and author of the *Shukra-niti*, the best known code of behavior and politics in all of Vedic literature. Venus has special powers of foresight as well as the capacity to restore life to those who have died. Venus is also the brightest object in the sky, after the Sun and the Moon, and as such joins the Sun and Moon as the three planets ruling over eyesight and vision. There are mythological reasons for this as well: Venus was blinded in one eye by Vishnu as a reprimand for overreaching his powers as a seer and in another story, Venus (Shukra) is blinded in one eye by Vishnu for creating obstacles toward the giving of charity. In both these myths in the *Puranas*, Venus was a man.

When Venus and Mars are intertwined in the chart and are also dominant through angularity or rulership of the Ascendant, it can indicate sexual energy that is either wasted or sublimated by the search for triumphant worldly power. There is an exaggerated need to fight and win battles, whether military, athletic, political, or social, including sexual conquests. By studying **Venus and Mars in Vedic mythology** we can understand this more deeply. The key manifestation of Venus in the form of a feminine deity is **Lakshmi**. She is associated with victorious kings, and attracted only to the bravest leaders who also win wars. She has both royal *and* spiritual authority. When Venus and Mars interact strongly there may be a willingness to sacrifice everything for a combination of power that brings both pleasure and greater control over life's circumstances. Venus is the giver of material abundance, as is Jupiter; but the difference from Jupiter is that Venus is also the deity ruling over many kinds of wealth: beauty, pleasure, fertility, the arts, full harvests, wealth of courage, of conveyances (cars or elephants), progeny, victory, knowledge, and monetary wealth.

The purpose of Mars, especially in the mythology of **Skanda**, is to have immense strength and be victorious in battle above all. Towards this end he remains forever young, vital, and unscathed by life's competitions. Most admired when he contains his sexual potency, he is associated with chastity, his only wife being the Army of the gods. In the Puranic literature, Skanda was born from the "spilling" of Shiva's seed that gets spread by the gods of Fire and Wind. He destroys the external enemies or the internal vices of his devotees. No woman was involved in Skanda's birth. Meanwhile, Lakshmi was born out of the ocean, emerging out of the purity of the lotus, unsullied by the mud in which the lotus grows. She is the Great Mother who fulfills desires, the water of fulfillment and the flowering of Divine grace and love.

With these qualities of both Lakshmi and Skanda, the close interaction of their planetary counterparts Venus and Mars in the astrological chart can bring an enormous quest for power. It may appear at times like the search for meaning within marriage or partnership, but more likely sexual energy will be either wasted or sublimated by the search for victory and worldly power. This dominates all other life themes, and the most important decisions are made in order to achieve it. If Venus and Mars are aligned with more purely ascetic *Moksha* influences, then themes of *Moksha* (spiritual liberation) will dominate the life. The whole chart would have to be considered. But **Venus is the planet of desire**, and the true spiritual master is free of desire, with no fear or anger over whether desires are met.

Nakshatras contained in Libra are: ***Chitra*** (starting from 23:20 Virgo, in the previous sign) 0 – 6:40 Libra (ruled by Mars), ***Swati*** 6:40 – 20:00 Libra (ruled by Rahu), ***Vishakha*** 20:00 – 30:00 Libra (ruled by Jupiter). The Libra Ascendant, or any planet situated in one of these three *nakshatras* is colored by being: 1) in Libra, ruled by Venus; 2) in one of these three *nakshatras*, ruled by Mars, Rahu, or Jupiter, respectively; and 3) by aspects to the Ascendant and Ascendant lord Venus.

Again, the best planets for Libra Ascendant (and also ***Nadi yogakarakas***) are: Saturn, Rahu, Mercury, and Moon, in that order. Saturn is exalted in Libra, maximum exaltation at 20:00 Libra. Sun is debilitated in Libra, maximum debilitation at 10:00 Libra.

Chart 28: Adolf Hitler

Birth data: Saturday, April 20, 1889, 18:30 LMT, Braunau am Inn, Austria, Long. 13E02 00, Lat. 48N15 00, *Lahiri ayanamsha*: -22:18:22, Class AA birth data. Zenit (Nov. 1993) quotes birth record. Birth and christening records kept at the local church in Braunau. Ascendant: 4:22 Libra.

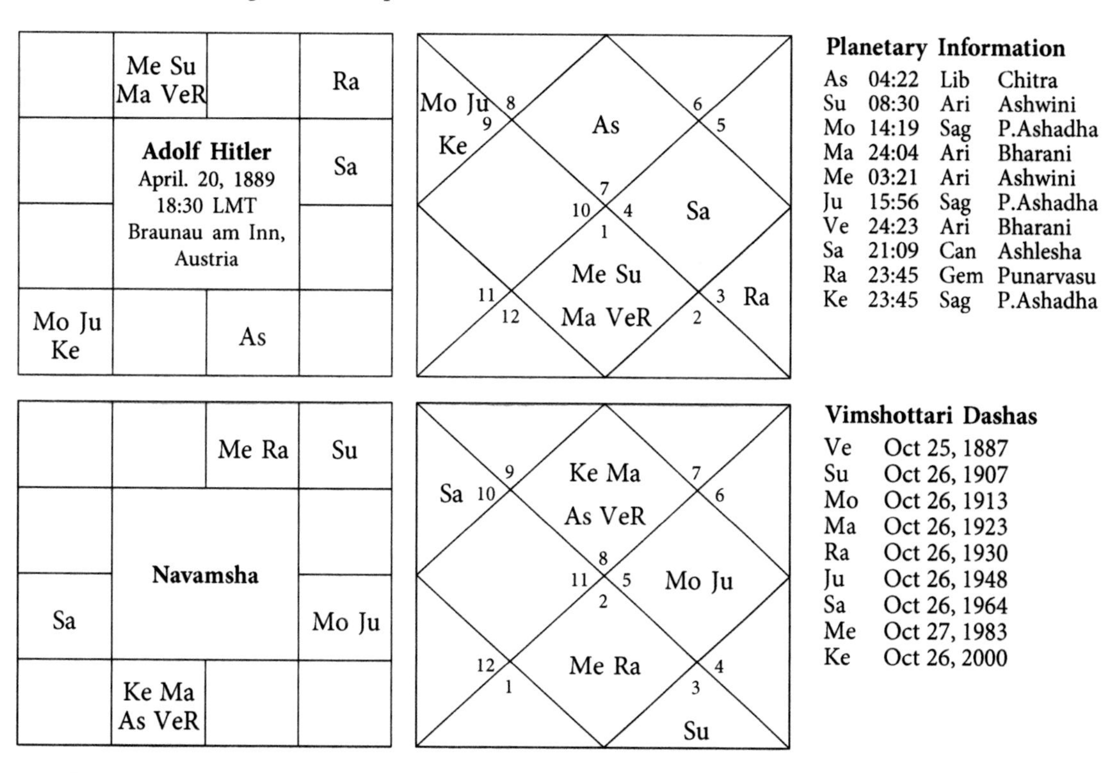

Planetary Information

As	04:22	Lib	Chitra
Su	08:30	Ari	Ashwini
Mo	14:19	Sag	P.Ashadha
Ma	24:04	Ari	Bharani
Me	03:21	Ari	Ashwini
Ju	15:56	Sag	P.Ashadha
Ve	24:23	Ari	Bharani
Sa	21:09	Can	Ashlesha
Ra	23:45	Gem	Punarvasu
Ke	23:45	Sag	P.Ashadha

Vimshottari Dashas

Ve	Oct 25, 1887
Su	Oct 26, 1907
Mo	Oct 26, 1913
Ma	Oct 26, 1923
Ra	Oct 26, 1930
Ju	Oct 26, 1948
Sa	Oct 26, 1964
Me	Oct 27, 1983
Ke	Oct 26, 2000

Brief biographical summary: The life story of Adolf Hitler and his era is a study in national pride and vengeance, of how vengeance begets more and more of the same on a grand scale, exploding among nations and races, highlighting his own as well as Germany's predicament from 1918 through 1945. This vengeance aims to protect citizens' honor, security and well being, but rips apart the very fabric of civilization. It dreams of restoring harmony, but assigns blame, exacts full punishment and seeks to eradicate and crush "the enemy" so they will never rise again.

Hitler served in the German military before becoming a political and military leader in his adopted country. He tried to overthrow the republican government in the famous

"Beer Hall Putsch" of Nov. 8 – 9, 1923. This resulted in a five-year prison sentence, of which he served nine months in 1924. Many assumed his political career was over, but Hitler rose again to become Chancellor of Germany at 11:15 AM on Jan. 30, 1933, ushering in the German Third Reich. On Feb. 28, 1933, he gained emergency powers through a presidential decree. He assumed both the Presidency and the Chancellorship, becoming Der Führer, or undisputed leader at 12 noon on Aug. 2, 1934 – a mere three hours after the death of the elderly President von Hindenburg. Hitler took the reins of power amidst the chaos of mass unemployment, ruinous inflation, and civil unrest. Germany was on the brink of bankruptcy, much of it resulting from the harsh terms of the Treaty of Versailles in 1919. From 1921, England and France demanded $33 billion in war reparations, a total that was negotiated down to $16.5 billion. An already weakened Germany paid off $4.5 billion between 1918 and 1932.

Hitler quickly turned the German economy around, by 1936 achieving nearly full employment. Very disciplined in many ways, he also had an exalted belief in his own abilities as a political and military leader. On Sept. 1, 1939, he directed German military forces to enter Poland – marking the start of World War II.

> "[By 1939 Hitler had created] the most effective fighting machine the world has ever seen."[227]

The war went relatively well for Germany up through spring 1942. But from 1943 onward, the Allied forces became unbeatable, in large part because with the U.S. entry into the war, they gradually possessed an overwhelming advantage over their foes in both manpower and resources. The Third Reich continued until Germany's official and unconditional surrender at 11:01 PM CET on May 8, 1945. To avoid capture by the Allied forces, Hitler committed suicide on April 30, 1945 in his Berlin bunker. As per his instructions, his body was cremated soon after.

FAMILY BACKGROUND & EDUCATION: Prior to his dramatic rise, Hitler started out weakly, even within his family lineage. His parents were Alois Hitler (1837-1903) and Klara Hitler (1860-1908). His father was a Customs officer in the villages of upper Austria from 1855 to 1895. Before their marriage (his third) on Jan. 7, 1885, they requested a special dispensation from the Pope to marry, having been rejected by the local bishop. Their family ties were close to incestuous as they were second cousins. Klara Hitler was 23 years younger, and called her husband "uncle" throughout their marriage. Adolf was the third child from this union. Four of Adolf Hitler's six siblings died in childhood or infancy. They were: Gustav (1885-1887), Ida (1886-1888), Otto (1887), and Edmund (1894-1900). One sibling was hidden from public view as an idiot, and another Paula (1896-1960), was mentally retarded. She was the only sibling who survived childhood. There was one half-brother, Alois (b. 1882) and one half-sister, Angela (1883-1949) both from his father's second marriage. Due to this troubled lineage and in-breeding, Adolf Hitler felt that his blood was weakened and regularly used leeches to purify his blood. His father was born Alois Schicklgruber, and changed his name to Hitler in 1877 at age 40. There were rumors that Adolf Hitler's paternal grandmother became pregnant by the Jewish son of her employer, as Alois Hitler was an illegitimate child. His surname, Schicklgruber, was that of his mother before taking his stepfather's name. Adolf Hitler kept this matter a closely guarded secret.

---

[227] Max Hastings, "How They Won," *The New York Review of Books*, Nov. 22, 2007, p. 24

Klara Hitler was indulgent, but Alois Hitler was a strict autocrat of a father. He wanted his son to do well in school and they quarreled violently over Adolf's desire to be an artist. He died when Adolf was age 13. Though he had done well in primary school and was good at gymnastics and drawing, Adolf was an undistinguished student and an under-achiever in secondary school (1900-1905). His main interests were history, geography and art. He left school at age 16 without completing his secondary education; he failed the exam and was unwilling to repeat the school year. He then tried to enter the famed Vienna Academy of Fine Arts to study painting, and was rejected twice, in Oct. 1907 and Oct. 1908. The second time his drawings were so poor, he was not even allowed to take the exam. Hitler explains in *Mein Kampf* that the Academy officials suggested he apply to the Vienna School of Architecture. However, he knew he lacked proper academic entry credentials to even consider this possibility. His mother died from breast cancer soon afterwards, on Dec. 21, 1908, and he never told her of these failures, which he found shattering. The next four years he lived a fairly aimless existence in Vienna, moving between various flophouses and hostels to avoid Austrian military service and any other serious life decisions. The tension was that he saw a grand destiny for himself, one as yet unmatched by reality. His mother left him a small stipend, and he made some small income from painting postcards. In *Mein Kampf*, Adolf Hitler's own account that he grew up in poverty is inaccurate. His mother continued to draw a pension and the family lived comfortably before and after the death of Alois in 1903.

RELATIONSHIPS: Hitler was secretive about his relationships with women and in this way he could continue being idolized by his German female public. When asked by his doctor why he did not marry, Hitler said:

> "Marriage is not for me and never will be. My only bride is my Motherland."

But he did not believe in either sexual or racial equality. His relationships also had a strange and rather sordid history. He preferred women much younger than himself whom he could dominate completely. His first love was the daughter of his half-sister Angela. Her name was Geli Raubal, then age 20. He set her up in a house they shared, and though he was in love with her, he severely limited her freedom of movement. She died of a gunshot wound, an apparent suicide in Sept. 1931, though the death was never fully solved. Local newspapers reported she had another violent quarrel with her uncle Hitler, and was beaten. Hitler wanted her to remain in his house, and she wanted to go to Vienna to be with her fiancé. From then on Hitler became a vegetarian, as eating meat was too painful a reminder of her corpse.

Eva Braun (b. Feb. 6, 1912) was a photographer's office assistant and model in Munich when she met Hitler in 1929. She became his mistress from 1932 through to the end of their lives, marrying at her request one day prior to their joint suicide April 30, 1945. Both families were against the relationship but as they never appeared together in public, most people were unaware of it. Even Hitler's intimates were unsure of the exact nature of their relationship. Braun first attempted suicide in Nov. 1932 and again in May 1935, finally getting her way with Hitler. She was jealous of his seeing other women, mostly actresses, including Renate Müller (1906-1937). Hitler, in turn, was jealous of Müller, who refused to end her relationship with her Jewish lover and soon died of an apparent suicide Oct. 1, 1937. It was unclear whether she threw herself out of her hotel window or whether the Gestapo was involved. She had declined to appear in Nazi propaganda films, and this caused problems in her relationship with Nazi leaders. Eva Braun was apolitical and remained unflaggingly loyal to Hitler to the end. She lived in ample luxury provided by Hitler, though mostly isolated from others, and spent much of her time in frustration, waiting for Hitler. As soon as guests arrived, he would banish her to her room. Hitler's staff called her "the girl in a gilded cage." In his memoirs *Inside the Third Reich,* Albert Speer, Hitler's closest friend, wrote: "Hitler kept his Eva like a puppet in a doll's house."

> "No strategist in history has been more clever in playing on the minds of his opponents – which is the supreme art of strategy."
>
> **Sir Basil Liddell Hart**, *The Other Side of the Hill*, 1948.

> "The masses find it difficult to understand politics, their intelligence is small. Therefore all effective propaganda must be limited to a very few points. The masses will only remember only the simplest ideas repeated a thousand times over. If I approach the masses with reasoned arguments, they will not understand me. In the mass meeting, their reasoning power is paralyzed. What I say is like an order given under hypnosis."
>
> **Adolf Hitler**, *Mein Kampf,* 1925.

> "The great masses of the people...will more easily fall victims to a big lie than to a small one."
>
> *Ibid.*, p. 10.

MILITARY & POLITICAL CAREER: Hitler moved to Munich, Germany in May 1913, both to avoid Austrian military service and due to his interest from 1908 in Pan-Germanism – a movement to unite all German and Dutch-speaking peoples across Europe. Pan-Germanism grew after German unification in 1871, but had more fervor by 1914. When the Great War broke out in Europe in late July 1914, Hitler's military career began within days. On August 3, 1914 he successfully petitioned King Ludwig III of Bavaria for special permission to join a Bavarian regiment. After years of wandering and relative laziness, life in the military gave him a sudden new sense of purpose.

He received several military honors for his courage in battle: the Iron Cross, Second Class December 1914, and the Iron Cross, First Class on Aug. 4, 1918. On Oct. 7, 1916 he was wounded in combat, suffering a grenade splinter in the leg, and on Oct. 13, 1918 a mustard gas attack blinded him for several weeks and also damaged his vocal chords, forcing him to learn to speak again. He was recuperating in the hospital in Nov. 1918 when he heard news of the unexpected German surrender. His shock at the news tempered the rest of his life and largely accounts for his overriding desire to return honor and prestige to Germany. While in the hospital he claims he also had a supernatural vision in which he suddenly saw his calling as a politician to free Germany. His mission was as Germany's savior, and he was sure it would lead to his deification in the nation's history.

From 1919 through 1930 Hitler worked to establish himself in German political life, though the years 1923-1930 were more trying, especially from Nov. 1923 with his failed coup d'état. He served nine months in prison and there he wrote the first volume of *Mein Kampf* (My Struggle), initially dictated to an assistant. Volume one was published in 1925 and Volume two in 1926. This became his manifesto and a kind of Nazi Bible. By May 1945, it was reported that over ten million copies had been sold or distributed, although up to 1930 the sales were very small. Noted Hitler biographer Joachim C. Fest describes *Mein Kampf* as:

> "... a mixture of autobiography, ideological tract, and theory of tactics; it also helped to complete the Führer legend ... [a] mythologizing self-portrait ... [but] behind the front of bold words lurks the anxiety of the half-educated author that his readers may question his intellectual competence."
>
> **Joachim C. Fest**, *Hitler*, 1973, English translation,1974, p. 202.

Hitler made up for any such fears by going on the offense, as his nature was imperious and impatient and he had high expectations. The Nazi party he helped to build gained steadily in membership by the end of 1931. "Nazi" was the familiar form of the German name for the NSDAP (National Socialist German Workers' Party), their official name from April 1920. Their favorite word was "dynamic" and their strategy was to promise everything to everyone. Hitler was a nationalist but never a socialist; and while the NSDAP was originally a working-class political party claiming to be socialist, from 1933 onward it targeted socialists and trade union workers as enemies of the state. Hitler became a German citizen on Feb. 25, 1932. Along the way he quickly rose in prominence through the Nazi party leadership, even with his distrust of argument and criticism. It so unnerved him that he developed a deep hatred of intellectuals and anyone capable of reasoning on their own, or disputing his ideas and principles. With this motivation he became a skilled orator and a superior political strategist, though his conversational style was chiefly as a giver of monologues on political or military topics that engaged or inflamed him. Totalitarianism was the perfect political modality for Hitler, as it abolishes plurality of opinion.

Hitler engineered several bloodless conquests in the late 1930s, undermining resistance to his larger schemes of expanding German living space, known as *lebensraum*, considered a major Nazi goal. This was in part to compensate for territories lost or surrendered at the end of WWI. Hitler's troops entered Poland on Sept. 1, 1939, and thus began the larger military war in Europe (World War II), 1939-1945. Two months later, there was a failed assassination attempt on Hitler November 8, 1939 at the Bürgerbräu Keller in Munich. All in all, there were 17 known assassination attempts on Hitler during his years in power. On July 20, 1944, one of them came close to succeeding, but only Hitler's arm was injured.

Though capable of histrionics and melodrama, he was considered a brilliant orator, very popular with the German public right up to the end of the Third Reich. His military decisions from 1942 became more flawed, but he was still viewed as a brilliant military strategist. Even so, his extreme views and obsession to seek revenge from Germany's enemies after World War I typically place him among the worst of the genocidal dictators, responsible for the Nazi Holocaust and deaths totaling approximately 12 million during World War II.

IDEOLOGY: Hitler wanted to restore the unity of the German people within a well-regulated, well-ordered society. Pan-Germanism had already contributed greatly towards this goal. To restore national power, he used the only established method for silencing detractors and uniting many divergent groups in a modern nation: preparation for war. The goal was also national expansion by war. Hitler sought to reduce the weakening effects of an entrenched class structure, to dignify work and the workingman, create low cost housing and the "folkswagon," among other things. He also wanted to remove the threat of aggressive godless Marxist Communism. Toward this end he employed many positive methods, but he is remembered for the most negative one – genocide: the systematic killing of certain groups of people who are perceived as weakening or threatening one's nation or culture. Toward this end, Nazi Germany targeted the Jews, Gypsies, homosexuals, the sick and infirm, and political and religious dissidents who could not otherwise be entirely controlled by the state. Upon acquiring Emergency powers in Jan. 1933, Hitler moved against the Social Democrats and their trade unions, seizing their headquarters in spring 1933. Along with socialists, liberals and trade unionists, more Communists were arrested and imprisoned as dangerous Leftists, and declared illegal and enemies of the people and of the state. They were the first to be imprisoned at Dachau in March 1933, establishing the first Nazi prototype for its prison camps.[228] (Trade unions are often the last organized force preventing total control of a nation by despots or corporate-controlled governments. If workers cannot organize and represent themselves politically as part of a modern democracy, it reflects the extent to which civil liberties may be reduced for all citizens of that nation. Prior to the start of the Third Reich on Jan. 30, 1933, Germany was a Parliamentary democracy since 1918.)

The premise of Nazi ideology was that Germany's plight derived from racial factors and could only be resolved by restoring German racial integrity. In *Mein Kampf* Hitler refers to the dangers he believed to be caused by Jewish citizenry in this regard, but it was not yet clear this would be later embraced as a major tactical aim. Derived from a traditional European brand of anti-semitism that had existed for centuries, Hitler's racism became more fanatic, as with Nazi ideology in its stated aims to purify and strengthen Germany.

Though Hitler ranted for many years about "the Jewish peril," some claim that the "Final Solution" (murder of the Jews) was not so much Hitler's real aim as the brainchild of others in his inner circle, notably Heinrich Himmler and Adolf Eichmann, and that they had more power than is generally known. At a conference on Jan. 20, 1942 held in a Berlin suburb called Wannsee (later called the **Wannsee Conference** at the Nuremberg Trials), top Nazi leaders gathered – though Hitler was not present. This meeting set a new and more radical Nazi policy to kill Jews and to proceed with their complete removal from the German-occupied territories. They would be deported from Europe and French North Africa to German-occupied areas in Eastern Europe, with Jews fit for labor used on road-building projects, after which they would die or be killed. When asked if Hitler approved these plans, his top officials said they assumed they had authority to go ahead with the new policy based on various prior statements by Hitler. One they cited in particular was Hitler speaking July 16, 1941 to a meeting of government ministers, including Hermann Goering and Heinrich Himmler, at which they discussed the administration of occupied Soviet territories. Hitler wanted to turn this area into a "German Garden of Eden."

> "[He said that] naturally this vast area must be pacified as quickly as possible; this will happen best by shooting anyone who even looks sideways at us."

In the end, approximately 5.5 million Jews were killed, along with some 3.5 million Soviet prisoners in German captivity, with an estimated total of 11 million noncombatants killed by Germany under Hitler's rule.[229]

Not a singular phenomenon, unfortunately, genocide is not exclusive to Nazi Germany, nor is eugenics as a philosophy. Under Mao Tse-tung's Communist regime, some 30 to 60 million

---

[228] Dachau prisoners were assigned numbers and colored triangles for identification in different categories: **Green**: Common criminals; **Black**: Communists, Socialists, Gypsies and Asocials; **Blue**: Slave laborers from occupied countries; **Purple**: Jehovah's Witnesses; **Pink**: Homosexuals; **Red**: Political prisoners; Two **Yellow** triangles making a six pointed Star of David: Jewish prisoners, who were also assigned a second triangle if they belonged to a second category. For example, many Jews were also Communists.

[229] This topic is also discussed in Chapter 1 under FIRE TO EARTH Mutation period, 1901-1961, citing statistics from historian Timothy Snyder's book, ***Bloodlands: Europe between Hitler and Stalin***, 2010. Snyder's figures are based on Eastern European archives opened in the 1990s and previously inaccessible.

Chinese perished as victims of famine, repression, or forced labor. Russian statistics are variable, but we know that Joseph Stalin executed at least 30 million or more Christians in Russia, and killed at least 7 million Ukrainians by starvation. Prior to 1914, the Belgians massacred between 10 and 12 million in the Congo, and the Turks massacred one million Armenians in 1915, to rid themselves of their "Armenian problem" – Christians in a Muslim nation. German-Jewish philosopher Hannah Arendt called it "the banality of evil" – which involves *not* distinguishing right from wrong.[230] Here is an example from a Hitler speech in Aug. 1939:

> "It was knowingly and lightheartedly that Genghis Khan sent thousands of men and women to their deaths. History sees in him only the founder of a state.... The aim of war is not to reach definite lines, but to annihilate the enemy physically. It is by this means that we shall obtain the vital living space that we need. Who today still speaks of the massacre of the Armenians?"
>
> **Adolf Hitler**, in a meeting with his military chiefs, August 1939, quoted in Samantha Power, *A Problem from Hell: America and the Age of Genocide*, 2002, p. 23.

As of the Tri-Partite Treaty of Sept. 27, 1940, Hitler had two other powerful allies – Italy and Japan, who also wished to expand their territories and restore national prestige. Hitler was inspired by America's expansion westward and its elimination of the Indians. His other models were the modern European empires that spread across Africa, Asia, and the Caribbean, and involved colonization, enslavement, and extermination. During the Nuremberg trials in 1945-1946 Hermann Goering said that Germany's persecution of the Jews was no worse than America's systematic persecution of the Native American Indian population.[231] The political principle of "necessity" was famously articulated by Italian political philosopher Nicolo Machiavelli, 1469-1527, in his classic book *The Prince* (1520).

In accord with Machiavelli's rules: Since moral law is no longer supreme, not all leaders get caught and vilified for abandoning it – especially not the victors. For instance, our view of Hitler and the Axis powers has been firmly established from history written by the victors, whose own atrocities and acts of violence before, during, and after WWII are rarely considered other than as necessities.[232] Even so, Hitler agreed with Machiavelli, whose ideas were encapsulated nearly 500 years later by South African author J.M. Coetzee:

---

[230] In an attempt to outlaw genocide, *The U.N. Convention on the Prevention and Punishment of the Crime of Genocide* was passed Dec. 9, 1948. But the U.S. did not ratify it for over forty years, and even then only with many opt-out clauses, rendering it largely symbolic. In spite of well-publicized accounts of genocide throughout the world, including among others – the Armenian genocide of 1915 and the plight of Jews in Nazi Germany – America has been mostly the bystander, with its long history of inaction through countless instances of genocide in the 20th century.

[231] The initial Nuremberg Trials (of 24 Major War Criminals) were held at the Palace of Justice in Nuremberg, Germany, from November 20, 1945 to October 1, 1946. Adolf Hitler, Heinrich Himmler, and Joseph Goebbels, committed suicide before the trials began. These trials pushed for a permanent International Criminal Court, which was finally established on July 1, 2002 with the Rome Statute. Its purpose is to prosecute genocide, crimes against humanity, war crimes and crimes of aggression; but it can do so only for crimes committed on or after that date. A growing number of member states (118 of them as of Dec. 1, 2011) are considered "States Parties to the Statute of this Court." Having "unsigned" ICC treaties, three states no longer have obligations to the principles of the Rome Statute: U.S., Israel & Sudan. Among United Nations member states, China & India are critical of the ICC, which does not have universal jurisdiction due chiefly to U.S. objections.

[232] Giles MacDonogh, ***After the Reich: The Brutal History of the Allied Occupation***, 2007.

> "Machiavelli says that if as a ruler you accept that your every action must pass moral scrutiny, you will without fail be defeated by an opponent who submits to no such moral test. To hold on to power, you have not only to master the crafts of deception and treachery but to be prepared to use them where necessary. Necessity, *necessita*, is Machiavelli's guiding principle. The old, pre-Machiavellian position was that the moral law was supreme. If it so happened that the moral law was sometimes broken, that was unfortunate, but rulers were merely human, after all. The new, Machiavellian position is that infringing the moral law is justified when it is necessary.
>
> ...The dualism of modern political culture...simultaneously upholds absolute and relative standards of value. The modern state appeals to morality, to religion, and to natural law as the ideological foundation of its existence. At the same time it is prepared to infringe any or all of these in the interest of self-preservation. Machiavelli does not deny that the claims morality makes on us are absolute. *At the same time* he asserts that in the interest of the State the ruler 'is often obliged... to act without loyalty, without humanity, and without religion.'"
>
> **J.M. Coetzee**, from *Diary of a Bad Year*, 2008, p. 17.
> Original source: **Niccolo Machiavelli**, *The Prince*, 1520, Chapter 18.

***Lebensraum* – expanded living space:** In his goal to achieve more "living space" for the Germanic peoples, Hitler saw himself as a world conqueror. In fact he came very close to conquering both Europe and Russia as of late 1941, and might well have done so if only he had coordinated more fully with his Japanese allies to put the squeeze on Russia from the East, and with India's rebel soldiers anxious to rid themselves of British rule. Hitler was well aware of the dangers of fighting on two flanks to the east and west of German borders. His famous predecessor, Otto von Bismarck (German Chancellor, 1871-1890) had warned against it, and this happened in both world wars, greatly contributing to Germany's loss in each case. Hitler tried to work around it by signing German-Russian economic and non-aggression pacts in Aug. 1939. His strategy was to defeat France first, and then invade Russia. The German invasion of Russia (June 22, 1941 – Feb. 2, 1943) was monumental in scale. Increasingly, the U.S. was steering supplies to the Soviet Union (and Britain), even though ideologically the U.S. was vehemently anti-Communist, as was Hitler. Hitler was both anti-Communist and anti-Jewish, while Stalin was anti-Jewish, anti-Christian and intolerant of most religions.

In the end, Germany was forced to retreat from Russia, but came very close to victory by Oct. 1941. Hitler's plans suffered from his racist assumptions of how easy it would be to defeat Slavs he regarded as "sub-human," led by "Jewish Bolshevics." Even so, only Stalin's willingness to suffer huge losses enabled Russia to prevail, paving the way for its alliance with Allied powers and uniting an unlikely trio – Stalin, Churchill, and Roosevelt. It was improbable that Hitler would defeat the combined forces of the British Empire, Russia and the United States, the latter just coming into its own militarily by 1942. Hitler was also unrealistic in assuming that with its **"triumph of the will"** Germany could win. Nor did he focus enough effort on protecting German cities from heavy bombing by Allied forces.

Hitler's rise to power began in late Feb. 1920 with his presentation of the 25-point program before the National Socialist Party (later known as Nazi party). In it he promoted the **separation of church and state**, as well as **"Positive Christianity,"** placing blame firmly on the Jews and Marxists not only for Germany's defeat in November 1918 but for huge economic hardships suffered in subsequent years. Many Marxists were also Jews, and leaders of the Russian Bolshevic revolution of Oct. 1917. Marxists also plotted to overthrow the German government, but their efforts were unsuccessful.

Rebuilding & rearming Germany: Hitler was very loyal to those who stood by him, including some of his original followers, such as Paul Goebbels, Hermann Goering, and Rudolf Hess. His supporters were fiercely united against the unfairness of the Treaty of Versailles, signed June 28, 1919 after six months of conferences, though never ratified by the U.S. Senate. By 1935, Hitler already disregarded the Treaty's refusal to allow German rearmament and militarization. He aimed for a higher standard of living as well as German military and industrial advantage, and merged state and economic power into one machine, as is the nature of Fascism. By September 7, 1937 he declared the end of the Treaty of Versailles. This enabled him to boost German economic progress and stability, in part by expanding the military and weapons production programs, as well as massively rebuilding the German infrastructure and outlawing labor unions. Citizens paid heavy taxes, more so than any other European nation.

The Olympic Games were held in Berlin in 1936. The Nazis used the Games to their advantage, showing themselves to be ahead of their times in use of the media, even experimenting with television technology, and broadcasting the international games on live radio – the first time ever for such events. Hitler encouraged family life, and urged women to stay at home to bear at least four children, focusing on their husbands and families and leaving more jobs for men. With this policy he wasted a valuable resource, as German women could have helped the hard-pressed armaments factories. Vast amounts of time and money were also wasted on the relatively useless German V-2 rockets.

THE MAJOR HISTORICAL DEBATE ON ADOLF HITLER: Since the end of World War II this debate has been chiefly between two British historians and their counterparts: **1)** Was Hitler an opportunistic adventurer totally lacking in ethics or beliefs? (Alan Bullock, 1914-2002) German playwright Bertolt Brecht also considered Hitler "a gangster in politics." **2)** Or was Hitler a man who did have beliefs – although repulsive ones – and those motivated his actions? (Hugh Trevor-Roper, 1914-2003) Others have argued as to whether Hitler sought global domination or only that of Europe. Trevor-Roper thought the only consistent objective Hitler sought was the domination of Europe. This seems odd, given Hitler's 18-month long attempt to conquer Russia – one of the largest and longest conflicts of the war. Trevor-Roper often participated in lively television debates with another British historian, A.J.P. Taylor (1906-1990). In his book, *The Origins of the Second World War* (1961), Taylor argues that:

> "... [Hitler's foreign policy] was that of his predecessors, of the professional diplomats at the foreign ministry, and indeed of virtually all Germans... to free Germany from the restrictions of the [Versailles] peace treaty, to restore a great German army; and then to make Germany the greatest power in Europe from her natural weight."

An astrological analysis supports the position that **Hitler had a strong set of beliefs** and a full-blown ideology, with its own broad internal consistency. He was also ambitious for power, and saw himself as a personal savior of the German people, sometimes working himself into a lunatic-like frenzy in his public speeches. His beliefs fuelled his actions through most of his life, igniting his sense of personal and even messianic destiny. His political passions can be documented from his late teens, when he became enthusiastic about Pan-Germanism. Anti-semitism was common in Germany since the time of Martin Luther in the early 16th century. It was also commonplace across Europe, especially in early 1900s Vienna, where many Jews converted to Catholicism to appear willing to assimilate. It would be easy to assume Hitler's anti-semitism derived from this period, but if it began then – it accelerated specifically due to Germany's defeat in Nov. 1918, the Treaty of Versailles in June 1919 and its aftermath, and his perception that Jews and Marxists were to blame for these events.

His beliefs: Hitler was a monomaniacal person who was intolerant of opposing views. Even so, his major gift in magnetizing people to him involved tuning in to the same grievances he shared with the German people since the conclusion of the Great War in Nov. 1918: **1)** restoring Germany to its rightful place in the world; **2)** assigning blame for Germany's demise squarely on the Jews and Marxists (often interchangeable at that time); and **3)** leveling fury at the blatant unfairness of the Treaty of Versailles, that aimed to crush Germany even further. Most of his voluminous speeches centered on these topics, unifying the country around a commonly perceived enemy – in this case an internal and external scapegoat. His beliefs also reflect his agreement with Machiavelli that the modern state and its rulers must at times "infringe on moral law in the interest of self preservation." For Hitler it was acceptable to engage in aggressive war and the killing of civilians for political purposes. The Nuremburg Trials in 1945 found this unacceptable, but many political leaders have carried on doing it anyway, in Machiavellian fashion.

Hitler admired those who confirmed his own ideas and beliefs, and read their works as much as possible: Jesus Christ, Mohammad, Moses, Martin Luther, Solon, Alexander the Great, Brutus, Caesar, Henry VIII, Frederick the Great, Napoleon, Oliver Cromwell, Otto von Bismarck, and Richard Wagner. He modeled himself on self-appointed dictators, especially Napoleon, in the way that "Napoleon was France, and France was Napoleon."[233]

> "It is the absolute right of the state to supervise the formation of public opinion."
>
> **Paul Goebbels** (1897-1945), Nazi Minister of Propaganda, Oct. 1933.

Lessons for modern times – Closing down an open society: Upon becoming Chancellor in Jan. 1933, Hitler moved quickly to quell the chaos and establish order. Many rulers in the same situation have declared martial law. Hitler consolidated his executive power, limited legislative and judicial powers, and destroyed constitutional freedoms formerly enjoyed by the German republic. In doing so, Hitler used classic methods for destroying constitutional freedoms in a democracy, invoking national honor and national security to advance his brand of despotism. At the Nuremberg Trials in Nov. 1945, when asked if it was true that the Nazis abolished democracy when they came to power, Hermann Goering replied: "We found it no longer necessary." Nazi propaganda asserted that in every region they conquered they would be greeted as liberators. But first, public opinion had to see the inevitability of war and the necessity for removing civil liberties for the sake of national security. American author Naomi Wolf identifies ten steps in this process of closing down an open society. They apply to Hitler as well as to some other contemporary regimes.[234]

1. **Invoke a terrifying internal and external threat.** (On Feb. 27, 1933, the Reichstag (German Parliament) was set on fire – now widely believed to be arson perpetrated by the Nazis

[233] Reference: Walter Langer, ***The Mind of Adolf Hitler: The Secret Wartime Report***, 1972.

[234] In her book ***End of America: Letter of Warning to a Young Patriot*** (2007), Naomi Wolf draws parallels between Hitler's methods in Nazi Germany and those used systematically by various dictators and despots around the world. She advises her readers to remember history, observing how those same techniques were being used by U.S. President George W. Bush (2001-2009) in his "War on Terror." In fall 2009, she was critical of U.S. President Barack Obama (2009-) for continuing many of Bush's policies along the same lines.

themselves in order to cancel most German civil and political liberties. Soon after, on March 23, 1933, a presidential decree gave Hitler emergency powers.)

2. **Establish secret prisons.** (Designed to intimidate the populace, such prisons are outside the rule of law. A fascist government can thus avoid prosecution, and imprison any of those designated as "state enemies.")
3. **Develop a paramilitary force.** (This is "a thug caste" used to terrorize ordinary citizens. The Brownshirts staged violent rallies throughout Germany.)
4. **Surveil ordinary citizens.** (People are encouraged to spy on each other.)
5. **Infiltrate citizens' groups.** (Any individual or group opposing government policies is a target for investigation or even imprisonment. With no *Habeas Corpus*, victims have no recourse to defend themselves.)
6. **Arbitrarily detain and release citizens.** (More intimidation techniques.)
7. **Target key individuals.** (Ditto.)
8. **Restrict the press.** (Persecute those attempting to report news in an independent manner. When citizens can no longer discriminate propaganda (fake news) from real news, they stop demanding accountability. Paul Goebbels, Minister of Propaganda, had complete control over the German press, radio, television, cinema, and theatre.)
9. **Cast criticism as "espionage" and dissent as "treason."** (This criminalizes a broad range of dissident voices. In April 1933, Goebbels purged the German Civil Service of attorneys not sufficiently pro-Nazi, paving the way for more brutal laws.)
10. **Subvert the rule of law.** (Any pretext is used to round up prisoners. *Habeas Corpus* is suspended.)

Building prejudice from religious and secular views: Among his large number of supporters were many German Christian churches and their pastors, who regarded Hitler as a strong leader. Nazi party and military leaders were frequently present in the churches. Though he grew up in a Roman Catholic household and was inspired by the Catholic liturgy to create ongoing political slogans, Hitler favored Protestant Christianity. The Nazis permitted various denominations, especially those more obedient to German nationalism and state control. Hitler abhorred both "Jewish materialism" and Marxist atheism. Stalin abhorred Hitler's Fascism and nationalism, though his racism was far less important to Stalin. Thus the German-Russian alliance was tenuous and brief prior to the German invasion of Russia, ending in February 1943.

While in prison for nine months in 1924, Hitler was greatly influenced by reading several books, including by Martin Luther (1483-1546), famous German leader of the Protestant Reformation. Though Luther was more tolerant of the Jews in his earlier years, he later became more and more pessimistic about converting them to Christianity and making them accept Jesus Christ as their Messiah. In his book ***On the Jews and Their Lies*** (1543), Luther denounced the Jews and urged their harsh persecution and even murder. Though in this regard, Luther's influence on the German public and on the Lutheran church for four centuries remains controversial, his influence on Hitler and Nazi Party philosophy is clear. In *Mein Kampf*, Hitler praised Martin Luther as a warrior, statesman, and theologian. Luther's 1543 book was used often as Nazi propaganda.

Europeans of many nationalities apparently needed little encouragement to turn against the Jews and participate willingly in their genocide – not only Germans, but Poles, Ukrainians, Lithuanians, French and Italians. Beyond any religious prejudices, they often saw Jews as interlopers, seizing the reins of power wherever they lived or moved. Martin Luther recommended that Jews be deprived of money, civil rights, religious teaching, and education, and forced to labor on the land. Otherwise, he said, they should be expelled from Germany and possibly even killed. Hitler was also greatly influenced by reading Henry Ford, the American car manufacturer. In his book ***The International Jew: The World's Foremost Problem*** (1922), Ford blamed Jewish industrialists for WWI. He issued an apology to the Jews in 1927, but later praised the growth of Nazi Germany under Hitler.

Racial purity: To achieve racial purity among the German people – one of the perceived solutions to Germany's problems – the Nazi regime persecuted those viewed either as enemies or *Lebensunwertes Leben* – "life unworthy of life" – later known as "ethnic cleansing." For centuries, many European countries had persecuted Gypsies (the Roma), but Hitler was the first to do so officially. He systematically persecuted and killed 1.5 million Gypsies, as well as various uncooperative or independent religious groups – such as the Jehovah's Witnesses and Freemasons, also homosexuals, political dissidents, the elderly and disabled. The Holocaust did not expand more fully until German expansion eastward into Russia began. Millions of Holocaust victims were non-Jewish, though some historians have focused mainly on the Nazi persecution of the Jews. As mentioned earlier, the latest figure cited is approximately 5.5 million Jews killed. According to a majority of witnesses, Jews were deported in the open, in full knowledge of the rest of the German public. Though Nazi leaders were mostly secretive about the concentration camps, their existence and purpose was well known outside of Germany. More and more German citizens became aware of them, as more and more Jews, among others, disappeared from daily life.[235]

Many puzzles remain. With the West so motivated to fight Hitler, especially for his persecution of the Jews, further questions arise as to why the Allied forces did not bomb railroads leading to the numerous concentration camps (this despite a huge number of Allied bombing raids on German domestic and military targets), or why so few Jewish refugees were allowed to immigrate to the USA during the war years. (This remains part of the long U.S. history of inaction to genocide.) Many other countries either expelled Jews or imposed strict immigration quotas at this time, which only increased the terrible plight and obvious martyrdom of the Jews. Up to 1918 many Russian Jews who fled persecution in Russia in the early 1900s had found Germany to be a welcoming and prosperous land. But from 1918 onward Palestine became the new hope.

BURIED HISTORY: The Führer – a product of his times? What is rarely discussed or assessed is why for so many years a large German public was vociferously receptive to Hitler's views. More often we are told that large numbers of German citizens were silenced because of Hitler's brutal totalitarian dictatorship or simply mesmerized by his hypnotic oratorical skills. Because of his views, Hitler's popularity and the broad sympathy for him among Germans is usually dismissed as outright racist hatred and anti-semitism, without any further foundation. In the Post-WWI years, why would ordinary Germans have any reason to seek vengeance against the Jews? Why

[235] United States Holocaust Memorial Museum. "The Holocaust." ***Holocaust Encyclopedia.*** **http://www.ushmm.org/wlc/en/index.php?ModuleId=10005143** (accessed Nov. 29, 2007).

not blame Germany's defeat in 1918 on the victors in WWI – the English, French, Italians and Americans? Why not especially blame the French? They never forgave Germany for defeating France in 1871. Or the English? They were chiefly responsible for the naval blockade of Germany, from 1915 through 1919. For greater historical perspective, we will look at some relatively buried history that might shed some light on this matter.

The Zionists: Though not all Jews were (or are) Zionists, Zionists were secular and religious Jews who aimed to create a Jewish homeland specifically in Palestine – the Holy Land they were forced to leave during the Roman conquest in 70 AD. As of 1918, Palestine was inhabited by some 750,000 Arabs, most of them Muslims. It was owned since 1516 by the Ottoman Empire (Turkey), an ally of the Germans in WWI, along with Austria-Hungary.[236] Due to the perilous situation of Jews in Russia, as well as in a newly nationalistic Europe, Zionists sought a refuge for the Jews. Since the 1890s, they had actively sought a way to promote mass migration of Jews to Palestine. They made no headway negotiating directly with the Ottoman Empire, though England had offered the Zionists land in East Africa (Uganda). Generally regarded as pro-German, Zionists comprised less than 1% of all Jews worldwide, as of 1913 statistics.[237] They had no real clout until 1915, by which time Jewish powerbrokers controlled key positions in global politics and finance, and most of them were also Zionists. They negotiated with both warring parties throughout WWI, but mostly played Berlin against London.

In the summer and fall of 1916, Germany had essentially won the war, especially up through spring 1917, and offered several reasonable peace proposals. The German people expected victory and were told it was coming. They also needed a speedy victory, as a naval blockade of Germany had been enforced since 1915. But in Dec. 1916, David Lloyd George was elected British Prime Minister (through 1922). This was a major turning point in the Great War: Lloyd George was against a British surrender to Germany and was staunchly pro-Zionist, having even litigated on their behalf in 1903, the year Britain gave its first official statement of guarded support to the Zionists.

> "...Palestine 'was to [David Lloyd George] the one really interesting part of the war.'"[238]

Lloyd George was reluctant to surrender British imperial interests, especially in the Middle East, and most notably Palestine. He saw an alliance with Zionists could greatly benefit the British Empire, and prevent Germany and the Turks from regaining control of the Middle East. The Zionists were equally eager to work with the British, to help them win the war and achieve their mutual goals.

Both the German and British governments wanted to win over the Zionists, but until the war was won, Germany was reluctant to promise the Jews mass migration to Palestine, especially since nearly one million Arabs already resided there and had done so for thousands of years. Muslims, Christians as well as Jews all claimed it as their Holy Land. While Germany awaited the imminent surrender of the Allies, Britain was still weighing its options based on some tempting offers from the Zionists. Few historians focus on how the prize of Palestine played such a key role in extending that war and changing its course irrevocably. Though both sides courted the

---

[236] For more material on Palestine, see Chapter 1 under National Independence, etc.

[237] David Fromkin, ***A Peace to End All Peace, Creating the Modern Middle East 1914-1922,*** 1989, p. 294.

[238] Ronald Sanders, ***The High Walls of Jerusalem: A History of the Balfour Declaration and the Birth of the British Mandate,*** 1984, p. 493, quoted in David Fromkin, ***A Peace to End All Peace***..., p. 287.

Zionists, Britain was willing to do far more. In exchange, what would the Zionists do for them? Most Zionists were pro-German and naturally backed a German victory in WWI. Would these same Zionists be willing to turn *against* Germany? One cannot say for certain whether they also promised to bring the U.S. into the war, but the anecdotal evidence is substantial that they probably did just that. American Zionists had positions of tremendous power and influence, especially on President Woodrow Wilson.

David Lloyd George, British Prime Minister (1916-1922), ***Memoirs of the Peace Conference, Volume II***, (New Haven, Yale University Press), 1939; Chapter. XXIII, pp. 737-738.

"**The Zionist leaders gave us a definite promise that**, if the Allies committed themselves to giving facilities for the establishment of a National Home for the Jews in Palestine, **they would do their best to rally to the Allied cause Jewish sentiment and support throughout the world**. They kept their word in the letter and the spirit, and the only question that remains now is whether we mean to honour ours. Immediately the declaration was agreed to, millions of leaflets were circulated in every town and area throughout the world where there were known to be Jewish communities. They were dropped from the air in German and Austrian towns, and they were scattered throughout Russia and Poland. I could point out substantial and in one case decisive advantages derived from this propaganda amongst the Jews. In Russia the Bolsheviks baffled all the efforts of the Germans to benefit by the harvests of the Ukraine and the Don, and hundreds of thousands of German and Austrian troops had to be maintained to the end of the war on Russian soil, whilst the Germans were short of men to replace casualties on the Western front. I do not suggest that this was due entirely, or even mainly, to Jewish activities. But we have good reason to believe that Jewish propaganda in Russia had a great deal to do with the difficulties created for the Germans in Southern Russia after the peace of Brest-Litovsk. The Germans themselves know that to be the case, and the Jews in Germany are suffering today for the fidelity with which their brethren in Russia and in America discharged their obligations under the Zionist pledge to the Allies.

... There is no better proof of the value of the Balfour Declaration *[described below]* as a military move than the fact that Germany entered into negotiations with Turkey in an endeavour to provide an alternative scheme which would appeal to Zionists. A German-Jewish Society, the V.J.O.D.[239] was formed, and in January, 1918, Talaat, the Grand Vizier, at the instigation of the Germans, gave vague promises of legislation by means of which "all justifiable wishes of the Jews in Palestine would be able to find their fulfilment."

... Since the Jewish opinion of the world is in favour of a return of Jews to Palestine, and inasmuch as this opinion must remain a constant factor, and further, as His Majesty's Government view with favour the realisation of this aspiration, His Majesty's Government are determined that in so far as is compatible with the freedom of the existing population, both economic and political, no obstacle should be put in the way of the realisation of this ideal.

The Arab leaders did not offer any objections to the declaration, so long as the rights of the Arabs in Palestine were respected. Pledges were given to the non-Jewish population of Palestine who constituted the great majority of its inhabitants, as well as to the Jews. These were the result of conversations which we had with such Arab leaders as we could get in touch with." *[NOTE: Palestinian Arabs were fighting against Britain in WWI, so were not contacted.]*

---

[239] Vereinigung Judischer Organisation en Deutschlands zur Wahrung der Rechte des Osten. (Alliance of the Jewish Organizations of Germany for the Safeguarding of the Rights of the Orient.)

Palestine – the prize: In reality, the Zionists wanted more than simply a mandate for the mass migration of Jews to Palestine. They wanted to establish a Jewish state, with the predominance of the Hebrew language and a Hebrew university, among other specific details. They were willing to give unprecedented support to whichever side of the Great War gave them what they wanted. The **Balfour Declaration,** signed Nov. 2, 1917, was at least the strongest sounding promise so far. Just prior to it was the **Bolshevic Revolution** in Russia Oct. 24-26, 1917, which changed the Russian power structure and eliminated the Czar. Zionists anticipated their Russian Jewish brethren, all key leaders of the Bolshevic movement, could later engineer a coup in Germany. These two events within ten days of each other signaled to Jews and Zionists worldwide that their support should now shift to Britain and its allies, and against Germany. This was an astounding reversal of allegiance, as up until then, most Jews, including Jewish immigrants to the U.S., were pro-German, and of German or Austrian ancestry. Many still lived in the German, Austro-Hungarian, or Ottoman empires. But Britain was promising more – if only on paper. And Palestine was the goal.

> "His Majesty's government view with favour the establishment in Palestine of a national home for the Jewish people, and will use their best endeavours to facilitate the achievement of this object, it being clearly understood that nothing shall be done which may prejudice the civil and religious rights of existing non-Jewish communities in Palestine, or the rights and political status enjoyed by Jews in any other country."
>
> Quoted in **David Fromkin**, *A Peace to End all Peace: Creating the Modern Middle East 1914-1922*, 1989, p. 297.

The Balfour Declaration: British Foreign Secretary Arthur Balfour wrote a short letter to Lord Edmond Rothschild, a prominent British Jew and Zionist. Known as "the Balfour Declaration," it was originally drafted in March 1916 and its veiled language obscured the intention of an eventual Jewish state. The final draft was signed on Nov. 2, 1917, announced in *The Times* a week later and celebrated at the London Opera House Dec. 2, 1917. In the letter, Britain promised Palestine to the Zionist Jews, even if Britain had no claim to Palestine at that time, and even if up to 1917 Britain was losing the war against Germany and its ally, the Ottoman Empire, who *did* own Palestine. The Balfour Declaration contradicted the earlier Sykes-Picot Agreement of 1915, and the Sykes-Picot-Sasanov Agreement of 1916, in which Britain promised its allies, France and Russia, to divide the Middle East and rule it jointly with them. The 1917 resolution also betrayed British promises of independence to the Arabs, whom they had encouraged to revolt against the Ottomans. Lloyd George referred to the Balfour Declaration as his **"contract with Jewry."** The Arabs refused to recognize it, and indeed its major beneficiaries were the Jews. By 1922 Britain was already defaulting on its reassurances to the Palestinian Arabs, having persuaded the League of Nations to give them a "mandate" to rule Palestine as a colony, encouraging Jewish immigration, and working to "secure the establishment of a Jewish national home." (British parliamentary papers, 1922)

Neither Britain nor the U.S. wanted to offend the influential Zionists, especially as they owned all the major international banking houses. And though sympathetic to Zionism, Woodrow Wilson was suspicious of Britain's motives regarding Palestine. Lloyd George saw the resolution as necessary to speed up U.S. intervention in the war. Nor could Britain survive the war without American financial aid, food supplies, and military intervention. Though the U.S. officially entered the war April 6, 1917, no American soldiers were at the front for months. Meanwhile, as Lloyd George writes in his *Memoirs*, the Zionists promised to persuade all Jews everywhere to turn against Germany economically, socially and politically. In exchange, the Zionists expected to be given

Palestine, or at least the promise of mass migration of Jews to Palestine. France was committed to keeping its own territorial hold in the Middle East, and was far less interested in Zionist plans for Palestine, especially as Zionists were seen as pro-German up to that time.

The Paris Peace conferences of 1919: Postwar Paris was probably a poor choice for the Peace conferences, with its already vengeful environment as a backdrop, and Premier Georges Clemenceau full of life-long hatred for Germany since its defeat of France in 1871. But he was not alone in his desire to force Germany to its knees. The Allied powers were unmerciful in continuing a naval blockade of Germany, even after all parties had signed the Armistice on Nov. 11, 1918. The Germans had hoped the Armistice would bring an end to the blockade, but it continued for another eight months, as the Peace conferences dragged on and on. In operation since 1915, the blockade had already caused the deaths of 763,000 Germans by starvation, according to the German National Health Office in Dec. 1918, and the deaths of two million civilians across Europe.[240] Some consider it a British war crime to have continued the naval blockade after the Armistice, as they were its chief enforcers. But as the German military staff was never forced to admit defeat on Armistice Day, there remained for some Allied nations the threat of further German military action.

Meanwhile, among some 10,000 attendees at the Peace conferences was a powerful delegation of Zionist Jews, including Bernard Baruch, American financier and economic advisor to President Woodrow Wilson. The Balfour Declaration had paved the way for this moment, as did the fulfillment of Zionist promises to the Allies. On Feb. 3, 1919 the Zionists gave their presentation to the Peace conferences, along with a proposed map of the new Zionist Palestine.[241] On March 4, 1919, a group of anti-Zionist American Jews presented their objections to "the segregation of the Jews as a nationalistic unit in any country." They were overruled by the more powerful contingent of American and British Zionists – including bankers, whose influence on the proceedings is far less known and understood as a major factor behind the events of the war itself, along with certain terms of the treaty, such as harsh war reparations demanded of Germany.

> "Seldom has such secrecy been maintained in any diplomatic gathering... The wishes, to say nothing of the interests, of the populations were... flagrantly disregarded.... Provinces and peoples were, in fact, treated as pawns and chattels in a game.... The territorial settlements in almost every case were based on mere adjustments and compromises between the claims of rival States."
>
> **Harold Nicholson**, *Peacemaking 1919*, 1933, p. 43.

> "The Treaty includes no provisions for the economic rehabilitation of Europe – nothing to make the defeated Central Empires into good neighbors, nothing to stabilize the new States of Europe... or to adjust the systems of the Old World and the New."
>
> **John Maynard Keynes**, *The Economic Consequences of the Peace*, 1920, p. 211.

The Treaty of Versailles was signed June 28, 1919 after six months of conferences. It was designed to spread the spoils among the victors and cripple the economic life of Germany. The result was prolonged famine and disastrous economic conditions in Germany, with a direct impact on the rise of a group such as the National Socialists (the NSDAP from April 1920, also known as

---

240 C. Paul Vincent, ***The Politics of Hunger: The Allied Blockade of Germany, 1915-1919***, 1985, p. 145.

241 Zionist proposals - Item #1: "The High Contracting Parties recognize the historic title of the Jewish people to Palestine and the right of Jews to reconstitute in Palestine their National Home."

the Nazi party). British economist John Maynard Keynes was correct in his predictions. The U.S. Senate never ratified the Treaty of Versailles, and Americans believed it was unfair and unjust in many ways, especially the punitive reparations and the **War Guilt Clause** placed on Germany. The British came around to the same conclusions and most Germans regarded the Treaty of Versailles as "a disgraceful peace" and unacceptable.

Worldwide boycott of German goods: Germany suffered poverty, civil war and chaos up through 1933, largely due to these harsh postwar conditions imposed on it. The loss and subsequent humiliation in 1918 was very degrading to Hitler and most Germans. It kindled their rage against the Jews in particular for their open declaration of war on Germany, even though in truth many parties were responsible for not providing famine relief for starving Germans after the Armistice, especially the British. But the Zionists had been trusted German allies, and they pushed for a world boycott of German goods that went on for years after WWI. It was especially effective in the USA, Britain, and Poland. The gradual escalation of this policy of confrontation on an international basis probably contributed to setting the stage for WWII, both before and after Hitler came to power in 1933. Newspaper transcripts and radio interviews document a consistent anti-Germany campaign by Jews and Zionists, a story which is still largely absent from history books.[242]

> "Enthusiasm for the boycott, led in America by the well-known New York attorney Samuel Untermeyer, and pressed elsewhere by an international Jewish congress held in Amsterdam and by the National Joint Council of the labor unions in England, was one of the ways in which liberal energies were mobilized against Hitler's Germany in the summer of 1933, in addition to the wide publicity given to atrocity stories, too often based on hearsay.... In no case did [a] German minority as small as the Jews gain the support the latter did in the American liberal press. It [i.e., the boycott and attendant hoopla] was probably the most remarkable campaign ever conducted in the interests of such a small minority group of citizens of a foreign state by any segment of American public opinion-making media in the nation's history."
>
> **J.J. Martin**, *The Nation*, Aug. 2, 1933, p. 223.

In truth, vengeance against the Germans was seen as excusable, and difficult to curb after years of anti-German propaganda from Western sources such as the U.S. Committee on Public Information (Apr. 1917 – Aug. 1919). President Woodrow Wilson appointed journalist and public relations maverick George Creel as head of CPI. Its purpose was to mobilize public support for America's entry into World War I – which Wilson optimistically called "a war to end all wars" and "a war to make the world safe for democracy." But CPI's raw, patriotic and militaristic materials were often complete fabrications, with images such as babies hoisted at the end of German

[242] This effectiveness of Zionist support for a Jewish state continues to the present. Israel proclaimed its nationhood May 14, 1948, and in 1953 America's pro-Israel lobby was founded: the American Zionist Committee for Public Affairs. Later renamed the American Israel Public Affairs Committee (A.I.P.A.C.), it is by far the most powerful lobby in Washington, D.C., with over 100,000 members. Congressional members and presidential candidates ignore AIPAC's agenda at their peril – such is the lobby's clout with Congress, the executive branch, and the mass media, impacting U.S. foreign policy and budget. Though a relatively wealthy nation, Israel receives over $3 billion annually in American military and financial aid, the largest amount of foreign aid to any nation. Perhaps M.J. Rosenberg said it best when he said: "Israel was established to have a secure refuge for Jews & financial aid.... [AIPAC's] only focus is on Israel... America is the best home that the Jews have ever had.... AIPAC has a chilling effect on debate [i.e., on the Israel-Palestine question, or on any matter questioning the correctness of Israeli or Jewish policies, past or present]." M.J.Rosenberg, Israel Policy Forum, quoted in an interview on *Bill Moyers Journal*, Public Broadcasting (PBS), Nov. 30, 2007.

soldiers' bayonets. Germans were portrayed as "evil monsters," "Huns" and sadistic killers. The effectiveness of this controversial propaganda campaign made subsequent anti-German material all the more acceptable.

Physical health & stamina: Adolf Hitler had remarkable powers of physical stamina and endurance. With seven planets in fire signs (all of them masculine signs), including the Ascendant lord Venus, Sun, and Mars, one is destined to have plenty of physical and masculine energy throughout the life, though it would not be fully galvanized until the 10-year Moon *Dasha*, from Oct. 26, 1913. This shows the power of the Moon as motivator and as *Nadi yogakaraka* planet for Libra Ascendant. Also, the Moon is lord of the 10th house of career, status and *karma*. (Six months into the Moon *Dasha* he moved from Vienna to Munich, and three months after that he was a soldier fighting for Germany.) All planets representing the physical, mental, and emotional characteristics are also in fire signs. From ages 25 to 29, during World War I he was a dispatch runner who took messages back and forth between the command staff in the rear and the fighting units near the battlefield. Always ready to volunteer for dangerous assignments, he often narrowly missed being hit by deadly explosions and was considered very fortunate to escape life threatening injury during his four years at the front. He recovered fully from war injuries sustained on two occasions: **1)** a grenade splinter in his leg (Oct. 7, 1916); and **2)** temporary damage to his eyes and vocal cords (Oct. 13, 1918). The 3rd house concerns short distance travel and travel on foot. It contains three planets (Moon, Jupiter, and Ketu). His war injury to his leg in Oct. 1916 occurred during his Moon-Jupiter *Dasha/Bhukti*. Jupiter is lord of the 6th house of conflict and health issues. The second injury in Oct. 1918 occurred during his Moon-Saturn *Dasha/Bhukti* (See below) and coincided with the Spanish Flu Pandemic that brought high fatality rates, especially in Germany and Austria. This pandemic targeted healthy young adults, but Hitler survived through that treacherous period.[243]

In 1932 (at age 42) he put in 24-hour workdays for weeks on end. He was considered "astonishingly brave." But as the assassination attempts on him grew in number (17 in all, most of them from Nov. 1939 onward), he demanded more and more security guards as well as personal physicians. His food and drink were usually tested for poisons. Hitler's physical health and safety were generally strong due to Ascendant lord Venus in a *Kendra* aspecting its own house and well placed in the *Navamsha* Ascendant. However, in both instances Venus is in the sign of Mars and in close contact to Mars, within one degree – in *Graha yuddha* (Planetary War), to be discussed below. Afflictions to Mars agitate the already *rajasic* (passionate) Venus, lord of the Ascendant. *Karakas* for physical strength and stamina are Sun and Mars. Sun at 8:29 Aries is within 1½ degrees of the maximum degree of exaltation at 10 Aries, bestowing excellent physical health and stamina as well as leadership ability. A planet will show exceptional qualities (related to what it signifies) when it is within such close range of its exact exaltation degree and approaching it. Also in Aries is Mars, ruler of the sign. In post-WWI years he sought very little physical exercise other than modest daily walks, but his basic constitution was excellent, especially up to the spring of 1942.

[243] In 1918-1919 a worldwide flu pandemic (also called the Spanish Flu) killed up to 50 million people, or around 3% of the world's population (then 1.8 billion). Most victims were healthy young adults. Some 500 million people, or 28% of the population, were infected. Germany and Austria had more fatalities than France and Britain, which may have been factors in WWI. The pandemic lasted from June 1917 to December 1920, and some consider it was worse than the Black Death of the 14th century in Europe.

Eyesight is ruled by Sun, Moon, and Venus – all strong in the chart; thus he made a full recovery from temporary blindness from a mustard gas attack on Oct. 13, 1918. At the time, malefics Mars and Rahu were transiting his 2nd house of eyesight, which bears the weakness of house lord Mars. Tr. Saturn in Leo closely aspected his Ascendant and he was in **Moon-Saturn** *Dasha/Bhukti* (Jan. 25, 1918 to Aug. 27, 1919). Natal Saturn is in the 8th house from natal Moon, indicating likely emotional loss during a period involving both Moon and Saturn, and difficulties pertaining to 3rd house matters, such as vocal chords. This is because natal Moon is situated in the 3rd house from Libra Ascendant.

Voice: The 3rd house and its lord (Jupiter) rule over the voice, as does Mercury as *karaka*. Mercury is also *Nadi Yogakaraka* for Libra Ascendant, as is the Moon, giving added power. In **Moon-Mercury** *Dasha/Bhukti* (Aug. 27, 1919 to Jan. 25, 1921), Hitler gave his very first public speeches at meetings of the National Socialist Party, later known as the Nazi party. From that time on he started to become known as a political orator: **1)** Oct. 6, 1919 (first-ever public talk, lasting 30 minutes); **2)** Feb. 24, 1920 (delivery of the party's 25 Point program). Natal Mercury in Aries is combust a strong Sun, within five degrees. Thus it can become overheated and feverish, but also inspired, lit by the fires of imagination. Mercury is also afflicted in the *Navamsha* chart by aspects from malefic planets: Mars, Rahu, and Ketu. Natal Jupiter is strong in its own sign of Sagittarius, though it suffers from contacts to Ketu and Rahu. When Hitler suffered injury to his vocal chords during the Oct. 1918 gas attack, it took him several weeks to learn to speak again. Mars is lord of the 2nd house of speech and is also combust the Sun. It suffers from being overheated by the Sun and overwhelmed by Venus in Planetary War. While delivering political speeches in later years, he pushed his voice, giving it a strange, high-pitched quality. But he was also famous for his fiery oratory, stirring crowds into a frenzy. His speeches averaged 2 1/2 to 3 hours, and were geared towards converting the audience to his ideas. He carefully memorized and rehearsed his speeches for maximum dramatic impact. In private conversation he delivered monologues that were often tirades, as if rehearsing for his next public speech.

Dietary habits: Hitler was a non-smoker and a vegetarian from Sept. 1931, one year into his Rahu *Dasha*. From the injury to his vocal chords in Oct. 1918 he could not tolerate food served at hot temperatures; thus all his cooked food was served lukewarm. He gave up most alcohol soon after his nine month prison term in 1924, but he occasionally drank beer or wine with a meal. Despite some biographical commentaries to the contrary, Hitler was not a complete teetotaler and was not opposed to alcohol, according to his personal valet.[244] But his lifelong taste for sweets was so extreme that it caused him many dental problems, even rotting teeth as well as major dietary imbalance. The 2nd house lord of food intake (also, tongue and teeth) is Mars, which is very weakened in the chart, as noted earlier, and overwhelmed by Venus's influence. Venus is sweet (taste), as is Jupiter, which aspects Mars and all planets in Aries. This caused him to gain weight in later years, though with Venus's natural attention to appearance he worked to keep his

---

[244] Source: Heinz Linge, ***With Hitler to the End: The Memoirs of Hitler's Valet***, 1980, 1st English language edition 2009, p. 57. Hitler was strongly opposed to eating meat and smoking tobacco. Due to his vocal chords he could not tolerate tobacco smoke around him. (Heinz Linge, b. March 23, 1913, d. March 9, 1980. Linge served as Hitler's bodyguard from 1933, household staff from 1935, and personal valet 1939 to 1945.) Some confusion on Hitler's teetotaler status, among other items, may come from Hitler himself, in the process of creating his desired image both for his public and for the Tax Collectors, who knew his spending far exceeded his income.

weight in check. Though he did not seek much physical exercise, he avoided beer completely after 1943. A radical reduction of his sweet intake would have modified his hot temperament and improved his overall health and digestion, changing his blood chemistry. In Ayurveda, excess sweet consumption ferments and acts like alcohol, exacerbating a fiery nature – which can tend to lose control emotionally.

Severe decline in physical health: This came from 1942 onward, as Germany was losing the war. From May 15, 1942 for the next three years, he was in Rahu-Venus *Dasha/Bhukti*. The sub-period of the Ascendant lord brought into sharp focus all of Venus's ability to outlast his enemy as well as Venus's excesses, the passionate attachment to stated goals, and with Rahu – a further expansion of any existing fanaticism. **Rahu-Venus** was an especially challenging period for a wartime leader due to Venus's conquest over Mars, planet of victory. (See upcoming section: *Venus and Mars at war.*) In those years he was severely sleep-deprived and suffered from grueling headaches, dizziness and insomnia. Four planets in Aries place focus on the head, especially as one of them is the Ascendant lord. The worsening war conditions probably contributed most to his severe decline in health, rather than some of the diseases attributed to him, including Parkinson's or even syphilis – which Hitler dreaded, as he considered it "a Jewish disease." As a hypochondriac, over the years he was fearful that he had syphilis, and despite medical evidence to the contrary he continued to seek treatment for it. In late 1942 his left hand began trembling and sometimes his left leg as well. Symptoms of muscular tremors, such as Parkinson's disease can come from an excess of fire in the body, though they are more often associated with excess air (vata). On Jan. 16, 1945 Hitler and his inner circle moved to their Berlin bunker to escape the bombing raids. Then as enemy troops approached their bunker, Hitler died at age 56 from a self-inflicted gunshot wound to the right temple, while Eva Braun swallowed cyanide. The two bodies were wrapped in blankets and immediately cremated, according to Hitler's instructions. Among witnesses to the burning of the bodies were Heinz Linge, his personal valet, Joseph Goebbels and Martin Bormann.

Drug use: Despite an aspect to 2nd house lord Mars from disciplinary Saturn, this Mars was impaired. Hitler used a wide range of drugs and medicines, especially as Germany's war traumas increased. Among his various doctors, Hitler preferred Dr. Morrell, whom he met in 1936 and who soon became a part of Hitler's inner circle. Dr. Morrell used treatments he claimed were holistic, unconventional, and alternative, though he did not often say what they contained, and opinions vary as to his correct use of medicines. Though generally healthy up to the spring of 1942, Hitler employed many drugs and medicines to keep himself energized and functioning, a pharmacopeia rivaling that of John F. Kennedy.

> "Over 70 'medicines' were given to Hitler by his doctor, Theodore Morell, widely known as a quack. Nick-named 'Meister-Jabber' by Hermann Goering, Dr. Morell recorded in his medical diaries 73 medications he administered to his famous patient, including a wide array of sedatives, hypnotics, stimulants, tonics, vitamins, and hormones. Included in the pharmacopoeia were Cardiazol, a cardiac stimulant, several variants of belladonna, barbiturates; cortisone; Orchikrin, a combination of male hormones; an extract of bull testes; hormones from the female placenta; an array of powerful narcotics; and laxatives...Morrell daily gave his own golden Vitamultin tablets to Hitler.... [containing] both caffeine and Pervitin... a form of amphetamine. He also regularly injected Hitler with ... Percodan, a narcotic of equivalent strength to morphine, for his abdominal pains."
>
> **Jerrold M. Post, M.D. and Robert S. Robins**, *When Illness Strikes the Leader*, 1993, p. 70.

After the assassination attempt on Hitler July 20, 1944, he also began receiving daily doses of intranasal cocaine for his chronic sinusitis, and this was to be the most powerful drug he used. (Cocaine was commonly used for medicinal purposes in Germany at this time.) From then on until his death some nine months later, Hitler increased his intake of cocaine, amphetamines and other stimulants. This probably also increased his outbursts of rage, irritability, distractibility, paranoia, intermittent euphoria, and erratic decision-making. As lord of the 2nd house of intake of food and drink, Mars tries to gain power and to galvanize his physical powers, but from such a weak position it can be excessively aggressive. (For more astrological details on the July 20, 1944 event, see the upcoming section on *Assassination attempts.*)

Appearance: The 2nd house also rules the face. Due to his poor teeth, he avoided smiling in public. Dark-haired, with medium height and build, his intense, flashing eyes and his mustache were his most notable features. Hitler was meticulous in his concern for his physical appearance. With Venus in the sign of militant Mars in both birth chart and *Navamsha*, he often wore the same style of suit or uniform during his leadership years. He was supremely confident of his popularity and of the powerful effect he could create on his public. He devoted much energy to his every public appearance. Venus cares about appearance and attractability, while natal Moon's close contact to Jupiter in both the birth and *Navamsha* charts brings buoyant optimism.

Fuel for discontent – Venus and Mars at war: The dominance of Venus over Mars and over the whole chart provides much underlying fuel for discontent, though it gives good physical health and stamina, with Venus as Ascendant lord. Since Mars is the classic planet of discontent, it delivers the most trouble when it *loses* a Planetary War. Meanwhile, Ascendant lord **Venus in Aries can be very aggressive in its pursuit of some ideal of harmony**. In this case, it was an ideal of the supremacy of the Aryan race and an expanded Germany, minus certain ethnic groups that were considered offensive. Venus can become very war-like here through taking on some of the energy of a vanquished Mars, sign lord of Aries – and a sore loser in *Graha yuddha*. Mars in this condition often feels wronged and is further catalyzed by the Moon (the nature of the mind) in righteous Sagittarius contacting Jupiter, Rahu and Ketu. This exaggerates an already self-confident and optimistic nature: a perfect setup for megalomania.

In the **Planetary War** between Venus and Mars, Venus Retrograde is at 24:23 Aries, Mars at 24:04 Aries. To qualify for *Graha yuddha*, the planets must be within one degree of each other in celestial longitude, even if in adjacent signs. Venus and Mars are located in his 7th house in Aries in the Venus-ruled constellation of *Bharani*. Venus wins primarily by its greater effulgence, but in this case Venus also happens to be situated further north by celestial longitude, declination and latitude. But Venus would win due to its effulgence alone, even if Mars were further north in longitude or declination.

Since the sign lord of Aries is destroyed in *Graha yuddha*, it drags down all four planets in Aries, even as Venus tries to lift them up and carry them all to victory. This creates enormous disequilibrium for the life and for the country that has this chart as a temporary national chart as long as Hitler is the country's leader. The two major *karakas* of military leadership, Sun and Mars, are both in Aries, with Mars combust and Sun approaching its maximum exaltation. If there were no detracting factors, we would expect this Sun to perform brilliantly. And it can do so for a while, though it suffers eventual defeat with its sign lord in defeat. As with combustion, creating a problem for the combust planet, the vanquished planet in *Graha yuddha* may well overreact, most especially Mars. It is fighting for its life due to Venus's overwhelming dominance and brilliance, temporarily hiding the presence of Mars in the heavens, which is already combust, or hidden by the Sun's brilliance. Mars also has extra heat to its nature being so close to a strong Sun. The effect is that the person with combust Mars also in *Graha yuddha* is always flexing the muscles of Mars to make sure he is there and that others recognize him as well. With motivational dynamics such as these, an overheated body and mind are predictable in Vedic terms. Further, Mars rules the blood; and Hitler used leeches regularly to purify his blood. His own assumption was that it was weakened due to a troubled lineage and in-breeding.

Hitler's four planets in Aries are affected not only by Venus's victory over Mars, but by her ownership of the *nakshatra* where house lord Mars resides. According to *Nadi* principles, Venus becomes the ultimate governing factor for 7 out of 12 houses, since a *nakshatra* lord can dominate over a sign lord. Seven house lords are situated in Venus-owned *nakshatras*: Moon, Mars, Venus, Jupiter and Ketu. Unfortunately, *Nadi* principles also consider Venus debilitated when situated in a *nakshatra* it rules, since it exaggerates the passionate, action-oriented *(rajasic)* qualities.

A *rajasic* Venus is overly attached to action, and with so many masculine, fiery planets the action is likely to be military or militant. The victory of *Lagna* lord Venus in *Bharani nakshatra* in the 7th house also accounts for Hitler's over-involvement in every aspect of World War II. It explains why his fortunes as well as those of Germany would go down in **Rahu-Venus** *Dasha/ Bhukti* (May 15, 1942 – May 15, 1945). Germany was largely victorious in WWII up through spring 1942. Hitler's joint suicide with Eva Braun, his bride of one day, occurred on April 30, 1945.

The steepest rise and fall of Hitler's personal and political fortunes occurred during his **18-year Rahu *Dasha***, from Oct. 26, 1930. (His *Navamsha* Rahu is in Taurus, considered by some Vedic astrologers to be in its sign of exaltation.) By contrast, Hitler held no overt political position of power during his **7-year Mars *Dasha*** (from Oct. 26, 1923), though his influence gradually spread. Shortly after the start of it he made an unsuccessful attempt to overthrow Bavaria's Republican government – the famous Beer Hall Putsch Nov. 8 - 9, 1923. Hitler drove to the event in his recently acquired red Mercedes, a luxury he could ill afford.[245] Sentenced to five years in prison, he only served from April 1 to Dec. 20, 1924. There he wrote Volume one of *Mein Kampf*, intended initially only for Nazi party members. The party claimed it was an immediate best seller; but Hitler biographer Joachim Fest says the book was not a success in the first few years and remained mostly unread. Even so, he began writing Volume two in 1925. Three planets are situated in Hitler's 3rd house of written communication (Moon, Jupiter, and Ketu). When *Mein Kampf* (Vol. 1) was published on July 18, 1925, he was 3 ½ months into his Mars-Jupiter *Dasha/ Bhukti*. Volume two was published on Dec. 10, 1926 during Mars-Saturn *Dasha/Bhukti*. Jupiter reflects his ongoing optimism and Saturn his persistence:

> "In daily life the so-called genius requires a special cause, indeed, often a positive impetus to make him shine... The hammer-stroke of Fate which throws one man to the ground suddenly strikes steel in another."
>
> **Adolf Hitler**, *Mein Kampf*, p. 293.

---

[245] Source: Joachim C. Fest, ***Hitler***, 1973; English translation 1974, p. 182-183; p. 251. There are conflicting accounts about what Hitler actually earned from *Mein Kampf*, some claiming that the royalties from the books were so large that the tax on them amounted to $8 million USD by 1933. But Fest's biography is considered authoritative, and he says that Hitler accumulated debts from heavy overspending through the 1920s and these were waived when he became the Führer in 1933. Again, confusion arises due to the very active German propaganda machine as well as Hitler's own tendency to mythologize himself.

Volume one sold nearly 10,000 copies in the first year but by 1928 dwindled to 3,015 copies for the two-volume set. However, by late 1930, with the start of Hitler's more favorable Rahu *Dasha*, the NSDAP rose in popularity and his book sales increased: 1930: 54,086 copies; 1931: 50,808 copies; 1932: 90,351 copies; 1933: over 200,000 copies.[246] The Nazi party alleged his total book sales in 1943 were 9.8 million copies, but this is probably exaggerated and sales figures tend to vary widely. In any case, many free copies were distributed each year, including to newly married couples and to soldiers at the front (1933-1945).

Hitler's publishing income was said to enable him to buy his mountain home in Obersalzberg, though his purchases during Mars *Dasha* far exceeded his income, including a six-seater open red Mercedes, bought in 1927 for 20,000 marks. As the 2nd lord of income, Mars destroyed in *Graha Yuddha* reflects the poor state of Hitler's finances during its *Dasha,* even though he continued to live well, mainly due to the protection of Venus. But during these years he wrote to the Tax Collectors of his inability to pay his taxes, and how he needed his red Mercedes for his political work. He also loved technology and public display. As the victor in the Planetary War, Venus also rules over automobiles; red is the color of Mars and the Sun, and all three planets are in fiery Aries.

Venus - planet of charisma: As the Ascendant lord and key *Dharma* planet for the chart, Venus encapsulates what this life is about. More powerful when situated in a *Kendra* and aspecting its own house, Venus brings a lot of personal command to this chart. It is a planet of charisma, in large part due to its ability to magnetize people when it is strong in the chart. Venus rules over politics and is considered to have invented it. Thus Venus is crucial for a politician. The attraction quotient is strong enough to overcome mere passing public interest. It brings in a component of sexual magnetism that often applies to male political leaders. A sub-topic of charisma, no less interesting to explore, is charisma that becomes tainted by history. In Hitler's case, he was a megalomaniac who loved to hypnotize others, especially through his oratory. But if anyone crossed him or disagreed with him, he became extremely angry. His volcanic eruptions were infamous, creating a cadre of advisors who avoided displeasing or angering him. Hitler did not permit himself any doubts, and had an exaggerated sense of his own infallibility and success.

Venus in Aries in the 7th house is aspected or contacted by five other natal planets: Sun, Mercury, Mars, Jupiter, and Saturn. Mars, the lord of the 7th house is greatly weakened by its defeat in a *Graha yuddha* with Venus. Mars is a very poor loser in *Graha yuddha,* railing against its loser status and causing much havoc. Meanwhile, Venus gains innate strength by being in its own *nakshatra,* and as we noted earlier is *Jiva,* or *nakshatra* lord of five planets: Moon, Mars, Jupiter, Venus, and Ketu. Consequently, Venus rules over the affairs of seven houses: Houses 1, 2, 3, 6, 7, 8, and 10. All four planets in Aries receive an aspect from classic benefic Jupiter in Sagittarius. But as ruler of the 6th house of conflict, **Jupiter** becomes an active temporal malefic and doubly so as Jupiter contacts Ketu in the same sign. It gives some protection but also excess - in this case an excess of enemies, as Jupiter rules the 6th house of enemies and 3rd house of neighbors. Jupiter also brings the desire to initiate action, a 3rd house matter.

Venus and Mars together in a fire sign are very sexually oriented, though frustrated by contact with neutral and asexual Mercury. Further emphasis comes from *Navamsha* Venus-Mars

---

[246] Joachim C. Fest, ***Hitler***, 1973; English translation 1974, p. 251, See also Note #35 on p. 251 of Fest's *Hitler*.

in Scorpio (also ruled by Mars), contacting Mercury, as well as Rahu and Ketu. Diminished by being combust and losing the Planetary War, Mars is bitter in defeat. As a planet of survival, Mars remains discontent and dangerous, especially when in an angle of the chart and in its own sign in both birth chart and *Navamsha*. It perceives a battle to be fought and is determined to win it. As with any prominent Mars, one is searching for the battle or the competition. Following Germany's bitter defeat in World War I, Hitler was among those Germans who were morbidly affected by that defeat. And this was his adopted country, to which he had moved in May 1913.[247] He personified the German national who was determined to seek revenge. As he lay recuperating in the hospital from temporary blindness and damage to his vocal chords from a gas attack, this was his reaction, upon learning of the German defeat in WWI, Nov. 11, 1918:

> "The more I tried to achieve clarity on the monstrous event in this hour, the more the shame of indignation and disgrace burned my brow. What was all the pain in my eyes compared to this misery? There followed terrible days and even worse nights — I knew that all was lost. Only fools, liars, and criminals could hope in the mercy of the enemy. In these nights hatred grew in me, hatred for those responsible for this deed."
>
> **Adolf Hitler**, *Mein Kampf*, 1925-1926, pp. 204-206.

Years later, he congratulated himself for what he had achieved following the Treaty of Versailles in 1919:

> "After fifteen years of work I have achieved, as a common German soldier and merely with my fanatical will power, the unity of the German nation, and have freed it from the death sentence of Versailles."
>
> **Adolf Hitler**, *Proclamation to the troops on taking over the leadership of the German armed forces*, Dec. 21, 1941.

Artistic ambitions thwarted: Hitler was twice rejected from the Vienna Academy of Fine Arts, in Oct. 1907 and in Oct. 1908, both shocking events for him as he was very confident of his artistic talents and unwilling to admit otherwise to anyone. As the planet ruling over the arts, **Venus** is the chief significator in the chart for success in this field. He was born early in the 20-year **Venus *Dasha***, and the *Dasha* ended when he was 18½ years old, on Oct. 26, 1907. During the same month he applied to the Vienna Academy of Fine Arts for the first time. Neither year was auspicious for someone intending to start an artistic career. Oct. 1907 marked the tail end of a 20-year *Dasha*, with its waning energy indicative of endings rather than beginnings. Also, this natal Venus is not well indicated for an artistic career. Its tendency is to be more martial in the fire signs, especially in Aries. Issues of aesthetics become less relevant than issues of victory and mastery over one's opponent. Venus has some extra power here as *Atmakaraka*, planet at the highest degree of celestial longitude. But its location in the sign of Mars in both birth chart and *Navamsha* chart promotes more aggressive life themes, such as a military and/or political life. Venus rules over the arts, but also politics.

---

[247] His strong interest in Pan Germanism had developed from his teens. Two days after Germany entered the war on Aug. 1, 1914 Hitler enlisted, having sought special dispensation to fight for Germany. His official German citizenship came on Feb. 25, 1932. Rahu *Dasha* began for him Oct. 26, 1930. (Rahu is the foreigner. In the birth chart it is situated in House 9 of foreign travel, and in the *Navamsha* House 7 of foreign residence.)

Hitler's second unsuccessful application to the Vienna Academy of Fine Arts in Oct. 1908 occurred one year into his 6-year **Sun *Dasha***. The Sun is better situated for a military career than an artistic one, but during its *Dasha* it suffers from being close to a weakened Mars, which rules the sign, and from being in the 8th house in the *Navamsha* chart. Hitler would continue to avoid military service in Austria during his entire Sun *Dasha*. Artistically, he might have fared better at the time by focusing on military subjects or monuments, or working with metals, hot materials, or anything involving a fiery process. Instead, he painted landscapes, mostly in watercolors. Hitler wrote that the Fine Arts academy that rejected him recommended he enter architecture school. But this was probably fabricated, as his drawings were considered poor the first year he applied and his application was rejected outright the second year. He also lacked the requisite diploma from secondary school that would qualify him to apply to architecture school.

Desire to create his own version of a harmonious world: Libra is concerned with harmony, equilibrium, and diplomacy, but here it is the Ascendant of one of history's most notorious and violent dictators. And though not an advocate of diplomacy, he was very artful (also Libra) in promoting his own version of a harmonious world. He was also relentless in his desire to create such a world. He was a man of action, not contemplation, and he pursued his convoluted ideas of German racial purity in a systematic way. His attempts to find fulfillment through artistic expression met with his father's strong objections early on and did not work out in the long run. His *Dharma* could be fulfilled only through some type of military or political activity, which he avoided until early Aug. 1914, at age 25, nine months into his Moon *Dasha*. Only then did he gain a sense of satisfaction and the successes he craved. The Libra Ascendant person may well create just the opposite of the harmony they seek. In Hitler's case this propensity is exacerbated by the *rajasic* condition of so many planets, including Venus.[248]

Fiery energy in the chart – desire to conquer the world: Seven out of nine planets in the chart are in fire signs (Aries and Sagittarius). All three fire signs express the royalty of the *Kshatriya* (warrior and politician), as well as being in odd (masculine) signs. So much masculine, fiery, and active energy can denote leadership, especially when this includes Ascendant lord (Venus), Sun and Moon. This energy also comes from the *Rajju yoga*, a planetary combination defined here by five planets in *Chara* (active) signs, and a *Chara* Ascendant. As a child, dreaming of heroic actions he would later perform, he liked to play military games in which he decided the moves and led the battles, similar to the young Winston Churchill, who insisted on being called "General" in childhood military games.

Five of the seven planets in fiery signs are located in Venus-owned *nakshatras*: Moon, Mars, Venus, Jupiter and Ketu. One is attached to passion and to action when the *rajasic* qualities of Venus are so featured. This combination alone can give compulsion for a very active life. As a counterweight, Saturn aspects the four planets in Aries and lends deep seriousness, consistency, and discipline to this life theme. The Mars-Saturn mutual aspects lend technical abilities, especially in his role as a military strategist.

---

[248] The *rajasic* condition of Venus (planet of the arts) in Hitler's chart also accounts for his tastes in music. His greatest passion was for the German operas of Richard Wagner (1813-1883) and also for the symphonies of Austrian composer Gustav Mahler (1860-1911), who in turn was strongly influenced by Wagner. Wagner popularized the Aryan myth in Germany and fostered Germanism as a form of national (and racist) superiority.

Life issues are increasingly intensified, the fewer houses planets occupy, notably three or less. In this case, all planets in the chart, excluding Rahu and Ketu (the shadowy planets), are in only three houses. This is called a ***Shoola yoga***, *shoola* meaning "thorn." (This same *yoga* occurs in Chart #27: Françoise Gilot, Chapter 8.) Add to this the dynamic of Hitler's *Graha yuddha* between Venus and Mars and you begin to understand his desire to conquer the world. Mars is the sign lord for the four planets in the 7th house, and when Mars loses a *Graha yuddha*, it does not accept defeat and may not know how to stop fighting, since it is the planet that rules over war. Furthermore, Mars in Aries and in a *Kendra* receives special distinction as a *Ruchaka Mahapurusha yoga*. The result is that the warrior planet gets a lot of attention, but at the same time it is highly afflicted. The cumulative combination of astrological factors helps to describe Hitler's ability to attract a public and to achieve results through what he called his **"fanatical will."**

The 7th house is also sometimes called "the house of war." It is where we deal with the other, in partnership, business or marriage, or as adversary. Classically it is also the battlefield, especially in worldly matters when assessing the chart of a national leader. Indeed, Hitler regarded Germany as "the world's battlefield." Mars in the 7th house brings allies but also enemies, and with so many planets in the 7th house, there is a huge interest in your partners or your enemies. With the Ascendant lord in the 7th house, one is inclined to go in search of one's partner. With four planets in the 7th house, including Ascendant lord Venus, Mars and an exalted Sun, there is an inordinate interest in projecting oneself out into the world in search of one's allies *and* enemies, all justified by the Libran sense of order and harmony. The aggression supplied by Aries tends to create a dynamic of impatience for harmony and order on command, deciding what that should be and going after it aggressively. But one's own version of a harmonious world can become perverted as one tries to force it on the rest of humanity. (Both Venus and Mars-ruled Ascendants must deal in some way with themes of disharmony or reclaiming harmony. When Venus or Mars rule the Ascendant, they also rule the more challenging 6th or 8th house.) Hitler's planetary combinations have the effect of demoting or diminishing a captain of war while simultaneously giving him command over impossibly large swaths of territory. Three planets in *Purva Ashadha* (Moon, Jupiter and Ketu) gave him "earlier victories."

> "Without consideration of traditions and prejudices, Germany must find the courage to gather our people and their strength for an advance along the road that will lead this people from its present restricted living space to new land and soil, and hence also free it from the danger of vanishing from the earth or of serving others as a slave nation."
>
> **Adolf Hitler**, *Mein Kampf, 1925-1926, p. 646.*

> "I shall give a propagandist cause for starting the war, never mind whether it be true or not. The victor shall not be asked later on whether he told the truth or not."
>
> **Adolf Hitler**. Quoted at The Nuremberg Trials (1945) by chief British prosecutor Sir Hartley Shawcross, in this opening speech (also describing "Hitler's gross disregard of international treaties and his intention of worldwide conquest").

Motivation – Sagittarian zeal and impaired Mars: The mental and emotional orientation for such an action-filled life comes from natal **Moon in Sagittarius** in the 3rd house, together with Jupiter and Ketu. Natal Moon on the Rahu-Ketu axis brings more emotional turbulence to the life, and more of it is involved in his motivations. This planetary combination in fiery Sagittarius is oriented towards finding the TRUTH. This is against the backdrop of already combustible astrological factors in Aries: Mercury, Sun, Mars, and Venus. Further, an afflicted

Moon in the 3rd house may be dissatisfied unless the life is action-filled and demands bravery and courage; but in close contact with Jupiter, Moon (the mind) is also self-confident and optimistic.

This may explain why Hitler entered the military and started to achieve his best career and personal results in the 10-year **Moon *Dasha***, from Oct. 26, 1913, especially with Moon-Jupiter very well placed in the *Navamsha* 10th house, and Moon being 10th lord in the birth chart. But his greatest triumphs were in the 18-year **Rahu *Dasha***, from Oct. 26, 1930, rising up in the first sub-period (Rahu-Rahu), and declining sharply in Rahu-Venus, from May 15, 1942. Rahu *Dasha* brings worldly activities and/or rewards, especially if Rahu is well placed in birth and *Navamsha* charts, in either *Kendra* or trinal houses. Rahu fulfills that requirement here. Venus sub-period brings extreme and decisive results, as it has overwhelming influence in the chart but is pulled down by defeated Mars, as Mars owns the sign where Venus resides.

The natural ethics of Jupiter in its own sign of Sagittarius can usually dominate over Ketu or Rahu. But we have noted that Venus brings much *rajasic* energy. Jupiter is in *Purva Ashadha nakshatra*, owned by Venus, and Jupiter aspects Venus. Mutual contacts of Jupiter and Venus often bring some form of excess, though natal Saturn provides an antidote. Saturn is Venus's great friend and *yogakaraka*, and aspects all four planets in Aries. Saturn provides powerful discipline to harness all the planets in fiery Aries, enabling Hitler to carry out his ambitious life plans, at least in the political and military arenas. From the time he joined the military in August 1914, he never stopped working towards his political and military ends: to defend and protect Germany. His shock at Germany's defeat in World War I changed his life. From Nov. 1918 this shock of defeat galvanized him and echoed his own defeated natal Mars.

Sagittarian motivation: Sagittarius is a sign of expansion, and applies to expansion both territorially and philosophically. It is the desire to expand one's horizons, or at the very least, one's sense of a larger universe, often through higher knowledge, higher education, or international travel. The same zeal that underlies Sagittarian search for the Truth, when misdirected, can also describe a search for vengeance. The **expansive nature of Sagittarius** desires to extend itself everywhere, especially in a trajectory outwards toward foreign places. Ideally, the *dharmic* motivation is to know more about the world, and its many cultures, languages, races, and beliefs in order to better understand it and interact with it. (Hitler did study other cultures and their authors, especially those of the American West, often in order to support his own ideas.) At their highest level of manifestation, key planets in Sagittarius seek to embrace everyone and everything, with the stated aims of making a better world. There is a quest for higher knowledge and of closeness to God. If operating at a lower, more negative level, a strong Sagittarius influence can spread its own beliefs at the expense of those of others. Since Jupiter rules Sagittarius, this type of expansiveness can be done with generosity and nobility, or at least with the appearance of it. With despotic leaders, nations, or empires, often their stated aims seem noble enough to inspire the populace, though the end results are quite different from what was initially stated. In the end, raw domination by one group, nation, or national leader over others may be exaggerated or minimized, depending on who writes the history.

For example, though Hitler's policy of *lebensraum* – expanded living space – has been vigorously condemned, most American history to date has presented its own expansionist policy as positive, justifiable, and even heroic. First officially enunciated by President James Polk in 1845 in enforcing The Monroe Doctrine of 1823, and later known as **American "Manifest Destiny"** – especially over the North and South American continents and their environs – the idea was that the U.S. had a duty and a right to extend its influence and territory throughout North America.[249]

Hitler was an expansionist in his philosophy. He greatly admired the American West and its "Manifest Destiny" and wanted to emulate it. Like so many others, Hitler romanticized America's expansion westward and all that went with it – including the messy business of having to exterminate the people who were there from previous millennia – in order to fully dominate the land and the continent. American colonials as well as British officers commonly referred to American Native Indians as "an execrable race" and "vermin" that needed to be removed, and that "had forfeited all claim to the rights of humanity."[250] Hitler had a similar philosophy in justifying the removal of undesirable people from Germany. In excess or denial, Sagittarius can find ways to justify unethical or dishonest expansiveness. In such cases, self-righteousness and dogmatism are typical Sagittarian traits, bound tightly together with religion and national identity and often resulting in a strong religious or nationalistic orientation. The Sagittarian warrior is unusually fierce, as he has the conviction God can only be on *his* side. Conservative Jupiter rules over Sagittarius. Dedicated to his priestly duties and rituals, at times he can overlook human levels of desire and emotion.

Hitler regarded himself as married to the Motherland, and spoke of himself as a savior of the German people. As mentioned earlier, his Sagittarian Moon in *Purva Ashadha nakshatra*, means **"the earlier victor."** Hitler fought in World War I and started World War II. Germany was winning both World Wars in the first 2 to 2½ years, but in both cases was ultimately crushed by overwhelming enemy forces.

Megalomania: Jupiter contacts to the Moon, especially the conjunction, can confer much optimism. But this optimism of Moon-Jupiter, when combined with Ketu, and opposite Rahu, is less benign due to its more extreme and turbulent nature. Moon and Jupiter occur together again in his *Navamsha* chart, though in Leo in the 10th house, an excellent position for career and public popularity. Hitler was supremely self-confident; it was his strength as well as his fatal flaw. He always believed his plans would work out with the help of what he called his "fanatical will." He also refused to see how they might *not* work out. Only in the last few months of the war did the German people begin to turn against him in larger numbers. By then they were living in misery due to so many bombings of German cities and factory sites. Most of Hitler's war effort went into external wars in order to extend the national "living space," and not enough into protecting German cities, and other civilian and military targets inside Germany.

---

[249] In an editorial written in the July/Aug. 1845 issue of *The United States Magazine and Democratic Review*, John L. O'Sullivan wrote that foreign powers opposed the U.S. annexation of Texas, then part of the sovereign state of Mexico, in order "to hinder the fulfillment of our manifest destiny to overspread the continent allotted by Providence for the free development of our yearly multiplying millions." The U.S. did annex Texas in 1845 and this led to the Mexican War, 1846-1848. Over 160 years later, with over 1000 military bases in some 130 countries, America's "Manifest Destiny" has continued unabated, justifying American domination and military involvement worldwide – also "pre-emptive wars" (illegal invasion and occupation of foreign countries), "enhanced interrogation" (torture), and "regime change" (overthrowing foreign governments). Despite the euphemisms, the American Empire, as well as Chinese and Russian totalitarian regimes, have relentlessly pursued the same "aggressive wars" and terror condemned and punished at the 1945 Nuremburg Trials. Other offenses at those Trials included violating the customs of warfare and committing crimes against humanity.

[250] **http://brneurosci.org/smallpox.html. http://www.nativeweb.org/pages/legal/amherst/lord_jeff.html. http://www.college.ucla.edu/webproject/micro12/webpages/indianssmallpox.html.**

Hitler was convinced of his own greatness. Not only was he supremely self-confident, but he hypnotized others into believing he was capable of anything. The German propaganda machine depicted him in the most glowing terms:

> "[He is]...the acme of German honor and purity; the Resurrection of the German family and home. He is the greatest architect of all time; the greatest military genius in all history. He has an inexhaustible fount of knowledge. He is a man of action and the creator of new social values. He is ... the paragon of all virtues."
>
> **Walter Langer,** *The Mind of Adolf Hitler: The Secret Wartime Report*, 1972, p. 53.

Hitler was even preoccupied with planning his own mausoleum, a great monument to be 700 feet high, and a kind of Mecca in Germany.

> "I know how to keep my hold on people after I have passed on. I shall be the Führer they look up at and go home to talk of and remember. My life shall not end in the mere form of death. It will, on the contrary, begin then."
>
> *Ibid.*, pp. 37-38. (Hitler quoted)

In the years leading up to his invasion of Poland in 1939, Hitler contacted Indian astrologer B.V. Raman, who had made some bold and accurate political predictions in the mid-1930s. Hitler asked Raman if he would win the war. When Raman wrote back from Bangalore that he would be defeated, Hitler ceased any further communication with Raman and burned any of his books on Vedic astrology in German translation or available in Germany. Raman shared this with family and close associates, but did not write about it in any of his many books. Nor did he share his astrological methods for making this prediction. But it is clear that within Hitler's sphere, any astrologer who was *not* predicting a German victory faced mortal danger.

More effects from an afflicted Mars – fighting to win but destined to lose: Lord of four natal planets in Aries, Mars is heavily afflicted in the chart.

> "When Mars is highly afflicted, the native may commit acts which may lead him to the scaffold."
>
> **B.V. Raman**, from his article "Astrology as an Aid to Medical Science," a 1991 lecture reprinted in *The Astrological Magazine*, Bangalore, India, Aug. 2007, p. 645.

Mars is born of fire and considered **god of War**. Hitler fought wars that were far grander than he could possibly win. Motivated by the expansionary combination of Moon-Jupiter-Ketu in Sagittarius in the 3rd house of courage, his warrior nature also derives from a powerful Venus in Aries, victorious over a heavily afflicted Mars also in Aries. As a planet of war, blood, courage, command, and competition, Mars wants victory above all. Its astrological condition echoes the fact that, in historical terms, any record of Hitler's heroism and military prowess is almost completely overshadowed by the memory of his losses and his bad deeds. Those are the only ones history remembers, and no one after 1945 wanted to be associated with his victories or any of his policies. His many supporters famously "evaporated."

Heated speech, anger and paranoia: Hitler's heated and dramatic delivery of speeches has been described, along with his damaged vocal chords and the feverish pitch of excitement his poltical speeches aroused in himself and his large audiences. Beleaguered Mars keeps trying to prove itself. As lord of the 2nd house of speech, defeated Mars is also likely to be angry, as Hitler was when things did not go his way, or when he heard news he disliked. The greater the stress, the more rigidly he held on to his views, raging at the objects of his wrath whenever possible and often frightening them a great deal. These were characteristics evident earlier in his career and greatly magnified with the use of drugs, from 1936 onwards and increasing through the war, especially after late July 1944. With at least 17 assassination attempts on him, his paranoia increased. When he caught anyone in a traitorous act, the suspected perpetrator was killed, along with their families and at times thousands of other associates. In April 1945, when his generals told him they were preparing to surrender, Hitler flew into a rage and accused them of betrayal and treason.

Drug use: See *Biographical summary.*

Hitler's lineage: See *Biographical summary.*

Partnerships with women: Since Mars loses the *Graha yuddha* and is also lord of the 7th house of partners, this brings generally unfortunate results for his life partners, or marriage and business partners. Most of his relationships involved women 20 to 30 years younger than himself, to whom he initially acted as "uncle" or chaperone. Hitler preferred women who were either apolitical or not involved in politics, without a strong intellect, and over whom he could exert his exclusive control, even while choosing to be with other women. His female partners were typically suppressed, overwhelmed, isolated and enslaved to his wishes, or possibly even murdered. His half-niece Geli Raubal and actress Renate Müller both died under suspicious circumstances. He had several girlfriends throughout his life, marrying for the first time at the request of his bride Eva Braun, one day prior to their joint suicide April 30,1945, and only two weeks prior to the end of his 3-year Rahu-Venus *Dasha.*

The sexual nature experiences some confusion and duality with defeated Mars as sign dispositor of Venus, planet of love and sexuality, and with Venus and Mars in contact with asexual Mercury. While imprisoned in 1924, Hitler was close to Rudolf Hess, a known transvestite. This has led to some speculation that Hitler had latent homosexual tendencies. Hess was Hitler's private secretary (Mercury again), transcribing and partially editing Hitler's *Mein Kampf.* Mercury is youth, and his lovers were considerably younger, replicating the relationship of his own parents, both in age gap and in close family ties. His first known partner was his half-niece.

Venus is the classical *karaka* (significator) for love matters, sexuality and marriage. Since it is retrograde in the sign of Mars, and contacting both Mercury and Mars in the birth chart and *Navamsha* chart, this Venus has some paradoxes. It desires physical passion, being closely associated with Mars, but creates some distance from the partner, either through speech or through the choice of partners much younger than himself. Saturn's aspects to Venus-Mars can bring perversion in sexual tastes, emphasized by the Mars-Saturn mutual aspect. Venus-Mars-Mercury contacts with Rahu-Ketu in the *Navamsha* chart bring further turbulence in love matters. Venus is victor, but defeated Mars owns the sign in which Venus resides. Thus Venus is pulled down as well, and cannot perform the usual Venusian functions. He had only one testicle, and a life-long fear of syphilis, which he considered a "Jewish disease." Though his doctors insisted he did not have it, he continued to request treatment.

Even so, the Venus factor drew women and crowds to him, and he was sexually magnetic in the way of many successful political leaders. During his leadership years, Hitler preferred to appear in public without a woman, so that German women would continue to regard him as their hero and savior.

Political allies: The 7th house rules over one's political alliances. As defeated 7th house lord and a planet of victory, Mars also represents Hitler's political allies. The weakness of Mars did not bode well for the ultimate military victory of Germany or its key allies, the Axis powers – Italy, Germany, and Japan. Italy was an ally since Oct. 1936; and Japan, from the signing of the Tripartite Treaty in Berlin Sept. 27, 1940. With the Tripartite Treaty, the intention was to keep the United States from entering the conflict in any arena. But Japan's attack on Pearl Harbor Dec. 7, 1941 did just that. Though enemies can also be associated with the 7th house, it is more appropriate in this chart to use the lord of the 6th house of conflict to represent **Hitler's enemies**. Jupiter is the 6th lord, and is strong in its own sign of Sagittarius in the 3rd house, aspecting 7 other planets. Only natal Saturn escapes its aspect, in the classical Parashari system. Therefore, enemies are stronger than allies, Jupiter being far stronger than Mars, lord of the 7th house.

Imprisonment, loss of power, and financial excesses during Mars *Dasha*: (See more details above under *Fuel for discontent – Venus & Mars at war.*) The *Dasha* of vanquished Mars – Oct. 26, 1923 to Oct. 26, 1930 – was generally a period of personal stagnation, with repeated setbacks and conflicts. Many of his competitors still dismissed him as "a talented agitator but not a politician." His party leadership was questioned and speeches by Nazi party leaders were banned for several years until spring 1927, when the ban imposed in Saxony and Bavaria was lifted. Volume one of *Mein Kampf* was published in 1925 and sold only 10,000 copies that year. Meanwhile, he overspent and tried to persuade a series of Tax Collectors he was not overspending and that each of his purchases (including a red Mercedes) was necessary in a life mostly devoted to political work. Mars as lord of the 2nd house of income loses the Planetary war to Venus – planet of luxuries, including vehicles.

Hitler's **Beer Hall Putsch** occurred Nov. 8-9, 1923. Here are some events leading up to it: In early 1923 Germany was unable to pay reparations. To force reparations payments, French and Belgian troops seized several towns in the Rhineland and occupied the mineral-rich Ruhr district (Jan. 1923 to Nov. 1924). Germany was forbidden to retaliate militarily by terms of the Treaty of Versailles, and on Jan. 21, 1923 German miners went on strike (for eight months) against the invasion. On Oct. 21, 1923, Bavaria and the Rhineland declared independence from Germany, which was in a potential state of dismemberment when Hitler and his followers attempted to overthrow the Bavarian government to avoid such a fate. He was tried Feb. 24, 1924 and sentenced to five years in prison, though imprisoned only nine months: April 1 to Dec. 20, 1924. In prison he was treated with lenience and even with admiration, as his actions – though illegal – were intended to restore German national unity. His quarters were comfortable and private and he was not required to fulfill work duties while a prison resident. He was also permitted to receive a stream of visitors (men and women) up to six hours daily and often copious gifts of food and flowers. By summer 1924 he had enough free time to begin work on his book. These details show some of the power of Venus to rescue Mars, even in defeat and in prison, and to bring more pleasant circumstances (flowers, women, sociability, good food and drink, leisure time – all Venusian). During Mars *Dasha* Hitler was mostly on the sidelines, including in semi-retirement at his newly rented home in the mountains.

Assassination attempts: Hitler's ability to survive in the midst of war and assassination attempts can be credited to a strong Ascendant lord. The sheer number of foiled assassination attempts – 17 – is another attribute of the weak Mars. Mars in the 7th house may cause great physical harm to the person, especially if natal Mars is powerful, as in the case of Robert Kennedy, Chart #2. Hitler was nearly assassinated **Nov. 8, 1939** at the Bürgerbräu Keller in Munich.[251] The most notable astrological components at the time were tr. Saturn at 3:24 Aries, and Ketu-Rahu at 5:21 Aries-Libra. The midpoint of the Saturn-Ketu-Rahu combination was almost exactly on Hitler's natal Ascendant at 4:22 Libra. If Saturn and Rahu were bad planets for Libra Ascendant, this could have been very dangerous for him. But they are excellent planets for the chart, even gaining *Nadi Yogakaraka* status. So instead of killing him, these planets gave him some added travel time due to weather conditions. He was comfortably in his limousine when the bomb went off. Tr. Mars was at 0:16 Aquarius, making no aspect to any of his natal planets. The *Dasha* period was Rahu-Mercury and the day was Wednesday, ruled by Mercury – a favorable planet for his birth chart.

Another assassination attempt occurred on **July 20, 1944.**[252] The event occurred within hours after a Solar eclipse at 4:17 Cancer, very close to the 10th house cusp of Hitler's natal chart, at 4:22 Cancer. Tr. Rahu was at 4:26 Cancer and tr. Sun at 4:29 Cancer. (Tr. Rahu is especially important to note as Rahu is *Dasha* lord.) Tr. Venus was combust in Cancer, and similarly important to note as his natal Ascendant lord as well as *Bhukti* lord. The day was Thursday, ruled by Jupiter, a problematic planet for Libra Ascendant. The *Dasha/Bhukti* was already dangerous, being Rahu-Venus, from May 15, 1942. As a reminder, natal Venus is tainted by its location in the sign of Aries, ruled by defeated Mars. From this event, Hitler suffered ruptured ear drums and injury to his right arm (both ruled by the 3rd house). Tr. Saturn at 11 Gemini was aspecting his 3rd house and three planets in it.

IMPORTANT TURNING POINTS IN HITLER'S MILITARY CAREER:

1. **Aug. 3, 1914, Munich, Germany – Joins military for the first time.** His 10-year **Moon *Dasha***, starting from Oct. 26, 1913, set the stage for this major life transition, and not Sun or Mars *Dasha* – as one might expect for military matters. But Moon is 10th house lord and *Nadi yogakaraka* for this chart. As 10th lord, a strong Moon can bring significant benefits of career and status to the life of this person, minus the frustrations of *Dashas* ruled by masculine planets too closely affected by the loser in the Planetary War. Natal Moon-Jupiter in Sagittarius blends the love for the military with the fervor of ideology. For an Austrian citizen with Pan-Germanic aspirations and no particular loyalty to Austria, being granted special permission to fight for Germany set the life purpose on fire.

   Since the start of Hitler's Moon *Dasha* nine months earlier, the first New Moon to occur in the heavens in Moon-ruled Cancer was on July 23, 1914 (in his 10th house of career and maximum visibility). This New Moon was at 6:43 Cancer, in *Pushya nakshatra*, ruled by Saturn,

[251] He began his speech at 8:00 PM, an hour earlier than scheduled, due to dense fog at Munich airport and the need to take an earlier train back to Berlin. He left the beer hall at 9:07 PM, and an explosive device detonated at 9:20 PM. Placed in a column directly behind the speaker's podium, it killed eight people and wounded 60 others, some seriously. Part of the ceiling even collapsed.

[252] Hitler's close associate Major Von Stauffenberg placed a briefcase with a bomb under the conference table inside Hitler's "Wolf's Lair"— his command post in Rastenburg, Prussia (now Poland). The bomb exploded at 12:42 PM, killing four officers and injuring Hitler's right arm and ear drums. Hitler later ordered the execution of around 5000 Germans, including Von Stauffenberg, who left shortly after placing the briefcase.

and opposite Jupiter debilitated in Capricorn, a very enthusiastic, but potentially misguided nationalistic combination. Tr. Jupiter was also in *Dhanishta*, *nakshatra* of "the wealthiest," hinting at how national loyalties could be corrupted by wealth and zeal for commercial advantage. This New Moon directly preceded the start of World War I in late July 1914, an event in turn preceded by the assassination on June 28, 1914 of Archduke Franz Ferdinand of Austria-Hungary and his wife Sophie, Duchess of Hohenberg, in Sarajevo. The Archduke was heir apparent to the Austro-Hungarian throne. Considered by most historians as the pivotal catalyst to WWI, the collapse of deep national ties once forged by royal bloodlines was soon to follow.[253]

Astrologically one could connect the gestation period from June 28, 1914 to the outbreak of the larger war one month later to the movements of tr. Venus and Mars. Both of them entered Venus-ruled *Purva Phalguni nakshatra* by July 20, 1914, giving overly *rajasic* or passionate qualities to this combination already destabilized on the Rahu-Ketu eclipse axis. Then, near midnight on August 3-4, 1914 transiting **Venus and Mars** entered a full four-day ***Graha yuddha*** in early Virgo in *Uttara Phalguni nakshatra*, ruled by the Sun. This continued the *rajasic* components and in a more quarrelsome sign. Mars again loses the battle, as in Hitler's natal chart, echoing the theme of a warrior fighting a battle he is determined to win but destined to lose. From Hitler's natal Libra Ascendant, this is located in the 12th house of hidden components and foreign residence – making him less aware of how unwinnable his wars and foreign adventures would be, after some initial victories. The transiting Venus-Mars *Graha yuddha* was a significant factor not only because Mars (planet of war) is more out of control when in *Graha Yuddha*, but also because it occurred prominently in his own natal chart. Transiting in Virgo, it was situated in the 10th house from natal Moon, then his *Dasha* lord, as Hitler's 10-year **Moon *Dasha*** began Oct. 26, 1913.

In general, **Saturn's transits** through Houses 9, 10 & 11 bring greater visibility. At the start of his military career on Aug. 3, 1914, tr. Saturn was in his 9th house at 5 degrees Gemini, in *Mrigashira nakshatra*, ruled by Mars. This was an important marker for the next Saturn return to this area of the zodiac, which occurred June 6, 1944, coinciding with D-Day. From this date, Germany faced the real possibility of defeat in WWII. (See item #6 below) Saturn is planet of *Karma*, and its return to either natal Saturn or its position during any important life event – can indicate times of reckoning.

At the start of his military career, tr. **Jupiter and Saturn** were **in mutual 6 - 8 house relationship,** the same pattern in his natal chart and the chart of Jan. 30, 1933, when he became Chancellor of Germany. This means no mutual classic aspect between these planets, and a conflict between the *Dharma* and *Karma* planets, even if out of aspect by a small orb. There is an inner and outer battle to achieve the *Dharmic* destiny, with obsession to overcome early defeats and meet expectations.

2. **Nov. 11, 1918, 11:00 AM, Compiègne, France (German/French border) – Cessation of hostilities on the Western Front. WWI ends.** Though his 10-year Moon *Dasha* was generally

---

[253] See material on World War I in Chapter 1: FIRE TO EARTH Mutation period (1901 to 1961).

favorable, the Moon-Saturn sub-period was the most difficult. This ran from Jan. 25, 1918 to Aug. 27, 1919. (The Treaty of Versailles was signed June 28, 1919. Hitler called its negative impact on Germany "a death sentence.") Tr. Saturn first reached its return position to the exact degree of Hitler's natal Saturn on Nov. 2, 1917 – the day the Balfour Declaration was signed, promising England's support of a Zionist Palestine, if the Zionists in turn would help ensure an Allied victory. This mutual promise had an enormous influence on the war's outcome, reversing the expected German victory. Hitler was also angry at the German politicians, "the November Criminals," who agreed to the surrender. After retrograding for 4 ½ months, tr. Saturn returned to the same degree in July 1918. It was a key turning point in Hitler's life, as his major war injuries occurred Oct. 13, 1918, shortly before the war's end. On Nov. 11, 1918, tr. Sun was debilitated at 25:21 Libra, tr. Venus combust at 22:24 Libra, opposite natal Venus and Mars at 24 Aries. The Aries-Libra axis is crucial for his life events, especially close to the exact degrees of natal planets.

3. **Nov. 8-9, 1923 Munich Beer Hall Putsch – Hitler's first insurrection fails and he is imprisoned.** On Oct. 26, 1923, Hitler entered the 7-year Mars *Dasha*, his most unfavorable *Dasha*, due to Mars losing the *Graha yuddha* to Venus. A Full Moon on Oct. 24,1923 at 7:23 Aries occurred near his natal Sun, Mercury, and Descendant, while the Putsch itself occurred within hours of a New Moon at 22:24 Libra, closely opposite his own Venus and Mars in *Graha yuddha* at 24 degrees Aries. Tr. Saturn and Mercury in Libra also opposed his natal Sun and Mercury in Aries. Saturn opposing the Sun challenges the status quo of the authority, though each was in its sign of exaltation. The New Moon in Libra was in his Ascendant, in Jupiter-ruled *Vishakha nakshatra*, on a Thursday (Jupiter's day). Jupiter is not favored as owner of the 6th house of conflict for Libra Ascendant. This was the first New Moon in his Mars *Dasha* – the waxing, still invisible Moon reflecting the premature actions of the future Führer, who lacked sufficient military and political support to carry off the coup. *Dasha* lord Mars was transiting 22 Virgo in his natal 12th house, a house of hidden activities. The Putsch was a failure, but Hitler celebrated this event every year on Nov. 8th at the same Beer Hall, and was nearly assassinated 16 years later on Nov. 8, 1939.

4. **Jan. 30, 1933, 11:15 AM, Berlin, Germany – Hitler becomes Chancellor of Germany. The Third Reich began from this moment.** This leap to power could not have happened during his 7-year Mars *Dasha*, which was detrimental to him due to his afflicted Mars. Hitler exited Mars *Dasha* and entered **Rahu *Dasha*** on Oct. 26, 1930. He became a German citizen sixteen months into his Rahu *Dasha*, on Feb. 25, 1932. As *Dasha* lord, **Rahu in transit** becomes more pivotal during its own *Dasha* period. In July 1932, when Hitler lost the General Election, tr. Rahu was in Aquarius in *Purva Bhadra nakshatra*, ruled by Jupiter (6th lord for Libra Ascendant chart). Rahu would be more beneficial to him when it transited *Shatabhisha nakshatra* (6:40 to 20:00 Aquarius), ruled by Rahu, a friend to Libra Ascendant. By late November 1932 tr. Rahu had entered the *Shatabhisha nakshatra,* where it remained through early Aug. 1933. On Jan. 30, 1933 tr. Rahu was at 16:47 Aquarius. A Solar eclipse on Jan. 26, 1933 at 12:38 Capricorn preceded Hitler's appointment by the reluctant President Hindenburg, who despised Hitler but was powerless by then to prevent his ascent to power. Hitler was absolute ruler of Germany for a full Jupiter cycle of 12 years. (On Jan. 16, 1945, he and his inner circle moved

permanently to their underground quarters in Berlin – "the Berlin bunker." German defeat was imminent when on March 13, 1945 tr. Jupiter reached the same exact degree as on Jan. 30, 1933.)

On Jan. 30, 1933 tr. Jupiter was at 29:33 Leo and tr. Saturn at 16:34 Capricorn, strong in its own sign. **Jupiter and Saturn** in **mutual 6 - 8 house relationship** repeat their mutual house positions in Hitler's natal chart, and the chart of his joining the military for the first time: Aug. 3, 1914 (see Item #1). Tr. Sun and Saturn within a few degrees of each other in Capricorn contrasts with the Sun-Saturn opposition Nov. 8-9, 1923 during the unsuccessful Beer Hall Putsch (see Item #3).

5. **Sept. 1, 1939, Poland – Germany starts WWII.** This occurred 2 ½ days after a Full Moon at 12:38 Aquarius, on a Friday, favorable for him, but with Venus closely combust the Sun in Leo, showing hubris, and in the sign of its enemy, the Sun, though in the Venus-owned *nakshatra, Purva Phalguni.* As the ruling planet in Hitler's chart, Venus is overly *rajasic*, a condition mirrored in this chart. Tr. Moon and Jupiter are close together in Pisces in *Uttara Bhadra nakshatra*, ruled by Saturn and considered fixed in its nature. That is, endeavors begun in a fixed *nakshatra* may endure. This Moon-Jupiter in Jupiter's sign echoes Hitler's birth Moon-Jupiter in Sagittarius. Tr. Saturn and Ketu close to 8 degrees Aries contacts Hitler's birth Sun and Mercury, similar to his natal Sun receiving an aspect from natal Saturn, and telling the story of self-discipline under duress while fighting off anyone else's version of authority. Tr. Mars exalted in Capricorn aspects five planets in this chart, and repeats in the *Navamsha* in Capricorn. Exalted Mars in Capricorn is considered one of the strongest possible positions of Mars: It has the best combination of physical strength plus intelligence and organizational abilities. British historians have praised Hitler's genius as a military strategist. Only towards the end of the war did he start making many military missteps – probably affected by the medications and drugs administered to him and unable to fully recover his strength and equilibrium in the aftermath of the assassination attempt July 20, 1944.
6. **June 6, 1944, Normandy coast – D-Day.** The mass landing of Allied troops on the Normandy coast was the largest seaborne invasion in history, extending for 24 days. Though there were significant losses on both sides, it marked the turning point in the war in favor of the Allies. Some of the same planetary patterns occurred at the start of Hitler's military career Aug. 3, 1914: notably – the **return of Saturn**, planet of *Karma* to the exact same degree: 5 Gemini, close to the 9th house cusp, and opposite Moon in *Purva Ashadha*, "the earlier victor." This is Hitler's natal Moon position, though tr. Moon was not opposite tr. Saturn until June 7-8, 1944. Moon-Saturn contacts were hard for him, as he disliked authority and only tolerated *being* the authority. Six weeks later Hitler barely escaped an assassination attempt, after which his health suffered a severe downturn and he became more dependent on his drugs and medicines. A Saturn return is a *karmic* return, and in this case, it is also approaching the natal 10th house position of Saturn, which foretells a possible fall from power.
7. **April 30, 1945, 3:30 PM, Berlin, Germany – Hitler & Eva Braun commit suicide.** It was a Monday, with Jupiter almost exactly rising – not the chart one would expect for a suicide, except that Jupiter is troublesome as 6th lord for the Libra Ascendant. Jupiter, planet of *Dharma*,

has a 12-year cycle. It was then at 24:44 Leo – having gone a full orbit around the Sun since Hitler became Chancellor. *Dharmic* results of his leadership were due. For extra emphasis, the Stationary Direct position of tr. Jupiter was 24:26 Leo on May 14, 1945. At the time of their joint suicide in the Berlin bunker, with enemy troops near the capitol, Leo was rising in *Purva Phalguni nakshatra* (ruled by Venus). Leo is the most royal of the fire signs. Tr. Moon was at 21:23 Scorpio, tr. Saturn and tr. Rahu in Gemini (closely opposite Hitler's natal Moon and Jupiter at 14 and 16 degrees Sagittarius, respectively).

On April 30, 1945 tr. Mercury (planet of communications) was defeated by Venus in *Graha Yuddha*. Weak and debilitated Mercury in watery Pisces on the 8th house cusp (from Leo Ascendant) reflects the amount of false information spread for many years about Hitler's death. Though they had conclusive dental evidence to the contrary, the Soviets who captured the Reichstag and the bunker denied his death. (In later years Hitler aides Otto Gunsche and Heinz Linge testified as witnesses to the event and Linge included his account in his 1980 book on Hitler.) Further contributing to the spread of false stories was Mercury's weakness as *Jiva* planet to transiting Moon, Venus and itself. Also, tr. Venus, Mars and Mercury in Pisces were in Hitler's 6th house of conflict in the sign of sacrifice and capitulation. Exalted Venus at 24:54 Pisces was close to its maximum degree of exaltation and in retrograde motion until May 6, 1945. Hitler was then in **Rahu-Venus** *Dasha/Bhukti*. He considered he had given everything to his country. The time had come to sacrifice himself, while encouraging his countrymen to continue fighting for a glorious Germany. But their unconditional surrender came soon after, on May 8, 1945. In his last recorded statement, Hitler urges German citizens to remain loyal to "the honor of the nation" above all else. One might have expected him to berate his military enemies, the Allied Powers. Instead, he continued to warn of the dangers of "International Jewry."[254]

Fall from power: Some Vedic astrologers at the time were predicting Germany's fall in World War II based on Hitler's 10th house Saturn. Saturn can fall far from grace when placed natally in the 10th house. It gives recognition and visibility for some time, but then can snatch it away. (U.S. President Richard Nixon had natal 10th house Saturn, as did the boxer Mohammad Ali.) When Saturn transits through the 10th house, it can bring the greatest amount of visibility. Saturn is the planet of *Karma* and the 10th house is the key house of *Karma*, especially in terms of public achievement, rank or position, recognition from the state, or whatever is most visible about a person. Natal Saturn in the 10th house may also put so much focus on ambition that it becomes an

---

[254] Hitler's Last Will and Political Testimony, 35½ hours prior to his suicide: Translated in the Office of United States Chief of Counsel for the Prosecution of Axis Criminality, *Nazi Conspiracy and Aggression*, Government Printing Office, Washington, 1946-1948, vol. VI, pp. 260-263. "Let them [the German people] be hard but never unjust, but above all let them never allow fear to influence their actions, and set the honor of the nation above everything in the world. Finally, let them be conscious of the fact that our task, that of continuing the building of a National Socialist State, represents the work of the coming centuries, which places every single person under an obligation always to serve the common interest and to subordinate his own advantage to this end. I demand of all Germans, all National Socialists, men, women and all the men of the Armed Forces, that they be faithful and obedient unto death to the new government and its President [Admiral Karl Dönitz, according to Hitler's wishes]. Above all I charge the leaders of the nation and those under them to scrupulous observance of the laws of race and to merciless opposition to the universal poisoner of all peoples, International Jewry." *Given in Berlin, this 29th day of April 1945, 4:00 A.M.* Adolf Hitler.

overwhelming issue and can bring about a fall. Hitler was born on a Saturday, adding to Saturn's power in the chart and its ability to pull him down.

In such an iconic chart, it is very important to study both natal and transiting Saturn, especially to its return position in the 10th house. This occurred for the first time on Nov. 2, 1917 – the exact date of the signing of the Balfour Declaration, in which England's promise to give Palestine to the Zionists insured that the Zionists would help the Allies deliver a resounding German defeat. After several months of retrograde motion, tr. Saturn returned to the same degree eight months later in July 1918, just prior to Hitler's injuries in WWI and the disgraceful end of the war for Germany. This was the raison d'être for the rest of his entire military career, just as the Pan-German movement inspired him to join the German military a few days into World War I. The second Saturn return to his natal Saturn came June 7-8, 1944, shortly after D-Day in World War II. (See items #2 and #6 above)

Hitler & the Third Reich – Legacy of genocide and defeat: Many nations and ethnic groups bear responsibility for racism and genocide, but just as many evade it, especially if they win the war.

> "There's a tendency for violence within all of us. There's even, I believe, a prehistoric desire for human sacrifice. You see it in all ancient cultures.... The Roman Coliseum used to be the cruelest place on earth....The real evil in the world is human cruelty. We all have a tendency for this that we're not willing to acknowledge....Human beings need to acknowledge human cruelty and renounce it.... All societies have a dream and a nightmare. ... [America's] nightmare has been ... our racism. We practically committed genocide on the people who were here – native Americans. We enslaved another race of people – the Africans. And then we dropped the Atom bomb on Asians. We would never have dropped that bomb in Europe, in my view. That proves the racism of it. That's the nightmare of America.... The dream of America is enunciated in the great speech by Martin Luther King: 'I have a dream.' ... It always remains a tension and a question as to which side of us, the good side or the bad side, will win out in the end. And I think that's true for every society."
>
> **Thomas Cahill**, American historian, on *Bill Moyers Journal*, Public Broadcasting – PBS, Nov. 9, 2007.[255]

Hitler's legacy as a genocidal dictator overshadows whatever good deeds and military brilliance he performed as "Germany's savior." His crushing defeat after an extraordinary rise to leadership is even more decisive to his **total loss of good reputation,** given that more atrocities are associated with him than with Allied nations, whose own atrocities were minimized in the writing of history. Victor nations often escape both vilification and accountability. As Hitler was well aware, hero status often bestows this type of immunity.

Hitler's chart indicates both the precipitous rise *and* crushing defeat of Germany during his twelve years of leadership, one complete Jupiter cycle. If natal Saturn in the 10th house indicates Hitler's likely defeat, defeated Mars in the 7th house indicates he would never stop fighting for his vision of a harmonious Germany (Venus), the adopted homeland he viewed as mightily wronged

[255] The Nazis were rightly vilified for their medical experimentation on Jews and non-Jews in the concentration camps. In her book ***Medical Apartheid: The Dark History of Medical Experimentation on Black Americans from Colonial Times to the Present*** (2007), Harriet Washington documents this long but little known history of ongoing unethical practices on Black Americans. However, most American perpetrators were never named or punished. Nor have such practices been completely terminated, according to Washington.

and in need of strength, honor and purification. Nor would he ever be exonerated or understood for any part of his actions. To defend anything about Hitler is generally regarded in the West as defending the indefensible, or "the Architect of Evil." He has become the accepted vessel for the world's outrage against most forms of violence, especially genocide.

In Vedic philosophy there is no evil or sin, only ignorance. If one commits violent deeds, one pays a *karmic* debt in this or future lifetimes. Such deeds are usually preceded by violent and ignorant thoughts. The birth chart of Adolf Hitler symbolizes not only his life and destiny, but that of Germany during the twelve years of his leadership. This was a nation initially struggling to survive, and later seeking to restore its former glory and rightful place in the world. In the process, it targeted and persecuted those identified as betraying the country from 1917 through 1945. The condition of Hitler's natal Mars and Saturn tends to erase any broader understanding of that reality, one likely to remain part of buried history. What is also buried and eviscerated is the very idea of a Germany that saw itself betrayed, attacked, and threatened. The leader of the Third Reich would be known mostly for killing Jews, long after the country itself was forgiven for such crimes.

Chart 29: Mahatma Gandhi

Birth data: Saturday, Oct. 2, 1869, 7:11:46 LMT, Porbandar, India, Long. 69E36 00, Lat. 21N38 00, *Lahiri ayanamsha*: -22:02:03, Class A birth data, from memory. "Three *Ghatis* and 12 *Palas* after sunrise." One *Ghati* = 24 minutes; one *pala* = 1/60th of a *ghati*, or 24 seconds. This is 76 minutes and 48 seconds after sunrise, which occurred at 5:54:58 AM LMT. Ascendant: 4:34 Libra.

Brief biographical summary: A lawyer, patriot, politician, and nation-builder, Mahatma Gandhi was above all a moral force. Millions of Indians followed him, insisting on calling him "Mahatma" (great soul) – despite his protests. His birth name was Mohandas Karamchand Gandhi, but Indian poet Rabindranath Tagore conferred this title on him in early 1915. Some controversy remains as to whether this sanctified title should have been given to a political figure with human failings, although masses of Indian people were eager to do so. Considered both a spiritual and political leader, one who held daily prayer meetings at dusk, he sought to foster peace and understanding between people of all races, creeds, and religions, especially between Hindus and Muslims in India. Gandhi's greatest achievements were in the field of social justice, especially civil rights, and rights for those whose creed, caste, or ethnic group was not in favor with the existing government. As an example, his ashram welcomed families of "untouchables," many of whose descendants had immigrated to South Africa.

> "The greatest fact in the story of man on earth is not his material achievements, the empires he has built and broken, but the growth of his soul from age to age in its search for truth and goodness. Those who take part in the adventure of the soul secure an enduring place in the history of human culture. Time has discredited heroes as easily as it has forgotten

	JuR		
	Mahatma Gandhi Oct. 2, 1869 7:11:46 LMT Porbandar, India		Ra Mo
Ke			
	Sa	As Me Ve Ma	Su

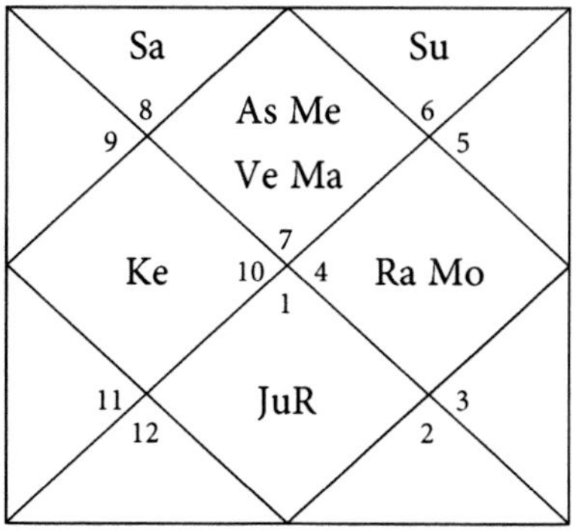

Sa Su
8 As Me 6
9 5
Ve Ma
7
Ke 10 4 Ra Mo
1
11 JuR 3
12 2

Planetary Information

As	04:34	Lib	Chitra
Su	16:53	Vir	Hasta
Mo	27:56	Can	Ashlesha
Ma	26:21	Lib	Vishakha
Me	11:44	Lib	Swati
Ju	28:08	Ari	Krittika
Ve	24:23	Lib	Vishakha
Sa	20:19	Sco	Jyeshta
Ra	12:08	Can	Pushya
Ke	12:08	Cap	Shravana

Mo	Ke	Ve Ma	Su
	Navamsha		
Sa Me			
JuR	As	Ra	

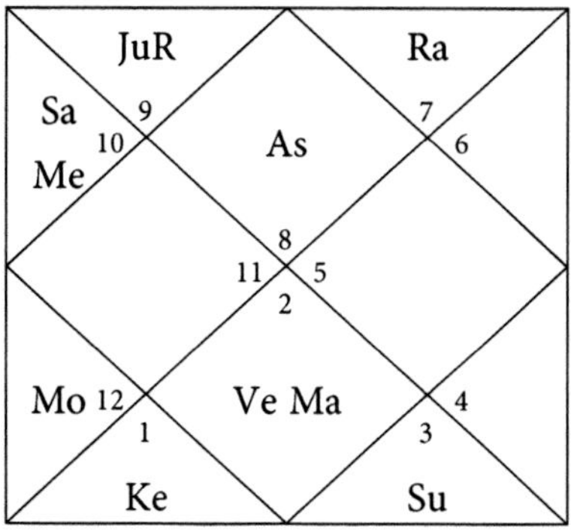

JuR Ra
Sa 9 7
10 As 6
Me
8
11 5
2
Mo 12 Ve Ma 4
1 3
Ke Su

Vimshottari Dashas

Me	May 20, 1855
Ke	May 19, 1872
Ve	May 20, 1879
Su	May 20, 1899
Mo	May 21, 1905
Ma	May 21, 1915
Ra	May 21, 1922
Ju	May 21, 1940
Sa	May 21, 1956

everyone else; but the saints remain. The greatness of Gandhi is more in his holy living than in his heroic struggles, in his insistence on the creative power of the soul and its life-giving quality at a time when the destructive forces seem to be in the ascendant."

S. Radhakrishnan (1888-1975), honoring the Mahatma on his 70th birthday, October 2, 1939.

Due to his relentless activism on behalf of Indians, initially in South Africa, then in his native country, he is credited with leading the movement that freed India from the shackles of British rule after several centuries. By 1915, on his return to India, Gandhi became convinced that British rule was enriching the Empire while impoverishing India, rendering its peoples politically and economically helpless. He vowed to change this and was uniquely equipped to do so. Though regarded by many as a saint – which left him open to vicious attacks as a political figure – he was a consummate politician, skilled lawyer and negotiator. He also correctly understood, earlier than most others, the strength of the Indian nationalist movement, along with the increasing unfeasibility of Britain retaining its hold on India. Britain was fatally sapped by each of the world wars.[256]

Gandhi's humility, tolerance, compassion and ability to forgive made possible his use of *Satyagraha* (passive resistance) and *ahimsa* (non-violence). This was crucial to his success and

[256] Gandhi had a sense of history, though not in the conventional sense. From 1919 onward, he understood the essence of what Queen Elizabeth remarked on years later at the American Bicentennial in Philadelphia, July 6,1976: **"We lost the American colonies because we lacked that statesmanship to know the right time, and the manner of yielding what is impossible to keep."** The same applies to Britain's reluctance to let go of India until after World War II. (Queen Elizabeth II is quoted in William L. Shirer, ***Gandhi: A Memoir***, 1979, p. 71.)

copied by other political movements, including the Civil Rights movement in the U.S. (1955-1968) led by Dr. Martin Luther King, Jr., as well as Nelson Mandela's anti-apartheid movement in South Africa.[257] By his own request, those who disagreed with, injured or betrayed Gandhi were always forgiven and never punished. To achieve his ends, both in South Africa and in India, he spent a total of 6½ years in prison. Meanwhile, he encouraged Indians to move toward self-improvement in many areas: in sanitary living habits, mutual cooperation, and knowing and defending their rights. From 1917 onward, Gandhi evolved in favor of Indians studying and learning their own languages. For a national language policy, he favored Hindi and Urdu (also called Hindustani).[258] He regarded language as a key tool in the struggle for India to free itself:

> "If we spend only half the effort we do in learning English in the learning of Indian languages, there will be born a new atmosphere in the country and a good measure of progress will be achieved..... The character of a people is evident in its language...."
>
> **Mohandas Gandhi**, 1909, member of the South African Delegation to a conference in London, England.

His legacy: When India gained its independence from Britain in August 1947, Gandhi's commanding presence loomed large, though he had been vehemently against partition of the country and was widely criticized for allowing it to occur. But over several decades – from his return to India in 1915 and his leadership of the Independence movement in April 1919 – his stature among Indians and his unique style of political activism contributed greatly towards a gradual conversion of British attitudes. This led to India's independence from Britain. In the process, Winston Churchill was his most formidable opponent in his own drive to preserve the British Empire.

Gandhi led the first nationwide campaign of civil disobedience and peaceful resistance. His desire was to lead a movement entirely based on "the truth" – *sat*, in Sanskrit. The movement would be known as "*Satyagraha.*" *Agraha* means insistence or persuading. Together the two Sanskrit words *Satya* and *Agraha* mean literally "insistence for truth." *Satyagraha* is also translated as "resting in the realm of truth," or also "those who hold to the truth." Gandhi's *Satyagraha* movement and philosophy are often regarded as his supreme achievements. He established the movement in 1907

---

[257] Mandela was born July 18, 1918, Umtata, S.A., time unknown. Recipient of the Nobel Peace Prize in 1993, he was the first black African to become a State President and regarded as one of the great moral and political leaders of his era. President of South Africa from 1994 to 1999, Mandela had become a leader of the African National Congress (ANC), a party founded in 1912 to fight injustice against all non-whites in S.A. This included the large population of immigrants from India arriving from 1860 onward. The ANC was banned for 30 years by the white South African government following the Sharpeville Massacre on March 21, 1960, when police fired on a crowd of unarmed blacks protesting apartheid Pass laws. Though Mandela had previously advocated non-violence, armed resistance to apartheid policies began from that date. It was a last resort against ongoing armed state repression of non-whites. For their armed resistance, the ANC was called "terrorist" by the S.A. govt. and only removed from USA terrorist lists in 2008. Mandela fought against apartheid his whole life, 27 years while in prison (Aug. 1962 to Feb. 1990). Before he was captured and imprisoned he was on the run for 14 months. He fought openly for his cause up to May 31, 1961 – when South Africa became a republic, separated from the British Commonwealth, and refused to break its apartheid policies, starting 33 years of international isolation. These policies were strongly in force in S.A. from 1948, and dominated ideologically from the time of the first Dutch settlements there in April 1652. The new nation – the Union of South Africa – came into being from legislation passed by the British Parliament on May 31, 1910.

[258] Hindi and Urdu are almost indistinguishable as spoken languages, but have different scripts, Urdu the Urdu script and Hindi the Devanagari (Sanskrit) script.

in South Africa. Ideally *Satyagraha* leads to the voluntary surrender of the opponent. Gandhi also headed many village improvement projects, including schools run by volunteers, drainage and cleanup projects, and modest clinics for medical treatment throughout India.

> "When I despair, I remember that all through history the way of truth and love has always won. There have been tyrants and murderers and for a time they seem invincible, but in the end, they always fall — think of it, always."
>
> Attributed to **Mohandas K. Gandhi**

Gandhi's philosophy and religion: Gandhi's philosophy centered around three concepts: ***Satya*** (truth), ***Ahimsa*** (non-violence), and ***Brahmacharya*** (abstinence). He grew up surrounded by the religious traditions of Hindus, Moslems, and Jains, who believed in *ahimsa*. Most of his family members were conservative and pious Hindus, and his mother was very influenced by the *Bhagavad Gita*. But Gandhi did not pay much attention to these influences until his second year in England, 1889-1890, where he was studying the law. He encountered some British Theosophists who asked him to teach them the *Bhaghavad Gita*. Thus began Gandhi's own lifetime study of this sacred work, of which he was to make his own translation later. It made a deeper impression on him through his contact with the West.

> "When a man ceases to have any attachment either for the objects of senses or for actions, and has renounced all thoughts of the world, he is said to have climbed to the heights of Yoga [union]. One should lift oneself by one's own efforts and should not degrade oneself; for one's own self is one's friend, and one's own self is one's enemy. One's own self is the friend of the soul by whom the lower self (consisting of the mind, senses and body) has been conquered; even so the very self of him who has not conquered his lower self behaves antagonistically like an enemy. The Supreme Spirit is rooted in the knowledge of the self-controlled man whose mind is perfectly serene in the midst of pairs of opposites, such as cold and heat, joy and sorrow, and honor and ignominy."
>
> ***Bhagavad Gita***, Ch. VI, v. 4-7. Translated by Jayadayal Goyandka, 2nd edition, 2002.

Gandhi became a devout if unorthodox Hindu who honored all religions. Even so, the *Bhagavad Gita* was to become the greatest single source of his inspiration. Though he read history, he preferred the *Mahabharata*, as he believed **"Truth transcends history."** He was also strongly influenced by the *New Testament* of the *Bible*, though he found the zealousness of Christian missionaries in India insulting, so determined were they to convert the local "heathens" to Christianity. Other influences came from authors Leo Tolstoy, Henry Thoreau (especially his essay on "Civil Disobedience"), and John Ruskin (notably his *Unto This Last*, which he read in 1904). From Ruskin, Gandhi learned about the moral dignity of manual labor and the beauty of community living on the basis of equality. In London he also began a serious study of vegetarianism.

Gandhi's study of the *Gita* and of the New Testament of the *Bible* forged the basis of his lifetime philosophy of non-violence and passive resistance, which he later renamed "*satyagraha*." Gandhi preferred the term ***satyagraha*** to "passive resistance" or "civil disobedience," though they are both associated closely with the concept. For Gandhi, living in the realm of Time was irrelevant compared to living in the realm of Truth, which he was always experiencing and reassessing, learning through trial and error. He thought that Truth should pervade all considerations of politics, ego, society and convention. Therefore, he did not define himself in the usual political terms as a pacifist, a socialist, etc. He looked to the pure, existing facts of life to make his decisions. Generally, he believed in non-violence, human freedom, equality and justice. If violence was sometimes necessary, then it is truthful to believe it. But he advised against Britain fighting the Nazis in World War II, and he had a radical suggestion for how to bring world attention to the genocide of the Jews.

> "What appalled him even more than the physical destruction of the Jews was the devastation of their self-respect. 'The Jews should have offered themselves to the butcher's knife,' he said. 'They should have thrown themselves into the sea from cliffs.... It would have aroused the world and the people of Germany.'"
>
> **M.K. Gandhi** quoted in George Woodcock, *Mohandas Gandhi*, 1971, p. 116.

Theme of self-restraint: Gandhi experimented with many levels of personal self-restraint and detachment, gradually simplifying his life and becoming an ascetic who lived in self-imposed poverty, especially as he became more involved as a leader of broad social movements. Taking a vow of *Brahmacharya* in June or July 1906, in part for birth control (though without prior consultation with his wife), he tried different forms of vegetarian diets to achieve this end more easily. He also fasted regularly and took periodic vows of silence, in one case for an entire year (1926), and for many years on Mondays – from midnight on Monday to midnight Tuesday. In Vedic tradition, an *upaya* (remedy) intended for Monday should begin at sunrise Monday and end at sunrise Tuesday. Though Gandhi was not using the purest Vedic timing, inadvertently he would have been placating his Moon, since the Moon is planetary ruler of Monday.

In the summer of 1914 he finally abandoned his European style of dress he had adopted since his days in England studying law, 1888 -1891. The same summer he achieved a big victory with the Indian Relief Act in South Africa, and vowed to live more and more simply, cutting his own hair, washing his own clothes, spinning cloth for his own garments, wearing only a *dhoti* (loincloth) and if needed, a shawl and sandals. His life as an ascetic had begun, enabling him to do his frequent fasts, both for personal and political reasons. For the British, Gandhi's political fasts created major problems, as they could not afford to let this widely acclaimed Indian hero die in prison.

Family background: He was born of distinguished parentage, and named Mohandas. His father's family was Hindu, of the *Vaishnava* persuasion and his mother's family was Pranami, a small sect that mingled Hindu and Moslem beliefs and stressed simplicity of living, strict vegetarianism, avoidance of alcohol and tobacco and periodic fasts. The Gandhis were of the *Vaishya* (merchant) class, and originally grocers. Even the family name signified grocer, though his grandfather, father, and uncle had served as prime ministers to the princes of Porbandar and other tiny Indian states in the Gujarat region. His father Karamchand Gandhi (1822-1885) was a member of the Rajasthanik Court, very influential at the time for settling disputes between chiefs and the fellow clansmen. He became Prime Minister in Rajikot and later in Vankaner. He married four times, having lost each wife by death. There were two daughters from his first and second marriages. With his last wife, Putlibai, he had one daughter and three sons: Mohandas was the youngest. Very close to both parents, he preferred their company to those of his school friends. His mother died while he was in England studying law, from ages 19 to 22. The skills he later exhibited as national negotiator were consistent with his paternal family tradition, though not apparent before his arrival in South Africa in 1893.

Marriage and family life: Gandhi was married for 62 years, until the death of his wife Kasturbai on Feb. 22, 1944. This left him in a state of deep sorrow. She died in his arms in a prison in Poona, both of them there for fighting for Indian independence. She had great mental and physical endurance but was always challenged to keep up with his world, in which she lacked intellectual or political understanding. She remained illiterate despite Gandhi's attempts to educate her. But she was strong, valiant and loyal, while he was often very hard on her. Common to Hindu tradition of the period, they were married at a very young age, when they were both 13 – in 1882. They had five children together, the first of whom died in infancy. Their four surviving sons were born between 1888 and 1900. At age 36, in 1906, Gandhi took a vow of celibacy, while still married and without consulting his wife, in part for birth control but mainly to improve and deepen his life through self-restraint. ***Brahamacharya*** involves celibacy as well as a strict vegetarian diet and abstinence from alcohol.

Some controversy surrounds Gandhi on the issue of *Brahmacharya*, as it became known that up through his 70s and for years prior to that, Gandhi slept with young women. He said it was in order to keep warm and also to test his vows of *Brahmacharya*. He only made it more public in his 70s, and explained it as part of his ongoing experiments with Truth. But many saw it as personal exploitation, especially as he refused to sleep in the same room with his wife (after 1906). Gandhi had an almost inexplicable dogmatism, insisting that those in his inner circle who were married should also observe *Brahmacharya*. Many of them, including Jawaharlal Nehru, objected strongly to having such dictatorial views imposed on them, even if they remained devoted to Gandhi in their common political aims. This was an area of Gandhi's life that contrasted sharply with his almost saintly status as spiritual and political leader of the Indian independence movement.

Education – the blend and clash of a Hindu in colonial India: A dull and timid student who did poorly in local secondary schools, Gandhi did not become fully self-confident for a number of years. In childhood, his shyness was so extreme that he concentrated mostly on his studies at school, returning home quickly to avoid contact with his fellow students. His father wanted him to follow in the tradition of local Prime Minister, but it was clear at that point that Mohandas was not up to it. The solution was to send him to London to study law at University College, as the exams were reported to be easy. On the eve of his departure, the elders of his subcaste, the Modh Banias, tried to forbid his journey because its rules prohibited voyages abroad, also eating and drinking with Europeans. Refusing to obey, Gandhi was excommunicated from his clan, who were forbidden to see him upon his return. This was his first major refusal to follow authority, and though it freed him from the more rigid Hindu ideas on caste, he still believed in *Varna* (caste) as a way of ordering Indian society. He set sail for England on Sept. 4, 1888, and for the first year was determined to become the perfect Englishman, dressing as a fop, taking lessons in dancing, elocution and the violin. He had promised his mother to avoid women, alcohol, tobacco and meat. And this he did, though staying away from women sexually would always be the most challenging for him.

After studying for several years he passed his law exams and was admitted to the British bar on June 10, 1891. Two days later, he set sail for India. He tried to establish a law practice in Bombay, but without much success, in part because he lacked training in the laws of his home country. But mainly, his extreme shyness was still a major problem for him. He collapsed in a Bombay courtroom in early 1893, unable to speak or to argue his case, his shyness was so overwhelming.

Physical health: Gandhi was not strong physically in childhood or adolescence, but in his 20s he gained strength, able to walk 10 to 12 miles easily. His daily routine included morning prayers followed by a vigorous walk at 5 AM as well as later in the day. He often outpaced those much younger than himself in later years. In March 1930 he was past 60 years old when he did the Salt March, a 24-day trek for 200 miles from his ashram to the sea. During his three years in England to study law (ages 19-21) he left behind his 16-year old wife and young son. He was often lonely and starving due to his dietary restrictions and lack of available vegetarian diet. But fortunately in London he met other vegetarians and began to study vegetarianism in a more scientific way. His 6 ½ years in and out of prisons were also very hard on his health, in addition to his many long fasts, some of them in prison.

Birth of a political activist – life-changing events: Being shy and insecure, Gandhi was unable to find steady work as a lawyer in India. Nor did he qualify to teach school. Even with a law degree from Britain he found it difficult to make a living for his young family in the early 1890s. But in spring 1893 a Muslim trading company with interests in Durban, South Africa hired him to work on a complex case. A few days after his arrival in late May 1893 in another mostly non-white country ruled by whites, his racial and political consciousness became galvanized. Asked to remove his turban in a Durban courtroom, he refused, exiting the premises. This caused an immediate stir and was reported the next day (May 29) in an article in the local newspaper, *The Natal Advertiser*: **"An Unwelcome Visitor."**[259] A few days later he began a trip to Pretoria that would be life-changing.

Gandhi started out on May 31, 1893 near the Natal/Transvaal border. It was a complicated trip via train and stage coaches, delaying his arrival in Pretoria until the evening of June 4. On May 31 he boarded a train which stopped at 9:00 PM at Pietermaritzburg (commonly known as Maritzburg, capital of Natal).[260] Though he held a First class ticket and was dressed impeccably in the Western style he had acquired while in England and befitting his position as a barrister – he was thrown off the train in Maritzburg (along with his luggage) for refusing to move to the train's Third class section when a white woman complained. It was winter in South Africa and colder on the high plateau. He shivered through the night in the unlit waiting room, his coat packed away in his luggage on the platform. The next morning (June 1) he sent telegrams reporting the incident to his new employer and to the railway's general manager. On the next segment of the journey by stagecoach he was beaten when he refused to stand on the outside footboard to make room for a European passenger. He was in a state of shock and debated whether to return to India. Instead he vowed to rectify this kind of social injustice for himself and for others. Out of necessity, he quickly overcame his shyness, as he was suddenly thrust into the role of spokesperson and mediator.

> "India was poor. The Indian settler went to South Africa in search of wealth, and he was bound to contribute part of his earnings for the benefit of his countrymen in the hour of their adversity.... Thus service of the Indians in South Africa ever revealed to me new implications of truth at every stage. Truth is like a vast tree, which yields more and more fruit, the more you nurture it. The deeper the search in the mine of truth the richer the discovery of the gems buried there, in the shape of openings for an ever greater variety of service."
>
> **M.K. Gandhi,** *An Autobiography: The Story of My Experiments with Truth*, 1957, Beacon Press edition, 1993, p. 218. (Translated from the original in Gujarati by Mahadev Desai.)

---

[259] Rajmohan Gandhi (grandson of Mohandas K. Gandhi), ***Gandhi: The Man, His People, and the Empire,*** 2007, 2008, p. 59.

[260] *Ibid.*, pp. 60-61.

South Africa (1893-1914): Gandhi's 21 years in South Africa began on April 19, 1893, the day he set sail from Bombay, arriving first in Zanzibar. After an 8 to 10 day stopover, he sailed from Zanzibar on May 14, arriving in the port of Durban on May 23. The province of Natal was then a colony of the British Empire and Durban's population was 30,000. The city would become Africa's largest port and among the world's busiest ports. Indians first arrived in Natal in 1860, brought in as indentured workers on the sugar plantations, replacing the Zulu workers who were considered less reliable. Indian traders and merchants soon followed, many of them Muslims from Gujarat, Gandhi's home region.

Muslim Indian traders (Dada Abdullah & Co.) hired Gandhi to be a link between the firm and its European lawyers, as they had no common spoken language. Gandhi understood he was being hired not so much as a barrister but as an interpreter and "a servant of the firm." However, his acceptance of a yearlong contract with Dada Abdullah provided a welcome turnaround from his failure to establish himself in India. He soon found his voice in South Africa, no longer allowing himself to be intimidated, as in India. The social injustices he saw all around him and experienced first hand served as a catalyst, and Indians residing there were a powerless alien minority. He finished his legal work and was about to return to India in 1894 when the Indian community in S.A. learned it was about to be disenfranchised completely. They begged Gandhi to stay and help their cause. This would lead to another twenty years in South Africa. Throughout this period he made his living as a lawyer working for the Supreme Court of Natal, refusing to accept any payment for his legal and political services to the Indian community. His wife and family joined him in 1896, and except for a few trips to India and England in the interim, they remained there until July 1914, when Gandhi considered his political work in South Africa had been achieved.

As he came into his own as a leader of the Indian community in South Africa, his attitudes towards racism and racial equality evolved. His awareness of the condition of all non-white races in Africa expanded beyond his initial prejudices against black Africans - common to the era and well beyond it. Even so, his major focus was to protect Indian immigrants, who suffered from the numerous indignities of racial apartheid, along with the black population. They were fingerprinted, forced to pay poll taxes, had severe curfews placed on them and were unable to own land. Also, Indian marriages were declared illegal and only Christian marriages were legal. Outlawing Indian marriages turned many Indian women into instant political activists, including Gandhi's wife Kasturbai. Victory came in July 1914 when South Africa passed **The Indian Relief Act**. It recognized Indian marriages, dropped poll taxes, softened immigration regulations, and started phasing out the indentured labor system. Though the passage of this law was a great personal triumph for Gandhi and the Indian community there, it was short-lived. The gains were soon lost when, after Gandhi's departure, no one could defend their interests and South Africa descended into more severe racial discrimination for many more decades up to the early 1990s. He had introduced his *Satyagraha* movement in South Africa in 1907 and it would turn out to be crucial for India and eventually for South Africa.

Indian independence movement (1915–1947): The struggle for *Swaraj* – independence of India from foreign domination – truly began in the mid 1700s, with the first resistance to British colonial rule through the expansion of the British East India Company. A year after the Indian rebellions of 1857, India came under direct rule from the British crown, which abolished the British East India company rule and declared Queen Victoria Empress of India. This was the British Empire's largest, most populous and profitable holding. Though Indians were promised equal treatment under British law, they were mistrustful of the British ever since the violence of the 1857 rebellions. In Victorian era-India, the English were the victorious Master Race, destined to rule over "lesser breeds," in the words of Rudyard Kipling. Into this colonial India Gandhi was born and lived until the age of 19. He then spent some 24 years abroad – over 3 years in England and 21 years in South Africa, prior to returning home to India in early Jan. 1915 at age 45. In South Africa he had learned how to stage and lead political movements and marches.

Prior to 1915 Gandhi had been an admirer of the British political system, until he reluctantly understood that British governance of India was exploitation that would never do more than drain the country of its resources and impoverish its peoples. One example of this was the thriving cottage industries, which had been spinning and weaving cloth with excellent results until they were ruined under British rule, leaving millions of Indians with little resistance to famines. Gandhi encouraged Indians to renew their cottage industries. And soon he was rarely seen without his spinning wheel, on which he made *khaddar* – handspun cloth. This was to become a major symbol for him personally and for his vision of an independent India.

For his political protests Gandhi was often thrown in prison, starting in South Africa in July 1907. He defended himself at numerous trials, starting in March 1922. He spent a total of 6 ½ years in prison in the two countries: 2089 days in India, and 249 days in South Africa. His fasts achieved various goals, including in his last year, a fragile peace from the slaughter between Hindus and Muslims after partition. In the last six months of his life he nearly died from several of these fasts and was deeply saddened by the ongoing violence between Indians. On his 78th birthday, Oct. 2, 1947, many congratulated him for liberating India.[261]

> "Where do congratulations come in?" he asked. "Would it not be more appropriate to send condolences? … There was a time when whatever I said, the masses followed. Today, mine is a lonely voice."
>
> **Mahatma Gandhi** quoted in William L. Shirer, *Gandhi: A Memoir*, 1979, p. 222.

Though he became a deeply religious man, Gandhi fought for decades for a secular India in which Hindus, Muslims and people of all faiths could live peacefully together. In spring 1919, he led the Hindus and the Indian National Congress in supporting the Muslim League, who sent a deputation to see British Prime Minister Lloyd George. Though he himself was a devout Hindu, Gandhi held out for a secular India that would tolerate all religions.

Muhammad Ali Jinnah, his Muslim counterpart and leader of the Muslim League, was a very secular Muslim nationalist who became more and more fanatical in his desire for a Muslim state.[262] Bombay lawyer Ali Jinnah became Gandhi's embittered opponent. Fearing that Muslims would be overwhelmed in an Indian state dominated by Hindus, he never backed down from his demand that India be partitioned. It could also be said that Gandhi did not go far enough

---

[261] There are conflicting dates for this quote. William Shirer and others date it from Gandhi's 78th birthday. But Narayan Desai (son of Mahadev Desai, Gandhi's personal secretary & biographer), in a lecture on Gandhi says this statement was given the day after India's independence. Asked to send a letter of congratulations, since he did not come to Delhi on Independence day, Gandhi replied in this way.

[262] Jinnah was born December 25, 1876 and died September 11, 1948. He suffered from tuberculosis through the 1940s, though few knew of it. Lord Louis Mountbatten, then Viceroy of India, said if he had known Jinnah was so ill he might have held off several months on independence, as Jinnah was "inflexible on Pakistan."

in trying to understand Jinnah, and seemed unable to tune into his fears and aspirations for the Muslim population. Gandhi was more skilled in his ability to understand the decency of the British and how they could be swayed. He was less accurate in estimating the capacity for violence on the part of his fellow Indians.

When he returned home from South Africa in Jan. 1915, Gandhi was a celebrated man in his own country. His mentor Gopal Krishna Gokhale advised him to spend his first several years educating himself about India's current needs. Gokhale was a leader for decades in Indian politics and had guided Gandhi for some years. But he died at age 48 on Feb. 19, 1915, just six weeks after Gandhi's return to India. Following Gokhale's advice, Gandhi stayed out of national politics until spring 1918. His political activities in India from 1915 to 1918 revolved around his ashram (founded in May 1915) and helping other Indian peasant farmers to assert their rights with mostly British landlords, backed by local administrations.

Amritsar, April 1919: In April 1918 Gandhi's national activism accelerated when he agreed to help the British enlist Indians to fight in WWI. For his efforts he was attacked by some for abandoning his vows of *ahimsa* (non-violence), but Gandhi hoped it might further the cause of Indian independence after the war. He was mistaken. In 1919 there was a major upsurge of discontent and militancy in India marking a turning point in the Indian independence movement, with greater activism by a larger segment of the population.

During the war, Indian Muslims were especially uneasy about England's fighting against the Turkish Caliphate. This was the center of spiritual leadership for the world's Muslims. But for both Muslims and Hindus in India, the British government had held out the promise of self-rule for India in return for India's supplying 1.3 million troops to fight in WWI. India suffered 106,000 casualties and contributed nearly $1 billion to the British cause. When the British showed no intention of following through on their promises, some 10,000 Sikhs, Hindus and Muslims gathered to protest on **April 13, 1919** at a peaceful meeting in the holy city of Amritsar. British soldiers opened fire, killing and wounding several thousand of the protesters. As a result of the British atrocities at Amritsar, Gandhi returned his two British war decorations and formally assumed leadership of the Indian nationalist movement. He was the logical person to do so, as his movement had been so successful in South Africa; but he refused to allow the participation of militant revolutionaries. Gandhi was also just recovering from a nervous breakdown and a bad case of dysentery. In the spring of 1919, the Spanish Flu pandemic had already killed 12 million Indians, and would kill up to 17 million.[263] Gandhi saw this as yet another reason the British were unfit to rule India.

He would dedicate the rest of his life to overthrowing British rule of India, and was at the center of negotiations on many occasions from 1919 onward. There was a serious meeting in 1931, in which he wanted *Purna Swaraj* (complete independence), but did not rule out an association with the British Empire to the mutual advantage of both parties, and with a right to secede. The first of his many imprisonments in India occurred in **March 1922**. Gandhi seized the chance to build bridges between Hindus and Muslims and to embarrass the British authorities, who were determined to maintain control of India and to continue giving false promises. The Indians used many means to make their point, some of them involving boycotting British goods.

---

[263] For Spanish Flu Pandemic, see under Hitler (Chart #28), and also Chapter 1, under Fire period: 1723-1921.

> "The coming of the second World War, in 1939, brought Gandhi once more back into politics and the struggle to free India of the British. As in 1914, the British government simply took India into the war by proclamation without consulting the Indians. Gandhi, preoccupied though he was by his work among the villagers, had seen the war coming. In 1938, revolted by the Nazi persecution of the Jews, he had written that 'if there ever could be a justifiable war in the name of and for humanity, war against Germany to prevent the wanton persecution of a whole race would be completely justified. But' – and here was the catch – 'I do not believe in any war.'"
>
> **William L. Shirer**, *Gandhi: A Memoir*, 1979, p. 211.

Because of his philosophy of non-violence, Gandhi did not support England in either World War. Nor did Gandhi support the partition of India, which was initially a British concept, and one fiercely defended by the Muslim leaders. As a way to divide and rule, the British had long fostered ongoing animosity between Hindus and Muslims. But Gandhi's greatest desire was for a unified and independent India, containing both Hindus and Muslims.

Salt March, March 12, 1930: This was a 200 mile trek from Gandhi's ashram at Ahmedhabad to Dandi, near the Gulf of Cambay: 24 days from start to finish, starting with 100 marchers and swelling to several thousand. To target just one of many examples of financial exploitation, Gandhi and his followers protested the British monopoly on salt. The British manufactured the salt, levied a tax on it, and prohibited Indians from making their own salt from the sea. Upon arriving at the sea, Gandhi made a small amount of it – an illegal act that was reported across India and around the world. Gandhi had broken the Salt Act. In this act of civil disobedience he was copied by thousands of Indians who swarmed to the beaches, gathered salt and sold it. Mass demonstrations followed, along with boycotts of British goods. A massive revolt thus began from a simple march and an attempt to break the British salt monopoly, just as American colonists broke the British monopoly on tea and that led to the American Revolutionary War and subsequent independence from the British Empire. Gandhi was imprisoned for this action, though released in1931.

Prologue to independence: The "Quit India" movement began in 1942, with many negotiations between Indians and British regarding India's participation in WWII. Many Indians hoped participation in the war would be rewarded by self-rule, but they had already been through many false promises after WWI. When World War II ended in 1945, it was clear that Britain was too depleted on all levels to hold off the fierce drive for Indian independence, begun so many decades earlier. The next two years were spent in negotiations as to how the Indians would govern themselves.

Although positive at one level, some of the developments within the country were troubling for Gandhi. He deplored the desire for greater power that saw no need to cultivate the peaceful village society he had envisioned. Even if his popularity was intact, his moral power was on the decline. He failed to stop the fighting between Hindus and Muslims, no matter how hard he tried everywhere: in Bengal, Bihar, Kashmir, Calcutta, and Delhi. He had some success, but failed to dissuade the militant traditionalists, one of whom would be his own assassin. When partition of India occurred, some never forgave Gandhi for not preventing it from happening. Though there were some inconsistencies in his policies, it is believed he never approved or accepted partition.

Assassination of Gandhi: There were six known attempts to assassinate Gandhi. The first was on June 25, 1934, then in May and Sept. 1944, on June 29, 1946, and twice in Jan. 1948. Finally on Jan. 30, 1948, at 5:17 PM in Delhi, India, he was assassinated on his way to a prayer vigil. A

Hindu fanatic shot him at point-blank range with three bullets. Gandhi died about half an hour later. His killer was Nathuram Godse, an activist member of the Hindu Mahasabha, a Hindu nationalist party that opposed a separate Muslim state, and became home for some of the more radical Hindu nationalists. They were angry at Gandhi for being too tolerant of the Muslims and hoped he would die during his last fast, begun Jan. 13, 1948, as he prayed for Indian unity and to calm the riots between Hindus and Muslims. Gandhi's fast ended after five days, encouraged by renewed promises of peace among Indians. When he was killed two weeks later by a Hindu and not a Muslim, some breathed a sigh of relief. But in spite of the larger victory of Indian independence, Gandhi's iconic role was such that some were bound to feel betrayed by his actions. Though he fiercely opposed partition, Gandhi was blamed by many Hindus and Muslims who distrusted his vision of Hindu-Muslim harmony. But even with so much bloodshed in the immediate aftermath of independence (with over half a million killed), history's longer view would honor Gandhi's role as crucial to Indian independence in 1947. His birthday was declared a national holiday that year and remains one of only three official national holidays.

Libra Ascendant: Making war or peace? : Libra is a *Chara* (active) sign, and the *Chara* emphasis gives an action-oriented life. Seven out of nine planets reside in *Chara* signs, and they are all in *kendras* (angles). The general shape of Gandhi's chart shows us his ability to inspire through leadership, with so much strength in the *kendras.* (Hitler has seven planets angular in his *Navamsha* chart, though in fixed signs.) *Chitra nakshatra* rises in the chart of both Gandhi and Hitler. *Chitra* is the shining jewel, and when rising in the chart, the person can glitter and shine, attracting with charm, elegance, and sexual power. These compelling qualities can depend on where the sign ruler and the *nakshatra* ruler are placed: Venus rules Libra, and Mars rules *Chitra nakshatra.* Gandhi's natal Venus and Mars are both placed in Libra in the Ascendant, whereas Hitler's natal Venus and Mars are in the 7th house in Aries within one degree, Mars afflicted by its loss to Venus in the *Graha yuddha.* By contrast, Gandhi's Venus is much stronger in the Ascendant in its own sign. His Venus and Mars are within two degrees, both in *Vishakha nakshatra* ruled by Jupiter and aspected by Jupiter, lord of the 6th house of conflict. Hitler's natal Jupiter also aspects his natal Venus and Mars. These shared components – Ascendant and *nakshatra* lords placed in *kendras*, aspected by 6th lord – give a destiny that must deal with major conflict. Recognition comes by presenting a compelling vision of harmony and a way to deal with the perceived conflict.

Hitler and Gandhi are major examples of what can happen either way – towards war *or* peace – with the Libra Ascendant. If Mars is the planet of war, there is no one planet of peace, although Venus and Libra are often assigned that role due to their association with equilibrium, and with a vision of harmony or a way of harmonious living. Venus and Libra rule over love and contractual relationships such as marriage. But Venus is also one's desire nature, and may tear down everything in its way in order to achieve the goal. Venus needs the discipline and detachment of Saturn, along with the tolerance and understanding of Jupiter in order to achieve true harmony. By itself, Venus and/or planets in Venus-ruled signs or *nakshatras* may be too full of desire to be capable of true harmony. We have already discussed Hitler's excesses. In the case of Gandhi, he may have clung too much to an anti-materialistic view of the future, one that – although true to deep spiritual values – may have been out of touch with modern materialistic "progress." Gandhi preferred his vision of the thousands of little self-sustaining and harmonious villages, enabling India to prosper in the best possible way.

His non-violent methods were in fact *not* totally in synch with the *Bhagavad Gita*, which says that if you are a *Kshatriya* (warrior) with a warrior's destiny, in certain circumstances you must fight. His one planet in a fire sign was Jupiter. Only Jupiter and his Ascendant *nakshatra* were ruled by fiery planets: the Sun and Mars respectively. In Hitler's case, fire was overwhelming in the chart, along with the *rajasic* strength of Venus. He led through fiery rhetoric and military prowess, in the traditional role of political and military leader. Gandhi's dual role as political and spiritual leader is less well charted in modern times. He wanted a politics completely suffused with spiritual principles. The problem was enacting *ahimsa* (non-violence) with a general public who had not been trained in such methods. His Holiness the **Dalai Lama XIV,** leader-in-exile of Tibet has had the same challenge, though his complete authority as a spiritual leader has helped him retain favor in the West.[264]

An urban sign, Libra understands the complexity of relationships in urban or heavily populated environments. It is concerned with social relationships and how they can work together effectively. As an Air sign, Libra is associated with the *Vaishya*, or merchant class. Gandhi came from a family of *Vaishyas*. When he moved back to Gujarat in 1915, he settled in Ahmedhabad – a brilliant choice for a popular base of support, with successful cooperation of merchants who provided most of his ashram's funding. When Venus is in Libra, especially in the Libra Ascendant, one can be a good mediator, especially if unafflicted. (Here we have two afflictions: the *Papa Kartari yoga* to the Ascendant, and an aspect from Jupiter, 6th house lord.) If Gandhi was overly attached to his peaceful village society model of India, he had the capacity to correctly reassess Britain's role in India over time. He began with an attitude of admiration and loyalty towards the British and moved through his periods of cooperation with British authorities, right up through 1918 and 1919, when he experienced a major shift due to the Amritsar massacre in April 1919. After that, he concluded that India would never have freedom and dignity while under British rule.[265]

The condition of Venus and Mars in this chart show passion, but the possibility of achieving greater balance with both planets in the sign of Libra, owned by Venus, the Ascendant lord. They recur in the *Navamsha* chart in Taurus, again owned by Venus – this time in the 7th house. Though unaspected by malefic planets, they are hemmed in by classic malefics Saturn and Sun (a *Papa Kartari yoga*), confirming that Venus-Mars have some initial difficulty expressing themselves. Lacking self confidence until age 24 when he arrived in South Africa, events demanded he either assert himself or submit to racist treatment he found intolerable.

Venus and the desire nature: The interplay between natal Venus and Mars shows the tendency for partnership and a sensual nature he insisted on taming through *Brahmacharya*. In the birth

[264] The Dalai Lama XIV was born July 6, 1935, 4:38 AM in Tengster Village, Tibet. He has Gemini Ascendant, with planets well placed in air signs: natal Jupiter in Libra and natal Saturn in Aquarius, and a total of five planets in air signs, along with Gemini Ascendant. Mercury *Digbala* in the Ascendant (especially in an air sign) helps one to be a skilled speaker and negotiator. Mahatma Gandhi shares this planetary signature. (For more on the Dalai Lama, see Chapter 1.)

[265] See Chapter 1, Fire to Earth Mutation period, 1901-1961. National independence can be dated from pivotal events which occur close to a JU-SA conjunction and/or a full Saturn cycle (28-29 years) earlier. The Amritsar Massacre was also some two months prior to the signing of the Treaty of Versailles, marking the end of WWI.

chart, Venus and Mars are more unbalanced by contacts from Mercury and Jupiter. Situated in Jupiter-owned *Vishakha nakshatra*, Venus and Mars are doubly affected by Jupiter's role as 3rd and 6th house lords, bringing the possibility of some conflict and/or excess in the matters affected by Venus and Mars.

Gandhi was married from the age of 13 and had five children, but took a vow of celibacy at age 36. His birth chart clearly shows that celibacy would be a major challenge for him, especially with such strongly passionate planets in the sign of Venus. Venus is in a dominant angular position in both birth chart and *Navamsha* chart. Mercury – nearby in Libra in the Ascendant of the birth chart – can neutralize planets of sexual desire. Mercury can have this effect when in combination with Sun and Moon (the marriage planets), or Venus and Mars (planets of sexual passion), but especially the latter. Though it can neutralize an overtly sexual combination, Mercury contacting either set of planets also tends to make one think about it, as Mercury is the communicator and the intellectual focus. In his case it is also the 9th lord of *Dharma*.

This combination may help to explain his sleeping with young Hindu women for some decades after he took his vow of *Brahmacharya* at age 36. Leveling with his followers, in his 70s Gandhi admitted this practice, saying his vows of *Brahmacharya* were only truly tested when the young women (some of them married) slept next to him naked and he was still free from sexual desire. Given that he refused to sleep with his wife from 1906 onward, claims of exploitation were reasonable and brought conflicts with his saintly image. The astrology illuminates the inner tension caused by a self-imposed discipline that went against his nature.

Astrological conditions in 1906: The *Satyagraha* movement Gandhi started in 1907 could not have happened without the events of 1906. During 1906 he was in the first year of his 10-year **Moon *Dasha***, which began May 21, 1905. Gandhi took his vow of *Brahmacharya* in June or July 1906, in his Moon-Mars period. Both Moon and Mars can be very emotional planets, especially the way they are configured in this birth chart. However, since natal Moon is in the 10th house, its *Dasha* is likely to be a period of leadership. And though the decision of *Brahmacharya* was made both for birth control and for spiritual practice, he may have chosen to place all priority on leadership, perhaps at the expense of his wife and family. Tr. Saturn in Aquarius was then in the 8th house from natal Moon. This is also the 5th house (of children) from the Libra Ascendant (indicating a theme of disciplining children or birth control). Called ***Ashtama Shani***, tr. Saturn in the 8th house from natal Moon or Ascendant can be one of the most challenging periods emotionally and mentally, and a time of loss and suffering or one that is geared towards *Moksha* (spiritual liberation). The two often go hand in hand.

In addition, tr. Rahu was in Cancer from Jan. 12, 1906 for the next 18 months. Echoing natal Rahu also in Cancer, it would have moved closely over the natal Moon for the first half of 1906, emphasizing the challenge Gandhi had earlier in his life to find his own inner equilibrium. The planetary backdrop (both transits and *Dasha*) helps to understand this decision, which though against his nature, was a way of attaining inner peace and freedom from his desire nature, especially sexual desires. Tr. Jupiter was in Taurus up through early July 1906 and thereafter in Gemini, aspecting Libra Ascendant but in the 12th house from natal Moon. Jupiter's influence as 6th house lord provided an inner level of conflict which mirrored the outer conflict he faced in various political situations in India and South Africa. Even so, going against his own sexual nature provided one of his major incongruities, and one that many people found difficult to accept.

Gandhi's **Ascendant lord Venus** is strong in the Ascendant in its own sign of Libra, and repeats in its own sign (Taurus) in the *Navamsha* chart in the 7th house. This gives him a good start in life and basic good health. The Venus position qualifies as a *Malavya Yoga*, one of the five *Mahapurusha yogas*, giving special significance to the life through the power of Venus. It should have given him more innate self-confidence from the start, but Venus and the Ascendant are hemmed in by malefic planets, and Jupiter's influence on Venus also created conflicts for him, as discussed. Up to age 24 he was timid but stubborn, and his real self-confidence did not come until 15 years into his 20-year Venus *Dasha*, during Venus-Saturn *Dasha/Bhukti* and Saturn's transit into Libra, its place of exaltation in his Ascendant. With so much interaction between natal Venus and Jupiter, accounting for the abundance of blessings *and* conflict in his life, we note that Gandhi passed his law school examinations on June 10, 1891 – during his **Venus-Jupiter** *Dasha/Bhukti*.

Venus is the major sign dispositor in this chart. The only planets that are not in this circuitry are Moon and Rahu. Moon is in its own sign of Cancer, and Moon conjunct Rahu is a *Raja yoga* in its own right, since Rahu is in a *Kendra* conjunct *Kendra* lord Moon in its own sign. In addition there is a powerful *Raja yoga* in the Ascendant formed by Mercury, Venus, and Mars in the Libra Ascendant. They are respectively owners of Houses 1, 7, and 9, two *kendra*s and a *trikona* house. Jupiter's aspect has a double impact: As 6th lord, it brings certain conflict, but it is also *guru* – the spiritual teacher. Jupiter contributes to deepening his spiritual life and making him a spiritual as well as political leader. This would embody so much of the life, thought, and actions of Gandhi, as he eventually took strong stands on social justice both in South Africa and in India.

From the Moon: Looking from the natal Moon in Cancer, Gandhi's planets are also powerfully placed. Jupiter as 9th lord is situated in the 10th – a *Raja yoga* for power and success, and good for the legal profession. Saturn is 7th lord placed in the 5th house, and 4th and 5th lords, Venus and Mars are in a *Kendra* from the Moon. Mercury detracts slightly as 12th lord (of foreign residence) from the Moon. This may have contributed to his living abroad for some 24 years, first to study law in England, then to practice law and take a position of community leadership in South Africa. There he attained so much fame for his political leadership that upon his return to India on Jan. 9, 1915 he was greeted as a hero and named a "Mahatma" soon after. This took place in the last few months of the 10-year **Moon *Dasha*,** ending May 21, 1915.

Jupiter is well placed in the 10th house from the Moon both in birth chart and *Navamsha* chart, forming a *Gaja Kesari yoga*, a yoga that can elevate a person in life, though Moon is somewhat afflicted by its contacts to Rahu and Ketu, as classic malefics. Thus, even though Jupiter is not a friend to Libra Ascendant, from the Moon it gives fame and visibility. Therefore, in **Jupiter *Dasha***, from May 21, 1940, Gandhi had increasing fame and power, but he faced increasing conflict from both the British and the Indians, as various sectors of Hindus and Muslims both opposed him for his goals. He survived less than half of Jupiter's 16-year *Dasha*. A Hindu fanatic killed him Jan. 30, 1948, in Jupiter-Ketu-Saturn period. Natal Jupiter and Saturn are both in *Maraka* houses, which can be death-producing, if all other factors coincide. In the last decade of his life Gandhi felt increasingly isolated in his quest for *Satyagraha* in India, despite the crowds who surrounded him. In 1940, he said:

> "Let no one say he is a follower of Gandhi.... It is enough that I should be my own follower."
>
> **Mahatma Gandhi** quoted in George Woodcock, *Mohandas Gandhi*, 1971, p. 100.

Natal Moon: Gandhi's Moon is in a powerful position for political leadership in the 10th house from Libra Ascendant. Moon is in its own sign of Cancer, and its *nakshatra* lord Mercury is well placed and has *Digbala* in the Ascendant. However, this Moon has several disadvantages. First, it is situated in the *nakshatra Ashlesha*, whose presiding deity is Sarpa, the divine serpent. *Ashlesha* represents snakes, and all those things associated with snakes – from special powers, including sexual powers, wisdom, and insight – to the poisonous ability to bite, torment, cause pain or death. A planet situated in *Ashlesha* can experience disruptions through the Serpent God, who is associated with intertwining, coiling, deception, painful separations, and misunderstandings. Gandhi experienced this on many levels, especially in the last years leading up to independence, when he could please neither Hindus nor Muslims. His assassin was a disgruntled Hindu who tried several times to kill him before succeeding.

At 27:56 Cancer, the Moon is situated in the last 3:20 of a water sign, considered *Gandanta*, and thus containing potential emotional difficulty. Also, though some 15 degrees wide from Rahu, any contact between Moon and Rahu in the same sign shows in the influence of a lunar eclipse.[266] Moon-Rahu also brings emotional turbulence and adds to the snake associations, as Rahu itself is the serpent's head with no body. (Ketu is its body with no head.) Gandhi went through a lot of suffering on an internal level regarding not only social injustices, but also taming his inner demons, mainly his sexual drive. His *Brahmacharya* vow came just at the start of his Moon *Dasha*, in Moon-Mars *Dasha/Bhukti*. Several months later he entered **Moon-Rahu** *Dasha/Bhukti* (Oct. 20, 1906 to April 20, 1908). During this period he was nearly killed (Feb. 1908) for his engagement in South African politics and his dealings with General Smuts regarding his criticism of the Registration Act for East Indians. Though his work eventually brought him national and international acclaim, he simplified his life more and more, demanding the same from those close to him. This was especially hard on his wife. From summer 1914 onward he wore nothing but a *dhoti* (loincloth) and at times a shawl and wooden sandals. He fasted regularly, at times for months in order to attain various political goals.

Eating: The 2nd house contains Saturn, which can show constraint in matters of eating and speaking. Physically, he was a slim and small man, whose vegetarian dietary habits and frequent fasting kept him extremely slight and at times threatened to kill him. He ate sparingly, mostly fruit and nuts while in South Africa, along with fruit juices, tea, and water. During most of his fasts he drank only water. Due to illness and weakness from so many long fasts and time in prison, he was forced to add goat's milk to his diet. He refused to drink cow's milk, a custom among some but not all Hindus.

He also maintained one day of silence every week (Monday – which appeased the condition of the Moon in his chart, whether or not he was aware of this Vedic custom). As lord of the 2nd house of speech, Mars creates some conflict, as it is in the Ascendant with Venus and Mercury. Though neutral to Venus, Mercury is enemy of Mars. Venus and Mars are both in *Vishakha nakshatra*, owned by Jupiter, owner of the 6th house of conflict. Even though Mars and Jupiter are mutual friends, any planet has some discomfort from Jupiter's aspect in this chart, given Jupiter's role as temporal malefic being owner of the 3rd and 6th houses.

Mars is the counterpart of Venus, being the 7th house lord. It is the OTHER. When the 7th house lord is in the Ascendant, oftentimes partners come to you. Thus, even though Gandhi was

---

[266] On Aug. 8, 1869 a Total Solar eclipse occurred at 23:19 Cancer, just 4½ degrees from his natal Moon.

a renunciant since age 36, many women would throw themselves at his feet, offering to massage him, bathe him or sleep next to him "for warmth."

Finding his voice – at age 24: With **Mercury** so well placed and having *Digbala* in the Ascendant, it might seem odd that Gandhi had a problem communicating until well into his 20s. However, there are several factors: Natal planets in the Ascendant are hemmed in by malefics (a *Papa Kartari yoga*). Also, *Navamsha* Mercury is with Saturn in Capricorn. **Saturn** can bring very careful speech and can delay its output. This is confirmed by Saturn's presence in the 2nd house of voice and speech, which can bring inhibitions, delayed or very serious speech. Since Saturn comes to maturity at age 36, one can reap the benefits of Saturn's delays after the age of 36 – in Gandhi's case after Oct. 2, 1905. He was in South Africa when he took a vow of *Brahmacharya* in early summer 1906. He always claimed it gave him greater depth and spiritual powers, even if he had much inner turmoil maintaining his chastity. In spite of that inner turmoil, his outer sense of authority and confidence began to flourish in this period.

> "I never had a brilliant career. I was all my life a plodder. When I went to England... I couldn't put together two sentences correctly. On the steamer I was a drone... I finished my three years in England as a drone."
>
> **M.K. Gandhi**, Speech at Law College in Travancore, March 1925.
> (Mahadev H. Desai, *Day-to-Day with Gandhi*, Volume VI, pp. 103-104.)

> "At school the teachers did not consider me a very bright boy. They knew that I was a good boy, but not a bright boy. I never knew first class and second class. I barely passed. I was a dull boy. I could not even speak properly. Even when I went to South Africa I went mainly as a clerk."
>
> **M.K. Gandhi**, Speech to Gandhi Seva Sangh, Hudli, April 17, 1937.
> *Collected Works*, Volume 65, pp. 100-101.

Writing his autobiography (1922-1926): Gandhi was asked to write an autobiography describing the background of his political campaigns. He started it in prison in spring 1922 and published it in book form in 1927. He worked on it four to five years and titled it *Gandhi – an Autobiography: The Story of My Experiments with Truth.* It covers his personal and political life from early childhood up to 1920. On May 21, 1922 his 18-year Rahu *Dasha* began. *Navamsha* Rahu is situated in the 12th house of foreign residence, or residence in ashrams, hospitals, or prisons. Due to the 12th house position of *Dasha* lord Rahu, its *Dasha* would have to start out with some disadvantages, though eventually bringing notoriety, as Rahu is placed with Moon in the prominent 10th house of the birth chart. The first two sub-periods produced the book: Rahu-Rahu and Rahu-Jupiter, up through June 27, 1927. The next sub-period (Rahu-Saturn) would be more successful in a worldly way due to Saturn's role as *yogakaraka* in the Libra Ascendant chart.

Other themes of personal restraint: Beyond his fasting and his vow of *Brahmacharya,* Gandhi kept a vow of silence on Mondays (from midnight Monday to midnight Tuesday). In 1926 he retired to his ashram for a year of silence. The following year, he toured India, expounding on his views of *Satyagraha*, and adding to it equality for women and for "untouchables" as well as abstinence from drugs and alcohol. Throughout 1926, Gandhi was in Rahu-Jupiter *Dasha/Bhukti.* Rahu was in Gemini in the 12th house from natal Moon, and tr. Jupiter was back and forth between Capricorn and Aquarius, signs ruled by the more constrictive Saturn. (Jupiter is also debilitated in Capricorn.) **Tr. Saturn** was 7 ½ months **in Scorpio** in its 2nd return of the life to its natal position (the last Saturn return for Gandhi), and 4½ months in Libra in the Ascendant aspecting the Moon. Though natal Saturn in the 2nd house may have a constraining effect on the voice, this is balanced out by an aspect of Jupiter to Mercury in the birth chart, typically bringing abundance of speech. But Saturn provided a delay factor in Gandhi finding his own voice, indeed any voice at all. In his first court cases in 1891-1893 he could not speak when required to do so as a trained lawyer in court. In one instance, he collapsed in the courtroom from the strain. Thus, most of his legal work in India in this period involved preparing briefs – a disappointment to his family.

Leading the Indian independence movement: The period from the end of WWI in Nov. 1918 to the bitter end of the long drawn-out Paris Peace Conference and the signing of the Treaty of Versailles June 28, 1919 was crucial as a backdrop to postwar Germany and to the political and personal motivations of Adolf Hitler. For countries such as Egypt and India and many others chafing under the rule of lumbering empires, the spring of 1919 also provided a major opening in the call to nationalism and independence. U.S. President Woodrow Wilson's Fourteen Points gave the nationalists even more ardent hope.

> "During the war [WWI], Mohandas Gandhi had arrived from South Africa with the tools of political organization and civil disobedience which he had perfected to transform the largely middle-class Indian National Congress into a formidable mass movement. Rapid inflation, the collapse of India's export trade and the revelations of how British military incompetence had wasted the lives of Indian soldiers in Mesopotamia disillusioned even those Indians who had thought that at least British rules provided good government. Although Britain promised a gradual move toward self-rule in 1917, it was being outflanked and out-witted. Indian nationalists noted President Wilson's talk of self-determination [in his Fourteen Points] with approval but at first they paid little attention to the [Paris] Peace Conference [of 1919]."
>
> **Margaret MacMillan,** *1919: Six Months that Changed the World*, 2004, p. 402.[267]

Gandhi developed his political and leadership skills chiefly during his **Moon *Dasha* (1905-1915)**. This is fitting, with his Moon in such a powerful position in its own sign in the 10th house of career status and leadership. In the house of highest visibility, this Moon also indicates that the people can push the individual into prominence, as they did do initially in South Africa. The people (Moon) are strongly represented in this chart, and for any politician, natal Moon in the 10th house is an excellent position. Gandhi was in Mars-Mercury *Dasha* from Nov. 20, 1918 to Nov. 17, 1919. These two planets straddle natal Venus in the Libra Ascendant of the birth chart, and they indicate he might have to speak (Mercury) forcefully (Mars) at this time. But as both planets are in Libra, and strongly impacted by the influence of Venus, he would strive to find a more elegant and harmonious way of having to speak strong words. Working in support of the Muslim League in spring 1919, it was through the spirit of cooperation (also Venus) with other Indian political movements that he would help pave the way towards Indian independence. That he could encounter plenty of obstacles and delays is shown through the *Papa Kartari yoga* to the Ascendant as well as the mutual positions of natal Jupiter and Saturn.

***Dharma* and *Karma* in the destiny:** Jupiter and Saturn in the birth chart indicate one's *Dharma* and *Karma*, respectively. In Gandhi's chart they are in 6 – 8 mutual house position, the same as

---

[267] This book was originally published in 2002 by Margaret MacMillan under the title ***Peacemakers: The Paris Peace Conference of 1919 and Its Attempt to End War.***

in Hitler's chart. In both cases there is unease in finding the true *Dharmic* role in one's life. It is not clear right away what one's path (including career path) will be, and there are some failures and ordeals along the way. We have seen how overly emotional factors fuelled Hitler's ideological point of view. Natal Jupiter and Moon are close together on his Rahu-Ketu axis. Extreme religious or ideological fanaticism can come with Moon-Jupiter-Rahu or Ketu, especially in either Sagittarius or Gemini. (They were transiting on this axis at the time of the attack on the World Trade Center in New York City Sept. 11, 2001.)

In Gandhi's case, since natal Jupiter and Saturn both improve in the *Navamsha* chart, the *dharmic* and *karmic* destiny also improves. Jupiter and Saturn are each in better sign positions (though neutral house positions): *Navamsha* Jupiter is in Sagittarius (own sign) in the 2nd house, *Navamsha* Saturn in Capricorn (own sign) in the 3rd house. The *Varga* chart specifically for career *(Dashamsha)* is not beneficial for Jupiter, with Jupiter debilitated in Capricorn in the 3rd house. Thus Jupiter *Dasha* would not be favorable for Gandhi, but especially because Jupiter is lord of the 6th house of conflict in the birth chart. (Jupiter's 16-year *Dasha* began for him on May 24, 1940.) Even so, Jupiter is also a great benefic, and can bestow strong spiritual influence - along with conflicts - as 6th house lord. As *Atmakaraka*, Jupiter is the planet at the highest degree of celestial longitude, and as such its ability to bestow good fortune is also present. It is just not enough to prevent his assassination.

Gandhi's natal Jupiter and Saturn are both in the sign of Mars, Saturn in Scorpio and Jupiter Retrograde in Aries. Mars is in the Ascendant in *Vishakha nakshatra*, and the Ascendant itself is in *Chitra nakshatra* ruled by Mars. Natal Mars is very influenced by all three classic benefics: Mercury, Venus, and Jupiter. Whereas Hitler's Mars was out of control due its condition in the birth chart, Gandhi's Mars is helped by being in the sign of Venus, whose position is strong in its own sign *(swakshetra)* and in the Ascendant.

Slow progress, many obstacles: The Libra Ascendant is hemmed in here by two classic malefic planets, Saturn and Sun - a ***Papa Kartari yoga.*** This can impede progress in life by causing the person to face many obstacles, often having to break out of barriers that have been constructed around him or her in various ways. For Mohandas Gandhi, his race and later his resistance to British rule (from April 1919) were the biggest barriers for him and ones he spent his life overcoming. Prior to that he faced other hurdles. When he sought to go to England to study law, the elders in his sub-caste forbade him to go, threatening excommunication by the clan if he did go. He took the risk, faced the punishment and went anyway. Up to the age of 24, he also suffered from debilitating shyness, which diminished his legal profession in India. Due to his political activism in South Africa and in India he suffered imprisonment as an ongoing factor, starting from July 1907 in South Africa. He spent a total of 6½ years in prisons. As with Hitler, Gandhi's most successful *Dashas* were those of the Moon and Rahu and to some extent Jupiter.[268] For Gandhi, none of these three planets are in the Ascendant and thus not subject to the constraints of the *Papa Kartari yoga*. Even so, Rahu *Dasha* (1922 to 1940) brought him many prison terms for all his political protest activities, though they were effective in the end and brought about a further erosion of British authority, as the Indian populace became more and more united under Gandhi in its desire for independence.

[268] Hitler's natal Moon and Rahu were free of influence from his afflicted Mars, at least in the birth chart.

IMPORTANT EVENTS IN GANDHI'S LIFE:

1. **Sept. 4, 1888** – Departure for England to study law. It was a Saturday (Saturn-ruled day), hinting at some of the hardships he would endure while in London. He was in Venus-Rahu *Dasha/Bhukti*. Tr. Venus was debilitated in Virgo on his 12th house cusp of foreign travel and residence. Tr. Rahu (the foreigner) was nearly 6:00 Cancer (the family), close to natal Rahu in Cancer and contacting natal Moon in late Cancer. Tr. Moon and Saturn were also in Cancer, amplifying this theme. Gandhi's first major act opposing authority figures was to defy elders from his sub-caste, who excommunicated him for travelling abroad. While creating a schism within the family, it freed him to oppose the rigidity of the Indian caste system. He was in the middle portion of his *Sade Sati* (first portion from Aug. 20, 1884). Tr. Saturn in his 10th house gave him visibility and responsibilities at age 19 for which he was not yet prepared, putting focus on his difficult Moon-Rahu in the natal 10th house. In 1885, when Gandhi was 15, his first child was born and died within a few days. His father died the same year, dealing a severe blow to the family finances. They borrowed money to send Mohandas to law school, with hopes of him rescuing the situation on his return. A second child was born in 1888 prior to his departure for England, leaving behind a young wife and child. He passed his law examinations and returned to India in June 1891.
2. **May 31, 1893 (9:00 PM, Maritzburg, S.A.)** – Thrown out of a 1st class compartment of a train in South Africa. **Tr. Ketu** (the outcaste or foreigner) was at 4:09 Libra, and almost exactly on his natal Ascendant at 4:34 Libra. This was a pivotal degree area that spring and for the next six months. A Total Solar eclipse April 16, 1893 preceded Gandhi's departure for South Africa on April 19th. The eclipse was at 4:26 Aries, just opposite his natal Ascendant. Five planets were in Aries: Sun, Moon, Rahu, Venus and Jupiter, with tr. Saturn at 16 Virgo very close to conjunct his natal Sun in the 12th house of foreign residence. A Total eclipse so close to his departure – with five planets opposing his Ascendant – wielded enormous influence in terms of upsetting important relationships and encounters. Partners, alliances, and also potentially life's battlefield comes in the 7th house. On May 31st at 9 PM tr. Moon had just entered his 3rd house of short distance travel. It was at the vulnerable *Gandanta* junction, at 0:37 Sagittarius. Tr. Mars opposed this Moon, heating it up, though tr. Jupiter also aspected it, easing the way in the end. He was in **Venus-Saturn-Ketu period**.
3. **May 25, 1915,** near Ahmedhabad, India (capital of Gujarat, and textile capital of India) – the Founding of the *Satyagraha* ashram. Just a few months earlier (Jan. 9, 1915) Gandhi returned to India as a hero, after some 21 years in South Africa. The ashram was founded four days into his 7-year Mars *Dasha*. Mars is the initiator, and *Dasha* lord Mars was defeated that day in a Planetary War with tr. Venus at 29 Pisces, this in his natal 6th house of service work and conflict. Tr. Venus was near its degree of exaltation in Pisces, but Venus-Mars close contacts denote major power struggles and Pisces denotes great sacrifice and potential for victimhood. (As a point of resonance, *Dasha* lord Mars suffered from its loss in this battle, but at least Venus is his Ascendant lord.) Five planets were at 29 degrees: Venus and Mars at 29 Pisces, Rahu at 29 Cancer, Ketu at 29 Capricorn, and Jupiter at 29 Aquarius. This can denote final stages of an enterprise. (The ashram lasted just two years in that location.) Tr. Rahu contacted his natal

Moon very closely, describing emotional and family chaos through this period. Though not as strong as his Moon *Dasha*, Mars *Dasha* gave him continued courage in his efforts to fight for more social and political justice around him. One of Gandhi's first important tests in this ashram was his acceptance of an "untouchable" family into the ashram. It caused uproar and an immediate loss of funding, though he soon found other funding. In civilian Hindu custom the "untouchable" caste was one below the *shudras*, putting 1/5th of Indians in a position of contempt.

The ashram was founded on a Saturday, the day after the New Moon at 29:35 Aries. Tr. Saturn (the servant) was in Gemini in his 9th house and tr. Jupiter in Aquarius. Both Jupiter and Saturn in commercial air signs reflect the ongoing support Gandhi received from local textile businesses, as the local bania were largely merchants, and Ahmedhabad had a long tradition of commercial viability and political self-assertiveness. Together with its financial, organizational, and cultural resources, Ahmedhabad and the Gujarat region served Gandhi very well in the Independence movement. (Gandhi was Gujarati, formerly from Rajkot.) On June 17, 1917 Gandhi relocated his ashram to a new site on the Sabarmati River, re-named the Sabarmati ashram, where it remained until he disbanded it in July 1933. After departing on March 12, 1930 for the Salt March, he never returned to the ashram. See Item #5.

4. **April 13, 1919**, Amritsar, India – British soldiers massacred several thousand Sikhs, Hindus and Muslims, among a total of 10,000 gathered at a peaceful meeting at the holy city of Amritsar. As a result, Gandhi returned his two British war decorations and assumed leadership of the Indian nationalist movement. The day was Sunday, with tr. Sun at 29 Pisces up to 11:10 PM that night. (Sun is the day lord and is much less stable in the *Gandanta* area at the junction of water and fire signs. Jupiter-ruled Pisces is a spiritual placement, but also denotes sacrifice and the potential for victimhood, as noted in Item #3.) Four planets were in the emotional water signs: Rahu, Sun, Mercury, and Saturn. The most prominent resonating astrological feature was tr. Saturn (retrograde) at 28:43 Cancer at noon on this day. In the 10th house of Gandhi's birth chart, it was close to his natal Moon at 27:56 Cancer. Not only is this in the midst of his *Sade Sati*, but a seminal moment for a public figure. The next time this occurred was 29 years later – a whole Saturn cycle, and preceded by a Mars-Saturn conjunction at 29:02 Cancer (Nov. 12, 1947). Gandhi's assassination was Jan. 30, 1948, when tr. Saturn was at 26:53 Cancer. Saturn in the 10th house can raise one to great heights and can also pull one down from those heights.

   Gandhi's **Moon-Rahu** in the 10th house gave more dramatic conditions for his status and visibility. Prior to the events of spring 1919, Gandhi suffered a protracted case of dysentery in July 1918. Moon is the stomach and rules the sign of Cancer. During their transit through Cancer, Saturn and Rahu can cause difficulties for the Moon, and extra volatility such as an intestinal disease when in *gandanta* degrees of a water sign. This is also *Sarpa Drekkana* when in the last 20 degrees of Cancer, and first 20 degrees of Scorpio. From early July 1918 to early June 1919 transiting Saturn was in Cancer while tr. Rahu was in Scorpio on Gandhi's natal Saturn in the 2nd house of eating (and speech). This period also coincided with the Spanish Flu pandemic of 1918-1919.

5. **March 1922:** Court case imprisoning Gandhi for the first of many times in India. This one was for encouraging Indians to boycott the arrival of the Prince of Wales at Bombay, Nov. 21, 1921. Those attending a reception for the Prince were attacked by a mob. A Solar eclipse at 14:17 Pisces occurred March 28, 1922 opposite Gandhi's natal Sun in the 12th house of imprisonment. This situation was compounded by tr. Jupiter and Saturn in Virgo, and other planets transiting the Virgo-Pisces axis. (The 20-year Jupiter-Saturn conjunction (3:49 Virgo Sept. 9, 1921) was one degree from Gandhi's 12th house cusp, the house of residence in prisons, hospitals, ashrams, or a foreign country.) In March 1922 Gandhi was in the last two months of his Mars *Dasha*, one of the worst *Dashas* for his health, and in Mars-Moon period. His 6-year prison sentence was shortened when he was operated on for appendicitis in Jan. 1924 and released in Feb. 1924. He started writing his autobiography in prison in 1922, giving the philosophy behind his political movements. Published in book form in 1927, it appeared in monthly installments before then in a magazine he ran.
6. **March 12, 1930**: The 200-mile Salt March from his ashram in Ahmedabad to the sea at Dandi, arriving on April 12, breaking the Salt Act by taking salt from the sea. Gandhi was imprisoned for his action but released in 1931. He declared he would not return to his ashram until India gained independence, and disbanded the ashram in July 1933. During the Salt March he was in Rahu-Saturn *Dasha/Bhukti,* with tr. Rahu across the Ascendant axis in 12:07 fiery Aries and tr. Saturn in galvanizing Sagittarius in the 3rd house of foot travel. *Navamsha* Saturn & Mercury in Capricorn are also in the 3rd house. Saturn's sub-period (June 27, 1927 to May 3, 1930) was generally very effective, as Saturn is *yogakaraka* in Gandhi's chart. His autobiography was published in 1927, initially in Gujarati.
7. **March 5, 1931**, New Delhi, India: Gandhi signed the Gandhi-Irwin Truce. Though the British conceded little, for the first time they were forced to deal seriously with the Indian revolution and the Indian National Congress that Gandhi dominated. The British still had not grasped the depth of the resurgent Indian nationalism. A Full Moon at 20:00 Leo *(Purva Phalguni)* the previous afternoon and tr. Jupiter and Saturn opposition from Gemini to Sagittarius highlighted a fiery confrontation of ideas and nationalist ambitions. But the Jupiter-Saturn conjunctions in 1940-1941 in Aries (*Bharani)* would give the independence movements far more weight. (See Chapter 1)
8. **Aug. 15, 1947**: India gained independence from Great Britain. This chart is set for Aug. 15, 1947, 00:00, New Delhi, India. Ascendant is at 7:43 Taurus, with Rahu close to the Ascendant and natal Sun at 27:59 Cancer, with a total of five planets in Cancer (family and tribes). There is a close connection to Gandhi's two planets in Cancer (Moon and Rahu) especially his natal Moon at 27:56 Cancer. Pakistan became a nation 24 hours earlier. Amidst the triumph there was much bloodshed between Hindus and Muslims in establishing the two new states. Gandhi had opposed the partition of India, but was sidetracked by his close aides and unable to stop it. (The India national chart is shown in the Appendix.)

Assassination of Gandhi: Jan. 30, 1948, 5:17 PM, Delhi, India. Gandhi was killed by Nathuram Godse, a Hindu fanatic who believed Gandhi was too tolerant of the Muslims and blamed him for the partition of India.[269] There are six known attempts to assassinate Gandhi from 1934 onward. Godse and his companions are known to have been involved in three of them and suspected of a fourth. They occurred in May 1944, on Sept. 9, 1944, and on Jan. 20, 1948. On June 25, 1934, a bomb was hurled at a car believed to be Gandhi's in a motorcade in Pune. As his car was delayed, Gandhi and his wife escaped unhurt, but 11 others (mostly police) were seriously injured. He was then in Rahu-Venus-Sun period. None of these planets were in *Maraka* houses (2 or 7) natally or by transit. Tr. Rahu was at 19 Capricorn in his 4th house. Tr. Sun was in Gemini in the favorable 9th house. Tr. Venus and Mars were in his 8th house in Taurus (their mutual contacts often producing a power struggle) opposite Moon in Scorpio, though softened by a Jupiter aspect. On Sept. 9, 1944, tr. Venus and Mars again in mutual contact were in a Planetary War in Virgo, his 12th house. Venus (his Asc. lord) won the war. No suspects were ever found for the June 29, 1946 train derailment near Bombay. Gandhi was the presumed target, as he was travelling on it.

On Jan. 30, 1948 Gandhi was on his way to a prayer vigil when he was hit by three bullets in the stomach (his vulnerable area) and chest, fired from a pistol at point blank range. (He died within half an hour.) Cancer was rising, with Saturn retrograde in the Ascendant opposite tr. Sun – confrontation with authority. Both tr. Saturn and tr. Jupiter (in Scorpio) were *gandanta* – in the last 3:20 of water signs. Though complementing each other in the same element, this degree area is a more extreme and emotionally driven placement for any planet. It repeats major themes from both Gandhi's birth chart and the India national chart. Tr. Saturn was close to Gandhi's natal Moon in the 10th house (27:56 Cancer) and the India national Sun (leader) position at 27:59 Cancer.[270] (He would never cease being identified as a leader of India.) Tr. Moon at his death closely conjoined his natal Sun at 16:53 Virgo in *Hasta nakshatra*. The assassination took place just 5½ months after India's independence from Britain, Aug. 15, 1947, and even closer to the **Mars-Saturn conjunction** at 29:02 Cancer on Nov. 12, 1947. Mars-Saturn can be a difficult and intense combination when in mutual contact, more so in the treacherous *gandanta* area. Amplifying the importance of this Mars-Saturn conjunction was a **Solar eclipse** 9½ hours later at 26:28 Libra, in *Vishakha nakshatra*, on the degree of Gandhi's Mars in his Ascendant and within two degrees from his natal Venus. At the time he was killed, Gandhi was in Jupiter-Ketu-Saturn period: Jupiter is the lord of his 6th house of conflict, and natal Jupiter is situated in a *Maraka* house in both birth and Navamsha charts, while natal Saturn is in a *Maraka* house in the birth chart.

---

[269] Nathuram Godse, b. May 19, 1910, Pune, India; d. Nov. 15, 1949 (by hanging). He plotted the assassination along with his brother Gopal Godse and six others. Nathuram possessed a dagger in two previous attempts to kill Gandhi, in both May and Sept.1944.

[270] The sectarian violence that rocked India at its independence and in subsequent periods is well described by the concentration of planets and/or affliction to planets in the late degrees of Cancer, the sign ruling over the family. It also describes family life – its culture, long-held traditions and continuity – as crucial factors in India, along with the typical reverence bestowed on ruling Indian political families. With five planets in Cancer in India's national independence chart, and Sun (the leader) at 27:59 Cancer, we note the continued confluence of key planetary positions in late Cancer in the charts of Indian political leaders: The Ascendant of Jawaharlal Nehru, first Prime Minister (1947-1964) is at 24:47 Cancer (natal Moon at 17:56 Cancer), using K.N. Rao's rectified time of 11:05 PM.. His daughter Indira Gandhi succeeded him in power, 1966-1977; 1980-1984. Her Ascendant is at 27:22 Cancer (natal Saturn at 21:47 Cancer). See: Chapter 6, Cancer Ascendant charts, Chart #18. She was assassinated on Oct. 31, 1984 by two of her own Sikh bodyguards.

The death charts of Mahatma Gandhi, Adolf Hitler, and John F. Kennedy all contain a close interaction of Venus, Mars, and Mercury. Venus is the planet of politics, and the inventor of politics. The Venus-Mars interplay can show power struggles and Mercury shows the news focus on all of this. At the time of Gandhi's assassination, tr. Mercury and Venus were in Aquarius, sign of philosophical idealism, opposite Mars in Leo, sign of royalty.

Legacy of a good reputation: Such a potent confluence of factors in 1947 into early 1948 shows that despite victory gained from India's independence from Britain, much work remained to be done and Gandhi would not necessarily be celebrated at this time. That his reputation would remain intact and grow is shown by a powerful 10th lord, along with strong Ascendant lord Venus, even if somewhat afflicted by aspects from Mars and Jupiter, especially Jupiter as 6th lord. This brought him many enemies, as would be expected from a prominent 6th house lord.

His 10th house lord is strong both in birth and *Navamsha* charts. Natal Moon in its own house in the birth chart gives a good career and reputation, especially when placed in the 10th house, with no malefic aspects, and Moon's *nakshatra* lord (Mercury) well placed: Mercury has *Digbala* (its best angular house) in the Ascendant. The 10th house position of Saturn can give the most visibility but also a fall from power, especially when natal Saturn is in the 10th. Though this is not the case for Gandhi, a fall from power is still possible when tr. Saturn is in the 10th house, even if not repeated in the birth chart. Wherever *Dharma* and *Karma* planets Jupiter and Saturn are transiting – their positions are vital for the economic, social and political situation. In the *gandanta* positions in water signs (as in Jan. 1948), there is great potential for a feverish emotional pitch, in this case directly related to the aftermath of Indian independence. This triumph after several hundred years of British rule was still overwhelmed with so many problems to be solved and so many still raw emotions.

The martyrdom of assassination can sometimes raise a tarnished reputation. Though Gandhi's reputation was not heavily tarnished, it had become bruised and battered in India's internecine battles for independence. He faced criticism from spiritual leaders such as Aurobindo Ghose, J. Krishnamurthi, Annie Besant, and others, though they recognized his greatness.

The Mahatma also had some startling idiosyncrasies, especially around his *Brahmacharya* vows. But they did not prevent him from retaining a legendary status akin to Jesus or Buddha. With the receding years, fewer people seem to know or remember some less saintly details. These are largely overlooked in light of his larger legacy, perhaps again a testimony to his powerful Venus and Moon, which gain strength in the *Navamsha* chart. His message of *Satyagraha* is still copied and his example of *ahimsa* (non-violence) endures as a bright light in an increasingly militaristic era. In 2007 – the centennial of his *Satyagraha* movement in South Africa – the United Nations declared his birthday henceforth as "International Day of Non-Violence." Sixty years earlier (from independence in 1947), Gandhi's birthday was established as a national holiday in India and continues as one of only three official national holidays. It honors a man still regarded as "Father of the Nation."[271]

[271] In addition to Oct. 2nd, the other two official Indian national holidays are: Aug. 15th (Independence day, 1947); and Jan. 26th, when India became a Republic in 1950 and established its own Constitution. Note also the important resonance of the 60-year cycle (five full Jupiter cycles and two full Saturn cycles), 1947-2007.

Chart 30: Ammachi (Mata Amritanandamayi)

Birth data: Sunday, Sept. 27, 1953, 9:10 AM IST, Vallickavu, India, Long. 76E31, Lat. 9N10, *Lahiri ayanamsha*: -23:12:54, Class B birth data: biography. Source notes: Linda Johnsen quotes 'Matruvani', printed in the Ashram at Amritapuri, India. Ascendant: 23:35 Libra.

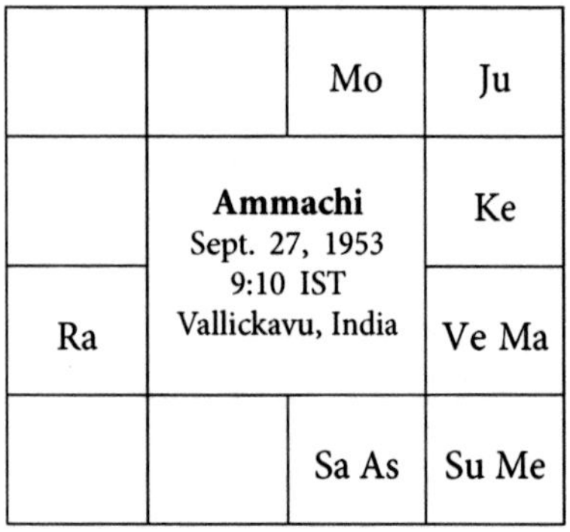

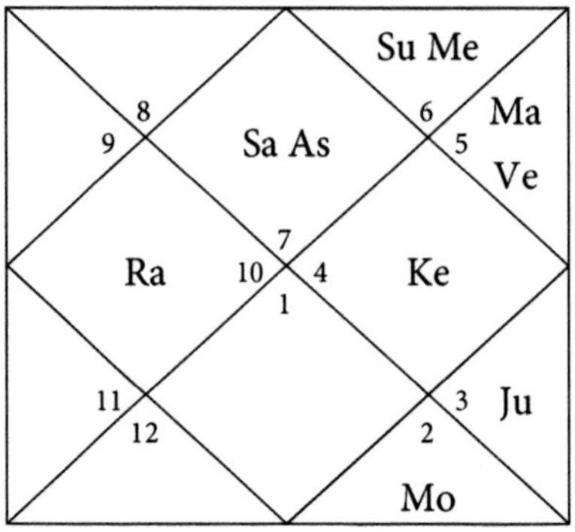

		Mo	Ju
	Ammachi Sept. 27, 1953 9:10 IST Vallickavu, India		Ke
Ra			Ve Ma
		Sa As	Su Me

Su Me
8 6 Ma
9 Sa As 5
Ve
7
Ra 10 4 Ke
1
11 3 Ju
12 2
Mo

Planetary Information

As	23:35	Lib	Vishakha
Su	10:31	Vir	Hasta
Mo	05:59	Tau	Krittika
Ma	14:36	Leo	P.Phalguni
Me	25:23	Vir	Chitra
Ju	02:43	Gem	Mrigashira
Ve	10:25	Leo	Magha
Sa	03:45	Lib	Chitra
Ra	06:36	Cap	U.Ashadha
Ke	06:36	Can	Pushya

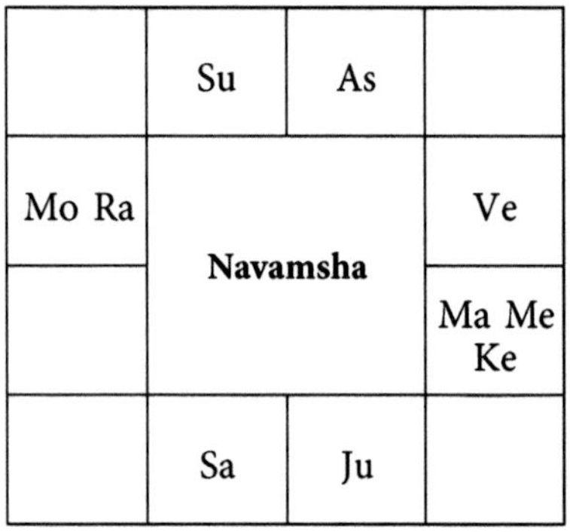

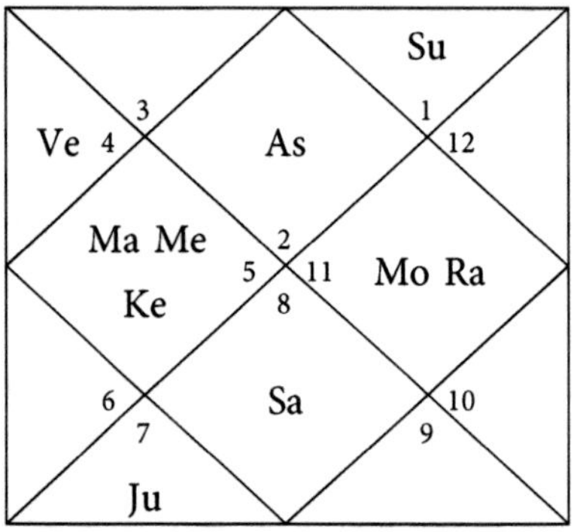

	Su	As	
Mo Ra	**Navamsha**		Ve
			Ma Me Ke
	Sa	Ju	

Su
3 1
Ve 4 As 12
Ma Me 2
5 11 Mo Ra
Ke 8
6 Sa 10
7 9
Ju

Vimshottari Dashas

Su	Jul 17, 1949
Mo	Jul 17, 1955
Ma	Jul 17, 1965
Ra	Jul 17, 1972
Ju	Jul 17, 1990
Sa	Jul 18, 2006
Me	Jul 17, 2025
Ke	Jul 18, 2042
Ve	Jul 18, 2049

Brief biographical summary: "Amma" or "Ammachi" means Holy Mother. Her full spiritual name, Mata Amritanandamayi (Mother of Immortal Bliss), was given to her by her monastic disciples in May 1981. She is widely acknowledged as another remarkable Incarnation of the Divine Mother. Popularly known as "the hugging saint," Ammachi gives *darshan*, or Blessing to millions of people around the world and has done so since 1972. As of October 2007, she had physically embraced more than 26 million people, blessing and consoling them. She continues to travel the world spreading her message of love. Ammachi describes the effect of her embrace, referring to herself always in the third person:

> "When Amma embraces someone, it is not just physical contact that is taking place. The love Amma feels for all of creation flows toward each person who comes to her. That pure vibration of love purifies people, and this helps them in their inner awakening and spiritual growth."
>
> **Ammachi** quoted in *Eye of Wisdom*, by Swami Ramakrishnananda Puri, 2007, pp. 160-161.

Selfless and tireless in this process, she also raises money for numerous charitable projects globally, as well as in India. They are designed to help those in need of health care, education and housing, and include hospitals, orphanages, secondary schools, universities and institutions of higher education, humanitarian aid to the homeless, widows, the elderly, and victims of numerous natural disasters in India and elsewhere. Ammachi personally oversees her many humanitarian projects even when travelling, though her base of operations is at her ashram in Amritapuri, Kerala, India.

FAMILY & EARLY CHILDHOOD: Her birth name was Sudhamani ("ambrosial jewel"). She was born into a poor fisherman's family of Hindu orientation that had lived on the Kerala coast for generations. Her father was industrious and her mother was known as a pious woman. It is said that Sudhamani was born fully enlightened; but her family was unaware of this, even with some remarkable omens. Both of her parents had strange visions while she was in the womb, some involving the Divine Mother, some Lord Krishna. Even so, her birth was treated as a non-event, and other relatives and friends were not informed.

> "The atmosphere that enveloped the child's birth was completely silent and peaceful. Damayanti [her mother] had not felt any discomfort... The babe had a beaming smile on her tiny face! The gaze of the child penetrated Damayanti's innermost heart and was never forgotten.... The parents were puzzled by the babe's dark blue complexion and the fact that the child lay in *padmasana*,[272] holding her fingers in *chinmudra*[273] with the tip of her thumb and forefinger touching to form a circle. They feared that this dark blue shade might be the symptom of some strange disease and that the peculiar posture might be due to abnormal bone structure or dislocation. Various doctors were consulted. [No such handicap could be found, but the dark skin could not be explained, as both parents were light tan.]
>
> **Swami Amritaswarupananda**, *Ammachi: A Biography of Mata Amritanandamayi*, 1994, pp. 14-15.

For six months after birth she retained the dark blue hue associated with Lord Krishna and Divine Mother Kali, but eventually it turned dark brown – except for those occasions later in her life when she began to submerge herself in the Divine Moods (*Bhavas*) of Krishna or Devi. This skin color was ironically the cause of much mistreatment by her family and relatives. They also had very little understanding of her devotional practices and her chanting, which began before she was two years old. Instead, her strange behavior caused alarm. She suddenly began to walk at the age of six months, soon after to run, and to speak in her native tongue of Malayalam. With no instruction from anyone, at age two she began to chant prayers and *bhajans* (devotional chants), composing many of her own. She has continued to sing *bhajans* throughout her life, a common Hindu devotional practice *(bhakti).*

Sudhamani was the fourth of thirteen children, five of whom died at birth or soon after. The remaining children were four daughters and four sons. Family members and relatives were disdainful of Sudhamani's dark complexion and even her hard-working ways. They treated her as a domestic servant and gave her more and more household chores, especially with the steady deterioration of her mother's health due to pregnancies. By the age of nine Sudhamani was up at three in the morning and worked until eleven at night. In this cruel and bleak environment, it was said that she lovingly performed her many duties. In the remaining time, she spent hours in her intense devotional spiritual practices, singing, praying and meditating. For this she was often fiercely harassed. Her family assumed she had gone mad, and her older brother Subhagan was especially intolerant. But other villagers began to see the true divinity in her and secretly tried to help her.

---

[272] Lotus posture in *Hatha yoga*.

[273] This symbolizes the oneness of the individual self with the Supreme.

EDUCATION: Sent to school at age five, Sudhamani showed intelligence and brilliance, with a sharp mind and memory. She immediately absorbed her lessons and could memorize very quickly, excelling well beyond her older brothers and sisters, who were favored to go to secondary school and/or university. Due to her onerous household duties, Sudhamani was barely able to attend 4th grade. Punished by her teachers for tardiness, she stood outside the classroom and paid close attention, enabling her to pass the 4th grade. This was the end of her formal education, as by the time she was ten years old there was no time for her to attend school. If not working at home, she was farmed out to other relatives. By the age of 16, if she had finished her chores, she was allowed to take sewing lessons one hour per day. She persisted and became very good at tailoring, though her father refused to provide her with her own sewing machine.

BEGINNINGS OF THE ASHRAM: Sudhamani began her *Krishna Bhavas* around May 1975 and her *Devi Bhavas* of the Divine Mother around Nov.- Dec. 1975. During these *Bhavas* she took on the Divine Mood of either Krishna or Devi, and gave *darshan* to those who attended. She learned how to become whatever form of god or goddess she contemplated, but it would be *Devi*, the Divine Mother that would endure. Thus began the pilgrimages to see her. Her first disciple resided on the property by 1976, and by late 1978, early 1979 a small group of well-educated young men and women became her first monastic disciples. Most of them would later become her senior swamis and swaminis. They began to reside permanently by her side at the family compound. Initially they lived outdoors much of the time, as the family quarters were already small and cramped. The ongoing presence of the young male monastics was especially problematic for her father Sugunandam, who still had three daughters to marry off. (He finally gave up on marrying off his eccentric daughter Sudhamani, but not without some high drama.) He viewed her disciples as single young men who were a threat to his sense of propriety.

In spite of his lack of funds, from Sept. 1980 through early 1981 Sugunandam was somehow able to marry off the other three daughters, with dowries and all the required gold jewelry. It is reported that Ammachi's "Divine Miracles" made all this possible, with checks showing up at the last minute as well as packets of the necessary jewelry. These three marriages cleared the way for the real beginnings of the ashram. A few thatched huts were constructed near Amma's family home, at the same location where the current ashram now houses 3000 residents who live, work and study there, with ongoing programs for local and international guests.

The legal founding of Mata Amritanandamayi Math and Mission Trust took place in Quilon, Kerala, India on May 6, 1981, with its registration under the Travancore-Cochin State Literary and Charitable Act of 1955. From this time onward, Ammachi officially adopted the name given her by one of the disciples: Mata Amritanandamayi. On Aug. 27, 1982, a *Vedanta Vidyalaya* (school of Vedantic knowledge) was started to teach the ashram residents Sanskrit and Vedantic knowledge. In 2005, the United Nations recognized M.A. Math in India as an NGO (non-governmental organization). Ammachi's now vast network of activities are documented in numerous videos and at her website: www.amma.org.

HER PHILOSOPHY & TEACHING: The basis for Ammachi's philosophy and teaching can be found in Hindu/Vedic texts and in the *Bhagavad Gita*. In this framework, there is one Supreme Reality, called by many names, and all souls ultimately realize Truth after reincarnating and evolving through many births until all *Karmas* are resolved. The universe undergoes endless cycles of creation, preservation, and dissolution. In the Hindu/Vedic tradition, a *satguru* is a spiritually awakened master who can quicken an individual's progress in finding the transcendent Absolute. Personal effort is also needed: self-discipline, good conduct, purification, self-inquiry and meditation.

> "Unless we have a certain degree of mastery over the mind, true peace is difficult to attain.... The purpose of one's life is to realize who we really are. By realizing our own Self we become full, with nothing more to gain in life. Life becomes perfect. ... The path of devotion *[bhakti]* and selfless-service *[seva]* is the safest and most conducive path for many people."
>
> **Ammachi** quoted at her website, www.Amma.org, 2008.

Amma herself is widely regarded as a *satguru*. She is open to all persons – in the Hindu tradition of accepting that no one religion has the monopoly on salvation and the experience of God's pure love. All life is considered sacred and to be loved and revered; thus *ahimsa* (non-violence) is practiced, as well as vegetarianism. Ammachi's ashram and events offer only vegetarian meals and snacks. Her organization also utilizes and offers Vedic astrology, or *Jyotish*.

The example of her own life is central to Ammachi's teaching: tirelessly and selflessly she offers comfort to suffering humanity. Her life purpose is to make spiritual education accessible to the world and to make the central principles of love and compassion an intrinsic part of *all* education and activity, whether social, economic, or political. In an era of expanding wars and domestic violence, she counsels men and women how to be more tolerant and loving towards each other, as well as towards all other sentient beings. She considers environmental concerns extremely urgent and tries to reawaken the once commonplace attitude of ancient times – that of protecting Nature and society because both are a reflection of God. And the Creator can be seen through the creation. She sees "the loss of Nature's harmony and our widening separation from Nature" as **the biggest threat to mankind**, more so than any third world war.

> "We should develop the awareness of a person at gunpoint. Only then can humanity survive. Life becomes fulfilled when humankind and Nature move together, hand in hand, in harmony."
>
> **Ammachi**, "Compassion: The Only Way to Peace," an address delivered at the 2007 Film Festival of Cinema Vérité, Oct. 12, 2007, Paris, France. (Translator: Swami Amritaswarupananda Puri, pp. 41-42.)

> "'This kind of service [with love and compassion, without expecting anything] will help one to lead a happy and balanced life in any circumstance even if one is starving. In reality spirituality is that which teaches us how to lead a perfect life in the world.... Everything depends on the mind. If the mind is calm and tranquil even the lowest hell will become an abode of happiness, but if the mind is agitated even the highest of heavens will become a place of tremendous suffering. That is what one gets from spirituality and spiritual masters, peace and tranquility without which one cannot live.'"
>
> **Ammachi** quoted in *Ammachi: a Biography of Mata Amritanandamayi*, by Swami Amritaswarupananda, 1994, pp. 327-328.

> "Our God-given abilities are a treasure that is meant for ourselves as well as for the entire world. This wealth should never be misused and made into a burden for us and for the world. The greatest tragedy in life is not death; the greatest tragedy is to let our great potential, talents, and capabilities be underutilized, to allow them to rust while we live. When we use the wealth obtained from nature, it diminishes; but when we use the wealth of our inner gifts, it increases."
>
> **Ammachi**, in her Keynote address, "May Peace and Happiness Prevail," at the Closing Plenary Session, 2004 Parliament of the World's Religions.

Assessing the chart of an Enlightened Being: Even for the most seasoned astrologer, it can be difficult to identify the birth chart of a person widely regarded as an Enlightened Being. This is true no matter at what stage of life they became enlightened. Stories circulate about such phenomena and they become mythical in stature, but not easy to repeat in practice. However, many of the same classic astrological rules still apply. The planets and their placements show us a lot about the life cycles and qualities of the person, even if there is much else that is beyond our capacity to understand.

For instance, **Saturn** is *shudra*, the servant, and it is the most prominent planet in this birth chart, placed in the Libra Ascendant and in the 7th house of the *Navamsha* chart – *Digbala* (best angular house for Saturn). For many years Ammachi was treated as a servant in her own family, and she willingly fulfilled obligations not required of her many siblings. She has also become a servant to the needs of many people around the world, blessing and helping them in both practical and spiritual ways. She is in service to "her children" of the world. She also stresses service orientation in her ashram, asking devotees to be mindful of how they can best serve others with *seva* (selfless service). Ammachi's many public programs depend on *seva* for their success and viability. She often participates, and if she sees a task that needs to be done, she starts off doing it. Soon many devotees follow suit and take it over. With a prominent Saturn, the individual often takes on many responsibilities. How effectively this operates depends on the power of its sign and *nakshatra* lords, as well as aspects to Saturn. In this case Venus and Mars are the sign and *nakshatra* lords, respectively, well placed in the 11th house of friends and financial gains.

Saturn in Libra in the Ascendant creates a *Shasha Mahapurusha yoga*, formed when Saturn is exalted or in its own sign in any *Kendra*. But Saturn is not easy in the Ascendant, considered its worst house placement. Even so, for Libra Ascendant, there are benefits, as Saturn is *Raja yoga karaka* planet, and being lord of the 4th and 5th houses gives both power and good fortune, along with the delays and hard work also associated with Saturn. As Saturn aspects the 3rd, 7th and 10th houses from itself (in clockwise motion in the South Indian chart, counter-clockwise in the North Indian chart), she would experience delays and obstacles in the areas of local travel, marriage, and status and visibility, respectively. This would be true for anyone with this birth chart. Ammachi traveled relatively little outside her birthplace until May 1987, when she started to travel widely on her world tours. Her parents tried to arrange to marry her off several times and each time she fought it, threatening in one case to kill off the prospective husband, and even dying in physical form for eight hours, as her father remained unconvinced of her adamant refusal to marry. This may have been partly Jupiter's role (as 6th house lord aspecting Saturn), bringing conflict to the idea of her playing the full traditional daughter's role in the family.

Though some neighbors and locals recognized the divinity in her from the time she was a very young child, it was not until 1975 (at age 22) that a larger public sought out Sudhamani. This forged a new situation with her family, who gradually moved from ignorance and resistance to acceptance, with the exception of her elder brother Subhaghan who resisted up to the time of his death in June 1978. This process would take another five years, from 1975 into late 1980. Sudhamani's decision to remain at home despite the enormous difficulties with her family is also a testimony to the perseverance of Saturn, helped by an aspect from Jupiter, planet of forgiveness. It is notable too that at the start of many of her talks, always given in her native language of Malayalam, Ammachi repeatedly invokes the Sanskrit name of Saturn: *Shani, Shani, Shani.*

Saturn delivers: Saturn is the planet of Truth, making it singularly capable of either elevating us or crushing us under its weight. Ammachi's deeper worth would be seen even more clearly at an international level during her 19-year **Saturn *Dasha***, starting from July 18, 2006. Due to its powerful status as *Yogakaraka* planet for Libra Ascendant, and its prominence in both birth chart and *Navamsha* chart, Saturn *Dasha* would be more beneficial for her than previous *Dashas*. In fact, it would be superior to all other *Dashas* in her lifetime. From July 2006, Ammachi's qualities and teachings were known to an increasingly broader public, and this brought her even greater responsibilities. Her 50th birthday celebrations in late Sept. 2003 brought millions of people to Cochin, India over a 4 to 5 day period.

Though she had many responsibilities unceasingly all throughout her life, the larger degree of public recognition brought the capacity to raise more funds and to initiate more and more humanitarian projects, both nationally and globally. Preceding her Saturn *Dasha* by six years was the Jupiter-Saturn conjunction in May 2000 in her 7th house. Situated in an angular house from her Ascendant, it denoted a more powerful 20-year period ahead for her personally. In Sept. 2004, her 7½ year *Sade Sati* ended. At the same time, tr. Saturn entered her 10th house of maximum visibility. This brought her hundreds of thousands more devotees in more countries. The French movie *Darshan: the Embrace,* by French director Jan Koenen, was released in 2005, and its DVD format on Nov. 14, 2006 – four months into her Saturn *Dasha*. This opened an even larger window on to Ammachi's extraordinary effect worldwide. Prior to this, in Dec. 2005, Amma's organization presented the Clinton-Bush Katrina Hurricane relief fund with $1 million, one of the largest donations from a non-governmental organization.

Magnetic attraction of the *Satguru*: Having resolutely rejected marriage, Ammachi instead receives the full spiritual devotion of her monastic disciples and her millions of devotees worldwide. Westerners are unused to connecting the idea of spiritual worship to the magnetic attraction of **Venus**. But it works much the same way as Venus attracts people to politicians or glamorous movie stars – except the motivation must be pure and of a higher order. If not, there can be negative results or none at all. In Vedic/Hindu thought, if you have great sincerity and persistence in propitiating a deity, a planet or a *satguru* – they have to pay attention to you in a sort of magnetizing process. By some accounts, results can take just moments or up to many long years, depending on one's *karmic* destiny. It may not even be in this lifetime. But by asking them to recognize your sincere state of longing, you may receive some kind of transmission, or some quality may now be more available to you if only because you are sincerely seeking it. The major goal is to bring certain divine qualities into your own human life. By establishing an inner intimacy (not to be confused with physical intimacy) with that deity or *satguru*, the aim is to have them keep you close in their mindfulness, which is considered infinite in any case. The practice of chanting Sanskrit mantras is also effective in this regard. They are given in many Vedic texts.

Ascendant lord Venus is well placed in the 11th house in Leo. In *Magha nakshatra*, meaning "the great one" and symbolized by the throne room, we see Ammachi, the embodiment of Love, sitting on her chair on the dais giving *darshan* for hour upon hour to tens of thousands of people. Venus is in Leo, the sign of royalty. If there were spiritual royalty, she would be the Queen Mother of Divine Love. Leo is owned by the Sun, which in turn is placed in the 12th house of *Moksha* (spiritual liberation) with exalted Mercury in Virgo. Venus is in *Magha nakshatra,* ruled by Ketu. Ketu is *Moksha* significator, which in turn is located in the 10th house of high visibility. The destiny is entirely towards *Moksha*, experiencing it, teaching it, and spreading that teaching.

Venus is with Mars, and when together in the same sign they form a powerful *Raja yoga* as lords of Houses 1 and 7 from the Libra Ascendant. They also form a *Dhana yoga* of wealth, as lords of Houses 1 and 2 together in a good financial house, House 11. The immense success of her fund-raising and her public good works is a testimony to the fine *yogas* in the 11th house of friends and financial gains.

As lord of the 2nd house, Mars rules over not only personal finances but the voice and speech. Venus and Mars together are usually associated with sexual passion, but in this case the fiery passion is transmuted into devotional love, or *bhakti*. This is expressed in a myriad of ways, but the **devotional music**, or *bhajans*, are among Ammachi's greatest spiritual inspirations and she personally leads or oversees most of them. In the familiar Hindu style of call and response, *bhajans* permeate most Ammachi events, in addition to visual presentations of her many talks around the world. Amma considers *bhajans* very effective for Westerners, who are largely mentally oriented. Music can pierce the heart often faster than spoken words, and in the case of devotional music and Sanskrit mantric chants – the vibrational power to elevate minds and hearts is enormously magnified. This concept is supported in Vedic texts.

Confirming the emotional power of her music and wise words is the placement of *Navamsha* Venus in Cancer in the 3rd house of music and communications, and Jupiter in Gemini aspecting its own house (the 3rd house) in the birth chart. Both natal Jupiter and Saturn in air signs confirm her *Dharma* as a communicator.

Spiritual orientation and the luminaries: Because of her unique destiny, some of the most important factors for her would relate to the *Moksha* houses (4, 8, and 12), and we expect them to be featured in some way. Natal Sun and Moon are both placed in *Moksha* houses, Sun in the 12th house and Moon in the 8th house in its sign of exaltation. Further, there is a *Nadi yoga* uniting the Sun and the Moon, as each of them is situated in the *nakshatra* of the other: Moon is in Sun-ruled *Krittika nakshatra* and Sun in Moon-ruled *Hasta nakshatra*. The exchange of *nakshatra* lords is considered very powerful, even more potent than an exchange of sign lords. The two royal planets are thus strongly linked, both from *Moksha* houses. The Sun in *Hasta nakshatra* ("the hand") hints at how she keeps a hand in all areas of her world. As always, the Moon's *nakshatra* placement is the most pivotal. The deity Agni, God of Fire rules over *Krittika nakshatra*. Its *shakti* (major energy or motivation) is to purify. When asked if she would like to enter politics, Ammachi said "no." Her function is that of a "sweeper" – to keep sweeping away the darkness and negative energies from human life, to make way for the life filled with light and for the spirit hidden in matter. She sees everything as saturated with one divine consciousness.

Key planets in *Moksha* houses can also cause many earthly problems for most mortals, especially during their *Dashas*. The most important and difficult of these *Moksha* houses are the 8th and the 12th houses, the 8th because it can involve ongoing suffering (or lingering illnesses), and the 12th because it can demand confinement in an ashram, a prison, hospital, or foreign country. The Moon is not well placed in the 8th house for mundane matters for most of us, as it tends to produce emotional turbulence rather than *Moksha*. But through deep suffering one can come to a place of the deepest understanding. Ammachi suffered greatly through her early years living amidst her own family and she overcame it through intense devotional practices. During most of her life she has slept and eaten very little.

Since 10th lord Moon is in the 8th house, causing a *Duryoga*, the native is "constantly away from home and living abroad." This has been more of a constant from spring 1987. Natal Sun is placed in the 12th house in both birth chart and *Navamsha* chart. In this case, we see the light shining on the inner world, on ashrams, hospitals, prisons, and on foreign residences. As a Spiritual Master, her inner world is reflected in her outer world and vice versa. There is no separation.

Natal Sun is surrounded by planets on houses either side of it. Called an *Ubhayachari yoga*, this *yoga* bestows many benefits – although it is preferable if the surrounding planets are classic benefic planets. This *yoga* is considered excellent for networking with other people, and always provides someone who can be helpful and influential in the life and usually quite a few such people. At the same time, the planets that create this *yoga* – in this case Mars on the one side and Saturn on the other – are both classical malefics. This in turn creates a *Papa Kartari yoga*, initially hemming in the planet and the house in question and requiring greater effort to break out of external barriers of the given life circumstances, especially during the *Dasha* of the hemmed in planet(s). The *Kartari yogas* usually apply to the Ascendant, but can also apply to any house or planet(s).

The Sun and the Moon are King and Queen, also father and mother. Since the Sun and the Moon are both situated in *Moksha* houses, both would bring her difficulty as well as spiritual growth. Read father also from 9th house lord and mother from 4th house lord. In earthly terms, since 9th lord Mercury is in the 12th house of loss and difficulties, with Sun located there as well, this confirms the many difficulties she experienced with her father. Natal Moon in the 8th house can present problems with one's mother. Both parents were relentless in their demands on their daughter. They lacked the ability to understand or appreciate her true nature, especially during the Sun and Moon *Dashas* and up until late 1980. At this point the triple Jupiter-Saturn conjunctions in Virgo (from Dec. 31, 1980 to July 23, 1981) eased the way for Ammachi's spiritual path. They brought her foreign exposure and the start of an ashram – both 12th house matters. The ashram was established legally on May 6, 1981. (See the upcoming section: *The roles of Jupiter and Saturn.*)

Thus even for an Enlightened Being such as Ammachi, the *Moksha* houses also caused her earthly problems. Because of the very nature of *Dashas*, these problems would be maximized during their *Dashas*: **Sun *Dasha*** from birth to July 17, 1955; 10-year **Moon *Dasha*** from July 17, 1955 to July 17, 1965. These were some of the worst years with her family and when she was the most vulnerable. The end of her formal education came in Moon-Venus and Moon-Sun *Dasha/ Bhuktis*: May 18, 1963 to July 17, 1965.

Difficulty in the early years – foreigner in the family: Young Sudhamani appeared strange to her parents and family from the time she was born. Starting from her birth, she was the foreign eccentric in many ways – being also several shades darker than her siblings. Natal Rahu (the foreigner) in the 4th house of family foundations can make one a foreigner in one's own family. The 4th house is also a house of education. As a separative planet, Rahu cut off her education, especially with *nakshatra* lord Sun in the 12th house. Sudhamani was treated as an outcast and a servant, and due to the increasing number of domestic chores demanded of her she was only allowed to attend school up to the 4th grade.

Her siblings, both older and younger, were given more education and privileges, while she experienced an extremely difficult time during her first 19 years. This is shown by the *Vimshottari Dasha* sequence. We have established the difficulties of the Sun and Moon *Dashas* up to July 18, 1965. This was followed by the 7-year Mars *Dasha*, which brought her a bit more freedom as a seamstress but still kept her very confined within the family. Mars is well placed house-wise, but its sign lord is in the 12th house. These years brought accelerated spiritual progress, but were truly the most grueling of her life.

There was little alleviation until she entered Rahu *Dasha* July 18, 1972. In 1972 she first started giving *darshan* to members of the public. Rahu is located in the 4th house in the sign of Capricorn. Since Rahu is in the sign owned by an exalted Saturn in Libra, Rahu *Dasha* shows an improvement for her in worldly terms. Rahu is also Saturn-like, and Saturn is an excellent planet for the Libra Ascendant chart.

There were bound to be some dangers for her during the sub-period of **Rahu-Jupiter** period (March 1975 to Aug. 1977), since Jupiter is 6th house lord for Libra Ascendant. Thus, natal Jupiter's aspect to the Ascendant brought ongoing conflicts, especially in its *Dasha* or sub-*Dasha* period. Sometime in 1975 (no date available), Sudhamani remained officially dead for eight hours before she came back to life, with no sign of physical damage. Her father had tried once again to marry her off and still insisted that as her father he had certain rights. He was strongly influenced by the local Rationalists, who maintained that Sudhamani could not possibly be a Divine Being. They kept trying to disrupt her events and causing her trouble. As she had threatened, her life force disappeared after he pressed to have his daughter back under his control. Seeing the results, he finally surrendered, weeping in desperation, and Sudhamani returned to life after eight hours.

On another occasion, her older brother Subhagan tried to have her murdered. He was ashamed and angry at having an unmarried sister who garnered all this attention he did not understand or respect. He also harassed her devotees and tried to dissuade them from going for *darshan.* Eventually he committed suicide by hanging on June 2, 1978, his death having been predicted several weeks earlier by Sudhamani. On that day, tr. Mars and Saturn were in early Leo in her 11th house of elder siblings. The 11th house lord Sun was transiting in her 8th house of loss and transition, aspected by tr. Saturn. A late waning Moon in Aries preceded the upcoming New Moon on June 6, 1978 at 21:11 Taurus in her 8th house, very close to her 8th house cusp. Subhagan's death removed the last major obstacle in the family and cleared the way for the arrival of more monastic disciples in late 1978. Sudhamani predicted Subhagan would be reborn within the family in just three years, and indeed there was the birth of a boy in just three years. He loved to spend time in the temple.

The Divine Mood, or *Bhava*: From late spring 1975 young Sudhamani began her *Krishna Bhavas,* and from late 1975 her *Devi Bhavas* of the Divine Mother, the embodiment of Universal Love. From the first *Devi Bhavas* she took on a new fierceness, fearlessness and protectiveness that helped to shield herself and her devotees from the local troublemakers, and from the obstacles caused by her father and brother Subhagan. Even so, Sudhamani had to deal continually with detractors such as the local Committee to Stop Blind Beliefs. They kept trying to set traps for her and each time she undid them. She had to enlist her father to keep handling the police and the courts, who received ongoing complaints from those who challenged her abilities to manifest her *Bhavas.* While in these *Bhavas,* she gave *darshan* to increasing numbers of people. This had the effect of inflaming her detractors, who were determined to expose her as a fraud.

So many people were now coming for *darshan* (and on their property) that her family also had to face certain realizations about Sudhamani's real power and stature. Tr. Saturn entered Cancer July 24, 1975 through Sept. 8, 1977, spelling several factors for her on a mundane and personal level. Cancer is her 10th house of maximum visibility and also the 3rd house from her Moon. It marked the end of her 7½ year *Sade Sati.* We know that *Sade Sati* can be an especially difficult period for anyone with natal Moon in the 8th house, even if exalted. Also tr. Saturn in the 10th is uniquely capable of giving maximum visibility. From the 10th house, it also aspects the 4th house of family foundations. Her parents and her family were the last to recognize who she was, but Saturn's opposition forced them to see.

Disciples and friends: The 5th house is disciples and 11th house is friends. Reading from the natal Moon in Taurus, the 11th lord is Jupiter – situated in the 2nd house from the Moon, creating wealth from friends. The 5th house from the Moon is well endowed, as it contains exalted Mercury and Sun in Virgo, both 4th and 5th lords from the Moon, forming a *Raja yoga*. Exalted Mercury also describes the native's high intelligence and speaking abilities, as well as that of the disciples. But initially there would be difficulties for these disciples that would only ease up after nine years. Most of the more highly educated group of spiritual disciples came in late 1978 and early 1979. From the time they first came to Ammachi, they faced ongoing harassment from their own families for leaving their worldly lives and going to reside at the ashram with a young woman. This was in addition to harassment from local troublemakers, who persisted in their efforts. But from Sept. 1980 through early 1981, Ammachi's father was finally able to marry off his other three daughters, clearing the way for the establishment of an ashram. This coincided with the first of the triple Jupiter-Saturn conjunctions in Virgo on Dec. 31, 1980.

Ammachi's ashram became a legal entity in spring 1981 during her **Rahu-Mercury** *Dasha/Bhukti*. Natal Mercury is beneficial as *Atma Karaka*, at the highest degree of celestial longitude. Mercury is a planetary friend of Venus, and both Rahu and Mercury are *Nadi yogakarakas* for Libra Ascendant. Natal Rahu is in the 4th house in the *nakshatra* of the Sun. The Sun, in turn, is situated with exalted Mercury in the 12th house. (The 12th house rules over ashrams and spiritual matters, but also losses and expenses in worldly affairs.) And though Mercury in its own sign is protective, this would reflect some of the worldly difficulties Ammachi still faced due to her increased prominence as a spiritual leader. Her leading disciples also had to deal with being branded eccentrics or worse. One of her most senior swamis, (formerly Balagopalan, or Balu) Swami Amritaswarupananda, experienced two final years of severe harassment from 1985 to 1987, when his father put out a search warrant for his arrest as a "lunatic." On Aug. 27, 1987 Swami A. appeared before the Kerala High Court, and with police protection. The case went in his favor, but his father was undeterred and soon appealed the decision in the Supreme Court of New Delhi. Ammachi urged her disciple to go to New Delhi, where he was finally cleared as a "lunatic" and allowed to pursue his spiritual path as both a Hindu and a devotee of Ammachi as his Spiritual Master. This last hurdle and *karmic* test with his family coincided with the start of Ammachi's world tours. He then accompanied her on all subsequent tours, and became her leading speaker, interpreter, writer and singer. For his extraordinary singing of *bhajans*, he is sometimes dubbed in the press as "the Hindu Pavarotti."

The roles of Jupiter and Saturn: The triple Jupiter-Saturn conjunctions of 1980-1981 in Virgo (*Hasta nakshatra*) marked the legal founding of her ashram in Kerala, India on Wednesday, May 6, 1981. These conjunctions occurred in the 12th house from her Ascendant, close to her natal Sun, and 5th house from natal Moon exalted in Taurus. The first of the conjunctions occurred on Dec. 31, 1980 at 15:55 Virgo. The next two would occur on March 4, 1981 and July 23, 1981. In early May 1981 transiting Jupiter and Saturn were close to the degrees where they would turn Stationary Direct in late May and early June 1981 respectively, giving them added potency. The New Moon prior to the legal founding was on May 4, 1981 at 20:01 Aries in *Bharani nakshatra*, just opposite Ammachi's Ascendant. Five planets were in Aries (Sun, Moon, Mars, Venus and Mercury) in her 7th house of partnerships and overseas residence, which can also be read as foreign travel. The confluence of all these astronomical events reinforces the astrological effect of their positions in the heavens.

The founding date on May 6, 1981 was also auspicious astrologically. The two luminaries, Sun (still in Aries) and Moon (now in Taurus) were each in the sign of their exaltation. There was a waxing Moon, a good *Tithi* (soli-lunar phase), with transiting Moon's *nakshatra* in *Rohini*, favorable to Ammachi's (the 2nd from her natal Moon *nakshatra* in *Krittika*), and an appropriate weekday, Wednesday – Mercury's day. It is possible Amma received astrological advice on selecting this date, although she is quite capable of dispensing such Divine Grace herself, even to her own organization. In any case, the basic principles of *Jyotish* help us to understand how an auspicious founding date blesses the activities of that organization or enterprise, allows it to grow, and minimizes the obstacles. Indeed, this was its primary use in ancient times: to time the sacred rituals in the Vedic civilization that would insure its continuity.

Air trine – communications highlighted: Natal Jupiter is in Gemini in the 9th house, and natal Saturn is in Libra in the 1st house receiving Jupiter's aspect. These are two of the best houses in the chart, giving more power to Jupiter and Saturn, the all-important *Dharma* and *Karma* planets. The trinal aspect is considered harmonious and stabilizing, and when it occurs between Jupiter and Saturn in the birth chart, it generally eases the way for the person to know their *karmic* and *dharmic* destiny and go towards it, having no inner conflict about it.

Although Jupiter's role as classical benefic and *Dharma* planet is still positive, as 6th lord for Libra Ascendant it brings outer conflict. This would show itself in the outer collision course Ammachi's spiritual life path created initially within the family and in the local environment. In the air signs, the emphasis is on communications and trade interests, as Ammachi's Krishna and Devi *Bhavas* brought many people to her. Confirming and enhancing the communications abilities, the final dispositor of the chart is Mercury, classic *karaka* for communications and publications. Mercury, in turn, is in its sign of exaltation in the 12th house of *Moksha*, or spiritual liberation. The earthy basis of her Sun and Moon gives a practical and physical modality to her message, while the air sign Ascendant, Jupiter and Saturn emphasize Ammachi's communications of love, spiritual upliftment and greater understanding between all peoples, regardless of race or creed. In addition to numerous humanitarian projects in India and around the world, Ammachi has originated a whole cottage industry of books, recordings (including a large collection of *bhajans*), and as of 2008 – Amrita TV. With only a 4th grade education, she wields the power of not only a Spiritual Master but the CEO of a very large international company. Such is the scope of all her operations, educational and philanthropic.

Chart 31: Jacqueline Kennedy Onassis

Birth data: Sunday, July 28, 1929, 14:30 EDT, Southampton, New York, USA. Long. 72W23' 24, Lat. 40N53'03, *Lahiri ayanamsha*: -22:52:13, Class A birth data, from memory. Frances McEvoy quotes her to mutual friends. Data not released until after her death in May 1994. Biography: Sarah Bradford, *America's Queen: The Life of Jacqueline Kennedy Onassis,* 2000. Ascendant: 25:06 Libra.

	Mo Ra	Ju Ve	
	Jacqueline Kennedy Onassis Jul. 28, 1929 14:30 EDT Southampton, NY		Me Su
			Ma
SaR		Ke As	

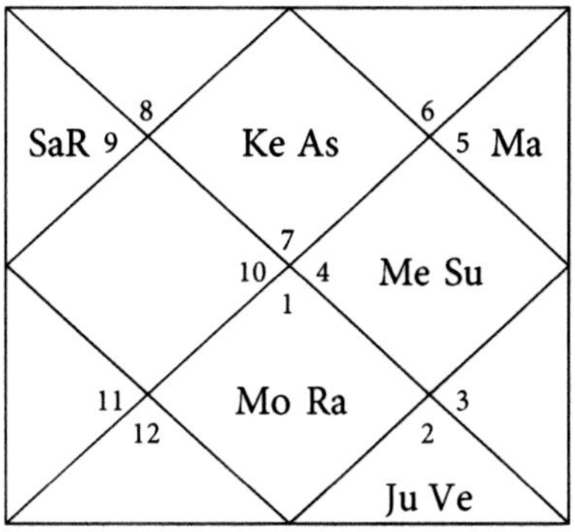

8 6
SaR 9 Ke As 5 Ma
7
10 4 Me Su
1
11 Mo Ra 3
12 2
Ju Ve

Planetary Information

As	25:05	Lib	Vishakha
Su	12:17	Can	Pushya
Mo	02:44	Ari	Ashwini
Ma	21:57	Leo	P.Phalg.
Me	09:31	Can	Pushya
Ju	16:42	Tau	Rohini
Ve	28:53	Tau	Mrigashi
Sa	01:46	Sag	Moola
Ra	24:20	Ari	Bharani
Ke	24:20	Lib	Vishakha

	SaR Mo	Ke As	Ju
	Navamsha		
	Ra	Ma Su	Ve Me

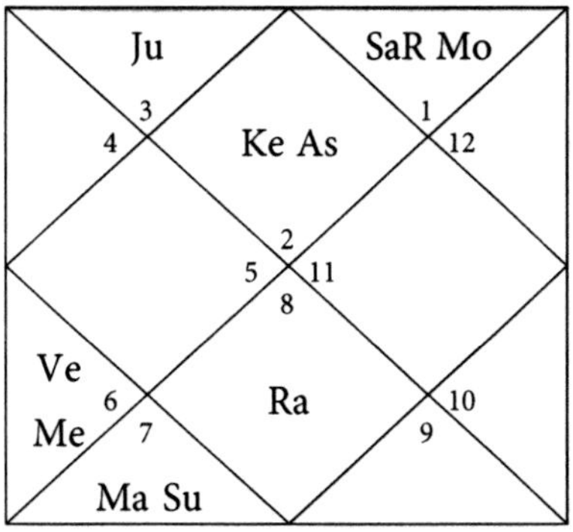

Ju SaR Mo
3 1
4 Ke As 12
2
5 11
8
Ve
6 Ra 10
Me 7 9
Ma Su

Vimshottari Dashas

Ke	Feb 20, 1928
Ve	Feb 20, 1935
Su	Feb 20, 1955
Mo	Feb 19, 1961
Ma	Feb 19, 1971
Ra	Feb 19, 1978
Ju	Feb 19, 1996
Sa	Feb 19, 2012
Me	Feb 19, 2031

Brief biographical summary: Jacqueline Bouvier Kennedy Onassis was among the most glamorous and cultured of America's First Ladies, known for her ability to be graceful and charming in good times and to stand strong in difficult times. She spoke four languages fluently (English, French, Spanish and Italian) and had a deep knowledge of the arts, literature, and history. She was greatly admired – as much for her beauty, intelligence, style and good taste as for her role as the widow of the martyred president and mother of his two children. Though she lived another 31 years after Kennedy's death and had various other partnerships – notably a second marriage to the billionaire Greek shipping magnate Aristotle Onassis – she is most remembered as the captivating wife and later grieving widow of John F. Kennedy, U.S. President 1961-1963.[274] Full of strength, grace and courage, she grieved not only for herself and her two children, but for a nation and a world in mourning for her late husband, slain in one of the most tumultuous political assassinations of the 20th century. Jackie's presence and noble demeanor were crucial to the healing of that wound. The years since November 1963 seemed to collapse instantly when at her death in May 1994 she was buried next to President Kennedy at the nation's capital.

At that moment, she was still largely a genuine American heroine, the carrier of the flame – though often unwillingly. Her desire to be an important part of history was matched equally by her resistance to the accompanying demands. She wanted to be "Queen of Camelot" but was far less willing to pay the social price. In a publicity-hungry era she refused to play by the rules, even in her years at the White House. Though she had to deal with the press, she barely tolerated their presence. She loved the sense of public power she commanded through her two successive

[274] For further material on John F. Kennedy, see Chapter 8, Chart #25.

husbands, but she also wanted a level of privacy that became almost impossible to maintain in the "Culture of American Celebrity," especially in some of her more reckless moments.

Her sense of history may have surpassed that of her slain husband, such was her sure sense of what rituals and symbols needed to be performed and seen by the nation. This was both during the presidency – with her detailed restoration of the White House, making it "a showcase for great American art and artists," with lavish dinners (including 16 State dinners), parties, concerts and performances at the White House – and also in late November 1963 in the days after her husband was killed. Though brother-in-law Sargent Shriver directed the funeral proceedings, Jackie was consulted on most details and credited with much of its success *and* flaws. (French President Charles de Gaulle thought it was over-produced and over-dramatized.) It was her idea to copy Lincoln's funeral, and to establish the eternal flame at Arlington National cemetery. She knew what she wanted written and insisted on it to such a degree that *Life* magazine ran the story she wished, under great pressure and expense, as the press deadline had to be extended. It would be about "the Camelot Presidency." She wanted her husband remembered as "a man of magic" and she was the chief purveyor of this myth. One week after the assassination she urgently summoned friend and journalist Theodore White to Hyannisport, Mass. Due to bad weather no planes were flying that night. He arrived by private limousine in a heavy rainstorm. After a long talk ending well after midnight, he wrote his article and phoned it in.

> "'... There's this one thing I wanted to say.... I'm so ashamed of myself.... When Jack quoted something, it was usually classical... I want to say this one thing, it's been almost an obsession with me, all I keep thinking of is this line from a musical comedy.... [It was *Camelot,* Jack's favorite]...'Don't let it be forgot, that once there was a spot, for one brief shining moment that was known as Camelot.'
>
> There'll never be another Camelot again.... Do you know what I think of history? When something is written down, does that make it history? The things they say! ... For a while I thought history was something that bitter old men wrote. But Jack loved history so... No one'll ever know everything about Jack. But history made Jack what he was ... Then I thought, for Jack history was full of heroes. And if it made him this way, if it made him see the heroes, maybe other little boys will see. Men are such a combination of good and bad....'
>
> And now she reverted to the assassination scene again, as she did all through the conversation, which had swung between history and death.... 'Then later I said to Bobby, What's the line between history and drama?'"
>
> **Jacqueline Kennedy** quoted in Theodore H. White's *In Search of History: A Personal Adventure,* 1978, Epilogue, pp. 678-679.

This now famous "Camelot interview" occurred on Nov. 29, 1963. It was one of the last interviews Jacqueline Kennedy would ever give. There were only two others: in April 1964 **William Manchester** interviewed her for his book *Death of a President, Nov. 20-25, 1963* (published 1967); and between March 2 and June 3, 1964 **Arthur Schlesinger, Jr.** interviewed her in seven recorded sessions. Those 8 ½ hours of tapes were intended to be released and published decades later, and only at the discretion of her two children Caroline and John Kennedy, Jr. In fall 2011 they were published in book and audio format as *Jacqueline Kennedy: Historic Conversations on Life with John F. Kennedy: Interviews with Arthur M. Schlesinger, Jr.* 1964, with an Introduction and Annotations by presidential historian Michael Beschloss, and a Foreword by Caroline Kennedy. This oral history of the Kennedy years was the first time since Nov. 1963 that her whispery voice would be heard publicly again, except informally in the context of her many artistic, philanthropic, or editorial projects over the next thirty years of her life. She would live an interesting life, always in the glare of the cameras but with the allure of mystery, soon embracing international society and leaving politics behind, aside from organizations and events supporting her late husband, including the presidential library project.

An aesthetic view of life: Kennedy loyalist and biographer Arthur Schleslinger observed that Jackie's view of the world was not intellectual or moralistic, but aesthetic. If she was a dilettante, she had a world view and always an eye to history, with a keen interest in the arts, artists, writers, and architects. Her strong interest in the preservation of cultural and architectural monuments led her to spearhead several major projects, starting with her renovation of the White House. This was a 13-month project which she led, both in fundraising and planning, to redecorate the White House with authentic furnishings and period pieces. Her televised *Tour of the White House* on Feb. 14, 1962 was seen by hundreds of millions of viewers in many countries, as it was rebroadcast and syndicated. Regular tours of the White House soon followed, as did Congressional recognition of its historical value. This was also the first major television documentary featuring a woman, even in the numerous scripted voice-overs. Jacqueline Kennedy's role was pivotal in leading the way to the greater recognition of women, both in traditional and non-traditional roles.

Jackie read voraciously and wrote well. Her fluency in several languages, notably French, dazzled the French public during her years as First Lady and only seemed to enhance her image of casual, contemporary chic. Far less known to her public was her biting and often derisive wit. (She referred to Lyndon and Lady Bird Johnson as "Colonel Cornpone and his little pork chop." Those who stared at her at Inauguration Balls she called "mesmerized cattle.")[275] In 1951, during her senior year at college she won *Vogue* magazine's Prix de Paris, a writing contest that awarded the winner an internship at Paris Vogue and six months in Paris. Asked to name the three men in history she would most like to have known, in her winning essay Jackie listed French poet Charles Baudelaire, Irish playwright Oscar Wilde, and Russian ballet impresario Sergei Diaghilev.

Her skill at managing her image and reputation: If her taste in artists and in men leaned towards "bad boys" and "pirates," she also understood that keeping a good reputation meant controlling her public image. Above all, Jackie was a major catalyst for the mythmaking of the Camelot presidency. She became increasingly skilled, even obsessed, with the management of public perception that would preserve her own place in history alongside that of Jack Kennedy.

> "It's not what you are that counts," Joe [Kennedy] liked to tell his children, "it's what people *think* you are."
>
> **Edward Klein**, *Just Jackie: Her Private Years*, 1998, p. 107.

At several key points, Jackie's public image suffered some reversals, much of it from October 1968 onward and perhaps inevitable after being on such a pedestal. According to Gallup polls Jackie was the "Most Admired Woman in the World" from 1962 to 1968. But she lost that position instantly with her October 1968 marriage to Aristotle Onassis. It damaged her reputation more than anything else in subsequent years: she went suddenly from Saint to Sinner. Headlines read:

---

[275] Jackie quoted by Larissa MacFarquhar, "The Importance of Being Jackie," *The New York Times*, Book Review section, Sept. 1, 1996.

"America has Lost a Saint," "Jackie Weds Blank Check," "Jack Kennedy Dies Today for a Second Time," and "The Reaction Here is Anger, Shock, and Dismay."[276] Her choice of partner was viewed as unseemly – a divorced billionaire 23 years her senior, and in business with a Greek military government. But he did offer enormous wealth, and Jackie's taste for the high life and reckless spending habits were just becoming more widely known.

> "My first impression [of Jackie], and it never changed...was that I was in the presence of a very great tragic actress."
>
> **William Manchester**, quoted in Edward Klein's *Just Jackie: Her Private Years*, 1998, photo inlay between pp.184-185.

Jackie also suffered some bad press in 1966-1967 with the flap over a book by William Manchester (1922-2004), historian and Kennedy biographer. In 1964, she commissioned Manchester to produce an account of the assassination. When he completed the book in 1966, she was not happy with the results, even though the publishers had edited and approved it. She tried to block the book and magazine serialization. Even the Kennedy family was distressed over Jackie's "sordid battle" and messy public feud with Manchester. To his own detriment, Manchester had signed a contract with the Kennedys giving them final editorial discretion over the contents of the book. After many delays a settlement was reached and *Death of a President, Nov. 20-25, 1963* was published in spring 1967. It was highly successful and still considered the definitive history of events by an historian who knew all the players, as did Arthur Schlesinger Jr. (1917-2007). Jackie also disliked Schlesinger's book, *A Thousand Days* (1965). Though both men were staunch Kennedy loyalists and both books won major prizes, Jackie had objections to both books. Though hard to pinpoint specifically, her objections appeared to center around putting into print a reality she had firmly in her own mind in mythical terms. Above all she wanted to control the message and the publicity. Pressure also came from the Kennedy clan, whose demands were due to further political ambitions of various family members, notably Bobby Kennedy.

Writer Norman Mailer referred to Jackie as a **"prisoner of celebrity."** For most of her life after becoming First Lady she was constantly followed by crowds and pursued by photographers, who often camped outside wherever she was residing, at home or on her travels. From 1964 onward, any potential suitor was also intensely scrutinized by the press. Though Jackie loved being photographed, she wanted to have control over the process. Among the most persistent and aggressive photographers was Ron Galella, who amassed a large number of photos of Jackie over the years. She initiated a lawsuit against him in 1972, and since the results left Galella feeling his First Amendment rights had been violated, he continued. In Dec. 1981, Jackie initiated a second lawsuit against him, when he clearly transgressed the 1972 injunction. She was satisfied with her victory, but was soon back in court with lawsuits against retailers whom she accused of using "Jackie Kennedy look-alikes." She found them distressing and disturbing to herself and her reputation. Her complaint:

> "I have never permitted the publicity value of my name, likeness or picture to be used in connection with any commercial products."[277]
>
> **Jacqueline Kennedy Onassis**

---

[276] Headlines respectively from: West German *Bild-Zeitung*, a British tabloid, Rome's *Il Messagero*, and *The New York Times*.

[277] Source: C. David Heymann, ***A Woman Named Jackie: An Intimate Biography of Jacqueline Bouvier Kennedy Onassis***, 1989, p. 611. Jackie Kennedy Onassis is quoted regarding her legal case against Christian Dior, Inc., Landsdowne Advertising Inc., photographer Richard Avedon, part-time model Barbara Reynolds and Ron Smith Celebrity Look-Alikes.

After Onassis died March 15, 1975, Jackie returned to Manhattan and began her new life. She also consciously set about rehabilitating her image from that of a voracious consumer to an honorable contributor to society. It took much effort, as her actions were closely followed and reported, including her purchase of 425 acres of land on Martha's Vineyard, and the building of her 19-room Cape Cod-style house on the island, completed in summer 1981. But in the end she largely succeeded. At the time of her death in May 1994 she was highly regarded.

Though she had not abandoned her rich tastes, she became known more as a gracious hostess and a patron of the arts, and less as an international jet setter and member of the idle rich. On Sept. 22, 1975 Jackie started her working life as a Consulting Editor at Viking Press, four days per week at a salary of $10,000 per year. In Feb. 1978, she started as Associate Editor at Doubleday – a job she held until her death in May 1994. She became their most aggressive chaser of celebrity biographies, among other projects. She started at three days per week at a salary of $20,000 per year – a big jump from her Viking salary, but a pittance compared to her $30,000 monthly allowance during her marriage to Onassis. She appeared more grounded and accessible as a working woman and as a selfless and devoted mother. The press, while still focused on her large expenditures, spoke well of her charitable endeavors and her projects to save local architectural treasures, such as Grand Central Station. Arts in the U.S had been treated like "a stepchild," she said, and needed more patronage, as in the European tradition.

By 1986, she reduced her workload at Doubleday to three half-days per week, spending as much time at the beauty parlor, as well as continuing as an active equestrian. During the 1980s she settled into a somewhat more private life, choosing not to remarry. Instead she had a discreet live-in relationship with financier Maurice Tempelsman. She sought independent status as a career woman and fundraiser, and this looked like a big change from her fame always being attached to a powerful man. The only real difference was that much of the public was unaware of Tempelsman's important role in her life, and still associated her closely with her two famous husbands. Their fame and wealth combined with her natural glamour gave her a kind of stardom in perpetuity, akin to prized royalty.

Jackie knew that John Kennedy's reputation could eventually be damaged by details of gross recklessness in his personal life, including his contacts with the Mafia. Along with the Kennedy clan, she sought to keep a lid on them for as long as possible, to "protect the children." However, from Dec. 1975 onward came the unleashing of information she most dreaded, as it had the power to diminish the image of the "Camelot presidency." This started with the Church Committee hearings in U.S. Congress in Nov. 1975, and blew up into sensational headlines on Dec. 17, 1975, identifying one of Jack Kennedy's many lovers – Judith Campbell Exner. Kennedy shared her with Mafia boss Sam Giancana from 1960 to 1962. This came as Congressional investigations revealed the Kennedy administration's plans to kill Fidel Castro and other "undesirable" world leaders, with help from Mafia leaders.

Childhood and early family life: Jackie was named after her father, Jack Bouvier, whose nickname for her was "Jacks" and whose exotic dark good looks she inherited. Her name first appeared in the newspapers with an announcement of her second birthday party, a society ritual followed 16 years later by the socialite rites of the debutante. In July 1947 she was introduced into society as a debutante and was even "Debutante of the Year" in fall 1947. She spent her early years between New York City and East Hampton, Long Island, and learned to ride almost as soon as she could walk, inheriting her mother's equestrian skills. Through her childhood and teenage years Jackie won many local equestrian prizes. With her mother's remarriage in June 1942, they settled into their stepfather's imposing Georgian mansion in Virginia, just across the Potomac River from Washington, D.C. Educated at the best private schools, she wrote poems and stories, drew illustrations for them and studied ballet. Her best athletic skill was horsemanship. Beyond that, she was not much of an athlete. She attended Vassar College for two years, the Sorbonne in Paris for one year and graduated from George Washington University with a major in French literature.

Sibling rivalry: Jackie and her younger sister Caroline Lee (b. March 3, 1933) were close, but very competitive and jealous of each other. Their father's obvious preference for Jackie did not help. The sisters had difficulty sharing the spotlight with each other and Jackie tended to gain most of it, much to Lee's chagrin. Lee had hoped to divorce her second husband and marry Aristotle Onassis, but after introducing him to Jackie in Oct. 1963 she soon found herself sidelined. They were mostly estranged during the last several decades of Jackie's life, as Lee never forgave Jackie for meddling in her life. Lee gained international attention at the start of her acting career in the 1960s, but it was cut short with poor reviews of her performances. Otherwise, she has lived an interesting life, with many financial and social privileges. She excelled at interior decoration and for years served as a public relations executive for the Italian fashion designer Giorgio Armani.

Turbulent heritage: Of Irish and French descent, Jackie's parents were married July 7, 1928, when Janet Lee was 20 and Jack Bouvier, a dashing stockbroker was 36. During the glittering society wedding celebrations, Jack clashed immediately with his new father-in-law, James Lee. The marriage would be continuously tense and perilous, threatened by the couple's deep lack of compatibility and an inability to live within their means. Other factors were Jack's flagrant womanizing, his drinking, gambling and unending failures in the stock market. He was also a profligate spender of his father-in-law's money and became increasingly dependent upon him for loan upon loan. The young family moved often and took lavish vacations.

Janet and Jack separated on Sept. 30, 1936. Jackie was then seven years old and her sister Lee was just three. Reconciling briefly in spring 1937, the couple separated permanently in Sept. 1937. Janet initiated divorce proceedings in Jan. 1940, and by the time it was finalized in July 1940, numerous allegations of Jack's adultery (later dismissed in court) were all over the newspapers. The result left Jack Bouvier's reputation in disrepute, causing shame for himself and his daughters for some years, and leading Jackie to seek respectability and the rehabilitation of the family's image. Her parents' bitter divorce also left young Jackie with a fear of poverty and a determination to marry well financially – a goal strongly reinforced by both parents. After marriage, and perhaps to ease the fear of poverty in their own minds, both sisters developed a penchant for extravagant and compulsive spending that far exceeded some of the wealthiest women in the world. Shopping for expensive items was a major preoccupation and one that needed to be financed.

At the time of her marriage to Jack Bouvier, Janet Lee was a slim, petite brunette with charm and a pretty face. Her friends described her as:

> "highly ambitious, smart, agressive as hell, a daredevil horseback rider who believed in hard work and self-reliance."[278]

[278] *Ibid.*, p. 11.

Her grandparents were Irish immigrants, and her parents were among the newly affluent through property and finance. Janet grew accustomed to a life of wealth and privilege, and sought continuity for this way of life. Her mishap was to marry a man who appeared wealthy, but whose family had more pretense to wealth and nobility than actually existed by the 1920s. Nor did Jack's nature help matters.

The couple shared a strong materialistic drive, but Janet was as pragmatic and proper as Jack Bouvier was decadent. Her ambitious nature and the failure of their marriage may well have exacerbated her quick temper and her tendency to be critical of others. She was both overly critical and insanely jealous of her daughters' love and preference for their father over their mother. She had ongoing animosity towards Jack Bouvier, whose bonds with their daughters continued to be a source of anger and conflict for her. Determined to cut him out of the social scene as much as possible, she did so even on Jackie's wedding day and the many gala events leading up to it. Janet's incessant criticism could be withering. Well into adulthood, Jackie said she found it easier to entertain kings and queens than to be in the presence of her mother. Janet was determined to marry upwards, once she was through trying to make the Bouvier marriage work. Her father operated a bank and provided an affluence she aimed to parlay into further upward social and financial mobility.

Janet was emotionally cool, shrewd and calculating. After her separation and divorce from Jack Bouvier, she dated numerous men. Some called it "the Great Manhunt." She finally settled on Hugh D. Auchincloss, heir to the Standard Oil fortune, and multimillionaire investment banker. They married June 21, 1942, and their union lasted until his death in 1979. **Janet Lee Bouvier Auchincloss** (b. Dec. 3, 1907, d. July 22, 1989) and her two daughters moved into their new life at Auchincloss's vast estates, one outside Washington, D.C. and a second in Newport, Rhode Island.

John Vernou Bouvier III (b. May 19, 1891, d. Aug. 3, 1957) earned the nickname "Black Jack" from both his dark good looks and his gambling habits. A stockbroker who gave lavish parties beyond his means, he claimed his heritage was French aristocracy. However, this was greatly exaggerated. (Jack's father had written a family genealogy full of errors and fabrications, but it perpetuated this myth.) In fact the Bouviers descended from "the petite bourgeoisie" of shopkeepers, farmers, tailors, etc. Though attracted to money, Jack lacked business sense, incurred mountains of debt and was far more adept at spending money than building and maintaining it. His chief assets were his looks, charm, and persuasiveness, especially with women or obtaining loans.

His personal story bore a striking resemblance to that of Jack Kennedy before and after his marriage to Jackie. Neither of them could deal realistically with money, Jack Kennedy far less so, though he was shielded by his father's wealth. With a lack of introspection and a propensity for a high-speed life of action, both men had zest for the sexual conquest of women as a major life activity. In turn, their wives hoped in vain to change this entrenched and decadent pattern. Even if initially bedazzled by the sheer bravado and hedonism of it, they had to live with the consequences. Judith Frame, a Bouvier family acquaintance, describes Jackie's father and "his compulsive lusting":

> "Black Jack was lethal, absolutely lethal. He was extremely handsome and knew it. He used to strut around the Maidstone [a Country Club in East Hampton, NY] like a barnyard rooster, fixing his hair, gazing in mirrors, parading for the women. He was the Don Giovanni type – seductive, charming, dangerous. He carried himself as if all eyes in the world were upon him. I don't know how tall he was. Six feet maybe, but he looked much bigger. It was the way he entered a room, the way he moved. Of course all this posturing

> may well have been a cover-up for some deep-seated insecurity on his part – it's hard to say. He used to go speeding up and down the roads in his convertible drawing attention to himself, disappearing in a haze of champagne and dust. It reached the point where Janet refused to be seen in public with her husband.... [Prior to their marriage] Janet knew about her husband's hyperactive premarital sexual adventures and found his appeal to other women tantalizing, [but] she had no intention of continuing to condone such behavior. Nor did her father.... The campaign to domesticate Black Jack soon faulted."
>
> **C. David Heymann**, *A Woman Named Jackie*, 1989, p. 34; 36.

In spite of his overly indulgent nature and all his womanizing, his daughters were the center of his life. After his 1940 divorce from Janet and their daughters' marriages, both in 1953, he never fully developed a life of his own. He became bitter and reclusive and died Aug. 3, 1957 of liver cancer (age 66). According to the attending nurse, his last word was "Jackie." Coming full circle, Jackie herself died on her father's birthday – May 19, 1994.

Jack Bouvier doted on his daughters and had an almost obsessive love for his daughter Jacqueline, sharing with her even his numerous sexual conquests. His pride in her was reciprocated by her devotion to him – whenever she could be pried away from the Auchinclosses, from 1942 onward. They offered her comparatively tempting choices prior to her 1953 marriage. Jack Bouvier often complained his daughters were so busy they neglected him. In this way, Jackie and her sister learned how to manipulate to their advantage the ongoing tug-of-war between their parents, as well as the competition between their impecunious father and their very rich stepfather. Ironically, their mother endured a very similar dynamic between her own parents, who were estranged for many years though not divorced. In another irony, during John F. Kennedy's State Funeral Procession there was a very spirited riderless horse. His name was Black Jack.

> "[Jackie]...was drawn to older men, piratical types with rampant sexual appetites. However, that was only part of the story. Black Jack Bouvier was Jackie's first and greatest mentor in the art of life. He had an eye for color, shape, and form, and he delivered lectures to his daughter on everything from architecture and art to antiques, interior decoration, and fashion. For Black Jack the most interesting art of all was the mating game between men and women. Pay attention to everything a man says, he told Jackie. Fasten your eyes on him like you are staring into the sun. Women gain power by affiliating themselves with powerful men."
>
> **Edward Klein**, *Just Jackie: Her Private Years*, 1998, p. 34.

Search for wealth *and* power: Jackie had a fiercely competitive edge, evident from childhood, though somewhat hidden under layers of charm. Her ambitious drive towards power and wealth was unquestionably influenced by both parents and reinforced from being raised and educated in an atmosphere of wealth and privilege. After her mother remarried in June 1942, she and her sister lived amidst the splendor of the Auchinchloss estates, but with such small allowances for their living expenses at college and elsewhere – that they felt poor. The stepsisters and brothers received much larger allowances. Jackie and Lee were not included in Auchinclosses' will, nor could they sustain their social position without relying heavily on favors from their stepfather or using their father's credit card without his permission. All this preceded 1953 – the year both sisters married. Much Bouvier family wealth was lost by the time Janet Lee married into it, and when Jackie's grandfather John Vernou Bouvier Jr. died Jan. 15, 1948, the expected paternal inheritance was reduced even further. The Bouvier fortune had been spent so lavishly over the previous 24 years that little remained for his heirs.

The goal – and the solution, as both parents emphasized – was to marry a rich man. Glory would be a plus but financial wealth was a must. With her marriage to Kennedy, Jackie managed to get both. And while appearing to offer impeccable social pedigree, her financial pedigree was almost non-existent – a fact much less well known to the world. But Jackie's inscrutable persona, obvious from childhood, became very useful in this regard. She presented herself successfully to the Kennedys as patrician to the core. She knew the sort of wife Joe Kennedy wanted for his son and that he could not achieve his goal of the Presidency without being married. Joe pushed Jackie as the ideal choice for Jack, and she knew she could play the role well.

> "According to Evelyn Lincoln [Jack's secretary], the marriage was forced on Jack Kennedy by his father. 'He was a politician who wanted to be president and for that he needed a wife. I am absolutely certain they were not in love. At least not at the time.'"
>
> **Christopher Andersen,** *Jack and Jackie: Portrait of an American Marriage,* 1997, pp. 117-118.

Capturing Jack Kennedy: By Jan. 1952 Jackie became engaged to the fledgling stockbroker John Husted, Jr. But after a few months the relationship cooled off. Her mother also considered Husted's financial wealth and social position insufficient. Not seeming to mind her mother's interference, Jackie was already looking elsewhere.

In May 1952 she met **Senator John F. Kennedy** at a dinner party and they began dating. Though she was warned of his rampant womanizing she had her sights set on him and decided that going after him was still worth the risks, especially if she could be at the center of events and have a decent role in the unfolding drama of his political life. He met her needs for a man with a keen mind, whose wealth and social position were substantial. Jackie aimed to capture Jack Kennedy, even if perhaps his heart could not ultimately be captured. He proposed in May 1953, the formal engagement was announced June 24, 1953 and the high society wedding took place on Sept. 12, 1953.

> "Gore Vidal [the famous author and Jackie's stepbrother] is right to the point: 'Jackie married Jack for money. Purely. There weren't that many other openings for her. Actually, if she hadn't married Jack she would have married someone else with money, although it wasn't likely she would have gotten someone as exciting as Jack in the bargain. When given a choice of glory or money, most people choose glory. But not Jackie. She also wound up with plenty of the latter, of course, but she didn't need that like she needed to be rich.'"
>
> *Ibid.*, p. 106.

Children: Jackie had four children with John F. Kennedy: Arabella (b. Aug. 23, 1956, stillborn), Caroline (b. Nov. 27, 1957), John Jr. (b. Nov. 25, 1960, d. July 16, 1999), and Patrick (b. Aug. 7, 1963, lived only 39 hours.) She had many difficulties bearing children, and suffered numerous miscarriages.

Partnerships after Kennedy – Aristotle Onassis:[279] The marriage to her second husband occurred on Oct. 20, 1968. Planned for earlier in the year, it was postponed at Bobby Kennedy's request during his Presidential campaign. When he too was assassinated in early June 1968, the pain of living in America became unbearable. Jackie escaped to Greece to marry her billionaire second husband.

---

[279] His date of birth is considered Class X data, i.e., unreliable. Onassis regularly changed the birth date and/or year. Born in Smyrna, Turkey, he listed his birth place as Salonika, Greece, when immigrating to Argentina as a teenager. One of the dates he gave is Jan. 20, 1906, but it could have also been 1900.

> "I don't think there are any men who are faithful to their wives." "The first time you marry for love, the second for money, and the third for companionship."
>
> **Jacqueline Kennedy Onassis**

To him the marriage gave new celebrity status. Though not at all handsome, he had a strong and magnetic charm. His long-time lover, Maria Callas, was heartbroken. Jacqueline herself suffered from adverse opinion from an American public reluctant to deliver their "Queen" into the hands of a Greek shipping magnate much older and shorter than she. But he offered her great wealth and above all the privacy that great wealth can afford. Even so, he himself loved publicity, as the more he had, the more confidence his bankers had in him. The couple honeymooned on his 325 foot yacht *Christina*, one of the world's largest privately owned yachts.

The marriage did not succeed for very long, with too many differences in style and taste. They lived mostly apart for the last several years until his death on March 15, 1975. As with her first husband, Onassis had a wandering eye. He resumed his affair with Callas not long after marrying Jacqueline. In 1974 and probably a year earlier, Onassis started to investigate divorce proceedings. By fall 1974, he was seriously unwell, and determined to extricate himself from the marriage, especially financially. He met several times with attorney Roy Cohn in Manhattan, who relates his conversations with Onassis:

> "'I am an enormously wealthy man,' he said, 'but still I find it hard to understand why I should receive a bill for 200 pairs of shoes. It isn't as though I don't provide her with a generous expense account – I do. What's more, the shoes are only one item. She orders handbags, dresses, gowns and coats by the dozen – more than enough to stock a Fifth Avenue specialty shop. This woman has no conception of when to stop squandering my money. I'm fed up with her, and I want a divorce.
>
> After a pause he came back to what I believed was the real reason behind his wanting a divorce – not her spending, which he could afford, but her coldheartedness. 'We socialize in different circles. Anywhere I am, she is somewhere else.... She wants my money but not me.'"
>
> **C. David Heymann**, *A Woman Named Jackie*, 1989, pp. 560-561.

When he died within six months, rumors still circulated about the pending divorce initiated by Onassis. But Jackie demanded that her stepdaughter Christina deny any such rumors to the press. After a bitter 18 month legal dispute Jackie was awarded $26 million from her late husband's $1 billion estate. She also had to relinquish any further claims to the Onassis estate. After taxes and fees, she netted around $19 million. But with her monthly expense account for jewelry, clothes, and travel, she had already received close to $42 million during her second marriage, or nearly $7 million per year from 1968 through 1975. Jackie returned to the 15-room Fifth Avenue apartment she had purchased in early 1964. It had remained her central residence among many, even through the years of extensive foreign travel and marriage to Onassis.

Partnerships after Kennedy – Maurice Tempelsman: Born in Antwerp, Belgium (Aug. 26, 1929), his home was Manhattan, NY, where he became a diamond dealer and only one of some 100 diamond traders dealing directly with the De Beers cartel and others in Africa, where he often traveled. For many years he had better connections in black Africa than most ambassadors, and this part of his life created a certain allure for Jackie. He was also well connected within the Democratic Party and knew Jackie for some years as a friend, including during her years in the White House. After the death of Onassis, he became her chief financial advisor, and as such, made her a very rich woman, increasing her holdings over ten-fold, beyond $200 million. He also finally moved out of the family apartment he shared with his wife Lilly.

> "Trapped in a sterile marriage, Tempelsman began looking around for female companionship. He was attracted to well-groomed, well-spoken, well-off women who moved gracefully in the highest levels of society. When it came to winning women, Tempelsman was not the equal of John Kennedy or Aristotle Onassis, but according to the testimony of several of his conquests, his old-world charm worked wonders."
>
> **Edward Klein**, *Just Jackie*, 1998, p. 306.

Tempelsman came from a trading family that understood the art of the deal, preferably a huge deal. Though not as wealthy as Onassis, they were alike in that way, as well as in their compulsion to better themselves, to break from their pasts and their fathers, and to be accepted by people of quality. An avid reader, traveler and art collector, Tempelsman began a discreet relationship with Jackie in the late 1970s, one that lasted until her death in May 1994. He shared her Fifth Avenue penthouse apartment from 1982 onward, though they decided against a formal marriage. He was at her side on May 19, 1994 when she died of non-Hodgkin's lymphoma, a form of cancer. After she was diagnosed in Jan. 1994 the disease moved rapidly, spreading to the membranes covering the brain and the spinal cord.

Private and mysterious nature – Ketu on the Ascendant: Jackie had a strong sense of her own stardom, but also an air of mystery about her, a loner quality with a tendency to be remote and secretive. These traits would grow more pronounced in adulthood. Her Sun near the top of the sky in the birth chart shows Jackie's ongoing visibility, while her paradoxical elusiveness is seen through both Ketu (the foreigner) on the Ascendant and Ascendant lord Venus in the mysterious 8th house. (Jack Kennedy also had his Ascendant Lord in the 8th house, though not in its own sign.) As if to confirm the qualities of Ketu and Venus, Jackie was born in **Ketu-Venus** *Dasha/Bhukti*.[280]

> "Jacqueline was a remarkable woman in many ways, but she was a very private person. She revealed little of her personal life, her thoughts, or feelings in either her writings or conversations, and she remained somewhat of an enigma to her family, friends, and the general public all her life. She once made an offhand remark to a friend that might offer something of a clue to this aspect of her personality: 'The trouble with me is that I am an outsider. And that's a very hard thing to be in American life.'"
>
> **John H. Davis**, *Jacqueline Kennedy, an Intimate Portrait*, 1996, p. xii.

Ketu is close to the exact degree of the Ascendant, so close that it repeats in the *Navamsha* chart, doubling its effect. Ascendant lord Venus placed in the 8th house adds to the magnetizing component she possessed. Both these factors also contributed greatly to her skill in managing the public's perception of her, a skill often associated with actors. Author William Manchester considered her "a very great tragic actress." Hounded by photographers from 1963 onward, she

[280] Her Ketu-Venus *Dasha/Bhukti* ended Sept. 17, 1929, six weeks prior to the NY Stock Market crash Oct. 29, 1929, severely impacting the family's finances. Both Jackie's grandfather "Major" Bouvier and her father suffered heavy and irrevocable financial losses. Her Ketu-Sun *Dasha/Bhukti* (Sept. 17, 1929 to Jan. 23, 1930) shows a big change, being in 6 – 8 house relationship from each other in the *Navamsha*, especially with debilitated Sun, owned in turn by a debilitated Venus. She (and her family of origin) would have to operate from some initial disadvantage.

refused all interviews from spring 1964 until her own death in May 1994. (Her last interviews in spring 1964 were with William Manchester for his 1967 book and with Arthur Schlesinger, Jr. Until fall 2011 the Schlesinger interviews were unknown to the public.)

Libra is an urban, social networking sign and Jackie lived most of her life in cities. For the last thirty years of her life she maintained a spacious 15th floor apartment in Manhattan. She could also hide in our midst, and this is described by Ketu in the Ascendant.[281] Even if not speaking in public she made an impressive appearance. And because of her iconic social position many public figures showed her extreme deference. She demanded special treatment and usually received it, though it was not always returned.

In the Planetary Cabinet, **Rahu and Ketu** are foreigners or outcasts, and associated with unpredictable behavior. **Ketu** is exotic, foreign and tends to disappear upon contact. Paradoxically Ketu is symbolized by a flag and thus draws attention to itself, even if as a persona it is largely unknowable. Though one of the most famous women in the world and among the most photographed and documented women of the modern era, Jackie too was largely unknowable. To discuss the life of someone with Ketu exactly on the Ascendant is like trying to break a code we are not supposed to know. Yet we are strongly tempted to know it, as we assume it will give us access to some valuable secrets. Or so we think.

Though **Ketu** flags our attention, we have to go inward to find it as Ketu is largely a *Moksha* indicator of spiritual liberation. As an eclipse planet Ketu obscures our ability to see what is directly in front of us, especially when in a prominent position in the chart. Placed almost exactly on the Ascendant, this Ketu enabled Jackie to obscure from the world and perhaps often from herself what was going on inside her heart and mind. From childhood, this enabled her to deal with her parents' painful marital situation, the ongoing domestic tension it created and the subsequent bitter and lengthy divorce. If she disliked what she was hearing, she had a habit of tuning it out or pretending it did not exist. Ketu in this position is capable of creating whatever image or mirage is desired, even more so when contacting the natal Moon, though by a wide margin here.

Ketu is usually associated with non-attachment to worldly benefits, with more attention to spiritual matters. But Jackie sought worldly wealth and power along with social respectability. She was determined to convey that image and to achieve it as a permanent reality. This more material orientation of Ketu is due to several factors here: both sign and *nakshatra* lords of Ketu are benefic planets (Venus and Jupiter, respectively), and placed in the 8th house, where benefic planets can be good for spiritual growth but can easily redirect their energies towards material growth and sensual pleasures. (Such is the paradox of the 8th house and the sign of Scorpio: sex and money – but also death and taxes. Money comes through inheritance or marriage.) This Ketu manifests in a different way due to the positions of Moon, Venus and Jupiter as well as the *Nadi yoga* between Venus and Mars. All these factors contribute towards attracting a partner and the wealth of a partner.

The houses of *Moksha* (spiritual liberation) are houses 4, 8, and 12. Ascendant lord Venus is placed in her 8th house and very close to the 8th house cusp. Normally the 8th house is a difficult placement for the Ascendant lord, creating an *Arishta yoga* that can indicate poor **physical health** and sometimes death as an infant. (Jack Kennedy's Ascendant lord on the 8th house cusp

---

[281] Ketu in the Ascendant is featured in Chart #15 (Francis Poulenc), also in the chart of Queen Elizabeth II, exactly conjunct her Sagittarius Ascendant.

created an *Arishta yoga*, with its close proximity to the 8th house cusp (less than one degree) as well as to the 8th house lord, thus far worse for his physical health and longevity.) In Jackie's case the 8th lord rules its own house, Taurus, creating a *Viparita Raja yoga* and also lending some protection. This *yoga* is buttressed by another *Viparita Raja yoga*: 6th lord Jupiter joins Venus in the 8th house. If other factors did not indicate such focus on material wealth, this could be a more spiritualized person. But Jackie has several notable *Dhana yogas* of wealth, bringing significant material wealth in her lifetime and the necessity of dealing with wealth. Further, Venus and Jupiter in mutual contact often bring excess in the material realm. The 8th house is also about political and sexual intrigue.

The sensual romantic: Jacqueline Kennedy Onassis is closely associated with many aspects of **Venus**: her physical beauty, personal refinement and sensuality; her support of the arts; her love and compassion, especially towards her children, and her fierce protectiveness of Jack Kennedy's legacy. The latter also reflects the *Nadi yoga* between Venus and Mars, to be described. There is a passionate yearning for fulfillment in marriage, and if that is not possible, then the achievement of power through marriage. Venus rules over sexual and love relations. When Jackie entered her 20-year **Venus *Dasha*** on Feb. 20, 1935 she came in touch early in life (at age 5½) with her own capacity for charm, style, and magnetism, encouraged by the ongoing adoration of her father and grandfather Bouvier. Venus *Dasha* also ushered her into a wider social arena as debutante, fiancée and bride.

> "My mother was a true romantic. She lived her life on a dramatic scale and responded to the poetry of love with a passionate intensity."
>
> **Caroline Kennedy** quoted in her anthology, *The Best-Loved Poems of Jacqueline Kennedy Onassis* (2001).[282]

Venus rules over fashion and jewelry. Jackie's keen sense of fashion and choice of ornaments is well known. As a fashion icon alone, she was a highly successful First Lady. Venus rules the chart and Moon is prominent in the birth chart. The gemstone for the **Moon** is pearl and for **Venus** – diamond. Jackie loved jewelry and had an extensive collection, including pearls and diamonds. In the early years, including at her 1953 wedding, she often wore strands of pearls. With her Moon in the 7th house of marriage, she might especially gravitate towards wearing pearls during the years she was formally married, notably during her first marriage, 1953 – 1963. With so much photo documentation available, one could study her use of pearls during her 10-year Moon *Dasha*, from Feb. 1961 thru Feb. 1971. At the time of her death at age 64, the photo circulating most widely showed her in her early 30s, coinciding with her **Moon *Dasha***. She is young and beautiful, clad in a light summer dress, with a triple strand of pearls around her neck, choker style. Conveying an atmosphere of luxurious ease, she leans her head and torso back against a white cushioned summer chair. This is an iconic photo of Jackie that triggers how we tend to remember her. It also highlights her natal Venus and Moon.[283]

---

[282] This anthology contains a wide variety of poems, from Robert Frost to William Shakespeare, and also four poems penned by Jackie herself, never before published. Her grandfather Bouvier introduced her to poetry and encouraged her love for it. A trial lawyer with a rich, booming voice, he often recited poems.

[283] This photo appears on the cover of Donald Spoto's book: ***Jacqueline Bouvier Kennedy Onassis: A Life***, 2000. Spoto provides no documentation to verify the date or year this photo was taken; but it was commonly seen in news media in the early 1960s, coinciding with Jackie's Moon *Dasha*.

Her contribution to the arts: Venus is the significator for artistic ability, fashion sense, love matters, and eyesight, among others. Given its strong position in her chart as Ascendant lord in its own sign Taurus on the 8th house cusp, we should note that while Jackie was artistic in many arenas (drawing, painting, writing poetry and decorating houses with her unerring fashion sense), she never sought to be a professional artist. Instead, her role in the arts was always as an aficionado, promoter, activist and philanthropist. She is known for her support of the performing arts, especially ballet. One result was the Jacqueline Kennedy Onassis School of Ballet at the American Ballet Theatre in New York City. She was a member of their Board for 25 years. She was also passionate about the visual arts and architectural monuments. Her preservation of the White House interior decor is a major project for which she is credited. She also actively promoted the protection of important American cultural and architectural monuments, notably Lafayette Square in Washington, D.C. and Grand Central Station in New York City. Her staunch support for arts that were both ephemeral (dance) and enduring as physical monuments (architecture) reflects two factors: First, Venus in the 8th house of the birth chart is more ephemeral. Secondly, Venus in earth signs in both birth chart and *Navamsha* chart is more enduring.

Glamour and magnetism: When the 8th house is strong, it can magnetize people to you, even bringing great fame. Venus rules the Ascendant of both birth chart and *Navamsha* charts, giving extra potency to the 8th house of the birth chart. Venus is also galvanized by its close interaction with Mars in a *Nadi yoga* (exchange of *nakshatra* lords). Venus is a great magnetizer, and if well positioned can strive for worldly power. Venus is better placed in the 5th house in the *Navamsha* chart, though in its sign of debilitation – Virgo. In the 5th house, it gives greater subtlety to her mind and creative instincts. In the 8th house of the birth chart Venus has special allure as planet of desire, and can represent that which is unattainable. Natal Venus in its own sign of Taurus is within four degrees of the 8th house cusp, adding the components of glamour, mystery, and unfathomable sexual allure. With his Ascendant lord on the 8th house cusp John F. Kennedy shared these traits, along with some other of the same astrological components.

Jackie's natal Venus is at 28:53 Taurus in *Mrigashira nakshatra*, ruled by Mars. Mars in turn is at 21:57 Leo in *Purva Phalguni nakshatra*, ruled by Venus. This creates a powerful *Nadi yoga*, an exchange between the two *nakshatra* lords, **Venus and Mars**. Jack Kennedy also had a *Nadi yoga* between Venus and Mars. Depending on the houses involved, this particular exchange of planetary energy can add to one's sexual passion. It can also increase one's desire for power in the world, a goal won chiefly through personal charm and magnetism.

If we review the **Vedic myths of Lakshmi and Skanda,** key deities representing Venus and Mars, respectively, we note that Lakshmi sought as her partners only the bravest and most victorious of leaders.[284] Skanda, in turn, had to show immense strength in battle. To maintain his eternal youth and vitality, he was revered for his ability to contain his sexual energy. When Venus and Mars are in mutual contact in the chart, and also dominate through angular house positions or through ruling the Ascendant, this can mean the search for triumphant worldly power causes sexual energy to be either sublimated or wasted. There is an exaggerated need to fight and win battles, whether military, athletic, political, or social, including sexual

[284] See Chapter 9 Introduction for a brief summary of the Vedic myths of Lakshmi and Skanda.

conquests. Everything may be sacrificed to this end. This may explain in part why Jackie was willing to overlook Jack Kennedy's stupendous level of sexual addiction in order to marry a powerful man, and why Jack in turn agreed to marry at all as a necessary step to gain the Presidency.[285]

Wealth and power through marriage: With uncanny accuracy, at age 14 Jackie jokingly pronounced that one day she would be

> "Queen of the Circus... [and marry the] Man on the Flying Trapeze."

She went on to marry two famous men, both of whom took enormous risks, most especially Jack Kennedy. He also put her at the center of world events. Likewise, she elevated the celebrity of her two husbands.

When she married Kennedy on Sept. 12, 1953 Jackie came into national and global public prominence. The event was dubbed "the Wedding of the Year." It occurred during Jackie's **Venus-Mercury** *Dasha/Bhukti* (Feb. 20, 1951 to Dec. 21, 1953). *Navamsha* Venus and Mercury are well-placed in the 5th house: This forms a *Dhana yoga* of wealth, combining *Navamsha* 1st and 2nd house lords, especially good financially. Since *Navamsha* Venus is debilitated in Virgo it means she starts out with some relative disadvantage in this regard. Placed in the 5th house of children, this Venus hints at the numerous problems she had bearing children.

Moon in the 7th house shows a certain mental and emotional focus on the partner, whether marriage or business partner, and also on the public. Because the Moon fluctuates, married life will go through more than average fluctuations.[286] There is the destiny to marry well financially, with two great benefic planets Venus and Jupiter in the 8th house. (The 2nd from the 7th house indicates wealth of the husband(s)). This is confirmed in the *Navamsha* chart by 11th lord Jupiter in the 2nd house aspecting the 8th house, which Jupiter also owns; also by *Navamsha* Venus and Mercury well placed together in the 5th house, two benefic planets, owning the *Navamsha* 1st and 2nd houses. This combination indicates the timing of the *karmic* destiny, and that as early as **Venus-Jupiter** *Dasha/Bhukti* but more likely **Venus-Mercury** *Dasha/Bhukti* could bring children and/or great wealth through marriage. (Neither Mercury nor Jupiter *Dasha* occurred in her lifetime.) In Sept. 1947, during **Venus-Jupiter period** (April 21, 1945 to Dec. 21, 1947) Jackie became "Debutante of the year," formally symbolizing her marriage eligibility. **Venus-Mercury period** (Feb. 20, 1951 to Dec. 21, 1953) brought at least two engagements as well as her first marriage. Mutual friends tried to introduce her to Jack Kennedy in 1948 when they attended the same party, but Jackie was preoccupied with another guest. She was then in Venus-Saturn *Dasha/Bhukti*, planets in 6 – 8 house relationship in both birth chart and *Navamsha*, thus not as auspicious as the next sub-period of Venus-Mercury, even if Saturn was *Darakaraka*, or spouse indicator. Meeting Jack Kennedy was to be further delayed.

---

[285] Either Venus-Mars or Mars-Venus *Dasha/Bhukti* would be most likely to manifest the fruits of this karmic destiny. Jackie was only 12-13 years old during her Venus-Mars *Dasha/Bhukti*. She was in Mars-Venus period from Jan. 15, 1976 through March 16,1977. The myth of the Camelot presidency first came crashing down in late 1975, with the Congressional testimony of ex-JFK lover Judith Campbell Exner. Newspaper revelations came on Dec. 19, 1975, also marking a full Jupiter cycle since JFK's death. See item #15 among Jackie's *List of major life events and dates.*

[286] For other biographical subjects in this book with natal Moon in the 7th house of the birth chart, see: Henry Miller (Chart #1); Marilyn Monroe (Chart #17); Indira Gandhi (Chart #18); and Woody Allen (Chart #22).

Personal happiness in marriage would be less likely due to several factors: **1)** Rahu-Ketu are exactly across the marriage (1-7 house) axis, giving turbulence and lack of marital stability, though it can also bring a foreign or exotic partner; **2)** Sun-Moon axis in the *Navamsha* is afflicted by contacts from Mars and Saturn, bringing older partners, difficulties with them and the tendency for widowhood; **3)** *Navamsha* Moon and Saturn are in Aries in the 12th house, a house of either foreign residence, or loss and suffering. Moon-Saturn in the 12th house evokes a woman alone, widowed, or even depressed, if not in spiritual practice); **4)** *Navamsha* Venus is debilitated, though corrected (*Neecha*) by its contact with exalted Mercury in Virgo. Natal Saturn on the marriage axis can bring **older partners** or separation from partners. From an early age Jackie's preference was for older men from whom she could learn. Favored by her father and grandfather Bouvier, in childhood she already preferred the company of adults to those her own age and she tended to prefer male companionship to that of females. Her mother (Moon in the chart) also married much older men, her first husband 16 years older, and her second husband 18 years older. Maurice Tempelsman was one month younger than Jackie and among the partners whom she did not marry, though it appeared to be the equivalent of a common-law marriage.

Jackie's legal marriages were tempestuous and difficult, though each of them brought her enormous financial rewards. Even Tempelsman, her longtime companion and live-in partner did extremely well in expanding her already sizeable fortune up to $200 million at the time of her death. Again, the astrological indicators are the two classical benefics, Venus and Jupiter placed in the 8th house of inheritance. Compared to her peers in the social registry, Jackie started with relatively little financially. But through marriage she surpassed the entire Bouvier family. By the time she entered the White House in Jan. 1961 she was spending $100,000 on her clothes budget alone. Both of her husbands complained of her expensive tastes in clothes and interior decoration, but each supported them in the end – Jack girded by his father's vast resources, and his promise to pay for Jackie's wardrobe while in the White House. Joe Kennedy, Sr. was her strongest ally in the Kennedy family. He suffered a stroke on Dec. 19, 1961, during her **Moon-Moon** *Dasha/ Bhukti*. Though still mentally intact, he lost all power of speech, and was confined to a wheelchair until his death Nov. 18, 1969.

Pregnancy and children: Jackie lost several children through either miscarriage or early death of the infant. The propensity for difficulties in this area is shown in the birth chart by several factors: The *karaka* (significator) for children is **Jupiter**, which resides in the potentially destructive 8th house in the sign of its enemy Venus and in the *nakshatra* of the Moon, which itself is very afflicted in the 7th house of partnership, alongside Rahu – the foreigner. Furthermore, classic malefic planets aspect **the 5th house of children**: from the Ascendant, Saturn and Mars both aspect the 5th house, and 5th lord Saturn is in *Mula nakshatra* owned by Ketu, giving Ketu power over 5th house matters. With its dissolving energy, Ketu does not tend to promote safe pregnancy or childbirth. When classic malefics Mars and Saturn aspect the 5th house, they have a separating effect and can cause separation from children, born or unborn. The sexual activities of the husband (Mars – 7th lord) did not help, and Mars in *Nadi yoga* with Venus hints at this outcome: Jack Kennedy had ongoing venereal diseases for which he took medication, but it was never enough to counteract the ongoing problem. His autopsy revealed he had chlamydia, a serious venereal disease that can cause reproductive disorders and problems with infants if a woman is infected while pregnant.

The third child born to Jack and Jackie Kennedy was Patrick Bouvier Kennedy (b. Aug. 7, 1963, 12:52 PM EDT, Otis Air Force Base, Mass.; d. Aug. 9, 1963, 4:04 AM EDT same location). He was six weeks premature, had Hyaline Membrane Syndrome and died of respiratory problems 39 hours after birth. Tr. **Jupiter** (significator of both children and husband) turned Stationary Retrograde on Aug. 9, 1963, just 7 hours after Patrick's death. Jupiter's Station was exactly on Jackie's 6th house of conflict and illness, magnifying its effect in her life. Tr. Jupiter received a close aspect from tr. Saturn, which turned Stationary Direct on Oct. 21, 1963 at 23:06 Capricorn, within two degrees of her 4th house of foundations. Saturn's passage through the 4th house of the birth chart can mark endings and new beginnings. Both parents were grief-stricken at the child's death, which shortly preceded John F. Kennedy's assassination and was like a precursor to it.

The only two children born to Jackie who lived to adulthood were Caroline and John, Jr., born during her Sun *Dasha*, in Sun-Jupiter and Sun-Venus periods. Her natal Sun is well placed and has *Digbala* in the 10th house; sub-lords Venus and Jupiter are well placed in the 11th house from *Dasha* lord Sun. *Navamsha* Venus is debilitated in the 5th house of children, but redeeming factors are: **1)** the good house placement of Venus; **2)** its sign lord Mercury being both a classic benefic planet and exalted in Virgo; and **3)** Mercury being a planetary friend to Venus and one of the *Nadi yogakarakas* for both Libra and Taurus Ascendants. For more on her pregnancies and childbirths see the list of Major Life Events at the end of this section.

Major influence of the parents: Parents have a larger than usual impact when both Sun (father) and Moon (mother) are in angles of the chart or in trinal houses. The negative effect of her parents' ongoing conflicts is shown here through Rahu's contact to the Moon and through the Sun and Moon house positions in the *Navamsha* chart, each receiving aspects from classic malefics Mars and Saturn, and also from temporal malefic Jupiter (*Navamsha* 8th house lord). Prominent in angular houses of the birth chart, Moon is in the descendant (7th house) and Sun is in the 10th house at the top of the sky. Sun has maximum visibility and reflects the enormous influence her father had over Jackie. However, her mother had more apparent control over details of Jackie's life through her ability to manipulate and through her marriage to her second husband, Hugh D. Auchinchloss, whose wealth far exceeded that of Jackie's own father, Jack Bouvier. The Moon's direct aspect to the Ascendant shows the mother's strong influence. Since Moon is with Rahu, there can be problems with the mother and/or a mother who is manipulative, teaching Jackie how to do the same, as Moon rules over one's own mind. Her mother changed the course of her life at several crucial points: her first engagement in early 1952 to John Husted, Jr. (not enough wealth or social position) and her Vogue prize in 1951 (Paris was too far away). She persuaded Jackie *not* to accept the Vogue prize, projecting her own insecurities on to her and convincing her the job was too much for her. She knew Jackie might well remain in Paris as she loved living there as a university student. But residence in foreign countries is more likely during the *Dasha* periods of key planets in the 7th or 12th houses. Jackie's *Navamsha* Moon is in her 12th house, but the 10-year Moon *Dasha* did not begin until Feb. 19, 1961. In 1951 she was in **Venus-Mercury** *Dasha/Bhukti*. Neither planet tenants these houses.

Houses ruling over parents are 4th house for mother and 9th and/or 10th for father. In this case, the 4th house lord (Saturn) is situated in the 12th house from itself, indicating problems with the mother. Father is ruled by 9th house (father as advisor, philosophical influence - South Indian), or 10th house (father as authority figure - North Indian). Mercury, the 9th lord is well placed in the 10th house, though the 9th house receives an aspect from malefic Saturn (mother - as ruler of 4th house). The 10th house lord is the Moon, coincidentally *karaka* for Mother, placed with Rahu in the 7th house. Both parents were extremely manipulative in competing for power, prestige and influence over their two daughters. Both were adamant their daughters should marry well, perhaps to avoid the financial woes they themselves suffered as a couple and as a family.

Sun and Moon positions in the *Navamsha* chart again reveal the difficulties Jackie's parents experienced together. And since Sun-Moon is also a marriage axis, its poor *Navamsha* house position and affliction by Mars and Saturn aspects shows that Jackie would suffer in the marital arena just as her own parents had suffered. The preference for older men is shown by Saturn's aspect to the Sun-Moon axis, though it can also show itself in the 7th house. Her preference for "Bad boys and pirates" as partners is reflected in some of the same combinations, including Rahu and Ketu across the marriage axis, delineating foreigners and/or eccentrics as partners and the possibility of some instability through them.[287] Since the destiny is to attract this sort of partner, either instability or foreignness in the partner is inevitable.

Sibling rivalry: The parents' message to compete for men was of such heightened importance that it eventually contributed to driving the sisters apart. Their competition became fierce to compete both for parental favors and for favoritism from men. Their father preferred Jackie, as did many men in their lives. In April 1953 Lee rushed into her first marriage (among three) in order to beat Jackie to the altar. Lee's 1974 marriage was cancelled by the groom five minutes before the ceremony, apparently due to pressure from Jackie's lawyers that he sign a pre-nuptial agreement guaranteeing Lee's financial security. (The bride-to-be had not requested this.) The younger sibling is shown by 3rd house lord and by *karaka*, Mars. Malefic Saturn is placed in the 3rd house, and 3rd lord Jupiter is situated in the unfavorable 8th house, bringing some loss and misfortune to Jackie's younger sibling. However, Lee's major misfortune seemed to be when in competition with her sister.

The nature of her mind is shown by the condition of **the Moon**. Moon at 2:44 Aries is in *Ashwini nakshatra*, symbolized by a horse's head. Its Vedic meaning centers upon owning horses, and its *shakti* (key energy) is speed. Aries also rules over the head and can make the person headstrong and strong-willed, as in this case. Jackie was a lover of horses and an accomplished equestrian who trained from the age of three, and who in turn taught her own children how to ride, just as her mother (Moon) – also a highly skilled equestrian – had taught her. Jackie's athletic abilities were mostly confined to horsemanship and she pursued it avidly throughout her life. Her equestrian talents were prominent throughout her early years, and by age 11 she was mentioned regularly in the newspapers of eastern Long Island, where she had already acquired some fame for her horsemanship. As a child, she was a unique combination of tomboy and "dream princess." In her early years, she was restless and easily bored at school. It was only when a perceptive headmistress talked with her, using the analogy of an untrained racehorse, that she started to grasp the deeper rewards of persistence and focus. Caring for her horses also provided a means of escape from a tension-filled domestic situation. Her favorite horse for many years was named "Danseuse," French for "the [female] dancer."

Her larger perspective and cultural breadth came from her considerable intellectual curiosity, love of books and learning. She was an avid reader of history, literature, philosophy and the arts. From an early age her teachers noted the extent of her **"inquiring mind."** Everything interested

[287] As a reminder, the marriage axis is identified as the 1-7 House axis; also the Sun-Moon conjunction or opposition. When planets are mutually angular on one or both of these axes, they impact the marriage destiny of the individual. Also, planets exactly on the Ascendant or Descendant can further describe marriage themes.

her up to a point, even politics, which she claimed to detest. But she studied it to keep up with Jack Kennedy, for whom it was a life passion. This is described astrologically by Jackie's Moon strong in Aries and *Vargottama* (repeating in the *Navamsha* chart in Aries). Aries Moon charges forward to discover new things in life, and being restless, has an interest in many aspects of life. It has a zest for competition and for taking the initiative, but prefers quick results – especially as the orientation of Moon in *Ashwini* is towards speed.[288] Moon is modified by its contact to Rahu in the birth chart. And in the *Navamsha* chart, Aries Moon is with Saturn in the 12th house, giving her more discipline, seriousness, and the need for privacy. Moon-Saturn can be hard working and conscientious, but when in the 12th house can also be prone to depression and even paranoia (areas ruled by Saturn and the 12th house). The Moon is the mind, Saturn is fear, and the 12th house is the past and the unknown – thus ungrounded fears. The mind is easily affected by its surroundings.

Yearning for attention *and* privacy: We pay special attention to Ketu, as owner of *Ashwini nakshatra* – where the Moon resides. Ketu has dominance in the chart, being right on the Ascendant, and Moon also has dominance being in an angular house. When the Moon is prominent above the horizon in the chart, there is an overt yearning for emotional connection. With a strong Venus as Ascendant lord – personal appeal and seductive power come into play. But Ketu is elusive and does not wish to be known. Even so, it draws you inward to discover what or who is there. These factors account for the paradox of her emotional appeal alongside her elusive nature, a woman who enjoyed adoring crowds but derided them for following her, in one case calling them "mesmerized cattle."

Moon and Rahu: Moon is located in the 7th house of partnership and the public. Since Rahu causes the eclipse of the Moon, Moon is compromised whenever Rahu is nearby, especially in the same house and sign. In this case they are 22 degrees apart, but still in the same sign of Aries, indicating the mental and emotional equilibrium (Moon) is disturbed by foreign elements (Rahu). There is also the tendency to magnetize this to you and to be attracted to the foreign or exotic. Jackie loved traveling abroad, but was often bothered by reporters and photographers, some of them doggedly relentless in pursuit of photographing her. This practice escalated since she gave no interviews after late spring 1964.

At the time of Jack Kennedy's assassination by bullets to the head, Jackie was in **Moon-Rahu** *Dasha/Bhukti*. (July 21, 1962 to Jan. 2, 1964), and narrowly missed being killed herself. Since natal Moon is in the 7th house, the 10-year Moon *Dasha* would highlight her marriage partnerships, though not in an easy way due to the afflictions to natal Moon. Moon *Dasha* brought both Jack Kennedy's murder and later on her second marriage to a foreigner. She married Aristotle Onassis during Moon-Ketu *Dasha/Bhukti*; Rahu and Ketu are both foreigners. The symmetry of the *Dasha* timing is arresting. Her own death of a brain tumor occurred on May 19, 1994 in **Rahu-Moon** *Dasha/Bhukti*. Moon in Aries is symbolic of the head. Located in the 7th house of marriage, the Moon-Rahu *Dasha/Bhukti* focused the symbols on her husband, who was killed by a bullet to the head. When she herself died of a brain tumor, again there was a foreign object (Rahu) lodged in the head (Moon in Aries).

---

[288] Of the three *nakshatras* situated in Aries, *Ashwini* is the most oriented towards speed, its primary *shakti*, or driving force. *Ashwini* means "owning horses," or "yoking horses," or "she who yokes horses."

In general, Moon-Rahu in the 7th house can show turbulence with the marriage partner. It can also indicate potential problems to the head, or to the brain, as in this case. (This is in addition to mental or emotional disturbance.) In the *Navamsha* chart Moon and Rahu are in 6 – 8 house relationship, a combination that can bring about turbulence or a big change during its period. This is the focus that the *Vimshottari Dasha* brings to the timing of the *karmic* destiny. Moon-Rahu issues come into sharp focus during their *Dasha/Bhuktis*, but they remain as life issues, and she would struggle throughout life to maintain a calm mind. She was a heavy smoker for much of her life, though she carefully hid this habit from her public. Only when daughter Caroline begged her to quit in Jan. 1994 did she do so. Jackie had just been diagnosed with cancer.

Moon-Saturn in Aries in the 12th house of the *Navamsha* chart describes Jackie's great yearning for privacy and the sadness that would precipitate that yearning. This combination can also show a tendency to live and reside abroad, or in pastoral places, though the Libran influence for urban living is also present. Fluent in four languages, Jackie would have traveled more as a young woman if not for the dictates of her mother and stepfather. They lobbied against her establishing herself abroad and dissuaded her from spending six months in Paris – her 1951 Vogue prize. Saturn creates delays and her desire to live abroad was delayed. Even so, her *Vimshottari Dasha* sequence confirms that foreign residence and/or travel would be more likely to occur during Moon and Rahu *Dashas*.

The time of greatest suffering: Because the Moon is afflicted and yet magnified by Rahu in this chart, we know that Jackie's 10-year **Moon *Dasha*** (Feb. 19, 1961 to Feb. 20, 1971) would bring the period with the most suffering along with some of the most responsibilities. It coincided with her ***Sade Sati*** (tr. Saturn in houses 12, 1, and 2 from natal Moon), April 9, 1966 to June 11, 1973. And though Saturn is a benefic planet for Libra Ascendant, its effect on this Moon would not be easy as Moon is afflicted in the chart.

Her Moon *Dasha* also brought the most eventful period, as Moon is in an angular house of the birth chart. One day prior to the start of her Moon *Dasha* was the seminal **Jupiter-Saturn conjunction**, at 1:52 Capricorn in her 4th house of foundations, reading from Libra Ascendant. This kicked off the Kennedy presidency. Reading from *Dasha* lord Moon, Jupiter and Saturn were in the 10th house, giving high status to herself and her partners. Kennedy's Inauguration was on Jan. 20, 1961. Though the timing of the Jupiter-Saturn conjunction was auspicious, it was not supported by the personal timing of the President and the First Lady. They were each at the waning end of personal *Vimshottari Dashas*. Jack Kennedy began his presidency in the last seven months of his 18-year Rahu *Dasha*, during his Rahu-Mars *Dasha/Bhukti*: not a good omen for the longevity of this presidency. Jackie, meanwhile, was in the last month of her Sun *Dasha*. And though Moon *Dasha* was only one month away, it had far more problematic indications than Sun *Dasha* due to the various afflictions to the Moon, as described.

Her first major loss in Moon *Dasha* was the active support and friendship of her father-in-law Joe Kennedy, Sr., with whom she had a very close relationship.[289] He succumbed to a stroke on Dec. 19, 1961, losing all powers of speech, though still mentally alert. She was still in **Moon-Moon period.** He died eight years later, on Nov. 8, 1969, during her **Moon-Venus period.**

---

[289] "Joe Kennedy's womanizing, while shocking to many, merely titillated Jackie, who had the experience of her own father against whom to measure the actions of her father-in-law. She and Joe established a closeness that she shared with few. They had a similar sense of humor…" (C. David Heymann, ***A Woman Named Jackie***, 1989. p. 142.)

Patrick Bouvier Kennedy was born six weeks early on Aug. 7, 1963 and he died after just 39 hours.[290] The event shattered both parents and brought them much closer together, a rare event in their marriage, thus magnifying the loss when Jack was killed a few months later. Jackie later considered the baby's death a precursor to her husband's death. She was in **Moon-Rahu period** from July 22, 1962 to Jan. 21, 1964. This sub-period would contain her deepest and most personal losses, but was closely matched by **Moon-Ketu period**, May 21, 1968 to Dec. 20, 1968, during which Jack's brother Bobby Kennedy was killed on June 5, 1968. (See Item # 11 in the list of Major Life events.)

Providing refuge during her first year of mourning for Jack Kennedy was Jackie's Fifth Avenue Manhattan apartment, which she purchased in January 1964. She was in **Moon-Jupiter** *Dasha/ Bhukti*. She lived there on and off for the next thirty years, more so from 1975 after the death of her second husband Aristotle Onassis.

Timing of the Camelot presidency: Rahu magnifies whatever it touches and in this case, Jackie sought to magnify - and many say exaggerate - the reputation and qualities of her first husband. Immediately after his assassination she set to work on this task. Thus it is fitting that all but one month of the Kennedy presidency occurred during her Moon *Dasha*, and her husband's assassination in **Moon-Rahu** sub-period. She wanted to guard the Kennedy legacy and only project it in a certain glorified way. But with the inevitable and continued spotlight on it by historians and investigative reporters, it would be impossible for Jackie to preserve only her version of the Camelot presidency. Even so, it is remarkable she was able to perpetuate it to the extent she did, fending off the darker truths to emphasize Kennedy's more "magical" qualities. But a series of credible books contributed to a different perception, among others Seymour Hersh's book, *The Dark Side of Camelot*, published in Nov. 1997. It unveiled more of John F. Kennedy's story to a larger public and magnified it, including many details still denied or minimized by Kennedy apologists. Such is the battlefield for dominance in the telling of history.

Reputation and image versus destiny fulfilled: In secular society, we tend to revere the maintenance of one's image and reputation, which may or may not be deserved. This can be assessed astrologically, but in Vedic terms - the greater emphasis is on whether a destiny has fulfilled its promise. This is not necessarily about the fullness or eventfulness of a life, though that plays a part. But it is more about the focus of the life. The key *Dharma* for Jackie as a Libra Ascendant person was to establish her version of a harmonious universe: she was active throughout her life to achieve this in an aesthetic way through her personal looks and charm, her family and relationships, the arts, and most of all - the "Camelot presidency." Otherwise, she would not have chosen to be buried next to a faithless husband who died 31 years earlier. This shows not only the tenaciousness of her vision and her dedication to Kennedy's memory but her desire to retain her own place of power in the nation's history, no matter how much she may have seen herself as an outsider to life in America, in her sardonic way.

The ***Navamsha*** (9th harmonic) chart shows how the *karmic* destiny bears fruit, as well as the direction of the soul's urge. The *Navamsha* chart is important to assess for ultimate results in the life, since it resonates with the future. If it is weaker than the birth chart in component after component, then the destiny will probably fail to bear the fullest fruit of what is promised in the

---

[290]For more on Patrick Bouvier Kennedy, see above under *Pregnancy and Children*.

birth chart.[291] The *Navamsha* chart shows where her destiny is going, and in this case shows quite a different story from the birth chart. Ketu repeats in the Ascendant, confirming the tendency to attract exotic, foreign or destabilizing partnerships, with Rahu and Ketu across the Ascendant (and marriage) axis. Both husbands were powerful and exciting men, both known for their sexual wanderlust. She had to accept these conditions to keep the marriages intact, and she herself could also exacerbate the situation by traveling often and living apart from her husband, as with her second marriage. This caused Onassis to start divorce proceedings in the fall of 1974.

In the *Navamsha* chart, other than Ketu and Rahu in the 1st and 7th houses, there are no other planets in *kendras*, so the worldly power declines except through partners. The trinal houses also remain unoccupied with the exception of two planets in the 5th house – Venus debilitated and Mercury exalted in Virgo. Since Venus is the Ascendant lord of both birth and *Navamsha* charts, its placement is extremely important. In the *Navamsha* chart, Venus is placed more auspiciously in a trinal house (as opposed to 8th house in the birth chart), and though debilitated in Virgo, it has *Neecha Bhanga Raja yoga*, corrected and uplifted by its contact to exalted Mercury also in Virgo. This gives her a good mind and a solid intellectual capacity, with talent for drawing and painting. The 5th house is the visual arts and also children. This gave her at least two children who grew to adulthood. (Her pregnancy problems are discussed elsewhere.) The *Navamsha* chart confirms the financial success given in the birth chart, both improving and amplifying these parameters. This is in part due to *Dhana* (wealth) *yoga* planet Mercury being exalted in the *Navamsha*, and contacting a debilitated Ascendant lord. It elevates her financially from disadvantageous circumstances. (Relatively speaking, in her society world she was at a lowly rung before marrying Kennedy.) Other *Dhana yogas* abound, including *Navamsha* 11th lord (Jupiter) in the 2nd house, aspecting the 8th house of partnership monies and inheritance, which Jupiter owns, protecting that area of life.

But *Navamsha* 9th and 10th lord Saturn falls in the 12th house of loss and expenditures, along with Moon. Birth chart 10th lord (Moon) also falls in the *Navamsha* 12th house.[292] Without spiritual discipline, this changes the outcome of the destiny and curbs personal happiness (Moon) and fulfillment in partnership or in career. (The 10th house is status and reputation.) This result is also due to the number of planets situated in Houses 6 and 12 in the *Navamsha* chart – a chart noteworthy for containing one exalted planet (Mercury in Virgo) and three debilitated planets: Venus in Virgo in the 5th house, Sun in Libra in the 6th house and Saturn in Aries in the 12th house. Debilitated planets rule Houses 1 and 6 (Venus), House 4 (Sun), and Houses 9 and 10 (Saturn). These are much weaker house and sign positions compared to the birth chart, although birth chart Ascendant lord in the 8th house does indicate some significant trials and tribulations in her life. In the planetary circuitry, debilitated Venus in the *Navamsha* chart owns both Sun and Mars in Libra, affecting their ability to function without some extraordinary outside help, which is sought and often found – if the planet is corrected through *Neecha Bhanga Raja yoga*. This occurs in every case here, since each of these planets is aspected by either its debilitation or exaltation lord.

[291] Very difficult or demanding destinies can be modified by spiritual practices, which though not removing the hurdles in the destiny, can minimize the challenges by making them easier to accept.

[292] The outcome of each area of life can be assessed through the position of house lord in the birth chart and *Navamsha* chart. If that house lord is located in House 6, 8, or 12 (*Dusthana* houses) in either birth or *Navamsha* charts, there are more obstacles to one's fulfillments, especially for reputation, with 9th or 10th lords.

With four *Navamsha* planets either exalted or debilitated, there would have to be some dramatic life situations, strong advantages and disadvantages – rising up and being cast down. There is continued focus on the need to better one's circumstances and upgrade one's life. This was a driving force from childhood, when Jackie first became aware of the financial insecurities of her family, who continued to live in luxury far beyond their means. The ongoing threat of losing these luxuries and the need to eliminate that threat in order to calm the mind – this is shown by Moon's condition in both birth chart and *Navamsha* chart. While no longer situated with destabilizing Rahu, the *Navamsha* Moon is now with debilitated Saturn in Aries in the 12th house of loss and expenditures. This may account in part for why she might seek ongoing fulfillment in the purchase of 200 pairs of shoes in one outing, along with many other luxury items. This Moon-Saturn configuration in the 12th house is less favorable for personal happiness without much time alone or ideally a spiritual orientation – which did not interest Jackie. She was more of a secular person who preferred the arts and literature. And though her Libran nature would gravitate towards the urban life, her chart shows she would tend to be happier in more private circumstances, more remote or pastoral places, including residence abroad (all 12th house arenas). This combination of factors helps us to understand why she gave no more interviews in her last thirty years and why in fact "America's Queen" may well have been happier living outside America.

MAJOR LIFE EVENTS & DATES:

1. **Parents separate first time (Sept. 30, 1936):** Jackie was 7 years old and in **Venus-Venus-Rahu period** (May 1, 1936 to Oct. 31, 1936). Her 20-year Venus *Dasha* ran from Feb. 20, 1935 to Feb. 20, 1955 and was a more problematic period for her parents. Venus in the 8th house of the birth chart already placed much of her own and her parents' attention on money or sex, causing many problems due to overemphasis and overspending. (Venus with Jupiter can bring excesses.) During Venus *Dasha* Jackie developed her own charms as well as her childhood fear of poverty. When her parents first separated, a Full Moon occurred in her 6th house of conflict, opposite transiting Mercury Retrograde, indicating a possible change of decision. Her parents did reconcile briefly in spring 1937, reversing the decision, but separated finally in one year. Venus-Venus-Rahu shows the family turbulence and the hijacked reputations Jackie worked so hard to restore throughout her life. Learning about domestic turmoil first hand, she sought to minimize its effect through her 8th house Venus. This brought a tendency to mythologize a life that is not how it appears to others.
2. **Parents separate final time (Sept. 1937):** Jackie was in **Venus-Venus-Saturn period.** With Venus and Saturn in 6 – 8 house relationship to each other in both her natal and *Navamsha* charts, a big change could happen then. Another catalyst was a **Total Solar eclipse on June 8, 1937** at 24:37 Taurus in Jackie's 8th house. It was close to the 8th house cusp (25:06 Taurus) and to her Ascendant lord Venus (28:53 Taurus), her *Dasha* lord at the time, thus receiving more emphasis. Tr. retrograde Mars at 28:53 Libra was aspecting tr. Venus in Aries as well as Jackie's natal Ascendant and five natal planets: Ketu, Rahu, Moon, Venus, and Jupiter. In a *Parivartana yoga* (mutual exchange) with each other, both tr. Jupiter in Capricorn and tr. Saturn in Pisces aspected the eclipse planets in Taurus. This eclipse impacts so many planets in Venus or Mars-owned signs, it would be likely to mark a seminal experience in her learn-

ing about life's power struggles, especially marital ones, and perhaps an early shattering of illusions about marriage and family life. (Jack Kennedy shared these *karmic* life themes. His natal Venus was at 24:02 Taurus and he had a *Nadi yoga* (*nakshatra* lord exchange) between Venus and Mars, the same as Jackie.)

3. **Parents' divorce initiated by mother (Jan. 1940, finalized July 1940):** The Jupiter-Saturn conjunctions of 1940-1941 in Aries marked a new 20-year era for everyone. The first one occurred on Aug. 7, 1940 at 21:26 Aries, closely opposite Jackie's Ascendant at 25:05 Libra. This was during her **Venus-Moon** *Dasha/Bhukti*. (Venus rules over marriage and Moon is mother: her mother initiated the divorce from Bouvier and aggressively sought a second marriage. Big money was a requirement and was likely to come, with the two great benefics in the 2nd house from Moon.) On June 21, 1942, her mother married Hugh Auchinchloss, the wealthy Standard Oil heir. This 2nd marriage took place during Jackie's Venus-Rahu *Dasha/Bhukti*, which was preceded by her Venus-Mars *Dasha/Bhukti*. Venus and Mars in a *Nadi yoga* indicated that during both **Venus-Mars** *Dasha/Bhukti* (Feb. 19, 1941 to April 21, 1942), and **Mars-Venus** *Dasha/Bhukti* (Jan. 15, 1976 to March 16, 1977), she would learn a great deal about gaining power through marriage and relationships.

4. **Marriage to John F. Kennedy (Sept. 12, 1953, 11:00 AM EDT, Newport, R.I.):** The marriage proposal and engagement occurred in spring 1953. The public announcement of the engagement on June 24, 1953 was during Jackie's Venus-Mercury-Jupiter period (Feb. 22, 1953 to July 10, 1953), with **marriage in Venus-Mercury-Saturn period** (July 10, 1953 to Dec. 21, 1953). Venus, Jupiter, and Mercury are each involved in the *Dhana yogas* of wealth in the *Navamsha* chart, reflecting the importance of financial wealth in this marriage to Jack Kennedy. At marriage, Saturn replaces Jupiter in the sub-sub period. Natal Saturn is *Darakaraka*, her spouse indicator (Jaimini *karaka* – see Glossary). Also, Saturn indicates how 12th house activities adversely impact marital happiness and prosperity. (The 12th house is both expenditures and pleasures of the bed. *Navamsha* Saturn in the 12th house aspects the 2nd house of income and family happiness.) A very visible or public marriage was evident in the marriage chart, with all physical planets above the horizon. Tr. Venus was at the top of the sky in Cancer on the Rahu-Ketu axis. Any planet on the eclipse axis is destabilized and in this case it is Venus, planet of love. Tr. Saturn's aspect to Venus shows the business angle to this marriage. Tr. Venus and tr. Moon in a *Parivartana yoga* (mutual sign exchange) amplify the number of female influences in the marriage. *Dharma* and *Karma* planets Jupiter and Saturn ushered Jackie into her life's most defining chapter, one that would shape all the rest. Tr. Jupiter was in early Gemini in her 9th house aspecting the Ascendant, while tr. exalted Saturn was in early Libra in her Ascendant. The AIR trine of Jupiter and Saturn was a positive indicator of the social, financial and aesthetic success the marriage brought to each of them. It also added some needed stability. Tr. Saturn conjoined tr. Moon in Libra – a serious moment and a formal commitment Kennedy had avoided up to then. For the bride, Moon-Saturn brought many new responsibilities, hard work, and also emotional distance and unhappiness, especially from Kennedy's ongoing and unstoppable promiscuity. Further, his medical needs, major back surgeries (1954 and 1955) and family and political obligations demanded a lot from her. (When Jack was killed, tr. Moon again joined tr. Saturn, though in Capricorn in her 4th house of endings.)

5. **Birth of daughter Caroline (Nov. 27, 1957):** Four months earlier Jackie's father "Black Jack" Bouvier died of liver cancer: Aug. 3, 1957. Both events occurred during Jackie's **Sun-Jupiter** *Dasha/Bhukti*. Sun is well placed in the birth chart in the 10th house (*Digbala* - best house for the Sun), though *Navamsha* Sun is debilitated in Libra and in the 6th house aspected by benefic Jupiter, also by Mars, Moon, and Saturn. The latter is problematic, as Saturn and Sun are enemies. Among *Dashas* during her fertile years, only Sun *Dasha* produced successful pregnancies. Sun is her *Jaimini karaka* for mother and motherhood (see Glossary), showing more likelihood of successful pregnancies during Sun *Dasha*. Her chart is not well disposed for childbirth and other *Dashas* suffered due to the house placement of the *Dasha* lord or malefic aspects to it. Planets in *Dusthana* houses (6, 8, or 12) can destroy the planned outcome through loss, disease, or some other *karmic* twist causing turbulence, especially when aspected by malefic planets: Sun, Mars, Saturn, Rahu or Ketu. But malefic planets in *Dusthana* houses can show progress with struggle. Jackie's Sun-Jupiter period facilitated a successful pregnancy due to harmonious mutual aspects between *Dasha-bhukti* planets in both birth chart and *Navamsha* chart (sextile and trine).

6. **Kennedy's Presidential Election & Inauguration (Nov. 8, 1960 and Jan. 20, 1961):** Both events occurred in Jackie's **Sun-Venus** *Dasha/Bhukti*. In the Inauguration chart (12 noon EST, Washington, D.C.) tr. retrograde Mars (aggression) closely aspects tr. Sun (the leader). In this case the Mars aspect was to the 8th house from itself, considered more challenging. Venus and Jupiter - planets of excess when in mutual association - were in a *Nadi yoga*, an exchange of *nakshatra* lords. Each of them was close to Jackie's 3rd and 5th house cusps, relating to communications (House 3) and children and the arts (House 5). Tr. Venus was on the tr. Rahu-Ketu axis, echoing the same configuration (though a different sign) as in her 1953 marriage to Kennedy. At the Inauguration, tr. Venus, Ketu and Moon were in Aquarius, the sign of idealism. Tr. Moon and Venus were both in *Purva Bhadra nakshatra*, whose *shakti* is to elevate consciousness. Tr. Moon at 29:58 Aquarius was about to leave the sign, within three minutes of clock time. In addition to the Mars-Sun contacts and tr. Moon about to shift signs, an ominous note is added by tr. Saturn's close aspect to tr. Moon (from Sagittarius) - the last aspect to the Moon, indicating that the enterprise starting at this moment could be cut short. With tr. Venus, Ketu, and Moon prominent we see charm, style and love matters magnified and causing turbulence (Rahu-Ketu). As First Lady Jackie initiated programs that gave more serious attention to the arts (Venus). After her husband's death, a National Cultural Center initiated by President Eisenhower in 1958 was re-named the Kennedy Center for the Performing Arts. This was intended as a living memorial to President Kennedy, who spoke often of "our contribution to the human spirit." It opened in Sept. 1971 in Jackie's Mars-Rahu *Dasha/Bhukti*. (In a *Nadi yoga* with Venus, her Mars gave her increased power while Rahu magnified it.)

7. **Birth of son John Jr. (Nov. 25, 1960, 0:22, Washington, D.C.):** He was born in Jackie's **Sun-Venus** *Dasha/Bhukti*, the last sub-period before the start of her Moon *Dasha* in Feb. 1961, which would not be as favorable for bearing children. His chart mimics hers in some ways. Instead of Ketu on the Ascendant opposite Moon-Rahu, he has Rahu exactly on the Ascendant

(Leo, in his case) opposite Moon and Ketu in Aquarius, giving him an exotic and handsome appearance, a tendency for celebrity status and changeability in relationships, most of them with actresses who failed to gain Jackie's approval. The theme of a mother's strong influence over the life continued. Jackie guided him against a career in the theatre, against learning to fly a plane, and towards law and real estate. Law was not easy for him; he failed the bar exam several times before passing it. Soon after her death in 1994 he learned to fly his own plane, started a political magazine (1995), and married his girlfriend (1996). His sudden death on July 16, 1999 came in a crash of his private plane into the ocean near Martha's Vineyard. He was the pilot and his wife and sister-in-law were passengers.

8. **White House Tour television documentary (Feb. 14, 1962):** Jackie was in **Moon-Mars** *Dasha/Bhukti*. Eighty million viewers tuned in that night. Five planets were transiting in Capricorn in her 4th house of home and foundations: Mercury, Mars, Jupiter, Saturn, and Ketu. Ketu and Jupiter were close to her 4th house cusp, amplifying the qualities of the historic place she then called home. These planets in Capricorn, though in the 4th house from her Libra Ascendant, were in the 10th house from her Moon, which as *Dasha* lord gave her maximum visibility. During her Moon *Dasha* tr. Saturn had a unique and dual role: On the one hand, reading from Libra Ascendant, Saturn marked the ending of a cycle; but reading from *Dasha* lord Moon in Aries as sub-Ascendant, it gave her enormous publicity, especially through her husband: Tr. Saturn was in Capricorn Feb. 1, 1961 to Jan. 27, 1964.

9. **John F. Kennedy assassination (Nov. 22, 1963, 12:30 PM CST, Dallas, TX):**[293] Jackie was in **Moon-Rahu** *Dasha/Bhukti*, both planets in her 7th house of partnership. Transiting Moon and Saturn were rising in Capricorn, while four planets were in Scorpio. The widow (Moon-Saturn) played a prominent role and people everywhere were grief-stricken (also Moon-Saturn). Venus and Mars were in a *Graha yuddha* (Planetary War) near 28 degrees Scorpio, a treacherous sector of the zodiac. Venus and Mars when together seek immense power. When Mars loses the war to Venus, violence can be a means to an end. Both tr. Mars and Venus opposed her natal Venus almost exactly, and her natal Jupiter more widely. She was almost killed herself. Tr. Saturn was in her 4th house, classically a low point in the life, and in the cycle of Saturn's orbit around the chart, especially so close to the 4th house cusp (bottom of the chart). Tr. Saturn closely aspected her Ascendant at 25:05 Libra from the 4th house of foundations, which were shattered at this moment. Saturn can create separations and ruptures. Kennedy's natal Mars was exactly opposite her Ascendant, showing the passion he evoked in her, as well as the life work he set up for her before and after his death. He inspired her to fight for his good reputation and a strong place in history.

10. **Publication of William Manchester's *Death of a President, Nov. 20-25, 1963* (spring 1967):** Jackie was then in **Moon-Mercury** *Dasha/Bhukti,* echoing a battle between her heart (Moon) and mind (Mercury). Tr. Jupiter was exalted in Cancer in her 10th house of maximum visibility. As lord of 6th house of conflict for Libra Ascendant, Jupiter can cause major problems

[293] For more analysis of John Kennedy's Assassination chart, see Chapter 8: Virgo Ascendant, Chart #25: John F. Kennedy, end of section.

at this time that cannot be hidden. Tr. Mars retrograded over her Libra Ascendant for much of spring 1967. As her 7th house lord Mars symbolized marriage or business partner(s): Anything she did or said was of enormous public interest, most especially in regard to President Kennedy in the months and years following his death. A Solar eclipse May 9, 1967 at 24:53 Aries was on her natal Rahu opposite natal Ketu, thus a return of the eclipse axis to its natal position, where it had a tendency to create both Jackie's expanded reputation as well as controversy around some of her decisions. She commissioned Manchester to write the book in 1964. When he finished it in 1966 she tried to block publication of the book and magazine serialization. (Mars with Rahu or Ketu can complicate and extend battles.) On April 9, 1966 she began her *Sade Sati* (7½ years of tr. Saturn in Houses 12, 1 and 2 from natal Moon). This denoted a period of heavier personal responsibilities: her afflicted Moon would suffer. There were legal settlements and Manchester published the book, which won prizes and widespread acclaim, though Jackie never liked it. Despite all the controversy, from 1962 through 1968 Jackie continued to be the "World's Most Admired Woman." She got a lot of slack as the widow everyone knew.

11. **Bobby Kennedy assassination (June 5, 1968):** This event had an enormous impact on Jackie, who was very close to Bobby. He had asked her to hold off marrying Onassis until after the Nov. 1968 election. After Bobby was killed in June 1968 she said:

> "I despise America, and I don't want my children to live here any more. If they are killing Kennedys, my kids are the number one targets."

She was then in **Moon-Ketu** *Dasha/Bhukti*. Several key planets were transiting across the Virgo-Pisces axis, which can bring themes of sacrifice and martyrdom: Saturn at 29:04 Pisces (a treacherous *gandanta* degree, within the last 3 degrees 20 minutes of a water sign); the eclipse planets Rahu and Ketu at 22:14 Pisces-Virgo; and Moon at 19 Virgo in *Hasta nakshatra*. On **April 12, 1968** there was a **Total eclipse** at 29:54 Virgo opposite Saturn, adding extra intensity, even more so and of a darker nature due to malefic planets transiting close to eclipse planets or degrees. When Jackie married Onassis 4 ½ months later, tr. Saturn returned to within one degree of its position on June 5, 1968 (at 28:14 Pisces). The familiar pattern of Moon-Saturn on the eclipse axis symbolized several themes, especially in houses 6 and 12 in her birth chart: sacrifice (6th house and Virgo), martyrdom (Virgo and Pisces, and houses 6 and 12), violent separation (Moon opposite Saturn on the Rahu-Ketu axis), and conflict through this event (planets in the 6th house or in Virgo). Jackie vowed to leave America and never return. She was more determined than ever to marry Onassis and escape from the United States.

12. **Second marriage (to Aristotle Onassis) (Oct. 20, 1968, 5:15 PM, Skorpios island, Greece):** Jackie was in **Moon-Ketu** *Dasha/Bhukti* up to Dec. 20, 1968. Venus *bhukti* started two months later and would have been more auspicious for marriage, as Venus rules over love matters. Moon-Ketu period emphasized the foreignness of her husband and the strained relationship that soon developed. True happiness seemed to elude them, as they lived far apart much of the time. In the marriage chart Pisces was rising in *Revati nakshatra* – a *nakshatra* of wealth,

with Saturn and Rahu in the Ascendant. Tr. Saturn and Rahu were opposite four planets in Virgo: tr. Jupiter, Ketu, Moon and Mercury retrograde. (Virgo was also Jackie's 12th house of foreign residence.) Transiting close to Ketu, Moon was in *Hasta nakshatra*, ruled by the Moon. Emotional distance (Moon-Saturn) and conflicts over money reigned for much of the marriage. In the marriage chart, Mars ruled two money houses (2 and 11), and was situated in the 6th house of conflict. This was a conflict that continued after Onassis died. For more details, see the next Item.

13. **Death of Aristotle Onassis (March 15, 1975):** By 1975, tr. Jupiter had completed one complete cycle (11.9 years) since the JFK assassination. Several key events occurred in 1975, starting with the death of Aristotle Onassis and culminating with the breaking of the "Camelot myth" and Judith Campbell Exner's Congressional testimony. (See item #15 below) Tr. Jupiter was in Pisces in Jackie's 6th house of conflict: from Feb. 20, 1975 to Feb. 26, 1976, with the exception of about 8 weeks (July 19, 1975 to Sept. 11, 1975). Tr. Saturn in Gemini was in her 9th house (of 2nd marriage) opposite her own Saturn, denoting hard work, endings or confrontations. Jackie was in **Mars-Mercury** *Dasha/Bhukti* (Mars-Mercury can be literally a "war of words.") Tr. Mars was in the 11th house of financial gains. Jackie entered a legal battle with stepdaughter Christina Onassis to claim her share of her deceased husband's estate. Christina demanded Jackie give up all claims to her father's estate and settle for $26 million. By Greek law, a widow was entitled to 12.5 % of his estate, approx. $125 million, in this case. Deciding against further legal battles, Jackie accepted the $26 million, perhaps to avoid more damage to her reputation via Christina, including any mention of a pending divorce. Mercury and Mars are in 2-12 relationship in both birth and *Navamsha* charts, reflecting issues of money, expenditures, and foreign interests.

14. **Jackie starts as consulting editor at Viking Press (Sept. 22, 1975, NYC):** Jackie did not need a job but she needed to re-ground her life and to refurbish her public image and reputation, both tarnished in the 1968-1975 period as Onassis's wife. Her *Dasha* and transit timing shows the focus of this two-pronged effort: Tr. Jupiter and Saturn were both in water houses: tr. Jupiter (and tr. Moon) in her 6th house of work and tr. Saturn in her 10th house of career (also status and reputation), the top of the cycle for Saturn and excellent for career visibility. She was in **Mars-Ketu** *Dasha/Bhukti*, with tr. Mars closely contacting her natal Venus at 28 Taurus, and tr. Ketu earlier in the sign. Tr. Venus in Leo contacted her natal Mars in Leo. Mutual Venus-Mars contacts indicate a quest for power and resonate with her Venus-Mars *Nadi yoga*. Tr. Jupiter opposite tr. Mercury on the Full Moon axis was auspicious for her entry into book publishing and as *Neecha Bhanga Raja yoga* (Mercury debilitated, with correction through Jupiter) shows that the initial disadvantage of inexperience would be corrected over time. Major assets she brought to the job were her celebrity (an asset for acquiring new projects), her fine intellect, her deep love of books and literature and her sincere respect for the writer. The relative stability of the tr. Jupiter-Saturn trine reflects her ability to carry on with this new part-time career for some years. Though the income was nominal in her upper echelon world, the work itself fed her intellectual and social needs. Though changing publishing houses two years later, she worked as part-time editor until near the time of her death in 1994. She believed in the world of books and their power to communicate important truths.

15. **"Camelot myth" cracks – Judith Campbell Exner in the headlines (Dec. 17, 1975):** The Congressional testimony of Judith Campbell Exner in Nov. 1975 broke into national headlines on Dec. 17, 1975. This was the first public revelation that Jack Kennedy was involved with the Mafia in plots to assassinate Fidel Castro, planned in part through Judith Campbell – Jack Kennedy's lover for two years, 1960-1962 while also linked with Mafia boss Sam Giancana. Jackie was then in **Mars-Ketu** *Dasha/Bhukti*. Tr. Mars was in late Taurus within half a degree of her natal Venus in the 8th house of scandal. (Venus-Mars combinations can evoke power struggles, and when in the 8th house – sexual or political scandals.) Tr. Moon was also in Taurus earlier in the sign, nearer to Jackie's natal Jupiter. Tr. Venus and Rahu were on her Libra Ascendant. Tr. Saturn in Cancer closely contacted her natal Mercury (depressing news). Tr. Sun was within one degree of her natal Saturn in Sagittarius in the 3rd house of news, doubling the impact. Tr. Saturn in the 10th house can bring one visibility that either elevates or brings one down suddenly. This event created the first major breaking of the "Camelot Presidency" myth Jackie had so carefully constructed.
16. **Death of Jackie's mother, from Alzheimer's disease (July 22, 1989):** Her mother died in Jackie's **Rahu-Ketu** *Dasha/Bhukti*. The eclipse planets fall on the same axis with Jackie's natal Moon (mother and memory, among other things). Afflictions of Rahu-Ketu to Moon in Aries indicate problems to the head or brain. Alzheimer's reduces the size of the brain and memory. Her death occurred just one week prior to Jackie's 60th birthday, culminating a period of return of both *Dharma* and *Karma* planets, Jupiter and Saturn, respectively. Tr. Jupiter had moved on to the next sign of Gemini, but tr. Saturn was still in its birth sign of Sagittarius, in a weak position for the mother, as it falls in the 12th house from the house it owns. This shows that though her mother was aggressive in forging a good life for herself and her family, she did so from a position of deep insecurity. Tr. Jupiter-Saturn in opposition across Jackie's 3 - 9 house axis indicates the possibility of some conflict at this time between Jackie and her sister Lee. (The 3rd house is younger sibling.) They were estranged for years in part due to their fierce competition for favors from their parents and from men.
17. **Jackie's death from non-Hodgkin's lymphoma (May 19, 1994, 10:15 PM EDT, Manhattan, NY):** She was diagnosed with the disease in Jan. 1994. Jackie died on her father's birthday, echoing his enormous influence in her life. She was in **Rahu-Moon-Saturn period** from March 24, 1994 to June 18, 1994. Her natal Saturn is in the 9th house (of father and legal matters) from her natal Moon and aspects the 9th house of father. Her last will and testament was dated March 22, 1994. She succumbed to a tumor on her brain, symbolized by Rahu (foreigner or foreign object) to the head (Moon in Aries). It eventually spread to her spine and liver, and on May 18, 1994 she chose no further treatment and went home from the hospital to die, slipping into a coma within hours. A memorial service followed on May 23rd at Saint Ignatius Loyola Roman Catholic Church in Manhattan, and a burial at Arlington National Cemetery next to Jack Kennedy and their two infant children, Patrick and Arabella. All of it was closely documented on national television. At her death, tr. Rahu, planet of celebrity, was at 29:56 Libra, having entered Libra earlier that same day and impacting her Ascendant. The backdrop for the outer planets shows that challenges to her health could start in fall 1993, or

even earlier, with tr. Saturn in Aquarius aspecting her Aries Moon from March 6, 1993. By Oct. 13, 1993, tr. Jupiter entered Libra, not favorably placed on her Ascendant as 6th house lord of health matters. Most important in confirming a challenging time for her health was the start of **Rahu-Moon period** on Aug. 3, 1993 (to Feb. 2, 1995). She was nearly killed in **Moon-Rahu period** in Nov. 1963, when bullets to the head killed her husband in front of the world. Rahu can amplify one's public (Moon). Moon and Rahu are also situated in a *Maraka* (potentially death-inflicting) house, though other factors have to be weighed. Rahu-Moon *Dasha/Bhukti* was as potentially dangerous for her as that of Moon-Rahu.

At Jackie's death, tr. Moon at 25:07 Leo was in *Purva Phalguni nakshatra*, ruled by Venus, and opposite tr. Saturn at 17:40 Aquarius. At many of her key life moments, tr. **Saturn** contacted tr. **Moon**. The two planets were conjoined (wedding to Kennedy and JFK assassination); tr. Saturn was in an exact sextile to tr. Moon and aspecting it (JFK Inauguration); and tr. Saturn opposed tr. Moon (wedding to Onassis and her own death). This resonates with Jackie's Saturn-Moon combination in the 12th house of her *Navamsha* chart opposite Sun-Mars. It shows a destiny to marry older men, to experience separation, distance, or loss through them and to know loneliness in both widowhood and celebrity. Moon-Saturn in the *Navamsha* 12th house also reflects her fierce need for privacy – both a response to her life circumstances and a means to find inner equilibrium. Even so, she was a chain smoker all her life, stopping only in January 1994 at the insistence of her daughter Caroline. Most of her public had no idea she smoked, but secrets are guarded in the 12th house, even if ultimately she could not suppress secrets about the Camelot presidency. At her moment of death it was still her father Black Jack's birthday and tr. Venus was in her 9th house of father. Tr. Moon was at 25:07 Leo, very close to the natal Moon of Jack Kennedy (24:30 Leo). She would soon be buried next to him at Arlington National Cemetery. Moon in *Purva Phalguni nakshatra* has the symbol of the conjugal bed and the desire "to have the best share." Tr. Venus was just setting on the Western horizon in Manhattan, New York as America was losing its "Queen."[294]

---

[294] Tr. Venus at 5:03 Gemini was also close to setting in the USA chart, using 8:59 Sagittarius Ascendant. Aside from this fitting symbolism, the month prior to Jackie's death and burial was breathtaking in its eventfulness, both nationally and internationally. Her memorial service was held the day before a Lunar eclipse, and was preceded by a Solar eclipse May 10, 1994 at 26:01 Aries – just opposite her Ascendant at 25:06 Libra, in the 7th house of partners, and on her natal Rahu (her current *Dasha* lord) at 24:20 Aries. Rahu has a magnifying effect on her partnerships and on her life in general. One month earlier, and two days after the previous Full Moon on April 25th, the funeral of Richard Nixon was held. Nixon had narrowly lost to John F. Kennedy in the Nov. 1960 Presidential election. With Jackie Kennedy's death so close on the heels of Richard Nixon's, it was said that "even in death Nixon was eclipsed by a Kennedy!" Nixon's funeral was held April 27, 1994. Then on May 9, 1994, ten days before her death, Nelson Mandela became the first black African to become a State President. A new six-color flag was raised in Pretoria to mark the birthing of a new democracy in South Africa. He had spent 27 years in prison (1962 to 1990) for his anti-apartheid activities, fighting up to May 1961, when South Africa began 33 years of international isolation, refusing to break its apartheid policies. President Kennedy did not fully understand how he reignited the black Civil Rights movement in the U.S., and how his stirring rhetoric emboldened both African Americans and black Africans. Jackie, in her unwavering dedication to John Kennedy's presidential legacy, would have been glad to see the end of apartheid and freedom for all South Africans, and for the first time since Europeans landed in Capetown in 1652. Thus did several seemingly unrelated endeavors – both begun in early 1961 – come to completion together. Their mutual resonance came in part from the Jupiter-Saturn conjunction Feb. 18, 1961 at 1:52 Capricorn, opposite the natal Sun of both Jackie and Nelson Mandela, the latter very closely.

REVIEW OF LIBRA ASCENDANT CHARTS:

Give the astrological reasoning in each case.

1. Why is the Libra Ascendant person often so driven to achieve his or her vision of harmony? Why is this often related to the interaction of Venus and Mars?
2. What factors in the chart of Adolf Hitler would make him particularly zealous in his desire to achieve his own vision of a harmonious universe?
3. What factors show his love for the military combined with the fervor of his ideology?
4. Why were Hitler's ambitions thwarted during his Mars *Dasha*? Why was he unable to realize his ambitions as an artist?
5. How did natal Venus and Mars in Libra contribute to Mahatma Gandhi's being an activist for peace, eschewing violence? Compare to Hitler's Venus and Mars in Aries.
6. What natal planet accounts for Gandhi's eventually "finding his voice" *outside* India? Why would this planet also give him the destiny to face conflict throughout his life?
7. Why was it more difficult for Gandhi to achieve his goals for India after May 1940?
8. How does Ammachi's chart show her destiny to be spiritually oriented? What is the astrological signature of her tireless capacity for hard work?
9. Why would Ammachi's early years bring her so many personal troubles? Why would Saturn *Dasha* bring her into greater prominence in the world?
10. What astrological components describe the mysterious nature of Jacqueline Kennedy Onassis? What accounted for the paradox of her wanting both public attention *and* tremendous privacy?
11. Why would her 10-year Moon *Dasha* be likely to be the most eventful of her life? Why would it also contain the most suffering?
12. Why was Jacqueline Kennedy Onassis likely to marry into great wealth?

Glossary of Sanskrit & Astrological Terms

Adhi yoga *(AH - dee - YOE - gah)* : *Adhi* = to add or receive; yoga = planetary combination. This yoga occurs when classic benefics (Venus, Jupiter, and Mercury) occupy the 6th, 7th, and/or 8th houses from the natal Moon (*Chandra-Adhi yoga)* or from the Ascendant *(Lagna-Adhi yoga)*. It brings great wealth and status, especially when all three classic benefic planets are involved. For full effect, the planets should be unafflicted, i.e., not combust or contacting classical malefic planets.

Agami Karma *(AH - GAH - mee - KAR - ma)* : the result of actions we perform in our present existence. (See also ***Karmas***)

Ahamkara (*Ah - ham - KAH - rah*) : the specific point in material reality when the consciousness becomes embodied, and considered for the birth of an individual astrologically as the Ascendant. As a *karaka* (significator) for the Ascendant, the Sun is also *karaka* for the *Ahamkara.* Neither entirely ego, nor personality – but encompasses the ego, and is the notion of individual existence. (See also **Ascendant**)

Amatyakaraka *(ah - MAH - tya - KAH - rah - kah)* : significator for the mind; also, one's confidant, as in the minister to the king. (See also ***Jaimini Karakas***)

Amsha (*AHM - sha*) : division, as in the divisional charts, such as *Navamsha*, 9th harmonic.

Arishta yoga *(ah - RISH - tah YOE - gah)* : a yoga of affliction, especially to one's physical health. It occurs with the mutual association of the Ascendant Lord and a *Dusthana* lord, i.e., ruler of House, 6, 8 or 12. If planets (2nd or 7th lords) are involved, the results are worse. An *Arishta yoga* can bring ill health, especially during the *Dasha/Bhukti* of the Ascendant lord and the associated *Dusthana* lord, and more so if the Ascendant is weak.

Artha *(AR - tha)* : material security, taking care of your physical and financial needs. It includes the survival needs – eating and nourishment on the physical level – a prerequisite to achieving status or recognition in the community and in the world. Artha is assigned to Houses 2, 6, and 10; also associated with the earth signs Taurus, Virgo, and Capricorn.

Ascendant: the point where the Eastern horizon intersects with the ecliptic, or the Sun's apparent path around the earth. At sunrise, the Sun is on the Ascendant to the exact degree. It takes on average two hours for a zodiacal sign to pass over the horizon. This varies according to the sign's ascension and the latitude in question. The destiny of any entity begins at the Ascendant, according to Vedic philosophy. (See ***Ahamkara***) With whole sign houses the Ascendant always falls in House 1. (See also ***Lagna***)

Ascension, Long: signs that take longer to rise over the eastern horizon: Gemini, Cancer, Leo, Virgo, Libra, Scorpio and Sagittarius in the Northern hemisphere. Capricorn to Taurus are of Long Ascension in the Southern hemisphere. (All of this applies to the sidereal zodiac.)

Ascension, Short: signs taking a shorter time to rise over the eastern horizon: Capricorn, Aquarius, Pisces, Aries, and Taurus in the Northern hemisphere (shortest are Pisces and Aries). In the Southern hemisphere signs of Short Ascension are Gemini to Sagittarius.

Ashtakavarga *(ah - SHTAH - kah - VAR - gah)* : eight sources of planetary energy, measured in *bindus* (points), indicating the strength of each planet in each house, especially relative to transiting planets. The total number of bindus accumulated from each planet in each house is the ***Sarva Ashtakavarga***, "*sarva*" meaning "all." One planet can contribute from minimum 0 to maximum 8.

Ashtama Shani *(AH - shtah - mah SHAH - ni)* : Transiting Saturn is in the 8th house from natal Ascendant or natal Moon, and considered the most difficult of Saturn transits, more so than *Sade Sati*, unless natal Moon is afflicted or located in House 6, 8, or 12, notably House 8. *Ashtama Shani* can bring losses, separations, or endings.

Aspects: Angular relationships between planets or in relationship to the Ascendant. (See ***Drishti***)

Atmakaraka *(AHT - mah - KAH - rah - kah)* : significator of the soul or self. (See also ***Jaimini karakas***)

Ayana *(ay - YAH - na)* : literally, "direction (moving); going, progress." In ancient Hindu astronomy it refers to the six month period when the Sun is travelling either on a northern path or a southern path. It is sometimes used to describe Solstice, half year or precession.

Ayanamsha *(ay - an - AM - sha)* : The difference between the tropical zodiac and the sidereal, or constellational zodiac. In Jan. 2010, the *Lahiri ayanamsha* was 24 degrees 00 minutes. To obtain the same position of a planet in the tropical zodiac, add the *ayanamsha* for the year in question to the position in the sidereal zodiac. (Lahiri advocated this *ayanamsha* in modern times. The full proper name would be *Chitra Paksha ayanamsha*, referring to *Chitra nakshatra*, as it falls in the middle of all the *nakshatras*.)

Bala *(BAH - la)* : strength.

Bhagya *(BHA - gya)* : fortune. The 9th house is a house of good *Bhagya*, more so if it is unafflicted and its house lord is unafflicted. A ***Bhagya yoga*** occurs in several ways, one of them when a strong and unafflicted benefic planet aspects both Ascendant and 9th house.

Bhanga *(BHA - nga)* : cancellation. (See also ***Neecha Bhanga Raja yoga***)

Bhava *(BHA – va)* : house, room, or attitude. Entering a new *Bhava* changes the way a planet operates, though not its basic nature.

Bhavat Bhavam *(BHA – vat BHA – vam)* : derived houses (not a literal translation). The house as far from the original house as its numerical value, i.e., counting one house from House 1 (comes to House 1), two houses from House 2 (comes to House 3), four houses from House 4 (comes to House 7), etc. If the lord of a house is in the *Bhavat Bhavam*, it is generally strengthening in a positive way to the affairs of the initial house, except for Houses 6, 8, and 12, and the 6th, 8th or 12th houses from their respective houses or from any house.

Bhratrikaraka *(BHRA – tri – KAH – rah – kah)* : significator for brothers, siblings. (See also ***Jaimini karakas***)

Bhukti *(BHUK – ti)* : sub-period, usually of a *Dasha*, thus *Dasha-Bhukti.*

Brahmins (*BRAH – mins*) : the priests, spiritual teachers, those seeking higher consciousness. Brahmins are associated with the water signs (Cancer, Scorpio, and Pisces), and include artists, notably musicians, especially when more planets occupy water signs, or Houses 4, 8, and 12. This may confer an emotional capacity to deliver a transcendent message to the public.

Buddhi *(BU – dhi)* : Mercury, also intelligence. It is also spelled ***Budha***. (See also ***Grahas***) As intelligence, it is associated with the ability to discriminate what is real and what is unreal. (See also ***Manas***)

Budha Aditya yoga *(BU – dah AH – DI – tya YOE – gah)* : If Mercury (*Budha*) is at least 10 degrees away from the Sun in the same sign, this condition produces a *yoga* of cleverness. If Mercury is unafflicted and at least 20 degrees away from the Sun in the same sign, the condition is greatly enhanced and gives mental brilliance.

Chaitra Shukla Pratipada (*CHAI – trah SHU – klah PRAH– ti – PAH – dah)* : ***Chaitra*** = "distinguished" (also refers to *Chitra nakshatra,* where the next Full Moon usually falls), ***Shukla*** = bright (i.e., bright half of the lunar month, *Shukla Paksha,* from New Moon to Full Moon), ***Pratipada*** = first *tithi* (lunar day, out of 30 *tithis* per lunar month). Sanskrit term for the Vedic New Year, considered generally very fortunate, and occurring every year with the New Moon in Pisces prior to the transiting Sun entering sidereal Aries. It falls somewhere between March 15th and April 15th each year. The chart of this event is often used to predict for the year, starting from this date until Vedic New Year the following year. It is set for the geographical location being studied, and most often the capital of a nation.

Chandra (*CHAHN – drah*) : Moon. (See also ***Grahas***)

Chara (*CHAH – rah*) : moveable, active, or in a state of motion; applies to Aries, Cancer, Libra, and Capricorn. Western tropical astrology refers to these signs as "cardinal." It is one of the three qualities of a sign. (See also ***Dwiswabhava*** and ***Sthira***)

Combustion: "burnt by the Sun's rays." A planet is compromised by being close to the Sun, especially if within five degrees either side of the Sun and behind the Sun in celestial longitude. Moon, Rahu and Ketu are excluded. The widest ranges of orbs used (on either side of the Sun) are these: Mars: 17 degrees; Mercury: 13 degrees; Jupiter: 11 degrees; Venus: 9 degrees; and Saturn:

15 degrees. (See also Chapter 7 Introduction)

Conjunction: Two or more planets are joined together in one sign of the zodiac. Those planets can affect each other more if they are in the same *nakshatra* (13 degree 20 minute segment), and even more intensely if within the same *Navamsha* (3 degree 20 minute segment). If they are within one degree, there are other conditions that can occur, such as **Combustion** and/or ***Graha Yuddha***. (See both Glossary entries) The conjunction of Jupiter and Saturn marks regular 20-year cycles of great importance.

Darakaraka *(DAH - rah KAH - rah - kah)* : Spouse indicator. (See also ***Jaimini karakas***)

Dasha (*DAH - sha*) : planetary period. There are at least 55 known *Dasha* systems of timing in Vedic astrology. (Alternative spelling is "*Dasa*," notably in South India.)

Dashamsha *(dah - SHAM - sha)* : the 10th harmonic chart, giving indications about 10th house matters, notably career and status. (All *Varga* charts are secondary in importance to the *Navamsha* chart. The *Dashamsha* chart is discussed only occasionally in this book.)

Debilitation: A planet at its lowest point by zodiacal sign, causing it to operate with some deficiency, especially noticeable during its *Dasha* or *Bhukti*. The severity of planetary debility can be lessened through other factors. See also **Exaltation** and ***Neecha Bhanga Raja yoga*** - giving the conditions under which a debilitation of the planet may be cancelled, though it still may express some initial disadvantage in some spheres of life.

Dhanu (*DHA - nu)* : Sagittarius, the 9th zodiacal sign; ruled by Jupiter.

Dharma (*DHAR - ma*) : the expression of one's true nature, shown by the Ascendant and the combination of planets from the Ascendant. Also, it is one's proper work in the world based on one's true nature. Assigned to Houses 1, 5, and 9, and associated with the fire signs - Aries, Leo, and Sagittarius. Collectively, this refers to right action - by thought, word, and deed - and to righteous duty as part of a moral code protecting the society from dissolution.

Digbala *(DIG - BAH - la)* : *dig* = direction; *bala* = strength; directional strength of a planet. (It is also spelled *Dikbala.*) *Digbala* assigns power based on angular house placement in a chart, and is perhaps the most important way of measuring a planet's strength. It is one of six components of the *Shadbala*. (See ***Shadbala*** in Glossary; ***Digbala*** in Chapter 2)

Drishti (*DRISH - ti)* : aspect. From the Sanskrit word root "to see." (See also **Aspects**) *Drishti* means the planet can "see" another planet or house. Therefore it has the capacity to influence the other, either one-way or mutually, depending on the rules established and used. The aspects are either between planets or in relationship to the Ascendant, always counted in forward motion in the zodiac, with the aspecting planet as starting point. (This is read clockwise in the South Indian chart, and counter-clockwise in the North Indian chart.) Classical Parashara aspects are used in this book. Jaimini aspects work differently.

Durudhara yoga *(duh - RUD - ha - rah YOE - gah)* : The Moon is surrounded by planets in Houses 2 and 12 from itself, giving many associations, and benefits from them. (Sun, Rahu and Ketu are excluded from this consideration.)

Dusthana** (duh - STAH - nah)* : Houses 6, 8, or 12 from the Ascendant, and considered the most difficult houses in Vedic astrology, notably the 8th house. The planet ruling the 8th house of an Ascendant chart will be problematic for that chart, unless it is also the Ascendant lord (as in Aries and Libra Asc.), also ruling over longevity. Though the Ascendant lord is always a benefic planet when ruling over the affairs of the first house (the physical body, and general well-being and good fortune), when the Asc. Lord is also 6th house lord (Taurus and Scorpio Asc.), as 6th house lord it can attract enemies. This is especially so for Venus, lord of Taurus Asc., more so than for Mars as lord of Scorpio Ascendant, as the nature of Mars is aggressive and competitive. It gravitates toward 6th house matters, whereas Venus tends to suffer some antagonism or sacrifice when associated with the 6th house or the 6th sign of Virgo, its sign of debilitation. (See also ***Viparita Raja yogas) The latter can protect or even enhance certain *Dusthana* house matters, though they do not apply to the health and well-being of the physical body.

Dwiswabhava** (DWIS - va - bha - vah)* : dual, or oscillating; applies to the signs of Gemini, Virgo, Sagittarius, and Pisces. Vedic texts also refer to it as a "common sign." It is one of the three qualities of a sign. (See also ***Chara and ***Sthira***)

Eclipse: The Earth, Sun, and Moon come into an alignment where either the Sun is obscured by the Moon from the viewpoint of the Earth (a Solar eclipse, occurring at the New Moon), or the Moon is obscured by the Sun from the viewpoint of the Earth (a Lunar eclipse, occurring at Full Moon). These occur several times a year, usually two in a row, after an interval of about six months. A Total eclipse (lunar or solar) is more intense in its effects, especially when it occurs at the solstice or equinox.

Exaltation: A planet is considered to be in its highest state in its sign of exaltation and will manifest in some extraordinary way, and becoming more remarkable the more planets are exalted (and/or debilitated). An exalted planet will operate best if its sign and *nakshatra* lords are strong, if it is well placed house-wise and receiving no afflictions from classical or temporal malefic planets. (See also **Debilitation**) The sign of debilitation is 180 degrees opposite the degree of exaltation. The same is true of maximum (or "deepest") degree of exaltation of each planet, given as follows:

Planet	Sign of Exaltation	Deepest degree	Sign of Debilitation
Sun	Aries	10:00	Libra
Moon	Taurus	3:00	Scorpio
Mars	Capricorn	28:00	Cancer
Mercury	Virgo	15:00	Pisces
Jupiter	Cancer	5:00	Capricorn
Venus	Pisces	27:00	Virgo
Saturn	Libra	20:00	Aries
***Rahu**	Taurus	20:00	Scorpio
***Ketu**	Scorpio	20:00	Taurus

*(Rahu and Ketu exaltation signs are not generally accepted by all astrologers, as Parashara does not include Rahu and Ketu when discussing this topic in the classic *Brihat Parashara Hora Sastra*, Vol 1, Chapter 3, v. 49-50.)

Gaja Kesari yoga *(Gah – jah KAY– sah – ree YOE – gah)* : literally, "the elephant – lion yoga." The person can walk tall in the jungle (of life) like an elephant, with the strength and daring of a lion. Natal Jupiter must be in a *Kendra* from natal Moon, and the results depend on how well Moon and Jupiter are situated by sign and *nakshatra*, and whether afflicted by malefic planets. If unafflicted, the person can enjoy many relations and have a good reputation that lasts even after death.

Galactic Center: The Sun of the Sun, or the center around which our Sun and our Solar system revolves. In celestial longitude it is situated at 2:35 Sagittarius, as of Jan. 1, 2011, using *Lahiri ayanamsha* (sidereal zodiac.)

Gandanta *(gan – DAHN – tah)* : *Ganda* = knot. *Anta* = the end. As a cyclical reference it is the end of one grouping and the start of another. Energy cannot flow smoothly through the obstruction of the knot. There are three types of *Gandanta*: *Rashi* (sign); *Tithi* (soli-lunar *cycle) and Nakshatra Gandanta.* This book does not cover *Tithi Gandanta. Rashi and Nakshatra Gandantas* refer to the last 3 degrees 20 minutes of a water sign, and first 3 degrees 20 minutes of a fire sign. These junctures are considered dangerous and critical because both the zodiacal sign and the *nakshatra* come to an end. However, there is more vulnerability at the end of water, and especially at the end of Scorpio. It is considered the most treacherous of the three junctures.

Graha *(GRAH – ha)* : planet. literally, "that which seizes you." (See also ***Grahas***)

Graha yuddha *(GRAH – ha YUDD – ha)* : Planetary War. Two planets are within one degree of each other in the same sign or in adjacent signs. Only five planets can qualify: Mercury, Venus, Mars, Jupiter, and Saturn. The winner takes on the energy of the losing planet, while overpowering it with its own energy and agenda. The victorious planet is chosen in this descending order: **1)** Size and influence of the planet (Saturn and its rings over Jupiter); **2)** Effulgence, or brightness; **3)** Speed of orbital motion; **4)** If the victorious planet is also north of the defeated planet by either celestial longitude or declination, especially the latter, then it is also usually the decisive winner. But the winner may also be to the south; **5)** Mars is a special exception, being generally the loser in *Graha yuddha*, even when situated further north, or within one degree of Mercury – the one exception in which the losing planet is also the larger planet. (See Index and definitions: Chapters 1 and 2)

Grahas: Among the nine classical planets in Vedic astrology, Rahu and Ketu are the only non-physical planets. *Grahas* are classically listed in their sacred order, coinciding with the order of the days of the week:

1. ***Surya*** *(SOO – ryah)* : Sun (108 names, but this is the primary one).
2. ***Chandra*** *(CHAHN – drah)* : Moon (108 names, but this is the primary one).
3. ***Mangala*** *(MAHN – gah – la);* ***Kuja*** *(KOO – ja);* ***Angaraka*** *(an – GAH – rah – kah)* : Mars (These are the three major names for Mars.)
4. ***Budha*** *(BU – dha) (BOO – dee)* : Mercury. (Also spelled ***Buddhi.***)
5. ***Guru*** *(GUH – ru); Brihaspati (bri – HAS – pa – tee)* : Jupiter (These are the two major names for Jupiter.)

6. ***Shukra*** *(SHOO – krah)* : Venus.
7. ***Shani*** *(SHAH – nee)* : Saturn.
8. ***Rahu*** *RAH – hoo)* : North Lunar Node.
9. ***Ketu*** *(KAY – too)* : South Lunar Node.

Guna *(GOO – nah)* : literally "rope" or "strand." All states of mind where there is an "I" or "mine" are the results of various combinations or permutations of the three *gunas*. The *guna* indicates one of three natures that human beings or human activities will possess to varying degrees.

Gunas: These are the three *gunas* in Vedic/Hindu philosophy:

1. ***Sattva*** *(SAT – vah)* is light, flowing, truthful, with divine energy.
2. ***Rajas*** *(RAH – jas)* is fiery and has ambition to achieve goals.
3. ***Tamas*** *(TAH – mas)* has inertia, dullness.

Guru *(GUH – ru)* : Jupiter. Literally, "the one who removes darkness." (See also ***Grahas***)

Jaimini *(JAY – mih – nee)* : A sage, and a great *Jyotishi* of ancient times. He may have been a contemporary of Parashara, who lived ca. 3100 B.C. (same era as Krishna and the *Mahabharata* War). His methods were originally taught orally, in the tradition of that era.

Jaimini Karaka: From the Jaimini system, temporary planetary significators indicated by the order of the degree of celestial longitude in the chart. During their *Dashas* or *Bhuktis* (sub-periods) there can be fruition of the area of life assigned to that planet, especially if other factors are in confluence, e.g., marriage when the *Darakaraka* is *Dasha* or *Bhukti* lord, career success when *Atmakaraka* is *Dasha* lord, etc. The *Atmakaraka* is often studied closely in the *Navamsha* chart, while the conjoining of AK and Amk forms a *Jaimini Raja yoga*. In descending order, from the planet at the highest degree (normally excluding Rahu and Ketu), the *Jaimini karakas* are:

1. ***Atmakaraka*** **(AK)** : the soul or self; can become a powerful indicator if unafflicted (related to House 1 significations and the Ascendant lord).
2. ***Amatyakaraka*** **(Amk)** : the mind; also, one's confidant, as in the minister to the king
3. ***Bhratrikaraka*** **(BK)** : brothers, siblings (related to House 3 significations)
4. ***Matrikaraka*** **(MK)** : mother, including one's own motherhood (related to House 4 significations)
5. ***Putrakaraka*** **(PK)** : children or creative enterprises (related to House 5 significations)
6. ***Jnatikaraka*** **(JK)** : cousins or relations; also distant relations, enemies, or diseases (related to House 6 significations).
7. ***Darakaraka*** **(DK*)*** : spouse; partnership (related to House 7 significations)

Jiva *(JEE – vah)* : life, or life-giver; living being. In *Nadi Jyotisha*, the *Jiva* is the *nakshatra* lord and considered to have dominance over the affairs of the planets situated in its *nakshatra(s)*, even more so than the sign lord. If it is a planetary friend to the Ascendant lord, the *Jiva* can be more beneficial. Its effectiveness manifests more clearly during the *Dasha-Bhuktis*.

Jnatikaraka *(NYAH - tee - KAH - rah - kah)* : significator for cousins or relations; also distant relations, enemies, or diseases. (See also ***Jaimini karakas***)

Jyotish *(JOE - tish)* : literally, "science of light." Vedic/Hindu astrology. *Jyotish* also refers to astronomy and meteorology. One of the six *Vedangas* (limbs) of the *Vedas*, and as "the eye of the Vedas" it was considered the most significant one. It was ***shruti*** - wisdom which was originally heard by the ancient seers and transmitted as oral knowledge. It was meant to be chanted and learned in verses called ***shlokas** or **mantras***. Not originally intended to be written or learned from written language, it was an oral tradition. Not until Parashara, ca. 3100 B.C. was it written in a codified version. (See also **Parashara**)

Jyotishi *(JOE - tih - shee)* : one who practices *Jyotish.*

Kala (*KAH - lah*) : time. (*Kala* is "black" in Hindi, but not in Sanskrit.) *Kala* is called "the Time-spirit" in the Vedic classic, *Srimad Bhagavata.*

Kala Purusha *(KAH - lah pu - ROO - sha)* : Time Personified. It extends across the signs of the zodiac and shows God, or the Divine force entering human life. God is inextricable from *Kala Purusha.*

Kali Yuga *(KAH - lee -YOO - gah)* : The "Iron age," lasting ca. 432,000 years. Civilization deteriorates more than usual in this darker period.

Kama (*KAH - mah)* : The desire nature, especially the desire to be in relation with others; with siblings and neighbors, those who are immediately accessible (Gemini), partners, either marriage or business (Libra), and friends, colleagues, and larger networks of people (Aquarius). *Kama* is assigned to Houses 3, 7, and 11; also associated with the air signs Gemini, Libra, and Aquarius. Venus is "the planet of desire" and rules a *Kama* house: Libra.

Kanya *(KAHN - yah)* : Virgo, the 6th zodiacal sign; ruled by Mercury.

Karaka (*KAH - rah - kah*) : significator. A *karaka* can be changeable or unchangeable and used alongside each other, with discrimination. In the Parashari system the *karaka* is usually unchangeable in any chart; e.g., Sun or Mars is a classic planetary significator for physical vitality. But *Jaimini karakas* are changeable and depend on the degree order in which they occur in the natal chart. (See also ***Jaimini karakas***)

Karakatwa *(kah - rah - KAT - wah)* : signification.

Karka, *(KAR - kah)*, or ***Kataka*** (*KAH - tah - kah)* : Cancer, the 4th zodiacal sign; ruled by the Moon.

Karma (*KAR - ma)* : our conscious and subconscious actions on an ongoing basis.

Karma phala (*KAR - ma PHA - la)* : the fruits *(phala)* of the karma; karmic fruit.

Karmas: These are the three *karmas* in Vedic/Hindu philosophy:

1. ***Sanchita Karma*** *(san - CHEE - tah KAR - ma)* : the totality of results emerging from our actions in all our previous lifetimes, pre-destined to some extent.
2. ***Prarabdha Karma*** *(pra - RABH - dah KAR - ma)* : the fruits of actions from previous lives that one is destined to experience in the present life.
3. ***Agami Karma*** *(ah - GAH - mee KAR - ma)* : the result of actions we perform in our present existence.

Kemadruma yoga *(Kay - mah DREW - mah YOE - gah)* : The Moon has no planets in adjacent houses. If it is not cancelled, it is one of the most severely negative yogas, as the person is unconnected to others through much of the life and does not get support from others. It can be corrected by a Jupiter aspect to the Moon or planets in *kendras* from the Moon. Even if the yoga is cancelled, the person can feel unsupported by others, even with appearances to the contrary.

Kendra *(KEN - drah)* : angle or quadrant. A *Kendra* house is House 1, 4, 7, or 10, usually from the Ascendant. (See also ***Digbala***)

Kendradhipatya Dosha *(ken - DRAH - dhee - PAH - tya DOE - shah)* : *Kendra* = angle or quadrant; *Adhhipati* = lord or owner; *Dosha* = flaw. Flaws (*doshas*) that cause planets owning *Kendras* to act in a different manner: Classical benefics lose their auspiciousness or have their beneficence reduced, while classical malefics become less malefic, though they are less affected by KD than are classical benefics, notably Jupiter and Venus. Since the Ascendant is both a *Kendra* and trinal house, the Ascendant lord is always auspicious. KD planets do not totally lose their essential nature, whether classically malefic or benefic, but they become neutral, especially during their *Dashas,* and this is more damaging to benefics. *Kendradhipatya Dosha* can produce a problematic situation, especially with the 7th house lord as *Maraka* planet. But benefic planets owning *kendras* can also be beneficial in some ways, including in various *Raja yogas* and the *Mahapurusha yogas.* (See Glossary) If the classical benefic also owns a trinal house it is protected, e.g., Venus for Capricorn Ascendant. Classic source: *Laghu Parashari (Jataka Chandrika),* chapter on "Combinations of Planets," verses on "Rules for *Kendradhipatya Dosha.*"

Ketu *(KAY - too)* : South Lunar Node. (See also ***Grahas.***)

Kshatriyas *(KSHAH - tree - yas)* : warriors, and/or political leaders. *Kshatriyas* are associated with the three fire signs - Aries, Leo, Sagittarius - also bestowing on them some royal status.

Kumbha *(KOOM - bha)* : Aquarius, the 11th zodiacal sign; ruled by Saturn.

Kuja Dosha *(KOO - ja - DOE - sha)* : "the blemish of Mars" that causes disturbance in marital harmony. This occurs when Mars is situated in House 1, 2, 4, 7, 8, or 12 from the Ascendant or Moon, magnified if it occurs in both birth and *Navamsha* charts. The condition is also called ***Mangalik*** *(MAHN - gah-lik.) Kuja Dosha* can be mitigated by marrying someone who also has *Kuja Dosha.* One tradition says that after age 28 (the maturity of Mars), the *Kuja Dosha* is less severe.

LMT (Local Mean Time): This is a time based on when tr. Sun crosses the Midheaven. It was universally used before Standard time was established Nov. 18, 1883, though also used long afterward in some locations.

Lagna *(LAHG - nah)* : the Ascendant. In any astrological birth chart it is where the Eastern horizon intersects with the ecliptic, or the Sun's apparent path around the earth. At sunrise, the Sun is on the *Lagna.* The destiny of any entity begins at the *Lagna.* (See also ***Ahamkara***)

Lagnadhipati *(lag - NAH - di PAH - tee)* : the Ascendant lord, or planet owning the sign on the Ascendant. It is usually favorable, as the Ascendant is both an angular and a trinal house; but its house position is crucial, along with the condition of its sign and nakshatra lords. *Lagnadhipati* wields its greatest influence during is own *Dasha* or *Bhukti* (i.e., *sub-Dasha). (**Lagna**:* Ascendant; ***Adhipati*** : lord or owner.)

Lahiri *(la - HEE - ree)* : the name of the astrologer whose *ayanamsha* is often used. (See also ***Ayanamsha***)

Maha *(MAH - ha)* : big, major; e.g., a ***Maha Dasha*** is a major *Dasha* period, with sub-cycles.

Maha purusha yoga (*MAH - ha puh ROO - shah YOE - gah*) : literally, "great soul." When a planet is in a *Kendra* from the Ascendant and in its own sign or exalted, there are five possible *Maha purusha yogas* involving only five planets: Mars (*Ruchaka yoga*); Mercury (*Bhadra yoga*); Jupiter (*Hamsa yoga*); Venus (*Malavya* yoga); and Saturn (*Shasha yoga*). This yoga can bring out the finer qualities of these planets in the chart of an individual, enhanced by being *Digbala* (best *Kendra* for that planet) and unafflicted. The yoga can also be diminished by various factors, including aspects or contacts from malefic planets.

Makara *(MAH - kah - rah)* : Capricorn, the 10th zodiacal sign; ruled by Saturn.

Manas *(MAH - nas)* : the Mind. The intelligent aspect of *Manas* is part of the higher, more divinely connected part of the individual, while the emotional aspect of *Manas* is part of the lower individual. Emotional aspects of *Manas* connect us to survival, earthly concerns, likes and dislikes, and human life. In Vedic thought, consciousness is reflected through *Manas*, and in Vedic astrology *Manas* is ruled by the Moon.

Mangala (*MAHN - gah - la*) : Mars. *(See also* ***Graha***; and ***Mangalik***, under ***Kuja Dosha****)*

Mantra *(MAHN - tra)* : sacred sounds. In the Vedic tradition they are in the Sanskrit language.

Maraka *(MAH - rah - ka)* : death-inflicting planet or house, referring to Houses 2 and 7, planets owning those houses or located there. Many other factors have to be considered, including the likely length of life and the strength of the Ascendant lord and the Ascendant in general. The *Dashas* of *Maraka* planets are especially significant. (See also ***Arishta yoga*** and ***Kendradhipatya Dosha***)

Matrikaraka *(MAH - tree - KAH - rah - kah)* : mother, including one's own motherhood. (See also ***Jaimini Karakas***)

Meena *(MEE - nah)* : Pisces, the 12th zodiacal sign; ruled by Jupiter.

Mesha *(MAY - shah)* : Aries, the first sign of the zodiac; ruled by Mars.

Mithuna *(mi - TOO - nah)* : Gemini, the third sign of the zodiac; ruled by Mercury.

Moksha *(MOE - ksha)* : liberation of the soul; spiritual liberation; the spiritual journey *Moksha* is assigned to Houses 4, 8 and 12; also associated with the water signs Cancer, Scorpio, and Pisces.

Muhurta *(muh - HOOR - tah)* : literally, the moment; also, auspicious time. A *Muhurta* is most often used to refer to the auspicious date/time/ and moment selected for the start of an undertaking. As a Vedic time unit, one *muhurta* = 48 minutes.

Mulatrikona *(MOO - la - TREE - koh - nah)* : literally, root trine. These are areas of the zodiac where a planet does especially well. For Moon and Mercury, the exaltation sign is repeated:

Sun : 4 to 20 degrees Leo

Moon : 4 to 20 degrees Taurus

Mars : 00 to 12 degrees Aries

Mercury : 16 to 20 degrees Virgo

Jupiter : 00 to 10 degrees Sagittarius

Venus : 00 to 15 degrees Libra

Saturn : 00 to 20 degrees Aquarius

***Nadi* astrology**: a form of Vedic astrology, with many components, among others giving greater importance to the *nakshatra* lord, and to each 3 degree 20 minute segment of the astrological chart (*pada*). In some parts of India, it refers to astrological information recorded on palm leaves.

Nadi yoga *(NAH - dee - YOE - ga)* : a mutual exchange of *nakshatra* lords, e.g., Moon in *Purnavasu nakshatra* ruled by Jupiter and Jupiter in *Rohini nakshatra* ruled by Moon. This *yoga* gives the two planets a strong influence on each other, regardless of whether there is any other mutual contact or aspect between them in the chart. *Nadi Jyotisha* considers the *Nadi yoga* more powerful than the *Parivartana yoga* (exchange of sign lords). (See also ***Jiva***)

Nadi yogakarakas *(NAH - dee - YOE - gah - KAH - rah - kahs)* : In *Nadi Jyotisha* , the best planets for each Ascendant chart, given in descending order. Mars is not included, either as Ascendant lord or as trinal house lord, while Rahu is elevated to *Nadi yogakaraka* staus for Venus-ruled Ascendants. Some of these differ slightly from the **Planetary friendship** list, where Jupiter is only neutral to the Moon. But as 9th house lord, Jupiter receives *Nadi yogakaraka* status for Moon-ruled Cancer Ascendant chart.

Aries : Sun, Jupiter, Moon

Taurus : Mercury, Saturn, Rahu

Gemini : Venus

Cancer : Jupiter

Leo : Jupiter

Virgo : Venus

Libra : Saturn, Rahu, Mercury, Moon

Scorpio : Sun, Moon, Jupiter

Sagittarius : Sun

Capricorn : Venus, Mercury

Aquarius : Mercury, Venus

Pisces : Jupiter, Moon

Nadis (*NAH - dees)* : channel systems or streams through which energy flows.

Nakshatra (*nah - KSHAH - tra)* : literally, "that which does not decay;" also called "lunar mansion." The 360 degree zodiac consists of 27 *nakshatras*, 13 degrees 20 minutes each, starting with zero degrees Aries and ending with 30 degrees Pisces. *Nakshatras* preceded use of the *rashis* (signs) in ancient Vedic literature.

1. ***Ashwini*** *(AH - shwee - nee).* Owning horses: 00:00 - 13:20 Aries.
2. ***Bharani*** *(BAH - rah - nee).* She who carries: 13:20 - 26:40 Aries.
3. ***Krittika*** *(KRIH - tee - kah).* Cutters: 26:40 Aries - 10:00 Taurus.
4. ***Rohini*** *(ROE - hee - nee).* Red, or growing: 10:00 - 23:20 Taurus.
5. ***Mrigashira*** *(MRI - gah SHEE - rah).* Deer's head: 23:20 Taurus - 6:40 Gemini.
6. ***Ardra*** *(AR - drah).* Moist, or moist one: 6:40 - 20:00 Gemini.
7. ***Punarvasu*** *(POOR - nar - VAH - su).* Good, or prosperous again: 20:00 Gemini - 3:20 Cancer.
8. ***Pushya*** *(PUH - shya).* Nourishing: 3:20 - 16:40 Cancer.
9. ***Ashlesha*** *(ah - SHLAY - shah).* Clinging, entwined: 16:40 - 30:00 Cancer.
10. ***Magha*** *(MAHG - ha). Great one:* 00:00 - 13:20 Leo.
11. ***Purva Phalguni*** *(POOR - vah phal - GOO - nee):* Former red one: 13:20 - 26:40 Leo.
12. ***Uttara Phalguni*** *(OOH - tah - rah phal - GOO - nee).* Latter red one: 26:40 Leo - 10:00 Virgo.
13. ***Hasta*** *(HAH - stah).* Hand: 10:00 - 23:40 Virgo.
14. ***Chitra*** *(CHEE - trah).* Bright, multicolored: 23:40 Virgo - 6:40 Libra.
15. ***Swati*** *(SWAH - tee).* Independent one: 6:40 - 20:00 Libra.
16. ***Vishakha*** *(vee - SHAH - khah).* Forked one, or Radha - delightful one: 20:00 Libra - 3:20 Scorpio.
17. ***Anuradha*** *(ah - nuh - RAH - dah).* Another Radha: 3:20 - 16:40 Scorpio.
18. ***Jyeshta*** *(jee -YESH - tah).* The eldest: 16:40 - 30:00 Scorpio.
19. ***Mula*** *(MOO - lah).* Root: 00:00 - 13:20 Sagittarius.
20. ***Purva Ashadha*** *(POOR - vah ah - SHAH - dah).* Early victory: 13:20 - 26:40 Sagittarius.
21. ***Uttara Ashadha*** *(OOH - tah - rah ah - SHAH - dah).* Later victory: 26:40 Sagittarius - 10:00 Capricorn.
22. ***Shravana*** *(SHRAH - vah - nah).* Ear: 10:00 - 23:40 Capricorn.
23. ***Dhanishta*** *(dhah - NEE - shtah).* Wealthiest: 23:40 - Capricorn - 6:40 Aquarius.
24. ***Shatabhisha*** *(SHAH - tah - BEE - shah).* 100 physicians: 6:40 - 20:00 Aquarius.
25. ***Purva Bhadra*** *(POOR - vah BHA - drah).* Front lucky feet: 20:00 Aquarius - 3:20 Pisces.
26. ***Uttara Bhadra*** *(OOH - tah - rah BHA - drah).* Rear lucky feet: 3:20 - 16:40 Pisces.
27. ***Revati*** *(RAY - vah - tee).* Wealthy: 16:40 - 30:00 Pisces.

Navamsha *(nah -VAHM - sha)* : 9th division, or harmonic; refers to the 3 degree 20 minute segment of each sign, with 9 *navamshas* in each sign (*rashi*). The first ninth of a sign is governed by the cardinal sign of the same element, followed by the rest of the signs from it in order through the zodiac: Aries 1st *navamsha* is Aries, Taurus 1st *navamsha* is Capricorn, Gemini 1st *navamsha* is Libra, etc. The 12 *rashis* contain a total of 108 *navamshas,* 108 being a sacred number in the Vedic system. A shift in *navamsha* (even by one minute of arc) changes the orientation of the planet within the *nakshatra* and the sign.

***Navamsha* chart:** 9th harmonic chart: It is considered the most important of the *Varga* (or harmonic) charts, especially in South India, and analyzed for delineation and timing alongside the birth chart.

Neecha Bhanga Raja yoga *(NEE - chah BHA - nga RAH - jah YOE - gah)* : *Neecha* = debilitated; *Bhanga* = cancellation. This planetary combination corrects and qualifies the yoga as a *Raja yoga* after some intitial setbacks. A debilitated planet can be corrected in this *yoga* in several ways: 1) the lord of the house where the debilitated planet is located is in a *Kendra* from the Ascendant or the Moon; 2) the debilitated planet is associated with or aspected by the debilitation or exaltation lord; 3) the debilitated planet exchanges signs (houses) with its debilitation lord; 4) the debilitated planet is conjunct an exalted planet; 5) the debilitated planet is exalted in the *Navamsha* chart.

Nimitta *(nih - MEE - tah)* : omen. In Vedic astrology, *Nimitta* is an entire branch of study involving the reading of omens. These are used alongside the planetary indicators, and sometimes by themselves, assisting the astrologer to judge the complexity of influences. The classic text *Prashna Marga* contains much information on *Nimittas.*

Nirbhagya yoga *(nir - BHA - gya)* : a remover of good fortune, occurring when the 9th lord is situated in a house of misfortune (6, 8, or 12 from the Ascendant, also called ***Dusthana*** houses). This yoga is made more complete when the 9th house is aspected by a malefic planet, but it can be modified when the 9th house is aspected by its own lord and/or classical benefic planets.

Pada *(PAH - dah)* : literally, a foot or segment. There are four *padas* in one *nakshatra*, equaling 3 degrees 20 minutes in each *pada.* The *nakshatra pada* determines the start of the *Vimshottari Dasha,* a *nakshatra*-based *Dasha* system.

Papa Kartari yoga *(PAH - pah KAR - tah - ree)* : *Papa* = "unfortunate," *Kartari* = "scissors." Classic malefic planets are in houses either side of any house, having the effect of hemming in a house or planet. Since it tends to block action from that house or planet, greater effort is required to counteract the malefic influences. Its effects are most notable when hemming in the Ascendant, and this may require breaking free of influences in one's immediate environment. One's own family, locale or even nation may not be favorable at first. (See also ***Shubha Kartari yoga***)

Parashara *(pah - RAH - sha - rah)* : A great sage and Jyotishi, who lived during the time of the Mahabharata War, ca. 3100 B.C. The great work *Brihat Parashara Hora Shastra* is attributed to him, but is probably a compilation of writings of various astrologers of Parashara's lineage. Parashara is usually the first to be credited with codifying Jyotish in written form, when previously it was always learned as ***shruti***, or wisdom that was orally transmitted. Though Jyotish is said to originate with Parashara, he was in fact preceded by other *rishis.* There is much evidence from the *Rig Veda* over 800 years earlier that Jyotish was already highly developed. (Also spelled Parasara, especially in South India.)

Parivartana yoga *(PAH - ree - VAHR - tah - nah YOE - gah)* : mutual exchange of sign lords. This can be a *Raja yoga,* especially if the exchange is between 9th and 10th house lords, or any combination of trinal and *kendra* lords.

Planetary friendships: The planets fall into two groups: 1) Sun, Moon, Mars, and Jupiter; 2) Mercury, Venus, Saturn, and Rahu. Though Rahu is Saturn-like and Ketu is Mars-like, they are not included in the usual list of planetary friendships. (See also ***Nadi yogakarakas*** and ***Dusthana*** houses)

PLANET	FRIEND	NEUTRAL	ENEMY
Sun	Moon, Mars, Jupiter	Mercury	Venus, Saturn
Moon	Sun, Mercury	Venus, Mars, Jupiter, Saturn	None
Mars	Sun, Moon, Jupiter	Venus, Saturn	Mercury
Mercury	Sun, Venus	Mars, Jupiter, Saturn	Moon
Jupiter	Sun, Moon, Mars	Saturn	Mercury, Venus
Venus	Mercury, Saturn	Mars, Jupiter	Sun, Moon
Saturn	Mercury, Venus	Jupiter	Sun, Moon, Mars

Planetary strength: A planet is strong in its own sign *(swakshetra)*, exalted, or *mulatrikona*. See these Glossary entries: **Exaltations** (which includes Debilitations), also ***Mulatrikona***, and ***Rashis***, listing the planetary owners of each sign.

Prakriti *(prah - KREE - ti)* : that which is seen. *Prakriti* is both matter and the stuff of experience. It is one of the two basic principles of *Sankhya*, the other being *Purusha*. Vedic texts (e.g., *Srimad Bhagavata Mahapurana*) describe how *Prakriti* has attributes in the form of the *gunas* (*Sattva, Rajas and Tamas*) and is thrown into action by *Kala* (the Time-Spirit). *Kala* in turn represents the creative will of the Cosmic being. (See also ***Gunas; Purusha.***)

Prana *(PRAH - na)* : the breath, or one respiration. Also, the time necessary to pronounce ten long Sanskrit syllables. *Prana* is the smallest time unit (4 seconds) usually considered in Vedic astrology, though astrological computer programs go to the second and sometimes beyond. (See *Surya Siddhanta*, Chapter 1, v. 11-12) The *Vedas* give a number of even smaller time units. Some of them are as follows: a **Paramáñu** = ca. 16.8 microseconds; a ***truti*** = 3 *trasarenus*, or ca. 1/3290th of a second; a ***vedha*** = 100 *trutis*; a **lava** = 3 *vedhas*; *a* ***nimesha*** *= 3 lavas, or a* blink of an eye (0.2112 seconds).

Prarabdha Karma *(pra - RAHB - dah KAR - mah)* : the fruits of actions from previous lives that one is destined to experience in the present life. (See also ***Karmas***)

Precession of the equinoxes: an astronomical phenomenon in which the point of the vernal equinox changes each year. Due to a slight wobble of the earth as it rotates on its axis, the earth moves backwards very slightly, about one degree of arc every 72 years against the backdrop of the fixed stars. As of Jan. 1, 2010, the two zodiacs are exactly 24 degrees apart. Depending on the *ayanamsha* used, it takes between 24,000 to 25,900 years for one complete revolution of the vernal equinox through the entire zodiac of constellations, called a **Precession cycle**. Though the Greek astronomer Hipparchus (2nd century AD) is credited with the discovery of precession, knowledge of precession preceded Hipparchus by millennia, as it was noted by the Egyptians,

Babylonians and Sumerians. India's *Rig Veda* contains many references to the *nakshatra* at the equinox and/or solstice.

Purana *(poo – RAH – nah)* : mythological literature, describing the structure of the universe, and the higher and lower worlds, all of it in a process of evolution, both physically and spiritually, through repeated incarnations, or manifestations. Viewed in modern times as mere religious belief, it bears examining in terms of scientific theory. Also narrated are the dynasties of rulers and clans of the great sages who lived on earth. The *Puranas* were produced in the Vedic culture over millennia, and through the many stories, astronomical phenomena in the remote past were often revealed. For instance, the *Vedas*, containing all the knowledge and scriptures of a civilization, were periodically stolen and hidden for safe keeping in the Milk Ocean, or Ocean of Milk (in Sanskrit, the *Ksheera Sagar*), also called the Cosmic Ocean, while civilizations are destroyed. Vishnu, the Preserver, would later come and retrieve them. Though commonly regarded as fiction, the *Puranas* may well describe events in the history of India, interspersed with cosmic events or events in the higher worlds.

Purusha *(poo – ROO – shah)* : the principle of pure consciousness, or the seer. *Purusha* and *Prakriti* are the two basic principles of *Sankhya.* (See also ***Kala Purusha** and **Prakriti**)*

Putrakaraka *(POO- trah KAH – rah- kah)* : significator for children or creative enterprises. (See also ***Jaimini karakas***)

Rahu *(RAH – hoo)* : North Lunar Node. (See also ***Grahas***)

Raja yoga *(RAH – jah YOE – gah)* : *Raja* = royal. A royal combination of planets, usually involving lords of both *kendras* (quadrant) and trinal houses. *Raja yogas* add benefit, unless weakened by other factors in the chart, such as the presence of classic malefics. *Raja yogas* involving the 9th and 10th houses are especially powerful, creating a ***Dharma Karma Adhipati** yoga.* These bring success, visibility, and fulfillment of life purpose.

Rajas *(RAH – jas)* : fiery and with ambition to achieve goals. (See under ***Gunas***)

Rashi *(RAH – shee)* : literally, "a heap." One of the 12 signs linked to the constellations. The *Rashi* chart is also the birth chart in its first harmonic. In South India, the spelling is "*rasi*," though pronounced the same. In general the "sh" is dropped in South India due to differences in the scripts from South to North.

Rashis: the 12 signs of the zodiac, in their natural sequence. Planetary ruler is given last.

1. ***Mesha*** *(MAY – sha)* : Aries, the 1st zodiacal sign. Mars.
2. ***Vrishabha*** *(vri – SHAH – bah)* : Taurus, the 2nd zodiacal sign. Venus.
3. ***Mithuna*** *(mi – TOO – nah)* : Gemini, the 3rd zodiacal sign. Mercury.
4. ***Karka,*** *(KAR – kah),* or ***Kataka*** *(KAH – tah – kah)* : Cancer, the 4th zodiacal sign. Moon.
5. ***Simha*** *(SIM – hah)* : Leo, the 5th zodiacal sign. Sun.
6. ***Kanya*** *(KAHN – yha)* : Virgo, the 6th zodiacal sign. Mercury.
7. ***Tula*** *(TOO – la)* : Libra, the 7th zodiacal sign. Venus.

8. ***Vrischikha*** *(vrih - CHEE - kha)* : Scorpio, the 8th zodiacal sign. Mars.
9. ***Dhanu*** *(DHA - noo)* : Sagittarius, the 9th zodiacal sign. Jupiter.
10. ***Makara*** *(MAH - kah - rah)* : Capricorn, the 10th zodiacal sign. Saturn.
11. ***Kumbha*** *(KOOM - bha)* : Aquarius, the 11th zodiacal sign. Saturn.
12. ***Meena*** *(MEE - nah)* : Pisces, the 12th zodiacal sign. Jupiter.

Rishi *(REE - shee)* : a sage, seer, wise man, also often an inspired poet. The purpose of a *rishi* is to impart spiritual wisdom and or training to the rest of mankind.

Sade sati *(SAH - day SAH - tee)* : 7 and 1/2. This is a 7½ year period on average for Saturn to transit House 12, 1 and 2 from the natal Moon. In Vedic astrology, it is a pivotal transit occurring every 22½ years. Saturn completes one full orbit around the Sun in 28-29 years.

Sankhya *(SAHN - khya)* : literally, "number." This refers to the number of things (*tattvas*) that comprise the philosophy of *Sankhya*, which is the most basic of all the systems of Indian philosophy, featuring the concepts of *Purusha* and *Prakriti*. (See also ***Purusha*** and ***Prakriti***)

Sanchita Karma *(san - CHI - tah KAR - mah)* : the totality of results emerging from our actions in all our previous lifetimes, pre-destined to some extent. (See also ***Karmas***)

Saraswati yoga *(SAH - rah - SWAH - tee - YOE - ga)* : Saraswati is the Hindu goddess of learning. For this yoga, Mercury, Jupiter, and Venus must be in an angle, trine, or 2nd house of the birth chart. Jupiter, planet of knowledge, must be in its own sign, exalted, or in the sign of a planetary friend (Sun, Moon, or Mars). One with this yoga is very learned and scholarly, and may receive fame because of it. It can give abilities as a writer or poet, or in higher mathematics or sacred scriptures.

Sarpa drekkana *(SAR - pah DREH - kah - nah)* : *Sarpa* is the snake; *drekkana* is a division of 10 degrees. These sectors are snake-like in quality; natal or transiting planets in this area can be volatile and unpredictable, and *sarpa* can bring deception. These occur only in water signs. The 2nd and 3rd *drekkanas* in Cancer are *sarpa*: 10:00 to 20:00 and 20:00 to 30:00 Cancer; in Scorpio, the first two *drekkanas*: 00:00 to 10:00 and 10:00 to 20:00 Scorpio; and in Pisces - the 3rd *drekkana*: 20:00 to 30:00 Pisces.

Sarva Ashtakavarga *(SAR - vah ah - SHTAH - kah - VAR - gah)* : The total number of bindus accumulated from each planet in each house, measuring a planet's strength in each sign by eight sources of energy; especially useful for assessing transits of planets. "*Sarva*" means "all." (See also ***Ashtakavarga***)

Sattva *(SAT - vah)* : light, flowing, truthful, with divine energy. (See under ***Gunas***)

Satyagraha *(SAT - yah GRAH - ha)* : Literally "insistence for truth." *Sat* = truth; *Agraha* = insistence or persuading. Also translated as "resting in the realm of truth," or also "those who hold to the truth." Not an astrological term. (See under Mahatma Gandhi, Chart #29)

Shadbala *(SHAD - BAH - la)* : six-fold strengths of a planet, as given in *Phala Deepika*, Chapter 4, Verse 1: "The strength of a planet is sixfold: These are (1) *Kalaja [Kala] Bala* (temporal); (2) *Chesta Bala* (motional); (3) *Ucchaja [Uchha] Bala* (due to exalted position); (4) *Dik [Dig] Bala* (directional); (5) *Ayana Bala* (due to north or south declination); (6) *Sthana Bala* (positional). Mars, the Moon and Venus are strong in the night time. Mercury is strong all the 24 hours of the day. The Sun, Jupiter and Saturn are strong during the day. The benefics are strong in the bright half of a month whereas the malefics are strong during the dark half of a month...." (Translation by S.S. Sareen, 1992)

Shakata yoga *(shah - KAH - tah YOE - gah)* : *shakata* = cart. The yoga is formed when natal Jupiter is in a *Dusthana* house (6, 8, or 12) from natal Moon. One goes through many ups and downs in life, like a cartwheel, causing toil and struggle due to fluctuating fortunes. The yoga is cancelled if Jupiter is in a *Kendra* from the *Lagna* and if Jupiter and Moon are both strong; it is also less severe when other factors coincide, such as a ***Durudhara yoga*** or an ***Adhi yoga.*** Otherwise, it can also bring miseries and poverty. (A different *Shakata yoga* occurs when all planets are located in either House 1 or 7. This brings extreme hardship, ill health, and poverty.)

Shakti *(SHAHK - ti)* : literally, "javelin." *Shakti* is the vibrational energy that infuses and empowers; the source of inspiration and motivation.

Shani (*SHAH - ni)* : Saturn. (See also ***Grahas***)

Shashtashtakha *(shah - SHTAH - shtah - kah)* : Literally, "six-eight." Planets are in mutual 6 – 8 house positions, creating tension and difficulty between them, and/or loss or major change during their respective *Dasha/bhuktis*, which could be positive or negative depending on other factors, including their *Navamsha* positions. (See also ***Shakata yoga***)

Shloka *(SHLOE - ka)* : verse in a Vedic text, usually giving astrological rules when in texts on *Jyotish*. In South India, the spelling is *sloka*, as their languages usually contain the "sh" sound within the letter "s," e.g., *navamsa, dasa, rasi.*

Shubha Kartari yoga *(SHOO - bah KAR - tah - ree)* : auspicious combination that occurs when classic benefic planets (Mercury, Venus, Jupiter, waxing Moon) are in adjacent houses to the Ascendant or any house. It is especially strengthening and protective for the Ascendant. (See also ***Papa Kartari yoga***)

Shudras *(SHOO - drahs)* : the servants and the laborers who get things accomplished in the material world. *Shudras* are associated with the earth signs: Taurus, Virgo, Capricorn.

Shukra *(SHOO - krah)* : Venus. (See also ***Grahas***)

Sidereal zodiac *(sigh - DEE - ri - al)* : from the Latin, *sidus* = star. A zodiac based on the relationship of the earth on its axis, against a backdrop of the Fixed stars.

Simha (*SIM - ha*) : the lion. The 5th sign of the zodiac; ruled by the Sun.

Sthira *(STEE - rah)* : fixed; applies to the zodiacal signs of Taurus, Leo, Scorpio, and Aquarius. It is one of the three qualities of a sign. (See also ***Chara*** and ***Dwiswabhava***)

Swakshetra *(swah - KSHEH - trah)* : own sign. Literally, *swa* = own; *kshetra* = field. A planet in its own sign has a natural strength, occupying the house it also rules.

Surya (*SOO - ryah)* : Sun. (See also ***Grahas***)

Tajika *(TAH - jee - kah)* : A branch of astrology, from Persia. Some of its components have parallels in Western tropical astrology, such as the use of the sextile.

Tamas *(TAH – mas)* : Inertia, dullness. (See under ***Gunas***)

Trikona *(TREE – koh – nah)* : trinal house. A trinal house is House 1, 5, or 9, House 1 being the Ascendant, and having special importance as both *kendra* and trinal house.

Tula *(TOO – la)* : Libra, the 7th zodiacal sign; ruled by Venus.

Ubhayachari yoga *(ooh – BHA – ya – CHAH – ree YOE – gah)* : Planets other than Moon, Rahu and Ketu occupy Houses 2 and 12 from the Sun, conferring on the native a large network of friends and supporters. But with no planets on either side, the Sun does not suffer.

Upachaya *(OOH – pah – CHAI – ya)* : Houses 3, 6, 10, 11, usually from the Ascendant. The affairs of these houses can improve over time, depending on the status of their respective house lords. The 11th house in particular can deliver financial gain, if well aspected and with a strong house lord.

Vaishyas *(VYE – shyahs)* : the mercantile or business class, who desire to make contact with other people and to make profits. *Vaishyas* are associated with the air signs: Gemini, Libra, Aquarius.

Vara *(VAH – rah)* : day, or 24 hour period. The Vedic astrological day begins at sunrise and extends until sunrise the following day, e.g., the Sun's *vara* begins at sunrise on Sunday and ends at sunrise on Monday, the Moon's *vara*. The calendrical day is noted, but the astrological day gives precedence to the energy of the planetary day lord and is considered in *Muhurta*, the astrology used for auspicious timing. The days of the week are in the sacred order of the *grahas* (planets). Rahu and Ketu rule no *varas*. (See also ***Graha***)

Varahamihira *(vah – RAH – ha – mih – HEE – rah)* : Great Jyotishi and sage who lived ca. 500 AD. He composed several works, including the *Brihat Samhita*, ca. 505 A.D. A court astrologer, he became famous for his accurate predictions combined with his fine intellect. Author of great Jyotish classics: *Brihat Jataka, Brihat Samhita, Swalpa/Laghu Jataka.*

Vargottama *(var – GOT – tah – mah)* : "the best division." A planet (or Ascendant of the birth chart) is in the same sign in the natal chart and in the *Navamsha* chart. The planet must be in the 1st *Navamsha* (3 degrees 20 minute segment) of *Chara* signs (Aries, Cancer, Libra, Capricorn), the 5th *Navamsha* of Fixed signs (Taurus, Leo, Scorpio, Aquarius); and in the 9th and last *Navamsha* of Dual signs (Gemini, Virgo, Sagittarius, Pisces). This is generally favorable and strengthening for the Ascendant, but it depends on how a *Vargottama* planet is configured in the chart and whether its emphasis is beneficial for the chart. (The principle of *Vargottama* may be applied to other divisional charts, though the emphasis is classically on the birth chart and *Navamsha* chart, as in this book.)

Varga *(VAR – gah)* : divisional, as in divisional chart. *Navamsha* is a 9th divisional chart.

Veda *(VAY – dah)* : Divine knowledge, from the Sanskrit root *vid*, to know or to see. The ***Vedas*** are a body of wisdom originally preserved through an oral tradition of verses *(mantras)* in ancient India. They are considered to contain all the knowledge and secrets of creation and life. They were revealed to the sages in a state of meditation in ancient times, and came from higher realms ruled by Brahma.

Vedangas *(veh – DAHN – gahs)* : limb; *Jyotish* is one of six *Vedangas*, or limbs, of the *Vedas* and given importance as "the eye of the Vedas."

Vedic *(VAY – dik)* : Adjective broadly referring to the literature and culture of the *Vedas* in ancient India. Using climate, agriculture, and genetics to study earliest civilizations, evidence suggests this dates the Vedic culture back as early as 7,000 or 8,000 B.C, with the origins of Indian culture over 50,000 years old. Some key components of their astrology, such as the Vedic *nakshatras*, appear in the *Rig Veda*, along with other sophisticated references to the 360 degree zodiac. Other numbers often repeated in the ancient *Vedas* are: 12, 30, 60, 180, and 360.

Vedic New Year: See ***Chaitra Shukla Pratipada.***

Vimshottari Dasha *(vim – SHOW – ta – ri DAH – sha)* : 120 year cycle; a system of timing encompassing 120 years, always in the same planetary sequence, and based on the *nakshatra* of the natal Moon in the birth chart. This is the most important of over 55 *Dasha* systems in Vedic astrology, as stated by sage Parashara: "*Vimshottari Dasha* is supreme in Kaliyuga.... [It is] the most appropriate for the general populace."

Viparita Raja yoga *(vee – pah – REE – tah RAH – jah YOE – gah)* : Literally a "crooked" *raja yoga*, occurring when the lords of the *Dusthana* houses (Houses 6, 8, and/or 12) are located in their own or each other's houses – from the Ascendant or the Moon. *(Harsha, Sarala, Vimala* refer to the individual *Viparita Raja yogas* pertaining to the lord of House 6, 8, and 12, respectively.) There is a rise in status, gain in wealth, and increased protection from adversities, especially if all three *yogas* are present from either Moon or Ascendant. They enhance material prospects but do not necessarily protect the physical body. Especially effective during their *Dasha* periods, planets involved in this yoga may indicate a sudden and unexpected rise in life. (See also ***Arishta yoga***)

Vrishabha *(vri – SHAH – bha)* : Taurus, the 2nd zodiacal sign; ruled by Venus.

Vrischikha *(vri – CHI – kah)* : Scorpio, the 8th zodiacal sign; ruled by Mars.

Yoga *(YOE – gah)* : literally, "union." In Vedic astrology, a yoga is a combination of two or more planets. The yoga indicates the effect of the planets on each other and/or in relation to the Moon or the Ascendant. In Vedic philosophy, yoga is a union of the individual soul with the cosmic spirit.

Yogakarakas *(YOE – gah – KAH – rah – kahs)* : These planets can deliver good results for the chart due to being a planetary friend of the Ascendant lord or ruler of a trinal house, especially during their *Maha Dashas* but also in the sub-periods as *Bhukti* lord.

Yuddha *(YUD – ha)* : fight or battle. (See ***Graha Yuddha,*** or Planetary war)

Yuga *(YU – gah)* : literally, "a yoke." Also, "an age." There are four *yugas* in the Hindu/Vedic system. In Earth years, they are: 1) *Satya Yuga* – 1,728,000 Earth years; 2) *Treta Yuga* – 1,296,000 years; 3) *Dwapara Yuga* – 864,000 years and *Kali Yuga* – 432,000 years. Their combined total: 4,320,000 years. The current *Kali yuga* is estimated to have begun ca. 3000 B.C.

Selected Bibliography

CLASSICAL TEXTS

- *Bhavartha Ratnakara,* by Sri Ramanujacharya (presumed author), Translated & notes by B.V. Raman, 1944, 10th edition, 1992.
- *Brihat Jataka,* by Varahamihira, Translated by Usha & Shashi, 1977.
- *Brihat Parasara Hora Sastra*, by Maharishi Parasara, Volumes 1 & 2, Translated & notes by R. Santhanam, 1989.
- *Brihat Samhita,* by Varahamihira, Translated by M. Ramakrishna Bhat, Part 1 & 2, 2nd revised edition, 1987.
- *Hora Ratnam,* by Bala Bhadra, Translated & notes by R. Santhanam, Part One, 1995.
- *Jatakalankara*, Translated & notes by Dr. K. S. Charak, 2007.
- *Jataka Parijata,* by Vaidyanatha Diskshita, Translated & notes by V. Subramanya Sastri, Volumes 1, 2 & 3, 1932.
- *Jataka Tatva,* by Mahadeva, Translated by S. S. Sareen, 1987.
- *Kalaprakasika,* Translated & notes by N. P. Subramania Iyer, 1991.
- *Laghu Parasari (Jataka Chandrika),* Translated & notes by O. P. Verma, 2002.
- *Phala Deepika,* by *Mantreswara,* Translated by S. S. Sareen, 1992.
- *Prasna Marga* (written ca. 1649 A.D. in South India by an anonymous Brahmin astrologer from Kerala, India), Parts 1 & 2, Translated & notes by B. V. Raman, 2nd revised edition 1991, 1992.
- *Saravali,* by Kalyana Varma, Translated by R. Santhanam, Volumes 1 & 2, 1996.
- *Surya Siddhanta: A Text-Book of Hindu Astronomy* (attributed by Al-Biruni to Lata), Translated & notes by Rev. Ebenezer Burgess, Edited by Phanindralal Gangooly, 1st edition 1860, 1997.
- *Uttara Kalamrita,* by Kalidasa, Translated by P. S. Sastri, 1994.

CONTEMPORARY TEXTS

- Bhasin, J. N., *Dictionary of Astrology*, Edited by Sharadendu, 1997.
- Bhat, M. Ramakrishna, *Fundamentals of Astrology*, 3rd edition, 1988.
- Braha, James, *Ancient Hindu Astrology for the Modern Western Astrologer*, 1986.
- Chadha, Shiv Kumar, *Significators in Astrology*, 1995.
- Charak, Dr. K. S., *Elements of Vedic Astrology*, Volumes 1 & 2, 1995.
- Charak, Dr. K. S., *Essentials of Medical Astrology*, 1994.
- Charak, Dr. K. S., *Yogas in Astrology*, 1995.
- Danielou, Alain, *The Myths and Gods of India*, 1991.
- deFouw, Hart and Robert Svoboda, *Light on Life: An Introduction to the Astrology of India*, 1996.
- deFouw, Hart and Robert Svoboda, *Light on Relationships: The Synastry of Indian Astrology*, 2000.
- Dowson, John (1820-1881), *A Classical Dictionary of Hindu Mythology and Religion*, 1998.
- Feuerstein, Georg, Subhash Kak, and David Frawley, *In Search of the Cradle of Civilization*, 1995.
- Frawley, David, *Astrology of the Seers*, 1990, 2000.
- Iyer, Sheshadri, *New Techniques in Prediction*, 1963.
- Kelleher, James, *Path of Light*, Volumes 1 & 2, 2006.
- Lal, Rattan, *Nadi System of Prediction (Stellar Theory)*, 1983.
- Meena (pen name of R. Gopalkrishna Rao), *Nadi Jyotisha or The Stellar System of Astrology*, 1st edition 1945, 2nd edition 1951, 3rd edition 1954.
- Mishra, Dr. Suresh Chandra, *Panchadhyaayee: A Compendium of Predictive Astrology*, Translated by Abhinav Mishra, 2005.
- Raman, B. V., *Graha and Bhava Balas*, 1979.
- Raman, B. V., *How to Judge a Horoscope*, Volumes 1 & 2, 11th edition, 1992.
- Raman, B. V., *A Manual of Hindu Astrology*, 1980.
- Raman, B. V., *Three Hundred Important Combinations*, 9th edition, 1983.
- Rao, K. N., *Astrology, Destiny and the Wheel of Time: Techniques and Predictions*, 1993.
- Santhanam, R., *Essentials of Predictive Hindu Astrology*, 1991.
- Svoboda, Robert, *The Greatness of Saturn*, 1997.

MUNDANE ASTROLOGY

- Campion, Nicholas, *The Book of World Horoscopes*, 3rd edition, revised, expanded and updated, 2004.

Appendix: Selected National Charts

The national chart can be assessed for numerous aspects of the nation's well-being, along with the Oath-taking of new leaders and the beginnings of any new legal or political entity as the nation evolves through time. The capital of the country is used when assessing important charts such as the Vedic New Year chart, *Chaitra Shukla Pratipada* (See Glossary).

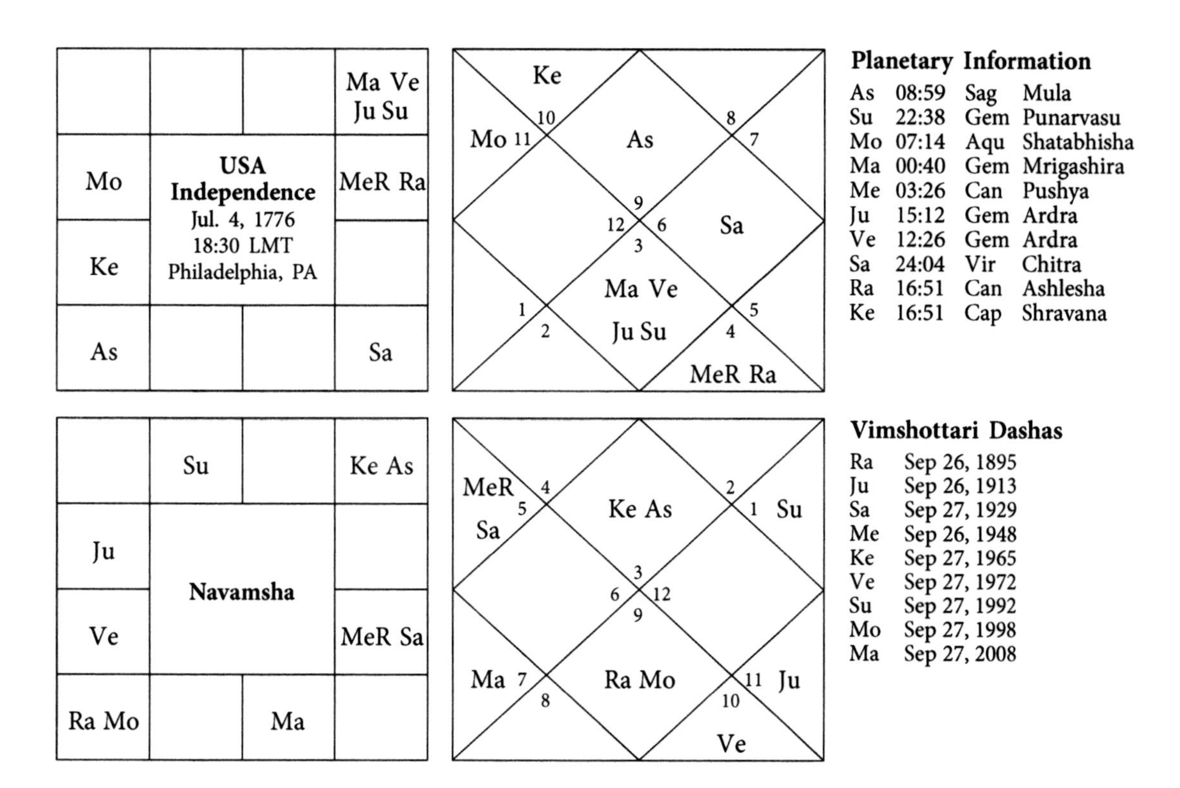

Planetary Information

As	08:59	Sag	Mula
Su	22:38	Gem	Punarvasu
Mo	07:14	Aqu	Shatabhisha
Ma	00:40	Gem	Mrigashira
Me	03:26	Can	Pushya
Ju	15:12	Gem	Ardra
Ve	12:26	Gem	Ardra
Sa	24:04	Vir	Chitra
Ra	16:51	Can	Ashlesha
Ke	16:51	Cap	Shravana

Vimshottari Dashas

Ra	Sep 26, 1895
Ju	Sep 26, 1913
Sa	Sep 27, 1929
Me	Sep 26, 1948
Ke	Sep 27, 1965
Ve	Sep 27, 1972
Su	Sep 27, 1992
Mo	Sep 27, 1998
Ma	Sep 27, 2008

This **USA chart** was rectified by Vedic astrologer and colleague James Kelleher in the late 1990s. I have used the chart continuously since Sept. 11, 2001, when I became satisfied with its accuracy.

See Nicholas Campion's ***The Book of World Horoscopes*** (3rd edition, revised & enlarged, 2004, pp. 348-376) for a lengthy discussion of the various historical charts used for the national chart of the USA. This reflects the amount of ongoing research and varying opinions there are on the subject. However, many astrologers, both Western and Vedic, tend to agree on a Sagittarius Ascendant for the USA chart. The second most popularly assigned Ascendant is Gemini, where there are four planets, sidereally. Vedic sages have tended to give the sign of Gemini to the USA and the sign of Capricorn to India, as general qualities of each nation. On the basis of this, some Vedic astrologers have watched for important astronomical events in these zodiacal areas. [Footnote #219] See also discussion of this USA chart in Chapter 8 (Virgo Ascendant), Chart #25, John F. Kennedy.

See my article: "Corporate Conglomerates vs. Nation-States: Which Nations will Survive and Thrive?" (Sept. 2010). It features the Feb. 18, 1961 Jupiter-Saturn conjunction in Capricorn, and how it works as a seminal global influence from that time up through April 2199, when the EARTH-dominant period ends. I devised a method for predicting national viability for a period of 238 years, described here. Available at my website: **www.edithhathaway.com.**

This chart documents the moment India achieved its legal independence from Britain on Aug. 15, 1947, New Delhi. Pakistan achieved legal independence at midnight on Aug. 14, 1947, Karachi, Pakistan, exactly 24 hours prior to India independence. Some astrologers prefer to use the chart of the republic of India, established on Jan. 26, 1950, New Delhi, with a proclamation at 10:15 AM

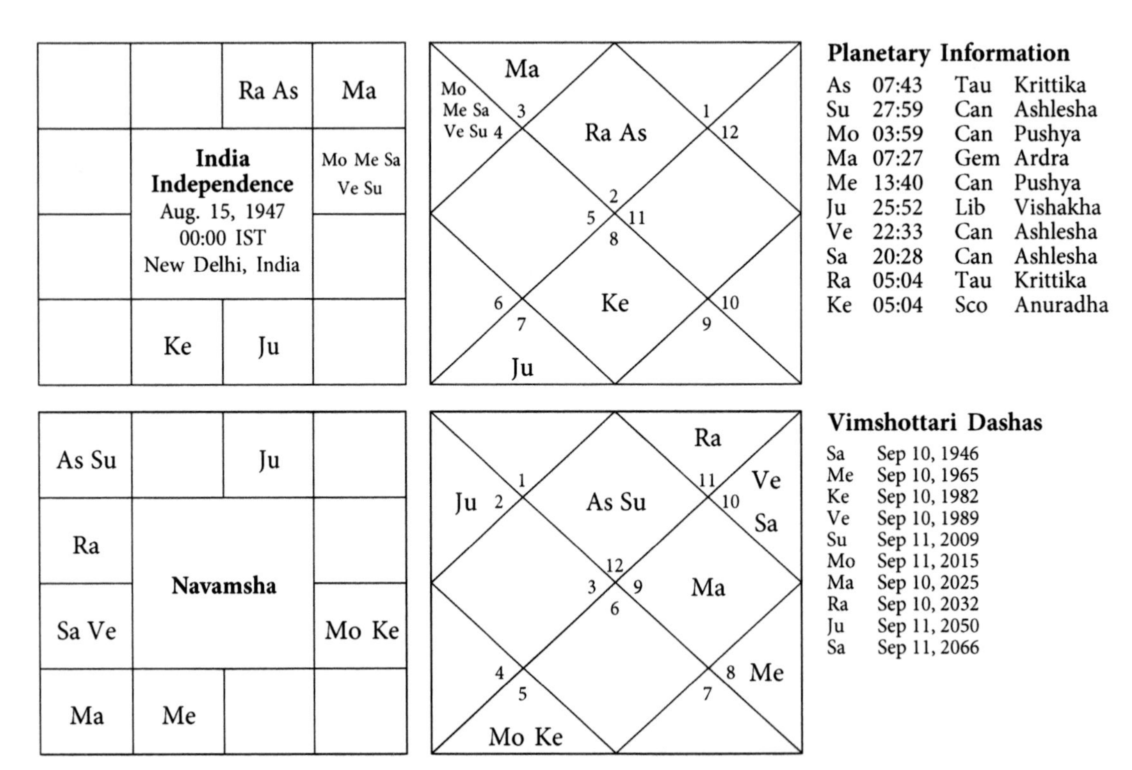

Planetary Information

As	07:43	Tau	Krittika
Su	27:59	Can	Ashlesha
Mo	03:59	Can	Pushya
Ma	07:27	Gem	Ardra
Me	13:40	Can	Pushya
Ju	25:52	Lib	Vishakha
Ve	22:33	Can	Ashlesha
Sa	20:28	Can	Ashlesha
Ra	05:04	Tau	Krittika
Ke	05:04	Sco	Anuradha

Vimshottari Dashas

Sa	Sep 10, 1946
Me	Sep 10, 1965
Ke	Sep 10, 1982
Ve	Sep 10, 1989
Su	Sep 11, 2009
Mo	Sep 11, 2015
Ma	Sep 10, 2025
Ra	Sep 10, 2032
Ju	Sep 11, 2050
Sa	Sep 11, 2066

that day. I prefer this India independence chart as an ongoing working chart for the national entity of India. The Ascendant is 7:43 Taurus, with Rahu close to the Ascendant and natal Sun at 27:59 Cancer, closely contacting Mahatma Gandhi's two planets in Cancer: natal Moon 27:56 Cancer and natal Rahu at 12:08 Cancer. (See Chapter 9, Libra Ascendant, Chart #29: Mahatma Gandhi) The cluster of planets in House 3 from Taurus Ascendant reflects the constant communications between family and/or tribal members, making India a nation especially ripe to succeed with the new Information Technology (IT) industries, burgeoning from the time of the 1980-1981 JU-SA conjunctions in Virgo.

The sectarian violence that rocked India at its independence and in subsequent periods is well described by the concentration of planets and/or affliction to planets in the late degrees of Cancer, the sign ruling over the family. It also describes family life – its culture, long-held traditions and continuity – as crucial factors in India, along with the typical reverence bestowed on ruling Indian political families. With five planets in Cancer in India's national independence chart, and Sun (the leader) at 27:59 Cancer, we note the continued confluence of key planetary positions in late Cancer in the charts of Indian political leaders: The Ascendant of Jawaharlal Nehru, first Prime Minister (1947-1964) is at 24:47 Cancer (natal Moon at 17:56 Cancer), using K. N. Rao's rectified time of 11:05 PM. His daughter Indira Gandhi succeeded him in power, 1966-1977; 1980-1984. Her Ascendant is at 27:22 Cancer (natal Saturn at 21:47 Cancer). See: Chapter 6, Cancer Ascendant charts, Chart #18. She was assassinated on Oct. 31, 1984 by two of her own Sikh bodyguards. [Footnote #270]

Index

A

Abdullah, Dada 437
Adams, Abigail 21
Adams, John 21
Addison's disease 349, 350, 351
Adelson, Merv 369
Adrenal glands 349, 351
Agent Orange 36
Age of Reason 18
Ahamkara 65, 277, 496, 497, 504
AIR period 7, 14-15, 41
Airplanes 23, 30
Alfred A. Knopf publishers 169
(See also **Knopf, Alfred A. publishers**)
Ali, Mohammad 80, 83, 428
Allen, Paul 33, 53, 209, 210, 212, 213, 214, 215, 216
Allen, Woody 119, 120, 122, 124, 125, 126, 137, 247, 297-308, 319, 320, 479
Altair 8800 33, 53, 209
Alter, David 23
American Ballet Theatre 478
American Revolutionary War 19, 20, 44, 45, 440
Ammachi 256, 454-464, 495
(See also **Mata Amritanandamayi**)
Amritsar massacre 27, 439, 442
Angelou, Maya 289-296, 300, 319, 320, 372
Angular house 67, 68, 499 (See also **Kendra**)
Andersen, Christopher 473
Anuradha 6, 8, 42, 43, 63, 105, 167, 204, 301, 303, 329, 499, 507
Apple Computer 33, 53, 56
Aquarius 3, 5, 6, 7, 8, 41, 42, 43, 60, 61, 62, 64, 66, 73, 497, 503, 504, 506, 507, 511, 512, 513
Archduke Franz Ferdinand 24, 48, 425
Ardra 8, 63, 94, 110, 177, 203, 217, 274, 507
Aries 2, 3, 4, 5, 6, 7, 13, 26, 28, 40, 44, 46, 48, 50, 54, 59, 60, 61, 63, 65, 66, 67, 69, 70, 73, 75, 76-80, 81-137, 497, 498, 499, 500, 504, 505, 506, 507, 510, 513
Artha 66, 138, 143, 278, 321, 496, 497
Asher, Peter 316
Ashlesha 6, 43, 63, 156, 157, 236, 237, 260, 264, 374, 445, 507
Ashtama Shani 144, 168, 191, 198, 293, 304, 318, 338, 380, 443, 497
Ashwini 63, 80, 97, 129, 249, 268, 271, 275, 289, 312, 316, 482, 483, 507
Aspects (planetary) 67, 68, 70 497, 499
Asperger Syndrome 187, 204
Assassination 24, 48, 81, 83, 93, 94, 95, 96, 102, 103, 105, 107, 109, 255, 265, 290, 345, 347, 349, 352, 360, 363, 366, 367, 397, 410, 413, 422, 423, 424, 425, 427, 440, 448, 450, 452, 453, 466, 468, 481, 483, 485, 490, 491, 492, 494
Assassination chart 94, 265, 367, 451-452, 490
Association Montessori Internationale (AMI) 239
Atlee, Clement 327
Atma Karaka 72, 497, 502 (See also **Atmakaraka**)
Atomic age 50, 288, 289, 337
Atomic bomb 26, 50, 184, 199, 327
Auchinchloss, Hugh D. 481, 488
Auchinchloss, Janet Lee Bouvier 470, 471, 472, 493
Automobiles 10, 22, 23, 26
Ayanamsha 2, 4, 59, 497, 501, 505, 509

B

Bacevich, Andrew 38
Bach, Johann Sebastian 18
Baldwin, James 290

Balfour Declaration 27, 48, 196, 405, 406, 407, 408, 426, 429
Ballmer, Steve 214
Bank bailouts 54, 55
Barenboim, Daniel 127, 128, 131, 132
Batista, Fulgencio 360
Bay of Pigs invasion 346, 358
Beck, Simone 159, 161
Beer Hall Putsch (Nov. 1923) 394, 414, 423, 424, 426, 427
Beethoven, Ludwig van 18
Bell, Alexander Graham 23, 238
Benz, Karl 23, 48
Berlin Wall 52, 54, 220, 347, 361
Bernays, Edward L. 12, 26
Bertholle, Louisette 159, 161
Bessemer, Henry 22, 47
Bessemer steel 18, 22, 47
Bhagavad Gita 1, 73, 79, 186, 281, 282, 433, 442, 456
Bhagya 70, 104, 113, 242, 248, 295
Bharani 7, 50, 63, 80, 122, 196, 337, 413, 414, 451, 463, 507
Bhava 61, 121, 212, 235, 462, 498, 516
Bhavartha Ratnakara 121, 305, 322, 515
Bill and Melinda Gates Foundation 210, 217
Bi-polar 331
Black Death 9, 16, 17, 410
Black Jack 471, 472, 489, 494
Bose, Jagdish Chandra 30
Bouvier, Caroline Lee 470, 482
Bouvier, Jack 469-472, 481
Bouvier, John Vernou 471, 472
Bowles, Chester 360
Brahmacharya 286, 433, 434, 435, 442, 443, 445, 446, 453
Brahmins 60, 178, 234, 498
Braque, Georges 378
Braun, Eva 395, 412, 414, 422, 427
Brecht, Bertolt 401
Bretton Woods 11, 50, 52
Brihat Parashara Hora Shastra 72, 74, 107, 174, 205, 350, 370, 508
Brihat Samhita 71, 513, 515
British Empire 4, 5, 19, 22, 25, 43, 45, 46, 47, 129, 327, 332, 335, 400, 405, 432, 437, 439, 440
British mandate (Palestine) 28, 405
Brokaw, Tom 219
Budha Aditya yoga 229, 278, 285, 312, 498
Buffet, Jimmy 313
Buffett, Warren 209, 210, 214, 217
Bullock, Alan xii, 401

C

Caesar, Sid 297
Cahill, Thomas 429
Calder, Alexander 152
Callas, Maria 134, 274
Camelot 344, 348, 360, 363, 364, 365, 465, 466, 467, 469, 479, 485, 492, 493, 494
Camelot Presidency 344, 467, 469, 479, 485, 494
Canby, Vincent 298
Cancer 3, 4, 5, 6, 13, 41, 43, 44, 60, 62, 63, 66, 73, 75, 234-237, 238-276, 497, 498, 500, 503, 505, 506, 507, 510, 511, 513, 519
Capricorn 3, 5, 7, 11, 13, 31, 33, 34, 36, 38, 39, 40, 51, 56, 60, 61, 62, 64, 66, 69, 73, 497, 498, 500, 504, 505, 506, 507, 511, 512, 513, 518
Capshaw, Kate 202, 207
Carter, Jimmy (U.S. President) 30, 35, 218
Casa dei Bambini 238, 241
Castro, Fidel 97, 346, 360, 368, 469, 493
Chagall, Marc 378
Chamara yoga 120
Chamberlain, Neville 327, 334, 337
Chappaquiddick 103
Chara 59, 60, 76, 77, 87, 498, 500, 512, 513
Chayefsky, Paddy 211
Chesta Bala 350, 370
Child, Julia 158-171, 172
Child, Paul Cushing 159-161, 164-169, 171
Chitra 7, 14, 56, 63, 93, 157, 323, 328, 332, 334, 336, 337, 382, 392, 441, 448, 497, 498, 507
Chomsky, Noam 83
Chopin, Frederick 19
Church Report, The (Nov. 1975) 349, 352, 360
Churchill Centre (Washington, D.C.) 331, 332
Churchill, Clementine Ogilvy Hozier 325, 343
Churchill, Randolph Spencer 325, 343
Churchill, Sir Winston 31, 50 188, 318, 322, 324-343, 346, 352, 357, 361, 362, 372, 387, 417, 431, 432
Citizens United (U.S. Supreme Court ruling) 11, 55

Civil Rights Act of 1964 12, 39, 80, 88, 89, 90, 92
Civil Rights movement 52, 80, 82, 89, 90, 91, 94, 432, 494
Claiborne, Craig 162
Clarke, Thurston 101
Climate change 13, 14, 35
Clinton, Bill (U.S. President) 40, 368, 375
Clinton-Bush Katrina Hurricane relief fund (2005) 459
Cluttering (speech defect) 331, 332
Cocteau, Jean 378
Coetzee, J. M. 399, 400
Columbus 16, 42
Colville, J. R. 328
Combustion 23, 48, 71, 278, 279, 413, 413, 498-499
Committee on Public Information 25, 409
Conant, Jennet 159
Copeland, Aaron 152, 157
Corporate conglomerates 11, 13, 37, 211
CPI 25, 26, 409
Creel, George 25, 26, 409
Cuban Missile Crisis (Oct. 1962) 358
Cugnot, Nicholas Joseph 22, 45
Czar Nicholas II 24, 25

D

Dada Abdullah & Co. 437
Daimler, Gottlieb 23, 48
Dalai Lama XIV 39, 442
Darakaraka 217, 231, 373, 479, 488, 499, 502
Dardanelles military campaign (March 1915) 322, 326, 330, 334, 335, 341, 342
Deepwater Horizon Oil explosion 39
De Feu, Paul 291
Denishawn 151, 152, 154, 155
Derived houses 70, 71, 498
Desai, Mahadev 436, 438
Desai, Morarji 255, 263
Desai, Narayan 438
Devi Bhava 436, 462, 464
DeVoto, Avis 167, 168, 169
Dhana yogas 104, 162, 206, 207, 213, 214, 215, 216, 227, 230, 339, 477, 486, 488
Dhanishta 7, 8, 64, 120, 191, 249, 250, 251, 293, 303, 304, 313, 367, 425, 507
Dharma 1, 2, 3, 11, 15, 58, 59, 62, 66, 73, 74, 77, 78, 85, 87, 88, 110, 116, 124, 149, 150, 169, 177, 180, 182, 188, 213, 215, 223, 231, 251, 263, 277, 278, 284, 288, 296, 303, 305, 312, 334, 338, 352, 353, 354, 360, 364, 373, 381, 415, 417, 425, 427, 443, 447, 453, 460, 464, 485, 488, 493, 499, 510
Digbala 68, 77, 80, 86, 87, 98, 101, 104, 107, 113, 120, 121, 123, 133, 150, 153, 154, 162, 168, 180, 186, 194, 206, 218, 221, 229, 230, 239, 258, 264, 269, 278, 284, 287, 301, 302, 311, 313, 314, 316, 337, 374, 380, 385, 442, 445, 446, 453, 458, 481, 489, 499, 504, 505
Dikötter, Frank 24
DiMaggio, Joe 246, 247, 249, 250, 252
Doumanian, Jean 307
Dred Scott decision 21, 47
Drug use 38, 143, 145, 318, 349
Dual signs 60, 75, 173, 174, 380, 513
Duchess of Cornwall 268, 274
Dukas, Helen 191, 192
Duke of Marlborough 325, 343
Dukes v. Wal-Mart Stores 39
Du Pré, Jacqueline 126-136, 137, 171, 180
Durudhara yoga 313, 499, 512
Dusthana house 68, 69, 70, 110, 114, 147, 150, 155, 204, 312, 391, 500, 512
Dwiswabhava 60, 498, 500, 512 (See also **Dual signs**)
Dyson, Frank Watson 185

E

EARTH period 7, 10-14, 18, 21, 26, 29-41
EARTH to AIR mutation period 14
East India Company 43, 46, 47, 437
Easton, Carol 128, 130, 132
Eclipse 49, 59, 68, 74, 91, 92, 93, 110, 124, 132, 136, 148, 170, 185, 195, 196, 268, 271, 273, 274, 275, 330, 339, 355, 356, 372, 381, 424, 425, 426, 445, 449, 451, 452, 476, 483, 487, 488, 491, 493, 494, 500
Eddington, Arthur Stanley 185
EEC (European Economic Community) 29, 52
Einstein, Albert x, 26, 49, 135, 174, 176, 180, 183-200, 224, 233, 243, 250, 332, 372
Einstein, Eduard 193
Einstein, Hans Albert 192, 193
Einstein, Maria (Maja) 192, 198

Einstein, Margot Löwenthal 192
Eisenhower, Dwight (U.S. President) 36
Emerson, Ralph Waldo 97
European Economic Community 29
Exner, Judith Campbell 349, 469, 479, 492, 493
Exxon-Valdez Oil spill 53

F

Farrow, Mia 73, 112, 118-126, 132, 137, 299, 300, 304, 306, 308
Farrow, Ronan Seamus 120, 299
Fauré, Gabriel 228
Feld, Bernard T. 184
Fest, Joachim C. 396, 414, 415
Finzi, Christopher (Kiffer) 130
FIRE period 6, 9, 16, 17, 18-23, 28, 44, 439
Fire to Earth Mutation period 4, 15, 23-28, 199, 398, 425, 442
Fixed signs 60, 513 (See also **Sthira**)
Ford, Henry 12, 23, 30, 48, 404
Ford Motor Company 33, 53, 360
Freed, Kayce 220, 222, 227
French Revolution 20, 45, 231
Fukushima 39

G

Gagarin, Yuri 31, 51, 357
Gaja Kesari yoga 221, 364, 444, 501
Galilei, Galileo 18, 43
Gandanta degrees 237, 264, 333
Gandanta, planets in 164, 237, 254, 260, 264, 279, 292, 333, 335, 445, 449, 450, 501
Gandhi, Feroze 254, 255, 256, 257, 258, 261
Gandhi, Indira 53, 253-265, 275, 276, 452, 479, 519
Gandhi-Irwin Truce 451
Gandhi, Karamchand 430, 434
Gandhi, Kasturbai 434, 437
Gandhi, Mahatma (Mohandas K.) xii, 28, 51, 73, 80, 82, 254, 258, 282, 327, 430-453, 495, 511, 519
Gandhi, Maneka 254
Gandhi, Rajiv 254, 255
Gandhi, Rajmohan 436
Gandhi, Sanjay 254, 255, 256, 258, 261, 263, 264
Gandhi, Sonia 254
Gandhi, Varun 254
Gates, Bill 33, 53, 70, 114, 175, 180, 208-218, 230, 233, 334, 371
Gates, Mary Maxwell 217
Gates, Melinda French 210, 215, 217
Gemini 3, 6, 7, 8, 18, 32, 42, 53, 60, 61, 62, 63, 66, 69, 73, 75, 173-177, 178-233, 497, 500, 503, 505, 506, 507, 510, 513, 518
General Motors 30
Genghis Khan 29, 399
George, David Lloyd (British Prime Minister) 196, 405, 406
Ghosh, Bhagabati Charan 280, 287
Gilot, Émile 377, 378
Gilot, Françoise 376-387, 388, 418
Glass-Steagall Act (U.S. Banking Act of 1933) 12, 49, 54
Glorious Revolution of 1688 16, 44
Godse, Nathuram 441, 451, 452
Goebbels, Paul 401, 402, 403
Goering, Hermann 398, 399, 401, 402, 412
Gokhale, Gopal Krishna 439
Gold-dollar link 52
Graham, Martha 151-158, 171, 172, 390
Graha Yuddha 71, 129, 130, 134, 135, 142, 162, 164, 301, 304, 305, 306, 307, 308, 370, 371, 372, 374, 379, 380, 383, 387, 410, 413, 414, 415, 418, 422, 425, 426, 441, 490, 501
Gray, Elisha 23
Guber, Lee 369
Gulf of Tonkin incident 37
Gulf of Tonkin Resolution 51, 82
Gunas 66, 67, 72, 391, 502, 509, 510, 511, 512
Gutenberg Bible 42
Gutenberg, Johannes 18, 42
Guthrie, Woody 309

H

Haiti 17, 28
Hammarskjold, Dag 52
Hansen, James 35
Hardy, G. H. 178, 179
Harrison, George 316
Hart, Sir Basil Liddell 396
Hartung, William 37
Harvard University 33, 37, 95, 100, 160, 209, 210, 212, 213, 214, 345, 346

Hasta 7, 14, 32, 53, 63, 163, 166, 214, 215, 264, 274, 294, 322, 323, 375, 379, 382, 385, 452, 460, 463, 491, 492, 507
Hebrew University 192, 196
Helú, Carlos Slim 209
Henny-Montessori, Marilenna 243
Henry the Navigator 17
Hersh, Seymour 348, 360, 363, 364
Hess, Rudolf 401, 422
Heymann, C. David 468, 472, 474, 484
Himmler, Heinrich 398, 399
Hitler, Adolf xii, 24, 49, 67, 104, 338, 341, 391, 393-430, 441, 442, 447, 448, 453, 495
Hitler, Alois 394, 395
Hitler, Klara 394, 395
Hoover, J. Edgar 94, 97, 100, 348
Hora Ratnam 58, 65, 515
Horst, Louis 152, 155
Houghton Mifflin (publishers) 163, 168, 169
House of Saxe-Coburg and Gotha 25, 48
House of Windsor 25, 48
Houses 60, 61, 65, 66, 67, 68, 69, 70, 71, 73, 74, 75, 496, 497, 498, 499, 500, 503, 504, 505,508, 509, 510, 512, 513, 514
Hundred Years' War 16, 42

I

IBM 33, 53, 209, 211, 214
IBM PC 33, 53, 209, 214
IMF 11, 50, 54
India 1, 4, 15, 17, 18, 19, 25, 27, 28, 31, 39, 41, 43, 57, 58, 59, 62, 64, 65, 71, 74, 76, 78, 79, 139, 159, 177, 178, 179, 182, 238, 239, 244, 248, 253-265, 280, 281, 282, 285, 286, 287, 288, 326, 327, 332, 335, 366, 369, 399, 400, 421, 430-453, 454, 455, 456, 459, 463, 464, 495, 499, 506, 508, 509, 510, 512, 513, 514, 515, 516, 518, 519
India national chart 254, 263, 264, 451, 452
Indian National Congress 254, 255, 438, 447, 451
Inquisition 4, 16, 18, 42, 43
Institute of Advanced Study (Princeton, N.J.) 186, 194, 195
Intergovernmental Panel on Climate Change 13, 14 (See also **IPCC**)
International Monetary Fund 11, 50, 54 (See also **IMF**)
Internet 11, 33, 34, 54
Irving, Amy 202, 207
Isaacson, Walter 184, 306
Isabella and Ferdinand 15, 16
Israel 27, 28, 51, 52, 127, 130, 196, 369, 399, 409

J

Jackson, Michael 368
Jacobson, Dr. Max 351
Jaimini 65, 72, 73, 123, 207, 268, 272, 287, 357, 488, 489, 496, 497, 498, 499, 502, 503, 505, 510
Janankananda, Rajarsi 286 (See also **Lynn, James**)
Jasanoff, Maya 19
Jataka Parijata 76, 174, 321, 515
Jefferson, Thomas 20, 44, 108
Jenkins, Roy 328, 334
Jennings, Charles 218, 220, 222
Jennings, Peter 66, 175, 176, 218-227, 229, 233
Jerome, Jennie 325
Jinnah, Muhammad Ali 438
Jiva 72, 104, 188, 240, 242, 243, 244, 249, 251, 260, 261, 262, 271, 289, 294, 295, 301, 303, 305, 311, 312, 332, 340, 352, 358, 371, 373, 380, 385, 390, 415, 428, 502, 506
Jobs, Steve 33, 53
Johnson, Haynes 98
Johnson, Lyndon B. (U.S. President) 82, 93, 95
Joling, Robert 109
Jones, Arthur Creech 28
Jovian year 3
Judt, Tony viii, x
Jupiter 2, 3, 4, 5, 11, 13, 21, 23, 27, 31, 32, 34, 38, 39, 40, 59, 60, 61, 63, 64, 67, 68, 69, 71, 73, 74, 496, 499, 500, 501, 502, 504, 505, 506, 509, 510, 511, 512, 518
Jupiter rocket 32
Jupiter-Saturn conjunction 2, 3, 200, 215, 223, 271, 288, 323, 451, 461, 463, 488
Jupiter-Saturn conjunction in Aries 4, 6, 7, 26, 28, 40, 46, 48, 50, 54, 170, 196, 214, 216, 295, 337, 451, 459, 488
Jupiter-Saturn conjunction in Capricorn 5, 7, 11, 13, 31, 33, 34, 36, 38, 39, 40, 51, 56, 88, 105, 108, 117, 118, 124, 132, 137, 163, 169, 251, 258, 307, 359, 484, 494
Jupiter-Saturn conjunction in Sagittarius 6, 7, 20, 21, 24, 25, 28, 44, 46, 48, 169
Jupiter-Saturn conjunction in Taurus 7, 14, 56

Jupiter-Saturn conjunction in Virgo 4, 7, 8, 9, 12, 14, 22, 24, 28, 29, 30, 32, 33, 40, 49, 50, 53, 56, 125, 200, 206, 215, 223, 229, 264, 271, 274, 288, 323, 461, 463

Jupiter-Saturn opposition 3, 29, 39, 148, 150, 262, 288, 493

Jyeshta 63, 180, 283, 507

K

Kael, Pauline 200, 201

Kaiser Wilhelm II 24, 25, 48, 49

Kaiser Wilhelm Institute 183, 184, 195, 196

Kala Purusha 57, 503, 510

Kalidasa 60, 71, 322, 515

Kali Yuga 2, 15, 67, 68, 503, 514

Kama 41, 66, 147, 173, 278, 343, 389, 503

Karakas 61, 134, 314, 410, 496, 505

Karma 1, 3, 58, 59, 61, 62, 63, 73, 74, 78, 85, 88, 110, 116, 117, 124, 133, 149, 150, 154, 169, 177, 182, 188, 213, 223, 231, 251, 284, 285, 288, 296, 305, 317, 334, 352, 373, 374, 381, 425, 427, 428, 447, 453, 464, 488, 493, 496, 503, 509, 510, 511

Karma phala 154, 503

Keaton, Diane 299, 300, 304

Kelleher, James 366, 516, 517

Kelley, William 22, 47

Kemadruma yoga 190, 194, 247, 382, 504,

Kempe, Frederick 347

Kendra 67, 69, 70, 73, 86, 87, 99, 100, 103, 104, 107, 113, 120, 122, 144, 145, 146, 150, 153, 156, 157, 162, 164, 167, 168, 175, 180, 188, 190, 194, 204, 205, 206, 213, 216, 241, 242, 243, 247, 250, 251, 256, 259, 260, 270, 305, 316, 323, 329, 333, 339, 356, 370, 371, 375, 380, 382, 391, 410, 415, 418, 419, 444, 458, 501, 504, 505, 508, 512

Kendradhipatya Dosha 69, 107, 176, 188, 323, 391, 504

Kennedy, Caroline 466, 477x

Kennedy, Edward (Ted) 102, 103, 106

Kennedy, Ethel Skakel 97, 101, 102, 104

Kennedy, John F. (Jack) (U.S. President) 12, 21, 31, 40, 51, 52, 95, 103, 344-367, 467, 469, 471, 473, 475, 476, 477, 478, 479, 480, 483, 484, 488, 492, 493, 494

Kennedy, John Jr. 466, 489-490

Kennedy, Joseph P. (Joe, Jr.) 355

Kennedy, Joseph P., (Joe, Sr.) 69, 94, 95, 101, 106, 348, 355, 363, 473, 480, 484

Kennedy, Patrick Bouvier 480, 485

Kennedy, Robert F. (Bobby) 12, 69, 87, 93, 94-110, 113, 114, 120, 125, 136, 137, 250, 293, 348, 352, 357, 424, 468, 473, 485, 491

Kennedy-Nixon debates 31, 344

Ketu 59, 60, 61, 63, 64, 67, 68, 69, 74, 229, 230, 231, 232, 475, 476, 480, 482, 483, 486, 487, 489, 491, 492, 493, 498, 499, 500, 501, 502, 504, 509, 513

Keynes, John Maynard 408

King George III 45

King George V 24, 25, 48

King, Martin Luther, Jr. 12, 21, 52, 80-94, 96, 97, 98, 101, 109, 110, 136, 137, 290, 292, 296, 429, 432

Kipling, Rudyard 438

Klein, Edward 467, 468, 472, 475

Klein, Naomi 37

Knopf, Alfred A. publishers 169

Koch Industries 35

Konigsberg, Ellen 297

Konigsberg, Martin 297

Konigsberg, Nettie 297

Krock, Arthur 362

Krushchev, Nikita 344, 347, 357, 359, 361

Kuja Dosha 78, 79, 504

L

Lache, Irene 378

Lafayette Square 478

Lahr, John 301

Lakshmi 115, 139, 392, 478

Lake, Carlton 376, 386, 387

Langer, Walter 402, 421

Lasser, Louise 298, 304

Lebensraum 397, 419

Leo 3, 5, 60, 62, 63, 66, 71, 73, 277-279, 280-320, 497, 499, 504, 506, 507, 510, 512, 513

Lepanto, Battle of 43

Lewinsky, Monica 368, 375

Libra xii, 3, 5, 60, 61, 62, 63, 66, 69, 70, 73, 389-393, 394-495, 497, 498, 500, 503, 506, 507, 510, 513, 519

Lindbergh, Charles 30, 49

Longevity 61, 62, 70, 71, 80, 85, 88, 97, 105, 112, 113, 139, 143, 153, 154, 156, 167, 168, 170, 171, 181, 229, 231, 257, 302, 329, 331, 350, 367, 370, 371, 372, 387, 388, 477, 484, 500

Löwenthal, Elsa 191, 192, 193
Luther, Martin 42, 401, 402, 403, 404
Lynn, James 286

M

MacDonogh, Giles 399
Machiavelli, Nicolo 399-400, 402
Macmillan, Harold 359
Macmillan, Margaret 447
Mafia 96, 98, 101, 348, 349, 360, 469, 493
Magha 6, 44, 63, 92, 162, 170, 222, 223, 269, 277, 279, 282, 283, 286, 291, 292, 300, 329, 343, 459, 507
Maha Bhagya yoga 113, 262
Maha Purusha Raja yoga 221, 250, 256, 505 (See also **Mahapurusha Raja yoga**)
Maha Samadhi 282, 289
Mahayuga 1
Mailer, Norman 112, 365, 468
Make, Vusumzi 290, 291
Malavya yoga 188, 444
Malcolm X 39, 80, 81, 290
Male client 141-145
Manchester, William 340, 466, 468, 475, 476, 490
Mandela, Nelson 432, 494
Mander, Jerry 33
Manhattan Project 184, 186, 198
Manic-depressive 331
Manifest Destiny 22, 46, 419, 420
Marable, Manning 39, 81
Maraka 69, 70, 176, 227, 302, 323, 367, 444, 452, 494, 504, 505
Marconi, Guglielmo 30
Marić, Mileva 186, 193, 243
Marriage chart 271, 274, 275, 488, 491-492 (See also **Wedding chart**)
Mars 14, 59, 60, 63, 64, 67, 68, 69, 70, 71, 73, 77, 78, 499, 500, 501, 503, 504, 505, 506, 509, 510, 511, 514
Marton, Kati 226
Mata Amritanandamayi 256, 454-464, 495 (See also **Ammachi**)
Matisse, Henri 376, 378, 381
Mayer, Jane 35
McCartney, Paul 316
McLean Psychiatric Hospital 310, 314
McLuhan, Marshall 32, 33, 34
McNamara, Robert S. 12, 36
Mehta, Zubin 134
Mein Kampf 395, 396, 398, 403, 414, 416, 418, 422, 423
Mercury 18, 31, 32, 34, 51, 59, 60, 61, 63, 64, 67, 68, 69, 70, 71, 73, 173, 174, 175, 176, 177, 321, 322, 323, 496, 498, 499, 500, 501, 503, 505, 506, 509, 510, 511, 512
Meyer, Mary Pinchot 349
Microsoft 33, 53, 114, 208, 209, 210, 211, 212, 214, 215, 216, 217, 233
Middleton, Kate 274
Miller, Arthur 245, 246, 248, 249, 251
Miller, Henry 110-118, 137
Millet, Kate 112
Mims III, Forest 33
Misdiagnosis 135
Moguls 29
Moksha 66, 106, 115, 145, 189, 232, 234, 241, 247, 248, 250, 251, 252, 278, 283, 284, 285, 287, 288, 289, 292, 293, 294, 314, 318, 331, 334, 391, 392, 443, 459, 460, 461, 464, 476, 505
Monroe Doctrine 22, 46, 419
Monroe, James (U.S. President) 22, 46
Monroe, Marilyn 98, 106, 165, 244-253, 275, 276, 348, 479
Montessori, Maria 235, 236, 237-244, 248, 269, 272, 275, 276
Montessori, Mario 239, 242, 243
Montgomery Bus Boycott 82, 83, 88, 90, 91
Moon 3, 57, 59, 60, 62, 63, 64, 65, 66, 67, 68, 69, 70, 72, 73, 74, 234-237, 496, 497, 498, 499, 500, 501, 503, 504, 505, 506, 508, 509, 510, 511, 512, 513, 514, 519
Morell, Dr. Theodore 412
Morris, Errol 36
Morse, Samuel 23
Mountbatten, Lord Louis 271
Mozart, Wolfgang Amadeus 18
Mrigashira 7, 8, 14, 56, 63, 84, 106, 140, 177, 241, 374, 425, 478, 507
Mula 6, 44, 64, 229, 232, 272, 285, 480, 507
Mulatrikona 104, 144, 509
Müller, Renate 395, 422
Multiple sclerosis 128, 131, 134, 135, 136, 137, 143
Murdoch, Rupert 33, 35

Murrow, Edward R. 220

Mussolini, Benito 239

Mutation period 3, 4, 9, 10, 14, 15, 18, 23, 24, 27, 199, 398, 425, 442

Muttit, Greg 37

N

Nadi Jyotisha 65, 66, 72, 73, 352, 390, 391, 502, 506, 516

Nadi yogakarakas 73, 140, 391, 393, 463, 481, 506, 509

Nadi yogas 72, 73, 120, 121, 222, 236, 241, 248, 249, 251, 259, 261, 271, 294, 295, 314, 351, 353, 355, 357, 365, 380, 382, 384, 385, 460, 476, 477, 478, 480, 488, 489, 492, 506

NASA 32, 35, 51

National Aeronautics and Space Administration 32, 35, 51 (See also **NASA**)

Nazi Germany 25, 26, 32, 184, 197, 198, 336, 340, 355, 397, 398, 399, 402, 404

Nazi party 49, 397, 400, 403, 409, 411, 414, 415, 423

Neecha Bhanga Raja yoga 126, 148, 149, 163, 165, 167, 187, 196, 213, 214, 241, 270, 304, 314, 383, 486, 492, 498, 499, 508

Nehru, Jawaharlal 254, 257, 258, 259, 263

Nehru, Kamala Kaul 254, 257, 258, 259

Nehru, Motilal 253, 254, 258

Nervous breakdown 128, 135, 193, 439

Nervous system 34, 61, 62, 129, 134, 174, 180, 211, 229, 232, 314, 323, 330

Newton, Isaac 44

Nicholson, Harold 408

Nin, Anaïs 111, 112, 115, 117

Nixon, Richard (U.S. President) 31, 52, 341, 348, 366, 368, 369, 428, 494

Nobel Prize 26, 89, 92, 186, 187, 190, 191, 193, 195, 199, 238, 328, 334, 339

Noguchi, Dr. Thomas 109, 246

Noguchi, Isamu 152, 157

Nuclear arms race 50, 359

Nuclear energy 10, 26, 39

Nuclear reaction 10

Nuclear weapons 10, 26, 36, 37, 50, 184, 186, 196, 200, 347, 348

Nupen, Christopher 129

Nuremberg Trials 398, 399, 402, 418

O

Obama, Barack (U.S. President) 35, 40, 55, 84

Ohga, Norio 33

Onassis, Aristotle 465, 467, 470, 473, 475, 483, 485, 491, 492

Onassis, Christina 492

Onassis, Jacqueline Kennedy 109, 163, 165, 232, 345, 349, 354, 464-494, 495

Oppenheimer, J. Robert 186, 197

Oratory 81-89, 96, 324, 327, 328, 332, 333, 339, 411, 415

Orwell, George 111

O'Sullivan, John L. 420

Ottoman Empire 42, 327, 330, 405, 407

P

Pakistan 19, 27, 51, 53, 255, 262, 438, 451, 518

Pakistan-India War 53, 255, 262

Palestine 27, 28, 48, 404, 405, 406, 407, 408, 409, 426, 429

Pan-Germanism 24, 198, 396, 397, 401, 416

Papa kartari yoga 113, 114, 116, 130, 132, 194, 203, 260, 442, 446, 447, 448, 461, 508, 512

Paramahansa Yogananda 277, 280-289, 292, 300, 319 (See also **Yogananda**)

Paris Peace Treaty, 1783 20, 45

Paris Peace Treaty, 1919 27, 408, 447

Parivartana yoga 113, 114, 121, 123, 130, 131, 134, 144, 145, 155, 156, 176, 177, 179, 182, 186, 188, 195, 203, 205, 222, 223, 227, 240, 241, 256, 262, 263, 302, 305, 329, 334, 338, 354, 373, 487, 488, 506, 508

Parker-Bowles, Camilla (Shand) 267, 268, 270, 271, 273, 274-275

Parks, Rosa 82, 90

Peace Corps 346

Pearl Harbor 50, 423

Pentagon Papers, The 348

Personal computers 53, 209, 211

Picasso, Pablo 376, 377, 378, 381, 382, 384, 387, 388

Pisces 3, 4, 5, 6, 17, 20, 39, 43, 44, 60, 61, 62, 64, 65, 66, 69, 73, 75, 492, 497, 498, 500, 505, 506, 507, 511, 513

Plessy v. Ferguson 21

Plutarch 41

Polio 118, 121, 133, 379

Polk, James (U.S. President) 22, 46
Polo, Marco 14
Pope Benedict XVI 275
Pope John Paul II 268, 274
Post, Jerrold M. 412
Poulenc, Francis 227-232, 233
Powell, Julie 161
Prana viii, 58, 66, 509
Precession of the Equinoxes 2, 59, 509
Previn, André 119, 122, 306
Previn, Soon-Yi 119, 120, 124, 298, 299, 302, 304, 306, 307
Prince Charles (of Wales) 33, 106, 265-275, 276
Prince Philip 265
Prince William 267, 274
Princess Diana (of Wales) 33, 249, 267, 268, 270, 271, 273, 274
Printing presses 18
Prostakoff, Theodore 146-150, 155, 156, 171
Prud'homme, Alex 160, 161, 171
Public transportation 10, 22, 30
Punarvasu 6, 8, 42, 63, 125, 130, 171, 177, 236, 241, 507
Puranas 62, 139, 236, 390, 392, 510
Purva Ashadha 6, 7, 46, 48, 64, 124, 129, 135, 169, 249, 261, 262, 304, 356, 418, 419, 420, 427, 507
Purva Bhadra 6, 43, 64, 89, 250, 311, 426, 489, 507
Purva Phalguni 6, 45, 47, 63, 222, 226, 261, 278, 279, 312, 355, 425, 427, 428, 451, 478, 494, 507
Pushya 6, 43, 63, 102, 148, 189, 236, 424, 507

Q

Queen Elizabeth II 31, 51, 232, 265, 268, 328, 431, 476
Queen Isabella 16, 42
Queen Victoria 25, 46, 48, 328, 437

R

Radhakrishnan, S. 431
Radio 30, 31, 32, 161, 218, 220, 222, 324, 366, 401, 403, 409
Rahu 59, 60, 63, 64, 67, 68, 69, 72, 73, 74, 498, 499, 500, 501, 502, 506, 509, 510, 513, 519
Railroads 4, 21, 22, 45
Rajas 67, 502, 509, 510
Raman, B. V. 421, 516
Ramanujan, Srinivasa 177-183, 185, 187, 190, 204, 232, 233
Rao, K. N. 516
Raubal, Geli 395, 422
Reagan, Ronald (U.S. President) 33, 35, 36, 37, 40, 53
Red Flag act 22
Reeves, Richard 351, 363, 364, 365
Revati 4, 6, 44, 64, 292, 295, 391, 491, 507
Rexroth, Kenneth 111, 115
Rig Veda 1, 59, 337, 508, 509, 514
Roberts, Ed 33
Robins, Robert S. 412
Rohini 7, 14, 56, 62, 63, 139, 140, 147, 148, 162, 168, 170, 294, 464, 506, 507
Roosevelt, Eleanor 26, 184
Roosevelt, Franklin D. (U.S. President) 50, 184, 196, 197, 198, 199, 233
Roque, Jacqueline 378, 385, 386
Rosen, Harlene 298, 304
Rostropovich, Mstislav 127
Ruiz-Picasso, Claude 380
Ruiz-Picasso, Paloma 378, 380, 385

S

Sade Sati 73, 74, 88, 92, 93, 114, 125, 136, 154, 190, 198, 224, 229, 251, 307, 317, 318, 339, 340, 341, 374, 375, 449, 450, 459, 462, 484, 491, 497
Sagittarius 3, 5, 6, 7, 20, 21, 24, 25, 28, 39, 40, 44, 46, 48, 60, 61, 62, 64, 66, 73, 75, 497, 499, 500, 501, 504, 506, 507, 510, 513, 518
Salk, Dr. Jonas 379, 385, 387
Santa Clara County v. Southern Pacific Railroad Company 21
Sanyasin yoga 381
Saraswati yoga 303, 511
Satguru Sri Sivananda Murty 2
Sattva 67, 502, 509, 511
Saturn 2, 3, 4, 5, 10, 11, 12, 13, 14, 27, 28, 31, 32, 36, 39, 40, 59, 60, 63, 64, 67, 68, 69, 71, 73, 74, 77, 78, 80, 497, 499, 500, 501, 502, 504, 505, 506, 509, 511, 512, 516, 518, 519
Saturn rocket 32
Satyagraha 431, 432, 433, 437, 443, 444, 446, 449, 453, 511

Satyagraha ashram 449
Saxe-Coburg and Gotha 25, 48
Scahill, Jeremy 38
Schlesinger, Arthur, Jr. 96, 97, 102, 348, 349, 361, 466, 468, 476
Scorpio 3, 5, 6, 8, 15, 41, 42, 43, 44, 60, 62, 63, 66, 70, 73, 497, 498, 500, 501, 505, 506, 507, 510, 511, 512, 513, 514
Selden, George 23
Self-Realization Fellowship 280
Shakespeare, William 43, 477
Shakti 62, 63, 64, 97, 99, 512
Shakti yoga 97, 99
Shashtashtaka 116, 257, 512
Shatabhisha 64, 93, 426, 507
Shepard, Alan 31, 51, 357
Shirer, William L. 431, 438, 440
Shravana 7, 13, 38, 56, 64, 89, 92, 179, 181, 340, 507
Shubha kartari yoga 117, 211, 225, 311, 332, 350, 371, 508, 512
Shudras 60, 512
Sibling rivalry 128, 130, 470, 482
Simon, Carly 308, 318, 319
Simon, Luc 378, 381, 385, 386, 387
Sinatra, Frank 119, 122, 123, 250
Sivananda Murty 2
(See also **Satguru Sri Sivananda Murty**)
Six-Day War (1967) 27, 52, 127, 130
Skanda 77, 115, 392, 478
Slave labor 10, 20, 56, 398
Slave trade 9, 17, 45
Sloan, Alfred P. 30
Smedwig, Caroline "Kim" 310
Smetana, Bedrich 19
Smith College 159, 160, 161, 169
Smith, Sidonie Ann 291
Snyder, Timothy 24, 398
Soma and Rohini (Puranic myth) 62
Soma and Tara (Puranic myth) 173, 236
Sorensen, Ted 348, 361, 362
South Africa 220, 430-433, 436-438, 442, 445, 446, 447, 448, 449, 453, 494
Space Race 31, 32, 357
Spanish Armada 43
Spanish flu 9, 191
Spanish Inquisition 16, 42
Speech disorder 331
Speech impediment 331, 333
Spielberg, Steven 200-207, 216, 233
Sputnik 32, 51, 357
Stalin, Joseph 335, 399
Stanley Steamer 23, 48
Stationary Direct 74, 84, 89, 91, 92, 93, 97, 98, 143, 169, 317, 350, 352, 353, 361, 365, 370, 428, 463, 481
Stationary Retrograde 74, 92, 93, 98, 110, 121, 136, 370, 374, 481
Steam engine 17, 22, 23, 44, 45
Steroids 349, 350, 351
Sthira 60, 139, 498, 500, 512 (See also **Fixed signs**)
Strasberg, Lee 244, 245, 246
Strategic Defense Initiative (SDI) 36
Stuttering 331, 332
Sudhamani 455, 456, 458, 461, 462
Sun 2, 3, 4, 5, 18, 57, 59, 60, 63, 64, 65, 67, 68, 69, 71, 72, 73, 74, 76, 277-279, 280-320, 496, 497, 498, 499, 500, 501, 503, 504, 506, 509, 510, 511, 512, 513, 519
Surya Siddhanta 1, 3, 58, 509, 515
Swami Amritaswarupananda 455, 457, 463
Swaraj 437, 439
Swati 7, 56, 63, 93, 114, 147, 149, 240, 392, 507
Syndicate, The 342

T

Talmey, Max (formerly Max Talmud) 185, 189
Tamas 67, 502, 509, 512
Taurus 3, 7, 14, 56, 60, 61, 63, 66, 70, 73, 75, 138-140, 141-172, 497, 500, 506, 507, 510, 512, 513, 514, 519
Taylor, A. J. P. 401
Taylor, Dr. Isaac 309, 315
Taylor, James 308-319, 320
Taylor, Kate 309
Taylor, Livingston 309
Tchaikovsky, Peter 18
Telegraph 22, 23, 30
Teller, Edward 186
Telephone 22, 23, 26, 30
Television 31, 32, 33, 53

Tempelsman, Maurice 469, 474, 480
Tesla, Nikola 30
Tevithick, Richard 22, 45
Theory of relativity 49, 183, 185, 187, 195, 198, 199
Third Reich 394, 396, 397, 398, 426, 429, 430
Thirty Years' War 43
Tiananmen Square protest (Beijing) 54
Treaty of Versailles 27, 49, 178, 394, 401, 402, 408, 409, 416, 423, 426, 442, 447 (See also **Paris Peace Treaty, 1919**)
Trevor-Roper, Hugh xii, 401
Trinal house 67, 69, 73, 88, 100, 123, 144, 170, 204, 206, 214, 241, 259, 283, 288, 311, 329, 486, 504, 506, 513, 514
Tripartite Treaty (Sept. 1940) 423
Triple conjunctions 4, 5, 6, 7, 8, 15, 26, 28, 32, 40, 41, 44, 50, 53, 463
Triplicities 3, 4, 41
Tse Tung, Mao 24

U

Ubhayachari yoga 87, 103, 269, 371, 461, 513
United Nations 13, 27, 28, 31, 51, 52, 83, 120, 327, 399, 453, 456
Upachaya house 70, 80
Upanishad 57, 140
U. S. Constitution 21, 28, 45
U. S. Presidential elections 40
USA national chart 40, 366, 517-518
Uttara Ashadha 7, 13, 38, 51, 56, 64, 87, 104, 105, 106, 107, 117, 118, 120, 169, 262, 385, 507
Uttara Bhadra 6, 44, 64, 102, 155, 287, 427, 507
Uttara Kalamrita 60, 71, 153, 322, 515
Uttara Phalguni 7, 49, 63, 126, 278, 279, 288, 323, 371, 425, 507

V

Vaishyas 60, 389, 442, 513
Van Praag, Phillip 109
Vargottama 4, 6, 7, 8, 13, 45, 46, 51, 56, 87, 93, 105, 111, 112, 113, 120, 143, 147, 149, 150, 154, 156, 157, 169, 203, 204, 206, 220, 221, 223, 225, 226, 242, 243, 261, 262, 301, 302, 304, 305, 329, 331, 334, 337, 338, 339, 350, 352, 353, 358, 370, 375, 483, 513
Venus xii, 59, 60, 63, 64, 67, 68, 69, 70, 71, 73, 77, 78, 138-140, 141-172, 389-393, 394-495, 496, 499, 500, 501, 502, 503, 504, 505, 506, 509, 510, 511, 512, 513, 514
Video News Releases 34 (See also **VNRs**)
Vietnam War 12, 31, 36, 37, 51, 81, 82, 83, 86, 89, 92, 93, 95, 97, 98, 99, 101, 104, 108, 109, 141, 143, 344, 348
Viking Press 469, 492
Vimshottari Dasha 65, 74, 154, 187, 195, 215, 272, 317, 353, 461, 484, 508, 514
Viparita Raja yogas 70, 110, 114, 139, 149, 213, 214, 215, 230, 243, 322, 500, 514
Virgo 3, 4, 7, 8, 9, 12, 14, 22, 24, 28, 29, 30, 32, 33, 39, 40, 49, 50, 53, 56, 60, 61, 62, 63, 66, 69, 73, 75, 321-323, 324-388, 497, 500, 503, 506, 507, 510, 512, 513, 518, 519
Vishakha 5, 8, 41, 63, 105, 181, 206, 250, 269, 332, 383, 392, 426, 441, 443, 445, 448, 452, 507
VNRs 34
von Bismarck, Otto 400, 402
Von Braun, Wernher 32
von Hindenburg, President 394
Von Stauffenberg, Major 424

W

Wagner, Richard 18, 402, 417
Walker, Kathryn 310, 315
Wallace, Mike 368
Walters, Barbara 98, 129, 188, 367-375, 379, 383, 387, 388
Wannsee Conference 398
Wars of the Roses 15, 16, 42
Washington, George (U.S. President) 20, 22, 45, 470
Washington, Harriet 429
WATER period 5
WATER to FIRE mutation period 9, 18
Watts, James 17
Wedding 33, 79, 101, 102, 127, 188, 210, 217, 267, 268, 271, 274, 275, 355, 387, 470, 471, 473, 477, 479, 494
Wedding chart 267, 268, 271, 274, 275, 488, 491 (See also **Marriage chart**)
Weizmann, Chaim 196
White, Theodore 466
White, William Allen 108

Wilson, Elizabeth 127, 128
Wilson, Woodrow (U.S. President) 23, 25, 406, 407, 408, 409, 447
Wilson, Woodrow 23, 25, 406, 407, 408, 409, 447
Wolf, Naomi 402-403
World Age 1
World Bank 11, 12, 36, 50, 360
World Trade Center, Attack 448
World War I (WWI) 9, 11, 12, 21, 23, 24, 25, 26, 27, 48, 49, 178, 184, 185, 193, 195, 196, 198, 199, 229, 281, 322, 326, 327, 335, 397, 404, 405, 406, 409, 410, 416, 419, 420, 425, 429, 439, 440, 442, 447
World War II (WWII) 4, 10, 11, 12, 24, 25, 27, 29, 30, 31, 32, 39, 50, 94, 100, 159, 160, 161, 184, 197, 199, 206, 220, 239, 244, 324, 326, 327, 338, 352, 355, 357, 394, 397, 399, 401, 409, 414, 420, 425, 427, 428, 429, 431, 434, 440
Wright brothers 23, 48

Y

Yama 63
Yogananda 277, 280-289, 292, 300, 319
(See also **Paramahansa Yogananda**)
Yogas, Jaimini 72
Yogas, Parashari 72
Yuga 2, 15, 67, 68, 503, 514
Yukteshwar 2, 281, 282, 284, 285, 286, 287, 288

Z

Zapruder, Abraham 347
Zionism 26, 184, 194, 196, 200, 407
Zionist movement 24, 196
Zuckerburg, Mark 209

CPSIA information can be obtained at www.ICGtesting.com
Printed in the USA
LVOW11s1923020813

346029LV00015B/299/P